D0816535

Glacier science and environmental change

Glacier science and environmental change

Edited by

Peter G. Knight

BLACKWELL PUBLISHING
350 Main Street, Malden, MA 02148-5020, USA
9600 Garsington Road, Oxford OX4 2DQ, UK
550 Swanston Street, Carlton, Victoria 3053, Australia

First published 2006 by Blackwell Publishing Ltd

2 2007

Library of Congress Cataloging-in-Publication Data

Glacier science and environmental change / edited by P.G. Knight.
p. cm.
ISBN-13: 978-1-4051-0018-2 (hardback : alk. paper)
1. Glaciology. 2. Global environmental change. I. Knight, Peter, 1961–
GB2403.2G53 2006
551.31—dc22
2005034252

A catalogue record for this title is available from the British Library.

Set in 9 on 11.5 pt Minion
by SNP Best-set Typesetter Ltd, Hong Kong
Printed and bound in Singapore
by Markono Print Media Pte Ltd

The publisher's policy is to use permanent paper from mills that operate a sustainable forestry policy, and which has been manufactured from pulp processed using acid-free and elementary chlorine-free practices. Furthermore, the publisher ensures that the text paper and cover board used have met acceptable environmental accreditation standards.

For further information on
Blackwell Publishing, visit our website:
www.blackwellpublishing.com

Contents

List of contributors

C. Acuña Centro de Estudios Científicos, Valdivia, Chile.

William George Adam School of Physical and Geographical Sciences, Keele University, Keele, Staffordshire ST5 5BG, UK.

Richard B. Alley Department of Geosciences and Earth and Mineral Sciences EESI, The Pennsylvania State University, University Park, PA 16802, USA.

Sridhar Anandakrishnan Department of Geosciences and Earth and Mineral Sciences EESI, The Pennsylvania State University, University Park, PA 16802, USA.

Suzanne Prestrud Anderson Institute of Arctic and Alpine Research, University of Colorado, Boulder, CO 80309-0450, USA.

John T. Andrews INSTAAR and Department of Geological Sciences, University of Colorado, Campus Box 450, Boulder, CO 80309-0450, USA.

Hernán De Angelis Department of Physical Geography and Quaternary Geology, Stockholm University, Stockholm, Sweden.

Greg Balco Quaternary Research Center and Department of Earth and Space Sciences, University of Washington, Mail Stop 351310, Seattle WA 98195-1310, USA.

Jonathan Bamber Bristol Glaciology Centre, School of Geographical Sciences, University of Bristol, University Rd, Bristol BS8 1SS, UK.

T. Bell Geography, Memorial University of Newfoundland, St John's, Canada.

Doug Benn School of Geography and Geosciences, University of St Andrews, Fife KY16 9AL, UK.

Richard Bintanja Institute for Marine and Atmospheric Research, Utrecht University, P.O. Box 80005, 3508 TA Utrecht, The Netherlands.

Gerard Bond Late of Lamont-Doherty Earth Observatory of Columbia University, P.O. Box 1000, 61 Route 9W, Palisades, NY 10964-1000, USA.

Ingmar Borgström Department of Physical Geography and Quaternary Geology, Stockholm University, Stockholm, Sweden.

Geoffrey S. Boulton School of Geosciences, The University of Edinburgh, King's Buildings, Edinburgh EH9 3JW, UK.

Tom Bradwell British Geological Survey, Murchison House, West Mains Road, Edinburgh EH9 3LA, UK.

Roger J. Braithwaite School of Environment and Development, University of Manchester, Manchester M13 9PL, UK

Chris Breemer GeoEngineers Incorporated, 15055 SW Sequoia Parkway, Suite 140, Portland, OR 97224, USA.

Tracy A. Brennand Department of Geography, Simon Fraser University, Burnaby, BC V5A 1S6, Canada.

Joseph Bulat British Geological Survey, Murchison House, West Mains Road, Edinburgh EH9 3LA, UK.

G. Casassa Centro de Estudios Científicos, Valdivia, Chile.

David Chandler Institute of Geography and Earth Sciences, University of Wales, Aberystwyth, SY23 3DB, UK.

Chris D. Clark Department of Geography, The University of Sheffield, Sheffield S10 2TN, UK.

Seth Cowdery Quaternary Research Centre and Department of Earth and Space Sciences, University of Washington, Mail Stop 351310, Seattle WA 98195-1310, USA.

K. M. Cuffey Department of Geography, University of California, Berkeley, 507 McCone Hall, Berkeley, CA 94720-4740, USA.

Paul M. Cutler Polar Research Board, National Academy of Sciences, 500 Fifth Street NW, Washington, DC 20001, USA.

Stephen Davison Department of Geology and Geophysics, University of Edinburgh, Grant Institute, West Mains Road, Edinburgh EH9 3JW, UK.

Julian A. Dowdeswell Scott Polar Research Institute and Department of Geography, University of Cambridge, Lensfield Road, Cambridge CB2 1ER, UK.

Paul Dunlop Department of Geography, NUI Galway, Galway, Ireland.

Paul Duval Laboratoire de Glaciologie et Géophysique de l'Environnement, B.P. 96, 38402 Saint Martin d'Hères Cedex, France.

Mark Dyurgerov INSTAAR, University of Colorado, Campus Box 450, Boulder, CO 80309-0450, USA.

David Elmore Purdue Rare Isotope Measurement Laboratory, Purdue University, West Lafayette, IN 47907, USA.

David J. A. Evans Department of Geography, University of Durham, Science Site, South Road, Durham DH1 3LE, UK.

Derek Fabel Research School of Earth Sciences, Australian National University, Canberra, ACT 0200, Australia.

David Fink AMS-ANTARES, Environment Division, ANSTO, PMB1, Menai, NSW 2234, Australia.

R. Finkel CAMS, Lawrence Livermore National Laboratory, Livermore, USA.

Urs H. Fischer Laboratory of Hydraulics, Hydrology and Glaciology, Swiss Federal Institute of Technology, CH-8092 Zurich, Switzerland.

Sean Fitzsimons Department of Geography, University of Otago, PO Box 56, Dunedin, New Zealand.

Kevin Fleming GeoForschungsZentrum Potsdam, Telegrafenberg, D-14473, Potsdam, Germany.

Sean W. Fleming Department of Earth and Ocean Sciences, University of British Columbia, 6339 Stores Road, Vancouver, B.C. V6T 1Z4, Canada.

David L. Goldsby Department of Geological Sciences, Brown University, Providence, RI 02912, USA.

J. C. Gosse Earth Sciences, Dalhousie University, Halifax, Canada.

J. T. Gray Geographie, Universitie de Montreal, Montreal, Canada.

Sarah L. Greenwood Department of Geography, University of Sheffield, Sheffield S10 2TN, UK.

G. Hilmar Gudmundsson British Antarctic Survey, High Cross, Madingley Road, Cambridge CB3 0ET, UK.

Wilfried Haeberli World Glacier Monitoring Service (WGMS), Department of Geography, University of Zurich, Irchel Winterthurerstrasse 190, CH-8057 Zurich, Switzerland.

Claus U. Hammer Niels Bohr Institute, University of Copenhagen, Geophysical Department, Juliane Maries Vej 30, DK 2100 Copenhagen, Denmark.

Jon Harbor Department of Earth and Atmospheric Sciences, Purdue University, West Lafayette, IN 47907, USA.

W. D. Harrison Geophysical Institute, University of Alaska, Fairbanks, Alaska 99775-7230, USA.

Clas Hättestrand Department of Physical Geography and Quaternary Geology, Stockholm University, Stockholm, Sweden.

Andy Hodson Department of Geography, University of Sheffield, Sheffield S10 2TN, UK.

G. Hoffmann Laboratoire des Sciences du Climat et de l'Environnement, UMR CEA/CNRS, Gif sur Yvette, France.

Alun Hubbard Department of Geography, University of Edinburgh, Edinburgh EH8 9XP, UK.

Bryn Hubbard Centre for Glaciology, Institute of Geography and Earth Sciences, University of Wales, Aberystwyth SY23 3DB, UK.

Philippe Huybrechts Alfred-Wegener-Institut für Polar- und Meeresforschung, Postfach 120161, D-27515 Bremerhaven, Germany.

Neal R. Iverson Department of Geological and Atmospheric Sciences, Iowa State University, Ames, IA 50011, USA.

T. H. Jacka Department of Environment and Heritage, Australian Antarctic Division, Channel Highway, Kingston, Tasmania 7050, Australia.

Krister N. Jansson Department of Physical Geography and Quaternary Geology, Stockholm University, Stockholm, Sweden.

Ian Joughin Polar Science Center, Applied Physics Lab, University of Washington, 1013 NE 40th Street, Seattle, WA 98105-6698, USA.

A. Kaab Department of Geography, University of Zurich, Irchel Winterthurerstr. 190, CH-8057 Zurich, Switzerland.

Georg Kaser Tropical Glaciology Group, Institute of Geography, University of Innsbruck, Innrain 52, A-6020 Innsbruck, Austria.

J. Klein Physics, University of Pennsylvania, Philadelphia, USA.

Johan Kleman Department of Physical Geography and Quaternary Geology, Stockholm University, S-106 91 Stockholm, Sweden.

Lisette Klok KNMI, Postbus 201, 3730 AE De Bilt, The Netherlands.

Debbie Knight School of Physical and Geographical Sciences, Keele University, Keele, Staffordshire ST5 5BG, UK.

Peter G. Knight School of Physical and Geographical Sciences, Keele University, Keele, Staffordshire ST5 5BG, UK.

Wendy Lawson Department of Geography and Gateway Antarctica, University of Canterbury, Christchurch, New Zealand.

C. F. Michael Lewis Geological Survey of Canada Atlantic, Natural Resources Canada, Bedford Institute of Oceanography, Dartmouth, Nova Scotia, Canada B2Y 4A2.

Liu Shiying Cold and Arid Regions Environmental and Engineering Research Institute, Chinese Academy of Sciences, Lanzhou Gansu 730000, China.

David Long British Geological Survey, Murchison House, West Mains Road, Edinburgh EH9 3LA, UK.

R. Lorrain Département des Sciences de la Terre et de l'Environnement, Université Libre de Bruxelles, CP 160/03, 50, avenue F.D. Roosevelt, B-1050 Brussels, Belgium.

Bryan G. Mark Department of Geography, Ohio State University, 1136 Derby Hall, 154 North Oval Mall, Columbus, OH 43210, USA.

Max Maisch Department of Geography, Glaciology and Geomorphodynamics Group, University of Zurich, Winterthurerstrasse 190, CH-8057 Zurich, Switzerland.

Silke Marczinek Hasseer Strasse 10, 24113 Kiel, Germany.

Shawn J. Marshall Department of Geography, University of Calgary, Earth Science 404, 2500 University Drive N.W., Calgary, Alberta, T2N 1N4, Canada.

Danny McCarroll Department of Geography, University of Wales Swansea, Swansea, UK.

Krista M. McKinzey Institute of Geography, School of GeoSciences, University of Edinburgh, Edinburgh EH8 9XP, UK.

Robert T. Meehan 86 Athlumney Castle, Navan, County Meath, Ireland.

Maurine Montagnat Laboratoire de Glaciologie et Géophysique de l'Environnement, B.P. 96, 38402 Saint Martin d'Hères Cedex, France.

Mandy J. Munro-Stasiuk Department of Geography, Kent State University, Kent, Ohio, USA.

Julian B. Murton Department of Geography, University of Sussex, Brighton BN1 9QJ, UK.

Renji Naruse Institute of Low Temperature Science, Hokkaido University, N19W8 Sapporo 060-0819, Japan.

Colm Ó Cofaigh Department of Geography, University of Durham, Science Site, South Road, Durham DH1 3LE, UK.

Byron R. Parizek Department of Geosciences and Earth and Environmental Systems Institute, The Pennsylvania State University, University Park, Pennsylvania, USA.

Frank Paul Department of Geography, Glaciology and Geomorphodynamics Group, University of Zurich, Winterthurerstrasse 190, CH-8057 Zurich, Switzerland.

Antony J. Payne Centre for Polar Observation and Modelling, School of Geographical Sciences, University of Bristol, Bristol BS8 1SS, UK.

Mark Person Indiana University, Department of Geological Sciences, 1001 East 10th St., Bloomington IN 47405, USA.

Erin C. Pettit Department of Earth and Space Sciences, Box 351310, University of Washington, Seattle, Washington 98195, USA.

Jean Robert Petit Laboratoire de Glaciologie et Géophysique de l'Environnement, 54, rue Molière—Domaine Universitaire—BP 96, 38402 Saint-Martin d'Hères cedex, France.

Jan A. Piotrowski Department of Earth Sciences, University of Aarhus, C.F. Mollers Alle 120, DK-8000 Aarhus C, Denmark.

Antoine Pralong Section of Glaciology, Laboratory of Hydraulics, Hydrology and Glaciology, Swiss Federal Institute of Technology, CH-8092 Zürich, Switzerland.

Pu Jianchen Institute of Tibetan Plateau Research, Chinese Academy of Sciences, Beijing 100029, China.

Brice R. Rea Department of Geography and Environment, Elphinstone Road, University of Aberdeen, Aberdeen AB24 3UF, UK.

Niels Reeh Oersted-DTU, Technical University of Denmark, Building 348 Oersteds Plads, DK-2800 Kgs., Lyngby, Denmark.

Eric Rignot Jet Propulsion Laboratory, California Institute of Technology, Mail Stop 300-319, Pasadena, CA 91109-8809, USA.

Ignatius Rigor Polar Science Center, Applied Physics Laboratory, University of Washington, Seattle, Washington, USA.

A. Rivera Department of Geography, University of Chile, Santiago, Chile.

Hazen A. J. Russell Geological Survey of Canada, 601 Booth Street, Ottawa, Ontario K1A 0E8, Canada.

M. Schwikowski Paul Scherrer Institut, Labor für Radio- und Umweltchemie, Switzerland.

Mark C. Serreze Cooperative Institute for Research in Environmental Sciences (CIRES), National Snow and Ice Data Center (NSIDC), University of Colorado, Boulder, Colorado, USA.

David R. Sharpe Geological Survey of Canada, 601 Booth Street, Ottawa, Ontario K1A 0E8, Canada.

John Shaw Department of Earth and Atmospheric Sciences, University of Alberta, Edmonton, Alberta, Canada, T6G 2E3.

Darren Sjogren Earth Science Program, University of Calgary, Calgary, Alberta, Canada.

Mike J. Smith School of Earth Sciences and Geography, Kingston University, Penrhyn Road, Kingston-upon-Thames, Surrey KT1 2EE, UK.

R. Souchez Département des Sciences de la Terre et de l'Environnement, Université Libre de Bruxelles, CP 160/03, 50, avenue F.D. Roosevelt, B-1050 Brussels, Belgium.

Chris R. Stokes Landscape and Landform Research Group, Department of Geography, The University of Reading, Reading RG6 6AB, UK.

Martyn S. Stoker British Geological Survey, Murchison House, West Mains Road, Edinburgh EH9 3LA, UK.

John O. H. Stone Quaternary Research Centre and Department of Earth and Space Sciences, University of Washington, Mail Stop 351310, Seattle, WA 98195-1310, USA.

Arjen P. Stroeven Department of Physical Geography and Quaternary Geology, Stockholm University, Stockholm, Sweden.

David Sugden School of GeoSciences, The University of Edinburgh, The Grant Institute, The King's Building, Edinburgh EH9 3JW Scotland, UK.

Shin Sugiyama Section of Glaciology, Versuchsanstalt für Wasserbau Hydrologie und Glaziologie, ETH, Gloriastrasse 37/39, CH-8092 Zürich, Switzerland.

Darrel A. Swift Department of Geographical and Earth Sciences, East Quadrangle, University of Glasgow, Glasgow G12 8QQ, UK.

James T. Teller Department of Geological Sciences, University of Manitoba, Winnipeg, Manitoba, Canada R3T 2N2.

Throstur Thorsteinsson Institute of Earth Sciences, University of Iceland, Sturlugata 7, 101 Reykjavik, Iceland.

Claire Todd Quaternary Research Center and Department of Earth and Space Sciences, University of Washington, Mail Stop 351310, Seattle WA 98195-1310, USA.

Martyn Tranter Bristol Glaciology Centre, School of Geographical Sciences, University of Bristol, Bristol BS8 1SS, UK.

M. Truffer Geophysical Institute, University of Alaska, Fairbanks, Alaska 99775-7230, USA.

Slawek Tulaczyk Department of Earth Sciences, University of California, Santa Cruz, Santa Cruz, CA 95064, USA.

David G. Vaughan British Antarctic Survey, High Cross, Madingley Rd., Cambridge, CB3 0ET, UK.

Andreas Vieli Centre for Polar Observation and Modelling, School of Geographical Sciences, University of Bristol, Bristol BS8 1SS, UK.

F. Vimeux Institut de Recherche pour le Développement, UMR Great Ice IRD, Paris, France.

J. L. Wadham Bristol Glaciology Centre, School of Geographical Sciences, University of Bristol, Bristol BS8 1SS, UK.

Richard I. Waller School of Physical and Geographical Sciences, Keele University, Keele, Staffordshire ST5 5BG, UK.

G. Yang Earth Sciences, Dalhousie University, Halifax, Canada.

Yao Tandong Institute of Tibetan Plateau Research, Chinese Academy of Sciences, Beijing 100029, China.

ONE

Glacier science and environmental change: introduction

Peter G. Knight

School of Physical and Geographical Sciences, Keele University, Keele ST5 5BG, UK

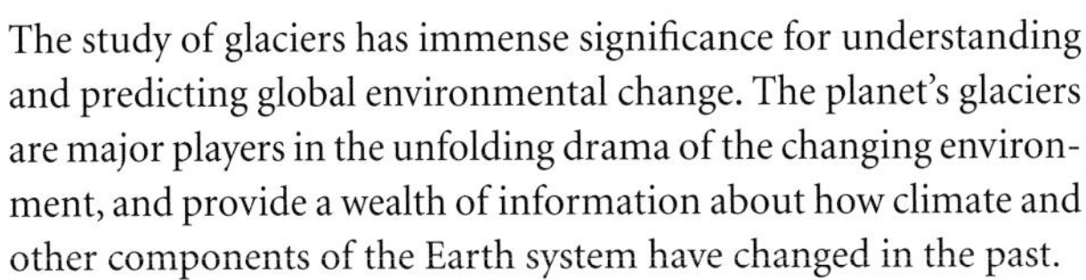

The study of glaciers has immense significance for understanding and predicting global environmental change. The planet's glaciers are major players in the unfolding drama of the changing environment, and provide a wealth of information about how climate and other components of the Earth system have changed in the past.

Scientists from different fields have begun to come together in their common interest in glaciers and the Earth's changing environment, and to recognize the increasing importance of interdisciplinary understanding in this area. The rate and scale of progress, however, has meant that researchers and students in fields such as glaciology, Quaternary studies, sedimentology and environmental science find it more and more difficult to keep abreast of the subject as a whole, and to recognize the key issues in areas outside their own specialism.

The purpose of this book is to provide a picture of current scientific understanding of key issues that relate the study of glaciers to the broader field of environmental change. The book presents not just the established common ground of the science, but also some of the uncertainty and controversy that accompanies progress in a dynamic and contested discipline. Contributors were invited to offer their personal perspectives on important topics, and where controversy emerged it has been highlighted as an indication of where the frontier of the science currently stands. For example, on controversial topics such as the role of meltwater in the creation of subglacial landscapes or the mechanisms of ice deformation within ice sheets, papers on the same topic by authors with contrasting views have been placed side by side so that the reader can judge the opposing arguments. Where authors, referees and editor have disagreed about particular issues, such as the naming of Antarctic ice streams or interpretations of data, the authors' preferences have wherever possible been allowed to prevail. This book has not been edited to present a consensus, but compiled to provide a snapshot of what different figures within the discipline consider to be important. The book thus provides:

1 an authoritative interdisciplinary compilation, accessible to both students and professionals, of the issues that are drawing together the research efforts of glaciologists and other scientists such as geologists, hydrologists, and climatologists who are concerned with how the global system responds to environmental change;
2 state-of-the-art reviews, and personal perspectives, from some of the world's foremost authorities, concerning key topics within glaciology and at the interface between glaciology and related disciplines in environmental change;
3 cutting-edge case studies by researchers from the scientific frontier where the conventional wisdom of current approaches comes face to face with unsolved problems.

Each of the book's five sections includes a keynote introduction, a series of articles reviewing particularly significant areas of the discipline, and a number of research case-studies relating to topics discussed in the review articles and keynote. The keynote introduction to each section is written by a senior figure within the discipline, providing a personal perspective on the fundamental issues that bring significance to the section and a broad context for the papers that follow. Each keynote is followed by a series of articles by leading authorities covering themes of major contemporary significance in the discipline. Distributed between these articles are shorter papers that provide research case studies that illustrate, or provide a counterpoint to, issues discussed or opinions promoted in the keynotes and reviews. Some of these short case studies are written by senior figures with established reputations, whereas others have been contributed by more junior researchers providing alternative perspectives on traditional approaches. Each section thus comprises a hierarchy of keynote, reviews and case studies, and a hierarchy of elder statesmen, established researchers and relative newcomers. The core of the volume is provided by the reviews, the integration of these reviews is achieved via the keynotes, and additional elaboration, illustration and debate is provided by the case studies. As well as the colour-plate section, selected figures are provided in colour at **www.blackwellpublishing.com/knight**

The papers in this volume are not intended to be exhaustive accounts, nor to reproduce introductory summaries of the subject that can be found in standard textbooks, but rather to present a picture of the issues that the discipline is currently engaged with, and to provide a starting point for further study. Most importantly this book provides a statement of what some of the leading figures in the field believe to be the most pressing issues in contemporary research in a discipline that concerns the planet's history, its present and its future.

PART 1

Glaciers and their coupling with hydraulic and sedimentary processes

TWO

Glaciers and their coupling with hydraulic and sedimentary processes

Geoffrey S. Boulton

School of Geosciences, University of Edinburgh, Kings Buildings, Edinburgh EH9 3JW, UK

2.1 Introduction

2.1.1 The sources of evidence

Understanding glaciers and their role in the Earth system demands both an understanding of the way in which their properties are organized in the modern time plane, and how they change through time. The time dimension is important because of the long lag times (10^2–10^3 yr) that may be required for dynamic changes to spread through the system and the possibility that there have been markedly different ice-sheet regimes through time (e.g. Clark, 1994).

In understanding the behaviour of glaciers in time and space, particularly the ice sheets that are by far the largest and climatically most influential part of the global glacier mass, it is important to combine evidence from two sources:

1 *from modern glaciers*, where we can directly measure properties and processes, and determine the magnitudes of pressures, forces and flow rates, but with the limitations that observations of time dependence have been restricted to a period of serious scientific study of little more than 50 yr, that vital processes at the bed are difficult to observe, except at the margin and through limited borehole tests, and that it is difficult to assess the representativeness of these latter observations;

2 *from deglaciated terrain*, where we can characterize the sedimentary and geomorphological character of former glacier beds, and can create a partial chronology of glacier variation, but with the limitations that we have to guess the processes, infer them from the sediments, or use analogues from modern glaciers to account for them. The strongly polarized debate between Shaw & Munro-Stasiuk on the one hand and Benn & Evans on the other (this volume, Chapter 8), about the origin of drumlins, exemplifies the problem when speculation is relatively unconstrained by definitive evidence.

2.1.2 Ice-sheet coupling to the Earth system

An ice sheet is coupled into the Earth system (Fig. 2.1a) across its interfaces with the atmosphere, the ocean and the lithosphere.

1 *Ice sheet–atmosphere coupling.* The atmospheric state (temperature, moisture content, energy transport) influences an ice sheet through:
- its impact on mass balance and ice temperature.

An ice sheet influences the atmosphere through:
- its high albedo;
- the deflection of atmospheric flow over the ice sheet, which influences the distribution of temperature, pressure and precipitation and thereby the mass balance distribution.

2 *Ice sheet–ocean coupling.* An ice sheet influences the ocean through:
- the discharge of meltwater (whether or not the glacier reaches the sea), which influences the temperature, salinity and turbidity of nearby ocean water;
- the discharge of icebergs from marine margins, which influences the temperature and albedo of the nearby ocean.

The ocean influences an ice sheet at marine margins through:
- water temperature, which influences ablation from ice shelves and tidewater margins;
- water depth and depth variation, which influence the buoyancy of a marine margin and subglacial water pressure in the terminal zone, and thereby its susceptibility to calving and to fast flow (marine drawdown and streaming);
- wave action, which influences calving through its influence on the extent of sea-ice and the action of waves against the ice front.

3 *Ice sheet–lithosphere coupling.* An ice sheet influences the lithosphere through:
- its mass, which is able isostatically to depress and flex the lithosphere as a consequence of flow induced in the Earth's mantle by differential loading;

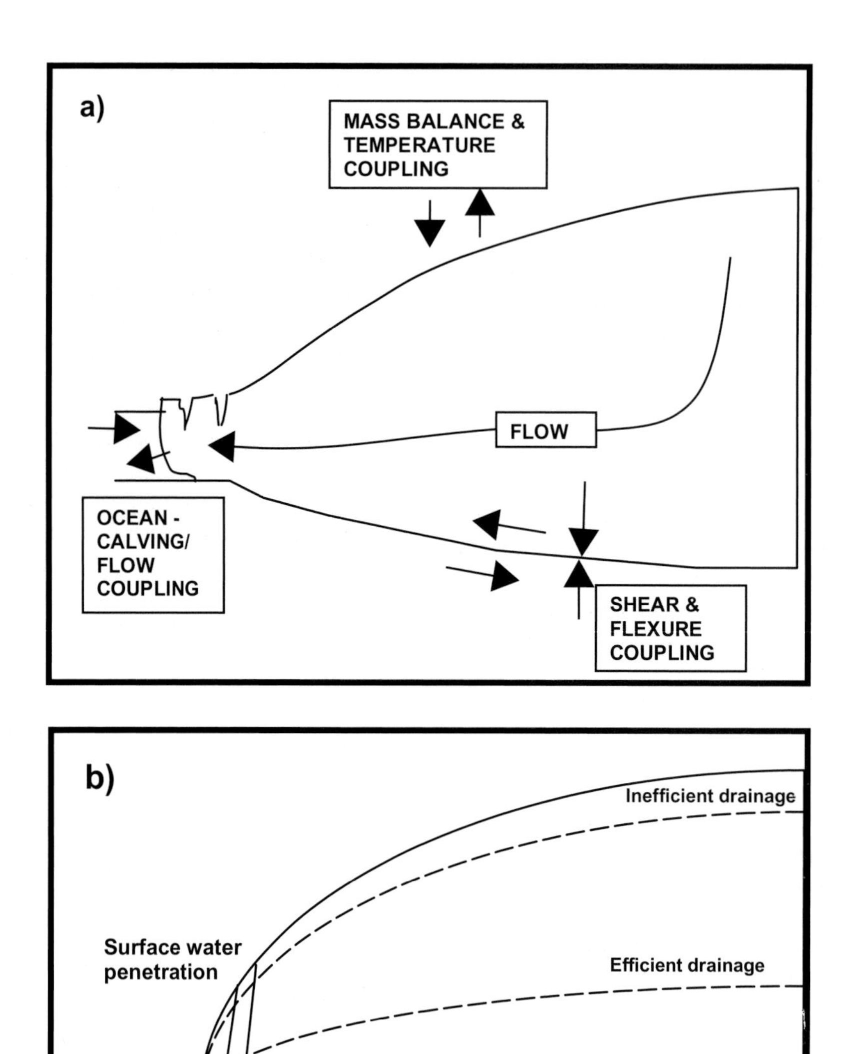

Figure 2.1 (a) Coupling at the surface, basal and marine boundaries of an ice sheet. (b) The ice-sheet drainage system, and the impact of drainage efficiency on generalized subglacial heads.

- meltwater, which is injected into the bed, so driving up pore fluid pressures;
- shear forces and flow that produce erosion of the bed, sediment transport over it, deposition on it, and deformation of bed materials;
- the effect on the thermal field in the lithosphere.

The lithosphere influences an ice sheet through:

- vertical movement of the ice-sheet mass through isostatic sinking or uplift with consequences for surface mass balance;
- the impact of water pressure (and therefore effective pressure) on friction, possible decoupling at the ice–bed interface and thereby on ice-sheet dynamics and form;
- the basal thermal regime, which influences the state of freezing or melting at the ice–bed interface and thereby basal friction and the nature of basal movement.

Ice-sheet surface processes, the transformation of snow to ice, the surface energy balance and mass balance in relation to surface meteorology are now well understood, quantified and used in a relatively sophisticated way in ice sheet models (e.g. Huybrechts, 1992; Payne, 1995) that have been successful in replicating current features of modern ice sheets and their future evolution (e.g. Huybrechts *et al.*, 1991b), although full time-dependent coupling between an ice sheet and a general circulation model of the atmosphere remains computationally taxing. Outstanding problems at the marine boundary include the development of a theory of iceberg calving and of ice shelves in understanding ocean–ice-sheet coupling, crucial to understanding the powerful iceberg fluxes associated with Heinrich events of the past and of the future evolution of the marine margins of modern ice sheets (e.g. Hindmarsh & Le Meur, 2001). The major problem of the basal boundary is the nature and magnitude of the coupling at the ice–bed interface determined by interactions between subglacial drainage, thermodynamics and basal friction, and how these processes are reflected in the subglacial sedimentary record.

2.1.3 The problem of the basal boundary

The nature of the coupling at the basal boundary is one of the major unresolved problems of glaciology. Ice, with a rheology defined by Glen's law (where the strain rate $\dot{E} = k\tau^n$, where k is a temperature dependent constant, τ is shear stress and n has a value of between 3 and 4), can be regarded as a perfectly plastic solid with a yield strength of about 100 kPa. If the ice–bed interface is a strong interface, the average shear stress (given by $\rho_i gh \sin\alpha$; where ρ_i is the density of ice, g the gravitational acceleration, h the ice thickness and α the surface slope) at the base of a glacier must have this latter value to drive the necessary flow. If, however, frictional interlocking at the ice–bed interface is weak, so that the frictional resistance mobilized at the interface is significantly less than 100 kPa, ice movement will be dominated by low stress décollement at the glacier sole, and there will be relatively little internal flow in the ice. Some modern ice streams flow continuously at shear stresses of as little as <15 kPa (Whillans & Van der Veen, 1997), and reconstructions of former Pleistocene ice sheets also indicate flow under similarly low shear stresses (Mathews, 1974). Currently, ice-sheet models define the basal condition in a way that does not depend upon physical understanding of the processes at the boundary. It is generally derived as a fitted parameter. Consequently, confidence is low that models have the capacity to predict, rather than merely explain, behaviour that depends on basal processes, such as streaming or surging.

A central issue for ice-sheet glaciology therefore is whether there is strong or weak coupling at the ice–bed interface, and the processes that determine this. The parallel issue for glacial geology is how the nature of coupling is reflected in the erosional or depositional products of glaciation. The degree of coupling and the magnitude of traction are determined by two components, the solid interactions that take place between the glacier ice and its bed, and the water pressures that are determined by the nature of drainage. If drainage is efficient, water pressures will be drawn down strongly, effective pressures will be high, and basal friction will tend to be at a maximum (Fig. 2.1b). If drain-age is inefficient, the converse will apply. Alternatively, the system may be self-organized in a way that favours a particular outcome.

The rest of this article is therefore concerned with processes at the basal boundary, and attempts to integrate modern glaciological evidence with the glacial geological and geomorphological evidence from the beds of former ice sheets.

2.2 The nature of the bed

Prior to the past 25 yr, glaciologists had implicitly assumed that glaciers flowed over hard, lithified and impermeable beds. Thus, any décollement at the ice–bed interface must be by ice sliding over a rigid, rock surface. Weertman (1957) suggested that sliding at the ice–bed interface must be associated with an intervening thin water layer, which would preclude tangential friction, and that the basal drag must be generated through movement by plastic flow and pressure melting/regelation around obstacles on the bed. Much of this conceptual frame has been confirmed by direct field observation (e.g. Kamb & LaChapelle, 1964). However, Boulton *et al.* (1979) conducted direct stress experiments beneath a glacier that showed that debris embedded in basal ice contributed a large, tangential, component of the drag at the base of the glacier (recent experiments by Iverson *et al.* (2003) have produced similar results). In this setting, water from basal melting is assumed to flow in a thin layer between the ice and the bed until it is able to reach subglacial drainage channels (e.g. Röthlisberger, 1972; Weertman, 1972). In 1979, Boulton & Jones showed that soft, unlithified sediments beneath a modern glacier were undergoing shear deformation at such a rate that they accounted for a large proportion of the forward movement of a glacier. Fischer & Clarke (1997) observed that slip at an ice–sediment surface occurred during periods of high water pressure, whereas shear movement was concentrated in the sediments during periods of lower water pressure. Boulton (1974) had drawn attention to the potential for groundwater flow to influence the effective pressure at the ice–bed interface, and Boulton *et al.* (1993, 1995a, 1996) and Piotrowski (1997a,b) showed that a large proportion of basal meltwater could be discharged in permeable subglacial beds rather than being restricted to flow in a water layer or in channels at the ice–bed interface. This intergranular drainage is clearly a vital process in determining effective pressures, the threshold for deformation in unlithified subglacial sediments, and shifts in the location of décollement between the ice–bed interface and underlying sediments. These insights are the basis for our understanding of sedimentary processes at the ice–bed interface and nature of the basal boundary condition.

The following framework is used to discuss processes at the ice–bed interface:

1 *Rock beds.* Basal ice, which may include sedimentary debris, is in direct contact with a hard, lithified rock surface. This is a common state in many glaciers in steep-sided mountain valleys, and has been observed on a number of occasions in artificial subglacial tunnels (e.g. Kamb & LaChapelle, 1964; Vivian & Bocquet, 1973).
2 *Thin till on rock.* Thin lenses or sheets of till (<10 m thick) are frequently found on predominantly rock beds in valley glaciers and over large areas of the Precambrian Shields that formed the core areas of the North American and European ice sheets.
3 *Thick unlithified sediment beds.* Beds of this type occur around the margins of the North American and European pre-Cambrian Shields, and on many low lying coastal plains onto which high-latitude glaciers flow, such as in southern Iceland and western Alaska.

Although we are currently only able to assess the nature of beds beneath the modern ice sheets of Greenland and Antarctica over a tiny proportion of their area, seismic surveys and drilling in the areas of a number of Antarctic ice streams have revealed an underlying soft bed (e.g. Alley *et al.*, 1987a—Ice Stream B, West Antarctica; Smith, 1997a—Rutford Ice Stream, Antarctica). These ice streams flow at high velocities under driving stresses as low as 10 kPa, and it is assumed that the soft bed, presumed to be till, is implicated in the fast flow process. The nature of these beds, and the mosaic of bed types that make up the whole bed of an ice sheet, almost certainly play a fundamental role in determining many large-scale dynamic aspects of ice-sheet behaviour. We now examine the role of the bed in relation to the key hydraulic processes which determine the effective pressure regime and the nature of the frictional coupling between glacier and bed.

2.3 Subglacial hydraulics

Water pressures at the base of a glacier are controlled by the rate of recharge of water into the subglacial system, and the efficiency with which it drains from it (Fig. 2.1b). Efficient drainage will be associated with relatively low water pressure gradients and low water pressures. Inefficient drainage will be associated with relatively high water pressure gradients and high water pressures.

2.3.1 Recharge to the bed

Basal water is derived from two sources (Fig. 2.1b): from melting of the glacier sole where this is at the melting point, at a rate determined by the heat supply from the geothermal flux and frictional heating by the moving glacier sole, and water draining from the surface down moulins, crevasses and other conduits. They are of very different magnitudes and distributions. The rate of basal melting is small, of the order of 10^{-3}–10^{-1} m yr^{-1}, but sustained through the year. A glacier 10 km in length would generate an insignificant two-dimensional, flowline discharge of about 3×10^{-6}–10^{-4} m^2 s^{-1}, producing a three-dimensional discharge of 0.03–3 m^3 s^{-1} from a catchment 10 km wide. However, an ice sheet with a 1000 km melting flowline would generate a two-dimensional discharge of about 3×10^{-4}–10^{-5} m^2 s^{-1} along a flowline and a three-dimensional discharge of about 3–30 m^3 s^{-1} from a catchment 10 km wide.

The flux to the bed from surface water can be very large. Boulton *et al.* (2001a) estimate a flux to the bed in the terminal zone of an Icelandic glacier of about 0.1 m day^{-1} in summer. If such a flux were only to penetrate to the bed of the glacier in the terminal 10 km, it would yield a two-dimensional flowline discharge of about 0.01 m^2 s^{-1} at the terminus and a three-dimensional discharge of about 100 m^3 s^{-1} from a catchment 10 km wide. This flux is generated only in the summer season. As a consequence, subglacial water pressures show a strong seasonal oscillation, in addition to a diurnal component.

A key question is how far from the glacier terminus can surface water penetrate to the bed, and therefore the extent of the zone in which annual fluctuations (and possibly diurnal fluctuations) of basal water pressure will occur. Arnold & Sharp (2002) assumed that surface meltwater was able to reach the bed of the European ice sheet over most of its area.

2.3.2 Discharge from the bed

Studies of modern glaciers suggest that by far the largest component of discharge of water from a glacier terminus is carried by turbulent streams that occupy so-called R-channels: channels that rest on the bed and arch up into the overlying ice (Röthlisberger, 1972). They are highly efficient means of discharging the subglacial water flux. Water pressure gradients along them are low, and water pressures are much lower than ice pressures (e.g. Fountain, 1994) because of the efficiency with which turbulent heating melts the tunnel walls and prevents ice flow towards the tunnel from strongly pressurizing tunnel water.

The key question is, how does water find its way into these efficient discharge conduits from the locus of melting or recharge of surface water to the bed, and how does pressure drop along the pathway?

2.3.3 A rock bed

Rock beds have been assumed to be impermeable. The classical view is that water will find its way into channels through a water film at the ice–bed interface driven by a pressure gradient that is given by the gradient of ice thickness (Walder, 1982). It has also been demonstrated that flow into major R-channels can take place through complex linked cavity systems that exploit the natural cavities in the lee of bedrock hummocks and can shrink or enlarge diurnally and seasonally to accommodate changing water fluxes (e.g Kamb, 1987). Any discontinuous lenses of till resting on the rock surface will have relatively little effect on the hydraulic system, and will have porewater pressures that are the same as the ambient pressures.

However, impermeable rock is a concept rarely realized in nature. Even the Precambrian rocks of the Fennoscandian shield commonly have hydraulic conductivities in the topmost 100 m or so of about 10^{-6}–10^{-7} m s^{-1} (largely determined by flow along frequent fractures), giving transmissivities of the order of 10^{-4}–10^{-5} m^2 s^{-1} (e.g. Gustafsson *et al.*, 1989). Comparing this with the basal melt fluxes in short valley glaciers (see section 2.3.1) suggests that in some cases groundwater flow can be an effective means of transporting at least the winter flux of basal meltwater towards major subglacial conduits.

2.3.4 Till on a rock bed

In cases when the rock is of low permeability compared with the till, the melt flux must be discharged through the till to the nearest low-pressure channel. Even tills in old, hard Shield areas, when in a non-dilatant state, tend to have hydraulic conductivities of the order of 10^{-6}–10^{-8} m s^{-1} (e.g. Engquist *et al.*, 1978). For a basal melt rate of the order of 10^{-9} m s^{-1}, a till 1 m thick with a conductivity of 10^{-7} m s^{-1} would be able to discharge the winter melt towards a low pressure channel with a head gradient low enough to ensure that pressures did not exceed ice pressures, provided there were efficient tunnel conduits spaced at about 2 km intervals. If the till was shearing and dilating to a depth of 0.5 m, with the conductivity increased by an order of magnitude in the dilatant horizon (see Piotrowski, this volume, Chapter 9), the spacing could be five times larger, and even larger if significant water fluxes were advected with the deforming till.

If significant quantities of summer meltwater were to reach the bed from the glacier surface, the necessary spacing of channels would be much reduced to the order of tens of metres. There is therefore a strong probability that an extreme oscillation of the system between winter and summer would produce an annual cycle of channel formation, from a winter condition of large, widely spaced channels, to a summer condition with a closely spaced network of channels. Boulton *et al.* (2001b) have suggested that beneath an ice sheet, the winter channels are stable, long lived R-channels that give rise to eskers. It is possible that the closely spaced network of summer channels could be the slowly flowing, high water pressure 'canals' of Walder & Fowler (1994), which would be short-lived, and thereby unable to integrate themselves and develop into an efficient low pressure channel system able to draw down water pressures over wide areas as well-developed R-channels. At locations remote from the glacier margin, they would be likely to be tributary to major R-channels.

The water pressure gradient in a till overlying low permeability bedrock will tend to be horizontal (Boulton & Dobbie, 1993), directed towards major low-pressure R-channels acting as sinks for groundwater flow. However, where there is a well connected fracture system in bedrock, such that water flow along interconnected fractures is able to draw down water pressures, there may be a strong, local downward water pressure gradient, sufficient to liquefy the till and force it into bedrock fractures. The results of such a process are commonly seen in bedrock areas, where till 'dykes' can penetrate many metres into bedrock (Fig. 2.2a). It is also suggested that this process of 'hydraulic intrusion' can lead to forcing apart of bedrock joint blocks by widening pre-existing or incipient fractures so as to break up the rock mass and make it more susceptible to erosion (Fig. 2.2b & c).

2.3.5 A thick unlithified sediment bed

Figure 2.3 shows an experimental site at the glacier Breidamerkurjökull in Iceland which was overrun by a 'minisurge' of part of the glacier margin (Boulton *et al.*, 2001a). The developing surge was anticipated some two months before it occurred, and prepared for by trenches dug in the forefield of the glacier to a depth of up to 2.5 m. They revealed a till unit 1–3 m thick overlying the sands and gravels of an extensive aquifer of up to about 80–90 m in thickness that occurs over a wide area in the vicinity of the glacier margin. Water pressure transducers were emplaced in the till and in the underlying aquifer, and monitored during the glacier advance. Figure 2.4 shows results from the transducers at 65 m along the transect. They indicate the prevalence of a strong, downwardly directed water pressure gradient, reflecting downward flow of water from the glacier sole through the till into the underlying aquifer. The maximum pressure drop across the till is about 90 kPa m^{-1}. The aquifer controls the water pressure at the base of the till, and the thickness and hydraulic conductivity of the till control the water pressure at the top of the till. It is clear that the greater the thickness of the till, the larger will be the water pressure and the lower the effective pressure in its topmost horizon for a given aquifer pressure. This has important implications for the evolution of mechanical coupling (see section 2.4).

In general, a thick bed of unlithified sediments, such as that described by Piotrowski (this volume, Chapter 9) from northern

Figure 2.2 Injection of till into jointed bedrock and incorporation of bedrock blocks into deforming till. (a) Till-filled bedrock joints at 6 m below the bedrock surface near Forsmark, Sweden. (b) Partially detached bedrock blocks in till at the bedrock surface, near Uig, Skye, Scotland. (c) Some detached bedrock blocks at the same locality are completely incorporated in the till as a result of till deformation. Others are partially detached.

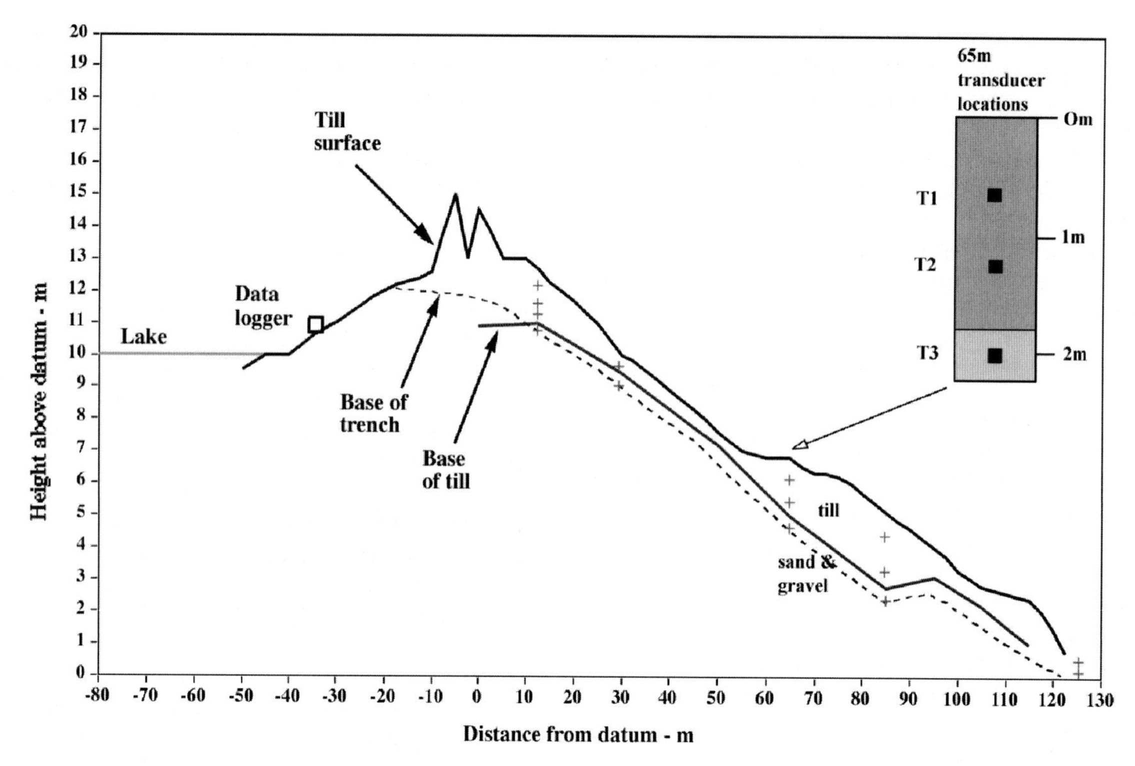

Figure 2.3 Site at the margin of Breidamerkurjökull, Iceland, overrun by a minisurge (Boulton *et al.*, 2001a). Water pressure transducers (marked by crosses) were emplaced in the walls of a trench dug prior to the minisurge. The glacier advance extended as far as the 0 m datum along the transect. Examples of the water pressure gradients at 65 m from datum are shown in Fig. 2.4.

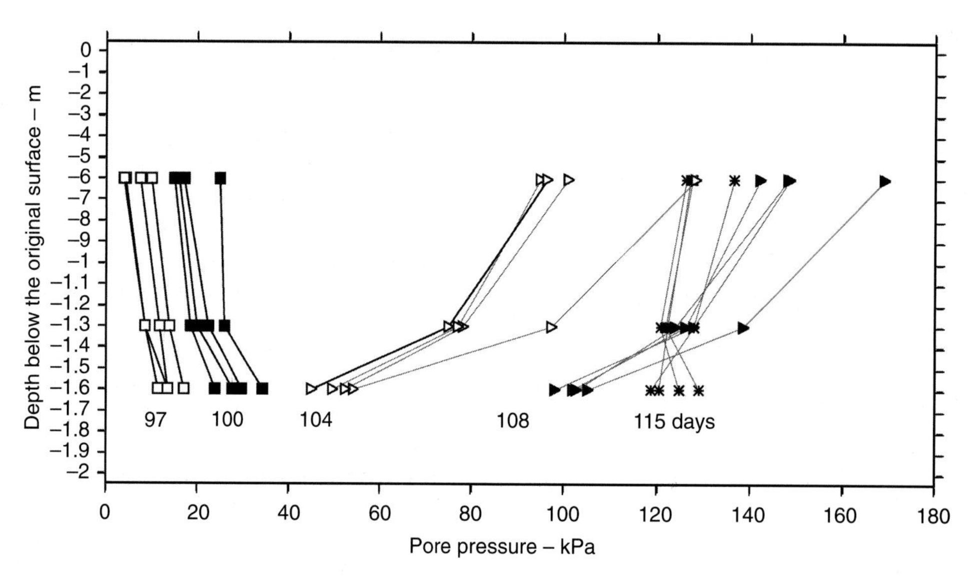

Figure 2.4 Water pressures in the till at 65 m along the transect (Fig. 2.3) at 6-h intervals on selected days. Early gradients (days 97, 100) reflect upward flow of water through the till just beyond and at the advancing glacier margin as water flowing outwards from beneath the glacier in the subtill aquifer wells upwards towards the surface. Later gradients (days 104, 108, 115) reflect downward movement of water through the till from the glacier sole. Some reversals of gradient occur on day 115 as aquifer pressures exceed till pressures. Maximum gradients tend to occur in the late afternoon during periods of peak glacier surface melting.

Germany, will contain both more and less conductive strata. In most cases, irrespective of the precise ordering of the pre-glacial strata, we would expect a relatively low conductivity till to be created by the glacier to overlie the pre-existing sequence, and that this till would have a significantly lower conductivity than those on the shield (clay-silt tills from lowland areas typically have conductivities in the range 10^{-8}–10^{-9} m s^{-1} compared with 10^{-7} m s^{-1} for the Iceland till). If such a till were to be inserted into the Icelandic setting, we would expect a pressure drop across the till that was one to two orders of magnitude greater (1000–10,000 kPa m^{-1}).

These results have several important implications.

1. The hydrological properties of a glacier bed and the operation of the subglacial hydraulic system must play a fundamental role in governing the frictional drag offered by the bed to glacier movement. Much existing work has focused on the role of channels and a water film. The role of groundwater flow is at least of similar importance.
2. If water is forced from the glacier sole through a till of very low hydraulic conductivity, a very large water pressure gradient will develop. Such a gradient may be large enough to create hydrofracturing, resulting in an increase in conductivity. Under these circumstances, conductivity is not an intrinsic property of the till, but a product of its hydraulic setting and granulometry.
3. As a till thickens as a consequence of deposition, the pressure difference across it will tend to increase, so that the effective pressure at the top of a till accumulating above a subglacial aquifer with a constant water pressure will tend to reduce. The till will therefore offer progressively less shearing resistance to the overriding glacier, thereby facilitating faster flow.
4. A very large flux of meltwater can drain as groundwater through a subglacial aquifer. Low pressures in the aquifer (strong pressure drawdown) can co-exist with water pressures at the top of the overlying till that approach the ice pressure.
5. Local variations in the effective pressure in a till are primarily dependent upon the drainage pathway and the water flux rate. The drainage pathway can change as a consequence of reorganization of channels that act as groundwater sinks, and the flux rate can vary diurnally and seasonally.
6. The maximum effective pressure that a till has suffered, without subsequent remoulding, determines its consolidation state. We expect this to reflect the periods of strongest drainage and smallest meltwater flux.
7. It is important to establish how far from an ice-sheet margin there are strong fluctuations in pressure generated by diurnal and annual recharge changes in surface water recharge to the bed.

2.4 Tractional processes

Because of the influence of basal friction on the large-scale dynamics of ice sheets (section 2.1.3), the strength of coupling between the basal ice and its bed is a key issue for glaciology. The parallel issue for glacial geology is how the state of traction is reflected in the erosional or depositional products of glaciation.

The traction at the base of a glacier depends upon the temperature at the ice–bed interface, the effective pressure at the interface, which determines the degree of interlocking between the glacier sole and the bed, and the strength of subsole materials, whether lithified or unlithified. These vary according to the lithology of the substratum, in which three principal types can be identified.

2.4.1 A frozen bed

Direct measurements and computer simulations of modern ice sheets show that the ice–bed interface tends to be at temperatures well below the pressure melting point in the central, divide regions, in the near terminal zone, where this terminates on land, in the summit regions of hills and mountains beneath the ice sheet where the ice is locally thin, and beneath slowly flowing ice-sheet zones between ice streams. Christoffersen & Tulaczyk (2003) have shown how the ice–bed interface can freeze when ice streams stagnate, potentially leading to processes of sediment incorporation by freezing in, as discussed by Weertman (1957) and Boulton (1972). Goldthwait (1960) produced evidence from northwest Greenland which shows that a frozen sediment bed can be largely uneroded by the glacier sole, and Holdsworth (1974) drew similar conclusions from study of a subglacial rock bed in Antarctica.

The rationale for non-erosion is provided by the results of Jellinek (1959), who showed that the adhesive strength of an ice–rock interface at temperatures well below the melting point was significantly greater than the maximum average shear stress that is normally generated at the base of an ice sheet. Under these conditions, we expect sliding to be inhibited at the ice–bed interface and the movement of ice sheets to be dominated by internal flow, with basal shear stresses of the order of 100 kPa. However, sliding can occur at temperatures lower than the ambient pressure melting point (Shreve, 1984; Echelmayer & Wang, 1987), and can locally abrade bedrock (Atkins *et al.*, 2002), leading to debris incorporation in basal ice (Holdsworth, 1974), and deformation of any frozen subglacial sediments either by brecciation and block incorporation (Boulton, 1979) or by ductile deformation (Davies & Fitzsimons, 2004). Although local stress–strength relationships may permit significant debris or sediment blocks to be torn away from the substratum, in general cold-bed erosion rates are likely to be very small compared with those achieved by a rapidly sliding glacier sole.

Very large frozen sediment and rock masses can be transported by ice if impeded, inefficient drainage beneath a subglacial frozen horizon creates high water pressures beneath a subglacial horizon. Provided that there is strong adhesion between the glacier sole and underlying frozen materials, high water pressures at the base of the frozen horizon can reduce the effective pressure and the friction to such a low value that sliding can occur along the frozen–unfrozen interface. It has been suggested that this process could permit very large sediment or rock masses to move with the glacier (Weertman, 1957; Mackay & Matthews, 1964; Moran, 1971; Boulton, 1972) and may explain some of the large, relatively undisturbed, pre-existing rock or sediment masses that are found in glacial till sequences (e.g. Christiansen, 1971) and many very large glacier tectonic structures. Permafrost effectively becomes part of the glacier, with the basal décollement beneath rather than above it.

2.4.2 An unfrozen rock bed

This is the classic domain for studies of glacier erosion because much early glaciological study was concerned with temperate, rock-floored, valley glaciers. The classic conception is of erosion processes dominated by 'plucking', where the ice prises away rock tools from the glacier bed, and abrasion in which these tools, embedded in the glacier sole, scratch and abrade the substratum to produce striated, smoothed and streamlined landforms. The debris generated by plucking and abrasion itself breaks down to produce a characteristic 'crushing distribution' of grain sizes (Haldorsen, 1981), and is transported away by inclusion in a basal ice layer by processes of relegation (Sharp *et al.*, 1989). In temperate glaciers, this debris-rich basal ice layer is typically of the order of centimetres in thickness, with concentrations of the order of 10–40% by volume (Table 2.1).

Implicit in the various approaches to abrasion of the bed by sliding ice is that there is a strong contrast between a state in which the glacier sole consists of a continuous debris carpet, and one where clasts are well separated in relatively clean ice. In the former case, a 'sandpaper' model is appropriate, where friction and abrasion rate are a function of effective pressure at the ice–bed interface, the driving stress and ice velocity (Drewry, 1986). Occasional large clasts that penetrate through the basal debris carpet because of the way in which they concentrate stress (Boulton *et al.*, 1979) will create larger deeper grooves in bedrock. In the latter case, the flow of ice around sparse, frictionally retarded clasts in traction over the bed will determine the abrasion rate (Boulton, 1974; Hallet, 1979). In the former mode, the tangential friction generated by a continuous basal debris carpet is a significant contribution to the resistance to glacier movement (section 2.2). The sandpaper mode is likely to produce greater rates of erosion than the sparse mode, generating even more debris, potentially clogging the basal transport system and thereby depositing subglacial till.

There has not, as yet, been an effective test for an abrasion law. Bedforms and glaciated valley profiles seem to provide only a weak test, as a wide variety of laws is able to generate typical erosional forms (Harbor, 1992). One of the difficulties in modelling valley glaciers and the overdeepening they produce has been the absence of negative feedback that would prevent the abrasion process from continuing to overdeepen the bed locally. Alley *et al.* (2003) have suggested that negative feedback can be provided by the triggering of a supercooling mechanism as overdeepened slopes increase, a mechanism that is suggested to inhibit erosion and promote deposition.

2.4.3 An unfrozen unlithified sediment bed—glaciological implications

A large proportion of the beds that directly underlay the soles of mid-latitude ice sheets of the last glacial period were composed of thick unlithified sediment sequences rather than rock. Soft beds have been the subject of much research in the past 25 yr since their significance was first recognized. They have proved to be phenomenologically rich, and as a consequence are still a very active source of research and debate. It is thus important to register what is known, what has been speculated and what the principal remaining problems are in determining their role in the traction at the bed of a glacier and in the generation of till. (The term till is frequently misused in the glaciological literature to refer to actively deforming subsole material. The term should be restricted to a deposited sediment, not to sediment in transport. Deforming subsole material is here referred to as deforming sediment, the deforming horizon or the deforming subsole nappe.)

1 On every occasion that boreholes through or excavations beneath modern glaciers have found unlithified sediment immediately below the glacier sole and measurements have been made that are able to detect deformation, they have done so. In the majority of cases, they report the thickness of the deforming horizon to be no greater than about 0.3–0.65 m (Boulton & Hindmarsh, 1987, 0.38–0.45 m; Blake, 1992, 0.3 m; Humphrey *et al.*, 1993, 0.65 m; Iverson *et al.*, 1994, 0.35 m;

Table 2.1

Glacier	Debris volume concentration (%)	Debris zone thickness (m)	Potential melt-out till thickness (m)	Source
East Antarctic margin	0–12	15	0–1.8	Yevteyev (1959)
Antarctic Byrd core	7	4.8	0.34	Gow *et al.* (1979)
Camp Century, Greenland	0.1	15.7	0.016	Herron & Langway (1979)
Nordenskiöldbreen Spitsbergen	40	0.4	0.16	Boulton (1970)
Barnes ice cap, Baffin	6–10	8	0.048–0.08	Barnett & Holdsworth (1974)
Breidamerkurjökull, Iceland	50	0.15–0.3	0.075–0.15	Boulton *et al.* (1974)
Breidamerkurjökull, Iceland	8–10	0.05–0.2	0.004–0.02	Boulton (1979)
Matanuska, Alaska dispersed facies	0.04–8.4	0.2–8	>0.0008	Lawson (1979a)
Matanuska, Alaska stratified facies	0.02–74	3–15	>0.006	Lawson (1979a)
Glacier d'Argentière, France	43	0.02–0.04	0.009–0.017	Boulton *et al.* (1979)
Myrdalsjökull, Iceland	15–31	2–5	0.3–1.55	Humlum (1981)
Bondhusbreen, Norway	0.39	5	1.95	Hagen *et al.* (1983)
Watts, Baffin Island	14–57	0.8–2.9	>0.4	Dowdeswell & Sharp (1986)

Based on data compiled by Kirkbride (1995)

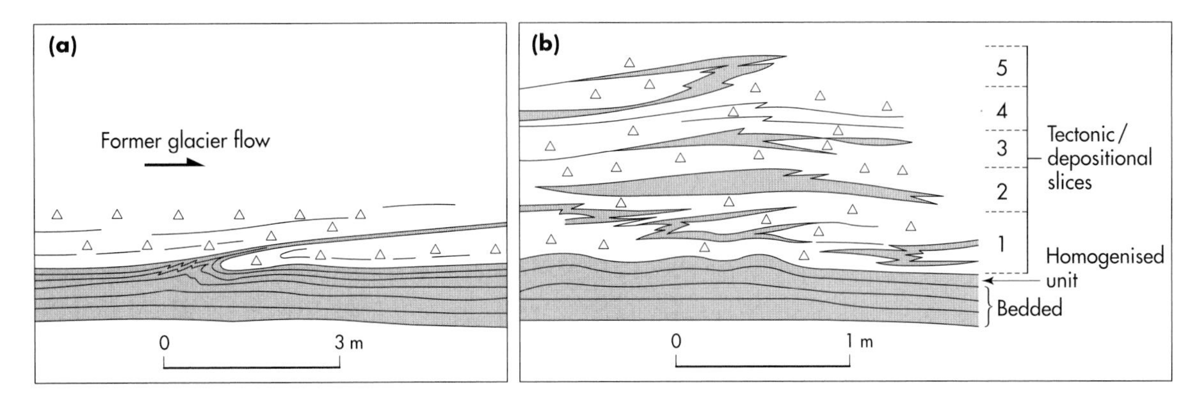

Figure 2.5 Examples of subglacial shear deformation. (a) Shear fold at the base of a thick till unit at Whitevale, Toronto, in which underlying sandy sediment have been folded into the till as a consequence of a local stress concentration that caused a fold that has subsequently been attenuated by simple shear strain. (b) Accumulation of a series of individual fold units (1–5) in a till near Sangaste, Estonia. The individual folds represent cumulative deposition of successive deforming horizons whose integrity and internal pattern of strain is reflected by individual folded units. Discontinuous shear planes occur between individual units. There is a thin zone of homogenized sediment at the base of the till, but below this, original sedimentary bedding has been relatively little disturbed.

Boulton *et al.*, 2001a, generally <0.5 m; Iverson *et al.*, 2003, >0.4 m). An exception to this has been the finding of Truffer *et al.* (2000) who demonstrated that a décollement surface in the deforming bed must lie at a depth of greater than 2 m below the glacier sole.

2 It is common to find shear and drag fold structures reflecting longitudinal shear deformation in sediments deposited beneath former glaciers and ice sheets (Hart, 1995b; Benn & Evans, 1998; Boulton & Dobbie, 1998). Some demonstrate subglacial shear deformation to depths of several metres (see Fig. 2.9). Such deep folding tends, however, to be localized, and may simply reflect local stress concentrations and blocking of shear movement that locally causes deformation to descend to greater than normal depths. It is more common to find structures in which fold packages are much thinner, with thicknesses similar to those of measured active deforming horizons (Fig. 2.5a). It is frequently found that folds in tills comprise a large number of such highly attenuated fold packages and boudins (Benn & Evans, 1996) that appear to have accumulated sequentially one above the other (Fig. 2.5b) rather than representing the 'freezing' of a single deforming horizon.

3 The roughness of the glacier bed is of fundamental importance to the décollement process. The effective roughness of a sediment bed is quite different from that of a rock bed. Whether décollement occurs by ice sliding over its surface or by internal deformation, the roughness on the surface of a shear plane is primarily at the millimetric or submillimetric scale of the grains and occasional metric scale of large clasts (if present) rather than at the 10–100 m roughness scale of smoothly eroded rock beds. As a consequence, the dominant mode of décollement is by ice sliding through regelation (Weertman, 1957) around individual grains, or by deformation within the sediment through grain against grain movement. Plastic flow of ice, that dominates on scales >10 cm to 1 m, will be relatively unimportant, in contrast to bedrock surfaces where it dominates. However, direct studies of the glacier sole in temperate valley glaciers (Kamb & LaChapelle, 1964; Boulton *et al.*, 1979) show, in most of the few cases studied, that the sole consists of a debris-rich horizon in the basal few centimetres, with the glacier sole forming a frozen sediment carpet. It seems in these cases that any sliding between the glacier sole and its bed does not occur at an ice–sediment contact but at a sediment–sediment contact. As in other granular sediments, water pressure will be a fundamental determinant of failure, either at the glacier sole or in the underlying sediments.

4 Fischer & Clarke (1997a) have demonstrated stick-slip behaviour at the base of a glacier in which slip occurs at the glacier sole during periods of the highest water pressures, with décollement being transferred down into the sediment bed as water pressures fall (see also Iverson *et al.*, 2003). Figure 2.6 shows the patterns of cumulative subglacial shear strain in 6-h increments recorded by strain markers (that are also water pressure transducers) in a subglacial sediment (Boulton *et al.*, 2001a). Strain is concentrated at the glacier sole (between transducers at 0 and 0.1) during water pressure peaks and at lower levels (between 0.1 and 0.3 m, or 0.3 and 0.5 m) during periods of lower water pressure. Figure 2.7 suggests how this might arise. Piotrowski & Tulaczyk (1999) and Piotrowski (this volume, Chapter 9) have suggested that there may also be a spatial variation in the partitioning between basal sliding and sediment deformation. Boulton (1987) suggested that sediment deformation would be minimized and friction would be maximized against the up-glacier parts of drumlins, which would be 'sticky spots' (Whillans, 1987) at the glacier bed, with easy deformation in interdrumlin zones.

5 The effective rheological behaviour of sediments deforming beneath a glacier is a matter of considerable debate. Boulton & Jones (1979) assumed a Coulomb failure criterion for mate-

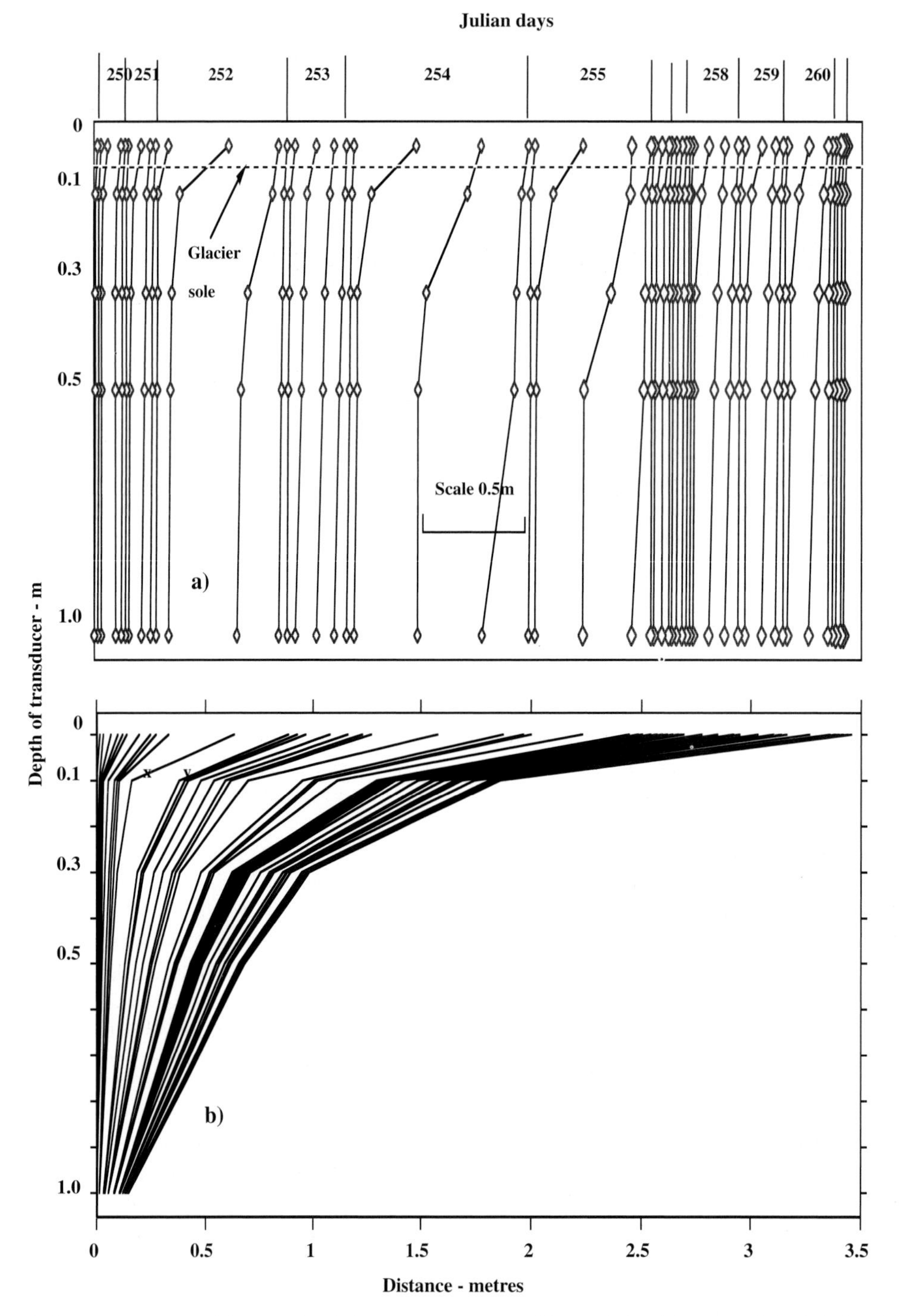

Figure 2.6 (a) Six-hourly patterns of longitudinal shear strain measured beneath Breidamerkurjökull, Iceland (Boulton *et al.*, 2001a). Strains are greatest on days 252, 254 and 255, which are also days of relatively high water pressure. Several patterns of strain occur. Most strain appears to be by basal sliding at 12.00 hours on days 252, 254 and 255 (periods of high and increasing water pressures), whereas at 18.00 hours on each of those days most strain appears to occur between 0.1 and 0.3 m depth. Significant strain occurs between 0.5 and 1.0 m on days 254 and 255. (b) Progressive net cumulative strain from days 250 to 261. Although detailed short-term patterns vary, as shown in (a), the net effect is a simple pattern, with about half the strain being taken up by sediment deformation and half by basal sliding. At *x* and *y*, for example, almost all net strain is by basal sliding.

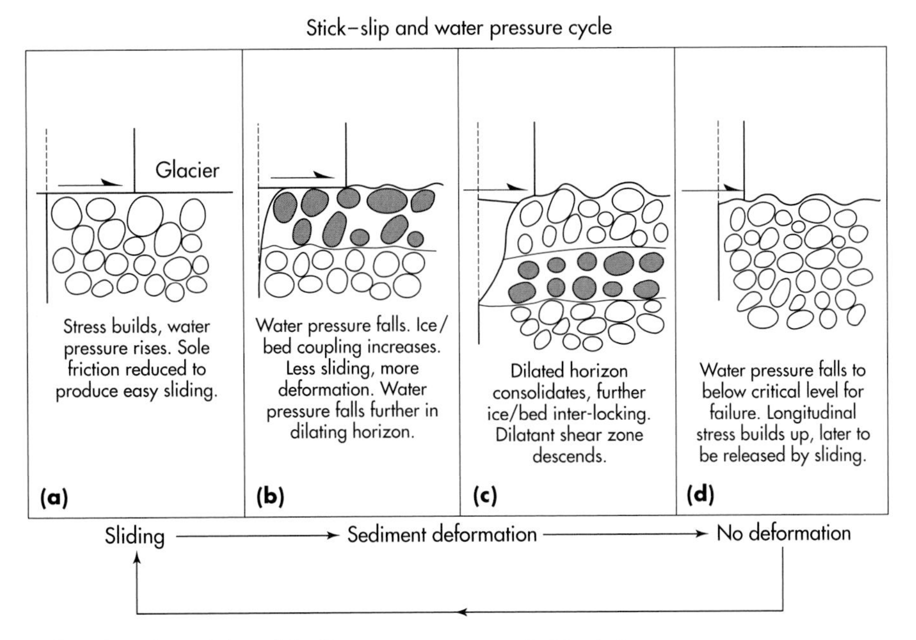

Figure 2.7 Suggested explanation of the stick-slip process apparent in Fig. 2.6.

rial in this setting. Boulton & Hindmarsh (1987) found that a non-linearly viscous law or a Bingham solid law would fit seven data points relating effective pressure to shear stress calculated from the average gravitational driving stress. A number of subsequent field experiments (Kamb, 1991; Hooke *et al.*, 1997; Tulaczyk *et al.*, 2000a) and laboratory experiments (Iverson *et al.*, 1998) have demonstrated that deforming sediments (in the former cases) and till (in the latter) show plastic behaviour that can be described using a Coulomb failure criterion. The anomaly is that such a failure criterion, if applied, for example, to the setting shown in Fig. 2.4, where effective pressure is least immediately beneath the glacier sole, predicts failure in a thin shear zone that the experiments of Hooyer & Iverson (2000b) suggest should not be more than 20 mm in thickness, immediately beneath the glacier sole. However, typical measured thicknesses of deforming horizons (**1** above) are more than an order of magnitude greater, and the vertical strain profile (Fig. 2.6b) is one characteristic of a viscous material. Clast interlocking (Tulaczyk, 1999) would create a shear zone of 10–15 times clast diameter, but even in most tills interlocking is only commonly likely between millimetric grains. Several suggestions have been made to reconcile these data:

- Hindmarsh (1997) has suggested that a plastic rheology may appear viscous at large scales, although failing to suggest the process by which small-scale plasticity is transformed into large-scale viscosity.
- Boulton & Dobbie (1998) and Iverson & Iverson (2001) have suggested that short-term water pressure fluctuations such as those shown in Fig. 2.4 could produce vertical variations in the location of Coulomb failure so that they aggregate to a time-integrated deformation profile of a viscous form as shown in Fig. 2.6b, as well as the stick-slip behaviour shown in Fig. 2.6a. Iverson & Iverson (2001) have simplified such a cumulative deformation to a law of the form:

$$\varepsilon = \frac{A}{P}\left[\left(\frac{S}{\mu N - \tau}\right) - 1\right]$$

where ε is the strain rate, A is a constant, P is ice pressure, S is sediment strength, μ is the coefficient of internal friction, N is effective pressure and τ is shear stress. Figure 2.7 also shows how localized failure and dilation could displace the location of failure without any external changes in water pressures.
- Fowler (2002) has drawn attention to a more fundamental problem: the unconstrained nature of the velocity field in perfectly plastic behaviour. The crucial issue remains therefore: how does a sediment bed generate resistance to glacier flow; what rheology and what flow law should be applied to sediment-floored ice sheets?

It is clear that the subglacial hydraulic regime, and its time dependence, are of fundamental importance to the behaviour of subglacial sediment beds. For a relatively fine-grained material, such as the clay–silt–sand matrix of a till, high porewater pressures that generate very low effective pressures are enough significantly to reduce interlocking and the strength of the till. Soft sediment beds are therefore fundamentally different from rock beds in that sustained high water pressures in them, resulting from poor drainage, can sustain a state in which interlocking is poor, the frictional resistance offered by the bed is perennially less than the yield strength of ice, and low shear stress flow can be sustained over long periods. Several of the active ice streams of the West Antarctic ice sheet appear to reflect this state (e.g. Alley

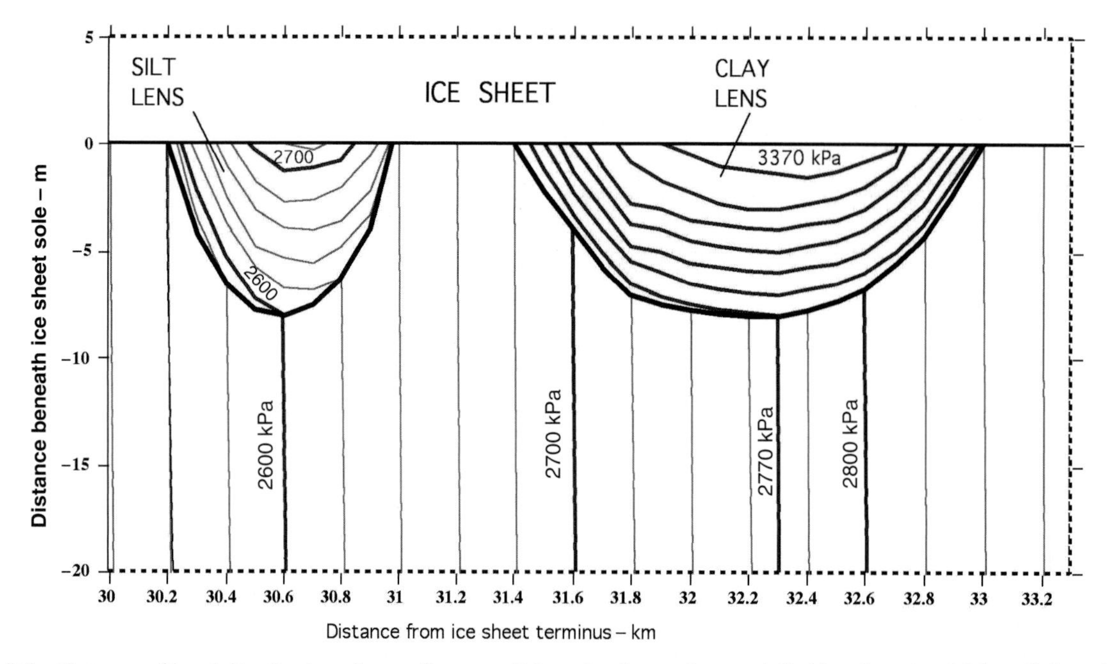

Figure 2.8 Patterns of head distribution along a flow-parallel section beneath a modelled ice sheet in which a silt lens (*k* about 10^{-8} m s^{-1}) and a clay lens (*k* about 10^{-9} m s^{-1}) overlie a thick sandy unit. There is a relatively small horizontal head gradient and insignificant vertical head gradient in the sand because of its high conductivity, but there are strong vertical head gradients in the clay and silt lenses as water drains through them from the glacier into the underlying sandy aquifer. At the ice–bed interface, effective pressures and friction will tend to be high along an ice–sand interface, lower at an ice–silt interface and even lower at an ice–clay interface, with strong areal variation in patterns of shear resistance and the nature of décollement at the ice–bed interface. Thick lines in the clay lens represent heads at 100 kPa intervals.

et al., 1987; Whillans & Van der Veen, 1997) and some parts of Pleistocene ice sheets seem to have done so (Mathews, 1974; Boulton & Jones, 1979). Time-dependent variations in water pressure may be the cause of apparently viscous behaviour of subglacial sediments (see Hindmarsh, 1997). It may also be the cause of partitioning of strain between the ice–bed interface and the sediment as a consequence of its effect on the degree of interlocking between the glacier sole and underlying sediments (Iverson *et al.*, 1995). It has been suggested by Piotrowski & Tulaczyk (1999) that a sediment-floored subglacial bed may be a patchwork of zones where slip is concentrated at the ice–bed interface and zones where a larger proportion of the forward movement of the glacier is accounted for by deformation in the sediment. Any such a patchwork is likely to change through time as a consequence of seasonal and diurnal changes, changes in the points of injection of surface water, and changes in the local hydraulic geometry of the bed that will change drainage pathways and effective pressures.

Figure 2.8 illustrates how the friction at the base of a glacier can vary as a consequence of varying geohydrological properties in the bed. The ice load and flux of water at the base of the glacier has been prescribed, as has the conductivity of subglacial beds and the dependence of permeability on effective pressure. The results indicate how fine-grained low-conductivity sediment masses overlying highly conductive strata can locally reduce effective pressures and frictional resistance on the bed.

2.4.4 An unfrozen, unlithified sediment bed—geological implications

Structures reflecting subglacial shear deformation of sediments, locally to a depth of metres, are widespread, although not universal, in both modern and formerly glaciated regions. Figure 2.9 shows an informative example, typical of many, observed on the Island of Funen in Denmark, during a 2001 field trip led by Jörgensen and Piotrowski in a zone of a former ice stream (Jörgensen & Piotrowski, 2003). Three zones can be readily distinguished, which are equivalent to the deformation zones of Boulton (1987).

1 *A lower zone* (C), in which proglacial fluviatile sediments are largely undisturbed;
2 *An intermediate zone* (B), in which strongly overturned shear folds occur in a sequence of till, sand and gravel, but which are either rooted in the sediments of zone C, or, if detached, can be recognized as derived from them. By allowing for tectonic thickening and thinning of these beds, but particularly by following the limbs of shear folds that have not been derooted, it is possible to reconstruct the approximate net strain in this zone. It suggests a tectonic transport at the top of zone B of about 70 m.
3 *An upper zone* (A), in which a diamicton containing numerous elongated lenses and wisps of sand, which are sometimes

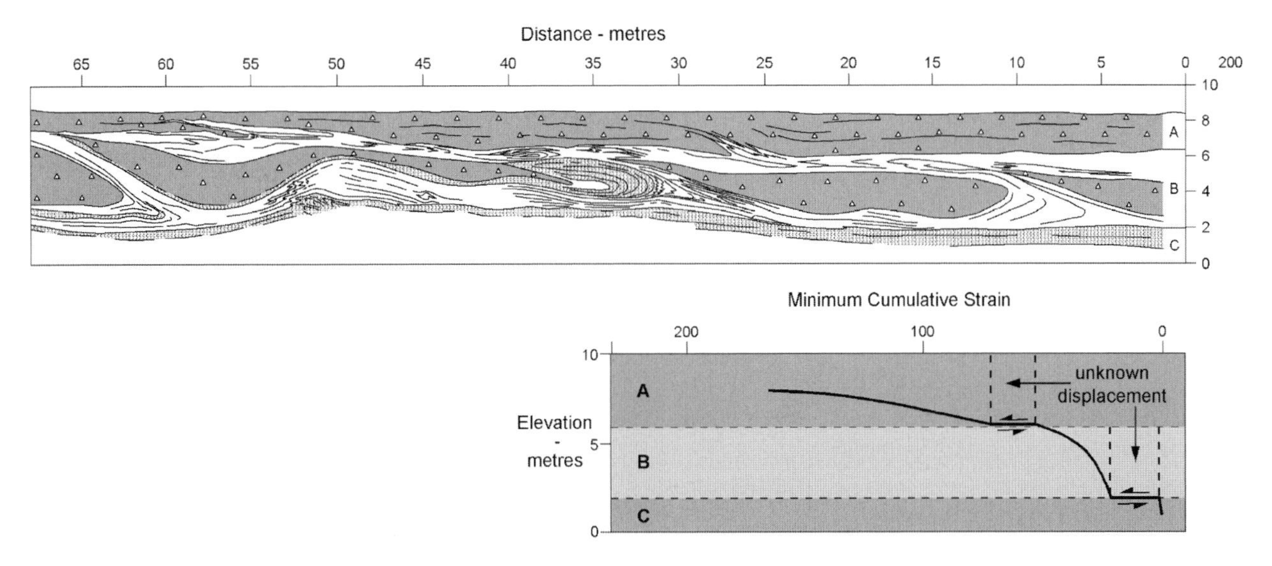

Figure 2.9 Deformation structures in a quarry at Davinde on the Island of Funen, Denmark in a section parallel to the direction of tectonic transport. The stippled ornament with triangles shows till, the unpatterned stratum is predominantly sand, in which lines show bedding planes. The circle ornaments show gravelly beds. Zone A, of the upper till and sandy masses that have been incorporated by folding, contains very highly attenuated folds reflecting the largest tectonic transport. Zone B is a zone of overturned folds reflecting lesser tectonic transport. Zone C is a zone of little deformation. The lower figure shows the estimated minimum strain in each of these three zones. There is décollement at the interfaces between the three zones, but the magnitudes cannot be determined. It is likely to be much larger at the A–B interface than at B–C.

folded, appears to lie unconformably on those of zone B. The estimated shear strain in zone A is an absolute minimum, derived by estimating the finite shear strain in individual isoclinal folds in this zone. However, as these folds are generally de-rooted or difficult to trace back to distant roots through extreme shear thinning, it is clear that the shear strain in zone A is far larger than the minimum.

The potential significance of this sequence can be best understood by referring to the pattern of monitored shear strain shown in Fig. 2.6. The cumulative shear strain in the uppermost 1.0 m is 3.5 m in 12 days (Fig. 2.6b). This approximates to a strain rate of 106 per year or 1060 in 10 yr, an extremely large finite shear strain. Applying this to the section illustrated in Fig 2.9, any sandy units from zone C or B, folded because of local stress concentrations (Fig. 2.5a) into a deforming mass such as that shown in Fig. 2.6b, would be enormously attenuated, recognized only as thin, subhorizontal sandy lenses or wisps. It is on this basis that the diamicton in zone A is suggested to be a deformation till that has formed in a zone of shearing such as that in the topmost 0.5 m in Fig. 2.6. Moreover, if local stress concentrations, or local increases in frictional resistance or consolidation on the surface over which it shears, produce folding, these laminae will tend to become progressively more strongly mixed into the deforming mass, to produce a homogenized sediment.

If such mixing by folding occurs, horizons that originally lay at the base of the rapidly shearing mass could be translated to the top, and vice versa. Under such circumstances, we could translate the average shear strain through the deforming bed in Fig. 2.6 into an average velocity of 53 m yr^{-1}. Clearly such a mass could travel far beyond the sediment source area from which it was derived, and would be added to by incorporated materials from further down ice. The unconformable relationship between the sediments in zone A and zone B would be a reflection of such strong de-rooted transport, particularly as the amount of time the site was last glaciated prior to final ice retreat in the last phase of late Weichselian glaciation in the area was probably about 2000–3000 yr (Houmark-Nielsen, 1999). This would be sufficient to permit material incorporated at this site at the time of initial glacier overriding to have been transported a very long distance, although the zone-A sediments at the site may themselves have been incorporated into the flow only recently, shortly before deglaciation. It is also possible that the A–B unconformity has been a surface of strong erosion from which sediment was incorporated into the shearing nappe above it. Given the potentially long period over which deformation may have occurred, the apparently large strains in zone B may reflect only very small strain rates, and may have lain in the lower zone of Fig. 2.6 in which the period of measurement was so short that only slight deformation was recorded.

Figure 2.10a & b shows a model of an advancing ice sheet with a deforming bed based on the theory of Boulton (1996b). In the zone up-glacier of the equilibrium line, the inner zone of accelerating flow, there is erosion of the bed. Down glacier of this zone, the outer zone of deccelerating flow produces a thickening till mass, which is itself eroded as the glacier advances over it, to produce an advancing wave of deposition (compression) and erosion (extension). As discussed above, a shearing nappe can be highly erosive through folding-incorporation of underlying sediments (Fig. 2.5a). Even if the deforming layer remains of

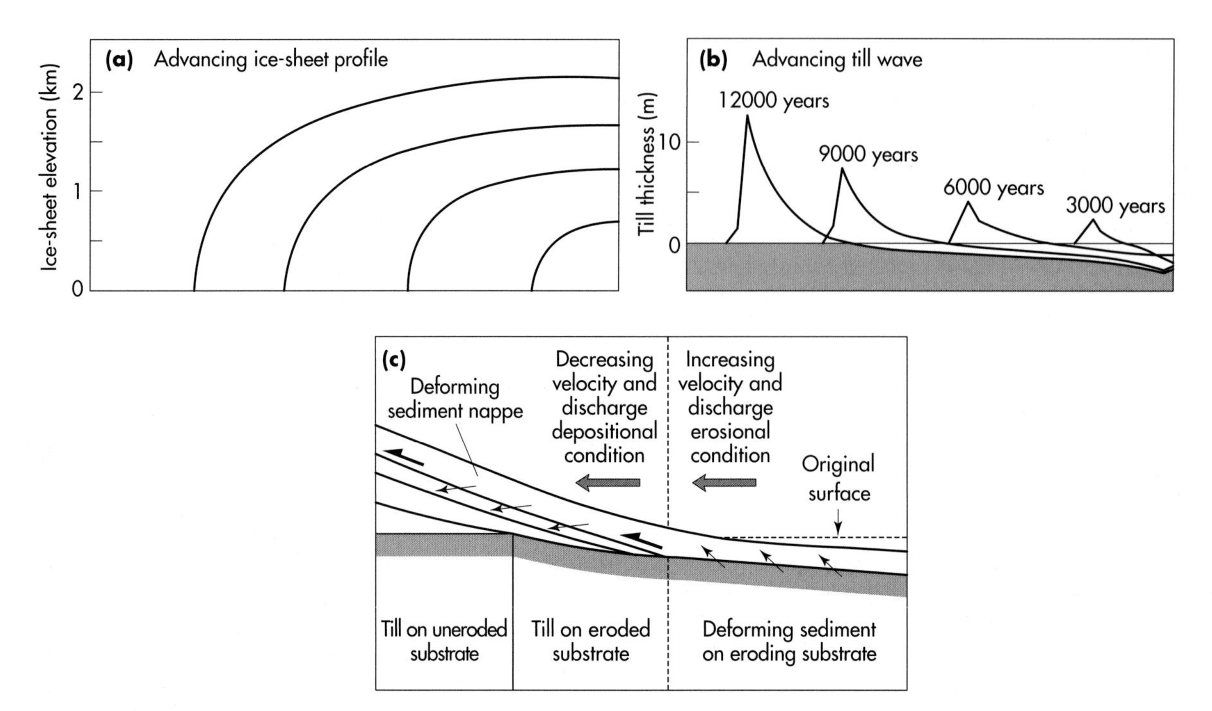

Figure 2.10 (a & b) A modelled till wave generated as the glacier advances. It is derived from the theory of till transport by subglacial deformation (Boulton, 1996b), but could apply equally to basal ice transport and lodgement. Zones of deposition and erosion extend outwards as the glacier advances. (c) A schematic diagram showing the zone of accelerating flow (extension) up-glacier of the equilibrium line, in which a thin deforming sediment nappe erodes the bed by incorporation of underlying sediment; and the zone of decelerating flow (compression) down-glacier of the equilibrium line which creates net deposition from the deforming nappe and thickening of the deposited till.

constant thickness, a down-glacier increase in velocity will permit the discharge of deforming sediment to increase, thus leading to incorporation (erosion) of yet more sediment in the deforming horizon (Fig. 2.10c). As a consequence, the sediments underlying a deforming sediment nappe may suffer aggregate deep erosion as successive deforming masses continuously move its surface, incorporating successive increments of sediment from it. As we pass into the terminal zone of a terrestrial glacier, where we expect basal velocities in general to decrease, the discharge of sediment in a deforming horizon of constant thickness will decrease and till will begin to be deposited from the base of the deforming horizon (Boulton, 1996b). From this point on, a thickening till stratum will form from successive increments of deposition from the base of the deforming horizon (Fig. 2.10c). Clark *et al.* (2003b) have suggested that because many tills are in excess of a metre in thickness and deforming horizons tend to be thin, that deformation cannot be an important source of erosion. This confuses the thickness of a deposited till with the thickness of the deforming horizon.

It is frequently observed that a sharp interface separates till and apparently undisturbed underlying sediment. This is most likely to be a product of an erosive deformation process. The existence of a soft, deforming till at the ice–bed interface, acting as a buffer between the glacier and an underlying stable bed, and able to incorporate irregularities that form local stress concentrations on the underlying surface by erosively folding them into the shearing nappe, is a means of creating an apparently undeformed, planar surface.

2.5 The origin of till and its properties

As with any other sediment, the thickness of a till is a product of the rate of transport into the zone of deposition and the period of time over which the rate is sustained. There are three principal modes of deposition of till: lodgement, deformation and melt-out.

2.5.1 Lodgement till

This is assumed to be deposited when the frictional drag between clasts transported in the basal ice and the bed is sufficient to halt the clasts against the bed. Lodgement is a cumulative process in which debris is continually imported into the region and progressively accumulated on the bed. In principle, a long period of till accumulation could produce a considerable till thickness. The till surface will bear streamlined features such as flutes and drumlins. Notwithstanding the many tills that have been ascribed to a lodgement process (e.g. Benn & Evans, 1998), we have no direct documentation of the process and no unequivocal demonstration of a lodgement origin for any deposited tills. Hart (1995b) has doubted that lodgement is a significant process by which till is finally deposited.

2.5.2 Deformation till

The way in which deformation tills might either be associated with underlying deformation structures or might overlie an undisturbed sediment across a planar interface as a consequence of erosion at its base has been described above. The till effectively acts to absorb stress at the base of the glacier, and can protect an underlying interface against deformation. If, for example, a melt-out till overlaid pre-existing sediments, the underlying surface would be a surface over which ice had flowed, and more deformation would be expected at the interface than in the case of a deformation till.

The till-creating potential of the deformation process does not depend upon the thickness of the deforming horizon, but on the sediment discharge in the deforming horizon. For example, a relatively thick (0.45 m) deforming horizon at Breidamerkurjokull had a two-dimensional discharge of about 25.7 $m^2 yr^{-1}$, whereas a thinner (0.3 m) deforming horizon at Trapridge Glacier had a discharge of 314 $m^2 yr^{-1}$ (Boulton *et al.*, 2001), simply because the flow velocity of the ice–sediment system is greater in the latter case. Even if the active deforming horizon is thin, the ultimate till that is deposited from it may be relatively thick. The ways in which the pattern of erosion and till depositon may vary through a glacial cycle based entirely on changes in the transporting power (Boulton, 1996b) are illustrated in Fig. 2.11. As in the case of lodgement till, a deformation till surface will be characterized by streamlined drumlin and flute forms.

2.5.3 Melt-out till

Melt-out till is the inevitable consequence of the slow melting out of debris-rich stagnant ice that is buried beneath a supraglacial sediment overburden. Simple thermodynamic considerations suggest that melting out will almost invariably be on the surface of the buried ice mass rather than beneath it. As this till represents the melting out of debris from a stationary ice mass, its ultimate thickness is limited by the mass of debris in a column of ice. It is not continuously being transported to the place of deposition as are the other two till types. This inevitably limits the thickness of melt-out tills, as the total debris content of a vertical column of ice is rarely enough to create more than a few decimetres and exceptionally metres of till (Table 2.1). However, their role in preserving buried stagnant ice, which then intercepts glacial drainage to create hummocky kame landscapes, is important, and some subpolar glaciers with relatively large debris loads, such as those of Spitsbergen, and some subpolar glaciers with relatively large debris loads can create melt-out tills with thicknesses in excess of a metre and potentially be a major source of supraglacial debris flows (flow tills) (van der Meer, 2004). Melt-out till deposition will tend to be associated with hummocky rather than streamlined glacial topography, although Monro-Stasiuk and Sjogren (this volume, Chapter 5) have suggested that 'hummocky terrain' can be of erosional origin, a puzzle that demands further analysis.

2.5.4 The state of consolidation of subglacial tills

It was formerly supposed that state of consolidation of tills and their tendency to overconsolidation was determined by the ice

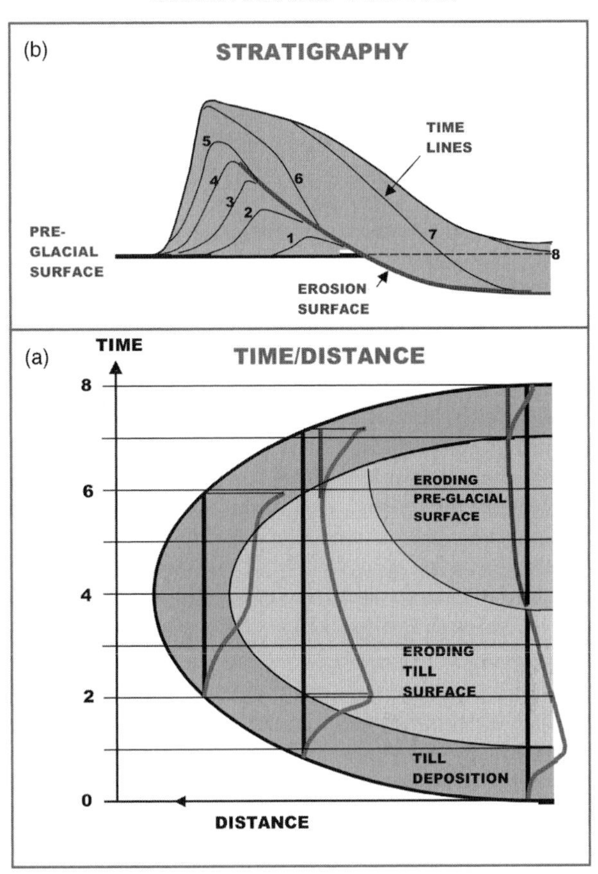

Figure 2.11 A schematic diagram of erosion/deposition through a simple glacial cycle. (a) Advance and retreat of an ice sheet in a glacial cycle. The longitudinal pattern of erosion/deposition along a specific timeline is as shown in Fig 2.10b. The three vertical lines show the sequence of events at specific locations through time. An early phase of till deposition (the 'till wave' as in Fig. 2.10b) is succeeded by a period in which this till is progressively eroded. Only at the right-most location does erosion occur for a period long enough to remove the earlier deposited till entirely and then to erode into the pre-till surface. Till is deposited on the eroded surface during the last phase of glacial retreat. At the middle location, earlier deposited till is eroded but not completely removed before the retreat-phase till is deposited, producing an erosion surface within the till, often marked by a boulder pavement and a lithological contrast (Boulton, 1996b). At the left-most location, till is deposited continuously. (b) The structure of the resultant till, including the location of internal erosion surfaces and timelines.

overburden pressure, and therefore that measured pre-consolidation values from tills could be used to infer former ice loads at the glacial maximum (e.g. Harrison, 1958). This involves three related assumptions: that ice load alone is the determinant of pre-consolidation pressure; that tills were necessarily present beneath the glacier at the maximum of glaciation to receive the imprint of contemporary pre-consolidation; and that the measured

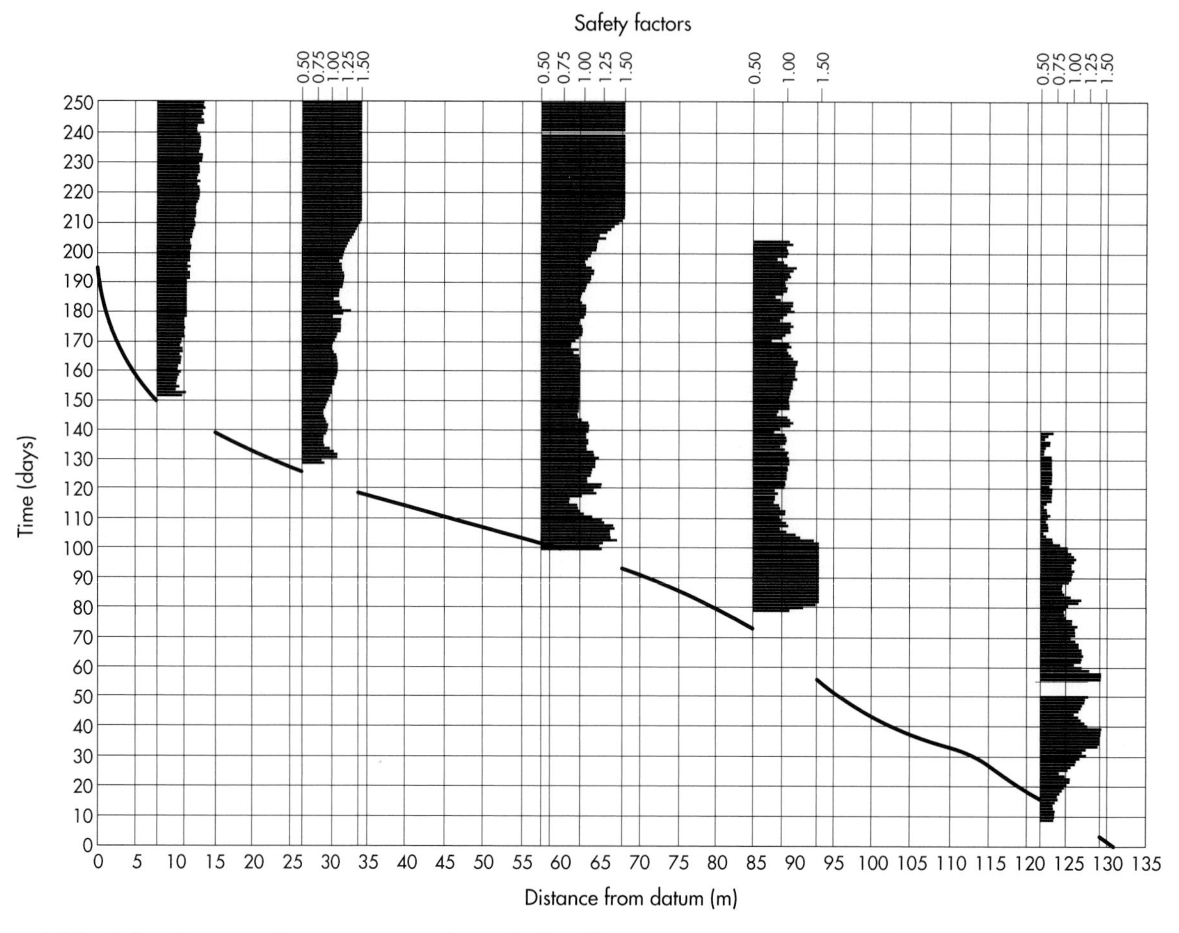

Figure 2.12 Safety factors at the topmost transducers in the till at 12 m, 30 m, 65 m, 85 m and 125 m along the transect shown in Fig. 2.3. The heavy line shows the glacier margin through time. The maximum ice thickness was achieved at about 195–200 days. The smallest safety factors (strongest deformation) occur early on during the advance; the largest safety factors (maximum effective pressures and pre-consolidation) occur after the period of maximum loading, when a more efficient drainage system had been established.

pre-consolidation reflects conditions at the glacial maximum. The first is very unlikely under any circumstances and there is no reason to believe that the second and third are commonly true.

The data in Fig. 2.4 show conditions that influence the state of consolidation of tills and associated subglacial sediments as they are overridden by a glacier. Using an assumed shear stress derived from the average gravitational shear stress, and assuming a Coulomb yield criterion, a safety factor (strength/stress) is calculated, in which values > 1 indicate stability, and values < 1 indicate failure. The results shown in Fig. 2.12 indicate some important conclusions.

1 The effective pressure varies both diurnally and seasonally, primarily in response to variations in recharge to the sub-till aquifer and to the top of the till from water draining to the bed from the glacier surface.
2 The maximum effective pressure (which would be the value of pre-consolidation recorded by the till provided that it is not remoulded by shear), which occurs after day 210, does not coincide with the maximum ice load, which occurs on about day 190. The maximum effective pressure is determined by late-stage drainage of the system.
3 Shearing in the till (safety factor < 1) occurs early during the glacier advance, whereas the maximum effective pressures occur later.

Measured pre-consolidation values large enough to inhibit failure cannot be used, as Hooyer & Iverson (2002) have done, to infer that the till could not have deformed. In the Breidamerkurjökull case, the period after the very active advance is a period of drainage reorganization that leads to a general fall in water pressures. A similar sequence is suggested to have occurred during the surge of Sefströmbreen in Spitsbergen in 1882–86 (Boulton *et al.*, 1996), where there is strong geological evidence and contemporary observation that the glacier surged forward on a deforming carpet of marine sediment and till, which would have required very low effective pressures in the sediment. Now, however, the sediments and tills are relatively heavily preconsolidated, which is suggested to have occurred when the glacier stagnated after the surge and a reorganization of drainage permitted water pressures to be reduced.

These observations reinforce the need to understand how far beneath an ice sheet surface water can penetrate to the bed. Is it merely in the terminal zone, or is it far from the margin as Arnold & Sharp (2002) suggest? If the latter, then a large variety of short-term, seasonal drainage effects would be important in driving highly variable subglacial processes. If the former, we would expect slowly varying hydraulic, geotechnical and depositional conditions beneath ice sheets.

2.6 Large-scale patterns of sediments and landforms and inferences drawn from them

An important current focus of glacial geological study is the reconstruction of the large scale properties of former ice sheets and the way in which they have evolved through a glacial cycle (e.g. Kleman *et al.*, this volume, Chapter 38). The advent of satellite imagery and broad swathe bathymetric devices has permitted coherent reconstructions to be made of landform systems that show very large scale patterns of distribution both on land and beneath the sea, rather than having to depend upon a fragmental patchwork of field surveys. There are currently three large-scale patterns that have been established for the European ice sheet, from which palaeoglaciological inferences can be drawn:

1 large scale drift lineations (drumlins and flutes);
2 relict landscapes;
3 esker distributions and tunnel valleys;

and a fourth one, the distribution of till thickness, which is less well known and possibly less diagnostic of origin.

In some cases, large-scale fossil features have been used to infer processes that occur beneath ice sheets, rather than merely being explained by reference to modern process studies.

2.6.1 Large-scale drift lineations

Figure 2.13 shows a compilation of the large-scale trends of lineations (primarily produced by drumlins) from the area of the last European ice sheet (Boulton *et al.*, 2001c). It reveals major crossing lineation sets, reflecting shifts in the centre of mass of the ice sheet through the last glacial cycle, and consequent changes in the pattern of flow. It shows that these lineations cross at particularly high angles in northern Sweden, in the ice divide zone, where Lagerbäck (1988) and Kleman *et al.* (1997) have shown that glacial geomorphological features have survived unaltered from the early part of the Weichselian glacial cycle. It shows the locations of former ice streams, particularly in the eastern and north-eastern area of the Fennoscandian Shield. Submarine bathymetric studies can reveal even more complete patterns, and have been used by Clark *et al.* (2003b) to demonstrate the megaflute lineations created by the ice stream that flowed along the Skaggerak and thence along the Norwegian channel towards the continental shelf edge during the Last Glacial Maximum. Large-scale lineation patterns are now a rich source of information about the locations of palaeo-ice streams in former ice sheets (see Stokes and Clark, this volume, Chapter 26).

In a series of papers over the past 20 yr, Shaw and collaborators (see Shaw, this volume, Chapter 4) have argued that drumlins are erosion marks created by subglacial water flow, and in the case of the North American Wisconsinan ice sheet were generated by very large outburst floods that, during one phase, are postulated to have generated a flow volume of 84,000 km^3. The discovery of very large lakes beneath the Antarctic ice sheet (Robin *et al.*, 1977; Kwok *et al.*, 2000) suggested how a possible source for such floods might have existed beneath the North American ice sheet. Although it would be premature, at this stage of our knowledge, to suggest that such floods cannot or did not happen, there are no characteristics of drumlin fields that have been described that cannot be explained by glacial transport processes that are known to occur. It would therefore also be premature to prefer a speculative flooding process for their origin.

2.6.2 Relict landscapes

The existence of well-preserved relict landscapes with early Weichselian eskers, frost shattered bedrock and the absence of erosional and depositional forms reflecting the flow of the late Weichselian ice sheet have been used by Kleman & Hätterstrand (1999) to map areas where the late Weichselian ice sheet was frozen to its bed and where there was therefore no sliding and little or no erosion (Plate 2.1c; see also Stroeven *et al.*, this volume, Chapter 90). The fact that erosion can occur below cold ice (section 2.4.1) does not affect the inference that the relict zone in Plate 2.1 was a cold ice zone. It could be that this zone was more extensive than mapped by Kleman & Hätterstrand (1999), but this seems unlikely because of the probable low rates of cold ice erosion. The data provide a strong constraint on palaeoglaciological reconstruction of the ice sheet, and particularly on the glacier–climate parameters that are used in ice-sheet simulation models. Plate 2.1a & b shows a simulation of the basal temperature distribution in the European ice sheet as it approached its maximum extent at the LGM. The simulations show that the areas of the ice sheet mapped by Kleman & Hätterstrand (1999) as having relict features coincides with an area that had persistent freezing conditions during the last glacial expansion, although they suggest that areas outwith this zone had chequered histories of basal thermal regime as the dynamic structure of the ice sheet evolved.

Plate 2.2 shows a time–distance simulation of the evolution of basal temperature in space and time along the given transect through the ice sheet. It shows that during the early part of the glacial cycle (100 ka) temperate bed conditions extended almost up to the ice divide, whereas during the last glacial maximum (LGM) there was a 250-km-wide zone of basal freezing in the divide zone. This matches well with the Kleman & Hätterstrand (1999) reconstruction of basal thermal regime for the LGM and the evidence of early Weichselian temperate conditions in the ice divide zone, under which eskers formed, but which were later preserved because of frozen bed conditions (Lagerbäck, 1988). It also suggests that during parts of the retreat from the LGM, the rate

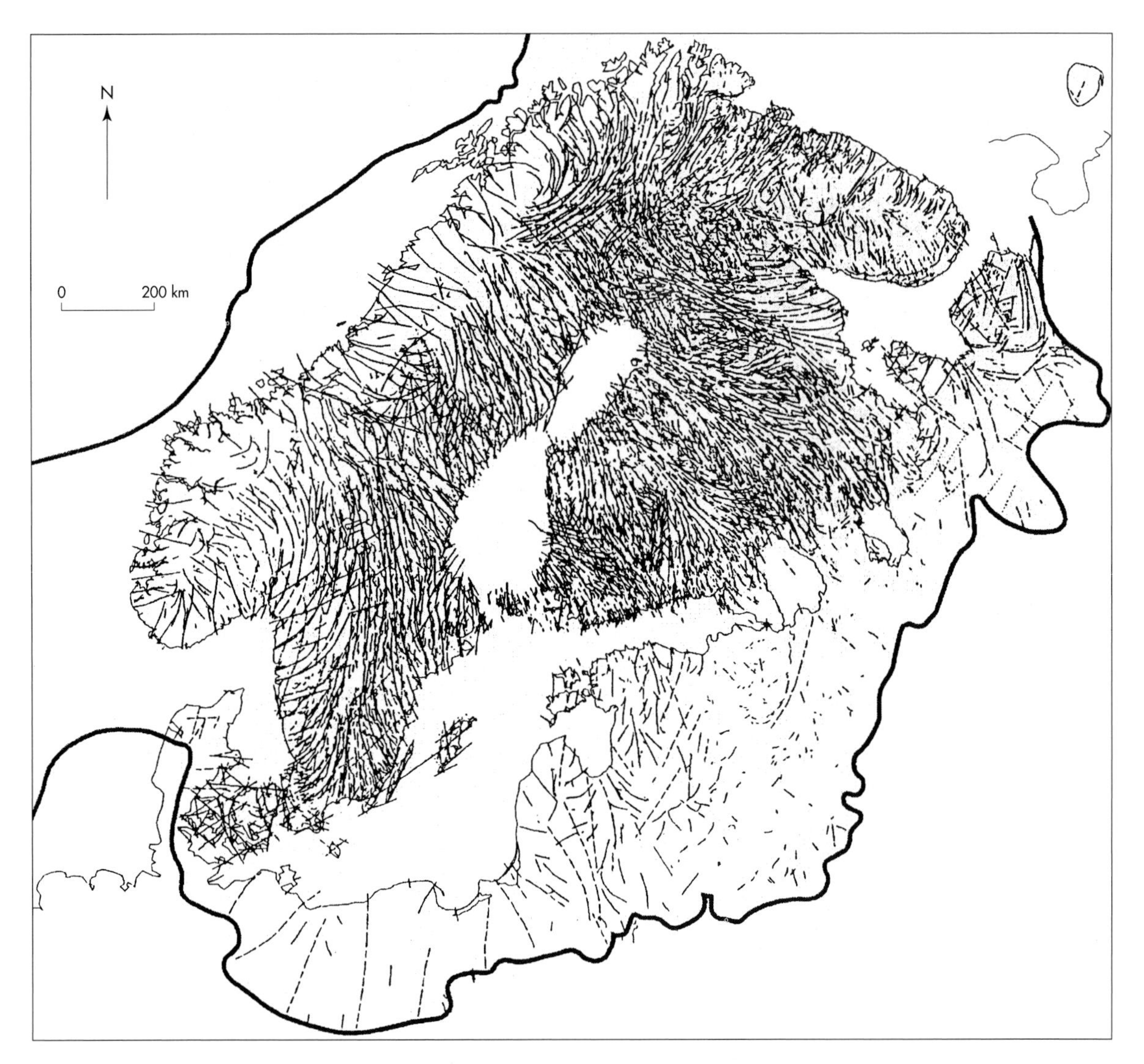

Figure 2.13 The large scale pattern of lineations lying within the area of the last European ice sheet (Boulton *et al.*, 2001c). Individual lines represent generalizations of more detailed mapping of lineations from Landsat images. The paucity of lineations in the southeastern region reflects the greater difficulty of resolving lineations in areas of arable agriculture.

of ice margin retreat was greater than the retreat rate of the junction between the outer temperate and inner cold ice-sheet zone, producing frozen bed conditions in the terminal zone (see also Boulton *et al.*, 2001c), a result consistent with the deduction of Hättestrand & Clark (this volume, Chapter 39) for part of the deglaciation of the Kola Peninsula.

2.6.3 Esker distributions and tunnel valleys

Plate 2.3 shows the distribution of eskers on the Fennoscandian Shield. Although eskers occur beyond the Shield, they are infrequent. A similar situation applies on and around the Laurentide Shield in North America. It is not entirely clear whether eskers are simply not so well preserved in the fringing soft-sediment areas, whether subglacial tunnel flow eroded deep channels into the substratum rather than being contained in R-tunnels, or whether eskers are largely replaced by tunnel valleys, or whether, as suggested by Clark & Walder (1994), channelled meltwater beyond the shield is discharged via high pressure 'canals' rather than low pressure R-tunnels.

Boulton *et al.* (2001b) have suggested that on the Shield, large R-tunnels that ultimately give rise to eskers occur where groundwater flow alone is unable to discharge the subglacial meltwater flux. They suggest that the spacing of eskers is that required to discharge the excess meltwater flux, a suggestion consistent with the observed increase in esker frequency with radial distance away from the ice divide, which would also be the groundwater divide. In any case, if eskers do represent the former locations of low pressure R-tunnels, they would inevitably act as sinks for groundwater, which would predominantly flow towards them. This pattern of flow is simulated in Fig. 2.14. This simulation demonstrates that the effect of tunnel drawdown would be to ensure that the

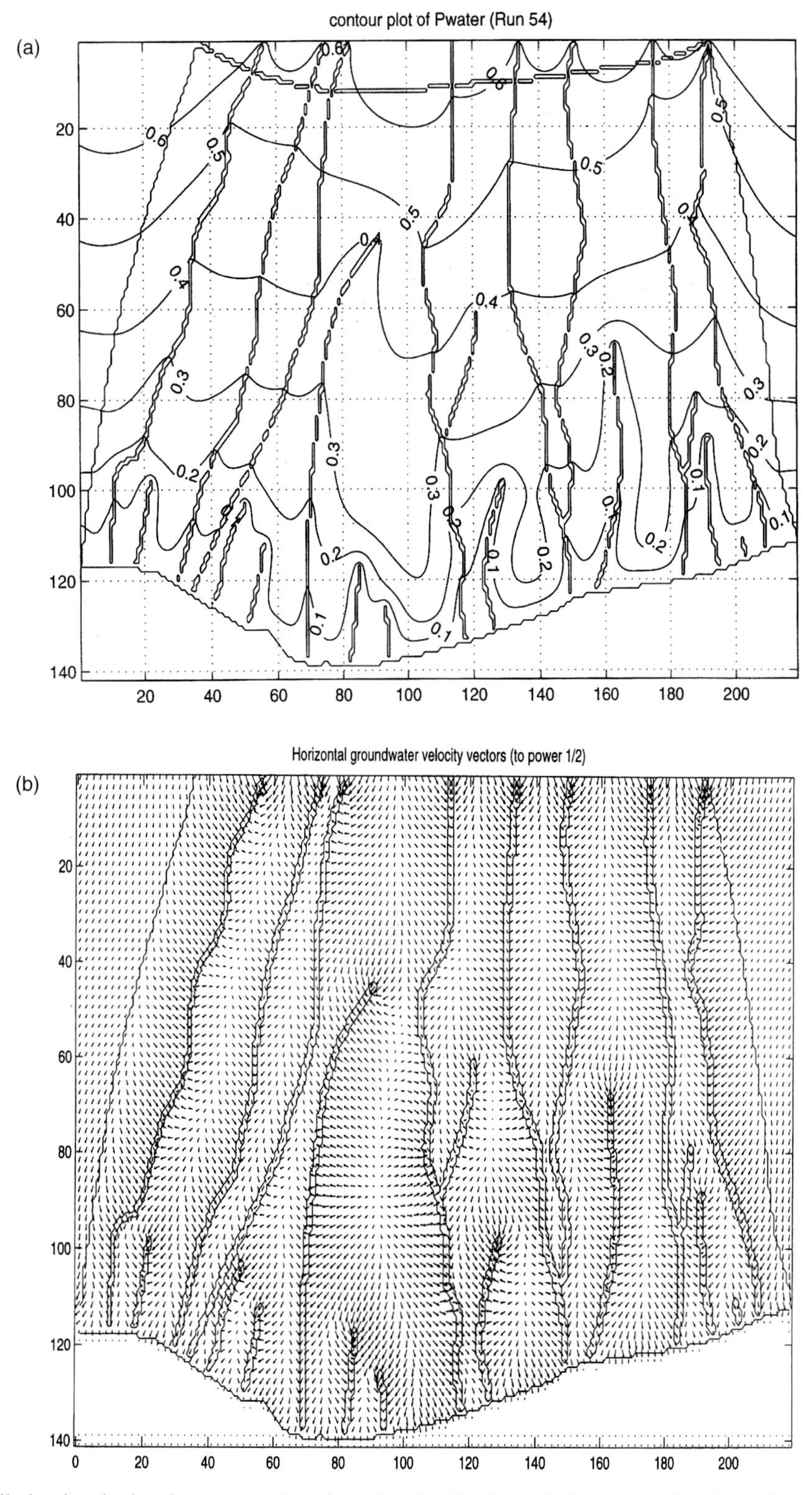

Figure 2.14 Modelled palaeohydraulic patterns based on the distribution of eskers (N–S lines) and bedrock conductivities in the area north of the Salpausselka moraines in Finland. Scale in kilometres. (a) Simulated water pressures in bedrock as a proportion of ice pressure. The pressure difference across a till lying above bedrock (see Fig 2.4) would produce larger water pressures at the ice–bed interface. (b) Groundwater flow vectors in bedrock. They are strikingly similar to the patterns that would be expected in unglaciated temperature regions.

dominant groundwater flow vector would be transverse to ice flow and not parallel to it as suggested by Boulton *et al.* (1993) and Piotrowski (1997b). Although I agree with Piotrowski and Piotrowski & Marczinek (this volume, Chapters 9 & 10) that in areas such as North Germany the transmissivity of the subsurface would have been inadequate to discharge even the basal meltwater flux alone to the margin via longitudinal flow, all meltwater can be discharged by groundwater provided the flow is transverse and towards esker/tunnel valley channels. The inferred hydraulic pattern would also play a major role in influencing ice-sheet dynamics through its influence on the effective pressure at the ice–bed interface.

So-called tunnel valleys that have long been regarded as products of subglacial fluvial erosion (Madsen, 1921) are common in the zone of sedimentary rocks that fringe the Shield area in Europe. Some prefer the term 'tunnel channel', that embraces both small channels and larger valleys, but I shall retain the term tunnel valley, in the recognition that they are large features that demand a special explanation. In Europe, features with this appellation are broad (0.2–5 km), deep (50–400 m), steep sided (up to 40° marginal slope) channels that can be up to 100 km in length. Unlike normal valleys, they rarely have till at their base, but tend to have sand and gravel fluvial sediments near their base, which are overlain by glaciolacustrine and marine sediments. Although many have been occupied during several glacial cycles, some were entirely eroded during the Weichselian. The volume of excavation that they represent would require very high rates of erosion. The larger ones in North Germany, if continuously eroded during the period of Late Weichselian glacier occupancy, would require a continuous sediment discharge rate of about $0.1\,m^3\,s^{-1}$, much larger than could be achieved if the water flux was derived from basal melting alone. It would imply either that they were eroded by short-period catastrophic floods (Wingfield, 1990) or that large quantities of surface meltwater found their way to the bed of the ice sheet and were channelled along the valleys. In the former case they may have formed by bankful discharges; in the latter a relatively small tunnel would have existed along the valley axis. Brennand *et al.* (this volume, Chapter 6) suggest that tunnel valleys in central Ontario are 'consistent with' an origin in which they, together with the regional drumlin fields, were eroded by a subglacial megaflood.

2.6.4 The distribution of tills

A scrutiny of geological maps showing the distribution of till shows that the till cover in central areas of the both the North American and European ice sheets is less extensive and thinner than in marginal areas. This may reflect the lesser erodability of shield rocks compared with fringing softer rocks, or it may reflect an almost inevitable consequence of outward transport and progressive till accumulation. Figure 2.11 shows a model of deforming bed transport by an advancing ice sheet and the net consequence of this mode of transport and deposition through an idealized glacial cycle (cf. Boulton, 1996b). A similar pattern would be produced by a lodgement mechanism but not by melt-out, which is not a continuous and cumulative process. Nor would we expect, if melt-out had been the dominant process of till deposition in Europe, to find the almost ubiquitous streamlining of drift surfaces that reflect active ice movement over the till surface, and which is reflected in Fig. 2.13. It is concluded therefore that the dominant process of till deposition must be a cumulative process of deposition beneath actively moving ice, implying either lodgement or deformation.

The keel-grooving mechanism for the creation of megascale lineations presented by Clark *et al.* (2003b) has been suggested by Tulaczyk *et al.* (2001b) not only to be a means of grooving pre-existing sediments, but also of transporting them. It is difficult to understand, however, how bedrock-created keels could be a major means of longitudinal transport rather than transverse transport due to keel grooving, unless longitudinal transport was really produced by subglacial shear deformation.

Acknowledgements

The assistance of and discussions with Magnus Hagdorn and Sergei Zatsepin are gratefully acknowledged.

THREE

Haut Glacier d'Arolla, Switzerland: hydrological controls on basal sediment evacuation and glacial erosion

Picture courtesy of Peter W. Nienow

Darrel A. Swift

Department of Geographical and Earth Sciences, University of Glasgow, Glasgow G12 8QQ, UK

At Haut Glacier d'Arolla, Switzerland (Fig. 3.1), suspended sediment transport during the 1998 melt season demonstrates the importance of subglacial drainage system morphology for basal sediment evacuation because it influences both the capacity of meltwater to transport basal sediment and the mechanisms by which sediment is accessed and entrained.

Early in the melt season, surface runoff enters a distributed subglacial drainage system (Nienow *et al.*, 1998) during which extreme increases in runoff stimulate periods of rapid glacier motion (termed 'spring events'; Mair *et al.*, 2003). Later in the season, removal of the surface snowpack from the ablation area results in increasingly peaked diurnal runoff cycles that promote the up-glacier extension of a hydraulically efficient network of subglacial channels (Nienow *et al.*, 1998). During 1998, two spring events occurred during steep rises in catchment discharge (subperiods 2 and 4, Fig. 3.2a & b; Mair *et al.*, 2003), the first coinciding with intense rainfall and the second with both widespread thinning of the snowpack and a rapid increase in the efficiency of meltwater routing to the glacier terminus (Swift *et al.*, 2005a). Dye tracer investigation demonstrated predominantly channelized subglacial drainage beneath the ablation area by late July, indicating rapid up-glacier extension of the channel network during subperiods 4 to 6 (Swift *et al.*, 2005a; cf. Fig. 3.2a).

Suspended sediment transport during 1998 was monitored in the proglacial stream draining the western subglacial catchment (Fig. 3.1), into which extraglacial sediment contributions were negligible. Hourly mean suspended sediment concentration (SSC) was obtained from a continuous record of proglacial stream turbidity calibrated using 1159 point-collected water samples (Swift *et al.*, 2005b). Catchment suspended sediment load

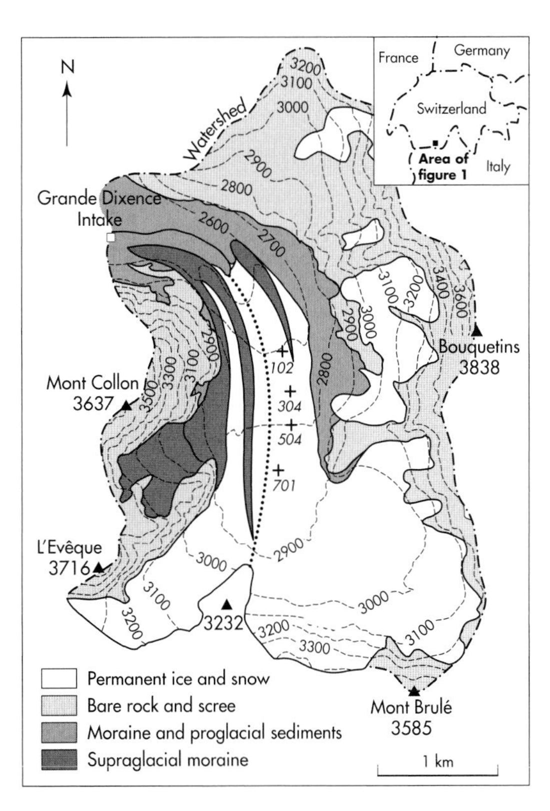

Figure 3.1 Haut Glacier d'Arolla, showing the supraglacial divide between the eastern and western subglacial catchments during 1998 (dotted line) and velocity stakes used in Fig. 3.2b (crosses). Contours (dashed lines) and elevations are in metres.

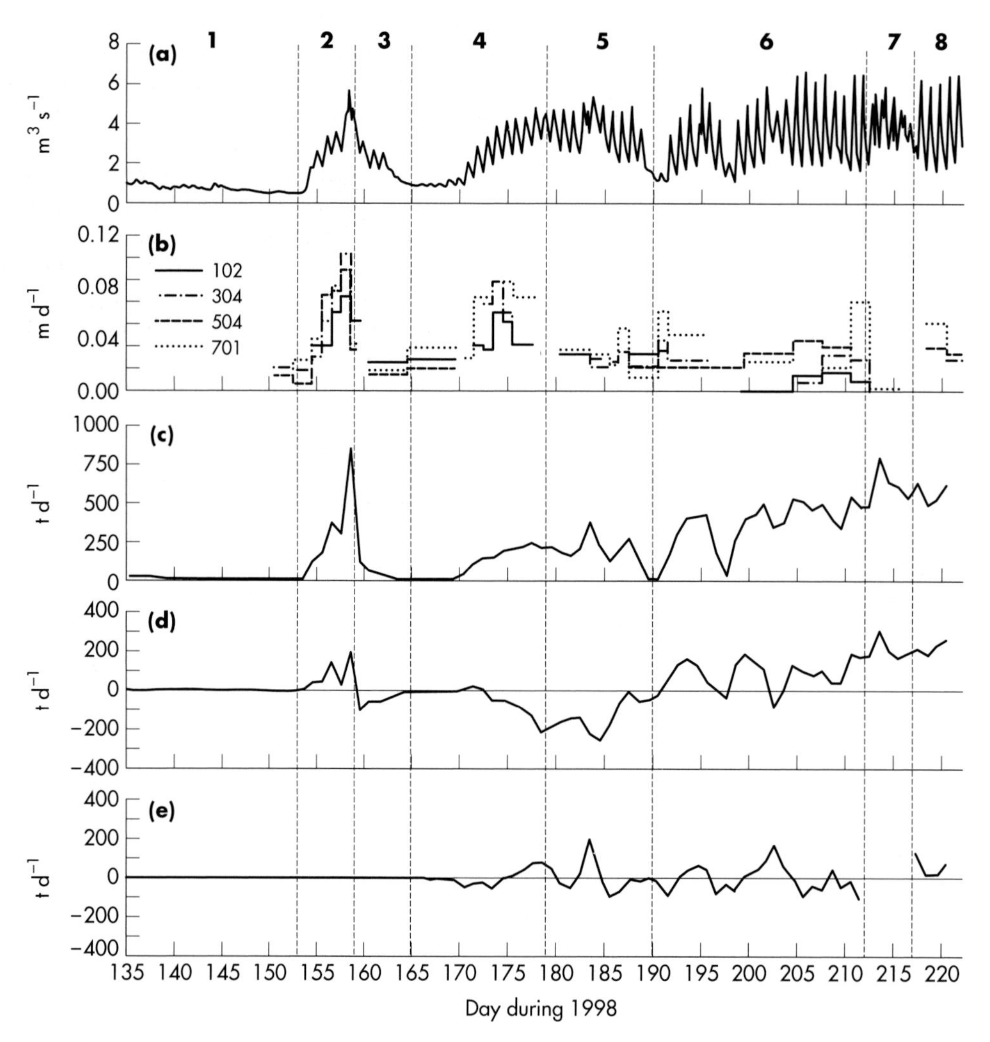

Figure 3.2 Catchment discharge, glacier velocity and sediment transport at Haut Glacier d'Arolla during 1998. (a) Hourly mean catchment discharge (Q) and subperiods of the melt season (numbered) used in Fig. 3.3; (b) glacier velocity at stakes 102–701 (Fig. 3.1; Mair *et al.*, 2003); (c) daily catchment suspended sediment load (SSL); (d) residual SSL from a log–linear relationship between SSL and Q; (e) residuals from a log–linear relationship between 'diurnal' SSL (see text) and Q diurnal amplitude.

(SSL; Fig. 3.2c) was calculated from SSC and hourly mean catchment discharge (Q) measured at the Grande Dixence S.A. intake structure (Fig. 3.1). Following log-transformation, SSL was highly correlated with Q ($r^2 = 0.93$); however, the residuals from this relationship (Fig. 3.2d) demonstrated subseasonal changes in the rate of sediment evacuation compared with discharge. Notably, the efficiency of sediment evacuation appeared to be lowest during subperiod 5 (Fig. 3.2d) when growth of the subglacial channel network was most rapid (Swift *et al.*, 2005a) and reached a maximum during subperiod 8 (Fig. 3.2d) when up-glacier extension of the network had largely ceased.

Relationships between SSC and Q (Fig. 3.3) for shorter periods of the melt season (cf. Fig. 3.2a) demonstrate that changes in the efficiency of basal sediment evacuation were controlled by both subglacial drainage system morphology and sediment availability (Swift *et al.*, 2005b). The gradient of the SSC versus Q graph (Fig. 3.3) is determined by the relationship between the sediment-transporting capacity of the flow, which scales linearly with flow velocity, and discharge. During subperiods 1–4, SSC $\propto Q^{-1.3}$ and therefore SSL $\propto Q^{-2.3}$ under predominantly distributed subglacial drainage conditions. However, during subperiods 5–8, SSC $\propto Q^{-2.2}$ and therefore SSL $\propto Q^{-3.2}$ under flow predominantly through hydraulically efficient channels. Flow velocity is therefore inferred to have increased more rapidly with discharge during subperiods 5–8, probably as a consequence of rapid discharge variation within subglacial channels under increasingly diurnally peaked surface runoff cycles (Swift *et al.*, 2005a,b). Relationship intercepts (Fig. 3.3) demonstrate changes in sediment availability under both distributed and channelized conditions, most notably a relative increase in availability between subperiods 5 and 8.

The limited availability of basal sediment during subperiod 5 (Figs 3.2d & 3.3) suggests that channelization confined meltwa-

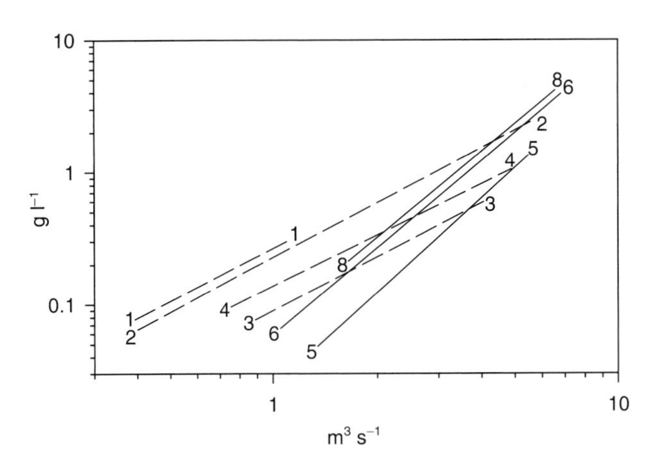

Figure 3.3 Relationships between SSC and Q plotted over the range of discharge observed for individual subperiods of the melt season (cf. Fig. 3.2); subperiod 7 has been excluded due to potentially high extraglacial sediment contributions during heavy rainfall. The most efficient evacuation of basal sediment occurred at flows >4 $m^3 s^{-1}$ during subperiods 6 and 8.

ter to areas of the bed from which sediment was rapidly exhausted. Thereafter, increasing sediment availability (Fig. 3.3), coupled with a strong increase in flow velocity with discharge, appears to have resulted in highly efficient sediment evacuation during the peak of the melt season (Fig. 3.2d). Strong increases within subglacial channels imply that water pressures also increased rapidly with discharge (Swift *et al.*, 2005b), suggesting that increasingly strong diurnal discharge variation may have increased sediment availability by encouraging local ice–bed separation, leading to extrachannel flow excursions, and/or a strong diurnally reversing hydraulic gradient between channels and the surrounding distributed system (Hubbard *et al.*, 1995). The importance of water pressure variation is supported by the absence of significant trends in the residuals from a relationship between 'diurnal' SSL (i.e. SSL calculated between diurnal discharge minima) and Q diurnal amplitude (Fig. 3.2e; Swift *et al.*, 2005b).

Whereas previous studies have generally emphasized declining sediment availability, these results demonstrate highly efficient sediment evacuation under channelized subglacial drainage conditions. Efficient flushing of basal sediment is critical in order to sustain the direct ice–bed interaction that is necessary for erosion (Alley *et al.*, 2003), and strong diurnal water pressure variation within subglacial channels may locally enhance glacier sliding. As a result, seasonal establishment of channelized drainage beneath temperate glaciers or ice caps has the potential to considerably elevate erosional capacity. Importantly, annual and glacier-to-glacier changes in the pattern and timing of subglacial drainage system evolution are likely to contribute significantly to variability in glacial sediment yield. Meaningful relationships between glacial sediment yield and surrogate indicators of erosional capacity are therefore unlikely to be found without explicit consideration of the hydraulics of subglacial drainage.

FOUR

A glimpse at meltwater effects associated with continental ice sheets

John Shaw

Department of Earth and Atmospheric Sciences, University of Alberta, Edmonton, Alberta, Canada T6G 2E3

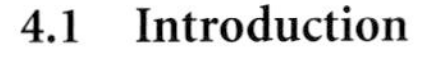

4.1 Introduction

This chapter covers some of the aspects of the subglacial landscape attributed to meltwater activities beneath the mid-latitude, Pleistocene ice sheets. A preliminary discussion deals with the sedimentary evidence for the presence of meltwater beneath their central parts. This is an essential step; without such evidence refuting the notion of cold-based ice-sheet centres, it would be

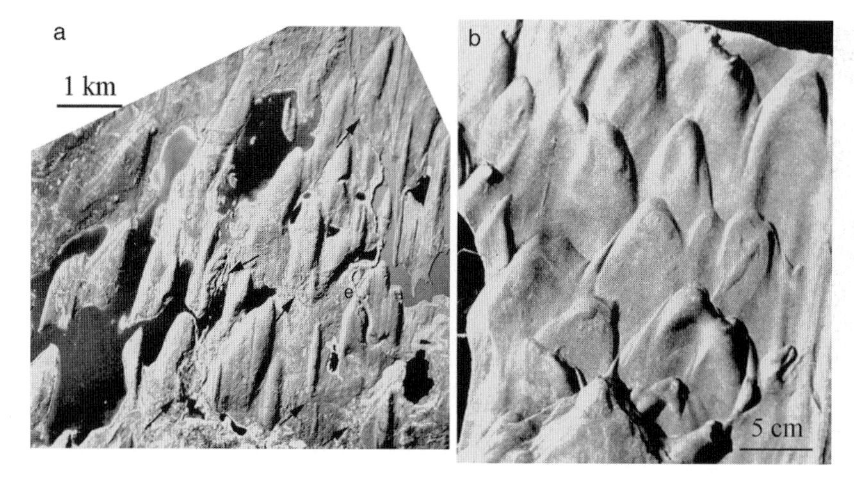

Figure 4.1 Form analogy. The aerial photograph (a) shows asymmetrical, parabolic drumlins. These bedforms are pointed up-flow and widen and grade gently into the surrounding surface distally. The asymmetry is about a dividing plane parallel to the flow and passing through the leading point. The sole marks or erosional marks (b) are orders of magnitude smaller in scale yet remarkably similar in form—asymmetrical, pointed up-flow, broadening and decreasing in depth down-flow (towards the bottom of the figure). Note the esker at e (Fig. 1a) and how the stream that formed the esker cut a tunnel channel over the drumlin. In the meltwater hypothesis, this combination represents a late stage of meltwater activity following the sheet flow that formed the drumlins.

difficult to argue for widespread meltwater activity extending to the so-called ice divides. Much of the discussion relates to subglacial landforms, both in surficial sediment and in rock. Finally, evidence for large-scale drainage paths and their broad significance is presented.

Subglacial landforms such as drumlins are commonly considered enigmatic because their formation has not been observed. We can, however, learn about the subglacial environment by studying its landforms. In the absence of actualistic observation, use of analogy is probably the best way to begin study of subglacial bedforms. By analogy, landforms and other features of known origin that resemble drumlins or other subglacial landforms are used to fill in the missing information about the subglacial processes that created immense fields of these streamlined landforms (Fig. 4.1).

The sets of analogues pairing subglacial bedforms with features created by turbulent fluids challenges modern drumlin studies holding that drumlins are formed by direct ice action on a deforming bed (Shaw, 1996). Erosional marks are formed by turbulent, low-viscosity flows that behave very differently than the high viscosity flows of ice and deforming sediment straining under stress applied by ice. The Reynolds number for water and wind analogues ($>10^4$) of subglacial forms compared with possible ice-flow analogies ($Re \sim 10^{-17}$) makes it extremely unlikely that actualistic wind and fluvial processes can be used as analogues for ice action and vice versa.

4.2 Evidence for abundant meltwater

If the meltwater explanation for drumlins and other bedforms is to be taken seriously, it is important to demonstrate that meltwater was available at the site of their formation. Some models of ice-sheet evolution include cold-bed conditions that preclude the meltwater hypothesis (Dyke, 1993; Kleman & Hätterstrand, 1999). The evidence for meltwater lies in sediments and geomorphology of the inner zones of the continental ice sheets, and the absence of melting would restrict the possibility of hydraulic connectivity between their inner and outer zones. The thermal and liquid water conditions are deduced from reconstruction of the processes that created the sediment and landforms. In this regard, the landforms and sediment more-or-less speak for themselves.

By interpreting the drumlins as meltwater forms, water necessary for their formation is invoked. Each future step in thinking carries the meltwater assumption and the observations to come must be compatible with water action if the hypothesis is kept. Should an observation flatly contradict this assumption, the hypothesis must be rejected. Thus, we are in search of the very evidence that will falsify the basic assumption of our hypothesis. The most significant question asks whether or not there was sufficient meltwater to form the subglacial landscape. The required magnitude of meltwater requires reservoir storage. The reservoirs and the source of water may have been subglacial and/or supraglacial. Supraglacial reservoirs may be favoured by abundant subglacial water. Reservoirs at the ice bed would cause flattening of the surface (Shoemaker, 1991) and, with climatic warming, the potential for supraglacial reservoirs in a greatly expanded ablation zone. Connection of a supraglacial lake with the bed (Zwally *et al.*, 2002a) might well have triggered megafloods (Shaw, 1996).

The stratigraphy in the inner zone points directly to meltwater at the ice-sheet bed (Bouchard, 1989). The bedrock around Lake Mistassini and Lake Albanel is striated and underlies beds of classic lodgement or, perhaps, deformation till. Deposition of these tills required melting. Associated stratified melt-out till with a high proportion of water-sorted beds, Sveg Till (Fig. 4.2a & b;

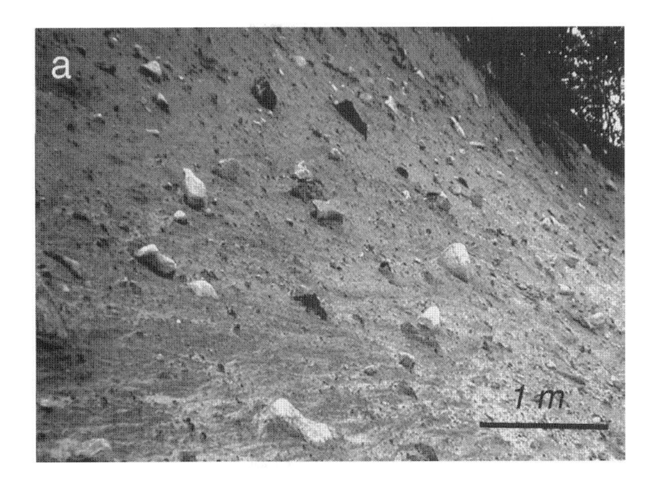

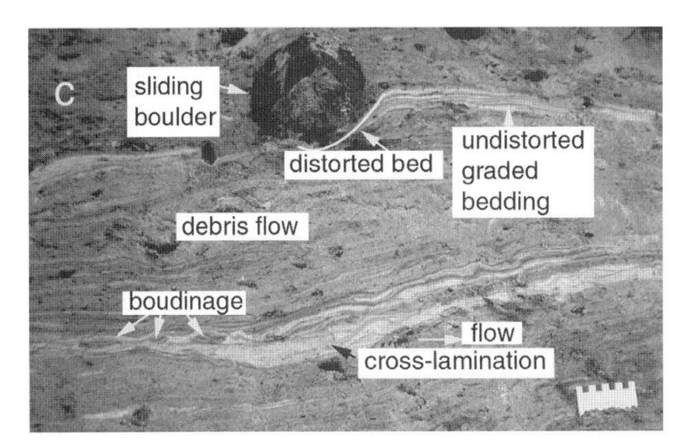

Figure 4.2 Stratified Sveg till (melt-out till, Shaw, 1979) and stratified sediment in Rogen moraine, Lake Albanel, northern Quebec. The bedded Sveg tills indicate abundant meltwater at the time of freezing-on and melting out. The draped, stratified and sorted sediment over the boulder (Fig. 4.2a) marks differential melting related to ice content. The deposits in the Rogen moraine show mainly graded deposits from suspension and minor deformation in boudinage where debris flowed in an extensional mode. A rock moving under gravity distorted primary graded bedding. This combination indicates deposition from suspension with minor sediment flow under gravity. A cavity environment with deposition of sorted sediment is suggested.

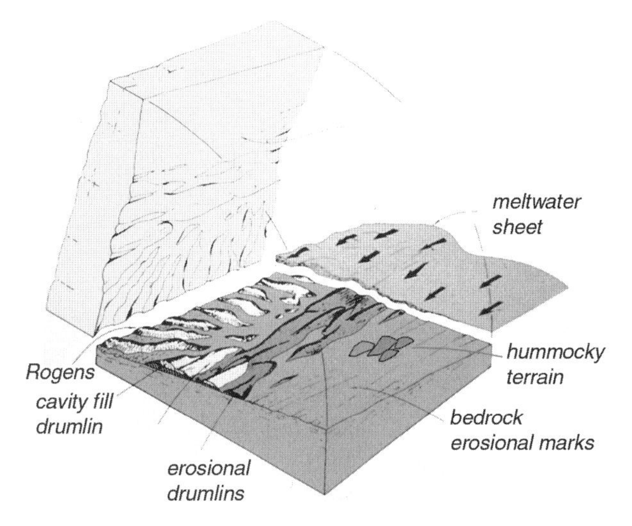

Figure 4.3 The meltwater model.

Shaw, 1979), rests on the fine-grained lodgement till. The Sveg Till marks an abrupt change in depositional regime. Its entrainment and deposition requires abundant meltwater for the freezing-on in the first place and subsequent melt-out of entrained basal debris (Shaw, 1979). The deposits of Rogen moraines are like the melt-out tills, relatively coarse, but they contain even more water-sorted sediment in thicker beds (Shaw, 1979; Bouchard, 1989; Fisher & Shaw, 1992).

The sedimentological and geomorphological evidence leaves no doubt that much of the inner zones of the Keewatin and Nouveau Quebec domes of the Laurentide Ice Sheet (Aylsworth & Shilts, 1989a; Bouchard, 1989) were warm based. Similarly, the interior of the Scandinavian Ice Sheet shows striations and lodgement tills and Rogen ridges there are composed of very local lithologies with ubiquitous sorted sediment indicating either formation by meltwater erosion and deposition or entrainment of sorted sediment prior to melt-out (Shaw, 1979).

The so-called Sveg tills (Fig. 4.2a & b) required meltwater at the bed during debris accretion and deposition. A deep investigation into the origin of Rogen moraine or Sveg tills is not necessary to support the view that there was abundant meltwater in the inner parts of the Scandinavian and Laurentide ice sheets. A cursory view of the sorting and stratification in the Sveg tills and associated Rogen moraines points to an abundance of subglacial meltwater at the time of their formation (Fig. 4.2).

4.3 Subglacial bedforms

Beyond the reasoning that the subglacial sediment indicates large quantities of meltwater in the inner zone (Fig. 4.3), the landforms have a story to tell. One of the most intriguing features of the Pleistocene ice sheets is the landscape they left behind. Much of this landscape appears almost synoptic: a vast, pristine, subglacial bed, although cross-cutting relationships show some time transgression (Shaw, 1996). The pristine nature of the landscape suggests that it was preserved with little alteration through deglaciation. Such subglacial landscapes were formed about the time of the late-glacial maxima of the ice-sheets; end moraines are surprisingly rare along the flow paths. This can only mean that recessional landforms were not formed over vast areas of the deglaciated landscape. We can speculate that meltwater storage and flow beneath the ice sheet caused flattening. When the ice returned to its bed, which sloped back up-glacier, there was insufficient slope to drive glacier flow. Melting under a regionally stagnant ice sheet is exactly the condition for the observed pristine subglacial ice-sheet bed. Nevertheless, there are arcuate end moraines associated with the Great Lakes, especially Erie and Michigan. There are also intriguing, nested, arcuate forms where the Bow Valley glacier spread out onto the Plains from its confined mountain and foothill valley. It is probable that these moraines represent local reactivation of the ice sheet following the last outburst. They are not to be confused with the giant ripples described by Beaney & Shaw (2000) further south in Alberta.

Bedforms of many types adorn the former bed of the Laurentide ice sheet. They indicate that for much of the landscape subglacial processes seem to have come to an abrupt halt. Assume the following for the sake of hypothesis testing: *the landforms considered here are created mainly by meltwater*. This is not to say that all subglacial landforms are fluvioglacial.

4.4 The meltwater hypothesis

Imagine a sheet flow of meltwater beneath an ice sheet on a scale of 100 km or so in width and a depth of a few tens of metres (Figs 4.3 & 4.4). The competence of this flow causes transport of all but the very largest boulders in its path. As well, cross-cutting assemblages of landforms may mark changes in current direction within broad flows. The geometry of the flow changes over the life of an event with an increasing formation of channels and a reduction in extent of sheet flow. Following an outburst, much of the glacier stagnates and melts *in situ*. The hypothesis explains the landforms at a basic level at which the hypothesis is expressed in the first instance. There would be no point in proceeding if this were not the case. What has been done over the past 20 yr or so is to test the hypothesis in terms of its ability to explain new observations.

4.4.1 Cavity-fill drumlins

The nature of the flows that formed drumlins in the form of infilled cavities (Fig. 4.3) can be determined only by deduction based on shape and composition of the drumlins themselves. The initial work in this regard was on the so-called Livingstone Lake drumlin field, northern Saskatchewan (Figs 4.1 & 4.2; Shaw, 1983; Shaw & Kvill, 1984). In the first instance, the drumlins in this field, when inverted, were seen to be identical in form to sets of erosional marks in areas of bedrock erosion by wind and water and, in particular, where turbidity currents scoured the ocean floor, generating sole marks (Fig. 4.1). Cavity-fill drumlins are positive forms on the landscape and are considered to be infills of cavities (inverted erosional marks) eroded upwards into the ice bed by broad sheets of meltwater (Fig. 4.3). The dominance of sorted sediment within the cavity fill drumlins and the predominance of

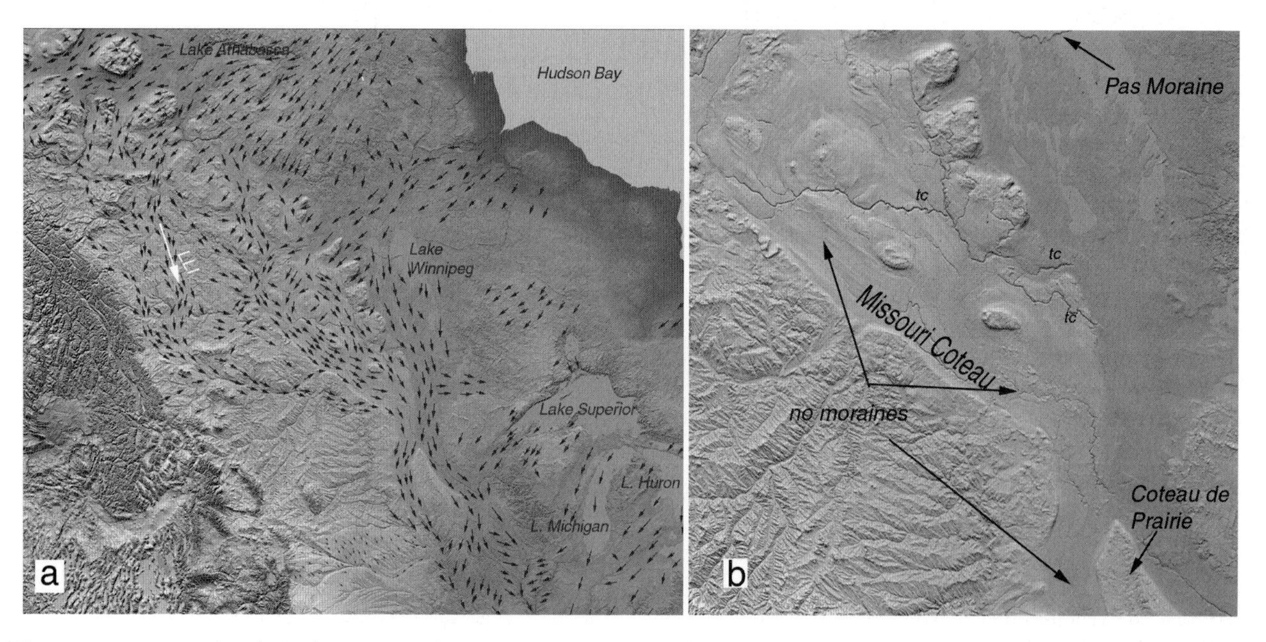

Figure 4.4 NASA Shuttle Radar Topography Mission (SRTM) hill shades. (a) The southwestern portion of the Laurentide Ice Sheet together with the extraglacial area. Arrows, indicating a synchronous flow of meltwater with strong channelling towards the west, mark flow paths. The northern part of the Livingstone Lake 'flood path' is a narrow filament in a broad flow. (b) Close-up of part of the flow with strong differential erosion demarking the Missouri Coteau and the Coteau du Prairie. The bedforms used for mapping flow directions in Fig. 4.4a are clearly visible in 'Mars like' channels. The Pas Moraine stands out, yet other tracts show no moraines over long stretches (Fig. 4.4b). In these stretches there is no overprint of glacial retreat. Rather, confluent flows eroded meltwater tracts which are preserved in the landscape.

subangular clasts with over 80% local bedrock lithologies indicate that the cavities were very efficient sediment traps. The sediment had been transported by water, but not very far (Bouchard, 1989; Shaw, 1989; Shaw *et al.*, 1989; Fisher & Shaw, 1992).

4.4.2 Erosional drumlins

These are probably the most common drumlins and their explanation is relatively uncomplicated. Each drumlin is a residual ridge left upstanding as the surrounding countryside is scoured (Fig. 4.5). The important problem here is to explain the preferential scour: it corresponds to erosion by horseshoe vortices (Fig. 4.5). Obstacle (drumlin) formation is self-perpetuating because the obstacle itself sets up the vortices. High Reynolds numbers are required because the extension of the vortices down-flow requires high inertial forces in the fluid. In the early stages of formation of these drumlins, hairpin troughs define the ridges. In more mature forms, these troughs are not so prevalent because the ridges become streamlined (Shaw, 1996). Figure 4.5 illustrates these points for drumlins, for hairpin scours in bedrock and for a model of the erosional processes and kinematics of horseshoe vortices. Tunnel channels, marking the passage from sheet flow to channelized flow, dissect extensive fields of these drumlins. Note that the configuration of the Kelleys Island scours and the drumlins are very similar (Fig. 4.5).

4.4.3 *S-forms*

Landforms and landform distributions produced in the hypothetical subglacial environment are indicated in a general model (Fig. 4.4). Like most bedforms, individuals may be transverse or longitudinal, depending on their alignment relative to the flow. Purely erosional landforms are scoured from hard bedrock and are referred to as *s-forms*, standing for sculpted forms (Kor *et al.*, 1991; Fig. 4.3). Ljungner (1930) conducted the first comprehensive study of *s-forms* on crystalline bedrock in southwestern Sweden. Hjulström (1935) and Dahl (1965) also argued that these forms are products of meltwater erosion. See Kor *et al.* (1991) for details.

4.4.3.1 Muschelbruche

Muschelbruche are scoop-like depressions in the rock, with sharp, parabolic upstream rims and a steep slope on the upstream side of the depression. In the down-flow direction the floors of depressions are gently sloping and merge imperceptibly with the rock surface. Experimental muschelbruche may be formed in flumes by running slightly acidic water over a plaster of Paris bed (Shaw, 1996). The forms of the flume muschelbruche and those in nature are identical. This experiment counters a common misconception regarding sheet flows and bedforms. After 30 h of flow there was no sign of channels forming in the bed, but muschelbruche continued to be created and to evolve beneath sheet or broad flows. There seems little point in raising theoretical objections to bed formation beneath broad flows when observations clearly show such formation.

4.4.3.2 Sichelwannen

Sichelwannen are erosional forms with sharp, crescentic upstream rims, and a crescentic main furrow or trough wrapped around a medial ridge (Ljungner, 1930). Some sichelwannen have lateral furrows alongside the main furrow (Shaw, 1996). The arms of the main furrow narrow downstream. Like muschelbruche, they form offsetting, *en echelon*, patterns.

Allen's (1982) sketch of the classic flute, an erosional mark formed by turbidity currents and commonly observed at the base of ancient turbidites (Fig. 4.1), is similar to sichelwannen. Identical forms are produced on the surface of glaciers where strong winds enhance ablation. As well, experimentally produced erosional marks show the same morphology as flutes and sichelwannen, including secondary furrows (Allen, 1982; Shaw, 1996). The form analogy between sichelwannen and forms generated by turbulent fluids is compelling.

4.4.3.3 Spindle form

Spindle form erosional marks are found in conjunction with sichelwannen and muschelbruche. They are narrow troughs with sharp rims, commonly asymmetrical about the long axis and pointed at the upstream ends (Allen, 1982; Shaw, 1996). Spindle marks are commonly curved and even sinuous in places, appearing to relate to coiled structures in the flow (Shaw, 1988). Spindle forms also have analogues with erosional marks observed in nature and those produced experimentally. They are particularly common as turbidite sole marks (Dzulinski & Walton, 1965) and as aeolian scours on resistant bedrock.

4.4.4 Associations of erosional marks

All three of the above erosional marks are found in association with other marks interpreted to be meltwater forms: potholes, furrows, troughs, transverse troughs, stoss-side-troughs, rat tails (Shaw & Sharpe, 1987; Murray, 1988; Shaw, 1988; Sharpe & Shaw, 1989; Kor *et al.*, 1991; Shaw, 1996; Sawagaki & Harikawi, 1997; Gilbert, 2000; Shaw, 2002). These forms are illustrated in the classification of Kor *et al.* (1991) and their morphology can be readily related to flow structures in turbulent and separated fluids (Shaw, 1996). Some details of rock surfaces further support meltwater formation. Wherever *s-forms* are pristine and unweathered, rock surfaces are highly polished by abrasion. Scanning electron microscopy (SEM) images of the surfaces indicate that detrital particles of fine silt and clay size were detached by abrasion.

4.4.5 Large-scale erosional marks

Large-scale erosional marks in bedrock might be loosely delimited as those with lengths >10 m. There are three main kinds: streamlined bedrock hills (Sawagaki & Hirakawa, 1997), crescentic scours and rock drumlins (Kor *et al.*, 1991), and furrows (Kor *et al.*, 1991). These large-scale forms all carry superimposed smaller forms.

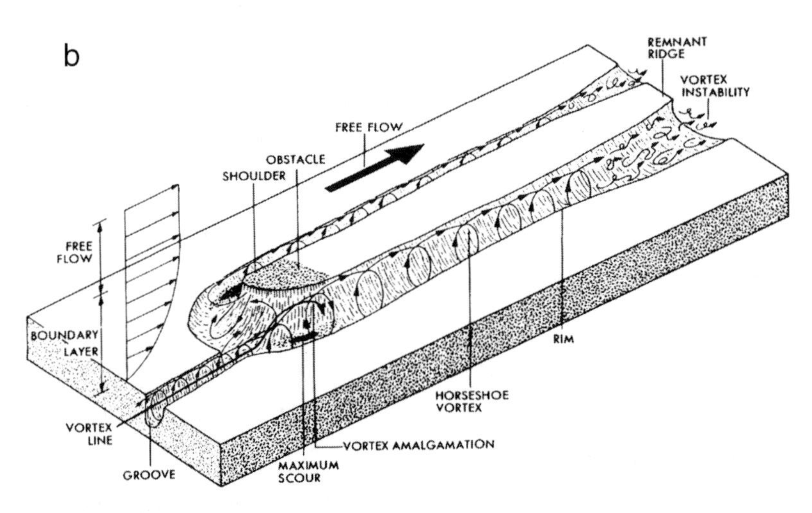

Figure 4.5 Erosional drumlins and horseshoe vortices. Drumlins near Prince George, British Columbia (a) are clearly defined by hairpin scours related to horseshoe vortices (b). The geometry of the horseshoe vortex and the amalgamation of vortices (b) explains the absence of cross-cutting hairpin scours in (a) and (c). Erosional marks on Kellys Island, Lake Erie (c, photograph by Mandy Munro-Stasiuk). These erosional marks show the integration of crescentic scour and linear furrows and several identical elements appear in (a) and (c), particularly hairpin scours and the way in which crescentic scours define the pointed, proximal parts of rock drumlins.

4.4.5.1 *Streamlined hills*

Streamlined hills in Antarctica are up to 300 m high and they are aligned with subglacial meltwater flow rather than ice flow (Sawagaki & Hirakawa, 1997). Similar streamlined hills on the Antarctic continental shelf have a distinctive crescentic trough wrapped around their proximal end and extending down-flow as two parallel furrows (Shipp *et al.*, 1999; Lowe & Anderson, 2003). A similar arrangement is observed at two classic erosional mark sites, Cantley near Ottawa (Sharpe & Shaw, 1989) and Kellys Island, Lake Erie (Fig. 4.5; Goldthwait, 1979).

4.4.5.2 *Crescentic troughs*

Crescentic troughs are wrapped around the upstream ends of rock bosses or rock drumlins. The knobs and drumlins usually carry stoss side furrows and the trough extends downstream as furrows before bifurcating in sichelwannen. The sichelwannen in turn are eroded into the residual rock, producing a long, tapered tail, such that the residual downstream from the crescentic scour takes on the form of a rock drumlin (Fig. 4.5). This arrangement of forms corresponds to erosion by horseshoe vortices (Fig. 4.5; Shaw, 1994) and is noted over a wide range of scales in areas of aeolian erosion, producing yardangs (Shaw, 1996). Consequently, these large-scale features are as expected for erosion of relief features submerged in a subglacial meltwater flow. As their morphology supports this explanation and there are analogies of similar scale in nature, application of the meltwater hypothesis to these large-scale crescentic scours and rock drumlins could only be denied if there was a strong argument showing that flows of the required scale could not exist.

4.4.5.3 *Furrows*

Furrows are large-scale subglacial features carved as troughs into bedrock. Furrow walls and floors are ornamented by all manner of erosional forms, which at a small scale replicate the furrows themselves. Probably the most spectacular example of furrow ornamentation is at Kelleys Island in Lake Erie (Fig. 4.5). This site illustrates the many nuances of meltwater erosion, particularly the importance of vortex interaction with bed features and the resultant crescentic scour/furrow sequences.

4.5 Bedform extent

The above brief commentary on bedrock features leaves unstated much of the detailed discussion of bedrock erosion by subglacial meltwater. At the same time it highlights the uncomplicated correspondence between analogous forms and, in a more sophisticated way, the correspondence between form and expected flow process in the meltwater hypothesis. In other words, the hypothesis is not simply based on form analysis. The brief comment on scale in the preceding section promises a telling test for the hypothesis. How can we determine the scale of the postulated meltwater flows?

Evelyn Murray (1988) mapped *s-forms* in the Kingston area and inferred a flood several tens of kilometres wide. Kor *et al.* (1991) used the same approach in the French River area, Georgian Bay. Ground mapping of small-scale features and aerial photograph analysis of large-scale features illustrate a flow at least 70 km wide. Subsequent mapping west of Kilarney increased the width to about 150 km. In the absence of cross-cutting relationships of the erosional marks, what we see is a synoptic view of an enormous meltwater flow. The width scale of this flow preempts the argument that meltwater floods on the scale of drumlin, Rogen and hummock fields are impossible.

4.6 Flow magnitude

From information garnered to this point we are close to being able to estimate the instantaneous discharge of the French River event. The relief of rock drumlins in the area is in excess of 20 m and the flow must have submerged these landforms. The mapped width of *s-forms* gives a minimum width of the flow. We take the conservative estimate of 70 km. It then remains to estimate the velocity of the flow to obtain the instantaneous discharge. Rounded boulders rest on the erosional surface and are found as boulder deposits in sheltered locations. Many of these boulders carry percussion marks indicating violent transport. Using the range of velocities required to transport such boulders (Kor *et al.*, 1991), we assume a conservative velocity of $10\,\mathrm{m\,s^{-1}}$. Thus we can obtain the instantaneous discharge using the continuity equation $Q = wdv$, where Q is discharge, w is width, d is depth and v is velocity. The estimated discharge of $1.4 \times 10^6\,\mathrm{m^3\,s^{-1}}$ would drain Lake Ontario (volume $1640\,\mathrm{km^3}$) in about 13 days.

4.7 Flow paths—a bigger picture

The existence of these enormous flow paths raises the obvious questions about their number and extent, the ways in which they affected the ice sheets, their timing and their extraglacial and climatic effects. Regarding climate, dramatic climatic change at the time of the Younger Dryas and the so-called 8.2 ka event recorded in the Greenland ice-cores are confidently attributed to meltwater outbursts from lakes (e.g. Clark *et al.*, 2003b), yet the potential importance of outbursts from beneath the ice itself are seldom considered. Blanchon & Shaw (1995) proposed that the sea-level and climatic changes around the time of the drainage of Lake Barlow Ojibway and Heinrich events H_0 and H_1 (ca. 12 ka and ca. 15 ka) were related to outburst floods from beneath the Laurentide ice sheet destabilizing ice grounded on continental shelves.

More recent work shows there is a time correlation between Laurentide and Cordilleran events. For example, there is a coincidence of the Laurentide outburst at ca. 15 ka and flooding from the Scablands at about 15.4 C^{14} kyr BP (Normark & Reid, 2003). The date is from mud that precedes deposition of a 57-m-thick turbidite said to have originated in a Lake Missoula flood. Until recently, the Cordilleran and Laurentide ice behaving synchronously would have been related to some external forcing such as climate. However, Shaw *et al.* (1999) suggested that the Scabland floods were connected to subglacial drainage from the Cordilleran Ice Sheet of interior British Columbia. Recent field work shows that the Cordilleran and Laurentide ice sheets were linked hydraulically during drumlin forming events. Consequently, linked, subglacial drainages involving the Cordilleran and Laurentide ice sheets could well have been responsible for the climate change and rapid sea-level rise event at about 15 ka (Blanchon & Shaw, 1995). As well, the meltwater events that triggered later, abrupt climatic change may have included a contribution from the Laurentide Ice Sheet itself. Shaw (1996) suggested that the sudden diversions and outbursts of Lake Agassiz might well have been a cascade effect, triggered by subglacial outbursts.

The NASA Shuttle Radar Topography Mission provides the evidence on hill shade maps, based on radar interferometry with a horizontal resolution of 938 m, for continent-wide, concurrent outbursts (Fig. 4.4). These images show distinct flow paths, marked by sharply defined erosional margins and streamlined bedforms. The paths are anabranching and the absence of cross-cutting relationships amongst the streamlined forms indicates that the flow patterns they represent are synoptic, that is the flows were part of a concurrent, continent-wide drainage system beneath the Laurentide Ice Sheet (Fig. 4.4). Figure 4.4 shows the flow paths as sets of arrows. The flow has been mapped directly from directional forms visible on the image.

In northern Alberta, the Livingstone Lake Event(s) scoured huge channels and left behind residual hills, several hundred metres above the channel level. These highlands stand out clearly on the hill shade, particularly the Caribou Hills and Birch Mountains (Fig. 4.4). Nevertheless, these hills were also overtopped by the enormous flows that sculpted their streamlined form. Trying to picture this is next to impossible; the scale is unimaginable! From the image we see that the flow to the east of the Caribou Mountains continued southwards and exited Alberta to the south of Calgary, east of the Cypress Hills. This flow is just part of the so-called Livingstone Lake Event (Rains *et al.*, 1993) with its source in the Keewatin Ice Divide zone (Shaw, 1996). It is a mere filament in a much wider flow.

The Livingstone event path is clearly identified and there is a wealth of detailed study supporting meltwater formation of features along this flood path [e.g., fluting (Shaw *et al.*, 2000); bedrock *s-forms*, cavity fill drumlins (Shaw & Kvill, 1984; Shaw *et al.*, 1989); sedimentary architecture and lithological composition (Shaw *et al.*, 1989), tunnel channels (Beaney, 2002), scablands and broad-scale erosion (Sjogren & Rains, 1995); hummocky terrain (Sjogren *et al.*, 1990; Munro & Shaw, 1997); lake systems (Shoemaker, 1991; Munro-Stasiuk, 2000)]. Shaw (1996) and Rains *et al.* (1993) present more general overviews which were designed to paint the bigger picture.

4.8 Earth system effects

The scale and coherence of the Livingstone Lake flow path are stunning, yet they pale in the larger scale image. The Shuttle Radar Topography Mission (SRTM) (NASA/JPL PIA03377) shows the true magnitude of these flows (Fig. 4.4). Probably the most exciting aspect of the flows is that, with few minor exceptions, where local relief dictates flow changes, the flows were *simultaneous*. The evidence for this is both simple and compelling. Rather than one set of forms cross-cutting another, the flow tracts merge and even carry interference patterns.

The Livingstone Lake event dominated the early thinking on flood tracts or paths, although it was clearly smaller than the drainage south of Winnipeg to the Mississippi. Leventer *et al.* (1982) provided independent evidence for just this kind of flood. Only the limitations of dating resolution prevented them from proposing extremely short-lived, high-magnitude drainage outbursts. With the continent-wide synoptic flow (Fig. 4.4a), the Livingstone Lake event pales to insignificance (Fig. 4.4a). We can make a rough estimate of the total drainage by extrapolating the flow estimates from French River (Kor *et al.*, 1991); the total discharge to the Gulf is about $2.7 \times 10^7\,m^3 s^{-1}$. The total volume of meltwater added to the Gulf of Mexico is more difficult to estimate. Taking the calculations for the Livingstone Lake event, $V = 84{,}000\,km^3$, where V is volume of flow (Shaw *et al.*, 1989), then extrapolating that figure to include the full width of flow, gives total rise in sea level of about 3.7 m attributed to water flowing to the Gulf of Mexico. This does not include contributing outlets via Hudson Strait, flows to the Arctic Ocean, or flow through the St Lawrence. The Arctic outlets must have been at least equivalent to those to the Gulf. As well, other continental ice sheets may have contributed meltwater directly to the Catastrophic Rise Event (CRE) at about 15 ka (Blanchon & Shaw, 1995). Consequently, the floods were capable of producing the rates and amounts of sea-level rise discussed by Blanchon & Shaw (1995). The predominant rise is at about 15 ka. The timing of this rise is close to that for the double peak of meltwater input to the Gulf of Mexico (Leventer *et al.*, 1982). In addition, sea-level rise might have destabilized ice resting on continental shelfs, causing the iceberg armadas of Heinrich events.

4.9 Conclusions

There is much more to write about the meltwater effects discussed here, both from the point of view of landscape and also from the large-scale, global effects of such high volumes of cold, sediment charged, freshwater. These global effects are expected to be extreme and the floods are expected to play an important part in explaining the various scales of abrupt climate change associated with Quaternary glaciations.

As many earth scientists do not consider the meltwater hypothesis credible, work must continue at the scale of landforms and landform associations. Although many are unlikely to be persuaded by such work, it must be done if the hypothesis is to be supported. The recent flurry of papers on Antarctic subglacial outbursts supports the concept of megafloods in the warmer ice sheets of the mid-latitude ice glaciers.

Acknowledgements

I am grateful to NSERC Canada for supporting this work from the beginning. Graduate students and colleagues have done much of the research cited here. I owe them an enormous debt. Above all, I could not have persevered in the face of often bitter criticism without the contribution and friendship of Bruce Rains. I am thankful to Peter Knight for his encouragement and generosity as this paper evolved.

FIVE

The erosional origin of hummocky terrain, Alberta, Canada

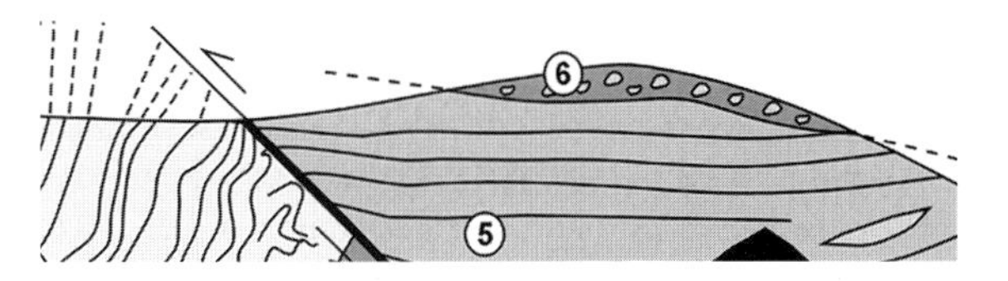

Mandy J. Munro-Stasiuk* and Darren Sjogren†

**Department of Geography, Kent State University, Kent, Ohio, USA*
†Earth Science Program, University of Calgary, Calgary, Alberta, Canada

Hummocky terrain is comprised of tracts of hummocks and depressions of various sizes and shapes that occur in formerly glaciated areas. Traditionally, this terrain is known as 'hummocky moraine', and is believed to represent deposition via letdown at, or near, the ice margins during ablation. Hummocks therefore have been used to delineate recessional stages of glaciation in many regions. For example, the four prominent north–south trending hummocky complexes (Fig. 5.1) in Alberta, Canada, are commonly identified as terminal or recessional moraines (e.g. Klassen, 1989) deposited by letdown during deglaciation. Observations in these hummocky zones, however, do not support the letdown theory. Other researchers have proposed alternative geneses for hummocky terrain in the region but we propose that these hummocks were formed by erosion and, more specifically, subglacial meltwater erosion. We also suggest that as such variation in hummock theory exists, the descriptive term 'hummocky terrain' should replace the genetic term 'hummocky moraine'.

The four major north–south trending hummocky belts in Alberta are known traditionally as (from west to east): the Duffield Moraine, the Buffalo Lake Moraine, the Viking Moraine and the Coteau Moraine (Fig. 5.1). Hummock form in these 'moraines' is typical of forms in most hummocky regions worldwide. We divide these forms into six types based on shape and pattern (Fig. 5.2; Table 5.1). Sediment in the Albertan hummocks, however, is atypical of letdown at the ice-margins. For instance, the 'Buffalo Lake Moraine' contains material including lodgment and melt-out till, *in situ* and disturbed lake sediments, and local *in situ* and thrust bedrock (e.g. Kulig, 1985; Tsui *et al.*, 1989; Munro & Shaw, 1997; Munro-Stasiuk, 2003). The 'Viking Moraine' contains pre-glacial lacustrine and glaciofluvial sediment, diamicton and *in situ* Cretaceous bedrock (Sjogren, 1999). The presence of *in situ* bedrock in some of these forms demonstrates that they are secondary; the product of erosion rather than deposition. In addition, exposures clearly show that intact regional lithostratigraphies and local sedimentary beds are truncated by hummock surfaces (Fig. 5.3). Thus, the hummocks are erosional, formed by excavation of the intervening depressions. Hummock surfaces are therefore representative of a landscape unconformity. This regional unconformity extends over a broad area at least as wide as the hummocky tracks (upwards of 50 km in places). The agent of erosion was one that involved movement, as simple observations note streamlining and transverse trends (type IV and V hummocks) towards the east-southeast and south-east(e.g. Munro & Shaw, 1997) (Fig. 5.2). These trends are similar to the other erosional fluted terrain observed in the region (Munro-Stasiuk & Shaw, 2002).

Sedimentary observations in the Buffalo Lake Complex point to a subglacial origin for the hummocks: subglacial eskers overlie the hummocks (Munro & Shaw, 1997), and the youngest recorded unit in the hummocks is a well-documented subglacial melt-out till (Munro-Stasiuk, 2000). Additionally, regionally consistent, strongly orientated clast fabrics in the till indicate ice movement towards the south-southwest, which is up to a 70° deviance from the surface trends noted in hummocks. Thus erosion of the hummock surfaces was therefore not contemporaneous with deposition of the underlying till; it occurred after till deposition.

We propose that the erosion was by subglacial meltwater and not by basal ice. Several lines of evidence support this: abrupt erosional surfaces are readily explained by fluvial erosion which removed sediment grain by grain, thus cutting into the underlying sediment but leaving beds undisturbed; surface boulders at many locations are best explained as fluvial lags resulting from lower flow competence in some areas; sorting of the lags suggests fluvial transport; many boulders are heavily pitted with percussion marks attesting to clast on clast collisions; type IV hummocks resemble fluvial bedforms and erosional marks produced on the underside of river ice (Ashton & Kennedy, 1972); and horseshoe-shaped troughs are wrapped around the upstream sides of some

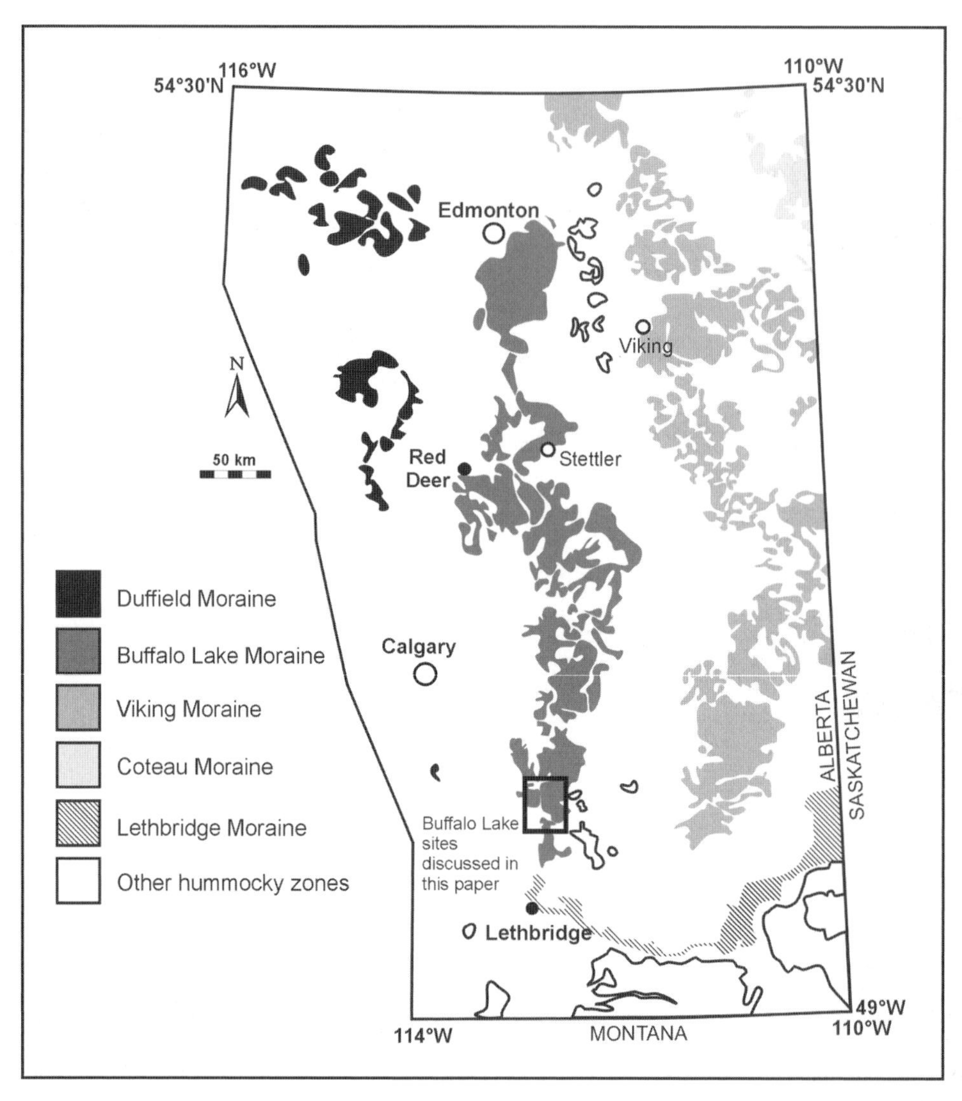

Figure 5.1 Distribution of hummocky terrain in central and southern Alberta. Names traditionally assigned to the 'moraine' belts are shown.

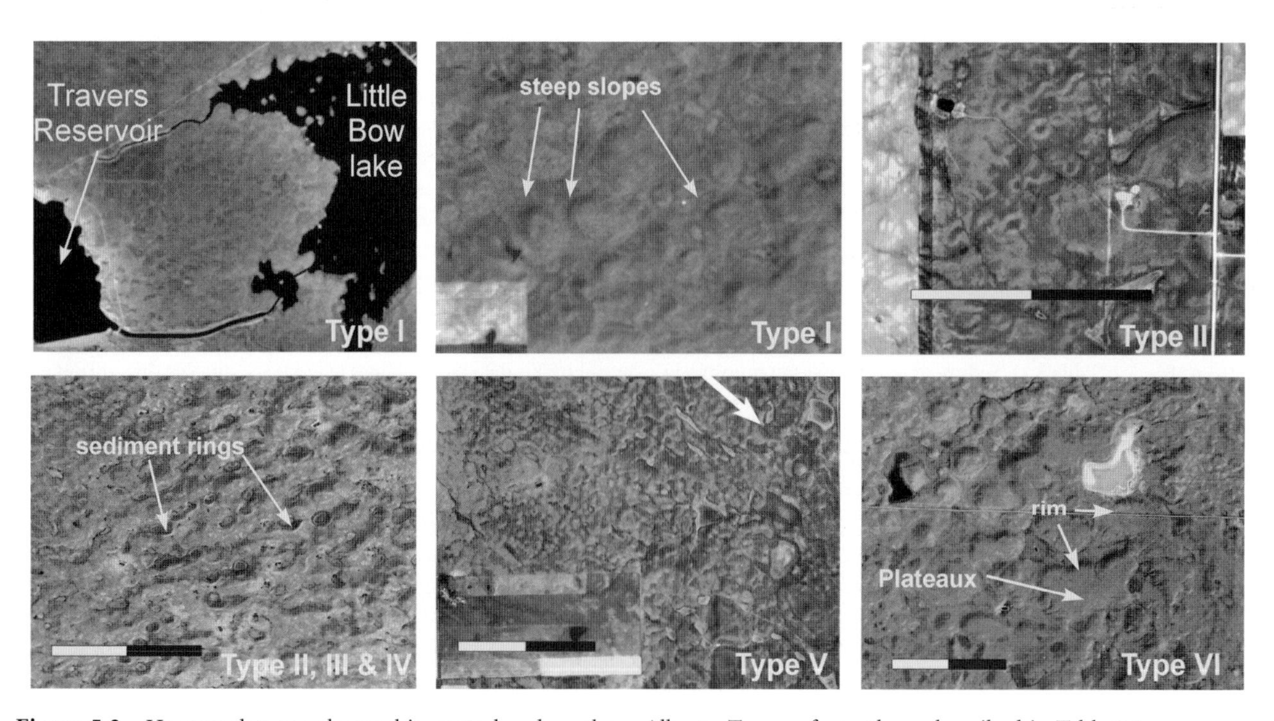

Figure 5.2 Hummock types observed in central and southern Alberta. Types refer to those described in Table 5.1.

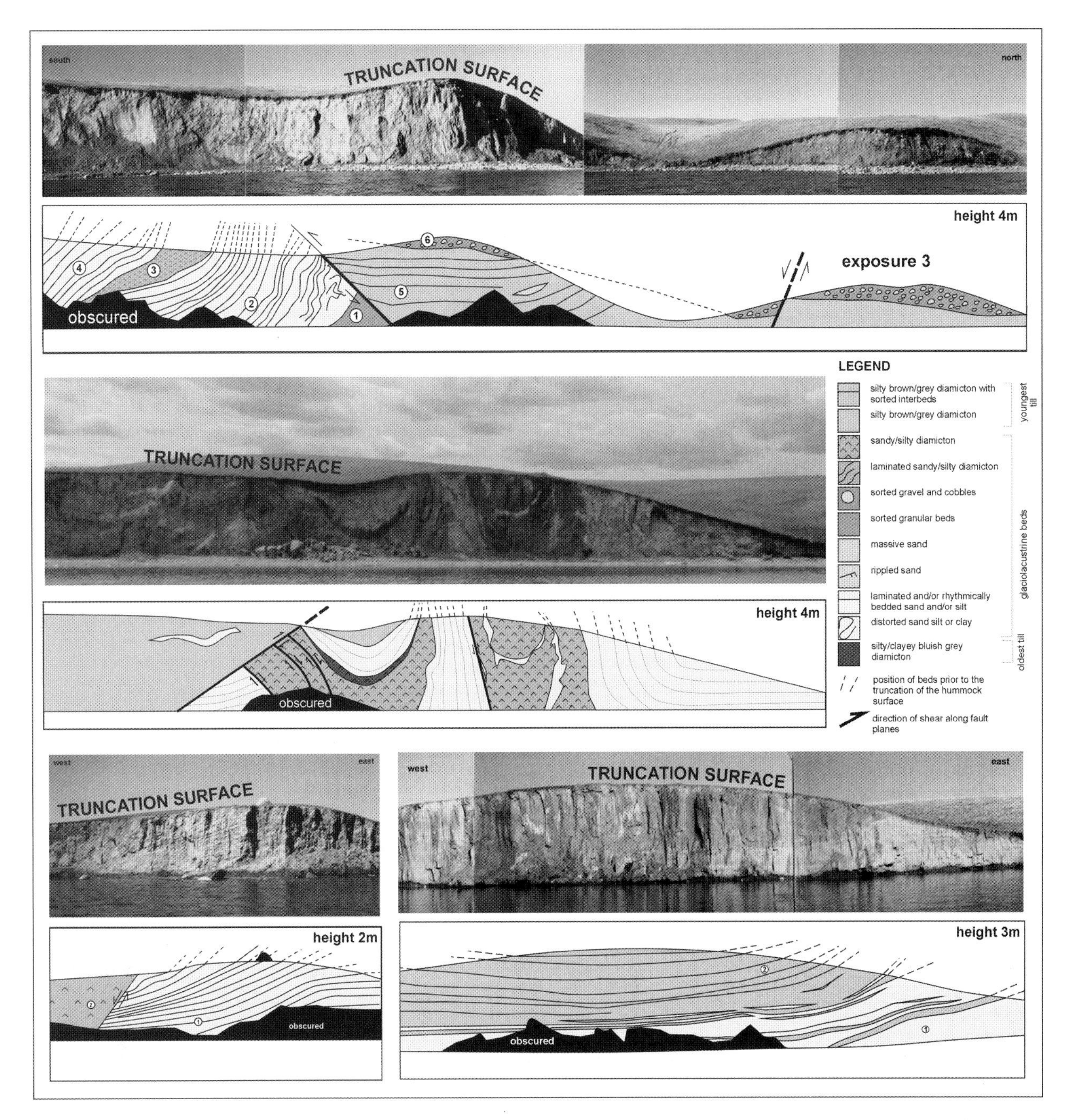

Figure 5.3 Representative exposures along Travers and McGregor Lake Reservoirs that illustrate the erosional nature of hummock surfaces.

Table 5.1 Hummock types and their description

Hummock type	Proposed name	Also known as	Description
Type I	Mounds with no discernible orientation or shape patterns	Stagnation moraine, disintegration moraine and uncontrolled moraine	Chaotically distributed mounds with varying size and height
Type II	Mounds with central depressions	Prairie doughnuts, rim/ring ridges and uncontrolled moraine	Mostly chaotically distributed mounds with minimal relief (<5 m) containing a shallow central depression
Type III	Linked mounds with central depressions	Hummock chains, rim ridge chains, donut chains and both controlled and uncontrolled moraine	Mostly chaotically distributed linked mounds with minimal relief (<5 m) containing a shallow central depression
Type IV	Ridged mounds	Cross-valley ridges, transverse ridges, transversal morainic hummocks and controlled moraine	Multiple semiparallel ridges resembling rogen moraine usually inferred to have formed at right-angles to ice flow direction
Type V	Elongate mounds	Drumlinized hummocky moraine, corrugated moraine, humdrums and controlled moraine	Mounds contain a distinct elongation in the shape and are usually inferred to have formed parallel to ice-flow direction; frequently have asymmetric shape and horseshoe shaped troughs around their steepest edges
Type VI	Moraine plateaux	Ice-walled lake plains and veiki plateaux	Larger and higher than surrounding mounds; can range from a few metres to several kilometres across; surface is generally flat to undulating, and is commonly surrounded by a discontinuous rim

elongate mounds (type V) and irregular shaped mounds (type I) suggesting scouring by horseshoe vortices generated at obstacles in the flow (e.g. Shaw, 1994).

While we propose that hummock formation was by subglacial meltwater erosion, in the absence of observations on hummock formation in the modern environment the agent and mechanics of hummock erosion are obviously open to debate. However, there is one major conclusion to be drawn from the Albertan observations: the Viking, Duffield, Buffalo Lake, and Couteau 'moraines' are *not* moraines. Consequently, reconstructions of Laurentide deglaciation in the prairies based on this assumption are misguided. We suggest that when relationships between underlying sediment and hummock surfaces are unknown, to avoid confusion and misinterpretation, the terms hummocky moraine, ice-disintegration moraine, stagnation moraine, and ice-stagnation topography should be abandoned in favour of 'hummocky terrain'.

Importantly, the interpretations presented here are based on specific field areas. As hummocks are known to form in subaqueous outwash at the margins of modern glaciers due to letdown and due to thrusting of debris bands, it is imperative that the morphology, sedimentology and structural relationships of hummocky terrain in each region of interest be studied thoroughly before determining landform genesis. Detailed descriptions and interpretations will lead to more accurate palaeoenvironmental reconstructions.

SIX

Tunnel channel character and evolution in central southern Ontario

Tracy A. Brennand*, Hazen A.J. Russell† and David R. Sharpe†

**Department of Geography, Simon Fraser University, Burnaby, BC V5A 1S6, Canada*
†Geological Survey of Canada, 601 Booth Street, Ottawa, Ontario K1A 0E8, Canada

6.1 Introduction

Valleys that truncate subglacial bedforms, contain eskers and follow upslope paths are the geomorphological expression of large subglacial channels—tunnel channels or valleys—that efficiently evacuated meltwater from beneath past ice sheets. In recent years there has been considerable debate as to the mechanism by which such large meltwater channels formed (e.g. Ó Cofaigh, 1996). Such debate is important as interpretations have a direct bearing on reconstructions of Late Wisconsinan ice-sheet dynamics and hydrology. Tunnel valleys may have formed at below bankfull conditions in a headward progression as saturated substrate dewatered and formed pipes at the ice margin or may have formed at bankfull conditions and filled rapidly over the course of a megaflood or jökulhlaup, draining a subglacial or supraglacial meltwater reservoir.

6.2 Background and methods

Central southern Ontario, Canada, exhibits a regional Late Wisconsinan unconformity that truncates Palaeozoic bedrock and a thick Quaternary sediment cover (e.g. Sharpe *et al.*, 2004; Fig. 6.1). This unconformity is composed of drumlins, s-forms and valleys (Fig. 6.2). Valleys were mapped and characterized using remote sensing, digital elevation models (DEMs) and field surveys (e.g. Brennand & Shaw, 1994; Russell *et al.*, 2003). Buried valleys were discovered and their geometry and sediment fill characterized by seismic reflection profiling and outcrop and drillcore sedimentology (e.g. Russell *et al.*, 2002).

6.3 Tunnel channel character

The valleys are assigned to five main classes based on their geomorphology, likelihood of breaching a regional till sheet (Newmarket Till), and probable depth of erosion (Figs 6.1 & 6.2). Structurally controlled, mainly bedrock valleys (class R) are steep-sided and form an anabranched NE–SW orientated system headward (north and east) of sediment-walled valleys. Bedrock valley walls are ornamented by s-forms (Shaw, 1988). Sediment-walled valleys continue the anabranched network and dissect a drumlinized terrain (Fig. 6.2). The largest sediment-walled valleys (class 1) trend NE–SW, are up to 40 km long and <7 km wide, have up to 50 m of topographic relief, and extend to bedrock at depths of >170 m. Between these large valleys are two systems of smaller, shallower (<20 km long, <2 km wide, <100 m deep), nested valleys. Deep channels (class 2) completely dissect the regionally

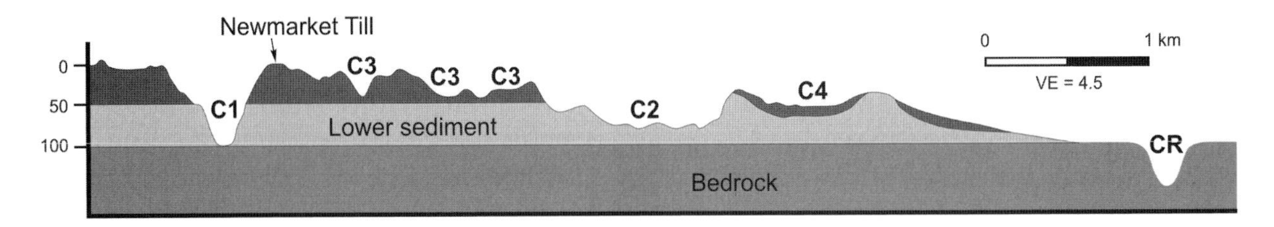

Figure 6.1 Simplified stratigraphy of central southern Ontario showing the relative depths of incision of five tunnel channel classes (C1, C2, C3, C4 and CR).

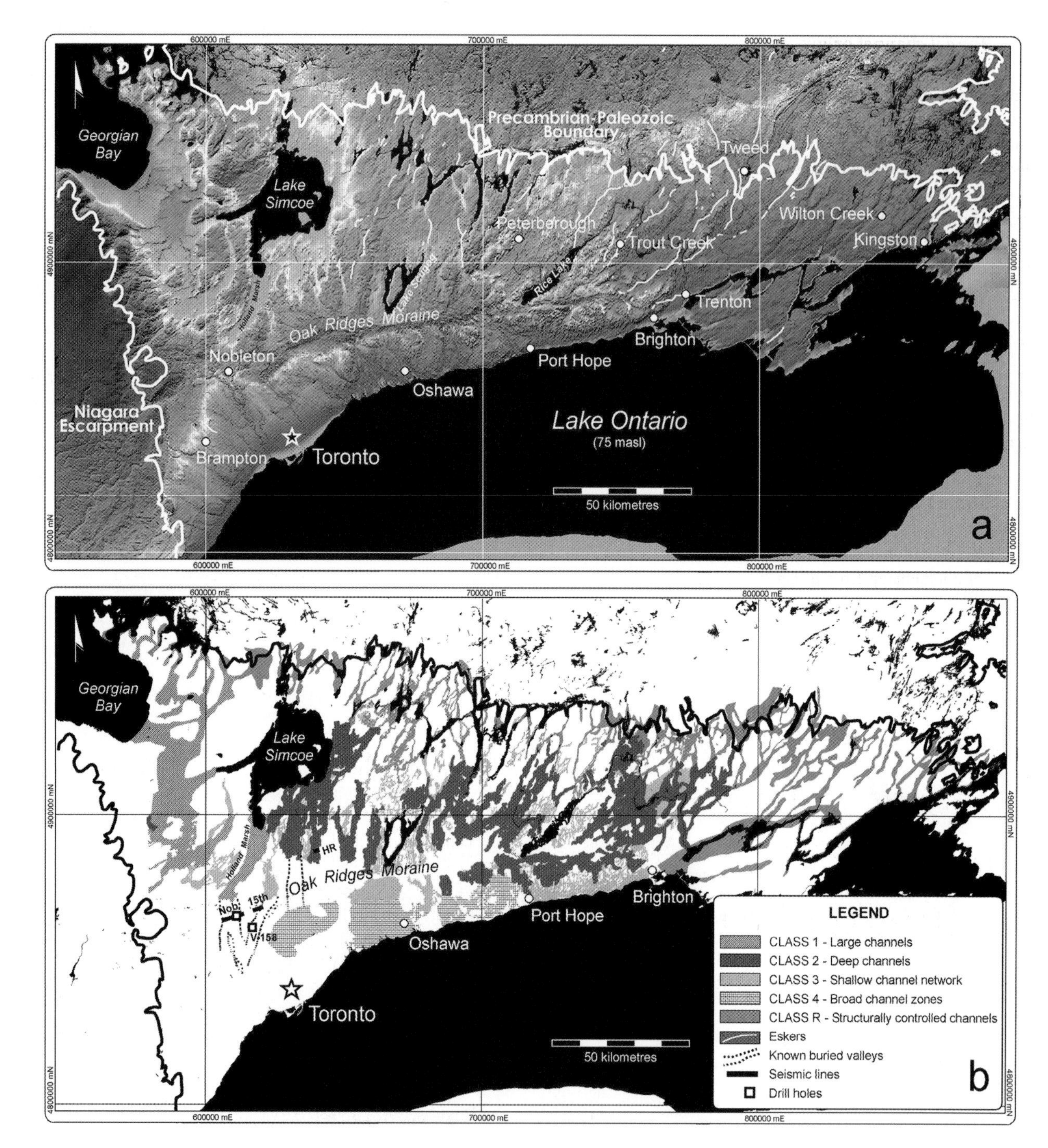

Figure 6.2 (a) Hillshaded digital elevation model (DEM) of central southern Ontario showing dissected drumlinized terrain, the Oak Ridges Moraine, escarpments and eskers. (b) Distribution of five tunnel channel classes, buried valleys, seismic lines and drill holes between the Precambrian–Palaeozoic boundary and the Niagara escarpment.

extensive Newmarket Till, whereas shallow channels (class 3) have a Newmarket Till substrate (Fig. 6.1). South of the Oak Ridges Moraine, broad, shallow erosional corridors (class 4) extend into Lake Ontario (Fig. 6.2).

Valley fills are up to ca. 150 m thick and include tunnel channel fills (20–60 m thick) in places overlain by ridge-building sediments of the Oak Ridges Moraine (<50 m thick), Halton Till (<30 m thick), deglacial lake sediments (<2 m thick) and/or Holocene fluvial and wetland sediments (<2 m thick). Tunnel channel fills often fine upward from gravel sheets, mesoforms (dunes) and eskers to beds of massive, graded and/or rippled sand to silt–clay rhythmites (e.g. Russell *et al.*, 2002).

6.4 Tunnel channel origin

This integrated, anabranched valley network is inferred to record a tunnel channel system that was produced and/or re-utilized by turbulent, subglacial meltwater flow released during outburst floods in the Late Wisconsinan, because valleys: (i) are incised into Late Wisconsinan, drumlinized till (Newmarket Till), are locally buried by Oak Ridges Moraine sediment and contain deglacial lake sediment; (ii) have undulating floors and upslope paths; (iii) locally contain eskers and are filled by sediments indicative of rapid sedimentation (e.g., sandy hyperconcentrated flow deposits); (iv) exhibit no evidence of convergent sediment deformation along their margins; (v) are cut to elevations below Lake Ontario base level and fail to terminate in deltas or fans at proglacial or modern shorelines; and (vi) contain modern underfit streams up to an order of magnitude narrower than the valleys (e.g. Brennand & Shaw, 1994; Russell *et al.*, 2002).

6.5 Tunnel channel evolution

The spatial variation in valley character records the temporal evolution of a jökulhlaup from a regional shallow channel network (class 3) to progressively fewer, larger channels (class 2 then 1) as flow concentrated and waned; the bedrock channels (class R) were probably antecedent and re-utilized. Bedrock structure, ice-bed gap width (Brennand & Shaw, 1994) and enhanced scour at thread confluences and hydraulic jumps (Russell *et al.*, 2002) controlled tunnel channel location. Erosion of sediment-walled channels was probably enhanced by groundwater flow and piping at depth (through sandy beds of the lower sediment, Fig. 6.1; Russell *et al.*, 2002). Channel fills record rapid and voluminous sedimentation during waning jökulhlaup flow (both fluidal and hyperconcentrated) followed by subglacially ponded sedimentation (e.g. Brennand & Shaw, 1994; Russell *et al.*, 2002).

This interpretation of central southern Ontario tunnel channels is consistent with the view that the subglacial landsystem (drumlins, valleys and s-forms) was eroded by a regional meltwater underburst—the Algonquin event—that unsteadily evolved from sheet to channelized flow (e.g. Shaw & Gilbert, 1990). As fan deposits are not observed at the southern ends of channels it is likely that channel formation was contemporaneous with an underburst event that eroded drumlins in Lake Ontario and swept away most sediment derived from channel erosion. Ice sheet thinning and flattening associated with underbursts facilitated deglaciation by regional downwasting and stagnation.

SEVEN

Glacial bedforms and the role of subglacial meltwater: Annandale, southern Scotland

Tom Bradwell

British Geological Survey, West Mains Road, Edinburgh EH9 3LA, UK

Since the advent of satellite imagery, geomorphologists have used remotely sensed images to gain a better understanding of large-scale landform assemblages. The main benefits of using satellite imagery are clear: the large field of view and the range of display scales, both allowing a higher speed of coverage. Many features undetectable on aerial photographs at large scales become readily apparent on LANDSAT images when viewed at a scale of 1: 100,000 or more. Unfortunately, few studies use both types of imagery in tandem. This study shows the advantages of combining modern satellite images with aerial photographs and traditional field survey techniques to address problems across a wide range of scales.

The geomorphology of Lower Annandale, between Lockerbie and Moffat, is very intriguing. When viewed from the ground or low-flying aeroplane the ground looks unremarkable—rolling farmland punctuated by occasional streams, typical of much of

Figure 7.1 Subset of a LANDSAT 5 image of Southern Scotland (bands 4, 5 and 7). Lockerbie is just off the southern edge of the image and Moffat is just to the north. The region is underlain by Silurian greywacke sandstones that strike NE–SW. The extent of the Permian sandstone in the south is also shown. Strong large-scale linear features cross-cut the geological 'grain'. The best developed, highly elongated, ridges and grooves are found in the central portion of the image, in the Annan Valley. The arrow shows the direction of palaeo-ice flow. B, Beattock; J, Johnstonebridge.

lowland Scotland. When viewed by satellite, however, 700 km above the earth, subtle patterns emerge. The LANDSAT 5 image in Fig. 7.1 is ca. 20 km across, has a pixel resolution of 30 m, and clearly shows the strong north–south lineations in the wide valley of the River Annan. The positive elongate features are in the same family of subglacial bedforms as drumlins: namely streamlined, elliptical, mounds of sediment or bedrock associated with ice-sheet flow. However, the large spindle-shaped landforms in Lower Annandale, typically 1–3 km long with length:width ratios up to 10:1, are probably best referred to as megadrumlins and megaflutes (*sensu* Rose, 1987). These features probably formed beneath the last ice sheet to cover the area, during the Late Devensian stage.

Using 1:24,000-scale aerial photographs, a more detailed inspection of the landform assemblage is possible. Further examination reveals a complex ridge-and-groove topography in Lower Annandale, with megaflutes and drumlins representing 'islands' defined by a large-scale network of interconnected grooves. Once recognized, these large grooves, often very shallow in form and gradient, can be mapped across the whole valley in detail. They occur on a range of scales (Figs 7.2 & 7.3), with spacings of tens to hundreds of metres, and cut through both bedrock and superficial deposits. The overall trend of the linear grooves in Lower Annandale is due north–south, whereas the strike of the steeply dipping greywacke sandstone is orientated NE–SW. A primary bedrock control on groove formation therefore can be ruled out.

Detailed walkover survey of the ground between Beattock and Johnstonebridge (Fig. 7.1) confirmed the existence of shallow, linear, depressions surrounding low, elongate, ridges. On the valley side, above ca. 200 m OD, topographic high points tend to be composed of bedrock (Fig. 7.2). On the valley floor, between the Rivers Annan and Kinnel Water (Fig. 7.3), hills are more commonly composed of glaciogenic sediments, typically lodgement till or coarse gravelly diamictons. The extensive intervening depressions are often gravel filled. These linear depressions, or grooves, are interpreted as glacial meltwater channels. The channels undulate in long profile, implying a subglacial origin. There is no possibility that these wide, largely dry, channels could have been cut during the post-glacial period. Furthermore, when mapped from aerial photographs, the channels represent an interconnected, anastamosing, network (Fig. 7.3). Given that the grooves cut into bedrock southwest of Beattock are subglacial meltwater channels, and that these same channels also cut through glacial sediments to define large drumlins and megaflutes (Fig. 7.2), it is argued that many of the streamlined landforms are erosional remnants.

It is the author's view that large volumes of debris-laden subglacial meltwater, acting under high pressure at the base of a mobile ice sheet, were responsible for carving the network of channels seen in Lower Annandale. The residual landforms are drumlins and megaflutes in bedrock and glacial sediments. The size and elongation of these streamlined forms increase gradually down-valley from Beattock to Johnstonebridge, possibly indicating increasing basal-meltwater flow linearity at higher velocities.

This case study echoes the conclusions made by Sharpe & Shaw (1989), working in the Canadian Shield, namely that the role of subglacial meltwater may be more significant than previously thought in shaping the glacier bed. It should be stated, however, that these findings are in sharp contrast to the more accepted theories of drumlin formation (cf. Boulton, 1987; Rose, 1987; Menzies *et al.*, 1997). Presently, there is no evidence to suggest whether the channel networks identified in this case study are the

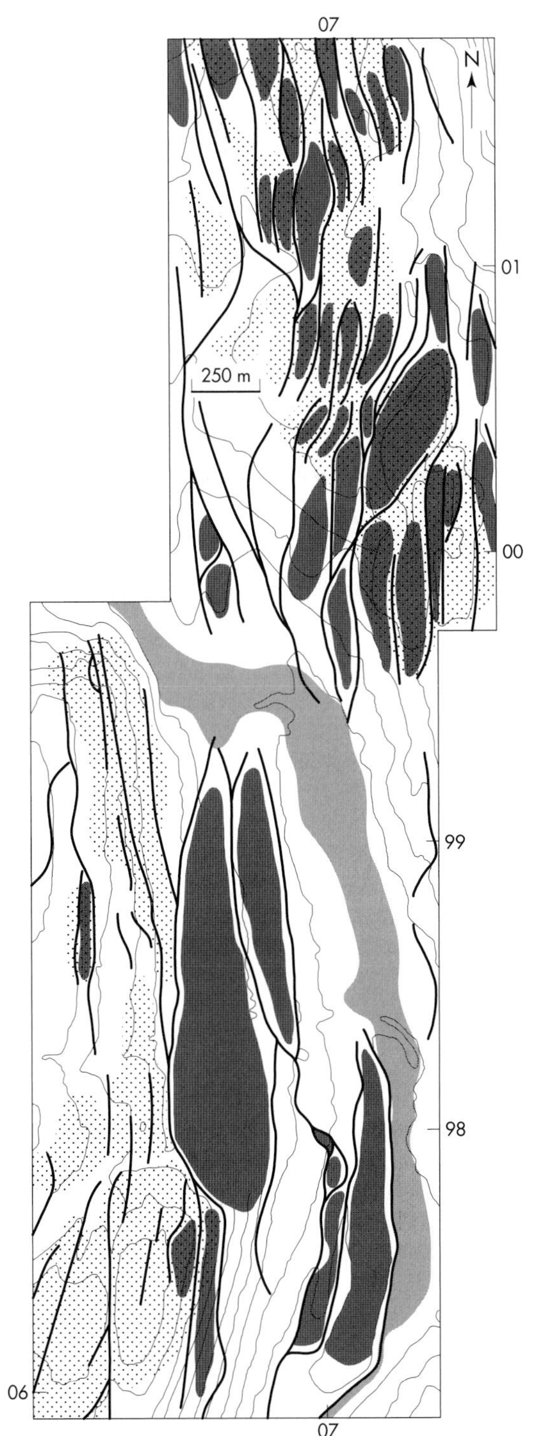

Figure 7.2 The geomorphology of the area immediately southwest of Beattock, Annandale. In the north of the area closely spaced subparallel channels define the outline of remnant, drumlinoid, bedrock hills. Individual channels can be traced over distances of 2–3 km. Further south these channels dissect glacial deposits and bedrock alike, resulting in elongate megadrumlinoid forms. Contours at 10 m vertical intervals; drumlinoid forms shown in dark grey; bedrock at surface shown as stipple; modern alluvium shown in light grey. See Fig. 7.1 for location. An aerial photograph is included at www.blackwellpublishing.com/knight

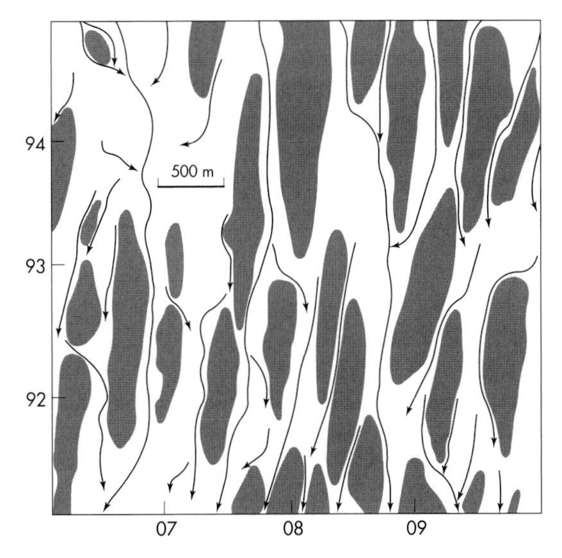

Figure 7.3 The geomorphology of the area north of Johnstonebridge, Lower Annandale. Subtle interconnected channels, cut into superficial deposits and bedrock, delimit the extent of streamlined drumlinoid ridges. These N–S glacial lineations are found widely across Lower Annandale. See Fig. 7.1 for location.

result of catastrophic subglacial flooding, as favoured by Shaw *et al.* (1989), or high volumes of basal meltwater associated with a fast-flowing ice stream, as found elsewhere by Ó Cofaigh *et al.* (2002a) and preferred by this author.

Finally, this case study highlights an example of equifinality in glacial geomorphology. Although landforms such as megadrumlins may appear similar, particularly when viewed on satellite imagery, they may be the result of very different geological processes.

Conclusions

1 A complex network of large-scale subparallel channels and streamlined ridges occur in Lower Annandale, southern Scotland.
2 The bedrock-cut channels are subglacial meltwater features aligned parallel to former ice-flow. They are not structurally controlled.
3 The existing landscape in Lower Annandale has been greatly influenced by subglacial meltwater erosion—many of the drumlinoid ridges and megaflutes being erosional remnants of bedrock and glacial deposits.
4 The nature and distribution of these landforms indicate that ice-sheet flow in this part of southern Scotland was convergent and faster flowing than in neighbouring areas, probably aided by high volumes of subglacial meltwater.

Acknowledgement

T. Bradwell publishes with permission of the Executive Director, British Geological Survey (NERC).

EIGHT

Subglacial megafloods: outrageous hypothesis or just outrageous?

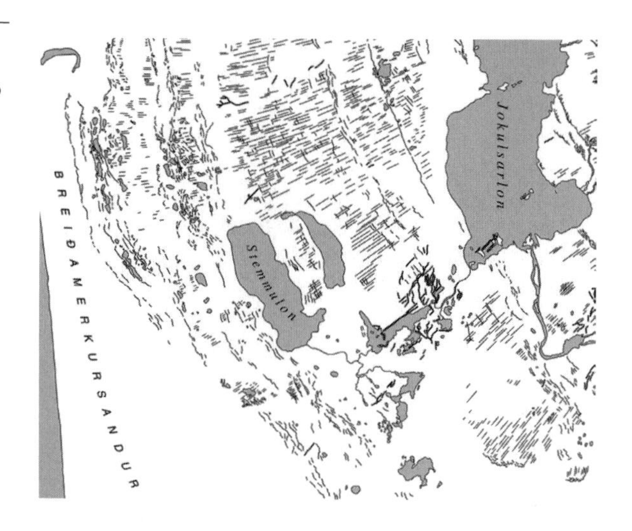

Douglas I. Benn* and David J.A. Evans†

**School of Geography and Geosciences, University of St Andrews, Fife KY16 9AL, UK*
†Department of Geography, University of Durham, Science Site, South Road, Durham, DH1 3LE, UK

Since the concept of drumlin formation by glacifluvial processes was first proposed by Shaw (1983), a large number of papers have been published interpreting a wide range of subglacial landforms in North America as products of subglacial megafloods (e.g. Shaw & Kvill, 1984; Shaw *et al.*, 1989, 2000; Shaw & Gilbert, 1990; Fisher & Shaw, 1992; Rains *et al.*, 1993; Brennand *et al.*, 1995; Sjogren & Rains, 1995; Shaw, 1996, 2002; Munro & Shaw, 1997; Beaney & Hicks, 2000; Beaney & Shaw, 2000; Munro-Stasiuk & Shaw, 2002). The sheer volume of peer-reviewed publications promoting the 'megaflood interpretation', and the fact that it has featured prominently in at least two recent compendia of earth science (Young, 2000; Brennand, 2004), may lend it an aura of respectability in the eyes of those unfamiliar with the evidence. However, most Quaternary scientists give little or no credence to the megaflood interpretation, and it conflicts with an overwhelming body of modern research on past and present ice-sheet beds. Despite this, there have been few attempts to systematically scrutinize the megaflood interpretation in print. In large part, this is because most working scientists are busy pursuing their own research programmes and are unwilling to invest time refuting ideas which are clearly incompatible with a huge body of mainstream research.

It is to the credit of journal editors that outrageous hypotheses (Davis, 1926) have been published. Science, after all, should proceed by the testing of ideas in the public domain, not by the censorship of papers simply because they are unorthodox. To become accepted, however, research should conform to the requirements of good science. Non-specialists can rarely be effective judges of whether this is or is not the case, and there is a risk that flawed science may appear to be mainstream in the eyes of a wider public. This is a common problem in the public perception of science, due to the unfortunate tendency for contentious theories to attract disproportionate attention. For this reason alone, the reluctance of many Quaternary scientists to publicly engage with the megaflood interpretation is regretable. This, combined with the fact that the megaflood interpretation has been seized by internet Creationist sites as 'proof' of the Noachian Flood, prompts us to undertake here the task of explaining, for the benefit of those unfamiliar with modern sedimentological literature, why the flood interpretation is unscientific, unnecessary and inconsistent with the evidence.

The megaflood interpretation is not a hypothesis in the Popperian sense of a provisional set of ideas that make clear, unambiguous predictions which can be objectively tested against new observations. The clearest example of this is the claim that megafloods can create drumlins by two different mechanisms, (i) by the infilling of subglacial cavities (cavilty fills) and (ii) by eroding away interdrumlin areas (erosional remnants). The cavity-fill interpretation was proposed by Shaw (1983) and Shaw & Kvill (1984), and visualizes drumlins as the infills of scours cut upward into the ice by turbulent waters below. This idea, which was based on the similarity of form between certain drumlins and sole marks below turbidites, and the presence of sorted sediments within some drumlins, makes the clear prediction that other drumlins should also be composed of sorted sediment (e.g. Sharpe, 1987). If this prediction is put to a Popperian test and a drumlin is found that does not contain sorted sediment, then the hypothesis should be rejected. However, this was not the option taken by Shaw *et al.* (1989, 2000) and Shaw (1993), who proposed that till-cored drumlins and flutings are the remnants of preexisting tills left behind when megafloods eroded the material between them. In other words, no matter what the internal composition, drumlins are interpreted as 'evidence' for megafloods. This being the case, the internal composition becomes irrelevant to the megaflood case, which is shown to rely exclusively on the perceived morphological similarity between drumlins and streamlined forms eroded by turbulent flows. Moreover, the megaflood interpretation apparently does not predict any systematic differences in the forms produced by these two mechanisms. If no such differences are expected, how can a single process (subglacial sheet floods) create two sets of erosional forms which are exactly equal and opposite in morphology? That is, why should moulds of erosional scours cut up into overlying ice look exactly like the remnants left behind by erosion of the substratum in other parts of the same flood? This difficulty is not experienced by Boulton's (1987) deformation model of drumlin formation or Tulaczyk *et al.*'s (2001) ploughing mechanism of substrate fluting (see also Clark *et al.*, 2003), for example, which potentially can explain all drumlins and flutings in terms of a single process, namely the streamlining of pre-existing bed materials be they composed of till, rock, or stratified sands and gravels. The deformation and ploughing models, moreover, make the clear prediction that the drumlins and flutings formed by such processes should be mantled by glacitectonite or till, a prediction that is borne out in our experience. These models do not, as Shaw and his co-workers repeatedly try to assert, champion the cause of pervasive deformation to the exclusion of all other processes.

By invoking the 'erosional remnant' mechanism to account for drumlins that cannot be explained by the 'cavity fill' process, Shaw is using a classic *ad hoc* protection device (Chalmers, 1976), the sole purpose of which is to remove difficulties encountered by the original theory. Any model that is protected from awkward evidence in this way is in effect unfalsifiable. This is not the hallmark of a hypothesis, but of a self-reinforcing belief system in which the interpretation of the evidence depends on pre-existing conclusions. Although many of the papers by Shaw and co-workers invoke the language of hypothesis testing, this does not stand up to scrutiny. For example, Shaw (this volume, Chapter 4) asks us to 'imagine' scenarios 'for the sake of hypothesis testing', then we find later in the paper that the imaginings are facts, which are then used to support further imaginings, and so on. Nowhere is there a serious and objective comparison of prediction with evidence.

As noted above, the megaflood interpretation relies very heavily, if not exclusively, on the morphological similarity between drumlins and certain kinds of small-scale erosional scours on the one hand, and between Rogen moraine and wavy fluvial bedforms on the other. There is indeed a superficial resemblance between drumlins and sole marks, as we illustrated in fig. 11.25 of *Glaciers and Glaciation* (Benn & Evans, 1998). However, this resemblance does not imply that they were necessarily formed by the same medium. The simple explanation is that obstacles below flowing media exhibit shadow effects, such that their presence influences patterns of erosion far down-flow. This effect applies not only to turbulent flows, but also to non-turbulent flowing media, such as ice. One need only to think of fluted moraines exposed on modern glacier forelands to see that this is so. Indeed, drumlins and megaflutings show a much stronger resemblance to streamlined subglacial landforms exposed by recent glacier retreat (in form and in scale) than they do to scours formed by turbulent media. The similarities are not merely superficial, but extend to numerous characteristics at a wide range of scales. Furthermore, historically produced fluting fields such as those in front of Breidamerkurjøkull in Iceland allow us to relate sediment and landform characteristics to genetic processes with a high degree of confidence (Evans & Twigg, 2002). At Breidamerkurjøkull, flutings are aligned parallel to known former ice-flow directions in slightly offset flow sets that terminate at moraines (Fig. 8.1). Tills in the flutings commonly have erosional lower contacts with glacitectonized or undisturbed outwash. Detailed process studies have demonstrated the role of subglacial lodgement and deformation in the origin of the tills and the flutings (e.g. Boulton & Hindmarsh, 1987; Benn, 1995; Benn & Evans, 1996; Boulton *et al.*, 2001). Eskers mark the location of channelized meltwater. This landsystem provides us with a clear process–form model and it is a small logical step to assume that it can be applied to ancient landform–sediment assemblages that have a wide range of closely similar characteristics.

To this end, landform assemblages illustrated through digital elevation models (DEMs) by Munro & Shaw (1997) have been remapped from aerial photograph mosaics, and have been shown to consist of discrete fields of glacially streamlined features (flutings) terminating at a series of inset transverse ridges (moraines) organized in broad arcuate bands (Fig. 8.2). More localized moraine arcs record topographically induced lobation of the ice margin during recession. Minor readvances of these lobes are documented by the localized superimposition of transverse ridges. This is a landform characteristic difficult to explain as a subglacial fluvial erosional ripple mark, the genesis of the transverse ridges suggested by Munro & Shaw (1997). Also evident are misaligned and cross-cutting flow sets, represented either by superimposed flutings or adjacent fluting fields with orientations that are significantly different and which cannot be explained by contemporaneous ice flow deviations. Juxtaposed fluting fields that display different orientations are convincingly interpreted as glacier flow sets (Clark, 1997) and the superimposition of flow sets is often identifiable in cross-cutting flutings (Dyke & Morris, 1988; Boulton & Clark, 1990; Clark, 1993). Flow sets have been interpreted as the subglacial imprint of fast glacier flow in an ice mass with shifting loci of ice dispersal and termination. Cross-cutting can be explained by the ice streamlining hypothesis but not the sheetflood hypothesis of fluting formation. In addition, evidence of subglacial and/or englacial meltwater activity is manifest in fragmented single and anabranched esker networks and occasional elongate water-filled depressions. This assemblage of landforms is similar in every respect to the glacial landsystem reported by Evans & Twigg (2002) from southern Iceland, characterized by inset sequences of integrated subglacial and ice-marginal landforms produced by lobate marginal recession of active temperate glaciers.

Given the availability of clear modern analogues for ice-sheet beds on modern glacier forelands, the case for the megaflood interpretation is seriously weakened, because turbulent flows are shown to be unneccesary to explain streamlined subglacial landforms. Historical jøkulhaups have occurred at Breidamerkurjøkull, but these were associated exclusively with small ice-dammed lakes at the western and eastern margins. Almost all of the glacier foreland (which is extensively fluted) has been unaffected by jøkulhlaups. If sheetfloods are unnecessary (indeed, impossible) as an explanation for streamlined subglacial landforms at Breidamerkurjøkull, why must we invoke them to explain closely similar sediment–landform associations in, say, Alberta? In the fourteenth century, William of Ockham wrote, '*pluralitas non est ponenda sine neccesitate*', which translates as 'entities should not be multiplied unnecessarily'. The principle that explanations of phenomena should not invoke any agencies not actually *required* by the evidence, led the way out of medieval superstition into scientific enlightenment. It still stands as a central tenet of science today. If we already have an ice sheet, and

Figure 8.1 Push moraines and flutings on the foreland of Breidamerkurjøkull, extracted from a 1998 map of the area by Evans & Twigg (2000). Each flow set of flutings is located between push moraines and when compared the flow sets record a slightly offset ice flow direction.

flowing ice can create streamlined landforms, why do we need a flood?

Thus, although the resemblances between streamlined subglacial landforms and features such as sole marks and sastrugi may appear significant, modern glacial landforms provide much stronger analogies. Attempts to boost the argument for turbulent flows by appealing to required Reynolds numbers are spurious. The Reynolds number *Re* is the ratio between inertial and viscous forces in a fluid. If it is assumed that streamlined subglacial landforms must have been made by turbulent flows (high *Re*), then of course it follows that ice was not the medium, since ice flow is not turbulent. But this is mere circular reasoning, in which the conclusion comes first, disguised as an argument.

The case for the megaflood interpretation would be strengthened if there were independent evidence that requires us to believe that large volumes of water were stored at the beds of former ice sheets. Shaw (this volume, Chapter 4) tackles this issue by arguing that the presence of tills implies that meltwater was 'abundant' in the inner parts of the Scandinavian and Laurentide ice sheets. Munro-Stasiuk (2000, 2003) uses the occurrence of stratified diamictons in an area of south-central Alberta to support the contention that englacial debris was melting out into large subglacial lakes beneath the southwest Laurentide Ice Sheet. The use of stratified diamictons as indicators of subglacial melt-out is a contentious one, but even if we accept that melt-out was the primary depositional process (and observations on sedimentation rates in modern subglacial lakes (e.g. Siegert, 2000; Siegert *et al.*, 2001) suggest that this is almost insignificant), the implied volume of water produced does not come anywhere close to the quantities required by successive manifestations of the megaflood interpretation. Consider the quantities involved. We are told that a megaflood draining into the Gulf of Mexico (only one component of the proposed cataclysm) was sufficient to raise global sea level by 3.7 m. This is about half the mass of the Greenland Ice Sheet. To claim that the occurrence of stratified tills supports the existence of enough water to supply floods of the required magnitude involves a logic jump that bypasses numerous stages of hypothesis testing. One obvious query relates to the nature of water release during melt-out till production—surely it must be released slowly and passively in order to preserve the delicate structures? Moreover, the melt-out process would have continued in areas of the bed of the former Laurentide Ice Sheet well after it had disappeared; it continues today in northern Canada (Dyke & Savelle, 2000; Dyke & Evans, 2003). The very nature of melt-

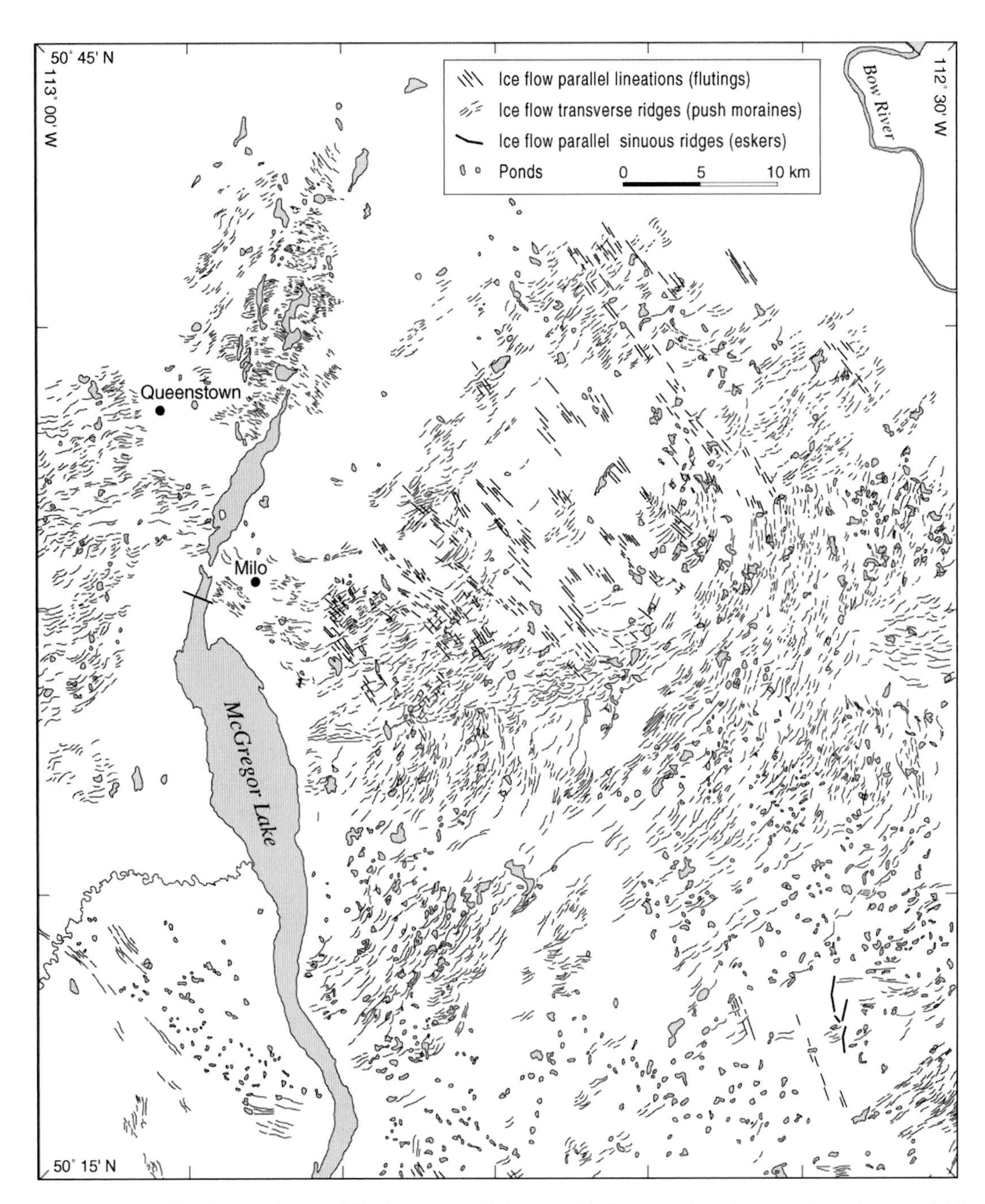

Figure 8.2 Sequences of push moraines and flutings mapped from aerial photographs of a part of south-central Alberta previously mapped from DEMs by Munro & Shaw (1997). Note the overprinting of push moraines with slightly offset alignments and the localized lobation that coincides with topographic hollows.

out till production precludes any central role it could have played in the production of catastrophic subglacial meltwater floods.

The idea of floods of such 'unimaginable' (Shaw, this volume, Chapter 8) dimensions is the outcome of taking flawed assumptions to their logical conclusion, a form of *reductio ad absurdum* in which the final absurdity is taken not as evidence of false premises but as fact. The initial flawed assumption, that the form analogy between drumlins and scours made by turbulent flows implies a common mode of formation, led to the conclusion that the Livingston Lake drumlin field in north Saskatchewan records a megaflood event (Shaw, 1983). Once this was accepted, then the close association between drumlins and other landforms (such as Rogen and hummocky moraine) was taken to imply that they too must have a flood origin (Munro & Shaw, 1997). As such landforms are very widespread inside the limits of the Laurentide Ice Sheet, this assumed origin in turn leads to the conclusion that 'unimaginable' megafloods must have occurred. Each stage of this chain of thought is deeply flawed, but has been woven into such a dense network of assumptions and conclusions that it appears solid to its proponents and to unwary onlookers. The apparently supporting 'evidence', such as widespread streamlined forms visible on DEMs (Shaw *et al.*, 1996; Beaney & Shaw, 2000), and the modelling work of Shoemaker (1992a,b) are no more than assumptions in disguise, which serve to perpetuate the myth in a self-reinforcing cycle in which the distinction between assumption and conclusion is constantly blurred.

This process is apparently so beguiling that the evidence itself becomes distorted to fit the interpretation. As an example, Shaw (this volume, Chapter 4) tells us that the landscape within the limits of the last Laurentide Ice Sheet represents a snapshot of the glacier bed at one moment in time, a pristine, synoptic assemblage preserved by continent-wide ice stagnation following a megaflood. The lack of recessional moraines or of cross-cutting relationships within drumlin fields are presented as facts, whereas several researchers have convincingly shown that the opposite is the case (see Fig. 8.2). Widespread cross-cutting streamlined landforms have been thoroughly documented for example by Dyke & Morris (1988), Boulton & Clark (1990) and Clark (1993). As was illustrated above with the Alberta case study, the very features used to support a subglacial megaflood (ripple marks of Munro & Shaw, 1997) in fact constitute the 'missing' recessional moraines that only become 'visible' when mapped objectively and systematically from aerial photographs.

It must be emphasized that our refutation of the arguments of Shaw and co-workers does not force us to conclude that no large subglacial outburst floods occurred during the lifetime of the great Pleistocene ice sheets. Indeed, evidence for such floods, in the form of tunnel channels for example, is found in many places (e.g. Brennand & Shaw, 1994; Patterson, 1994; Ó Cofaigh, 1996; Clayton *et al.*, 1999; Cutler *et al.*, 2002). Subglacial reservoirs, mostly of the order of 10 km across, are now known to be widespread below the Antarctic Ice Sheet (Siegert, 2000), and some such reservoirs appear to have drained catastrophically in the past (Shaw & Healy, 1977; Denton *et al.*, 1993). Tunnel channels and anastomosing channel networks such as the Labyrinth in the Antarctic Dry Valleys provide compelling evidence for catastrophic drainage events, but acknowledgement of this does not lead logically to the conclusion that landforms such as drumlins, flutings and hummocky moraine record vastly larger floods. The juxtaposition of demonstrably glacifluvial landforms (channels) with flutings or other landforms does not imply that all were formed simultaneously by the same mechanisms, as consideration of the landform associations at Breidamerkurjøkull makes clear (Evans & Twigg, 2002). To make the step from local, albeit large floods to events of Biblical proportions falsely polarizes the situation into an 'all or nothing' scenario, which actually distracts attention away from the important business of determining the true importance of catastrophic discharges from former ice sheets.

Reply to Benn and Evans by John Shaw and Mandy Munro-Stasiuk

This response to the comments of Benn and Evans is divided into three parts: fact, omission and philosophy.

Fact

Benn and Evans argue that the basis for the interpretation of the Livingstone Lake drumlins was the form and the presence of sorted sediment in the drumlins. This is a simplification: the interpretation was based on the form of drumlins, the *style*, not the *presence*, of sediment, clast lithology and rounding, landform association and landform sequence.

They also write that the meltwater hypothesis does not predict any systematic difference in the morphology of cavity fill and erosional drumlins. In fact, Shaw (1996) gives the two types of drumlins two different names because they are distinctly different in form. Shaw (1983) had previously drawn attention to the remarkable difference in form between the Livingstone Lake drumlins and classic drumlins. It was this difference that prompted the meltwater hypothesis.

As far as the Livingstone lake drumlins are concerned, Boulton's (1987) paper is contradicted by the field evidence; where he shows attenuation of fold limbs, the sediment is undeformed (Shaw *et al.*, 1989). Tulaczyk *et al.* (2001) specifically state that their ploughing mechanism does not explain drumlins. How could ploughing explain crescentic scours around the upstream ends of drumlins? As well, Tulaczyk *et al.* (2001) point out that the deep, pervasive deformation required by Boulton to explain drumlins is unlikely in light of observations beneath modern glaciers and experiments on deformation. Clarke *et al.* (2003) write on landforms without any observations on their internal structure. They deal exclusively with form and ignore our field observations on structure.

Benn and Evans state incorrectly that the megaflood hypothesis relies very heavily, if not exclusively, on morphological similarity with other forms. In addition to morphology we studied landform pattern, and detailed sedimentology including sedimentary architecture, landform associations, clast lithology and roundness (Shaw *et al.*, 2000), computational fluid dynamics (Pollard *et al.*, 1996), stone lags (Rains *et al.*, 1993; Munro & Shaw, 1997; Beaney, 2002), hydraulic modelling (Beaney & Hicks, 2000), and valley profiles (Rains *et al.*, 2002; Beaney, 2002).

As an example, they state that Munro & Shaw (1997) interpret Rogen and hummocky moraine as flood landforms because hummocks are associated with other landforms interpreted as flood forms and therefore assumptions become based upon assumptions. This is misleading because Benn and Evans are forcing the reader to believe that Munro & Shaw (1997) presented form analogy as their only evidence. They do, in fact, miss the point and ignore the main evidence presented: hummocks are truncated at their surfaces regardless of internal structure and without deformation of the immediate underlying sediments (see Munro-Stasiuk and Sjogren, this volume, Chapter 5, for images), and boulder lags heavily pitted with percussion marks sit on the surface of many hummocks. Although form was the starting point in formulating the megaflood hypothesis and remains an essential element, our hypothesis testing has clearly become more sophisticated.

Benn and Evans have used aerial photograph mosaics to contradict the morphological work of Munro & Shaw (1997) in south-central Alberta. It should be noted that mosaics provide less information on form than stereo aerial photographs or DEMs because they do not give a three-dimensional view of the landscape. It is also surprising then that Benn and Evans criticize us for using form analogy because they also use analogy. When they compare landforms around McGregor Lake Reservoir in south-central Alberta with forms in the forefield of Breidamerkurjökull, they give no indication of internal sediment, or sediment–landform relationships. Instead, they ignore the sedimentary evidence presented by Munro & Shaw (1997) (see previous paragraph). Here we present an aerial photograph with an interpretive map (Fig. 8.3) that lies within the mapped region presented by Benn and Evans. Clearly there are a number of different 'transverse' landforms on this photograph, yet Benn and Evans show all ridges as push moraines. They map undifferentiated mounds, ridged mounds (some with central depressions), linear ridges and some eskers as push moraines. The eskers show a flow direction towards the southwest, towards McGregor Lake Reservoir (transverse to the main regional flow direction). Of all the features on Fig. 8.3, few resemble the features on the forefield of Breidamerkurjökull (compare Fig. 8.3 with fig. 7 in Evans *et al.*, 1999). It is possible that the linear transverse ridges are moraines, as they are quite similar to Icelandic features. There are other features like these in southern Alberta only 20 miles from the McGregor Lake Reservoir near the town of High River that strongly resemble many of the moraines around Vatnajökull in Iceland. However, all other features are dissimilar and require an alternative interpretation. We have tried to provide an alternative explanation based on the similarity of these forms with large-scale ripple marks, *and* based on sedimentology (Munro & Shaw, 1997). It appears that Benn and Evans are confusing landforms. There are some broad arcuate ridges that have a very restricted distribution in Alberta. Although they may represent the ice margins they cannot be used as evidence for ubiquitous ice recession on the Canadian Prairies because they are not present everywhere. On the other hand the transverse features described by Munro & Shaw (1997) as hummocky terrain and the giant ripples described by Beaney & Shaw (2000) are distinctly different. The giant ripples are sinusoidal, and in places show rhomboidal patterns. The forms described by Beaney & Shaw (2000) are eroded from undeformed bedrock so how could they be push moraines?

Benn and Evans refer to the use of Reynolds numbers in our hypothesis as spurious. Considering erosional marks in bedrock (Kor *et al.*, 1991), we find rounded, lag boulders, over 1 m in diameter and with distinctive percussion marks. These boulders rest on erosional marks that are identical in form to those inter-

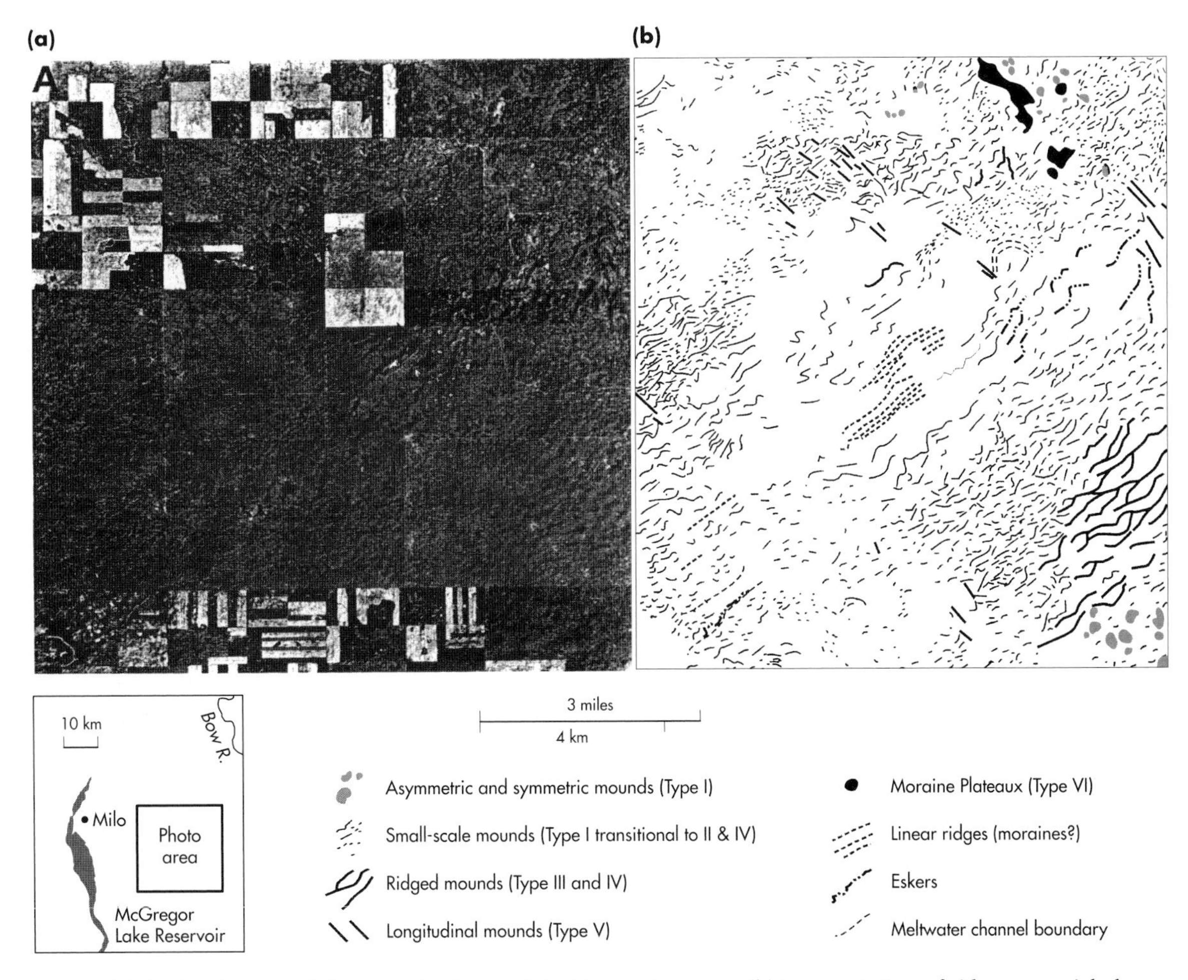

Figure 8.3 (a) Aerial photograph from the McGregor Lake Reservoir region. (b) Interpretation of ridges on aerial photographs based on photograph analysis and field work. Small map inset at base shows the position of the photograph relative to the area presented in Fig. 8.2.

preted by Allen (1982) as products of turbulent, separated flow. It is literally impossible for laminar flow to have transported the boulders and formed the erosional marks. Contrary to the opinion of Benn and Evans, Reynolds number calculations follow from the field observations; the flow depths and velocities required for these calculations are not assumed.

Benn and Evans provide an extensive discussion on the source and amount of water for the floods. They miss the point. We do not argue that melt-out produces *all* the water for floods. Munro-Stasiuk (2000) specifically noted that the melt-out processes (associated with small volumes of water) preceded the flood events and that the water for the major flood was part of the Livingstone Lake Event and was derived far to the north of her study site. She also provided extensive evidence for sedimentation into subglacial reservoirs that occupied the local pre-glacial valley system (Munro-Stasiuk, 2003). The sedimentary facies representative of the reservoirs consists of a range of subaqueous deposits (mostly gravity flow deposits and *not* melt-out deposits) that are chaotically deposited and range from entirely undisturbed to pervasively deformed (interpreted as ice-reactivation). Although Benn and Evans go on at length arguing that the subglacial reservoirs discussed by Munro-Stasiuk (2003) were not large enough to provide the volume of water required for a megaflood, Munro-Stasiuk noted that the reservoirs were 'small', never contended that the reservoirs were the source of the megaflood waters, and clearly stated that the presence of reservoirs was followed by till deposition and then by the event responsible for creating the erosional landforms. In this volume, Shaw (Chapter 4) refers to abundant meltwater to refute the notion of frozen bed conditions near the centre of the ice sheet and to account for hydraulic connectivity between the central zone of the ice sheet and the margins. Shaw (1996) proposed a supraglacial origin for the meltwater, a suggestion that is well supported by observations on modern glaciers and by modelling of past ice sheets (Marshall *et al.*, 2002; Zwally *et al.*, 2002a). This proposal clearly establishes that melt-out is not considered as the primary source of meltwater for the flood events.

Benn and Evans argue that modern glacial environments provide all the necessary analogues for past glacial landscapes. This claim is unlikely. For example, there are no known modern glacial landforms resembling Rogen moraine, nor any that show the sedimentary and morphological characteristics of hummocky moraine on the Western Plains of Canada. Furthermore, the streamlined, loess hills of the unglaciated Channelled Scablands in Washington are identical in form to drumlins 100 km or so to the north. Also, the fluted bedrock above Dry Falls in the Scablands is identical to similar bedrock fluting in glaciated areas. Both of these Scabland landforms required immense water sheet floods for their formation. They were formed *beyond* the ice limits and it is therefore impossible that these could have been created by glacial processes. At the same time they provide powerful analogues for drumlins and erosional marks. Obviously, we need something other than analogues from modern glacial environments to explain these non-glacial landforms. Thus, it is reasonable to use Scabland flood landforms as part of a hypothesis on subglacial megafloods.

Shaw never claimed that drumlins and fluting do not show cross-cutting relationships. In fact, Shaw (1996, fig. 7.39) presents a map clearly showing such cross-cutting. We cannot imagine why megafloods could not cause cross-cutting. All that is required is a variation in flow direction. Indeed, Shaw & Gilbert (1990) differentiated the Algonquin and Ontarian events on the basis of the cross-cutting of one flow path by the other. The Algonquin came first. By contrast, there are vast (hundreds of kilometres long) tracts of drumlins and flutings, the Livingstone Lake flow path for example, that do not show cross-cutting relationships, nor do they display systematic superimposed forms. It is our view that these tracts represent pristine subglacial surfaces.

In summary, Benn and Evans state that the megaflood case 'is shown to rely exclusively on the perceived morphological similarity between drumlins and streamlined forms eroded by turbulent flows'. As discussed above, the megaflood case is grounded in sedimentary, hydrological and glacial theory, and is supported by extensive morphological and sedimentary observations and interpretations.

Omission

As is commonly the case with those who review our work, Benn and Evans fail to relate bedrock erosional marks to the meltwater hypothesis. Such erosional marks in granite and gneiss in the French River area of Georgian Bay provide very strong evidence for broad, catastrophic subglacial floods (Kor *et al.*, 1991). There is no possibility of interpreting them otherwise (Shaw, this volume, Chapter 4). The meltwater hypothesis is well supported by evidence from the streamlined hills and fluted bedrock of the Scablands and from the French River erosional marks, all of which require extensive, turbulent sheet floods for their formation.

Benn and Evans also ignore some of the main evidence for large-scale erosional events such as erosion into undisturbed preglacial gravels in the Blackspring Ridge flute field in south-central Alberta (Munro-Stasiuk & Shaw, 2002). They make the blanket statement that glacitectonite and till should appear in these landforms. They also ignore the observation that many landforms have truncated surfaces representative of a landform unconformity (e.g. Munro & Shaw, 1997; Munro-Stasiuk & Shaw, 2003, Sharpe *et al.*, 2004).

Philosophy

Benn and Evans seem to be intent on discrediting the meltwater hypothesis on the basis of questionable reasoning and philosophy. We would like to comment critically on their approach.

They state that most Quaternary scientists give little credence to the megaflood hypothesis and that non-specialists need to be protected from its flawed science. Did they conduct a poll to determine that most Quaternary scientists do not believe the hypothesis? Hardly: there are many Quaternarists who embrace these ideas. Surely Benn and Evans are not suggesting that these researchers are 'unfamiliar with the evidence' or that their research is also 'unscientific, unnecessary and inconsistent with the evidence'. As well, is there really a wider public that needs to be protected against our unscientific thoughts? The alternative to Benn and Evan's view is that the megaflood hypothesis is not really flawed; rather it challenges and threatens establishment research. It would not be the first time that establishment figures have railed against a fruitful hypothesis because they found it repugnant.

The reasoning on the dual interpretation of drumlins misrepresents our work and the work of Popper. We do not hold that there is a need for two types of drumlins *because* there is sorted sediment in one and not in the other. We hold this view *because* the sedimentary structure and architecture, clast shape and clast lithology in one type are so different from those in the other type. As well, the forms of cavity fill drumlins, spindle, parabolic and transverse asymmetrical, are so unlike classic drumlins that we felt obliged to give them different names—Livingstones and Beverleys (Shaw, 1996). Beverleys, with troughs wrapped around the proximal ends, are almost certainly erosional, but we do not, as Benn and Evans assert, base this argument 'exclusively' on perceived morphology. We consider morphology related to the action of horseshoe vortices, turbulent structures at forward-facing steps, truncation of internal structure, stone lags on the erosional surface and landform associations (e.g., Shaw *et al.*, 2000). We are well aware that the internal structure and composition is less important in deducing the formation of erosional landforms, compared with that of depositional forms. In the case of landforms such as the French River erosional marks cut in granite and gneiss, this point is so obvious it is taken for granted. By contrast, we go to great lengths in the case of hummocks and large-scale fluting to demonstrate that internal structure and surface form are largely independent (Munro & Shaw, 1997; Shaw *et al.*, 2000; Munro-Stasiuk & Shaw, 2002). We certainly do not accept that sedimentary structure, architecture and clast lithology are irrelevant to understanding subglacial landforms in the megaflood hypothesis. Why would much of our work involve sedimentology if it were so evidently irrelevant to the meltwater hypothesis? Erosional drumlins may contain sorted and stratified sediment. Benn and Evans appear to have overlooked this point and insist that erosional drumlins are till cored. Hence their confusion on the duality of drumlins and their assertion that the megaflood hypothesis for drumlins is unfalsifiable. In reality, details of form, architecture, sedimentology and lithology allow us to distinguish cavity fill drumlins, Livingstones, from erosional drumlins, Beverleys. There are very specific predictions on the characteristics of erosional and cavity fill drumlins and we continue to use these predictions when interpreting landforms. We wonder why Benn and Evans make such a fuss over a matter that we have treated exhaustively in a number of papers.

We do not make our hypothesis unfalsifiable by protecting it from awkward evidence: the hypothesis is easily falsifiable. For example, the hypothesis would be rejected if the properties of the sediment in depositional landforms contradict the specific predictions for cavity fills (e.g. it is aeolian or marine or is deformed into the shape of the landform) or, for the case of erosional landforms, the hypothesis is rejected if the patterns of defining erosional troughs show cross-cutting rather than bifurcating and merging relationships. Rather, we have amended the hypothesis as it became apparent that it was contradicted by certain observations. Thus, we introduced an erosional version because drumlins in bedrock, for example, demanded it. Benn and Evans argue that this is sleight of hand. We suggest that it is sensible. It seems illogical to argue that erosional drumlins and cavity fills cannot both exist. Sedimentologists and geomorphologists describe some bedforms as erosional and some as depositional without concluding that hypotheses on their genesis are unfalsifiable (Allen, 1982). Since Benn and Evans fail to recognize, rather they ignore, certain observations (e.g. the truncation at the land surface, the presence of boulder lags over erosional landforms, the origin of bedrock forms), they are the ones using the *ad hoc* protection device they accuse us of using. They are protecting their own models from the 'awkward' evidence we present. For instance, they state that in their experience all flutings and drumlins are mantled by glacitectonite or till and thus they can be easily explained by ploughing or deformation processes. They ignore observations that contradict their prediction (e.g. Rains *et al.*, 1993; Munro & Shaw, 1997).

Shaw (this volume, Chapter 4) does not present imaginings as fact as Benn and Evans assert. More than any other modern hypothesis on subglacial bedforms, our work is grounded on fieldwork, experiment and imagery. We believe that whereas our facts are based on observation, much of the literature on subglacial deformation is based on modelling. We also believe that the most powerful explanations of these bedforms will come ultimately from a combination of both approaches.

William of Ockham's tenet—'entities should not be multiplied unnecessarily'—is very helpful in this discussion because the multiplication of entities has been necessary. Analogues from modern glaciated areas cannot explain Rogen moraine. Nor can they explain streamlined forms in bedrock and loess in the Scablands, lying beyond the limits of the Cordilleran Ice Sheet. Obviously we need additional entities to those from modern glacial environments. Hence, the Scablands streamlined forms are excellent analogues for erosional drumlins and fluting in bedrock and it is proper to use them in a hypothesis stating that *some* subglacial landforms resulted from megafloods.

The processes Benn and Evans champion as generally applicable fail to explain the characteristics of many bedforms. For example, Tulaczyk *et al.* (2001) point out that their ploughing method does not explain drumlin patterns. We invite Benn and Evans to explain the topology of a ploughing mechanism that bifurcates around the upstream ends of drumlins and does not cross-cut downstream crescentic troughs. It is topologically impossible that ploughing can behave in this way. Indeed, Tulaczyk *et al.* (2001, p. 64) write: 'Whilst this (*the carving of intermediate grooves by ploughing*) does not explain the pattern of all bedforms such as drumlins and Rogen moraine, it may explain the observed form of megalineations'. Italics added. Chris Clark freely admits that this is a problem for their hypothesis and is the reason why they only deal with megalineations and not with drumlins. It is worth noting that Tulaczyk *et al.* (2001) argue very strongly against the ideas of Boulton (1987), yet Benn and Evans present the contradictory views of these authors as supporting their conclusions. Such inconsistency detracts from their arguments.

The modern glacial analogue approach is incapable of explaining either Rogen moraine or the scale of megalineations. Ploughing by ice keels cannot explain the form and pattern of drumlins. The megaflood hypothesis is attractive because it is not faced by these difficulties. More to the point, such difficulties make it *necessary* to introduce new entities; the megaflood hypothesis presents such new entities. There is no quarrel with William of Ockham here.

The comments on melt-out till and the generation of meltwater for megafloods set up a red herring. We have been at pains to point out that the melt-out till precedes the megaflood and is commonly a remnant within erosional drumlins (e.g. Shaw *et al.*, 2000). To demonstrate this, we present some of the most detailed field sketches of fluting sediment together with structural and fabric data. A stone lag, interpreted to have been produced by flood erosion, lies on an erosional surface, truncating melt-out till and diapiric mélange. It is this erosional surface that defines the fluting. The melt-out till preceded the megaflood that eroded the drumlin and meltwater involved in the formation of the till was probably of little consequence to the flood which originated far to the north. Of course, as pointed out since 1982, the water for melt-out till was released slowly; we present estimates of thousands of years for melt-out till formation (Shaw *et al.*, 2000). Once again, we are misrepresented and the arguments we make for a supraglacial origin for the megaflood discharge are overlooked.

Obviously, there is a major difference in perception between Benn and Evans and us. They consider that the megaflood hypothesis flies in the face of a huge body of mainstream research. In our defence, there is no known observation that contradicts the hypothesis. Nor does this hypothesis violate any fundamental principle in science. It might be incompatible with mainstream research, but the same can be said of any new paradigm. In answer to their assertion that our work is unscientific and unnecessary, our only response is—we do not think so. It would help if Benn and Evans were more specific where they write that our work is inconsistent with the evidence. Again, there are no known observations that *contradict* the megaflood hypothesis.

NINE

Groundwater under ice sheets and glaciers

Jan A. Piotrowski

Department of Earth Sciences, University of Aarhus, C.F. Møllers Allé 120, DK-8000 Århus C, Denmark

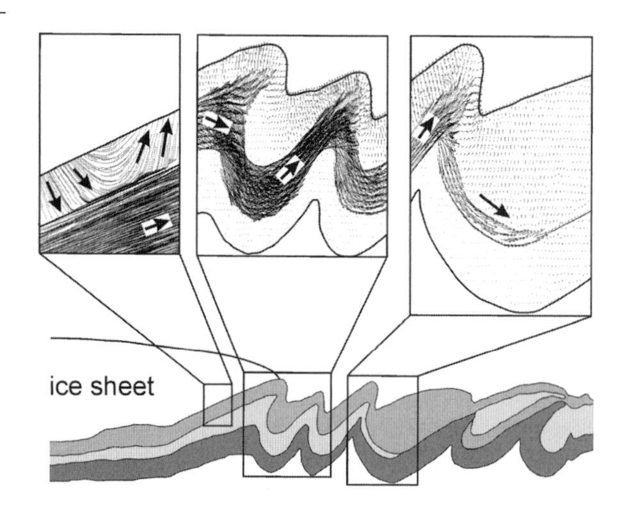

9.1 Introduction

It has been realized only recently that groundwater under ice sheets and glaciers is an important, integral part of the hydrological system in environments affected by glaciation. The late start in research and the resulting scarcity of published work on subglacial groundwater was caused by its position in the no-man's land between glaciology and hydrogeology, despite its relevance for both. It is now recognized that water in permeable rocks and soft sediments overridden by glaciers, through a system of feedbacks, influences glacier stability, movement mechanisms, sediments and landforms. Water discharged from melting ice contributes to the renewal of groundwater resources.

Large-scale groundwater circulation patterns and dynamics experience fundamental changes in glacial–interglacial cycles subjected to repeated loading and relaxation by kilometre-thick ice. Old glacial groundwater trapped in low-permeability areas yields important information about past environmental changes, and modifications of future groundwater flow dynamics in areas likely to be affected by prospective ice sheets must be considered in disposal strategies of toxic waste. It is therefore clear how important the impact of glaciation on groundwater is, which explains the recent interest in this field.

9.2 Water source and drainage systems

Subglacial meltwater originates from a range of sources, mainly from melting of ice by geothermal heat trapped at the glacier sole and by the frictional heat caused by ice movement past the substratum. These two sources yield up to some 100 mm yr^{-1} of water. Close to the ice margin, in the area where englacial conduits extend to the bed, surface ablation water may reach the ice sole with recharge several orders of magnitude greater than the basal meltwater alone. It is difficult to estimate how wide this area is, but as deeper conduits tend to close under cryogenic pressure or they bend horizontally towards the ice margin, it is probable that ablation water would reach the bed only where the ice thickness is less than about 100–200 m (Reynaud, 1987; Fountain & Walder, 1998). Under polythermal conditions, such as those that characterized continental Pleistocene ice sheets, the outermost marginal zone and the ice divide zone may be cold-based and do not contribute to the basal water recharge.

Subglacial water can be evacuated to the ice margin through high-discharge drainage systems consisting of channels incised into the glacier sole (R-channels; Röthlisberger, 1972) and channels carved into the bed (N-channels; Nye, 1973). R-channels typically form on bedrock, whereas N-channels tend to occur on soft beds. This is shown by the distribution of landforms across different substrata overridden by the Fennoscandian and Laurentide ice sheets where on bedrock areas eskers dominate, whereas in areas of sedimentary basins away from glaciation centres meltwater channels and tunnel valleys prevail (Clark & Walder, 1994).

Another type of subglacial drainage is the low-discharge, distributed system operating at the ice–bed interface or in the rocks below. A linked-cavity network consists of broad and shallow water lenses connected by orifices, favoured by rapid sliding and high bed roughness (Nye, 1970), which originally was suggested for hard bedrock areas by Lliboutry (1976) and Kamb (1987). A modification of this drainage mechanism was proposed by Walder & Fowler (1994) and Clark & Walder (1994) for soft, deformable beds. It consists of broad, shallow 'canals' interconnected in a non-arborescent system without orifices, capable of evacuating more water than the linked cavities. Yet another mechanism is the subglacial water film, typically about 1 mm thick, generated by regelation at the ice–rock interface (Weertman, 1972). Under special circumstances, parts of a glacier may be lifted by pressurized water leading to short-lasting subglacial sheet-floods of high magnitude.

9.3 Basic laws of groundwater flow through the subglacial sediment

If a glacier or an ice sheet rests on a permeable bed, be it rock or soft sediment, a part of the subglacial meltwater will enter the bed and be evacuated as groundwater flow, governed by the same physical rules as groundwater flow in confined aquifers outside the glaciated areas. The major differences, however, are that (i) the flow is driven by hydraulic gradient imposed by ice overburden, and (ii) some groundwater may be advected within the sediment if it deforms in response to glacier stress.

The basal meltwater will enter the substratum if the pressure at the ice–bed interface is greater than the pressure within the bed. If the bed is a porous medium such as glacial deposits and most other sedimentary rocks, water flow will be governed by the Darcy law

$$Q=\left(\frac{KA}{\rho_w g}\right)\left(\frac{dh}{dl}\right)$$

where Q is the water flux, K is the hydraulic conductivity, A is flow cross-sectional area, ρ_w is the density of water, g is the acceleration due to gravity, h is the hydraulic head and l is the flow length. The flow is driven by the hydraulic gradient $(1/\rho_w g)(dh/dl)$ determined largely by the ice-surface slope (Shreve, 1972; Fountain & Walder, 1998). The potentiometric surface of groundwater confined by an ice sheet runs approximately parallel to the ice surface at a depth corresponding to the water pressure (Fig. 9.1A). At the ice margin the pressure in the aquifer is atmospheric, thus the groundwater changes from a confined to unconfined flow. Under conditions of high meltwater recharge and low drainage capacity of both the bed and the drainage systems at the ice–bed interface, the groundwater pressure may be elevated to the vicinity of the ice overburden pressure (glacier flotation point). Under such conditions groundwater recharge ceases because of equilibrated pressures and lack of head drop into the sediment.

Groundwater flow under polythermal glaciers and ice sheets will be affected by the distribution of frozen ground in ice-marginal areas (e.g. Haldorsen *et al.*, 1996; Cutler *et al.*, 2000) and under central parts of ice sheets where basal freezing may occur due to advection of cold ice (Fig. 9.1B). Because frozen soil is orders of magnitude less permeable than the same soil in unfrozen state (Williams & Smith, 1991), permafrost will act as a confining layer. Pressurized groundwater thus will be forced under the permafrost, and the confined drainage system will first terminate outside the permafrost zone or at large discontinuities in the frozen ground such as taliks under rivers, lakes and over salt domes.

Pressurized, often artesian (Haldorsen *et al.*, 1996; Flowers & Clarke, 2002b; Flowers *et al.*, 2003) groundwater at a glacier margin can cause hydrofracturing of sediment and rocks. Boulton & Caban (1995) suggested that such fractures may occur to a depth of about 400 m, and the zone affected by hydrofracturing can extend many tens of kilometres into the ice foreland. They gave examples of small-scale sediment dykes filling hydrofractures in Spitsbergen, England and Sweden (see also Larsen & Mangerud, 1992; van der Meer *et al.*, 1999; Grasby *et al.*, 2000). Plate 9.1 shows a hydrofracture of similar origin in northwest Germany.

Pressurized groundwater may also facilitate formation of push moraines, diapirs and large-scale extrusion moraines (Boulton *et al.*, 1993; Boulton & Caban, 1995), typically by sediment liquefaction. Artesian groundwater behind an ice front is capable of lifting loosened, thick rock slabs (Bluemle & Clayton, 1984; Pusch *et al.*, 1990), which may then be redeposited by ice as allochthonous rafts and megablocks within glacial sediments (Aber *et al.*, 1989). Water pressure increase will also occur in subglacial aquifers that wedge out in the direction of glacier flow. Such groundwater traps were an important cause of glaciotectonism in northwestern Germany during the last glaciation (Piotrowski, 1993), and facilitated glaciotectonism along the Main Stationary Line in Denmark (Piotrowski *et al.*, 2004; Fig. 9.2). Deformation also may be caused by fast ice retreat where a swath of land with pressurized groundwater is exposed, causing sediment blow-ups before pressure equilibrium is established.

Groundwater flow dynamics, especially the flow field, velocity and rate, strongly depend on the substratum porosity and hydraulic conductivity, which act as 'knobs that open and close subglacial drainage valves' (Flowers & Clarke, 2002a). In the case of sorted granular materials such as sand and gravel, these parameters do not differ much from those under non-glacial conditions. More difficult to estimate are hydrogeological parameters of subglacial tills, yet this sediment type is of particular impor-

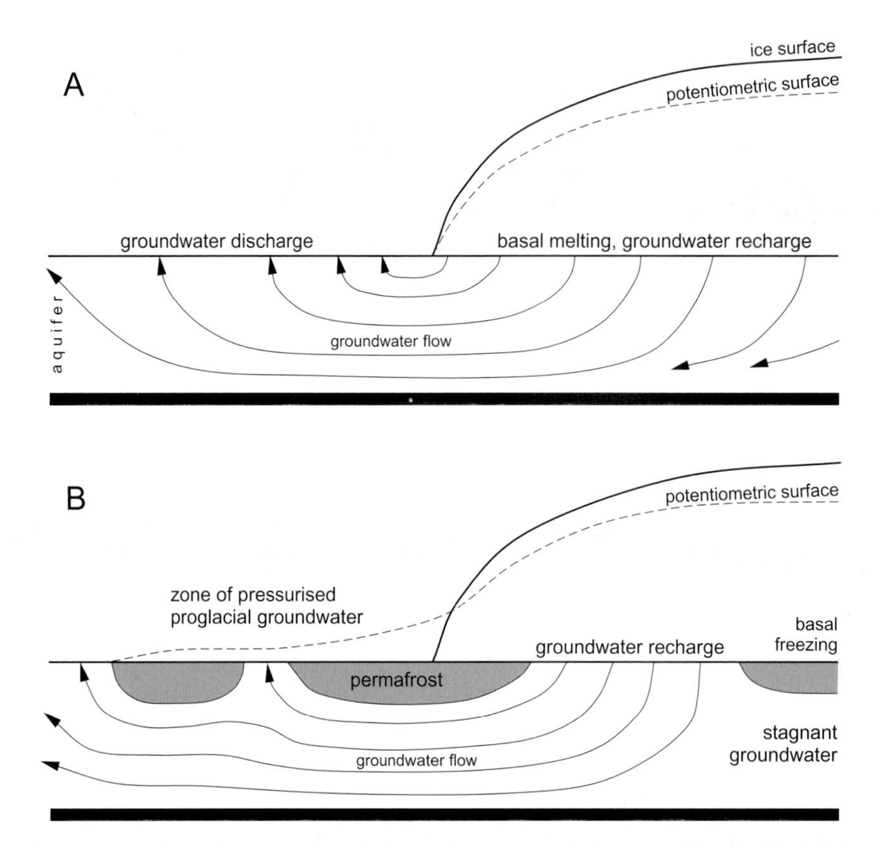

Figure 9.1 Schematic representation of groundwater drainage through a subglacial aquifer under a glacier resting on unfrozen (A) and partly frozen (B) bed.

tance because it occurs immediately under most glaciers and is the first sediment on the pathway of the subglacial groundwater into the bed.

When a basal till is dilated due to shear deformation and high porewater pressure, its porosity is around 0.4 (Blankenship *et al.*, 1986; Engelhardt *et al.*, 1990b), whereas non-dilated tills appear to have porosities around 0.25–0.3 (Fountain & Walder, 1998). The difference corresponds to an increase in flow velocity of up to about 10 times. Hydraulic conductivity of tills from past glaciations varies over several orders of magnitude, typically between 10^{-1} and 10^{-7} m s^{-1} (Freeze & Cherry, 1979), with an extreme effect on groundwater fluxes (see the Darcy equation).

Conductivity of a till under a glacier can be expected to vary accordingly, but its direct determination is complicated owing to limitations in the accessibility of glacier beds. Reliable estimates are, therefore, scarce. Examples of *in situ* estimates using different methods, and some laboratory tests on till samples collected from under modern glaciers and ice sheets, are given in Table 9.1 and show the extreme spread of values between some 10^{-2} and 10^{-12} m s^{-1}. Worth noting is that *in situ* measurements tend to yield higher conductivities than laboratory tests, owing to large-scale non-Darcian flow pathways and large-scale till heterogeneities (e.g. Gerber *et al.*, 2001), such as sand and gravel lenses, known to occur in nature but not accounted for in laboratory tests.

Besides the textural composition, also compression and shear deformation influence the conductivity of subglacial tills. Both result in highly anisotropic distribution of till properties with horizontal conductivity up to two orders of magnitude greater than the vertical one owing to particle alignment, as shown by laboratory experiments (e.g. Murray & Dowdeswell, 1992). Till deformation in response to glacier stress may increase the conductivity (in the case of dilation; Clarke, 1987b) or decrease it (in the case of compaction), as demonstrated experimentally by Hubbard & Maltman (2000). They also showed that hydraulic conductivity in tills is inversely related to the effective glacier pressure because pressurized porewater prevents sediment compaction and closure of drainage pathways.

If till undergoes pervasive deformation, substantial volumes of groundwater may be advected toward the ice margin within the high-porosity deforming layer (Alley *et al.*, 1986a, 1987a). Such deformation may destroy drainage paths at the ice–bed interface (Clarke *et al.*, 1984), but it will enhance drainage through the bed (Murray & Dowdeswell, 1992). Despite our still fragmentary knowledge about hydrogeological properties of tills under glaciers, one can safely conclude that these properties vary extremely both in space and time, influenced by the nature of the source material and stresses applied by the overriding glacier.

9.4 Subglacial groundwater in past and modern environments

As ice sheets grow and expand over permeable rocks, groundwater flow evolves from a subaerial, precipitation-fed system con-

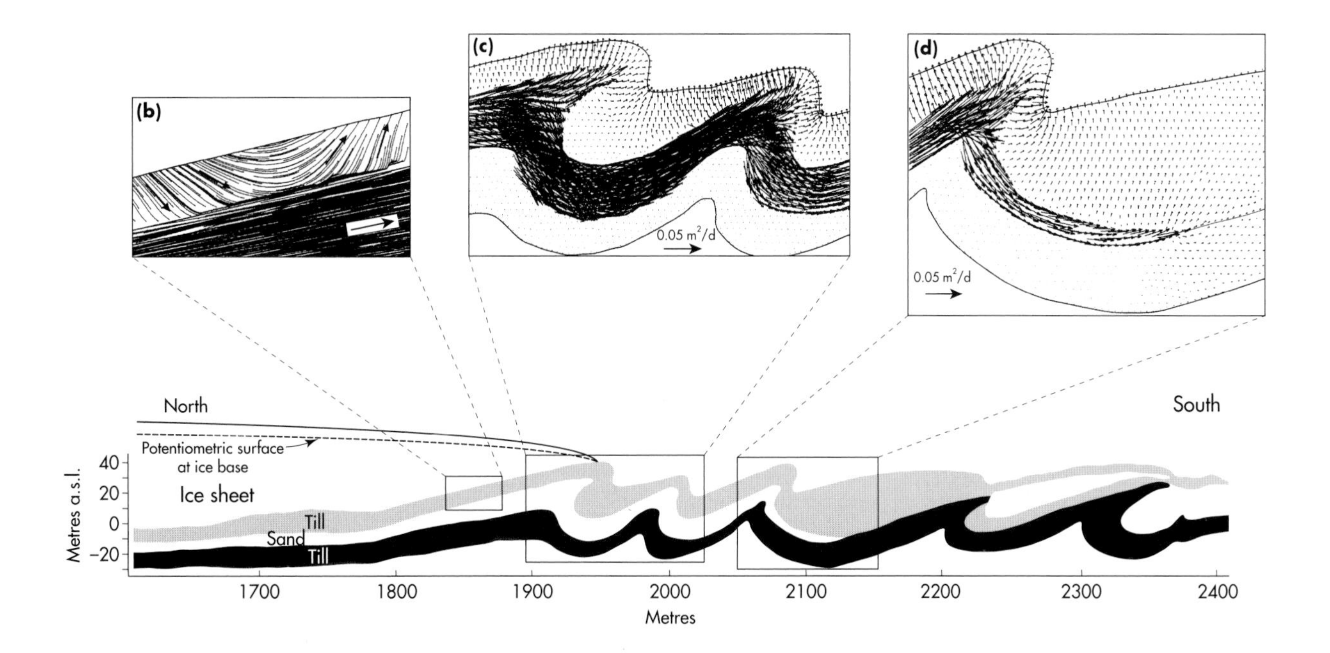

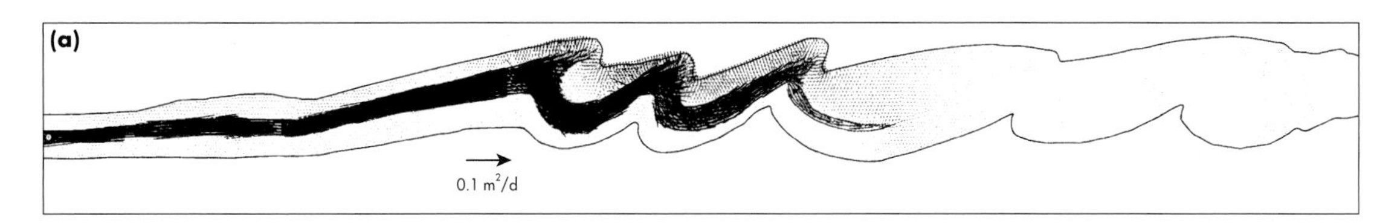

Figure 9.2 Modelled groundwater flow pattern under the margin of the Weichselian ice sheet at its maximum extent at the Main Stationary Line in Denmark (Bovbjerg). Large glaciotectonic folding was facilitated by high porewater pressure in the low-transmissivity bed, partly due to thin aquifers wedging out in the direction of groundwater flow. Note that the transition from groundwater recharge to discharge still occurs under the ice sheet (b), caused by the low hydraulic conductance of the bed not capable of evacuating all meltwater to the ice foreground (Piotrowski *et al.*, 2004).

Table 9.1 Hydraulic conductivities of some subglacial tills derived from *in situ* and laboratory measurements

Location	Hydraulic conductivity (ms^{-1})	Reference
Bakaninbreen	3×10^{-7} to 8×10^{-8}	Porter & Murray (2001)
Bakaninbreen	8×10^{-3}	Kulessa & Murray (2003)
Midre Lovénbreen	2×10^{-5}	Kulessa & Murray (2003)
Trapridge Glacier	5×10^{-4}	Stone *et al.* (1997)
Trapridge Glacier	$1–2 \times 10^{-9}$	Waddington & Clarke (1995)
Trapridge Glacier	2×10^{-8}	Murray & Clarke (1995)
Trapridge Glacier	1×10^{-8}	Flowers & Clarke (2002a)
South Cascade Glacier	10^{-7} to 10^{-4}	Fountain (1994)
Haut Glacier d'Arolla	10^{-7} to 10^{-4}	Hubbard *et al.* (1995)
Haut Glacier d'Arolla	8×10^{-12} to 9×10^{-9}	Hubbard & Maltman (2000)
Breidamerkurjökull	$1–2 \times 10^{-6}$	Boulton *et al.* (1974)
Breidamerkurjökull	6×10^{-7} to 4×10^{-4}	Boulton & Dent (1974)
Storglaciären	10^{-7} to 10^{-6}	Iverson *et al.* (1994)
Storglaciären	10^{-9} to 10^{-8}	Fischer *et al.* (1998)
Storglaciären	10^{-7} to 10^{-6}	Baker & Hooyer (1996)
Gornegletscher	2×10^{-2}	Iken *et al.* (1996)
Ice Stream B	2×10^{-9}	Engelhardt *et al.* (1990b)
Ice Stream B	10^{-9}	Tulaczyk *et al.* (2001a)

trolled by landscape topography, to a system pressurized by and recharged from the overlying ice. Studies reconstructing water flow in subglacial sediments are very scarce, but they all show that flow pattern and velocity differed substantially from the interglacial situation in the same areas. Owing to the complexity of the glacial and groundwater systems, numerical methods have been applied following the pioneering study of Boulton & Dobbie (1993). They have demonstrated that under glaciers resting on aquifers with high hydraulic transmissivity, all basal meltwater can be evacuated through the bed and no other drainage systems at the ice base need to form. If, on the other hand, a glacier is underlain by fine-grained sediment, its conductance may be insufficient triggering formation of more efficient drainage pathways, such as channels. This conclusion is important for inferences on ice movement mechanisms and for interpreting the origin of eskers and tunnel valleys, which, accordingly, should indicate areas with excess of basal meltwater.

9.4.1 Did all basal meltwater drain through the bed?

Boulton & Dobbie (1993) suggested that areas in Holland, during the Saalian glaciation, had sufficient transmissivity to drain all meltwater from the ice sheet base. This inference was based on calculations with very low basal melting rates of 2–3 mm yr^{-1}, but claimed valid also in subsequent simulations with melting rates of 20 mm yr^{-1} (Boulton *et al.*, 1993). In a numerical model of groundwater flow under ice sheets of the last two glaciations along a transect from the ice divide in Scandinavia to the ice periphery in Holland, Boulton *et al.* (1995) used a melting rate of 25 mm yr^{-1} with a similar conclusion. In their model, however, the entire Quaternary sequence was lumped together into one aquifer with conductivity of 3×10^{-4} m s^{-1}, corresponding to the conductivity of sand. This value is clearly too high, neglecting the true nature of Quaternary strata characterized by interlayered sediments of much lower bulk transmissivity. The hypothesis of basal water drainage entirely through the bed is also inconsistent with widespread sediment deformation postulated by Boulton (1996a), because such deformation will occur only if porewater pressure is at least 90–95% of the overburden pressure (Paterson, 1994, p. 169), that is when the bed capacity to absorb meltwater is reached. As pointed out by Arnold & Sharp (2002), the modelling of Boulton and co-workers is seemingly contrary to most geological evidence. More recently, Boulton *et al.* (2001b) proposed a revised model of basal hydraulic regimes that considers a zone where the meltwater flux is too large to be discharged by groundwater flow alone, and in which tunnels form.

In a three-dimensional model of groundwater flow under the Weichselian ice sheet in northwestern Germany that accounts for the heterogeneity of Quaternary sediments, located close to Boulton's transect, Piotrowski (1997a) showed that only about 25% of meltwater could have been evacuated through the bed, and the rest was drained in spontaneous outburst episodes through subglacial channels. This is supported by the occurrence of tunnel valleys, some up to ca. 80 m deep (Piotrowski, 1994), abundantly found in this area. Modelling the adjacent area around the Eckernförde Bay showed that about 30% of basal meltwater drained as groundwater flow, and the rest through channels (Marczinek, 2002; Marczinek & Piotrowski, this volume, Chapter 10). It should be noted that these simulations were made with conservative basal melting rates of 36 mm yr^{-1} not considering surface ablation water, which probably recharged the bed close to the ice margin. Accounting for this additional water source would further substantiate the conclusion about the insufficient drainage capacity of the bed, consistent with the formation of tunnel valleys as high-discharge drainage pathways. It also should be stressed that if a permafrost wedge under the ice-sheet margin is considered, the bed conductivity will be yet lower (Fig. 9.3), further increasing the likelihood of tunnel valley formation.

It is tempting to suggest that similar hydraulic deficiency of the bed could have initiated tunnel valleys elsewhere across the Central European Lowland and at the bottom of the North Sea, where they represent the largest glacial features of the Elsterian and Weichselian glaciations, some over 500 m deep and hundreds of kilometres long (e.g. Huuse & Lykke-Andersen, 2000). If tunnel drainage is predicted to have extended up to 150 km from the ice sheet margin (Arnold & Sharp, 2002), then some of these tunnel valleys possibly formed time-transgressively during deglaciation.

Because subglacial channels form in response to the excess of water at the ice sole, a succession of drainage mechanisms operating in cycles can be envisaged. Each cycle starts with low water pressure in the bed and groundwater recharge, followed by gradual increase of water pressure, formation of a basal water layer or a lake at the ice flotation pressure, succeeded by spontaneous drainage through channels, lowering the water pressure and ending with basal recoupling (Piotrowski, 1997a). Formation of the channels would thus act as a stabilizing feedback preventing widespread basal decoupling and catastrophic collapse of an ice lobe. Origin of subglacial lakes owing to meltwater ponding in low-transmissivity areas of northern Germany was also considered by van Weert *et al.* (1997), and in marginal zones of the Laurentide Ice Sheet by Cutler *et al.* (2000), in accordance with the considerations of Shoemaker (1991) who gave a physical framework for the formation of such lakes.

Most reconstructions of groundwater flow under past ice sheets in North America and Europe indicate a deficiency of bed materials to evacuate all the incoming meltwater, which is interesting, bearing in mind the diversity of beds overridden by ice sheets on the two continents. Brown *et al.* (1987) showed that only a fraction of basal meltwater could have been evacuated through a soft substratum under the last glacial Puget Lobe of the Cordilleran Ice Sheet. They envisage a widespread basal decoupling by water pressurized to the ice flotation level, indicated among other things by lenses and layers of sand in till, and low overconsolidation of subglacial clays. The same line of sedimentological and geotechnical evidence was used by Piotrowski & Kraus (1997) and Piotrowski & Tulaczyk (1999) to suggest bed drainage deficiency in northern Germany. Shoemaker (1986) demonstrated that even under an ice sheet resting on a metre-thick gravel bed, channels must develop to keep porewater pressures below the ice overburden pressure, which was later confirmed by Ng (2000a). Two recent studies along transects from the Hudson Bay to the southern margin of the Wisconsinan ice sheet independently show that subglacial aquifers were not capable of evacuating the basal meltwater, possibly due to a combination of low bed-permeability and permafrost, compensated by channelized drainage (Breemer *et al.*, 2002) or subglacial water storage (Cutler *et al.*, 2000). Under

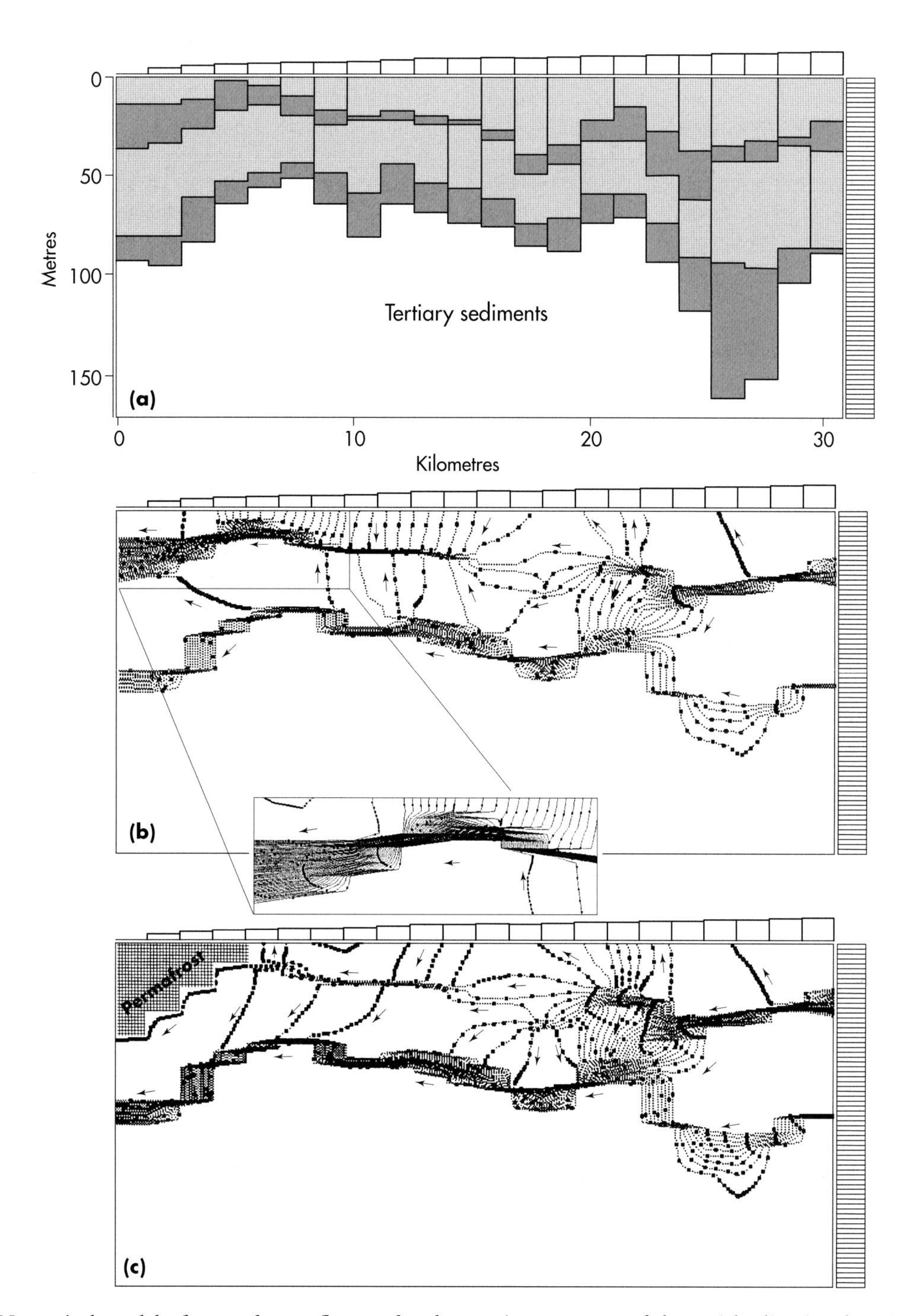

Figure 9.3 Numerical model of groundwater flow under the maximum extent of the Weichselian ice sheet in northwestern Germany along an ice-flow parallel transect between the ice margin (0 km) and the present Baltic Sea coast (30 km). Geology (a) is generalized into two aquifers (dark grey) and two aquitards (light grey) resting on impermeable Tertiary sediments. Groundwater flow lines are shown with time markers and direction arrows. Scenario in (b) assumes non-permafrost conditions, and scenario in (c) considers a permafrost wedge under the ice margin. Note that in (c) the whole upper aquifer is frozen, so that (i) more groundwater drains through the lower aquifer, (ii) less water recharges the bed from the ice sole, and (iii) more water is forced upward toward the ice sole from the bed. (From Piotrowski, 1997b.)

the Green Bay Lobe, only raising the conductivity values by three orders of magnitude would have allowed all basal melt to be evacuated as groundwater (Cutler *et al.*, 2000). A comprehensive numerical study of the hydrology under the largest European ice cap, Vatnajökull, showed that buried aquifers may only evacuate up to ca. 30% of subglacial water (Flowers *et al.*, 2003). Under Trapridge Glacier, Yukon, the aerial extent of aquifer saturation is 70–90%, with the groundwater drainage system at the edge of its capacity beneath the ablation zone (Flowers & Clarke, 2002b).

Also worth noting is that data from beneath the Whillan's Ice Stream (formerly Ice Stream B), Antarctica, indicate the substratum inefficiency in capturing basal meltwater. Lingle & Brown (1987) suggested that the subglacial water discharge mechanism is advection in the deforming layer, but in the light of a much thinner zone of deformation than originally assumed (Engelhardt & Kamb, 1998) this mechanism is not likely to be efficient. Subglacial ponding and localized ice decoupling are likely, among other effects, owing to the postulated upwards-directed groundwater flow in parts of the bed (Tulaczyk *et al.*, 2000b). Likewise, marine-based ice sheets terminating on shelf sediments that tend to be fine-grained could not have discharged all the melt through the bed (Hindmarsh, 1999).

9.4.2 Interaction between groundwater and water in subglacial channels

Where glaciers rest directly on aquifers or are separated from them only by a thin layer of low-permeability sediment, the groundwater is in contact with the water in channels at the ice–bed interface. Channel formation reduces the water pressure and generates a hydraulic gradient, which will drive groundwater from the surrounding sediment into the channel and create a catchment area along the channel. This was suggested independently by Shoemaker & Leung (1987) and Boulton & Hindmarsh (1987), and recently measured in piezometers nested around a channel under Breidamerkurjøkull in Iceland (Boulton *et al.*, 2001b).

Radial groundwater flow to the channel can fluidize the sediment, followed by its injection into the channel and subsequent removal by water flow. In a steady-state situation, this mechanism could produce large incisions resembling tunnel valleys by erosion in relatively narrow R-type channels serving as sediment sinks for the surrounding aquifers (Boulton & Hindmarsh, 1987). Possible capturing of groundwater by tunnel valleys was also suggested for one prominent tunnel valley tract in northern Germany (Piotrowski, 1994). However, it was postulated that an increase of effective pressure at the channel flanks would strengthen the sediment there and, in particular, prevent pervasive deformation of the glacier bed.

Modelling by Boulton *et al.* (2001b) and Fleming & Clark (2000) shows a progressively damped pressure wave propagating away from a channel into the sediment for tens of metres, consistent with some field measurements (e.g., Fountain, 1994; Hubbard *et al.*, 1995; Murray & Clarke, 1995). Boulton *et al.* (2001b) predicted a bulb of low pressure in a shallow zone beneath the channel. Occurrence of this low-pressure zone is confirmed by Piotrowski *et al.* (1999), who documented a soft-sediment diapir under a subglacial channel of Saalian age in eastern Germany. The diapir demonstrates sediment creep into the channel from below, and yields support for theoretical considerations of Boulton & Hindmarsh (1987) and Shoemaker & Leung (1987) suggesting a sediment mobilization into a channel and erosion rates higher than could be estimated from channel dimensions alone.

Coupling of water flow in channels and in the bed also implies that, under specific circumstances such as a wave of ablation water reaching a subglacial channel or channel blockage by sediment injection or roof collapse, a water pressure increase will propagate into the aquifer. This will happen with delay, caused by hydraulic resistance of the sediment, particularly in low-conductivity, fine-grained aquifers. The result will be a reversal of hydraulic gradient, now driving the water from the channel into the aquifer. Frequent shifts in groundwater flow directions close to channels would be expected where the subglacial hydrology is dominated by diurnal or seasonal ablation cycles, i.e. close to a glacier margin.

9.4.3 Groundwater flow dynamics under past ice sheets and preservation of old glacial groundwater

Numerical modelling shows that groundwater flow velocities and hydraulic heads in northwestern Europe, especially in the relatively shallow aquifers, were significantly higher under ice sheets than they are at present. For an area bordering the Baltic Sea in Germany, Piotrowski (1997b) estimated flow velocities in the upper aquifer under the last ice sheet as about 30 times higher than at present. Furthermore, a reversal of flow direction occurred. At present, the groundwater drains to the Baltic Sea; during the glaciation it was forced in the opposite direction by the ice sheet advancing out of the Baltic Sea basin. A corresponding flow reversal was also determined by Boulton *et al.* (1996) and van Weert *et al.* (1997) in a large-scale, vertically integrated palaeoflow model (Fig. 9.4), by Glynn *et al.* (1999) in a transect between the ice-sheet centre in Scandinavia and its periphery in Poland, by Grasby *et al.* (2000) in a transect from South Dakota to Manitoba during the last glaciation, and by Marczinek & Piotrowski (this volume, Chapter 10) in a coastal area of northwest Germany. Maximum flow velocities in upper subglacial aquifers in northwestern Europe were estimated at $20\,m\,yr^{-1}$ (Boulton *et al.*, 1995), $200\,m\,yr^{-1}$ (Boulton *et al.*, 1996) and over $100\,m\,yr^{-1}$ (van Weert *et al.*, 1997).

Given the high flow velocities and hydraulic heads, groundwater can penetrate deep into the subglacial sediments and rocks. Because of the typically layered structure of sedimentary beds, the flow pattern is highly anisotropic with preferential flow approximately horizontal in the transmissible, coarse-grained sediments and vertical in aquitards. Most studies indicate that upper aquifers with residence times in the order of several thousand years were completely flushed out during glaciations and the groundwater derived from precipitation during the interglacials was replaced by glacial meltwater. In Holland the entire upper (Quaternary) aquifer system, about 300 m thick, and a large part of the lower (Mesozoic) aquifer, up to 1500 m thick, are believed to have been flushed under the Saalian ice sheet, whereas the Tertiary aquitard in between was partly penetrated by meltwater (Boulton *et al.*, 1993). Deep meltwater circulation is also suggested for northwestern Germany, where during the Weichselian glaciation the flow field in the whole Quaternary sequence consisting of two major aquifers and two aquitards up to ca. 200 m thick was reorganized by ice overriding (Piotrowski, 1997a,b). Similar conclusions were drawn for the Illinois basin, USA where the several-kilometres-thick drainage system of superposed aquifers and aquitards was significantly modified by the last ice sheet, as modelling (Breemer *et al.*, 2002) and hydrochemical analyses (McIntosh *et al.*, 2002) indicate. Under the Des Moines lobe in Iowa, subglacial meltwater penetrated over 300 m down into the

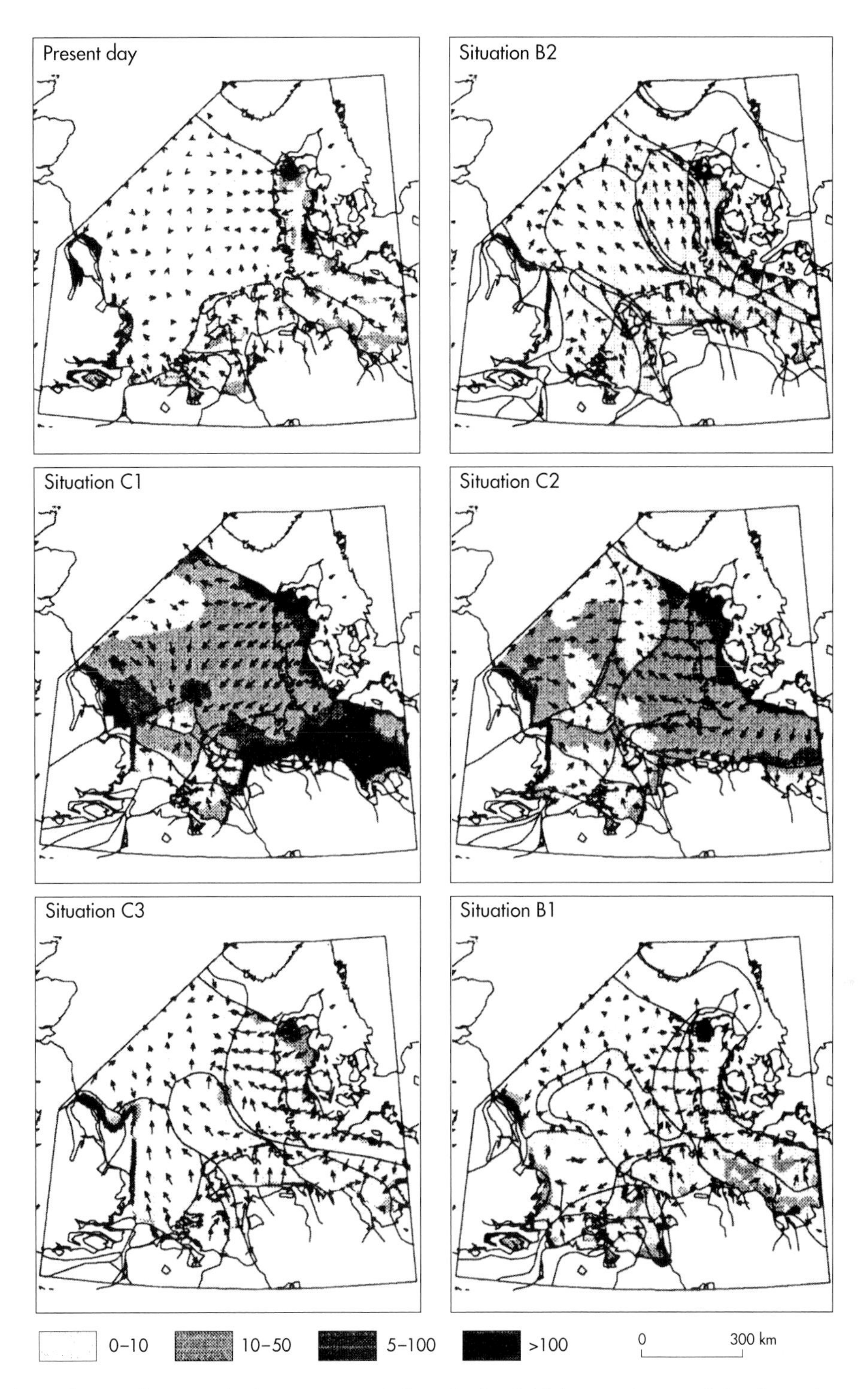

Figure 9.4 Horizontal groundwater velocity vectors and the velocity field (ranges in $m\,yr^{-1}$) in the major regional aquifer in northwest Europe at present and during the expansion of Fennoscandian and Scottish ice sheets (different palaeogeographical scenarios with maximum ice extent in C1). Note the reorganization of flow patterns under glacial conditions as compared with the present time. (From van Weert *et al.*, 1997.)

bed (Siegel, 1991), and the hydrogeology of the Atlantic continental shelf off New England experienced short-lived but dramatic reorganization during the Last Glacial Maximum, including large-scale disturbances in freshwater/seawater equilibrium (Person *et al.*, 2003). Pressurized glacial meltwater penetrated the bedrock in Northwest Territories, Canada, to depths over 1.6 km (Clark *et al.*, 2000). Under some circumstances, however, a glacier may seal the substratum preventing groundwater recharge, as was the case in one Alpine valley covered by the last ice sheet (Beyerle *et al.*, 1998).

Admittedly, numerical models of past groundwater flow are notoriously difficult to validate, and the range of possible solutions vary with the conceivable range of input parameters and boundary conditions, such as the ice sheet thickness and glacier persistence in a certain area, lithospheric response to loading, and spatial variability of hydraulic conductivities. Heuristic studies are promising but scarce (e.g. Fleming & Clark, 2000). A semi-quantitative test of model results may be provided by the isotope composition of deep groundwater, preferably in aquitards where the preservation potential of old water is high. In most cases direct dating with radiocarbon is not possible due to its relatively fast decay, but unstable isotopes with half-lives comparable to glacial cycles may be used (see below).

Indeed, several studies have shown that glacial meltwater flushed through and remained trapped in fine-grained sediments since the last glaciation in lowland areas of North America and Europe. Remenda *et al.* (1994) reported old groundwater in shallow tills and glaciolacustrine deposits from sites spanning a distance of 2000 km along the margin of the Last Glacial Maximum ice sheet in North America, identified by $\delta^{18}O$ values around −25‰, much lower than the modern precipitation of around −13 to −14‰ (see also Stueber & Walter, 1994; Clark *et al.*, 2000). Similarly, Marlin *et al.* (1997) have found old glacial porewater depleted in ^{18}O and ^{2}H in thick clay sediment inside the extent of the last glaciation in northern Germany. In another area in northern Germany at Gorleben, deep, saline water on top of a salt dome was formed in a cold Pleistocene climate, very likely under glacial conditions (Schelkes *et al.*, 1999). Glacial meltwater has also been documented in several places, notably down to depths of at least 500 m, in fractured rocks of the Fennoscandian and Canadian shields (Wallin, 1995; Smellie & Frape, 1997; Tullborg, 1997a; Glynn *et al.*, 1999; Laaksoharju & Rhén, 1999) and in deep Cambrian aquifers in Estonia (Vaikmäe *et al.*, 2001). These case studies show that, under favourable hydrogeological conditions, glacial meltwater injected into low-conductivity rocks under high pressure can remain there for thousands of years after deglaciation, and can be used as a proxy of global climatic changes of the past. In many other places groundwater recharged during cold periods of the Pleistocene also has been documented, but it may not necessarily be water derived from melting of glacier ice (e.g. Zuber *et al.*, 2004).

9.5 Hydrochemical and environmental aspects

In a cycle comprising a glaciation and two bracketing interglacial periods, the predicted changes in groundwater chemistry will be as profound as the changes in groundwater-flow dynamics (Fig. 9.5). Worth noting is isostatic rebound that may cause fracturing and matrix expansion, thereby increasing the rock permeability on a regional scale and enabling migration of new fluids into the substratum after deglaciation (e.g. Weaver *et al.*, 1995). Glacial meltwater recharging the subsurface has certain characteristics that distinguish it from waters derived from other sources and under warm climate conditions, which help its identification and facilitates reconstruction of groundwater circulation under ice sheets (Smellie & Frape, 1997; Glynn *et al.*, 1999). Such features are:

1 Low ^{18}O and ^{2}H contents. Glacial groundwater may have $\delta^{18}O$ values as low as ca. −30‰. Most old deep glacial groundwaters, however, show slightly higher signatures, probably due to mixing with heavier, pre-glacial water. Low $\delta^{18}O$ is also found in minerals precipitated from subglacial meltwater, e.g. in some calcites and iron oxyhydroxides on bedrock surfaces in Scandinavia.

2 High dissolved oxygen content. During ice formation, oxygen is trapped in air bubbles that will later dissolve under the pressure of the overlying ice and enrich the groundwater. Using O_2 concentrations measured by Stauffer *et al.* (1985) in the Greenland ice sheet, Glynn *et al.* (1999) concluded that O_2 content in subglacial meltwater may be three to five times greater than the normal concentration at equilibrium with the

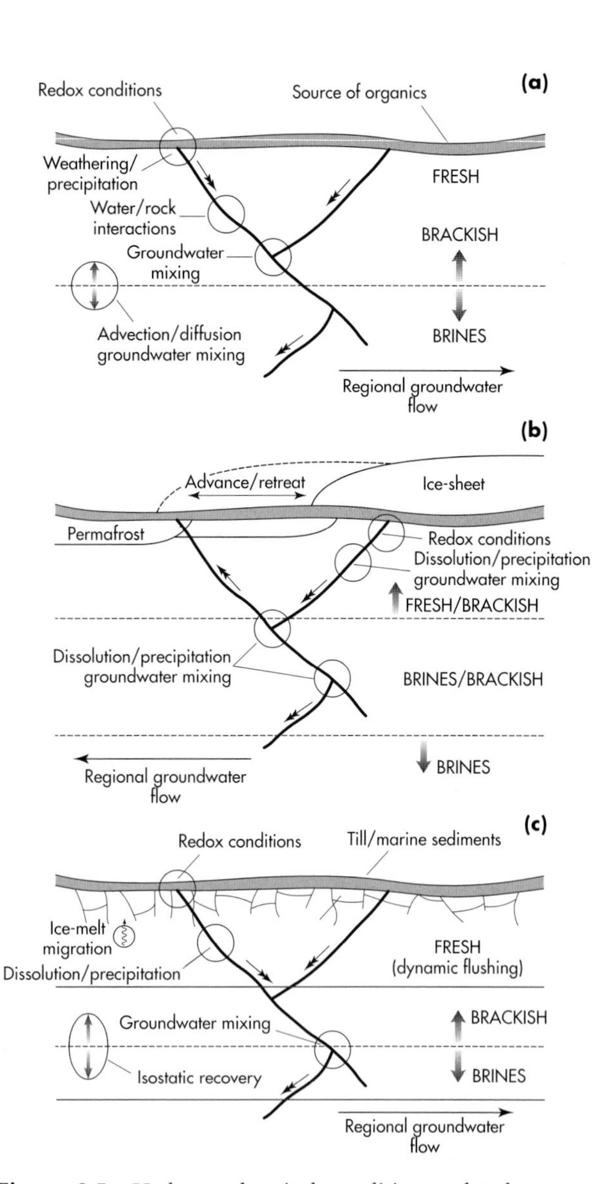

Figure 9.5 Hydrogeochemical conditions related to pre-glacial (a), glacial (b) and post-glacial (c) periods in fractured bedrock areas. Circled are aspects sensitive to surface conditions that may provide signatures of glaciations or interglacials. (Adapted from Smellie & Frape, 1997.)

atmosphere at 0°C. This can lead to the formation of non-hydrothermal iron oxyhydroxides such as those coating fractures in bedrock at a depth of several hundred metres in Sweden (e.g. at 900 m at Äspö in southeast Sweden; Glynn *et al.*, 1999). Up to a depth of 120 m, occurrence of iron oxyhydroxides coincides with the absence of pyrite and calcite (Tullborg, 1997b). Meltwater rich in oxygen would also lead to sulphide oxidation and characteristic $\delta^{13}C$ values (Wallin, 1992).

3 Lack of pyrite along the conductive fractures. Borehole cores from crystalline bedrock around Äspö reveal absence of pyrite in hydraulically active fractures, whereas this mineral occurs in closed fractures and in the rock matrix (Tullborg, 1989; Smellie & Laaksoharju, 1992). This possibly is due to the penetration of oxidizing glacial groundwater during the last glaciation, which did not affect the low-conductivity matrix and fractures outside the active groundwater system.
4 High values of $\delta^{13}C$ in calcites. Values around −5‰ and higher (together with low $\delta^{18}O$ contents) commonly occur in fracture fills in the Fennoscandian shield, e.g. at 600 m depth at Laxemar, possibly indicating glacial meltwater as the precipitant source (Wallin, 1995).
5 Compositional zoning of fracture minerals. Smellie & Frape (1997) pointed out that minerals may reflect more than one glacial cycle as discrete zones, whereas water composition will be a hybrid resulting from mixing, dominated by the most recent events. They give examples of zoned calcites from Sellafield, UK with depleted, low temperature fluid inclusions often interpreted as a glacial signature, dated to around the end of the last glaciation. Also at several sites in Finland and Sweden multiple generations of mineral precipitation were documented, interpreted as reflection of repeated cold and warm climatic events (Smellie & Frape, 1997).
6 Low dissolved organic carbon concentration. Water released from the melting glacier base would typically infiltrate the bed without passing through the organic-rich soil zone, and thus have bicarbonic acid concentrations two orders of magnitude lower than found in interglacial meteoric water. A low ^{14}C content (<12 ppm C) is a strong indication of glacially derived groundwater (Tullborg, 1997a).
7 Low total dissolved solids contents. Studies in Iceland and Spitsbergen show that modern glacial groundwater is characterized by a non-equilibrium (undersaturation) with respect to most constituents otherwise derived from host rocks, indicating efficient flushing, short residence time and low degree of water–rock interactions (Sigurdsson, 1990; Haldorsen *et al.*, 1996). Dissolved solids are expected to rise after ice retreat when the groundwater flow slows significantly.
8 Uranium-series disequilibrium (USD). Penetration of oxic meltwater should affect the U/Th ratio in minerals along groundwater flow passages because uranium is much more mobile than thorium under oxidizing conditions. This can be useful in assessing the role of glacial meltwater in the substratum during the late Pleistocene glaciations. Indeed, USD determined in bedrock in Palmottu, southeastern Finland indicates clear periodicity in uranium mobilization several times during the past 300,000 yr, correlated with major glaciation phases (Suksi *et al.*, 2001; Rasilainen *et al.*, 2003). At Kamlunge, northern Sweden the USD indicating depletion of uranium in the past 500,000 yr was determined to a depth of about 500 m in high-conductivity zones of bedrock (Smellie, 1985).

Remnants of glacial meltwater with these characteristics are found in numerous localities in crystalline shield of Scandinavia and North America as well as in some places in soft sediments south of the crystalline basement. Documentation of such remnants at depths of at least 400 m is convincing. Bearing in mind the likelihood of the next glacial maxima at ca. 5, 20, 60 and 100 ka from now as predicted by the Central Climate Change Scenario (King-Clayton *et al.*, 1997), the old glacial groundwater found at such depths and numerical models suggesting penetration of subglacial water down to several kilometres must be of concern in designing radioactive waste disposal strategies (Talbot, 1999). Repositories are considered, for example, in Sweden and northern Germany, well within the range of prospective glaciations. The major problem is the high dissolved oxygen content of glacial groundwater, which would increase the solubility and mobility of many radionuclides (U, Pu, Tc, Np) by several orders of magnitude and may put the waste canisters stability at risk due to oxidization (Glynn *et al.*, 1999). A special hard rock laboratory on the island of Äspö, southeast Sweden serves as a test and research facility to study the suitability of crystalline bedrock for hosting radioactive waste repositories (SKI, 1997). The laboratory, situated 450 m under the ground surface has served, among others, to evaluate the impact of groundwater flow under a perspective ice sheet on the repository's safety. The majority of the results show that the site would be affected by enhanced, oxidizing groundwater penetration under a future ice sheet. Because the stability of nuclear waste repositories must be ensured for tens to hundreds of thousands of years (e.g. in USA for 10,000 yr, in Switzerland and France for 1,000,000 yr), groundwater stress periods imposed by glaciations must be taken into account.

9.6 Summary

Aquifers overridden by ice sheets and groundwater recharged into permeable beds from the melting ice base are important components of the hydrological cycle in regions affected by glaciations. Research on subglacial groundwater is very scarce and often disputable, but the available data allow the following generalizations:

1 Glaciers pressurize groundwater and impose regional head gradients driving groundwater toward the ice margin where it is released into subaerial drainage systems. If the glacier margin rests on permafrost, the zone of pressurized groundwater may extend for tens of kilometres into the ice forefield.
2 Groundwater systems are significantly if not entirely reorganized during glaciations as compared with the non-glacial conditions. Local topographic catchments are replaced by large, ice-shape controlled regional catchments that extend from glacier termini to the ice divides or to cold-based inner parts of ice sheets. Groundwater flow velocities increase several times, and flow depth may reach several kilometres flushing deep aquifers and aquitards. Non-glacial meteoric

groundwater may be replaced completely by glacially fed groundwater.

3 Drainage capacity of the bed influences the ice movement mechanism by controlling the sediment strength and the extent of basal coupling. A substratum of low hydraulic conductivity facilitates fast ice flow by bed deformation and/or basal sliding on a thin water sheet. Indirectly, the drainage capacity of the bed influences glacier transport processes ranging between the subglacial transport in the deforming bed and englacial transport when the deforming bed is absent.

4 Large Pleistocene ice sheets overrode beds whose transmissivity was typically insufficient to evacuate all the subglacial meltwater as groundwater flow. The system responded by formation of channels, the remnants of which, in the form of tunnel valleys (some over 500 m deep and hundreds of kilometres long) and eskers, occur abundantly across Europe and North America.

5 Pressurized groundwater may trigger profound disturbances in sediments and rocks such as hydrofracturing and glaciotectonic thrusting and folding, and it may create specific landscapes of soft-sediment extrusion.

6 Glacial groundwater is characterized by a distinct chemical composition, notably low ^{18}O and ^{2}H content and high content of dissolved oxygen, and rocks affected by glacial meltwater may show specific mineralogical and chemical signatures. Old glacial groundwater trapped in aquitards has been documented from several bedrock and sediment areas in Europe and North America, down to depths of several hundreds of metres and more.

7 The predicted reorganization of groundwater flow in areas affected by future glaciations, especially its increased flow dynamics, deep penetration and highly oxidizing chemistry, is of major concern for planning repositories of radioactive waste because many radionuclides have half-lives well outside the next glacial cycle.

TEN

Groundwater flow under the margin of the last Scandinavian ice sheet around the Eckernförde Bay, northwest Germany

Silke Marczinek* and Jan A. Piotrowski†

**Hasseer Str. 10, 24113 Kiel, Germany*
†Department of Earth Sciences, University of Aarhus, C.F. Møllers Allé 120, 8000 Aarhus C, Denmark

Northwestern Germany experienced several ice advances and retreats during the Last Glacial Maximum. This part of the Scandinavian Ice Sheet, located at the margin of the Baltic Sea basin, was dominated by a land-based but highly dynamic Baltic Ice Stream that terminated about 30 km southwest of the study area at its maximum extent. Abundant tunnel valleys, drumlins, and low preconsolidation ratios of tills suggest fast ice flow caused by some combination of enhanced basal sliding and deformation of soft sediment in this and adjacent areas (Piotrowski & Kraus, 1997; Piotrowski & Tulaczyk, 1999), both indicative of subglacial water pressure elevated to the vicinity of ice flotation point. This study was conducted to evaluate the capacity of the glacier bed

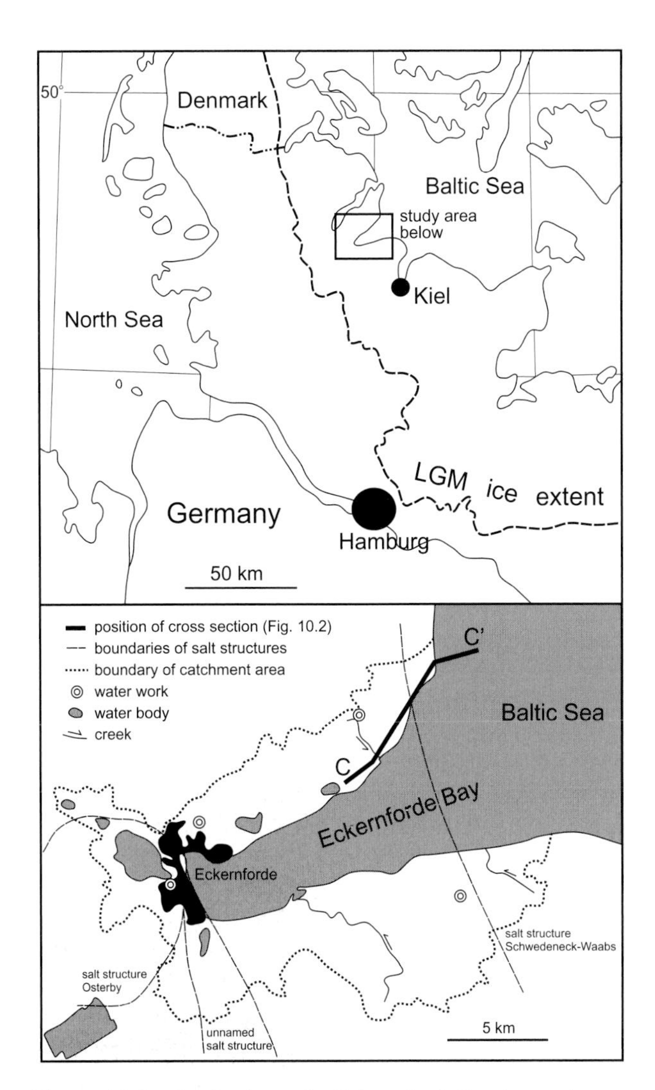

Figure 10.1 Study site at the Eckernförde Bay in northwestern Germany.

around the Eckernförde Bay (Fig. 10.1) to drain basal meltwater as groundwater flow, and constrain conditions favourable for generation of tunnel valleys, which are the most prominent features of the glacial landscape in northwestern Germany.

A hydrogeological model of the Eckernförde Bay was made using the available borehole data, geological maps at the scale of 1:25,000 and field mapping of coastal cliff sections. The model comprises the entire Pleistocene succession and a part of the Tertiary, down to the first widespread aquitard considered impermeable and thus hydrologically inactive. The sediment succession was generalized into five major units as follows (Fig. 10.2): unit 1, Miocene lignite sands; unit 2, Miocene mica clays; unit 3, Pleistocene lower till complex; unit 4, Pleistocene sand horizon; unit 5, Pleistocene upper till horizon. These units represent two aquifers (1 and 4) and three aquitards (2, 3 and 5). Owing to the discontinuous distribution of the aquitards, in some places the two aquifers are in direct hydraulic contact.

Hydraulic conductivities were estimated from material description of the borehole data or directly derived from pumping tests. They range from ca. 5×10^{-5} to $5 \times 10^{-4}\,\mathrm{m\,s^{-1}}$ for aquifers and from ca. 5×10^{-8} to $1 \times 10^{-6}\,\mathrm{m\,s^{-1}}$ for aquitards. Spatial interpolation of thicknesses and conductivities of the hydrogeological units was carried out with a fuzzy-kriging procedure (Piotrowski *et al.*, 1996; Marczinek & Piotrowski, 2002).

A three-dimensional, steady-state groundwater flow model for the catchment area of the Eckernförde Bay was created for the modern day conditions with the Finite Difference code MODFLOW (Plate 10.1A & C). The model was validated using water level data from 63 wells screened in both aquifers with a satisfactory result corresponding to the model of Kaleris *et al.* (2002), which shows a confluent discharge pattern from the surrounding aquifers into the bay, a regional groundwater sink at present. Subsequently, steady-state groundwater flow was simulated for full glacial conditions with the ice margin situated at a still-stand line several kilometres southwest of the study area. The lateral boundaries of the model running parallel to the ice-flow direction were taken as no-flow boundaries, the up-ice and down-ice boundaries were open to facilitate flow, and a prescribed head boundary was assigned to the uppermost layer, with head values corresponding to the maximum water pressure determined by the ice thickness. The ice thickness was calculated using a parabolic formula adjusted for soft beds and warm ice (cf. Piotrowski & Tulaczyk, 1999), in accord with palaeoglaciological data and theoretical considerations. The calculated ice profile rises from southwest to northeast and the ice thickness reaches about 250 m in the northeast corner of the model area.

The simulation under the ice sheet cover shows a fundamental reorganization of the groundwater flow field as compared with the modern-day (interglacial) situation. Driven by heads diminishing towards the ice margin, the groundwater in both aquifers discharges in the opposite direction, i.e. away from the Eckernförde Bay and the Baltic Sea basin (Plate 10.1B & D). Under glacier coverage, hydraulic heads are between ca. 85 and 185 m, whereas at present they are between ca. 2 and 14 m. Subglacial groundwater flow is faster by a factor of 30 than under non-glacial conditions, and the discharge along the down-ice boundary is ca. $4\,\mathrm{m^3\,s^{-1}}$.

In consecutive runs of the model we considered scenarios for both frozen and unfrozen ice margins, as well as channelized drainage along the ice–bed interface (Table 10.1). A single subglacial channel with water pressure just slightly below the glacier flotation pressure would increase the total discharge by about one-third. Permafrost under the ice margin would reduce the discharge by about 8% whereby the bulk drainage would occur through the Miocene aquifer. If both permafrost and the channel are considered (an unlikely scenario), then the discharge at the ice margin would increase by about one-quarter with respect to the basic model.

A conservative assumption of basal melting ($36\,\mathrm{mm\,yr^{-1}}$, Piotrowski, 1997a) as the only source of water at the ice sole gives recharge of ca. $14\,\mathrm{m^3\,s^{-1}}$ in the model area and its upstream catchment area. This is several times greater than the discharge at the down-ice boundary in all scenarios, which shows that the hydraulic capacity of the substratum was by far insufficient to evacuate all the basal meltwater to the ice margin as groundwater flow, similar to the situation in other parts of northwestern

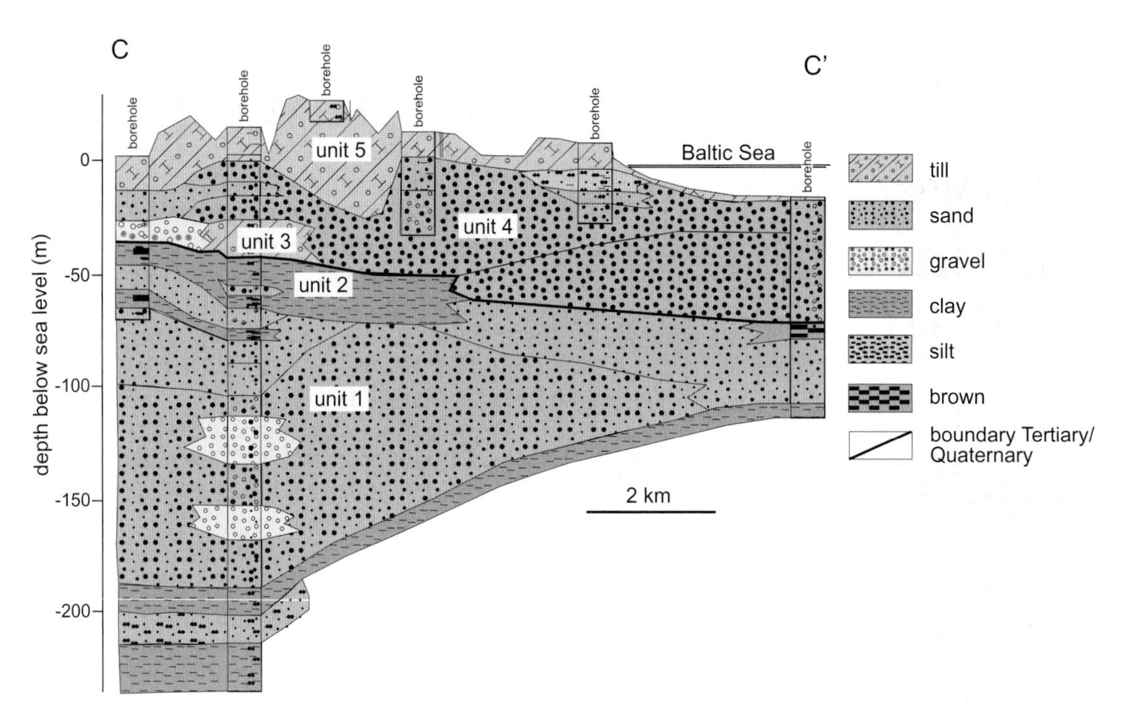

Figure 10.2 Geological section showing the major hydrogeological units 1–5. Location of the section in Figure 10.1.

Table 10.1 Water discharge at the ice-sheet margin for different scenarios including permafrost and a subglacial channel

Scenario	Discharge at ice margin ($m^3 s^{-1}$)
Basic model without permafrost and without subglacial channel	4.0
Without permafrost and with subglacial channel	5.3
With permafrost and without subglacial channel	3.7
With permafrost and with subglacial channel	4.9

Germany modelled by Piotrowski (1997a,b). This in turn implies a likelihood of hydraulic decoupling, and formation of more efficient drainage mechanisms for the surplus meltwater, such as subglacial sheet floods or channelized flow. Tunnel valleys, occurring abundantly across the entire central European lowland, yield convincing support for the latter. We therefore suggest that tunnel valleys formed to secure the stability of the ice sheet by reducing the water pressure at the ice–bed interface, thereby preventing catastrophic surges and ice-sheet collapse.

ELEVEN

Simulation of groundwater flow and subglacial hydrology, Late Pleistocene Lake Michigan Lobe, Laurentide Ice Sheet

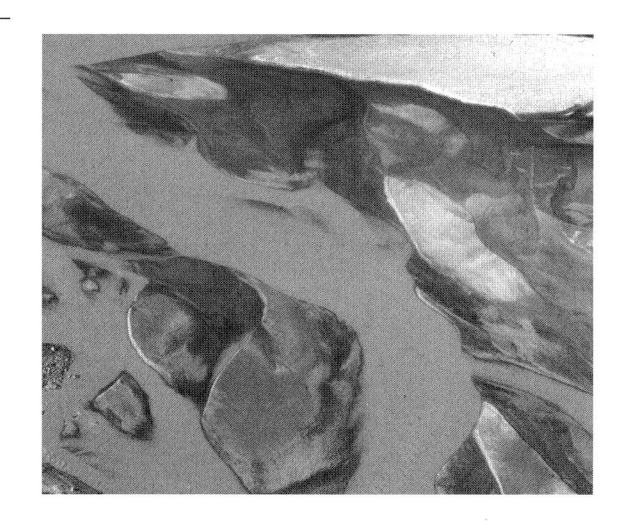

Chris Breemer

GeoEngineers, Incorporated, 15055 SW Sequoia Parkway, Suite 140, Portland, Oregon 97224, USA

11.1 Introduction

The Lake Michigan Lobe was a dynamic feature of the Laurentide Ice Sheet during the last glaciation. The lobe had an extremely low surface profile, yet sustained fast ice flow, suggesting that the lobe flowed over a low-friction bed. We used the groundwater model MODFLOW (McDonald & Harbaugh, 1988) to estimate the water pressure at the base of the Lake Michigan Lobe that would have influenced mechanisms of fast ice flow during the last glaciation.

11.2 The groundwater model

Our simulation is based on a two-dimensional flow line parallel to ice flow (Fig. 11.1). The model includes a hydrostratigraphic description of six regional aquifer units and four regional confining units that were present below the Lake Michigan Lobe (Fig. 11.2). Hydraulic characteristics of the hydrostratigraphic units are based on regional estimates (e.g. Mandle & Kontis, 1992). A summary of hydrostratigraphic units and hydraulic characteristics is presented in Table 11.1. We assume a conservative melt rate ($6\,mm\,yr^{-1}$) along the entire ice-covered portion of the flow line and we assume that all of the meltwater generated along the flow line north of Lake Superior arrived at the ice–bed interface. We validated the model by performing simulations using modern non-glacial boundary conditions. Simulated steady-state head values and surface-discharge volumes were close to measured modern values, suggesting that the model functions as a reasonable approximation of the modern groundwater flow system.

11.3 Results

Our primary goal was to examine those conditions by which the effective pressure (ice pressure minus water pressure) at the ice–bed interface remains greater than zero, and thus prevents ice flotation. The likely ice thickness at Lake Superior may have ranged from 1200 to 2000 m (Licciardi *et al.*, 1998); we refer to these endpoints as 'thin-ice' and 'thick-ice' conditions, respectively.

Under boundary conditions of the Last Glacial Maximum with hydrogeological properties set as for the validation model, the water-pressure head beneath the Lake Michigan Lobe ranged from 50,240 m at the northern boundary of the model to 2,292 m at the southern terminus, indicating that simulated water pressure at the ice–bed interface would have far exceeded the ice overburden pressure for any reasonable Lake Michigan Lobe thickness.

By increasing the hydraulic conductivity of the glacial drift to $3.7 \times 10^{-2}\,m\,s^{-1}$, which is a value three to nine orders of magnitude greater than typical till in Illinois, the maximum simulated subglacial head is reduced to 3498 m, a value close to, but still greater than, the likely ice-overburden pressure. To test the sensitivity of the model to bedrock *K*, we increased the hydraulic conductivity of all bedrock units by two orders of magnitude. Subglacial head reached a maximum of 816 m at the northern upper boundary and 359 m at the ice margin. These results suggest that the subglacial aquifers could not evacuate the meltwater while maintaining subglacial water pressure at a level less than the ice-overburden pressure. Therefore, some type of drainage system may have developed at the ice–bed interface. Such a drainage system may have consisted of a meltwater film or a network of canals.

To simulate the effects of a subglacial water film, we redefined the upper layer of the model to simulate a water film. The transmissivity of the film was determined by the cubic law (Romm, 1966). A 7 mm film reduces maximum subglacial head to 1447 m and an 8 mm film reduces head to 1099 m, suggesting that a water film with a thickness between 7 and 8 mm is sufficient to reduce subglacial head to values close to or less than the ice-flotation level for thin- and thick-ice conditions. Our simulations are based on the assumption that a water film occupies 100% of the ice–bed interface. Only a very small increase in film thickness is needed, however, to decrease the subglacial drainage area. Thus, water may have drained through some type of canal-type system that occupied less than 100% of the bed.

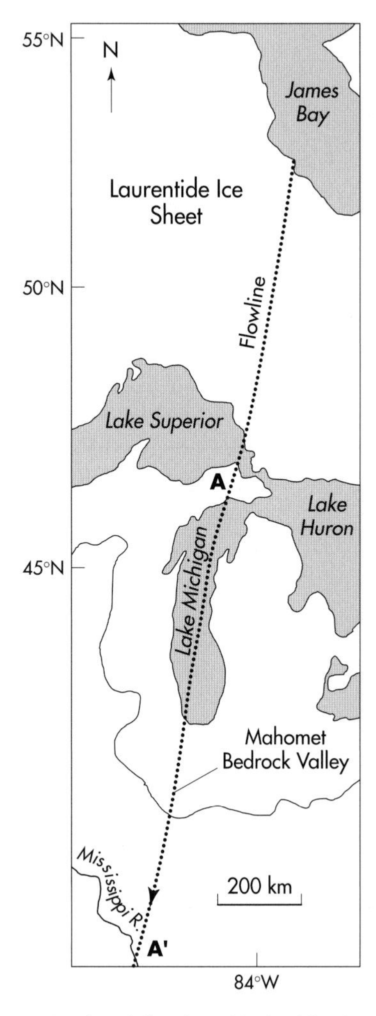

Figure 11.1 Simulated flowline (dashed line). The ice limit is shown as the dark grey line and the ice cover is represented by the grey pattern. A and A' show the boundaries of the finite-difference grid. The approximate location of the Mahomet Bedrock Valley is shown where it is crossed by the simulated flowline.

To investigate the possible effects of permafrost near the ice-sheet margin, we altered the model boundary conditions by defining drift and water film cells within 50 km of the ice margin as impermeable to water. Simulations based on these conditions generate a maximum head of 1418 m, suggesting that permafrost had very little effect on subglacial water pressure. The relatively small impact of permafrost probably reflects the influence of the Mahomet bedrock-valley system on the subglacial hydrology. The Mahomet bedrock-valley system lies more than 50 km upstream of the lobe terminus and is thus beyond the range of our assumed extent of permafrost. Because the Mahomet bedrock-valley system is filled with thick layers of highly permeable sediment and extends far beyond the ice margin (Kempton *et al.*, 1991), it had the capacity to divert water away from the permafrost, thus preventing subglacial water pressure from substantially increasing.

When permafrost is simulated covering the Mahomet bedrock valley, the model generates subglacial heads of more than 50,000 m, further illustrating the importance of an unfrozen Mahomet valley in maintaining low subglacial water pressure. If the valley was frozen, the mechanism that could reduce the subglacial water pressure is unclear.

11.4 Glacial and non-glacial groundwater flow comparison

Under non-glacial conditions, groundwater is typically recharged in topographically high areas and discharged from topographic lows. Our simulations indicate, however, that the Lake Michigan Lobe altered or reversed topographically driven pressure gradients, resulting in groundwater flow patterns and velocities that differed substantially from those of current conditions. The non-glacial (current conditions) simulation shows a flow pattern consisting of recharge in the high-elevation areas north and south of Lake Michigan and upward discharge through aquifers underlying the lake. South of Lake Michigan, groundwater velocity is substantially reduced, and flow directions are variable.

Under glacial conditions, simulated groundwater velocities are high, and flow is consistently directed toward the southern end of the flow line. Along the northern two-thirds of the flow line, groundwater has a relatively strong downward component in the drift aquifer. Along the southern one-third of the flow line, flow in the drift aquifer is negligible, owing to flow through the Mahomet bedrock valley and discharge at the ground surface. South of the ice margin, groundwater in deep bedrock aquifers is directed upward and discharges to regions that currently serve as recharge zones.

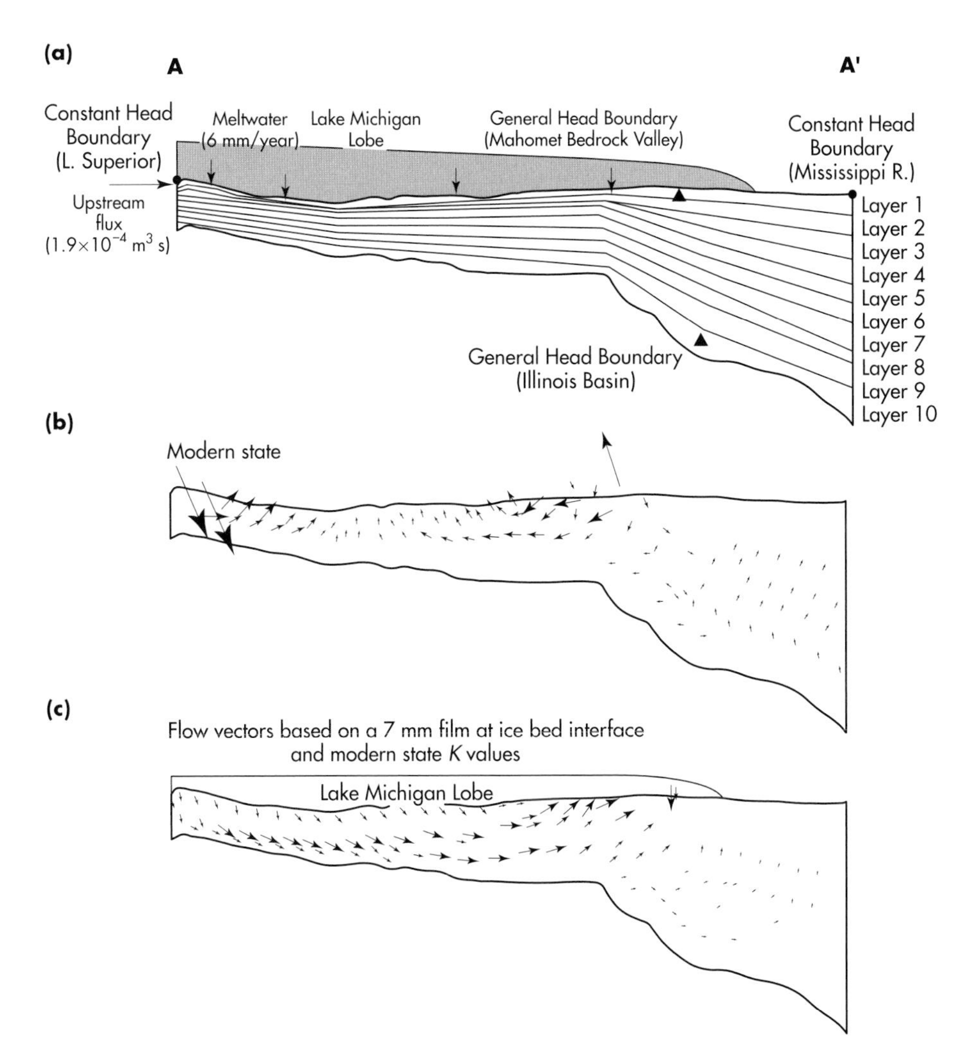

Figure 11.2 (a) Generalized model geometry and boundary conditions along flowline A–A′ (see Fig. 11.1). Horizontal length is 1040 km. Vertical exaggeration is 100×. Illustrated stratigraphy shows ten modelled hydrostratigraphic units (not to scale) (see Table 11.1 for a description of hydrostratigraphic units). (b) Groundwater flow vectors under simulated modern conditions. (c) Groundwater flow vectors based on modern *K* values, with Lake Michigan Lobe present and a 7-mm-thick film at the ice–bed interface. The vector scale varies between illustrations.

Table 11.1 Model hydrostratigraphy and hydraulic conductivity values in Illinois

Model layer	Hydrostratigraphic description	*K* ($m\,s^{-1}$) assigned to the validation simulation*	Geometric mean	−1? ($m\,s^{-1}$)	+1? ($m\,s^{-1}$)	Total observations
1	Quaternary Drift 1	3.7×10^{-7}		N.A.		
2	Quaternary Drift 2	3.7×10^{-7}		N.A.		
3	Pennsylvanian through to Middle Devonian rocks	7.6×10^{-11}		N.A.		
4	Silurian–Devonian carbonates	1.4×10^{-5}	9.0×10^{-5}	4.6×10^{-5}	1.6×10^{-4}	1816
5	'Maquoketa Shale, Galena, Dolomite, and the Decorah, Platteville, and Glenwood Formations'	1.8×10^{-10}		N.A.		

Table 11.1 *Continued*

Model layer	Hydrostratigraphic description	K (m s^{-1}) assigned to the validation simulation*	Geometric mean	−1? (m s^{-1})	+1? (m s^{-1})	Total observations
6	'St Peter Sandstone, Prairie du Chien Group and the Jordan Sandstone'	1.5×10^{-5}	1.0×10^{-5}	1.6×10^{-6}	6.4×10^{-5}	539
	'St Peter Sandstone, Prairie du Chien Group, and the Jordan Sandstone (southern third of the flow line)'	6.1×10^{-5}	1.0×10^{-5}	1.6×10^{-6}	6.4×10^{-5}	539
7	St Lawrence and Franconia Formations	2.4×10^{-9}		N.A.		
	St Lawrence and Franconia Formations (southern third of the flow line)	3.0×10^{-11}		N.A.		
8	Ironton–Galesville Sandstones	2.1×10^{-4}	4.3×10^{-5}	1.2×10^{-5}	1.5×10^{-4}	58
	Ironton and Galesville Sandstones (southern quarter of the flow line)	1.5×10^{-4}	4.3×10^{-5}	1.2×10^{-5}	1.5×10^{-4}	58
9	Eau Claire Formation	3.0×10^{-10}		N.A.		
	Eau Claire Formation (southern third of the flow line)	2.4×10^{-10}		N.A.		
10	Mount Simon Sandstone and Elmhurst Sandstone	1.1×10^{-4}	1.8×10^{-5}	2.4×10^{-6}	1.4×10^{-4}	99

*All K values except those of Quaternary drift are based on the best estimates of the Northern Midwest Regional Aquifer System Assessment (Mandle & Kontis, 1992). Standard deviation data are available only for the bedrock aquifer units. K values assigned to the confining units are close to the values estimated by Walton (1960) and Young (1976). Quaternary drift K is one order of magnitude less than Walton's (1965) estimate for diverse drift deposits in Illinois.

11.5 Conclusions

The simulations indicate that, under much of the Lake Michigan Lobe, subglacial aquifers did not have the capacity to transmit the estimated meltwater flux. Thus, water may have been present at the ice–bed interface at pressures that could contribute to fast and unstable ice flow. An unstable Lake Michigan Lobe may have led to climatic instability and may have caused significant variability in the locations and quantities of groundwater recharge and discharge.

TWELVE

Modelling impact of glacier–permafrost interaction on subglacial water flow

Paul M. Cutler

Polar Research Board U.S. National Academy of Sciences, 500 Fifth Street NW, Washington, DC, 20001, USA

12.1 Introduction

Permafrost was a dynamic and prominent element in the forefield of and under the southern margin of the Laurentide Ice Sheet (LIS) during the last glacial cycle. As it waxed and waned in response to changing climate, it had an impact on groundwater flow and the subglacial thermal regime and, consequently, the extent, form, and dynamics of ice lobes and the landforms they produced (Mickelson *et al.*, 1983).

In one of the first examinations of time-dependent groundwater–permafrost–glacier interactions, Cutler *et al.* (2000) investigated the influence of permafrost on subglacial conditions using a coupled numerical ice-flow–permafrost model. The model was applied to the Green Bay Lobe, which flowed into the Great Lakes region and terminated in Wisconsin. Experiments focused on conditions during ice advance, as it is during such conditions that subglacial permafrost was likely most extensive and most influential to subglacial drainage, ice motion and landform evolution.

We know little about the time-transgressive nature of ice–permafrost interaction and the likely thickness and horizontal extent of pro- and subglacial permafrost. Consequently, Cutler *et al.* (2000) addressed three fundamental questions:

1. How deep was ice-marginal permafrost under the Green Bay Lobe?
2. How wide was the subglacial frozen zone (measured from the ice margin and extending upglacier) during ice advance?
3. How quickly does subglacial permafrost thaw once overridden?

12.2 Methods

Cutler *et al.* (2000) used a time-dependent, two-dimensional, thermomechanically coupled finite-element flowline model. The model domain extends 1000 m into the subsurface to accommodate permafrost dynamics and groundwater flow.

Permafrost growth and decay is modelled by solving the heat flow equation for mixtures of parent material, water, and ice and accounting for latent heat from phase changes. Subglacial groundwater flow, fed by basal melting, is calculated from the Darcy equation. At a given node, the available vertical extent of permeable substrate—if any—for meltwater drainage varies through time as permafrost evolves. The model tracks the volume of subglacially stored water.

Ice-sheet evolution, driven by changes in mass balance, is forced by air temperature and precipitation at the ice surface between 55 and 21 kyr before present. Their temporal variation was estimated from palaeoclimate records. Because heat flow is modelled in the bed as well as the ice lobe, air temperature fluctuations also drive permafrost dynamics. The impacts of geothermal heat and potential energy released from groundwater are also modelled.

12.3 Results

The results of 21 model runs that tested model sensitivity to input parameters enabled Cutler *et al.* (2000) to respond in general terms to the three fundamental questions posed at the outset.

How deep was ice-marginal permafrost under the Green Bay Lobe? At its maximum, permafrost thickness may have approached 200 m—the average thickness of the aquifer in Wisconsin. Its thickness was always at least a few tens of metres up to the Last Glacial Maximum (LGM).

How wide was the subglacial frozen zone during ice advance? The frozen zone upstream from the ice margin extended 60 to 200 km at the LGM.

How quickly does subglacial permafrost thaw once overridden? The answer depends primarily on the thickness of the permafrost before it is overridden. In the case of the Green Bay Lobe, it lasted for centuries to a few thousand years under advancing ice.

12.4 Discussion

Given the magnitude of simulated permafrost thickness, horizontal extent and duration, it seems likely that permafrost had a significant impact on subglacial water flow and processes at the ice–bed interface.

Because permafrost was at least a few tens of metres thick up to the LGM, water flow through unlithified sediments was probably blocked and channelized flow was absent. As the Green Bay Lobe approached its maximum extent, the combination of the available potential gradient, the probable hydraulic conductivity (ca. 10^{-6} m s^{-1}), and the thickness of the remaining unfrozen lithified aquifer (ca. 100 m) was inadequate to convey all basal meltwater from thawed-bed areas upstream. Indeed, only by raising simulated hydraulic conductivity to close to that of gravel (10^{-3} m s^{-1}) could all subglacial meltwater drain. The presence in Wisconsin of tunnel channels and large boulders in tunnel channel fans has been linked to release of subglacially stored water (e.g. Cutler *et al.*, 2002). The model results support this linkage. The absence of such features in landsystems of the neighbouring Lake Michigan Lobe (Colgan *et al.*, 2003) suggests that permafrost was less important there. This may have been due to the presence of proglacial lakes.

By locally preventing permafrost development, proglacial lakes had an impact on the basal heat balance, subglacial drainage and, consequently, lobe dynamics. Cutler *et al.* (2001) added a proglacial lake parameterization to the model and qualitatively matched geomorphological evidence of differences in ice extent, timing of ice advance, LGM ice-surface profile, and likely subglacial hydrological conditions between the Green Bay Lobe and its neighbouring lobes.

The foregoing discussion illustrates the value of time-dependent numerical models in the glacial geologist's toolkit. Models complement other investigative techniques, enabling researchers to test and quantify (perhaps with order-of-magnitude-scale uncertainty) the role of potentially important parameters in glacial environments.

THIRTEEN

Pleistocene hydrogeology of the Atlantic continental shelf, New England

Mark Person

Department of Geological Sciences, Indiana University, 1001 East 10th Street, Bloomington, IN47405, USA

During the 1960s and 1970s, scientific drilling campaigns along the Atlantic continental shelf offshore New England, USA (Hathaway *et al.*, 1979) revealed that the freshwater–saltwater interface is far out of equilibrium with modern sea-level conditions (Fig. 13.1). Aquifer salinity levels are less than 5 parts per thousand (5 ppt) over 100 km offshore of Long Island (wells 6009, 6011 and 6020, Fig. 13.2). In addition, salinity levels within confining units beneath Nantucket Island are 30–70% of seawater levels and exhibit a parabolic profile consistent with ongoing vertical diffusion (well 6001, Fig. 13.2). Analytical models of vertical solute diffusion for the Nantucket confining units suggest that flushing of aquifers beneath Nantucket began in the late Pleistocene between about 195 and 21 ka assuming a diffusion coefficient of 3.0×10^{-11} $m^2 s^{-1}$ (Person *et al.*, 2003).

Kohout *et al.* (1977) proposed that the presence of unusually fresh water within the permeable units of the Atlantic continental shelf could be attributed to meteoric recharge during Pleistocene sea-level low stands. Marine isotopic records and precise dating of shoreline facies and corals suggest that Pleistocene sea-level varied by about 120 m with a period between 40 ka and 100 ka. Average sea-level for the Pleistocene was about 40 m below modern. During sea-level low stands, large portions of the continental shelf

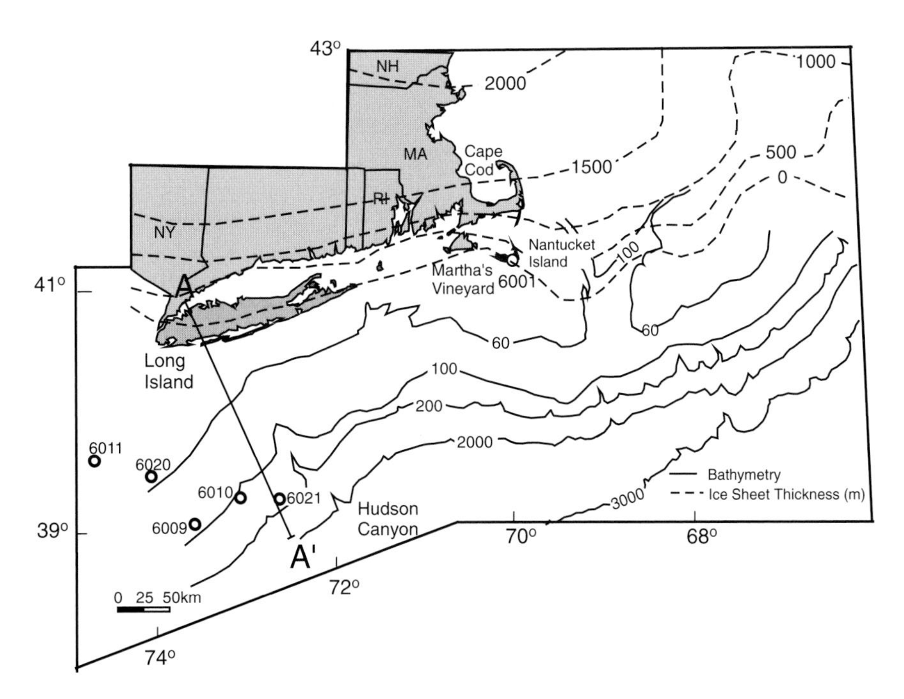

Figure 13.1 Bathymetric map (solid contour lines in metres) of the Atlantic continental shelf, New England. The dashed contours are maximum reconstructed Laurentide Ice Sheet thickness (in metres) for 21 ka (after Colgan *et al.*, 1981; Hathaway *et al.*, 1979). The circles indicate the location of exploration boreholes used to construct the hydrostratigraphic cross-sections presented in Fig. 13.3.

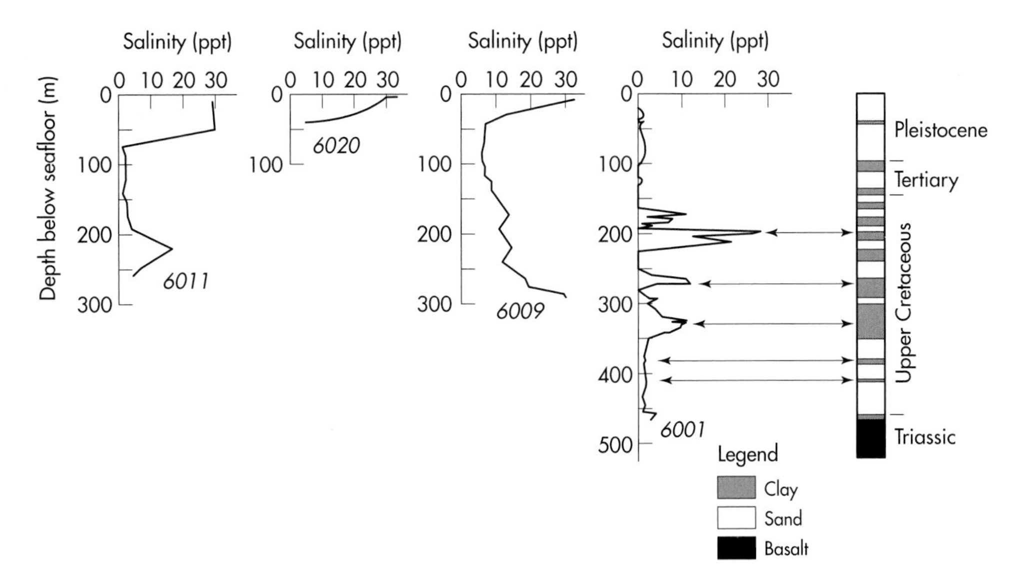

Figure 13.2 Salinity profiles and stratigraphy of Pleistocene and Cretaceous continental shelf sediments below Nantucket Island, MA (well 6001) and offshore Long Island (wells 6009, 6011, 6020; no stratigraphy plotted). The parabolic salinity profiles within clay-rich confining units in well 6001 suggests ongoing vertical diffusion (after Folger *et al.*, 1978; Hathaway *et al.*, 1979). The locations of the wells are listed in Fig. 13.1. The well numbers are in italics at the bottom of each scattergram.

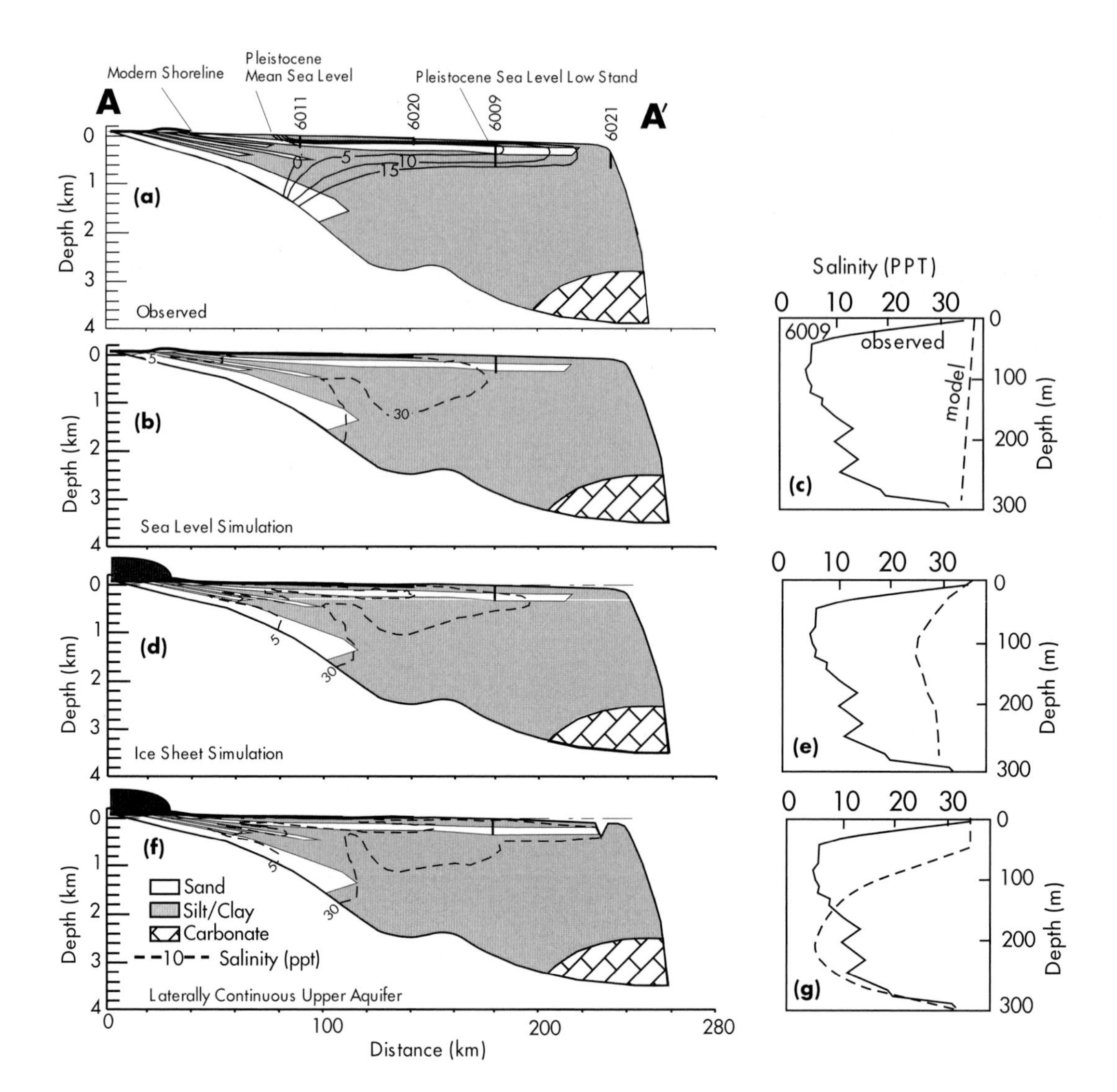

Figure 13.3 (a) Observed (solid lines, in parts per thousand) salinity contour map for the Atlantic continental shelf sediments offshore Long Island, NY. (b) Computed salinity contours applying sea-level boundary conditions (20 sea-level cycles about the Pleistocene mean sea-level of 40 m below modern using a period of 100,000 yr with an amplitude of 120 m). (d) Computed salinity contour map (dashed lines, ppt) for the Atlantic continental shelf sediments off shore Long Island, NY applying ice-sheet recharge and (f) ice-sheet model the same as (d) but allowing groundwater in shallow aquifer to discharge along submarine canyons. Both ice-sheet simulations used the computed salinity from the Pleistocene sea-level simulation shown in Fig. 13.3b as an initial condition. Both ice-sheet simulations are presented following 2000 yr of imposed ice-sheet boundary conditions. A comparison between simulated and observed concentrations profiles for well 6009 for b, d & f are shown in c, e & g, respectively. The location of the cross-section is listed in Fig. 13.1 (after Person *et al.* 2003).

were exposed to meteoric recharge. However, numerical models of Person *et al.* (2003) which represented variable-density groundwater flow and solute transport could not reproduce the relatively low-salinity groundwaters observed off Long Island by applying boundary conditions consistent with Pleistocene sea-level fluctuations (Fig. 13.3a–c). These researchers also considered the effects of subice sheet recharge between about 21 and 18 ka when the Laurentide Ice Sheet advanced out onto the continental shelf forming the islands of Nantucket, Martha's Vineyard and Cape Cod. Including the effects of ice-sheet recharge (i.e. applying a specified head boundary condition for 2000 years equal to 9/10th the ice-sheet height across much of Long Island) helped to drive the freshwater–saltwater interface much farther out onto the continental shelf (Fig. 13.3d & e). Observed salinity conditions were most closely matched by also allowing groundwater to discharge from Miocene/Pliocene aquifers along submarine canyons near the continental slope (Fig. 13.3f & g). Simulated recharge induced by Laurentide Ice Sheet meltwater was short lived (<2000 yr) but, on average, about two to ten times greater than modern subaerial levels.

FOURTEEN

Glacial chemical weathering, runoff composition and solute fluxes

Martyn Tranter

Bristol Glaciology Centre, School of Geographical Sciences, University of Bristol, Bristol BS8 1SS, UK

14.1 Introduction

Glaciers are the largest freshwater reservoir on Earth (Knight, 1999), containing some $28 \times 10^6 km^3$ of water (Table 14.1). Glaciers are acknowledged as powerful agents of physical erosion (Paterson, 1994; Hallet *et al.*, 1996). Globally significant volumes of sediment may be transported and deposited from glacial runoff (Syvitski *et al.*, 1987). By contrast, glacial chemical erosion is less well documented (Sharp *et al.*, 1995; Anderson *et al.*, 1997), but globally significant quantities of solute may be transported by glacial runoff in relatively brief periods during deglaciation (Tranter *et al.*, 2002a).

Chemical erosion in glaciated regions proceeds at rates comparable to those of temperate catchments with comparable specific runoff (Anderson *et al.*, 1997). The factors usually associated with elevated rates of chemical weathering in other environments, such as the continual presence of water, soil and vegetation (Drever, 2003), are not definitive features of glacial environments. Glaciated regions are largely frozen for significant periods each year, the residence time of liquid water in the catchment is low (Knight, 1999), there are thin, skeletal soils at best and vegetation is either absent or limited (French, 1997). Even so, chemical erosion rates in glaciated terrain are usually near to or greater than the continental average (Sharp *et al.*, 1995; Wadham *et al.*, 1997, Hodson *et al.*, 2000). This is because glaciated catchments usually have high specific runoff, there are high concentrations of freshly comminuted rock flour, and adsorbed organic matter or surface precipitates that may hinder water–rock interactions are largely absent (Tranter, 1982). In short, the rapid flow of water over fine-grained, recently crushed, reactive mineral surfaces maximizes both the potential rates of chemical weathering and chemical erosion.

14.2 Glacial chemical weathering

The principal reactions that comminuted bedrock undergo in glaciated terrain are summarized below. We assume that the bedrock is primarily composed of silicates and aluminosilicates.

Glacial comminution crushes bedrock and exposes the trace reactive components within crystal aggregates more rapidly than would be the case in temperate and tropical soils, where new minerals are ultimately accessed via solubilization of aluminosilicate lattices. Hence, glaciers are effective at promoting the solubilization of trace reactive components in the bedrock, which include carbonates, sulphides and fluid inclusions. Laboratory experiments and direct sampling of waters from the glacier bed (Tranter *et al.*, 1997, 2002b) show that the initial reactions to occur when dilute snow and ice melt first access glacial flour are carbonate and silicate hydrolysis (Equations 1 & 2). These reactions raise the pH to high values (>9), lower the PCO_2 (to ca. 10^{-6} atms) and maximize the water's potential to adsorb CO_2. Carbonate hydrolysis produces a solution with a Ca^{2+} concentration of ca. $200\,\mu eq\,L^{-1}$, with HCO_3^- the dominant anion.

$$\underset{\text{calcite}}{Ca_{1-x}(Mg_x)CO_3(s)} + H_2O(l) \leftrightarrow (1-x)\,Ca^{2+}(aq) + xMg^{2+}(aq) + HCO_3^-(aq) + OH^-(aq) \quad (1)$$

$$\underset{\text{K-feldspar}}{KAlSi_3O_8(s)} + H_2O(l) \leftrightarrow \underset{\text{weathered feldspar surface}}{HAlSi_3O_8} + K^+(aq) + OH^-(aq) \quad (2)$$

The relatively dilute meltwater in contact with fine-grained glacial flour promotes the exchange of divalent ions from solution for monovalent ions on surface exchange sites. Hence, some of the Ca^{2+} and Mg^{2+} released from carbonate and silicate hydrolysis is exchanged for Na^+ and K^+.

$$\text{glacial flour} - Na_{(2-z)},K_z\,(s) + (1-x)\,Ca^{2+}(aq) + xMg^{2+}(aq) \leftrightarrow \text{glacial flour} - Ca_{(1-x)}, Mg_x(s) + (2-z)\,Na^+(aq) + z\,K^+(aq) \quad (3)$$

The high pH derived from hydrolysis enhances the dissolution of aluminosilicate lattices, as Al and Si become more soluble at pH > 9. Hydrolysis of carbonates results in a solution that is near

Table 14.1 The concentration of major ions in glacial runoff from different regions of the world (after Brown, 2002). Concentrations are reported in $\mu eq\,L^{-1}$

Region	Σ^+	Ca^{2+}	Mg^{2+}	Na^+	K^+	HCO_3^-	SO_4^{2-}	Cl^+	Source*
Canadian High Arctic	280–3500	260–2600	21–640	1–190	0.1–39	210–690	59–3900	—	1
Antarctica	550–3100	72–1300	120–336	360–1400	0.8–110	91–1600	34–1200	0.6–1000	2
Svalbard	330–1900	120–1000	99–540	110–270	5.1–41	110–940	96–760	5–310	3–5
Canadian Rockies	1300–1500	960–1100	290–310	3.7–36	5.8–9.2	890–920	380–520	1.7–25	6
Iceland	170–960	110–350	30–120	30–480	2.8–12	190–570	26–130	30–87	7,8
Himalayas	130–940	75–590	6.6–230	25–65	22–51	200–730	160–410	1–22	9
Norway	20–930	8.8–623	1.6–66	8.3–210	1.0–29	1.4–680	7–140	0.9–190	10
European Alps	37–910	20–640	6–140	4.9–92	5.9–33	11–400	10–240	0.9–92	11–13
Alaska	670	550	36	25	61	430	260	2	14
Greenland	280–387	130–170	68–98	78–110	5–9	220–340	90–200	16–30	15
Cascades	56–150	35–80	8.3–20	2.5–17	9.7–37	83–100	7.9–29	—	16
Global mean runoff	1200	670	280	220	33	850	170	160	17

*1, Skidmore & Sharp (1999); 2, De Mora *et al.* (1994); 3, Hodgkins *et al.* (1997); 4, Hodson *et al.* (2000); 5, Wadham *et al.* (1997); 6, Sharp *et al.* (2002); 7, Raiswell & Thomas (1984); 8, Sigurður Steinþórsson & Óskarsson (1983); 9, Hasnain *et al.* (1989); 10, Brown (2002); 11, Brown *et al.* (1993); 12, Collins (1979); 13, Thomas & Raiswell (1984); 14, Anderson *et al.* (2000); 15, Rasch *et al.*, 2000; 16, Axtman & Stallard, 1995; 17, Livingstone (1963).

saturation with calcite and aragonite. It is only in these types of waters that aluminosilicate dissolution is greater than carbonate dissolution. The influx of gases (including CO_2 and O_2), either from the atmosphere or from basal ice, and CO_2 produced by microbial respiration (see below) both lowers the pH and the saturation with respect to carbonates. In addition, sulphide oxidation produces acidity (see below). Hence, almost all subglacial meltwaters are undersaturated with respect to calcium carbonate. The rapid dissolution kinetics of carbonates with respect to silicates means that carbonate dissolution continues to have a large impact on meltwater chemistry, despite carbonates being present often in only trace concentrations in the bedrock. For example, Haut Glacier d'Arolla has a bedrock which is composed of metamorphic silicate rocks. Carbonates and sulphides are present in trace quantities in bedrock samples (0.00–0.58% and <0.005–0.71% respectively). There are also occasional carbonate veins present in the schistose granite. Despite the bedrock being dominated by silicates, sulphide oxidation in subglacial environments dissolves carbonate to silicate in a ratio of ca. 5 : 1 (Tranter *et al.*, 2002b), compared with the global average of ca. 1.3 : 1 (Holland, 1978).

The acid hydrolysis of silicates and carbonates (Equations 4 & 5) that arises from the dissociation of CO_2 in solution is known as carbonation. Carbonation occurs in a restricted number of subglacial environments because ingress of atmospheric gases to these water-filled environments is restricted. It largely occurs in the major arterial channels at low flow, particularly near the terminus, and at the bottom of crevasses and moulins that reach the bed. Fine-grained sediment is flushed rapidly from these environments, and there is little time for the formation of secondary weathering products, such as clays. Hence, silicates dissolve incongruently, as crudely represented by Equation 4.

$$\underset{\text{anorthite}}{CaAl_2Si_2O_8(s)} + 2CO_2(aq) + 2H_2O\ (l) \leftrightarrow Ca^{2+}(aq) + 2HCO_3^-(aq) + \underset{\text{weathered feldspar surfaces}}{H_2Al_2Si_2O_8(s)} \quad (4)$$

$$\underset{\text{calcite}}{Ca_{1-x}(Mg_x)CO_3(s)} + CO_2(aq) + H_2O\ (l) \leftrightarrow (1-x)Ca^{2+}(aq) + xMg^{2+}(aq) + 2HCO_3(aq) \quad (5)$$

There is a limited body of evidence which suggests that microbial oxidation of bedrock kerogen occurs (Wadham *et al.*, 2004), and if this is the case, carbonation as a consequence of microbial respiration may occur in debris-rich environments, such as in the distributed drainage system and the channel marginal zone.

$$C_{org}(s) + O_2(aq) + H_2O(l) \leftrightarrow CO_2(aq) + H_2O(l) \leftrightarrow H^+(aq) + HCO_3^-(aq) \quad (6)$$

The dominant reaction in subglacial environments is sulphide oxidation, because, following hydrolysis, this is the major reaction which provides protons to solution, so lowering the pH, decreasing the saturation index of carbonates, so allowing more carbonate dissolution (Equation 7). Sulphide oxidation occurs predominantly in debris-rich environments where comminuted bedrock is first in contact with water. It is microbially mediated, occurring several orders of magnitude faster than in sterile systems (Sharp *et al.*, 1999). It consumes oxygen, driving down the pO_2 of the water.

$$\underset{\text{pyrite}}{4FeS_2(s)} + 16Ca_{1-x}(Mg_x)CO_3(s) + 15O_2(aq) + 14H_2O(l) \leftrightarrow 16(1-x)\,Ca^{2+}(aq) + 16xMg^{2+}(aq) + 16HCO_3^-(aq) + 8SO_4^{2-}(aq) + \underset{\text{ferric oxyhydroxides}}{4Fe(OH)_3(s)} \quad (7)$$

Earlier studies suggested that the limit on sulphide oxidation was the oxygen content of supraglacial melt, because subglacial supplies of oxygen are limited to that released from bubbles in the ice during regelation, the process of basal ice melting and refreezing as it flows around bedrock obstacles. Studies of water samples from boreholes drilled to the glacier bed, however, show that the SO_4^{2-} concentrations may be two or three times that allowed by the oxygen content of supraglacial meltwaters (Tranter *et al.*,

2002b). This suggests that oxidizing agents other than oxygen are present at the glacier bed. It seems very likely that microbially mediated sulphide oxidation drives certain sectors of the bed towards anoxia, and that in these anoxic conditions, Fe(III), rather than O_2, is used as an oxidizing agent (Equation 8). Sources of Fe(III) include the products of the oxidation of pyrite and other Fe(II) silicates in a previous oxic environment, as well as that found in magnetite and haematite.

$$FeS_2(s) + 14Fe^{3+}(aq) + 8H_2O(l) \leftrightarrow 15Fe^{2+}(aq) + 2SO_4^{2-}(aq) + 16H^+(aq) \quad (8)$$

Support for anoxia within subglacial environments comes from the $\delta^{18}O$-SO_4, which is enriched in ^{16}O when sulphide is oxidized in the absence of oxygen (Bottrell & Tranter, 2002).

The realization that there is microbial mediation of certain chemical weathering reactions in subglacial environments (Sharp *et al.*, 1999; Skidmore *et al.*, 2000, Bottrell & Tranter, 2002) has resulted in a paradigm shift, as the types of reactions that may occur in anoxic sectors of the bed include the common redox reactions that occur, for instance, in lake or marine sediments (Drever, 1988). A key difference in glacial systems is that the supply of new or recent organic matter is limited to that inwashed from the glacier surface, such as algae, insects and animal faeces, or overridden soils during glacier advance. By contrast, the supply of old organic matter from comminuted rocks is plentiful. Given the thermodynamic instability of organic matter in the presence of O_2 or SO_4^{2-}, it seems likely that microbes will have evolved to colonize subglacial environments and utilize kerogen as an energy source. The first data to support this assertion is stable isotope analysis from Finsterwalderbreen, a small polythermal-based glacier on Svalbard that has shale as a significant component of its bedrock (Wadham *et al.*, 2004). The $\delta^{18}O$-SO_4 of waters upwelling from subglacial sediments are very enriched in $\delta^{18}O$; the $\delta^{34}S$ is enriched in ^{34}S, which suggests that cyclical sulphate reduction and oxidation has been occurring. The $\delta^{13}C$ of DIC (dissolved inorganic carbon) is negative, consistent with the assertion that organic matter has been oxidized. Mass balance calculations suggest that a possible source of organic matter is kerogen, but the necromass of dead bacteria cannot be discounted. Whatever is the source of organic matter, sectors of the bed at Finsterwalderbreen are so anoxic that sulphate reduction is occurring (Equation 9).

$$\underset{\text{organic carbon}}{2CH_2O(s)} + SO_4^{2-}(aq) \leftrightarrow 2HCO_3^-(aq) + H_2S(aq) \quad (9)$$

It is possible that methanogenesis occurs under certain ice masses, because methanogens have been isolated from subglacial debris (Skidmore *et al.*, 2000). The low $\delta^{13}C$-CH_4 and high concentration of methane found in gas bubbles within the basal ice of the Greenland Ice Sheet are consistent with there being methanogenesis within the basal organic-rich palaeosols.

The colonization of subglacial environments by microbes suggests that both energy and nutrient sources are readily available. Energy sources, such as sulphides and kerogen, have been discussed above. Comminuted bedrock may also provide a source of nutrient. Average crustal rock contains 1050 ppm of P. Typically, this is contained in sparingly soluble minerals such as apatite, and calcium, aluminium and ferrous phosphates (O'Neill, 1985). Comminuted bedrock and basal debris provides a renewable source of P on mineral surfaces, and it is likely that uptake of P by microbes maximizes the extraction of P from these activated surfaces. Hodson *et al.*(2004) suggest that 1–23 $\mu g\,P\,g^{-1}$ is present as readily extractable P on the surface of glacial flour. Sources of N also may be derived from comminuted rock (Holloway & Dahlgren, 2002). The N content of rocks is typically 20 ppm (Krauskopf, 1967), but may exceed 1000 ppm in some sedimentary and metasedimentary rock (Holloway & Dahlgren, 2002). For example, bedrock has been shown to be a source of NH_4^+ from schists in the Sierra Nevadas, California (Holloway *et al.*, 1998), and there may be appreciable concentrations of NH_4^+, which substitutes for K^+, in biotite, muscovite, K-feldspar and plagioclase (Mingram & Brauer, 2001). It follows that glacial comminution of bedrock and basal debris maximizes the likelihood that N-producing surfaces are exposed to meltwaters and microbes, and, given that bedrock in the Sierra Nevadas can act as an N source, it is likely that comminuted glacial debris is also a potential source of N.

The predominance of carbonate hydrolysis, carbonation and sulphide oxidation in subglacial weathering reactions on aluminosilicate/silicate bedrock is also found on carbonate bedrock. The balance between carbonate dissolution and sulphide oxidation, however, depends on the spatial distribution of sulphides in the bedrock and basal debris (Fairchild *et al.*, 1999). Non-congruent dissolution of Sr and Mg from carbonate is also observed in high rock:water weathering environments, such as the distributed drainage system, in which water flow is also low (Fairchild *et al.*, 1999).

To date, there are few studies of glacial chemical weathering on bedrock with a significant evaporitic content. Work at John Evans Glacier in the Canadian High Arctic has shown that gypsum is dissolved in some areas of the bed, and that mixing of relatively concentrated Ca^{2+}–SO_4^{2-} waters with more dilute Ca^{2+}–HCO_3^-–SO_4^{2-} waters results in $CaCO_3$ precipitation owing to the common ion effect (M.L. Skidmore, personal communication, 2002). Kennicott Glacier, Alaska, is underlain by a sabkha facies limestone, which contains trace quantities of halite. Waters accessing sites of active erosion readily acquire Na^+ and Cl^- (Anderson *et al.*, 2003).

A key feature of the above chemical weathering scenarios is that relatively little atmospheric or biogenic CO_2 is involved. Hence, whereas ca. 23% and ca. 77% of solutes, excluding recycled sea salt, found in global mean river water is derived from the atmosphere and rock respectively (Holland, 1978), atmospheric sources account for a maximum of 3–11% of solute in glacial runoff (after Hodson *et al.*, 2000).

14.3 Chemical composition of glacial runoff

The chemical composition of glacial runoff from ice sheets, ice caps and glaciers around the world is shown in Table 14.1 (after Brown, 2002; Tranter, 2003), which also includes the composition of global mean river water for comparative purposes. Sea salt is a variable component of glacial runoff, and the dominant non-sea-

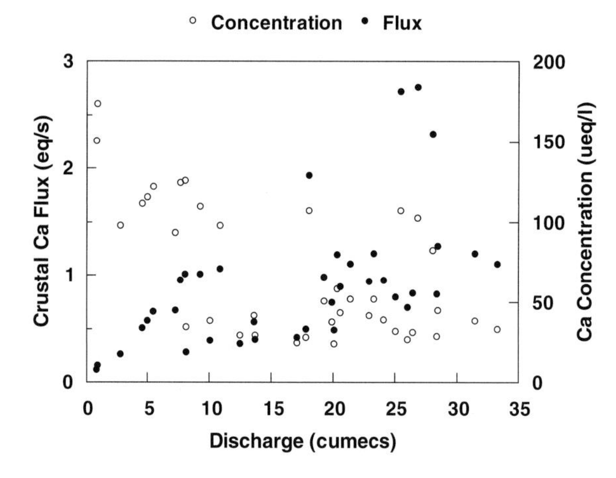

Figure 14.1 Scatterplot of crustal calcium flux and calcium concentration versus discharge at Manitsoq Glacier, southwest Greenland.

salt ions are Ca^{2+}–HCO_3^-–SO_4^{2-}–Mg^{2+}. The concentration of glacial runoff is usually inverse to discharge, so that a rough rule of thumb is that low discharge waters are concentrated, whereas high discharge waters are dilute. The concentration of low discharge waters approaches ca. 1 meq L^{-1} (of positive charge) in temperate, lower latitude glacial runoff, and ca. 3 meq L^{-1} at higher latitudes, presumably as a consequence of freezing effects. Low discharge waters do not usually make a significant contribution to the annual solute flux (Sharp *et al.*, 1995; Tranter, 2003). Paradoxically, the concentration of the dilute, high discharge waters is more important, because high discharge transports significantly more solute than low discharge (see Fig. 14.1). This is because the increase in discharge over the ablation season may be of the order of one to threee orders of magnitude, whereas the dilution of the solutes is less, being up to an order of magnitude less.

The sum of cation equivalents in glacial runoff ranges from ca. 10 to 3500 μeq L^{-1}. Glacial runoff is usually more dilute than global mean river water, and usually contains more K^+ and less Si for a given specific runoff (Anderson *et al.*, 1997). The Ca^{2+} : Si and HCO_3^- : SO_4^{2-} ratios of glacial meltwaters are high and low respectively, when compared with the principal world river water, making glacial runoff an end member of global riverine water (Tranter, 2003). This is because glaciers preferentially weather carbonates and sulphides from the bedrock.

Raiswell (1984) showed that the base cation composition of glacial meltwaters does not reflect that of the lithology of the bedrock. The predominant cation is always Ca^{2+}, even on acid igneous and metamorphic bedrocks. This is because the dissolution kinetics of Ca^{2+} from trace carbonates, which are ubiquitous in most bedrocks, and from aluminosilicates are more rapid than those of monovalent ions. Hence, Ca^{2+} may be a relatively minor base cation in the bedrock, but becomes the dominant base cation in solution (White *et al.*, 2001).

Holland (1978) showed that specific annual discharge is the most significant control upon chemical erosion in temperate catchments, and the same is true in glacierized basins (Anderson *et al.*, 1997; Hodson *et al.*, 2000). The lithology of the catchment is an important secondary control on chemical erosion rates, with carbonate-rich and basaltic lithologies exhibiting the highest chemical weathering rates. Hodson *et al.* (2000) show that there is year on year variability in specific runoff, and therefore on cationic denudation rate. Currently, there are more studies of chemical erosion rates in the glacierized basins of Svalbard than in other region (Table 14.2). Crustally derived solute fluxes from 10 basins are equivalent to a mean cationic denudation rate of 350 $\Sigma meq^+ m^{-2} yr^{-1}$ (range: 160–560 $\Sigma meq^+ m^{-2} yr^{-1}$), which lies within the global range of 94–1650 $\Sigma meq^+ m^{-2} yr^{-1}$ for the other 15 glacier basins in the Northern Hemisphere (Hodson *et al.*, 2000). The mean value for Svalbard is close to the continental average of 390 $\Sigma meq^+ m^{-2} yr^{-1}$ (Livingstone, 1963).

Table 14.2 Specific runoff and cationic denudation rates for glaciers in different regions (after Hodson *et al.*, 2000)

Region	Specific runoff ($m\,yr^{-1}$)	Cationic denudation rate ($\Sigma meq^+ m^{-2} yr^{-1}$)
Svalbard	0.4–1.5	190–560
European Alps	1.4–2.3	450–690
North America	0.7–7.7	94–1600
Iceland	1.8–2.1	650–1100
Asia	1.1–3.5	460–1600
Continental average	0.31	390

14.4 Glacial solute fluxes

Approximately 10% (ca. 15.928 × 10^6 km^2; Knight, 1999) of the Earth's surface is presently covered by glaciers, with most being concentrated in Antarctica and Greenland. Estimates of the current annual runoff from ice sheets range from 0.3 to 128 × 10^{12} $m^3 yr^{-1}$ (Jones *et al.*, 2002), of which ca. 0.328 × 10^{12} $m^3 yr^{-1}$ is from Greenland (Oerlemans, 1993a) and 0.4–5028 × 10^9 $m^3 yr^{-1}$ is from Antarctica (Jacobs *et al.*, 1992). The amount of runoff from smaller terrestrial ice masses is not well known. A crude illustration of potential values can be obtained from the average residence time of ice in the smaller ice masses. Table 14.3 shows that the current volume of terrestrial ice that exists in locations other than the Antarctic and Greenland ice sheets is of the order of 0.228 × 10^{15} m^3. If the average volume-weighted residence time of ice in these glaciers that is lost as runoff is of the order of 10^2–10^4 yr, then runoff from these smaller ice masses is of the order of 0.02–2.028 × 10^{12}. Hence, glaciers may contribute ca. 0.7–7% to current global annual runoff (ca. 4628 × 10^{12} $m^3 yr^{-1}$; Holland, 1978).

The volume of terrestrial ice approximately trebled at the LGM, when ca. 22–29% of the present-day land surface was covered by glaciers. Table 14.3 shows that much of the additional ice volume was in the two great ice sheets of the Northern Hemisphere, the Laurentide Ice Sheet that grew over North America and the Scandinavian Ice Sheet that grew over Europe. The average runoff from ice sheets during the past 100 kyr was 1.328 × 10^{12} $m^3 yr^{-1}$, about three times that of current ice-sheet runoff (Tranter *et al.*,

Table 14.3 Volumes of terrestrial ice contained in ice sheets and other ice masses (after Knight, 1999)

Region	Current Volume (10^6 km^3)	Volume at LGM (10^6 km^3)
Antarctica	26	26
Greenland	2.6	3.5
Laurentide		30
Cordilleran		3.6
Scandinavian		13
Other ice masses	0.2	1.1
Total	28.4	77.2

2002). Average global glacial runoff might be double this value, dependent on the volume-weighted average residence time of ice in smaller ice masses. There were several occasions between 80 and 10 ka when ice-sheet runoff approached 20–30% of current global annual runoff to the oceans, but the most sustained period of relatively high ice-sheet runoff was between 15 and 5 kyr BP, when ice-sheet runoff was of the order of 2–1028 × 10^{12} m^3 yr^{-1}. The impact of runoff from smaller ice masses has not been documented yet.

Glacial chemical weathering does not remove large quantities of CO_2 from the atmosphere. This is because rock components are mostly solubilized in the absence of free contact with the atmosphere. Instead, glacial chemical weathering may result in a source of CO_2 to the atmosphere. The Ca^{2+} and HCO_3^- dissolved by sulphide oxidation (Equation 7) is ultimately deposited as carbonate in the oceans, and this gives rise to a net release of CO_2 to the atmosphere (Equation 10).

$$Ca^{2+}(aq) + 2HCO_3^-(aq) \leftrightarrow CaCO_3(s) + H_2O(l) + CO_2(g) \quad (10)$$

The impact of major ions in glacial runoff on global geochemical cycles is presently very limited. This is because the runoff is dilute when compared with riverine concentrations, and because the volume of glacial runoff is small when compared with riverine discharge to the oceans. However, during deglaciation, when global glacial discharge is believed to have equalled global riverine discharge to the oceans, there may have been short periods when glacial solutes made similar contributions to riverine fluxes of major ions (Tranter *et al.*, 2002a).

To date, modelling that attempts to determine the impact of glacial chemical weathering on atmospheric CO_2 concentrations has not found any significant perturbations (Ludwig *et al.*, 1998; Jones *et al.*, 2002; Tranter *et al.*, 2002a). However, linkage of terrestrial chemical erosion models to ocean carbon cycle models is still at an early stage of development. It may well be the case that the next generation of ocean carbon cycle models will be able to explore in greater detail the changes in dissolved inorganic carbon species at depth that arise from changes in ocean circulation (Broecker, 1995; Dokken & Jansen, 1999). Atmospheric CO_2 perturbations from changing terrestrial chemical erosion may be amplified as a consequence, as it is anticipated that enhanced terrestrial chemical erosion during times of reduced sea level is less well buffered by slower deep water turnover (Jones *et al.*, 2002).

FIFTEEN
Solute enhancement in the proglacial zone

J.L. Wadham

Bristol Glaciology Centre, School of Geographical Sciences, University of Bristol, Bristol, BS8 1SS, UK

15.1 Introduction

The proglacial zone is an area of potentially high geochemical activity, because it contains a variety of comminuted glacial debris, is subject to glaciofluvial reworking and can be colonized by vegetation (Anderson *et al.*, 2000; Cooper *et al.*, 2002). Chemical weathering reactions resemble those in subglacial environments (Anderson *et al.*, 2000; Wadham *et al.*, 2001a), with sulphide oxidation/carbonate dissolution dominating in recently deglacierized moraines, and carbonation of silicates becoming more important in older moraines as the reactive sulphides and carbonates are exhausted (Anderson *et al.*, 2000). Significant differences between subglacial and proglacial environments are that the surface of the proglacial zone freezes, thaws and dries on an annual basis, and that it is a deposition site for snow and rain. Further, there may be ingress of atmospheric gases through the dry surface sediment. These processes perturb the composition of groundwaters within the proglacial zone, for example, there is concentration and recycling of salts via evapoconcentration and freeze concentration and oxic and anoxic conditions may prevail in groundwaters under different conditions of saturation.

15.2 Case study: Finsterwalderbeen, southwest Svalbard

Finsterwalderbreen is a polythermal, 44 km², High Arctic glacier. Subglacial waters drain seasonally via a subterranean upwelling (Wadham *et al.*, 1998) and outburst events (Wadham *et al.*, 2001b). The proglacial zone is characterized by an outer zone of push moraines resulting from surges (Hart & Watts, 1997) and an inner zone of smaller moraines and widespread fluvial deposits (Fig. 15.1). Most of the proglacial zone (4.2 km²) has been exposed over the past 100 yr at a rate of 10–40 m yr^{-1}. Proglacial sediments contain material from all elements of the catchment lithology (Precambrian carbonates, phyllite and quartzite, Permian sandstones, dolomites and limestones and Triassic to Cretaceous siltstones, sandstones and shales).

15.3 Proglacial hydrology

The proglacial hydrology is relatively complex owing to several water inputs, an irregular topography and 200–300 m of per-

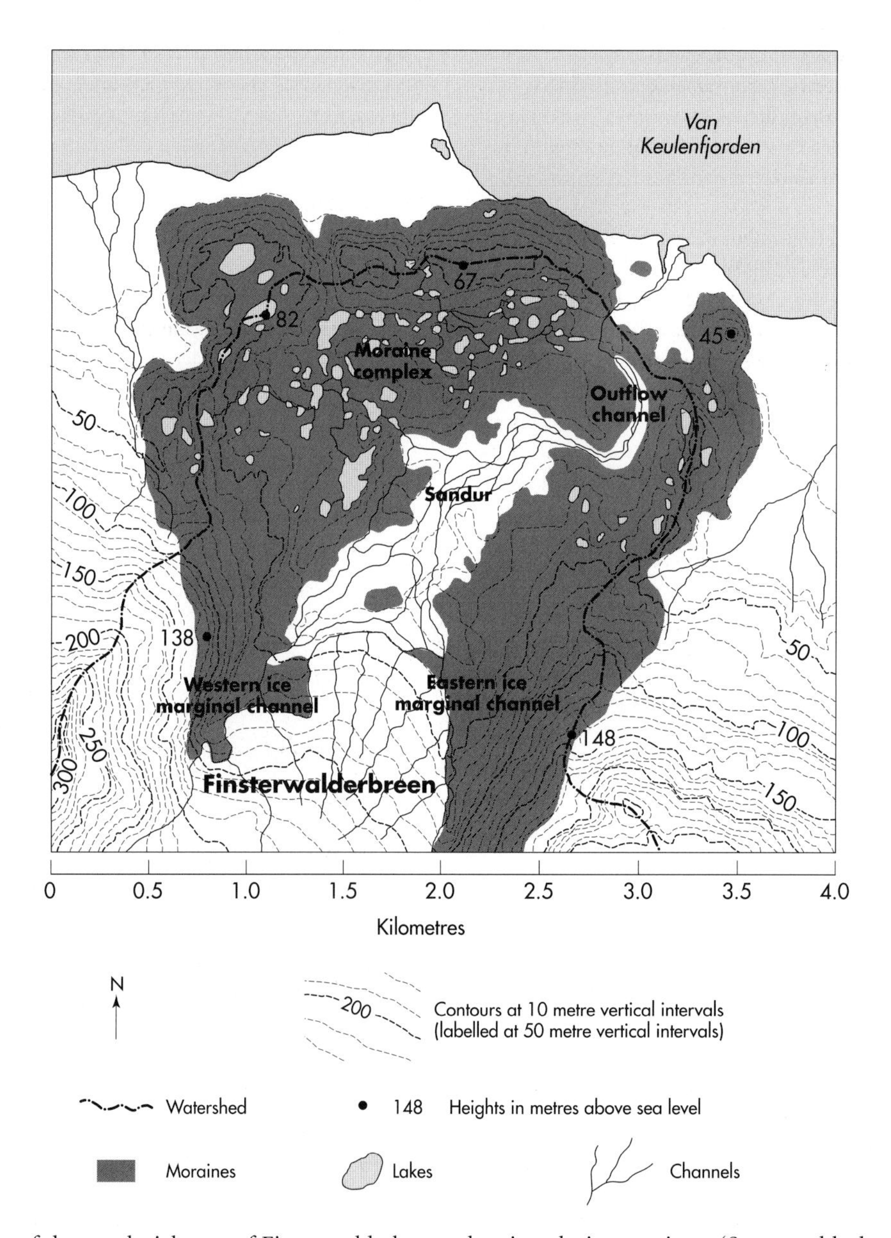

Figure 15.1 Map of the proglacial zone of Finsterwalderbreen, showing glacier terminus. (See www.blackwellpublishing.com/knight for colour version.)

mafrost. Main water inputs derive from snowmelt, rainfall and active layer melt. These accumulate in active layer sediments, forming lakes in topographic depressions in the moraine (5% of the proglacial area). These groundwaters drain through the summer along a topographic gradient to the proglacial floodplain via ephemeral streams. The streams dry out and water takes a subsurface routing as the active layer deepens and with progressive evaporative loss through the summer. Evaporative losses represent the greatest output from the proglacial hydrological system, accounting for 70% of water inputs from snow and rain.

Glacial bulk meltwaters act as a significant water input locally in near-channel environments. They flow through the proglacial zone in two major channels originating on the eastern and western glacier margins (eastern and western ice-marginal channels: Fig. 15.1). Together, they form a braided stream network that floods a significant area of the proglacial zone during summer. This network finally reforms as a single channel, discharging meltwaters to the fjord ca. 1.5 km from the glacier terminus. The glacial bulk meltwater stream and the proglacial active layer groundwater system are closely coupled in the near-channel zone. Glacial meltwaters are forced out into fluvial sediments and moraines as bulk meltwater discharge rises and drain back to the main channel as discharge falls. These channel and active-layer groundwater interactions are evident in stage records of wells emplaced in active-layer sediments bordering the main channel (Fig. 15.2). Periods of high bulk meltwater discharge give rise to high groundwater levels, with a decreasing effect with distance from the main channel (Wells 1 to 3). This channel–groundwater coupling contrasts with observations in some other proglacial environments where the lack of an active layer, less complex topography and smaller groundwater reservoirs lead to very little interaction between glacial meltwaters and proglacial sediments (Fairchild *et al.*, 1999).

15.4 Proglacial solute fluxes and chemical weathering rates

Solute fluxes from the proglacial zone at Finsterwalderbreen are dominated by HCO_3^-, SO_4^{2-}, Ca^{2+} and Mg^{2+}, of which the snowpack contributes by ca. 7% in total (Table 15.1). The negative silica flux is within the error margin of calculations, and indicates that there is very little Si discharged from the proglacial zone. This proglacial solute flux represents an enhancement of ca. 30% over glacial solute fluxes. A chemical weathering rate three times that for the glacial part of the catchment also signifies the proglacial zone as highly geochemically reactive. The elevated rates of chemical weathering can be attributed to the relatively young age of proglacial material (<100 yr) and the close coupling between glacial bulk meltwaters and proglacial sediments, achieved by an extensive active layer groundwater system and associated ephemeral streams.

15.5 Chemical weathering mechanisms and sources of solute

High proglacial chemical weathering rates at Finsterwalderbreen can be explained only by the chemical weathering of rock material in the proglacial zone. Increases in Ca^{2+}, Mg^{2+} and HCO_3^- signify the importance of limestone and dolomite dissolution in contributing solute to meltwaters. The ultimate source of the SO_4^{2-} is oxidation of sulphide minerals, although the dissolution

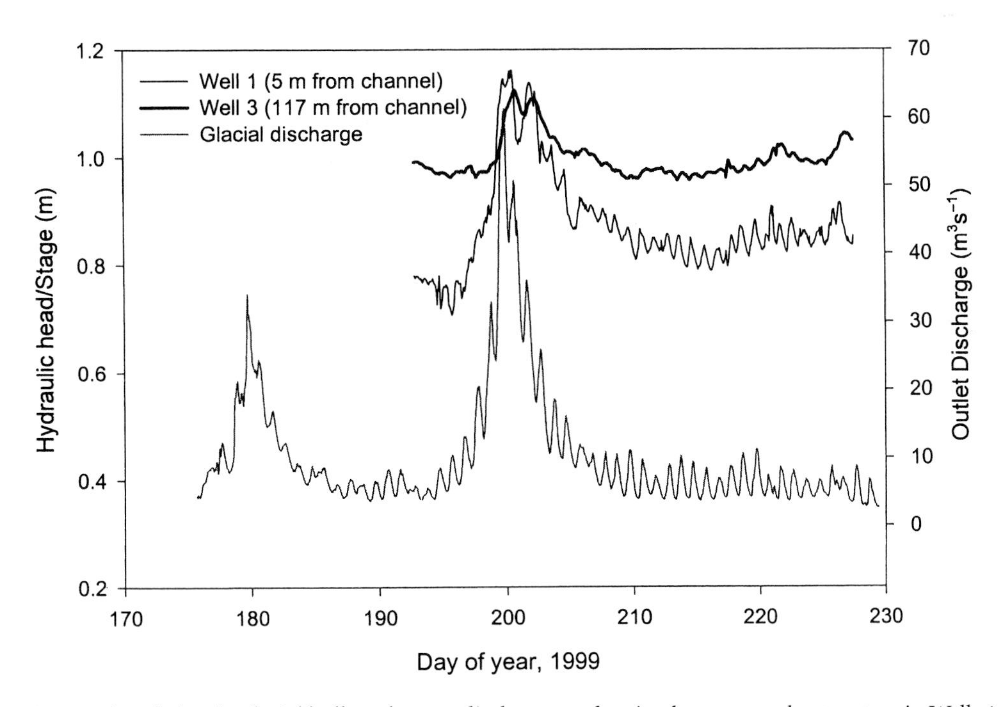

Figure 15.2 Temporal variation in glacial bulk meltwater discharge, and active-layer groundwater stage in Wells 1 and 3 during the 1999 melt season at Finsterwalderbreen.

Table 15.1 Glacial and proglacial solute fluxes and chemical weathering rates during the 1999 melt season

Species	Glacial solute flux (kg)	Proglacial solute flux (kg)	Proglacial/glacial flux ×100
Cl^-	110,000	5200	0
HCO_3^-	1,200,000	310,000	26
SO_4^{2-}	680,000	250,000	47
Na^+	83,000	7500	9
K^+	29,000	2800	10
Mg^{2+}	120,000	35,000	29
Ca^{2+}	470,000	160,000	34
Si	14,000	−800	0
Total	2,700,000	770,000	29
Water	4.9×10^7	-5×10^6	−13
Chemical weathering rate (meq $\Sigma m^{-2} yr^{-1}$)	790	2,600	329

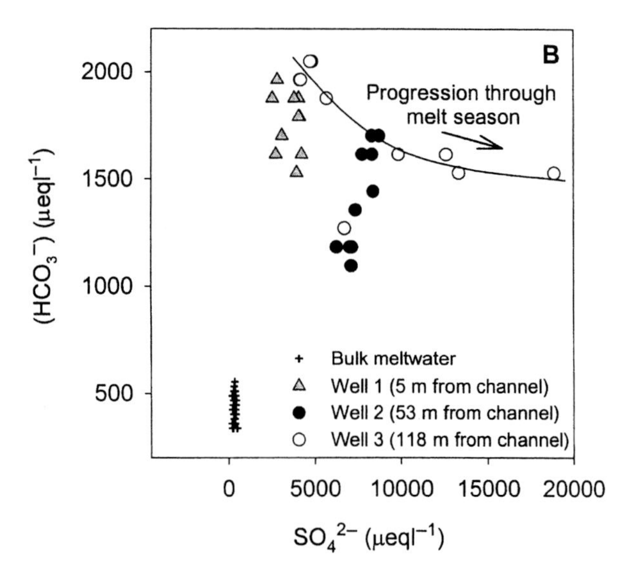

Figure 15.3 Association between SO_4^{2-} and HCO_3^- for active layer groundwaters from Wells 1, 2 and 3 and bulk meltwaters sampled during the 1999 melt season in the Finsterwalbreen proglacial zone.

of secondary sulphate salts is also important locally. In-channel weathering of suspended sediment contributes some Ca^{2+} and HCO_3^- to glacial meltwaters, but most proglacial solute derives from weathering of material in the active layer. Hydrochemical trends in active layer groundwaters can be explained by four geochemical processes: (i) sulphide oxidation, (ii) carbonate dissolution, (iii) precipitation of calcite and Mg-Ca sulphate salts and (iv) dissolution of the secondary calcite and sulphate salts (Fig. 15.3). At distance from the glacial bulk meltwater channel, there is little direct mixing of channel and active-layer groundwaters (although variations in stage are forced by channel discharge fluctuations). Here (Well 3), waters evolve by carbonate dissolution/sulphide oxidation, together with the dissolution of sulphate salts. The latter form by freeze concentration in winter and evaporative concentration in summer, and give rise to high Ca^{2+} and SO_4^{2-} concentrations that decrease through the melt season. Close to the glacial channel (Wells 1 and 2), any salts formed on the proglacial surface are rapidly removed by glacial meltwaters during frequent channel flood. Hence, carbonate dissolution coupled to sulphide oxidation is the dominant geochemical weathering process, with trends of dilution being evident as sediments are flushed by dilute channel waters (Fig. 15.3).

SIXTEEN

Impact of mineral surface area on solute fluxes at Bench Glacier, Alaska

Suzanne Prestrud Anderson

Institute of Arctic and Alpine Research and Department of Geography, University of Colorado, Boulder, CO 80309-0450, USA

Temperate glaciers are noted for exceptionally high erosion rates, which should promote chemical weathering because rates of physical and chemical denudation are generally coupled. Yet fluxes of dissolved cations from glaciers are indistinguishable from non-glacierized basins, and fluxes of dissolved silica are about an order of magnitude lower than from non-glacierized basins (Anderson *et al.*, 1997, Hodson *et al.*, 2000). In this study, I model the solute flux from a glacier to gain insight into this apparent paradox.

The study site is Bench Glacier, a 7.5 km^2 temperate glacier in south-central Alaska (Anderson, 2005). The underlying bedrock is metasedimentary; carbonate constitutes <1 wt% of the bedrock. Solute fluxes, sediment yields and glacier dynamics have been documented (MacGregor, 2002, Riihimaki, 2003, Anderson *et al.*, 2004).

16.1 Glaciers as flow-through reactors

To first order, the subglacial geochemical environment consists of finely crushed rock bathed in snowmelt, analogous to flow-through reactors used to measure mineral weathering rates. The solute flux due to weathering, Q_{solute}, out of a flow-through reactor is given by

$$Q_{solute} = C^* Q_{water} = r\beta\lambda S_{geo} \qquad (1)$$

where C^* is solute concentration in the outflow corrected for inputs (mass/volume), Q_{water} is the water discharge (volume/time), r is the weathering rate constant (moles/mineral surface area/time), β is a stoichiometric coefficient (mass/moles), λ is a dimensionless surface roughness parameter, and S_{geo} is the geometric mineral surface area (based on grain size).

16.2 Mineral surface area

I assume that the geometric mineral surface area at the glacier bed is equivalent to the mineral surface area produced by glacier erosion. This implies that there is little storage of fine-grained sediment at the bed of Bench Glacier, an assumption supported by video camera observations of the bed (Harper *et al.*, 2003), year-to-year consistency in sediment yield, and dynamics of the glacier.

The surface area production rate, S_{geo}, is given by $S_{geo} = n_{gr}\, a_{gr}$, where a_{gr} is the surface area per grain, and n_{gr} is the number of grains produced per unit time, given by $n_{gr} = EA/v_{gr}$, where E is the erosion rate (length/time), A is the basin area, and v_{gr} is the volume per grain. For a mixture of grains of simple geometry (either spheres or cubes)

$$S_{geo} = \sum_i 6 f_i \frac{E}{D_i} A \qquad (2)$$

where D_i is the particle size of grain size class i (length) and f_i is the fraction of the sediment yield represented by grain size class i. Bench Glacier sediment yield of 1.5 mm yr^{-1} falls in the middle of the global range for glaciers. I assume a sediment size distribution of 50% coarse material (with insignificant surface area), 25% coarse silt (62 μm) and 25% coarse clay (2 μm). This yields S_{geo} = 8100 km^2 yr^{-1} for Bench Glacier.

Natural mineral surfaces are rough compared with S_{geo}. The true reactive surface area is approximated with S_{BET}, the surface area determined from gas adsorption isotherms. A roughness parameter λ, defined as $\lambda = S_{BET}/S_{geo}$, has values commonly 10 or greater. Surface roughness tends to increase as mineral surfaces age (White & Brantley, 2003), empirically following

$$\lambda = 13.6 t^{0.20} \qquad (3)$$

where t is the exposure age of the mineral surface in years.

Table 16.1 Regressions of r versus time from White & Brantley (2003). The regressions are of the form $\log r = a + b \log t$, where t is time in years, and r is the mineral weathering rate ($\text{mol}\,\text{m}^{-2}\text{s}^{-1}$)

Mineral	a	b
Plagioclase	−12.46	−0.56
K feldspar	−12.49	−0.65
Hornblende	−12.67	−0.62
Biotite	−12.32	−0.60

16.3 Weathering rate constants

Mineral weathering rate constants, r, depend on temperature and on mineral surface age. I use expressions for $r(t)$ for major silicate minerals (Table 16.1), and the carbonate (calcite) weathering rate constant, from Morse & Arvidson (2002), is kept constant in time. The rate constants are adjusted from laboratory temperatures (20°C) to $T = 0$°C with the Arrhenius relationship

$$\frac{r}{r_0} = \exp\left[\frac{E_a}{R}\left(\frac{1}{T_0} - \frac{1}{T}\right)\right] \qquad (4)$$

where r is the rate at temperature T (in kelvin), r_0 is the rate at temperature T_0, E_a is the activation energy of the reaction (energy/mol), and R is the gas constant (energy/mol/T). The apparent activation energy (E_a) used for plagioclase, K-feldspar and hornblende is $55\,\text{kJ}\,\text{mol}^{-1}$, for biotite is $27\,\text{kJ}\,\text{mol}^{-1}$ (White *et al.*, 1999), and for calcite is $9\,\text{kJ}\,\text{mol}^{-1}$ (Morse & Arvidson, 2002). Solute fluxes are computed over a range of sediment ages (t), which affect values of r and λ.

16.4 Results

Table 16.2 shows calculated solute fluxes for Bench Glacier compared with measured solute fluxes from Bench Glacier. The model considerably overpredicts cation fluxes from carbonate dissolution, probably because of the effect of saturation on dissolution rate. Subglacial waters are often saturated with respect to calcite, which reduces to zero the solute flux from this reaction. For silicate weathering reactions, the best model fit is for $t = 1$ yr. The measured silica fluxes are matched, whereas predicted cation fluxes are lower than measured fluxes by a factor of about five. The low value of the best-fitting t supports the assumption that little fine-grained sediment is stored subglacially. Underprediction of the cation flux from silicate weathering in the model arises entirely from underprediction of K. Very high K concentrations commonly observed in glacier runoff have been attributed to physical shearing of biotite grains during glacial abrasion (Blum, 1997). Mechanical release of cations is not characterized by the dissolution rate constants, r, used in this model.

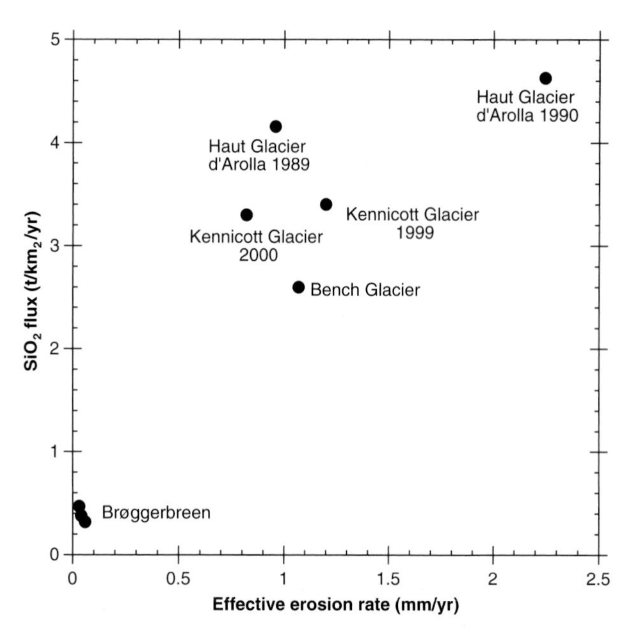

Figure 16.1 Annual silica flux as a function of the annual suspended sediment yield from active glaciers. Haut Glacier d'Arolla is an $11.7\,\text{km}^2$ glacier in Switzerland, on metamorphic and igneous rocks (Sharp *et al.*, 1995). Kennicott Glacier is a $420\,\text{km}^2$ glacier in Alaska, on volcanics, limestone and greenstone rocks, and Bench Glacier is a $7.5\,\text{km}^2$ glacier in Alaska on metasedimentary rock (Anderson *et al.*, 2003). Brøggerbreen is a $32\,\text{km}^2$ polar (cold-based) glacier in Svalbard on carbonate-rich sedimentary rocks (Hodson *et al.*, 2000).

Table 16.2 Glacier chemical denudation rates: calculated and measured

Composition	Glacier erosion rate ($\text{mm}\,\text{yr}^{-1}$)	t (yr)	Cation flux from calcite ($\text{t}\,\text{km}^2\text{yr}^{-1}$)	Cation flux from silicates ($\text{t}\,\text{km}^2\text{yr}^{-1}$)	SiO_2 flux ($\text{t}\,\text{km}^2\text{yr}^{-1}$)
Metagreywacke	1.5	1	1330	0.99	2.41
(Bench Glacier)		10	2110	0.40	0.97
45% plag, 20% kspar,		100	3340	0.16	0.39
13% hbld, 2% bio, 20%		1000	5300	0.07	0.16
qtz, 0.1% calcite		10,000	8400	0.03	0.06
Bench Glacier—measured	1.5		22.4	5.6	2.70

16.5 Erosion rate and subglacial weathering

For silicate weathering, which in dilute glacial systems is unaffected by saturation state, Equation (1) predicts a direct dependence of weathering solute fluxes on mineral surface area produced by erosion. Data from the few glaciers where erosion and solute fluxes have been measured simultaneously appear to show an increase in dissolved silica flux with glacier erosion rates (Fig. 16.1). This correlation suggests that erosion and mineral surface area production exert important control on the silicate weathering fluxes from glaciers. However, reduction in rate constants due to low temperatures keep silicate weathering fluxes from glaciers lower than other environments.

Acknowledgements

Teams anchored by C.A. Riihimaki and K.R. MacGregor collected data at Bench Glacier. I am grateful to Art White and Alex Blum for discussions on glacial erosion and weathering. The National Science Foundation funded this work through grant numbers 9818251, 9812945, 9912129 and 0345136.

SEVENTEEN

Phosphorus in glacial meltwaters

Andy Hodson

Department of Geography, University of Sheffield, Sheffield S10 2TN, UK

The liberation of phosphorus (P) from rocks comminuted by intense physical erosion is likely to be an important factor in the dynamics of young, post-glacial ecosystems. Phosphorus release from such sources is also likely to be highly sensitive to climate change, and P fluxes from glacial environments are likely to respond directly to increased meltwater runoff.

17.1 Determining phosphorus in glacial meltwaters

Phosphorus is present in a number of phases that are operationally defined. Dissolved P is usually that which passes though a 0.45 μm grade filter and so colloidal forms may be included inadvertently (these have yet to be assessed in glacial settings). The dissolved P pool is partitioned into reactive and non-reactive phases following the quantitative measurement of a reduced phosphomolybdate complex in filtrate and digested filtrate respectively (e.g. Tockner *et al.*, 2002). Digestion of an unfiltered sample may also enable estimation of particulate P, which may be further characterized using sequential extraction. These methods are not without their problems, the most important being that concentrations in glacial meltwaters are often below conventional detection limits (usually 1–3 $\mu g\,P\,L^{-1}$). Recent work by the author has therefore developed a novel analytical procedure (pioneered for marine analyses by Karl & Tien, 1992) for detecting dissolved reactive phosphorus at nanomolar concentrations. This involves the co-precipitation of dissolved P with brucite ($Mg(OH)_2$) after spiking water samples with $MgCl_2$ and NaOH. The dissolved P content of large samples thus becomes adsorbed to a small quantity of precipitate, which is then separated by centrifugation and dissolved during the acidification step of the phosphomolybdate colorimetric method. In this way detection limits may be decreased to levels 25 times lower than those achieved with conventional methods.

17.2 Phosphorus in the snowpack

Dissolved forms of P are present in snowpacks of the Arctic (this paper), the European Alps (Tockner *et al.*, 2002), and the Himalayas (Mayewski *et al.*, 1984) (Table 17.1). Mayewski *et al.*'s (1984) study of a firn/ice core suggests that P adsorption onto wind blown Fe-oxide particles might be a major precursor to P deposition at high-altitude continental sites. Otherwise, winter

Table 17.1 Published phosphorus data for glacier basins. Average concentrations are given as $\mu g\,P\,L^{-1}$ for snow and runoff, and yields (values in parentheses) are in $kg\,P\,ha^{-1}yr^{-1}$. P_{ads} denotes adsorbed P that is base extractable from suspended sediment (see Hodson *et al.*, 2004)

Site	Dissolved reactive phosphorus		Dissolved unreactive phosphorus		Particulate phosphorus	
	Snow	Runoff	Snow	Runoff	Snow	Runoff
Midre Lovenbreen, Svalbard*	0.43 (0.33)	1.05 (1.8)	—	—	<3.0 (–)	516 (530) P_{ads}: 5.0 (22)
Austre Brøggerbreen, Svalbard†	<3.0 (–)	<3.0 (–)	—	—	<3.0 (–)	82 (64) P_{ads}: 0.61 (0.48)
Val Roseg, Switzerland‡	10 (–)	2.0 (3.0)	8.0 (–)	5.0 (5.0)	12 (–)	97 (420)
Sentik Glacier, India§	11 (–)§	—	—	—	—	—
Manso Glacier, Argentina¶	—	4.3 (–)	—	—	—	190 (–)
Castano Overo, Argentina¶	—	21 (–)	—	—	—	51 (–)

*Unpublished data for 2003; †Hodson *et al.* (2004, 2005); ‡Tockner *et al.* (2002); §Mayewski *et al.* (1984), average values from snow/ice core; ¶Upstream proglacial sites only ($n = 3$), Chillrud *et al.* (1994).

snowpack data show that the mean dissolved reactive P content of snow is ca. $10\,\mu g\,P\,L^{-1}$ in the European Alps but only ca. $0.5\,\mu g\,P\,L^{-1}$ in more remote Arctic environments (see Table 17.1). The first fractions of snowmelt P, however, may be up to 12 times more concentrated than the surrounding snowpack and so elution has the capacity to generate runoff more concentrated in dissolved P than might otherwise be expected.

17.3 Phosphorus reactions at mineral surfaces

Hosein (2002) showed how the etching of apatite mineral surfaces confirms that dissolution is a source P to glacial meltwaters. As the concentrations of P in snow, however, are broadly equivalent to those in bulk runoff (Table 17.1), the fate of P liberated by rock weathering in glacial environments is unclear. Chillrud *et al.* (1994) have argued that pyrite oxidation exerts a stoichiometric control upon this P in the absence of strong redox gradients within the glacial system. Thus dissolved reactive P liberated by dissolution becomes adsorbed to the Fe-oxyhydroxide minerals produced following pyrite oxidation (which is a major reaction in high rock–water contact environments of the glacial system: Tranter, this volume, Chapter 14). However, the assumption that strong redox gradients do not exist is no longer supported by studies of subglacial environments (e.g. Tranter, this volume, Chapter 14) and so P binding to such surfaces may be reversed during cyclic oxidation/reduction. A further problem with Chillrud *et al.*'s (1994) reasoning is that particulate P is thought to be dominated by Fe-oxyhydroxide-bound P, which is clearly not the case when the composition of particulate P in turbid glacial runoff is examined (see below).

17.4 Particulate phosphorus

Sequential extraction of glacial sediments show that the overwhelming majority (ca. 95% in the case of two Svalbard glaciers) of the particulate P phase is associated with unweathered, acid-extractable fractions (e.g. Hodson *et al.*, 2004). This largely recalcitrant fraction is often assumed to be 'apatite/calcite-bound'. Exchangeable P (or loosely sorbed and oxyhydroxide-bound P), which may be liberated by simple base extraction, has been found to represent just 1–23 µg P/g or up to 3.4% of the particulate P phase of a wide range of glacial sediments (Hodson *et al.*, 2004). Thus, unlike many non-glacial environments, particulate P is poorly representative of P adsorbed to reactive mineral surfaces. However, glacial environments may have high sediment yields, and so the small exchangeable fraction of P may indeed represent a significant P yield (see below).

17.5 Phosphorus yields

The above scheme suggests that estimates of P budgets are needed that distinguish between reactive and non-reactive forms of P in the dissolved phase and then between exchangeable and acid-extractable P in the particulate phase. The notion that secondary carbonates sequester P during precipitation has not been supported by field data in Svalbard (Hodson *et al.*, 2004) and so chemical weathering probably represents all P except the acid-extractable particulate P phase (after atmospheric inputs have been accounted for). With this provenance model, the data in Table 17.1 suggest that the liberation of P by dissolution is restricted in glacial settings. This probably reflects the fact that rock–water contact within the glacial system is limited by short residence times, especially where delayed flow pathways through the subglacial environment are absent. Thus the most significant impact of glacial erosion upon the P cycle probably involves the generation and redistribution of freshly comminuted apatite mineral surfaces which do not undergo significant dissolution until deglaciation has occurred. Significant lags might therefore characterize the changing impact of glaciation upon this global biogeochemical cycle.

EIGHTEEN

Glacial landsystems

David J.A. Evans

Department of Geography, University of Durham, Science Site, South Road, Durham DH1 3LE, UK

18.1 Introduction

The landsystems approach is a holistic form of terrain evaluation, linking the geomorphology and subsurface materials in a landscape and relating them genetically to process–landform studies. Early developments of landsystems-type investigations in glaciated terrains include those of Speight (1963), who mapped and grouped landforms according to common origin and age, and Clayton & Moran (1974) who linked process and form in assessing the spatial distribution of glacial features. Glacial landsystems *per se* were first compiled by Fookes *et al.* (1978) to simplify the geomorphological and sedimentary complexity in glaciated basins and provide engineers with genetic classifications of glacigenic landform–sediment assemblages. They recognized the 'till plain landsystem', 'glaciated valley landsystem' and 'fluvioglacial and ice contact deposit landsystem'. The component parts of landsystems are land units (e.g. drumlin or fluting field, Rogen moraine belt, suite of recessional push moraines, etc.) and land elements (e.g. individual drumlins, flutings, eskers, kame terraces, moraines, etc.), the landsystem then being a recurrent pattern of genetically linked land units. Eyles (1983a,b), Eyles & Menzies (1983) and Paul (1983) offered a new approach by incorporating into landsystems all of the landform and sediment types associated with particular glaciation styles (e.g. subglacial, supraglacial and glaciated-valley landsystems). Glaciated terrain could then be mapped according to the dominant landsystems (Eyles & Dearman, 1981; Eyles *et al.*, 1983a), thereby providing a predictive tool for rapid assessments of subsurface materials in resource and engineering project management.

The glacial-landsystem approach has prioritized the assessment of groups of landforms, rather than individual forms in isolation, and the recognition of process continuums. The most recent developments of the glacial landsystem concept (e.g. Evans, 2003a) have stressed the complexity of glacial depositional environments and the fact that variability in landform–sediment assemblages is dictated not only by the location of deposition but also by the 'style' of glaciation. Glaciation 'styles' are a function of climate, basement and surficial geology and topography (Fig. 18.1) and consequently a wide range of glacial landsystems have been compiled for different ice masses and dynamics (Benn & Evans, 1998; Evans, 2003a). Moreover, spatial and temporal variability in predominant landsystems results in the addition of glacial features to a landscape as a series of layers (stratigraphy). This acknowledges the complexity of glacial depositional systems and highlights the fact that spatially coherent landform–sediment assemblages can be superimposed in a landscape and that glacial landscapes are just as much a palimpsest as any other terrain type (Boulton & Clark, 1990a, b; Dyke *et al.*, 1992; Kleman, 1992, 1994; Kleman *et al.*, 1994; Punkari, 1995).

18.2 Landsystems as modern analogues

The landsystems approach is given more credibility and applicability wherever landscape evolution can be monitored, thereby providing modern analogues for the interpretation of ancient glaciated terrain (e.g. Price, 1969; Gustavson & Boothroyd, 1987; Kruger, 1994; Dyke & Savelle, 2000; Kjær & Krüger, 2001; Evans & Twigg, 2002). This type of research on modern glaciers has led to the identification of landform–sediment suites indicative of specific styles of glaciation (e.g. plateau icefields (Rea *et al.*, 1998; Rea & Evans, 2003) and arid polar glacier margins (Fitzsimons, 2003)) and ice dynamics (e.g. surging glaciers (Evans & Rea, 1999; 2003) and ice streams (Stokes & Clark, 1999, 2001; Clark *et al.*, 2003a)). Once a landform–sediment suite pertaining to a single period of glacier occupancy or activity can be identified, it often becomes possible to differentiate overprinted signatures (e.g. Dyke & Morris, 1988; Clark, 1993, 1999; Krüger, 1994) and landscape palimpsests.

Examples of terrestrial landsystems models that have recently elaborated upon earlier landsystems classifications by incorporating modern process research include those of the glaciated valley (Spedding & Evans, 2002; Benn *et al.*, 2003) and active temperate glacier margin (Evans & Twigg, 2002; Evans, 2003b). These landsystems represent particular styles of glaciation according to glacier morphology and environmental controls. For example, it has been recognized that the details of glaciated valley landsystems will reflect the relative relief, bedrock, climatic regime and debris supply and transfer rates of the mountainous terrain in which they are located (e.g. Boulton & Eyles, 1979; Owen &

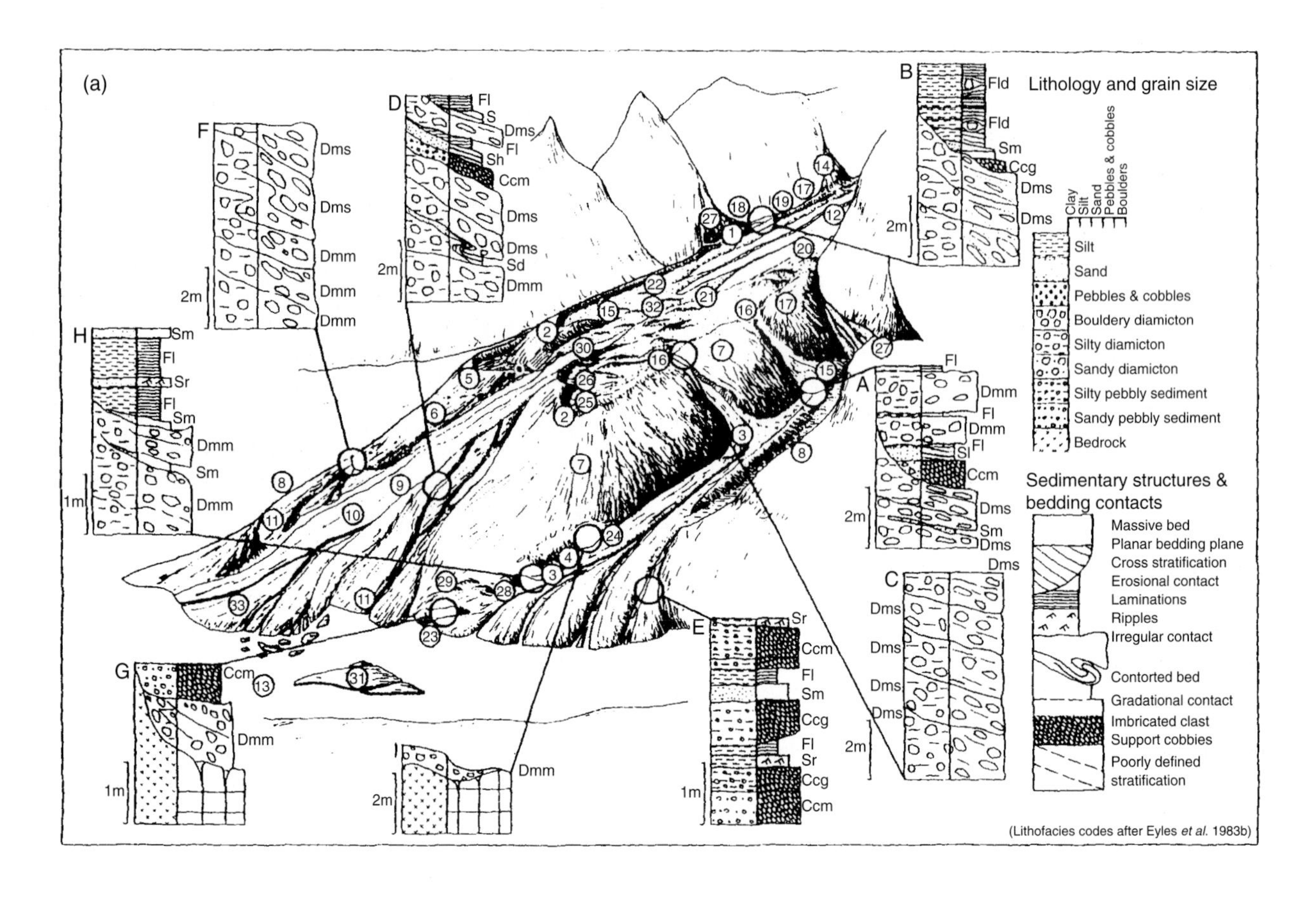
(a)
Lithology and grain size
Clay
Silt
Sand
Pebbles & cobbles
Boulders
Silt
Sand
Pebbles & cobbles
Bouldery diamicton
Silty diamicton
Sandy diamicton
Silty pebbly sediment
Sandy pebbly sediment
Bedrock
Sedimentary structures & bedding contacts
Massive bed
Planar bedding plane
Cross stratification
Erosional contact
Laminations
Ripples
Irregular contact
Contorted bed
Gradational contact
Imbricated clast
Support cobbies
Poorly defined stratification
(Lithofacies codes after Eyles et al. 1983b)
A
B
C
D
E
F
G
H
Dms
Dmm
Fl
Fld
Sm
Sr
Sl
Sh
Sd
S
Ccm
Ccg
1m
2m

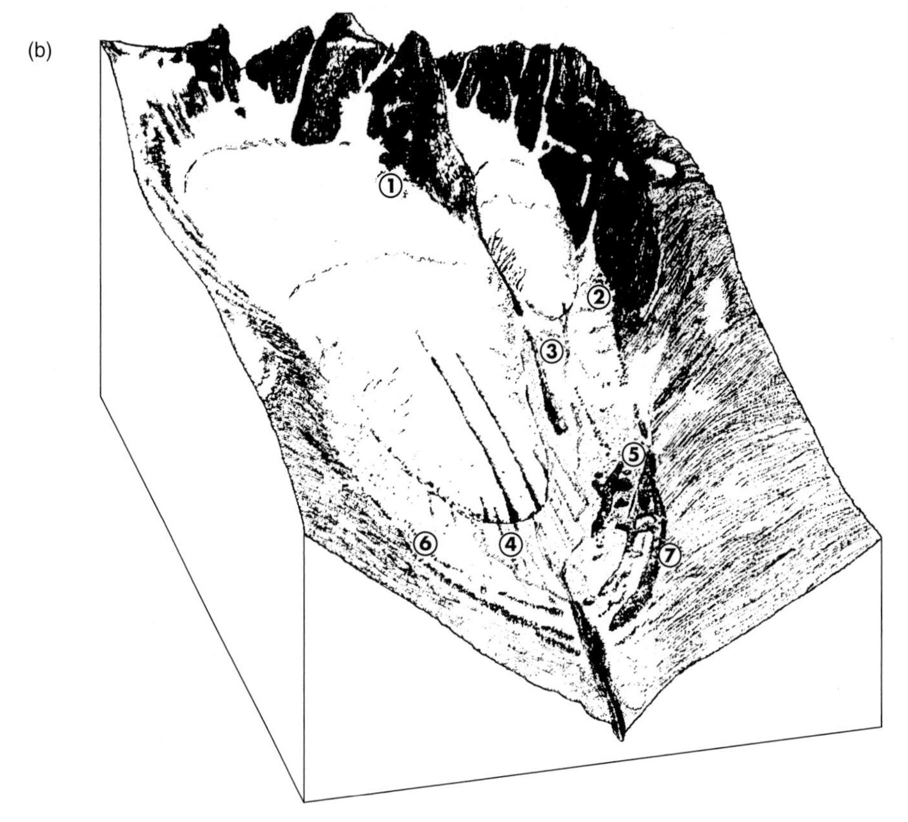
(b)

Derbyshire, 1989; Evans, 1990; Spedding & Evans, 2002; Benn *et al.*, 2003). These interrelated controls are emphasized by Benn *et al.* (2003) in the development of a conceptual continuum of glaciated valley landsystems (Fig. 18.2). Specifically, Benn *et al.* (2003) identify distinct landsystem associations that are defined by the degree of coupling between glacial and proglacial environments via the glacifluvial system. Water discharges from a glacier are a function of catchment area, climatically determined catchment water storage and mean precipitation. Therefore, although there is a climatic significance attached to the glacial geomorphology, it is modulated by catchment size and topography. Coupled and decoupled glacier snouts are thereby defined respectively as those with efficient and inefficient glacifluvial transport between glacial and proglacial systems. At coupled glacier margins, due to the effective sediment transfer from the snout to the fluvial system, moraine development is limited and large amounts of sediment pass through the proglacial zone. This gives rise to the dominance of the outwash head/aggrading sandur landsystem. At decoupled glacier margins there is insufficient meltwater discharge to transfer sediment away from the snout, giving rise to the build up of debris around the glacier perimeter to form moraines. The repeated superposition of moraines around debris-covered glacier margins leads to the construction of giant bounding moraines and moraine-dammed glaciers and also rock glaciers in some settings.

The active temperate glacial landsystem (Evans & Twigg, 2002; Evans, 2003b) is a development of the most prominent terrestrial component of the subglacial landsystem of Eyles & Menzies (1983) in that a modern lowland glacier snout, Breidamerkurjokull in Iceland, overlying a varied substratum is used as a central case study. Recent research on process–form relationships at modern glacier margins highlights three depositional domains (Fig. 18.3). First, the marginal morainic domain consists of extensive, low-amplitude marginal dump, push and squeeze moraines derived largely from material on the glacier foreland. These moraines often record annual recession of active ice but some may be superimposed during periods of glacier stability (Price, 1970; Sharp, 1984). Subglacial fluting and push moraine production is genetically linked wherein the deformation/ploughing of subglacial material into flutings results in the advection of subglacially deforming sediment to the ice margin where squeezing and pushing then combine to create moraines. Larger push moraines are constructed by stationary glacier margins involving either the stacking of frozen sediment slabs, the prolonged impact of dump, squeeze and push mechanisms at the same location, or the incremental thickening of an ice-marginal wedge of deformation till. Overridden push moraines are recognizable as arcuate, low-amplitude ridges draped by flutings and recessional push moraines. The paucity of supraglacial sediment in active temperate glaciers generally precludes the widespread development of chaotic hummocky moraine, although low-amplitude, bouldery hummocks are produced by the melt out of medial moraines and by the melting of debris-charged glacier snouts in settings where marginal freeze-on produces debris-rich ice facies. Second, the subglacial domain includes assemblages of flutings, drumlins and overridden push moraines on surfaces that lie between ice-marginal depocentres. Subglacial materials often display a vertical continuum comprising glacitectonized stratified sediments capped by subglacial till. The flutings are traditionally explained as the products of till squeezing into cavities on the down-glacier sides of lodged boulders. Larger drumlins are explained as the streamlined remnants of coarse-grained sandur fans (Boulton, 1987). Third, the glacifluvial and glacilacustrine domain is characterized by sandur fans (both ice-contact and spillway fed), ice-margin-parallel outwash tracts and kame terraces, topographically channelized sandar, pitted sandar (ice-marginal and jökulhlaup types), and eskers of single and more complex anabranched forms. Although hummocky terrain located at receding glacier margins is often referred to as 'kame and kettle topography' it can evolve through time due to melt out of underlying ice into complex networks of anabranched eskers. The clear landform–sediment signatures of active temperate glacier recession have been recognized in some ancient glaciated terrains (e.g. Evans *et al.*, 1999), thereby demonstrating the potential of the landsystems approach for deciphering palaeoglacier dynamics and their linkages to climate change.

Examples of landsystems that incorporate the landform and sediment suites indicative of particular glacier dynamics are the palaeo-ice stream (see Stokes & Clark, this volume, Chapter 26) and surging glacier landsystems (Evans & Rea, 1999, 2003). These landsystem signatures are contained within the larger imprints of ice sheets or icefields. For example, Evans & Rea (1999, 2003) recognize a suite of landforms and sediments produced by modern glacier surging at the margins of upland icefields in Iceland and

Figure 18.1 Examples of glaciated valley landsystems. (a) High-relief terrain, from Owen & Derbyshire (1989) based upon glaciers located in the Karakoram Mountains where rates of debris supply are extremely high. Number codes are: (1) truncated scree; (2) termino-lateral dump moraine; (3) lateral outwash channel; (4) glacifluvial fan; (5) slide moraine; (6) slide-debris flow cone; (7) slide-modified lateral moraine; (8) lateral outwash fan; (9) meltwater channel; (10) meltwater fan; (11) abandoned meltwater fan; (12) bare ice; (13) trunk valley river; (14) debris flow; (15) flow slide; (16) gullied lateral moraine; (17) lateral moraine; (18) ablation valley lake; (19) ablation valley; (20) supraglacial lake; (21) supraglacial stream; (22) ice-contact terrace; (23) lodgement till; (24) roche moutonnée; (25) fluted moraine; (26) diffluence col; (27) high-level till remnant; (28) diffluence col lake; (29) fines from supraglacial debris; (30) ice-cored moraines; (31) river alluvium; (32) supraglacial debris; (33) dead ice. (b) Low-relief mountain terrain, from Benn & Evans (1998) based upon northwest Europe where supraglacial inputs of debris are relatively low. Number codes are: (1) supraglacially entrained debris; (2) periglacial trimline above ice-scoured bedrock; (3) medial moraine; (4) fluted till surface; (5) paraglacial reworking of glacigenic deposits; (6) and (7) lateral moraines, showing within-valley asymmetry.

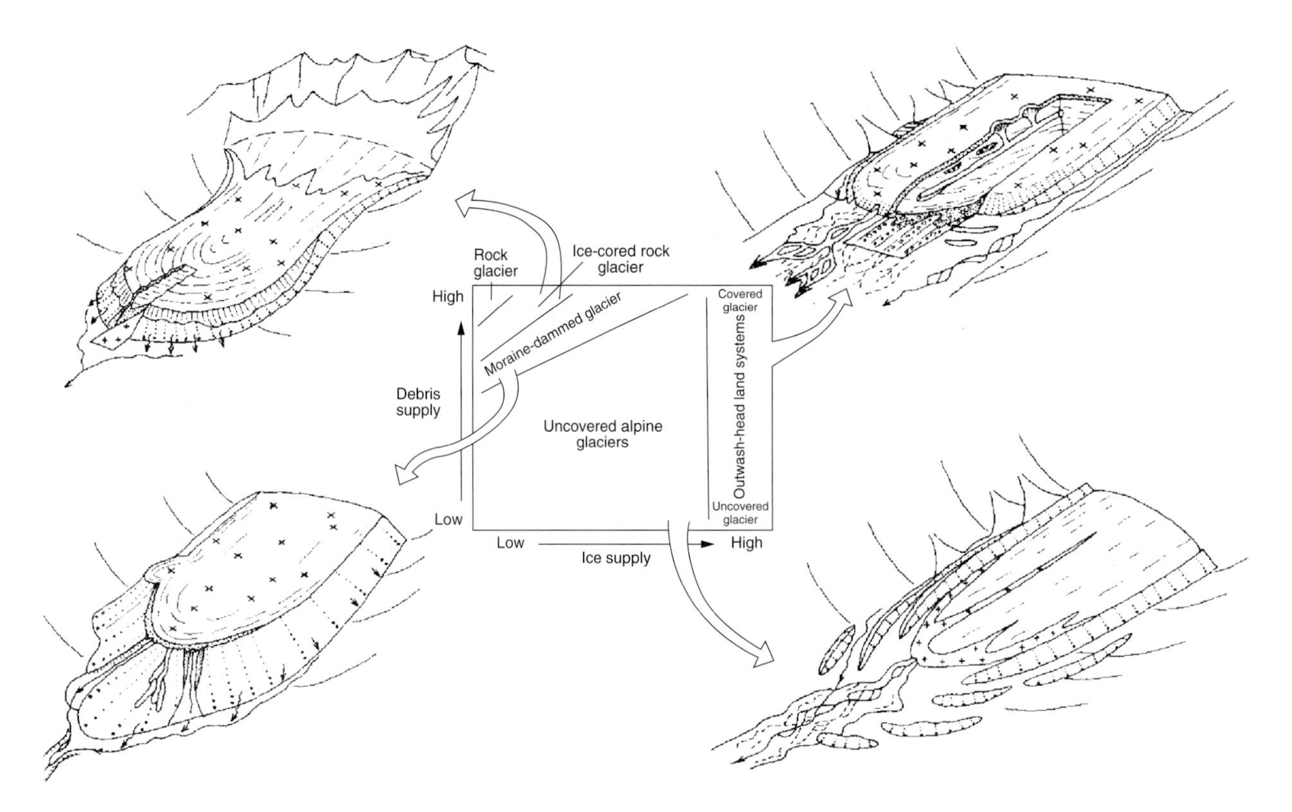

Figure 18.2 Constraints on landsystems development around valley glaciers, showing development pathways to four landsystem associations (from Benn *et al.*, 2003).

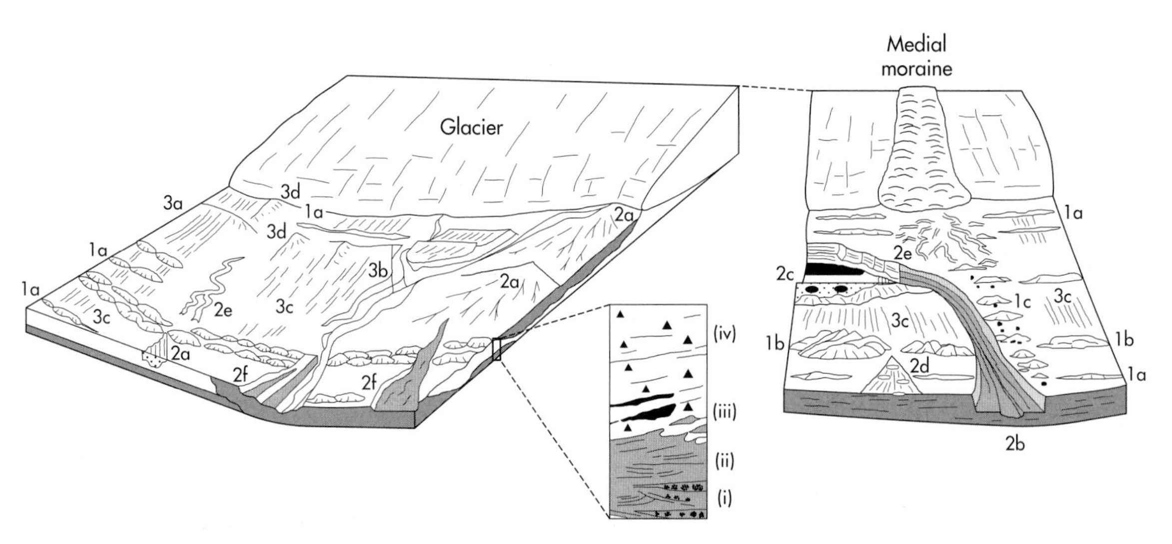

Figure 18.3 The active temperate glacial landsystem (after Krüger, 1994; Evans & Twigg, 2002; Evans, 2003b). Landforms are numbered according to their domain (1, morainic domain; 2, glacifluvial domain; 3, subglacial domain): (1a) small, often annual, push moraines; (1b) superimposed push moraines; (1c) hummocky moraine; (2a) ice-contact sandur fans; (2b) spillway-fed sandur fan; (2c) ice-margin-parallel outwash tract/kame terrace; (2d) pitted sandur; (2e) eskers; (2f) entrenched ice-contact outwash fans; (3a) overridden (fluted) push moraines; (3b) overridden, pre-advance ice-contact outwash fan; (3c) flutes; (3d) drumlins. The idealized stratigraphical section-log shows a typical depositional sequence recording glacier advance over glacifluvial sediments, comprising: (i) undeformed outwash; (ii) glacitectonized outwash/glacitectonite; (iii) massive, sheared till with basal inclusions of pre-advance peat and glacifluvial sediment; (iv) massive sheared till with basal erosional contact.

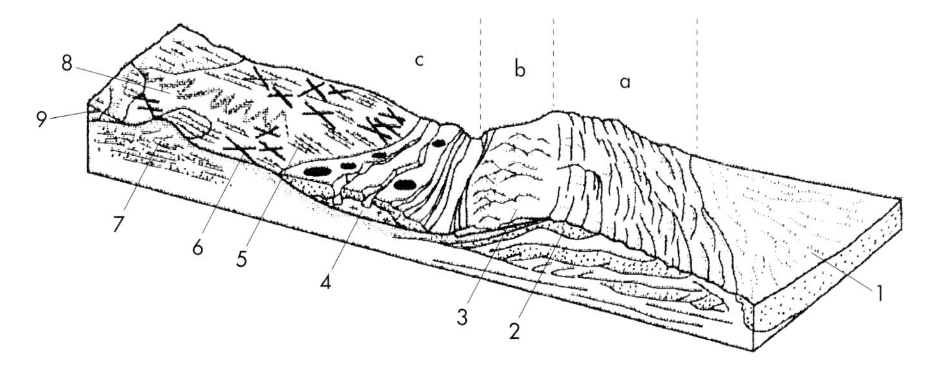

Figure 18.4 Landsystems model for surging glacier margins (after Evans *et al.*, 1999; Evans & Rea, 1999, 2003): (a) outer zone of proglacially thrust pre-surge sediment which may grade into small push moraines in areas of thin sediment cover; (b) zone of weakly developed chaotic hummocky moraine located on the down-ice sides of topographic depressions; (c) zone of flutings, crevasse-squeeze ridges and concertina eskers; 1, proglacial outwash fan; 2, thrust-block moraine; 3, hummocky moraine; 4, stagnating surge snout covered by pitted and channelled outwash; 5, flutings; 6, crevasse-squeeze ridge; 7, overridden and fluted thrust-block moraine; 8, concertina esker; 9, glacier with crevasse-squeeze ridges emerging at surface.

Svalbard (Fig. 18.4). This includes an outer zone of thrust-block moraines and push moraines produced by rapid ice advance into proglacial sediments and the consequent failure and stacking of large contorted and faulted blocks due to high proglacial and submarginal compressive stresses. Evidence of earlier surges is often manifest in the form of overridden thrust-block moraines. These are recognizable as ice-moulded (fluted) hills on the down-ice side of topographic depressions from which they were displaced by thrusting. Inside the thrust-block moraines of an individual surge lies a zone characterized by 'concertina' or 'zig-zag' eskers, crevasse-squeeze ridges, flutings and pockets of hummocky moraine and ice-cored outwash. The zig-zag eskers were produced englacially or supraglacially by meltwater that exploited the extensive network of crevasses created during the surge. The crevasse-squeeze ridges are the product of subglacial sediment squeezing into the widespread cross-cutting basal crevasses produced during the surge. The intersection points of flutings and crevasse-squeeze ridges show that the sediments of both landforms were displaced upwards into crevasses, demonstrating that the flutings are also surge features. Moreover, the great length of the flutings verifies that they were produced by a fast-flow event. The hummocky moraine is produced by the overriding, overthrusting and incorporation of debris-rich stagnant ice dating to a previous surge. Additionally, thrusting, squeezing and bulldozing of proglacial lake sediments and outwash lying over pre-existing stagnant ice also occur. The post-surge evolution of such supraglacial lake and outwash sediments is manifest in tracts of ice-cored outwash fans and glacilacustrine sediment bodies with surfaces that become increasingly kettled and incised through time, exposing stagnant ice in the sides of ice-walled channels. The surging glacier landsystem has been applied to the regional geomorphology of western Canada (Evans *et al.*, 1999; Evans & Rea, 2003) by highlighting the spatial association of thrust-block moraines, crevasse-squeeze ridges and long flutings produced by an ice stream within the receding Laurentide Ice Sheet.

18.3 Subaqueous landsystems

Subaqueous depositional environments were recognized as extensions of the primary landsystems presented by Eyles (1983c; e.g. Eyles & Menzies, 1983). Models of the wide spectrum of glacimarine processes and depositional features have appeared in reviews of offshore environments (e.g. Powell, 1984; Eyles *et al.*, 1985), but subaqueous landform–sediment assemblages have only recently been given separate treatment in a landsystem context (e.g. Benn & Evans, 1998; Powell, 2003; Teller, 2003; Vorren, 2003). This is exemplified by Vorren's (2003) compilation of the landforms and sediments on glaciated passive continental margins (Fig. 18.5). Glaciation of the shelf is recorded by huge erosional troughs containing glacially streamlined forms and recessional ice marginal depocentres. Grounding-zone wedges and trough-mouth fans mark former glacier margins at the shelf edge. Although the glacial bedforms are draped by extensive glacimarine sediments and forms associated with glacier-margin, ice-shelf and iceberg processes, the imprint of former glacier flow is often impressive and can be readily used in palaeoglaciological reconstruction (e.g. glacier surges (Solheim, 1991), palaeo-ice streams (Canals *et al.*, 2000; Ó Cofaigh *et al.*, 2002a; Ottesen *et al.*, 2002; Sejrup *et al.*, 2003), glacitectonic thrusting (Sættem, 1990), trough-mouth fan construction (Vorren & Laberg, 1997; Ó Cofaigh *et al.*, 2003)). Sedimentary sequences and depositional architecture in former subaqueous environments (e.g. Powell, 1981, 2003; Lønne, 1995, 2001) also provide valuable information for reconstructions of glacier marginal stability and oscillations, and feedback into reconstructions of palaeoglaciation style along glaciated coasts.

18.4 Conclusion

Palaeoglaciological reconstructions, at scales ranging from local to continental scale, play a significant part in the development of

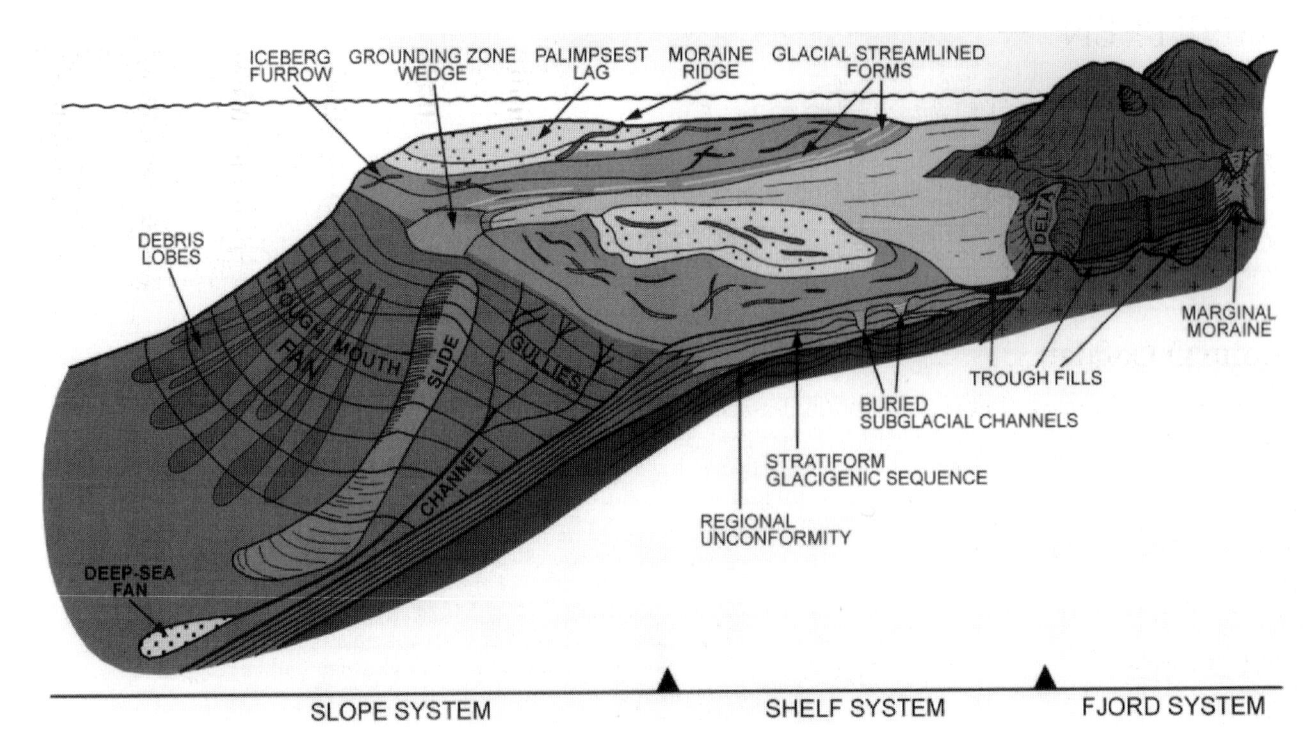

Figure 18.5 Landsystems model of the main glacigenic features and sediments on a passive continental margin, exemplified by the northern Norwegian shelf (from Vorren, 2003).

models of global palaeoclimate. However, the great spatial and temporal complexity of glacial processes and forms have necessitated the compilation of process–form models or landsystems that relate to various glaciation styles and systems, which may then be applied to specific physiographic and climatic settings. A number of contemporary glacierized landscapes have been used as modern analogues for the landform–sediment assemblages produced by different glaciation styles and ice dynamics in differing climatic, geological and topographic settings. The identification of modern landsystems, for example those pertaining to active temperate glacier snouts, surging glaciers, ice streams, valley glaciers, subaqueous depocentres, arid polar glaciers and plateau icefields (Evans, 2003a), provide glacial geomorphologists with the criteria to decipher palaeoglacier dynamics and former glacier–climate relationships in ancient glaciated terrain.

NINETEEN

The subpolar glacier landsystem of the Canadian High Arctic

Colm Ó Cofaigh

Department of Geography, Durham University, Durham DH1 3LE, UK

19.1 Glaciers

The Canadian High Arctic comprises the Queen Elizabeth Islands (QEIs), located north of Parry Channel (Fig. 19.1). Present-day ice cover ranges from large ice caps that feed outlet glaciers and piedmont lobes, to small plateau icefields. The smaller islands of the central and western QEIs are characterized by lowland topography, and glacier cover is restricted to scattered, small ice-masses. Subpolar glaciers of the QEIs are characterized by a frozen marginal zone that passes up-glacier into warm-based ice (Blatter, 1987; Skidmore & Sharp, 1999). Glaciers with a relatively high-mass turnover and/or strong converging flow will contain the most extensive zones of warm-based ice. This thermal regime means that subpolar glaciers can be characterized by both subglacial and supraglacial/lateral meltwater systems, and by significant compressive stresses where the zone of warm-based ice passes into the frozen bed at the margin. These characteristics are critical for debris entrainment and transport and are reflected geomorphologically and sedimentologically in the subpolar glacier landsystem (Fig. 19.2).

19.2 Debris entrainment

Debris is eroded and transported from warm-based zones towards the frozen glacier margin, where debris-rich basal ice is produced by net adfreezing, and thickened by compressive flow. Marginal debris extruded from the glacier surface, dry-calved ice blocks, buried glacier ice and alluvium may also be entrained during glacier advance through apron entrainment. The resulting end products are the thick debris-rich basal ice sequences that are observed at the margins of most subpolar glaciers in the Canadian High Arctic. Supraglacial debris cover is most extensive at the snout where debris-rich basal ice is exposed by glacier marginal thinning. In contrast, outlet glaciers emanating from plateau icefields are generally deficient in supraglacial debris.

19.3 Glacial geomorphology and geology

19.3.1 Glacial debris release and moraine deposition

Exposure of debris-rich basal ice in many Canadian High Arctic subpolar glaciers during snout downwasting results in the formation of controlled moraine (Fig. 19.2). Preservation potential of these moraines is low, however, owing to sediment redistribution during melt-out. Hummocky till veneers interspersed with glacifluvial outwash occur on valley floors where piedmont glaciers have receded onto surrounding uplands, leaving buried glacier ice at lower elevations. Where debris accumulates by rockfall below bedrock cliffs, this process may produce supraglacial lateral moraines, medial moraines and occasional ice-cored rubble cones.

At the margins of plateaux icefields and upland outlet glaciers, debris turnover is low and moraines are rare. This is reflected by the occurrence of trimlines formed of lines of boulders or rubble veneers. However, the major outlet glaciers that occupied the fjords and valleys of the region during the last glaciation deposited till sheets and extensive lateral moraines. This is predominantly a function of the thermal characteristics of larger glaciers, the soles of which reached pressure melting point.

Thrust-block moraines are formed by proglacial glacitectonic thrusting of glacifluvial, glacilacustrine, and emergent glacimarine sediments on valley floors where compressive stresses in the glacier snout are transmitted to unconsolidated sediments (Fig. 19.2). Thrust-block moraines of the region typically comprise relatively intact blocks of sediment displaced in *en echelon* arcs. Individual blocks are commonly tens of metres high, hundreds of metres wide, and, en echelon, may be hundreds of metres long. Bedding in the displaced blocks generally dips back towards the glacier snout, suggesting that they are imbricately stacked blocks partially rotated during thrusting. However, some moraines comprise blocks with bedding dipping away from the snout, indicating that the glacier was responsible for deep-seated wedging of proglacial material (Evans & England, 1991). Thrust blocks can

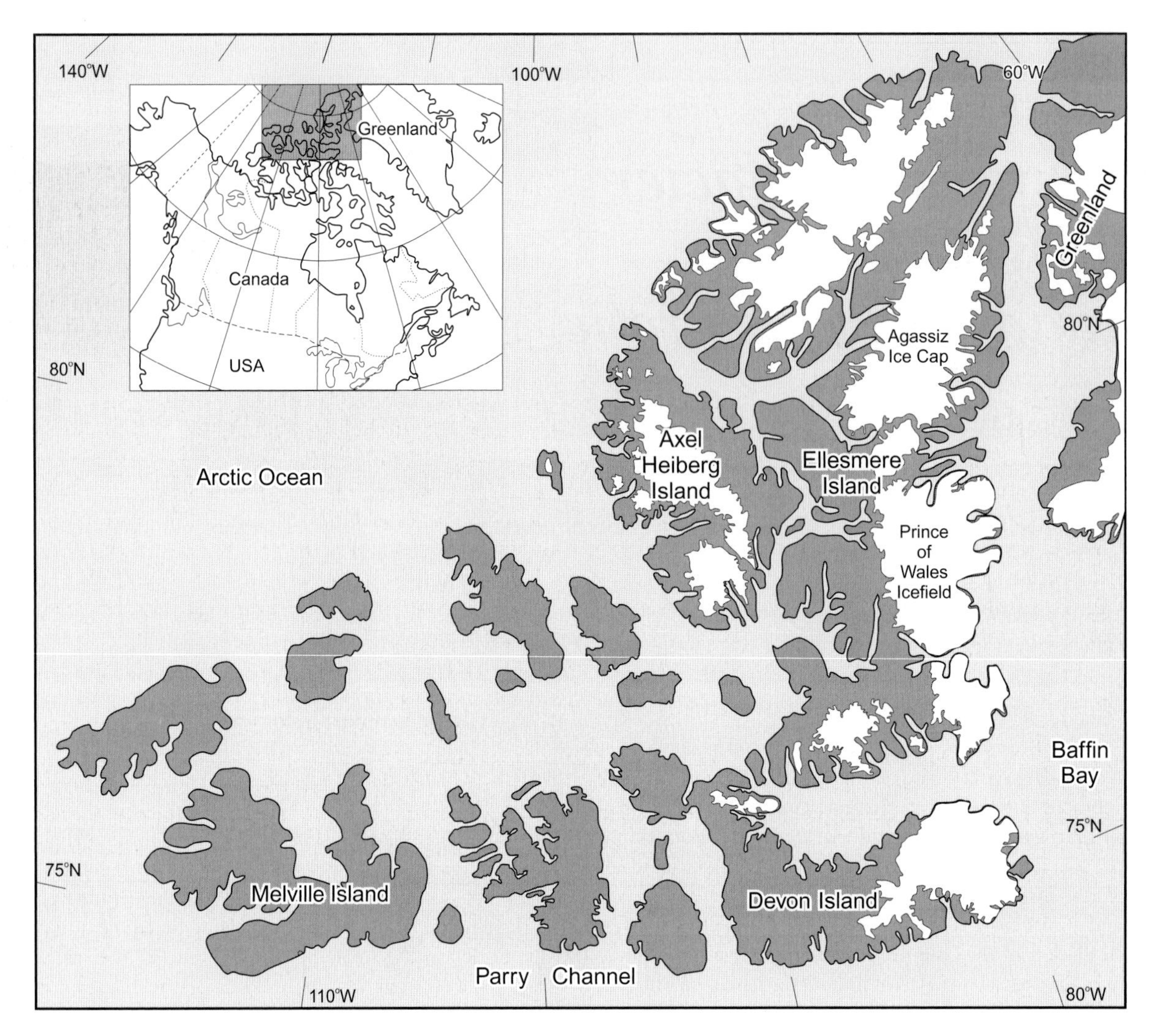

Figure 19.1 Location map of the Queen Elizabeth Islands, Canadian High Arctic and contemporary ice cover.

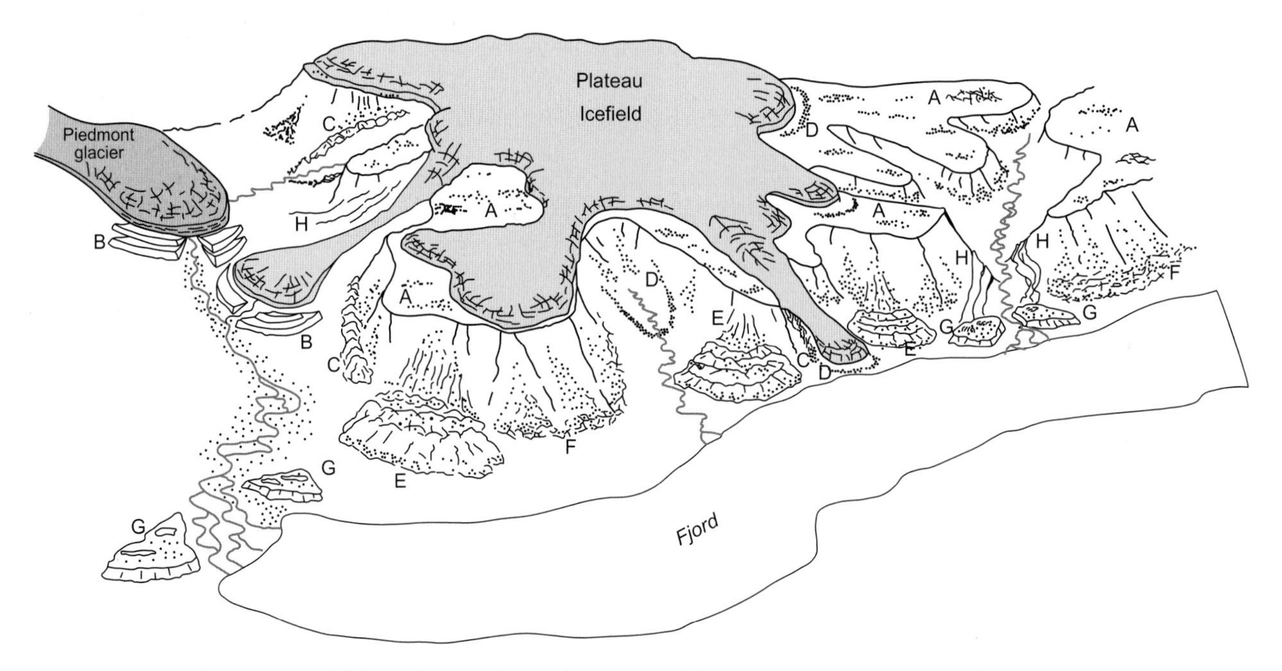

Figure 19.2 Landsystems model for subpolar glaciers (plateau icefields and piedmont lobes) of the Canadian High Arctic: (A) blockfield/residuum; (B) thrust-block moraine; (C) ice-cored lateral moraine; (D) trimline moraine; (E) glacier-ice-cored protalus rock glacier; (F) periglacial protalus rock glacier; (G) raised, former ice-contact deltas; (H) lateral meltwater channels.

act as a source for later sediment entrainment during glacier overriding. This often gives rise to controlled moraine ridges being superimposed on the inner thrust blocks during glacier downwasting.

19.3.2 Glacifluvial processes and landforms

Because meltwater is commonly routed along the frozen lateral margins of subpolar glaciers, lateral meltwater channels eroded into bedrock or sediment during this drainage are common (Fig. 19.2). The nested patterns of such channels document successive recessional positions of glacier snouts confined by topography (Dyke, 1993; Ó Cofaigh *et al.*, 1999). Although subglacial meltwater has been reported from subpolar glaciers in the region, eskers are rare. This probably reflects the restricted nature of subglacial drainage and the sparsity of debris available for meltwater transport. Because large volumes of meltwater are directed along frozen glacier margins, sediment carried by such flows can be deposited as kame terraces or deltas.

19.3.3 Rock glaciers

Talus-foot or valley-side rock glaciers are common, and have been subdivided by Evans (1993) into glacier ice-cored and permafrost-related categories (Fig. 19.2). The former represent the palaeo-margins of outlet glaciers. The lateral moraines of former outlet glaciers can form discontinuous rock glaciers when the valley or fjord becomes deglaciated. Paraglacial activity can also contribute to the production of rock glaciers in such settings where talus buries parts of ice margins during glacier downwasting.

19.3.4 Ice-contact glacimarine and glacilacustrine landforms

Information on glacimarine and glacilacustrine sedimentation from subpolar glaciers in the Canadian High Arctic is largely based on investigations of emergent Holocene sediments (e.g. Ó Cofaigh *et al.*, 1999; Smith, 2000). Where subpolar glaciers terminate in marine environments, englacial and subglacial meltwater emanating from these ice-masses forms subaqueous grounding-line fans, ice-contact deltas and morainal banks. Spatially, the location of these subaqueous depocentres exhibits a strong relationship to fjord bathymetry. Grounding-line fans form where sediment-laden meltwater enters deep water from englacial or subglacial conduits. Morainal banks are deposited along grounding lines during glacier stillstands. They range in height from about 5 to 30 m, and their elongate morphology reflects deposition from point sources along the ice-front, as well as ice-marginal fluctuations that bulldoze and squeeze sediment. Where grounding-line fans and morainal banks aggrade to sea level, they form marine limit deltas (Fig. 19.2). Distinguishing characteristics are steep ice-proximal slopes and kettled surfaces due to melt-out of buried ice.

TWENTY

Plateau icefield landsystem

Brice R. Rea

Department of Geography and Environment, Elphinstone Road, University of Aberdeen, Aberdeen AB24 3UF, UK

20.1 Ice dynamics

The dynamics of plateau icefields are controlled by the ice provenance and thermal regime experienced during either extensive regional to full glacial conditions where the topography is submerged and ice-sheet-surface slopes control ice-flow directions, or local- to regional-scale glaciation when ice accumulation is centred on plateaux, occurring during interstadials/interglacials. With the onset of full glacial conditions newly forming or expanding plateau icefields are likely to be, or become, cold-based, thereby 'freezing' and thus preserving the plateau surface. As ice masses thicken, plateau ice will likely remain cold-based, whereas ice in surrounding valleys may reach the PMP (pressure melting point), overdeepening existing valleys and perhaps cutting through-valleys, in the classic style of 'selective linear erosion' (Sugden, 1974).

During local- to regional-scale glaciation, plataeu ice thicknesses are reduced and controlled by the plateau dimensions (Rea

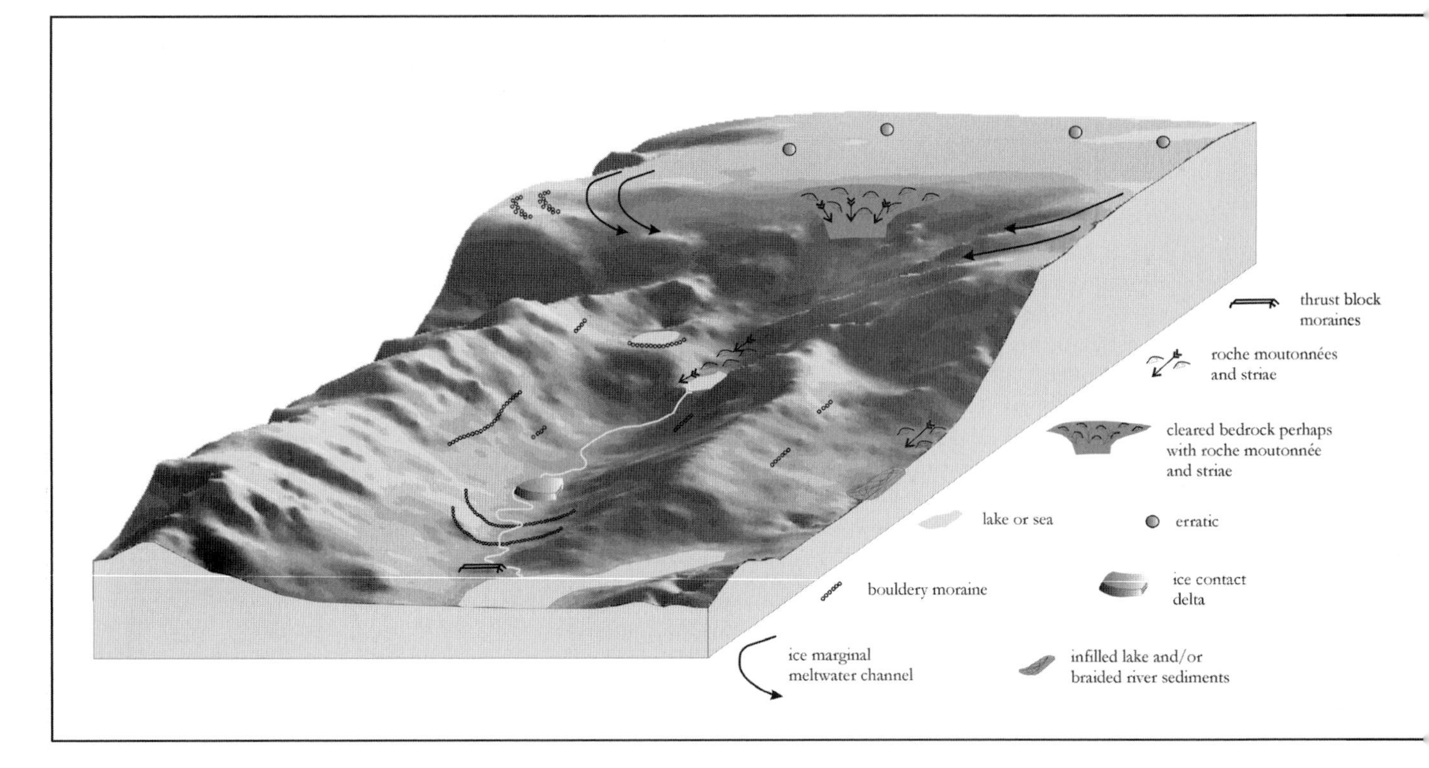

Figure 20.1 Land elements and units that comprise the plateau icefield landsystem. Plateaux surfaces that show no signs of glacierization must be evaluated for a potential ice cover by an area altitude approach. Reprinted by permission of Hodder Arnold.

et al., 1998). Simplistically, below a critical summit size, the smaller the plateau the higher it must be above the regional firn line to support an icefield, and so the lower will be the ice temperature. Thus, the potential for producing a clear geomorphological signature varies directly with plateau size (thicker warmer ice) and inversely with plateau altitude above the regional firn line (colder protective ice). In polar regions even large icefields are unlikely to reach the PMP. If present, plateau icefields may contribute mass to valley glaciers, by direct connection or avalanching, thus having an impact upon climate reconstructions based on glacier equilibrium line altitudes (ELAs).

20.2 Plateau icefield landsystem

Ice sheet imprints may be recorded as major landscape features such as overdeepened/widened and through-valleys transecting plateaux. If they exist, erosional forms on plateau tops will be aligned parallel/subparallel to main and through-valleys. Exotic erratics further demonstrate ice-sheet glaciation and ice-flow directions. Ice-sheet-produced land units and elements must be distinguished from the regional- to local-scale glaciation land units and elements, which comprise the plateau icefield landsystem shown in Fig. 20.1 (Rea & Evans, 2003) and discussed below.

20.2.1 Moraines

If plateau ice was cold-based, lateral and frontal moraines will be found in the valleys only, but if some plateau ice was warm-based, moraines may be found on top of, or leading onto, the plateau from the surrounding valley-head outlets. Valley moraines tend to be dominated by large, cobble- to boulder-size, angular material, indicating that the dominant debris source is rock fall. Where considerable thicknesses of valley floor sediments are available, outlet glaciers may construct large end-moraine sequences, especially where proglacial thrusting takes place. This appears to be most effective where retreat-driven isostatic rebound allows aggradation of permafrost, creating a 'potential' shallow décollement surface within the valley floor sediments.

20.2.2 Sediments

Sediments found on the plateau are most likely to be allochthonous or autochthonous blockfields/felsenmeer/detritus, and have been interpreted as the product of long-term weathering (Weichselian to pre-Pleistocene). On plateaux, tills tend not to be formed, owing to the dynamics of the ice cover. In the valleys thin, generally patchy tills are found and proglacial reworking of sediments by large braided river networks may occur. Lake sediments

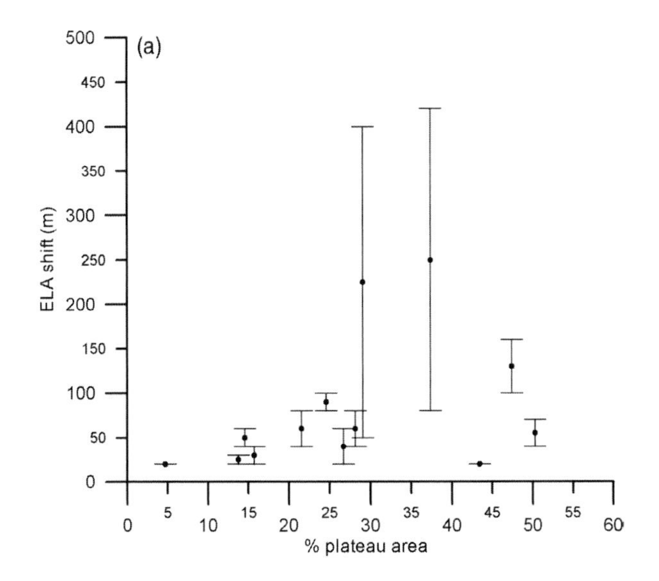

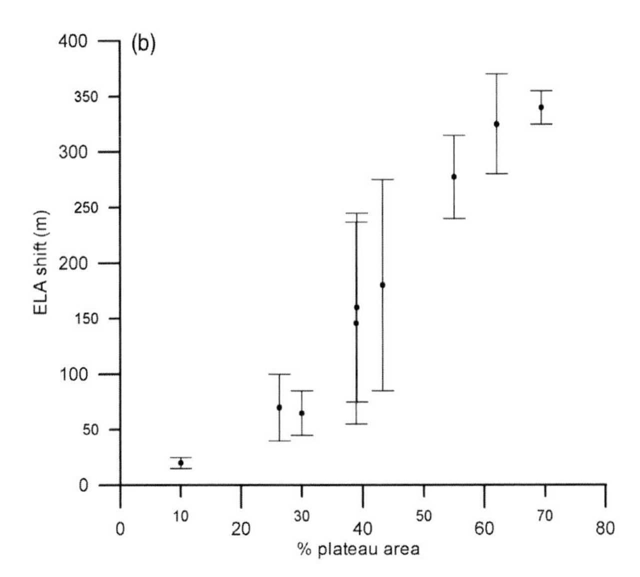

Figure 20.2 Equilibrium line altitude (ELA) shifts for the inclusion of contributing plateaux to coupled plateau–outlet-valley-glacier systems for (a) the Little Ice Age maximum in the southern Lyngen Peninsula, North Norway, and (b) the Younger Dryas reconstruction of Øksfjordjøkelen, North Norway. Equilibrium line altitudes were calculated using the accumulation area ratio (AAR) method. Bars indicate the limits of the shift for AARs of 0.5 and 0.8.

may be deposited in ice or moraine-dammed lakes and in overdeepened basins. When outlet glaciers terminate in standing water, ice-contact deltas and/or grounding-line fans indicate ice-marginal locations (Evans *et al.*, 2002).

20.2.3 Meltwater channels

For cold-ice cover, if sufficient meltwater is produced, lateral meltwater channels will typically mark the retreat of outlet glaciers onto plateaux. In higher relief landscapes, where plateaux tend to be smaller/more dissected, steep valley-head outlets, direct calving of plateau edges and low melt volumes tend to preclude the cutting of lateral meltwater channels. Larger (lower) icefields tend to have outlets with basal ice at the PMP, producing subglacial meltwater and allowing surface meltwater to percolate to the bed. The subglacial bedrock topography and ice-surface slope act to channel the meltwater towards the outlets, forming channelized subglacial and ice-marginal meltwater streams. If the water drains directly off the plateau edge steep alluvial fans will form at the base of plateau edge gullies.

20.2.4 Glacial erosion

Thermal regime controls the erosion potential of the ice. Plateau surfaces above valley-head outlets, cleared of blockfield and with striae-adorned bedrock, indicate warm-based plateau ice cover. More extensive blockfield stripping, bedrock abrasion and roche moutonnée formation indicates ice cover with a larger and more persistent warm-based component. Extensive evidence of erosion may fingerprint the provenance of ice as local (erosional elements indicating radial ice flow), or ice sheet (erosional elements are parallel/subparallel across plateaux). In the valleys erosional elements, for example trimlines, abraded bedrock, roches moutonnées, may be present.

20.2.5 Bedrock weathering zones

Provided ice retreat was 'slow enough', for example punctuated by periodic still stands, it may be possible to identify a progressively 'younging' zonation towards the plateau centre, using lichen and/or rock hardness proxy dating (Gellatly *et al.*, 1988) or cosmogenic exposure dating.

20.2.6 Erratics

Erratics from outwith the plateaux margins indicate ice cover and fingerprint the ice provenance as ice sheet rather than local. Local, within-plateau erratic transport must be treated with caution as it may indicate either ice sheet or plateau-centred ice cover. Only in favourable situations would the erratic dispersal pattern indicate a radial drainage pattern and thus local ice cover. It is more likely that the orientation of erratic trains have to be compared to other ice flow indicators in order to establish the ice source.

20.2.7 Plateaux with no evidence of ice cover

No land elements appearing on a plateau poses the question, 'cold-based ice, or, no ice?', the answer being very important for glacier–climate reconstructions. If land elements and units of the plateau icefield landsystem are found in outlet valleys, an area altitude relationship can be used to evaluate if plateau ice existed (Rea *et al.*, 1998, 1999). If it did, then it must be factored into any ELA reconstruction for a coupled plateau-icefield–outlet-glacier system.

20.3 Impact of plateau-ice contribution on reconstructed ELAs

Ignoring plateau icefield contributions to coupled plateau-icefield–outlet-glacier systems can seriously affect ELA reconstruction. Figure 20.2 plots the plateau area as a percentage of the coupled plateau-icefield–outlet-glacier system against the altitudinal shift in ELA resulting from inclusion of the plateau icefield area. Figure 20.2a shows results from the reconstructed LIA maximum glaciers in the southern Lyngen Peninsula, Norway, where ice supply from icefield to valley was mainly by avalanching. The ELA shifts shown in Fig. 20.2a reduced the equilibrium winter accumulation rates by up to 20% (Rea *et al.*, 1998). Figure 20.2b presents evidence from the Younger Dryas reconstruction of Øksfjordjøkelen, North Norway. The plateau contributing areas are larger, resulting in greater ELA shifts. It is obvious from Fig. 20.2 that climate reconstructions ignoring plateau icefield contributions become increasingly erroneous as the plateau contribution increases.

PART 2

Glaciers, oceans, atmosphere and climate

TWENTY-ONE

Glaciers, oceans, atmosphere and climate

John T. Andrews

INSTAAR and Department of Geological Sciences, Box 450, University of Colorado, Boulder, CO 80309, USA

I have yet to see any problem, however complicated, which, when you look at it in the right way, did not become still more complicated. Poul William Anderson

21.1 Introduction

This section of the volume deals with 'Glaciers, oceans, atmosphere and climate'—a truly daunting task to attempt any synthesis of such a large and complex subject. Nevertheless, in the past several decades, and largely post-World War II, many in the international earth science research community have been intrigued by the various feed-backs that link glacier variations (mainly ice sheets from a global mass (water) balance perspective) to various forcing mechanisms that exist within the Global Climate System (Fig. 21.1). It is worth reiterating that the focus of this chapter, and indeed of this volume, is the 'glacier' part of the system; hence variations in the oceans, atmospheres and biosphere that occur on 'climate' time-scales (>10^1 yr) are of prime interest. 'Changes in climate' are also caused by continental-scale tectonics that have changed the geography of the planet (Ruddiman & Kutzbach, 1991), and even the composition of the atmosphere (Royer *et al.*, 2004), on 10^6 yr time-scales. Some even call upon variations in cosmic ray flux as a cause of Phanerozoic climate change (Shaviv & Veizer, 2003), although this argument is disputed (Rahmstorf *et al.*, 2004).

In addition to the conventional view that the atmosphere and ocean boundaries are important controls on glacier response there is now increased recognition that the 'climate' at the bed of the ice sheet or glacier is also fundamental in controlling or explaining ice-sheet dynamics. Examples of such a control are the North Atlantic Heinrich events (MacAyeal, 1993a; Alley & MacAyeal, 1994; Clarke *et al.*, 1999), although there are arguments as to how far these massive glacial outbursts represent ice-sheet instabilities and drive climate or, alternatively, are a response to climate. The critical questions are: what controls the basal thermal regime of ice bodies and over what time-scales can changes in these basal conditions be propagated? Boulton (Boulton & Jones, 1979; Boulton, 1996b) amongst others examined this problem. The temperature at the base of a glacier/ice sheet is controlled primarily by the mean air temperature (MAT°C) at the surface of the ice mass, ice thickness, ice advection and the geothermal heat flux. Only in those rare cases (e.g. Iceland) where ice lies over mantle plumes is the geothermal heat flux highly variable. For areas such as the Laurentide Ice Sheet (LIS), or the British Ice Sheet, it is reasonable to consider the geothermal heat flux as constant. For many glaciologists it is difficult to understand how an atmospheric climate perturbation can be transmitted rapidly to the base of an ice stream given the poor conductive properties of ice (Drewry, 1986; Marshall *et al.*, 2000; Marshall & Clark, 2002). The central question is then over what time-scales does the basal temperature regime function and is it essentially decoupled from the atmospheric climate?

21.1.1 Scope and goals

The concept that glaciers and ice sheets have different responses to ocean and atmospheric climate, depending on a specific time-scale, is an appropriate but not the only way to organize this chapter. On the decadal to century scales, such as the Little Ice Age (LIA), then the changes in glacier extent are largely a function of changes in glacier mass balance, which in turn are linked with changes in ocean and atmosphere climates (Fig. 21.2). On the 10^4–10^5 yr scale (the Milankovitch or orbital time-scales), however, changes in glacier or ice-sheet extent are influenced by other, non-climatic factors, such as glacial isostatic depression of the crust, which can have a major effect on ice-sheet mass balance, especially if they terminate in the oceans. Furthermore, with new data on the changes in the greenhouse gases CO_2 and CH_4 during several glacial cycles (Ruddiman, 2003a,b) then it is imperative that the role of the biosphere be an integral part of the total system (Fig. 21.1). Indeed, there are some who argue that variations in CO_2 are either the primary driver in climatic change (Royer *et al.*, 2004) or an important amplifier (Shackleton, 2000).

The crux of the issue that I address is what 'climatic' information can be drawn from the land and marine stratigraphical records that archive histories of ice-sheet advance and retreat, and what are the feed-backs between the ice sheets and the rest of the

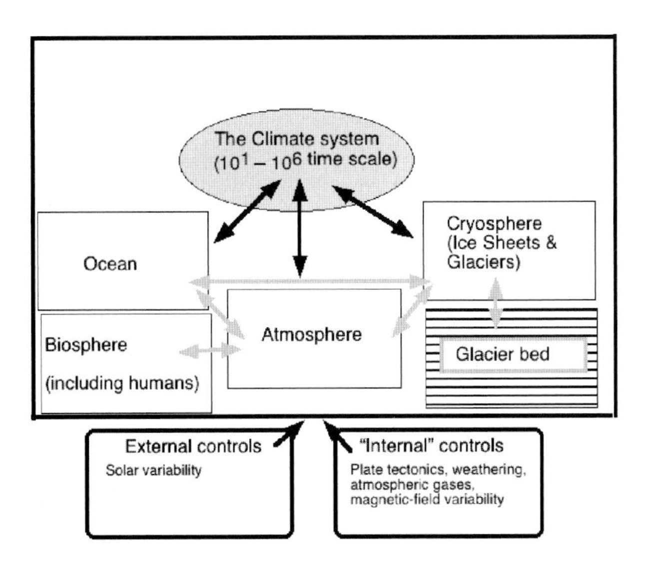

Figure 21.1 The global climate system as it pertains to glacier responses to and with climatic change.

Earth's climate system? The term 'climate' is usually interpreted to represent long-term, integrated records of temperature, precipitation, cloudiness, etc. In reality most climatic interpretations of glacial responses are couched in terms of temperature changes, although in reality the ice body responds to changes in mass balance, the product of both temperature and precipitation (Meier, 1965). Modelling the response of ice sheets to various climatic scenarios (see Marshall, this volume, Chapter 32) has great promise to extend our knowledge of the relative importance of the various forcing functions as they vary temporally and spatially.

This chapter evaluates the evidence for recent and future glacier–climate relationships and interactions, and then examines data associated with the Little Ice Age, Holocene glacial intervals, Heinrich and Dansgaard–Oeschger oscillations, and some comments on issues associated with the orbital forcing of global glacial cycles. Earlier surveys of the topic include the papers by Meier (1965) and Porter (1981), and the various editions of Paterson's book are an invaluable source of information (Paterson, 1969, 1981, 1994). A concise review of the issues involved in modelling the dynamics of ice sheets is given by Hindmarsh (1993). A conscious effort is made to include a broad temporal range of references.

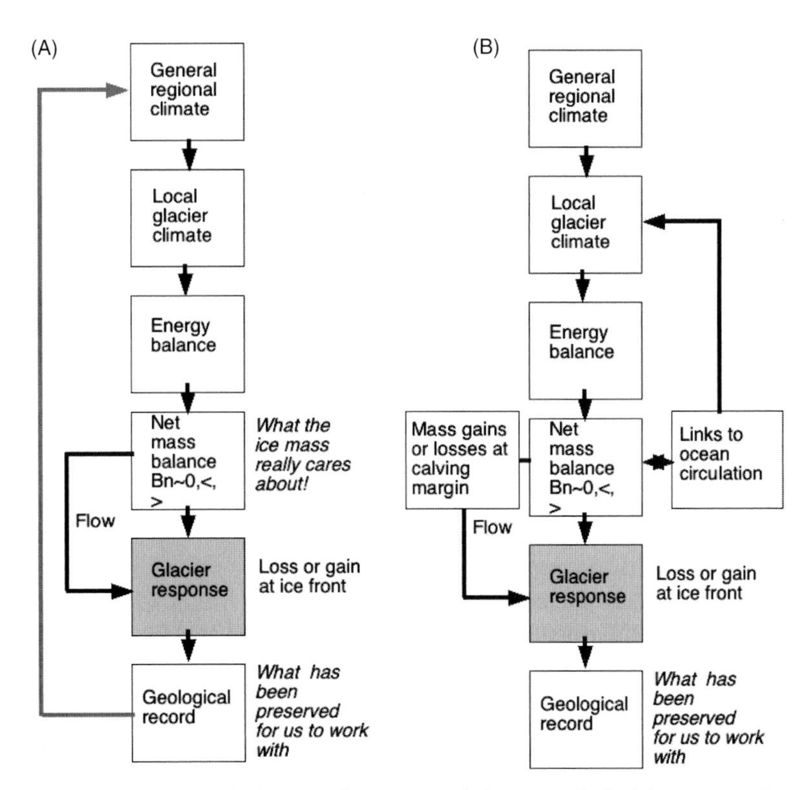

Figure 21.2 Relationship between regional climate, glacier mass balance and glacial response for terrestrial and tidewater margins. The solid grey line that leads from the evidence of a glacial event/response back to the climate box is the usual intellectual pathway but this ignores the intermediate elements in the climate/glacier interactions. (From Meier, 1965; Andrews, 1975.)

21.2 Glaciers and climate—how simple is the relationship?

The cryosphere is an integral part of the climate system, thus emphasis is rightly placed on the response of ice bodies to changes in the biosphere, atmosphere and ocean climate. There is also a positive feed-back (Fig. 21.1) between the extent, thickness and spatial geometry of the large ice caps and ice sheets, especially on the atmospheric circulation and the waveforms in the upper atmospheric circulation (Felzer *et al.*, 1994). At the simplest level, a change in the extent of snow and ice over an area of the LIS (ca. $12 \times 10^6\,km^2$) would have an impact on the hemispheric surface reflectivity and on the net radiation budget, whereas changes in the height of ice sheets have the possibility of affecting the geometry of the upper atmospheric waves. Global circulation models (GCMs) are being used to evaluate the importance of ice sheets on the climate of the planet during glacial intervals (Felzer *et al.*, 1994; Pollard & Ingersoll, 1980; Calov *et al.*, 2002).

Another important feed-back is the impact of glacial meltwater on the global ocean circulation (Broecker & Denton, 1989; Broecker, 1997). The outburst of large volumes of glacial meltwater from the retreating LIS into the North Atlantic (Teller, 1995b; Barber *et al.*, 1999) is theorized to curtail the thermohaline circulation (THC) and trigger abrupt cold intervals. Less attention, however, has been given to the impact of freshwater during intervals of glacial build-up. One would be forced to assume that as the ice sheets grew across the Northern Hemisphere the amount of freshwater reaching the Arctic Ocean (presently around $1700\,km^3\,yr^{-1}$) would be reduced, as would the freshwater entering Hudson Bay and the Labrador Sea. In addition, as global ice volume increased then the inflow of relatively fresh Pacific Water into the Arctic Ocean would be reduced and would cease when sea level fell to about −50 m (Elias *et al.*, 1996). However, the global atmospheric and ocean circulation changes associated with the build-up of ice sheets are not greatly commented upon (Johnson, 1997).

Glacial geologists and geomorphologists frequently equate the retreat or advance of ice bodies with specific climatic parameters, especially summer temperature, and the notions of 'glacial intervals = cold and interglacial intervals = warm' are instinctive (Knight, 1999) interpretations, or so it appears. It is worth redrawing Meier's (1965) diagram (Fig. 21.2A), however, which shows the steps in the chain of processes between changes in climate and changes in the response (advance, retreat, or still-stand) of the glacier (Porter, 1981; Knight, 1999). This diagram shows a unidirectional cascade of relationships, but there are feed-back loops between the different elements. For example, the overall energy balance is partly related to position of the equilibrium line altitude (ELA), which is a function of the net mass balance. It is clear from Fig. 21.2 that the 'regional' climate is archived by an ice mass through a series of elements. First is the 'local' climate for that particular cirque or valley glacier, where elements such as orientation, direction related to dominant winter snow storms, shading from summer insolation, etc., come into play (Williams, 1975; Porter, 1981). The critical mass balances (b_n) affecting a glacier on a yearly or longer cycle are a product of the energy balance on the glacier's surface, where for the majority of the world's ice bodies (i.e. those with surface characteristics that are classified as either subpolar or polar) the critical issue in summer losses is not temperature *per se* but the balance of net radiation (Paterson, 1981).

The position of the glacier's frontal margin is controlled on a year-to-year basis by mass loss, the result of a particular summer's climate, and on the forward motion of the glacier, which represents the integrated flow response to events in the accumulation and ablation zones over some interval of time (Knight, 1999). The flow component has a time-constant that depends on the mass exchange of the glacier (which is largely a function of whether it is temperate, subpolar, or polar (Andrews, 1975; Knight, 1999) and the flow law, especially the temperature sensitive coefficient which varies by two orders of magnitude between ice at 0° and −20°C (Paterson, 1981). This parameter is balanced by the mass turnover of the glaciers—the accumulation at the ELA in polar glaciers is around 0.1 m whereas it can reach 10 m in some temperate areas. Thus the response of a glacier's margin to a change in climate is not immediate and will lag by years, decades or even centuries. The response of the margins of polar, subpolar and temperate glaciers to a given climatic perturbation might not be synchronous because of these factors.

Meier's diagram (Fig. 21.2A) of the links between climate and glacier response is designed primarily to reflect the realities of the situation for terrestrial mountain glaciers. The situation is dramatically different when glaciers terminate in marine or lacustrine basins (Meier & Post, 1987; Meier *et al.*, 1994) (Fig. 21.2B). There is now an important, non-climatic, control on the mass balance and resultant glacier response (Mercer, 1961; Brown *et al.*, 1982; Mann, 1986; Alley, 1991a). Most frequently the effect on glaciers or ice streams that terminate in fresh or seawater is an additional mass loss to the glacier system by the mechanical removal of icebergs by calving—it must be admitted that the controls on calving are not well known (Thomas, 1977; Warren, 1992; Kenneally & Hughes, 1995–96). However, in the right circumstances investigations in Antarctica have shown that mass can be gained by the freezing-on of seawater to the base of the ice shelf. In East Greenland, Syvitski *et al.* (1996) noted that the ocean dynamics in Kangerlussuaq Fjord (68°N) lead to the importation of relatively warm, salty water at depth to balance the surface freshwater outflow from the summer ablation. There is thus a positive feed-back mechanism between the atmospheric climate and the oceanic estuarine circulation (Fig. 21.2B). If, for example, cold summers resulted in reduced meltwater this would dampen the importation of warmer waters from the deep shelf trough (ultimately being fed by Atlantic Water in the Irminger Current (Malmberg, 1985; Azetsu-Scott & Tan, 1997)), which would then reduce basal melting at the ice front.

During intervals of positive mass balance, tidewater glaciers face the problem of advancing along fjords with water depths between 100 and 1000 m (Syvitski *et al.*, 1987). As the ice advances into deep water there is a tendency for calving to increase (Brown *et al.*, 1982) and thus there is a question as to how outlet glaciers advance along deep-water fjords. Observations in the temperate Alaskan fjords indicate that the advancing ice constructs a shoal (moraine) by transporting overrun fjord sediments to the ice front (Alley, 1991a). In subpolar and polar areas Andrews (1990) suggested an alternative mechanism based on the rate of calving being retarded by the formation of a siqussaq (a mélange of sea ice and icebergs) that extends down fjord and anchors on the sill. In all areas, however, once tidewater glacier retreat starts it can be irreversible unless there are changes in the three-dimensional geometry of the fjord (Mercer, 1961).

Terrestrial and marine studies by earth scientists focus on the sedimentary records for glacier response to climate (Fig. 21.2), and records need to be interpreted in ways that clearly identify what is known or what is assumed about depositional mechanisms (Porter, 1981). Take, for example, the issue of dating moraines from the temperate glacier forelands of Norway versus the subpolar areas of the Canadian Arctic. In the first case the moraines are built-up by sediments entrained in the lower traction zone with some additional materials supplied from rockfall. On an annual basis the margins retreat somewhat during the summer months when ablation is high and then readvanced during the winter when the forward motion is not counterbalanced by surface melting. When the summer ablation is much greater than the winter forward movement the moraines are isolated and lichens could colonize the surface boulders (Matthews, 1977). In contrast in the subpolar landscape, sediments are brought to the surface of the ice sheet, and a process of thermal protection and sediment redistribution produces ice-cored moraines (Boulton, 1972). In such an environment the linkages between moraine formation and 'climate' are even more attenuated than in the temperate glacier situation, and the established ages of the moraines are times of ice-core stabilization (Davis, 1985)—which lag the timing of the climate change by some decades or even centuries (Østrem, 1965).

The discussion of glaciers and climate (Figs 21.1 & 21.2) would not be complete without some mention of the role of surging glaciers and the forcing of glacier response by non-climatic controls. Surging glaciers and glaciers that end in marine or lake waters (Fig. 21.2B) can have responses that are out-of-phase with climate changes. Mercer (1961) showed that changes in the three-dimensional geometry of fjords determine where an ice margin can be stable or can be in retreat. Hillaire-Marcel *et al.* (1981) also argued that large moraines associated with the LIS can be formed in association with the marine limit because of the reduction in the ice sheet's mass balance as the margin moved from a calving-dominated regime to one where summer conditions forced mass loss (Andrews, 1987b). Surging glaciers occur today in most glacierized areas of the world (Meier & Post, 1969; *Canadian Journal of Earth Sciences*, 1969) and are typified by quasi-periodic periods of rapid flow following a longer interval of slow flow and build-up. Their contorted medial moraines identify them but it is unclear whether past surges leave any distinctive signature in terms of depositional architecture or characteristic sediments. Surging glaciers respond to some self-triggering mechanism(s) and the relationship to climate and changes in mass balance is thus tenuous.

An important, but climatically ambiguous element in the reconstructing of past climates is the determination of the elevation of former ELA (sometimes erroneously referred to as the 'snowline') (Østrem, 1964; Andrews & Miller, 1972; Porter, 1981; Furbish & Andrews, 1984). On past and present mountain glaciers the ELA can be defined as the upper limit of lateral moraines of a particular age. Changes in elevation of the ELA through time have been measured by a number of different techniques (Furbish & Andrews, 1984). Whereas the ELA is glacier-specific, the delimitation of the glaciation threshold (GT) or limit is a method for determining the regional climatic controls on the distribution of ice bodies (Porter, 1977; Williams, 1978b; Østrem *et al.*, 1981). In most cases, however, changes in the elevation of an ELA or GT are expressed as a change in temperature; this approach does not attach any importance to the effect of winter accumulation on a glacier's mass balance.

There are relatively few efforts to link glaciological indexes of climate, such as changes in b_n, in a regional synthesis. One fundamental question is: how far do the changes in ELA or mass balance from a single glacier correlate with glaciers in a 10, 50, 100, 1000 km or more radius (Dugdale, 1972; Cogley *et al.*, 1995)? With the interest in the recent increase of temperatures on small glaciers this has become an important issue (Dyurgerov & Dwyer, 2001; Dyurgerov & Meier, 2000). Cogley *et al.* (1995) presented an important synthesis (and data set, their appendix F) of existing northern high-latitude glacier mass balance data and the degree to which the 50 or so glaciers showed similar mass balance trends since ca. AD 1960. Their focus was on the long glacier–climate records from the White Glacier, Axel Heiberg, Arctic Canada (Fig. 21.3). Although the lengths of the individual data sets are varied an important finding (their fig. 4.7) is that glaciers more than 500–700 km apart show little temporal correlation in their mass balances. This suggests some spatial limit to the 'regional climate' in Fig. 21.2.

Diagrams such as Figs 21.1 & 21.2 are useful conceptual tools but in terms of glacial history these diagrams need to be adjusted for changes in the time domain. Thus the next endeavour is to consider the glacier climate system at successively longer time-scales, starting with multidecadal to century scale oscillations, such as the Little Ice Age, to millennial-scale features such as Heinrich (H-) (Heinrich, 1988), Dansgaard–Oeschger (D–O) (Johnsen *et al.*, 1992), and Younger Dryas-like events (Peteet, 1995), and finally to orbital-scale Milankovitch events (Hays *et al.*, 1976; Weertman, 1976; Imbrie *et al.*, 1992; Ruddiman, 2003b).

A repetitive question through all temporal scales of glacier–climate relationships is what do we mean by synchrony in response? Are correlations between glacial events synchronous if they differ by ±10, ±50 and ±100 yr? Although this question sometimes gets involved in semantics there has to be some reason or rule for correlating a glacial event within a region, hemisphere or globally. Thus the designation of glacial events in New Zealand as 'Younger Dryas' (Denton & Hendy, 1994) raises a question because the type events in Norway and Scotland are 400–600 yr younger. This issue becomes even more important with the current interest in 'see-saw' climate systems, whether it is the 'Greenland above/Europe below' model operating at North Atlantic Oscillation (NAO) time-scales (multiyearly) (Rogers & van Loon, 1979) (Fig. 21.3) or the bipolar oscillation (Broecker, 1998; Rind *et al.*, 2001b), which operates on longer (millennial) scales.

It is important, especially for researchers starting in the field, to realize that the human species is endowed with both a facility and need for 'pattern recognition'. The reader of the literature on the 'correlation of glacial events,' on all spatial and temporal scales, will invariably find that the paper in question does indeed

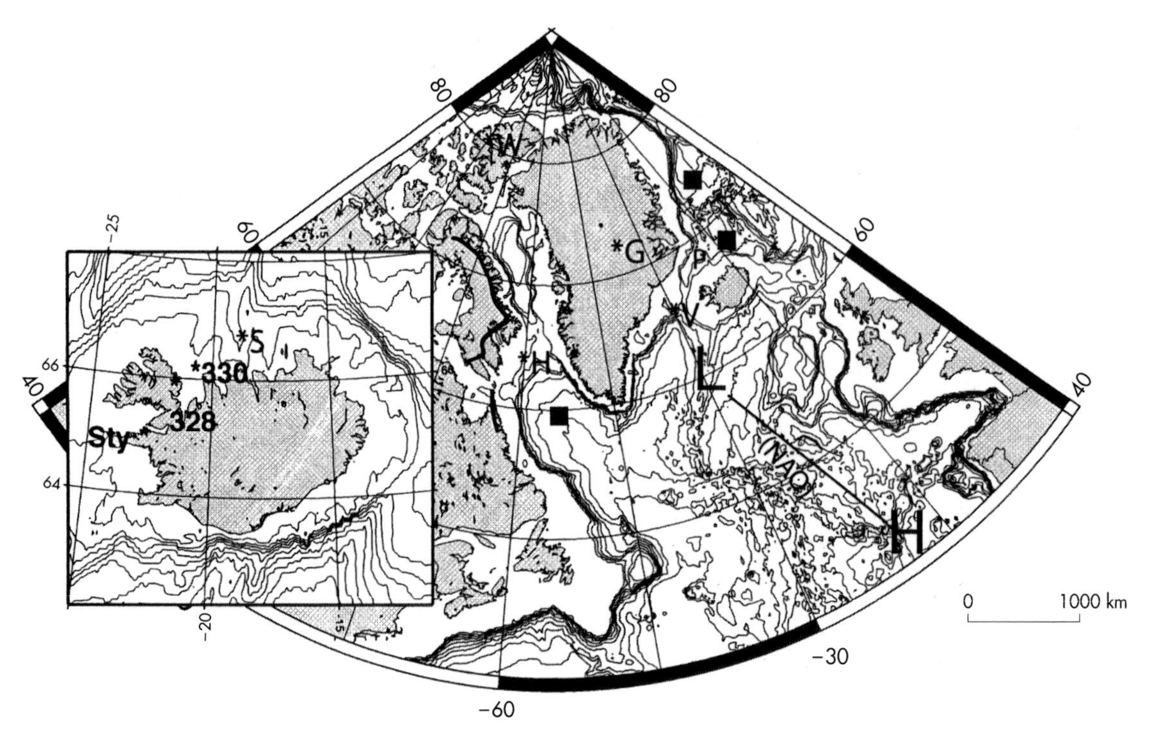

Figure 21.3 Map showing North Atlantic sites (Iceland as an insert—Siglunes = *S, Stykkisholmur (Sty) B997-328 and -330): W, White Glacier; *H, core HU87-009; *V, core V28–14; *G, Greenland Summit ice cores; P, core PS2644; solid black line = Cockburn Moraines; black squares = convection sites; L and H, the general locations of the Iceland Low Pressure system and the Azores High Pressure cell, which define the North Atlantic Oscillation.

find a correlation between those records and others in the literature! Very, very rarely does a researcher go out of his or her way to argue that they have a glacial record that has no correlative! The term 'correlative' is of course the key. At what time-resolution, given the complexities of the glacier/ice sheet's response to the integrated climate system, are glacial responses correlative? There is no simple answer to this question that I am aware of, but I suspect that many correlations of glacial events are forced into an existing paradigm.

21.3 Glacier–climate interactions at different time-scales

In the sections below I will address what we know of these interactions on different time-scales. As we will see the actual 'glacier/ice sheet' relationships to climate, through mass balance (e.g. Figs 21.2 & 21.4), become increasingly tenuous so that in many cases the 'glacier response' is interpreted by proxies that do not delimit a specific glacier or ice sheet, such as the $\delta^{18}O$ of benthic foraminifera, which records an integrated global ice volume signal (Shackleton & Opdyke, 1973). It is necessary to re-emphasize that the mass balance is the product of a winter accumulation minus summer mass losses. Hence, although the literature on past glacial changes tends to focus on inferred temperature changes this should not hide the fact that changes in accumulation must also be taken into account.

In a similar vein, the use of ice-rafted debris (IRD) as an indicator of climatic change during the Holocene in marine sediments is increasingly being used, although there are significant caveats as to what interpretations can be placed on an increase in a particular sand-size fraction (Andrews, 2000). A first question is whether the IRD event represents the transport and deposition of sand-size particles from icebergs or from material entrained in sea-ice? If IRD increases at a site it may be that icebergs have changed their tracks or that the entrainment of sediment in outlet glaciers has changed (Warren, 1992). It is usually assumed that the only reason for changes in IRD is a change in the iceberg flux and that this is related to climate. As noted earlier, however, the response of tidewater glaciers to climate is not simple and involves glacier dynamics (Mann, 1986).

In the examples below my own biases are evident with a focus on the area from Iceland to the Canadian mainland. Iceland is an important area within the North Atlantic as it is strongly influenced by changes in marine conditions such as the Great Salinity Anomaly (GSA) (Dickson *et al.*, 1988), including the extent of sea-ice and changes in the thermohaline circulation (Malmberg & Jonsson, 1997; Jónsson, 1992). Iceland's climate also reflects the atmospheric NAO (Hurrell, 1995; Rodwell *et al.*, 1999), as one end member of the NAO is the strength of the Iceland Low.

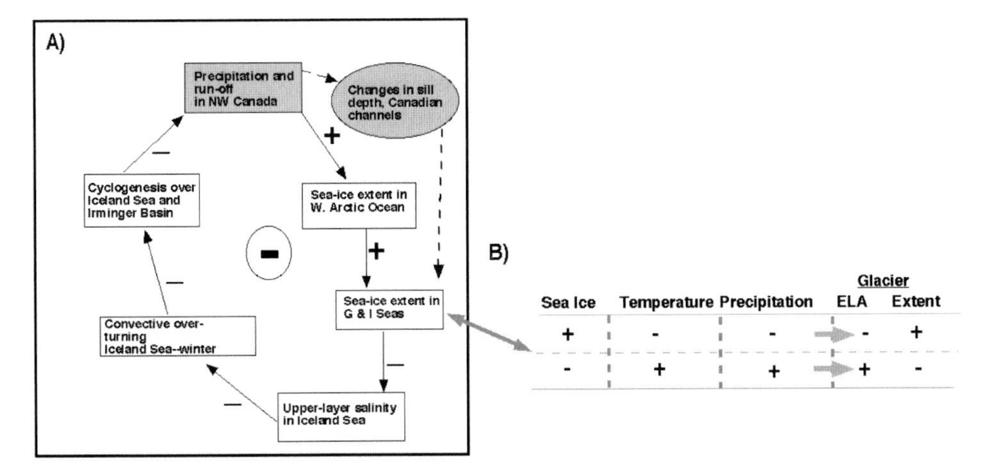

Figure 21.4 (A) Feed-back loops associated with the effect of increased runoff and freshwater export from the Arctic Ocean (from Mysak & Power, 1992). (B) Feed-backs between ice extent, climate, and glacier response in north Iceland (Stotter *et al.*, 1999).

21.3.1 Present and future glacier–climate interactions

In the past decade there has been a tremendous increase in the interest in the glacier–climate relationships because of the link to the measured rise of sea level (Meier, 1984; Arendt *et al.*, 2002; Meier & Dyurgerov, 2002). Several studies have shown that mid-latitude glaciers receded extremely rapidly during the past two to three decades, a period when concerns for the anthropogenic influence on climate became a scientific focus (Mann, 2000; King, 2004). The link between climate, glacier recession and changes in sea level is strong, but the effect of changes of climate on the mass balance of the large ice sheets, Greenland and Antarctica, is difficult to gauge because of their sheer size (Reeh, 1985). If, as it appears so, global warming consists of an increase in winter temperatures then a corollary is that winter accumulation of snow should increase at high latitudes and the Greenland or Antarctic ice sheets thus might have a positive mass balance (Zwally *et al.*, 1989; Krabill *et al.*, 2000; Thomas *et al.*, 2000; Joughin & Tulaczyk, 2002).

In an analysis of Northern Hemisphere mass balance data from AD 1960 to 1991, Cogley *et al.* (1995) concluded that it was 95% certain that 16 glaciers were probably shrinking, 12 glaciers were in approximate balance, and only 3 glaciers were 'growing'. An additional nine glaciers were difficult to categorize. Dyurgerov (2002), in a compilation of current mass balance data, demonstrates that the winter balance of glaciers has been increasing at an average rate of $1.7\,m\,yr^{-1}$, which is approximately balanced by the increased rate of meltwater production. In addition, the large Russian rivers that drain to the Arctic Ocean, and contribute some $1400\,km^3\,yr^{-1}$ of freshwater to this ocean, have shown an increase in discharge over the past two to three decades (Peterson *et al.*, 2002). This freshwater is exported from the Arctic Ocean as sea ice (Aagaard & Carmack, 1989), and has influenced the rate of convection (i.e. the thermohaline circulation) in the Arctic Ocean (Meincke *et al.*, 1997) and Nordic Seas, as well as overall climate (Smith *et al.*, 2003).

The notion of a warming world, especially at high north latitudes, carries with it the seeds of a significant negative feed-back loop associated with the increased export of freshwater into the Nordic Seas, largely as sea ice (Smith *et al.*, 2003), and its impact on the thermohaline circulation (Meincke *et al.*, 1997; Rahmstorf, 2000; Fichefet *et al.*, 2003) (Figs 21.3 & 21.4A). Several authors have warned of the potential devastating impact of export of this freshwater on the production of North Atlantic Deep Water (NADW) (Fichefet *et al.*, 2003). The shutting down of the ocean conveyor system would result in a major decrease in temperatures over northwest Europe associated with the reduction in the flux of the North Atlantic Drift (Aagard *et al.*, 1985; Broecker, 1997; Rahmstorf & Ganopolski, 1999; Delworth & Dixon, 2001). A foretaste of such an event occurred in the late 1960s associated with the GSA (Dickson *et al.*, 1988; Rogers *et al.*, 1998) (Fig. 21.4A).

If the climate is warming, glaciers melting and sea level is rising, then the concerns about the collapse of the 'marine based' West Antarctic Ice Sheet, which were an ongoing topic of debate starting in the 1960s and 1970s (MacAyeal, 1992a; Mercer, 1978; Hughes, 1992), are more than doom-day scenarios. In a later section I will show that the marine based LIS suffered frequent (in a geological sense) collapses during the last glacial cycle.

In a regional example, studies of changes in the length of 27 Icelandic non-surging glaciers (Sigurdsson & Jonsson, 1995) show that '. . . the advance/retreat records of non-surging glaciers show a clear relationship to climate.' About half the glaciers have been readvancing after a cool interval that started in the 1940s, with a substantial change in the mid-1960s. This analysis of the Iceland glacier data indicates that there is only a 10-yr lag between a climate shift and the terminus showing a response.

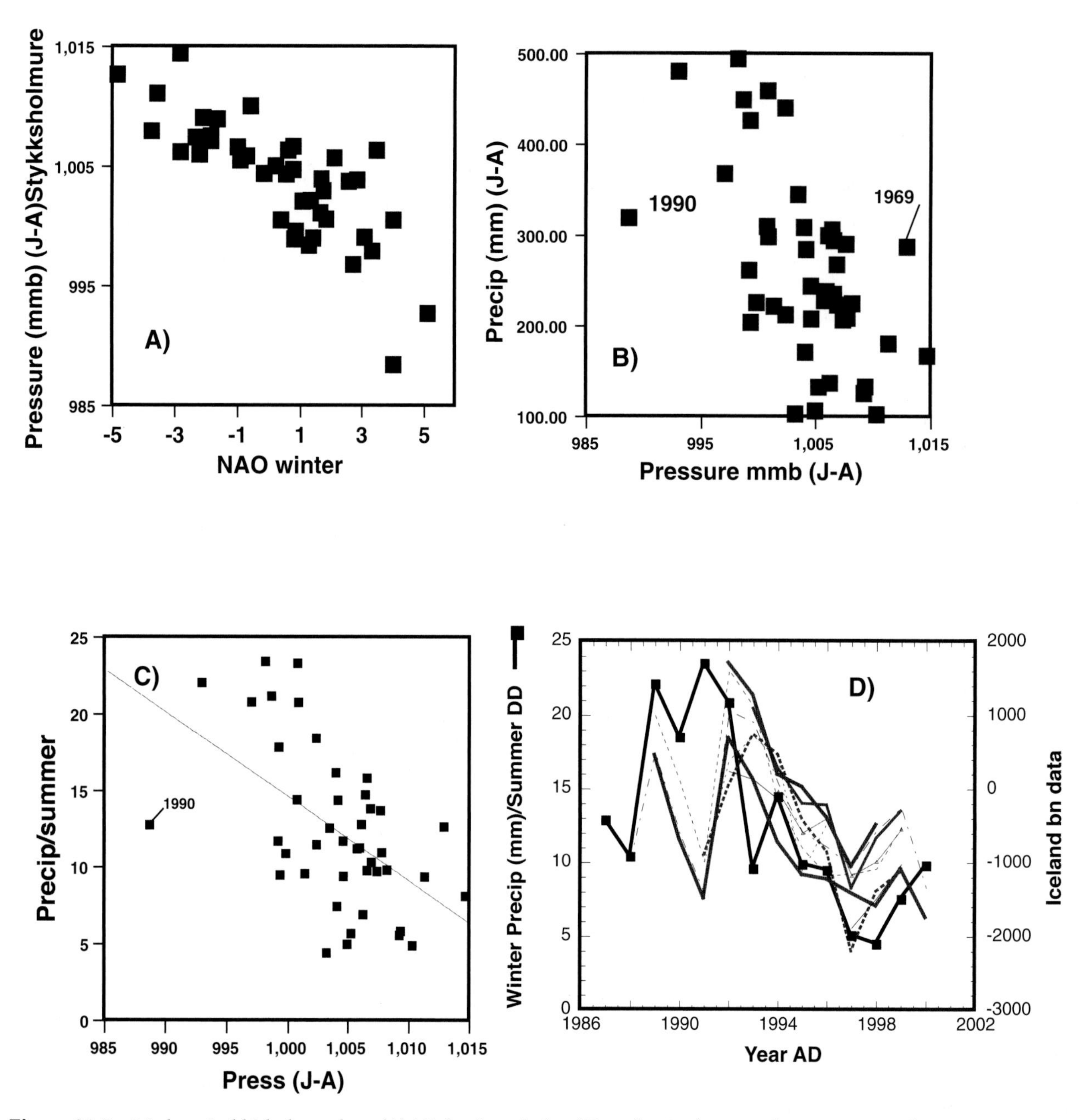

Figure 21.5 Modern Stykkisholmur data. (A) NAO winter index (Hurrell, 1995) versus the January–April average pressure at Stykkisholmur (Fig. 21.3). (B) Winter pressure versus winter precipitation (mm) at Stykkisholmur. (C) 'Net mass balance index' (winter precipitation (mm)/summer degree-days versus Stykkisholmur winter pressure). (D) The 'net mass balance index' (C) versus net mass balance data from Icelandic glaciers (Dyurgerov, 2002) for the period AD 1987–2002.

How far can weather station data predict regional glacier variations? Monthly climatic data (temperature, precipitation, pressure) from Stykkisholmur in northwest Iceland (from www.vedur.is) (Fig. 21.3) from AD 1960 to 2002 were compiled for the winter accumulation season (taken as January to April) and the monthly degree-days for the summer ablation season (June, July, August). Barlow (2001) discussed the agreement between this temperature time-series and central Greenland ice-core isotope data and showed a 64% decadal-level agreement in the sign of the signal for the period AD 1830–1980. The explained variance between the average winter pressure values at Stykkisholmur and the NAO winter index (Fig. 21.3) (Hurrell, 1995; Rodwell *et al.*, 1999) is $r^2 = 0.61$ indicating that the pressure variations at Stykkisholmur are closely coupled, as expected, to NAO variations (Fig. 21.5A). The winter pressure variations correlate with changes in the winter precipitation (Fig. 21.5B), with intervals of low pressure tending to be winters with higher accumulation. The data show large variations in precipitation (261 ±

105 mm) whereas summer degree-days have little variability (21 ± 2 degree-days). A simple index for the mass balance of glaciers in the region can be computed as: total winter precipitation/summer monthly degree-days (units mm/degree-day). This index is strongly driven by the variations in winter accumulation and mirrors the winter NAO (Fig. 21.5C). Sigurdsson & Jonsson (1995) used a similar but more complex approach to compute from climatic data a 'snow-budget' index for a glacier in southern Iceland. They show that a change in both the index and termini variations occurred around AD 1970, the time of the Great Salinity Anomaly off north Iceland (Fig. 21.4A) (Malmberg, 1985; Olafsson, 1999). It coincided with an extreme negative NAO index (Fig. 21.5C) and indicates a coupling between the oceans and atmosphere, although this coupling is not simple (Rogers *et al.*, 1998; Olafsson, 1999).

There are few long-term glacier mass balance observations from Iceland (Dyurgerov, 2002), although there is a long history of measuring the movement of termini (Sigurdsson, 1991; Sigurdsson & Jonsson, 1995). When the Stykkisholmur 'mass balance index' (Fig. 21.5B) is examined against the net mass balance data from nine glaciers starting in AD 1987 (Dyurgerov, 2002) there is a clear parallelism in the records, with a rapid increase in negative mass balance starting in 1992 to 1998 with a small upturn since then (Fig. 21.5D). A Stykkisholmur index of ca. 14 appears to delimit the move from positive to negative mass balance on the glaciers surveyed. This agreement (Fig. 21.5D) suggests that changes in winter accumulation are the dominant element in the sign of the mass balance, because the variations in summer (J, J, A) temperatures are slight.

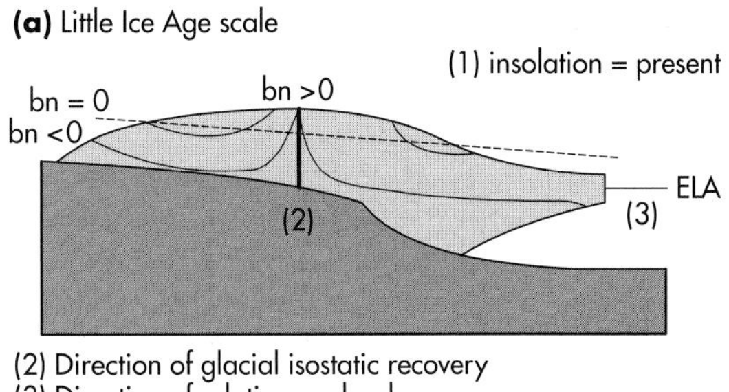

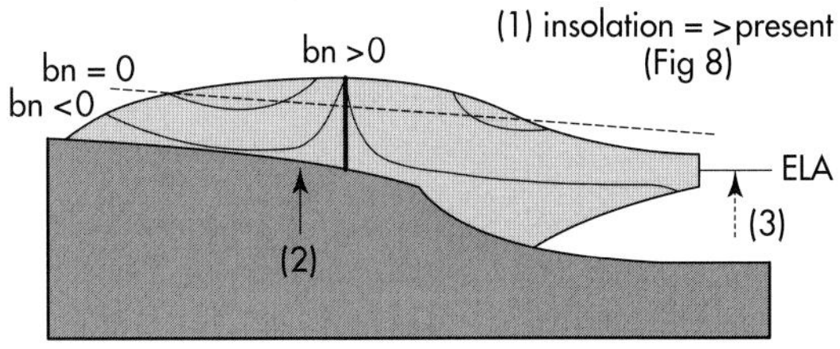

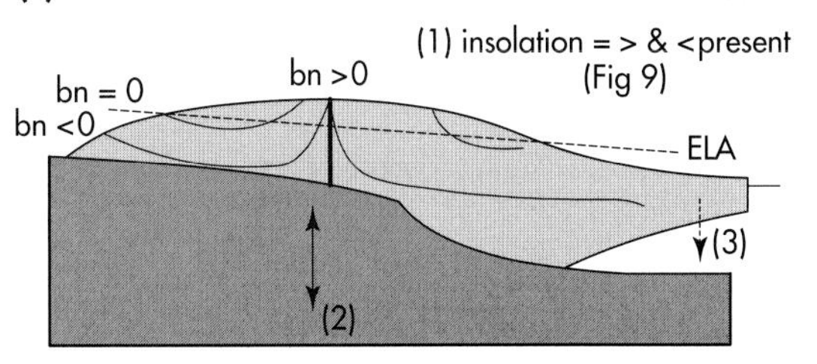

Figure 21.6 Sketch of ice sheet with terrestrial and marine margins and controls on (a) Little Ice Age (LIA) and (b) last glacial cycle.

21.3.2. Glacier–climate relationships: Little Ice Age (LIA) scale

'The term "Little Ice Age" relates to the behaviour of the glaciers not directly to climate' (Grove, 2001, p. 76; after Luckman). What was once a well accepted, if poorly understood, interval of Earth history is now in debate, as is its companion, the Medieval Warm Period (MWP) (Bradley & Jones, 1993; Hughes & Diaz, 1994; Broecker, 2001). Instrumental data for climatic variables do not exist prior to the 17th century, therefore the atmospheric and ocean climates (Figs 21.1 & 21.2) have to be inferred from a variety of proxy data. With few exceptions the evidence for glacial response in this time period is no longer couched in terms of mass or energy balances but usually in climatic terms (Fig. 21.2). Within the LIA period, or approximately the past 600 yr, the primary record of glacial response to climatic perturbations are end moraines that are located down-valley from present-day glaciers. It appears from the literature (Grove, 2001) that the LIA was underway by the 13th and 14th centuries. The boundary conditions for LIA glacier–climate systems are simpler than the response and interactions on longer time-scales (Fig. 21.6a versus 6c). On the longer time-scales, even during the Holocene, the numbers of tidewater glaciers and marine-based margins with their particular dynamics are more evident. On LIA time-scales the dynamics of change rest largely within the atmospheric climate system, and includes the oceans (Fig. 21.6a). On these time-scales tectonic movements, including glacial isostatic depression or recovery, are small, probably exist (Clark, 1977), but do not force a glacier response. Changes in relative sea level (RSL) are also small. We will consider the LIA history and neoglaciation of Iceland as representing a climate system that should integrate a substantial hemispheric signal because of its location in relationship to the NAO, its proximity to sites of deep sea convection in the Iceland and Greenland Seas (Fig. 21.3) (Broecker, 1997; Malmberg & Jonsson, 1997; Delworth & Dixon, 2001), and the known effect of sea ice on the climate (Fig. 21.4A & B) (Bjornsson, 1969; Chapman & Walsh, 1993; Deser *et al.*, 2000; Ogilvie & Jónsson, 2001). Barlow (2001) noted that the Greenland isotopic records for the period AD 1400–1980 might be negatively correlated with European temperature anomalies because of the climate 'see-saw' (Rogers & van Loon, 1979).

Interactions between glacial history and the surrounding land/oceans has been evaluated by Mysak & Power (1992) in terms of the feed-backs associated with the GSA of the late 1960s (Fig. 21.4A) and on a more regional scale by Stotter *et al.* (1999) for North Iceland (Fig. 21.4B). Although there is some dispute on the origin of the salinity anomaly in the Nordic Seas, what is not in dispute is that it probably originated 1000s of kilometres away within the borders of the Arctic Ocean, either as an excursion in the flow of the McKenzie River or by the increased transport of sea ice through Fran Strait associated with atmospheric (wind) processes (Mysak & Power, 1992; Serreze *et al.*, 1992). An additional forcing mechanism, and one not explicitly considered by these authors, is the influence of changes in solar activity on the climate, a topic which has received considerable attention or re-emphasis in the last decade or so (Harvey, 1980; Stuiver *et al.*, 1991; Bond *et al.*, 2001; Shindell *et al.*, 2001; Lean *et al.*, 2002).

The Mysak & Power's (1992) negative feed-back loop indicates that there is some level of self-regulation in at least part of the climate system (Fig. 21.4A). Hydrological changes on the Canadian Arctic margin are considered to propagate into the Greenland and Iceland Seas and cause a freshening of the surface waters, which reduces cyclogenesis, thus decreasing precipitation and reducing the freshwater anomaly. An additional box has been added to this model to interface with longer Holocene time-scales. On longer time-scales, changes in bathymetry of the key Canadian Arctic Channel sills are associated with glacial isostatic recovery and changes in the global ice volume (Figs 21.4A & 21.6b).

Changes in sea ice have an impact on the temperature and precipitation over north Iceland (Ogilvie, 1997) so that an increase in sea ice is associated with a decrease in temperature, a decrease in precipitation, a fall in the ELA, and an increase in glacier extent (Fig. 21.4B) (Stotter *et al.*, 1999). Although there are no long-term mass balance measurements on northern Iceland, the sense of this association is that the decrease in temperature, associated with the increase in sea ice off the coast, is more important than the decrease in precipitation (Fig. 21.4B). This assertion does not appear to explain the trends of the past 40 yr, however (Fig. 21.5D).

In Vestfirdir (northwest Iceland) there are numerous semi-permanent snowbanks and evidence for LIA and neoglacial moraine-building intervals (Eythorsson, 1935; Thorarinsson, 1953). The Dragnajokull Ice Cap lies on the Tertiary basalt surface with an ELA around 570 m asl. However, the hypsometry of Vestfirdir is such that an ELA lowering of only 100 m would increase the glacierization of the area by 50% (Principato, 2003). The ages of LIA moraines on Iceland indicate major intervals of moraine formation over the past three centuries (Grove, 2001; Wastl *et al.*, 2001) although there is also evidence for a glacial moraine deposition around 700 yr BP.

Figure 21.2 represents a conceptual view of the glacier–climate system. It is useful in terms of present-day interactions but how well can we move backwards and infer climate from the glacial record? Figure 21.7 illustrates some of the available data, which can place glacial response on Iceland into the context of hemispheric, regional and local changes of climate. It shows the estimated hemispheric temperature anomalies for the past 600 yr (Mann *et al.*, 1999), the changes in sea ice off Iceland since the Settlement in ca. AD 870 (Bergthorsson, 1969; Björnsson, 1969; Ogilvie, 1991, 1992; Ogilvie *et al.*, 2000), and inferred changes in seafloor temperature from a small North Iceland fjord (Andrews *et al.*, 2001; Castaneda *et al.*, 2004) (Figs 21.3 & 21.7B). Mann *et al.* (1999) (Fig. 21.7A) examined which climatic forcings were most important in determining climate change prior to human intervention (but see Ruddiman, 2003a). Variations in solar activity were most strongly associated with the 600 yr record (Lean *et al.*, 1995).

For the past 600 yr (Fig. 21.7A) the hemispheric picture is dominated by temperatures below the mean of the reference series. A pronounced cold interval appears around AD 1450 and other prolonged cooler intervals are centred around AD 1600, 1700 and between ca. 1800 and 1900. How far does climate data from Iceland mimic this record? The data from Stykkisholmur, northwest Iceland (Fig. 21.3), extends back to AD 1820. The winter (J, F, M) and summer (J, J, A) trends (Fig. 21.7C) indicate an interval of both cold winter and summer temperatures centred on AD 1860 followed by a rather dramatic increase in temperatures, led by winter values, starting ca. AD 1900. The correlation between winter and the subsequent summer temperatures is quite evident ($r^2 = 0.48$); the correlation between the Mann *et al.* (1999) series and Stykkisholmur is only $r^2 = 0.14$.

A reconstruction of changes in the Iceland Low pressure system has been presented on the basis of the Greenland ice-core data (Meeker & Maywekski, 2002). This reconstruction suggests that a fundamental mode change occurred around AD 1400 when overall high-pressure conditions changed abruptly to lower pressures, suggesting an intensification of the Iceland Low (Fig. 21.3). Based on the recent data from Stykkisholmur (Fig. 21.5) this might imply that, starting between AD 1400 and 1450, winter accumulation would have increased over Iceland.

In attempting to extend our knowledge of Iceland climate and glacial variations over the past millennium a critical question is how far changes in the offshore sea temperatures are correlated with temperatures on land, and with glacier response (Sigurdsson & Jonsson, 1995)? Iceland is only ca. 100,000 km^2 and studies of climatic data show a very strong correlation between weather stations across the island (Einarsson, 1991). The 50-yr time series from the Siglunes hydrographic CDT data, off north Iceland (Fig. 21.3) (Olafsson, 1999), shows a significant correlation with the Stykkisholmur MAT data ($r^2 = 0.55$) (Fig. 21.7D), hence it is reasonable to postulate that changes in the marine realm can be used to hindcast temperature conditions on land.

A proxy for marine temperatures in the late Holocene is the stable $\delta^{18}O$ of foraminifera from marine cores, as it has been shown (Smith *et al.*, 2005) that these primarily reflect changes in temperature. Figure 21.7B shows a reconstruction of sea-ice severity off Iceland versus the $\delta^{18}O$ record in core B997-328PC from the head of a small fjord in Vestfirdir (Fig. 21.3) (Castaneda *et al.*, 2004). The agreement is substantial since the Settlement ca. AD 870. There was a $\Delta\delta^{18}O$ change of ca. 0.6‰ between the MWP (Hughes & Diaz, 1994) and the LIA. This corresponds to a temperature change of ca.3°C, which is within the temperature variations caused by the GSA of the late 1960s when water-column temperatures fell 5°C (Malmberg, 1985). Adjustment in the chronology of B997-328, given the errors in ^{14}C dating, would easily result in an even stronger association. The $\delta^{18}O$ results indicate a prolonged interval of cold marine conditions between AD 1500 (500 yr BP) and ca. AD 1900.

21.3.3 LIA-like cycles in the Holocene?

As the search for forcing mechanisms to explain glacial history on Holocene time-scales intensifies there is a need to ask: are there

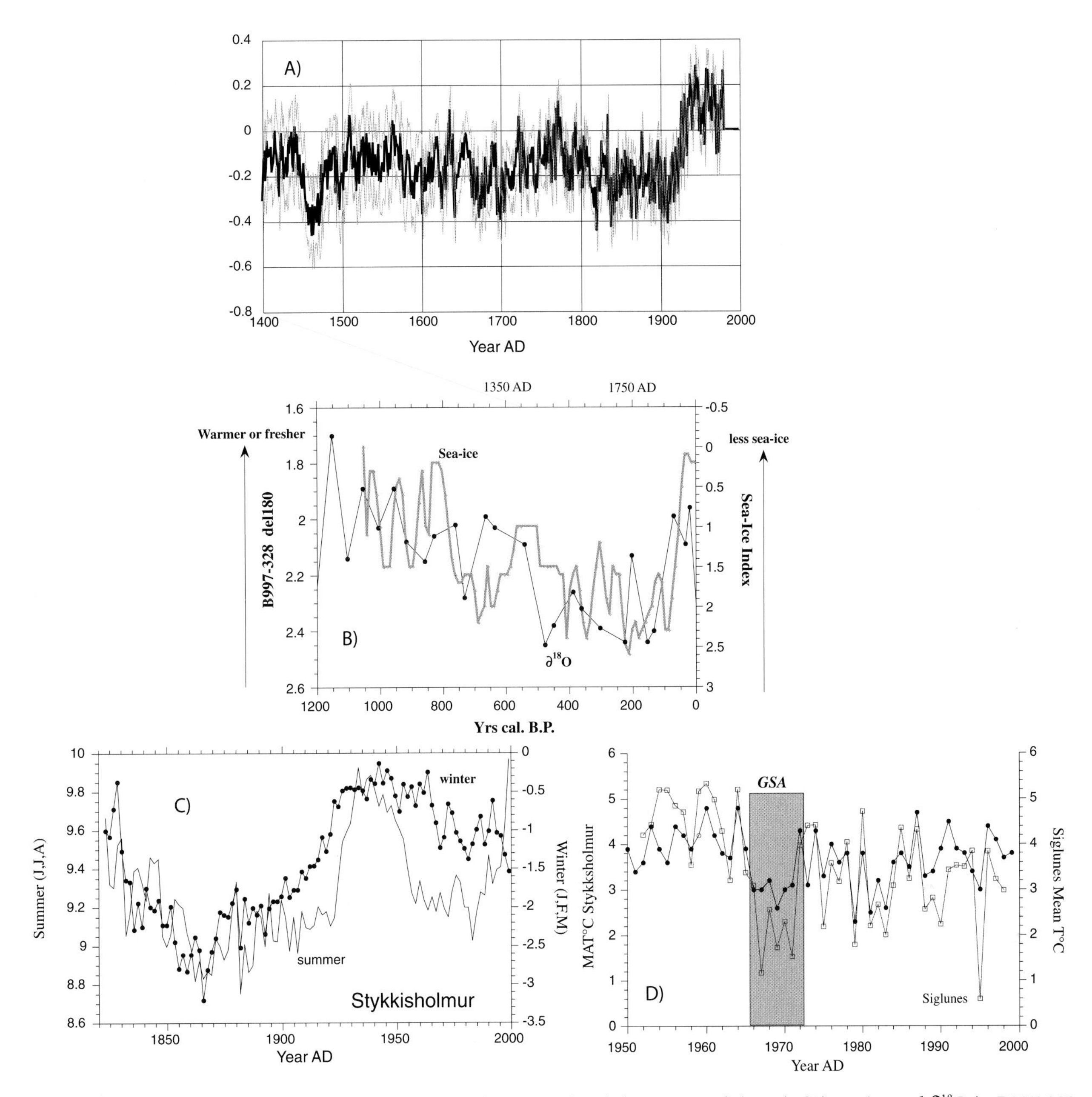

Figure 21.7 Data from Iceland on Little Ice Age (LIA) time-scales. (A) Mann *et al.* (1999), (B) sea ice and $\delta^{18}O$ in B997-328, (C) Stykkisholmur winter and summer temperature trends, (D) Siglunes water column temperature and Stykkisholmur mean annual temperature (MAT).

periodic LIA-like events within the Holocene (Bond *et al.*, 1999, 2001), and are there recent analogues for the LIA within the instrumental record? Lamb (1979), for example, suggested that the GSA of the late 1960s off Iceland and elsewhere in the North Atlantic (Dickson *et al.*, 1988; Belkin *et al.*, 1998) was an analogue for the climate of the LIA. This suggestion clearly advocates changes in the ocean, and specifically in sea-surface temperatures, as a potent force in regional climate changes and associated glacier response. Changes in ocean salinity (Fig. 21.4A) in key areas north of Iceland have the potential to disrupt the global THC (Jónsson, 1992; Clark *et al.*, 2002b) (Fig. 21.3; such changes are frequently called on to explain abrupt climate change on the H- and D–O (= longer) time-scales.

However, a persistent problem with the analogue approach to retrodicting past climatic conditions (e.g. Lamb, 1979) is the issue of 'persistence'. How does the climate system switch from decadal-scale oscillations, such as the NAO in either positive or negative mode, to series that have a century- or millennial-scale persistence?

Once we move in time over 1000 yr then the glacier–climate interactions become more complex to interpret because we enter a period when boundary conditions change, such as rates of glacial isostatic recovery, relative sea level or solar insolation, and the picture (Fig. 21.6b) becomes increasingly removed from the application of modern analogue conditions (Lamb, 1979). Summer insolation was higher than present at northern latitudes in the early to mid-Holocene (Berger & Loutre, 1991; Kutzbach *et al.*, 1996) but these conditions were opposed by the melting of the LIS and the export of freshwater to the North Atlantic (Giraudeau *et al.*, 2000; Teller *et al.*, 2002). For example, between 7 and 12 cal. kyr the retreating LIS (Dyke & Prest, 1987a; Dyke *et al.*, 2002) was sending copious volumes of meltwater into the North Atlantic via Gulf of Mexico, Hudson Strait or the Arctic Ocean (Teller, 1995a; Barber *et al.*, 1999; Fisher *et al.*, 2002; Teller *et al.*, 2002; Aharon, 2003; Nesje *et al.*, 2004). In addition, the depths of the Canadian Arctic channels, leading from the Arctic Ocean into Baffin Bay, were 100 m or more deeper due to glacial isostatic depression, hence the 'freshwater' outflow from the Arctic Ocean to the North Atlantic would be affected by this time-dependant relaxation (Williams *et al.*, 1995; Dyke & Peltier, 2000) (Fig. 21.4A). During the early Holocene in particular, as ice sheets retreated, large tidewater glaciers (Fig. 21.6b) lay within most fjords in the Northern Hemisphere and their response to changes in mass balance (see earlier) may not always be equated with climate.

The 'natural' climate of the early and mid-Holocene is now being questioned (Ruddiman, 2003a); comparisons between atmospheric greenhouse gas interglacial records contained in ice cores and the Holocene suggest that both CO_2 and methane records reflect human occupation. Thus the natural decrease in solar insolation (Fig. 21.8A) and the onset of neoglaciation might have been mitigated by the rise in greenhouse gases 5–8 cal. kyr (Ruddiman, 2003a). In effect, Ruddiman (2003a) argued that the severity of neoglaciation was reduced by anthropogenic changes to the climate system.

The 'climate' part of the glacier relationship (Fig. 21.2) for the past 12,000 yr is based on a variety of proxy data from ice, lake and marine cores. The 'glacier' response is often archived in a series of terminal moraines at a variety of distances from present-day glacier margins, changes in the past ELAs of glaciers (Porter, 1981), or variations in ice-rafted debris (IRD). Thus changes in sediment type, such as grain-size, in both lake (Leonard, 1986; Nesje *et al.*, 1991) and marine settings (Andrews *et al.*, 1997; Bond *et al.*, 1999; Andrews, 2000) are also used in this interval to deduce 'glacial' activity. In reality there is little specific 'climate' data that has been developed that applies to variations in glacier mass balances during the past 12,000 yr. The most direct estimates are those derived from ice cores where changes in accumulation (Meese *et al.*, 1994), stable isotopes of the precipitation (assumed to reflect largely air temperature) and borehole temperatures (Cuffey & Clow, 1997; Dahl-Jensen *et al.*, 1998) (which reflect a long-term integrated measure of air temperatures at ca. 3 km over central Greenland) have been derived.

Dating of glacial moraines in front of existing valley and cirque glaciers in the Northern Hemisphere led to the concept of 'neoglaciation' (Porter & Denton, 1967; Davis, 1985; Karlen, 1988; Karlen & Denton, 1976), an interval of renewed glacial expansion, starting 5–6 cal. kyr after a 'thermal optimum' (Kaufman *et al.*, 2004) (Fig. 21.8B). This was most frequently associated with the orbitally driven reduction in summer insolation at high northern latitudes (Berger & Loutre, 1991) (Fig. 21.8A). During this same interval there was a noticeable increase in IRD on the East Greenland margin, suggesting increased calving from Greenland tidewater outlet glaciers (Andrews *et al.*, 1997; Jennings *et al.*, 2002b). Inferences on the climate of the neoglacial interval over Iceland over the past 5–6 cal. kyr (Gudmundsson, 1997; Stotter *et al.*, 1999) are drawn from intervals of moraine formation (Fig. 21.8C), although these may underestimate the real number of glacier advances because of overriding of older moraine systems (Schomacker *et al.*, 2003).

At present, the most direct proxy climate record that exists for north Iceland is the $\delta^{18}O$ record from benthic foraminifera from cores B997-328 and -330 (Fig. 21.3) (Castaneda *et al.*, 2004; Smith *et al.*, 2005). An error of $\pm 0.5°C$ has been inferred based on a Monte Carlo simulation on coefficient errors in the calibration equation (Shackleton, 1974). The seafloor temperature data (Fig. 21.8B) show a peak in warmth ca. 7 cal. kyr, and there are marked oscillations in temperature, especially over the past 6 cal. kyr. Previously it was shown that the record from B997-330 has a significant correlation with changes in ΔC^{14} (Andrews *et al.*, 2003), suggesting a link with changes in the strength of the THC (Clark *et al.*, 2002b). A 1400–1500-yr climate oscillation has been detected in several Holocene records and also in the GISP2 ice-core data (Stuiver *et al.*, 1991; Bianchi & McCave, 1999; Bond *et al.*, 2001) (but see Wunsch, 2000). The cause of the 1500-yr oscillation is not known with certainty (Schulz *et al.*, 1999). It has been attributed to stochastic resonance (Alley *et al.*, 2001); attention has also been drawn to a 1470-yr lunar cycle (Berger *et al.*, 2002); and it has also been attributed to aliasing (Wunsch, 2000). A 1470-yr filter was fitted to the B997-330 seafloor temperature data (Andrews & Giraudeau, 2003; Smith *et al.*, 2005) (Fig. 21.8B). This suggests 'LIA-like' events off north Iceland occurring ca.500, 2000, 3500 and 5000 cal. yr BP. However, there is little evidence in the temperature series for older LIA-like events; there is some evidence to signal the 8.2 cal. kyr cold event (Alley *et al.*, 1997b; Barber *et al.*, 1999; Andrews & Giraudeau, 2003).

Changes in the flux of IRD (either contributed from sea ice and/or icebergs) into the northern North Atlantic (Bond *et al.*, 1999, 2001) might imply climate–glacier feed-backs sketched on Fig. 21.4A &B. However, Wastl *et al.* (2001) recognized six neoglacial advances in north Iceland (Fig. 21.8C), but these do not fit neatly into a ca.1500-yr oscillation, nor is there a clear temporal correlation between moraine formation and north Iceland seafloor temperature, nor the variations in haematite-stained quartz in core V28-14 just to the west in Denmark Strait (Figs 21.3 & 21.8C) (Bond *et al.*, 1999). In-point-of-fact, neither the north Iceland temperature record nor the Denmark Strait data (Fig. 21.8C) bear a close temporal correspondence to the north Iceland moraine intervals, which appear to be grouped between inferred 'cold' events.

The 'glacial' origin of distal Holocene IRD sediments in the North Atlantic is difficult to prove versus an alternative hypothesis that they represent sand-size particles carried on or within sea-ice. There is some correlation between the distal IRD events and the IRD record closer to the East Greenland coast (Jennings *et al.*, 2002b), but it is far from clear why the delivery and transport of Icelandic volcanic shards to sites to the south and east of Iceland is not primarily a wind-driven process that is independent of sea

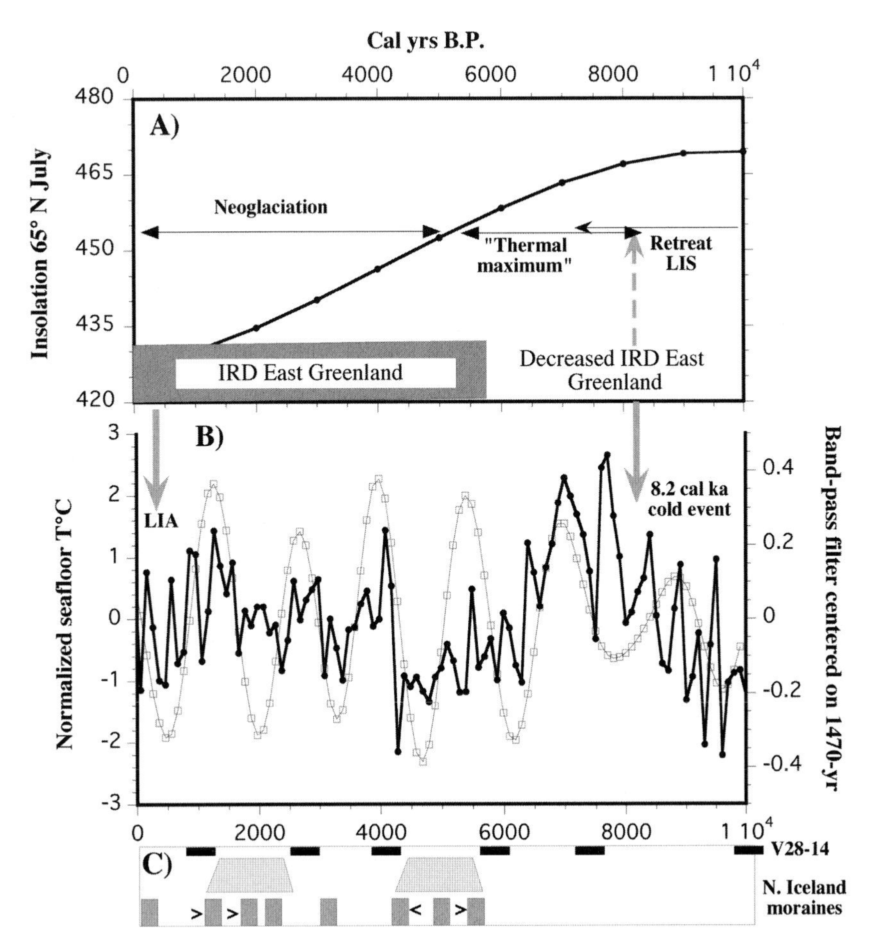

Figure 21.8 (A) Solar insolation and IRD history of East Greenland shelf, (B) 330 temp estimates and a 1470 filter response (note difference in scales), (C) peaks in haematite grains in core V28–14 from Denmark Strait (Bond *et al.*, 1999) (Fig. 21.3) and intervals of moraine formation around north Iceland glaciers (Wastl *et al.*, 2001). The rectangular grey areas represent groups of moraines between inferred cold events.

ice as a transport agent (Lacasse, 2001). There is increasing evidence for the presence of Icelandic tephras in lake and peat sites on Ireland, Britain, Norway and Sweden (Bjorck & Wastegard, 1999; Lacasse, 2001; Hall & Pilcher, 2002; Wastegård, 2002) as well as in the Greenland ice core (Zielinski *et al.*, 1994; Grönvold *et al.*, 1995), and sediments on the east Greenland shelf (Jennings *et al.*, 2002a). Studies are required of the continuous flux of Icelandic glass shards at several sites around Iceland so that the variability in tephra production can be factored out of the possible IRD contribution. Such a study is not yet available.

21.3.3.1 Meltwater pulses and deglaciation

The evidence for late marine isotope stage (MIS) 2 and 1 'global meltwater pulses', as deduced from the relative sea-level records from sites well-distant from the ice sheets (Fairbanks, 1989; Blanchon & Shaw, 1995; Aharon, 2004), does not coincide with the evidence for 'meltwater events' deduced from the planktonic isotopic evidence of foraminifera ocean-core records for the northern North Atlantic (Andrews *et al.*, 1994; Jones & Keigwin, 1988; Hald & Aspeli, 1997). The former show two massive meltwater pulses that are associated with rapid rises in global sea level. Surprisingly, given the wealth of data that exists for the interval of deglaciation (19–10 cal. kyr), no-one has been able to identify the ice sheet(s) associated with MWP1a and Antarctica has been called on to supply the required sea-level rise (Clark *et al.*, 1996a, 2002a; Weaver *et al.*, 2003), although that scenario is not without problems (Licht, 2004). The global meltwater pulses presumably represent increased melting on the ice sheets associated with the rise in summer insolation (Fig. 21.9), although the pulse-like nature of the process indicates other processes must have dominated at times, such as calving into pro-glacial lakes (Andrews, 1973). The planktonic isotope data from the northern North Atlantic indicates that surface melting on the LIS, Greenland and Fennoscandian ice sheets preceded MWP1a and may reflect the collapse and rapid retreat of ice across glacierized continental margins (Jennings *et al.*, 2002a), probably associated with rapid calving of tidewater margins.

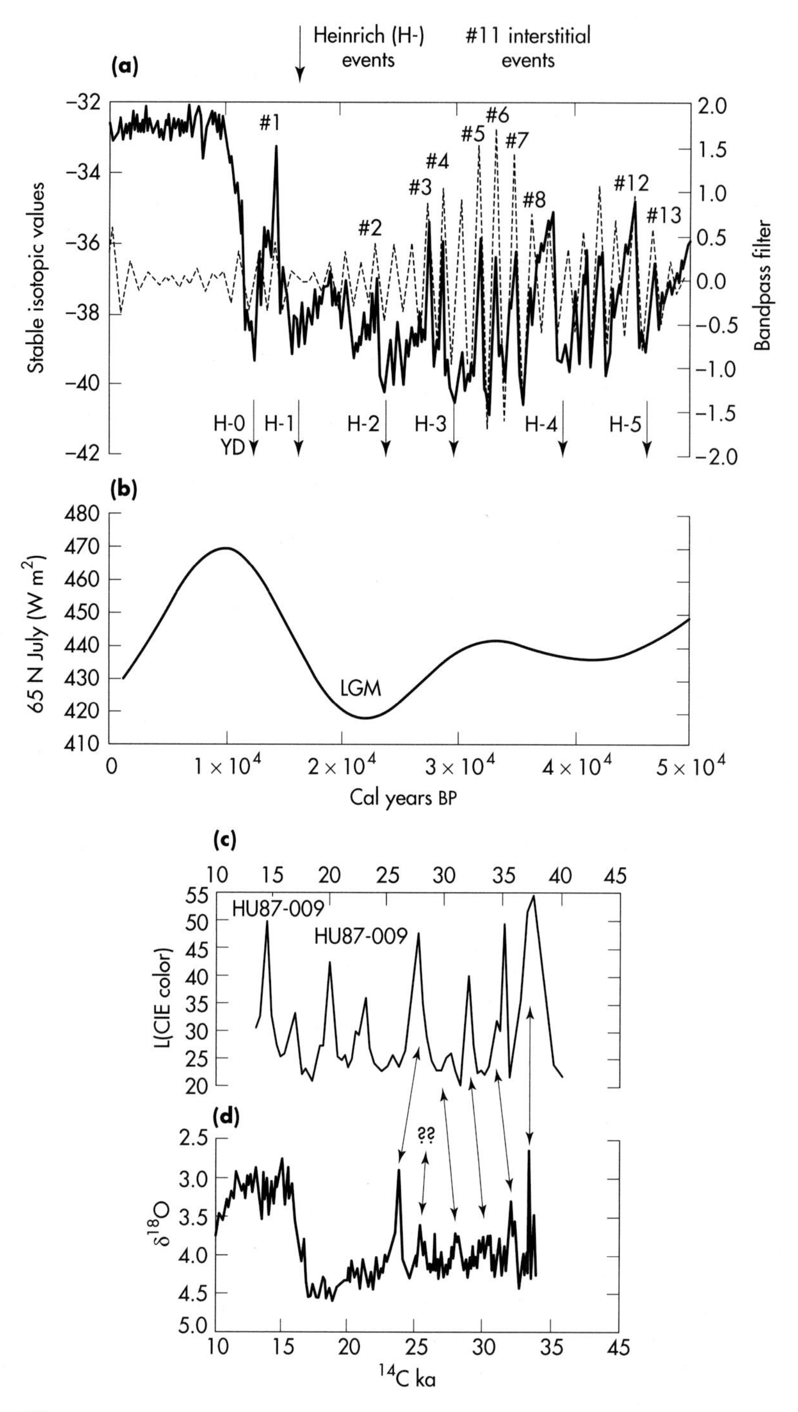

Figure 21.9 (a) GISP2 $\delta^{18}O$ data, with 1470-yr band-pass filter. (b) Solar radiation at 65°N in July. (c) Multitapered (MTM) spectra. Data for (c) and (d) from Blosseville Basin and Labrador Sea—D–O events. Notice the different age scales. The arrows in (c) and (d) link possible correlative events (see Fig. 21.3 for location of cores).

21.3.3.2 The Cockburn enigma

One of the world's longest moraine complexes runs from the northern tip of Labrador northward across the fjord heads on eastern Baffin Island, on to Melville Peninsula, and thence across the Arctic mainland where it lies close to the marine limit (Falconer *et al.*, 1965; Blake, 1966; Andrews & Ives, 1978; Andrews, 1989) (Fig. 21.3). The moraine complex is dated by marine shells found in associated raised glacial marine deltas (Andrews, 1989, fig. 3.75, p. 294) at around 8 ^{14}C kyr BP and are referred to as the Cockburn Moraines (Ives & Andrews, 1963). The Cockburn Moraines post-date the last major readvance of Labrador ice across Hudson Strait during the Noble Inlet advance 8.4–8.9 ka

(MacLean *et al.*, 2001; Jennings *et al.*, 1998; Manley & Miller, 2001) and they appear to pre-date the 8.2 cal. kyr cold event associated with the final break-up of the LIS in Hudson Strait and Hudson Basin (Barber *et al.*, 1999; Clarke *et al.*, 2004). The Cockburn readvance represents a positive change in the northeast LIS's mass balance, which appears to be coeval with the introduction of a 'warm' mollusc fauna into Baffin Bay (Andrews, 1972; Dyke *et al.*, 1996). However, the Cockburn Moraines invariably lie close to a marine limit or at a fjord head, so the apparent positive change in mass balance might be caused by the reduction in calving (mass loss) as the ice sheet retreated from a tidewater- to land-based situation (Fig. 21.2B) (cf. Hillaire-Marcel *et al.*, 1981). Hence the lack of a correlation with any changes in temperature or accumulation on the nearby Greenland Ice Sheet (Fig. 21.3) might argue for a non-climatic but dynamic driving force.

21.4 The last glacial cycle: abrupt glaciological responses to climate?

Research on marine cores from the 1970s onward have shown that the isotopic changes in foraminiferal $\delta^{18}O$ have variations that match those predicted from calculations of the variations in the Earth's orbit (Hays *et al.*, 1976). In particular, peaks in spectral analysis with periodicities of ca. 22, 41 and 100 kyr match variations in obliquity, precession and eccentricity, and it was widely argued that changes in solar insolation, at high northern latitudes, forced changes in the global ice-sheet inventory (Budd & Smith, 1981).

Broecker & Van Donk (1979) drew attention to the 'saw toothed' nature of the global ice volume changes, with the glacial intervals being abruptly terminated after a slow and progressive build-up of the global ice volume. The abrupt terminations of the glacial cycles suggest that causes other than the predictable changes in orbitally driven solar insolation provide positive feedback during the deglaciation processes. During glacial stages many continental ice-sheet margins fronted tidewater (Fig. 21.6c) and one explanation for the abrupt terminations is based on the collapse of the Northern Hemisphere's marine-based ice sheets (Denton *et al.*, 1986; Andrews, 1991).

When we consider glacier–climate relationships (Fig. 21.1) on time-scales of 10^4 yr we are now dealing with the growth and destruction of large ice sheets (Drewry & Morris, 1992) (Fig. 21.6c). The interaction between the height and albedo of large ice sheets and the climate system is important, as is the impact of freshwater inputs to the ocean and the effect on the THC. In addition on these time-scales, changes in the elevation of ice sheets and relative sea level will occur at the margins of ice sheets as the relaxation time for the process of glacial isostatic adjustment is of the order of 2–3 kyr for the LIS (Dyke & Peltier, 2000) (Fig. 21.6c). The massive Northern Hemisphere ice sheets over northern North America (LIS) and Fennoscandia were centred spatially over the large interior seas of Foxe Basin/Hudson Bay, and the Baltic/Gulf of Bothnia, hence are 'marine-based' (Denton & Hughes, 1981). Owing to the weight of the ice sheets and subsequent glacial isostatic depression of the crust, a large fraction of the beds of these ice sheets was well below sea level during glacial maxima (Peltier & Andrews, 1976; Peltier, 1994).

The modern analogue for these ice sheets is the West Antarctic Ice Sheet; theoretical arguments for the sensitivity of this ice sheet to future, and past, climate changes have been advanced (Hughes, 1977; Mercer, 1979; MacAyeal, 1992a; Anderson, 1999). It came as a surprise, however, that the evidence for massive and abrupt changes in ice-sheet discharge (from studies of marine cores in the North Atlantic) was associated with periodic collapse of the LIS that are now referred to as Heinrich or H-events (Andrews, 1998; Bond *et al.*, 1992; Broecker *et al.*, 1992; Broecker, 1994; Heinrich, 1988). In the original paper, however, Heinrich (1988) envisioned a link with insolation forcing, which was the ruling paradigm at that time, and only with the 1992 publications (Andrews & Tedesco, 1992; Bond *et al.*, 1992; Broecker *et al.*, 1992) did the nature of the glaciological response take centre stage. Thus, in the 1990s the focus on ice-sheet–climate shifted to take into account evidence for abrupt changes in the cryosphere–climate system at much higher periodicities than solar insolation changes related to the Earth's orbit around the sun (Fig. 21.9a & b).

Over the past decade or more the literature on 'glacial' events dated between ca. 10 and 50 cal. kyr is dominated by the results from the study of marine sediments. The evidence on land for glacial H- or D–O events is not particularly compelling. The marine studies highlighted a number of dramatic glacial responses, largely viewed from the perspective of changes in the ice rafted component (IRD) of deep-sea sediments, and the evidence for significant additions of meltwater to the surface ocean. During major glacial cycles the ice sheets thickened and extended outwards. In many areas this has resulted in ice sheets extending beyond the coastline onto the continental shelf, often to the shelf break (Piper *et al.*, 1991; Vorren *et al.*, 1998) (Fig. 21.6c). During the advance phase of the glacial cycle, the crust will be undergoing glacial isostatic depression and the relative sea level at the seaward margin of the ice sheet will, probably, be rising (this depends on the balance between global ice-sheet growth, which extracts water from the ocean, and the regional rate of ice-sheet growth, which results in isostatic depression (Peltier & Andrews, 1976; Lambeck, 1990; Peltier, 1996; Lambeck *et al.*, 2000)). Because of the density differences between ice and marine water (900 versus ca. 1028 kg m^{-3}) there is a buoyancy force working to lift the ice sheet from its bed. This works to reduce friction at the bed and to cause an acceleration of glacier flow, which could not be matched by an increase in accumulation, hence leading toward a collapse and rapid retreat of ice margins on glaciated continental margins. This scenario is also enhanced by the fact that deep troughs which lead toward fjords cross most glaciated continental shelves (Holtedahl, 1958). As ice retreats, glacial isostatic recovery would cause uplift at the sea floor and a marine regression. Thus there are hints here of a self-regulating cycle of glacial response where an increase in the regional mass balance results in an ice advance onto the continental margin, but thereafter the termination of the advance and the retreat might be caused by rapid calving driven by changes in water depth (Fig. 21.6c). This is a fundamentally different set of responses from those climate–mass-balance interactions that control the activity of a margin on land.

The initial evidence for abrupt changes in climate, but not a priori glacial response, during MIS 2 and 3 (or ca. 13 to 50 cal.

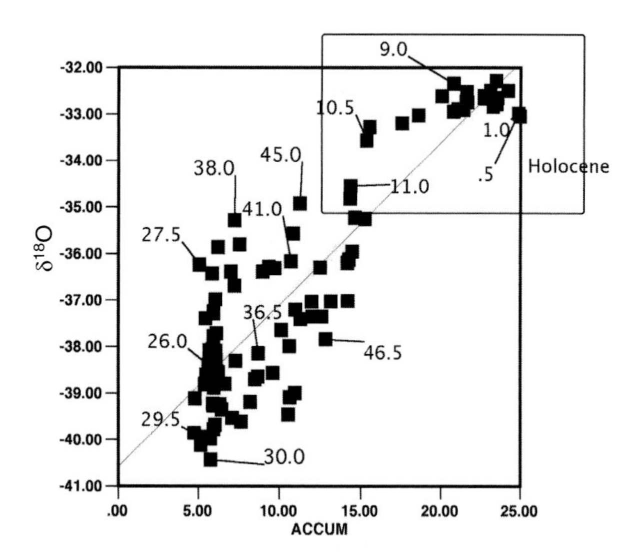

Figure 21.10 GISP2 data—accumulation (cm yr^{-1}) versus δ^{18}O at 500-yr intervals for the past 50,000 yr. The figures represent the estimated age (cal. kyr) for a data point. The line is the best fit for the correlation between the two variables.

kyr) was highlighted by the data from the Greenland Ice Sheet (Fig. 21.9a), which indicated very abrupt changes in the isotope (climate) data (Johnsen *et al.*, 1992)—the question is whether these were reflected in a glacial response? Data from marine sites close to the margins of the Greenland Ice Sheet (Voelker, 1999, 2000; van Kreveld *et al.*, 2000) and the eastern LIS (Andrews & Barber, 2002) indicated that in the interval 12–50 cal. kyr there were coeval changes in the marine environments (Fig. 21.9c & d) on either side of Greenland (Fig. 21.3). The climate for the region during this interval (Fig. 21.10) (Meese *et al.*, 1994) indicates that colder temperatures occurred with lower accumulation ($r^2 = 0.77$). The points in the upper right quadrant of Fig. 21.10 are the Holocene values and the graph shows that there is an overall relationship between accumulation and δ^{18}O (which is generally considered to be linearly related to temperature (Dansgaard *et al.*, 1969)). This linkage persists in the time domain, with interstadials being times of higher net accumulation on the summit of the ice sheet. Of concern in this chapter are the interrelationships between the climate–ocean variables and the feedbacks between ice sheets and glaciers (Figs 21.1 & 21.2) (Sakai & Peltier, 1997).

The data from the Greenland Ice Sheet reflect changes in climate at around 3 km in the atmosphere; the marine data show a series of parallel changes in a variety of proxies (Fig. 21.9), but how do these link to actual changes in the extent and position of ice margins during the D–O or H-events? Heinrich events represent large-scale collapse of the LIS with a 5–7 kyr periodicity, and possibly linked to events around other ice sheets (Bond & Lotti, 1995). Dansgaard–Oeschger events represent climate cycles present in ice-core and marine records (van Kreveld *et al.*, 2000; Andrews & Barber, 2002), with a series of them forming a saw-toothed pattern that leads to an H-event (Moros *et al.*, 2002); these have been termed 'Bond cycles' (Broecker, 1994). Note, however, that the glaciologically massive H-events are not dramatically different in their isotopic composition from the D–O stadial events (Fig. 21.9a). Furthermore, as noted by several authors (Mayewski *et al.*, 1997), a ca. 1500-yr cycle is embedded in the Greenland ice-core data; hence Bond *et al.* (1999) argued that the IRD events within the Holocene are a continuation of this fundamental cycle (e.g. Fig. 21.8B). This periodicity is seen in the band-pass filter on the GISP2 δ^{18}O data (Fig. 21.9a) and on the multi-tapered (MTM) spectra (Mann & Lees, 1996) from 100-yr resolution GISP2 data (not shown).

Heinrich events must represent major changes in the dynamics of the LIS but whether these massive collapses of the ice sheet, which may be associated with rapid changes in sea levels (Andrews, 1998; Chappell, 2002), are driven by internal glaciological dynamics, 'climate' at the bed–ice interface (Fig. 21.1), or by surface climate, is debatable (Alley *et al.*, 1999; Clarke *et al.*, 1999; van Kreveld *et al.*, 2000). The 'binge–purge' model for the origins of H-events (MacAyeal, 1993a,b) is similar to some theories for the origins of surging glaciers (*Canadian Journal of Earth Sciences*, 1969). In these cases the links with atmospheric or ocean climate are not direct and the events are driven by changes in the conditions at the bed–ice interface, particularly temperature and water. Not included within the 'binge–purge' model is the possibility that large subglacial lakes may have existed under the LIS and H-events might be associated with 'outburst' floods (Shoemaker, 1992a,b; Hesse & Khodabakhsh, 1998). Because H- and D–O oscillations are associated with ice-sheet behaviour largely associated with marine (tidewater) margins, many of the caveats associated with the relationship (or lack thereof) between glacier response and climate (e.g. Figs 21.2B & 21.6b) at LIA and Holocene time-scales (Mann, 1986) apply on these longer time-scales. In particular, the presence on the Canadian shelf of a large 600 m deep basin seaward of the Hudson Strait sill (Andrews & MacLean, 2003) may do much to explain the massive collapse of the LIS called for in H-events.

As noted above, much of our information on abrupt glaciological variability in ice sheets comes from marine cores (Jennings *et al.*, 1996; Stoner *et al.*, 1996; Scourse *et al.*, 2000; Groussett *et al.*, 2001; Hemming *et al.*, 2002b), so are there coeval responses of land-based margins (Mooers, 1997; Kirby & Andrews, 1999; Rashid *et al.*, 2003)? Margins that descend to sea level or lake level (tidewater and calving margins) are intuitively more prone to be unstable (Thomas, 1977, 1979a; Broecker, 1994). However, one of the dramatic revolutions in the past several years has been the recognition of fast-flowing ice streams and margins that are situated on soft sediments, which deform rapidly (Clark & Walder, 1994; Ó Cofaigh & Evans, 2001; Dowdeswell *et al.*, 2004a). This was partly predicted in the 1970s when Matthews (1974) documented very low gradients on marginal lobes of the LIS across the Canadian Prairies. These low gradients were compatible with shear stresses of 4–10 kPa, an order of magnitude lower than from most glaciers (Paterson, 1981); similar shear stresses were also measured on moraine segments from the highly lobate margins of the southwest and northwest LIS (Beget, 1987; Clark, 1994; Clark *et al.*, 1996b). Soft sediments also probably lay under the large ice streams that drained the LIS, and especially Hudson Strait (Andrews & MacLean, 2003). However, D–O events are characteristic of MIS 3 (Johnsen *et al.*, 1992) (Fig. 21.9a & b) and, at

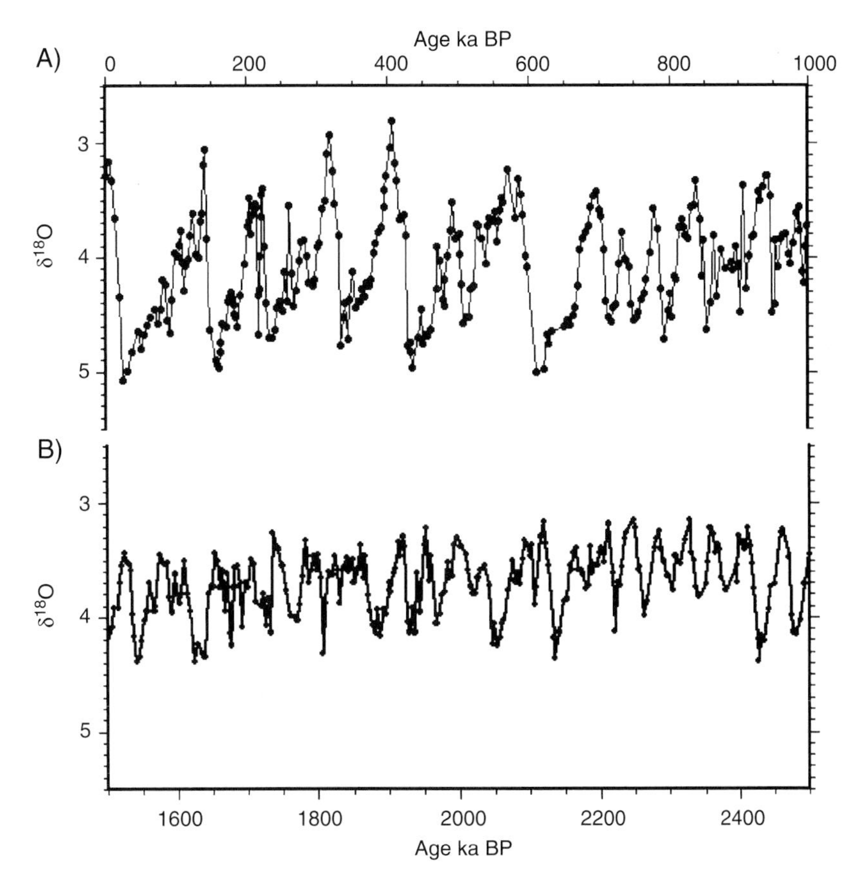

Figure 21.11 (A) $\delta^{18}O$ records on benthic foraminifera for the past 0–1 Myr, and (B) 1.5 to 2.5 Myr (see text for source).

present, it appears that the LIS did not extend onto the Canadian Prairies during this period of time (Andrews, 1987b, Fulton, 1989; Dyke *et al.*, 2002). Thus although the southwest margin of the LIS was very dynamic during MIS 2 (Lowell *et al.*, 1995) it appears (= no evidence?) that this sector of the ice sheet was not involved in the D–O interstadial–stadial responses coeval with those in the Greenland ice-core record (Fig. 21.9a) or seen in the North Atlantic sediment records (van Kreveld *et al.*, 2000; Moros *et al.*, 2002). It is probably no accident that H- and D–O events are recorded offshore from the margins of ice sheets terminating in the sea, hence involving glacier dynamics associated with changes in water depth driven by a combination of glacial isostatic forcing and exchange of freshwater between oceans and ice sheets.

On D–O time-scales there is actually a significant disconnect between the spatial continuity of ocean and climate evidence and the glacial record. For example, although the margins of the Younger Dryas (YD) (H-0, Fig. 21.9a) ice sheets are relatively well-mapped in Fennoscandia (FS) and Great Britain, there is no comparable moraine system around the margin of the Greenland Ice Sheet (it might be the Milne Land stadial? (Funder, 1989; Kelly *et al.*, 1999)). Furthermore, although the margins of the YD time period are reasonably well constrained for the LIS (Dyke, 2004) there does not appear to be the same continuity of moraine expression as noted for the FS ice sheet, although some moraine sequences have been linked to the YD. There are indeed coeval moraine intervals to some of the younger H-events (Lowell *et al.*, 1995) but there is not a 1:1 match, and the link between the marine-based H-events and the history of the terrestrial LIS ice extent over the past 50,000 yr is not easy to decipher (Stoner *et al.*, 1996; Kirby & Andrews, 1999; Rashid *et al.*, 2003;). Exactly what the terrestrial glacier/ice sheet responses were to D–O events is uncertain in terms of the creation of a geological record, such as an end moraine or a till stratigraphy.

21.5 Orbitally driven glacier–climate interactions

As we move back in time the connection between individual glacier or ice-sheet response and the climatic forcings becomes unknown. Indeed, the history of Quaternary 'glacial' cycles (Fig. 21.11) is not based on glacial geological evidence *per se* but on changes in the isotopic composition of foraminifera (Shackleton, 1973; Shackleton *et al.*, 1973), which reflect the global hydrological balance.

The advent of extensive marine coring expeditions after World War II, combined with the development of ^{14}C dating, $\delta^{18}O$ measurements of foraminifera and palaeomagnetic chronology, eventually led to the recognition that the global hydrological balance in terms of ice-sheet growth and retreat had significant similarities with the calculated changes of solar insolation associated with the orbital or Milankovitch variations (Hays *et al.*, 1976). Although the association is widely accepted it is still unclear how the relatively small changes in insolation trigger such massive changes in global ice volume. Some non-linear feed-backs are required in the glacier–climate system (Birchfield & Weertman, 1982), with one being the feed-back between changes in extent of snow cover with its high albedo. With new data on the phasing of ice volume and greenhouse gas content over the past 0.5 Ma or so (Ruddiman, 2003b, 2004) the role of the biosphere is becoming more important. The dramatic terminations may be best explained by the rapid loss of mass of those ice-sheet margins, which extended onto the continental shelves (Andrews, 1987a). This process could result in a rapid deglaciation and a positive feed-back as ice sheets collapsed and global sea level rose.

The major change that occurred on Quaternary time-scales is the switch about 0.7–0.8 Ma from the previous domination of the 42 kyr periodicity to the subsequent ca. 100 kyr periodicity (Fig. 21.11A & B) (Ruddiman *et al.*, 1989; Raymo *et al.*, 1989; Raymo, 1992). There has been speculation and theories as to what mechanism or mechanisms were responsible for this dramatic shift in glacial response (Clark & Pollard, 1998) but it is obvious from Fig. 21.11 that the response mode to orbital forcing changed drastically around 0.8 Ma ago.

Ruddiman (2003b, 2004) provided the latest assessment of the role of solar insolation in orbital-scale climatic change. However, the change in stable isotopes of oxygen reflects an integrated global assessment of the addition and subtraction of ice on land, and hence does not specifically address the issue of the histories of individual ice sheets and certainly is totally unable to resolve the record of mountain glaciations, except by inference. One would assume that on Quaternary time-scales the LIS is still the major component of the $\delta^{18}O$ signal in foraminifera but exactly how the history of this ice sheet relates to the major changes in $\delta^{18}O$ is uncertain. Ruddiman (2003b, 2004) used the 'greenhouse gas' data from the Greenland and Antarctica ice cores to derive a model for the response of the global ice signal to a combination of changes in insolation and changes in CO_2 and CH_4. Explicit in this argument is the recognition that ice-sheet growth and decay lag the forcing functions. Furthermore, the emphasis is squarely on the importance of summer insolation (specifically mid-July) operating at the 41 and 22 kyr periods in the high-latitude Northern Hemisphere to change the rate of ablation of the Northern Hemisphere ice sheets.

Although the importance of the orbital forcing on the geological records of the past 2 Ma is nearly universal, Wunsch (2004) has presented an important critique of the paradigm. A major conclusion is (p. 1001) that 'The fraction of the record variance attributable to orbital changes never exceeds 20%.' Wunsch (2004) then proposes that the records are consistent with '. . . stochastic models of varying complexity' with only a small-scale response associated with obliquity. Note that somewhat similar arguments have been advanced for the 1/1.5 kyr cycle in late Quaternary records (Alley *et al.*, 2001; Rahmstorf & Alley, 2002).

There are (at the moment) no regional glacial chronologies with time-scales that can be matched simply with the global ice-volume record archived in the benthic isotope data (Roy *et al.*, 2004). Thus the history of the continental ice sheets is largely inferred from the marine records and there is, for example, no glacial stratigraphy on land which preserves all the 22, 41 and 100 kyr cycles that are inferred from marine records, hence the 'glaciology' of the climate response is unclear for individual ice sheets and glaciers. Some researchers have modelled ice-sheet variations that are forced by orbital changes (Budd & Smith, 1981; Marshall & Clark, 2002) and this would appear to be the most productive method for understanding ice sheet/climate interactions.

21.5.1 Ice-sheet growth

One of the significant problems in ice-sheet ↔ climate interactions is the rate of growth of large ice sheets. Previous studies have tended to reaffirm Weertman's (1961a) conclusion that ice-sheet size is very sensitive to mass balance distribution (Hindmarsh, 1993). The Broecker & Van Donk (1979) saw-tooth model implies a slow build-up of the global ice volume from say MIS 5d at 115 cal. kyr peaking in MIS 2 at ca. 20 cal. kyr. However, studies of coral reef sequences indicate that sea level fell rapidly during the onset of the last glacial cycle in MIS 5d. In Barbados, recent studies (Speed & Cheng, 2004) suggest a fall of ≥37 m or 3 to 4 m kyr^{-1}! This is about half the volume of the LIS during MIS 2 or the equivalent of building a present-day Greenland Ice Sheet, or at least half an ice sheet, every 1000 yr. Looking at the temperature–precipitation relationships from the GISP2 data (Fig. 21.10) the paradox between the rate of growth and climate is apparent. If glacial cycles primarily reflect changes in temperature then colder conditions are equated with lower accumulation—thus how are fast rates of growth maintained? An early three-dimensional glaciological modelling exercise on the 'instanteous' growth of the LIS (Ives, 1962; Andrews & Mahaffy, 1976) pushed an increase in mass balance to the limit of around three times the present (equilibrium line altitude placed at sea level and mass accumulation increased to 0.9 $m\,yr^{-1}$) and managed to account for a 20 m lowering of sea level in 10,000 model-years. The patterns of ice-sheet inception and subsequent growth in this early model are similar to more recent glaciological simulations (Marshall & Clarke, 1999), which, however, do not fully capture the rate of ice build-up.

The critical question is what atmospheric and oceanographic conditions are required to produce such a rapid and positive increase in the net mass balance of the incipient LIS? Is it a function of colder temperatures or must it involve an increase in winter accumulation, or some elements of both? Given the clear positive correlation between the length of the open-water season and snowfall along the coast of Baffin Island the growth of the LIS might be linked to suggestions for a more open water (i.e. less sea ice) in the Labrador Sea and Baffin Bay during glacial inception (Johnson, 1997; Johnsen & White, 1989). These arguments have in part been based on the presence of 'warm water' molluscs in glacial marine sediments associated with the onset of the last glaciation (Andrews *et al.*, 1981). Arguments such as these have raised the possibility that global warming could result in renewed glacierization of Arctic Canada because of increased winter precipitation (Miller & de Vernal, 1992; Glidor, 2003) but Alley (2003) argued forcefully against such a scenario.

21.6 Conclusions

In many ways the assessment of the interrelationships between the factors illustrated on Figs 21.1, 21.2, 21.4 & 21.6 has been rather gloomy. In point of fact, although a figure such as Fig. 21.2 is easy to draw there are few, if any, studies on present-day glaciers that mimic this degree of investigation on all aspects of a glacier, and such an integrated system certainly is not available for the present-day large ice sheets—our knowledge of present-day large ice sheets in terms of the parameters on Fig. 21.2 is sketchy at best. Hence the application of the axiom 'the present is the key to the past' should excite a high-level concern from earth scientists interested in examining past glacier–climate interactions, because there is so much to learn about modern day conditions never mind the inferences about the past.

It is clear that the specific glaciological controls on the response of ice sheets and glaciers becomes increasingly uncertain as we step back from the past two to three decades, to the LIA, the Holocene and then into the last glacial cycle, and finally to the Quaternary glacial cycles. A key resource, however, which will serve to link these various time-scales of glacier response, is that of modelling glacier/ice-sheet–climate interactions. Starting with Mahaffy's efforts in the mid-1970s (Mahaffy, 1976) a variety of three-dimensional glaciologically driven models have been integrated, to a greater or lesser extent, with 'climate' to predict the responses of individual glaciers or large continental-scale ice sheets (Marshall *et al.*, 2000; Siegert *et al.*, 2002) and their contributions to deep-sea sediments (Dowdeswell & Siegert, 1999; Clarke & Prairie, 2001). It is, however, fair to say that no model yet incorporates the full complexity of the interactions illustrated in Figs 21.1 & 21.2, although they are increasingly adopting more and more complex interactions (Vettoretti & Peltier, 2004). One linking theme in this chapter has been the argument that glaciers and ice sheets at tidewater margins drive much glacial history because many of our data sets are derived from an interpretation of records archived in marine sediments rather than terrestrial glacial deposits.

We can observe the present-day climate–glacier mass balance interactions that are leading to a significant reduction in glacier mass and a rise in global sea level, but as we go back in time the problems increase in quantifying these interactions and usually the simple glacier response ↔ climate loop is taken (Fig. 21.2A). One example of the problems that arise farther back in time as we seek to understand the Earth's climate system is that, because of glacial erosion, the landscape of glaciated areas is to a significant degree unknown once we move back into the mid- or early Quaternary. We have seen the importance of the Hudson Strait on the stability of the LIS during the last glaciation, but what was the form of this Strait 1 or 2 Myr ago, and how has the landscape at the bed of LIS changed over this interval as a result of glacial erosion (Bell & Laine, 1985)? During glacial periods much of the Canadian Shield was scoured by ice (Sugden, 1978) so that literally 10,000s of small to medium lakes dominate the landscape in the subarctic and arctic areas. A fact of life is that snow and ice clears significantly later off lakes than land (Wynne *et al.*, 1996), hence the progressive erosion of the shield and the development of the 'ice scoured' landscape would progressively increase interglacial summer albedo on geological time-scales. This is but one small example of the complexities involved in fully understanding and predicting glacier–climate interactions.

Acknowledgements

I would like to thank Peter Knight for asking me to contribute a chapter to this volume. I would like to thank my colleagues and graduate students at the University of Colorado for 30+ years of stimulation, and the National Science Foundation for their grant support to a variety of programmes. Dr Maureen Raymo kindly provided the data for Fig. 21. 11. Two reviewers made suggestions for changes and requests for clarity.

TWENTY-TWO

A multidisciplined approach to the reconstruction of the Late Weichselian deglaciation of Iceland

Alun Hubbard

Department of Geography, University of Edinburgh, Edinburgh EH8 9XP, UK

Examining changes in past ice cover across Iceland is enlightening owing to the many feedbacks that are apparent between ice, atmosphere, ocean and the lithosphere. Unravelling this interaction presents a tough job but is important from three related perspectives:

1 Iceland is influenced by an extreme maritime climate which yields ice-caps with large turnover and outlet glaciers that are both sensitive and responsive. Hence, changes in long-term North Atlantic circulation will have an immediate and significant impact on glacier dynamics and extent.
2 Its location astride the mid-Atlantic ridge means that Iceland experiences extreme geothermal conditions, in certain cases leading to intense subglacial volcanism and concomitant meltwater flooding or Jökulhlaups. This is not only a contemporary concern but has been a key control on past ice dynamics (Jull & McKenzie, 1996).
3 Sediments originating from Iceland have been found in the lower (i.e. earlier) sequences of Heinrich IRD layers, suggesting that large and critically located pulses associated with the collapse of the ice sheet during the Late-glacial may have affected North Atlantic Deep Water (NADW) formation (Grousset *et al.*, 2000). Rapid isostatic rebound, which promotes enhanced mantle upwelling, means that volcanic activity may have been a hundredfold greater during glacial unloading compared with the present (Maclennan *et al.*, 2002) and therefore has the potential to trigger widespread ice-sheet collapse.

Investigating these interactions is a complex and multidisciplined task involving empirical data from a variety of sources, but is one in which numerical modelling potentially plays a pivotal role. One significant obstacle to a comprehensive Late-glacial reconstruction is chronological uncertainty in the sedimentary and geomorphological records. It is generally believed though that the Weichselian glaciation most likely culminated in its maximum extent at ca. 21 kyr BP (the Last Glacial Maximum—LGM) and terminated ca. 10 kyr BP (Norddahl, 1991). Strong evidence for the extent, thickness and flow of the former ice sheet certainly persists but apart from a handful of key ocean cores, the pertinent evidence (in particular the terrestrial geomorphology) is invariably undated. Furthermore, the piecemeal offshore record (summarized in Fig. 22.1) remains equivocal. Limits are identified either from bathymetric or seismic surveys or are inferred from radiocarbon dated basal diamicton from recovered cores. Dates from Marine Oxygen Isotope Stages (MIS) 2 and 3 have been obtained from sediments recovered from Húnaflóaáll (north Iceland), Djúpáll (northwest Iceland) and Látra Bank (west Iceland) (Andrews *et al.*, 2000; Andrews, in press). Furthermore, one core at Djúpáll, 50 km off the northwest, indicates that the site has been unglaciated for 30.9 ^{14}C kyr BP and thus provides a definitive constraint on LGM extent in this sector. Elsewhere, the absolute extent of the LGM ice sheet remains ambiguous but is placed at the shelf edge in the south and west. Grímsey Island, off northern Iceland shows extensive evidence of ice erosion, hence the LGM limit is inferred to be more extensive in this sector also.

Here, a three-dimensional coupled ice-flow–degree-day model is used to provide an experimental framework by which the interactions between ice–atmosphere–lithosphere can be investigated, validated and refined against available observation data. Such modelling provides the classic link between form (the erosional and depositional record) and process (climate change, glacier-meteorology, geothermal activity and ice-dynamics). Specifically,

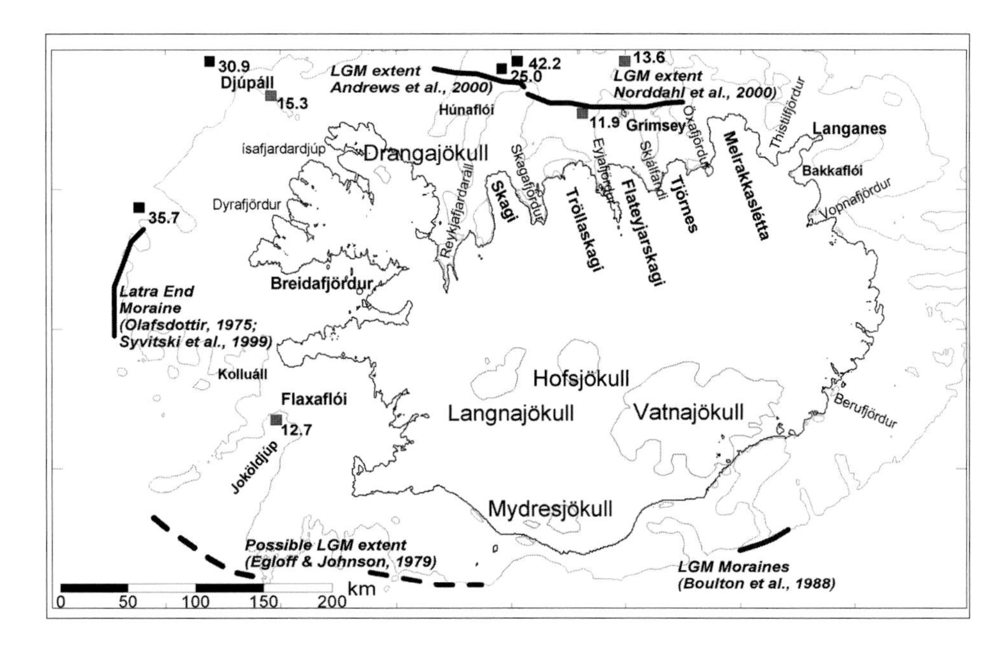

Figure 22.1 Map of Iceland showing key locations referred to in the text. Also shown are critical basal marine core dates, evidence of former ice limits and the main ice caps in existence today. The 200 m bathymetric contour is marked for reference and all dates are in k ^{14}C yr BP. (See www.blackwellpublishing.com/knight for colour version.)

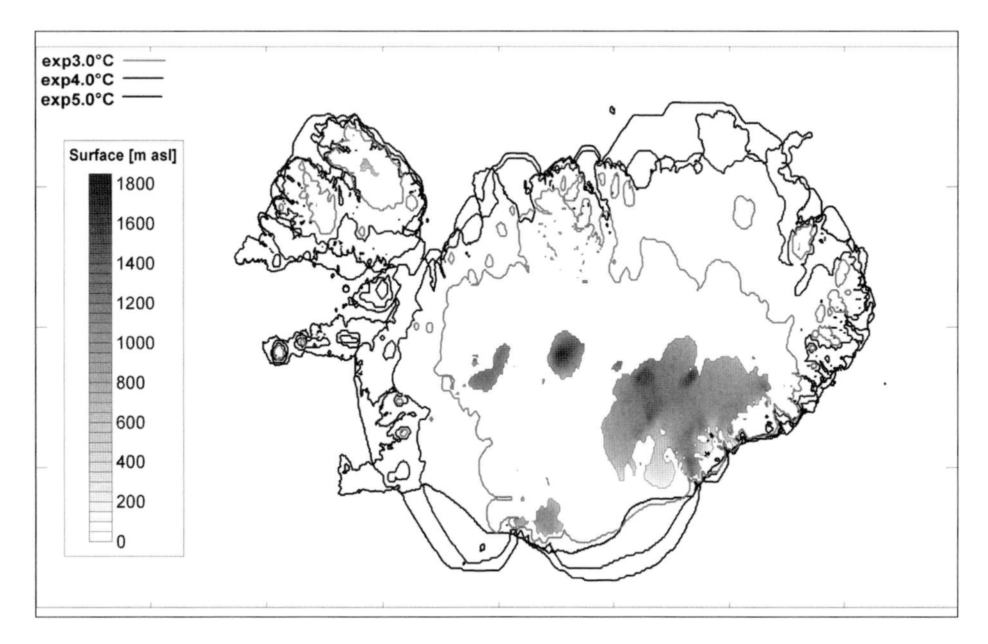

Figure 22.2 Modelled present-day ice surface using a two-stepped cooling of 2°C for 1000 yr followed by 200 yr at −1°C. Also shown are the modelled ice-sheet extents associated with 2000 yr of cooling of 3, 4 and 5°C under the present precipitation regime and no sea-level change. (See www.blackwellpublishing.com/knight for colour version.)

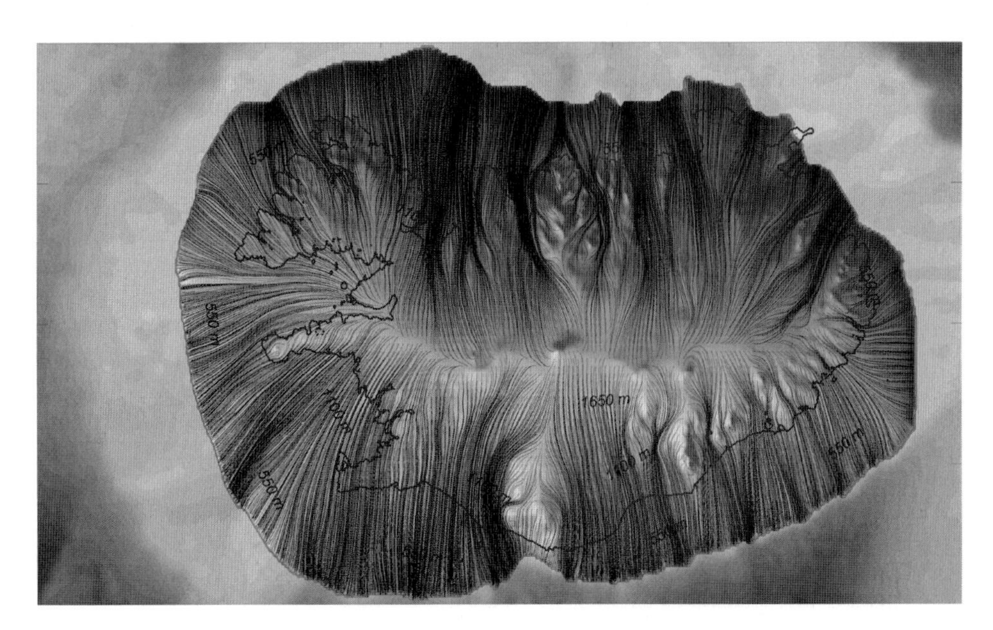

Figure 22.3 The optimum LGM modelled ice-sheet surface and its flow regime most compatible with the available offshore evidence corresponding to a cooling of 12.5°C and a 35% decrease in precipitation with an additional 30% suppression applied north of the 65th parallel. (See www.blackwellpublishing.com/knight for colour version.)

the time-dependent, thermomechanical model is applied to a 2-km grid and requires distributions of topography, geothermal heat flux, temperature and precipitation. It enables the variables of ice thickness, bed adjustment, flow and temperature to interact freely and caters for the dynamics of ice-shelves, calving and thermally triggered basal sliding through longitudinal coupling. It is forced through perturbations from present by changes in sea level, precipitation and temperature and yields space–time distributions of ice thickness, isostatic response, stress, velocity, temperature, melt and iceberg flux. As an initial test, Iceland's contemporary ice cover is well replicated from ice-free initial conditions given 1000 years of 2°C cooling followed by 200 yr at 1°C (Fig. 22.2). Experiments also reveal that given its present rainfall regime, Iceland is highly susceptible to widespread glaciation, with an ice sheet advancing to the southern coastline given just 3°C annual cooling and beyond it in many quarters with 4 to 5°C cooling (Fig. 22.2).

Modelling the LGM presents a challenge though, given that virtually no local palaeoclimatic data exist. Assuming though, that the form, if not the magnitude, of the forcing climate signal took that of the GRIP oxygen isotope curve, then a 'shot-gun' approach may be adopted. Fifteen experiments were initiated from 24 kyr BP for 3 kyr using the GRIP isotope curve scaled for maximum cooling at 21 kyr BP ranging from 5 to 15°C in 2.5°C increments, each with a corresponding 20, 35 and 50% suppression of today's precipitation. Two of these experiments generally matched the offshore limits but all overran appreciably to the north. The optimal LGM configuration required 10 to 12.5°C of cooling with 35% overall precipitation reduction, with a further 30% enhanced aridity across the north. This results in a large, offshore ice sheet with an area of $3.29 \times 10^5 km^2$ and a volume of $2.38 \times 10^5 km^3$. Sensitivity experiments highlight the central role played by geothermal activity in activating extensive zones of basal melting, yielding a highly dynamic and low-aspect ice sheet with a mean thickness of ca. 800 m drained by numerous ice streams extending far into the interior and breached by numerous nunataks (Fig. 22.3).

Norddahl (1983) uses trimlines and the upper limit for evidence of glacial erosion and deposition on nunataks to reconstruct the LGM ice-sheet surface in north Iceland across the Tröllaskagi and Flateyjarskargi mountains. The high level of correspondence observed between the modelled profiles down both Eyjafjördur and Bárdardalur ice streams and those reconstructed by Norddahl (1983) indicates not only that the modelled LGM is coherent with this evidence but that the model is also replicating the dynamics of these outlet ice streams satisfactorily (Fig. 22.4). The low-lying aspect of the modelled LGM is also consistent with the profile of the 'Pleistocene ice sheet' reconstructed by Walker (1965) from the summit altitudes of palagonite tuff-breccia (Móberg Mountains). These 'table-mountain' volcanoes form subglacially, but summit capping of subaerial lavas indicates that they topped-out at the level of the palaeo-ice-sheet surface.

Comparison between the orientation of ice-contact features indicating past ice flow with basal velocity vectors from the modelled LGM is also encouraging, even if the result is not entirely unambiguous (Fig. 22.5a). The main issue is whether the features observed can be considered chronosynchronous with the LGM. However, assuming that ice thickness, volume, extent and hence

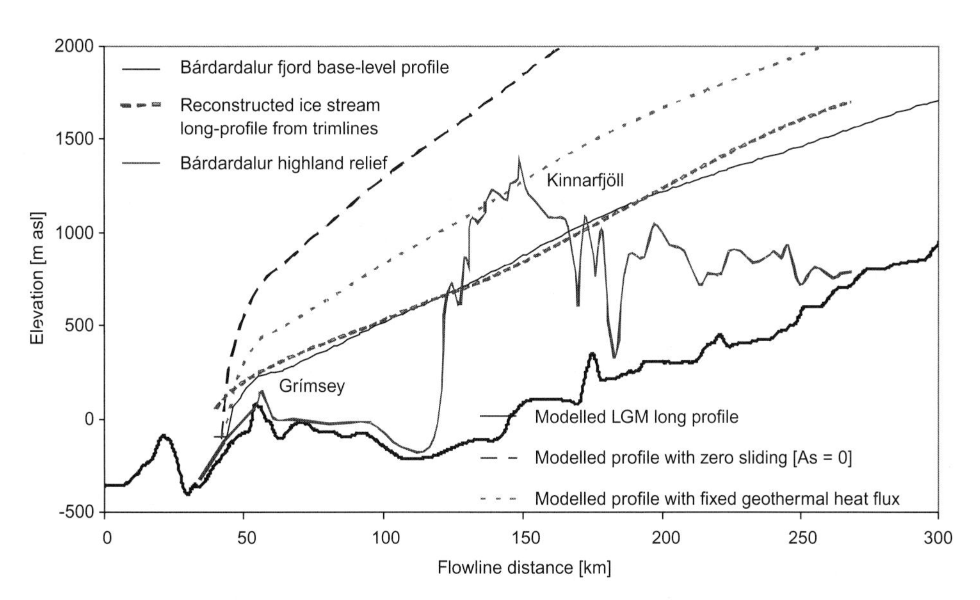

Figure 22.4 The long profile of Bárdardalur fjord showing the basal topography, its mountain relief, the modelled optimum LGM ice stream and that reconstructed by Norddahl (1991) from trimline evidence. Superimposed are the two long-profiles modelled using the standard value for geothermal heat flux ($G = 54.2\,\mathrm{mW\,m^{-2}}$) applied across the model domain and zero basal sliding ($A_s = 0$). (See www.blackwellpublishing.com/knight for colour version.)

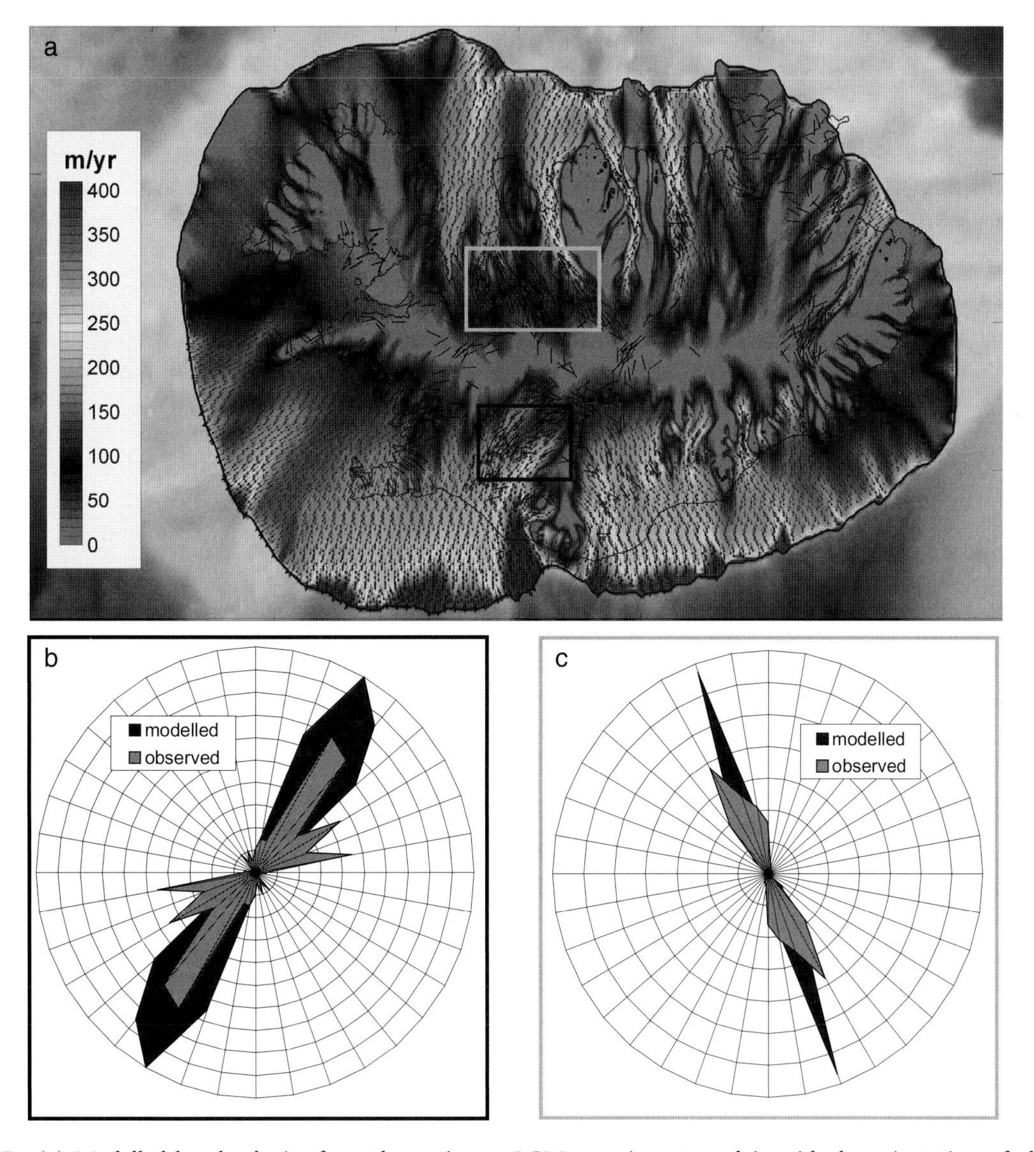

Figure 22.5 (a) Modelled basal velocity from the optimum LGM experiment overlain with the orientation of observed ice directional features (adapted from Bourgeois *et al.*, 2000) superimposed with the direction of the corresponding modelled basal vector for the same location. (b & c) Rose diagrams of modelled basal velocity vectors and observed orientations of ice directional features for the two locations indicated. (See www.blackwellpublishing.com/knight for colour version.)

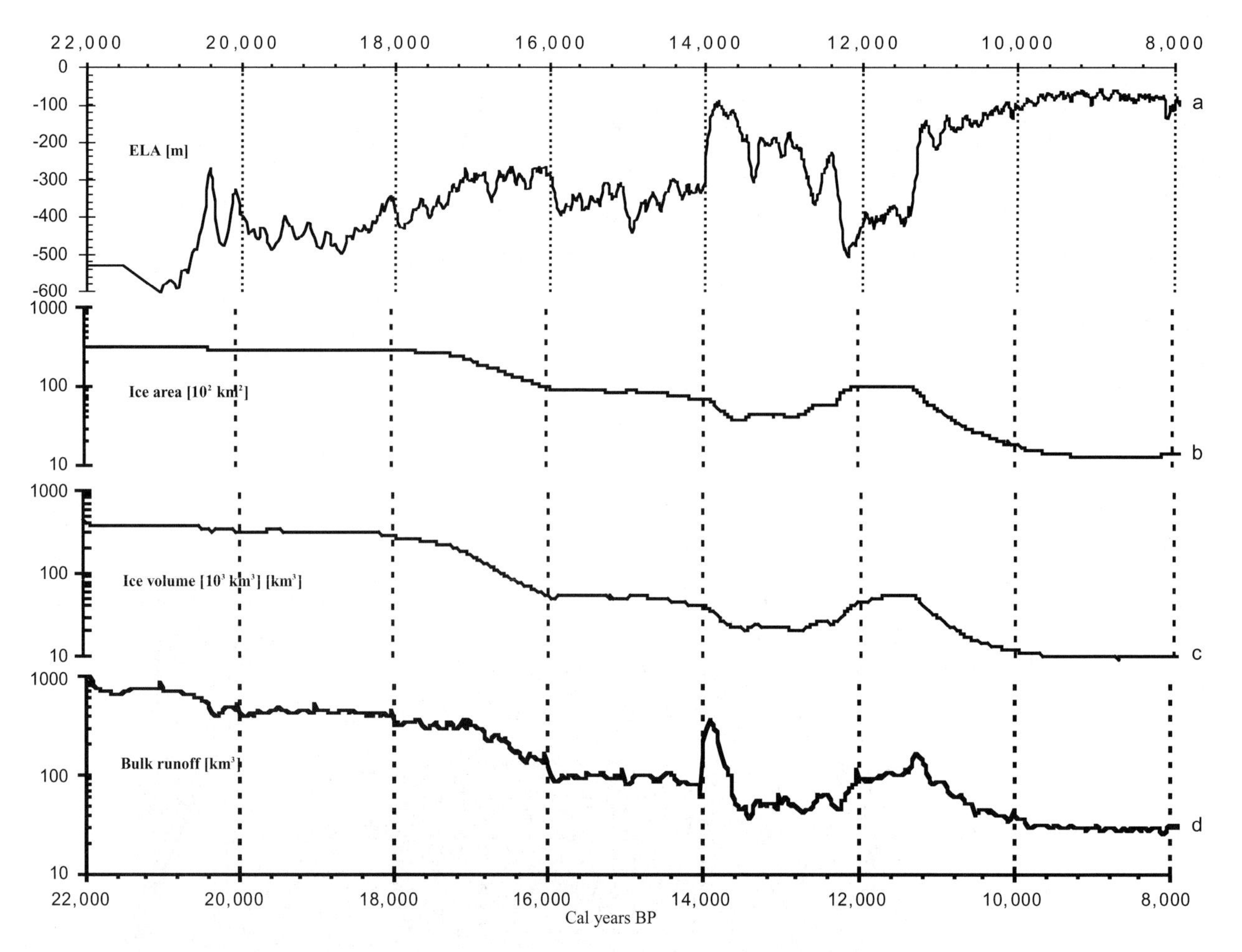

Figure 22.6 Modelled time-series from LGM through to early Holocene of: (a) predicted mean ELA, (b) ice sheet area, (c) ice volume and (d) bulk freshwater runoff.

potential 'work-done' on the substrate were maximized during the LGM, then it may be argued that a significant proportion of these features should have formed during this time. A broad comparison reveals numerous areas displaying a high level of consistency, but also some specific areas under the ice divide that are divergent, or where there is no basal sliding component modelled. Detailed comparison of model and observed orientations in two specific localities, within the northern fjords and the southern central lowlands, are, though, encouraging (Fig. 22.5b & c). In both cases modelled vectors not only correspond but are 'tighter' than observed orientations, reflecting the alternative flow regimes associated with a subsequent, less extensive ice sheet during deglaciation.

From this optimum LGM configuration the model is integrated forward in time using the scaled GRIP record from 21 kyr BP to the early Holocene. Output time-series (Fig. 22.6b, c & d) reveal a highly dynamic ice sheet characterized by large and rapid fluctuations in area, volume and bulk runoff (melt and calving flux). A most notable pulse of up to 300 $km^3 yr^{-1}$ occurs at ca. 14 kyr BP, for a period of ca. 100 yr during a period of ice-sheet retrenchment onshore which is synonymous with the Heinrich 1 event. Such an event will have been amplified with the onset of volcanism due to magma upwelling in the crust after the initial deglaciation episode up to 16 kyr BP. By 12 kyr BP though, the ice sheet readvances in the Younger Dryas episode during which ice expands to the present-day coastline (Fig. 22.7a), although the ice cap over the northwest remains independent of the main ice and many areas, especially the main peninsulas and northern highlands, escape inundation.

Comparison of modelled ice-extent with mapped shorelines and end moraines in the north, northeast and southeast (Fig. 22.7b, c & d) demonstrates that the model accurately creates a coherent three-dimensional reconstruction of the Younger Dryas, which matches available observation data and which, given the freely determined nature of the simulation, corroborates its output and robustness. To summarize it is hoped that this modelling effort demonstrates that the unique climatic, volcanic and oceanic setting of Iceland had a major impact on the Late-glacial ice sheet rendering it highly dynamic with significant potential for rapid collapse, which through the resulting salinity changes of the Nordic Sea could have had substantial impacts extending well beyond Iceland's immediate coastline.

Coupled, time-dependent modelling is one means of unifying the reconstruction of the late Weichselian Icelandic ice sheet,

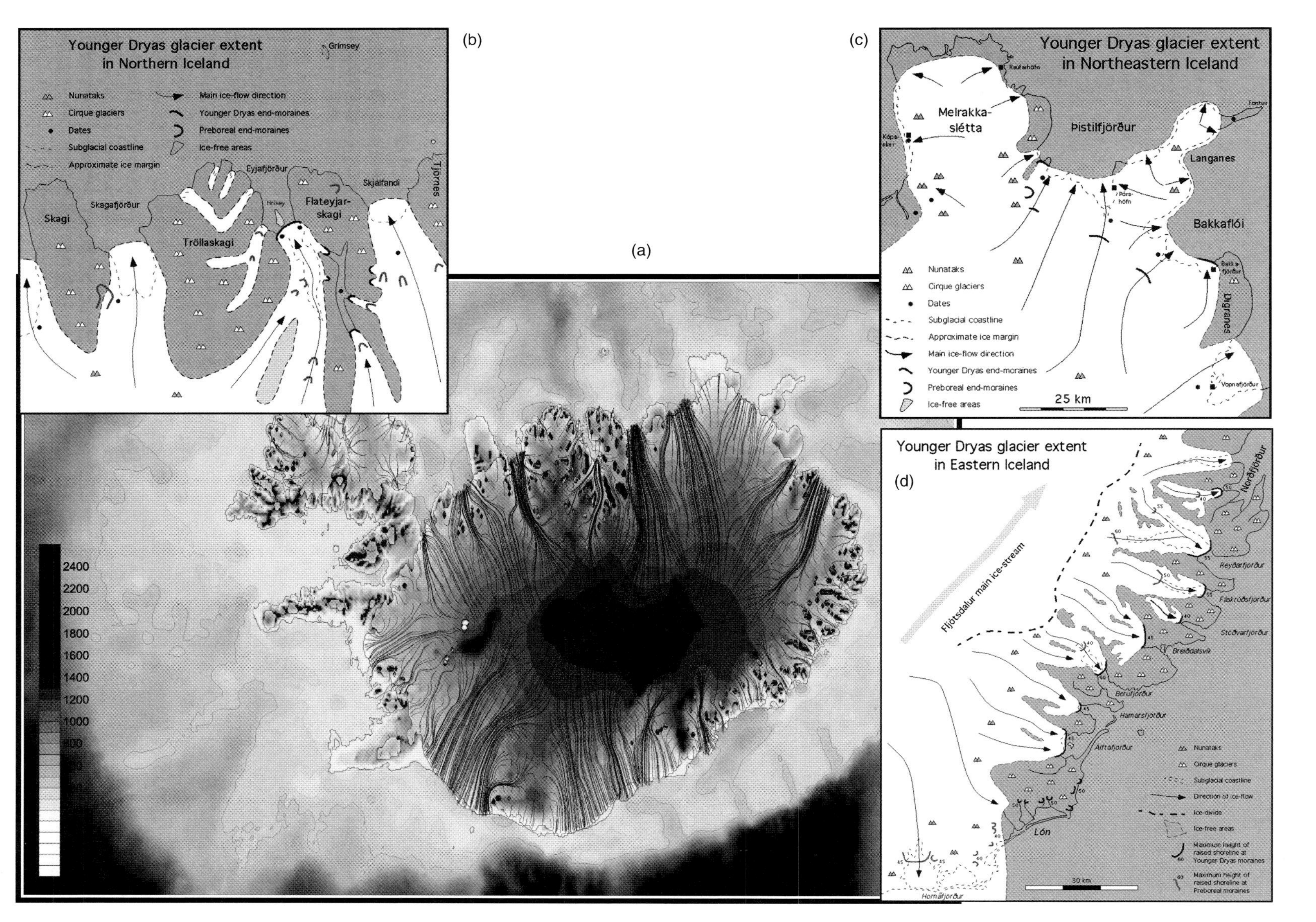

Figure 22.7 (a) The modelled Younger Dryas ice-sheet geometry and associated flowlines compared with empirical reconstructions for: (b) the northern fjords, (c) the northeast and (d) the southeast based on geomorphological mapping and dating of end moraines, trimlines and raised shorelines (Norddahl & Pétursson, in press). (See www.blackwellpublishing.com/knight for colour version.)

providing a coherent framework by which the underlying link between form and process can be explored and understood. Specifically, it allows for the controlled and systematic perturbation of an envelope of external environmental forcing variables along with those internal parameters governing ice dynamics, to determine the sensitivity and response of the climate–ice-sheet–landscape system. Significantly, the approach enables the trajectory of multiple virtual ice sheets to be investigated and, ultimately, guided to be optimally compatible with the on- and offshore glacial records. Here, such a framework has been specifically used to: (i) investigate the palaeoclimatic change required to yield an LGM ice sheet which is optimal with the empirical record; (ii) elucidate the mechanisms that control its response; and (iii) provide insight into the limits and geometry of the palaeo-ice sheet in areas where field evidence is deficient.

TWENTY-THREE

The cryosphere and climate change: perspectives on the Arctic's shrinking sea-ice cover

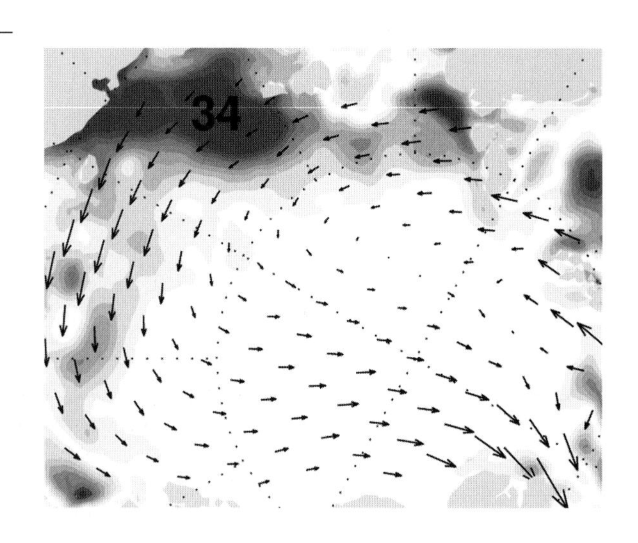

Mark C. Serreze* and Ignatius Rigor†

**Cooperative Institute for Research in Environmental Sciences (CIRES), National Snow and Ice Data Center (NSIDC), University of Colorado, Boulder, Colorado, USA*
†Polar Science Center, Applied Physics Laboratory, University of Washington, Seattle, Washington, USA

23.1 Introduction

The Arctic environment is in the midst of significant change (Serreze *et al.*, 2000). The past several decades have seen pronounced warming in northern high latitudes, especially for winter and spring. Over land, there are indications of a shift from tundra to shrub vegetation, as well as regional warming of permafrost. The mass balance of small glaciers in the Arctic has been generally negative (Dyurgerov & Meier, 1997), paralleling a global tendency. Based on data from 1979–1999, Abdalati & Steffen (2001) show that the area of the Greenland ice sheet exhibiting summer melt has increased. It also seems that the lower coastal areas of the Greenland ice sheet thinned in the 1990s, with the coastal ice losses contributing to sea-level rise (Thomas, 2001). Of all the changes that have recently been observed in the Arctic, the decline in sea-ice cover stands out prominently. To complement the other chapters in this book it is useful to review the changes in Arctic sea ice.

Satellite passive microwave imagery has allowed for detailed assessments of Arctic sea-ice extent and concentration. Coverage from October 1978 through to 1987 is provided by the Nimbus-7 Scanning Multichannel Microwave Radiometer (SMMR) and since 1987 from the Defense Meteorological Satellite Program (DMSP) Special Sensor Microwave/Imager (SSMI). Figure 23.1 gives the time series of Arctic sea-ice extent through 2002 from the combined satellite record as normalized monthly anomalies. Sea-ice extent is defined as the region with an ice concentration of at least 15%. There is considerable month-to-month variability in normalized departures as well as evidence of a multiyear

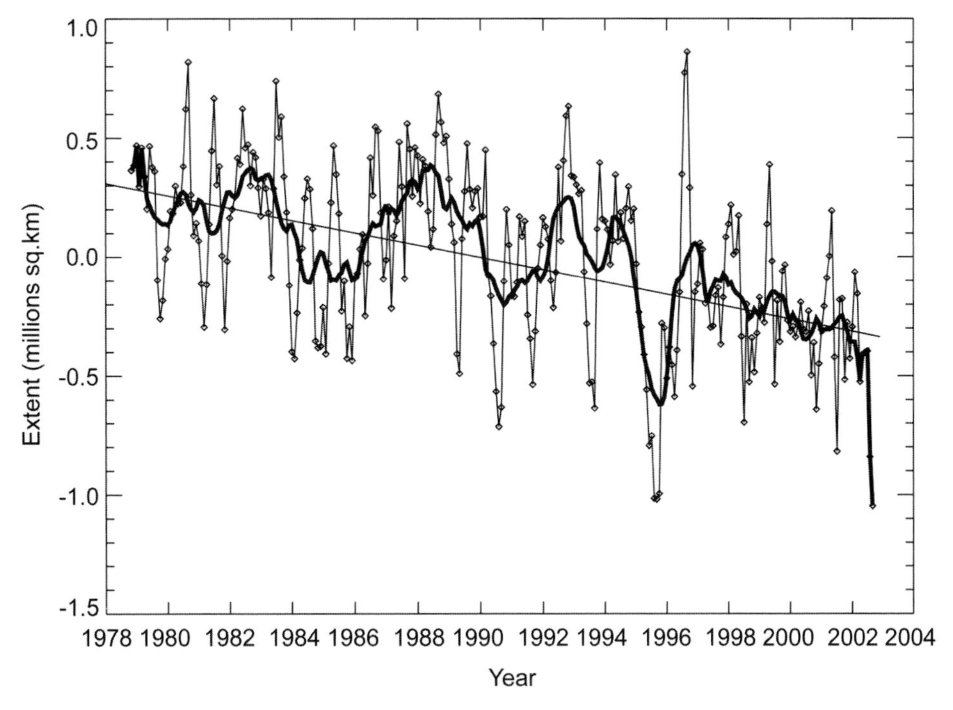

Figure 23.1 Time series of Arctic monthly sea-ice extent anomalies, 12-month running anomalies and least-squares linear fit for the Arctic basin, based on data from 1979 to 2002. (Courtesy of National Snow and Ice Data Center, Boulder, CO.)

cycle. These features are in turn superimposed upon a statistically significant downward trend. Using data though 1997, Cavalieri *et al.* (1997) calculated a trend of 2.9 ± 0.4% per decade. Closer inspection of the record indicates that the downward tendency is driven primarily by sea-ice reductions in late summer and early autumn. Extreme minima were observed in September of 1990, 1993, 1995, 1998, 2002 and 2003. September of 2002 set a new record low for the passive microwave era (Serreze *et al.*, 2003). Comparisons with earlier, albeit less reliable data, based on visible band satellite imagery, aircraft and ship reports, suggest that ice extent in September 2002 was the lowest in at least the past 50 years. Conditions were nearly as extreme in September 2003 (Plate 23.1). From satellite time series updated at the National Snow and Ice Data Center (NSIDC) in Boulder, Colorado (http://nsidc.org/data/seaice_index/), 2004 was another low sea-ice year. Records from ship reports and other sources provide some evidence of century-scale reductions over the North Atlantic (e.g. Walsh *et al.*, 1999), but the quality of these data is open to question.

There is additional evidence of attendant thinning of the perennial ice pack in the central Arctic Ocean (Rothrock *et al.*, 1999). This is the part of the ice cover that is present year-round, comprised of ice floes typically 3–5 m in thickness that have survived at least one melt season (known as multiyear ice). This contrasts with first-year ice, which forms in areas of open water during the autumn freeze up and winter. Through the use of upward-looking sonar, submarines can provide information on sea-ice draft—the fraction of the total ice thickness (about 90%) that projects below the water surface. Comparisons between sea-ice-draft data acquired during submarine cruises between 1993 and 1997 with earlier records (1958–1976) indicate that the mean ice draft at the end of the summer melt period decreased by 1.3 m in most of the deep-water portions of the Arctic Ocean. Questions have arisen regarding how much of this thinning represents the effects of melt/reduced ice growth, a wind-driven redistribution of thicker ice towards the coast of the Canadian Arctic Archipelago (Holloway & Sou, 2002), or a flux of ice out of the Arctic basin via Fram Strait (between Svalbard (Spitzbergen) and northern Greenland) (Rigor & Wallace, 2004). Although all of these processes may be at work, the latter seems quite important (see later discussion). An updated analysis by Rothrock *et al.* (2003), based on both submarine observations and models, gives further supporting evidence for thinning in the late 1980s through to 1997, with some indication of recent recovery.

How can we explain these ice losses? A number of studies (e.g. Maslanik *et al.*, 1998; Serreze *et al.*, 2003) have addressed some of the recent large anomalies (e.g. 1998, 2002) as case studies. These studies document the importance of anomalies in the atmospheric circulation that alter the distributions of surface air temperature (hence ice melt/growth) and the wind-driven circulation of the sea ice cover (e.g. promoting ice drift away from the coastlines). Thermodynamic and dynamic forcing tend to be closely intertwined. Although such studies have certainly been useful, a more basic framework is required.

One possible unifying framework is greenhouse-gas warming. A common feature of climate model projections is that, largely due to albedo feed-backs involving snow and sea ice, the effects of greenhouse gas loading will be observed first and will be most pronounced in the Arctic region (Holland & Bitz, 2003). Sea-ice loss is a near universal feature of these projections. Although the rate of decline varies widely between models, some predict a complete loss of summer ice cover by the year 2070. Given that the Arctic has warmed in recent decades, it is certainly tempting to view the observed ice losses as an emerging greenhouse signal. As reviewed below, however, the more obvious explanations involve changes in the atmospheric circulation. These changes account not only for much of the warming, but have led to important impacts on the sea-ice circulation. Might the circulation changes themselves be at least partly a response to changes in atmospheric trace gas composition? Although there is no firm consensus on this issue, there is growing evidence that this may be the case.

23.2 The North Atlantic and Arctic Oscillations

The circulation changes of primary interest here involve the North Atlantic Oscillation (NAO) and the Arctic Oscillation (AO), the latter also known as the Northern Hemisphere Annular Mode (NAM). The NAO represents covariance between the strength of the Icelandic Low, an area of mean low pressure at sea level centred near Iceland, and the Azores High, an area of mean high pressure at sea level centred near the Azores. In the positive mode, both are strong. Thompson & Wallace (1998) argued that NAO should be placed in the more general framework of the AO. The AO framework describes an oscillation of atmospheric mass between the Arctic and middle latitudes. In the positive mode, pressures are low over the Arctic, most strongly expressed in the vicinity of the Icelandic Low.

Although debate has ensued as to which framework is most appropriate, both on statistical and physical grounds, for many purposes, including the present discussion, we can consider the NAO and AO as different expressions of the same basic phenomenon. They can be identified at any time of the year, but are best expressed from late autumn into early spring. The strength and phase of the NAO and AO can be described by simple indices. As evaluated using data for December through to March (Fig. 23.2), the period from about 1970 onwards has been characterized by an overall trend toward the positive modes of the NAO/AO. There has certainly been high variability over this period. Note in particular that the NAO/AO was especially positive from about 1989 to 1995, but has subsequently regressed toward a more neutral state.

The characteristics of recent Arctic warming have been widely examined (see the review of Serreze & Francis, in press). Most studies have focused on changes in surface air temperature (SAT) over land areas where the longest time series are available, some of which extend into the 19th century. Computed trends are sensitive to the period examined, the data source, and the way in which the data are analysed. However, it is quite clear that land areas have experienced warming since about 1970, at rates larger than that for the Northern Hemisphere as a whole. Although warming is evident in all seasons, it has been most pronounced over the Eurasian and North American sub-Arctic in winter and spring (Serreze & Francis, in press). As outlined in a number of studies (Hurrell, 1995, 1996; Thompson & Wallace, 1998, 2000; Thompson *et al.*, 2000; Hurrell *et al.*, 2003), this winter and spring warming is consistent with the change to the positive NAO/AO mode. The basic reason is that the positive mode of the NAO/AO

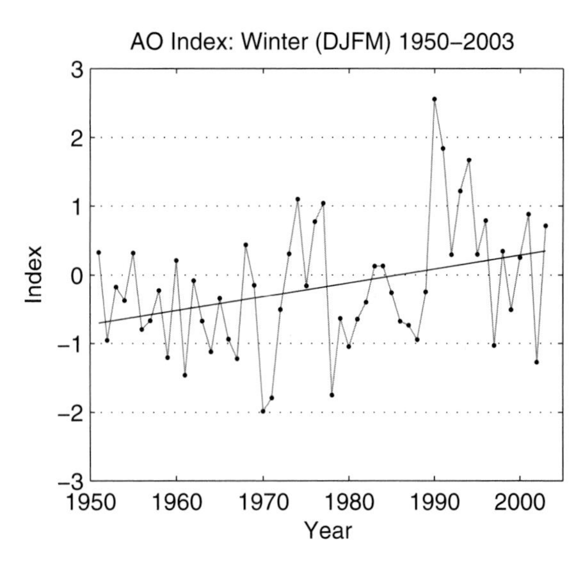

Figure 23.2 Time series of the winter (December through to March) index of the Arctic Oscillation from 1950 to 2003 and linear least-squares regression line.

promotes advection of relatively warm air across large parts of the Arctic and sub-Arctic land area.

Trends have also been observed over the Arctic Ocean. Information comes primarily from the Russian 'North Pole' measurements for 1950–1991, collected at a series of manned drifting camps (providing only one to four monthly means per year), since 1979 from arrays of drifting buoys maintained by the International Arctic Buoy Programme (IABP) and from the late 1970s or early 1980s onwards from various satellite retrievals. An analysis of the Russian North Pole records (1961–1990) by Martin *et al.* (1997) showed a significant increase in Arctic Ocean SAT during May and June (0.89 and 0.43°C per decade). Rigor *et al.* (2000) assessed trends for the period 1979–1997 north of 60°N via gridded fields that combine land-station records with data from the North Pole stations and the IABP. In basic agreement with the Martin *et al.* (1997) study, positive ocean trends are most pronounced and widespread during spring. The subsequent effort by Comiso (2003) made use of clear-sky surface temperature retrievals from Advanced Very High Resolution Radiometer (AVHRR) satellite data for 1981 onwards. Based on updates using improved algorithms, the Arctic Ocean has experienced warming during winter, spring and autumn, with greatest increases in spring and autumn.

All of these studies point to an earlier seasonal melt onset and lengthening of the melt season. A longer melt season would favour less total ice cover at summer's end. Once the ice cover begins to melt, the reduction in the surface albedo provides a feed-back to foster further melt. Comiso (2003) calculates an increase in the melt season of 10–17 days per decade. Further support for extended melt comes from Belchansky *et al.* (2004), who base their independent analysis on passive microwave satellite retrievals. Comiso's (2003) lack of summer SAT trends is not surprising over the Arctic Ocean. Presumably, his summer trend includes the effects of an earlier June transition to the melting point.

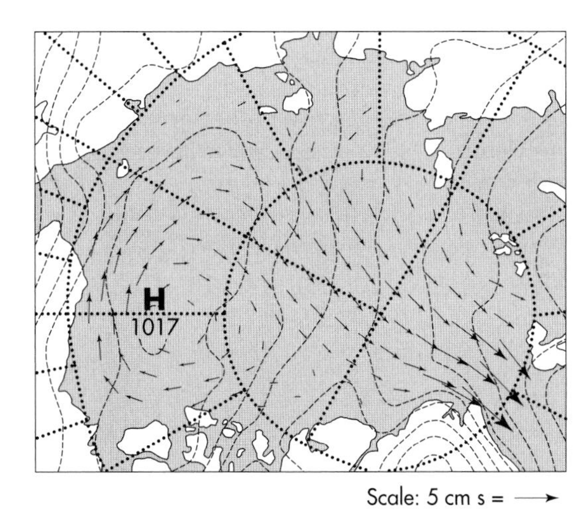

Figure 23.3 Mean annual circulation of the Arctic sea-ice cover, based on data from the International Arctic Buoy Programme, with overlay of sea-level pressure from 1979 to 2002. Contours are shown every 1 hPa.

As discussed, the pronounced winter and spring SAT changes over land areas reflect the direct advective effects of the change in the NAO/AO. It might be tempting to attribute spring warming over the Arctic Ocean and earlier melt onset to this same process. Indeed, one finds that the largest reductions in sea-ice extent have occurred along the coasts of Siberia and Alaska, where the rises in spring SAT have been most pronounced. Temperature advection probably is playing some role. A closer analysis forces us to conclude, however, that although the ice reductions are closely tied to the NAO/AO, the primary links are actually more 'indirect'.

23.3 Impacts of the sea-ice circulation

Except for areas of landfast ice near the coast, the Arctic ice pack is in near constant motion. The mean annual circulation has two major features, a clockwise motion in the Canada Basin known as the Beaufort Gyre, and a motion of ice from the Siberian coast, across the pole and through Fram Strait, known as the Transpolar Drift Stream (Fig. 23.3). Most of the ice that leaves the Arctic on an annual basis exits through the Fram Strait (between Spitzbergen and Greenland). This flux is primarily of thick, multiyear ice. The long-term mean ice circulation is determined by roughly equal contributions of the surface wind field and surface ocean currents. However, the ice motion exhibits pronounced variability, which is primarily wind driven. Depending on the wind field and internal stresses in the ice cover (floe-to-floe interactions), the ice motion may be divergent, decreasing ice concentration, or convergent, increasing ice concentration. If ice concentration is already 100%, convergence will raft ice floes atop one another (part of the sea ice deformation process). Divergence leads to more open water. In autumn and winter, the open-water areas will quickly freeze, forming areas of new, thinner first-year ice. In summer, the exposure of dark open-water areas associated with divergence promotes strong absorption of solar radiation, increasing melt. Understanding the observed loss of Arctic Sea ice requires that we address such factors.

The NAO/AO link seems to involve several related processes. The first two can be understood with the aid of Plate 23.2, adapted from the study of Rigor *et al.* (2002). The top panel shows observed trends in summer sea-ice concentration (based on passive microwave records) along with observed trends in sea-ice motion for the previous winter (based on IABP data), both computed over the period 1979–1998. Following earlier discussion, the trends in summer ice extent have been largest along the Siberian and Alaskan coasts. Ice concentration trends largely mimic the trends in extent. The trends in winter ice motion indicate a change to a more cyclonic (counterclockwise) pattern. Care should be exercised in the interpretation—although the actual winter ice motion remained generally anticyclonic (broadly like that in Fig. 23.3), the trend has been toward more cyclonic conditions. This manifests itself as a shrinking of the Beaufort Gyre, and an increase in the size of the Transpolar Drift Stream during high NAO/AO conditions. The bottom panel shows the components of summer ice concentration and winter ice motion for each year regressed on the winter AO index. The key point is that the patterns are very similar—most of the change in summer ice concentration and winter ice motion relates to the trend in the winter AO.

The relevant processes are as follows. Changes in the winter wind field associated with the general upward trend in the winter AO changed the sea-ice motion field. This change features enhanced transport of ice away from the Siberian and Alaskan coasts and more ice divergence (or less convergence). Both processes lead to openings in the ice cover, which quickly refreeze to form thin, first-year ice. By spring, there is an anomalous coverage of this thin ice. With more thin ice, less energy is needed to completely melt the ice in summer. Unless there is a compensating motion of ice into the region, there will be less ice at summer's end. A second effect is that the thinner ice also leads to an earlier melt onset, further favouring ice losses. Except in summer, sea ice insulates the cold atmosphere from the relatively warm underlying ocean. With thinner ice in spring, the insulating effect is reduced. There are strong heat fluxes to the atmosphere, manifested as a positive SAT trend. The warmer air temperatures, together with the fact that the thinner, first-year ice melts at a lower temperature, fosters early melt (salt depresses the freezing point, and first-year ice is saltier than multiyear ice). Hence, the SAT rise in spring favouring early melt is not so much a direct advective effect of the NAO/AO, but rather a result of wind-induced changes in the ice cover.

The Rigor *et al.* (2002) model does not explain all aspects of the recent ice trend, such as the extreme minima of 2002 and 2003. As noted, the AO, although showing a positive trend since about 1970, and very positive from about 1989–1995, has in recent years retreated toward a more neutral state (Fig. 23.2). Serreze *et al.* (2003) highlight the very strongly cyclonic circulation of the atmosphere during the summer of 2002. A manifestation of the sea ice momentum balance is that ice drift tends to be to the right of the surface geostrophic wind (the wind parallel to contours of sea-level pressure, representing a balance between the pressure gradient and Coriolis forces). As a consequence, when the atmospheric circulation (hence sea-ice circulation) is cyclonic, the ice cover will tend to diverge, resulting in more open water. Although by itself, ice divergence in summer will spread the existing ice over a larger area, Serreze *et al.* (2003) argued that enhanced absorption of solar energy in the resulting dark open-water areas between floes would lead to much stronger melt, reducing ice extent. For each month from June through to August 2002, the sea-level pressure fields featured mean closed lows over the Arctic Ocean, an extremely unusual, stormy pattern. Further analysis shows the summer of 2003 as also characterized by an unusually cyclonic pattern.

The subsequent analysis of Rigor & Wallace (2004) gives a different view. Using a simple sea-ice model, they show that these recent extreme sea-ice minima may have been induced dynamically by changes in surface wind field over the Arctic associated with the previous very positive AO state from about 1989 to 1995. Put differently, the sea-ice system seems to have a memory of the previous high AO state. Plate 23.3 shows the estimated age (in years) of sea ice for 1986 and 2001 based on the model. The obvious difference between the two years is the much younger age of ice in 2001. In the 1980s, winds associated with primarily negative AO conditions favoured a larger Beaufort Gyre, which trapped sea ice in the central Arctic Ocean for over 10 years. New ice that was formed along the coast was advected away from the coast towards Fram Strait. For example, in the top panel we mark areas of sea ice that formed in the coastal Beaufort and Chukchi seas in the autumn of 1981 and 1984 which are drifting towards Fram Strait in the Transpolar Drift Stream. By comparison, sea ice returning to the Alaskan coast along the eastern flank of the Beaufort Gyre has been drifting in the central Arctic Ocean for many years. From 1989 to 1991, the average age (thickness) of sea ice in the Arctic Ocean decreased rapidly with the step to high NAO/AO conditions. The areal extent of older, thicker ice declined from 80% of the Arctic Ocean to 30%. Although some of this loss may represent melt, it seems that the old ice was primarily exported out of the Arctic via Fram Strait. Several investigators have documented an upward trend in either ice volume or area fluxes through the Fram Strait from the late 1970s through much of the 1990s (e.g. Kwok & Rothrock, 1999). The overall effect is that sea ice stayed in the Arctic Basin for a shorter period, meaning less time for the ice to ridge and thicken.

The AO then declined from its high positive state. During the summers of 2002 and 2003, the anomalously younger, thinner ice was advected into Alaskan coastal waters where very pronounced ice melt occurred. For example, in the bottom panel of Plate 23.3, we show the recirculation of the much younger ice formed in the Beaufort and Chukchi seas in the autumns of 1993, 1997 and 1999 back towards the Alaskan coast instead of leaving the Arctic through Fram Strait. This argument helps to explain why the past several years have seen large ice reductions, despite the decline in the winter AO.

Rigor & Wallace (2004) show that, statistically, ice divergence associated with the cyclonic summer pattern noted by Serreze *et al.* (2003) would actually increase, rather than decrease, ice extent (the ice would be spread over a larger area). By this argument, the more dominant role of the cyclonic summer circulation was probably its contribution to the unusually diffuse ice cover in the Beaufort Sea seen in September of both years (Plate 23.1). Another remarkable aspect of September 2002 and 2003 (again see Plate 23.1) was the virtual absence of sea ice in the Greenland Sea (off the east coast of Greenland). Serreze *et al.* (2003) argued that the cyclonic summer circulation led to a wind field that limited ice transport through Fram Strait and into the Greenland Sea. In the Rigor & Wallace (2004) framework, however, the Fram Strait flux was probably composed of unusually thin ice, such that ice entering the Greenland Sea rapidly melted away. A reasonable hypothesis is that both processes were at work.

23.4 Ocean influences

A mean vertical profile of temperature and salinity over the Arctic Ocean reveals several features. There is a low salinity surface layer, with temperatures near the salinity-adjusted freezing point. Below this surface layer, extending to about 200–300 m depth, is a rapid increase in salinity. This is attended by an increase in temperature to maximum (and above-freezing) values at around 300–500 m depth. Although temperature falls off at greater water depths, from about 400 m downward salinity stays nearly uniform at 34.5–35.0 psu (practical salinity units). The layer of rapid salinity increase is termed a halocline. The layer of rapid temperature change is termed a thermocline. Over most of the global ocean, a stable upper-ocean stratification (less dense water at the top) is maintained by higher temperatures closer to the surface. The situation in the Arctic is quite different. At the low water temperatures of the Arctic, the density is determined by salinity. Consequently, the halocline is associated with a strong increase in density with depth (a pycnocline). This means that the upper Arctic Ocean is very stably stratified. The warmer water at depth, if brought to the surface, would quickly melt the sea-ice cover. Suppression of vertical mixing by the 'cold Arctic halocline' is one of the key features of the Arctic (along with low winter air temperatures) that allows sea ice to form and persist.

The fresh surface layer is maintained primarily by river runoff, the influx of relatively low salinity waters from the Pacific into the Arctic Ocean through the Bering Strait and net precipitation over the Arctic Ocean. It is also influenced by the growth and melt of the sea-ice cover. As sea ice forms, brine is rejected. The density of the surface layer increases as does the depth of vertical mixing. As ice melts in summer, the surface layer becomes fresher and vertical mixing is inhibited. The temperature maximum layer at about 300–500 m depth manifests the inflow of Atlantic-derived waters. This is provided by two branches, one west of Spitzbergen (the West Spitzbergen Current) and one through the Barents Sea (the Barents Sea Branch). It appears that this Atlantic inflow has changed.

As summarized by Dickson *et al.* (2000), results from a number of oceanographic cruises indicate that, in comparison with earlier climatologies, the Arctic Ocean in the 1990s was characterized by a more intense and widespread influence of Atlantic-derived waters. The Atlantic-derived sublayer warmed 1–2°C compared with Russian climatologies of the 1940s to 1970s and the subsurface temperature maximum shoaled (to about 200 m in from some observations). The boundary between waters of Atlantic and Pacific origin spread west, resulting in an extension of the

Atlantic water range by nearly 20%. Evidence also arose of a weakening of the cold halocline in the Eurasian Basin (Steele & Boyd, 1998).

The change in the cold halocline is attributed to eastward diversion of Russian river inflow in response to changes in the atmospheric circulation. The Atlantic layer changes are attributed primarily to changes in the Atlantic inflow in the early 1990s, and some warming of this inflow. These changes have been modelled successfully, and are viewed as a response to changes in surface winds associated with increasing dominance of the positive phase of the NAO/AO. As argued by Maslowski *et al.* (2000), this enhanced inflow of warm, Atlantic-derived waters promotes a stronger upward oceanic heat flux, which may be contributing to the observed sea-ice decline. Subsequent work (Maslowski *et al.*, 2001) suggests an additional role of a warmer inflow of Pacific waters through Bering Strait. Whether changes in the cold halocline may also be involved is not clear.

23.5 Discussion

Do the observed changes in Arctic sea ice, temperature and atmospheric circulation represent natural variability or is there an anthropogenic influence? As evaluated over the 20th century, the Arctic has exhibited considerable variability on decadal and multidecadal time-scales. Although a great deal of emphasis has been placed on the recent warming, available records, albeit primarily from inland and coastal stations, document a rise in surface air temperatures from about 1920 to 1940 that was just as large, followed by a period of cooling. As frequently pointed out (e.g. Polyakov & Johnson, 2000; Polyakov *et al.*, 2002; Semenov & Bengtsson, 2003) such low-frequency variability in the Arctic can make it very difficult to separate natural climate fluctuations from those due to trace-gas loading. The earlier and most recent warmings are certainly quite different. The earlier warming was largely restricted to Northern high latitudes, with complex seasonal and spatial expressions (Overland *et al.*, 2004). By sharp contrast, the post-1970s warming is essentially global. That it is most strongly expressed in northern high latitudes reflects the influence of the NAO/AO. This includes both the direct advective influences, and, as discussed here, impacts of the declining ice cover itself.

An ongoing puzzle concerns the cause of the NAO/AO trend that seems so strongly linked with the sea-ice decline. The NAO/AO is a natural atmospheric mode. Decadal variability in the NAO/AO is a fundamental characteristic of internal atmospheric dynamics, with lower-frequency variability involving ocean coupling. A growing weight of evidence, however, suggests that the recent change is not entirely natural. Several studies point to links with changes in the stratosphere. Cooling of the stratosphere, in response to greenhouse gas (GHG) loading or losses of stratospheric ozone, can lead to 'spin up' of the polar stratospheric vortex, resulting in a positive shift in the NAO/AO (e.g. Gillett *et al.*, 2003). Such ideas are being tested in a wide variety of model simulations, but no consensus has been reached. Another idea, outlined by Hoerling *et al.* (2001) and expanded in subsequent efforts (Hurrell *et al.*, 2004; Hoerling *et al.*, 2004), is that an observed slow increase in sea-surface temperatures in the tropical Indian Ocean have helped to 'bump' the NAO/AO toward a preferred positive state. They further argue that the ocean warming may itself represent a non-linear response to GHG loading. Links with ENSO (Cassau & Terray 2001) and feed-backs on the circulation associated with factors such as sea-ice loss (e.g. Semenov & Bengtsson, 2003; Delworth & Knutson, 2000; Deser *et al.*, 2004) also require further investigation.

Could changes in terrestrial ice volume have an impact on the future state of the sea-ice cover? Taken by itself, an enhanced discharge of meltwater from glaciers and ice sheets might lead to greater upper-ocean stability, increasing sea-ice production in winter. Although, presumably, this effect would be overwhelmed by a reduction in winter ice growth and greater summer melt due to warmer climate conditions, at least one model simulation suggests a strong and abrupt weakening of the Atlantic thermohaline circulation at the end of the 21st century, triggered by enhanced freshwater input arising largely from melting of the Greenland Ice Sheet. This simulated event caused marked cooling over eastern Greenland and the northern North Atlantic (Fichefet *et al.*, 2003), to which the ice cover would presumably respond. Although perhaps an unlikely scenario, it forces us to acknowledge that the Earth's future climate may hold surprises.

Acknowledgments

This study was supported by NSF grants OPP-0229651, OPP-0242125, NASA contract NNG04GH04G, NOAA grant NA17RJ1232, and ONR grant N00014-98-1-0698.

TWENTY-FOUR

The interaction of glaciers and oceans in the context of changing climate

Gerard Bond[1]

Late of Lamont-Doherty Earth Observatory of Columbia University, P.O. Box 1000, 61 Route 9W, Palisades, NY 10964-1000, USA

24.1 Introduction

In the context of changing climate, interactions between glaciers and the oceans occur mainly within two bands. One includes the Milankovitch oscillations of 20,000 to 400,000 yr and the second contains millennial-duration changes of between 1000 and 10,000 yr. Although both are currently of great interest to the climate community, I focus here on the millennial band where rapid transients in forcing mechanisms led to complex non-linear responses in the ice–ocean–climate system. It is within this band that dramatic, abrupt climate changes were first recognized and that possible scenarios for future abrupt events first emerged.

24.2 Background

During the last glacial cycle, Earth's climate was punctuated by two series of abrupt changes of millennial duration. One of these, known as Dansgaard–Oeschger cycles, contains oscillations of 1000 to about 5000-yr duration that are best developed in the North Atlantic in Marine Isotope Stage 3 (Bond *et al.*, 1993; Broecker, 1994). As revealed in records from Greenland ice and North Atlantic sediments, each D–O cycle began with a remarkably abrupt warming of several degrees, probably occurring within decades, if not several years, followed by increasing amounts of ice-rafted debris (IRD) and a gradual cooling to a full stadial (Figs 24.1 & 24.2).

The second series, known as Heinrich events, consists of cycles that occurred every 7000 to 10,000 yr. As best documented in records from the glacial North Atlantic (Fig. 24.2), Heinrich events occurred suddenly at the end of cooling ramps containing D–O cycles, and were accompanied by marked increases in ice-rafted debris (IRD) and cold sea-surface and land temperatures that exceeded those of D–O cycles (Heinrich, 1988; Bond *et al.*, 1992; Hemming, 2004). Each Heinrich event was followed by an exceptionally abrupt warming that raised sea-surface temperatures by several degrees to nearly Holocene values.

Both the D–O cycles and Heinrich events are thought to have been part of nearly global reorganizations of ocean–atmosphere climates. Within the limits of dating error, apparent correlatives of both series have been found in marine and terrestrial records from high northern latitudes, through the tropics, and into at least subtropical southern latitudes (Voelker, 2002). In deep-sea cores from the southeast Atlantic Ocean, Kanfousch *et al.* (2000) found apparent correlatives of Heinrich events in records of IRD discharge from Antarctica. Distinct millennial temperature variations also occur in Antarctic ice but whether these lead, lag or are in phase with Greenland's D–O cycles is widely debated owing to uncertainties in ice-core age models (e.g., Steig & Alley, 2002, Roe & Steig, 2004, Shackleton *et al.*, 2004).

24.3 The 'conventional view'

Although there is no agreement on the nature of glacier–ocean interactions during D–O and Heinrich events, one widely held view maintains that ocean–climate changes were forced by oscillatory internal dynamics of glaciers flowing into the northern North Atlantic. Recurring surges or collapses of those glaciers are thought to have caused repeated and rapid increases in the flux of icebergs (i.e. freshwater) into the ocean. The sudden flow and subsequent melting of the icebergs in convecting regions north and south of Iceland is presumed to have lowered ocean surface salinities enough to force ocean circulation across a threshold, induce a hysteresis behaviour, and abruptly reduce or shut down North Atlantic Deep Water (NADW) formation (e.g. Broecker,

[1] Sadly, Gerard Bond passed away during the production of this volume. We are extremely grateful for his contribution and proud to be able to publish one of the last writings of a great scientist who has contributed so much to the field.

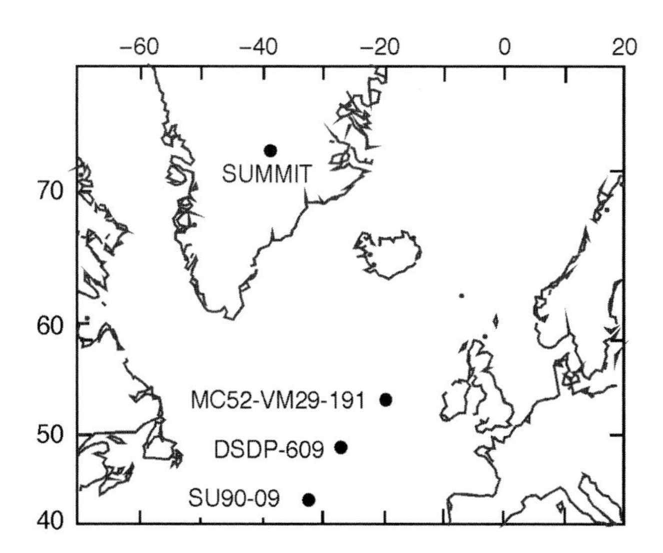

Figure 24.1 Locations of cores mentioned in the text.

1994; Ganopolski & Rahmstorf, 2001; Rahmstorf, 2002). Results of modelling experiments suggest that an increase in freshwater flux from 0.03 Sv (D–O cycles) to 0.15 Sv (Heinrich events) would have been sufficient to alter deep circulation and lower temperatures in the northern North Atlantic and surrounding regions by a few to several degrees C (Ganopolski & Rahmstorf, 2001; Weaver *et al.*, 2001). The modelled temperatures are broadly consistent with those of palaeoclimate records from the glacial North Atlantic (Cacho *et al.*, 1999; van Kreveld *et al.*, 2000; Landis *et al.*, 2004).

The key pieces of evidence that formed the basis for the glacial instability concept came from North Atlantic deep-sea records of IRD concentrations and surface salinities. Marked increases in IRD concentrations accompany the cold phases of each D–O cycle and even larger increases in IRD coincide with Heinrich events (Fig. 24.2). Petrological and geochemical tracers in IRD indicate that Heinrich icebergs were produced mainly by massive collapses of Laurentide ice in the Hudson Bay region (Grousset *et al.*, 2001; Hemming, 2004; Figs 24.2 & 24.3). For the D–O cycles an Icelandic source is well documented and geochemical fingerprinting may implicate sources in Europe as well.

Large reductions in surface salinity during the IRD maxima in both series are suggested by planktonic and dynocyst census counts, measurements of planktonic Mg/Ca ratios, and measurements of planktonic oxygen isotopes. Although estimating salinity anomalies is notoriously difficult, values of up to −4 per mil have been suggested for Heinrich events and −1 to −2 per mil for cold phases of D–O cycles (Cayre *et al.*, 1999; de Vernal & Hillaire-Marcel, 2000, van Kreveld *et al.*, 2000). Based on results of modelling experiments, those values would have been sufficient to reduce or shut down NADW formation (e.g. Ganopolanski & Rahmstorf, 2001).

To explain the oscillatory dynamics required by the glacier instability concept, MacAyeal (1993a) proposed a free oscillation mechanism that has come to be known as the binge–purge model. A large glacier or ice sheet frozen to bedrock will slowly build up during the binge phase. The purge phase (i.e. the Heinrich event) occurs when geothermal heat melts the basal sediment and produces a lubricated discharge pathway. The model produced massive collapses of the ice every 7000 yr, agreeing reasonably well with the observed timing of Heinrich events. In a more elaborate development of the model, Greve & MacAyeal (1996) found that free oscillations from 1000 to 10,000 yr could occur within a large ice sheet, thereby potentially providing an explanation for both Heinrich events and the faster-paced D–O cycles.

24.4 The chicken or egg problem

The elegant glacier-centred hypothesis, regarded as a boon to the field of glaciology, was called into question, however, by the results of efforts to test two of its corollaries. If the hypothesis were correct, the IRD increases in both series should occur at the same time as, or slightly lead, the ocean–climate response, and synchronicity of discharges from different glaciers would be unlikely owing to the vagaries of internal glacier dynamics. Instead, it was found in some high-resolution records from the North Atlantic that the onset of ocean surface coolings in both series actually preceded, by at least several hundred years, the IRD increases (e.g. Bond & Lotti, 1995; Bond *et al.*, 1999). Even more troubling was evidence from records of petrological and geochemical tracers that the IRD in D–O cycles was discharged from different glaciers at the same time (Bond *et al.*, 1999; Grousset *et al.*, 2001; Figs 24.2 & 24.3).

Probably the most surprising discovery, though, was that each Heinrich event appeared to have been immediately preceded by an increase in IRD with a different composition to that of the IRD in the overlying Heinrich layers. High-resolution petrological and geochemical analyses focused on the precursory layers demonstrated that the IRD recorded the same simultaneous discharges of icebergs from different sources that were found in D–O cycles (Figs 24.2 & 24.3). The sequence of events was documented in records from the same cores, thereby removing any question about dating error. It appeared then that Heinrich events were immediately preceded by the onset of IRD build-up in a D–O cycle.

Those findings raised the spectre of the chicken or egg problem. The lead of ocean surface coolings and the synchronous IRD discharges from different glaciers during D–O events seemed best explained by the effects of a climate–ocean mechanism that operated upon glaciers flowing into the North Atlantic. The observation of D–O-like IRD discharge immediately preceding each Heinrich event implied that even Heinrich events were triggered by the same climate mechanism.

If true, then the glacier-centred hypothesis for iceberg discharges during D–O cycles and Heinrich events is incorrect and must be replaced by a climate-centred hypothesis in which climate triggers the glacial instabilities. It should be emphasized that the impact of the iceberg discharges on formation of NADW and the subsequent large and geographically widespread climate responses would be the same for either hypothesis. In the case of the climate-centred hypothesis, though, the iceberg discharges and reduction in NADW production operate as non-linear feedbacks that amplify the climate mechanism and transmit large, abrupt coolings into and well beyond the North Atlantic region.

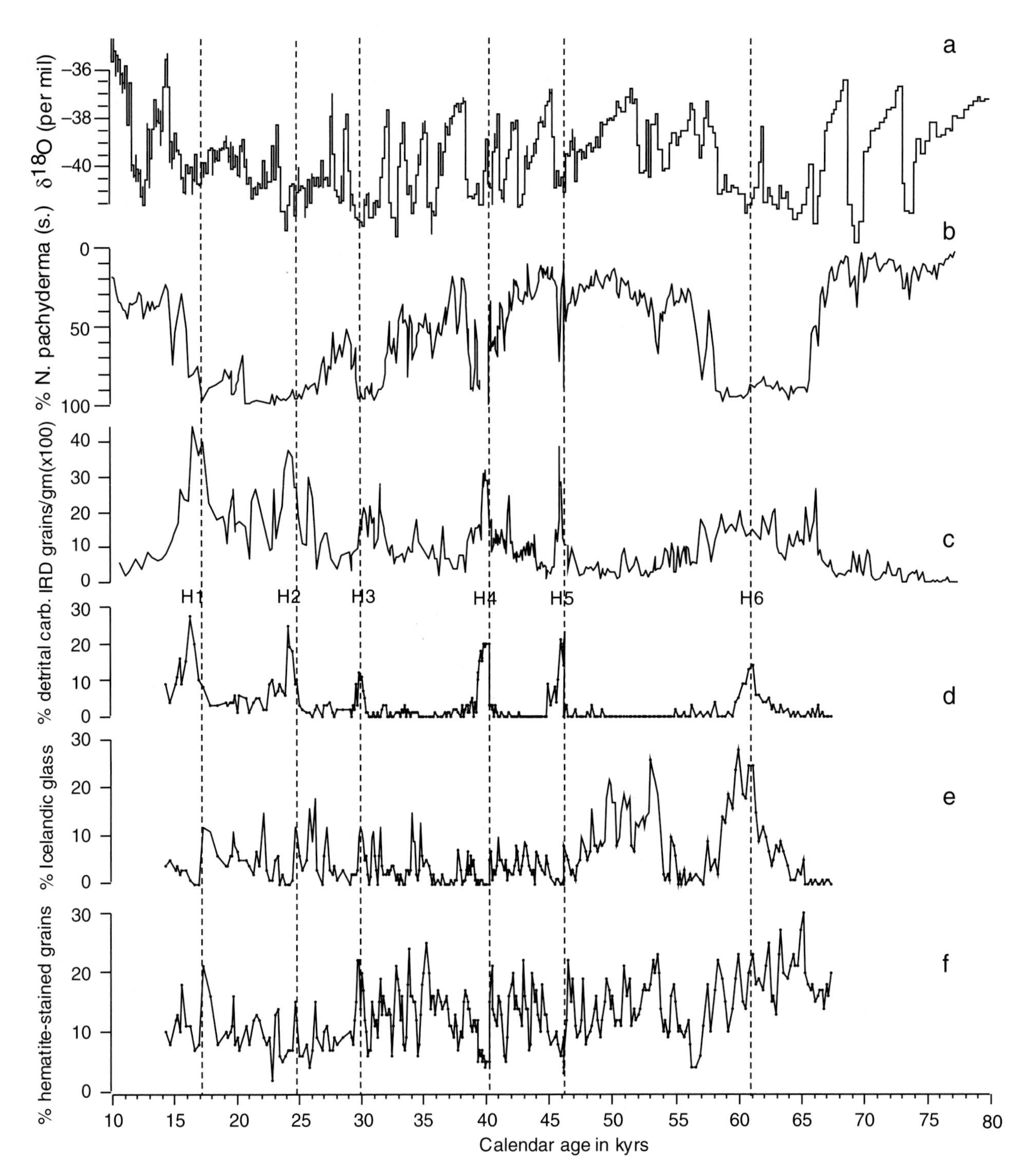

Figure 24.2 Records of climate, IRD and petrological components in IRD: (a) oxgen isotope composition in GISP2 ice core at Summit Greenland; (b–f) climate and IRD records from DSDP site 609. H refers to Heinrich event. Note that in (b) cold peaks point down; in (c–f) cold peaks point up. Dashed lines mark start of Heinrich events as defined by increases in detrital carbonate. Precursory events can be seen in both haematite-stained grains and Icelandic glass. See text and Bond *et al.* (1999) for details.

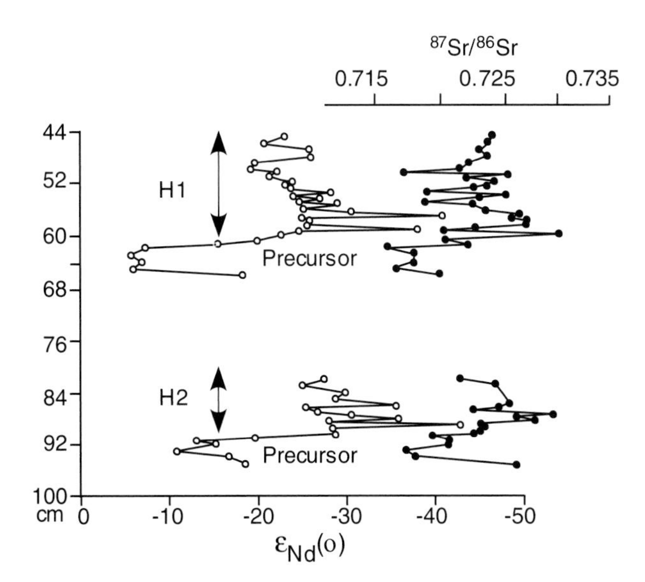

Figure 24.3 Sr and Nd isotope composition of IRD in core SU90-09 as fingerprints of IRD sources just before and during Heinrich events 1 and 2. Compositions of layers just below each Heinrich event are precursors and consistent with European sources, including Iceland. Compositions of Heinrich layers are consistent with a Laurentide source. See Grousset *et al.* (2001) for details.

Attempts to identify the climate triggering mechanism have thus far failed to produce a consensus. Broecker *et al.* (1990) proposed a salt oscillator in which a climate-driven salt build-up in northward flowing strands of the Gulf Stream strengthened NADW formation, warming the ocean and melting glacial ice. The flow of meltwater and icebergs into the ocean weakened NADW formation and cooled the ocean, leading to glacier regrowth and reduced freshwater discharge. That in turn strengthened NADW formation and initiated the next cycle. The difficulty with that concept is that the relationship of peak IRD (freshwater) discharge to ocean temperature is the opposite of what is observed. Alley *et al.* (2001) argued that a weak periodic forcing of the climate system combined with noise from ice-related events could, through stochastic resonance, produce the observed increased iceberg discharges and mode switches of D–O cycles. The model required specifying the period of the weak forcing, which they argued was the presumed 1500-yr pacing of the D–O events. What would cause a pacing with that period, and indeed whether the D–O pacing is really periodic are open questions. Hulbe *et al.* (2004) suggested that Heinrich events were preceded by growth of a large, unstable ice shelf in the Labrador Sea, which, in response to brief atmospheric warmings, broke up and triggered collapse of massive Laurentide ice in Hudson Strait. Evidence for abundant planktonic foraminifera in the Labrador Sea during Heinrich events (Andrews *et al.*, 1998b), however, argues against a large ice shelf in the Labrador Sea, and there is no evidence that atmospheric warmings preceded Heinrich events.

The tropics also have been invoked as possible drivers of the millennial variability in the glacial North Atlantic. Clement *et al.* (2001) suggested that orbital variations coupled with El Niño oscillations lock the seasonal cycle in the tropical Pacific into a La Niña state that persists for several centuries, thereby cooling the Earth and triggering the cold phases of D–O cycles and associated iceberg discharges. Stott *et al.* (2002) and Visser *et al.* (2003) argue that oscillations of a 'super El Niño' caused large, recurring ocean surface salinity changes in the tropical Pacific that appear to correlate with the D–O variations in the North Atlantic. They suggested that the tropical salinity changes altered the greenhouse gas concentration of the atmosphere, thereby changing the Earth's temperature and triggering the D–O variations in the North Atlantic. The critical piece of evidence that would place the tropics in the driving role, a lead of the tropical signal relative to that of the D–O variations, however, is lacking (see also Broecker, 2003).

Some have even questioned the basic tenant of the climate-centred hypothesis, arguing that there is no simple relation between climate, glacial advance and discharge of icebergs from tidewater glaciers. Discharges of icebergs from a glacier in an outlet are known to occur simultaneously with decreases of iceberg discharge from an adjacent outlet, for example. Even if climate drives glacial advances and retreats, the timing of the response at the terminus may vary with the size of the glacier or ice sheet. Ice-rafted detritus may reflect changing debris loads in the basal layers of the ice and whether increases in IRD flux are due to ice-margin advances or retreats is not clear (Clark *et al.*, 2000). In fiords, sea ice slows iceberg transit and IRD in basal layers of the ice may melt before icebergs reach the ocean (Reeh *et al.*, 1999).

In an interesting new approach to the problem, however, Kaspi *et al.* (2004) suggested from modelling results that all of the observations pointing to a climate trigger are consistent with an elaborate version of the binge–purge model. They argue that it is not reductions in NADW formation that force the large climate responses. Rather, they found that the strong climate response could be due mainly to the albedo effects of sea ice that would build up in the northern North Atlantic in response to even small reductions of NADW formation. The sea-ice cover created a strong atmospheric cooling which then cooled the ocean surface enough that icebergs drifted farther from their sources and spread IRD far into the North Atlantic, thereby explaining the synchronous changes in IRD from different sources during D–O cycles without invoking climate forcing of glaciers and ice sheets.

If more than one ice sheet or glacier was undergoing a binge–purge type of oscillation, each with a different frequency, the Kaspi model predicted that after a few binge–purge cycles the sea ice-induced atmospheric coupling of the different glaciers caused them to become phase locked, thereby explaining the synchronicity of iceberg discharge observed in the sediment record. (Fig. 24.4a). Their model also predicted that smaller glaciers or ice sheets could have produced precursor events because under certain conditions, even though the phase locking causes the smaller and larger ice sheets to oscillate with the same frequency, the smaller ones will tend to reach the collapse phase first by several hundred years (Fig. 24.4b).

The Kaspi version of the binge–purge model is compelling because potentially it explains all of the observational evidence in hand surrounding D–O cycles and Heinrich events. The massive build-up of sea ice that is fundamental to the model, however, is difficult to test owing to the lack of reliable proxies for sea ice in

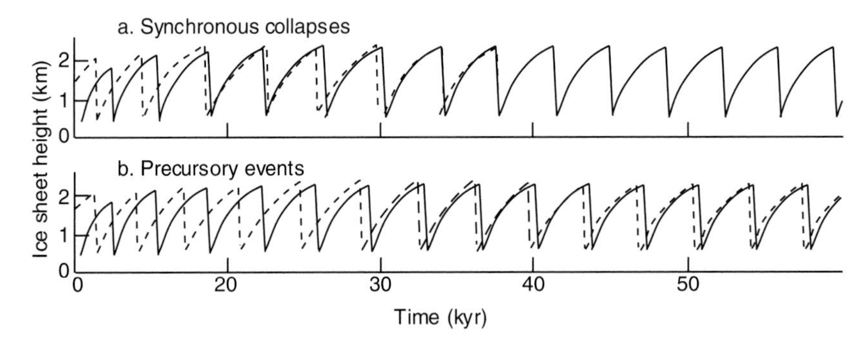

Figure 24.4 Results from the Kaspi *et al.* (2004) model suggesting that, owing to an atmospheric coupling of ice sheets as sea ice cools the North Atlantic, millennial oscillations eventually will become phase locked, producing (a) synchronous ice-sheet growth and decay. For certain model parameters (b) ice sheet growth and decay have the same periodicity, but smaller ice sheets lead larger ones, thereby producing precursory-like events seen in the North Atlantic marine records.

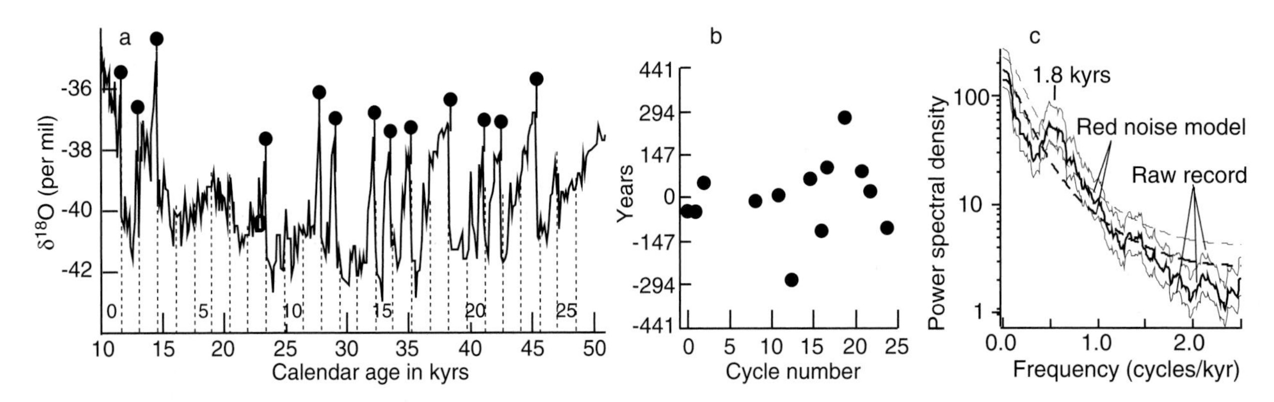

Figure 24.5 (a) The precise 1470-yr timing of D–O events from Rahmstorf (2003). Dashed lines mark out exact 1470-yr intervals. Markers with solid dots denote onset of D–O cycles as defined by warming above a threshold defined by amplitude and rate. (b) Deviation in years of onset of D–O cycles from the exact 1470-yr pacing. (c) Spectral analyses (multi-tapered (MTM) method) of haematite-stained grain cycles in Fig. 24.2. The heavy line in raw data is the average spectrum; light lines are the upper and lower 90% confidence limits. The heavy dashed line is the mean AR1 (red noise) spectrum; the light dashed line is the upper 90% confidence limit. The prominent peak centred on 1800 yr indicates a broad-band process (not periodic) that is different from red noise. The process could be periodic, however, and is distorted in the record by errors in the marine age model.

palaeoclimate analyses. Such proxies are being developed and a rigorous testing of the model should be possible in the near future.

24.5 The enigmatic '1470-yr cycle'

Any effort to explain ocean–ice interactions during the last glaciation must take into account the puzzling and widespread ca. 1470-yr cyclicity found in glacial records from ice cores and in marine and terrestrial sediments in both hemispheres (see Alley *et al.*, 2001). In records of $\delta^{18}O$ from Greenland ice, the cycle appears as a sharp peak in spectral energy that suggests a truly periodic process with a period of about 1470 yr (Mayewski *et al.*, 1997; Grootes & Stuiver, 1997). Recently, Rahmstorf (2003) demonstrated that the spacing of D–O cycles in the GISP2 ice core from Greenland was exactly 1470 yr or exactly even multiples of 1470 yr, thereby tying the D–O oscillations to the 1470-yr cycle (Fig. 24.5a & b). Rahmstorf argued from those findings that D–O events are discrete events paced by the 1470-yr cycle rather than being discrete cycles themselves. He concluded that internal processes in a complex non-linear system such as the Earth's climate system would not likely produce such a regular oscillation.

If Rahmstorf (2003) is correct, the implication is that a regular, external process must underlie D–O cycles. Orbital cycles would meet that criterion, but there is no known orbital cycle with a period of 1470 yr. Regardless of the mechanism, however, the regularity of the 1470-yr cycle could provide the weak periodic forcing required by stochastic resonance models, such as proposed by Alley *et al.* (2001). It should be emphasized though that ice-core age models are undergoing revision, and improved chronologies in the future may reveal a much less regular oscillation within the 1000 to 2000 yr band.

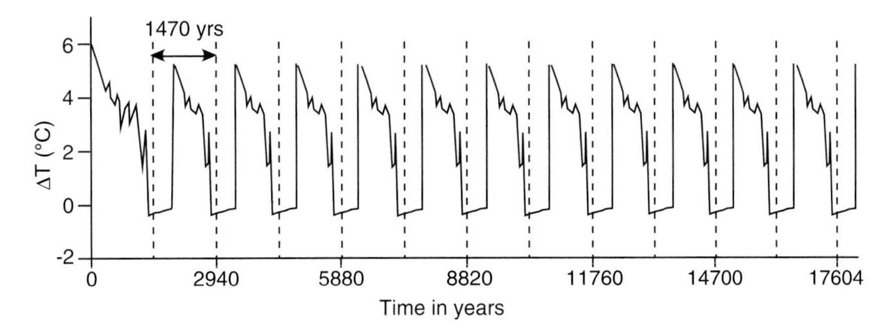

Figure 24.6 The record of haematite-stained grains from the Holocene in cores MC52 and VM29-191. Vertical lines mark out an exact 1470-yr cycle. The dashed line is the result of filtering the haematite-stained grain record with a Gaussian filter centred on 1470 yr and bandwidth of 1000 to 2000 yr. Note that although the Holocene cycle pacing is not exactly 1470 yr, it falls within the same range as its glacial counterpart in Fig. 24.5c.

Further insight into the nature of the 1470-yr cycle came from efforts by Bond *et al.* (1999) to produce a detailed record of IRD variations during the entire glacial cycle using measurements of two of the petrological tracers of IRD, haematite-stained grains and Icelandic glass (Fig. 24.2). They found that increases in both tracers (and, hence, increases in IRD) were coherent and occurred in an irregular non-periodic cycle with a mean value of 1470 yr and a dispersion around that mean (at 1 σ) of ± 500 yr (Fig. 24.5c). A true 1470-yr period could be present, though, and obscured by the large errors in constructing age models for the marine records. The cycle was not confined to just the cold phases of D–O events; rather, it punctuated the entire glacial record, including the LGM and the interstadial phases of the D–O cycles. It would appear then that the IRD maxima were not tied to D–O cycles, but instead could be a manifestation of the underlying 1470-yr cycle identified by Rahmstorf (2003).

24.6 The Holocene dilemma

Once regarded as a long period of climate stability, it is now clear from marine and terrestrial records that the entire 12,000 yr of the Holocene is punctuated by a series of robust cycles (not necessarily periodic) of millennial duration. Many of those records demonstrate that the cycle pacing occurs within a band of 1000 to 2000 yr. Curiously though, ice-core records reveal few climate shifts within the Holocene, the main events occurring at about 8200 yr and within the past 2000 yr where the most recent cycle corresponds to the Medieval Warm Period–Little Ice Age oscillation.

Upon extending their analyses of the two petrological tracers (haematite-stained grains and Icelandic glass) into the Holocene, Bond *et al.* (1997, 1999, 2001) made an unexpected discovery. They found that the robust variations of the last glacial extended through the entire Holocene with a pacing that fell within the 1470 ± 500-yr pacing of the glacial cycles (Fig. 24.6). Given the absence of large glaciers and ice sheets during most of the Holocene, the mechanism for the persistent and simultaneous increases in both tracers within the range of the 1470 ± 500-yr could not have been synchronous increases in iceberg discharge from more than one source as previously thought. Bond *et al.* (1999) concluded that it was more likely that recurring southward advections of cooler surface waters from the seas north of Iceland carried ice with IRD rich in both haematite-stained grains and Icelandic glass, implying that the cycle reflected a mechanism operating within the ocean–atmosphere system. Given the similarity in composition and pacing of the Holocene and glacial petrological cycles, it seemed possible that the glacial petrological cycles might be a manifestation of the same climate mechanism that operated during the Holocene.

Hence, the petrological record of the 1470-yr cycle suggested that whatever its fundamental origin, the cycle gave rise to oscillations in the ocean–atmosphere system that controlled deposition of at least some of the IRD associated with D–O cycles in the North Atlantic. That conclusion would seem to lend further support to the climate-centred hypothesis for glacier–ocean interactions. The argument has been criticized, however, because the Holocene cycles do not have the precise 1470-yr periodicity of the D–O oscillations (e.g. Schulz *et al.*, 2002; Fig. 24.6).

Braun *et al.* (2004, 2005) have proposed an intriguing solution to this problem. Building on evidence in Bond *et al.* (2001) that the Holocene cycles may have been forced by variations in solar irradiance, Braun and his colleagues made the remarkable discovery that their climate system model CLIMBER-2, when forced by the DeVries solar cycle (210 yr) and the Gleissberg solar cycle (88 yr), can produce robust D–O events with an exact period of 1470 yr under glacial boundary (Fig. 24.7) conditions but not under Holocene boundary conditions. Their surprising results have far reaching implications for the fundamental mechanisms that underlie abrupt climate change, and they clearly deserve much further investigation and rigorous testing.

24.7 Summary

Views on what causes ocean–glacier interactions during D–O cycles and Heinrich events fall into two camps. One is glacier-centred and assumes that free oscillations in glaciers and ice sheets cause recurring discharges of icebergs into the ocean. The iceberg

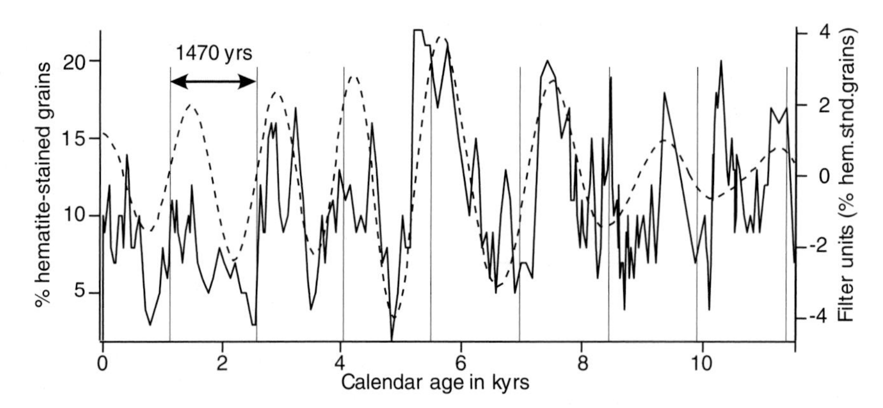

Figure 24.7 Model results from Braun *et al.* (2005) using the CLIMBER-2 model forced with the DeVries solar cycle (210 yr) and the Gleissberg solar cycle (88 yr) under glacial boundary conditions. Vertical dashed lines demark an exact 1470-yr pacing. Even though weak, the solar forcing in the model produces ocean surface-temperature anomalies of several degrees C with an exact pacing of 1470 yr.

freshwater injections force recurring changes in the North Atlantic's deep-water circulation that in turn produce a series of robust abrupt climate changes with a global or nearly global footprint. The other is climate-centred and assumes that a climate mechanism of unknown origin forces increases in iceberg discharge either through direct effects on glacier mass balance or through changes in the North Atlantic deep circulation. In this case the glacier activity and its effects on NADW production operate as a non-linear positive feed-back that amplifies the climate signal and transmits a series of abrupt climate shifts well beyond the North Atlantic region. Both views have been criticized and neither can account for the striking 1500-year pacing of the D–O cycles and the presence of similar cycles in the Holocene when the large glaciers and ice sheets were absent.

TWENTY-FIVE

Northern Hemisphere glaciers responding to climate warming by increasing their sensitivity and their contribution to sea-level rise

Mark Dyurgerov

INSTAAR, University of Colorado, Campus Box 450, Boulder, CO 80309-0450, USA

Glaciers are becoming 'warmer', responding faster to ongoing change in climate, contributing more to the global water cycle and receding at an accelerating rate. Glacier sensitivity is the key parameter which links changes in climate, glacier volume and the eustatic component of sea-level rise. To show the relation between these changes in the Northern Hemisphere, where such data are more complete and accurate, I used observational data of air temperature and glacier mass balance over the years 1961–1998. The last decade of the twentieth century was the warmest over the previous hundreds of years (Mann & Jones, 2003). The global water cycle has accelerated in terms of increases in precipitation rate and rise in sea level (RSL) (Church *et al.*, 2001b). The eustatic component of RSL has increased and the source of this must be continental (Miller & Douglas, 2003). We have attributed this continental source to the ongoing process of volume losses by mountain and subpolar glaciers (Meier *et al.*, 2003). We need to gain better knowledge on what has caused a large increase in glacier wastage on a hemispheric scale.

The annual-balance sensitivity to temperature ($\partial b/\partial T$: where b is the annual mass balance and T is the air temperature in °C) is used for most projections of glacier wastage. Mass balance sensitivity calculated here uses all available time series for Northern Hemisphere glaciers in respect to observed Northern Hemisphere annual air temperature. Glacier sensitivity to climate depends on mass turnover (Oerlemans & Reichert, 2000). In regions with dry climate conditions and small mass turnover, summer temperature is the major driver of glacier mass balance change. In wetter climates with larger mass turnover, glaciers' mass balance is very sensitive to change in the amount of precipitation. This suggests a different glacier response by regions with climate warming, such as:

1 in the high Arctic, and the Canadian Archipelago, ice caps are disappearing fast because they are very sensitive to change in equilibrium line altitude (ELA), due to a glacier's surface topography—i.e. a small increase in ELA results in an enormous decrease in accumulation area;
2 everywhere in continental mountains a small warming results in a decrease in the snow/rain ratio, thus there is a decrease in surface albedo, and the increased ablation rate causes more negative mass balance—glaciers with a summer peak of precipitation are more sensitive to air temperature change (Naito *et al.*, 2001);
3 large tidewater glaciers in Alaska and the Arctic, and individual ice caps in Greenland and other parts of the Arctic, may have been especially sensitive to an increase in sea-level and the direct impact of oceanic water—i.e. as a result of wave erosion, propagation of the grounding line towards the land and acceleration of bottom melting (Pfeffer *et al.*, 2000; Zwally *et al.*, 2002a; Thomas *et al.*, 2003; Steffen *et al.*, 2004).

Different responses resulted in huge variability in glacier mass balance sensitivity over 1961–1987 (Fig. 25.1). This has changed sharply. Since 1988 $\partial b/\partial T$ decreased (more negative mass balance) enormously from 0.017 m °C^{-1} yr^{-1} in 1961–1987 to −0.734 m °C^{-1} yr^{-1} in 1988–1998 (the warmest period); less change in summer temperature is required to cause the same glacier wastage. Variability in sensitivity reduced enormously (Fig. 25.1).

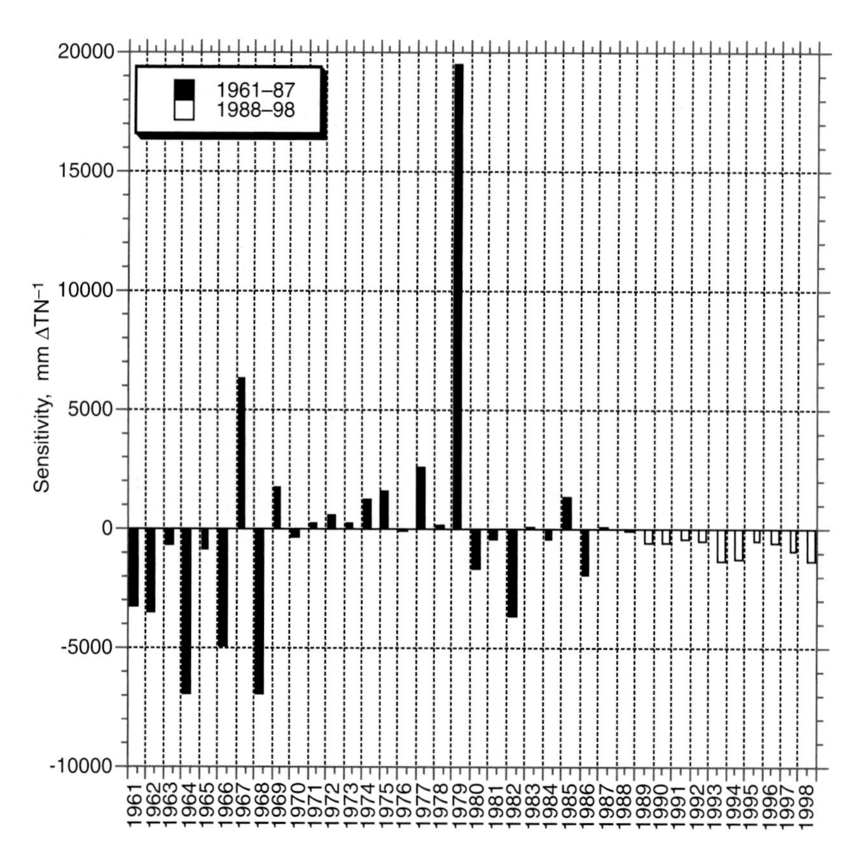

Figure 25.1 Northern hemisphere glacier mass balance sensitivity to annual air temperature. Long-term annual mass-balance time series averaged for about the same 40 bench mark glaciers have been used to calculate averages (Dyurgerov, 2001). Note, this is a different measure of sensitivity than that used by the IPCC (Church *et al.*, 2001b), which involves a change between two steady states.

Such changes in $\partial b/\partial T$ have never previously been observed. These hemispheric-scale changes remain unexplained. They may be due to the complex combination of large-scale atmospheric circulation patterns, which requires further analyses. Before this is accomplished, the simple hypothesis is that these changes are due to a hemispheric increase in air temperature.

Realistic knowledge of glacier regimes and their involvement in global processes depends on data from direct observations. This is more true now than it was 10–20 yr ago. Direct observations cannot be substituted by modelling results. Observational data have revealed previously unknown processes and changes, such as changes in sensitivity and its variability. Knowledge of these two is crucial in order to predict glacier response to climate change and glacier contribution to sea-level rise.

TWENTY-SIX

Influence of ice streaming on the ocean–climate system: examining the impact of the M'Clintock Channel Ice Stream, Canadian Arctic Archipelago

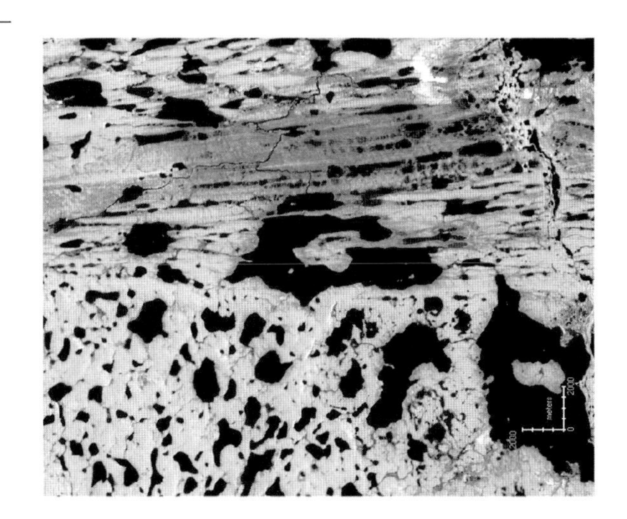

Chris R. Stokes* and Chris D. Clark†

**Landscape and Landform Research Group, Department of Geography, The University of Reading, Reading RG6 6AB, UK*
†Department of Geography, The University of Sheffield, Sheffield S10 2TN, UK

Ice streams not only exert a profound influence on ice-sheet mass balance and stability but they can also influence the ocean–climate system through profligate iceberg discharge and subsequent melting. In recent years, palaeo-ice stream imprints have been identified with a greater degree of confidence than ever before (see Stokes & Clark (2001) for a review) but further progress requires data to constrain ice-stream timing and their long-term evolution. One way to achieve this goal is to link the known location of palaeo-ice stream imprints (source) to the ocean sedimentary record (sink). Given an intact ice-stream imprint, it is possible to provide an estimate of the potential ice flux and the range of lithologies that should pass through the system, which can then be constrained and possibly dated by the ocean sedimentary record. Such a link has been made for the large Hudson Strait Ice Stream (Laurentide Ice Sheet) which is thought to have operated episodically throughout the last glacial cycle, contributing to Heinrich events and profoundly influencing North Atlantic ocean circulation and climate (Andrews & Tedesco, 1992). However, the extent to which other Laurentide ice streams influenced ocean circulation and climate is largely unknown.

One location where a large marine-terminating ice stream is known to have existed (at the very least during Late-glacial times) is in the M'Clintock Channel at the northwestern margin of the Laurentide Ice Sheet (Hodgson, 1994; Clark & Stokes, 2001). The reconstructed location of the Late-glacial ice stream imprint is shown in Plate 26.1 with an example of the remarkable bedform imprint that it inscribed on Victoria Island (Plate 26.1b). The ice stream cut down through soft carbonate sediments and was comparable in size (ca. 770 × 140 km) to the Hudson Strait Ice Stream, with a surface area of ca. 162,000 km^2, a catchment area of around 400,000 km^2 and an estimated cross-sectional area in excess of 100 km^2 (cf. Clark & Stokes, 2001). The final activity of the ice stream (ca. 10,400–10,000 ^{14}C yr) is constrained to around the same time as Heinrich Event-0 (Hodgson, 1994; Clark & Stokes, 2001) but whether it was part of a pan-ice-sheet response or acted independently is unknown. It is also unknown whether the ice stream operated at other times, prior to deglaciation, although the bedform record strongly supports an enlarged ice stream during full glacial conditions (Plate 26.1a).

More recently, marine sedimentary evidence has been documented which may shed some light on the long-term history of the M'Clintock Channel Ice Stream and its influence on Arctic oceanography and climate. Of major significance is that Arctic Ocean circulation was markedly different from present-day patterns (see Plate 26.2). During glacial intervals, Bischof & Darby

(1997) suggest that icebergs issued from the vicinity of Victoria and Banks Island are able to exit the Arctic Ocean directly through Fram Strait and down the east coast of Greenland, without the multiple rotations in the western Arctic Ocean which characterize present-day conditions (the Beaufort Gyre). The significance of this is that the transport and subsequent melting of icebergs from the Canadian Arctic Archipelago could have played a key role in reducing surface water salinity and North Atlantic Deep Water formation in the Greenland–Iceland–Norwegian Seas (cf. Bischof & Darby, 1997). Indeed, evidence from a sediment core taken from Fram Strait (Plate 26.2) suggests that there were four major iceberg discharge events between 12 and 31 ka (^{14}C yr) (Darby *et al.*, 2002). These events were of relatively short duration (<1–4 kyr) suggesting rapid purges of ice through Fram Strait. The composition of these iceberg rafted debris (IRD) events exactly matches a till sourced from Victoria Island, and we suggest that an enlarged M'Clintock Channel Ice Stream was the major contributor.

The far reaching effects of the ice stream are also evidenced in sediment cores from the Mendeleev Ridge in the western Arctic Ocean which also show peaks in carbonate IRD layers sourced from Victoria and Banks Island (Polyak *et al.*, 2004) and major meltwater influxes (Poore *et al.*, 1999). It is also interesting to note that, despite chronological uncertainties, the IRD events issued from the north-western margin of the LIS appear to immediately precede Heinrich events (Darby *et al.*, 2002). Although very tentative, this suggests that the activity of the MCIS may have important implications for a lead-lag relationship whereby the northern margins of the LIS disintegrated prior to the eastern margins that produced the Heinrich events (Darby *et al.*, 2002).

In addition to the IRD and meltwater input, the ice stream may also have influenced the formation of a large Antarctic-type ice shelf in the western Arctic Ocean, up to 1 km thick and covering hundreds of kilometres (Polyak *et al.*, 2001). Evidence for such an ice shelf includes submarine bedforms indicative of glacial scouring and moulding of the sea floor around the Chukchi plateau in the western Arctic Ocean (Plate 26.2). The orientation of bedforms points to a source emanating from the broad straits of the Canadian Arctic Archipelago, suggesting possible interactions with the ice stream (cf. Polyak *et al.*, 2001). However, the evolution and size of this Arctic ice shelf are yet to be constrained. Given the strong evidence for several phases of iceberg delivery from the ice stream, it is likely that icebergs were able to bypass this ice shelf en route to the Fram Strait. Further work to determine the age and extent of an Arctic Ice Shelf will help to elucidate the pos-sible interactions between the MCIS, iceberg delivery and Arctic Ocean IRD events. These developments may feed into our knowledge of the interactions between West Antarctic Ice Streams and the ice shelves they currently nourish.

TWENTY-SEVEN

Influence of ocean warming on glaciers and ice streams

Eric Rignot

Jet Propulsion Laboratory, California Institute of Technology, Mail Stop 300-319, Pasadena, CA 91109-8099, USA

Thinning of floating ice shelves with resulting unpinning and reduced buttressing of continental ice and grounding line retreat is potentially the largest contributor to sea-level change (Weertman, 1976; Thomas, 1979b; van der Veen, 1986). Ice-shelf thinning could occur from enhanced melting at the underside of ice shelves, which could result from changes in properties and circulation of the ocean caused by climate change (Jacobs *et al.*, 1992). Melt rates near the grounding lines of deep-draft outlet glaciers are particularly relevant to ice sheet mass balance because continental ice discharge is controlled by the channelized flow of these ice streams into the ocean. If these regions are the locus of high basal melting, the potential exists for substantial ocean control over ice shelf, if not ice sheet, mass balance. Model simulations indicate that for moderate global warming, resulting in

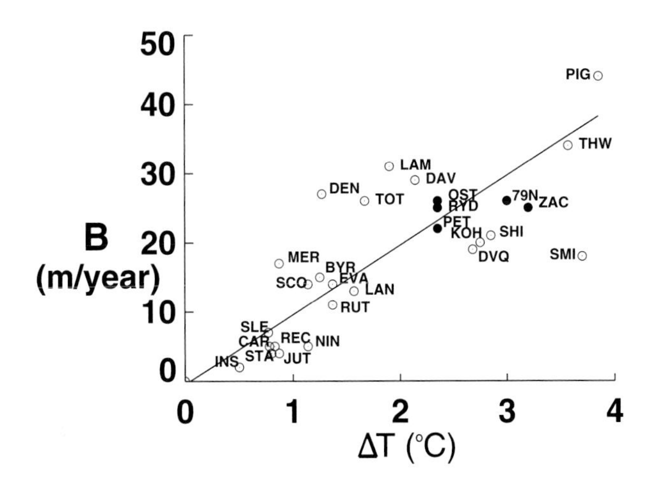

Figure 27.1 Basal melt rates, *B*, seaward of Antarctic (hollow circle) and Greenland (solid circle) grounding lines, versus thermal forcing, ΔT, from the ocean, which is the difference between the nearest *in situ* ocean temperature data and the seawater freezing point at a depth of 0.88 times the maximum grounding line ice thickness. The regression indicates that a 1°C increase in effective ocean temperature increases melt rate by 10 m yr^{-1}. PIG, Pine Island; THW, Thwaites; SMI, Smith; KOH, Kohler; DVQ, DeVicq; LAN, Land; BYR, Byrd; DAV, David; NIN, Ninnis; MER, Mertz; TOT, Totten; DEN, Denman; SCO, Scott; LAM, Lambert; SHI, Shirase; JUT, Jutulstraumen; STA, Stancomb-Wills; SLE, Slessor; REC, Recovery; INS, Institute; RUT, Rutford; CAR, Carlson; EVA, Evans; OST, Ostenfeld; PET, Petermann; RYD, Ryder; ZAC, Zachariae Isstrom; 79N, 79 north glaciers.

increases in accumulation, surface melt and bottom melting, the large increase in bottom melting is the dominant factor in the long-term response of the Antarctic Ice Sheet (Warner & Budd, 1998; Huybrechts & De Wolde, 1999).

Unlike melting under a grounded ice sheet, processes beneath floating glaciers are governed by the transport of ocean heat and by the seawater freezing temperature dependence on pressure (Robin, 1979). This allows sensible heat to be obtained from the cold dense shelf waters resulting from sea-ice formation, as well as warm deep water that intrudes onto the continental shelf and flows into ice-shelf cavities. Bottom melting freshens and cools the seawater, adding buoyancy that drives upwelling as the ice shoals seaward. In some regions, the rising seawater–meltwater mixture drops below the *in situ* freezing point to form marine ice that can comprise a substantial part of ice-shelf volume (Oerter *et al.*, 1992). Area-average melt rates for the large ice shelves are about 40 cm yr^{-1} (Jacobs *et al.*, 1992). Much of the actual melting occurs in the deepest parts of the subice-shelf cavities (Plate 27.1), however, where direct measurements are difficult to obtain.

Basal melt rates near grounding zones, calculated using remote sensing observations of Antarctic and Greenland glacier velocity and thickness distribution, and applying principles of mass conservation, range from less than 4 m yr^{-1} for several glaciers that flow into the Filchner–Ronne Ice Shelves to greater than 40 m yr^{-1} for Pine Island Glacier (Fig. 27.1). The wide range is consistent with earlier studies of several of the individual glaciers, using a variety of techniques (Potter & Paren, 1985; Smith, 1996; Rignot, 1998; Lambrecht *et al.*, 1999; Rignot & Jacobs, 2002), and stems from differences in grounding line drafts, seawater temperatures, and ice topographies and velocities. Nevertheless, most of the melt rates calculated near grounding lines exceed the area-average rates for the largest ice shelves by one to two orders of magnitude.

The largest thermal forcing in the Antarctic, 4°C above the *in situ* melting point, is associated with Pine Island, Thwaites and Smith Glaciers that flow into the Amundsen Sea. This results from the nearly unaltered Circumpolar Deep Water that extends southward across the floor of the Antarctic continental shelf in Pine Island Bay (Jacobs *et al.*, 1996). Modified Circumpolar Deep Water is known to upwell at other locations along the East Antarctic continental shelf, e.g. Shirase Glacier, and will increase grounding-line melt rates where its density allows it to intrude along the sea floor.

The potential impact of basal melting on short-term (<100 yr) ice-sheet stability is greatest in regions where deep water has direct access to glacier grounding lines. Ocean temperatures seaward of Antarctica's continental shelf break have risen about 0.2°C over recent decades, sufficient to increase basal melting by 2 m yr^{-1} where that change has reached vulnerable grounding lines. This may account for the rapid thinning of ice shelves in the western Amundsen Sea (Rignot, 1998), which would explain the observed acceleration of their nourishing glaciers (Rignot *et al.*, 2002) and the resulting negative mass balance of the entire basin (Rignot *et al.*, in press).

Submarine melting is also significant near the terminus of tidewater glaciers, which control the mass budget of coastal Alaska, Patagonia and Greenland, with summer rates reaching 12 m day^{-1} at the terminus of LeConte Glacier, Alaska, or 60% of the estimated mass loss at the front (Motyka *et al.*, 2003). High bottom melting could explain the high correlation between calving rate and water depth because the subglacial area in contact with the ocean increases in deeper waters. Even in the absence of large floating sections, a warmer ocean may therefore have a major influence on the evolution of glaciers and ice streams.

TWENTY-EIGHT

Glacial runoff from North America and its possible impact on oceans and climate

C.F. Michael Lewis* and James T. Teller†

**Geological Survey of Canada Atlantic, Natural Resources Canada, Bedford Institute of Oceanography, Dartmouth, Nova Scotia, Canada B2Y 4A2*

†Department of Geological Sciences, University of Manitoba, Winnipeg, Manitoba, Canada R3T 2N2

28.1 Introduction

At the Last Glacial Maximum (LGM), about 18–21 ka (21,000–25,000 cal. yr), ice over Canada and the northern USA formed a nearly continuous cover which merged with the Greenland Ice Sheet (e.g. Dyke & Prest, 1987a; Dyke *et al.*, 2003, Dyke, 2004). Beyond this area, glaciers expanded over the higher mountains in the western USA, Alaska and the Yukon Territory. Contiguous ice sheets included the Cordilleran (mostly over British Columbia, and in southern Yukon and Alaska at its maximum about 16.5 ka (19,700 cal. yr)), Innuitian (in Arctic Canada), Laurentide (mostly over the remainder of Canada and adjacent northern USA) and Greenland (Fig. 28.1). The largest and most variable of these was the Laurentide Ice Sheet (LIS), which has been estimated to have accounted for 50% of the increased ice volume on Earth during the LGM (Boulton *et al.*, 1985). By about 5 ka (5700 cal. yr), ice remained only over Greenland, the highlands of Baffin and Queen Elizabeth islands of Arctic Canada, and the highest elevations of the western Cordillera.

The baseline surface runoff from precipitation and meltwater from the southern margin of the LIS had profound effects on North America and on the North Atlantic Ocean and climate (Teller, 1990a,b; Licciardi *et al.*, 1999). Supplementing this runoff were periodic outbursts of water stored in ice-marginal lakes and, possibly, in subglacial reservoirs (Shaw, 1989, 2002; Teller *et al.*, 2002). Iceberg production also added freshwater to the oceans, with the most intense periods known as Heinrich events, based on ice-rafted detritus found in Atlantic Ocean sediments (Ruddiman, 1977; Heinrich, 1988; Bond *et al.*, 1999; Clarke *et al.*, 1999). In this chapter, we review the history of surface runoff from the southern margin of the LIS during the last period of deglaciation, with particular reference to major abrupt changes in runoff magnitude and routing and their possible relationship to changes in thermohaline circulation (THC) and the North Atlantic Deep Water (NADW) formation. Recent advances in modelling ice–ocean–atmosphere and lake–ice–atmosphere feed-back processes are also discussed in the context of their importance in forcing ice-marginal fluctuations and diversions of runoff routing. Finally, the sensitivity of North Atlantic THC to freshwater runoff during the last deglaciation is explored with particular reference to events of abrupt climate change.

28.2 Controls on North American runoff magnitude and routing

Runoff from the melting Laurentide Ice Sheet (LIS) and precipitation, plus abrupt releases of stored water in proglacial lakes along the southern margin of the LIS, discharged at different times to the Gulf of Mexico, North Atlantic Ocean, western Arctic Ocean and finally to Hudson Bay. A primary control on the variability in runoff routing stems from the changing configuration of hydrological catchments and drainage basins during deglaciation, which developed by the combination of location of the ice margin, relief on the ice sheet and the deglaciated continental surface, and differential isostatic rebound. The latter raised the land surface more in the northern and northeastern regions, and shifted overflow from proglacial lakes from one outlet to another.

In this chapter, the four main glacial drainage basins (Fig. 28.2) that directed runoff to widely separated continental margins will be discussed. These basins are:

1. the Mississippi River Valley (draining to the Gulf of Mexico);
2. the St Lawrence and Hudson River valleys (draining to the North Atlantic Ocean);

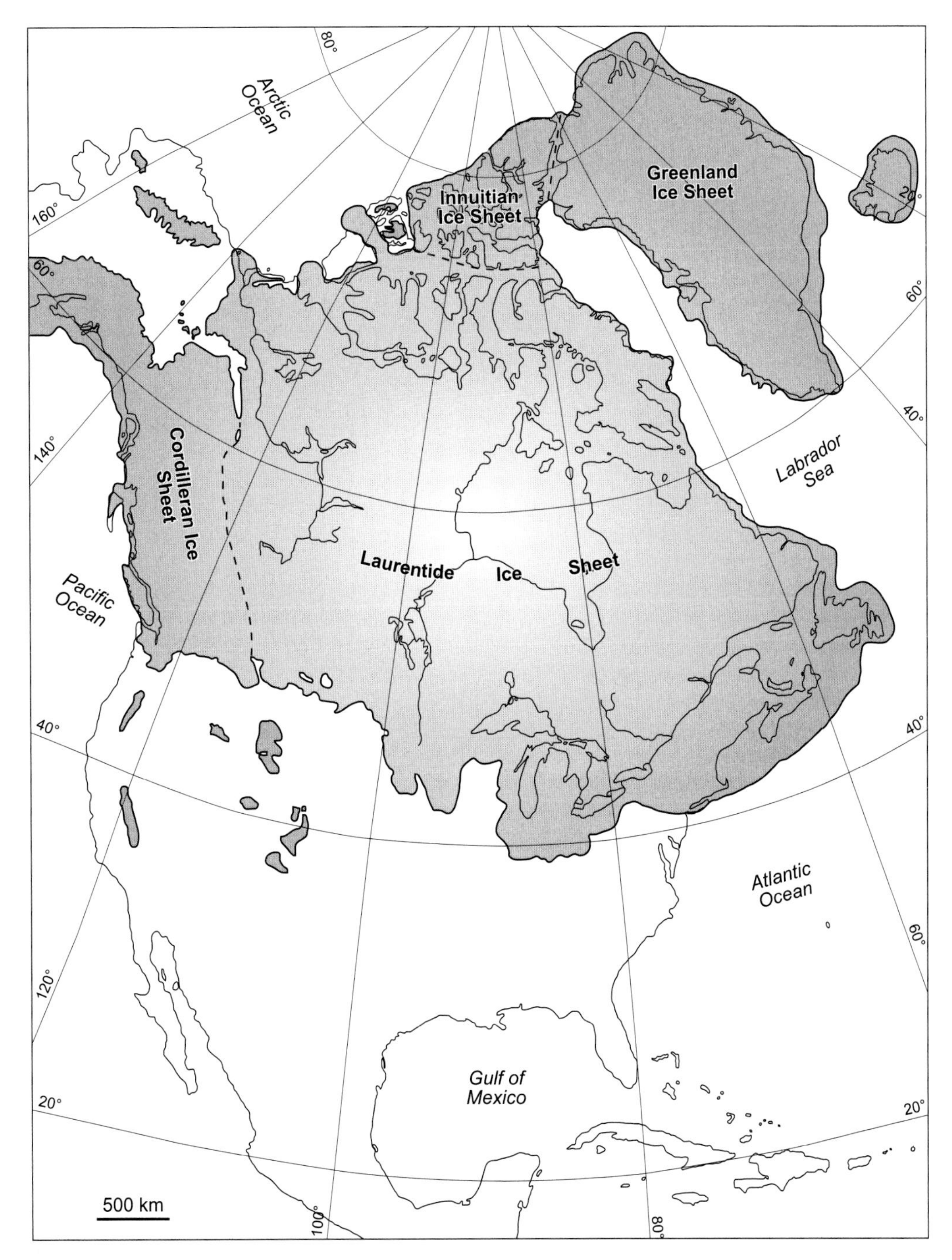

Figure 28.1 Map of North America and Greenland showing four ice sheets at their maximum extent in the last glaciation. Oscillations in the various routings of precipitation and ice-melt runoff from the southern part of the Laurentide Ice Sheet between 18 ka and 7 ka and their oceanic impact are the subject of this chapter. (See www.blackwellpublishing.com/knight for colour version.)

3 the Athabasca–Mackenzie River Valley (draining to the Beaufort Sea and western Arctic Ocean);
4 Hudson Bay watershed (draining via Hudson Strait to the western Labrador Sea and the northern North Atlantic Ocean).

Growth of the LIS blocked the northward-draining basins. This blockage, together with the great elevation of ice, ranging up to 2–3 km over Hudson Bay and adjacent regions, drastically changed the configuration of continental surface slopes. In addition, the Earth's crust was isostatically depressed hundreds of metres by the ice load (Peltier, 1998). The surface configurations of the ice sheet and the continental surface beyond the ice margins defined new 'cryohydrological' drainage basins for continental runoff (Teller, 1990a,b; Licciardi *et al.*, 1999). At the LIS maximum, about 21–18 ka (25,000–21,000 cal. yr), all drainage

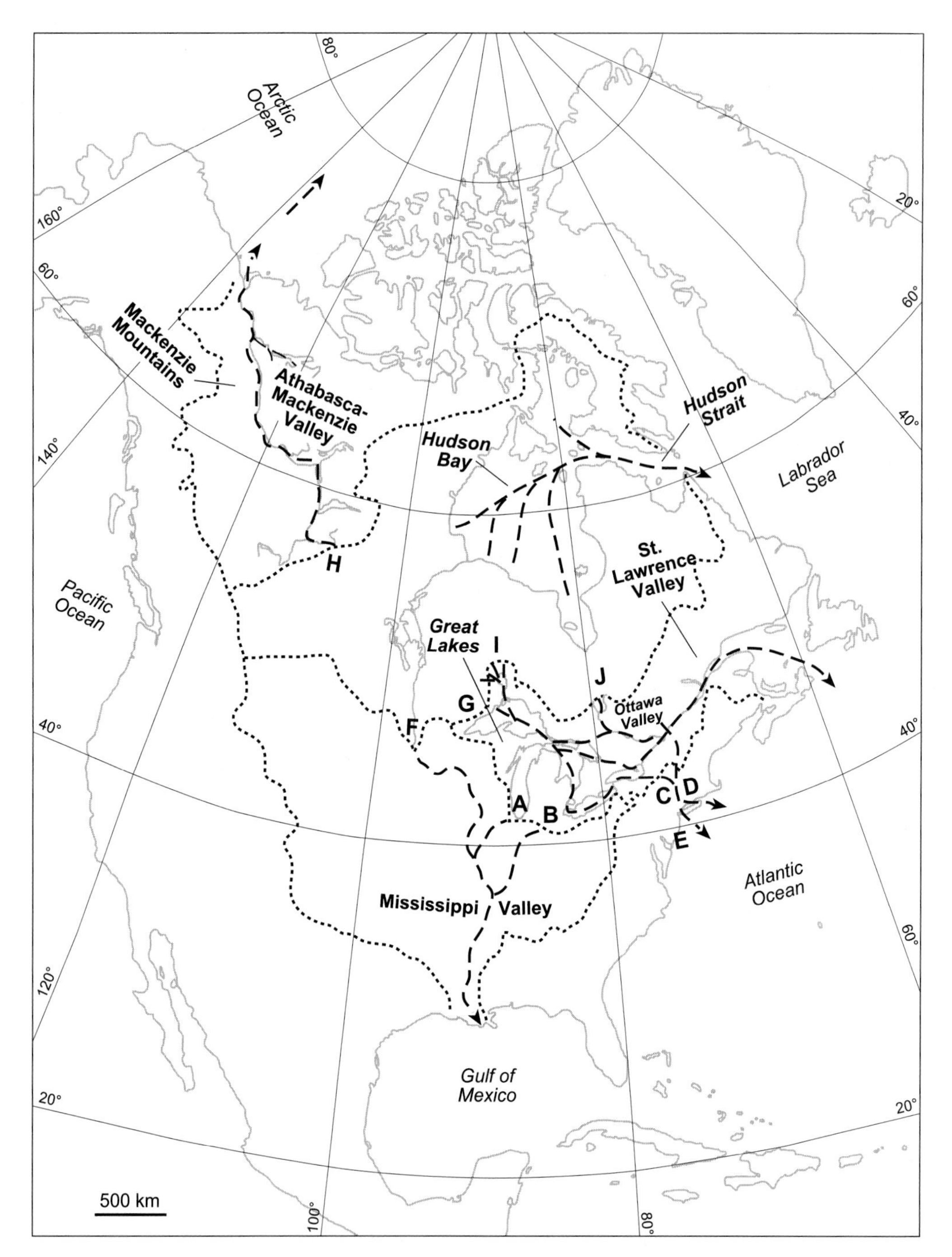

Figure 28.2 A similar view as in Fig. 28.1 showing four continental watersheds which played significant roles in directing runoff to the oceans. These watersheds are the Mississippi Valley catchment draining to the Gulf of Mexico, the Great Lakes–St Lawrence Valley catchment discharging to the North Atlantic Ocean, the Athabasca–Mackenzie Valley catchment draining to the western Arctic Ocean, and the Hudson Bay catchment discharging via Hudson Strait to the northern Atlantic Ocean (western Labrador Sea). Outlets that were significant controls on runoff routing include Chicago (A) from southern Lake Michigan basin, Wabash Valley (B) from western Lake Erie basin, Mohawk Valley (C) connecting Lake Ontario basin with Hudson River Valley, Hudson River Valley (D) draining to Long Island Sound and the Atlantic Ocean, Hudson Shelf Valley (E), a possible extension of the Hudson Valley route (R. Thieler, personal communication, 2004), southern Lake Agassiz outlet (F), first eastern Agassiz outlet (G), northwestern Agassiz outlet (H), second eastern Agassiz outlet (I), and Kinojevis outlet (J) to Ottawa and St Lawrence River valleys. See Fig. 28.4a for names and locations of Great Lakes basins. (See www.blackwellpublishing.com/knight for colour version.)

from the southern portion of the LIS was directed south to the Gulf of Mexico or east to the Atlantic Ocean (Figs 28.1 & 28.2). For the next 13–10 kyr (16,500–12,500 cal. yr ago), watershed configurations were dynamic, controlled by shifting ice divides as the LIS retreated and downwasted. The routing of runoff repeatedly changed, at times abruptly, whenever the retreating ice margin uncovered alternative lower drainage pathways to different sectors of the continental margin (Licciardi *et al.*, 1999).

The routing of runoff was also influenced by isostatic recovery of the crust, as ice thinned and retreated. The amplitude of recovery was greatest where ice loads had been thickest. Although some recovery was instantaneous upon reduction of the ice load (elastic crustal response), much uplift was influenced by redistribution of viscous mantle material beneath the lithosphere, which caused crustal recovery or vertical uplift to proceed on a millennial timescale, generally at a decaying exponential rate (Peltier, 1998). Recovery amplitudes diminished toward the outer limits of the former ice sheet, and this effect imparted a differential aspect to the isostatic recovery. As a result, former horizontal surfaces, for example glacial-lake shorelines, are now uplifted more in the north than in the south (e.g. Teller & Thorleifson, 1983; Larsen, 1987; Lewis & Anderson, 1989).

Precipitation runoff and melting of the ice sheet during deglaciation yielded annual baseline flows of freshwater to the adjacent oceans that significantly exceeded postglacial discharges (Teller, 1990a,b; Licciardi *et al.*, 1999) (Fig. 28.3a). An important increment to the baseline flow arose from the repeated rapid drawdown of lakes that were impounded between the downslope ice margin and the upslope continental surface (Teller *et al.*, 2002; Teller & Leverington, 2004) (Fig. 28.3a). These outbursts channelled large volumes of water through newly opened outlets for short periods, and were superposed on the baseline flow of precipitation and meltwater runoff (see Teller & Leverington, 2004, fig. 5). Outlet erosion, especially during catastrophic outbursts from lakes when large flows entrenched spillways, also affected the subsequent routing of runoff from the lakes.

28.3 History of runoff diversions, outburst floods and proglacial lakes during deglaciation

28.3.1 Introduction

Diversions of glacial runoff from one entry point to the ocean to another resulted in freshening of surface seawater in widely separated regions. The history of the glacial palaeohydrology of North America is closely linked to the history of ice-marginal lakes. As will be discussed later, these changes are thought to be of great significance in influencing THC and climate. The following simplified account of the history of baseline deglacial runoff is tied to changes in proglacial lakes, and is structured to identify and characterize diversions of continental-scale runoff. Other, more detailed discussions of the history of North American glacial drainage and proglacial lakes can be found in publications such as Teller & Clayton (1983), Teller (1987, 1990a, 2004), Karrow & Calkin (1985), Fulton (1989), Teller & Kehew (1994) and Licciardi *et al.* (1999), as well as in many of the references cited in this chapter. In this account, the baseline runoff was periodically augmented by large discharges arising from lake outburst floods when ice retreat opened new lower outlets from the glacial Great Lakes between about 16.5 and 11.2 ka (19,700–13,100 cal. yr) and from glacial Lake Agassiz between 10.9 and 7.7 ka (12,900 and 8450 cal. yr) (Licciardi *et al.*, 1999; Teller *et al.*, 2002; Teller & Leverington, 2004) (Fig. 28.3). Releases of subglacial meltwater (Shaw, 1989, 2002; Flower *et al.*, 2004), the timing and magnitude of which are poorly known, would have supplemented the surface runoff described in this chapter.

From the LGM to the Erie Interstade about 16.5 ka (19,700 cal. yr), drainage from the entire southern margin of the LIS between the Appalachian (east) and the Cordilleran (west) mountains flowed to the Gulf of Mexico as baseline runoff (Figs 28.1 & 28.4a). This runoff would have consisted of large baseline flows of annual precipitation and meltwater modulated by seasonal variation. Runoff from eastern and northern sectors of the LIS would have gone to the North Atlantic and Arctic Oceans, respectively.

28.3.2 The Great Lakes drainage basin

After 16.5 ka (19,700 cal. yr), the southern LIS margin receded into the basins of the southern Great Lakes (Erie Interstade) (Fig. 28.4a & b) (Barnett, 1992). For several hundred years the Ontario–Erie and Huron lobes of the LIS retreated enough to switch 0.038 Sv of drainage from the Gulf of Mexico to the North Atlantic Ocean via the Mohawk and Hudson valleys (Licciardi *et al.*, 1999; Lewis *et al.*, 1994) (Figs 28.2 & 28.4b, outlets C and D). (1 Sv = $10^6\,m^3\,s^{-1}$, or about the present combined flow of all of Earth's rivers.) Runoff from the central LIS, from Michigan basin (outlet A) west, continued south through the Mississippi Valley to the Gulf of Mexico (Figs 28.2, 28.3c & 28.4b, outlet A). The eastward drainage diversion was switched back after 15.5 ka (18,500 cal. yr) to the Mississippi Valley and Gulf of Mexico when the ice margin readvanced (Port Bruce Stade) to cover all the Great Lakes basins again (Barnett, 1992; Lewis *et al.*, 1994) (Fig. 28.4c).

After 14.5 ka (17,400 cal. yr), a second recession of the Ontario–Erie, Huron and Michigan ice lobes again brought the ice margin back into the basins of the southern Great Lakes where large proglacial bodies of impounded meltwater were established. These lakes first continued to drain south to the Mississippi Valley, then switched to the Hudson Valley and North Atlantic route again when ice recession during the Mackinaw Interstade opened the Mohawk Valley (Fig. 28.2, outlet C) about 13.5–13.1 ka (16,200–15,800 cal. yr) (Barnett, 1992; Lewis *et al.*, 1994; Licciardi *et al.*, 1999). This opening would have reduced the level of glacial lakes in the Ontario, Erie and southern Huron basins. The draw down of these lakes may have resulted in outburst floods to the North Atlantic Ocean superposed on the 0.056 Sv increase in baseline runoff from the southern Great Lakes region east of the Michigan basin (Fig. 28.3b) (Teller, 1990a; Licciardi *et al.*, 1999).

Readvance of the southern LIS to moraines of the Port Huron Stade about 13.0–12.8 ka (15,600–15,400 cal. yr) again closed the Mohawk Valley route to the Atlantic Ocean. Glacial runoff from the eastern Great Lakes basins was switched again back to the Gulf of Mexico, and Licciardi *et al.* (1999) calculated that this resulted in an increase in runoff through that route by 0.059 Sv (Fig.

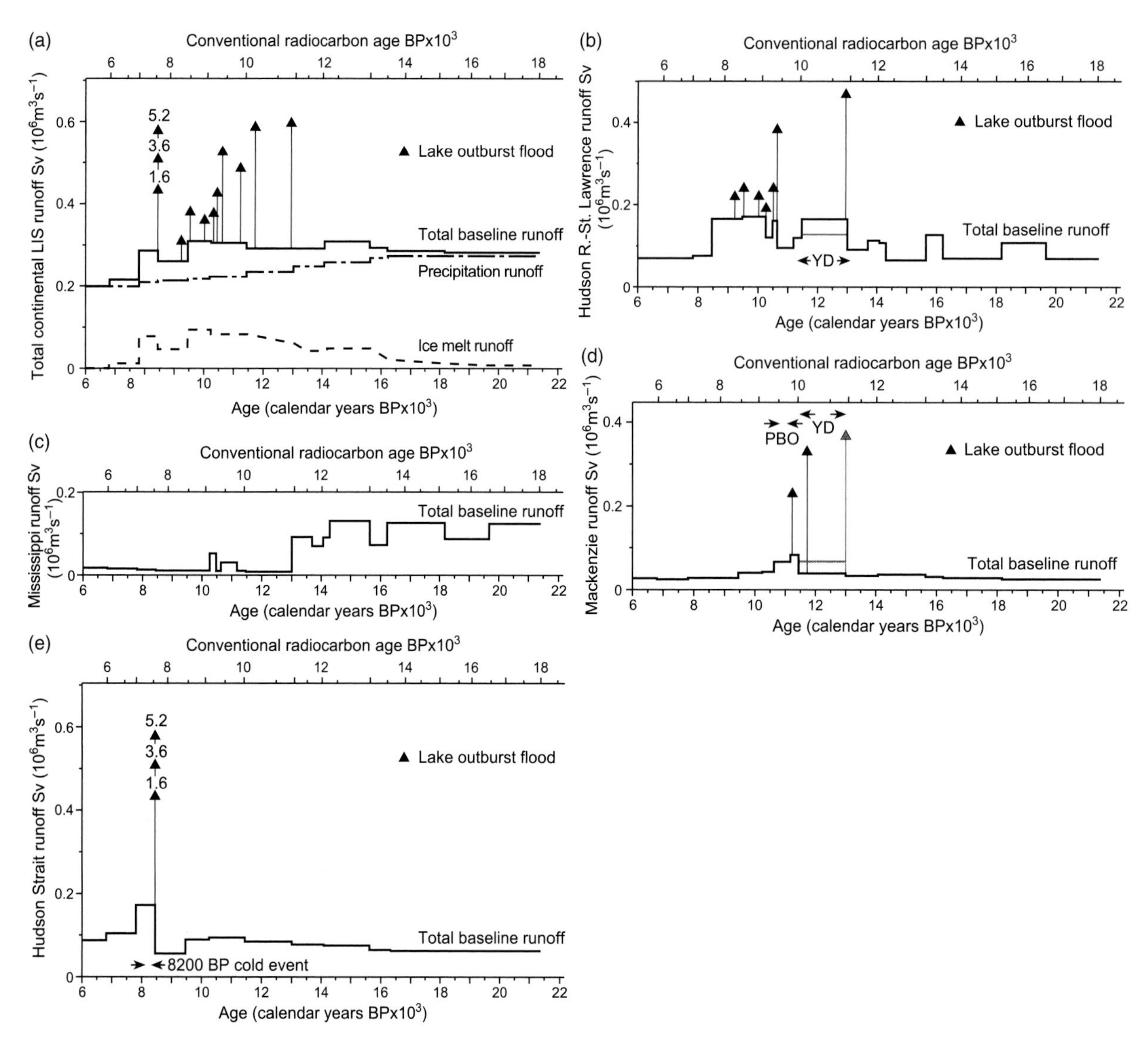

Figure 28.3 History of Lake Agassiz baseline runoff and outburst floods to oceans from southern LIS after Licciardi *et al.* (1999) and Teller *et al.* (2002). Triangles illustrate Lake Agassiz outburst flood fluxes assuming each lake draw down was completed in 1 yr after the opening of a lower outlet by ice retreat. Flood discharges are added to baseline runoff. (a) Total runoff showing ice melt and precipitation components plus outburst floods through all routes. (b) Total (ice melt + precipitation) runoff plus outburst floods (triangles) to the North Atlantic Ocean via first the Hudson River and after 11 ka (13,000 cal. yr) the St Lawrence Valley routes: YD and horizontal arrows indicate period of the Younger Dryas cold event. In an alternative scenario the large Agassiz flood shown at the onset of the Younger Dryas event would not have discharged by this route, and baseline runoff would have been reduced (thin grey line) (see also Fig 28.3d). (c) Total (ice melt + precipitation) runoff via the Mississippi Valley route to Gulf of Mexico. Note the strong antiphased relationship in discharge with that of the route via Hudson River–St Lawrence Valley. (d) Total (ice melt + precipitation) runoff to western Arctic Ocean. The two outburst floods (black triangles representing floods after 12,000 cal. yr) through Lake Agassiz's northwest outlet are thought to have triggered the Preboreal Oscillation (PBO) cool event. An alternative scenario, shown by the first outburst flood (grey triangle at 13,000 cal. yr) and enhanced baseline flow (thin grey line), illustrates flux for the first Agassiz diversion beginning about 11 ka (13,000 cal. yr) if it went via the Athabasca–Mackenzie Valley rather than the Great Lakes–North Atlantic route. (e) Total (ice melt + precipitation) runoff to Atlantic Ocean via Hudson Bay and Hudson Strait. The final demise of the LIS in Hudson Bay led to one huge outburst flood of 5.2 Sv, or two or more closely spaced floods of about 3.6 Sv and 1.6 Sv. Horizontal arrows indicate the period of the 8200 yr BP cold event as recorded in Greenland ice cores.

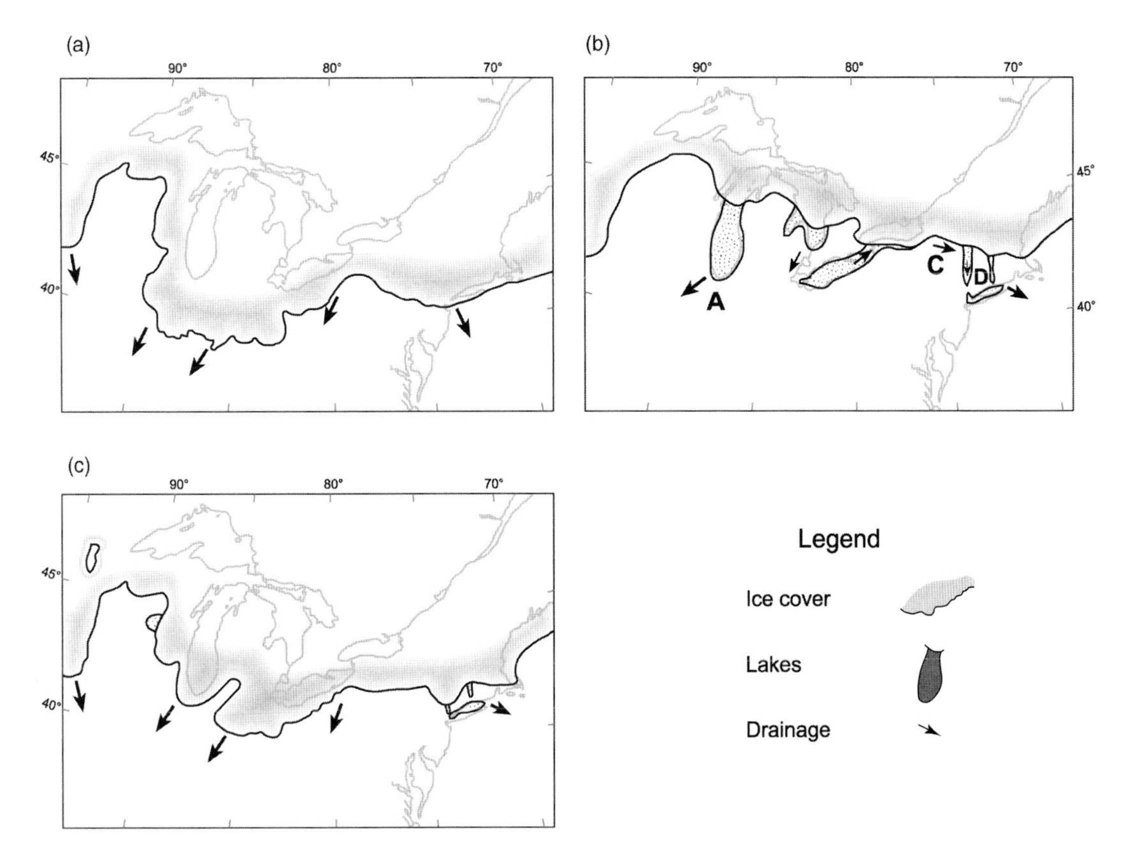

Figure 28.4 Ice-marginal positions during the first oscillation of runoff between Mississippi and Hudson River valleys, after Licciardi *et al.* (1999) and Dyke *et al.* (2003). Base image from Figs 28.1 and 28.2. (a) The ice margin at the LGM (about 18–21 ka; 21,000–25,000 cal. yr) covered the Great Lakes basins (named) and Hudson River Valley. Runoff west of the Appalachian Mountains (dotted) drained to Gulf of Mexico, and runoff to the east of the mountains went to the North Atlantic Ocean. (b) From about 16.5–15.2 ka (19,700–18,200 cal. yr) ice had retreated enough to switch runoff generated in the eastern Great Lakes region from Gulf of Mexico to the Atlantic Ocean via the Mohawk (C) and Hudson (D) valleys. Runoff from southern Lake Michigan basin continued to drain via the Chicago outlet (A) to Gulf of Mexico. (c) By 15 ka (17,900 cal. yr) ice advance had diverted the Mohawk–Hudson drainage back to the Mississippi River and Gulf of Mexico. (See www.blackwellpublishing.com/knight for colour version.)

28.3c). This runoff was discharged by overflow from glacial Lake Whittlesey in the southern Huron and Erie basins to glacial Lake Glenwood II in southern Michigan basin, and thence to the Mississippi Valley (Barnett, 1992; Lewis *et al.*, 1994).

Following the Port Huron Stade, ice margins were in a sensitive position relative to outlets in the southern Great Lakes region; their variations caused lake basin drainage to oscillate between the Mohawk–Atlantic and Mississippi–Gulf-of-Mexico routes. From about 12.8–12.6 ka (15,400–14,600 cal. yr) Erie basin overflow (Lake Wayne) was briefly routed east to the Atlantic Ocean via the Mohawk–Hudson Valley, then switched to a westward route into the Michigan basin, which overflowed into the Mississippi River system (Lake Warren II). About 12.5 ka (14,500 cal. yr), overflow returned to the Mohawk–Hudson route during the Lake Grassmere to Early Lake Erie stages (Calkin & Feenstra, 1985). By 12.2–12.0 ka (14,200–14,000 cal. yr), with recession of ice from the Michigan basin during the Two Creeks Interstade, waters of the Michigan, Huron and Ontario basins were all draining to the North Atlantic via the Mohawk and Hudson valley routes (Fig. 28.3b) (Hansel *et al.*, 1985; Calkin & Feenstra, 1985; Lewis *et al.*, 1994). With the Greatlakean (formerly Two Rivers) advance about 11.8 ka (13,600 cal. yr) into the Michigan basin, overflow there was diverted south into the Mississippi Valley and Gulf of Mexico (Calumet Lake phase) (Hansel *et al.*, 1985), while the eastern basins continued draining to the Atlantic (Lewis *et al.*, 1994). Ice retreat from the northern flank of the highland east of the Ontario basin about 11.3 ka (13,200 cal. yr) (Fig. 28.4b) released glacial Lake Iroquois down the Hudson River Valley into the Atlantic Ocean where it is thought to have suppressed oceanic thermohaline circulation and induced the Intra-Allerod cold period (Donelly *et al.*, 2005). Continued ice retreat switched discharge routes and released a glacial lake outburst flood and subsequent drainage down the St Lawrence Valley to the Atlantic Ocean about 11.1 ka (13,010 cal. yr) (Richard & Occhietti, 2005) close to the initiation of the Younger Dryas cold event. After recession of Greatlakean ice about 11.2 ka (13,100 cal. yr), waters from the Michigan–Huron and eastern basins returned to their eastward routing (Hansel *et al.*, 1985; Barnett, 1992; Lewis *et al.*, 1994). At this time, the switching of overflow out of the Great Lakes, back and forth from east to south, between the North

Atlantic and Gulf of Mexico, ended. Additional details and insight of the above switches in continental runoff have been obtained through studies of proxies of enhanced freshwater inflow in marine sediments of the Gulf of Mexico (Brown & Kennett, 1998; Brown *et al.*, 1999; Aharon, 2003).

After about 10.2 ka (11,900 cal. yr), water in the Michigan and Huron basins went to the St Lawrence Valley when ice retreat opened a lower overflow route via the Ottawa Valley (Fig. 28.2). This overflow continued until about 5.5 ka (6300 cal. yr) when differential rebound raised the Ottawa Valley outlet region to the elevation of the southern rim of the basin. Southward overflow at Chicago (Fig. 28.4b, outlet A) from the Michigan basin and from the southern end of the Huron basin at Port Huron (Fig. 28.4b) continued until about 4.7 ka (5400 cal. yr) when the northern outlet to Ottawa Valley emerged above lake level owing to its faster uplift (Lewis, 1969). By about 3.4 ka (3600 cal. yr) outlet erosion lowered the Port Huron channel. This lower channel captured all overflow from the upper Great Lakes, directing it through the Erie and Ontario basins (Fig. 28.4b) to the St Lawrence River (Fig. 28.2), a routing that continues today (Larsen, 1985; Blasco, 2001).

28.3.3 The Lake Agassiz basin

During the first six millennia after the LGM, glacial meltwater and precipitation runoff west of the Michigan basin passed without interruption down the Mississippi River Valley, as the LIS margin retreated toward the northern limit of the Gulf of Mexico watershed. By 11.7 ka (13,600 cal. yr), the margin had retreated north of the continental divide, and had begun to impound water in proglacial lakes which drained south over the divide (Fig. 28.5a). Continued retreat of the ice margin expanded the area and depth of these lakes, particularly Lake Agassiz in the Red River basin (Clayton & Moran, 1982; Fenton *et al.*, 1983; Teller, 1987; Kehew & Teller, 1994; Teller & Leverington, 2004), as well as that of glacial lakes Duluth and Algonquin in the Superior basin (Farrand & Drexler, 1985).

Shortly after 11 ka (13,000 cal. yr), an outlet opened about 110 m below the early Agassiz spillway that had carried overflow to the south. All published interpretations have concluded that this new outlet was east into the Great Lakes (Figs 28.2 & 28.5b, outlet G). An outburst flood of 9500 km^3 resulted, with a flux of about 0.3 Sv, assuming the drawdown of the lake occurred in 1 yr (Leverington *et al.*, 2000; Teller *et al.*, 2002). The new outlet also diverted baseline runoff estimated at 0.075 Sv from the Gulf of Mexico to the Great Lakes and the Atlantic Ocean (Fig. 28.3b) (Licciardi *et al.*, 1999). Recent evaluation of deglaciation, however, suggests that the eastern outlet area may not have been open at 11 ka (13,000 cal. yr), and that the Agassiz outburst flood may have discharged via a northwestern route to the Arctic Ocean (Teller *et al.*, 2005).

Once lower northern outlets were opened by ice retreat in the more deeply depressed parts of the basin, differential glacial rebound began to play an important role in controlling the relative elevations of outlets, and in transferring lake discharge from one outlet to another. The interaction of ice retreat and outlet elevation with differential rebound for the Agassiz basin is illustrated in Fig. 28.6. A similar situation applied during ice retreat through the Great Lakes basin. In the Lake Agassiz basin successively lower outlets were uncovered in the direction of ice retreat and more rapid rebound. When ice retreated (box 1) in the lake's northern sector, a lower outlet was commonly opened (box 2). The lake then abandoned its beach and was drawn down to a lower elevation with smaller surface area (box 3). Ongoing differential rebound raised this northern outlet relative to the lake's southern shore (box 4), causing Lake Agassiz to transgress its basin as it rose and expanded in surface area (box 5) (Teller, 2001). Whether the rise of the lake reached the elevation of a previous outlet farther south, or not, determined the routing of lake drainage (circle 6). While the water level remained below the previously higher outlet, Lake Agassiz continued to overflow through its more northern outlet, i.e. by either the eastern or northwestern routings to the Atlantic or Arctic oceans, respectively (box 7). If the transgression of the lake level reached the previously higher outlet elevation, overflow was diverted, either to the Mississippi River valley from the eastern or northwestern outlets, or to the Great Lakes and Atlantic Ocean from the northwestern outlet (box 8).

The diverted baseline flow and initial outburst flood at about 10.9 ka (13,000 cal. yr) may have gone eastward via the Great Lakes to the North Atlantic Ocean (Teller, 1987, 1990a). Alternatively, the 11–10 ka outflow (approximately 13,000–11,400 cal. yr ago) may have escaped via the northwestern Agassiz outlet to the Arctic Ocean (Figs 28.2 & 28.5c, outlet H). Teller & Leverington (2004) and Teller *et al.* (2005) describe a possible scenario where all overflow from Lake Agassiz between about 10.9 and 10.1 ka (12,900–11,700 cal. yr) may have been entirely through the northwestern outlet into the Athabasca–Mackenzie River Valley and to the Arctic Ocean. This routing difference (into the Arctic Ocean versus into the North Atlantic Ocean) is significant because of the potential role that Agassiz freshwater overflow played in changing ocean circulation at this time (see later discussion). Glacial readvance around 10 ka (11,400 cal. yr) would have closed the northwestern outlet and forced Lake Agassiz to overflow briefly to the Mississippi Valley before being rerouted to the northwestern outlet for the next 500–600 yr. Floods associated with draw downs at these times of overflow diversion are estimated at 0.3 and 0.19 Sv, and were superposed on baseline runoff of 0.047 to 0.034 Sv (Teller *et al.*, 2002).

Rapid differential rebound then raised the northwestern outlet, and by about 9.4 ka (10,600 cal. yr) the lake had transgressed to the level of an eastern col on the divide between Lake Agassiz and the Great Lakes watershed. Overflow from the 2 million km^2 Agassiz watershed was then diverted into the Nipigon basin northwest of Lake Superior, where overflow was then into the Great Lakes–North Atlantic system, with an initial outburst of 7000 km^3 (0.22 Sv, if it occurred in 1 yr) superposed on a 0.034 Sv baseline flow (Fig. 28.2, outlet I) (Licciardi *et al.*, 1999; Teller *et al.*, 2002). Only briefly did overflow return to the southern outlet, before returning to the eastern outlets for the next 1400 yr. A succession of outbursts from Lake Agassiz punctuated the retreat of the LIS, as progressively lower spillways were deglaciated (Teller & Thorleifson, 1983; Leverington & Teller, 2003). The lake phases associated with these discharges are recorded in the Agassiz basin by a series of beaches, each representing the transgressive maximum of the lake (caused by differential rebound) reached before a lower outlet was opened (e.g. Teller, 2001). These draw downs ranged from 8 to 58 m and generated flood fluxes of about 0.06 to 0.26 Sv from flood volumes of 2000 km^3 to 8100 km^3

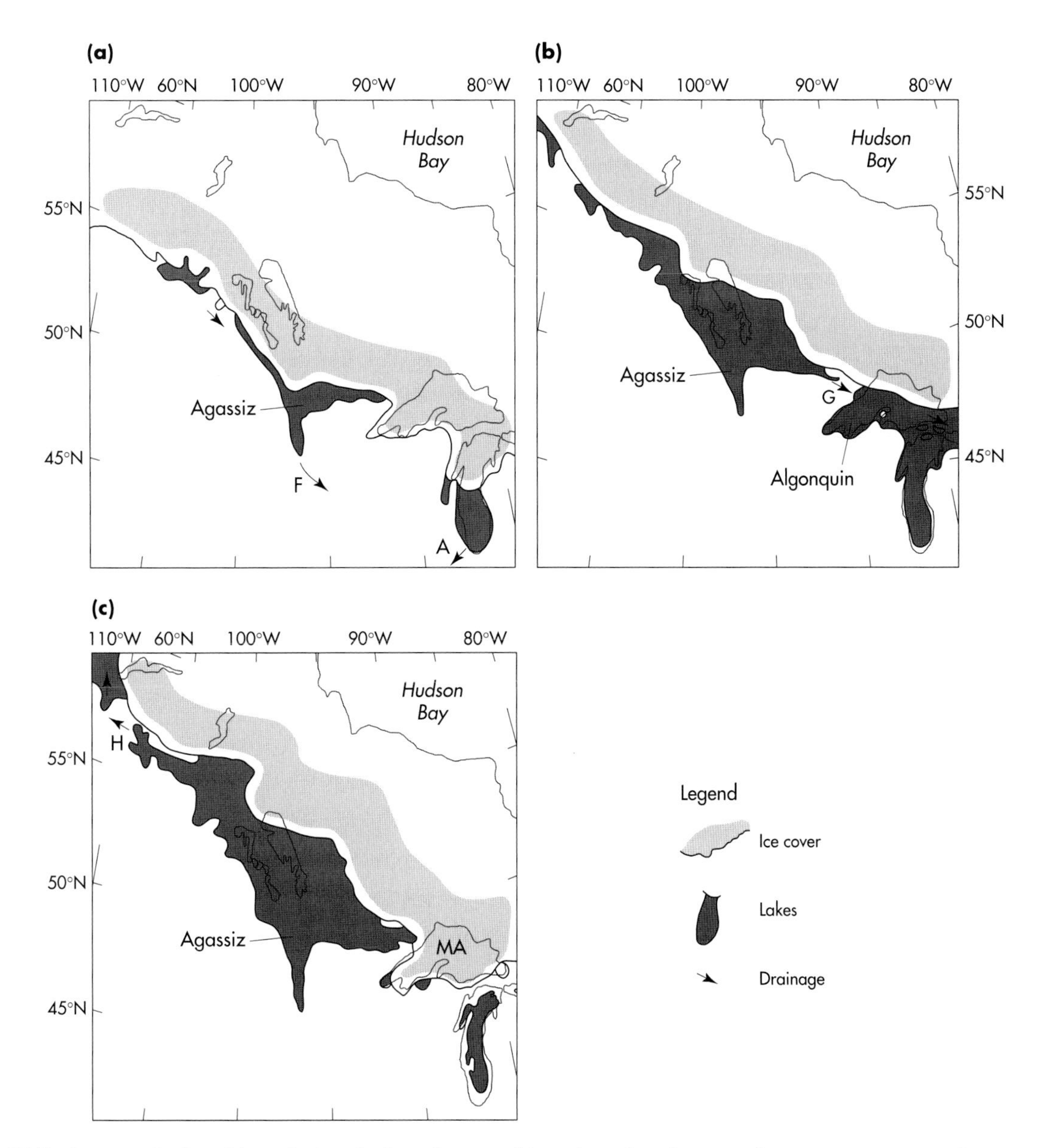

Figure 28.5 Ice-marginal positions during the first changes of Agassiz outlets, from southern to eastern to northwestern outlets. (a) Just prior to 11 ka (13,000 cal. yr), Lake Agassiz drained south at (F) to the Mississippi Valley and Gulf of Mexico. (b) Ice retreat about 11 ka (13,000 cal. yr) uncovered first eastern (lower) outlet (G) to Superior basin in the Great Lakes, thereby initiating a large outburst flood as the lake was drawn down to the outlet sill. The flood discharge and Agassiz baseline runoff were diverted to the North Atlantic Ocean via the Great Lakes–St Lawrence route. Alternatively, if ice retreat was sufficient, Agassiz discharge may have switched first to the northwestern outlet as in Fig. 28.5c. (c) By 10 ka (11,400 cal. yr), advance of the Superior ice lobe (Marquette advance, MA) had closed the first eastern Agassiz outlet and diverted the Agassiz baseline runoff briefly back to the Mississippi Valley route to Gulf of Mexico. Then, with opening of the northwestern outlet (H) to the Athabasca–Mackenzie Valley as ice retreated along the southwestern margin of LIS, runoff and a draw-down outburst flood were directed to the western Arctic Ocean. (See www.blackwellpublishing.com/knight for colour version.)

(Leverington & Teller, 2003), superposed on baseline runoff flows ranging 0.034–0.050 Sv (Teller *et al.*, 2002). Following overflow via the Nipigon basin to the Great Lakes, ice retreat farther east may have briefly opened outlets directly to Lake Superior (Thorleifson, 1996).

About 8 ka (8900 cal. yr), Lake Agassiz merged with glacial Lake Ojibway to the east, forming a huge lake of >800,000 km^2 (Leverington *et al.*, 2002), more than three times the area of the modern Great Lakes, which bordered the entire southern LIS south of Hudson Bay. Drainage from these combined glacial lakes bypassed the Great Lakes basin down the Ottawa and St Lawrence valleys to the North Atlantic Ocean (Fig. 28.2, outlet J).

In addition, beginning about 10 ka (11,400 cal. yr), another great proglacial lake developed along the northwestern part of the LIS in the Mackenzie River system, Lake McConnell, which overflowed to the western Arctic Ocean. This lake eventually covered an area of 215,000 km^2 in the isostatically depressed region on its eastern shore adjacent to the retreating LIS. This lake changed depth and

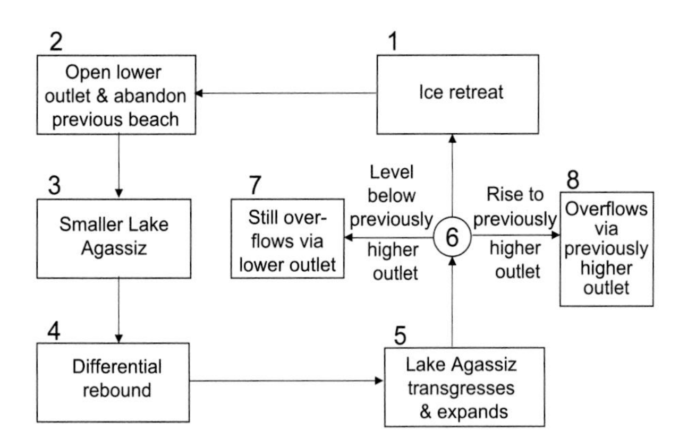

Figure 28.6 Interaction of ice retreat, opening of lower outlets, and differential rebound resulting in the switching of Lake Agassiz discharge from one outlet to another (after Teller *et al.*, 2001). See text for explanation.

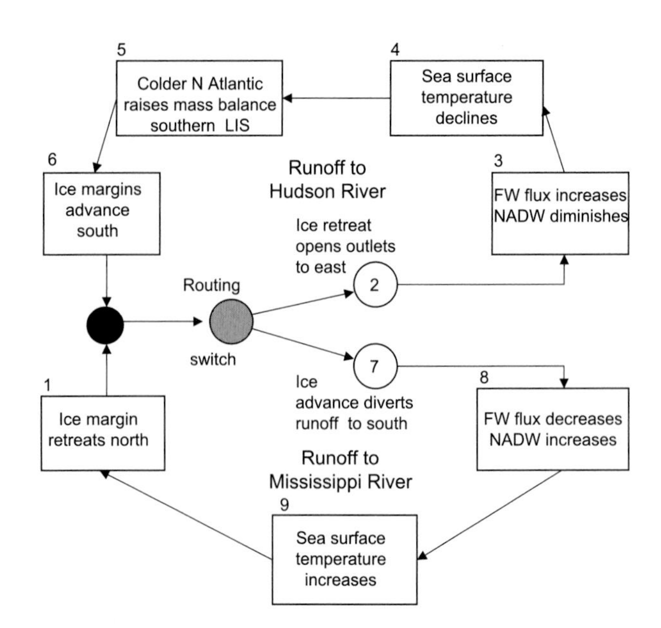

Figure 28.7 Schematic diagram of ocean–ice interaction and oscillatory switching of Mississippi runoff to and from Hudson River Valley (after Clark *et al.*, 2001). See text for explanation.

configuration in response to ice-marginal retreat and isostatic rebound, while it expanded downslope until it drained by 8.3 ka (9300 cal. yr) (Lemmon *et al.*, 1994; Smith, 1994). Lake Agassiz discharged to Lake McConnell when it overflowed through the northwestern outlet into the Athabasca River valley. As noted before, this may have been only from around 10 ka until 9.4 ka (11,400 until 10,600 cal. yr) (Smith, 1994), or could have spanned a longer interval between 10.9 and 9.4 ka (12,900 and 10,600 cal. yr).

Downwasting and thinning of the residual ice sheet in Hudson Bay significantly reduced its weight relative to buoyancy forces of the sea entering Hudson Strait to the north and those of the deep (>400 m) proglacial lakes to the south (Leverington *et al.*, 2002). By 7.7 ka (8450 cal. yr), the merged Lake Agassiz–Ojibway breached the LIS in the Hudson Bay basin (Barber *et al.*, 1999) and drained catastrophically, initially passing beneath the remaining ice in Hudson Bay, and then through Hudson Strait into the North Atlantic Ocean. Teller *et al.* (2002) calculated that this event released 163,000 km^3 of water from the lake (5.2 Sv, if released in 1 yr). Alternatively, the final release may have occurred in two steps, with an initial flood of 113,100 km^3 or 3.6 Sv, and a subsequent flood of 49,900 km^3 or 1.6 Sv a short time later (Teller *et al.*, 2002). Clarke *et al.* (2003, 2004) modelled this final drainage of the world's largest lake, concluding that, once initiated, complete drainage would have occurred in only 6 months. This terminated overflow of the Hudson Bay drainage basin through the St Lawrence system, which had been occurring since 9.4 ka (10,600 cal. yr), and established modern runoff patterns on the continent.

28.4 Models of oscillations in runoff diversion: feed-back interactions of ice–ocean/lake–atmosphere

28.4.1 Interaction of the southern LIS and North Atlantic Ocean: oscillations of the Mississippi–Hudson valley routings 16.5–11.5 ka (19,700–13,500 cal. yr)

Switches in runoff routing direction, early in deglaciation, between south to the Gulf of Mexico via the Mississippi River Valley, and east to the Atlantic Ocean via the Hudson River Valley, may be a manifestation of an ocean–ice feed-back oscillator (Clark *et al.*, 2001). This system depended on feed-backs from the North Atlantic to the mass balance of the southern Laurentide Ice Sheet. Clark *et al.* (2001) model an oscillatory system for ice cover in eastern North America during early phases of LIS retreat, as illustrated on Fig. 28.7. They note that when the LIS retreated into the Great Lakes region, its margin, which comprised thin lobes of ice, overlay a soft deformable substrate. Under these conditions, the ice margin was inherently unstable and dynamically responsive to changes in mass balance induced by feed-back from changes in temperature and sea-ice cover of the adjacent North Atlantic Ocean. As global climate warmed after the LGM, the LIS retreated north into the Great Lakes basins (box 1). This retreat (e.g. from the Ontario basin) opened the Mohawk River Valley outlet to the Hudson River Valley and the North Atlantic Ocean (circle 2). As indicated in many coupled ocean–atmosphere models (e.g. Manabe & Stouffer, 1988; Stocker & Wright, 1991; Weaver & Hughes, 1994; Ganopolski & Rahmstorf, 2001, Rind *et al.*, 2001a), the increased freshwater flux would cause a reduction in NADW production (box 3) and a consequent decline in sea-surface temperature (box 4). The colder North Atlantic would promote a mass balance increase in the LIS (box 5) and a readvance of the southern LIS (box 6). The advance would then close the Hudson River Valley outlet and divert glacial runoff back to the Mississippi River (circle 7). The resulting reduction of inflow to the North Atlantic Ocean would lead to a resumption of, or increase in, deep water production (box 8), an increase in sea-surface temperatures (box 9) and a warmer North Atlantic Ocean. This would decrease the mass of the LIS; causing a northward retreat in the ice margin (box 1). The oscillation may then have begun again.

28.4.2 Lake Agassiz, ice retreat, differential rebound and climatic feed-back effects: oscillations of St Lawrence, Mackenzie and Mississippi valley runoff routings

In the later part of deglaciation of middle North America, after 11.7 ka (13,600 cal. yr), the retreat of ice under the influence of relatively high summer insolation led to the formation of proglacial Lake Agassiz. The lake, which was impounded by the retreating ice margin on its northern side, initially overflowed the continental divide to the south. During its lifespan, Lake Agassiz enlarged significantly and varied its overflow among southern, eastern and northwestern outlets, as described above. The variations in outlet use are understood to have resulted from a combination of oscillations in ice-margin positions and changes in relative outlet elevation by differential glacio-isostatic uplift. Three possible explanations or hypotheses for ice-margin and outlet changes that may have influenced Agassiz overflow routings are described below in terms of external insolation forcing, regional climatic feed-backs, and differential rebound.

1 Numerical modelling studies by Hostetler *et al.* (2000) addressed the early history of Lake Agassiz (11–9.5 ka; 13,000–10,700 cal. yr) when southern overflow was first diverted eastward through a lower outlet opened by ice retreat. The overflow was subsequently diverted away from the eastward outlet by the Marquette ice advance across the Lake Superior basin (Fig. 28.5c). Comparisons of the lake–atmosphere–ice system in this region at about 11 ka (13,000 cal. yr), with and without Lake Agassiz, showed that the lake induced a zone of reduced precipitation along its adjacent ice margin, including the ice that dammed the eastern outlets of the lake. When waters of Lake Agassiz were present, the lake–ice–atmosphere interactions may have set up oscillations in the ice margin which induced rerouting of overflow and outburst floods when the lake fell to the level of a new lower outlet. The climatic feed-back effects of Hostetler *et al.* (2000) are outlined in Fig. 28.8a. Overall, the deglaciation is assumed to have been driven by high summer insolation (Kutzbach *et al.*, 1998) which caused long-term, progressive ice retreat and growth of the impounded lake surface (boxes 1 and 2). Based on the numerical experiments of Hostetler *et al.* (2000), low evaporation from the cold lake surface, combined with anticyclonic air flow, blocked atmospheric moisture from reaching the lake and adjacent LIS (box 3). This feed-back would have reduced mass balance of the adjacent LIS (box 3), leading to ice retreat (box 4). The retreat would have increased the lake area (return to box 2), and/or may have opened a lower outlet, in this case to the east (box 5). A rapid draw down of the lake to the sill elevation of the lower outlet would have produced a reduction in lake size (box 6). Crustal rebound (box 7) would have enlarged the lake through time. Simultaneously, according to the Hostetler *et al.* (2000) model, climatic feed-back from the reduced lake surface area would have enhanced atmospheric moisture and glacial mass balance (box 8), leading to advance of the ice margin (box 9). If the advance overrode an outlet (box 10), the lake would refill to the next higher outlet level. The larger lake (box 2), arrived at by either an ice advance, or differential uplift, or both, would start the cycle again.
2 In contrast, Teller (1987, fig. 24) suggested that an increase in lake size would lead to more evaporation and *growth* of the ice sheet. The presence of a large ice-marginal lake along the LIS would also encourage brief rapid expansions of the ice margin by surging, although subsequent rapid calving of the ice margin would occur, and that would expedite the demise of the ice sheet (Teller, 1987). In this conceptual model (Fig. 28.8b), external forcing by relatively high summer insolation (Kutzbach *et al.*, 1998) drove general ice retreat as before (box 1). The ice retreat could result in a larger lake surface (box 2). Alternatively, if retreat opened a lower outlet (box 3), a smaller lake (box 4) would result after a draw down and outburst flood through the new lower outlet. Considering the first scenario, the resulting larger lake (box 2) would have led to more atmospheric water vapour by evaporation from the enhanced water surface (box 5). In turn, more vapour would have increased precipitation on the adjacent ice surface, and enhanced glacial mass balance (box 6). This increased nourishment of the ice sheet would have led to an advance of the ice margin (box 7), and a consequent reduction in lake surface area (box 4). This smaller lake would have induced climatic effects (boxes 8–10) which, along with differential uplift (box 11), would have resulted in enlargement of the proglacial lake. Continued ice retreat as a result of high summer insolation (box 1) could have reinforced the lake enlargement (box 2), or have deglaciated a new lower outlet (box 3), starting the cycle again.
3 Krinner *et al.* (2004) modelled the lake–ice–atmosphere interaction in relation to the big Eurasian glacial lake complex, concluding differently to Hostetler *et al.* (2000). Specifically, they found that, although a large ice-marginal lake would bring about cooling and less snowfall, there was a significant reduction in summer melting of the ice sheet that resulted in readvance (or slowed retreat) of the ice margin. The application of this alternative interpretation to Lake Agassiz is shown in Fig. 28.8c. As before, the general deglaciation and ice retreat is assumed to follow from an increase in insolation (box 1), which would enlarge the lake surface area (box 2), or open a new lower outlet (box 3). A lower outlet would reduce the lake area after a draw down and outburst flood (box 4). The larger lake would induce cooling (box 5), reduce summer ice melt (box 6) and thereby increase glacial mass balance (box 7). Either this ice advance or a draw down through a lower outlet (box 3) could lead to a smaller lake (box 4). As before, the climatic effects of the reduced water surface (boxes 8–10) and the parallel effect of differential uplift (box 11) would tend to increase the lake area (box 2) and start the cycle again.

28.5 The land–ocean interface: sensitivity of ocean thermohaline circulation to freshwater runoff

Surface ocean currents bring relatively warm water to northern latitudes of the Atlantic Ocean from equatorial and temperate regions where it has gained heat (Broecker, 1987, 1997; Clark *et al.*, 2002b; Rahmstorf, 2002). As the surface water cools in

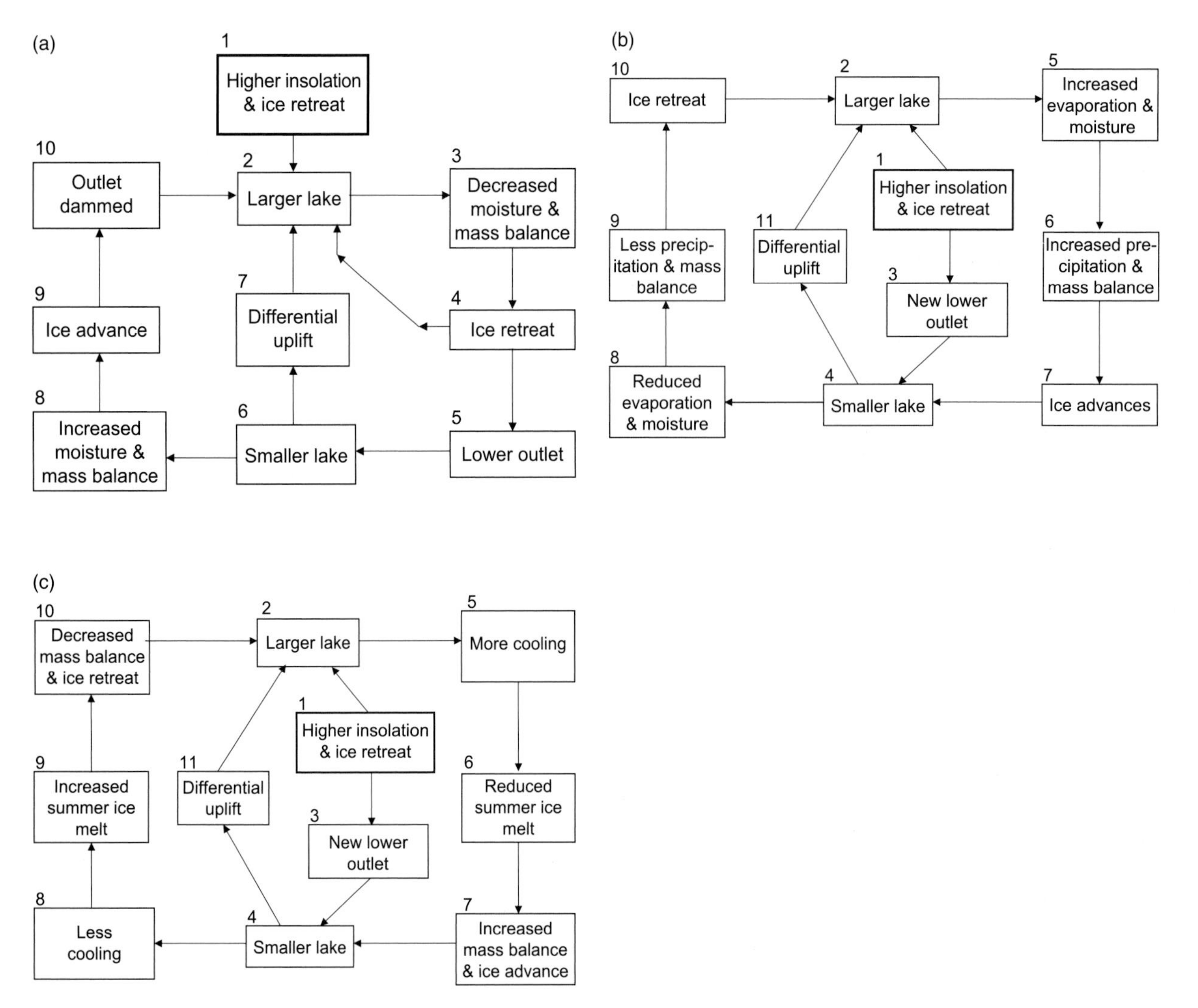

Figure 28.8 Schematic diagrams of oscillations of Lake Agassiz area, level and outflow routings as a result of ice retreat, climatic feed-backs and differential rebound. In these diagrams, the external forcing of relatively high summer insolation which drove general ice retreat (Kutzbach *et al.*, 1998) is represented in box 1 with the heavy border. (a) Lake–atmosphere–ice interactions suggest decreased precipitation over the adjacent ice sheet as the lake enlarges, based on numerical modelling by Hostetler *et al.* (2000). This model accounts for ice retreat and the first opening of eastern outlets following early growth of Lake Agassiz. (b) Following concepts in Teller (1987), a large proglacial lake possibly provided more moisture and nourished growth of the adjacent ice sheet. See text for further explanation. (c) Possible feed-backs and effects using the modelling results of Krinner *et al.* (2004) for Eurasian glacial lakes. Large lakes induced cool climate that suppressed summer ice melting, and thereby enhanced growth of the adjacent ice sheet. Positive feed-backs to lake enlargement as in Figs 28.8B and 28.8C may account for ice readvances such as the Marquette Advance (Fig. 28.5c) which closed the eastern outlet about 10 ka (11,400 cal. yr).

northern North Atlantic seas between Greenland and Norway, and off Labrador, it imparts enormous quantities of heat to the atmosphere and gains density by further evaporation and cooling. This heat warms the climate of Europe and the North Atlantic basin. The cooled salty surface water, too dense to be buoyed up by underlying water masses, sinks to great depth and forms North Atlantic Deep Water (NADW). This deep water circulates southward and eastward, forming a Great Conveyor Belt of THC circulation, merging with Antarctic deep water masses in the southern ocean, and ultimately entering the Indian and Pacific Oceans. This Great Conveyor Belt is an important planetary mechanism for distributing heat to northern Atlantic regions and salinity to other oceans (Broecker, 1987, 1997; Clark *et al.*, 2002b; Rahmstorf, 2002).

Being a density-driven flow, the conveyor belt is sensitive to variations in surface salinity at sites of deep water formation, for example, in the North Atlantic Ocean (Rooth, 1982). Increases in freshwater input, as from deglacial continental drainage switches and lake outburst floods, will reduce salinity and density of surface water, and thus will reduce or even stop the production of deep water, thereby inducing abrupt climate change (cooling) (Broecker, 1987, 1997; Broecker *et al.*, 1990, Clark *et al.*, 2002b;

Rahmstorf, 2002). Thermohaline circulation and atmospheric warming may increase elsewhere, for example in Antarctic seas, when NADW formation is reduced (Broecker, 1998; Rahmstorf, 2002). Thermohaline circulation and the rate of NADW formation is non-linear with respect to freshwater input and reduction of surface salinity (Broecker *et al.*, 1985; Stocker & Wright, 1991; Manabe & Stouffer, 1995; Rahmstorf, 1995a,b; Rind *et al.*, 2001a). At higher salinity, THC is vigorous, but below a 'threshold' salinity value, THC is sluggish. In between these two states, THC is very sensitive to changes in surface salinity and may shift from high to low rates of NADW formation (or *vice versa*) with a relatively low freshwater forcing, as shown and described in Fig. 28.9. This may help explain why the relatively small outburst of Lake Agassiz overflow around 11 ka (13,000 cal. yr) (9500 km^3) could have initiated the Younger Dryas cooling (Broecker *et al.*, 1988), whereas the introduction of >160,000 km^3 of freshwater when Lake Agassiz completely drained appears to have had only a small impact on climate around 7.7 ka (8200 cal. yr) (Teller *et al.*, 2002). During the Younger Dryas, the roughly 1000-yr long shift of baseline overflow from the 2 million km^2 watershed of glacial Lake Agassiz, from the Gulf of Mexico to the North Atlantic, may have sustained the relatively small Agassiz perturbation to THC and the associated cooling (Teller *et al.*, 2002).

Numerical models simulating North Atlantic climatic and oceanographic processes generally confirm the sensitivity of THC to freshwater inputs as noted in Fig. 28.9, although there are differences in detail. The freshwater input required to affect THC has been explored by numerous experiments in numerical modelling, and some results are reviewed in Teller *et al.* (2002). These experiments have shown that different stable modes of THC are possible, and that various North Atlantic responses may occur for a range of modelled influxes that overlap with the reconstructed influxes shown here (e.g. Manabe & Stouffer, 1988, 1995, 1997; Stocker & Wright, 1991; Weaver & Hughes, 1994; Tziperman, 1997; Ganopolski & Rahmstorf, 2001; Rind *et al.*, 2001a). For example, Manabe & Stouffer (1997) found a reduction of THC and a decline of sea-surface temperature of about 6°C in less than 100 yr for an input of as little as 0.1 Sv. Larger freshwater additions generated more abrupt rates of change. An increase of 0.06 Sv for a few hundred years achieved shutdown of NADW formation in modelling by Rahmstorf (1995a). A larger flux of 0.16 Sv directly inserted into the North Atlantic for just 4 yr could achieve a similar change (Rahmstorf, 1995b).

Fanning & Weaver (1997) studied the effects of geographical and temporal variation in freshwater influx. They concluded that when pre-Younger Dryas runoff to the Gulf of Mexico had reduced salinity in the North Atlantic, a major slow down in NADW production could be brought about in 200 yr by a freshwater inflow of only 0.026 Sv directly through the St Lawrence Valley. Rind *et al.* (2001a) found that a flux of 0.12 Sv through the St Lawrence Valley reduced normal NADW formation by 32% in 100 yr, or by 48% if the THC was weaker; a large sustained freshwater flux of 0.53 Sv would reduce normal THC by 98% in 100 yr. As expected, the impact on NADW production varies with the magnitude and duration of freshwater flux, and the strength of ocean circulation. Resumption of THC occurred with cessation of freshwater inflow if NADW production were only reduced, but did not occur after complete NADW shutdown (Rind *et al.*, 2001a). Teller *et al.* (2002) conclude that large draw-down events of Lake Agassiz would likely have triggered a change in THC, especially when the influx occurred under weakened THC conditions during the glacial–interglacial transition, and when it was followed by a diversion and increase in baseline freshwater inflow.

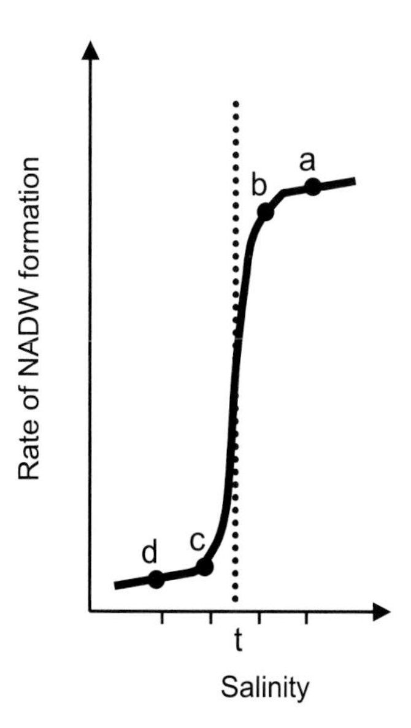

Figure 28.9 Curve illustrating the response of North Atlantic Deep Water (NADW) formation to the introduction of an arbitrary volume unit of freshwater that reduces surface salinity at a time when there is a specific 'Rate of NADW formation' (after Stocker, 1996; Bond *et al.*, 1999). For example, if salinity was reduced by 1 unit on the 'Salinity' scale, when NADW formation was high (point a), there would have been only a small response in NADW formation rate (point b). This situation is like the present interglacial mode, which is relatively insensitive to freshwater additions. If salinity was reduced by 1 unit when NADW formation was low and salinity was also relatively low (point c), as probably occurred during a full glacial, there would only be a small response in the rate of NADW formation (point d). In contrast, if freshwater additions reduced the salinity by 1 unit when salinity was near the threshold value *t* at point b, it would reduce the rate of NADW formation dramatically to point c. This probably is what the North Atlantic Ocean was like during the transition from glacial to interglacial conditions, and may have led to the large change in NADW formation associated with the Younger Dryas that is attributed to the 9500 km^3 outburst from Lake Agassiz.

The subsequent Preboreal Oscillation (PBO) at 9.6 ka (10,900 cal. yr) and the climatic event called the 8.2 cal. yr cold event, which are registered in Greenland ice cores and terrestrial records in Europe, are also attributed to diversions and outburst floods of freshwater from the North American continent. The PBO, known from vegetation shifts in Europe and eastern North America, has been linked to the opening of the northwestern outlet of Lake Agassiz around 9.9 ka (11,300 cal. yr) (Fisher *et al.*, 2002; Teller *et al.*, 2002). Fisher *et al.* (2002) suggest that the diversion of runoff and lake outburst floods down the Mackenzie Valley to the Arctic Ocean may have induced sea-ice thickening there and greater outflow of sea ice and freshwater into the North Atlantic Ocean, inhibiting THC. Additional freshwater contributions to the North Atlantic about this time came from draining of the Baltic Ice Lake in Europe (Björck *et al.*, 1996; Hald & Hagen, 1998). The relatively short duration (about 200 yr) and limited impact of the PBO may be a consequence of Agassiz outflow to the Arctic Ocean (versus the North Atlantic Ocean), a more robust phase of THC, and a lack of previous preconditioning freshwater outflow to the Gulf of Mexico (Teller *et al.*, 2002).

The 8200 cal. yr cold event (Alley & Agústsdóttir, 2005) lasted only 100 or 200 yr, and is manifested in vegetative, limnological and speleothem proxy records in Europe (Klitgaard-Kristensen *et al.*, 1998; von Grafenstein *et al.*, 1998; Tinner & Lotter, 2001; Baldini *et al.*, 2002), depressions of air temperature over Greenland (Alley *et al.*, 1997), and by increased wind speed in the Atlantic tropics due to high-latitude cooling (Hughen *et al.*, 1996). This event is attributed to THC slowdown related to the enormous final discharge of merged glacial lakes Ojibway and Agassiz (Fig. 28.3e; 5.2 Sv for one flood or 3.6 and 1.6 Sv for two closely spaced drainages). This drainage is associated with the collapse of the LIS about 8400 cal. yr and the onset of Hudson Bay watershed runoff to the North Atlantic via Hudson Strait (Barber *et al.*, 1999; Clarke *et al.*, 2003, 2004). Teller *et al.* (2002) suggest that the relatively small THC impact of this huge influx, which preceded the 8200 cal. yr cooling by several hundred years, can be explained by the fact that the North Atlantic Ocean was in a more stable mode of circulation (Fig. 28.9). Dean *et al.* (2002) offer the view that the 8.2 ka cold event resulted from a reorganization of atmospheric circulation related to marked changes in the proportion of land, lake and glacial ice in North America at this time, the most important of which was the drainage of Lake Agassiz and disappearance of the central LIS over Hudson Bay. These changes, they argue, may have produced a perturbation in climate over Greenland that resulted in the brief cold pulse detected in ice cores and elsewhere.

28.6 Summary

During the deglaciation of North America after the Last Glacial Maximum, the position of the southern margin of the Laurentide Ice Sheet (LIS) was characterized by millennial-scale oscillations (northward retreat followed by southward advance and subsequent retreat). These oscillations, related to ice–lake/ocean–atmosphere interaction, together with differential isostatic rebound and palaeotopography, caused the drainage of North American runoff to be periodically rerouted to different oceans during deglaciation. Oscillations of the ice margin in the Great Lakes basin between 16.5 and 11.5 ka (19,700 and13,500 cal. yr) led to at least two switches of ice melt and precipitation runoff between the Gulf of Mexico and North Atlantic Ocean. After 11.7 ka (13,600 ka), ice margin oscillations in the Agassiz basin caused switches of drainage among routes to the Gulf of Mexico, North Atlantic and Arctic oceans, ending with a final large discharge to Hudson Bay and Hudson Strait about 7.7 ka (8450 cal. yr). Normal baseline runoff, due to precipitation and melting of the LIS, was supplemented on numerous occasions by the release of stored water in ice-marginal lakes, including the proglacial Great Lakes and Lake Agassiz. The timing, magnitude, duration and location of these catastrophic releases, and the associated baseline freshwater flows appear to have influenced North Atlantic thermohaline circulation and associated climatic excursions, namely as related to the Younger Dryas, Preboreal Oscillation, and 8.2 cal. yr cold events recorded over large areas of the circum-Atlantic region.

Acknowledgements

We thank Drs Ann Miller and Thymios Tripsanas for constructive review comments, and Alan Bateman, Ken Hale and Gary Grant for preparation of the illustrations. This is Geological Survey of Canada Contribution No. 2004116.

TWENTY-NINE

Impacts of climatic trends upon groundwater resources, aquifer–stream interactions and aquatic habitat in glacierized watersheds, Yukon Territory, Canada

Sean W. Fleming

Department of Earth and Ocean Sciences, University of British Columbia, 6339 Stores Road, Vancouver, B.C. V6T 1Z4, Canada

This case study briefly turns the spotlight upon a topic with a broader significance that may prove to belie its perhaps modest appearance: differences in the winter groundwater responses of glacierized and glacier-free watersheds to climatic warming. Glacial runoff could be altered significantly by projected anthropogenic climate warming, and observed long-term trends in surface water resources under historical warming conditions can be dramatically different between glacierized and nival catchments. Virtually no attention has been devoted, however, to potential differential trends in groundwater availability. Such differential trends, if present, may be particularly important from an ecological perspective. Groundwater resources control quality and quantity of cold-regions winter aquatic habitat, largely by sustaining the (typically very low) winter baseflows of freshet-driven rivers, with powerful and direct impacts upon fish populations and biogeography (see Reynolds, 1997; Cunjak *et al.*, 1998). Temporal trends in winter groundwater supplies, and interbasin variability in such trends arising from the presence or absence of watershed glacial cover, would thus strongly imply parallel fluvial hydrological and ecological effects.

Consistent multidecadal piezometric measurements from aquifers undisturbed by human utilization are very rare, but usably long streamflow records from sufficiently pristine watersheds are relatively common. As winter baseflow is a direct reflection of hydrogeological conditions, such data offer a surrogate measure of groundwater resource availability and also 'speak' directly to riverine impacts. Time series of annual minimum daily discharge were therefore constructed from historical streamgauge records (Environment Canada, 1999) for five glacierized and four nival basins in southwest Yukon and northwesternmost British Columbia (Fig. 29.1). These watersheds possess a number of properties advantageous to an empirical analysis of the effects of watershed glacierization upon water resource responses to historical climatic warming, and have been used previously for the purpose. Fleming & Clarke (2003) provide a detailed description of the study area and data; two key points are that the region is experiencing a climatic warming trend, and that systematic differences in the hydrological responses of glacierized and nival catchments are indeed directly attributable to presence or absence of glacial cover. Note that surface and subsurface hydrological catchment areas appear to correspond reasonably well within the study area (see Owen, 1967). Each of the nine time series was normalized by its own maximum value to facilitate interbasin comparisons, and winter baseflow trends, β, were determined for each

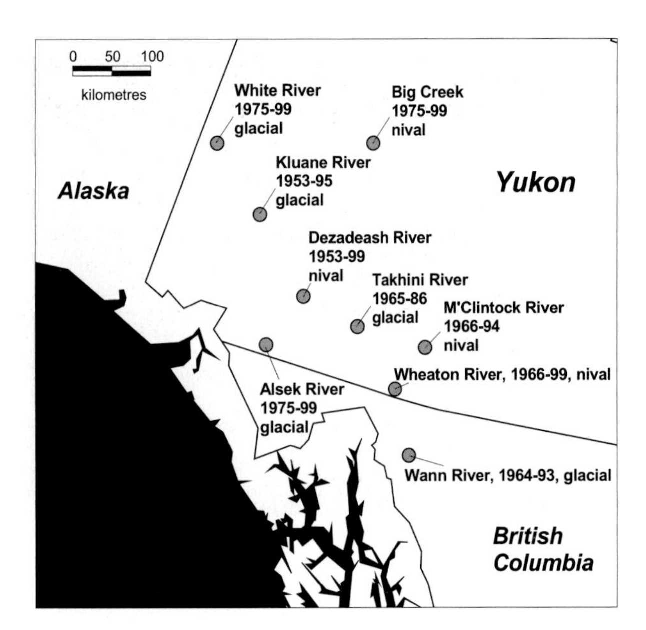

Figure 29.1 Gauge locations and corresponding hydrological record durations.

catchment using a rank-based approach resistant to outliers (Theil, 1950)

$$\beta = \text{median}\left[\frac{x_j - x_k}{j-k}\right] \forall k < j \tag{1}$$

where x_i, $i = 1, N$ is the annual time series of length N. Results are shown in Table 29.1. The median trend value for glacierized catchments is somewhat higher than that of the nival basins, but the difference is not statistically significant ($P > 0.05$, 2-tailed Mann-Whitney test; see also Fleming & Clarke, 2003). The ranges are also similar. However, the overall trend pattern appears to reveal a stark contrast: glacierized catchments yield uniformly positive trends, whereas the various nival watersheds exhibit a mixture of both progressively increasing and decreasing winter baseflows.

The likelihood that this pattern arises from chance trends in random time series is assessed using a non-parametric Monte Carlo randomization technique. In a given realization, random baseflow records are generated by scrambling the temporal order of the observations in each of the nine original time series. Trends are then estimated using Equation (1). The trend pattern thus generated is taken to match the observed pattern if the following condition is satisfied

$$\{\beta_l > 0 \text{ all } l = 1, N_g\} \text{ and } \{(\beta_m > 0 \text{ any } m = 1, N_n) \text{ and } (\beta_m < 0 \text{ any } m = 1, N_n)\} \tag{2}$$

where N_g and N_n are the number of glacierized and nival basins, respectively. A total of 10^5 such realizations are synthesized. The probability, P, that the observed trend pattern occurs by chance is the proportion of the ensemble that satisfies Equation (2). By this technique, $P = 0.02$; the observed pattern is statistically significant.

Although streamflow data are utilized, the baseflow trend pattern is reflective of changes in hydrogeological conditions and must be interpreted in that light. I posit that the observed pattern arises from systematic basin-to-basin variability in the net balance of trends in aquifer recharge and aquifer properties. Fleming & Clarke (2003) demonstrated that freshet magnitudes here are increasing in glacierized watersheds but decreasing in nival watersheds. This essentially results from the relative importance of trends in temperature and precipitation and their effects upon evapotranspiration and glacial melt production, and implies increased (decreased) summertime aquifer recharge in glacial (nival) watersheds. However, aquifer properties are also believed to be changing due to the permafrost impacts of historical climatic warming. These watersheds lie in areas of sporadic (valley bottoms) to continuous (alpine zones) permafrost, which exerts powerful control over hydrogeological properties. Permafrost may be degrading regionally in various fashions under warming conditions, which has been hypothesized to increase the connectivity and storage capacity of aquifers, resulting in greater winter baseflow (Woo, 1990; Michel & van Everdingen, 1994). Warming-induced trends in aquifer recharge and aquifer storage capability are in the same direction for glacierized catchments, resulting in uniformly positive baseflow trends. In contrast, such hydrogeological trends lie in opposite directions for nival catchments, so that the net direction of groundwater resource impacts is sensitive to interbasin variability in basic geological characteristics, leading to a mixture of positive and negative baseflow trends in different nival watersheds. Implications for basin-to-basin consistency in the direction of climatically induced changes in aquifer-stream coupling and winter aquatic habitat availability are clear.

The above interpretation, although plausible, is by necessity conjectural. Two important steps toward a deeper understanding of glacial influences upon groundwater responses to climatic shifts are to replicate the analyses presented here and by Fleming & Clarke (2003) in other regions, and to obtain year-round piezometric measurements in undisturbed fluvial aquifers to better constrain aquifer–stream interactions in glacial and permafrost environments, enabling more effective and thorough use of available long-term baseflow datasets.

Table 29.1 Estimated baseflow trends (10^{-3}yr^{-1})

Glacial watersheds					Nival watersheds			
Wann	Alsek	Kluane	White	Takhini	Big Creek	M'Clintock	Dezadeash	Wheaton
1.2	5.2	9.7	6.2	4.5	−0.2	0	4.6	−3.3

THIRTY

Ice sheets and marine sedimentation on high-latitude continental margins

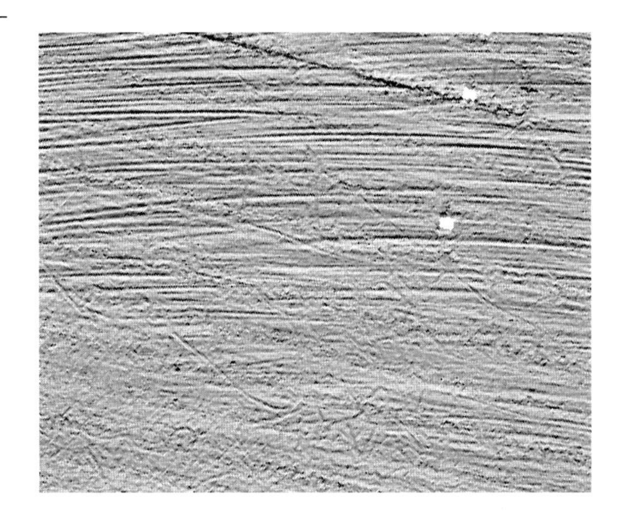

Julian A. Dowdeswell

Scott Polar Research Institute and Department of Geography, University of Cambridge, Cambridge CB2 1ER, UK

30.1 Introduction

Ice sheets cover about 16 million km^2 of the Earth's surface today. During the last glacial maximum, about 20,000 yr ago, ice covered about double this area, contributed to mainly by the growth of the North American and Eurasian ice sheets and the advance of the Antarctic and Greenland ice sheets across the adjacent continental shelves (Clark & Mix, 2002). Significant portions of the margins of these contemporary and past ice sheets terminate in marine waters. From this ice–ocean interface, icebergs, meltwaters and the sediments they transport, are released into the oceans. About 10% of modern oceans are affected directly by glacier-derived sediments and sedimentation, and this area approximately doubled at the last glacial maximum.

The continental margins of high- and mid-latitudes, that is the continental shelves, slopes and deep-sea basins beyond (Fig. 30.1), therefore contain a record of both present and past glacimarine processes and products. Here, we describe the large-scale sedimentary architecture of these glacier-influenced margins, together with inferences that can be made from this evidence concerning the nature of past ice-sheet flow and environmental change.

30.2 Background: the flow of modern ice sheets

The dynamics of modern ice sheets, and the nature of their spatial and temporal variability in flow, are important to our understanding of the high-latitude marine sedimentary record. This is because the flux of ice, and any associated meltwater and sediment load delivered to the adjacent ocean, will vary significantly depending on the ice-flow regime. The large-scale form and flow of contemporary ice sheets and large ice caps have been examined using a variety of satellite and airborne remote-sensing systems. Ice-surface features indicative of flow dynamics can be observed at the resolution of a few tens of metres over vast areas of ice-sheet interiors (e.g. Bindschadler *et al.*, 2001b). Satellite radar altimeters can measure ice-sheet surface elevation to a few tens of centimetres, and SAR interferometry allows the calculation of ice velocity, also at synoptic scales (e.g. Wingham *et al.*, 1998; Joughin *et al.*, 1999).

When ice-surface topography, surface features and velocity structure are mapped over large areas of Antarctica and Greenland, it has been shown that the ice sheets exhibit two basic modes of flow. A smooth and unbroken ice surface is indicative of slow ice flow, whereas rougher and often heavily crevassed ice with abrupt lateral margins defines areas of fast flow. These fast-flowing features, known as ice streams, are up to tens of kilometres in width and penetrate from the coast up to several hundred kilometres into the ice-sheet interior. Altimetric data show that they drain very large interior basins of up to 10^4 km^2 in Greenland and 10^6 km^2 in Antarctica (Joughin *et al.*, 1997; Bamber *et al.*, 2000a). Ice-surface velocities range from hundreds to thousands of metres per year in ice streams, whereas the smooth topography beyond the boundaries of fast flow often has velocities of only a few metres per year. These two modes of flow are typical not only of the great ice sheets, but also of the many large ice caps in the island archipelagos of the Arctic (e.g. Dowdeswell *et al.*, 1999a, 2002a).

The ice streams of the West Antarctic Ice Sheet are perhaps the most thoroughly investigated fast-flowing ice masses (Alley & Bindschadler, 2001). Each of six ice streams is separated by a ridge of ice that flows only very slowly, and almost all the mass transfer in the region occurs through the ice streams. Interestingly, one ice stream, C, has switched from fast to slow flow during the past 140 yr (Whillans *et al.*, 2001). Thus, it appears that, in addition to the spatial variability described above, there is also temporal switching in ice-flow regime. In Greenland too, over 80% of mass loss from the ice sheet is through about ten ice streams, each of which drains through the peripheral mountains of the island as a fast-flowing outlet glacier. Switching between slow and fast flow

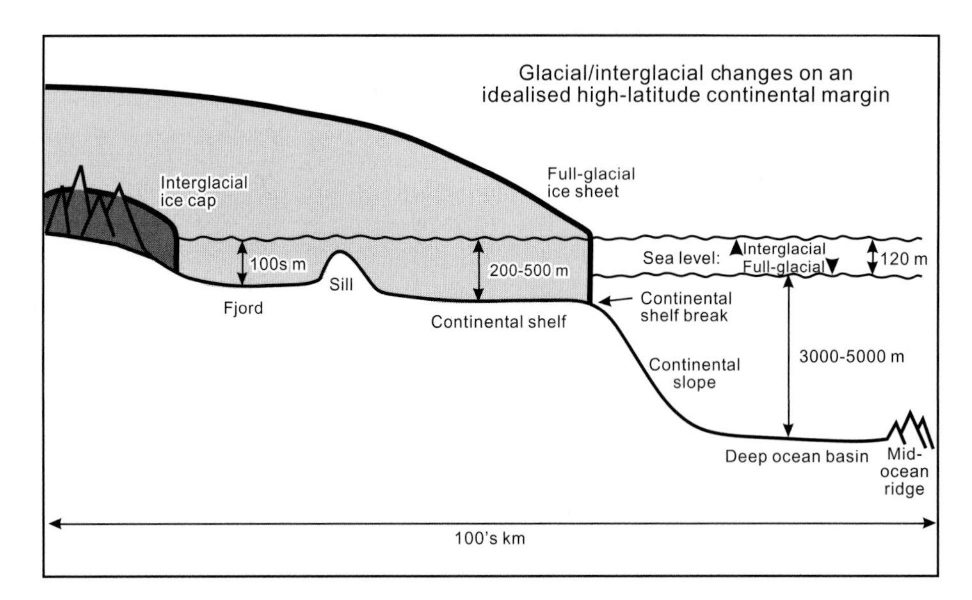

Figure 30.1 Schematic diagram of a glacier-influenced continental margin. (Modified from Dowdeswell *et al.*, 2002b.)

also has been observed in Greenland outlet glaciers, such as Storstrommen (Reeh *et al.*, 1994). In this case, glacier surging appears to have taken place, whereby a few years of fast flow punctuate much longer periods of stagnation.

The nature of the ice-sheet terminus where it reaches high-latitude seas also varies. The margin may be either grounded or floating. Mass loss from grounded margins (known as tidewater margins) is in the form of iceberg production and the release of any meltwater at terminal ice cliffs. By contrast, floating margins, known as ice shelves (or ice tongues where they are long relative to their width), allow the penetration of marine waters into the cavity beneath. The cavity can be hundreds of metres deep and extend from tens to hundreds of kilometres upstream to the grounding zone, where ice rests directly on a sedimentary or bedrock substrate. The grounding zone, and the base of floating ice shelves particularly close to this zone, are important areas of mass loss and sediment delivery by melting, in addition to the calving of icebergs at the ice-shelf edge. Large ice shelves, exemplified by the Ross, Ronne and Amery ice shelves of up to 10^5 km^2, are present today only in Antarctica; ice is about 300 m thick at the ice-shelf edge and often over 1000 m thick at the grounding zone. Smaller fringing ice shelves and glacier tongues also occur elsewhere in Antarctica and at the margins of fast-flowing outlets of the Greenland Ice Sheet; several Greenland outlet glaciers have floating tongues of about 500 m in thickness (e.g. Olesen & Reeh, 1969; Mayer *et al.*, 2000). Such variations in the nature of the ice–ocean interface contribute to the observed patterns of glacier-influenced sedimentation on high-latitude margins.

30.3 Marine geological and geophysical records on high-latitude margins

Continental margins extend up to hundreds of kilometres from landmasses, which are often mountainous, to continental shelves with water depths typically of a few hundred metres, the outer limit of which is defined by a shelf break usually at about 500 m (Fig. 30.1). Beyond the shelf break, the margin continues into deep water, via a continental slope (of between about 1° and 10°), to a deep-ocean basin of very low gradient at abyssal depths of several thousand metres. In the deep sea, the outer boundaries of an ocean basin are normally defined by tectonic features, such as spreading centres and fracture zones, that rise above the abyssal plain (Fig. 30.1).

30.3.1 Continental-shelf morphology and sediments

High-latitude continental shelves range from a few tens to several hundred kilometres in width. The shelves of both the Arctic and Antarctic are typically made up of relatively deep cross-shelf troughs and intervening shallower banks. The Norwegian shelf provides a Northern Hemisphere example (Plate 30.1), as does the Ross Sea in Antarctica (e.g. Anderson, 1999; Shipp *et al.*, 1999; Ottesen *et al.*, 2002). Many troughs are linked to relatively narrow deep-water fjords which dissect the mountainous hinterland of adjacent landmasses (Syvitski *et al.*, 1987). Ice has advanced across most of these shelves to reach the shelf edge during successive full-glacial periods, retreating again in subsequent interglacials (Fig. 30.1).

Some Arctic continental shelves, in particular the Laptev, East Siberian and Beaufort shelves which fringe the Arctic Ocean, are very shallow at tens of metres deep. They are supplied with sediments mainly by major Arctic river systems, rather than from glacial delivery of debris. They were emergent during the lower sea levels of full-glacial intervals. By contrast, in Antarctica the surrounding shelves often deepen landward, from about 500 m at the shelf edge to sometimes in excess of 1000 m close to the margins of the Antarctic Ice Sheet (Anderson, 1999). This is a result of isostatic loading of the crust beneath Antarctica by ice that averages about 2.5 km in thickness over the continent (Drewry *et al.*, 1983), combined with glacial erosion over the late Cenozoic (Anderson, 1999).

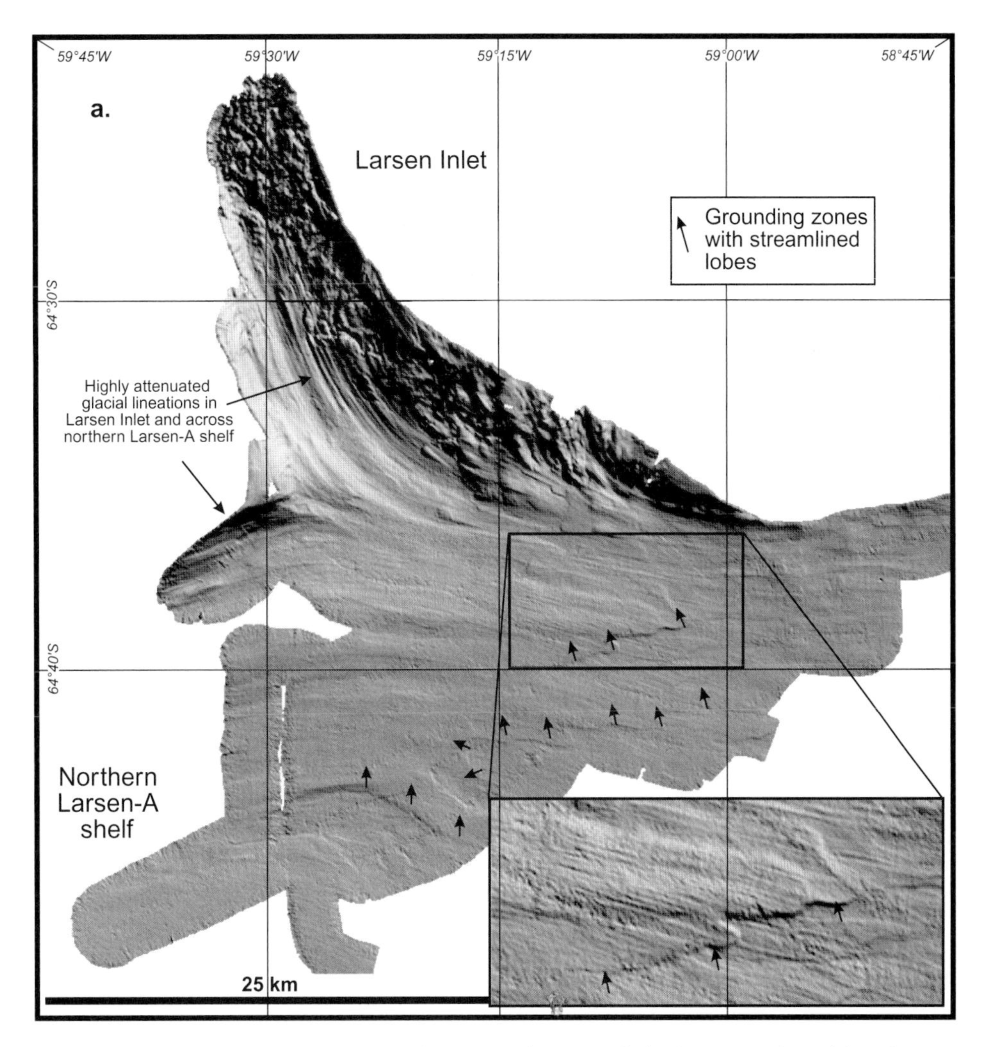

Figure 30.2 Morphology of high-latitude continental margins from swath-bathymetric data. (a) Sediment scarp marking a palaeo-grounding line formed during the retreat of ice offshore of the Larsen Ice Shelf, Antarctica (arrowed) (modified from Evans *et al.*, 2005).

The more detailed morphology of polar continental shelves reveals a series of characteristic sea-floor features (Fig. 30.2). These submarine landforms include moraine ridges and grounding-zone deposits marking the former extent of ice, a variety of streamlined glacial landforms indicating past ice-flow directions and dynamics, and ice-keel scours resulting from the grounding of floating icebergs and sea ice (e.g. Ottesen *et al.*, 2005).

Submarine moraines are sedimentary ridges, orientated parallel to the shelf edge, marking either the maximum extent of past ice sheets or still-stands during retreat (e.g. Anderson, 1999; Shipp *et al.*, 2002; Ottesen *et al.*, 2005). Grounding-zone sedimentary wedges, identifiable on acoustic stratigraphical records and sometimes with surface morphological expression, indicate the past locations of the zone where ice begins to float and form an ice shelf (e.g. Evans *et al.*, 2005) (Fig. 30.2a). The development of both moraines and grounding-zone wedges usually requires that ice is stable on the shelf for some time, in order for significant sediment build-up to take place. Past locations of the ice front and grounding zone thus can be reconstructed and, if sediment cores are available, dated to provide a chronology for ice-sheet retreat across high-latitude shelves (e.g. Svendsen *et al.*, 1992; Domack *et al.*, 1999).

Elongate, streamlined submarine landforms, such as megascale lineations (Clark, 1993) and both sedimentary and rock drumlins (Fig. 30.2b), have been used to infer the direction of former ice flow on glaciated continental shelves (e.g. Lowe & Anderson, 2002; Evans *et al.*, 2004, 2005; Ottesen *et al.*, 2005). In addition, megascale glacial lineations have been observed widely in high-latitude cross-shelf troughs (e.g. Shipp *et al.*, 1999; Canals *et al.*, 2000; Wellner *et al.*, 2001; Ó Cofaigh *et al.*, 2002a), formed in soft, deformable diamictic sediments a few metres thick (Dowdeswell *et al.*, 2004a; Ó Cofaigh *et al.*, 2005). These lineations, which have wavelengths and amplitudes of hundreds of metres and less than about 5 m, respectively, are often kilometres in length. They are interpreted to indicate the presence of former ice streams, which drained large interior ice-sheet basins to the shelf edge during full-glacial conditions (Stokes & Clark, 1999).

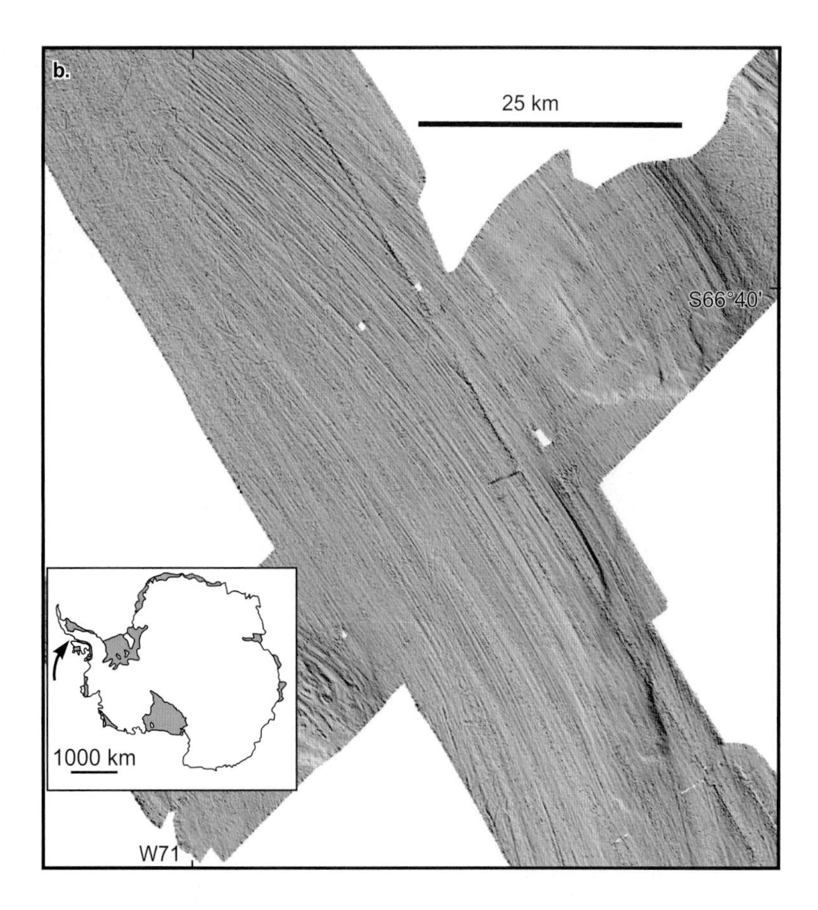

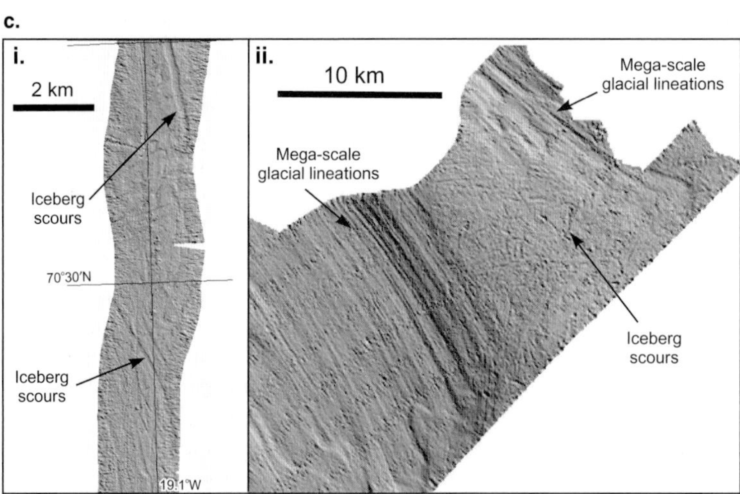

Figure 30.2 *Continued* (b) Megascale glacial lineations in Marguerite Trough, Antarctic Peninsula (modified from Dowdeswell *et al.*, 2004a). (c) Irregular pattern of scours produced by the keels of drifting icebergs impinging on sea-floor sediments. (i) Iceberg scours on the East Greenland continental shelf offshore of the Scoresby Sund fjord system. (ii) Iceberg scours on the continental shelf west of the Antarctic Peninsula. Note the irregular pattern of the scours in plan view and the morphological contrast with the subglacially produced, megascale glacial lineations (arrowed) to either side.

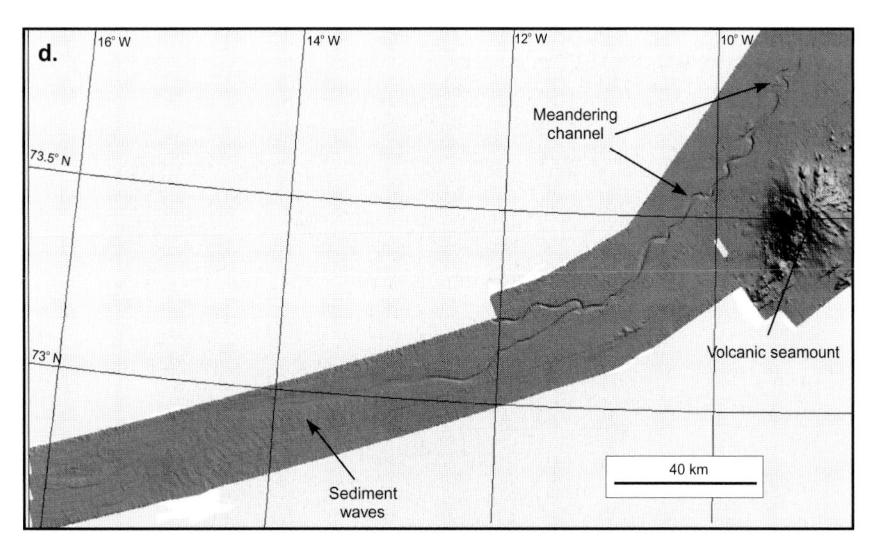

Figure 30.2 *Continued* (d) Submarine channels in the Greenland Basin, east of Greenland at 73–76°N. (Modified from Ó Cofaigh *et al.*, 2004.)

Scouring of high-latitude shelves by the intermittent grounding of iceberg and sea-ice keels produces very large numbers of irregular grooves on the sea floor ranging from metres to tens of metres in width, up to several metres deep and from tens to thousands of metres in length (e.g. Belderson *et al.*, 1973; Barnes, 1987; Dowdeswell *et al.*, 1993) (Fig. 30.3c). The keels of modern icebergs can reach up to 500–600 m and, as a result, most polar shelves are heavily scoured, leading to extensive reworking of seafloor sediments and the destruction or severe disturbance of the stratigraphical record (Vorren *et al.*, 1983; Dowdeswell *et al.*, 1992, 1993, 1994; Ó Cofaigh *et al.*, 2002b). Sea-ice keels are usually <10 m in depth, although pressure-ridging can produce occasional keels up to about 20–30 m in thickness. Even so, the shallow shelves fringing much of the Arctic Ocean, the Beaufort, Laptev and East Siberian shelves are severely scoured by sea ice (Barnes *et al.*, 1982; Reimnitz *et al.*, 1994).

Typical sedimentary facies can be found on high-latitude continental shelves over which ice advanced at the last glacial maximum and retreated during deglaciation (e.g. Svendsen *et al.*, 1992). Erosion during ice advance produces a sharp erosional contact overlain by diamictic sediments with a relatively high shear strength deposited at the ice-sheet base. If a fast-flowing ice stream develops, a soft deforming diamict layer of high porosity and low shear strength forms (Dowdeswell *et al.*, 2004a; Evans *et al.*, 2005; Ó Cofaigh *et al.*, 2005). As ice retreats across the shelf during deglaciation, grounding-zone deposits comprising interlaminated gravel, sand and finer grained layers are produced during still-stands in retreat, followed by increasingly fine-grained silts and clays with progressively fewer iceberg-rafted pebbles as the shelf becomes more distal to the retreating ice front. The whole sequence may be capped by relatively organic-rich interglacial muds (Domack *et al.*, 1999; Evans & Pudsey, 2002). Where the shelf is shallow enough for iceberg keels to rework sediments, massive diamict is common (Dowdeswell *et al.*, 1994). Current reworking may lead to the removal of fine material and armouring with a lag made up of shells and iceberg-derived pebbles (e.g. Andruleit *et al.*, 1996). In the coldest Antarctic waters, close to the freezing point of saline ocean water at −1.8°C, no meltwater-transported sediment is delivered. Icebergs traverse the area without melting to release their debris load and slow biogenic sedimentation dominates (Domack, 1988).

30.3.2 Continental-slope and ocean-basin morphology and sediments

The angle of the continental slope offshore of high-latitude shelves varies from <1° to >10°. The gradient of about 13,000 km or 70% of Arctic continental margins was analysed by O'Grady & Syvitski (2002). They found that lower slope angles were associated with convergent modern and full-glacial ice flow and, by implication, with more rapid rates of sediment delivery (Dowdeswell & Siegert, 1999). Side-scan sonar and seismic-reflection surveys of Arctic and Antarctic continental slopes have shown that large submarine fans formed from glacier-derived debris are present beyond many, but not all, cross-shelf troughs (e.g. Bart & Anderson, 1996; Dowdeswell *et al.*, 1996; Vorren *et al.*, 1998; Ó Cofaigh *et al.*, 2003). Seismic stratigraphical investigations show that such margins are usually prograding; that is, the continental shelf edge has built out seaward through hundreds of thousands to millions of years, during successive periods of high full-glacial sediment delivery (e.g. Larter & Vanneste, 1995; Vorren *et al.*, 1998; Anderson, 1999).

These fans, forming cone-shaped depocentres beyond the shelf break (Fig. 30.3), have particularly low slope angles (0.5 – 2°). They have areas of 10^4–10^5 km^2 and are made up largely of stacked series of debris flows, derived from diamictic sediment delivered by fast-flowing ice streams to the shelf break and upper slope

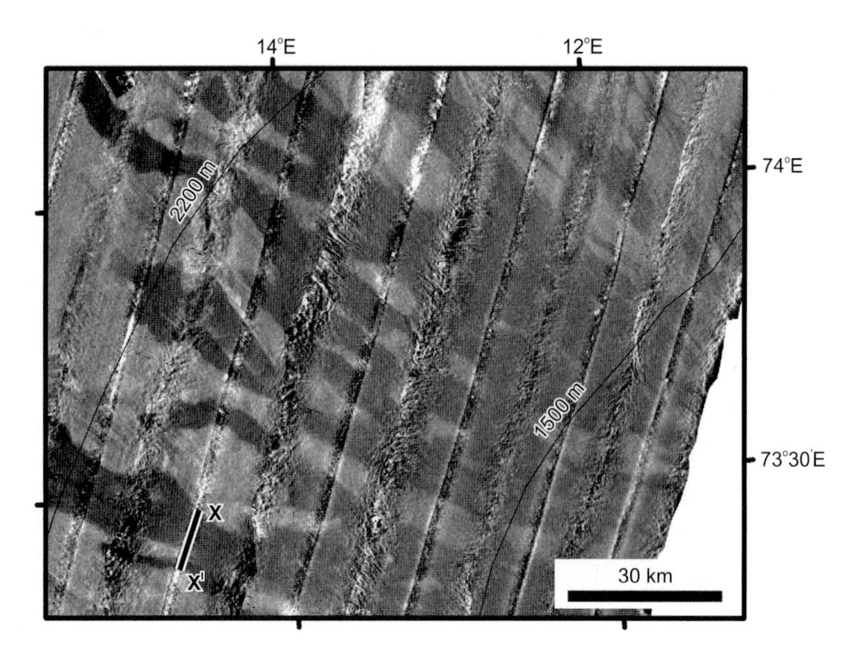

Figure 30.3 Oblique view of the Bear Island Fan, Norwegian–Svalbard margin, with debris flows observed on 6.5 kHz long-range side-scan sonar imagery shown. (Modified from Taylor *et al.*, 2002.)

during full-glacial conditions (e.g. Vorren *et al.*, 1989; Dowdeswell *et al.*, 1996, 1997; King *et al.*, 1996). Packages of debris flows are separated by thin layers of fine-grained hemipelagic debris, representing successive interglacials or interstadials (King *et al.*, 1996). Acoustically laminated silts and clays with a substantial ice-rafted component also contribute to full-glacial sedimentation on some fans, suggesting that ice streams may shift their position between successive full-glacials (Taylor *et al.*, 2002). Sometimes, large slope failures or slides are also present, either on the fans themselves or on adjacent parts of polar continental slopes (e.g. Laberg & Vorren, 1993; Dowdeswell *et al.*, 1996; Vorren *et al.*, 1998). Characteristic features include failure scars on the upper slope, and zones of deposition of the failed material, often associated with large blocks, on the lower slope and beyond (e.g. Dowdeswell *et al.*, 1996, 2002b).

Between these high-latitude fans, the continental slope is sometimes relatively steeper. Canyons are occasionally present (Taylor *et al.*, 2000), especially where the rate of glacial sediment delivery is low. On the Lofoten margin, for example, a series of canyons is located seaward of what was probably a relatively restricted full-glacial drainage basin of low ice flux, located between two much larger, fast-flowing systems (e.g. Ottesen *et al.*, 2005). In general, canyons are much less common on glacier-influenced margins than at lower latitudes, probably because glacial sediments are delivered from line sources (i.e. the seaward ice-sheet margin), rather than from the quasi-point sources provided by lower-latitude riverine activity (e.g. Dowdeswell *et al.*, 1996).

However, rates of full-glacial sediment delivery directly to the upper slope vary considerably with the dynamics of the ice margin (Dowdeswell *et al.*, 1996, 2002b). Fast-flowing ice-stream margins, often tens of kilometres in width, with an ice flux several orders of magnitude greater than slower moving margins, release considerably greater quantities of icebergs, meltwater and sediments to the upper slope (e.g. Dowdeswell & Siegert, 1999; Dowdeswell & Elverhøi, 2002; Siegert & Dowdeswell, 2002). Thus, submarine fans represent major sedimentary depocentres on high-latitude ice-influenced margins.

By contrast, relatively steeper slopes of 5–10° are also found offshore of some Antarctic cross-shelf troughs (Ó Cofaigh *et al.*, 2003; Dowdeswell *et al.*, 2004b), which are known to have contained major fast-flowing palaeo-ice streams (Ó Cofaigh *et al.*, 2002; Dowdeswell *et al.*, 2004a). In these locations, such as the Marguerite Bay trough and slope system west of the Antarctic Peninsula, sediments are inferred to be delivered rapidly from a full-glacial ice stream. Debris largely bypasses the relatively steep slope to be deposited as sediment drifts, and more distally as turbidites, in the deep-ocean basin beyond (e.g. Rebesco *et al.*, 1996; Pudsey, 2000; Dowdeswell *et al.*, 2004b).

Marine geophysical studies have identified several further submarine landforms associated with ice-influenced continental slopes and the deep-sea basins beyond. Systems of submarine channels are well-developed in some high-latitude settings (e.g. Mienert *et al.*, 1993; Hesse *et al.*, 1997; Dowdeswell *et al.*, 2002b, 2004b). In the 250,000 km^2 Greenland Basin, for example, a branching network of channels is present on the upper slope, coalescing into a few major meandering channels on the abyssal plain beyond (Fig. 30.2d). The channels reach several kilometres in width, up to 100 m in depth, and are several hundred kilometres in length (Mienert *et al.*, 1993; Dowdeswell *et al.*, 2002b; Ó Cofaigh *et al.*, 2004). Beyond the channel margins, acoustically laminated levees, related to successive downslope-flow events, are built up by turbidity-current transport and deposition of debris during overbank flow events (Ó Cofaigh *et al.*, 2004). In the most distal and lowest gradient parts of the Greenland Basin, the channels lose their definition and deposit extensive lobes of relatively sandy sediment (Dowdeswell *et al.*, 2002b).

Extensive areas of sediment waves and well-sorted contourite drifts are also clearly defined submarine landforms in some high-latitude slope and basin settings, particularly in interfan areas, where along-slope transport of fine-grained sediments is the dominant transport process (e.g. Kenyon, 1986; Kuvaas & Leitchenkov, 1992; Rebesco *et al.*, 1996; Pudsey, 2000). In the series of sediment drifts at the base of the continental slope on the west side of the Antarctic Peninsula, individual sediment drift mounds cover thousands of square kilometres and rise several hundred metres above the general elevation of the sea floor. The drifts are usually separated by channel systems draining from the slope to the abyssal basin beyond (Plate 30.2).

30.4 Inferences on past ice-sheets and environmental change

The sedimentary stratigraphy of high-latitude continental margins, where it is undisturbed by processes of reworking (e.g. Dowdeswell *et al.*, 1993; Anderson, 1999; Ó Cofaigh *et al.*, 2002b), provides a quasi-continuous record of environmental change over time-scales ranging from 10^3 to 10^7 yr. However, much of the sedimentary record on polar continental shelves is removed by renewed ice-sheet advance across the shelf in successive glacial periods; shelf sediments, therefore, may contain a record only of retreat from the last glacial maximum, albeit at a relatively high temporal resolution. Where submarine fans are present on the continental slope, records may be derived of ice advances to the shelf edge in a series of past glacial episodes over the past 2 Myr or so through periods of enhanced sedimentation (e.g. Andersen *et al.*, 1996; Vorren *et al.*, 1998). Finally, the deep-sea record in polar abyssal basins extends over tens of millions of years and may contain both sedimentary, isotopic and biogenic evidence on the inception of the late Cenozoic Ice Age (e.g. Anderson, 1999).

The long-term record of environmental change at high latitudes is held mainly in deep-sea sediments and large submarine fans. For example, an important indicator of the initiation of continental-scale glaciation to sea level at the beginning of the Cenozoic Ice Age is the first occurrence of coarse-grained iceberg-rafted debris within fine-grained abyssal muds; thus, major ice-sheet growth began about 35–40 Ma in Antarctica, 7 Ma in Greenland and about 2 Ma in Fennoscandinavia (e.g. Jansen & Sjøholm, 1991; Larsen *et al.*, 1994; Barrett, 1996). Similarly, the variability in grain size of debris in finely laminated drift sheets on high-latitude margins demonstrates the changing velocity and direction of major ocean-current systems over glacial–interglacial intervals (e.g. Pudsey, 1992). Such changes are important to global-scale transfers of mass and energy in the Earth system.

More recently, the six sandy iceberg-rafted or 'Heinrich' layers in North Atlantic sediments demonstrate that the 2 million km^2 Hudson Bay–Hudson Strait drainage basin of the North American ice sheet underwent a series of major collapses during the past 60,000 yr that affected hemispheric climate (e.g. Bond *et al.*, 1992; Dowdeswell *et al.*, 1995). The icebergs calved during each collapse event transported debris over 3000 km across the Atlantic from their source (Dowdeswell *et al.*, 1995). The layers are an indicator of ice-sheet instability rather than a direct response to climate change. In the same way, less extensive layers of ice-rafted debris reported from the Norwegian–Greenland Sea may also reflect the behaviour of individual drainage basins of the full-glacial Greenland and Eurasian ice sheets rather than providing a direct record of regional climatic changes (e.g. Dowdeswell *et al.*, 1999b, 2001).

At shorter time-scales, the timing of ice-sheet advance and retreat across continental shelves, its maximum extent and retreat chronology in particular, can be reconstructed from shelf sediments together with moraine-ridge landforms and grounding-zone wedges (e.g. Domack *et al.*, 1999, 2001; Shipp *et al.*, 2002; Evans *et al.*, 2005). The nature of past ice-flow dynamics also can be inferred from sedimentary bedforms. Megascale glacial lineations and other streamlined landforms indicate past ice-flow speed and direction, and whether ice flow is convergent or divergent. The sedimentary facies on high-latitude shelves allow inferences to be made about past sediment-delivery processes and environments, for example, whether iceberg–ice-shelf systems or meltwater systems dominated deposition (Dowdeswell *et al.*, 1998; Evans & Pudsey, 2002). Shallow acoustic stratigraphy and sediment rheology have also been used to infer the nature of processes taking place at the former ice–bed interface, an area that is difficult to access beneath modern ice sheets (Dowdeswell *et al.*, 2004a; Ó Cofaigh *et al.*, 2005).

30.5 Conclusions

The sedimentary record of high-latitude margins has been influenced strongly by the presence of ice sheets during the late Cenozoic. These ice sheets were the dominant suppliers of debris to the shelf–slope–basin system during successive glacial periods. Repeated ice-sheet advance and retreat shifted the focus of sediment delivery from fjords and mountainous hinterlands during interglacials, to the shelf edge during glacials (Fig. 30.1). In addition to this temporal variability, ice sheets are also variable spatially in terms of their flow regime. Fast-flowing ice streams and outlet glaciers, tens of kilometres wide at their seaward margins and fed from huge interior drainage basins, punctuate slower flowing ice. Thus, during full-glacials, the rate of sediment delivery to the shelf edge and upper slope is highly variable along a given marine ice-sheet margin. Huge submarine fans, built of glacier-derived sediments delivered to the mouth of ice streams traversing the continental shelf, are a clear product of these processes (Fig. 30.3). Beyond high-latitude shelves, where glacial erosion has removed or reworked much of the record of earlier glaciations, the sediments in fans and deep-sea basins contain a long-term archive of past ice-sheet and wider environmental change at high latitudes.

Acknowledgements

Geophysical data used in the production of figures for this paper were acquired on cruises of the RRS *James Clark Ross*, funded by UK NERC grants GR3/8508, GST/02/2198, GR3/JIF/02, NER/T/S/2000/00603 and NER/T/S/2000/00986. Colm Ó Cofaigh and Jeff Evans provided helpful comments and prepared the figures.

THIRTY-ONE

Seismic geomorphology and Pleistocene ice limits off northwest Britain

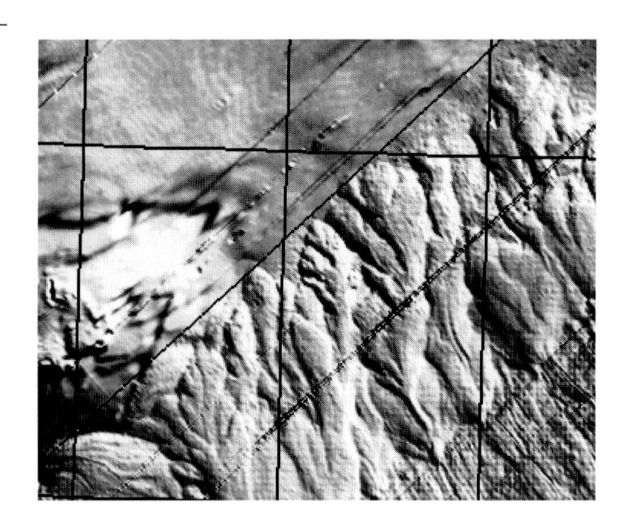

Martyn S. Stoker*, David Long*, Joseph Bulat* and Stephen Davison*†

**British Geological Survey, Murchison House, West Mains Road, Edinburgh, EH9 3LA, Scotland, UK*
†Department of Geology and Geophysics, University of Edinburgh, Grant Institute, West Mains Road, Edinburgh, EH9 3JW, Scotland, UK

During the past decade, insights into the history and extent of the former British Ice Sheet (BIS) off northwest Britain have been revealed by the integration of geomorphological information from geophysical data with stratigraphical and lithological data from sea-bed cores. In particular, the emergence of three-dimensional seismic acquisition as a tool for regional reconnaissance has resulted in near complete coverage of the outer continental shelf and slope in areas of active oil exploration. The first signal returns from three-dimensional exploration surveys can be utilized to create a regional image of the sea-bed morphology. Such an image has been developed for the outer continental shelf and slope west of Shetland (Bulat & Long, 2001) (Fig. 31.1). When combined with more traditional two-dimensional seismic reflection profile data, this three-dimensional imagery reveals the footprint left by an expansive BIS.

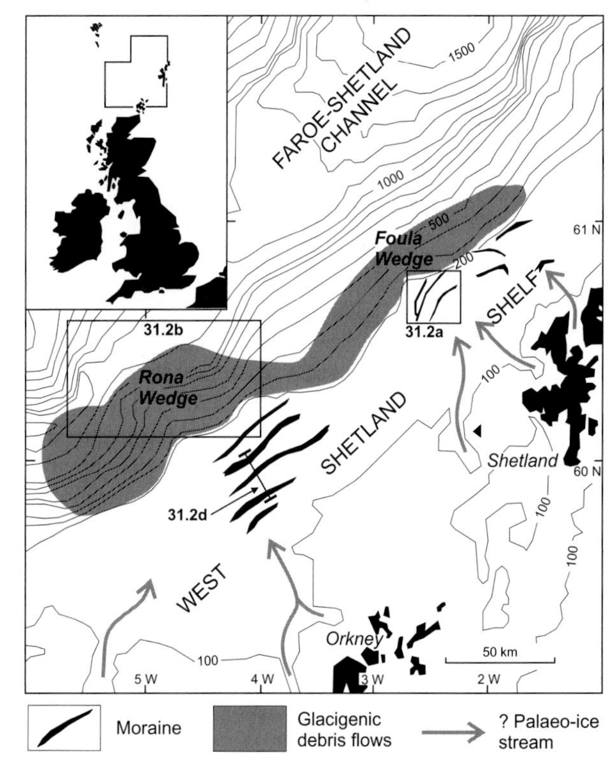

Figure 31.1 Location of the West Shetland margin; bathymetric contours in metres.

The Pleistocene succession off northwest Britain is preserved as part of a seaward-thickening wedge of sediment that has prograded into the adjacent deep-water basins since the Pliocene (Stoker, 2002). West of Shetland, the shelfbreak has migrated seaward by up to 50 km concomitant with this progradation of the continental margin. In general terms, the prograding sediment wedges can be subdivided into two sequences separated by an irregular, erosional unconformity informally termed the 'glacial unconformity' (GU) (Fig. 31.2). Although the underlying, prograding, Pliocene to middle Pleistocene sediments preserve a record of ice-rafting since the late Pliocene (about 2.48 Ma), the development of the GU marks the onset of extensive shelf glaciation off northwest Britain, instigated during the early mid-Pleistocene (about 0.44 Ma: marine isotope stage (MIS) 12), and repeated on several occasions during the mid- to late Pleistocene (Stoker, 1995). In places the GU represents a composite unconformity, reoccupied and/or modified during each subsequent

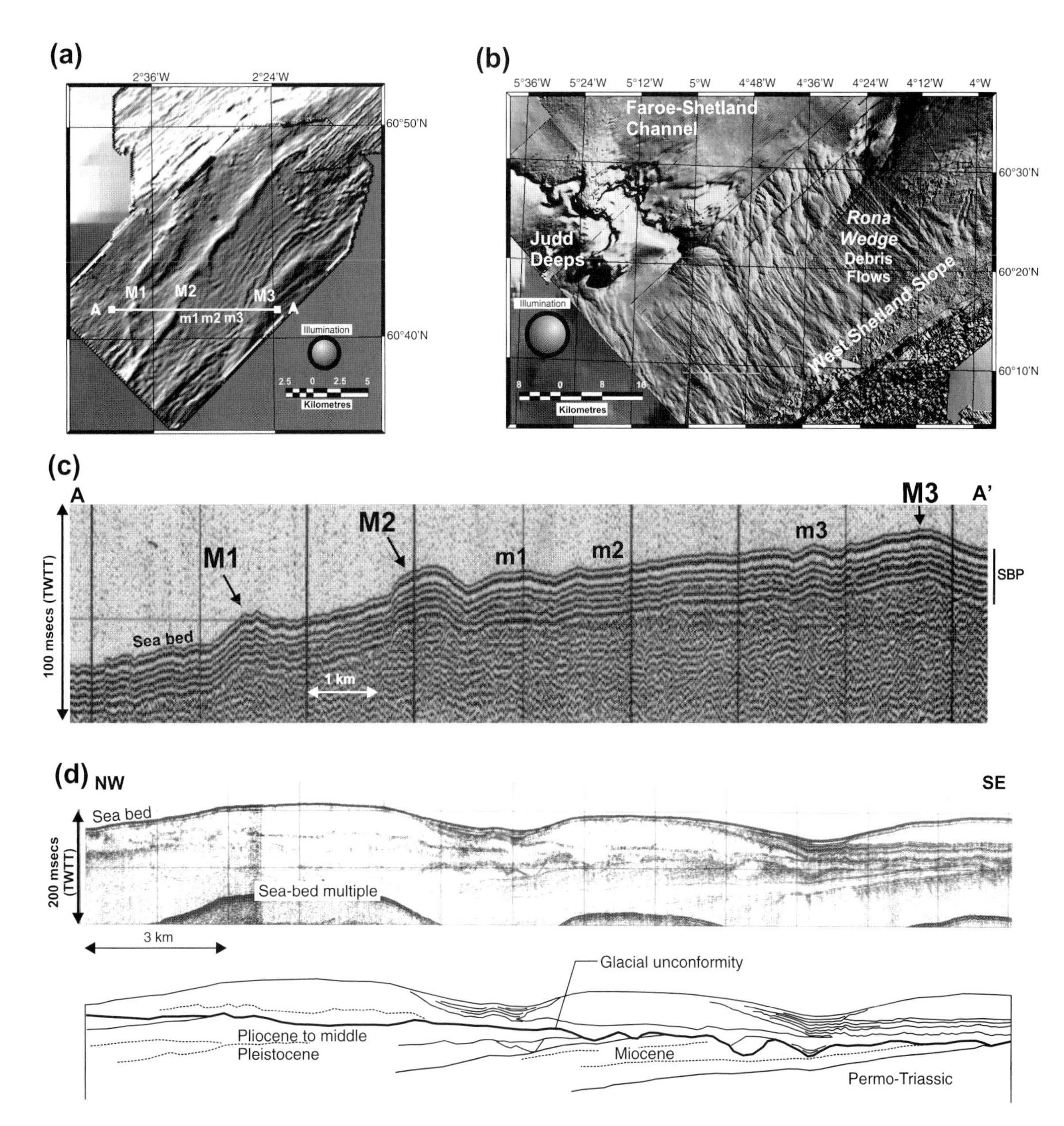

Figure 31.2 (a) Sea-bed relief image of the outer shelf illuminated from the northwest illustrating the northern area of morainal ridges of various scales; (b) sea-bed relief image illuminated from the northeast over the southern end of the West Shetland Slope and Faroe–Shetland Channel, illustrating the debris flows of the Rona Wedge, and the smooth to irregular (Judd Deeps) floor of the Channel; (c) BGS high-resolution seismic (1 kJ sparker) profile (79/14-23) across the large (M1–M3) and small (m1–m3) morainal ridges illustrated in (a); (d) BGS high-resolution seismic (1 kJ sparker) profile (83/04-31) and interpreted line drawing of the morainal ridges in the southern area, highlighting their younging to the southeast, and their relation to the acoustically well-layered, ponded, sediments. Images and profiles in (a), (b) and (d) located in Fig. 31.1. Abbreviations: TWTT, two-way travel time; msecs, milliseconds; SBP, sea-bed pulse.

expansion of the BIS. West of Shetland, the present-day sea-bed morphology is considered to largely reflect the last (late Devensian, MIS 2) expansion of the BIS (Stoker, 1995; Hall *et al.*, 2003).

Geophysical data have revealed two main geomorphological features that indicate a shelf-edge terminal position for the BIS during peak, late Devensian, glaciation, west of Shetland: (i) a series of end moraines preserved on the outer shelf, beyond which lie (ii) the glacially fed sedimentary fans of the Rona and Foula wedges, on the West Shetland Slope (Fig. 31.1). The end moraines behind the Foula Wedge are clearly evident on the sea-bed image, which shows three arcuate moraines (M1–M3) that can be traced for up to 30 km (Fig. 31.2a). At the southwest edge of the image, M1 and M2 appear to coalesce. A high-resolution seismic reflection profile crosses the moraines indicates that they are up to 1.5 km wide and 20 m high (Fig. 31.2c). The crests of the moraines currently lie between 110 and 145 m water depth. The area between M2 and M3 preserves similar, albeit smaller, features (m1–m3) that have a relief of about 5 m or less. Unfortunately, the internal structure of all the moraines is obscured by the seismic artefact of a thick sea-bed pulse.

A second area of end moraines occurs behind the Rona Wedge (Fig. 31.1). Although these are less well defined on the sea-bed image, in profile they are individually much larger than those to the north, being up to 50 m high, 8 km wide and traced laterally for up to 60 km (Stoker & Holmes, 1991) (Fig. 31.2d). The crests of these moraines are currently at a water depth of 140–150 m. Internally, the moraines are largely acoustically structureless, although hyperbolic reflections are present, most probably originating from cobbles and boulders on the ridge crests and from sporadic bedding surfaces. The northwest flank of each moraine interdigitates with acoustically well-layered, ponded, strata. This indicates both a coeval relationship between the moraines and ponded sediment, and the systematic younging of the glacial succession towards the land (to the southeast).

Linkage between the two areas of moraines is unclear owing to incomplete coverage of the sea-bed image on the outer shelf. We interpret the end moraines in both areas to be recessional ice-sheet moraines. In the northern area, the smaller moraines (m1–m3) may represent short-term halts in the retreat of the BIS, whereas the larger moraines mark a relatively long halt, with time enough for the localized deposition of the ponded sediments in the southern area. The observation that the most landward moraine imaged in Fig. 31.2d overlies layered strata, which post-dates the middle moraine, may reflect a considerable readvance of the BIS during the general phase of overall recession.

On the West Shetland Slope, stacked, glacigenic, debris-flow deposits comprise the glacially fed Rona and Foula sediment wedges, which are up to several hundred metres in thickness (Stoker, 1995; Davison & Stoker, 2002). The sea-bed image highlights the Rona Wedge, and reveals the characteristically elongate and lobate form of the debris flows, together with indications of linear channelling in the mid-slope region (Fig. 31.2b). Seismic profile data show that individual lobes commonly range from 5 to 20 m in thickness, and the sea-bed image indicates that their downslope extent can be measured in kilometres. On the Rona Wedge, the debris-flow deposits extend the length of the slope, and are gradually transgressing across the floor of the Faroe–Shetland Channel, infilling the major deep-water erosional scours of the Judd Deeps (Fig. 31.2b). In contrast, the Foula Wedge is less extensively developed, being essentially restricted to the upper slope (Fig. 31.1). This lateral variation in slope-apron architecture is well illustrated on the sea-bed image by the relatively sharp northeastern edge, at about 60°30′N, of the Rona Wedge. This implies that the glacially fed fans represent zones of discrete, focused, deposition on the West Shetland Slope.

The association of the fans and the moraines is consistent with a late Devensian ice terminus for the BIS, which locally reached the edge of the West Shetland Shelf. The glacially fed fans developed during peak glaciation, whereas the moraines represent stages in its recession. It is becoming increasingly apparent that glacially fed fans (also referred to as 'trough mouth fans') are important indicators of former ice streams (Stokes & Clark, 2001). It may be no coincidence that a number of cross-shelf bathymetric troughs converge on the outer West Shetland Shelf adjacent to the fans (Fig. 31.1). Preliminary indications suggest that a zone 40–60 km wide on the outer shelf may have fed the Rona Wedge. An ice stream of this width is within the dimensions of those identified elsewhere by Stokes & Clark (2001).

Acknowledgement

This paper is published with the permission of the Executive Director, British Geological Survey (NERC).

THIRTY-TWO

Modelling glacier response to climate change

Shawn J. Marshall

Department of Geography, University of Calgary, Calgary AB, Canada

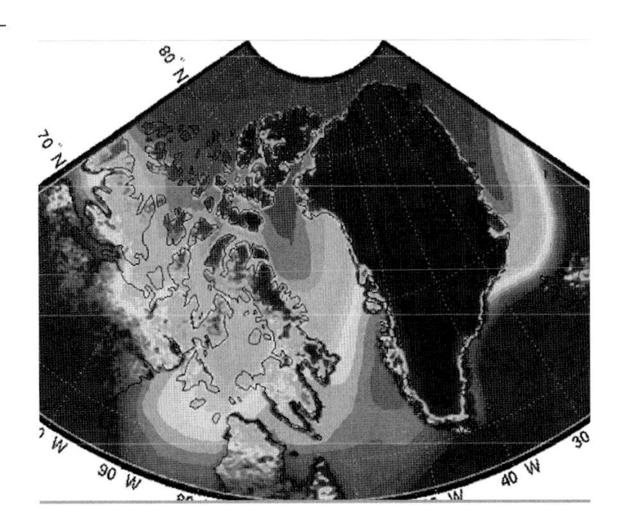

32.1 Introduction

This chapter focuses on the coupling of glacier and climate models for simulation of glacier and icefield response to climate change. Sections 32.2 and 32.3 summarize glacier mass balance and ice dynamics considerations, concentrating on concepts that are central to understanding of global-scale glacier and ice-sheet evolution. Sections 32.4 and 32.5 discuss the coupling of climate and glaciological models, along with some of the important limitations and uncertainties in glacier–climate forecasts. The discussion in this review is necessarily selective, with a focus on the current state-of-the-art in glacier–climate modelling. For a more comprehensive examination of glacier–climate interactions, mass balance modelling, and the response of glaciers and ice sheets to climate change, interested readers are referred to the excellent texts of Oerlemans (1989, 2001) and Paterson (1994). Chapters in this volume by Braithwaite, Haeberli, Reeh, Vaughan and Huybrechts also expand on these topics.

I give a broad treatment of continental ice masses, from alpine glaciers through to continental-scale ice sheets in Greenland and Antarctica. Although these ice masses have different climate-change response times and mass balance regimes, the same basic processes govern their behaviour and their expected response to ongoing climate change. Because ice masses in Greenland and Antarctica represent almost 97% of the glaciated area and an estimated 99.8% of ice volume in the world (Ohmura *et al.*, 1996), the impact of climate change on glacier mass balance in Greenland and Antarctica is of primary importance for sea-level change. Intermediate-scale icefields and ice caps, such as those in Patagonia, Iceland, Alaska and the Arctic, also present an interesting case. These ice masses are small enough to be significantly impacted by decade- and century-scale climate change, but are large enough to be important to eustatic sea level (Arendt *et al.*, 2002; Meier *et al.*, 2003). In addition to questions of global sea level, changing glacial environments in the 21st century are also expected to produce significant climate feed-backs in all of the world's glacierized regions.

32.2 Glacier mass balance

Glacier mass balance, $\dot{b}$, is a measure of the net accumulation minus ablation of snow and ice, $\dot{b} = \dot{a} - \dot{m}$. The most common convention is to consider net annual mass balance, where $\dot{a}$ represents the total annual snow/ice accumulation and $\dot{m}$ is the total annual snow/ice loss through ablation. Sources of accumulation are primarily meteoric, although 'internal accumulation' occurs through the refreezing of surface meltwater that percolates into the snow or firn. Internal accumulation represents an important mass balance term above the glacier or ice-sheet equilibrium line in subpolar and polar regions, where up to 100% of summer meltwater can percolate and refreeze. Snow and ice ablation occur through iceberg calving, surface melting, sublimation and basal melting. Note that these are all represented as annual rates. Each variable is typically measured in $\mathrm{m\,yr^{-1}}$ ice- or water-equivalent, although this can also be expressed as $\mathrm{kg\,m^{-2}\,yr^{-1}}$.

In practice, the glacier mass balance year runs from ca. September (start of the snow accumulation season) to the following August (end of the melt season) in the Northern Hemisphere, but this varies regionally and is an idealization, because summer snowfall and non-summer snow melt occur in most glacial settings. In glacier–climate modelling studies, it is more common to consider calendar years, with net mass balance estimated from monthly or annual climate fields. In all cases, net annual mass balance is a direct measure of thinning or thickening of an ice mass. Measurements are local—at a particular location on an ice mass—and must be spatially integrated to provide an estimate of the large-scale mass balance of a glacier, icefield or ice sheet.

The following subsections discuss the different components of mass balance and the methods being used to estimate mass balance for glacier–climate modelling. I restrict the discussion to contemporary ice masses, for both present-day reconstructions and century-scale climate change scenarios. I focus on the modelling of spatially integrated mass balance, at scales ranging from individual glaciers to continental ice sheets. Mass balance models for local- or regional-scale glacier and icefield complexes commonly make use of observational climatology (networks of local monitoring stations). For global-scale icefield analyses or climate change scenarios, mass balance fields are typically estimated from modelled or reanalysed climatology (e.g. Kalnay *et al.*, 1996). Extrapolation of climate data from monitoring stations and interpolation of climate fields from models pose similar challenges for glacier–climate modelling, discussed further in section 32.5.

32.2.1 Snow accumulation

Annual snow accumulation, $\dot{a}(\lambda, \theta)$, is determined by the local precipitation rate, $P(\lambda, \theta)$, and an estimate of the fraction of precipitation to fall as snow, $\phi_S(\lambda, \theta)$: $\dot{a}(\lambda, \theta) = \phi_S(\lambda, \theta)P(\lambda, \theta)$, where spatial locations are denoted by their longitude–latitude co-ordinates (λ, θ). Given spatially distributed precipitation rates at a particular time, an arbitrary air temperature cutoff is normally used to dictate whether local precipitation falls as snow, $T < T_t$, for a threshold temperature T_t. Operationally, T_t is typically taken to be 0°C or 1°C. This is a poor treatment if monthly or mean annual climatology are used to estimate snowfall, as the cutoff temperature prohibits snowfall at mean values slightly above the threshold, and diurnal and synoptic variability will obviously give extensive periods with subfreezing temperatures. Degree-day methodology offers a physically based alternative to estimate the fraction of precipitation to fall as snow, ϕ_S. If observational weather data are available at frequent (subdiurnal) time intervals, ϕ_S can be estimated from the fraction of time below T_t during a precipitation event.

Using monthly or annual climatology, simple statistical models can be introduced to parameterize temperature variability. For instance, temperature can be assumed to have a Guassian distribution around mean monthly temperatures T_m, with standard deviation σ_m. Under this assumption, ϕ_S is calculated from the integral

$$\phi_S = \frac{1}{\sigma_m \sqrt{2\pi}} \int_{-\infty}^{T_t} \exp\left[\frac{-(T-T_m)^2}{2\sigma_m^2}\right] dT \quad (1)$$

The standard deviation σ_m is derived ideally from hourly temperature data, in order to capture the diurnal temperature cycle. For annual climatology, a sinusoidal temperature distribution is typically assumed to represent the seasonal temperature cycle (Reeh, 1991)

$$T_d(t) = \bar{T} - (T_{max} - \bar{T})\cos(2\pi t/\tau) \quad (2)$$

where t is time in days, τ is the length of the year (365.24 days), T_d is the daily temperature, T_{max} is the summer temperature maximum, and $\bar{T}$ is the annual average air temperature.

January 1 is taken to be $t = 0$ in Equation (2), although a time lag can be built in to incorporate the effects of seasonal surface temperature lags, a function of local surface heat capacity and the radiative environment. The equivalent integration to Equation (1) is then

$$\phi_S = \frac{1}{\sigma_d \sqrt{2\pi}} \int_{-\infty}^{T_t} \frac{1}{\tau} \int_0^{\tau} \exp\left[\frac{-(T-\bar{T})^2}{2\sigma_d^2}\right] dt\,dT \quad (3)$$

where σ_d represents the standard deviation in daily temperature, calibrated to include both diurnal and synoptic temperature variability. In model implementations, atmospheric temperature lapse rates become important, as monthly or annual temperatures need to be interpolated or extrapolated over the landscape for application in Equations (1)–(3).

32.2.2 Surface ablation

Surface ablation is the removal of surface snow or ice from a site. It occurs through sublimation, melting and, on a local scale, wind-scouring. Surface melting is the dominant ablation mechanism for all global ice masses with the exception of the Antarctic Ice Sheet, where temperatures are too low to produce large-scale melting. Summertime surface melting occurs on all major Arctic and alpine icefields. Melting is limited in interior and northern regions of the Greenland Ice Sheet, but a narrow ablation zone is present around the entire ice-sheet periphery (Abdalati & Steffen, 2001; Plate 32.1). Surface melting in ablation zones of southern and southwestern Greenland is extensive, up to a few metres per year. Melt rates as high as 10 m yr^{-1} occur in maritime icefields in Iceland, Svalbard, and western North America (e.g. Björnsson, 1979). Similar annual melt rates are found in the ablation zones of mid-latitude alpine ice masses (e.g. in the Andes, the Alps and the Rockies).

The case study by Bintanja (this volume, Chapter 33) discusses the importance of sublimation in Antarctica, where mass is lost to the atmosphere through both direct surface sublimation and, under sufficiently windy conditions, sublimation of windblown snow. The latter case can be a very effective ablation mechanism where loose surface snow is available, as winds can move the snow above the near-surface boundary layer, which quickly becomes saturated by surface sublimation. Simulations of Greenland Ice Sheet mass balance suggest that surface- and blowing-snow sublimation may contribute up to 15% of the total ablation from the ice sheet (Box *et al.*, 2004), suggesting that these processes are also important to glacier mass balance in Arctic regions. One outstanding question is whether the net effect of sublimation is similar to wind scour, serving to redistribute snow rather than remove it from the system.

The physics of snow and ice melt are well understood; melt rates $\dot{m}$ are governed by local surface energy balance

$$\rho_{S/I} L\dot{m} = Q_V(1-\alpha) + Q_{IR}^{\downarrow} - Q_{IR}^{\uparrow} + Q_H + Q_L - Q_C + Q_P - Q_R \quad (4)$$

where $\rho_{S/I}$ is the snow or ice density, L is the latent heat of fusion for water, α is the surface albedo, Q_V is incoming solar radiation, $Q_{IR}^{\downarrow}$ and $Q_{IR}^{\uparrow}$ are incoming and outgoing infrared radiation, Q_H and Q_L are the sensible and latent heat fluxes, Q_C is the flux of energy conducted into the snow or ice, Q_P is the advective energy delivered by precipitation, and Q_R is the advective energy associated with meltwater runoff from the snow or ice surface. All fluxes have units W m^{-2}, and fluxes are defined to be positive for heat transfer from the atmosphere to the snow/ice. Hence $Q_{IR}^{\uparrow}$, Q_C and Q_R are energy losses from the surface–atmosphere interface. Latent heat exchange occurs through sublimation, deposition and evaporation of surface water during the melt season. The precipitation source term, Q_P, accounts for heat transferred to the snowpack or ice surface as rainwater of temperature T_P cools to 0°C

$$Q_P = \rho_W c_W P T_P \tag{5}$$

where ρ_W and c_W are the density and specific heat capacity of water. Rainfall temperature T_P is measured in °C.

Further elaborations of the terms in the energy balance can be found in Arnold *et al.* (1996), Cline (1997a, b), and Marks *et al.* (1999). Klok & Oerlemans (2002) discuss the spatial variability of energy balance terms on Morteratschgletscher in the Swiss Alps, and Denby *et al.* (2002) introduce a boundary layer model for simulation of energy balance at ice-sheet scales. Case studies and chapters in this volume from Braithwaite, Klok and Bintanja provide further insight into the energy balance terms and their relative importance for snow and ice ablation.

Equations (4) and (5) assume that the snow/ice surface is at the melting point, such that all available energy at the snow/ice surface is dedicated to meltwater production. In most glacier environments, however, overnight radiative cooling leads to refreezing and cooling of the snow/ice surface during the ablation season, leading to refreezing of both ponded surface water and near-surface snowpack porewater. This delays the onset of 'new' melt until the surface warms to the melting point and refrozen water thaws the next day. If significant refrozen water is present this delay can be significant, delaying melting until several hours after local sunrise on both snow and ice surfaces. This effect can be incorporated through the Q_L term above, but it is not generally accounted for as it requires a separate parameterization for the amount of free surface water.

In the most cold glacial environments (polar regions and extremely high altitudes), the snowpack remains below the melting point throughout the ablation season, and surface meltwater percolates into the snowpack, where it will refreeze and release latent heat. This also occurs in alpine snowpacks in springtime, limiting the amount of net mass loss via runoff, $\dot{m}$, and having an impact on the snowpack energy balance. This refreezing is difficult to quantify, as a proper treatment of the process requires two- or three-dimensional modelling of surface runoff, meltwater percolation in the snowpack, and the detailed vertical thermodynamic evolution of the snow and firn. The customary approach in ice-sheet modelling is to crudely estimate a fraction of the total melt that refreezes, of the order of 60% in the ice-sheet accumulation area. Total refreezing quantities are much less in ablation zones or in alpine settings, of the order of 2–5% (Jóhannesson *et al.*, 1995).

A study by Janssens & Huybrechts (2000) explored explicit modelling of the refreezing process in Greenland, concluding that differences from the assumption of 60% refreezing had only minor impacts on the modelled mass balance. The extent of refreezing, however, will vary in different climatic and topographic settings, as well as temporally during the melt season, so physically based models of the process are clearly preferable for application in different environments (e.g. ice-sheet ablation versus accumulation zones; polar versus mid-latitudes; steep, confined glacier outlets versus gently sloping lobes). Internal snowpack thermodynamical models are well-developed in avalanche, sea-ice and snow-hydrology research and can be expected to become standard in glacier–climate modelling. In addition to improving the representation of meltwater refreezing, internal snowpack thermodynamics and hydrological modelling will allow consideration of important energy balance feed-backs on Equation (4), including thermodynamically explicit modelling of internal heat conduction, Q_C, advective heat transport via precipitation, Q_P, and meltwater heat transport, Q_R. All three of these terms are better-quantified as volume rather than surface processes.

The controls of surface melt are amenable to field measurement, making them reasonably well understood, but they are difficult to quantify in a spatially distributed model owing to their dependence on local meteorological conditions. The turbulent heat fluxes, Q_H and Q_L, for instance, are governed by local wind, humidity, and surface roughness properties, which exhibit substantial spatial and temporal variability. Other processes such as albedo evolution and snowpack hydrology are spatially complex and difficult to quantify in large-scale models.

The meteorological data demands and the spatial variability of governing processes make it difficult to apply a rigorous energy balance model to surface-melt modelling in ice-sheet models. As a gross but tractable simplification, temperature-index models are widely used for estimation of surface melt (e.g. Huybrechts *et al.*, 1991; Jóhannesson *et al.*, 1995). These models make use of observations that air temperature is a strong indicator of radiative and sensible heat energy available for melting, and parameterize meltwater generation over time interval τ, $m(\tau)$, as a function of positive degree days, $PDD(\tau)$,

$$m(\tau) = d_{S/I} PDD(\tau) \tag{6}$$

where $d_{S/I}$ is the degree-day melt factor for snow or ice, an empirically determined coefficient that differs for snow and ice to reflect the higher albedo of snow. Degree-day factors have units of metres water-equivalent melt production per °C per day. The *PDD* is a measure of the integrated heat energy in excess of the melting point over the time interval of interest. Surface melt estimation from Equation (6) is convenient because temperature (hence, *PDD*) is the only governing meteorological variable and is relatively amenable to spatial interpolation/extrapolation. Temperature fields from climate models or point measurements at automatic weather stations can be distributed over the landscape through application of atmospheric temperature lapse rates (e.g. Giorgi *et al.*, 2003).

In glacier and ice-sheet modelling, where decadal to millennial time-scales are of interest, either monthly or mean annual temperatures are applied to estimate net annual melt via temperature-index models. Where monthly temperature fields are available, monthly mean temperatures are used to calculate PDD_m (Braithwaite & Olesen, 1989; Braithwaite, 1995),

$$PDD_m = \frac{\tau}{\sigma_m \sqrt{2\pi}} \int_0^\infty T \exp\left[\frac{-(T - \bar{T}_m)^2}{2\sigma_m^2}\right] dT \tag{7}$$

where τ is the length of the month in days and σ_m is the standard deviation in monthly temperature, as in Equation (1). The primary objective of the statistical distribution of temperatures is to capture the critical influence of daily temperature maxima on meltwater production.

If annual mean climatology is applied, using the Reeh (1991) sinusoidal function as in Equations (2) and (3), net annual *PDD* are calculated from

$$PDD_a = \frac{1}{\sigma_d \sqrt{2\pi}} \int_0^\infty \int_0^\tau T \exp\left[\frac{-(T-\overline{T})^2}{2\sigma_d^2}\right] \mathrm{d}t\,\mathrm{d}T \quad (8)$$

Monthly or annual *PDD* go first towards melting of this year's snow accumulation, where available. Once there is no remaining snow from the annual accumulation, it is assumed that melting has penetrated to the previous year's firn or ice surface, and any remaining melt energy is directed to ice melt (cf. Marshall & Clarke, 1999). Degree-day factors $d_{S/I}$ most commonly used in ice-sheet modelling are taken from Greenland Ice Sheet studies (e.g. Braithwaite & Zhang, 2000), although it is recognized that these values may not be appropriate for the extremely different radiation regime of mid-latitude and tropical glaciers. Even within Greenland, different sets of degree-day factors appear preferable in different geographical sectors of the ice sheet (Bøggild *et al.*, 1996; Lefebre *et al.*, 2002), possibly related to local meteorological conditions and snow/ice surface properties (albedo, roughness).

The technical step of improving snow-hydrology and melt-rate calculations in ice-sheet models is within reach, and more physically based process models are likely to become the glaciological modelling standard over the next few years. This advance will be driven in part by the current effort to improve coupling between sophisticated atmospheric and ice-sheet models. It should be possible to move towards a more physical energy balance approach, including snow hydrology, following similar developments in atmospheric general circulation model (AGCM) simulation of present-day seasonal snowpack evolution (e.g. Marshall & Oglesby, 1994; Yang *et al.*, 1999).

An intermediate step towards the goal of more physically based melt models that do not have prohibitive data demands is to include the effects of solar radiation in heat-index modelling (e.g. Cazorzi & Dalla Fontana, 1996; Hock, 1999). This can be done through a parameterization of the form

$$m = d_{S/I} PDD + f_{S/I}(\lambda, \theta, t) Q_V \quad (9a)$$

or

$$m = d_{S/I} [PDD + f(\lambda, \theta, t) Q_V] \quad (9b)$$

where Q_V is incoming solar radiation and $f_{S/I}(\lambda, \theta, t)$ is a radiation index for snow/ice melt. It is spatially and temporally variable to allow the effects of changing surface albedo to be parameterized (e.g. Shea *et al.*, in press).

This relationship can be calibrated empirically through field observations, and it has important advantages over pure temperature-index modelling. In particular, incorporation of incoming solar radiation allows the effects of latitude and time of year (length of day, solar zenith effects), as well as terrain effects (shading, aspect) to be explicitly and objectively built into the melt model. For distributed modelling, where spatial estimation of variables is required over a much larger area than is amenable to monitoring, Hock (1999) used potential direct solar radiation for Q_V, with a univariate regression equation of the form in Equation (9b). Potential direct radiation can be calculated for any place and time, with shading effects incorporated through digital elevation models (DEMs); the relationship is purely geometric and straightforward to implement.

As it is surface radiation (rather than potential direct radiation) that matters for snow and ice melt, additional data are required to apply radiation-temperature melt models to a broad region, including parameterizations of diffuse versus direct radiation and atmospheric transmissivity, which includes the important influence of varying cloud cover. Given sufficient insights about actual meteorological conditions or statistical cloud distributions in a region, the radiation index $f_{S/I}(\lambda, \theta, t)$ could be made more complex to parameterize the effects of cloud cover as well as changing albedo. This has not yet been attempted in glacier–climate modelling, but it is a promising avenue to pursue until a full energy balance becomes practical. Statistically based insights into regional and seasonal cloud conditions can be derived from regional-scale climate modelling, or they can be based more crudely on local/regional precipitation rates.

32.2.3 Marine ablation

In present-day Antarctica, the dominant ice-sheet ablation mechanisms are iceberg calving and basal melting beneath floating ice shelves, as air temperatures are too cold to produce surface melting on most of the continent. The controls of ice-shelf wastage through both calving and subshelf melting are complex and are not well quantified. Calving involves fracture generation and propagation processes and is known to be dependent on ice thickness and tensile strength (hence temperature), water depth, tidal forcing, coastal/embayment geometry, and the flux of ice across the grounding line. There also appears to be a close relationship with spring/summer air temperatures, via the extent of the summer melt season (Doake & Vaughan, 1991; Vaughan & Doake, 1996). Scambos *et al.* (2000) demonstrate the role of water-filled crevasses in weakening overall ice-shelf competence, through forcing of vertical crack propagation.

Melting beneath ice shelves is also difficult to quantify and is known to be spatially variable, ranging from centimetres per year to several metres per year. Basal accretion occurs beneath some Antarctic ice shelves. Melt (or accretion) rates are controlled by ocean water temperature, density and subshelf bathymetry, with much of the spatial variability associated with convection cells in subshelf waters (Jacobs *et al.*, 1992, 1996; Jenkins *et al.*, 1997).

Ice shelves that drain the Greenland Ice Sheet and Arctic icefields in Canada and Russia are modest in scale relative to their Antarctic counterparts, but behave in similar fashion, calving large tabular icebergs. Other marine-based outlets of the Greenland Ice Sheet and the high Arctic islands are more akin to tidewater valley glaciers, calving blocky icebergs from floating glacier tongues or at the marine grounding line. Fracture propagation is still the primary control on the calving process in this case, but the flux of inland ice across the grounding line is the main determinant of calving rates. Approximately 50% of Greenland's total ablation is believed to occur through iceberg calving (Paterson, 1994). Recent interferometric studies in northern Greenland ice shelves indicate that basal melting beneath ice shelves is an important ablation mechanism in northern Greenland (Rignot *et al.*, 1997, 2000), with inferred melt rates as high as $20\,\mathrm{m\,yr^{-1}}$.

Current continental-scale ice-sheet models do not have the technical capacity for explicit representation of the processes of iceberg calving, ice-shelf breakup and subshelf melting. The governing mechanisms in each case have a finer scale than present model resolution (ca. 20 km), and the governing physics for calving and ice-shelf breakup are not fully understood and may not be deterministic. Explicit modelling of ice-shelf basal melting requires coupling with a regional-scale ocean model (e.g. Hellmer & Olbers, 1989; Beckmann *et al.*, 1999). This is technically feasible but has not yet been attempted in the ice-sheet modelling community.

From the perspective of modelling ice-sheet mass balance, it may be sufficient for many studies to neglect the details of marine-based ablation and accept that all ice to cross the grounding line ablates—the mechanism is unimportant. This is probably acceptable for first-order ice-sheet reconstructions, and it is essentially valid if one is not interested in the detailed timing of ice loss in coastal regions, as the marine ablation mechanisms discussed above will effectively dispose of ice to cross the grounding line within a time frame of years to centuries. However, the details of marine ablation processes become very important for questions of marine-triggered ice-sheet instability (e.g. tidewater glacier or ice-shelf collapse). This is therefore an important area of uncertainty for decade- and century-scale forecasts of West Antarctic ice-sheet response to climate change, ocean warming and sea-level change.

Zweck & Huybrechts (2003) give a summary of current methods of portraying marine ablation mechanisms in continental ice-sheet models. In the simplest treatments, all ice to cross the grounding line or a predefined bathymetric contour (e.g. 400 m water depth) is simply removed. Zweck and Huybrechts introduce a slight variation on modern bathymetric controls, allowing time-varying water depth controls on calving rates. This builds in both the bathymetric and sea-level influences on marine ice extent. Other studies parameterize calving losses, $\dot{m}_C$, as a function of water depth, ice thickness and ice temperature, proxy for ice strength/stiffness (e.g. Marshall *et al.*, 2000):

$$\dot{m}_C = k_0 \exp[(T - T_0)/T_C] H H_W \tag{10}$$

where H_W is the water depth, H is the ice thickness and T is ice temperature. The parameters k_0, T_0 and T_C establish calving vigour and the reduction in calving rates with decreasing temperature. The value for T_0 is typically set to 273.16 K, such that calving rates are maximal for isothermal ice ($T = 273.16$ K; $\dot{m}_C = k_0 H H_W$) and $\dot{m}_C$ exponentially decreases for colder ice. This crudely mimics the difference in calving rates observed in mid-latitude tidewater environments versus polar environments, where calving rates are low enough to permit ice-shelf development.

This calving model permits floating/shelf ice to expand over the continental shelf or shelf break, but it is not necessarily closer to the truth; effects of oceanic circulation, air temperature (crevasse-forced fracture propagation) and the geometry of marine embayments are not captured. Pfeffer *et al.* (1997) explored a calving parameterization similar to that above, but including an explicit treatment of fracture propagation. High-resolution ice-shelf models that include the physics of ice-shelf deformation (longitudinal stress/strain and horizontal shear stress) are better able to simulate these processes and controls (MacAyeal *et al.*, 1995). These models are difficult to couple with inland ice models, and fully articulated ice-shelf models have not yet been coupled with continental-scale ice-sheet models. This technical step is imminent, however, with considerable recent progress focused on the West Antarctic Ice Sheet (Payne, 1998; Hulbe & MacAyeal, 1999).

32.3 Glacier, icefield and ice-sheet dynamics

Glaciers, icefields and ice sheets respond to climate change on a time-scale determined by (i) the size of the ice mass, (ii) the dynamic and thermodynamic regime of the ice mass, (iii) the topographic environment (i.e. slope and hypsometry of the glacier bed) and (iv) the general climatic environment (e.g. maritime versus continental). All ice masses experience instantaneous areally averaged thinning or thickening in response to negative or positive mass balance. The four factors noted above govern the longer term ice-volume response that is witnessed at the ice-sheet margin: changes in glacier length and area.

Rigorous quantification of this response requires consideration of ice dynamics and glacier-specific topographic and climatic conditions. However, very instructive simple models have been developed to estimate glacier response times to step changes in climate (Nye, 1960; Jóhannesson *et al.*, 1989; Harrison *et al.*, 2001). These studies do not explicitly simulate ice dynamics, but apply a scaling relationship for the area and volume responses of an ice mass to climate (or mass balance) changes. For a characteristic thickness scale H representing the region in the ice mass most affected by the climate change, the simplest assumption is that rates of volume and area change are constant and proportional

$$\frac{dV}{dt} = H \frac{dA}{dt} \tag{11}$$

or $dV = H dA$ (Jóhannesson *et al.*, 1989; Harrison *et al.*, 2001).

With this assumption, Harrison *et al.* (2001) consider the long-term volume response of an ice mass to a change in mass balance with respect to an initial 'reference' geometry (elevation and area distribution). This provides a framework for explicit consideration of the effects of changing surface elevation as a glacier responds to climatic changes. Harrison *et al.* (2001) introduce these effects through the parameter $\dot{G}(z) = d\dot{b}(z)/dz$, which represents the vertical mass balance gradient (see also Elsberg *et al.*, 2001). An effective glacier-wide vertical mass balance gradient, $\dot{G}_e$, is calculated through a weighted averaged of $\dot{G}(z)$ over the ice mass. The predicted volume response time to a mass balance perturbation, τ_V, can then be estimated from

$$\tau_V = \frac{1}{-\dot{b}_e / H - \dot{G}_e} \qquad (12)$$

where $\dot{b}_e$ is the mass balance rate at the glacier terminus (Harrison *et al.*, 2001). This relationship provides significant insight about the relaxation response time of a retreating or advancing glacier and its memory of past mass balance perturbations.

More complicated numerical models explicitly simulate glacier dynamics, discretizing a glacier, icefield, or ice sheet into one-dimensional, two-dimensional or three-dimensional control elements (or volumes). For valley glaciers or major icefield outlets, cross-sectional models that resolve vertical and longitudinal (down-glacier) two-dimensional elements are common, whereas ice caps and ice sheets are typically simulated using either vertically-integrated, plan-view two-dimensional models or fully-resolved three-dimensional control volumes. Finite-element and finite-difference techniques are both common for glacier and ice-sheet discretization.

Consider a point (λ, θ), with glacial ice cover of thickness $H(\lambda, \theta, t)$. For surface topography $h_s(\lambda, \theta, t)$ and subglacial (bed) topography $h_b(\lambda, \theta, t)$, it follows that $H(\lambda, \theta, t) = h_s(\lambda, \theta, t) - h_b(\lambda, \theta, t)$ for grounded (non-floating) ice. Conservation of mass at this point gives the dynamical evolution equation

$$\frac{\partial H}{\partial t} = -\frac{\partial}{\partial x_j}\left(\bar{v}_j H\right) + \dot{b} \qquad (13)$$

where $\dot{b}$ is the ice-equivalent mass balance rate, as defined in section 32.2, and $\partial_j(\bar{v}_j H)$ denotes the horizontal divergence of ice flux, where $\bar{v}_j$ is the vertically integrated horizontal velocity. The vertically integrated ice flux in Equation (13) is estimated as a function of ice-sheet geometry using a constitutive equation that relates strain rates (hence, velocity) to the stress field in the ice. Stresses that induce internal, creep deformation in ice masses are associated primarily with gravitational normal and shear stresses, and are a non-linear function of ice thickness and surface slope. Additional, generally second-order, stresses arise as a result of velocity gradients in the ice (compressive or extensional flow), topographic channelization (e.g. flow constrictions or convergence of tributary glaciers), or gradients in friction at the bed and the glacier margins (e.g. side drag from valley walls).

Glen's flow law (Glen, 1958; Paterson, 1994) is the prevailing ice rheology used in glacier flow modelling. Glen-flow-based ice-sheet models have had good success in simulating the internal shear deformation that dominates flux in ice sheets that are well-coupled with their bed (e.g. East Antarctica, Greenland), but they are not adequate in ice masses where ice flux is dominated by basal flow. In the West Antarctic Ice Sheet, basal sliding and/or plastic failure in a thin layer of subglacial sediment are responsible for most of the motion in fast-flowing ice streams that drain over 90% of the ice sheet (Paterson, 1994; Tulaczyk *et al.*, 2000a).

Modelling of basal flow remains a challenge, as the governing physics involve subglacial hydrology, roughness elements (basal pinning points or 'sticky spots'), and sediment dynamics (Alley *et al.*, 1987a; MacAyeal *et al.*, 1995; Tulaczyk *et al.*, 2000a,b). These controls are complex, difficult to observe, and are subgrid-scale in most modelling studies. Even in West Antarctica, one of the most well-studied ice masses in the world, continental-scale ice-sheet models do not yet give good simulations of ice dynamics. Subglacial geology and basal ice thermal regime play important roles in dictating where basal flow is possible, but the primary control is believed to involve the subglacial water system. Simulation of subglacial hydrology is another emerging focus in glaciological modelling, and the first coupled models of ice dynamics and subglacial hydrology are just becoming available (Arnold & Sharp, 2002; Johnson & Fastook, 2002; Flowers & Clarke, 2002a; Flowers *et al.*, 2003). Although parameterically unwieldy because of the uncertainties in subglacial hydrological processes, this development should provide improved representation of the controls of basal flow and the dynamical response of glaciers to climate change.

32.4 Modelling glacier response to climate change

Different approaches have been taken to estimate glacier, icefield and ice-sheet response to climate change. Haeberli (this volume, Chapter 84) summarizes present-day glacier and ice-sheet changes. I discuss the methods being used to estimate glacier sensitivity to climate change. A variety of techniques have been used to provide forecasts of glacier and ice-sheet response to the anticipated 21st century climate change, at both local and global scales.

32.4.1 The response of individual glaciers to climate change

For individual ice masses, ground- and satellite-based meteorological and snowpack observations can be combined to provide quantitative insights about the climatic controls of mass balance. Field-based mass balance measurements are frequently co-ordinated with energy balance studies at a site (cf. case studies of Klok and Bintanja, this volume, Chapters 33 and 34). Over a number of years, this gives insights about the relative influence of precipitation versus temperature in governing interannual mass balance variability. To illustrate this, Fig. 32.1 plots the winter, summer and annual mass balance record of the Peyto Glacier, Canadian Rockies, for the period 1966–1995 (Demuth & Keller, 1997). Linear correlation analyses on the winter and summer versus net annual mass balance indicate that both the melt season and the accumulation season are important for net annual balance at this site ($r = 0.77$ for the summer balance and $r = 0.67$ for the winter balance).

This gives evidence of the important roles of both maritime influences (precipitation) and continental influences (the length and strength of the summer melt season) on the Peyto Glacier. Numerous in-depth mass balance/climate studies relevant to this site and to western North America have been carried out (e.g. Yarnal, 1984; Letréguilly, 1988; Demuth & Keller, 1997; Hodge *et al.*, 1999; Bitz & Battisti, 1999). Studies of this sort have been done in most of the world's glaciated regions, providing a large knowledge base on local-scale meteorological controls of mass balance for a small number of ice masses that are possible to monitor. Climate-change sensitivity can be estimated reasonably

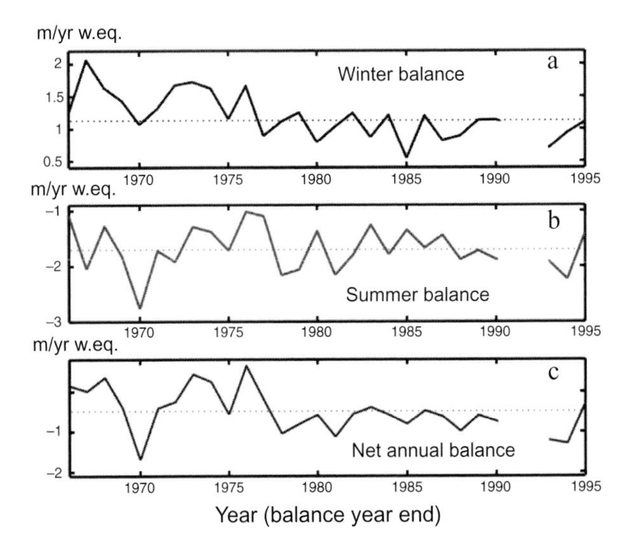

Figure 32.1 Mass balance observations on the Peyto Glacier, AB, 1966–1995, adapted from Demuth & Keller (1997). Winter balance is the net accumulation of snow during the autumn, winter and spring seasons and summer balance is the total amount of summer mass loss due to surface melt. Net annual balance is the sum of the winter and summer balance. All variables are averaged over the area of the glacier and are expressed in metres of water equivalent. Dashed lines indicate the 30-yr means. No data were collected in 1990–91 and 1991–92.

well for these ice masses, because the impact of (for instance) a 1°C warming or cooling or a 20% change in precipitation can be translated directly into glacier-wide mass balance impacts (e.g. Fleming *et al.*, 1997).

Long-term simulations at a given local study site need to consider the feed-backs associated with changing glacier elevations. Glacier hypsometry (the area–elevation distribution) dictates how responsive the glacier-wide mass balance is to sustained climate perturbations (Furbish & Andrews, 1984). In a climate-warming scenario, for instance, a 0.5°C warming may correspond to a 200-m increase in the snowline at a particular site. Low-sloping ice masses will be heavily influenced in this situation, as an increase in snowline (the elevation at which ablation is equal to accumulation) will result in a larger area of the glacier surface that is converted to the ablation zone and is now subject to net annual mass loss. Cumulative long-term mass-balance changes therefore depend on an individual glacier's hypsometry, which is controlled by bedrock topography, ice thickness, ice rheology and ice dynamical regime. Cold, stiff ice that is frozen to the bed, for instance, tends to have a steeper profile than ice that experiences significant basal flow.

Long-term studies also need to include glacier dynamical responses associated with changes in the flux of ice from the accumulation area to the ablation area of an ice mass (e.g. Oerlemans *et al.*, 1998; Albrecht *et al.*, 2000). The dynamical response-time dictates the response of glacier area and terminus position to temporally and spatially integrated mass-balance changes. Using a sophisticated three-dimensional model of glacier dynamics, Schneeberger *et al.* (2001) simulated the response of Störglaciaren, Sweden to 2 × CO_2 climate-change forecasts derived from high-resolution AGCM simulations. A doubling of CO_2 is expected by close to 2050, so these simulations represent ca. 50-yr forecasts of glacier length, area and volume response to potential climate change. Schneeberger *et al.* have extended these simulations to 50-yr forecast scenarios for several well-studied glaciers in Europe, Asia and North America, and these are among the most definitive modelling studies to date of individual glacier response to climate change.

32.4.2 Global-scale mass balance impacts of climate change

The methods introduced above are applicable to larger-scale ice masses and continental to global-scale climate change scenarios, but few ice masses are as well-studied and thoroughly understood as Störglaciaren or the Peyto Glacier. Mass balance records have been made on a total of roughly 280 glaciers over the past 60 yr (Dyurgerov, 2002), out of an estimated 160,000 glaciers worldwide. Long-term records are available from less than 25% of these sites, with data from only 44 glaciers available for the 1964–1995 global mass-balance assessment of Dyurgerov (2002). In addition, and of greater importance for decade- and century-scale sea-level forecasts, large-scale icefields and ice sheets have too great an area to fully assess with mass-balance field campaigns. Satellite techniques (e.g. Zwally & Giovinetto, 1995; Abdalati & Steffen, 2001) and laser altimetry (e.g. Arendt *et al.*, 2002) are creating new possibilities to quantify interannual mass balance variability, and show promise for glacier-change monitoring.

For model-based estimates of global-scale glacier and icefield response to climate change, two principal methods are possible. One approach that has been insightful involves sensitivity tests of different glacial systems to a hypothetical step change in climate. A 'reference' climatology representing present-day conditions can be taken from observational reconstructions (e.g. Legates & Willmott, 1990a,b), reanalysis fields (Kalney *et al.*, 1996) or AGCM simulations. It is common to take the reference state as a 30-yr average of observed, reconstructed or simulated climate, such as the 1961–1990 climate normals. Global-scale glacier mass-balance fields (accumulation and ablation rates) are then estimated for this reference climatology, as illustrated in Plate 32.1 for the western Arctic.

Using the same schemes for mass balance calculation, climate sensitivity can then be tested through a step change in the reference climatology, such as a 1°C warming or a 10% increase in precipitation (e.g. Braithwaite & Zhang, 2000). This kind of experiment illuminates the sensitivity of different glacial systems to temperature versus precipitation shifts, and it provides quantitative insight into the mass balance (hence, ice volume) response to climate change in different geographical regions (Gregory & Oerlemans, 1997). The seasonality of potential climate change is also important and is variable in different glacial environments (Oerlemans & Reichert, 2000). Numerous ice-sheet-scale investigations of mass-balance sensitivity have also been carried out using AGCM-derived climate change scenarios. For instance, Ohmura *et al.* (1996) and Thompson & Pollard (1997) assessed the impacts of doubled CO_2 on mass balance in Greenland and Antarctica, without regard to ice dynamical feed-backs.

At present, global-scale studies of the mass-balance impacts of a given climate change scenario can only be explicitly coupled with ice dynamics models for large-scale ice caps and ice sheets, such as those in Iceland, the Arctic islands, Greenland and Antarctica. As discussed in section 32.5, the complex relief in mountain regions is too poorly resolved in global gridded observational, reanalysed or modelled climatology to provide meaningful mass-balance information for individual glaciers or alpine icefields. General insights can be attained, however, through large-scale hypsometric characterization of glacierized areas (e.g. Marshall & Clarke, 1999). Knowing the distribution of areas in different elevation bands, generalized regional-scale mass balance impacts can be assessed for seasonal or annual climate perturbations (Marshall, 2002).

32.4.3 Coupled glacier and climate models

Glacier and ice sheet sensitivity experiments such as those described above highlight regions of the world that are most sensitive to expected changes in temperature or precipitation. More realistic, geographically and temporally explicit scenarios are required for quantitative forecasts of glacier and icefield response to climate change. Studies of this type require coupling of glacier and climate models, and have been carried out to assess the transient response of various ice masses to model-based climate change scenarios (e.g. Schneeberger *et al.*, 2001; Huybrechts *et al.*, 2002). Studies to date have all involved one-way or 'offline' forcing, where time-dependent glacier or ice-sheet simulations are driven by reanalysed or modelled climate fields. Future forecasts use monthly, annual or decadal average climatology to estimate mass-balance fields, and integrate the cumulative mass balance and ice dynamics response for a given ice mass.

Numerous studies have analysed the likely response of the Greenland Ice Sheet to future climate change using this methodology (e.g. Huybrechts *et al.*, 2002). Plate 32.2 presents an illustrative simulation for the year 2200, using climate forecast fields from a coupled ocean–atmosphere simulation with the NCAR CCSM v2.0 (B. Otto-Bliesner, personal communication, 2003). These climate fields have been used to drive a three-dimensional dynamical ice-sheet model through a 200-yr transient run, using an ice sheet model that is spun up through a prior simulation of the last glacial cycle (Marshall & Cuffey, 2000). Differences from the reference climatology of Plate 32.1 are plotted to illustrate the spatial patterns of ice-sheet response to the simulated climate change.

The degree-day mass balance model used to generate these results is oversimplified and the AGCM representation of modern-day temperature and precipitation patterns is poor in some regions of the ice sheet (cf. Ohmura *et al.*, 1996; Fig. 32.1). There are therefore significant uncertainties in this type of forecast. High-resolution AGCM simulations focusing on the polar regions show continually increasing promise for modelling snow accumulation patterns over ice sheets (Bromwich *et al.*, 1995, 2001; Ohmura *et al.*, 1996; Wild & Ohmura, 2000; Wild *et al.*, 2003). With appropriate 'downscaling' or interpolation strategies to link between AGCM and ice-sheet model grids (of the order of 20 km for continental ice sheets), ice-sheet-scale mass balance can be simulated with reasonable accuracy (Pollard & Thompson, 1997; Glover, 1999). There are nevertheless many outstanding challenges in climate field downscaling, particularly for regional-scale icefields. I discuss this further in section 32.5.

Coupled glacier and climate models that address future climate change impacts require an initial glacier distribution to represent present-day conditions. This initial distribution shapes the surface topography and it provides the reference mass balance state. The resulting mass-balance sensitivity tests should be considered only as snapshots of glacier distribution and mass balance for a particular climatic and glacierized state. This neglects ice dynamical feed-backs and mass balance feed-backs that will result from changing glacier geometry (hence surface topography and local energy balance/boundary-layer meteorology feed-backs).

These studies also implicitly treat the initial glacier distribution as if it is in an equilibrium state with the reference climatology, which certainly is not the case. Even small glaciers take decades to respond to climatic changes, while continental ice sheets take tens of thousands of years. Many contemporary ice masses are in a state of negative balance for modern reference climatologies (e.g. 1961–1990), and a long integration with a perturbed climate scenario would compound the effects of both the disequilibrium reference state and climate change scenario. These simulations can be run out to equilibrium for pedagogic purposes, with or without glacier dynamics feed-backs, but some of the glacier distributions that are predicted in such a future scenario will simply reflect the disequilibrium of modern-day ice masses with the reference climatology. Experiments of this type are therefore best interpreted as 'instantaneous' mass balance perturbation experiments, with the resultant mass balance fields differenced from those of the reference state.

One tactic to develop improved glacier and ice sheet initial conditions is to run global-scale 'spin-up' simulations using historical or modelled climatology. For alpine glaciers and icefields (small glacial systems) or those with a rapid climatic/dynamic response time (e.g. Icelandic ice caps), a simulation from 1800 to 2000 would give a reconstruction of present-day ice masses that is reasonably in tune with the historical climatic forcing. Polar icefields and continental-scale ice sheets have much longer dynamical response times. The Greenland and Antarctic Ice Sheets are still responding to late Pleistocene climate and the Pleistocene–Holocene transition, with englacial temperatures and impurity concentrations reflecting Pleistocene conditions. This influences ice rheology and the basal temperature distribution of the ice sheets (hence, areas where basal flow is possible). For these ice masses, spin-up simulations typically extend for at least 120 kyr, encompassing the last glacial cycle (e.g. Huybrechts *et al.*, 1991, 2002).

32.4.4 Insights from glacier–climate modelling

Studies to date all indicate the dominant sensitivity of glacial systems to changes in temperature; ablation rates are much more variable than accumulation rates and are the primary control on glacier mass balance and long-term glacier distributions (Oerlemans & Reichert, 2000; Marshall, 2002). On the Greenland Ice Sheet, for instance, a 1°C warming translates to roughly a 50%

increase in present-day melt area extent (Abdalati & Steffen, 2001). Temperature changes have a generally greater impact than precipitation changes because of compounding influences; a warming, for instance, will expand the ablation area, lengthen the melt season, increase the extent of melt at a given site, and increase the proportion of precipitation that falls as rain rather than snow.

The seasonality of temperature changes is important for these influences. Winter warming, for instance, has little effect on annual ablation and does not have an impact on rain versus snow events in most glacierized regions, as it is too cold in winter. In this situation, warming has the potential to increase winter moisture supply, hence net accumulation, although this depends on the synoptic-scale controls of precipitation in a region.

There are local exceptions to the general rule of temperature influences winning out over precipitation variability. Maritime ice masses experience more net accumulation and greater interannual variance in precipitation/accumulation rates. Precipitation changes in wet or dry years can be tens of per cent in coastal settings, whereas interannual temperature variability can be buffered by the ocean. This gives precipitation variability a larger role in mass balance fluctuations. In some regions, such as Norway, the strong positive phase of the North Atlantic Oscillation in the 1980s and 1990s has increased regional precipitation and resulted in growth or stability of many glaciers, bucking the global trend in this period (Dyurgerov & Meier, 1997). Even in this situation, however, recent warming has overtaken the precipitation increases in a number of regions (e.g. Iceland) and glacial systems have moved into a negative mass balance regime. Recent warming in many other maritime regions (e.g. Alaska, coastal North America, Patagonia, the high Arctic) has also resulted in near-unanimous ice retreat (Dyurgerov & Meier, 1997; Arendt *et al.*, 2002; Meier *et al.*, 2003).

The most important broad exception to the general temperature control of glaciation is in Antarctica, where it is too cold for significant melt over most of the ice sheet. In this setting, large-scale changes in snow accumulation have an impact on mass balance over a very large area, whereas a modest warming or cooling has very little effect on ice loss. Warming-induced moisture increases in interior Antarctica may increase mass balance in the decades ahead. The Antarctic Peninsula is an exception (see the review by Vaughan, this volume, Chapter 42), as annual temperatures are significantly warmer on the Peninsula relative to the interior plateau of the continent.

For past or future climate-change scenarios, AGCMs can be used to simulate changes in precipitation patterns associated with changing temperatures and surface boundary conditions. In contrast, stand-alone glaciological models that explore temperature perturbation scenarios generally adopt present-day spatial precipitation patterns, using either modelled or observational climatology. A common treatment is to assume that climate warming/cooling will be accompanied by increases/decreases in moisture availability

$$P(\lambda,\theta,t) = P(\lambda,\theta,0)\exp[\beta_P(T(\lambda,\theta,t) - T(\lambda,\theta,0))] \tag{14}$$

where β_P parameterizes the temperature–humidity relationship in accord with the Clasius–Clapeyron relationship. Time t = 0 refers to present-day climatology (precipitation rates and temperature).

The validity of this assumption of a globally consistent positive correlation between increasing temperature and precipitation is not certain, as changes in moisture supply are more complex than a simple thermodynamic control. The variable feed-backs associated with shifts in storm tracks and orographic forcing under different synoptic and topographic conditions are important in long-term glacier–climate forecasts.

Globally, temperature and precipitation are positively correlated (Fig. 32.2a), but the relationship is weak or even negative in some regions. Figure 32.2b plots the zonally averaged correlation between surface temperature and precipitation calculated from the global gridded observational climatology of Legates & Willmott (1990a,b), at 0.5° resolution (V2.01 of global precipitation data and V2.02 of surface temperature data). The average global correlation is 0.52 for annual mean temperature and precipitation, but the relationship weakens at regional scales (see e.g., Leung & Qian, 2003, fig. 9) and at subannual time-scales. The average monthly correlation of the Legates & Willmott data is 0.42. There is a slightly stronger relationship in high latitudes and at cold temperatures, well-illustrated in a continental-scale subset of this data shown in Fig. 32.2c. This figure plots mean climatological (1971–2000) temperature versus precipitation for all available climate stations in Canada (N = 1077; Environment Canada, 2003). The linear correlation value of this long-term annual data is 0.48, indicating a positive relationship but also illustrating that much of the variance in precipitation rates is due to factors other than temperature.

I have downplayed the importance of precipitation, as typical climate change scenarios for the decades and centuries that lie ahead predict global temperature changes of a few °C, which will be much more important for glacier mass balance than the predicted precipitation changes, of the order of 10%. However, long-term mass balance forecasts need to consider changes in both precipitation and temperature, as the climatic controls of glacier mass balance differ in all regions and the patterns and rates of anticipated climate change will differ significantly between regions (Church *et al.*, 2001).

32.5 Climate-field downscaling for mass-balance modelling

There are unique challenges of scale for efforts to estimate climate change impacts on the world's glacial systems. For mountain glaciers, regional-scale icefields and ice-sheet margins, temperature and precipitation fields need to be downscaled to the scale of relevance for glacier mass balance, of the order of 1–10 km. A full treatment of energy balance on glaciers and icefields also requires wind, humidity and radiation balance terms at the 1–10 km scale. This poses a significant challenge at the resolution of current operational climate models, ca. hundreds of kilometres.

Pollard & Thompson (1997) discuss many of the issues and strategies involved in downscaling of AGCM climate fields for glaciological modelling. The main challenge in simulating mass balance is adequate resolution of the topographic detail. Altitudinal controls on temperature govern both surface melt rates and

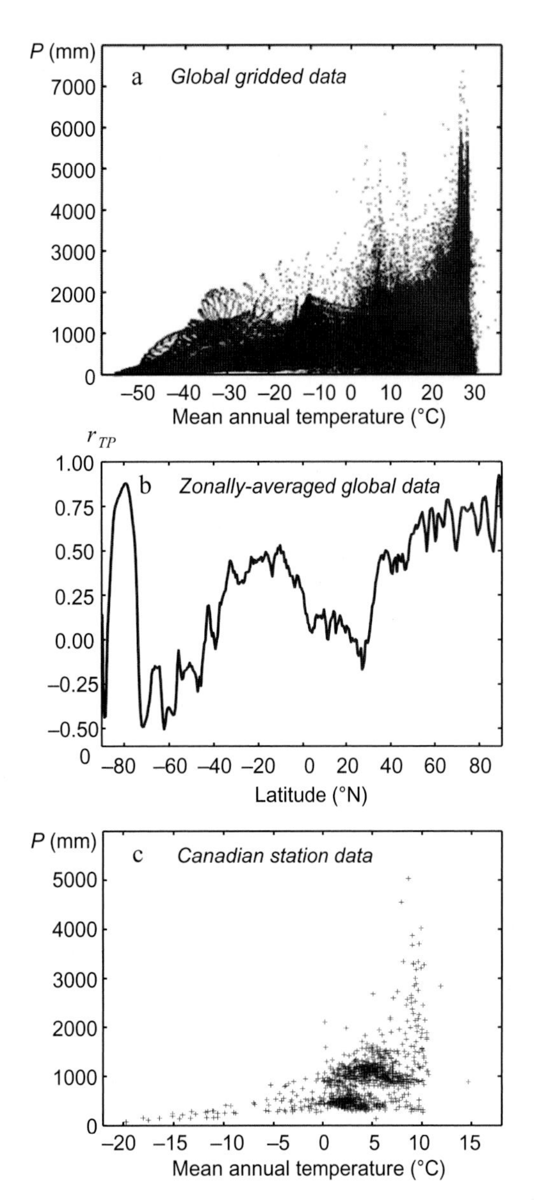

Figure 32.2 Relationship between annual gauge-corrected precipitation and mean annual surface temperature. (a) Global gridded observational climatology (Legates & Willmott, 1990a,b), for all data points in 0.5° by 0.5° latitude–longitude grid cells. (b) Zonally averaged linear correlation coefficient r_{TP} for surface temperature and precipitation. (c) Precipitation versus mean annual temperature for all available 1971–2000 climatic average station data in Canada (N = 1170), illustrating the relationship for a mid- and high-latitude continental environment (Environment Canada, 2003).

the critical transition between liquid and solid precipitation. If subgrid topographic distributions are known, atmospheric (temperature–elevation) lapse rates can be used to estimate surface temperature at a mosaic of subgrid cells (e.g. Thompson & Pollard, 1997; Giorgi *et al.*, 2003) or over a statistical distribution of subgrid terrain elevations (Walland & Simmonds, 1996; Marshall & Clarke, 1999). Surface temperature lapse rates (vertical gradients of surface temperature) are governed by local surface energy balance rather than free-air processes, and are typically less than free atmosphere lapse rates. Surface lapse rates also vary spatially and temporally, so there are significant uncertainties in temperature downscaling from direct altitude adjustments. Simple statistical or dynamical methods are needed to estimate surface lapse rate variability and its controls in different environments.

In addition to the direct effects of altitude on temperature, spatial accumulation rate patterns have a complex dependence on orographic forcing. The intrinsic spatial variability of precipitation is typically much less than the resolution of either modelled or reanalysed climatology, with significant spatial gradients of snow accumulation on scales of hundreds of metres to several kilometres (e.g. Koerner, 1977; Erxleben *et al.*, 2002). At regional scales, this complexity in snow accumulation patterns is largely governed by the interactions of topography with synoptic systems. For instance, orographic forcing of frontal cyclonic systems is responsible for most of the moisture deposition on the Greenland Ice Sheet (Chen *et al.*, 1997). At smaller scales, spatial snowpack gradients are associated with wind redistribution and local topographic influences.

This introduces the requirement to downscale precipitation to subgrid scales. In practice, the most common approach is to distribute resolvable (i.e., frontal) AGCM grid-scale precipitation uniformly across subgrid terrain features (Giorgi *et al.*, 2003), although it is recognized that subgrid orographic controls will give highly non-uniform distributions in complex terrain. Physically based statistical or dynamical models of precipitation variability need to be applied to generate a more detailed approximation of spatial distributions (e.g. Daly *et al.*, 1994).

Mesoscale climate and forecast models are now approaching the resolution of interest for regional topographic detail and precipitation processes (Giorgi *et al.*, 1994, 1996, 2003; Colle *et al.*, 1999; Leung & Qian, 2003). Regional-scale models simulate atmospheric dynamics over a high-resolution domain that is nested within a global-scale model, with boundary conditions coming from the global model. Studies of the Pacific northwest USA by Colle *et al.* (1999) explored grid cells as fine as 4 km, but with little improvement in precipitation forecasts relative to simulations at 12-km resolution. This points to the fact that model performance is limited by physical process understanding as well as resolution of topographic detail and land-surface heterogeneities. Parameterizations of subgrid physics need to be retuned at different resolutions (Giorgi & Marinucci, 1996), and parameterizations may not be generalizable to different regions.

Limits to resolution-derived improvements are also expected in association with model nesting strategies, the coarse resolution of boundary forcing in nested regional models, and the lack of mesoscale data that is available to constrain and test mesoscale models. Leung & Qian (2003) explored two-way nesting in regional climate model experiments over complex terrain, with promising results in simulation of precipitation distribution at 13-km resolution. They found marked improvements over simulations at 40-km resolution, although modelled snow distribution at both 40- and 13-km scales is poor. Climate models may require additional subgrid-scale precipitation physics to give improved, dynamically based estimates of precipitation in complex terrain (e.g. Hulton & Sugden, 1995; Chen *et al.*, 1997; Roe, 2002).

Mesoscale model development is an active research frontier, with significant progress to be expected on this front. As subgrid physical processes become better understood and modelled, dynamical downscaling may become the standard for simulating the mass balance of alpine glaciers and icefields. It is nevertheless practical to assume that global-scale climate models will not reach the necessary resolution for some phenomena of interest in the foreseeable future (e.g. kilometre-scale hydrological and land-surface processes; mountain glacier distributions). A combination of physically based downscaling strategies and regional/high-resolution climate models therefore needs to be pursued.

32.6 Summary

Glacier mass balance is monitored at numerous sites worldwide, and remote sensing methods for monitoring snow accumulation and snow/ice surface melting are rapidly improving. The latter offers the potential to monitor glacier, icefield and ice-sheet changes with unprecedented spatial coverage, although field-based calibration and validation studies are still needed at present. Mass balance processes are relatively well-understood and are amenable to modelling, with the notable exception of marine ablation through iceberg calving and basal melting beneath ice shelves. At sites where extensive meteorological and snow survey data are available, successful mass balance models have been developed. That is to say, with knowledge of snow distribution and spatial–temporal energy balance components, surface melt rates and net annual mass balance can be well quantified through theoretical energy balance modelling.

Unfortunately, detailed snowpack and energy balance data are not available for most ice masses, nor are they well-predicted by global- or regional climate models at the scale of relevance for glaciers and icefields. This has led to the development of reduced models for snow and ice melt, parameterizing melt rates as a function of cumulative positive degree-days and/or incoming solar radiation. Temperature is generally considered to be more easily downscaled or interpolated over the landscape than other terms in the local energy balance (e.g. winds, relative humidity), and (potential direct) solar radiation can be theoretically predicted at a site. Hence, these simplified parameterizations of snow and ice melt permit distributed modelling in a region where ground observations are scarce. Despite their simplicity, these models have been extremely successful at melt modelling. Degree-day based mass-balance models are presently used by most glaciological models, at scales that range from individual glaciers to continental-scale ice sheets.

Mass balance sensitivity to climate change has been explicitly simulated for the Greenland and Antarctic Ice Sheets, based on either simple sensitivity tests (e.g. a 1°C warming) or climate change scenarios generated by climate models (e.g. 2 × CO_2 experiments). Similar mass-balance sensitivity tests have been carried out for a number of individual glaciers, with explicit representation of the glacier geometry. This is not possible on a global basis, but regional-scale numerical experiments have investigated the general mass balance sensitivity of most of the world's glacierized regions, based on regional climatic and topographic environments. Collectively, these numerical experiments give a broad idea of regional mass-balance regimes, glacierized regions that are expected to be most sensitive to climate change, and the global-scale mass balance impact of a given climate change scenario.

Long-term simulations of glacier and icefield response to climate change require consideration of ice dynamics. Changing glacier geometries, through both glacier dynamics and cumulative mass-balance response, introduce feed-backs that will alter future mass balance as well as local and regional climate. Numerous studies have included glacier dynamics and simple elevation-based mass balance feed-backs in simulations of glacier response to climate change, but climatic feed-backs that arise due to changing albedo and topographic forcing conditions are seldom included. There is a need for improved two-way coupling in glacier–climate modelling, where changing surface boundary conditions are fed back into the climate model. It is generally believed that most climatic feed-backs will be positive (e.g. melting increases surface albedo which begets greater melting; melting lowers the surface which increases temperatures which increase melting; decreased surface slopes will lower orographic forcing, diminishing precipitation, and so on). Hence, two-way coupling may be important in some regions, and in long-term forecasts it is expected to improve quantitative estimates of glacier response to climate change.

Coupled modelling of glacier–climate evolution is rapidly developing, driven in part by intense efforts to improve the representation of the hydrological cycle in weather forecast and climate models. Advances are being made across the board, including model physics, technical capabilities, limits of resolution and the sophistication of climate system coupling (e.g. atmosphere–ocean–land-surface schemes). With the current emphasis on improving regional-scale precipitation and climate change forecasts, via both subgrid physical parameterizations and climate downscaling strategies, glacier–climate models can be expected to evolve and improve for the foreseeable future.

Acknowledgements

I thank Peter Knight for the opportunity to contribute to this volume, and for his patience in awaiting late manuscript drafts. Research support for the author is provided by the Natural Sciences and Engineering Research Council (NSERC) of Canada, the Canadian Institute for Advanced Research, and subcontract OPP-0082453 from the U.S. National Science Foundation. Additional thanks to Bette Otto-Bliesner of the U.S. National Center of Atmospheric Research for the NCAR CCSM v2.0 climate fields used for the Greenland Ice Sheet simulations.

THIRTY-THREE

Energy and mass fluxes over dry snow surfaces

Richard Bintanja

Institute for Marine and Atmospheric Research Utrecht, Utrecht University P.O. Box 80005, 3508 TA Utrecht, The Netherlands

Over snow and ice, the surface mass balance determines whether the surface loses mass to or gains mass from the atmosphere. The mass balance is the sum of accumulation and ablation. In the case of dry snow surfaces, melt and runoff can be ignored, which means that ablation can take place only through sublimation and wind erosion. Surface sublimation represents the transport of water vapour directly from the snow surface into the atmosphere. Wind erosion occurs when winds are stronger than the threshold for snowdrift initiation, and snow particles become mobile. The snow taken up by the wind either can be redeposited downflow (in which case there is no net effect on the overall mass balance) or be sublimated when in the air. The latter is commonly referred to as snowdrift sublimation. The sum of both sublimation components represents the net ablation, with a magnitude that is strongly related to the near-surface atmospheric profiles of wind, temperature and especially humidity. I will illustrate this relation for a case study in Antarctica.

In 1997–98, a meteorological expedition to Dronning Maud Land, Antarctica, studied the details of the Antarctic boundary layer (Bintanja, 2001a). An extensive array of techniques was used at various locations, of which just one will be described here. Wind speed, temperature and relative humidity were measured at five levels (0.5 to 9 m above the surface) at a horizontally homogeneous location on the Antarctic ice sheet, near Swedish research station Svea (74°11′S, 10°13′W, 1150 m above sea level). This site is subject to katabatic winds and prevailing synoptic easterlies, with an unobstructed fetch of at least 10 km. The measurements took place from late December 1997 to early February 1998, with mean values of wind speed, temperature and relative humidity of 4.3 m s^{-1}, −10.2°C and 70%, respectively. Sensors were sampled every 2 min, and half-hourly means were calculated. For the purpose of this paper we will focus on a period with particularly strong winds (8–10 January 1998).

Most snowdrift-related processes are of highly non-linear nature. The amount of particles in the air increases exponentially with wind speed (e.g. Kobayashi, 1979), and Bintanja *et al.* (2001) showed that snowdrift transport rates at the study site concur with this. The floating particles are continuously subject to snowdrift sublimation, which is found to be a very efficient way to pump moisture into the atmosphere (Schmidt, 1972). This is because the total exposed surface area of all snowdrifting particles is many orders of magnitude larger than that of the surface. Hence, once particles are swept into the undersaturated atmosphere, snowdrift sublimation starts, which decreases their size and mass. The lowest atmospheric layers quickly become saturated, because these are already close to saturation. Hence, snowdrift sublimation decreases again, because its value is proportional to the undersaturation of the ambient air (note that snowdrift sublimation rates are maximum near the surface, where most particles reside). This constitutes a negative feed-back (e.g. Déry *et al.*, 1998). Figure 33.1 illustrates this effect. It shows observed profiles of relative humidity in a stormy case with snowdrift and in a quiet case without snowdrift. In the snowdrift case, snowdrift sublimation has significantly moistened the lowest layers, with levels below a certain height (z_{sat}; here about 12 cm) being entirely saturated. Thus, the lowest 12 cm of the atmosphere were saturated completely through the action of snowdrift sublimation. This means that surface sublimation (which is proportional to the vertical moisture gradient at the surface) is essentially shut off.

The temporal variation of z_{sat} and other variables over a 5-day period including the strong wind period is shown in Fig. 33.2. During 3 days (8–10 January) winds are sufficiently strong to generate drifting and blowing snow with peak transport rates of 2000 kg m^{-1} s^{-1}. Snowdrift sublimation steadily increases to more than 100 W m^{-2} on 8 January, after which it drops because relative humidities increase and a saturated layer of several centimetres thick is established (notice that surface sublimation vanishes simultaneously). In the early morning of 9 January, advection of relatively dry air enables snowdrift sublimation rates to increase to 250 W m^{-2} (equal to 8 mm of ice per day). Also, the stronger winds lift the particles to higher levels where the air is still undersaturated and snowdrift sublimation can occur. The increased moisture flux to the atmosphere enhances the thickness of the saturated layer to its peak value of almost 18 cm (in a similar study at coastal station Halley during winter, saturated layers of several metres have been observed during strong wind events (Mann *et al.*, 2000)). When wind speeds drop, relative humidity decreases and surface sublimation becomes non-zero again. Over the 5-day

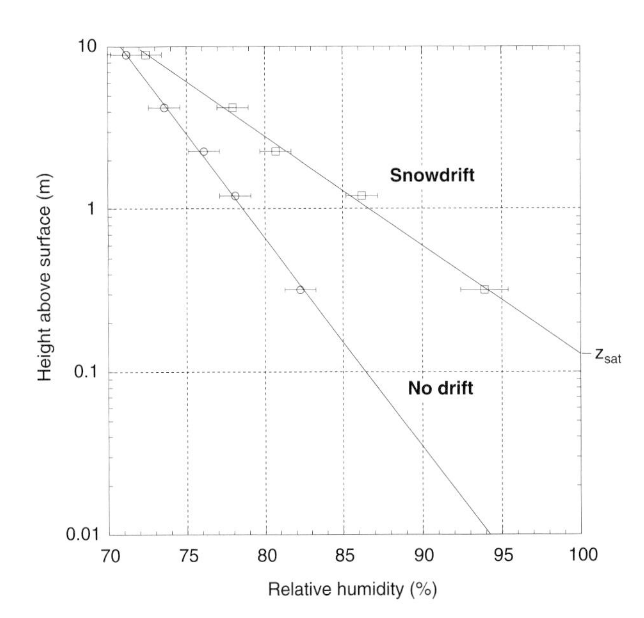

Figure 33.1 Vertical profiles of relative humidity. Measured values represent half hourly averages. The line labelled 'Snowdrift' was taken at 9 January, 1800–1830 hours, during the peak of the storm. The other, labelled 'No drift', was taken at 11 January, 2030–2100 hours, during a period with weak winds and no snowdrift. The level z_{sat} is indicated at the right vertical axis for the 'Snowdrift' profile.

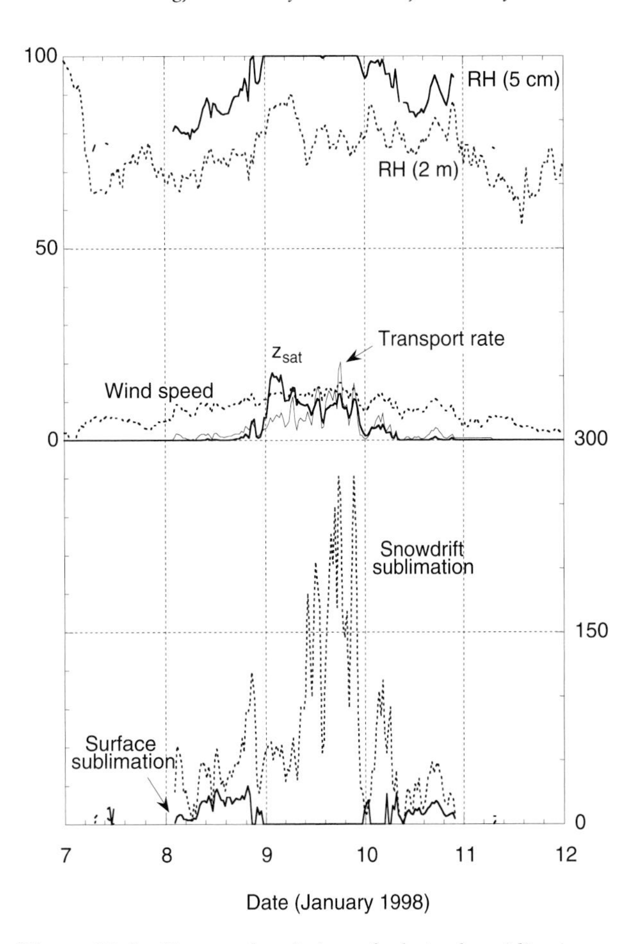

Figure 33.2 Temporal variation of relative humidity, in per cent (at 5 cm and at 2 m), wind speed, in m s^{-1} (2 m), z_{sat}, in cm, transport rate, in 10^2 kg m^{-1} s^{-1}, snowdrift and surface sublimation, in W m^{-2}, during a 5-day period in January 1998.

period, average values of surface and snowdrift sublimation are 7.8 and 44.1 W m^{-2}, respectively, illustrating the importance of blowing snow sublimation. Note that verification of snowdrift sublimation estimates remains one of the great challenges of snowdrift research (Pomeroy & Essery, 1999; Bintanja, 2001a).

This case study highlights the complex interaction between surface sublimation, snowdrift sublimation and moisture budget of the lower atmosphere. The key observation is that the effectiveness of the snowdrift sublimation process limits its own value as well as that of surface sublimation. In high winds, the total moisture flux into the atmosphere is dominated by snowdrift sublimation, as shown in Fig. 33.3. For winds stronger than about 13 m s^{-1}, the formation of a saturated layer close to the surface effectively diminishes surface sublimation, leaving snowdrift sublimation as the only source of moisture to the atmosphere. Whether or not snowdrift increases the total moisture flux over a situation without loose surface particles (with only surface sublimation) depends strongly on the prevailing atmospheric conditions. In a katabatic wind region, horizontal advection and vertical entrainment of dry air continuously supply the lowest atmospheric layers with relatively dry air, rendering the negative feed-back process less effective. The feed-back is more effective and sublimation values are more reduced if the moisture produced by snowdrift sublimation is somehow able to remain near the surface (Mann *et al.*, 2000). In such situations, model experiments demonstrate that the total moisture flux attains maximum values at intermediate winds (Bintanja, 2001b).

Averaged over the entire measuring period of 37 days, surface and snowdrift sublimation rates were about equal at this

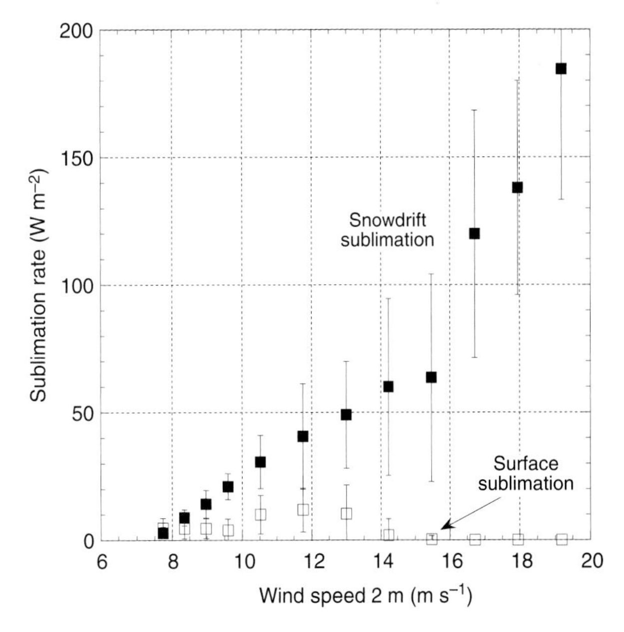

Figure 33.3 Dependence of snowdrift and surface sublimation rates on wind speed (2 m). Symbols represent grouped data, and the error bars represent the standard deviations of the average values.

location (6.4 and 6.7 W m^{-2}, respectively). This is because in below-threshold winds (which occurred 79% of the time) snowdrift sublimation rates are obviously equal to zero whereas surface sublimation is non-zero. In windier regions, snowdrift sublimation will be more important. Sublimation rates peak in summer owing to warm conditions, in spite of the relatively modest winds. As a rough estimate of annual values we assume that the calculated rates are valid for the three summer months, and zero for the rest of the year. In that case, sublimation would remove 20–30% of the annual snow accumulation. This value may change considerably over the Antarctic continent, depending on the prevailing climate conditions (wind, temperature, humidity). Even though the magnitude of wind erosion is unknown, it seems likely that sublimation is the primary ablation mechanism at this site. This contrasts sharply with the situation in Greenland and other glacier environments, where ablation is dominated by surface melt. This case study demonstrates that for a proper evaluation of the surface mass-balance of dry snow regions it is imperative that the contribution of snowdrift sublimation is taken into account, and to do so one needs to consider explicitly the *interaction* between snowdrift and the lowest atmospheric layers.

THIRTY-FOUR
Energy fluxes over Morteratschgletscher

Lisette Klok

KNMI (Royal Netherlands Meteorological Institute), P.O. Box 201, 3730 AE De Bilt, The Netherlands

34.1 Morteratschgletscher

Morteratschgletscher is a valley glacier located in southeast Switzerland (46°24′N, 8°02′E). The glacier mainly faces north and is surrounded by high mountains. Its altitude ranges from 2000 to 4000 m a.s.l. The glacier area is about 17 km^2 and Morteratschgletscher is currently 7 km long. An automatic weather station (AWS) is located on a relatively flat part of the glacier snout at an altitude of 2100 m a.s.l. It measures the four radiation components, air temperature, relative humidity, wind speed and wind direction, all at 3.5 m above the glacier surface. Data from this AWS are useful for understanding the energy fluxes at the glacier surface, and also for developing and validating parameterizations used in energy- and mass-balance models.

34.2 An analysis of the AWS-derived surface fluxes

For the year 2000, all components of the surface energy balance are calculated from the AWS data (Fig. 34.1). Daily mean incoming shortwave radiation shows a clear annual cycle. The day to day variation is due to clouds. In winter, reflected shortwave radiation almost equals incoming shortwave radiation owing to the high surface albedo, which drops in May when snow is starting to melt and ice appears at the surface. The mean annual albedo is 0.53. Incoming longwave radiation is generally less than outgoing longwave radiation. On cloudy days, however, it sometimes exceeds longwave radiation emitted by the glacier surface. In summer when the surface is at melting point, fluctuations in outgoing longwave radiation are small.

The sensible and latent turbulent heat fluxes are estimated from measured air temperature, humidity and wind speed using a bulk approach and a constant turbulent exchange coefficient of 0.00153 (Oerlemans & Klok, 2002). The sensible turbulent heat flux is positive throughout the entire year because mean daily air temperature always exceeds the surface temperature at our site. It is largest in summer when the surface temperature is limited to the melting point and the vertical temperature gradients are largest. This also generates a strong katabatic flow, which in turn enhances the turbulent exchange. The latent turbulent heat flux is small and on average positive, indicating that normally condensation or riming takes place. This adds about 2 cm water-equivalent per year to the glacier surface at the AWS site. Between

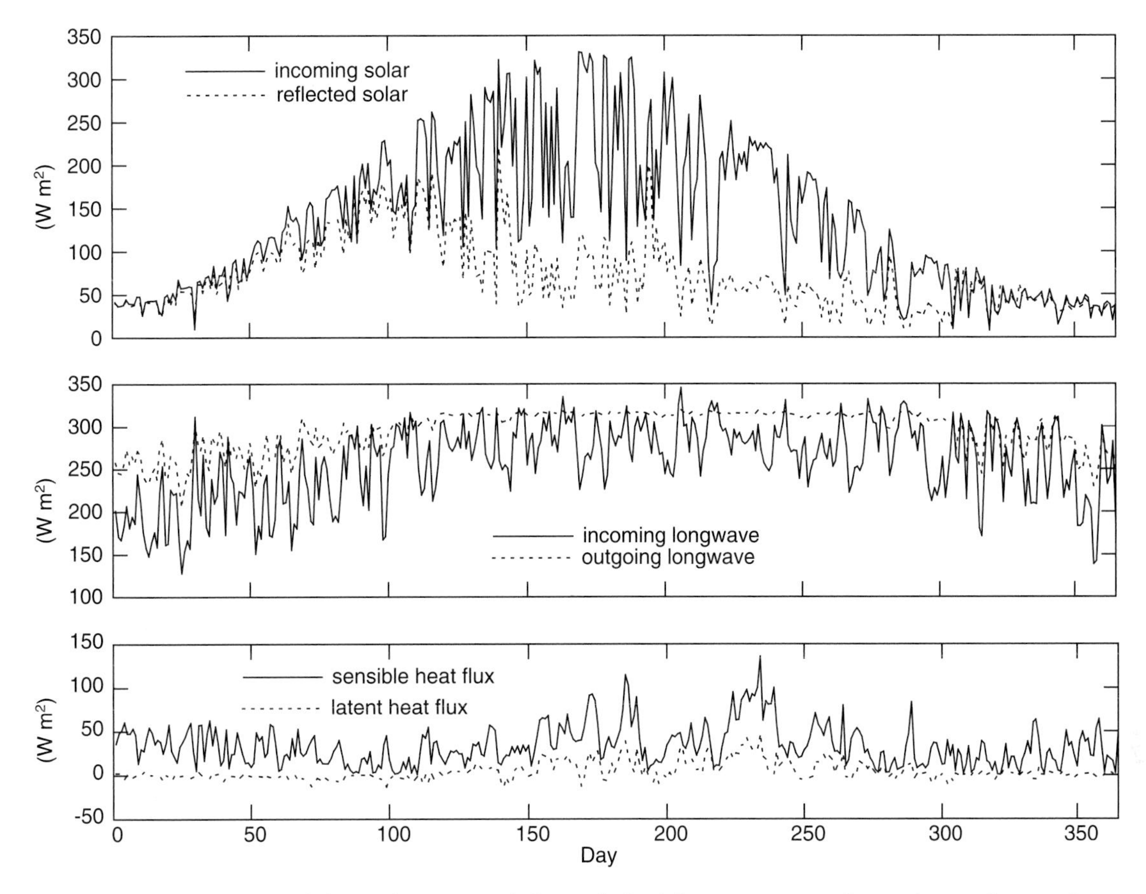

Figure 34.1 Daily mean values of the surface energy balance derived from an automatic weather station on the tongue of Morteratschgletscher for the year 2000.

day 127 and 303, when the AWS site was free of snow, 7.3 m ice melted according to the surface energy balance calculations. Altogether, 75% of the energy for this melt was supplied by net solar and longwave radiation and 25% by the turbulent heat fluxes.

34.3 Energy and mass fluxes calculated from a two-dimensional model

The altitudinal variation in the energy balance components of the year 2000, shown in Fig. 34.2, is derived from a two-dimensional model (Klok & Oerlemans, 2002). This model is driven by meteorological input from synoptic weather stations located in the vicinity of Morteratschgletscher. Its spatial resolution is 25 m and half-hourly time steps are used. The model accounts for the effects of shading, surface orientation, obstruction of the sky and reflection from the surrounding slopes on the incoming shortwave radiation. The albedo of each grid cell is estimated with the method of Oerlemans & Knap (1998). The parameterization of the turbulent heat fluxes follows the model of Oerlemans & Grisogono (2002). The conductive heat flux into the glacier is estimated from a simple two-layer subsurface model.

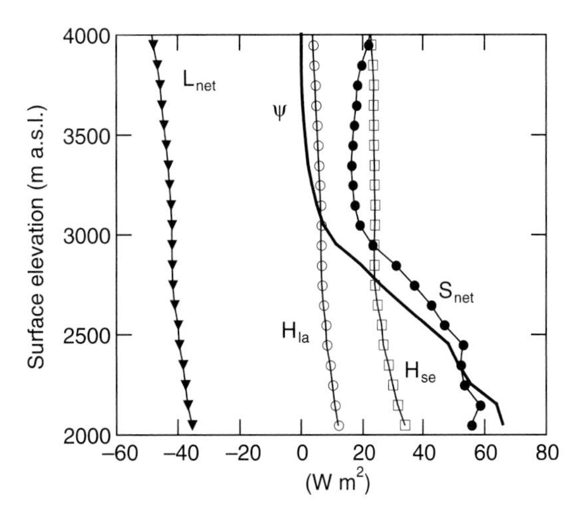

Figure 34.2 Mean modelled annual surface energy flux (χ), net shortwave radiation (S_{net}), net longwave radiation (L_{net}), sensible (H_{se}) and latent (H_{la}) turbulent heat flux averaged over 100 m height intervals for the year 2000.

Averaged over the year 2000, net longwave radiation becomes more negative at higher altitudes. Likewise, the turbulent heat fluxes decrease with altitude. These features can be explained by the decrease in temperature with elevation. Net shortwave radiation is at minimum at 3300 m a.s.l. and at maximum at 2100 m a.s.l. This variation is mainly due to variations in the surface albedo, but also due to the orientation of the surface, shading, sky obstruction and reflection from the surrounding slopes. Averaged over the entire glacier, these topographical effects reduce shortwave reception by 37%. The surface energy flux is approximately zero at the very high altitudes, implying that almost no melting takes place and the glacier surface does not gain or lose heat on an annual basis.

The total melt as calculated from the surface energy balance together with the estimated snowfall and the final mass balance are plotted in Fig. 34.3. Snowfall was computed from daily precipitation data measured at close to Morteratschgletscher and a precipitation gradient of $0.5\,\mathrm{mm\,yr^{-1}\,m^{-1}}$ (Schwarb, 2000). The equilibrium line is located at an altitude of about 2900 m a.s.l. The calculated mean specific mass balance of the glacier for the year 2000 is 0.23 m water-equivalent, which reflects the healthy accumulation area of Morteratschgletscher. Sixty-four per cent of the glacier area is above the 2900 m altitude. The average snow accumulation amounts to 1.82 m water-equivalent, which means that on average 1.59 m water-equivalent ablated from the glacier surface. Figure 34.3 also shows the mass balance components when the air temperature is raised by 1°C. This mainly affects the melt rate through changes in longwave radiation and the turbulent heat fluxes, enhanced by the albedo feed-back mechanism. Snow accumulation decreases little when the temperature increases. Altogether, the modelled mean specific mass balance of 2000 decreases by 0.70 m water-equivalent for a 1°C change in air temperature and becomes negative.

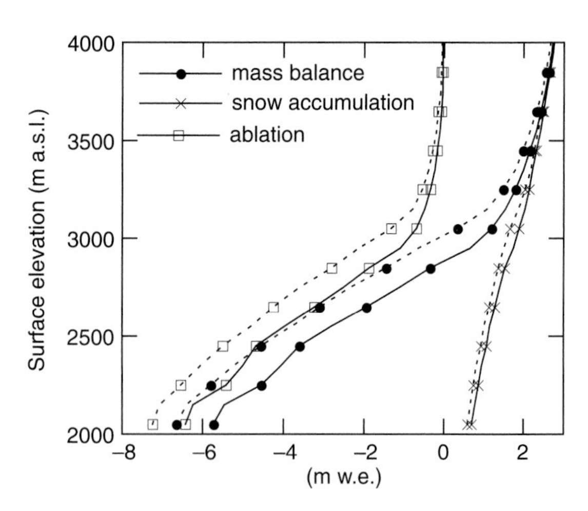

Figure 34.3 Modelled mean specific mass balance, ablation and snow accumulation for 100 m height intervals for the year 2000 (solid) and for a 1°C increase in air temperature (dotted).

THIRTY-FIVE

The environmental significance of deuterium excess in meteoric and non-meteoric Antarctic ice

R. Souchez and R. Lorrain

Département des Sciences de la Terre et de l'Environnement, Université Libre de Bruxelles, CP 160/03, 50, avenue F.D. Roosevelt, B-1050 Brussels, Belgium

35.1 Introduction

Stable isotopes of the water molecule have been widely used for palaeoclimatic reconstructions. Most of the studies were concerned with the oxygen isotopic composition but high precision deuterium measurements allowed co-isotopic studies to be performed, both in $\delta^{18}O$ and in δD. $\delta^{18}O$ is defined as the relative range between the $^{18}O/^{16}O$ ratio in a sample and the $^{18}O/^{16}O$ ratio of standard seawater, expressed in parts per mil. δD is similarly defined as the relative range between the $^{2}H/^{1}H$ ratio in a sample and $^{2}H/^{1}H$ ratio of standard sea water, also expressed in parts per mil. ^{2}H is also called D (for deuterium). The plotting of δD against $\delta^{18}O$ opens up the possibility to visualize important processes. Deuterium and oxygen-18 concentrations in precipitation (rain or snow) are linearly related. The equation expressing this fact is very well obeyed. Thus has been defined the meteoric water line, usually expressed as $\delta D = 8\ \delta^{18}O + d$, where the independent term d is called the deuterium excess. The value of d for a sample is thus its δD value minus eight times its $\delta^{18}O$ value. It is clear that for samples aligned along a meteoric water line (MWL), d is constant (actually equals 10 on the Global Meteoric Water Line). There are, however, situations where the slope of the linear relationship between δD and $\delta^{18}O$ is different from 8. In such cases, d is not constant along the line. If the slope is less than 8, the deuterium excess is usually correlated with δD or $\delta^{18}O$, decreasing with increasing δ values and *vice versa*.

Meteoric ice can be defined as ice resulting from the metamorphism of snow or more generally of solid precipitation. Such ice is by far the most important constituent of glaciers and ice sheets. A meteoric water line is usually displayed for such ice with a well-defined value of deuterium excess. By contrast, samples of ice resulting from water freezing can be aligned in a δD–$\delta^{18}O$ diagram on a straight line with a slope lower than 8. This slope is the signature of a freezing process and will be considered later in this paper. Such ice not directly derived from solid precipitation is called here non-meteoric ice. The significance of the deuterium excess concept in non-meteoric ice is thus completely different from that in meteoric ice. The purpose of this paper is to review the environmental significance of deuterium excess in both situations. The first one is developed in the case of the Vostok and EPICA deep-ice cores. The second one is illustrated in different case studies where ice is formed by water freezing.

35.2 Origin of precipitation, moisture source temperature and deuterium excess in meteoric ice

Isotopic fractionation occurs at phase changes of water in the atmosphere. Two types of isotopic fractionations are involved. First, because the saturation vapour pressures of HDO and $H_2{}^{18}O$ water molecules are slightly lower than that of $H_2{}^{16}O$, the condensed phase is enriched in heavy isotopes with respect to the vapour phase. During evaporation of sea water, the δD and $\delta^{18}O$ values of the vapour will be more negative than those of the liquid phase. This is an equilibrium fractionation. Independently, a kinetic effect results from differences in molecular diffusivities in air of these molecules. The molecular diffusivity in air of HDO or $H_2{}^{18}O$ is lower than that of $H_2{}^{16}O$. This kinetic effect, which is independent of temperature, applies to non-equilibrium processes of evaporation and condensation. Both isotopic effects—equilibrium and kinetic fractionation—are eight times higher for HDO than for $H_2{}^{18}O$.

The initial isotope concentration in an air parcel can be expressed, with certain simplifying assumptions, as a function of sea-surface temperature, relative humidity and wind speed (Merlivat & Jouzel, 1979). Models have been constructed by Jouzel & Merlivat (1984) taking into account the kinetic fractionation associated with snow formation by inverse sublimation in a supersaturated environment and later by Ciais & Jouzel (1994) allowing the coexistence of supercooled liquid droplets and ice crystals in mixed clouds between −11°C and −30°C with different saturation conditions over water and ice. These models are able to reproduce the basic behaviour of δD and δ^{18}O in precipitation at mid- and high latitudes. Such works have outlined the fact that deuterium excess *d* in polar precipitation is largely determined by the source conditions and this has motivated attempts to use *d* as a proxy of ocean surface conditions. Petit *et al.* (1991) have pointed out that the sea-surface temperature largely determines deuterium excess values in Antarctic snow. Warmer sea surface at the source of the vapour produces a higher *d* in snow. Based on GCM results, the role of expected changes in relative humidity is less influential in determining the *d* value.

It is therefore possible to infer from the deuterium excess in ice the temperature of the moisture source and, to some extent, the region of origin of the vapour producing the snow. Such an approach is used to reconstruct palaeoenvironmental situations from the study of deuterium excess in ice cores, as developed in the next paragraph.

35.3 Recent deuterium excess ice-core studies

The Vostok ice-core has provided a great deal of information about environmental changes in the past 420,000 yr (Petit *et al.*, 1999). In a δD–δ^{18}O diagram, the ice samples are beautifully aligned on a regression straight line with a slope of 7.94. Vimeux *et al.* (1999, 2001) have interpreted Vostok deuterium excess variations as depending on fluctuations of the temperature of the oceanic moisture source only, by using a relationship between relative humidity of the air and sea-surface temperature. The variation of the ocean isotopic composition has been removed from the deuterium excess profile by using the marine isotope record. This correction does not modify the long-term variations but significantly affects the amplitudes. Central to the analysis of Vimeux *et al.* (1999, 2001) is the physically based assumption that the temperature difference between the vapour source and the precipitation site principally controls the δD of Antarctic snow, because net isotopic distillation during atmospheric transport from the source to the precipitation site is driven by fractional reduction of the air-mass water content. New information on source-region temperatures can be obtained by measuring the deuterium excess in precipitation.

Deuterium excess in the most recent 250,000 yr of the Vostok ice-core is dominantly modulated by the obliquity periodicity. During periods of low obliquity, the annual mean insolation at high latitudes is low and the latitudinal insolation gradient between 20°S and 60°S is maximized. This increases evaporation at low latitudes and latitudinal atmospheric moisture transport. The latter enhances the contribution of remote oceanic moisture sources and reduces the contribution of local cooler sources. All this together acts to increase the deuterium excess. An anticorrelation has in fact been observed between the obliquity and the deuterium excess. It is interpreted in terms of the relative contribution of low and high latitudes to the precipitation at Vostok. The Vostok data allow all the interglacial periods and glacial inceptions in the past 420,000 yr to be examined. A constant relationship can be observed between deuterium and deuterium excess during these periods. The deuterium excess starts to increase during the warmest period and reaches a maximum value at the beginning of the next cold stage. Then the deuterium excess decreases through the glacial period. In all cases, the glacial inceptions occur when the obliquity is low and the relative contribution of low latitudes to Vostok precipitation is maximized. This suggests that, at glacial inceptions, the temperature of the oceanic surface at low latitudes remains at its interglacial level for some time after the high latitudes have abruptly cooled. Prior to 250,000 yr, there is a lack of correlation with obliquity, probably due to a remote origin for the deep ice at Vostok station because of the ice flow.

The Vostok core also shows a correlation of CO_2 and δD for the past 150,000 yr, the strength of which is $r^2 = 0.64$. This strength is limited primarily by the rapid decrease of δD during and immediately following the last interglacial. Such a large temperature drop is puzzling. Cuffey & Vimeux (2001), by using measurements of deuterium excess for the temperature reconstruction, were able to show that the relative mismatch of the CO_2 and deuterium records is an artefact caused by variations of climate in the water vapour source regions. Using a model correcting for this effect, the co-variation of CO_2 and temperature is clearly improved for the past 150,000 yr ($r^2 = 0.89$). This excellent correlation strongly supports the role of carbon dioxide as a forcing factor of climate.

Below a depth of 3310 m, where the age of the ice is about 420,000 yr and above accreted ice from subglacial Lake Vostok at 3538 m depth, there is evidence of complex ice deformation resulting in folding and intermixing of ice at a submetric scale and, for the upper part of this sequence (3310–3405 m), in interbedding of ice layers from distinct origins at a larger scale. Souchez *et al.* (2002) have used deuterium and deuterium excess properties of the ice to document the build up of these highly deformed basal ice layers. First, there is a damping of δD variations with depth from top to bottom of this core section. Second, the plotting of ice samples on a *d*–δD diagram (Fig. 35.1) shows striking features. The two oldest glacial–interglacial climatic cycles are displayed. They represent ice between 2755 and 3108 m depth (black circles) and between 3109 and 3309 m depth (open circles) respectively. The dot distribution takes the shape of a ring. Now, samples of ice below 3405 m depth (crosses) are also plotted in this figure. The crosses representing these ice samples cluster inside the ring. Moreover, if a vertical line (dotted) is drawn from the centre of the ring, splitting into two equal parts the maximum δD range for the climatic cycles displayed, it can be seen that most of the sample points from the two oldest climatic cycles are within the part containing the more negative values. As one sample was measured per metre of core, this indicates that the colder periods are more developed in terms of ice thickness than the warmer ones. Now, the crosses representing ice samples below 3405 m depth are all included within this part containing the more negative values. Such characteristics point to the occurrence of a folding/mixing process for the basal ice. Indeed folding/mixing

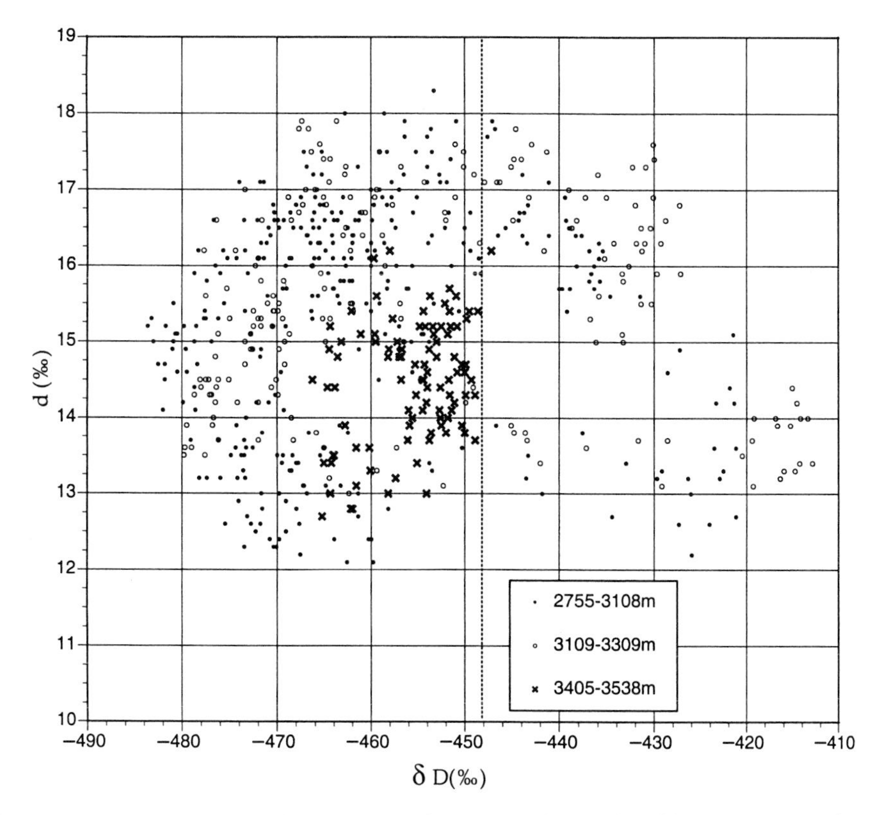

Figure 35.1 *d*–δD diagram in meteoric ice from the Vostok ice-core. (Reproduced by permission of American Geophysical Union from Souchez *et al.* (2002). Copyright American Geophysical Union.)

produces ice having isotopic compositions intermediate between those of the ice layers involved, before they were deformed, hence the clustering of the sample points in the hole of the ring. There is a higher probability that the intermediate isotopic compositions produced result from folding/mixing of ice from colder periods, more frequently present at depth. The distribution within the part containing the more negative δD values thus can be understood. Within this context, the reduction in amplitude in the δD variations with depth can be viewed as implying more complex ice deformation at depth, a very likely situation if one considers the uneven bedrock topography where the ice is grounded upstream from Vostok station. Deformation more complex than simple shear is most pronounced close to the bed, where ice viscosity is reduced by higher temperatures.

In the EPICA ice-core at Dome C in East Antarctica, deuterium excess study of the ice from the last deglaciation period reveals the timing and strength of the sea-surface temperature changes at the source regions for Dome C precipitation (Stenni *et al.*, 2001). It can be shown that an Oceanic Cold Reversal took place in the southern Indian Ocean 800 yr after the Antarctic Cold Reversal, a cold period of the last deglaciation. During this deglaciation period, the temperature gradient between the oceanic moisture source and the Antarctic continent shows a temporal trend similar to the Dome C sodium profile, illustrating the strong link between this temperature gradient and the strength of the atmospheric circulation.

35.4 Deuterium excess in ice formed by water freezing

Heavy isotopic water molecules are preferentially incorporated into the growing ice so that the solid is enriched in deuterium and oxygen 18 compared to the water. Equilibrium fractionation always occurs at the ice–water interface but the observed fractionation between ice and bulk water can be lower, depending on the isotopic concentration in the water at the interface. Souchez & Jouzel (1984) have demonstrated that, by partial freezing of an open or a closed system, samples of ice plot in a δD–δ^{18}O diagram on a so-called freezing slope different from the meteoric water line. The equation of the slope *S* is given by:

$$S = \frac{\alpha_D(\alpha_D - 1)(1000 + \delta_i D)}{\alpha_{18}(\alpha_{18} - 1)\left(1000 + \delta_i^{18}O\right)}$$

where α_D and α_{18} are the equilibrium fractionation coefficients for deuterium and oxygen-18 respectively, and $\delta_i D$ and $\delta_i^{18}O$ are the δ-values for deuterium and oxygen-18 respectively of the initial water at the onset of freezing. The distribution of sample points along the freezing slope is dependent on several factors: percentage of freezing in a closing system with progressive disappearance of the liquid phase, variations of the isotopic composition of the liquid at the interface due to diffusion and to convection, freezing rate and trapping of unfractionated water pockets during ice

Figure 35.2 *d*–δD diagram in lake ice from Taylor Valley, Antarctica. Numbers are increasing with depth. (Reproduced by permission of American Geophysical Union from Souchez *et al.* (2000). Copyright American Geophysical Union.)

accretion that freeze completely afterwards. In such a freezing process, deuterium excess is inversely related to the δ values. The significance of deuterium excess in such a non-meteoric ice is related to the distribution of sample points on the freezing slope, thus to the factors cited above. Variations in deuterium excess are in this case the result of the conditions prevailing during freezing. This is in strong contrast with the general acceptance of deuterium excess in precipitation. In Fig. 35.2 is shown the very good inverse linear relationship between *d* and δD values of samples of lake ice retrieved from a small lake frozen to the bottom, adjacent to Suess Glacier in Taylor Valley, South Victoria Land. In basal ice reached in a tunnel cut at the front of Suess Glacier, a freezing slope is also well displayed in a δD–δ^{18}O diagram, indicating the role of water freezing in its formation in a subglacial environment where the basal temperature is several °C below the pressure-melting point. The deuterium excess profile in this basal ice is a mirror image of the δ^{18}O profile (Sleewaegen *et al.*, 2003). This inverse relationship between *d* and δD results from the fact that the freezing slope is lower than 8.

The study of deuterium excess in ice formed by water freezing can reveal specific processes involved as developed in the two cases described below.

35.5 Deuterium excess in ice formed at the grounding line

In Antarctica, special conditions must be met for water freezing at grounding lines. Thermohaline circulation frequently exists under ice shelves and brings water above its pressure melting point near the grounding line. Efficient tidal mixing in this area also promotes melting. Only in areas sheltered from this deep thermohaline circulation is freezing possible. A double-diffusion process within the pores of subglacial sediments is worth considering in this respect (Souchez *et al.*, 1998). At some distance upstream from the actual separation of the glacier from its bed,

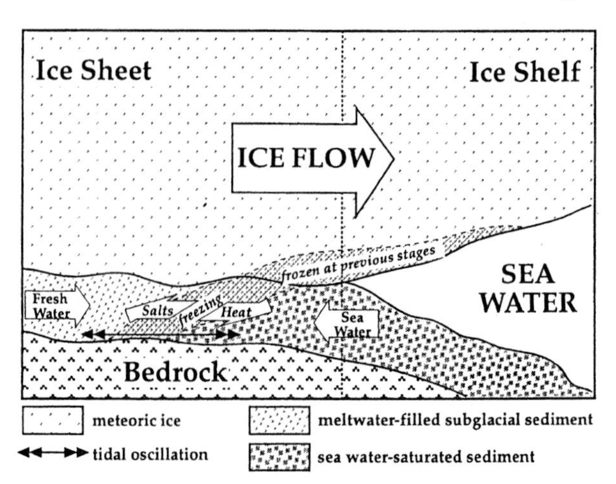

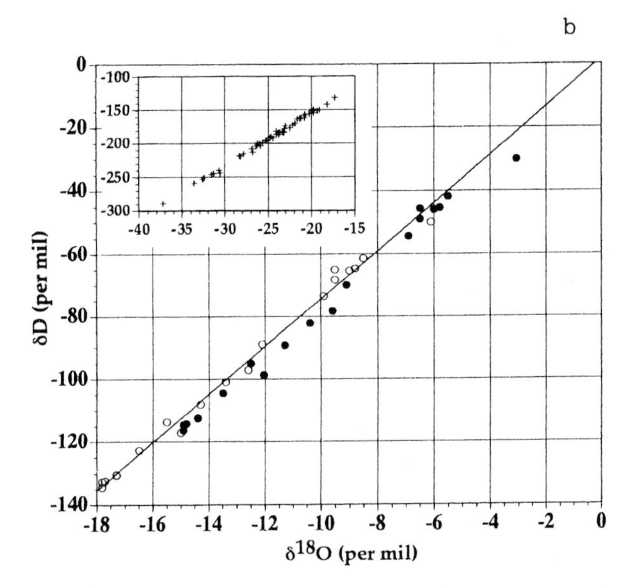

Figure 35.3 Double-diffusion freezing at the grounding line in Terra Nova Bay. (a) Sketch of the suggested mechanism. (b) δD–δ^{18}O diagram of the ice samples. Black circles, ice samples; open circles, initial water samples computed from δ values of ice samples and equilibrium fractionation coefficients. The straight line represents the best-fit line for the waters. Insert shows a diagram for glacier ice samples in the area. (Reproduced by permission of International Glaciological Society from Souchez *et al.* (1998). Copyright International Glaciological Society.)

sea water seeping through the sediment comes into contact with continental meltwaters (Fig. 35.3a). Because the freezing point of sea water is lower than that of continental meltwater, heat diffusion occurs. If both water bodies are at their respective freezing point, heat diffuses through both liquid and solid fractions from upstream to sea water. On the other hand, salinity of sea water being higher than that of continental meltwater, salt diffuses from the seawater-saturated sediment into the meltwater-filled sub-

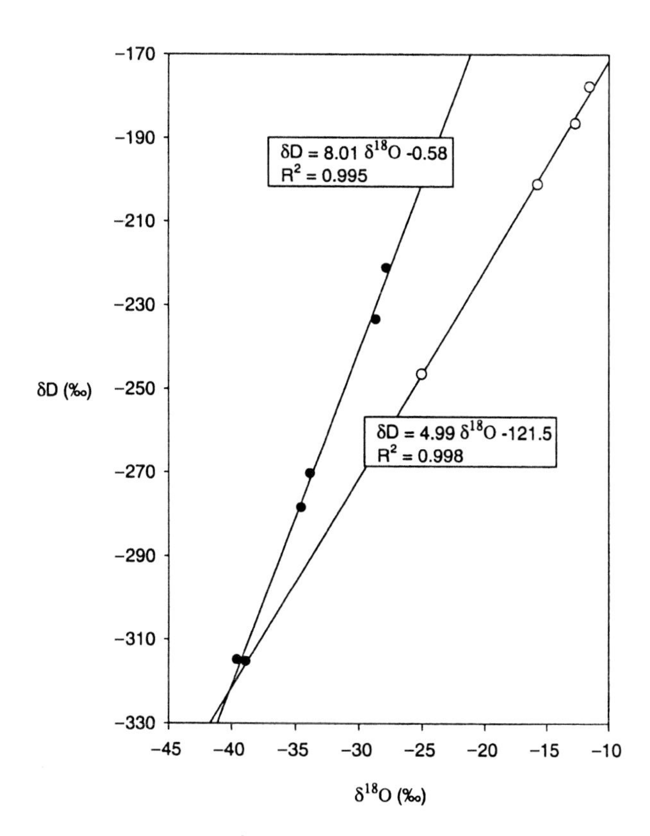

Figure 35.4 δD–δ^{18}O diagram for different ice types in Beacon Valley. (Reproduced by permission from Marchant *et al.* (2002). Copyright Geological Society of America.)

glacial sediments. Salt can diffuse only through liquid, unlike heat. Thermal diffusivity is also an order of magnitude higher than salt diffusion. The meltwater-filled sediment then loses heat more rapidly than it gains salt, and freezing occurs, welding the sediments to the bottom of the ice. As this mechanism occurs within the subglacial sediment, rapid mixing between continental meltwater and sea water is precluded. The isotopic composition of ice supposedly formed at the grounding line in the region of Terra Nova Bay in East Antarctica provides evidence that such a process really occurs.

Glacier ice samples in the area considered are aligned in a δD–δ^{18}O diagram on a straight line with equation δD = 7.91 δ^{18}O + 2.59. The δ values of all the samples are more negative than −17‰ in δ^{18}O and −130‰ in δD. By contrast, the debris-rich ice layers thought to have been formed at the grounding line have δ^{18}O values between −15‰ and −3‰ and δD values between −120‰ and −30‰ (black dots in Fig. 35.3b). Such samples are also aligned on a straight line in the diagram with a slope of 7.7 (Fig. 35.3b); with thus an inverse relationship between δD and deuterium excess, as explained above. Such characteristics are the result of a mixture, in various proportions, between meltwater from glacier ice and sea water. If, as indicated above, glacier-ice meltwater within the subglacial sediment enters into contact with sea water at the grounding line, diffusion does occur. Like salt, heavy isotopes of oxygen and hydrogen diffuse from sea water, when they are less impoverished, to continental meltwater, where they are more depleted. Diffusion coefficients of stable isotopes and of salt in liquid water have the same order of magnitude, so that the double-diffusion mechanism described above also leads to isotopic diffusion. Therefore, the isotopic composition of the debris-rich ice layers formed by freezing depends on the magnitude of the diffusion process prior to freezing. Isotopic composition probably is a better indicator of this diffusion process than salt because the different isotopes are within the ice lattice. They undergo only solid-state diffusion at very long timescales. From the isotopic composition of the ice layer formed, it is possible to reconstruct the approximate composition of the initial water. Subglacial freezing at or near the grounding line is most probably very slow, so that equilibrium fractionation can be considered. Therefore, taking into account the equilibrium fractionation coefficients for deuterium and oxygen-18 and the respective δ values of the ice, it is possible to compute the isotopic composition of the water that was later partially frozen (open circles in Fig. 35.3b). The best fit for these waters is close to points representing δ values of sea water in the area and δ values of glacier ice and meltwaters derived from it. Deuterium excess is thus in this case totally dependent on the position of a sample on a diffusion line connecting continental meltwater and sea water in the area.

35.6 The case of the very old ice from Beacon Valley

Buried glacier ice from Beacon Valley has been studied by Sugden *et al.* (1995) and claimed to be about 8 million yr old. ^{40}Ar/^{39}Ar analyses of single volcanic crystals within *in situ* air-fall ash deposits within the drift above the buried ice provide the geochronological control. These results have been criticized (Hindmarsh *et al.*, 1998) on the basis that sublimation over such a long period would have resulted in the complete disappearance of this buried ice. The buried ice is, however, protected from intense sublimation by the protective cover of the drift. Furthermore, this drift is partly composed of ice-cemented sands. The co-isotopic analysis, both in δD and δ^{18}O, of this ice cement provides clues about its formation. In a δD–δ^{18}O diagram, the samples of the ice cementing the deposit are aligned along a slope of ca. 5, considerably lower than the slope corresponding to the buried glacier ice and modern snow (Fig. 35.4), which is a meteoric water line (slope of ca. 8). The intersection point between this precipitation slope and the line on which the ice that cements the sands is aligned is very close to modern snow samples, suggesting that this snow could be at the origin of the cementing ice. The deuterium excess values of the cementing ice (−85‰ to −45‰) are surprisingly extremely negative and much more negative than in modern snow samples (−4‰ to +2‰). This effect is best explained if the source of the cementing ice has undergone strong evaporation before freezing. Although the deuterium excess of ice can be lowered by partial freezing following melting, the change in deuterium excess never exceeds a few parts per mil. Conversely, evaporation of snow also produces a shift off the precipitation slope but with a larger decrease in the deuterium excess. The simplest scenario in agreement with the isotopic data is that snow has accumulated, has undergone significant evaporation loss and partial summer melting, and has percolated into permeable sand layers where refreezing above the buried glacier ice has occurred. Small amounts of snow melt on sunny summer days have been observed when radiative heating was sufficient to overcome subzero air and ground temperatures. Downward percolation of meltwater is probably facilitated by the depression of the freezing point by the high salt content of the Dry Valleys soils (Bockheim, 1990). Depressions of the ground surface (polygon troughs for example) trap windblown snow and initiate a negative feed-back that slows further sublimation of the underlying glacier ice (Marchant *et al.*, 2002). This negative feed-back involves the development of secondary-ice lenses within the covering sands. Such a process acts as an effective protective cap against sublimation for the underlying buried glacier ice, allowing its long time preservation. The highly negative values of deuterium excess, as encountered in this study, are thus in strong support of the proposed scenario.

35.7 Conclusion

The environmental significance of deuterium excess in ice is dependent on the mechanisms of ice formation. In any case, it reflects physical conditions existing in the environment where isotopic fractionation of hydrogen and oxygen is occurring during phase changes. As such, the study of deuterium excess in ice opens to us the possibility of environmental reconstructions of prevailing conditions in the past.

Acknowledgement

The Belgian Antarctic Programme (OSTC) is gratefully acknowledged for supporting the work.

THIRTY-SIX

Deuterium excess in Antarctica: a review

G. Hoffmann* and F. Vimeux†,*

**Laboratoire des Sciences du Climat et de l'Environnement, UMR CEA/CNRS, Gif sur Yvette, France*
†Institut de Recherche pour le Développement, UMR Great Ice IRD, Paris, France

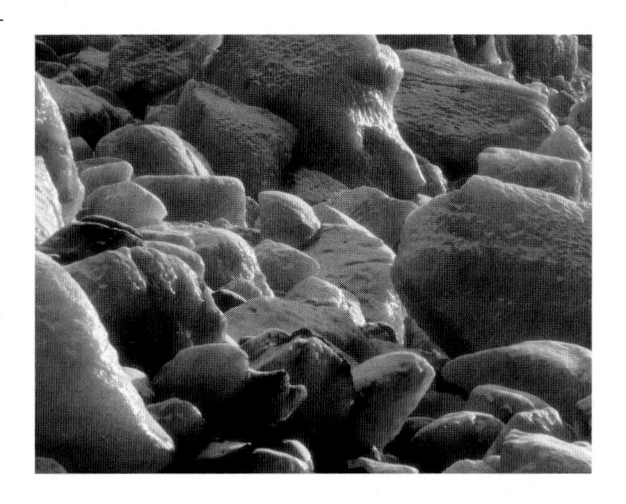

For more than 40 yr the stable water isotopomers, $H_2{}^{18}O$ and $HD^{16}O$, have provided *the* principal means of deriving past temperatures from ice-core drillings in polar regions. Their concentration relative to the main isotopomer $H_2{}^{16}O$ ($H_2{}^{18}O/H_2{}^{16}O$ and $HD^{16}O/H_2{}^{16}O$) is expressed as a δ value ($\delta^{18}O$, δD), that is, as a deviation from the Standard Mean Ocean Water—SMOW (Craig, 1961b), usually given in parts per mil. In many studies the isotopic composition of the ice was linearly related to past site temperatures using a calibration based on the modern spatial δ-value–temperature relationship (Johnsen *et al.*, 1989; Jouzel & Koster, 1997; Jouzel *et al.*, 2003a). This empirical approach is supported theoretically by Rayleigh distillation models (Dansgaard, 1964; Ciais & Jouzel, 1994; Jouzel & Merlivat, 1984). On their way from low to high latitudes water vapour masses become isotopically more and more depleted (that is the δ value becomes more negative) owing to a successive rainout process that continuously transfers the heavier isotopomers into the liquid phase during condensation. The amount of vapour lost as precipitation is controlled by condensation temperatures, which is why the water isotopes eventually depend on temperatures.

Even in the early years of isotope geochemistry (Craig, 1961a; Dansgaard, 1964), however, it was noted that when calculating the scaled differences between the two isotopomers the result should be independent of the described first-order processes, that is, the dominating temperature control, and second-order processes should be detectable. The slope of the global meteoric water line, i.e. the relation between δD and $\delta^{18}O$ in global precipitation, is close to 8. Therefore the deuterium excess was defined originally (in Craig, 1961a) as $d = \delta D - 8\delta^{18}O$. Several non-equilibrium processes influence the deuterium excess d.

1 Evaporation from the ocean surface into the free atmosphere is a non-equilibrium process. After equilibrium evaporation into a microskin layer above the sea surface the turbulent/diffusive transport into the atmosphere is affected by kinetic diffusion. The latter depends strongly on the saturation of the atmospheric planetary boundary layer, that is, principally on temperature and relative humidity. This is the reason for using deuterium excess records from polar ice to reconstruct past source conditions (Merlivat & Jouzel, 1979; Jouzel *et al.*, 1982).
2 Further non-equilibrium processes take place during evaporation of rain drops falling through an undersaturated atmosphere (Stewart, 1975) and during the formation of ice crystals at very low temperatures (Jouzel & Merlivat, 1984).

In addition to these actual kinetic processes the relationship of the equilibrium fractionation coefficients of $H_2^{18}O$ and $HD^{16}O$ is temperature dependent, which is why site temperature has an additional influence on the deuterium excess.

Here we report on the major progress that has been made in recent years both in understanding the kinetic processes at the sea surface during evaporation and in interpreting deuterium excess records from Antarctica in terms of climatic variability during the past 400 kyr.

36.1 Fractionation physics during evaporation

An important reassessment of the physics of non-equilibrium fractionation has been made in Cappa *et al.* (2003). Under non-equilibrium conditions molecular diffusion of the different water isotopomers produces greater fractionation of the lighter water molecules and explains the global deuterium excess value of about $d = +10‰$ (if only equilibrium processes controlled evaporation, global excess would equal $d = 0\%$). Mathematically this process was first described in Merlivat & Jouzel (1979): isotopic fractionation during evaporation is controlled by an effective fractionation comprising an equilibrium and a non-equilibrium (kinetic) part. The latter is affected by the relation of the molecular diffusivities between the different isotopomers. In kinetic gas theory molecular diffusivity is a function of the mass and the collision parameter of the respective molecules. Mainly focusing on the global deuterium excess value, Merlivat & Jouzel (1979) assumed a significantly different collision parameter for the $HD^{16}O$ molecule than for the $H_2^{18}O$ molecule, basing their argument on the broken symmetry of the deuterized water molecule. With this *ad hoc* assumption on molecular collision parameters the deuterium excess could be described successfully in the model of Merlivat & Jouzel (1979).

Cappa *et al.* (2003) carefully conducted laboratory experiments to measure the influence of non-equilibrium evaporation on the effective fractionation. They controlled the relative humidity of an air stream over an evaporation chamber and measured finally the isotopic composition of the vapour and the liquid phase. In their interpretation no different collision parameters for the different isotopomers have to be assumed. In order to explain the relation between bulk water temperature and observed evaporation fractionation, however, they give a major role to an evaporative cooling affecting a skin layer on top of the water surface. Depending on relative humidity this evaporative cooling varies and leads to typical cooling of 1–3°C in this skin layer. Recomputing with these modified temperatures the corresponding effective fractionations, the authors could not only successfully explain their own measurements but also reinterpret former experiments. Their laboratory experiment waits now for a global *in situ* evaluation by measuring the isotopic composition of the first vapour formed over the ocean surface, which astonishingly is still unknown even after decades of using water isotopes. If the model of Cappa *et al.* (2003) is confirmed this has major consequences both for the description of the water isotopes in global atmospheric models (Jouzel *et al.*, 1987; Hoffmann *et al.*, 1998; Kavanaugh & Cuffey, 2003) and for the interpretation of the deuterium excess in palaeorecords. For today's conditions they predict, for example, a deuterium excess of the evaporative flux a couple of parts per mil larger than former estimates.

36.2 Results from Vostok

However, when interpreting palaeo-excess records there are even more ambiguities to consider than just evaporation conditions and their exact influence on the double isotopic composition. Currently Antarctic deep-ice drillings cover several glacial–interglacial cycles. The Vostok ice-core provides us with a deuterium excess record spanning now the past ca. 400,000 yr (Vimeux *et al.*, 1999, 2001, 2002). Fully interpreting this long-term record, Vimeux *et al.* (2002) came to an important re-evaluation of the Antarctic excess, which includes the combined effects of local site temperature and source condition changes. As mentioned before the deuterium excess is influenced principally by climate conditions prevailing in vapour source regions of the corresponding precipitation site. This feature led to a quantification of polar excess signals exclusively in terms of sea-surface temperature (SST) changes (Johnsen *et al.*, 1989). Using simple single-trajectory distillation models the sensitivity of the excess to changes in SSTs in the corresponding source areas was estimated to 0.7–0.8°C $‰^{-1}$. Such an approach has several fundamental premises. The most important one is probably the assumption that there is just one single and geographically stable source region to which deduced SST changes can be associated. Moreover this method assumes that the relative humidity—like the SST influence on evaporating conditions—remains unchanged even under drastically different climate conditions such as the last glacial maximum. These premises can be fully addressed only in future numeric simulations with general circulation models (GCMs) equipped with water isotope diagnostics (Hoffmann *et al.*, 1998).

However, another important mechanism affecting synchronously the water isotope signal ($\delta^{18}O$, δD) and the deuterium excess (d) is now included in the interpretation. Obviously SST changes influence the temperature gradient between the evaporation and the condensation site. This temperature gradient is the leading climate control on the water isotopes (and not just site temperature) and changes at the evaporation site are at least second order for the isotopic composition of precipitation. On the other hand condensation temperatures affect the $\delta^{18}O$ /δD

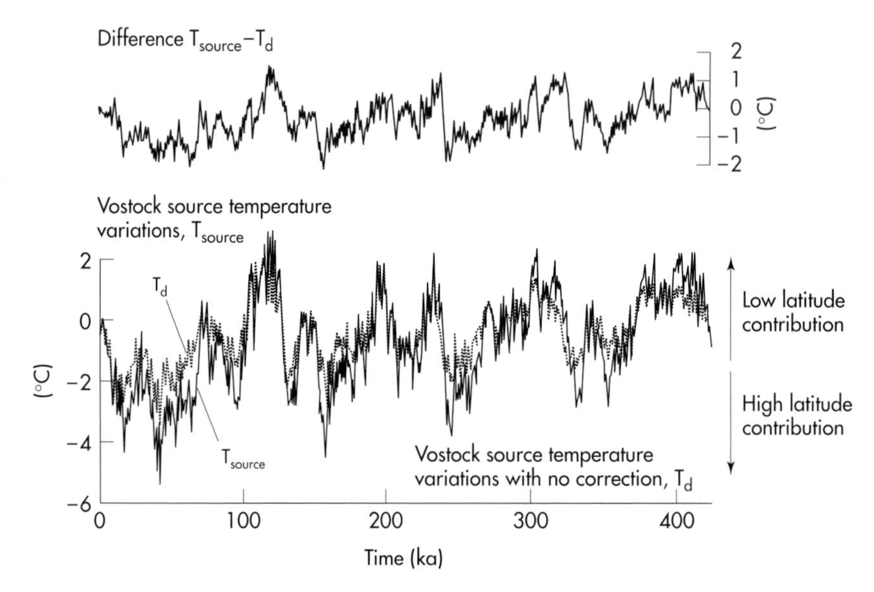

Figure 36.1 Vostok source region temperatures derived from the 400-kyr-old East Antarctic ice-core record (lower panel). Following the classic approach for interpreting the deuterium excess, T_d assumes that the effective source temperature can be deduced from the excess alone. However, T_{source} takes into account the combined effects of condensation-site temperature changes *and* source temperature changes. In the upper panel the difference between the two computations is shown indicating that these differences are growing and shrinking following the glacial–interglacial 100 kyr cycle.

slope and therefore the deuterium excess. A physically satisfying interpretation of the water isotopes ($\delta^{18}O$, δD) and the deuterium excess therefore needs a synchronous computation of SST and site-temperature changes. In fact we are dealing with two unknowns (SST and site temperature) and two measured variables ($\delta^{18}O/\delta D$ and d) in a weakly non-linear system. The system is in fact only weakly non-linear because SST changes are second order for the interpretation of the water isotope records and *vice versa* site temperature (T_{Site}) changes are second order for the interpretation of the deuterium excess in terms of SST changes.

Figure 36.1 shows the intriguing consequences of the simultaneous computation of SSTs and T_{Site} by using a linearized version of a Rayleigh distillation model (Vimeux *et al.*, 2002). The deuterium excess and therefore the source temperatures deduced from the excess are dominated by a strong obliquity periodicity of about 40 kyr. Vimeux *et al.* (2002) argue that obliquity insolation modulates the temperature gradient between high- and mid-latitudes and affects not only directly the SSTs but also the intensity of meridional vapour transport. The 'classic' calibration of source temperatures, T_d, assumes that evaporation conditions can be deduced from the deuterium excess alone $T_d = d/1.3°C\ ‰^{-1}$. Conversely T_{Source} includes the effects of changing condensation site temperatures (see Fig. 36.1). The corrections due to the latter effect amount to up to 50% of the original signal (thus in fact more than just second order) and increases T_{Source} changes during the last glacial to as much as 6°C. Such large changes are very unlikely for a single individual source oceanic region and are not observed in marine sediments for the time period covered by the Vostok core. Therefore the authors rather argue for insolation driven geographical redistribution of source areas (high versus mid-latitudes) combined with real SST changes ending up in an effective temperature change of 6°C.

Antarctic long-term excess records put strong constraints on the hydrological cycle in the Southern Hemisphere and its functioning in the past. They represent therefore a valuable and critical test for our understanding of past climate changes.

PART 3

Changing glaciers and their role in earth surface evolution

THIRTY-SEVEN

Changing glaciers and their role in earth surface evolution

David Sugden

Institute of Geography, School of GeoSciences, University of Edinburgh, Edinburgh EH8 9XP, UK

37.1 Introduction

A report card on the progress of glacial studies over the past 40 yr or so might read 'encouraging progress, but surprisingly large gaps in knowledge remain'. There has been a host of activity in a number of disciplines studying glaciers and many important discoveries that have changed the way we view the world. However, we still cannot tell a politician whether global warming will lead to continuous warming or plunge us into a new Ice Age, or whether or not it will cause the West Antarctic Ice Sheet to disappear. We are equally uncertain about the role of glaciers on other planets and moons in our solar system.

To practitioners that cut their teeth in the 1960s the overriding concern at the time was the lack of communication between glaciologists, with their stress on theory, and field geomorphologists/geologists with their focus on description and evolution. Paterson laid down the challenge to fieldworkers with his famous quote that 'a mere handful of mathematical physicists, who may seldom set foot on a glacier, have contributed far more to the understanding of the subject than have a hundred measurers of ablation stakes or recorders of advances and retreats of glacier termini.' (Paterson, 1969, p. 4). Clarke (1987c) encapsulated the problem when he compared the changing fortunes of papers published in the *Journal of Glaciology* that contained only equations with those relying on maps. After decades of separation, he was able to identify the first paper with both! The danger of separation was the underlying rationale for the writing of *Glaciers and Landscape* (Sugden & John, 1976), as could be seen if a reader used the index to look up 'intrepid fieldworker' or 'armchair theorist'. The problems of separation were obvious to all. Strange glaciological processes were invoked to explain landforms, the historical Davisian paradigm dominated field interpretations with the cycle, *corrie glaciers–ice sheet–corrie glaciers,* seen as an agent of landscape denudation (Linton, 1963), while theoreticians lacked a body of quantitative empirical data with which to constrain theories.

37.2 Background to recent progress

The substantial progress that has been made in subsequent decades is well illustrated in the papers in this volume. One can recognize that progress has been stimulated by both technical developments that have increased the nature and areal coverage of data and the means to analyse them, as well as the emergence of the new theoretical structures of earth system science with its stress on interconnectedness.

Remote sensing has opened up new horizons on a revolutionary scale. Whereas fieldworkers in the 1960s relied on aerial photographs, often patchy in coverage, local and different in scale from place to place, it is now possible to view landscapes anywhere in the world at an unparalleled level of detail. We can monitor mass balance over the Antarctic and Greenland ice sheets (Vaughan and Reeh, this volume, Chapters 42 & 44) and on mountain glaciers (Naruse and Yao Tandong, this volume, Chapters 46 & 55). Such real-time monitoring allows current trends of glacier health to be assessed, for example, in the dynamic parts of West Antarctica (Wingham *et al.* 1998; Shepherd *et al.*, 2002). We can also test hypotheses on the beds of former mid-latitude ice sheets. For example, it is possible to ascertain the pattern of bed streamlining or transverse moraines on a continental scale (Andrews and Clark *et al.*, this volume, Chapters 40 & 50).

The advances in computing power are so complete that the full significance is best appreciated from a perspective of several decades. When running a model of the Laurentide Ice Sheet in 1975 in Boulder, Colorado, the requirement was an armful of punch cards and the key concern was not to trip up as you went to the computer centre each evening to load up a single run of the model. In the 1980s I can remember research students, relieved of punch cards, but still orientating their life around the night run of the model. Now we can run infinitely more complex models from the desk top in a matter of minutes. Although there are sceptics, one hopes that progress in understanding matches the technical advances. Perhaps the real value is that it is now possible to simulate ice sheets on the large spatial and temporal scales that are necessary to understand their behaviour. Indeed, it seems the only way in which one can carry out experiments on ice sheet behaviour, as is well illustrated in the case of Iceland (Hubbard, this volume, Chapter 22).

Advances in dating have revolutionized our subject. Our understanding of current ice sheets and their link to climate change has been transformed by the results of the analysis of ice cores and comparisons with ocean cores. And we are in the throes of another revolution, namely the use of cosmogenic nuclides in exposure age dating. For the first time it is possible to establish the rate of weathering of a rock surface, how long it has been exposed to the atmosphere, how long it has been buried by ice without disturbance and, indeed, even when it was covered. Such studies are revealing their potential in understanding the history of the East Antarctic Ice Sheet (Ivy-Ochs *et al.* 1995; Summerfield *et al.*, 1999), and the thickness and trajectory of the West Antarctic Ice Sheet over the last glacial cycle (Stone *et al.* 2003). In formerly glaciated areas, such studies are establishing the depth of erosion accomplished by valley glaciers in a glacial cycle and the great age and complex history of surfaces that have survived multiple glaciations (Fabel & Harbor, 1999; Fabel *et al.*, 2002). There is now the prospect that the former beds of mid-latitude ice sheets hold a detailed record of changes in past glacier thickness, ice dynamics and thermal regime.

The emergence of earth system science has stressed the importance of the interactions between the cryosphere and the lithosphere, atmosphere, biosphere and oceans. Feed-back between the various subsystems leads to non-linear behaviour and has brought a change in the approach to understanding. The profundity of this change is once again reinforced by a longer perspective. Classic theories, such as John Glen's Flow Law and John Nye's analysis of the mechanics of glacier flow, were a demonstration of the power of deterministic models in understanding glacier behaviour (Nye, 1952; Glen, 1955). They formed the basis of an approach which stressed the importance of equilibrium, whether it was glacier form and extent or the shape of a drumlin or glacier trough. So pervasive was the concept that I completely failed to recognize that it underpinned my attempt to model the Laurentide Ice Sheet, until it was pointed out to me by a kindly referee (Sugden, 1977). Appreciation of the importance of non-linear behaviour reduces the primacy of equilibrium as a concept. It draws attention to the importance of system trajectory and the possible limitations of assumptions of steady state. It stresses the importance of probabilistic as well as deterministic approaches. It highlights the importance of accurate measurements of the current state of a system and the need to identify thresholds of instability in system behaviour. Above all, the recognition of non-linear behaviour has changed the balance between fieldwork and theory. Whereas Paterson saw fieldwork as a side show, it is now clear that models need to be tightly constrained by empirical data if model predictions are to carry weight. At the very least a model needs to be based on well-constrained empirical data and to be able to replicate present and past behaviour.

Such arguments are exciting and humbling. They open up new vistas but also draw attention to how much we need to know. We need to know about system trajectory on a variety of time-scales, from tens of millions of years in the case of the Antarctic Ice Sheet, through 100,000 yr glacial cycles, millennial-scale Dansgaard–Oeschger events, tidal effects on glacier flow, to seconds in the case of stick-slip basal sliding. We also need to acknowledge the remarkable spatial variability in glacier behaviour. This might be the contrast between an ice stream and the adjacent ice sheet, or between mountain glaciers in the tropics and those in mid-latitudes (Kaser, this volume, Chapter 53). Or we can exploit the remarkable spatial heterogeneity of the beds of former mid-latitude ice sheets to understand so much more about their behaviour. This variability occurs at different scales and is potentially a rich archive reflecting past ice-sheet growth and decay. Finally, we can exploit the contrasts between glaciers on Earth and those on different planets and moons.

37.3 Variability over time

Progress in unravelling ice-sheet evolution can be illustrated by the observation that in the 1960s it was assumed that the Antarctic Ice Sheet was a child of the Pleistocene. Now it is believed that extensive ice built up in Antarctica as long as 34 Myr ago at the end of the Oligocene and that in the ensuing 14 Myr there was a phase of progressive global cooling culminating in a full-bodied Antarctic Ice Sheet in the middle Miocene around 15 Ma. The period of ice build-up began in cool temperate climates with Croll–Milankovitch cyclic glacier fluctuations coinciding first with beech forests and subsequently with tundra vegetation (Naish *et al.*, 2001; Sugden & Denton, 2004). The maximum ice sheet advanced to the outer continental shelf but subsequently, around 13.6 Ma, it withdrew to its present extent with limits in East Antarctica approximately on the present coast. The history of the ice sheet since the mid-Miocene to the present has been the centre of debate. Some have argued for a dynamic history of ice sheet growth separated by episodes of massive deglaciation (Webb *et al.*, 1984). Others have argued for essential stability of the ice sheet for over 13 Myr with only coastal glaciers and ice shelves responding to changing global sea levels (Denton *et al.*, 1993). The latter view is consistent with a view that the ice sheet has influenced local climate and ocean circulation to such an extent that the system has achieved a new state of stability (Huybrechts, 1993). Warmer temperatures are unable to increase melting significantly since overall temperatures are too far below the melting point; instead, warmer air brings more snowfall and causes glaciers to thicken. The case of the West Antarctic Ice Sheet may differ, because much of it is grounded below sea level and is thus susceptible to fluctuations in global sea level. Although there is no firm evidence that it has disappeared since it too formed in the early Miocene, there has been progressive thinning throughout the past 10,000 yr, apparently in response to sea-level rise following the loss of Northern Hemisphere ice sheets at the end of the last glaciation (Conway *et al.* 1999; Stone *et al.*, 2003).

In spite of the long history of study since the mid-19th century, the form and behaviour of the Northern Hemisphere ice sheets has been the centre of active debate. A landmark publication was *The Last Great Ice Sheets* in which a minimum and maximum model were put forwards (Denton & Hughes, 1981). This stimulated a new attack on the problem. It now appears that the Laurentide and Greenland ice sheets were extensive and covered North America on a scale equivalent to that of Antarctica today. The Scandinavian Ice Sheet too was extensive, covering the Barents Sea and extending as ice shelves into the Arctic. Ice in eastern Siberia, however, was limited by a lack of snow and glaciers were centred on mountain massifs. Overall, ice build-up, most of which was in the Northern Hemisphere, was equivalent to a lowering of global sea level by 120 m. Although mountain glaciers have been present in the Northern Hemisphere for millions of years, cycles of massive ice-sheet growth and decay every 100,000 yr are characteristic only of the past 500,000 yr or so.

Elaboration of this story of ice-sheet evolution over long timescales has raised many challenging questions. What was the distribution of land masses and ocean currents necessary to initiate ice sheet growth in Antarctica? What were the feed-back relationships between the growing ice sheet, ocean circulation and climate that caused ice to grow to the outer continental shelf? Why did it then retreat from its maximum extent to the present coast? Could it be that it eroded its offshore bed to such an extent that, during higher-frequency cycles of growth and decay, it could no longer advance into deep offshore water? Why did massive ice-sheet growth in the Northern Hemisphere occur only in the past 500,000 yr? What controlled the pace of ice-sheet growth and decay? Given the close correlation between insolation changes in the Northern Hemisphere and global ice-sheet growth and decay, what mechanisms transmit the signal worldwide?

A feature of recent decades has been the appreciation of the extreme variability of climate on time-scales finer than that of ice-sheet growth and decay. The ice-core records show abrupt changes in climate, especially Dansgaard–Oeschger events which represent a sharp warming of up to 10°C followed by a slower phase of cooling that recur every few thousand years (Hammer, this volume, Chapter 78). Such variations have drawn attention to possible mechanisms of change and the importance of establishing records of glacier fluctuations over the globe. Some workers argue that glacier fluctuations, including those on a millennial scale, are synchronous over the world (Denton *et al.*, 1999), thus arguing for a rapidly changing driver of change, such as the levels of carbon dioxide or water vapour in the atmosphere. Others stress that glaciers display antiphase behaviour on a millennial scale, thus arguing for oceanic drivers of change, such as the bipolar seesaw (Broecker, 1998; Clark *et al.*, 2002b). This latter mechanism involves the thermohaline circulation whereby warm conditions in the North Atlantic (as occurs now) draw heat from the Southern Hemisphere oceans. Conversely, a cold North Atlantic suppresses the thermohaline circulation and leads to warmer conditions in the Southern Ocean. Both situations can be driven from conditions in the north or south and imply antiphase behaviour of glaciers on a millennial scale. The debate is still alive and its firm solution, at least for the last glaciation, awaits new dating methods that have a resolution sufficient to unpick fluctuations on time-scales of centuries. In this context, there is much to be gained by the use of different approaches, for example, glacio-isostasy (Fleming, this volume, Chapter 45), or study of more recent glacial episodes, such as the Little Ice Age, where other dating methods are available (McKinzey, this volume, Chapter 54).

37.4 Spatial variability

There has been a tendency to see glacial erosion as a uniform process that can be compared, for example, with fluvial erosion. One of the realizations from the use of remote sensing over the beds of former ice sheets is just how spatially variable glacial erosion is. There are now numerous accounts of surfaces, commonly marked by river valley networks, tors and blockfields, which have survived inundation by an ice sheet without significant modification, apparently as a result of cold-based glacier ice that is unable to slide at the ice–rock interface. The distribution of such preserved land surfaces in the Transantarctic Mountains, northern Canada, east and northern Greenland and northern Scandinavia points to the importance of cold continental climatic conditions in creating conditions suitable for ice accumulation at temperatures below the pressure melting point (e.g. Kleman & Hättesrand, 1999; Sugden, 1978). Such a link with cold-based ice is reinforced by more local patterns whereby the subaerially weathered landscapes are preferentially preserved on uplands where the presence of thinner, diverging ice favours cold-based ice. This is particularly well displayed on Baffin Island, Arctic Canada, and in Marie Byrd Land in West Antarctica where erratics of the last glaciation overlie bedrock surfaces that originated much earlier and actually lie in relict weathering pits and on the surface of tors (Briner *et al.*, 2003; Sugden *et al.*, 2005). In these cases the use of cosmogenic isotope analysis reveals the contrasting age of erratics and bedrock.

Such protected rock surfaces are often cheek by jowl with landforms that reflect glacial moulding. The latter are most common at lower elevations, reflecting thicker, converging ice and are assumed to be a result of glacial erosion beneath warm-based ice that can slide at the ice–rock interface. The most extreme version of this landscape contrast is the landscape of selective glacial erosion that characterizes the plateau scenery of eastern Baffin Island and Labrador, eastern Greenland and the higher eastern hills of the former Scottish and Scandinavian ice sheets. In some situations the contrast is so sharp that one can sit on a relict surface and dangle one's feet over the glacially eroded trough.

In many low-lying shield areas of former ice sheets and in the warmer maritime sectors the landscape is worn extensively by glacier action. Such landscapes of areal scouring are common in western and central Scandinavia, western Scotland and in much of Canada and western Greenland. The blend of bare rock outcrops, moulded into streamlined hills or roches moutonnées, and bedrock basins is the result of glacier sliding, favoured by ice at the pressure melting point that permits sliding and erosion at the ice–rock interface. There has been a long debate as to how much erosion is achieved under such conditions. In support of relatively limited rates of erosion, one can point to the preservation of pre-glacial river valleys and patches of weathered bedrock, for example in the shield areas of Arctic Canada and Scandinavia. Indeed, old forestry maps of the river routes used to float logs in a glacially modified part of eastern Sweden clearly show the survival of the pre-existing river network, as well as locally disrupted areas where glacial erosion has achieved more of a transformation. On the other hand, analysis of the volume of offshore sediments of glacial erosion points to a greater depth of erosion by ice sheets (Andrews, this volume, Chapter 40).

One of the exciting developments in formerly glaciated areas resulting from the analysis of satellite imagery at the scale of an ice sheet is the recognition of contrasts in regional patterns of depositional landforms. Such studies in Arctic Canada and Scandinavia have revealed assemblages of streamlined landforms that mark the presence of former ice streams and they reveal remarkably abrupt margins (Stokes & Clark, 2001; Boulton *et al.*, 2003). By looking at cross-cutting relationships between different assemblages of features and their relationship to preserved landscapes, it is possible to reconstruct the flow dynamics of an ice sheet and some insight into basal thermal regime during build up and decay. As this line of enquiry is developed and combined with surface exposure dating, there is the prospect of a record of change over time that will be invaluable for constraining models of former ice sheet behaviour. Another exciting step is to link the terrestrial record of ice streams with side-scan radar surveys of their offshore continuations. The offshore evidence of glacial streamlining marking the presence of former ice streams is remarkable for its clarity and resolution (Canals *et al.*, 2000).

There is a stepped change in our potential to exploit the spatial variability of glacier forms and processes by engaging in extraterrestrial geomorphology. Studies of Mars have already established a theoretical flow model of current ice sheets (Nye, 2000), and high-resolution imaging shows rock glaciers and a variety of meltwater landforms (Baker, 2001; Lucchitta, 2001). Indeed there is a remarkable similarity between the rock glaciers of the Dry Valleys, Antarctica and those on Mars. New vistas open up with every new mission that deploys a new instrument and these present clear opportunities to apply and refine our terrestrial expertise.

37.5 Grand challenges

There is a risk of pomposity and arrogance in even contemplating a list of grand challenges. Everyone will have their own list. However, in the belief that it is sometimes useful and humbling to highlight what we do not know, here is my list.

1 The West Antarctic Ice Sheet stands out as a potential threat in that if it approaches a threshold of collapse, it could raise global sea level by 6 m and change the ocean and atmospheric circulation in the Southern Hemisphere. The signs that it has been on a trajectory of decline throughout the Holocene and that other parts of the ice sheet are thinning today could be significant. Immediate questions are: How secure and widespread is the evidence of Holocene decline? Is the Weddell Sea sector involved? How has the rate of change varied during the Holocene? What evidence is available from nunataks and the immediate offshore zone to improve our level of understanding? Is there a threshold of collapse? All these questions are answerable by a combination of field work and modelling.

2 The North Atlantic thermohaline circulation plays a fundamental role in initiating and/or modulating climate change. It is possible that global warming will lead to continued warming of the North Atlantic. It is also possible that an increase in the surface water flux in the Arctic is sufficient to close off the circulation and lead to cooling and a further Ice Age. Crucial unknowns are the volume of fresh water supplied by the Greenland Ice Sheet and the sensitivity of the locations at which it is incorporated in the thermohaline circulation. Here is a clear glaciological task that can be accomplished by a concerted effort.

3 Ice-sheet models are a sophisticated way of representing the behaviour of ice sheets and could be used for prediction if we had faith in them. At present there seem two major impediments. One is the lack of an effective sliding theory. Some would argue that the parameter is so important and knowledge is so limited that the predictive power of any current model is questionable. Another major problem is the lack of data about the morphology of the bed of current ice sheets. There are parts of Antarctica the size of western Europe with only a handful of depth soundings. Bearing in mind the importance of topography in determining the flow and thermal regime of an ice sheet, it is not easy to have faith in model predictions based on such sparse information, however sophisticated the glaciological components of the model.

4 There is a wealth of data on past glacier behaviour at the time the world changed from a glacial to interglacial mode between the onset of deglaciation ca. 17,500 yr ago and the start of the Holocene ca. 11,500 yr ago. High-resolution dating for this time span is hindered by the error margins associated with radiocarbon and surface-exposure age dating. In particular it is difficult to obtain firm evidence on which to test rival theories about leads, lags and synchrony of glacier fluctuations. Higher resolution dating over this period would firm up our understanding of the mechanisms of abrupt climate change.

5 The interdisciplinary approach that brought together glaciological theory and empirical evidence has proven its value over the past decades. Is it possible to detect a weakening of the link, perhaps because we are in danger of taking it for granted? There are currently two areas where some in the field community are using arguments that are controversial among glaciologists, namely the importance of thrusting in creating marginal moraines and the assumption that lateral meltwater flow indicates a cold-based ice margin. One suspects that discussion among the wider cross-disciplinary community would lead to deeper understanding on such issues.

6 Finally, are we ready to contribute fully to the interpretation of remote sensing data obtained by new extraterrestrial probes? Already it is clear that there is an exciting and challenging range of new glacial forms and processes. Not only are they fascinating in their own right, but they add a humbling perspective to our earth-bound glaciers.

THIRTY-EIGHT

Reconstruction of palaeo-ice sheets—inversion of their glacial geomorphological record

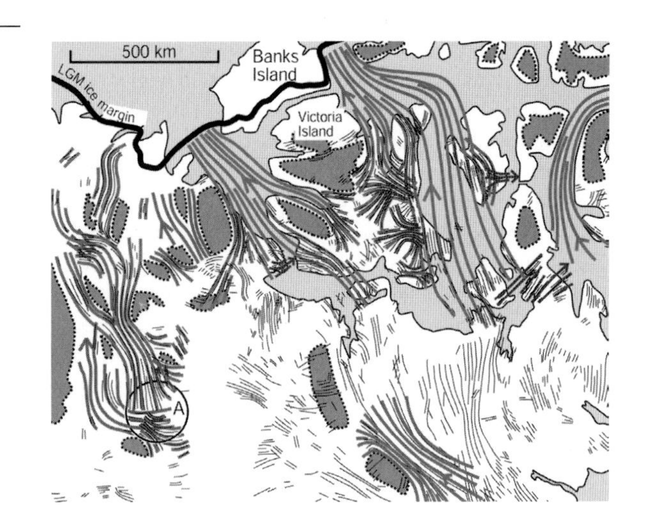

Johan Kleman, Clas Hättestrand, Arjen P. Stroeven, Krister N. Jansson, Hernán De Angelis and Ingmar Borgström

Department of Physical Geography and Quaternary Geology, Stockholm University, Stockholm, Sweden

38.1 Introduction

The former presence and extent of North American and Fennoscandian ice sheets was first documented in the middle-to-late 19th century using field evidence such as striae, marginal moraines, erratics and the extent of till sheets (Torell, 1873; Chamberlain, 1895; Tanner, 1914). A second phase, using landform mapping from aerial photographs as a prime tool, started in earnest in the 1950s and provided the spatial information for a reasonably robust picture of post-LGM (Last Glacial Maximum) evolution and the deglaciation pattern (Prest *et al.*, 1968). In a third phase, advancement of radiocarbon dating led to a refinement of the temporal framework of the ice sheets (Denton & Hughes, 1981; Dyke & Prest, 1987a). This work has since progressed further so that we today have detailed information of the deglaciation pattern and chronology for many formerly glaciated areas (Ehlers & Gibbard, 2004).

Parallel to the field-based ice-sheet reconstructions, analyses using numerical ice-sheet modelling, following the explosive development of computer facilities in the 1970s, have led to a major leap in our understanding of ice-sheet dynamics and the formulation of new research directions (e.g. Budd & Jenssen, 1975; Andrews & Mahaffy, 1976; Oerlemans, 1982; Huybrechts & T'siobbel, 1995; Clark *et al.*, 1996b; Payne & Dongelmans, 1997; Marshall *et al.*, 2000; Boulton *et al.*, 2001c). Modelling efforts rely on geomorphologically based reconstructions for a validation and evaluation of simulated ice-sheet topology and history. Modellers typically ascertain model performance and plausibility by crudely checking model output against published map data. A similar need for geomorphologically based spatial models exists for the interpretation of the multitude of dates related to ice-sheet extent. Stratigraphical (point) data can be used efficiently only when analysed in a spatial context provided by ice-flow indicators, marginal moraines, etc.

It is of crucial importance, in this respect, that the geomorphology of a particular region has been analysed in a conceptual framework embracing recent advances in ice-sheet dynamics, particularly with regard to basal thermal zonation, subglacial hydrology, evidence for fast ice flow (palaeo-ice streams), and an appreciation of the length of time represented by the glacial landform record. We refer to this conceptual framework as the process of inversion (Kleman & Borgström, 1996), i.e. extraction of ice-sheet properties from glacial geomorphology and geology.

Here we discuss the nature of the glacial inversion problem, which is concerned with the deciphering of ice-sheet evolution through time and therefore quite different from the sphere of the genetic problem, i.e. deciphering of the processes by which particular landforms are created. In essence the inversion problem can be considered as being five-dimensional, i.e. composed of three-dimensional space, time and process components. A glacial inversion model is a set of assumptions, a genetically based classification scheme for landform systems that allows palaeoglaciological inferences from the glacial landform archive, and a procedural outline for a stepwise and internally consistent data handling, as well as the integration of absolute chronological data.

The purpose of this article is to discuss the inherent nature of the ice-sheet-geomorphology inversion problem and, based on recent advances in the interpretation of glacial landforms and landscapes, suggest a new working model for palaeoglaciology on the ice-sheet-scale.

The presented inversion model rests on a set of assumptions concerning the time–space domains in which various glacial landforms are created. We emphasize two cases that previously have not received attention in proportion to their importance, frozen-bed preservation of older landscapes, and the recent discovery of major ice-stream networks in previously glaciated areas. The main tool in the inversion models is the recognition of *swarms*, which are the cartographical representation of coherent directional landform systems that can be traced to particular formative conditions. We recognize three event or flow types as being the minimum necessary to capture real-world complexity. These are the deglacial envelope, event swarms and ice-stream swarms.

Meltwater landforms (eskers, meltwater channels, traces of glacial lakes) form a data layer that can and should be used independently of glacial lineations for tracing the retreat pattern.

Two examples, from northern Fennoscandia and Keewatin, show that the outlined inversion approach can be applied successfully, and in both cases reveal landform systems that can constrain modelling of glacial events far pre-dating the LGM.

38.1.1 Methodological approaches

Previous attempts to use an inversion protocol to deduce ice-sheet properties at the ice-sheet scale reveal important differences in methodology and underlying genetic assumptions. Here we review four studies to illustrate these differences (Plates 38.1 & 38.2).

The Boulton *et al.* (1985) study (Plate 38.1a) is fundamentally a geomorphologically driven reconstruction of the Laurentide Ice Sheet, with an important modelling component. It is based on the assumption that coherent ice-sheet-wide patterns of glacial lineations were formed during the decay of the ice sheet because they reflect near-marginal ice-flow conditions. Pockets of deviating and sometimes cross-cutting glacial lineations were noted, but were disregarded as anomalies in the actual reconstruction. This pioneering study, however, does not fully recognize the age span of the landform record and force-fits apparently relict lineation patterns into the retreat pattern, with the important consequences of a partly corrupt decay pattern and the discard of important information pertaining to pre-deglacial ice-sheet configurations.

Dyke & Prest (1987a) present a combined geomorphological and dating-driven reconstruction of the post-LGM evolution of the Laurentide Ice Sheet (Plate 38.1b), and it is still a benchmark paper. Their chronology was based on the radiocarbon dates available at the time and underscores the importance of the publication. However, the robustness of the palaeoglaciological reconstruction, i.e. the evolution of flow pattern and the locations of ice divides and dispersal centres, is more debatable. Their use of glacial lineation patterns closely resembles that of Boulton *et al.* (1985) and is adequate when tracing the decay pattern of ice sheets. In contrast to the three other papers discussed here, neither geomorphological method nor inversion assumptions were clearly stated.

The Boulton & Clark (1990a,b) approach differed fundamentally from Boulton *et al.* (1985), because the glacial lineation record (Plate 38.1c) is seen as an amalgamation of discrete patches of glacial lineations (flow sets) formed at different times during the last glacial cycle. Deciphering the relative age of cross-cutting glacial lineations, using Landsat MSS imagery as the primary data source and aerial photographs as complementary information at key sites where flow sets intersect, allowed Boulton & Clark (1990a,b) to reconstruct the Laurentide Ice Sheet evolution as a stack of events, marked by major shifts in the location of dispersal centres. No effort was made to systematically trace the deglaciation pattern, and meltwater landforms (eskers, marginal channels and glacial lake traces) were not used in the reconstruction.

Kleman *et al.* (1997) present a geomorphology-driven reconstruction of the Fennoscandian Ice Sheet evolution through the last glacial cycle that is methodologically based on the conceptual inversion model of Kleman & Borgström (1996). It essentially combines the approaches of Boulton *et al.* (1985) and Boulton & Clark (1990a,b) with the addition of two important new elements: the explicit consideration of basal thermal zonation and the treatment of lineation patterns and meltwater patterns as two separate entities (Plate 38.1d).

Plate 38.2 visualizes the differences between the four works discussed, focusing on particular time–space domains and data types used in the inversion procedures. From Plate 38.2 it is apparent that no study yet published fully utilizes the three prime data sources of extramarginal datings, landforms reflecting the decay stage and overprinted glacial lineations (and other stratigraphically derived directional indicators) that shed light on older glacial events. A shortcoming shared by all four works is that little attention is paid to ice streams, and the rapid (order of magnitudes faster) configuration changes that may occur in ice-stream networks, as compared with sheet-flow situations.

38.1.2 The nature of the inversion problem

The glacial inversion problem is quite different from the genetic problem, i.e. deciphering of the processes by which, and the conditions under which, particular landforms are created (Plate 38.3). Most of our direct process knowledge pertains to marginal landforms such as end moraines, meltwater channels, proximal glaciofluvial deposits, etc. Observations of subglacial processes are restricted to marginal or thin-ice (<100 m) situations or rely on borehole or geophysical data where either the three-dimensional spatial context or the information content and resolution are extremely restricted. Hence, the landforms that are the most important in the inversion context, i.e. glacial lineations and ribbed moraine, are the least understood in terms of genetic processes. Yet, answers to genetic questions are necessary as basic assumptions in inversion models.

The formative processes for individual landforms have attracted considerable attention. For example, linkages between particular marginal landform assemblages and glacier dynamics have been established (e.g. Sharp, 1988; Krüger, 1993; Hart, 1995a; Evans *et al.*, 1999; Glasser & Hambrey, 2002). The understanding of linkages between interior ice-sheet dynamics and particular subglacial landform assemblages, which is critical for ice-sheet-scale palaeoglaciological reconstructions, has also experienced significant progress in the past decade (e.g. Kleman, 1992; Dyke, 1993; Clark, 1993, 1994, 1999; Kleman & Borgström, 1994; Hättestrand & Kleman, 1999; Kleman & Hättestrand, 1999).

The problem of genesis of glacial landforms can, at least in theory, be approached in a manner similar to other laboratory sciences, with the main problems being related to practical and logistical considerations and the extreme hostility and inaccessibility of the subglacial environment. The glacial inversion problem, on the other hand, is a reconstruction problem based on proxy data of phenomenal richness and complexity. The sheer size of glaciated areas requires that literally hundreds of independent data sources be utilized, if credibility in terms of space, time and glaciological inference is to be achieved. These circumstances require that any full-scale ice-sheet reconstruction has to be performed in a stringent manner in order to be trustworthy.

38.2 Key considerations in designing an inversion model

Six aspects are critical in geomorphology-based reconstructions at the ice-sheet scale. This is because they place different and higher demands on the methods used than those required in local to regional-scale reconstructions:

38.2.1 Data volume and data reduction

When working at the ice-sheet scale, a necessary step in the analysis is data reduction, meaning aggregation of thousands of individual landforms, striae and till-fabric analyses into a modest number of coherent 'packages', where each package reflects a flow system that is spatially and temporally coherent. The data reduction is far more critical in subcontinental-scale reconstructions than in local–regional reconstructions, owing to the amount and complexity of the data, and also the presence of significant within-system age gradients, which increase in importance with the size of the system studied. The most appropriate way to achieve data reduction is through a landscape-level classification scheme, where main types of glacial landscapes are classified on the basis of genetic conditions. These landscapes can then be grouped together based on linkages in the time–space domain. Only if such a classification system is based on genetic conditions and time–space domains can inversion from geomorphology and geology into palaeoglaciological parameters, such as basal thermal regime, flow pattern and configuration (palaeogeography), be performed.

38.2.2 Mapping practice

The basic mapping that is typically available as input for work at the ice-sheet scale comprises information at different levels of resolution. Eskers are generally marked on maps as individual landforms, whereas glacial lineations usually are marked in a generalized way emphasizing the orientation, to a lesser extent density, and only rarely the size of individual glacial lineations. Marginal moraines are likewise generalized, with discontinuous moraine belts or closely spaced ridges aggregated to generalized map symbols. Marginal meltwater channels represent a special problem, in that they are often of small size and occur in complex systems. Hence, they are often not represented on small-scale or overview maps. A proper analysis of deglaciation pattern based on channels will also require a detailed topographic rendition or digital terrain model to provide the necessary topographic context. The same holds true for glacial-lake shorelines in topographically complex terrain (Jansson, 2003). In reality, ice-sheet-scale inversion projects will to some extent rely on diverse mapping, with inherent differences in classification and cartographical principles. Any realistic inversion model will have to embrace a varying standard of map input and such differences have to be identified, carried through and compensated for in the entire reconstruction procedure.

38.2.3 Time and temporal gradients

When the size of the area analysed increases, the importance of intrasystem age differences increases dramatically. In local to regional-scale (<50 km) glacial geology there is usually little reason to consider this time-transgressiveness, and glacial history is typically described in terms of a relative age-sequence of flow directions (event sequence). However, for continental-scale, apparently continuous, glacial lineation systems it is reasonable to assume that the glacial lineations were formed relatively close inside the retreating margin. They witness, consequently, massive reorganizations of the ice-sheet margin and the location of its dispersal centre. The logical conclusions are that a 'continuity line' that traces glacial lineations in the up-glacier direction does not necessarily reflect a true flowline at a specific time and that realistic inversion models need to include a deconvolution procedure for time-transgressive systems. The latter conclusion follows from the argument that time-transgressive map representations, for example a 1000 km-long deglaciation landform system, cannot be used for validation of time-slice output from numerical ice-sheet models.

38.2.4 Genetic conditions of different landform system types

The purpose of Plate 38.4 is to visualize some of the complexity and variation that exists on the landscape level, which necessitates the formulation and implementation of a glaciologically sound classification scheme at this level in inversion procedures.

Any credible landscape classification system has to address adequately the very wide range of real-world glacial landscapes illustrated above, and make the correct linkages to the glaciological conditions (or evolutionary chains) responsible for a particular landscape type.

We also note that the time perspective will differ fundamentally between landscape types, with a long time axis being available in frozen-bed core-area landscapes and deposition-dominated marginal landscapes. On the other hand, the time perspective will be short in erosion-dominated landscapes, and particularly in ice-stream corridors.

38.2.5 Cold-based glaciation

Two major problems in ice-sheet reconstruction are how to interpret glaciated landscapes that show little or no direct evidence for glaciation and how to treat landscapes which contain old glacial landforms but little or no evidence for younger glacial events. It is now generally acknowledged that dry-bed conditions (ice sheet frozen to substrate) are the probable explanation for the inability of ice sheets to produce glacial lineations and for the preservation of older large- and small-scale morphology despite prolonged periods of ice cover (Plate 38.5; Sollid & Sørbel, 1984; Lagerbäck & Robertsson, 1988, Kleman, 1992, 1994; Dyke, 1993; Kleman & Borgström, 1994; Kleman & Stroeven, 1997; Hättestrand & Stroeven, 2002). Although cold-based glaciers are known to transport subglacial debris and are suggested to be capable of eroding their beds (Cuffey *et al.*, 2000b), the rate of erosion/deposition is many orders of magnitude lower than for warm-based glaciers. However, cold-based ice marginal areas still produce abundant meltwater during decay phases. These conditions set the stage for the formation of a conspicuous landform assemblage entirely dominated by meltwater landforms related to marginal and extramarginal drainage, but lacking subglacial meltwater traces such as eskers. The tracing of the spatial pattern of deglaciation over such 'relict surfaces' has to be based on the elevation and location of marginal channels, ice-dammed lakes and spillways (Borgström, 1989; Jansson, 2003), because 'conventional' glacial landforms may be lacking altogether or, even more problematic, be misleading because they formed long before final deglaciation.

We also note that landscapes of areal scouring, which are generally considered to indicate conditions of fast and/or highly erosive ice flow, may be of widely varying age. There is no justification for assuming a priori that they reflect such conditions under the last ice sheet. Most classic scouring landscapes occur in Precambrian bedrock areas, where resistant bedrock, a glacially deranged fluvial drainage pattern and a thin soil cover, all contribute towards an almost stagnant morphological development under subaerial conditions. Hence, very little morphological change through basin in-fill and fluvial erosion is likely to occur during a typical 10 kyr interglacial period. It will, consequently, be very difficult to distinguish morphologically a recently formed scouring landscape from a scouring landscape formed during an earlier glacial cycle and which experienced interglacial conditions and intervening dry-bed glacial conditions. Such discrimination will require field inspection and, probably, will involve the application of cosmogenic dating (Fabel *et al.*, 2002; Stroeven *et al.*, 2002c).

38.2.6 Integration of chronological data into geomorphology-based inversion models

Thousands of radiocarbon datings collected over the past 50 yr form the backbone of our dating constraints on ice-sheet evolution during the last glacial cycle. This data base is phenomenally rich (e.g. Denton & Hughes, 1981; Dyke & Prest, 1987a; Ehlers & Gibbard, 2004) and well constrains the post-LGM shrinkage of the Laurentide Ice Sheet and the Fennoscandian Ice Sheet and, with less precision, the Cordilleran and British ice sheets.

However, in light of the aim to resolve glacial evolution through a full glacial cycle, there are substantial inherent limitations, primarily the ca. 40 kyr maximum limit for reliable ^{14}C dating and the fact that extremely few datings give any direct evidence about subglacial events. In essence, the information provided directly by a radiocarbon date is that ice was not present at a specific site at a specific point in time (Plate 38.6). All other inferences, such as proximity to the ice margin, result from stratigraphical interpretation of the sites, and inferences made from that.

38.3 Ice streams

Areas dominated by sheet flow, e.g. the southwestern shield-area parts of the Keewatin and Quebéc domes, display 'classic' glacial landscapes comprising abundant glacial lineations, eskers aligned to lineations, ribbed moraine and a scarcity of sharp lateral contrasts in lineation length and development. Older ice-flow directions may be manifested by overridden large 'ghost' drumlins or older cross-cut striae.

Landscapes that probably are sites of former ice-stream webs or networks show fundamental differences from sheet-flow landscapes (Clark, 1999; Stokes & Clark, 2001) in that sharp lateral boundaries of lineation swarms are common, eskers are few and often not aligned to flow traces, and ribbed moraine is lacking or present only near former ice-stream heads. Older flow directions are mainly manifested as beheaded lineation patches *outside* ice-stream corridors (Plate 38.7). Topographical control on flow direction generally appears to have been strong in this type of landscape. In analogy with evidence from the present-day ice streams at Siple Coast in Antarctica, we interpret such landscapes to reflect networks of ice streams. Typical for these are that they display upstream tributaries into sheet-flow areas, frozen-based interstream ridges, head convergences and variability in length, width and velocity on short time-scales (Hodge & Doppelhammer, 1996; Bamber *et al.*, 2000a; Gades *et al.*, 2000).

Reconstruction of event sequences in such areas necessitates an approach significantly different from that used in sheet-flow areas. In sheet-flow situations, regional cross-cutting and overprinting of old lineation systems is seen as reflecting slow migration of dispersal centres, in response to climate-controlled mass-balance distribution changes on the ice-sheet surface. In ice-stream-dominated landscapes, on the other hand, disjunct and beheaded lineation patches probably reflect rapid onset of ice streaming at a particular location, and consequent drawdown in individual ice drainage basins. The time-scale for major directional changes may be in the order of a few hundred years. Inferences about overall ice configuration changes must in such areas be based on the collective evidence from several ice streams. For instance, the onset of the four major Finnish ice streams (Kleman *et al.*, 1997; Boulton *et al.*, 2001c) in Late-glacial time most probably caused a major westward shift in the ice-divide position of the Fennoscandian Ice Sheet. This conclusion would be rather uncertain if based only on the evidence from one ice-stream corridor.

Our investigations of palaeo-ice streams, hitherto, suggest that the term *ice stream*, although coherent and meaningful from a process point of view, in reality embraces a range of different fast-flow situations. This situation necessitates a subdivision into functionally different types of ice streams in the inversion context. We have observed and mapped the traces of a large number of palaeo-ice streams, varying widely in size and glaciological context, and with topographic guidance varying between strong and insignificant. In addition, we have uncovered evidence for small ephemeral ice streams in the Canadian Arctic during the deglaciation. They indicate flow patterns that are very different from older apparently semi-stable ice streams in the respective areas. We hold the view that a relevant classification should comprise at least three different types of ice streams.

1 *Topographically governed ice streams* are constrained by topography and fixed in space but variable in time (Marshall *et al.*, 1996; Stokes, 2000; Kaplan *et al.*, 2001). The Hudson Strait, Laurentian Channel and Norwegian Channel ice streams are prime candidates for this type.
2 *Transient rigid-bed ice streams* form without topographical or substratum control when thawed spots start to develop under a largely cold-based ice sheet, which then finds itself with a steeper profile than the reduced bed traction can sustain. The Dubawnt ice stream (Stokes, 2000) is the prime candidate for this type.
3 *Ephemeral ice streams* probably develop in response to a rapid calving and break-up of ice in adjacent marine areas. Prime candidates are the east-trending ice stream on Prince of Wales Island (Dyke *et al.*, 1992) and the small Cap Krusenstern ice stream, the traces of which overprint large glacial lineations of the Amundsen Gulf ice stream of probable LGM age.

The development of appropriate inversion methods for areas dominated by ice streams is only in its infancy, although important progress has been achieved (Stokes & Clark, 2001).

38.4 An inversion model

The procedure described here to reconstruct past ice sheets from the glacial geomorphological record, combined with the stratigraphical record, builds on previous inversion models of Kleman & Borgström (1996) and Kleman *et al.* (1997). The inversion model comprises a classification system for glacial landform assemblages and a stepwise deciphering procedure. It should be noted that it has to be adapted to the particular area and time frame in question. The following *assumptions* are used in the model (see further discussion under the key considerations described above):

1 the basic control on landform creation, preservation and destruction is the location of the phase boundary between water and ice, separating frozen from thawed material, at or under the ice-sheet base (i.e. basal temperature);
2 basal sliding requires a thawed bed;
3 glacial lineations can form only if basal sliding occurs;
4 glacial lineations are created aligned parallel to local ice-flow directions and perpendicular to the ice-sheet surface contours at the time of creation;
5 frozen-bed conditions inhibit the reshaping of the subglacial landscape;
6 regional deglaciation is *always* accompanied by the creation of a spatially coherent but metachronous system of meltwater features, such as meltwater channels, eskers and glacial lake traces;
7 eskers are formed in an inward-transgressive fashion inside a retreating ice front (Norman, 1938; De Geer, 1940; Hebrand & Åmark, 1989; Bolduc, 1992);
8 meltwater channels will form the major landform record during frozen-bed deglaciations, whereas eskers are typically lacking under these conditions.

The main analytical components used in the inversion model are called *swarms*. These are temporary tools in the inversion model and represent glacial landform systems, or landform sets, with similar morphological characteristics. The use of *swarm* serves the purpose of data reduction, because each swarm is a simplified and spatially delineated map representation of many individual landforms. Therefore, they allow relative chronologies to be applied to a manageable number of cartographical units. Coherent swarms are defined on the basis of spatial continuity of landforms in a landform system, and the resemblance to a glaciologically plausible pattern, i.e. a minimum-complexity assumption.

38.4.1 Landscape-level classification

The swarm types we recognize are classified as follows (Plate 38.8).

1 The deglacial envelope. This swarm type subdivides into two components.
   - Wet-bed swarms defined by flow traces (typically drumlins and flutes) with aligned eskers. If the wet-based zone has resulted from thawing of previous cold-based conditions, these swarms will also hold ribbed moraines. These 'classic' swarms are interpreted to represent inward-transgressive formation of flow traces (Boulton *et al.*, 1985; Kleman *et al.*, 1997), which become preserved as new areas are successively deglaciated.
   - Frozen-bed deglaciation swarms. In these, the landform record consists solely of a see-through pattern of meltwater traces overprinted on a relict surface. Marginal channels are dominant and eskers are small or lacking. The relict surfaces may be former subaerially developed non-glacial landscapes or may contain (usually non-aligned) flow traces from an older glacial event.

   Note that in addition to the deglacial envelope from the last ice sheet there can exist older deglacial swarms in a particular area. This will be the result if the last deglaciation event was characterized by frozen-bed conditions, preserving any pre-existing landscapes, or only experienced a short period of warm-based conditions that did not completely reshape previous deglacial swarms.
2 'Event' swarms. These swarms are defined by landform systems with abundant flow traces (typically drumlins and flutes) but lacking aligned meltwater traces. In some cases they can be interpreted as the sites of former ice streams; in other cases they may have formed by slow sheet-flow far inside the margin. If such a swarm is defined by glacial lineations lacking a later overprint, the termination of lineation creation was probably caused by change to a frozen bed. If a swarm is defined by a low-frequency but regional occurrence of older

striations, no inferences can be made regarding basal temperature during subsequent events. On the spatial scale of individual roches moutonnées, lee-side protection and preservation is operational. Hence, old striae can be preserved, despite sustained wet-based ice flow from other directions.

3 Ice-stream swarms. These swarms represent events of enhanced ice flow, draining considerable amounts of ice. They typically have a strongly convergent head zone, probably reflecting the transition from sheet flow to stream flow. Ice-stream swarms associated with land-terminating flow often have a distinctive bottleneck pattern, with a divergent terminal zone, whereas water-terminating swarms lack the divergent-flow zone (Stokes & Clark, 1999). At present we use only one class for ice-stream landscapes, but as understanding progresses, it is likely that a more elaborate classification will have to be developed to adequately cover functionally different ice-stream types, as described above in the section on key considerations.

Landform systems may be formed during one single event or formed in a time-transgressive fashion. A mapped swarm is therefore the orthogonal projection of a system that may be sloping in the three-dimensional time–distance domain.

38.4.2 Deciphering procedure

As a first step in the deciphering procedure, swarms are spatially delineated and classified (Plate 38.9) according to the morphological criteria described above.

Where swarms cross-cut, relative chronologies are established, using striae and till fabric, as well as cross-cutting glacial lineations. The swarms are locally sorted into relative-age stacks, according to the relative chronologies, with the first result being the reconstruction of the deglacial envelope. Event swarms and fragments of old deglacial envelopes are then aggregated into groups forming glaciologically plausible, coherent flow patterns. For the reconstruction of time-slice flow patterns, the deglacial envelope is deconvoluted in terms of the changing ice sheet configurations during the deglaciation. This is done by going up-swarm and relating successively younger parts of the swarm to successively younger dispersal centre locations.

The time-slices are distributed into stadials on the basis of correlations with stratigraphical sequences of regional significance. Of particular use in this context are dated interstadial/interglacial sequences bracketing individual glacial sediment units that can be directly correlated to specific swarms. Examples of such glacial sediment units are glaciofluvial deposits (e.g., in eskers forming part of a deglaciation swarm), and till beds that have till fabric directions that can be directionally correlated with glacial lineations (drumlins, flutes) in a swarm.

38.4.3 The treatment of residuals and conflicting evidence

Residuals and unresolvable conflicts between different input data are inevitable given the scope of ice-sheet-scale reconstructions. Clusters of conformable observations, for example regarding relative age, are generally to be trusted, and given preference over individual observations. One of the crucial points is the data reduction: can we know for certain that, for example, spatially separated patches of glacial lineations that indicate roughly conformable ice-sheet patterns really pertain to the same event? Caution has to be exercised, because there is a significant chance that roughly similar flow patterns are repeated. Iteration is necessary; if conflicts arise it is necessary to backtrack and test another slightly different data reduction.

38.5 Examples of inversion model application

38.5.1 Northern Keewatin

The shield area west of Hudson Bay comprises a huge radial swarm of eskers that for the most part are directionally conformable with the glacial lineations in the area. It has long been recognized that Keewatin was one of the two primary retreat centres of the Laurentide Ice Sheet (Tyrell, 1897; Lee, 1959). However, there are at least two additional glacial landform systems that can shed light on the evolution of this sector of the Laurentide Ice Sheet. A 400-km-long divergent–convergent system of extremely elongated glacial lineations trending west-northwest occurs northwest of Dubawnt Lake (Aylsworth & Shilts, 1989b; Kleman & Borgström, 1996; Stokes, 2000). This lineation system disturbs the seemingly simple pattern of deglacial lineations and eskers, and has been interpreted as indicating a surge (Kleman & Borgström, 1996) or the site of an ice stream (Stokes, 2000; Stokes & Clark, 2001, 2003b). The flow direction of this swarm, the Dubawnt Ice Stream, is in general agreement with the deglacial flow pattern, suggesting that it was active during a brief phase during final deglaciation of the area.

A striking older landform stratum is formed by disjunct patches of degraded glacial lineations trending NE–SW in an arc-shaped corridor stretching northwest from Dubawnt Lake (Aylsworth & Shilts, 1989b), at right angles to the deglacial lineations and eskers. These lineations are discordant to maximum-stage and deglacial flow patterns and indicate an older dispersal centre in northern Keewatin or the central Arctic. The areal extent of evidence for this flow is actually larger, because observations by Lee (1959) of preserved striae from the north extend well into the area east of Dubawnt Lake. These striae are invariably older than those formed by Late Wisconsinan ice flow from the Keewatin Dome (or from a southern extension of the M'clintock Ice Divide (Dyke *et al.*, 1982)). Glacial lineations in the eastern part of the corridor are aligned with the westernmost north-striations reported by Lee (1959) and we interpret these to reflect the same ice-flow event. Aylsworth & Shilts (1989b) speculated that these anomalously orientated glacial lineations might reflect preservation in frozen-bed patches. Kleman & Hättestrand (1999) drew similar conclusions, based on a wider regional context. Plate 38.10 shows the three main landform systems in Keewatin.

38.5.2 Northern Fennoscandia

Three distinctly different glacial landscapes can be recognized in northern Fennoscandia. Because they cross-cut it is also possible to resolve the relative age chronology (Plate 38.11). Swarm 1, the oldest, is a classic deglaciation swarm, and includes drumlins, eskers and end moraines, formed by southeastward ice flow emanating from the Scandinavian Mountain Range. This landscape is locally overprinted and reshaped by later ice flow from the Bothnia Bay to the south/south-southeast (Swarm 2). Swarm 2 consists exclusively of glacial lineations, and hence classifies as an event swarm, and is interpreted to have formed during the LGM (Kleman *et al.*, 1997). A deglacial envelope (Swarm 3) overprints both Swarm 1 and 2, and forms a radial pattern leading in towards a final deglaciation centre in the eastern part of the Scandinavian Mountain Range. Most of this deglacial envelope is defined by wet-based flow traces (drumlins) and eskers. However, much of the eastern sector of the deglacial envelope essentially lacks warm-based ice-flow indicators, and lateral meltwater channels form a coherent pattern fanning out towards the east. This indicates that this sector was dominated by frozen-bed conditions during the last deglaciation.

38.6 Conclusion

1. The glacial geomorphological inversion problem is extremely complex and requires a systematic approach including landscape-level classification based on current glaciological understanding of formative conditions for different glacial landform assemblages. In essence this inversion problem is five-dimensional, i.e. composed of three-dimensional space, time and process components.
2. We recognize that three event or flow types are rquired as a minimum in order to capture real-world complexity. These are the deglacial envelope, event swarms and ice-stream swarms.
3. Meltwater landforms (eskers, meltwater channels, traces of glacial lakes) form a data layer that can and should be used independently of glacial lineations for tracing the retreat pattern.
4. Ice-stream webs also appear to have been extremely important in the former mid-latitude ice sheets. Deciphering of such landscapes involves the same basic techniques as sheet-flow landscapes, but the time-scale for major flow-pattern changes is much shorter than in sheet-flow landscapes.
5. Palaeo-ice streams may provide the best chronological constraints for subglacial events and evolution, because they are the synchronously formed vehicles that allow extramarginal datings on sediments or landforms to be linked to flow traces in the interior of the ice sheet.
6. Relict landscapes, i.e. landscapes lacking a clear glacial imprint or composed of ancient glacial landforms, are important components in formerly glaciated areas. These landscapes offer important insight into former basal conditions, and may preserve direct evidence of ice-sheet configurations and flow-patterns much older than the last glacial maximum.
7. Two examples, from northern Fennoscandia and Keewatin, show that the outlined inversion approach can be applied successfully.
8. The full potential of geomorphological inversion can be realized only when geomorphologists (three-dimensional space) directly collaborate with dating experts (time) and numerical ice-sheet modellers (glaciological processes). Only thereby are all five relevant dimensions of the problem fully addressed.

THIRTY-NINE

Reconstructing the pattern and style of deglaciation of Kola Peninsula, northeastern Fennoscandian Ice Sheet

Clas Hättestrand* and Chris D. Clark†

**Department of Physical Geography and Quaternary Geology, Stockholm University, Stockholm, Sweden*
†Department of Geography, The University of Sheffield, Sheffield S10 2TN, UK

The last deglaciation of the Fennoscandian Ice Sheet is fairly well constrained, particularly around the southern and eastern margin. The deglaciation of the northeastern sector of the ice sheet, including the Kola Peninsula, is much less well understood. The area is of particular interest for ice-sheet reconstructions, however, because of its position between three ice masses: the Fennoscandian Ice Sheet, the Barents Sea Ice Sheet and the White Sea ice lobes.

Kola Peninsula displays a rich and complex pattern of glacial landforms, and yet the ice-sheet history remains elusive. Even the main deglaciation pattern is uncertain, and little or no attention has been paid to prior glacial phases. Deglaciation reconstructions range from a coherent ice-marginal retreat towards the central part of the peninsula (Kleman *et al.*, 1997), to almost complete deglaciation of the central part of the Kola Peninsula while ice lobes still flanked it (Niemelä *et al.*, 1993). A third alternative has been the possible existence of a more or less independent Ponoy Ice Cap, centred on eastern Kola Peninsula in Late-glacial time (e.g. Ekman & Iljin, 1991). These discrepancies result partly from an incomplete coverage of detailed geomorphological maps over the area, which would allow a reconstruction of the spatial patterns of deglaciation, and partly from the lack of geochronological data, which would allow the advancement of time constraints on the ice-flow events.

To reconstruct the pattern and style of deglaciation, we have focused specifically on mapping the distribution and direction of meltwater channels and eskers (Fig. 39.1a). Cold-based ice sheets, which were common in arctic regions during the last glaciation, leave no subglacial landform record (e.g. Dyke, 1993), and only a sparse and patchy sedimentary record. However, all ice sheets leave a meltwater record during deglaciation. In areas where warm-based conditions dominate, eskers will constitute the main record of deglaciation, whereas lateral and marginal meltwater channels will form in areas where the ice sheet was predominantly cold-based.

In this study, we have used Landsat 7 ETM+ satellite images to map the glacial geomorphology over Kola Peninsula and adjacent areas in northwestern Russia. This imagery has a spatial resolution of 15 m, which allows both eskers and meltwater channels to be mapped accurately (Fig. 39.1b), at least in areas without extensive forest canopy. For some regions, primarily in and around the central Kola mountains, the satellite-image-based mapping has been complemented with interpretation of aerial photographs (ca. 1 m resolution) and field-based mapping in 2001 and 2002.

The distribution of eskers and meltwater channels is shown in Fig. 39.1a. Eskers are common along the peripheral and western parts of the peninsula, whereas meltwater channels are frequent in most areas, although particularly in the east-central Kola Peninsula and generally in the highlands. The directional pattern of these meltwater landforms provides a uniform picture of the ice retreat, indicating an ice sheet with a lobate ice margin that retreated westwards across the peninsula, without major deviations (Fig. 39.2). It also appears that the highest part of the ice

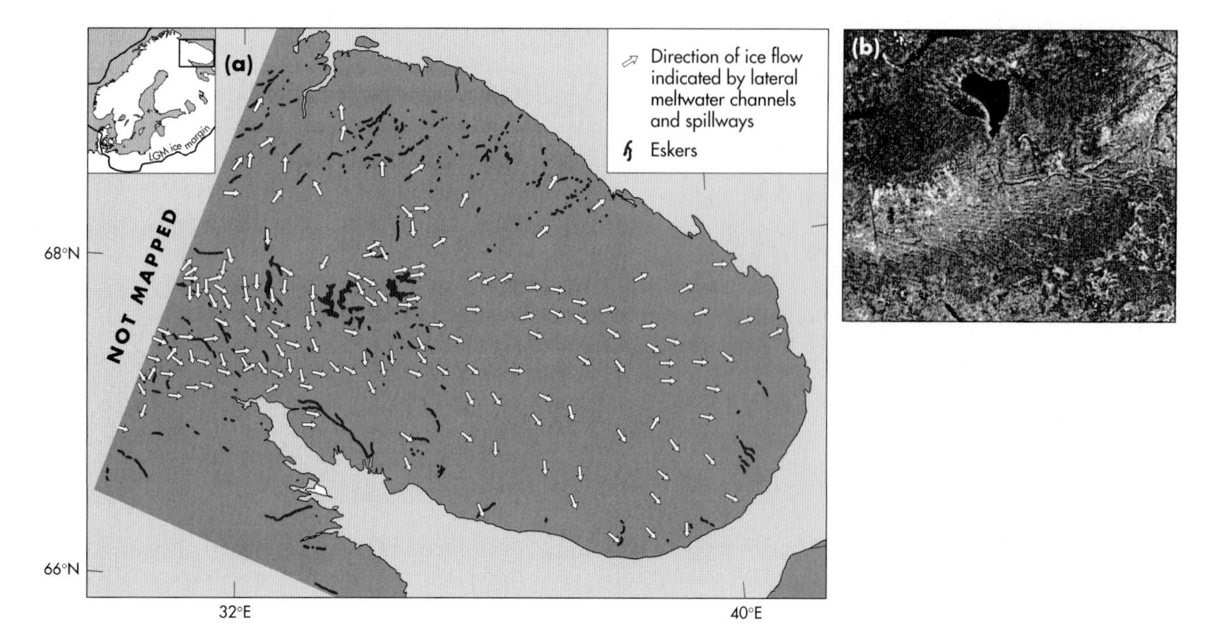

Figure 39.1 (a) Meltwater landforms on Kola Peninsula. The central Kola mountains are shown as dark areas in the terrain model. (b) Satellite image of a suite of lateral meltwater channels on the southern slope of Lovozero Mountains, central Kola Peninsula, indicating that the latest ice flow was towards the east, and that the margin successively backstepped from northeast to southwest. North is towards the top.

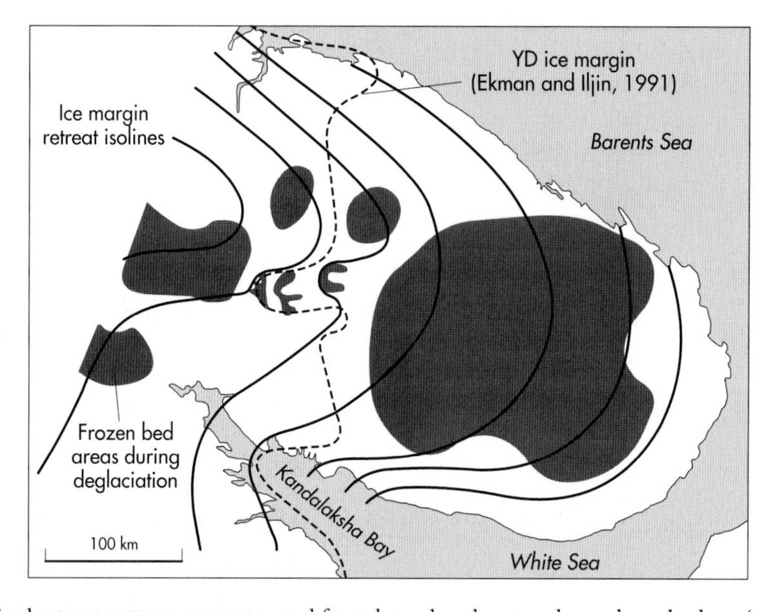

Figure 39.2 Ice-marginal retreat pattern reconstructed from lateral meltwater channels and eskers (see Fig. 39.1a). Frozen-bed areas are reconstructed based on the distribution of (i) lateral meltwater channels, indicating little or no penetration of meltwater through the ice, and (ii) subglacial landforms (e.g. eskers, drumlins), directly marking warm-based conditions.

mass was centred along the elevation axis of the peninsula, and ice flowed out radially to the east. Kandalaksha Bay probably acted as a calving bay, draining ice from the interior of the ice sheet, which caused an embayment in the ice margin.

There are no lateral meltwater channels, eskers, or marginal moraines, indicating ice flow from east to west on the central part of the peninsula. There are some (<10) extramarginal spillway channels going in a southwesterly direction just east of the central Kola mountains (Fig. 39.1a). Because they are not marginal, they do not by themselves prove ice-flow direction, and when set against the large number and consistent pattern of lateral channels documenting systematic west to east retreat, we relegate the significance of these and tentatively presume that they relate to local-scale intricacies of the ice margin. Hence, it appears very unlikely that an independent Ponoy Ice Cap existed during Late-glacial time. Such an ice sheet, even if cold-based (which would explain the lack of subglacial ice flow indicators, such as striae and drumlins, pertaining to this hypothetical ice cap), probably would have left at least some meltwater landforms when it melted away. We do not challenge the existence of ice cover over eastern Kola Peninsula in Late-glacial time, but demonstrate that ice in this area was part of the ordered retreat of the Fennoscandian Ice Sheet (Fig. 39.2) rather than an ice mass of its own (i.e. the hypothesized Ponoy Ice Cap).

In the westernmost part of the region, all meltwater channels indicate general ice flow towards the east during deglaciation, but we note an unusual complexity in the pattern. There appear to be two sets of channel direction: one generally towards the northeast and one towards the southeast. These may represent two different deglaciations, where the southeast channels probably relate to the last deglaciation, because of their directional alignment with eskers and the wider pattern of channels. Alternatively, they all belong to the last deglaciation and just reflect local topographically induced ice flow variations when the ice sheet thinned.

The spatial relationship between the distribution of lateral meltwater channels against eskers and other subglacial landforms also provides information on former subglacial temperature conditions. The shaded areas in Fig. 39.2 are areas with extensive sets of lateral meltwater channels and with a complete lack of subglacial landforms (eskers, drumlins, ribbed moraine, etc). We therefore interpret these areas as being dry-, and hence, cold-based during the last deglaciation.

Although the *spatial* deglaciation pattern can be deduced from the landform record, less is known about the *timing* of events. Between the central Kola mountains and the northern coast, a series of end moraines running north–south have been mapped (e.g. Ekman & Iljin, 1991; Niemelä *et al.*, 1993). These moraines have been suggested to be of Younger Dryas (YD) age, and based on these, Ekman & Iljin (1991) reconstructed an ice margin for that stage, running more or less north–south across the peninsula (Fig. 39.2). We note that their YD ice margin fits reasonably well with the ice margin reconstructed from the meltwater landform record in some places (in the coastal areas), whereas it is almost perpendicular in other areas. We argue, however, that any credible ice-marginal reconstruction has to be compatible with the meltwater landform record. Although a YD age of these moraines seems plausible from a theoretical point of view (the spatial continuity with the Karelian Salpausselkä ridges; Niemelä *et al.*, 1993), there is very little direct evidence. As this is the only major sector along the Fennoscandian Ice Sheet margin where the YD ice margin is not well established, resolving this issue is an important task for the understanding of YD ice-sheet dynamics.

In conclusion:

1 there was no separate Ponoy Ice Cap in Late-glacial time;
2 the Fennoscandian Ice Sheet in the northeastern sector, over Kola Peninsula, accomplished deglaciation as a simple ordered ice-margin retreat westwards with a large ice lobe centred along the elevation axis of the peninsula;
3 Kola Peninsula was largely deglaciated under cold-based conditions.

FORTY

The Laurentide Ice Sheet: a review of history and processes

John T. Andrews

INSTAAR and Department of Geological Sciences, Box 450, University of Colorado, Boulder, CO 80309, USA

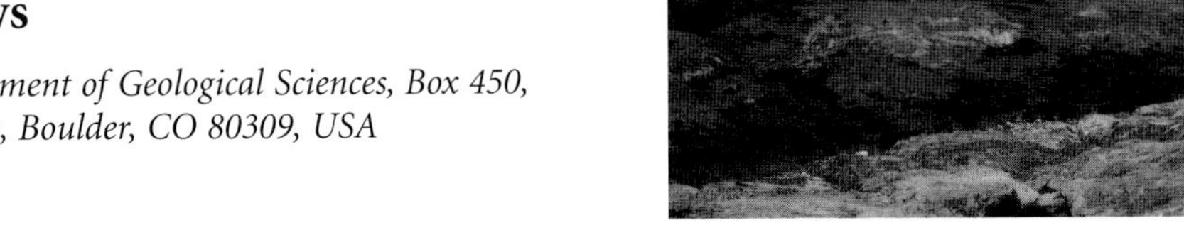

40.1 Introduction

For most individuals the Laurentide Ice Sheet (LIS) refers to Antarctic-size ice sheets that extended from the general area of the Great Lakes in the USA northward to the Canadian Arctic coast, and westward from the uplands and fjords of Baffin Island and Labrador to the foot of the Rocky Mountains (Fig. 40.1). However, the late V. Prest insisted that the term was applicable only to the ice sheet during the last glaciation. In this section I will adopt the more general usage for an ice sheet that had an

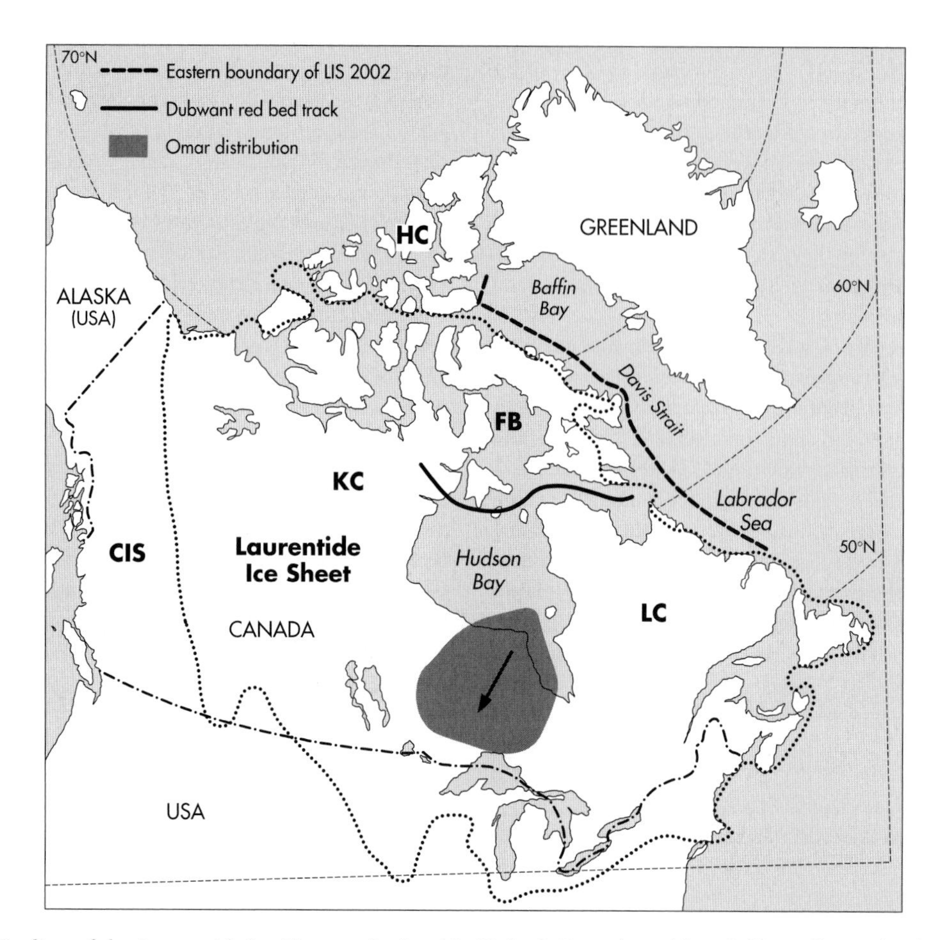

Figure 40.1 Outline of the Laurentide Ice Sheet as depicted in Dyke & Prest (1987b) and the outline today (Dyke *et al.*, 2002). The large bold lettering refers as follows: CIS, Cordilleran Ice Sheet; IIC, Innutian Ice Sheet; FB, Foxe Basin ice divide; LC, Labrador ice divide; KC, Keewatin ice divide. The positions are approximate and shifted through time. Also shown are examples of long distance erratic plumes (see text for discussion), specifically the Dubwant red bed sandstones and the 'omars' (Prest, 1990) of northern Ontario (oolitic jaspers and greywackes).

area of approximately $12 \times 10^6\,km^2$, a central thickness estimated to be in the range of 3–4 km, and sufficient mass to cause a worldwide fall in sea-level of around 70 m (Denton & Hughes, 1981; Paterson, 1972). Given the estimated volume of the ice sheet, computed in various ways but invariably resulting in an estimate ca. 70±m of equivalent sea-level, then the global history of sea level deduced from the variations in the $\delta^{18}O$ (Shackleton, 1973; Shackleton & Opdyke, 1973) must be dominated over the past 2.5 Myr by the growth and retreat of this ice sheet. The Antarctic Ice Sheet is about the same size as the former LIS but all indications are that it has been much more stable and has certainly not participated in the dramatic 100 and 41 kyr glacial cycles that are characteristic of the global ice-volume record (Raymo, 1992) (see Andrews, this volume, Chapter 21). Thus the LIS has to be considered a major component in the Cenozoic glaciations of our planet.

In this section: (i) I will briefly outline the history of exploration and thought that resulted in the concept of a vast North American Ice Sheet, the major unit of which is the LIS. I will then go on to address some critical issues in respect to this ice sheet, namely: (ii) the erosional history of the ice sheet; (iii) the nature of the ice sheet's bed and evidence for glacial transport; (iv) evidence for a complex growth and retreat. Because this is a brief survey readers should be aware of several major compilations that should be consulted for additional details. In particular the 1989 compendium *Quaternary Geology of Canada and Greenland* (Fulton, 1989) covers many regional and ice-sheet wide topics. Recent compilations of note include those by Dyke *et al* (2002) and Dyke (2004); these are especially concerned with the chronology of the last glaciation and deglaciation.

Enormous strides have been made in understanding the spatial and temporal variability in glaciological processes under and marginal to the LIS since my involvement, which started in 1959. In a general chronological sequence the major elements aiding this have been:

1 the aquisition of 1 : 50,000 aerial photographs and the provision of 1 : 250,000 topographic maps, starting in the late 1940s onward;
2 the development and usage of radiocarbon dating in the late 1950s;

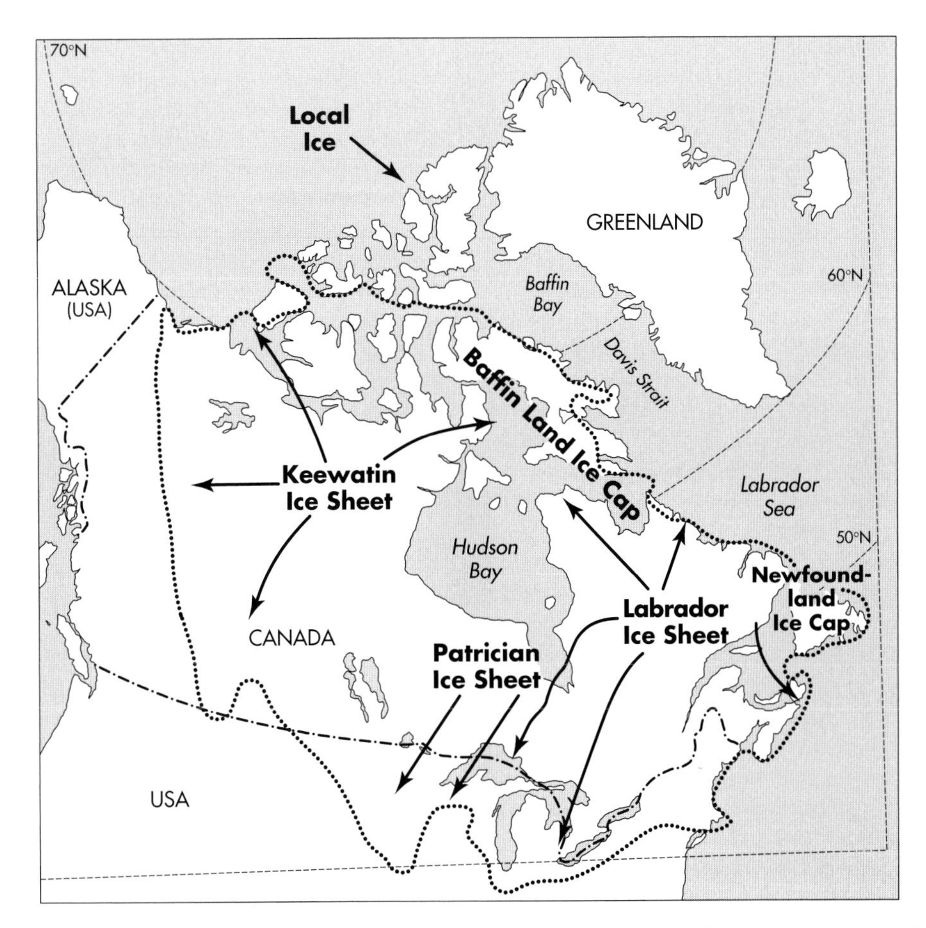

Figure 40.2 Redrawn map showing Tyrrell's concept of the ice sheet at the turn of the last century (1898). Note the overall similarities with the more modern concepts on Fig. 40.1.

3 the military and national infrastructure in the Canadian north (e.g. DEW Line and Weather Stations), which enabled field parties to be supported;
4 the LANDSAT imagery of the 1970s;
5 the interaction, starting in the late 1970s, between glacial studies on land and offshore marine studies on glaciated margins and adjacent deep-sea basins, largely promoted by the Bedford Institute of Oceanography;
6 the recent use of cosmogenic exposure dating and isotopic studies on provenance to aid in understanding ice-sheet chronology, erosion and transportation.

40.2 A brief history of changes in the concept of the LIS

By the 1890s a good deal of research had been carried out in the glaciated area of the USA and in southern Canada. State and Provincial Geological Surveys undertook much of this research. Of particular note were the long journeys (over several months) undertaken by Canadian geologists, often members of the Geological Survey of Canada. In retrospect the work of individuals such as Low, Tyrrell, Bell and others was amazing in its perception (Bell, 1884, 1898; Tyrell, 1898; Low, 1893; and see review in Prest, 1990), given the fact that much of the area north of 50°N latitude was scientifically largely unexplored. In 1898 Tyrrell produced a map of the former North America ice sheets (Fig. 40.2). This map is remarkably similar in many ways to present-day concepts of the LIS, that is, an ice sheet that like the Antarctic Ice Sheet today had several dispersal centres. Even 110 yr or so ago he showed that the ice sheet consisted of ice divides over Labrador, Foxe Basin, Keewatin and the Patrician Centre off southwest Hudson Bay. This notion prevailed until the late Professor Flint (1943) published a seminal paper in which he argued for the growth of an ice sheet from the mountains of the eastern Canadian Arctic to finally form a massive single-domed (~ ridge) ice sheet centred over Hudson Bay. Flint's model was not based *per se* on any specific field evidence but rather on a model for the growth and development of the ice sheet. This model continued to have adherents into the 1970s and 1980s (Denton & Hughes, 1981), although it is important to note that there is no erratic evidence on the east side of Hudson Bay for a flow from a supposed Hudson Bay centre (Figs 40.1 & 40.3).

An alternative model to that proposed by Flint was developed in the 1960s and onward. It was based initially more on a different view of how the onset of glacierization occurred and to a large degree it was offered on the basis both of field work and, as importantly, the first topographic maps (1:250,000 to

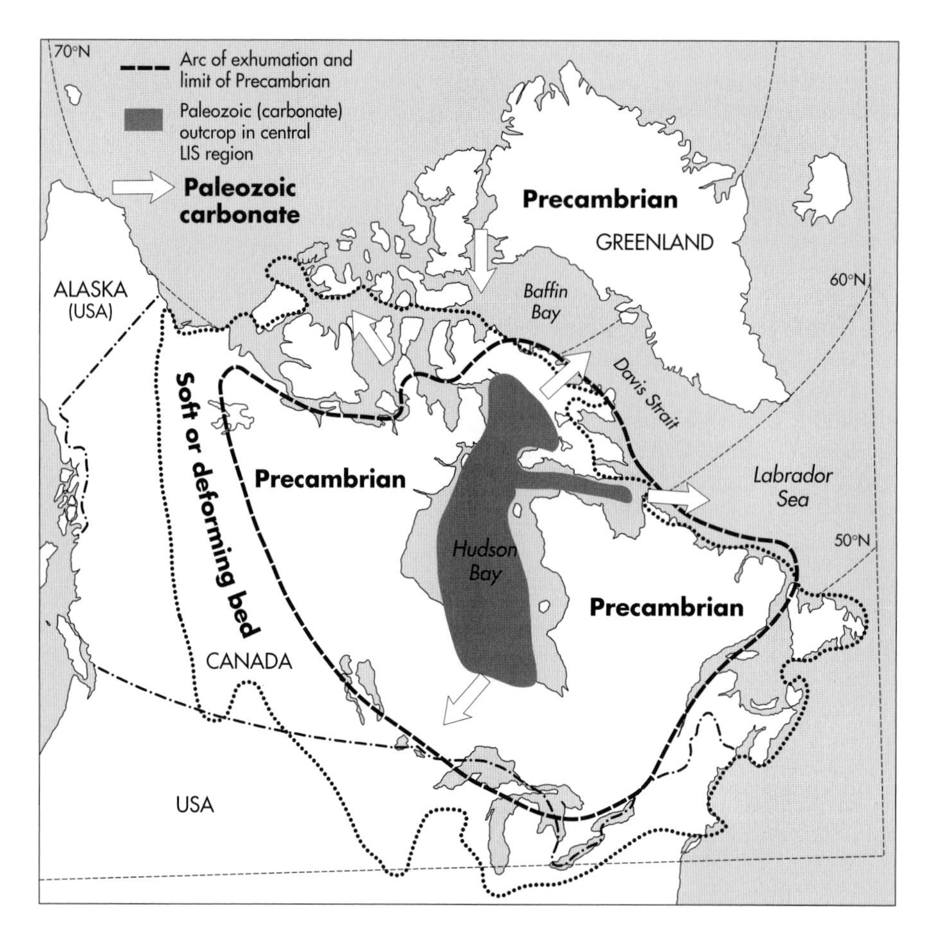

Figure 40.3 Map of the extent of the Precambrian shield rocks under the Laurentide Ice Sheet and the arc of exhumation (White, 1972) that lies along the contact between the 'hard' shield rocks and the onlapping sedimentary rocks. This extends along the Labrador coast as a coast-parallel trough, but this is not noted along the Baffin Island coast. Arrows indicate areas of erosion and transport of Palaeozoic carbonates (calcite and dolomites) onto land or to marine sites. The distribution of Palaeozoic carbonates in the High Canadian Arctic and North West Greenland is too complicated to show on this map but basically crop out on the floors of the major channels. Some of the major transport directions for the carbonates are shown (note the lack of carbonates from Hudson Bay on the Labrador shield region).

1:500,000) of large areas of the Canadian Arctic. What struck Ives (1957, 1962) and earlier writers (Brooks, 1926) was that the eastern Canadian Arctic consisted of the uplifted rim of the Precambrian shield, so that there are vast areas of high, rolling uplands >500 m a.s.l. Based on field and imagery observations in Labrador and Baffin Island (Ives, 1962; Andrews *et al.*, 1976) the concept on 'instanteous glacierization' was proposed and examined by some of the first energy-balance and three-dimensional glaciological models (Andrews & Mahaffy, 1976; Williams, 1978a). Coeval with this conceptual work on the distribution of raised marine beaches, now datable by radiocarbon dating, came arguments that the glacial rebound of the LIS also suggested the presence of several rebound centres (Andrews, 1970). The inception of LIS glaciation on the unscoured uplands of the Eastern Canadian Arctic, first examined rigorously in the 1970s, is supported by later, more sophisticated three-dimensional glaciological reconstructions (Kleman *et al.*, 2002; Marshall & Clark, 2002).

The debate between a single-domed ice sheet and a multidomed LIS then became involved with the issue of the extent of the ice sheet, especially along its eastern and northern margins, and the debate between 'big and little ice or minimum and maximum ice sheets' was born. The two sides of the debate were effectively discussed by Denton & Hughes (1981) in their book *The Last Great Ice Sheets*. I was one of the 'minimum ice sheet' crowd primarily for two reasons:

1 radiocarbon dates from the margin of the Eastern Canadian Arctic invariably (and still to this day) gave a series of results ≤9000 ^{14}C yr BP and often associated with a massive moraine complex called the Cockburn Moraines, located near the fiord heads, whereas on the outer coast Late-glacial/Holocene marine limits were often low and there were complex stratigraphical exposures with the uppermost sediments dating >35,000 to >54,000 ^{14}C yr BP (Andrews, 1980);
2 the evidence for the maximum model was frequently based on geomorphological arguments with no temporal control, e.g. shelf troughs were cut by outlet glaciers of the ice sheet, therefore the ice sheet extended along the trough during the Last Glacial Maximum (LGM).

The nail in the coffin of the absolute minimum model was delivered by the work of Anne Jennings in Cumberland Sound, Baffin Island (Jennings, 1993), who showed unequivocally that the Sound, shown as largely unglaciated in Dyke & Prest (1987a), had been filled with a major ice stream that had not retreated from the outer basin until ca. 10,200 ^{14}C yr BP. Since her work there has been a substantial revision of the glacial history based in large measure on research into the stratigraphical record on adjacent shelves, slopes and deep-sea basins (Jennings *et al.*, 1996; Andrews *et al.*, 1998b), and on the application of cosmogenic exposure age dating (Steig *et al.*, 1998; Marsella *et al.*, 2000; Briner *et al.*, 2003), the results of which have tended to confirm some of the earlier suspicions articulated by Sugden & Watts (1977) and has resulted in 'just the right ice' model for the LGM (Miller *et al.*, 2002), with ice reaching the coastline in some areas but extending well seaward in other areas (Fig. 40.1), such as Cumberland Sound (Jennings *et al.*, 1996) and Hudson Strait (Hesse, 1995; Hesse & Khodabakhsh, 1998).

40.3 Erosion by the LIS

In many ways our concept of glacial erosion is governed by data from present-day, often small, mountain glaciers (Hallet *et al.*, 1996). Thus when considering glacial erosion these data usually are considered as more efficient per unit area than fluvial erosion. However, we need to consider present-day rates in the longer Quaternary context of the LIS. Thus a rate of 0.5 mm yr^{-1} indicates the removal of 500 m of bedrock in 1 Myr or nearly 1500 m since the generally accepted date of Northern Hemisphere general glaciation ca. 2.6 Ma.

Flint (1971) attempted to compute the volume of glacial sediments from various papers and reports and suggested an average lowering of 10 m (or 0.001 mm yr^{-1}) but the topic of how much erosion had occurred under a 12×10^6 km^2 ice sheet did not attract much attention until 1972 when White (1972) published an article that claimed evidence for 1000 m or so of erosion in the centre of the ice sheet, that is Hudson Bay. The paper provoked a series of sharp exchanges, most of which were, rightly so, extremely critical (Gravenor, 1975; Sugden, 1976), but one observation that White made is important and that is he noted that along the contact between the Canadian shield (mainly granites and gneisses) and the onlapping younger and less resistant bedrock lay 'the arc of exhumation' or a series of large lake basins, assumed to be the result of glacial erosion. The location of these basins, fringing the Precambrian Shield, can be carried seaward along the Labrador margin where a deep, coast-parallel trough lies along the contact between the shield rocks and the softer sediments of the continental shelf (Holtedahl, 1958) (Fig. 40.3). The contrast between the 'hard bed' of the Canadian Shield and the 'soft bed' that lies toward the margins of the former ice sheet is now considered to be fundamental in terms of glaciological processes.

Erosional products from terrestrial sites, given enough time, will eventually be transported seaward and their final point of deposition will be the deep-sea basins. Bell and Laine (Laine, 1980; Bell & Laine, 1985) examined the offshore evidence for the amount of sediment stored in the basins flanking the eastern sector of the LIS. The evidence consisted of a series of deep-sea drilling sites, the lithostratigraphy of the cores, and the interpretation of seismic profiles. With these data they were able to compute the volume of sediments associated with the eastern sector of the LIS. They arrived at a value of around 100 m of erosion in the past 2.6 Myr or so, or a long-term average erosion rate of ca. 0.04 mm yr^{-1}.

The next issue then is how this average rate might apply to the bed of the ice sheet. How variable was erosion and what were its spatial characteristics?

40.4 Geomorphological expression of glacial erosion, or lack thereof

The years following World War II saw earth scientists exploring a wide range of glacier environments from Antarctica to mountain glaciers and ice caps lying at the Equator. This geographical coverage combined with recruitment into the new field of 'glaciology' of physicists and mathematicians led to the development of models of temperature conditions at the bed of glaciers and the large ice sheets. Sugden (1977, 1978) wrote two seminal papers that first developed an estimate for the temperature at the bed of a North American Ice Sheet (the ice sheet was not temporally fixed), proposed how this might affect glacial erosion or glacial protection, and then used maps and aerial photographs to produce a map of the bed of the LIS showing areas of scour, selective linear erosion (e.g. fiords), alpine glaciation, and areas showing little evidence for glacial erosion. As a dramatic example of the protective nature of an ice sheet where the bed is frozen there is a thin (few centimetres thick) Paleocene (60–70 Ma) deposit sitting on a hilltop a few tens of kilometres from the Barnes Ice Cap (Andrews *et al.*, 1972), which in turn is a relict of the LIS (Hooke & Clausen, 1982)!

In terms of the work of Bell & Laine (1985) a key feature of Sugden's map, and subsequent studies (Andrews *et al.*, 1985a,b), was that the 100 m of erosion for the eastern sector of the LIS must have varied spatially from ca. 0 to >>100 m. Because large areas of the eastern Canadian Arctic show little evidence of glaciation (Sugden, 1978) then the rate of erosion in the areas of scour or selective linear erosion must be moderately high. However, it would be a mistake to adopt another of White's arguments (White, 1988) and conclude that the deep sounds and channels that are a feature of the eastern and northeastern margin of the LIS (e.g. Hudson Strait, Frobisher Bay, Cumberland Sound, Lancaster and Jones sounds) are the product of long-term glacial erosion. Undoubtedly ice streams have flowed along these features but in all these cases there is undisputable geological evidence that they are fault-bounded and, moreover, have 'soft' sedimentary rocks lying at the seafloor (MacLean, 1985; MacLean *et al.*, 1986; MacLean, 2001a,b), these would include mainly Palaeozoic carbonates but in the case of Cumberland Sound, Cretaceous mudstones. Thus the large-scale features that fringe the margin of the LIS and often lead into its interior are grabens, probably date to the Tertiary break-up of Canada and Greenland. This being the case then the question has to be revisited—what are the origins of the fiords that are such a dramatic feature of the Labrador and Baffin landscape, or, more specifically, what fraction of their current volume is associated with the removal of rock by glacial erosion and how much might be allocated to the creation of space by faulting (Dowdeswell & Andrews, 1985)? This is an unanswered question and probably both processes explain some fraction of the fiord volumes. At ODP Site 645, off the coast of Baffin Island and on the floor of Baffin Bay (Hiscott *et al.*, 1989; Srivastava *et al.*, 1989), the rate of sediment accumulation is ca. 0.13 m kyr^{-1} and the rate of accumulation at other sites within Baffin Bay are also relatively modest at ca. 0.1 m kyr^{-1} (Andrews *et al.*, 1998a), especially given the fact that the Bay was, and is, surrounded by large ice sheets and glaciers.

How much erosion has occurred on the scoured landscapes of the Canadian Shield is difficult to estimate but along the eastern sector it is worth noting that the erosional products from the scoured areas have to transit the uplands, which show little evidence of active glaciation (Andrews *et al.*, 1985a), before converging into fiord outlet glaciers and in some cases ice streams. On the Baffin margins these ice streams are marked by plumes of carbonate-rich till (Tippett, 1985), which indicate transport from the Palaeozoic outcrop around and in Foxe Basin toward the eastern coast of Baffin Island (Fig. 40.3). Suprisingly, the major trough that leads to Cumberland Sound does not contain such a plume (Andrews & Miller, 1979).

In other parts of the LIS field observations and glaciological modelling indicate that the LIS, like the Antarctic and Greenland ice sheets today, contained fast-flowing ice streams (Marshall & Clarke, 1996) that characterized the southern, eastern and northern margins. As in Baffin Island the paths of these former ice streams can be delineated by bedforms and frequently by the composition of tills, often detectable from satellite imagery (Dyke & Morris, 1988), especially when carbonate-rich tills are spread across the rocks of the Canadian Shield. Other distinctive rocks are used for tracers of glacial transport, however, such as the red bed sandstones of Keewatin that were transported toward and along Hudson Strait (Shilts, 1980; Laymon, 1992; Aylsworth & Shilts, 1991) and a unique volcanic rock from the Belcher Islands, Hudson Bay, that Prest (1990) traced for thousands of kilometres to the south and west toward the LIS margin (Fig. 40.1).

Fisher *et al.* (1985) modelled the impact of a deforming bed on the shape and volume of the LIS. The notion of a deforming bed was then quite new and was based in part on measurements from beneath Icelandic glaciers (Boulton & Jones, 1979) and theory and observations from Antarctic ice streams (Hughes, 1977; Alley *et al.*, 1994; Weertman & Birchfield, 1982). There are two geographical areas of deforming beds under the LIS—one is linked with the ice streams that follow the structural channels around the eastern and northern margins of the ice sheet (e.g. Hudson Strait, Cumberland Sound, Lancaster Sound, etc), whereas the broader concept involves the ring of softer sedimentary rocks that lie to the south and west of the Precambrian shield (Fig. 40.3) and White's (1972) arc of exhumation (but see Stokes & Clark, 2003a). Clark & Walder (1994) showed that the distribution of eskers at the bed of the LIS is largely confined to the hard-bed of the shield, whereas on the fringing area of sedimentary rocks (Fig. 40.3) there are few eskers, but it is in this area where Mathews first identified outlets of the LIS with extremely low gradients (Matthews, 1974), and this work has been extended along this fringe. The presently accepted notion is that these outlets were lying on deforming sediments (Clark, 1994; Clark *et al.*, 1996b) but there are challenges to this inclusive notion (Stokes & Clark, 2003a).

One final aspect of issues pertaining to the rates of glacial erosion is the tremendous potential for the application of cosmogenic exposure age dating to not only issues of glacial chronology but also to the fundamental question of the rate of glacial erosion. Near the southern margin of the LIS Colgan *et al.* (2002) collected 22 samples from five striated rock outcrops comprising granites, metarhyolites and quartzite. In two of the outcrops nuclide abundances were consistent with accumulation since deglaciation, implying an erosion of ca. 2 m or more of rock. However, in three outcrops the nuclide abundances '. . . were up to eight times higher than predicted by the radiocarbon chronology'. For these data minimum limiting glacial erosion rates of 0.01–0.25 mm yr^{-1} were estimated (Cogley *et al.*, 2002, p. 1581). Clearly it will be extremely interesting to obtain cosmogenic exposure age dates from the unscoured uplands of Baffin Island and Labrador—this is currently being investigated (Briner & Miller, personal commununication, 2004).

40.5 Evidence of changes in ice-sheet geometry and abrupt changes in mass balance

If the LIS had been erosive at all points under its bed then the amount of information on past flow directions would be nonexistent. Far travelled glacial erratics (see above) would provide some information but these are not imprinted on the bed of the ice sheet, hence relative changes in flow regimes could not be ascertained. With the advent of LANDSAT and other high-resolution imagery several research groups have developed procedures for mapping the relative ages of different flow directions on the bed of the former ice sheet (Boulton & Clark, 1990; Clark, 1993, 1997; Kleman & Hattestrand, 1999). These papers suggest a more dynamic LIS than might be judged from the relatively uncomplicated pattern of ice retreat since the LGM, which shows a retreat of the LIS towards the north and east, thus towards the ancestral homeland on the uplands of the eastern Canadian Arctic (Andrews, 1973). Indeed, the Barnes Ice Cap is a relic of the LIS. The mapping of ice-flow directions from imagery is also being combined with fieldwork to verify the interpretations, however some of the interpretations, especially in the area of Labrador–Ungava, are controversial (Veillette *et al.*, 1999).

Up until the very early 1990s researchers often talked about the possible past and future collapse of the West Antarctic Ice Sheet (WAIS) (Mercer, 1978; Thomas *et al.*, 1979; Weertman & Birchfield, 1982; MacAyeal, 1992a). However, in 1988 (Heinrich, 1988) and subsequently (Andrews & Tedesco, 1992; Andrews, 1998; Bond *et al.*, 1992; Broecker *et al.*, 1992; Hesse & Khodabakhsh, 1998) it was shown that massive collapses of the LIS had occurred during the Wisconsinan glaciation in what are now referred to as Heinrich (H-) events. These have been labelled and dated at ca. 13 (H-0), 16.5 (H-1), 24 (H-2), 29 (H-3) and 40 (H-4) ka, with two older events, H-5 and H-6, at 46 and 60 ka (Chapman & Shackleton, 1999; Bond *et al.*, 1999). These events were centred on Hudson Strait (MacAyeal, 1993a), and largely involved the transport of glacially eroded detrital carbonate from the floors of Hudson Strait, and possibly Hudson Bay, to sites as far distant as the deep-sea basins off Ireland and even south to near Portugal (Lebreiro *et al.*, 1996). The cause(s) of H-events is not known with certainty but may have involved a 'binge–purge' cycle associated with changes in ice-stream thickness, activity, and basal temperatures (MacAyeal, 1993a, Alley & MacAyeal, 1994; Clarke *et al.*, 1999). Although the literature most often stresses the iceberg rafted (IRD) component, defined as lithics >125 μm, it is critical to note that in the ice-proximal area below the shelf-break off Hudson Strait the sediments are fine-grained, often laminated and produced by turbidites (Hesse *et al.*, 1997, 2004). The evidence suggests that massive outburst floods were part and parcel of the conditions associated with H-events.

A variety of tracers are now being used to better understand where the sediments associated with H-events came from on the bed of the LIS. The tracers include $^{40/39}$Ar dates on individual hornblende grains (Hemming *et al.*, 2000, 2002a), the isotopic composition of the <63 μm decalcified sediment fraction (Farmer *et al.*, 2003), the petrology of the sand fraction (Bond & Lotti, 1995), the rare earth composition of the sediments (Benson *et al.*, 2003), and rock magnetic properties (Stoner *et al.*, 1996; Stoner & Andrews, 1999). The recent work of Farmer *et al.* (2003) from the eastern margin of the LIS was able to characterize the glacially transported sediments in a series of offshore marine cores and link them with source areas. An important aspect of this work was the demonstration that the bulk of the sediment was derived from the adjacent continent. Down-core studies (Hemming *et al.*, 2002a; Groussett *et al.*, 2001) indicate that these various approaches will capture changes in the flow directions of the ice sheet and hence offer the possibility of linking the geomorphological bed record, which is on a relative dating scale, with a more numerical age scale based on the rate of sediment accumulation in marine cores. An exciting prospect!

40.6 Retreat of the LIS from the Last Glacial Maximum

The advent of radiocarbon dating rapidly resulted in compilations of data for the deglacial history of the LIS. This exercise was initiated in 1969 by the publication of two series of maps from different authors but with rather similar results (Bryson *et al.*, 1969; Prest, 1969). These data were used in several publications dealing with LIS volume changes and patterns and causes of retreat (Paterson, 1972; Andrews, 1973). A major revision was undertaken in time for the INQUA meeting in Canada in 1987, including a discussion and a series of maps showing retreat every 1 kyr or so (Dyke & Prest, 1987a,b). In the past decade or so, additional radiocarbon dates have been obtained and the outline of the margin has changed, especially in the eastern Canadian Arctic, including the High Canadian Arctic and the Maritimes of Canada, as workers wrestled with the ice extent during the LGM and Marine Isotope Stage 2 (Dyke *et al.*, 2002; England, 1999; Miller *et al.*, 2002; Clark *et al.*, 2003). The outline of the Precambrian shield (Fig. 40.3) broadly mimics the outline of the LIS at about 10,000 ^{14}C yr BP and there is a significant asymmetry in the isochrons of deglaciation with hundreds of kilometres of retreat along the deformable bed margins versus tens of kilometres of retreat along the Labrador and Baffin Island margins (Andrews, 1973).

An important question in terms of the dynamics of ice sheet build-up and global sea level is the extent of the LIS during Marine Isotope Stage 3. The Canadian data, reviewed in Andrews (1987), suggests that the ice sheet had retreated on to its hard Precambrian Shield bed (Fig. 40.3), and thus there is presently no evidence for ice across the western sectors of the LIS, nor indeed within the region of the Cordilleran Ice Sheet. Heinrich events 3, 4 and 5 occurred within this interval, however, indicating that a dynamic ice sheet still extended across the region with a large drainage basin draining ice from Hudson Bay through Hudson Strait.

40.7 Conclusions

The bed of the LIS, and other former ice sheets, represent a rosetta stone for deciphering past changes in ice-sheet dynamics, including changes in thermal conditions, rates of erosion, and transport vectors. New tools and new approaches have characterized our efforts to learn more about the LIS during the late Cenozoic Ice Age. Observations and fieldwork can be used to provide tests of new and sophisticated three-dimensional ice sheet models (Marshall & Clark, 2002; Marshall *et al.*, 2002; Hildes *et al.*, 2004) and this interaction will surely increase our understanding of the tantalizing fragments of evidence that ice sheets leave, hence allowing researchers to better understand bed conditions at the base of large ice sheets.

Acknowledgements

This paper reflects materials that I used in teaching Glacial Geology (GEOL 4360/5360) at the University of Colorado between 1968 and 2003. I thank Peter Knight for giving me an opportunity to sound off on a topic of long-standing interest. These materials were used in a 2-week intensive course at the University of Nottingham in the autumn of 2003, arranged through the efforts of Dr R.E. Dugdale.

FORTY-ONE

What can the 'footprint' of a palaeo-ice stream tell us? Interpreting the bed of the Dubawnt Lake Ice Stream, Northern Keewatin, Canada

Chris R. Stokes* and Chris D. Clark†

**Landscape and Landform Research Group, Department of Geography, The University of Reading, Reading RG6 6AB, UK*
†Department of Geography, The University of Sheffield, Sheffield S10 2TN, UK

Rapidly flowing ice streams exert a profound influence on ice-sheet configuration. It is, therefore, essential to incorporate the spatial and temporal activity of ice streams in order to reconstruct accurately the evolution of an ice sheet through time. Recently, considerable progress has been made in identifying palaeo-ice stream imprints in formerly glaciated terrain and we now have a large (>50) population of palaeo-ice stream tracks to investigate (Stokes & Clark, 2001). Their identification has refined our reconstructions of ice-sheet dynamics but they also provide an unprecedented opportunity to advance our understanding of ice-stream behaviour and stability. For example, once a palaeo-ice stream track has been identified confidently, its 'footprint' may provide answers to several important glaciological questions such as:

1 How big was the ice stream and what was its likely catchment area?
2 What factors triggered its location within the ice sheet?
3 What was its likely flow mechanism?
4 For how long did it operate and did it operate more than once?
5 What factors led to its shut-down?
6 What was its wider impact on the ice sheet?

Given that the bedform record inscribed by an ice stream is intimately linked to its activity then it is possible to use inversion techniques to answer such questions. This case study demonstrates how the bedform record of the Dubawnt Lake Ice Stream in North Keewatin (Canada) can be used to glean pertinent information about the functioning of ice streams.

The Dubawnt Lake Ice Stream represents a >450-km-long bottleneck flow pattern of lineations, trending in a west-northwest direction northwest of Dubawnt Lake, Keewatin. In Kleman & Borgström's (1996) inversion model, it was highlighted as a surge fan generated rapidly during deglaciation of the Keewatin Sector (see also this chapter). More recently, detailed mapping of the flow-set and surrounding areas has established its prominence as one of the best-preserved terrestrial ice-stream imprints available for scrutiny (Stokes & Clark, 2003a).

The location of the flow-set is shown in Plate 41.1, which also illustrates typical examples of the size and pattern of subglacial bedforms in the onset zone, main ice-stream trunk and terminus. Detailed mapping reveals that bedforms get longer and more closely packed together in the main trunk of the ice stream and that the overall pattern of bedform elongation (Plate 41.1c) mimics the expected velocity field for a terrestrial ice stream. The sharp southern margin of the flow-set is another reliable indicator of an abrupt change in ice dynamics, marking the boundary between the rapidly flowing ice stream and the neighbouring slow-flowing ice (and the probable boundary between the fast warm-based ice and slower cold-based ice). The northern margin of the flow-set is less clear but suggests that the ice stream was around 140 km wide at its narrowest point, significantly wider than contemporary ice streams in the Antarctic and Greenland. The length is reconstructed at 450 km and the ice stream had an estimated catchment area of around 190,000 km^2.

The location of the ice stream on the relatively flat Canadian Shield conflicts with the paradigm that ice streams require topographic funnelling or soft subglacial sediments for their initiation (Stokes & Clark, 2003a). Although softer sediments interrupt the characteristically hard crystalline bedrock on the Canadian Shield in this location, it is unlikely that they were in sufficient quantity to 'trigger' the ice stream. Rather, it is speculated that the evolution of large (>3000 km^2), deep (ca. 120 m) proglacial lakes impounded at the ice sheet margin may have been important in initiating ice stream activity by inducing calving and taking the system beyond a threshold that was sufficient to trigger fast ice flow. Evidence of these lakes suggests that the only location where they existed for a significant length of time exactly matches the location of the ice-stream flow pattern (see Craig, 1964; Stokes & Clark, 2004).

Elucidating the flow mechanisms beneath ice streams is a major scientific challenge and we argue that the bedform record of palaeo-ice streams can shed some light on this problem. Rather than viewing the lineaments on the Dubawnt Lake Ice Stream bed as streamlined ridges, it is suggested that a better description may be that of a highly grooved till surface (see Plate 41.1b). On the basis of this observation and investigations of megascale glacial lineations that characterize other ice-stream beds (Clark *et al.*, 2003b) our hypothesis is that the ice stream flowed by 'groove-ploughing'. Under this mechanism, large keels at the base of the ice stream plough through sediments, carving elongate grooves and deforming material up into intervening ridges. If correct, this mechanism holds important implications for ice-stream mechanics, not least of which is that it suggests that they may be able to widen under steady state conditions and increase ice discharge (Clark *et al.*, 2003b). This may partly explain why many palaeo-ice streams that operated during deglaciation are wider than their contemporary counterparts.

In places, the regular parallel pattern of ice-stream bedforms has been modified by the presence of ribbed moraines that clearly developed after the ice-stream bedforms were generated. Plate 41.2 shows an example of this superimposition in the ice-stream onset zone. Given that ribbed moraines are not associated with fast ice flow, we argue that they represent areas of high basal drag (sticky spots) that developed either during or in response to ice-stream shut-down. One possibility is that they resulted from basal freeze-on and are a manifestation of shearing and stacking of debris and/or fracturing of cold-based ice (see Hättestrand & Kleman, 1999), or some other unknown mechanism associated with a reduction in ice velocity.

The ice stream imprint is orientated approximately parallel to the overall deglacial direction and comparison with retreat patterns (isochrons) suggests that it operated for only a brief phase (<500 yr?) during deglaciation, likely to be around 8.2 kyr BP (Stokes & Clark, 2004). Its activity would have had a profound affect on the Keewatin Sector of the Laurentide Ice Sheet, considerably lowering its surface profile and accelerating its demise at the end of the last glacial cycle.

FORTY-TWO

The Antarctic Ice Sheet

David G. Vaughan

British Antarctic Survey, High Cross, Madingley Road, Cambridge CB3 0ET, UK

42.1 Introduction

Antarctica (Fig. 42.1) has been described as 'a continent for science'; it is certainly a continent of superlatives—the coldest, highest, driest, and certainly the least known and least understood. Just 30 yr ago, vast tracts were unmapped and unexplored, and were represented by largely featureless maps showing only the routes of the handful of expeditions to have crossed the interior of the continent. The mapping of the continental ice sheet was, however, greatly improved in the 1970s as long-range airborne survey became practical, and as a result the first convincing maps of the interior of the continent were published (e.g. Drewry, 1983a). But despite the cooperative efforts of many nations, the coverage obtained from aircraft remained far from complete, and only with the advent of imaging satellites in the 1970s did we gain a truly continental perspective on the ice sheet. In the following two decades, the increasing supply of satellite data allowed high-resolution mapping of the elevation and flow of the entire ice sheet. In the past half-decade, simple mapping has ceased to stretch our capabilities, and measurements of change in the ice

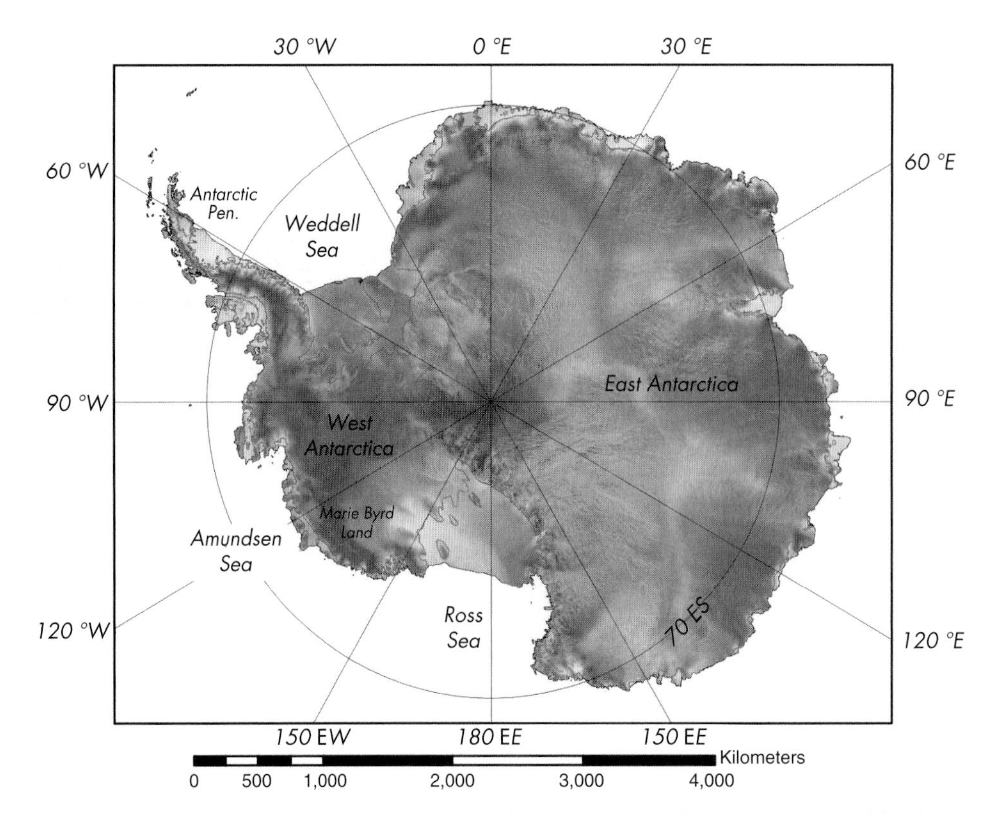

Figure 42.1 The Antarctic Ice Sheet today. The image is a synthetic aperture radar mosaic acquired by the Canadian satellite Radarsat. (Reproduced courtesy of Radarsat and Ken Jezek, Byrd Polar Research Institute.)

sheet are now possible and we have moved forward into a new realm of investigation—strong evidence is emerging that the Antarctic Ice Sheet is changing. In this chapter, we shall show that some of these changes are gross whereas others are surprisingly subtle, and that they operate over a range of time-scales with a variety of causes and varying degrees of predictability.

Thirty years of research in Antarctica have taken us from an almost blank map to a growing understanding of a constantly changing ice sheet of enormous complexity. Understanding this complexity and the interplay between drivers of change is a huge task that could take many more years. Unfortunately, we probably cannot afford to take that long, because it is also increasingly clear that changes in the Antarctic Ice Sheet will be felt across the world, sooner rather than later.

42.1.1 The imperative of ice-sheet research

For 2.6 Myr, glacial cycles have altered every aspect of the Earth's environment. The rapidity of climate change through regular glacial cycles drove pre-humans into new habitats and is thought to have provoked an evolutionary response that coincided with the earliest evidence of tool-making. Proxy records from the late Quaternary (approximately the past 1 Myr) show that in each cycle, deglaciation was far more rapid than the onset of glacial conditions. During the most recent period of deglaciation (since the Last Glacial Maximum (LGM) ca. 17 ka) two 1000-yr periods of ice-sheet retreat caused global sea level to rise at rates of ca. 1 m per century (Fairbanks, 1989), which is approximately ten times the rate of sea-level rise during the 20th century. Recent 'sea-level fingerprinting' has been used to argue that much of the rise measured for the first of these events resulted from deglaciation in Antarctica (Clark *et al.*, 2002a).

This history clearly shows us that Antarctica and similar ice sheets have had the potential to raise global sea level rapidly and to cause significant changes in the Earth System, but we should also ask if that potential still exists today? We live in an interglacial period, but despite the persistence of interglacial conditions for more than 10 kyr the Earth appears to be only partially deglaciated. There are still glaciers and ice caps on every continent, and the vast ice sheets covering Greenland and Antarctica are probably still extensive compared with previous interglacials. This relative abundance of glaciers, compared with previous interglacials, results from a modern climate that is cooler than that at the height of two of the last three interglacials (Petit *et al.*, 1999). However, these slightly cooler conditions may not continue: despite the scare stories prevalent in the media during the 1970s that predicted an imminent descent into a new glacial period ('The New Ice Age'), there is actually little evidence that our interglacial is coming to an end. Predictions based on the eccentricity in the Earth's orbit suggest that interglacial conditions could continue for a further 5–50 kyr (Berger & Loutre, 2002), and what is more, there is a strong likelihood that the anthropogenic greenhouse effect may drive further widespread warming in the next few decades (IPCC, 2001): perhaps such warming will take us back to conditions in previous interglacials. Together, the abundance of glaciers and the likelihood of future warming suggest that we could be on the verge of a period of further glacial retreat; perhaps this could be widespread enough to be termed deglaciation.

In fact, it is possible that such a period of deglaciation has already begun. Throughout this volume there are documented examples of dramatic glacial retreat in tropical, temperate and polar regions, and in this chapter we will add more examples from Antarctica. But simply cataloguing glacial change as it happens is a very poor tool with which to make predictions, and there is a clear burden on scientists to make realistic and supportable predictions for the future. Every episode of Quaternary deglaciation caused turmoil throughout the Earth System and even minor global deglaciation today would have profound economic and social consequences. For example, currently around ten million people each year suffer from coastal flooding, and accepting the best prediction of sea-level rise (ca. 44 cm, Church *et al.*, 2001a), this number is likely to increase to ca. 200 million by 2080 (DETR and The Meteorological Office, 1997). It is, however, arguable that the most potent potential contributor to sea-level rise, the Antarctic Ice Sheet, was underestimated in those predictions of sea-level rise and that any increased contribution from Antarctica would further increase the population at risk.

Thus the imperative for understanding the present and future changes in the Antarctic Ice Sheet is clear. In this chapter we will review some of the reported observations of change and discuss them in terms of their likely causes, and the time-scales over which they may act. We will discuss direct and rapid changes in response to contemporary climate change and the changes that have occurred over longer time-scales; and finally, we turn to observed changes, the causes of which we do not yet understand, but which could have an impact that would eventually dwarf all others. We begin, however, with a discussion of the steady-state assumption that we apply to ice sheets and the methods of measuring deviations from the steady-state.

42.2 Measuring change in an ice sheet

With a surface area of around 12 million km^2, the grounded Antarctic Ice Sheet is the single largest control on world sea level. The East Antarctic Ice Sheet holds the equivalent of 52 m of global sea-level rise, and the West Antarctic Ice Sheet, which rests on rock that lies considerably below sea level and is inherently less stable, contains the equivalent of 5 m of sea-level rise (Lythe *et al.*, 2001).

More than 50% of the outflow from the continental ice sheet passes through the largest 40 outlet glaciers and ice streams (hereafter collectively termed ice streams). It is these ice streams that are the most important control on the dynamics and volume of the ice sheet. Assuming that ice flow is generally in a downslope direction, it is relatively easy to define the portion of the ice sheet that feeds each glacier, i.e. its drainage basin (Fig. 42.2). And as each drainage basin operates largely independently of its neighbours, it makes sense to try to understand the evolution of the ice sheet in terms of these basins.

Our understanding of the evolution of the glacier basins in the Antarctic Ice Sheet (indeed any ice sheet, ice cap or glacier) is built around the concept of steady-state ice flow. Ice is driven to flow downslope by gravity, and this gravitational driving stress is controlled by surface slope and ice thickness. Thus as thickness increases so does ice flow. An ice sheet is said to have reached steady-state when it has thickened enough to cause sufficient flow at every point to remove all the snow accumulating between that point and the ice divide. After that, so long as the accumulation rate remains the same, and the relationship between driving stress and ice-flow velocity remains constant, a drainage basin should achieve something close to a steady-state. In reality, it is implausible that any major glacier drainage basin will ever reach a precise steady-state, because long before it is reached, some change (perhaps in snowfall or temperature) will have occurred, altering the accumulation rate or the way in which ice flow responds to stress. Having said this, that a true steady-state is implausible, we find that most drainage basins are actually surprisingly close to steady-state, and imbalances (known as the mass balance of the basin) have proved difficult to measure. Because any deviation from steady-state implies an immediate contribution to sea-level change, however, this remains an important measurement for glaciologists to make.

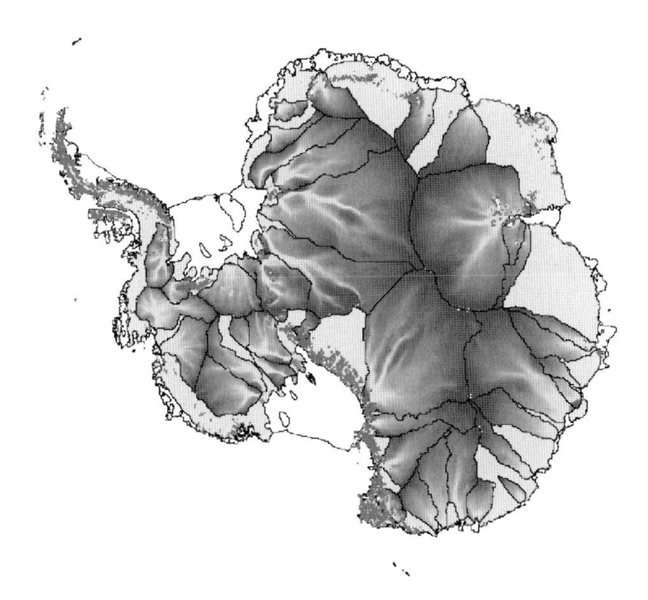

Figure 42.2 Drainage basins for 30 of the most significant glaciers in Antarctica. Within the basins for the significant glaciers the tone indicates the calculated balance flux (taken from Bamber *et al.*, 2000), with the lighter tones indicating faster flow. Areas of the ice sheet not drained by these significant glaciers are shaded grey and ice shelves are unshaded.

Estimates of the mass balance of the Antarctic Ice Sheet and its basins began very soon after the basic techniques for measuring accumulation rate, thickness and ice-flow velocity were established. These early assessments tended to use a credit/debit approach, in which the amount of snow accumulating over a particular domain, usually a complete drainage basin, was estimated, together with the flux of ice leaving this domain. Many such calculations have been published, but until quite recently, every one has been hindered by an unfavourable error budget. After all, the credit/debit approach relies on taking the difference between two

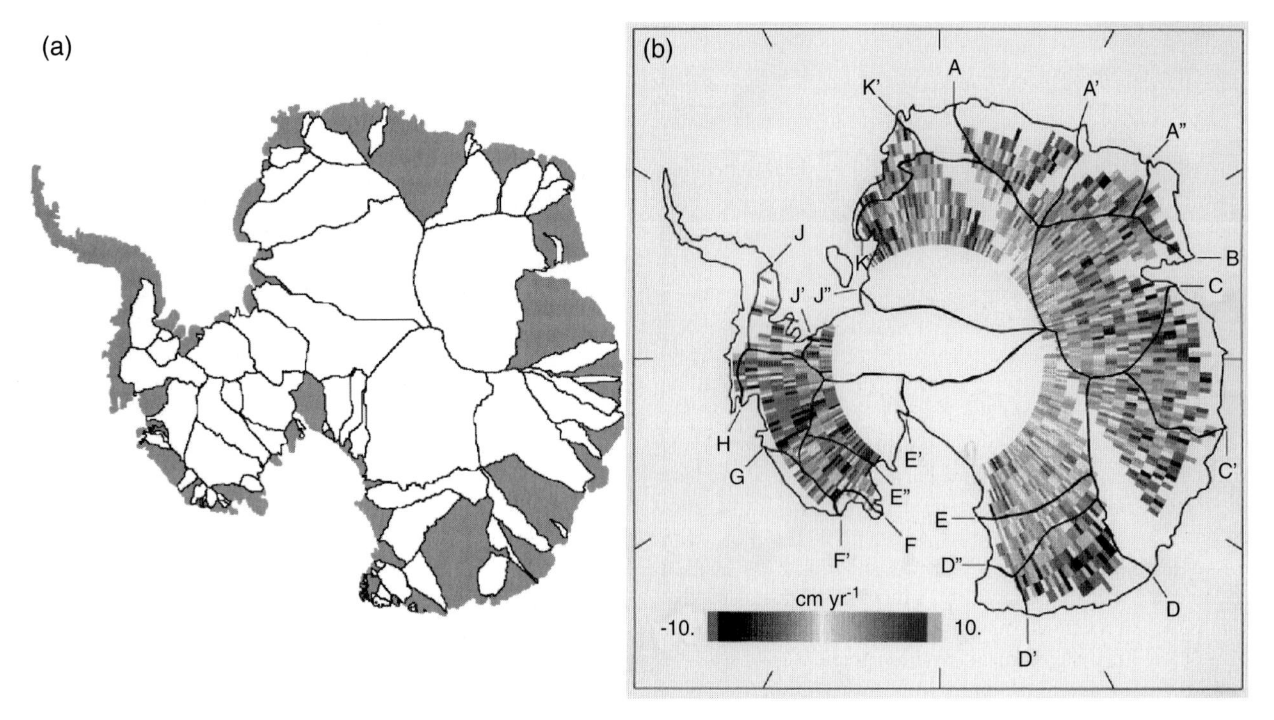

Figure 42.3 (a) The drainage basins of the principal outlet glaciers and ice streams in Antarctica, derived from a digital elevation model (Bamber & Bindschadler, 1997). (b) Map of surface elevation change in the period 1992–1996 derived from ERS-1,2 satellite altimetry. (Reproduced from Wingham *et al.* (1998), courtesy of American Association for the Advancement of Science.)

large numbers that contain substantial uncertainty. A 10% uncertainty in the mean accumulation rate across the basin (which would be a particularly precise estimate) means that even with ice flux known precisely the mass balance can only ever be found to ±10%. Not surprisingly, the credit/debit approach has rarely resulted in a statistically significant measurement of imbalance.

So although many, including myself, have tried to use the credit/debit method to measure change in the Antarctic Ice Sheet, these efforts were mired in uncertainty and the credit/debit approach has failed to provide a useful answer for most of the basins in the ice sheet (e.g. Jacobs *et al.*, 1992). This impasse remained until satellite altimetry finally came of age.

The idea has existed for many years to use satellite altimetry data to make a different kind of assessment of mass balance in ice sheets: a direct measurement of the rate of change in the surface elevation over a sufficiently long period to be able to observe mass imbalance directly. Notable efforts in both Antarctica and Greenland using several satellite altimeters (Zwally *et al.*, 1989; Lingle & Covey, 1998) were published, but only as improved satellite orbit determinations became available in the late 1990s did the technology finally come of age and creditable assessments become possible. The first entirely convincing assessment was published by Wingham *et al.* (1998; see Fig. 42.3). Although satisfactory altimetry could still not be retrieved from steep coastal zones, the area beyond the satellite's orbital range (south of 81.5°S), or those portions of the ice sheet where the satellites were out of range of ground receiving stations, this study produced estimates of change over 63% of the continental ice sheet. Somewhat surprisingly, it showed that between 1992 and 1996 there was little evidence for any surface elevation change exceeding ca. 10 cm yr^{-1} over the majority of basins. Owing to the limitations of the technique, small-scale changes could not be ruled out but most of the individual basins appeared to be close to balance. There was, however, one substantial region that showed a spatially coherent change: the drainage basins feeding Pine Island, Thwaites and Smith glaciers in West Antarctica, which appeared to be thinning at a rate of >10 cm yr^{-1}. Given the high snowfall rates in these basins, it was not possible for Wingham *et al.* (1998) to be certain whether this change was due to unusual snowfall or a dynamic imbalance in the glacial flow, but it was clear that unique and substantial changes were occurring in this area.

Later in this chapter we will return to the successful credit/debit assessments (section 42.4.2) and to glacier basins identified by Wingham *et al.* (1998) (section 42.5), but first we will discuss changes that have been observed elsewhere in Antarctica, which illustrate some of the causes of change that could lead to basin-wide imbalance.

42.3 Rapid response to contemporary climate change

It can be argued that the majority of the Antarctic Ice Sheet may be uniquely insensitive to small changes in atmospheric climate change. This is because it is so large that the time-scale of the dynamic response is measured in 10 to hundreds of thousands of

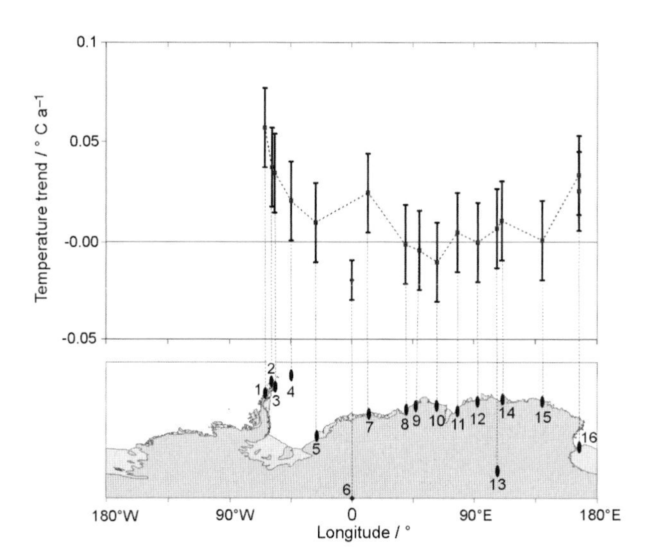

Figure 42.4 Long-term (>30-yr) trends in mean annual temperature measured at meteorological stations around Antarctica. Note the absence of trend data from the sector 70°W to 170°E. (Reproduced from Vaughan *et al.* (2003a), courtesy of Kluwer Academic Publishers.)

years, and it is sufficiently cold that a small warming is unlikely to provoke significant melting. But the climate of Antarctica does appear to be warming (Fig. 42.4) and the literature contains many small-scale exceptions to this generality which we will not discuss, rather we will focus on two particular areas of change that have the potential to make a noticeable impact on sea level.

42.3.1 Warming on the Antarctic Peninsula

In many respects the Antarctic Peninsula is quite different from the rest of the continental ice sheet and arguably shares greater similarity with areas such as coastal Greenland, Svalbard, Patagonia and Alaska. In contrast to the homogeneous ice sheet covering most of Antarctica, the Peninsula consists of a system of over 400 largely independent mountain glaciers draining into ice shelves or marine tidewater glaciers, with just a handful terminating on land. It represents only ca. 2% of the total area of the grounded Antarctic Ice Sheet but receives ca. 7% of the snowfall, equivalent to ca. 0.37 mm yr^{-1} of global sea-level change. Meteorological data from the Antarctic Peninsula show ca. 2°C warming since the 1950s, substantially faster than elsewhere in Antarctica (Vaughan *et al.*, 2001), and probably more sustained than in Greenland or Alaska (Hansen, 2003). Owing to the region's strong climatic gradients (with mean annual temperatures falling from 0°C to −17°C over a distance of 1000 km (Morris & Vaughan, 2003)), however, this exceptional rate of warming has resulted from only a modest geographical migration in the climate patterns.

The sensitivity of the Antarctic Peninsula ice sheet to contemporary climate warming is confirmed by the many observations of recent glacial change, retreating glaciers (Fig. 42.5, and examples given by: Splettoesser, 1992; Morris & Mulvaney, 1995; Smith *et al.*, 1999), reduction of permanent snow cover (Fox & Cooper, 1998), thickening of the ice sheet at high altitude (Morris & Mulvaney, 1995) and a lengthening melt season (0.5 ± 0.3 days yr^{-1} over 20 yr) (Torinesi *et al.*, 2003). Furthermore, the retreat of ice shelves, a long-predicted consequence of warming (Mercer, 1978), is well underway (Vaughan & Doake, 1996): nine ice shelves have retreated during the latter part of the 20th century (Fig. 42.6, and examples given by: Doake & Vaughan, 1991; Ward, 1995; Rott *et al.*, 1996; Cooper, 1997; Luchitta & Rosanova, 1998; Skvarca *et al.*, 1998; Scambos *et al.*, 2000; Fox & Vaughan, in press). It is also clear that climate warming is causing acceleration of glacier flow (Rott *et al.*, 2002; De Angelis & Skvarca, 2003), either directly though enhanced lubrication similar to that observed in Greenland (Krabill *et al.*, 1999; Zwally *et al.*, 2002) or indirectly as ice shelves are lost (Rott *et al.*, 2002).

So, although the Antarctic Peninsula is a small fraction of the entire continent, its contribution to sea-level rise could be rapid and substantial. The 'largest glaciological contribution to rising sea level yet measured' originates from the 90,000 km^2 of Alaskan glaciers. Over the period from the mid-1990s to 2000, changes in Alaska probably contributed 0.27 ± 0.10 mm yr^{-1} to sea-level rise (Arendt *et al.*, 2002). For comparison the Antarctica Peninsula supports 120,000 km^2 of grounded ice sheet with lower reaches that suffer substantial melt, and 45,000 km^2 lie at less than 200 m above sea level. Although at the time of writing there is no coherent assessment of the magnitude of change on the Antarctic Peninsula, the likelihood is that if atmospheric warming continues, the contribution from the Antarctic Peninsula will be significant.

42.3.2 Climate change and snowfall

Long before anthropogenic warming became a significant issue, let alone gained common acceptance, it was suggested that if climate warms over Antarctica, warmer air will be able to carry more moisture over the ice sheet and this will increase the precipitation (Robin, 1977). This simple argument is based on the increase in saturation vapour pressure, a measure of the ability of the air to carry moisture, with temperature. This is a very potent effect: at −20°C the saturation vapour pressure increases by around 10% per degree centigrade.

Several authors have invoked this effect to calculate the increase in accumulation due to particular warming scenarios (e.g. Fortuin & Oerlemans, 1990). Most recently, van der Veen (2002) showed that this argument implies that climate change could actually cause an increase in Antarctic snow accumulation equivalent to between 3.0 and 14.8 cm of global sea level by AD 2100, in part compensating for sea-level rise due to melting of non-polar glaciers, thermal expansion of the oceans and changes in terrestrial water storage. To cite this figure alone would, however, be to misrepresent van der Veen's study, since he argued that these models based on saturation vapour pressure arguments have not, in any objective sense, been verified as an accurate representation of reality, and that the level of uncertainty in their predictions remains extremely high, encompassing both a positive and negative contribution to sea-level change. He argued cogently that models based on such simple parameterizations of saturation vapour pressure ignore much more important mechanisms for changing accumulation rate, such as changes in the mean patterns

(a)

(b)

Figure 42.5 The retreat of ice around the British Antarctic Survey summer-only air facility, Fossil Bluff, has been continuing at least since the mid-1980s and resulted in around 10 m depression of the snow surface: (a) 1985–1986 (D.G. Vaughan, BAS) and (b) 1995–1996 (Peter Bucktrout, BAS). Note that between the two dates the main hut was rebuilt and extended, however the position of the white Stevenson screen in front of the hut remains unchanged. (See www.blackwellpublishing.com/knight for colour version.)

of atmospheric circulation and changes in the frequency with which cyclones penetrate onto the continent to produce more frequent snowfall events.

A more valuable approach will be to use general circulation models (GCMs) of climate change to investigate the likely increase in precipitation associated with particular climate scenarios. For the present, only a few such studies exist. Wild *et al.* (2003) used a coupled atmosphere–ocean model running at relatively high resolution to suggest that doubled carbon dioxide would lead to a warming over Antarctica and increased precipitation. They suggested that the warming would be insufficient to cause significant melt and that the dominant effect would be to compensate sea-level rise at a rate of ca. 0.86 mm yr^{-1}.

Whichever approach turns out to be most reliable, it seems that at present, estimates from both the simple saturation-vapour-pressure and GCM approaches are in the same ballpark. Collectively, they suggest that this effect may be significant. Conceivably it might be the dominant change in the overall mass balance of the Antarctic Ice Sheet, and it could be sufficient to compensate for a notable fraction of the predicted sea-level rise that will result from melting of non-polar glaciers and ice caps, thermal expansion of the oceans and changes in terrestrial water storage. Indeed,

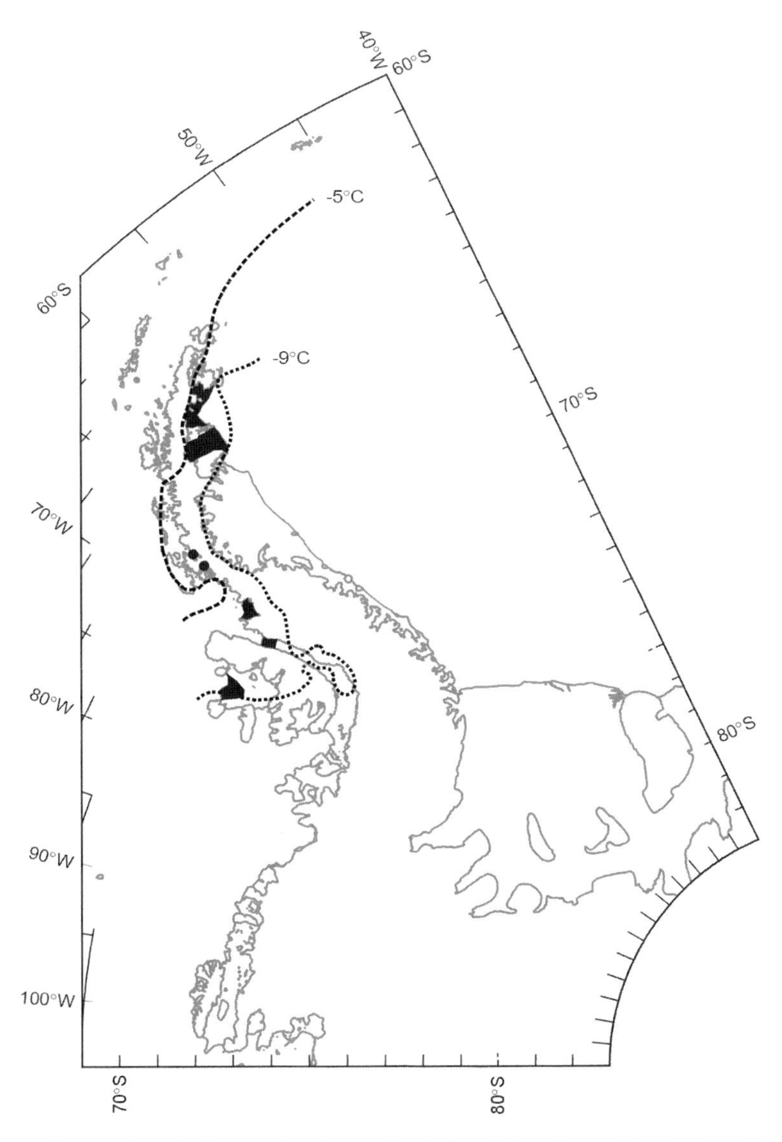

Figure 42.6 Distribution of retreating glaciers on the Antarctic Peninsula for the period for which observational and satellite records exist (i.e. the past 50–100 yr). Note that all ice shelves that show retreat exist between the –5°C and –9°C contours of mean annual temperature. (Reproduced in modified form from Morris & Vaughan (2003).)

this effect may turn out to be the single largest mitigator of sea-level rise, but even adding a healthy uncertainty to the present estimates, there appears to be little chance that this will amount to more than a compensating effect.

42.4 Long-term responses to Holocene climate change

These rapid responses to contemporary climate change discussed in the previous section are not the whole story. The scale of East and West Antarctic Ice Sheets is such that their dynamic response to any change might take in the order of 10,000 and 100,000 yr respectively, thus we should consider any rapid response to contemporary climate change as overprinting on the long-term trend of an ongoing response to millennial-scale changes. If we are unfortunate, the two effects will interact in a complicated manner and the prediction of future change will be considerably more difficult as a result. Either way, we need to establish the pattern of long-term change in the ice sheet as a separate phenomenon and some recent studies are now providing direct evidence of such changes.

42.4.1 Continuing Holocene retreat in West Antarctica

There are several methods available that allow us to determine the past configuration of ice sheets; these are primarily marine geophysical and geomorphological techniques, but one relatively new technique is substantially improving our understanding of the history of West Antarctica and is worthy of special mention. Using isotopic analysis, it is now possible to determine how long rocks have been exposed to cosmic radiation. When this technique is applied to glacial erratics deposited on the slopes of nunataks

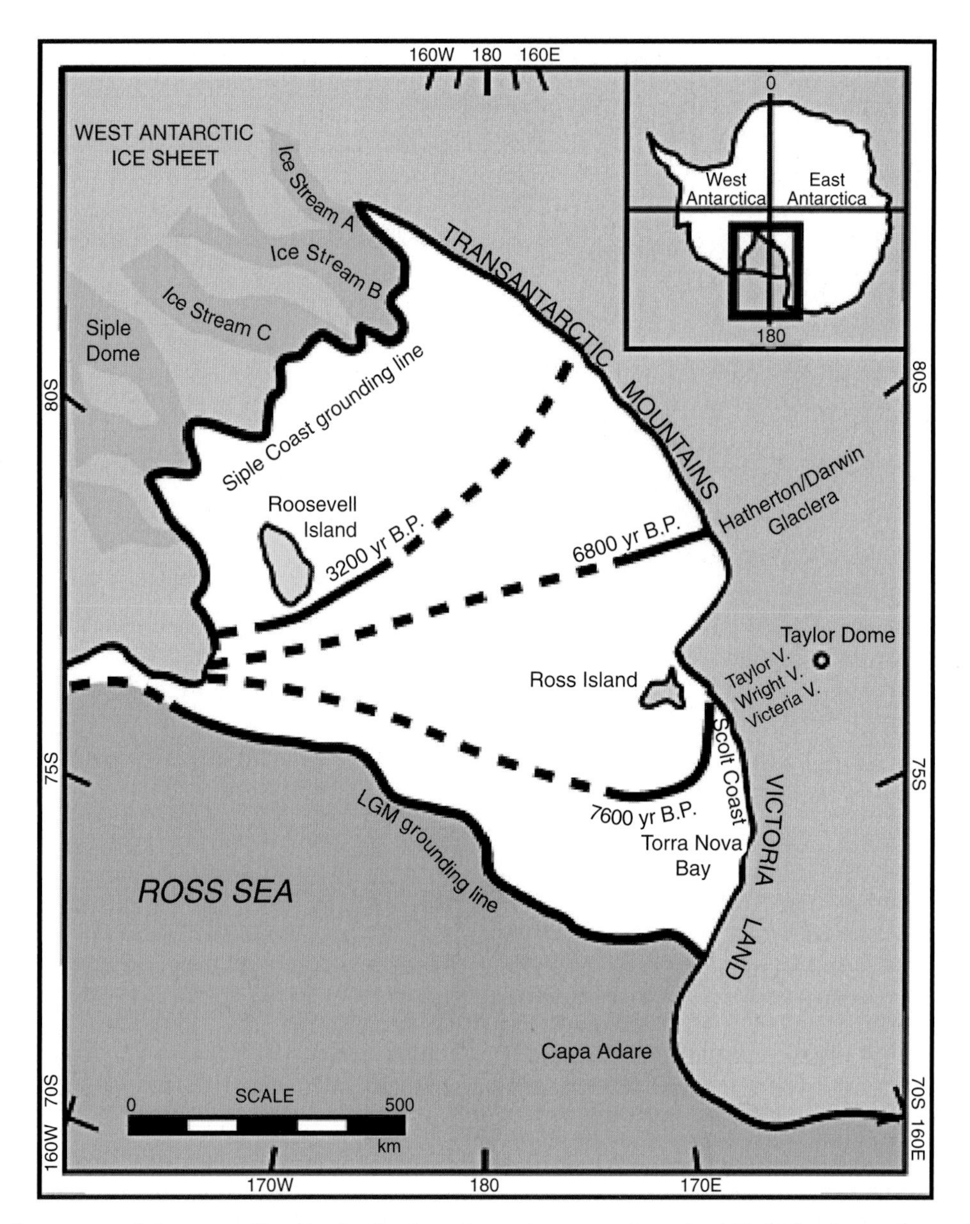

Figure 42.7 The retreat of the grounding line in the Ross Sea embayment since the LGM. The figure shows that the retreat was substantially later than the post-LGM rise in global sea level, and is perhaps still ongoing. (Redrawn from Conway *et al.* (1999), courtesy H. Conway, University of Washington.)

during the last deglaciation, we obtain dates for when those erratics were last covered by a significant thickness of ice, i.e. the timing of deglaciation at that point.

This method of cosmogenic dating has been applied in several areas of Antarctica (Brook *et al.*, 1993; Ackert *et al.*, 1999; Tschudi *et al.*, 2003), but one study provides a detailed insight into the continuing response of the ice sheet to changes after the LGM. Stone *et al.* (2003) used rock-exposure dating to provide evidence of more than 700 m of thinning of the ice sheet in Marie Byrd Land over the past 10 kyr. This provides direct evidence of prolonged retreat of the ice sheet in this area in response to changes after the LGM. Indeed, their measurements show no evidence that the Holocene thinning of the ice sheet has slowed in the past few millennia, and it is entirely possible that Holocene deglaciation of Antarctica may not yet be complete, but continuing at a similar rate to the last few millennia.

A similar type of signal has also been noted in the retreat of the grounding line across the Ross Sea onto the Siple Coast (Fig. 42.7, reproduced from Conway *et al.*, 1999). Conway *et al.* used a variety of glaciological and geological sources to map the retreat of the grounding line in the Ross Sea embayment since the LGM. Although there is uncertainty over the precision of each of the dates, they found reason to believe that the retreat since the LGM was linear, and is still ongoing.

Although Stone *et al.* (2003) measured ice thickness and Conway *et al.* (1999) measured ice-sheet extent, both studies came

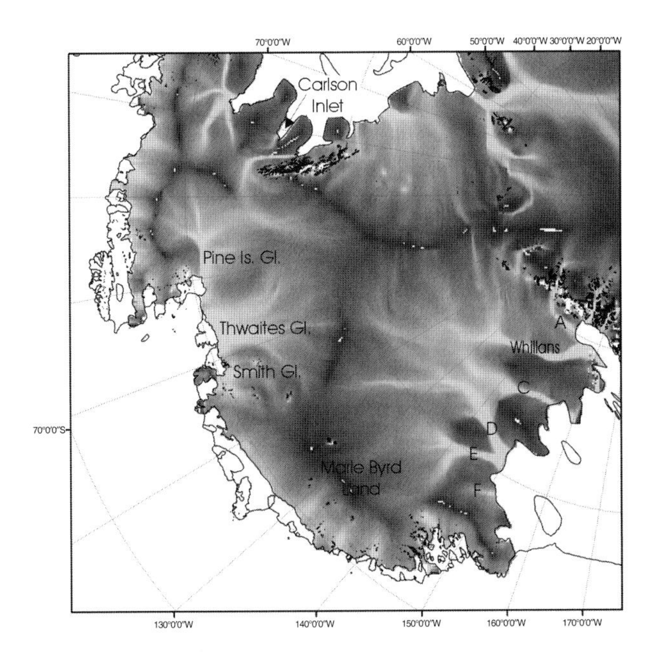

Figure 42.8 Major glaciers in West Antarctica.

to the same conclusion, that the post-LGM response of the ice sheet in West Antarctica was not tied to periods of rapid change in sea level, but has been prolonged and may be ongoing. The mean rate of ice-sheet thinning noted by Stone *et al.*, ca. 7 cm yr^{-1} is of sufficient magnitude that eventually it could become measurable using satellite altimetry, but the likelihood is that it will take many decades and huge improvements in our understanding of snowfall variability before we can be confident in assigning some portion of modern elevation change to the long-term trend, without further evidence from cosmogenic dating.

42.4.2 Ice-stream instability

Despite the shortcomings noted in an earlier section, two recent assessments using the credit/debit approach (Joughin & Tulaczyk, 2002; Rignot & Thomas, 2002) found one additional basin that is significantly out of balance. Both studies considered the part of the West Antarctic Ice Sheet that drains through the five major ice streams of the Siple Coast (Fig. 42.8), an area beyond the coverage of current satellite altimetry. Although, the studies disagree in the detail of their assessments and their estimated uncertainty, taken together, these studies indicate that most of the ice streams (A, Whillans, D, E and F) are not unambiguously out of balance at this time. Both studies do, however, agree that the basin of Ice Stream C is thickening at a mean rate of around 14 cm yr^{-1}— a significant fraction (38%) of the total annual accumulation in this basin.

Although this represents a fairly gross imbalance, it was not entirely unexpected. Evidence of buried crevassing from the downstream portion of Ice Stream C, which is clearly visible in ice-penetrating radar data, suggests that the ice stream effectively shut down around 140–150 yr ago (Retzlaff & Bentley, 1993; Anandakrishnan *et al.*, 2001). The upstream parts of the glacier, which were once tributaries to the main glacier, are still active and the log-jam between active and stagnant ice flow is likely to be causing an area of considerable thickening, perhaps approaching 1 m yr^{-1}. The apparent rapidity with which flow terminated has been a puzzle to glaciologists for some time and the observation of this single glacier has been most influential in driving the study of basal conditions beneath ice streams. Anandakrishnan *et al.* (2001) usefully summarized the several mechanisms that have been proposed to explain the stagnation of Ice Stream C, and although several of these mechanisms can now be discounted, it appears that some combination of diversion of subglacial water, evolution of thermal conditions at the bed, and possible changes in the distribution of subglacial 'sticky-spots' was responsible.

Although Ice Stream C remains the only example of a dated ice stream switch-off, there are other glaciers (e.g. Carlson Inlet, Doake *et al.*, 2001) that may well have behaved in a similar fashion. Even if it were alone, the significance of the stagnation of Ice Stream C is undeniable; it is evidence that an ice-stream basin, once close to balance and which might have looked near to steady-state, changed rapidly and in a direction away from balance. This implies that 'non-linearity' in the internal dynamics of the ice sheet was sufficiently strong to cause reorganization of flow away from the stable state. Such non-linear behaviour (i.e. a response that is not proportional to its cause) is the first requirement for a system to become chaotic. This potential for chaotic behaviour is important because it implies that the apparent proximity to balance, which we see in much of the ice sheet, cannot be taken to imply that rapid and unpredicted changes in flow will not occur.

42.5 Changes in the Amundsen Sea sector

The shutdown of Ice Stream C is clear evidence that non-linear behaviour is possible in particular glacier basins, but the discovery by Wingham *et al.* (1998) of surface elevation change in the Amundsen Sea sector of West Antarctica revitalized the debate about a much more serious type of non-linear behaviour that may govern the fate of the West Antarctic Ice Sheet as a whole.

Even when the maps of Antarctica were still blank, Weertman (1974) presented a theoretical analysis of the junction between an ice shelf and an ice sheet grounded below sea level. This suggested that the grounding line could never be stable, but should always be in the process of migrating, either seaward towards the continental shelf edge or inland. Inland migration, which caused acceleration and thinning of the grounded ice, could lead eventually to complete collapse of the ice sheet, as increasingly large areas thinned sufficiently to float and calve off as icebergs. As much of the West Antarctic Ice Sheet is grounded on rock substantially below sea level, it was argued that this ice sheet was most vulnerable to collapse (Thomas *et al.*, 1979). This vulnerability can be seen in the map of the hydrostatic overburden (Fig. 42.9), which indicates the thickness of ice required to keep the ice sheet in contact with the bedrock for each portion of the ice sheet resting on rock below sea level. In particular, the portion of the West Antarctic Ice Sheet feeding the Amundsen Sea is identified as being particularly vulnerable to collapse because it has a particularly low divide resting on bedrock several thousand metres below sea level (Hughes, 1981) and has only narrow ice shelves to buttress the ice sheet.

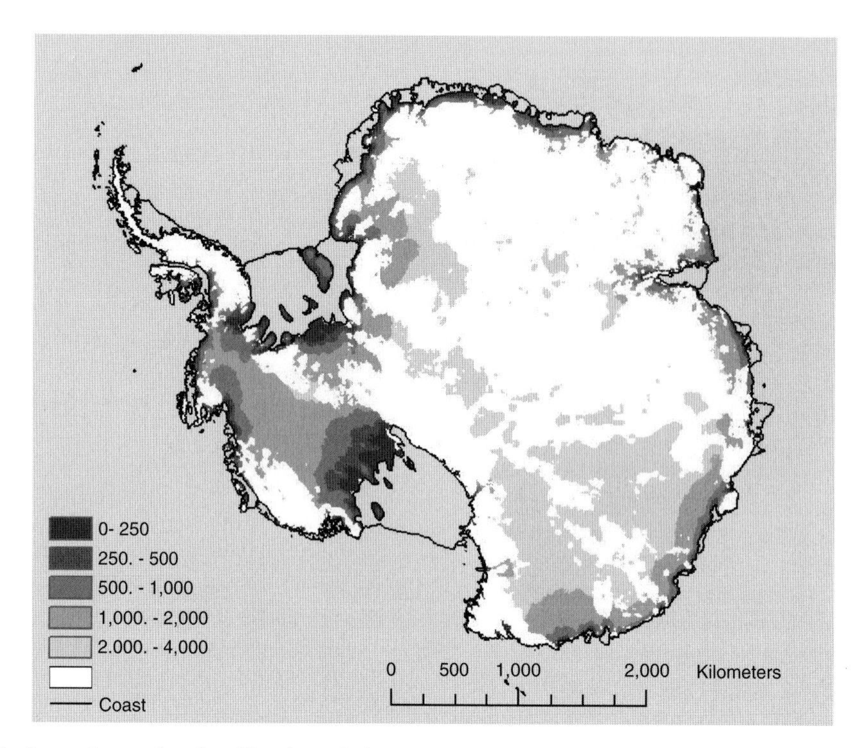

Figure 42.9 The hydrostatic overburden (free-board) for areas of the Antarctic Ice Sheet grounded on rock that is currently below sea level. Unshaded areas of the ice sheet indicate areas grounded above sea level. In simple terms, the freeboard in any particular area indicates the thickness change required before the ice sheet begins to float. (See www.blackwellpublishing.com/knight for colour version.)

Table 42.1 Reported basin-scale changes in the Amundsen Sea sector of the Antarctic Ice Sheet

Reference	Data source	Observation	Period of observation	Conclusion
(Wingham *et al.*, 1998)	ERS-1, 2 altimetry	Mean surface lowering of 0.12 ± 0.01 m yr^{-1}, for the non-coastal parts of the basins from Pine Island Glacier to Smith Glacier	1992–1996	Apparent basin-scale imbalance possibly resulting from variability in snowfall or possibly from dynamic change
(Shepherd *et al.*, 2001)	ERS-1, 2 altimetry	Confirmed the mean thinning over the Pine Island Glacier basin as 0.11 ± 0.01 m yr^{-1}	1992–1999	
(Shepherd *et al.*, 2002)	ERS-1, 2 altimetry	Mean surface lowering of basins from Pine Island Glacier to Smith Glacier, at rate of 0.09 ± 0.02 m yr^{-1}. But this signal dominated by lowering > 1 m yr^{-1} of fast flowing ice streams and ca. 50 cm yr^{-1} on their tributaries, and insignificant change on interior of basins	1991–2001	Correlation of surface lowering with fast flow indicates dynamic change in all three glaciers

Although the evidence for potentially rapid ice-sheet retreat, which exists in the record of global sea level, is undeniable, the general instability of marine ice sheets is not universally accepted. For example, it has been shown that Weertman's (1974) original analysis did not include elements such as ice streams that may act to allow the grounding line to achieve a neutral equilibrium, in which collapse was not inevitable (Hindmarsh & Le Meur, 2001). Although the surface lowering of Pine Island and Thwaites basins was taken by some as evidence for emergent collapse, Wingham *et al.* (1998) were rightly cautious in their interpretation. They noted that the elevation change was 'not unusual in comparison with the expected snowfall variability' over the period and that a 'snowfall variation is certainly implicated in the volume reduction'.

The debate remains unresolved but the results of Wingham *et al.* (1998) provided the first clear evidence that a significant portion of the West Antarctic Ice Sheet is thinning at a measurable rate. This, together with the observation by Rignot (1998) that a portion of the grounding line of Pine Island Glacier was retreating, spurred many researchers to focus on this area and investigate the changes, and their causes, in greater detail. The results for the basin-scale imbalance and the particular changes in Pine Island Glacier are summarized in Tables 42.1 & 42.2, and the pattern that has emerged is coherent, if not yet fully interpretable.

Table 42.2 Recently reported changes on Pine Island Glacier

Reference	Data source	Observation	Period	Conclusion
(Rignot, 1998)	ERS-1 InSAR	Retreat of $1.2 \pm 0.3\,km\,yr^{-1}$ in the central portion of the grounding line, implying thinning of $3.5 \pm 0.9\,m\,yr^{-1}$	1992–1996	First indication of non-steady behaviour on recent time-scales
(Shepherd *et al.*, 2001)	ERS-1, 2 altimetry	Mean thinning of $0.75 \pm 0.07\,m\,yr^{-1}$, on the trunk, with greatest rates of thinning nearest to the grounding line, and with surrounding areas showing only minor changes	1992–1999	Glacier thinning too large to be due to snowfall variability, indicating probable dynamic origin
(Rignot *et al.*, 2002)	ERS-1, 2 InSAR	An increase of 18 ± 2% in speed of the downstream part of the glacier over 8 yr, with increase in the rate of acceleration in 1996–2000, over 1992–1996	1992–2000	Confirmation of sufficient dynamic variability to implicate acceleration as a cause of thinning
(Rignot, 2002a)	Aerial photography, various satellite imagery	No discernable migration of the ice front of the floating portion of Pine Island Glacier, although most recent images show cracks further inland than previously noted.	1947–2000	Floating portion of the glacier is not retreating dramatically, although continued thinning is noted. Suggests that a significant oceanographic change is possible
		Progressive retreat of adjacent ice-shelf front	1947–2000	
		Further retreat of central portion of the grounding line, implying thinning of 21 m over 8 yr	1992–2000	
(Bindschadler, 2002)	Landsat imagery	Progressive widening (ca. 5 km) of the floating portion of Pine Island Glacier into adjacent ice shelf	1973–1997 (possibly since 1953)	Confirms thinning of ice shelf adjacent to Pine Island Glacier and likely oceanographic change
		Thinning of up to 134 m of the ice shelf, adjacent to north of Pine Island Glacier	1973–2001	
		New areas of crevassing close to the grounding zone of Pine Island Glacier	1973–2001	
(Rabus & Lang, 2003)	ERS-1, 2 SAR Feature-tracking	A progressive increase in velocity of Pine Island Glacier of ca. 12% over 8 yr	1992–2000	Independent confirmation of earlier result (Rignot *et al.*, 2002)
(Joughin *et al.*, *et al.*, 2003b)	Landsat feature-tracking, ERS-1, 2 InSAR	Two roughly equal periods (1974–1987, and 1996–2000) of acceleration (totalling 22%) on the floating and grounded glacier. These periods of acceleration separated by a period of steady flow	1974–2000	Refinement of timing of glacier acceleration and indication of stepped change

In summary, it appears that much of the floating portions of Pine Island Glacier and adjacent ice shelves have been thinning for many decades. This thinning appears to be broadly consistent with the rate of grounding-line retreat. There have been some periods of substantial acceleration of the lower reaches of the glacier that are unlikely to be due to a change in snowfall rate, or even a shift to general erosion, and so must be due to dynamic change. Its neighbour, Thwaites Glacier, appears not to have accelerated but to have increased its flux by widening (Rignot *et al.*, 2002), and both Pine Island Glacier and Thwaites have reached a quite distinctly negative mass balance. The most recent estimate for the mean thinning across the entire Amundsen Sea sector suggests a contribution of ca. 0.04 mm of sea-level rise. This is, however, dominated by thinning in the coastal parts of the glaciers, and it is possible that the interior is not thinning at all (Shepherd *et al.*, 2002). This, combined with the fact that changes in the velocity of Pine Island Glacier occurred quite recently and over a very short period, should cause us to be extremely cautious in predicting that this negative mass balance will persist long enough to have a significant impact on sea level.

Thinning is concentrated in the downstream portions of the glaciers and is similar across several neighbouring basins, which implies a cause other than an Ice Stream C-type instability. It is tempting to suspect as a root cause a change in the oceanic boundary condition, such as an increase in the supply of warm water to the ice sheet, which would act regionally at the glacier fronts (Payne *et al.*, 2004; Shepherd *et al.*, 2004). Such a hypothesis would fit in with the thinning of ice shelves adjacent to Pine Island Glacier. We have, however, no time-series of oceanographic measurements that could prove this, and it is difficult to rule out the possibility that the change is part of an ongoing, stepwise Holocene retreat of the ice sheet, or even a sign of emergent collapse. The causes of change in the Amundsen Sea sector remain uncertain, but our increasing understanding of the many potential factors that could cause such change should now make us cautious of interpreting this as evidence of imminent ice-sheet collapse.

42.6 Summary

The preceding sections suggested that Antarctica's potential to protect us from the hazard of sea-level rise is considerably less than its potential to make the problem much worse. Predicting the future of the great ice sheets is likely to become increasingly important as global climate change is finally accepted at a governmental level. The present state of change in the Antarctic and Greenland ice sheets was succinctly summarized by Rignot & Thomas (2002): 'As measurements become more precise and more widespread, it is becoming increasingly apparent that change on relatively short time-scales is commonplace: stoppage of huge glaciers, acceleration of others, appreciable thickening and far more rapid thinning of large sectors of ice sheet, rapid breakup of vast areas of ice shelf and acceleration of tributary glaciers, surface melt-induced acceleration of ice sheet flow [although probably only in Greenland], and vigorous bottom melting near grounding lines. These observations run counter to much of the accepted wisdom regarding ice sheets, which, lacking modern observational capabilities, was largely based on 'steady-state' assumptions.'

My opinion is similar, if less succinctly stated. There is now a growing raft of evidence that large changes are afoot in the major glaciers in the Amundsen Sea sector of West Antarctica. Despite this evidence and the long-lived speculation that this portion of the West Antarctic Ice Sheet could be particularly prone to collapse, however, it is too early to predict disaster. Indeed, the overall opinion of a panel of experts brought together to assess the risk of collapse of the West Antarctic Ice Sheet was that there remains a 5% probability that this sheet will make a 1 $cm\,yr^{-1}$ contribution to sea-level rise, and a 30% probability of a 2 $mm\,yr^{-1}$ contribution over the next 200 yr (Vaughan & Spouge, 2002). As with all assessments of risk, though, the likelihood of a particular hazard occurring must be balanced by its potential to cause harm, and even a relatively small addition to predicted sea-level rise could have huge social and economic consequences—the lack of certainty over the likelihood of collapse cannot be taken as good reason to reduce the relatively modest efforts in trying to make reliable and defensible predictions of change in general. It is also the case that we cannot afford to focus all our effort on this one area, because observational evidence of change elsewhere in the Antarctic Ice Sheet is also building; we have observed local changes on the Antarctic Peninsula that are probably a direct and immediate response to the recent rapid regional atmospheric warming and we have strong evidence that ongoing Holocene ice-sheet thinning may be persistent across much of West Antarctica. Finally we should bear in mind that the stagnation of Ice Stream C demonstrates the potential for highly non-linear and rapid change in the ice streams which are so significant in controlling the rate of Antarctic ice loss.

Without ruling out surprises, there are five mechanisms of change that may become important in the Antarctic Ice Sheet over the predictable future.

1 In the warmest areas, on the Antarctic Peninsula and perhaps in some coastal areas of East and West Antarctica, contemporary climate warming may cause increased melting and promote immediate glacial retreat. This is likely to be particularly noticeable where ice shelves are driven to retreat.
2 Subtle changes in the temperature of ocean water in contact with the ice sheet around the grounding line may drive increased melt and grounding-line retreat. We must work to understand if such changes have the potential to interact with glacier acceleration to amplify retreat or even to lead to collapse of whole drainage basins.
3 There is a high likelihood that climate change will cause increased precipitation over much of the interior of the Antarctic Ice Sheet. This could cause thickening of the ice sheet and act to compensate to some extent for other sources of sea-level rise.
4 A continued retreat in response to long-term Holocene warming may continue to cause continued background thinning of the ice sheet, especially in West Antarctica.
5 Internal instability, or non-steady behaviour of ice streams and outlet glaciers, may cause rapid, hard to predict changes, especially in West Antarctica.

Each of these mechanisms could have a substantial impact on the ice sheet and hence on future sea levels.

The clear task facing researchers is to begin to understand each of these mechanisms, their likely longevity and predictability. Given the theoretical and proven potential of ice sheets to exhibit non-linear responses, if several of these effects operate together we may find the pattern a complex one to unravel, let alone predict with confidence. We will require much more data describing contemporary and past changes and an improvement in our understanding of the basic mechanics of glacial flow and restraint even to attempt the task.

Acknowledgement

My thanks go to Hamish Pritchard for his thorough comments on the manuscript.

FORTY-THREE

Antarctic Ice Sheet reconstruction using cosmic-ray-produced nuclides

Greg Balco, Seth Cowdery, Claire Todd and John O. H. Stone

Quaternary Research Centre and Department of Earth and Space Sciences, University of Washington, Mail Stop 351310, Seattle WA 98195-1310, USA

43.1 The problem: finding, mapping and dating Antarctic ice-marginal deposits

The Antarctic Ice Sheets are the largest extant ice masses on Earth, and understanding their history is relevant not only to past environmental changes but also to ongoing changes in global climate and sea level. The glacial-geological record in Antarctica provides a means of reconstructing this history, but the unique features of the Antarctic environment present several challenges that do not arise in more temperate latitudes. In addition to the basic fact that the Antarctic continent is nearly completely covered in ice, leaving few exposed surfaces on which glacial deposits might be preserved, much of the ice in Antarctica is frozen to its bed. It transports little sediment to the terrestrial ice margins that do exist, and may advance and retreat repeatedly without appreciably disturbing the landscape. Antarctic moraines and glacial drift are usually very thin, often consisting only of scattered cobbles on an otherwise bare bedrock surface. The combination of cold-based ice and extraordinarily slow rates of subaerial erosion during ice-free periods means that the deposits of both recent and long-past glacier advances and retreats may not only be found together, but be nearly indistinguishable. Finally, not only is it difficult to identify and correlate ice-marginal deposits, but there exist few ways to date them. Organic material that could be radiocarbon dated is rarely found outside of coastal areas, and waterlain sediment suitable for optical dating techniques is equally unusual.

43.2 A solution: exposure-age dating with cosmic-ray-produced nuclides

The chief recent advance in understanding the history of the Antarctic Ice Sheets, therefore, has been the development of a dating technique that is perfectly suited to the Antarctic landscape: exposure-age dating with cosmic-ray-produced nuclides. This relies on the measurement of rare nuclides such as ^{10}Be, ^{26}Al and ^{3}He, which are produced within mineral grains by cosmic-ray bombardment of rocks exposed at the Earth's surface. These nuclides are useful for dating ice-marginal deposits because nearly all cosmic rays stop within a few metres below the rock (or ice) surface, so any clast that is quarried by subglacial erosion at the bed of the ice sheet and brought to the ice margin arrives there with a negligible nuclide concentration. The surface production rate of these nuclides varies in a predictable way with altitude (Stone, 2000), and once this is determined, the nuclide concentration in an erratic cobble or boulder is related only to the duration of subsequent surface exposure, that is, the time since the ice margin lay at that position. Thus any erratic, lying on any previously glaciated surface, is a record of the past ice-sheet configuration. Inferring deglaciation ages from nuclide concentrations in erratics relies only on the two assumptions that the rock samples of interest have been emplaced with zero nuclide concentration, and that they have not been eroded, moved, or covered with a significant thickness of soil or snow since exposed by ice retreat. In principle, these assumptions could be true of bedrock surfaces eroded subglacially and exposed by deglaciation as well as erratic clasts. The important difference is that, in practice, there is generally no assurance that subglacial erosion was sufficient to remove any nuclide inventory that might date from a previous period of exposure (e.g. Briner & Swanson, 1988). Cosmic-ray-produced nuclide concentrations in bedrock surfaces can provide some information about glacial history (e.g., Stroeven *et al.*, this volume, Chapter 90); however, they are hard to interpret. For Antarctic erratics, on the other hand, both assumptions are nearly always met. First, the extreme scarcity of exposed rock, and near absence of supraglacial debris, in most of the continent means that any glacially transported clast was almost certainly derived

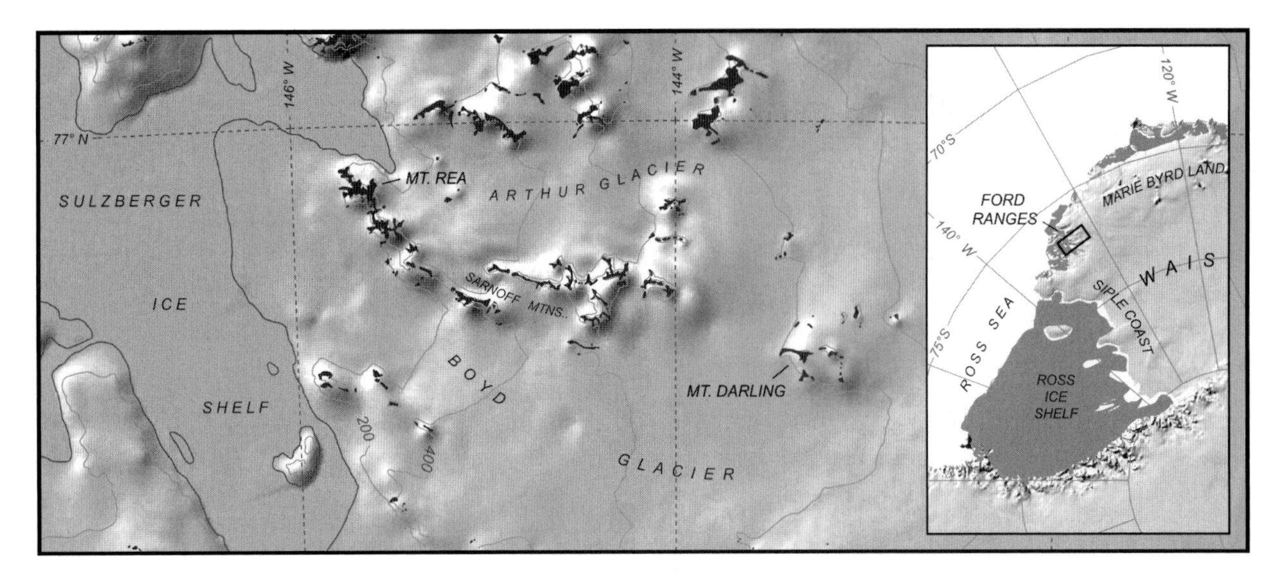

Figure 43.1 Map of part of the Ford Ranges of Marie Byrd Land, West Antarctica. Shaded-relief background from RAMP-DEM version 2; vector data from the Antarctic Digital Database.

from the glacier bed and thus will have negligible inheritance. In the case of erratics of lithologies that do not crop out at all above the ice surface, this assumption is always true. Second, once such a clast arrives at the ice margin, the typical occurrence of only thin and patchy glacial deposits, and the windy, arid climate of most nunataks, mean that it is unlikely to be covered by soil or snow, and the extremely slow rates of erosion ensure that it is preserved unmodified for very long periods of time.

43.3 An example from the Ford Ranges of West Antarctica

The Ford Ranges of West Antarctica consist of scattered nunataks that separate large outlet glaciers draining the interior of the West Antarctic Ice Sheet into the Sulzberger Ice Shelf (Figs 43.1 & 43.2). Ice-free surfaces in the Ford Ranges consist mostly of steep slopes of bare granite and phyllite bedrock or locally derived talus and blockfields. Striated surfaces at all elevations indicate past overriding by the ice sheet. Ice retreat is recorded only by a few patches of thin till and by erratic cobbles scattered on rock surfaces (Fig. 43.2). We collected such erratic cobbles from a range of elevations on several nunataks, measured ^{10}Be concentrations, and found that they yielded a precise record of continuous Holocene lowering of the ice-sheet surface and consequent exposure of the peaks (Stone *et al.*, 2003; Fig. 43.2). Analyses of adjacent cobbles from a few sites agreed within analytical uncertainty, and cobbles very close to the ice margin had exposure ages of only a few hundred years, reinforcing the idea that these cobbles arrived at the ice margin with no inherited nuclide inventory. Sets of cobbles from single nunataks yielded smooth deglaciation histories, and adjacent nunataks produced similar results, indicating that the cobbles had not been significantly covered by snow or sediment, and had not moved since they were emplaced. A few nunataks, however, yielded more complicated results. At Mount Rea (Fig. 43.2) we found that samples at lower elevations, as well as the youngest sample near the summit, recorded the same Holocene deglaciation as on nearby nunataks. At higher elevations, we found many erratics that were physically indistinguishable from Holocene erratics, but had very much greater exposure ages. Some of these old erratics also had ^{26}Al/^{10}Be ratios below the surface production ratio of 6.1, indicating that they had been covered by ice for long periods (Cowdery, 2004; also see Stroeven *et al.*, this volume, Chapter 90). It appears that these erratics were deposited during long-past episodes of ice retreat, and remained undisturbed by subsequent ice advances: their ^{10}Be concentrations integrate several periods of exposure in addition to the most recent one. These repeatedly exposed erratics persist through multiple glacial–interglacial cycles, not only because of the ineffectiveness of weathering processes that might degrade or destroy them during ice-free periods, but also because of the common occurrence of cold-based ice at higher elevations where overriding ice was thinner. Although some parts of Antarctica (the McMurdo Dry Valleys in particular, where surfaces have been exposed for millions rather than thousands of years, and weathering and periglacial disturbance are correspondingly more important) present additional complexities in interpreting cosmogenic-nuclide measurements, much of Antarctica is similar to this example from the Ford Ranges. The near absence of ice-free areas that might supply pre-exposed erratics, the minimal ice-marginal sediment accumulations, and the very slow rates of weathering provide the best possible environment for reconstructing ice-

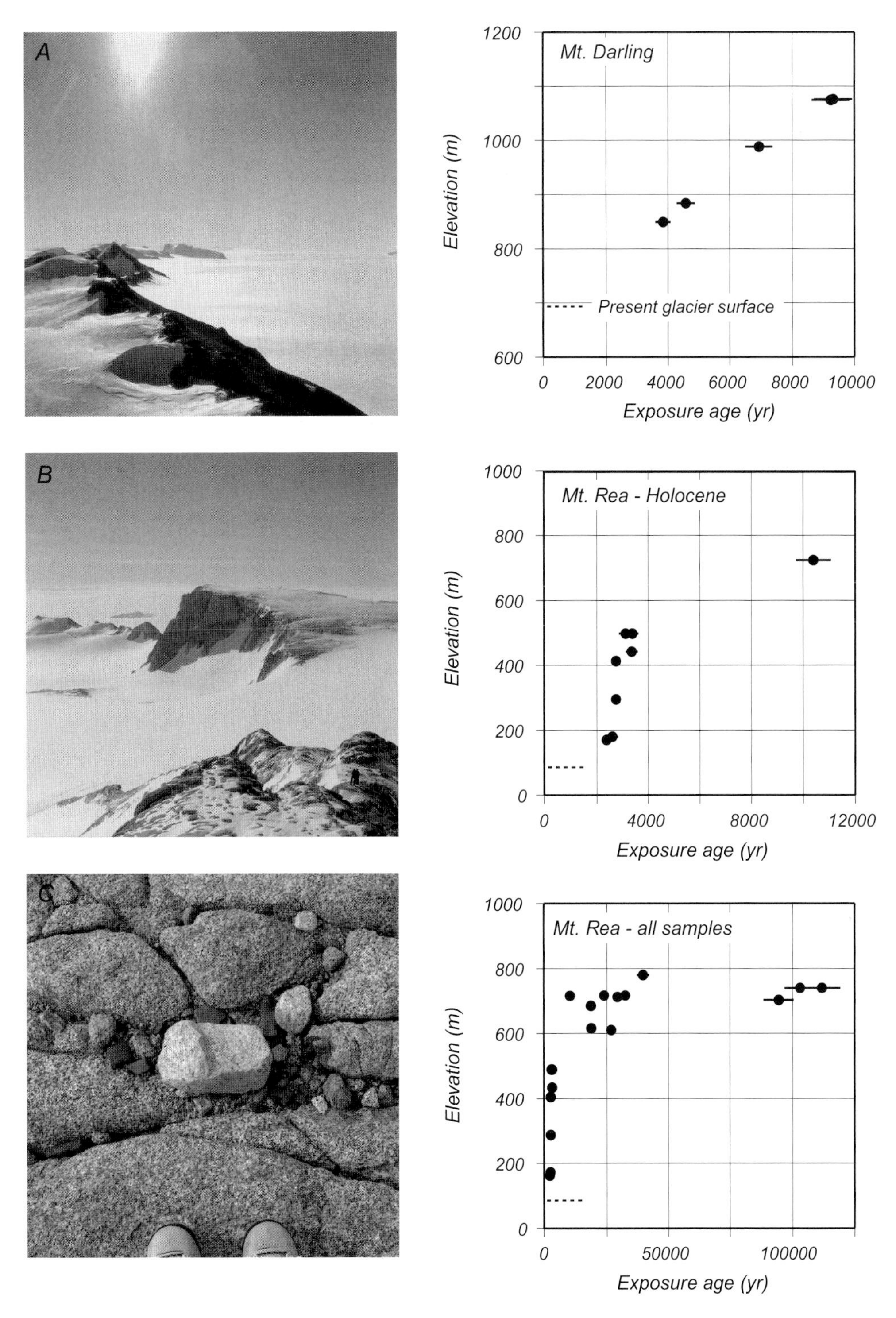

Figure 43.2 Exposure ages of glacial erratics from the Ford Ranges. Photographs: (A) View of the western Sarnoff Mountains and the Mount Rea massif from the east. (B) View of Mount Darling seen from the south. Ice flow is from right to left. (C) Typical granite erratic, resting on granodiorite, in the eastern Sarnoff Mountains. Panels to right: ^{10}Be exposure ages of erratic cobbles from Ford Ranges nunataks. These data were previously published in Stone *et al.* (2003) and Sugden *et al.* (2005).

sheet history by exposure-age dating. If the goal is to accurately reconstruct the most recent deglaciation, it is important to keep in mind that glacial erratics may survive many ice advances and retreats without being disturbed. On the other hand, these persistent erratics are a unique source of other useful information: evidence of the past subglacial temperature distribution, and a potential record of not one but many past ice advances and retreats.

FORTY-FOUR

Current status and recent changes of the Greenland Ice Sheet

Niels Reeh

Ørsted-DTU, Technical University of Denmark, Building 348
Ørsteds Plads, DK-2800 Kgs., Lyngby, Denmark

44.1 Introduction

The past decade has seen major advances in our knowledge of the mass balance of the Greenland Ice Sheet (PARCA, 2001). Based on this new knowledge, Rignot & Thomas (2002) concluded that, presently, the Greenland Ice Sheet is losing mass at a rate sufficient to raise sea level by 0.13 mm yr^{-1} because of rapid near-coastal thinning. However, it is not known whether this change is part of a long-term trend, or whether it reflects short-term temporal variability of accumulation and melt rates. So, despite many decades of intensive field studies and more recent air- and satellite-borne remote sensing measurements, we still do not know whether, in a longer term view, the Greenland Ice Sheet is increasing or decreasing in mass.

Two main factors contribute to the difficulty of measuring total ice-sheet mass balance:

1 short-term (interannual to decadal scale) fluctuations in accumulation and melt rate cause variations in surface elevation that mask the long-term trend;
2 climate changes that occurred hundreds or even thousands of years ago still influence ice flow and, therefore, current mass changes, as do more recent climate changes.

The total mass balance of the ice sheet can be determined by two entirely different methods: (i) direct measurement of the change in volume by monitoring surface elevation change, and (ii) the budget method, by which each term of the mass balance budget (accumulation, runoff, iceberg calving) is determined separately. The strengths and limitations of the two methods in the context of modern, remote-sensing observation techniques are discussed by Reeh (1999).

44.2 Surface elevation change

The glacial-geological record and the historical record show that, generally, the marginal zone of the ice sheet has thinned and retreated over the past 100 yr (Weidick, 1968). Whether this mass loss was compensated, partly or fully, by thickening in the interior is unknown. Although several expeditions have crossed the ice sheet since the late nineteenth century, the earliest data of sufficient precision to permit calculation of surface-elevation change appear to be those of the British North Greenland Expedition (BNGE) that crossed the ice sheet in 1953–1954. Comparing these data with modern surface elevations measured by satellite radar altimetry and airborne laser altimetry shows that between 1954 and 1995, ice thickness along the BNGE traverse did not change much on the northeast slope, whereas the northwest slope thinned at a rate of up to 30 cm yr^{-1} (Paterson & Reeh, 2001), see Fig. 44.1. Repeated height measurements in 1959, 1968 and 1992 along another profile across the ice sheet in central Greenland showed thickening on the west slope between 1959 and 1968, but subsequent thinning between 1968 and 1992 (Möller, 1996), probably reflecting decadal-scale fluctuations in accumulation rate.

Since 1978, satellite-borne radar altimeters have allowed precise monitoring of the surface elevation of interior regions of the ice sheet south of 72°N. In contrast to the previous traverse data, these measurements provided extensive area coverage. Other satellite-borne radar altimeters extended the area coverage of surface-elevation measurements to almost the entire ice sheet, showing the interior regions to be close to balance (within 1 cm yr^{-1}) for the past few decades. However, local areas showed quite rapid thickening or thinning (Thomas *et al.*, 2001) (see Plate 44.1b), probably resulting from short-term variations of snow accumulation (McConnell *et al.*, 2000b). Airborne laser-altimeter measurements from the period 1994–1999 confirmed this result for the interior ice-sheet region, but showed substantial thinning along the periphery of the ice sheet (Krabill *et al.*, 2000). Thinning rates of more than 1 m yr^{-1} were measured in many outlet-glacier basins in southeast Greenland, in some cases at elevations up to 1500 m (see Plate 44.1c), possibly indicating dynamic thinning caused by increased glacier velocities resulting from increased meltwater supply to the glacier bottom.

In order to interpret volume changes derived from repeated accurate measurements of surface-elevation change in terms of mass changes—the relevant quantity influencing global sea

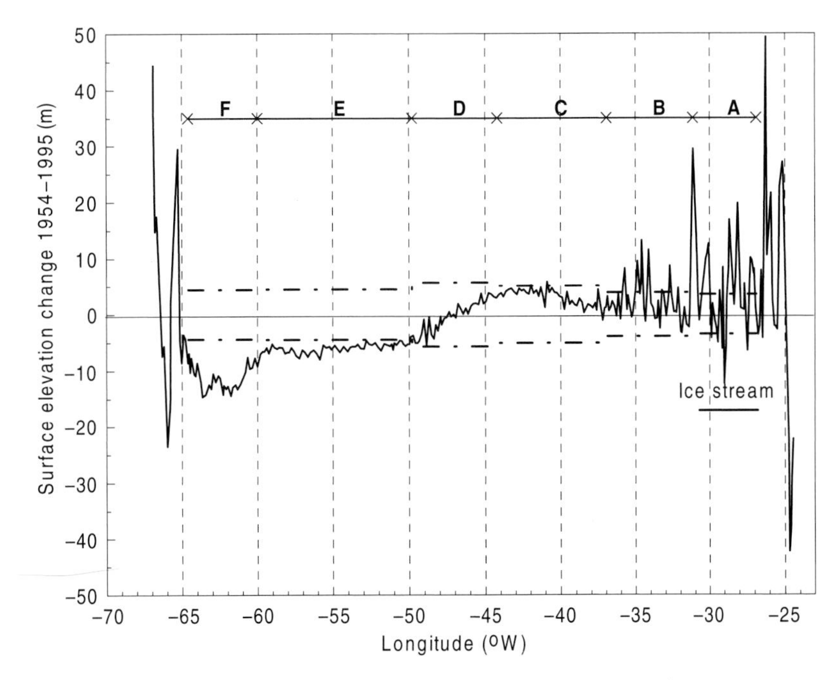

Figure 44.1 Measured changes in surface elevation between 1994 and 1995 along the traverse route of the British North Greenland Expeditions. For location, see map in Plate 44.1. (From Paterson & Reeh, 2001.)

level—temporal changes of specific mass balance and near-surface temperature, and their influence on surface density and hence ice-sheet volume change, must be understood and accounted for. Changing snow fall over the central accumulation region and changing ice-melt rates in the marginal ablation area will result in immediate surface-elevation (volume) changes that can be translated directly into mass changes by using the density of firn or ice. In the accumulation zone, temporal variations of surface energy balance and accumulation rate will, in addition, cause temporal changes of the rate of densification of the near-surface layers, giving rise to a surface-elevation (volume) change, not accompanied by a corresponding change of mass. Model studies of firn densification (Arthern & Wingham, 1998; Cuffey, 2001) analysed this effect in the dry-snow region of ice sheets where surface melting/refreezing is insignificant. Arthern & Wingham (1998) showed that, in central Greenland, the maximum expected rate of change of surface-elevation due to time-dependent density changes is $3\,\mathrm{cm\,yr^{-1}}$, $1.5\,\mathrm{cm\,yr^{-1}}$ and $0.5\,\mathrm{cm\,yr^{-1}}$ for, respectively, a 10% step-change in surface density, a 10% step-change in accumulation rate, and a 1 K step-change in surface temperature. These surface-elevation changes are significant, but small. Much larger surface-elevation changes without a corresponding change of mass may occur in the percolation and wet-snow zones of the ice sheet (for definition of zones in an ice sheet, see Paterson (1994, chapter 2)). Here, the density–depth profile is to a large extent controlled by melting–refreezing processes, by which low-density surface snow is transformed into high-density ice lenses deeper in the snow pack (Paterson, 1994, chapter 2). A temperature change—or rather a change of the energy balance at the surface—will change the amount of surface melting and the subsequent refreezing of ice lenses or formation of superimposed ice. The resulting density change will cause an immediate change of surface elevation, the magnitude of which in the wet snow zone may reach values of 0.1–$0.2\,\mathrm{m\,yr^{-1}}$. As first pointed out by Braithwaite *et al.* (1994), this has important consequences for how observed changes of ice-sheet surface elevation should be interpreted: a change of surface elevation does not necessarily mean that the mass of the ice sheet has changed. Part of the elevation change, or all of it, may be due to a change of surface-layer density, and therefore may merely reflect a change of ice-sheet volume without a corresponding change of ice-sheet mass. This is illustrated in Plate 44.2, which, for a surface-temperature increase of 1 K over the Greenland Ice Sheet, shows the calculated increase of ice-lens formation (Plate 44.2a), and the part of the surface-elevation change not contributing to the mass change (Plate 44.2b). The accumulation rate is assumed to be unaffected by the temperature change. Moreover, the small effect on snow density of the temperature change considered by Arthern & Wingham (1998) is neglected. The increase of ice-lens formation is derived by means of the melting–refreezing model presented by Reeh (1989). For the total Greenland Ice Sheet, the increased volume loss resulting from a 1 K change of surface temperature amounts to $128\,\mathrm{km^3\,yr^{-1}}$, whereas the increased mass loss by runoff amounts to only $96\,\mathrm{km^3\,yr^{-1}}$ of ice equivalent, showing that only 75% of the volume change for a 1 K warming actually represents a change of mass.

44.3 Mass budget

The mass-budget method was applied by Thomas *et al.* (2000b) to estimate the mass balance of the interior region of the Greenland Ice Sheet (Plate 44.1a). They compared the total snow accumulation over Greenland ice-sheet drainage basins with the flux out of them. The study included determination of surface ice motion by using repeated GPS measurements at 30 km intervals along the 2000-m elevation contour around the whole ice sheet. The ice thickness along the same line was measured with airborne ice radar. The ice flux across the 2000-m elevation contour was calculated with due account of the difference between surface velocity and depth-averaged velocity. This method should measure the dynamic response of the ice sheet on a longer time-scale, because flow velocity is believed to be relatively unaffected by moderate changes of surface slope, ice thickness, internal ice temperature and basal conditions that might occur on short time-scales. Fast-flowing ice streams embedded in the ice sheet, for example the ice stream feeding the Jacobshavn Glacier in West Greenland (Echelmeyer & Harrison, 1990) and the Northeast Greenland ice stream (Fahnestock *et al.*, 1993), however, may undergo significant short-term changes of ice dynamics. Apart from problems caused by possible unsteady ice-stream motion, the method should provide a reasonably accurate estimate of the long-term average mass flux out of the central Greenland region. In order to assess the mass balance of the region, however, this mass flux must be compared with the long-term average input to the ice-sheet area above the 2000-m contour line. As mentioned in the following section, a reliable dataset providing such information is still missing, causing some uncertainty of the mass-balance estimate. The derived average thickness changes for different regions of the ice sheet are shown in Plate 44.1c. Whereas the total interior ice sheet region appears to be nearly in balance (thickness change = $0 \pm 7\,\mathrm{mm\,yr^{-1}}$), some individual regions appear to be undergoing large changes. In particular, a large area of the ice sheet in the southeast has thinned significantly ($261 \pm 52\,\mathrm{mm\,yr^{-1}}$) over the past few decades in accordance with the short-term changes inferred from radar- and laser-altimeter measurements.

The individual terms of the mass budget of the Greenland Ice Sheet: accumulation rate (520 ± 26), runoff (329 ± 32), and iceberg calving (235 ± 33) are also encumbered with a large uncertainty. Numbers are from Houghton *et al.* (2001) in $10^{12}\,\mathrm{kg\,yr^{-1}}$.

44.3.1 Accumulation rate

Accumulation rate compilations are based on spatially and temporally inhomogeneous information from pits and ice cores. From these data, the average accumulation rate over the ice sheet is estimated to be ca. $0.30\,\mathrm{m\,yr^{-1}}$ (H_2O) with a standard error of 0.02–$0.03\,\mathrm{m\,yr^{-1}}$ (Bales *et al.*, 2001). Plate 44.3 shows the distribution of accumulation rate based on the recent compilation by Bales *et al.* (2001).

44.3.2 Runoff

Most measurements of surface melt and runoff have been concentrated in southwest Greenland with a few scattered observations from other regions (Weidick, 1995). The results, obtained by the traditional method of repeated reading of stakes drilled into the ice, have been extended to the entire ablation region of the ice sheet by degree-day or energy-balance modelling (Reeh, 1989; Van de Wal & Oerlemans, 1994). Records of annual melt duration derived from satellite microwave radiometric data also have been used to regionalize *in situ* point measurements of melt rate (Mote, 2000). This study indicates large interannual variability of the runoff. For the total ice sheet, for example, the annual runoff varied by a factor of five within the period 1988–1996, with a minimum of $115\ 10^{12}\,\mathrm{kg\,yr^{-1}}$ in 1992 and a maximum of $590\ 10^{12}\,\mathrm{kg\,y^{-1}}$ in 1995. A key issue for improving estimates of ice-sheet runoff is to utilize satellite remote-sensing information to obtain a better understanding of spatial and temporal variations of albedo on the ice margin (Stroeve, 2001). Another promising method is to increase the density of observationally based ice-sheet melt rates by combining space- and airborne remote sensing measurements of ice motion, ice thickness and surface-elevation change (Reeh *et al.*, 2002)—see section 44.4 on ice dynamics.

44.3.3 Calving of icebergs

Iceberg calving is the term for the Greenland Ice Sheet mass budget that is still encumbered with the largest uncertainty. Within the past few decades, Satellite Synthetic Aperture Radar (SAR) Interferomery has revolutionized measurement of ice-sheet flow, providing area distributions of surface velocity which, when combined with ice-radar measurement of calving-front thickness, allows derivation of iceberg calving fluxes. However, until now the new technique has been applied only to north and northeast Greenland (Rignot *et al.*, 2000, 2001). Observations of calving rates by traditional methods, i.e. repeated surveying from fixed rock points and repeated aerial photography, are summarized by Reeh (1994) and Weidick (1995). Table 44.1 shows measured calving rates from Greenland glaciers. The total measured calving flux amounts to ca. $211\,\mathrm{km^{-3}\,yr^{-1}}$ of ice equivalent corresponding to ca. $193\ 10^{12}\,\mathrm{kg\,yr^{-1}}$. Still, however, many calving glaciers, particularly in northwest and southeast Greenland, have not been studied. Hence, only about 80% of the estimated total calving flux of $235\ 10^{12}\,\mathrm{kg\,yr^{-1}}$ from Greenland is based on observations of calving-front thickness and velocity. Moreover, most of the calving-front velocities used for calculating the calving fluxes shown in Table 44.1 are 'snap shot' measurements over short periods (a few days or at best a few weeks) performed at different times during the past 50 yr. Summing up, application of SAR interferometry and other remote-sensing techniques has, until now, not substantially reduced the uncertainty of the estimate of the total iceberg calving loss from the Greenland Ice Sheet.

44.3.4 Basal melting

For ground-based ice-sheet regions, basal melt/freeze-on rates are normally in the order of a few millimetres per year and, therefore, in general are neglected in the mass balance budget. An exception is reported by Fahnestock *et al.* (2001) who derived basal melt rates of up to $15\,\mathrm{cm\,yr^{-1}}$ beneath an ice stream and adjacent areas in northeast Greenland, probably caused by higher than normal geothermal heat flow. On the other hand, for the extended floating glacier tongues in north and northeast Greenland, basal melting averaging up to $10\,\mathrm{m\,yr^{-1}}$ is the dominant ablation mechanism (Reeh *et al.*, 1999; Mayer *et al.*, 2000; Rignot *et al.*, 2001). At the grounding line, peak values of basal melt rate up to $40\,\mathrm{m\,yr^{-1}}$ are derived (Mayer *et al.*, 2000; Rignot *et al.*, 2001). The

Table 44.1 Measured calving fluxes in ice equivalent from the Greenland Ice Sheet

Region	Latitude band (°N)	Calving flux ($km^{-3} yr^{-1}$)	Grounding-line flux ($km^{-3} yr^{-1}$)	Reference
South and southwest	61.0–61.3	7.2		Weidick (1995)
Central west	68–72	90.2–101.6*		Reeh (1994)
Northern west	72–75	14.1		Weidick (1995)
Northwest	75–78	67.1		Weidick (1995)
North (Humboldt Gletscher)	79–80	7.7		Weidick (1995)
North	80–82	5.6	31.0	Weidick (1995), Rignot *et al.* (2002)
Northeast	76.5–80	2.0	10.8	Rignot *et al.* (2002)
Central east	70–72	17.7		Reeh (1994)
Total		211.6	41.8	

*The two different numbers are based on aerial photographs taken in 1957 and 1964, respectively.

difference of ca. 35 $km^{-3} yr^{-1}$ of ice between grounding-line flux and calving flux for north and northeast Greenland glaciers derived from Table 44.1 is due mainly to bottom melting.

It is apparent from Table 44.1 that the major part of the Greenland icebergs is produced from glaciers in south, central, and northwest Greenland. The calving fronts of these glaciers are located at or within a few kilometres of their grounding lines, and iceberg calving occurs relatively frequently. In contrast, the release of icebergs from the north and northeast Greenland floating glaciers occurs at much larger intervals. The terminal parts of these glaciers disintegrate into kilometre sized, relatively thin, 'ice islands' that during long periods are hindered from drifting away by semi-permanent land-fast ice in the fjords or on the inner shelf. Changing the ocean conditions underneath the floating glaciers or increasing summer air temperatures is likely to mean relatively fast collapse of the floating tongues (Reeh *et al.*, 1999; Rignot *et al.*, 2001), and consequently a retreat of the glacier front to a position near the grounding line. As presently in south, central and northwest Greenland, iceberg calving would take over as the dominant mass loss, and, as a consequence, the concentration of drifting icebergs in the East Greenland Polar Current would increase.

44.4 Ice dynamics

Besides being used for determining calving fluxes, velocities derived by SAR interferometry have been used to detect glacier surges (Joughin *et al.* 1996; Mohr *et al.*, 1998), and to infer grounding-line positions and their migration. Rignot *et al.* (2001) report grounding-line retreat of most north and northeast Greenland floating glaciers between 1992 and 1996. The retreat rate varies from several hundred metres per year up to 1 $km yr^{-1}$. Rignot *et al.* (2001) argue that the corresponding glacier thinning is unlikely to be explained by enhanced surface melting but must be of dynamic origin.

Another application of satellite SAR interferometry is to derive specific mass balance, particularly ablation rate, by combining surface velocities derived from SAR interferometry, airborne radar measurement of ice thickness and surface-elevation change determined by repeated laser-altimetry (Reeh *et al.*, 2002).

Considering the very long time-scales involved in ice-sheet evolution, it seems unlikely that the Greenland Ice Sheet has adjusted completely to its past mass-balance history, as demonstrated by the model studies of Huybrechts (1994b) and Huybrechts & LeMeur (1999); see Plate 44.4. These studies simulated the evolution of the ice sheet during the last glacial–interglacial cycle, and subsequently derived local thickness changes for the present time. The model results displayed in Plate 44.4 are not necessarily comparable with regional surface-elevation changes derived from observational data (Weidick 1991; Thomas *et al.*, 2000b) because of the different time-scales (decadal or shorter mean values for the observational data versus 200 yr mean values for the model).

Plate 44.4 confirms that large spatial differences of surface-elevation change can be expected. There are many reasons for these differences. To mention a few, they could be caused by regional differences in past climate history, or by regional differences in accumulation rate and flow dynamics leading to different warming rates of the basal ice layers. Plate 44.4 suggests that the present rate of change of ice-sheet volume or mass has a component that is determined by pre-anthropogenic forcing operating on century or millennial time-scales. Therefore, in order to understand the mechanisms behind the current rate of change of ice-sheet mass, which is important for predicting future changes, modelling of the ice-sheet evolution over the last glacial–interglacial cycle (the past 100,000 yr) is required. This emphasizes the need for investigating past climate forcing over such long periods (to be obtained, for example, from deep ice-core records) in addition to studying present ice-sheet dynamics and mass balance.

44.5 The future

Continued monitoring of the ice sheet, for example by using existing and future space-borne laser- and radar-altimeters (ICESAT/GLAS; CRYOSAT), is likely to improve our knowledge of the current volume change of the Greenland Ice Sheet. Combining this information with high-accuracy space-borne measurements of changes of the gravity field (GRACE) should allow discrimination between ice-sheet volume change and ice-sheet mass change.

In order to improve estimates of future mass-balance changes, the following studies should also be given high priority: (i) better understanding of albedo changes and feed-back mechanisms; (ii) studies of outlet glacier dynamics with emphasis on their potential for triggering persistent, fast changes in ice-sheet volume; and (iii) improving ice-dynamic models for determining the long-term response of the ice sheet to past climate change.

It is important to stress that although the use of remote-sensing measurements for studying changes of the Greenland Ice Sheet will increase in importance in the future, there will still be a large demand for *in situ* measurements for calibration/validation purposes. It also should be emphasized that future mass-balance changes are strongly dependent on future changes of climate. As a consequence, our ability to predict future mass-balance changes of the Greenland Ice Sheet is closely linked to the ability of general atmosphere and ocean circulation models to predict future changes in regional climate over Greenland. Recent GCM model runs, for example, predict a larger increase in accumulation rate over Greenland associated with a temperature increase than did previous studies (Van de Wal *et al.*, 2001). If true, this will to a large extent compensate for the increased runoff resulting from the predicted climate warming.

FORTY-FIVE

The impact of ice-sheet fluctuations and isostasy on sea-level change around Greenland

Kevin Fleming

Department 1, Geodesy and Remote Sensing, GeoForschungsZentrum—Potsdam, Potsdam, D-14473, Germany

45.1 Glacial-isostatic adjustment and sea-level change around Greenland

Present-day sea-level change around Greenland is dominated by glacial-isostatic adjustment (GIA), which is the Earth's viscoelastic response to surface-load changes associated with the redistribution of continental ice and ocean water. The most prominent changes occurred after the Last Glacial Maximum (LGM, ca. 21 kyr BP), which involved the melting of ice equivalent to ca. 130 m of water when distributed evenly over the present-day ocean surface (Lambeck *et al.*, 2000). We consider separately the changes in the Late Pleistocene ice sheets located outside of Greenland, and those involving the Greenland Ice Sheet (GIS) itself. The GIS's fluctuations are in turn divided between those following the LGM, and more recent changes such as the neoglaciation, when the ice sheet in some areas retreated behind its current margin before readvancing from ca. 4 kyr BP (Weidick, 1996), and present-day changes.

Figure 45.1 presents modelled predictions of the GIA contribution to sea-level change around Greenland. The computation programs and input models are from the Research School of Earth Sciences, the Australian National University (e.g. Lambeck *et al.*, 1998, 2000). The ice models are realistic spatial and temporal descriptions of the North American, European, Antarctic and Greenland Ice Sheets. A three-layered earth model is used, consisting of an elastic lithosphere of thickness h_L, an upper mantle of viscosity h_{UM} extending to the 670 km seismic discontinuity, and a lower mantle of viscosity h_{LM} extending to the core–mantle boundary, the mantle being treated as a Maxwell-viscoelastic body. The earth-model parameter values used have been

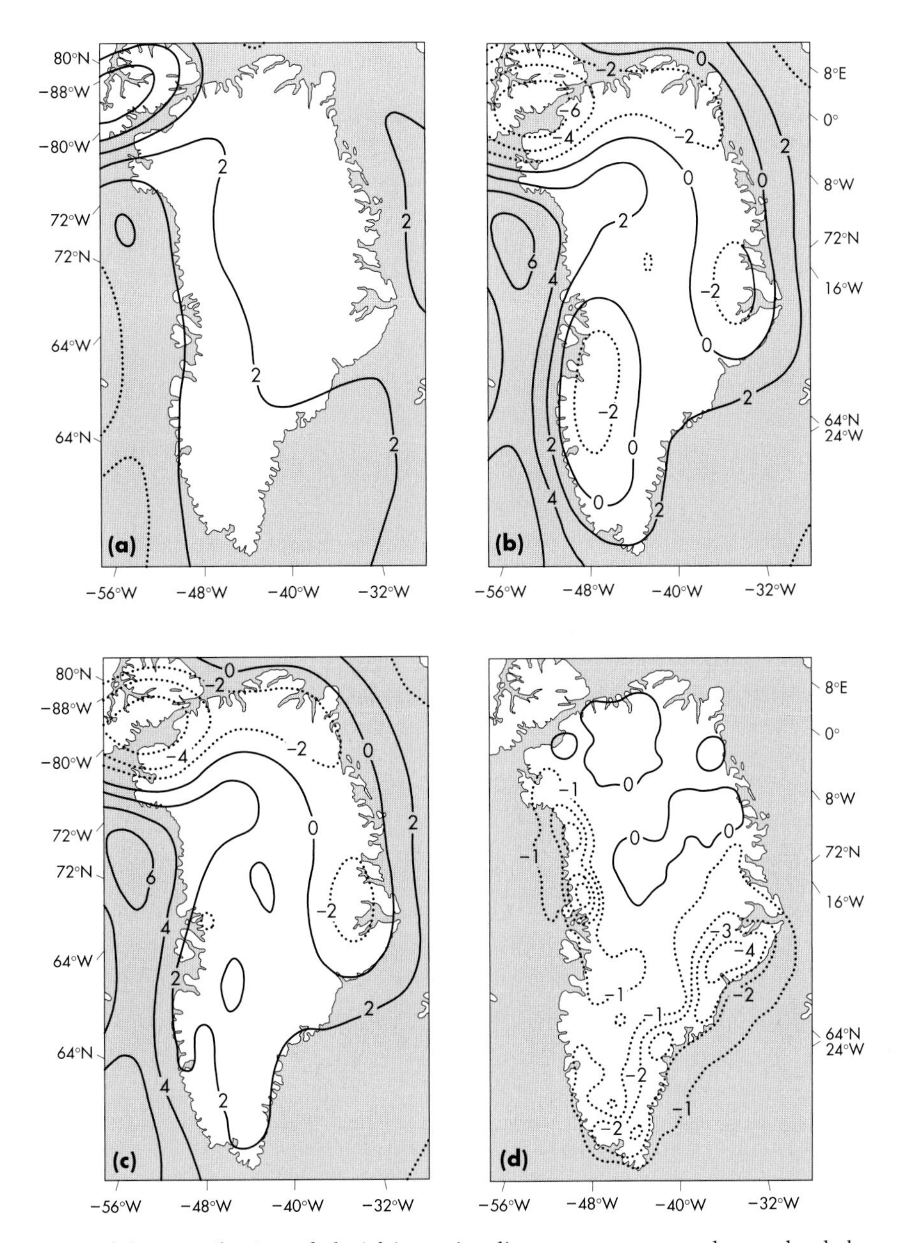

Figure 45.1 Predictions of the contribution of glacial-isostatic adjustment to present-day sea-level change about Greenland (mm yr^{-1}, positive indicating rising sea levels): (a) the contribution from changes in the Late Pleistocene ice sheets located outside of Greenland; (b) the total contribution from changes in the Late Pleistocene ice sheets, including Greenland (neglecting the neoglaciation); (c) the same as (b) but including a neoglacial part in the southwest Greenland Ice Sheet; (d) the contribution from present-day changes in the Greenland Ice Sheet. (See www.blackwellpublishing.com/knight for colour version.)

inferred from previous GIA studies and are appropriate for regions such as Greenland that are made up of older cratonic units (e.g. Mitrovica, 1996; Lambeck *et al.*, 1998). These values are $h_L = 80$ km, $h_{UM} = 5 \times 10^{20}$ Pa s and $h_{LM} = 1 \times 10^{22}$ Pa s.

Figure 45.1a presents the GIA contribution from changes in the ice sheets located outside of Greenland following the LGM. The result is rising sea levels of the order of several millimetres per year, decreasing from west to east (Tarasov & Peltier, 2002). This is largely due to Greenland's location on the collapsing forebulge that surrounds the former North American ice sheets. The falling sea level over Ellesmere Island (northwest corner) comes from the ongoing uplift following the deglaciation of the Innuitian Ice Sheet. An important point is that if the spatial description of the Innuitian were modified, it may significantly alter the results for northwest Greenland.

Figure 45.1b presents the contribution from past changes in all ice sheets, including the GIS. The GIS model (Fleming & Lambeck, 2004) defines the retreat of the expanded GIS to have been completed by 7.5 ka, ignoring in this case the neoglaciation. Much greater spatial variability is observed, mirroring the observed marine limits (e.g. Funder & Hansen, 1996). The more-rapid sea-level changes occur where the GIS had expanded most

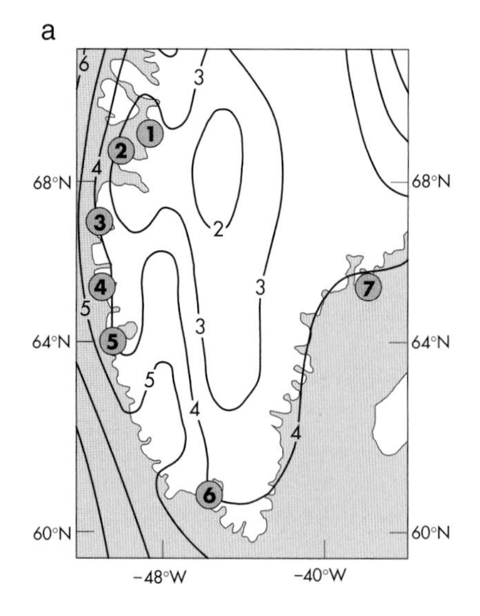

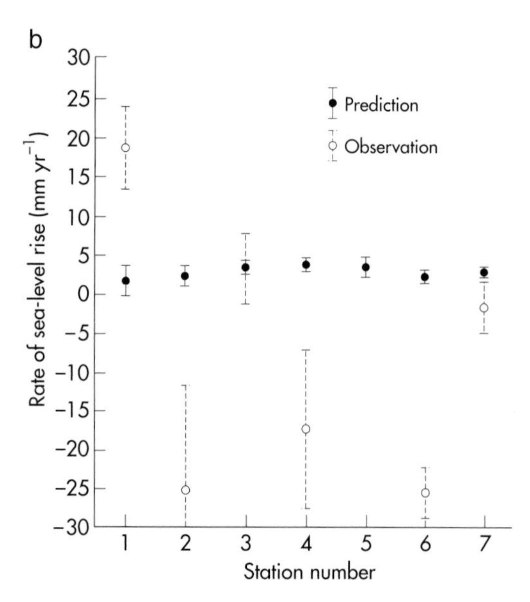

Figure 45.2 (a) Predictions of present-day sea-level change in South Greenland (mm yr^{-1}). The numbered circles correspond to the tide-gauge stations (see Table 45.1). (b) Comparing modelled predictions of present-day sea-level change with estimates derived from the corrected tide-gauge time series. (See www.blackwellpublishing.com/knight for colour version.)

extensively relative to the present day, specifically the southwest, east and northwest. The effect of a neoglacial contribution is demonstrated in Fig. 45.1c, where a simple model defining the GIS to have retreated behind the current margin by 40 km before readvancing over the past few thousand years is incorporated into the southwest GIS (e.g. van Tatenhove *et al.*, 1996; Wahr *et al.*, 2001). A substantial modification to the GIA signal occurs, up to 6 mm yr^{-1} in some areas, with a reversal in sign sometimes observed.

The predicted GIA response from changes in the present-day GIS is shown in Fig. 45.1d. The ice model is based on airborne-altimetry data (Krabill *et al.*, 2000). The largest changes occur in the lower parts of the ice sheet (<2000 m elevation), especially along the ice margin and outlet glaciers in the southeast and the southwest. Over yearly time-scales, the Earth essentially reacts elastically to these changes, the GIA signal in the areas where they are greatest being of the order of several millimetres per year. However, there are also large stretches of the coastline where the predicted response is < 1 mm yr^{-1}.

45.2 Comparing predictions and observations

We now compare our predictions with sea-level-change trends inferred from tide-gauge data (Fig. 45.2). We use monthly-average values as supplied by the Permanent Service for Mean Sea Level (PSMSL, e.g. Woodworth & Player, 2003). Details of the tide-gauge stations are listed in Table 45.1. Most stations provide less than 10 yr worth of data, the exception being Nuuk (5), which has unfortunately ceased operation. In fact, if the criteria listed by, for example, Douglas (1997) were applied, then none of these time series would be considered usable. However, we will still make use of them to obtain preliminary estimates for present-day sea-level change in south Greenland. We first correct the time series by removing the annual and semi-annual cycles. A linear trend is then fitted to the corrected values, the resulting gradient being the local rate of sea-level change.

Figure 45.2a presents the total predicted present-day sea-level-change rate in south Greenland, combining the results from Figs 45.1c & 45.1d, and a eustatic, or global average, contribution (1.9 mm yr^{-1}, Douglas, 1997). Sea level is predicted to be rising all along the southern coast, with millimetres per year spatial variability along the lengths of the larger fjords (hundreds of kilometres). The predicted values for each tide-gauge station are compared with the values inferred from the time series in Fig. 45.2b. The uncertainty range in the predictions is a result of the use of a range of realistic earth-model parameter values, and models of the ice sheets located outside of Greenland (90 to 110% of the nominal ice volume change, Fleming & Lambeck, 2004). Three stations show relatively good agreement between predictions and observations; Sisimiut (3), Nuuk (5) and Ammassalik (7). On the other hand, the values inferred for Ilulissat (1), Aasiaat (2), Maniitsoq (4) and Qaqortoq (6) are quite unrealistic, being of the order of several centimetres per year, the result of the too-short time series that do not accommodate well decadal fluctuations in sea level, leading to large discrepancies between the predicted and observed rates.

45.3 Conclusions

The GIA contribution to sea-level change around Greenland arising from changes in the ice sheets outside of Greenland following the LGM, and the equivalent signal from the GIS, are of a similar magnitude, although differing in their spatial form. We also see how the manner in which the recent (few thousand years)

Table 45.1 Details of the tide-gauge stations and the time series used in this work. The station numbers correspond to those in Fig. 45.2: FRV, The Royal Danish Administration of Navigation and Hydrography; DMI, Danish Meteorological Institute

Station and authority	Location	General
(1) Ilulissat (FRV)	63°13′N/51°06′W	1992–1997. No longer operational
(2) Aasiaat (FRV)	68°43′N/52°53′W	1997–1999
(3) Sisimiut (FRV)	66°56′N/53°40′W	1991–1999
(4) Maniitsoq (FRV)	65°25′N/52°54′W	1997–1999
(5) Nuuk (Godthaab) (FRV, DMI)	64°10′N/51°44′W	1958–2001. No longer operational. Only data from 1970 is used since the station was moved in 1969
(6) Qaqortoq (FRV)	60°43′N/46°02′W	1991–1999
(7) Ammassalik (FRV)	65°30′N/37°00′W	1990–1999

history of the GIS is treated is very important. The corresponding response from present-day changes in the GIS is generally smaller in magnitude and more restricted in extent, although the rates in areas experiencing the largest response are still of the order of millimetres per year. Unfortunately, the available tide-gauge time series are generally not very useful owing to their short duration. Therefore, to constrain the load history of the GIS, other observations will need to be considered, such as crustal uplift as measured by GPS (e.g. Wahr *et al.*, 2001), as well as the further analyses of geomorphological observations.

FORTY-SIX

The response of glaciers in South America to environmental change

Renji Naruse

Institute of Low Temperature Science, Hokkaido University, Sapporo, Japan

46.1 Distribution of glaciers in South America

A number of glaciers exist in the Andes, extending from the Equator to 55°S on the western side of South America. They are in the forms of small hanging or cirque glaciers, valley glaciers, ice caps and vast ice fields. In terms of glacier area, about 65% of the total is made up by two separate icefields in Patagonia (45°–53°S), with the rest located mostly at high altitudes in the Andes from 10°N to 45°S (Williams & Ferrigno, 1998), and in Tierra del Fuego and other small islands in southern Patagonia (Holmlund & Fuenzalida, 1995; Casassa *et al.*, 2002b).

The distribution of glaciers and the total glacierized surface area in each region of South America are summarized in Table 46.1. The methods and the standards used to compile glacier areas vary between the regions (or authors). Some include only inventoried glaciers, whereas others include estimates of uninventoried glaciers. If we simply take a sum of the areas, the total glacierized area in South America yields 26,100 km^2, compared

Table 46.1 Glacier areas in South America

Country	Region	Total glacier area† (km^2)
Venezuela	Sierra Nevada de Merida	*3
Columbia	Sierra Nevada de Santa Marta and others	*111
Ecuador	Cordilleras Oriental and Occidental	*120
Peru	Cordillera Blanca and others mountains	**2042
Bolivia	Cordilleras Oriental and other mountains	*566
Chile	Andes other than Patagonia Icefields	**2015
Argentina	Andes other than Patagonia Icefields	*1385
Chile	Northern Patagonia Icefield	***4200
Chile–Argentina	Southern Patagonia Icefield	***13,000
Chile–Argentina	Tierra del Fuego	*2700

†*Sources:* * IAHS(ICSI)–UNEP–UNESC (1989); **Casassa *et al.* (1998); ***Aniya (1999).

with an area of 25,900 km^2 reported by IAHS(ICSI)–UNEP–UNESCO (1989).

The north–south distribution of the present equilibrium line altitude (ELA) along the Andes has been roughly estimated based on aerial photographs by Nogami (1972). The ELA rises gradually from the Equator to around 25°S, which corresponds to the Atacama Desert in the northern Chile, and lowers sharply from 30°S to 42°S in northern Patagonia. This abrupt descent in the ELA is attributed to the prevailing westerlies south of 35°S, which carry moisture from the Pacific Ocean (Ohmura *et al.*, 1992). More detailed estimates have been made in the central Andes using photographs, topographic maps and Landsat imageries. In the Cordillera Blanca (7°–10°S), the mean ELA is located around 5000 m a.s.l., whereas it is significantly lower at about 4400 m on the eastern side (Amazon Basin) of the Cordillera Oriental (Rodbell, 1992). This trend is interpreted to be due to the effect of precipitation distribution resulting from vapour advection from the Amazon basin in the east (Casassa *et al.*, 1998). The major feature of the present ELA in the tropical Andes is the general rise from northeast to southwest: the ELA rises westward to 5100 m in central Peru and rises southwestward to the extremely high altitudes >5800 m in the western cordillera along the Bolivia–Chile border (Klein *et al.*, 1999).

In South America, tropical glaciers experience strong solar radiation and very dry conditions. The climate in the Cordillera Blanca is characterized by relatively large daily and small seasonal temperature variations as well as by a distinct succession between dry and wet seasons (Kaser *et al.*, 1990). Although only a few studies have been performed on mass balance and variations of glaciers in the tropical and central Andes, it is known that recently most glaciers have retreated. For example, drastic shrinkages of small glaciers in the Cordillera Real (16°S) in Bolivia have been observed during the past two decades, and a probable extinction of these glaciers in the near future could seriously affect the hydrological regime and the water resources of the high-elevation basin (Ramirez *et al.*, 2001). In northern and central-south Chile (18°–41°S), the 13 glaciers studied have all receded during the past 40 or 50 yr (Rivera *et al.*, 2002).

In contrast to glaciers in the high Andes, glaciers in Patagonia (south of 40°S) are located at lower altitudes and typified by temperate, maritime glaciers with high accumulation and ablation throughout the year. In Patagonia, the ELAs of most glaciers range from about 900 m to 1300 m (Aniya, 1988; Aniya *et al.*, 1996). In the following sections, we concentrate only on the characteristics and behaviour of Patagonian glaciers.

46.2 Icefields and outlet glaciers in Patagonia

Two large ice-covered regions occur in Patagonia. The smaller one is called Hielo Patagónico Norte (Northern Patagonia Icefield: NPI) and is about 100 km in length and 50 km wide, located around 47°S (Fig. 46.1). The larger one is called Hielo Patagónico Sur (Southern Patagonia Icefield: SPI), which extends for about 350 km from 48°20′ to 51°30′S along 73°30′W (Fig. 46.2). The total surface areas of the NPI and the SPI were first estimated as 4400 km^2 and 13,500 km^2 (Lliboutry, 1956), and were amended later to be 4200 km^2 and 13,000 km^2, respectively, based on the 1974–1975 aerial photographs (Aniya, 1988) and the 1986 Landsat TM images (Aniya *et al.*, 1992). Elevations of the central plateaux of the icefields range from about 800 m on the western side and about 1500 m on the eastern side. Several high nunataks (altitudes greater than 3000 m) soar above the icefields, namely Monte San Valentín and Cerro Arenales (NPI), and Volcán Lautaro (SPI).

A number of glaciers discharge from the icefields in all directions. In the NPI, a total of 28 outlet glaciers were inventoried. Thirteen of these calve into proglacial lakes, and only San Rafael Glacier calves into a fjord. Among 48 outlet glaciers inventoried in the SPI, all but two glaciers calve into lakes on the east side and into fjords on the west side (Aniya *et al.*, 1996).

46.3 Changes in glacier length and ice-thickness during the past 50 yr

Changes in frontal positions and surface areas of outlet glaciers during the past 50 yr were clarified by analysing satellite data (Landsat, Spot, and others), aerial photographs and topographic maps with field survey data (Aniya *et al.*, 1992, 1997). Variations

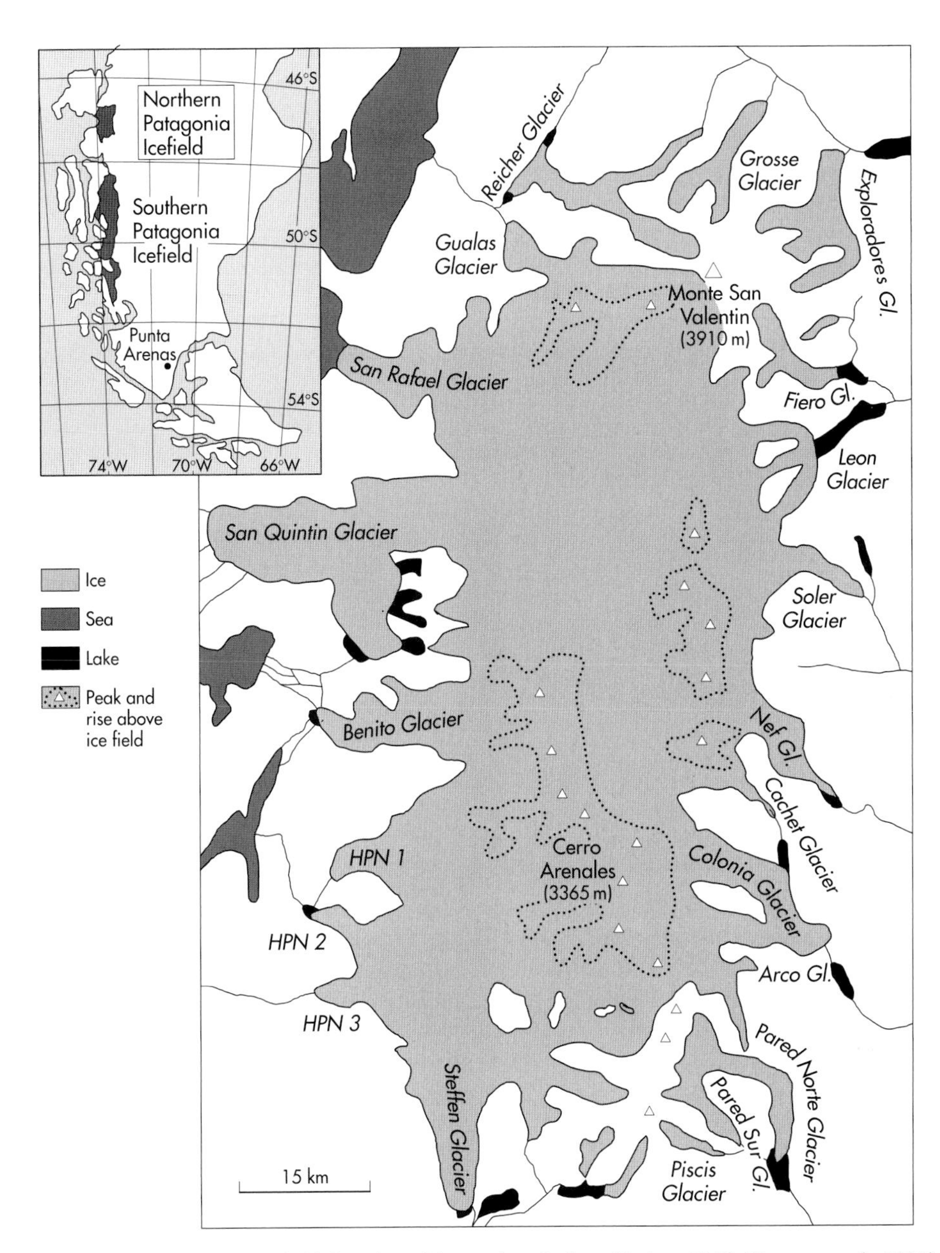

Figure 46.1 The Northern Patagonia Icefield (NPI) and its outlet glaciers (Aniya, 1988; Naruse *et al.*, 1995). Reproduced by permission of the International Glaciological Society.

in frontal positions of eight major calving glaciers in the SPI since 1945 are shown in Fig. 46.3. At first sight, a large retreat is evident at O'Higgins Glacier, a northeastern outlet glacier, which showed a 13 km retreat during 41 yr, the mean being 330 m yr^{-1}. This rate is about one order of magnitude greater than the mean retreat rate in the NPI (about 30 m yr^{-1}). Whereas Upsala and Tyndall Glaciers in the southern SPI have retreated at relatively high rates, the other three glaciers in the northern SPI, Jorge Montt, Occidental and Viedma, have been retreating gradually.

A second feature to note is the peculiar behaviour of the advancing Pío XI (or Brüggen) Glacier, the southern calving tongue of which has advanced 9 km during 31 yr from 1945 and the northern tongue is still advancing. A possibility of surges in 1976 and 1992–1994 is suggested (Rivera & Casassa, 1999). In contrast, Perito Moreno Glacier has been almost in a steady state. In the NPI, most of 22 outlet glaciers retreated at an increasing rate from 1945 to 1990; however, the retreat has slowed down since then (Aniya, 1999).

Surface elevations of bare-ice in ablation areas were measured several times with the conventional survey method at Soler Glacier in the NPI, and Tyndall, Moreno and Upsala Glaciers in the SPI since 1983 (Naruse *et al.*, 1995). From the elevation data measured in different years, changes in ice thickness were obtained (Fig. 46.4). Large thinning rates of ice, from 3 to 5 m yr^{-1}, were measured at Soler and Tyndall Glaciers; especially, Upsala Glacier has thinned markedly with a rate of 11 m yr^{-1}. Compared with the thickness change data at 42 glaciers compiled by IAHS–UNEP–UNESCO (1993), the rate of Upsala Glacier is amongst the largest and one order of magnitude greater than the mean rate of receding glaciers in the world.

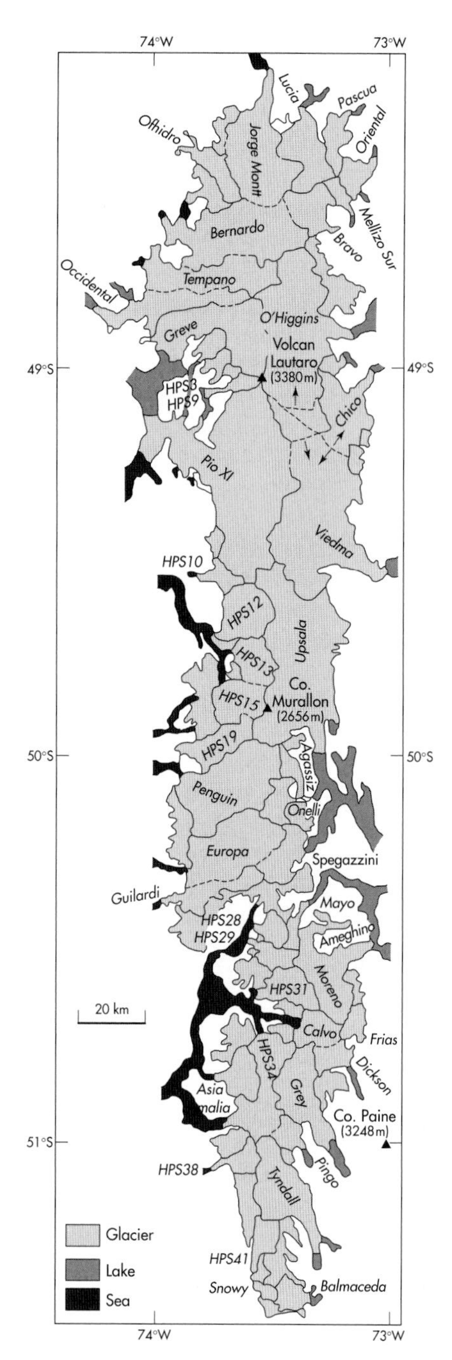

Figure 46.2 The Southern Patagonia Icefield (SPI) and its outlet glaciers. Determined (or inferred) ice divides are shown by solid (or broken) lines (Aniya *et al.*, 1996, 1997).

In contrast, the ice thickness of Perito Moreno Glacier has been almost unchanged between 1990 and 2000. Considering this with the glacier-front positions, the glacier is regarded to be in a steady state at present, which is a rather peculiar behaviour in Patagonia.

Aniya (1999) roughly estimated the total ice-volume loss due to area reduction and ice thinning of Patagonian glaciers during the past 50 yr to be $825 \pm 320\,km^3$, which corresponds to a global mean sea-level rise rate of $0.038\,mm\,yr^{-1}$. If we take a median

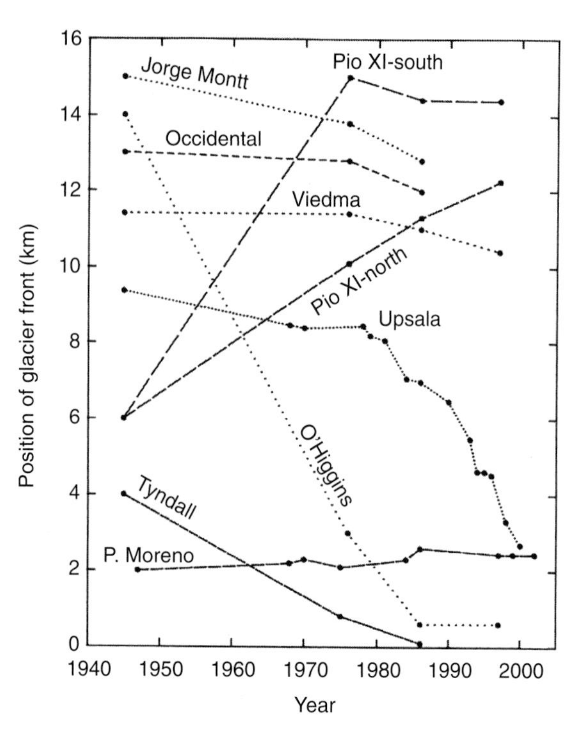

Figure 46.3 Fluctuations in frontal positions of eight major calving glaciers of the SPI. Positive gradients indicate advances and negative gradients indicate retreats. Jorge Montt, Occidental and Pío XI-south (southern tongue) glaciers calve into fjords, and the other five glaciers and Pío XI-north (northern tongue) glacier calve into lakes. (Compiled using various sources: Aniya *et al.*, 1992, 2000; Aniya & Skvarca, 1992, Naruse *et al.*, 1995; Naruse & Skvarca, 2000; Skvarca *et al.*, 2002.) Reproduced by permission of the International Glaciological Society and the Instituto Antárctica Argentina.

value of estimates, $0.3\,mm\,yr^{-1}$, for the contribution of melting of mountain glaciers and ice caps to sea-level rise (IPCC, 2001a), Patagonian ice contributes to 13% of that value, although Patagonian ice occupies only 3.2% of the total surface areas of mountain glaciers and ice caps. From this argument, we can recognize how extensively the Patagonian glaciers have been shrinking.

46.4 Recent trend of climate change in southern South America

When glacier variations in Patagonia are discussed from a climatic point of view, meteorological data at Puerto Aisén and Punta Arenas are often cited, from which we cannot identify any significant trends in air temperature and annual precipitation (Warren & Sugden, 1993). Although these stations provide relatively long, continuous records from around Patagonian glaciers, they are located far from glaciers, the former station being about 150 km to the north of the northern margin of the NPI and the latter being about 250 km to the south of the southern margin of the SPI.

Here, we select two minor meteorological stations near the SPI, namely the Station Islote Evangelistas on a small island about 150 km to the southwest of the southern end of the SPI and the

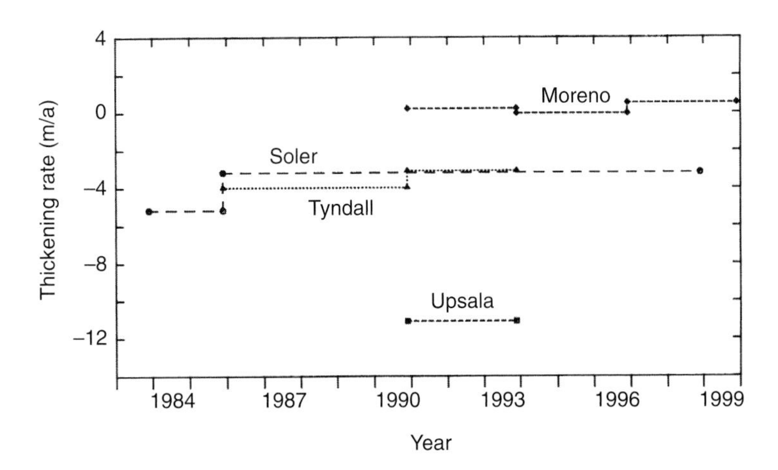

Figure 46.4 Annual thickness change in ablation areas of Upsala, Perito Moreno and Tyndall Glaciers (SPI), and Soler Glacier (NPI), which terminated on land before around 1992 and now calves into a newly formed lake. (Sources: Naruse *et al.*, 1995; Skvarca & Naruse, 1997; Naruse & Skvarca, 2000; Skvarca *et al.*, 2004).

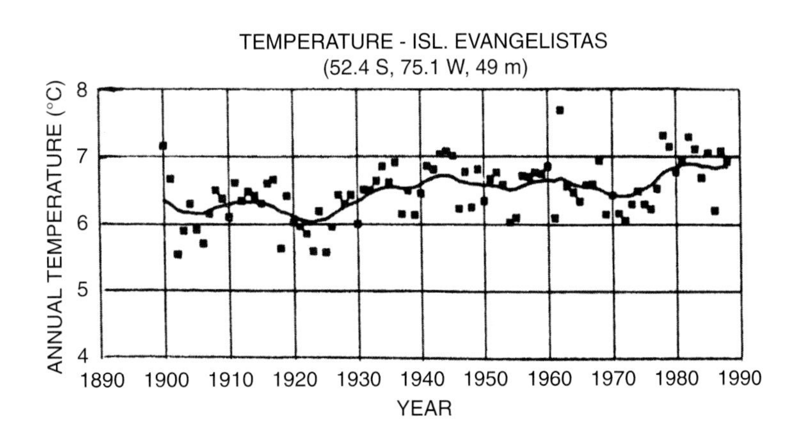

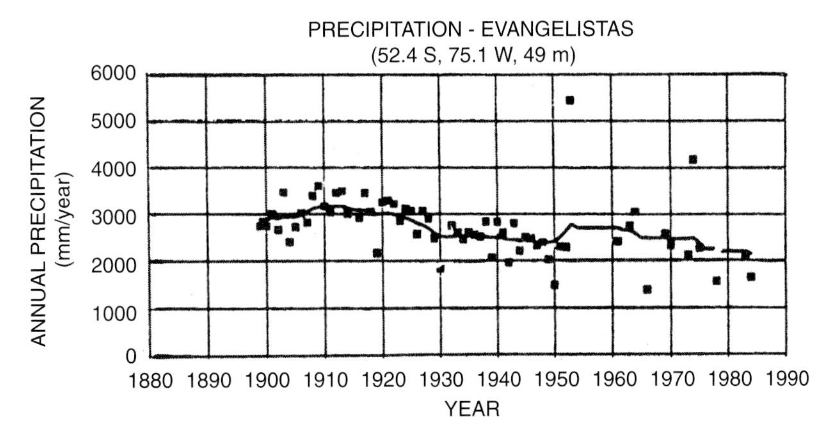

Figure 46.5 Annual mean air temperature and annual precipitation between 1900 and 1988 at the Station Islote Evangelistas, Chile (Rosenblüth *et al.*, 1995). Reproduced by permission of the Bulletin of Glacier Research.

Station Lago Argentino at Calafate about 60 km from the terminus of Perito Moreno Glacier. Fluctuations in annual mean temperatures and annual precipitations at these two stations are shown in Figs 46.5 & 46.6. Although year-to-year variations are considerable, we can notice a slight rising trend in temperature during the 20th century or the past 50 yr. Rosenblüth *et al.* (1995) stated that, south of 46°S a definite warming trend appears in the records, both on the Pacific and the Atlantic coast, especially in the past two decades. As to annual precipitations, weak decreasing trends are recognized at both stations.

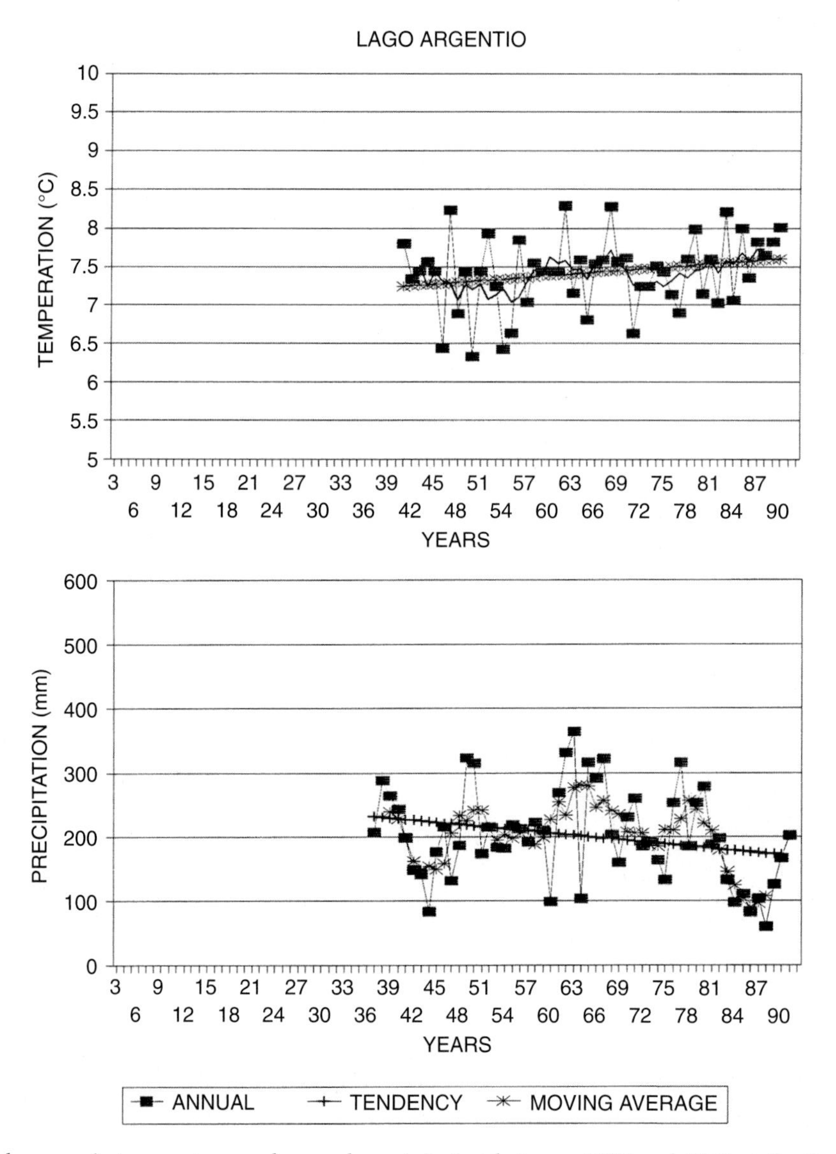

Figure 46.6 Annual mean air temperature and annual precipitation between 1937 and 1990 at the Station Lago Argentino, Argentina (Ibarzabal y Donangelo *et al.*, 1996). Reproduced by permission of the Bulletin of Glacier Research.

An analysis based on the NCEP Reanalysis data (Climate Data Center, National Oceanic and Atmospheric Administration) indicated that air temperature at 850 hPa over southern South America rose with a mean rate more than 0.05 K yr^{-1} between 1948 and 1989 (K. Kubota, unpublished data). This result, as well as the trends of temperature and precipitation at the above two stations, seems to support the interpretation of recent recessions of most glaciers in Patagonia.

At Perito Moreno Glacier, based on the ablation measurements in the summer of 1993–1994 and a degree-day calculation, mean annual ablation rates over the past 30 yr were estimated as about 14 ± 2 m in ice thickness (Naruse *et al.*, 1997). Thus, a range between the maximum and the minimum of the annual ablation caused by year-to-year temperature variations can be considered as about 4 m in thickness, which is much smaller than the measured annual ice thinning. It was concluded that the change (rise) in temperature alone cannot explain the thinning rate of 11 m yr^{-1} at Upsala Glacier. Therefore, the recent recession of the glacier must be caused by a non-climatic effect, that is, the glacier dynamics.

46.5 Calving dynamics and glacier variations

Studies on calving dynamics of glaciers mostly have been made at fjords in Alaska and the Arctic. It was known that although some Alaskan tidewater glaciers were retreating catastrophically, others were oscillating or advancing slowly (Mercer, 1961). Change in the terminus position of a calving glacier can be expressed as

$$dL/dt = U_i - U_c \quad (1)$$

where L is the terminus position, U_i is ice velocity at the terminus and U_c is the calving rate (speed), which is defined by the volume rate of iceberg discharge (including the frontal melting) divided by the cross-sectional area of the terminus.

Calving rate is often represented empirically with a linear relation of water depth at the terminus. In Fig. 46.7, several empirical relations are shown. It is noted that calving rates of tidewater glaciers are more than five times larger than those of freshwater calving glaciers. This is due to the effects of tide, water current, salinity, and so on. These calving relations clearly demonstrate the typical characteristics of tidewater glaciers: when the terminus of a glacier retreats into deeper water from the moraine shoal or the bed rise, the rate of calving increases with the relation (Fig. 46.7), which may lead to the further recession of the glacier by Equation (1).

It is sometimes pointed out that back-stress (or back-pressure) arising from the shoal or islands is crucial to the dynamics or stability of a glacier. Meier & Post (1987) explained rapid disintegration of grounded tidewater glaciers owing to a feed-back process including back-pressure: that is, retreating→decrease in back-pressure→longitudinal stretching→ice thinning→decrease

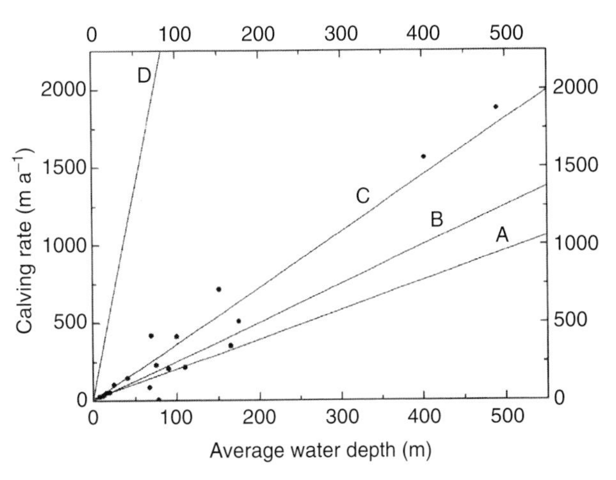

Figure 46.7 Relationships between calving rate and water depth near the glacier termini. Dots and their regression line C are obtained from freshwater calving glaciers in Patagonia (Skvarca *et al.*, 2002), the lines A and B are empirical relations for fresh water proposed by Funk & Röthlisberger (1989) and Warren *et al.* (1995), respectively, and the line D is derived for tidewater glaciers in Alaska (Brown *et al.*, 1982).

Figure 46.8 The frontal part of Upsala Glacier calving into the western arm of Lago Argentino, in November 1993 (upper photograph) and March 1999 (lower photograph), viewed from the control station of field surveys on the eastern (left-hand side) bank. Width of the glacier is about 3 km. The tributary glacier flowing from the western valley is Bertacchi Glacier, and the snow-covered, pyramidal peak is Cerro Cono (2440 m). (Both photographs taken by P. Skvarca.) (See www.blackwellpublishing.com/knight for colour version.)

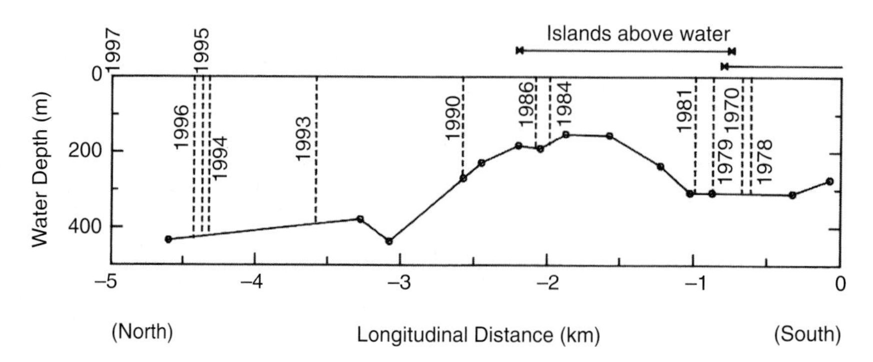

Figure 46.9 Longitudinal profile of the bedrock in the lower reaches of Upsala Glacier. Horizontal distance is measured northward from 50°S. Small islands are exposed to the south of the 2-km position in the western part of the lake. Glacier front in 1997 is approximate, and the fronts in 1998 and 2000 are located at about 6 km and 6.7 km, respectively (Naruse & Skvarca, 2000; Skvarca *et al.*, 2002). Reproduced by permission of the International Glaciological Society and the Instituto Antárctica Argentina.

in subglacial effective pressure→furtherstretching→calving→further retreating.

By analysing data from Columbia Glacier, it was shown that, during the retreat, the thickness at the terminus appears to be linearly correlated with the water depth (Van der Veen, 1996). This is the 'height-above buoyancy model', which states that the position of the terminus is controlled by such geometric factors as ice thickness and water depth in order for the thickness in excess of flotation at the terminus to become about 50 m. However, the physical basis of this model has not been fully clarified.

Recent fluctuations of Upsala Glacier are worthy of note. From 1945 to 1978 the glacier had been almost stable, and the retreat started in 1978 (Fig. 46.3). A remarkable change in the front part of the glacier from 1993 to 1999 clearly can be seen in the photographs (Fig. 46.8). The width-averaged front positions of Upsala Glacier are indicated in Fig. 46.9. The front retreated remarkably in the periods of 1981–1984, 1990–1994 and 1996–1998.

Measurements of ice-thickness with ice radar have not been conducted successfully at this glacier owing to heavy crevasses and a large amount of water within the glacier. Instead, water depth surveys carried out in 1997 at the proglacial lake made it possible to reveal the subglacial topography under the glacier terminus in the period from 1970 to 1996 (Fig. 46.9). Bedrock rises were found to spread out from the exposed islands in the western part of the lake. Naruse & Skvarca (2000) proposed the following model. Between 1978 and 1990, the western half of the glacier terminus was located at the bedrock rises, which suggests that the front fluctuations were strongly regulated by the basal drag. During 1990–1993, the glacier terminus shifted upstream from the bedrock rises, and then the glacier was stretched longitudinally (with a strain rate of $0.22\,yr^{-1}$ deduced from the continuity consideration), which resulted in ice thinning. This process may cause extensive calving, so that the glacier retreats further.

At San Rafael Glacier, Warren (1993) stated that, on annual to decadal time-scales up to about three decades, glacier oscillations may not be related directly to climate change. When we consider the glacier fluctuations at the longer time-scales of more than several decades or a century, however, climatic effects become more dominant. Thus, fluctuations in the calving terminus of glaciers over a short time-scale, such as a few decades or so, are considered to be controlled by non-climatic effects, specifically topographic, dynamic and subglacial hydrological conditions near the glacier terminus.

FORTY-SEVEN

Glacier mass-balance data for southern South America (30°S–56°S)

G. Casassa*, A. Rivera† and M. Schwikowski‡

**Centro de Estudios Científicos, Valdivia, Chile*
†Department of Geography, University of Chile, Santiago, Chile
‡Paul Scherrer Institut, Labor für Radio- und Umweltchemie, Switzerland

Southern South America (SSA), i.e. south of 30°S, comprises an estimated glacier area of ca. 27,500 km^2 (Fig. 47.1), which represents ca. 89% of all Andean glaciers. The glaciers in the region show a generalized retreat and thinning (Naruse, this volume, Chapter 46). In spite of their importance in water resources within a region affected by climate changes, glacier mass balance is to a large extent largely unknown. This summary aims at reviewing the few glacier mass-balance studies available for SSA, obtained by means of firn coring and by the traditional stake method.

Data for 15 glacier sites with annual mass-balance observations are presented in Table 47.1. Data obtained from modelling studies, climatological extrapolations or seasonal (and intraseasonal) stake observations are not included in this work. Firn core data include seven sites, covering periods from 1 yr to 68 yr; multistake mass-balance studies which cover complete glacier basins exist for five sites and three sites show mass-balance data from single or multiple stakes covering limited portions of glaciers.

The annual values of accumulation, ablation and net balance have been plotted for the above glacier sites in Fig. 47.2. Data representation is limited due to the restricted spatial coverage of the mass-balance data. In general both accumulation and ablation values increase southward, consistent with the meridional gradient in precipitation driven by the southern position of the storm tracks controlled by the westerlies, as opposed to the location of the stable Pacific anticyclone to the north which results in very dry conditions. This north–south gradient is clearly seen in the data, with net balance data in the upper accumulation areas ranging from 30 cm water equivalent (w. eq.) at Cerro Tapado, largely controlled by sublimation (Ginot *et al.*, 2002), to a record value of 1540 cm w. eq. at Glaciar Tyndall in Patagonia (Shiraiwa *et al.*, 2002; Kohshima *et al.*, accepted for publication).

A clear west–east gradient is observed in the mass balance south of 33°S, particularly in Patagonia, which is affected by an enhanced westerly circulation with precipitation occurring largely on the windward side of the Andes. For example, on the upper accumulation area of the Southern Patagonia Icefield a minimum net balance of 31 cm w. eq. has been recorded at 2300 m a.s.l. on Cerro Gorra Blanca (Schwikowski *et al.*, 2003), located on the eastern margin, in contrast to the record value of 1540 cm w. eq. obtained near the ice divide at Glaciar Tyndall.

The low values of net accumulation measured on the leeward-facing glaciers of Patagonia are not regarded as representative for the accumulation area, and are much smaller than expected from hydrological models (e.g. Escobar *et al.*, 1992), precipitation estimates from neighbouring meteorological stations (Carrasco *et al.*, 2002), *in situ* precipitation observations (Fujiyoshi *et al.*, 1987; Casassa *et al.*, 2002a), and glacier dynamics models (e.g. Naruse *et al.*, 1995), which estimate that an annual accumulation of 5–10 m w. eq. is needed in the upper accumulation areas for maintaining a steady-state at SPI, in order to balance the large annual ablation values that occur at lower elevations, such as those measured at Glaciar Moreno (Stuefer, 1999) and Glaciar Lengua (C. Schneider, personal communication, 2004).

Strong interannual variations are found within the mass-balance records of Glaciar Piloto Este and Glaciar Echaurren (ca. 33°S), closely related to ENSO events, with dry La Niña years and wet El Niño years. At Glaciar La Ollada, Cerro Mercedario, interannual variations in snow accumulation related to ENSO events are not obvious in the record. After 1988 both mass-balance series of Glaciares Piloto Este and Echaurren Norte show a clear negative trend, consistent with the retreating trend of glaciers in the region, which probably is largely due to a regional warming (Rosenblüth *et al.*, 1997). A negative trend is also observed in the short mass-balance record of Glaciar Martial Este.

Table 47.1 Glacier mass-balance data. The mass-balance data represent mean annual values for the indicated period. In the case where mass balance is monitored with a stake array, minimum and maximum represent the altitudinal range of the glacier. For firn core and single stake sites, minimum and maximum represent the altitude of the site. The altitude of the firn core site at Glaciar Moreno is taken from Godoi *et al.* (2001). At Glaciar Tyndall the mass-balance data is an average value from multiproxy analysis based on stable isotope data and microalgae concentrations

Glacier	Latitude (°S)	Longitude (°W)	Location*	Altitude (m a.s.l.)		Ablation (cm w. eq.)	Accumulation (cm w.eq.)	Balance (cm w. eq.)	Period (year)	References
				Minimum	Maximum					
Cerro Tapado†	30.13	69.92	E	5550	5550	24	54	30	1920–1998	Ginot *et al.*, 2002
La Ollada†	31.97	70.12	E	6100	6100			45	1986–2002	Bolius *et al.*, 2004
Piloto Este‡	32.50	70.15	E	4185	4740	110	77	−33	1979/80–1996/97	Leiva, 1999
Echaurren Norte‡	33.55	70.13	W	3650	3880	289	269	−20	1975/76–2003/04	Escobar *et al.*, 1995; F. Escobar, personal communication, 2004
Mocho‡	39.92	72.03	W	1603	2422	346	258	−88	2003/04	Rivera *et al.*, 2005
San Rafael†	46.73	73.53	W	1296	1296			345	1984	Yamada, 1987
Nef†	46.93	73.32	E	1500	1500			220	1996	Matsuoka & Naruse, 1999
Cerro Gorra Blanca†	49.13	73.05	E	2300	2300			31	1995–2001	Schwikowski *et al.*, 2003
Chico§	49.18	73.18	E	1444	1444			57	1994/95–2001/2002	Rivera, 2004
De los Tres‡	49.27	73.00	E	1120	1830	225	232	7	1995/96	Popovnin *et al.*, 1999
Moreno§	50.50	73.00	E	365	365	1025			1998–1999¶	Stuefer, 1999
Moreno†	50.63	73.25	E	2000	2000			120	1980/81–1985/86	Aristarain & Delmas, 1993
Tyndall†	50.98	73.52	W	1756	1756			1540	1998/99	Shiraiwa *et al.*, 2002; Kohshima *et al.*, accepted for publication
Lengua§	52.80	73.00	W	450	450	640			2001–2004**	C. Schneider, personal communication, 2004
Martial Este‡	54.78	68.40	E	1000	1180	86	81	−5	2000–2002	Strelin & Iturraspe, accepted for publication

*Location of the glacier site, E(W) indicates that the glacier is located east(west) of the main Andean range.
†Firn core data.
‡Net mass-balance data from stake array.
§Net balance data from one single stake.
¶20 March 1998 to 7 March 1999, average for three stakes.
**Annual average for three stakes.

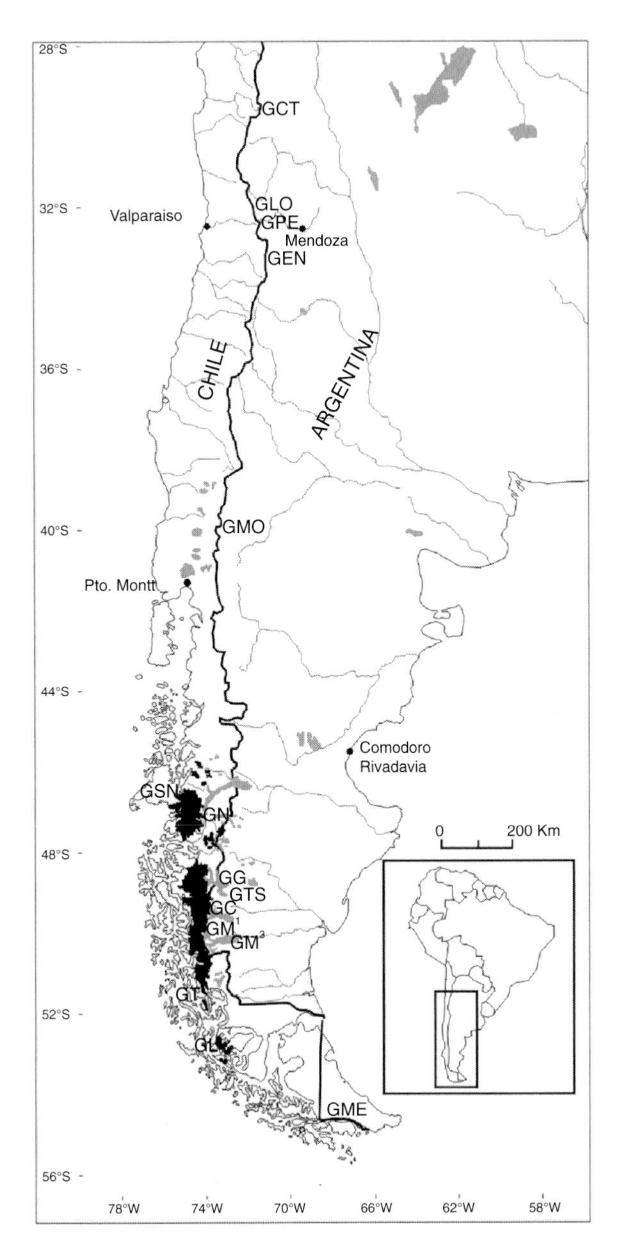

Figure 47.1 Map of southern South America showing the glacier sites mentioned in the text: GCT, Glaciar Cerro Tapado; GLO, Glaciar La Ollada; GPE, Glaciar Piloto Este; GEN, Glaciar Echaurren Norte; GMO, Glaciar Mocho; GSN, Glaciar San Rafael; GN, Glaciar Nef; GG, Glaciar Cerro Gorra Blanca; GC, Glaciar Chico; GTS, Glaciar de los Tres; GM[3], Glaciar Moreno, ablation stakes site; GM[1], Glaciar Moreno, firn core site; GT, Glaciar Tyndall; GL, Glaciar Lengua; GME, Glaciar Martial Este.

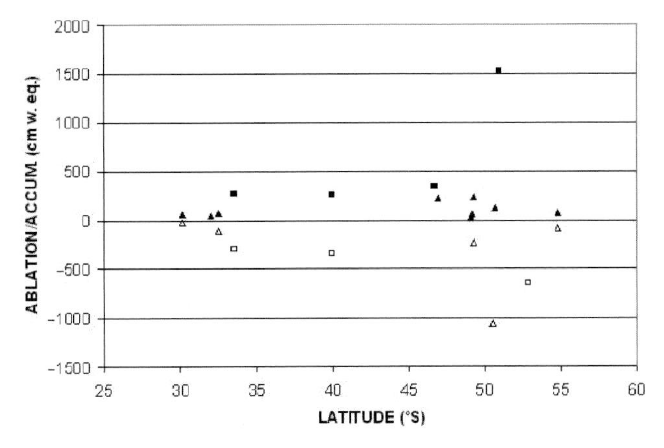

Figure 47.2 North–south transect along the Andes showing annual ablation and accumulation values, as quoted in the text. Solid squares and solid triangles represent accumulation values for western Andean glaciers and eastern Andean glaciers respectively. Hollow squares (triangles) represent ablation values for western (eastern) glaciers. Values increase southward to a maximum at ca. 50–51°S.

FORTY-EIGHT

Quantifying the significance of recent glacier recession in the Cordillera Blanca, Perú: a case study of hydrological impact and climatic control

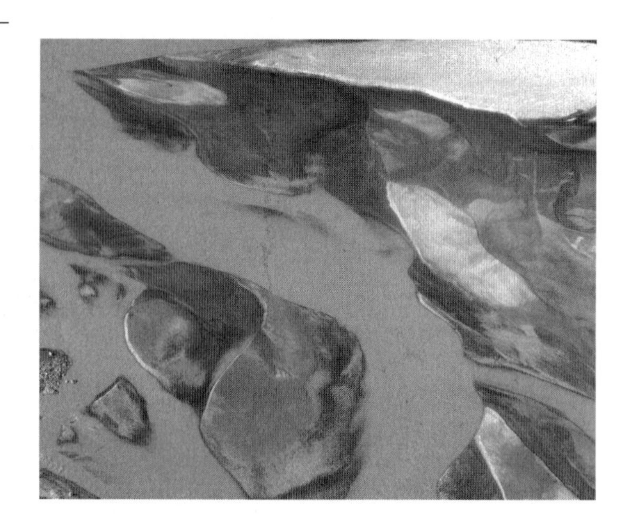

Bryan G. Mark[1]

Department of Geography, The Ohio State University, Columbus, OH 43210, USA

48.1 Introduction

The Peruvian Cordillera Blanca is the most extensively glacierized mountain range in the tropics, and an important location to study the practical impact and climatic control of ongoing glacier volume loss. Draining most of the glacierized area in the Cordillera Blanca, the Río Santa (Fig. 48.1a) maintains the second largest and least variable annual discharge of all rivers in Perú flowing to the Pacific Ocean, from which the economically developing population derives water. If the glaciers disappear, as has been predicted for other tropical regions, important climatic archives will be lost, and the region could face a future water supply crisis. A programme of routine surveying was established in 1968, and four target glaciers were selected to monitor the magnitude of changes to these frozen reservoirs, providing a particularly important geographical database (Ames, 1998). Moreover, gauges in tributary basins discharging to the upper reaches of the Río Santa, known as the Callejon de Huaylas (Fig. 48.1a), provide a 50-yr record of runoff and precipitation. This case study focuses first on one of these target glaciers, Glaciar Yanamarey, where the magnitude of recession is known, then quantifies the downstream impact of meltwater.

[1]Formerly at Max Planck Institute for Biogeochemistry, Jena 07745 Germany and Department of Geography & Geomatics, University of Glasgow, Glasgow G12 8QQ, UK

48.2 Quantifying 50 yr of net glacier recession: Glaciar Yanamarey

Data from GPS surveys of Glaciar Yanamarey extend the record of mapped terminus positions to over three decades at this site, where early quantitative evaluations of modern tropical glacier volume loss were made (Hastenrath & Ames, 1995a,b) (Fig. 48.1c). Although the terminus experienced some small advances, the late 20th century featured extensive glacier recession, consistent with other tropical glaciers and pervasive wastage of glaciers globally. Recession rates also increased, as in other Andean locations. These length variations express a rapid response to mass-balance changes, captured by the basin hydrological balance (Kaser *et al.*, 2003). Using a simplified annual mean budget based on a decadal-scale velocity and net-balance data, Hastenrath & Ames (1995b) estimated that about 50% of the water discharging from Glaciar Yanamarey was not renewed by precipitation, but provided by progressive thinning, and that the receding glacier

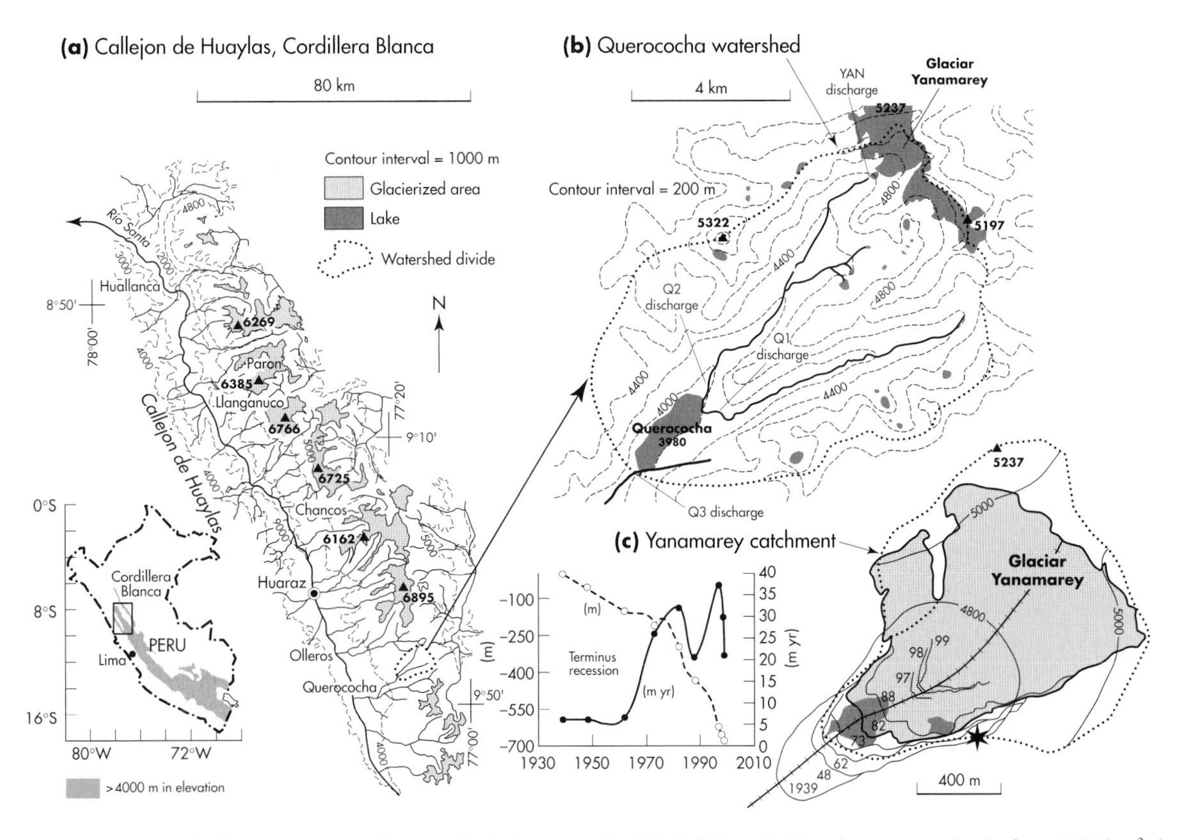

Figure 48.1 Case study location maps of successively larger scale: (a) Callejon de Huaylas, a watershed of ca. 5000 km^2 draining the Cordillera Blanca, Perú, to the upper Río Santa. Stream-gauge locations mentioned in the text are identified; (b) Querococha watershed, 60 km^2, showing the discharge and water sampling points mentioned in the text; (c) Yanamarey catchment, 1.3 km^2 between 4600 m and 5300 m, 75% of which is covered by glacier ice. The shaded region shows the outline of Glaciar Yanamarey in 1982, with contours and a centre-line to show distance from the headwall at 100-m intervals (after Hastenrath & Ames, 1995a). Terminus positions are mapped onto a common datum, based on surveys for 1939, 1948, 1962, 1973, 1982, 1988, 1997, 1998 and 1999. The latter three positions were mapped using differential GPS. The cumulative terminus recession from the 1939 position is shown (m) on the inset graph as a solid line, with solid rectangles for years with corresponding terminus position mapped (data from A. Ames, personal communication, 1998), along with average recession rate between years with mapped termini (in m yr^{-1}). Asterix marks the location of a weather station, where daily temperature and monthly precipitation were recorded discontinuously from 1982.

would survive another 50 yr in the present climate. Thus, a quantitative estimate of how much the downstream hydrology is impacted by glacier wastage is required for water resource planning.

48.3 Quantifying downstream impact of glacier meltwater: Callejon de Huaylas

Monthly observations of specific precipitation and discharge (P and Q_t, respectively) were collected with hydrochemical samples over the 1998–1999 hydrological year at the Yanamarey glacier catchment (YAN) and in the larger Querococha watershed downstream, where a confluence of glacierized (YAN, Q2) and non-glacierized (Q1) streams forms a tributary stream to the Río Santa (Q3) (Fig. 48.1b). Maximum Q_t precedes the peak in average P by 4 months at YAN, whereas Q_t is diminished and closely correlated in time to P at Q3 (Fig. 48.2). A simple water-balance calculation shows that the maximum in specific melt occurs in October for Glaciar Yanamarey, and the April minimum is negative, representing net accumulation (Fig. 48.2a). However, melt contributes a maximum relative percentage of the monthly Q_t during the dry season months (June–September). During this period of little to no precipitation, glacier melt contributes up to 100% of Q_t, thereby buffering the downstream flow. Assuming that the loss in glacier storage is exclusively by melting, then glacier meltwater comprises 35–45% of the total annual stream discharge from YAN. The Querococha watershed is a good analogue for the entire Callejon de Huaylas, as both are about <10% glacierized. Here, the relative influence of glacier meltwater to the annual runoff regime diminishes downstream of YAN at Q2, to become precipitation dominated at Q3, as in the non-glacierized Q1 (Fig. 48.2b). A volume-weighted hydrochemical mixing model (using dissolved anions and cations) revealed that YAN contributes about one-third of the discharge at Q3 over an annual cycle (Mark & Seltzer, 2003). By analogy, the larger Río Santa watershed thus

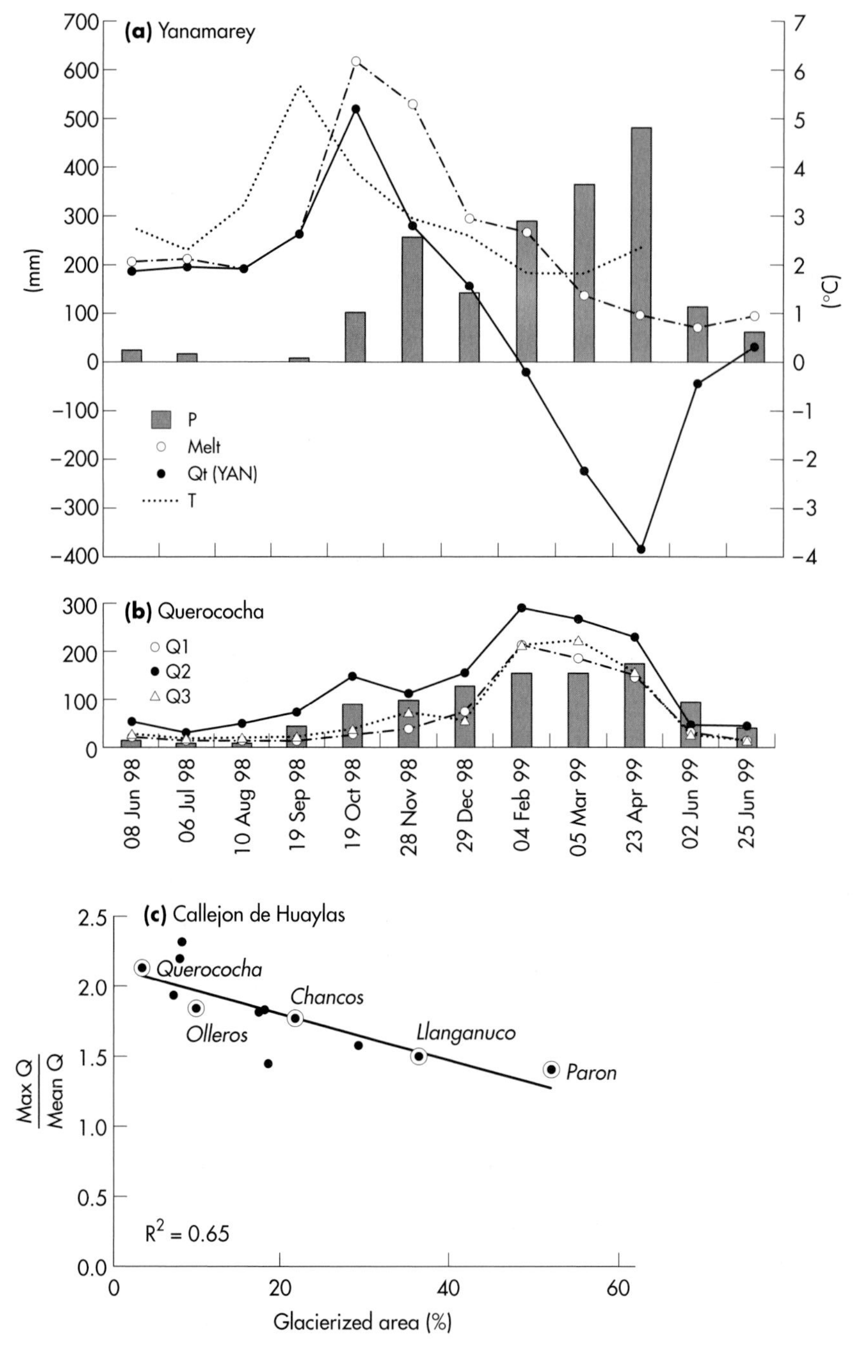

Figure 48.2 Hydrological and climatological data from the successively larger catchments of the case study (see Fig. 48.1): (a) observational data from the Yanamarey glacier catchment, including monthly measurements of specific discharge (Q_t) (mm) from YAN plotted with the monthly precipitation totals (*P*) (mm) and monthly average temperature (*T*) (°C) sampled over the 1998–1999 hydrological year, plotted with the glacier melt (*Melt*) calculated from a simplified hydrological mass balance; (b) specific discharge data from locations in the Querococha watershed plotted with monthly precipitation at the Querococha gauge (both in mm), on the same scale as (a); (c) magnitude and variation of annual stream discharge with percentage of glacierized area in the Río Santa tributaries, shown by ratio of maximum monthly discharge to mean monthly discharge (max *Q* / mean *Q*); labelled data points correspond to gauge locations shown in Fig. 48.1a.

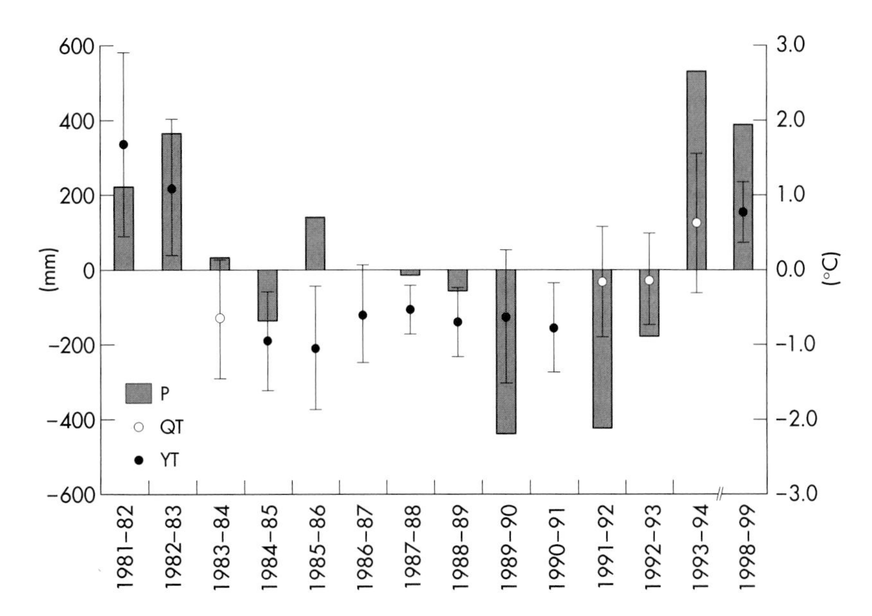

Figure 48.3 Total annual precipitation (*P*) (in mm) and mean annual temperature (in °C) anomalies for the Yanamarey catchment during years with complete data (1981–1994, 1998–1999). Monthly measurements were aggregated over the course of a hydrological year (June–May). Temperature data are shown with one standard deviation error bars based on monthly means from the weather station in Yanamarey catchment (*YT*) where available, and corrected using lapse rate computed from 1956 to 1997 time series measured at the Querococha stream gauge (*QT*). Observations are incomplete after 1994, and only re-established in 1998, as denoted by the break in the *x* axis.

receives a significant amount (10–20%) of its annual discharge from melting glacier ice. Historical records of discharge from tributary basins of the Callejon de Huaylas confirm that watersheds feature enhanced mean annual discharge and less variable runoff in proportion to glacierized area (Fig. 48.2c).

48.4 Insights into climatic control

The tropical austral spring climate during maximum glacier discharge is conducive to melting ice, emphasizing the hydrological and climatological significance of these transitional months between the late dry and early wet seasons. Mass balance and basin hydrology are tied most closely to atmospheric conditions during these spring–summer months, when a lower surface albedo absorbs more long-wave energy and an increase in humidity shifts the latent energy balance at the glacier surface towards producing more melting than sublimation/evaporation. Thus cloudiness, humidity and precipitation probably have more control on melt in the short term (Francou *et al.*, 2003). This confirms a combination of climatic variables originally hypothesized to have forced 20th century recession at Glaciar Yanamarey (Hastenrath & Ames, 1995b). Recent research on tropical glaciers and runoff in Bolivia and Ecuador (e.g. Francou *et al.*, 2000) has shown that total net all-wave radiation, not temperature, is the main factor controlling ablation. However, because temperature is strongly interconnected with these variables, it remains an important indicator of longer term glacier evolution linked to large-scale climate processes such as ENSO (e.g. Vuille *et al.*, 2003; Kaser *et al.*, 2003). Observational evidence from Andean meteorological stations shows a strong positive trend in temperature (Mark, 2002), and model results confirm that the century-scale Andean glacier recession is best explained by increased temperature and humidity (Vuille *et al.*, 2003). Temperature measurements from the Yanamarey catchment show that discharge is actually better correlated with temperature than precipitation (Fig. 48.2a). Moreover, discontinuous time series show an intriguing association of higher temperature with increased precipitation during at least the 1982, 1994 and 1997–1998 El Niño events (Fig. 48.3), opposite from the expected deficit of precipitation driving enhanced ablation during El Niño further south along the Andes (e.g. Wagnon *et al.*, 2003). Downstream, the seasonal contrast in discharge due to a strong amplitude precipitation signal is mitigated by glacier melt, distinct from mid-latitude glaciers responding to strong seasonal temperature contrasts (Mark & Seltzer, 2003).

FORTY-NINE

Glacier variations in central Chile (32°S–41°S)

A. Rivera*,†, C. Acuña* and G. Casassa*

**Centro de Estudios Científicos, Valdivia, Chile*
†Department of Geography, University of Chile, Santiago

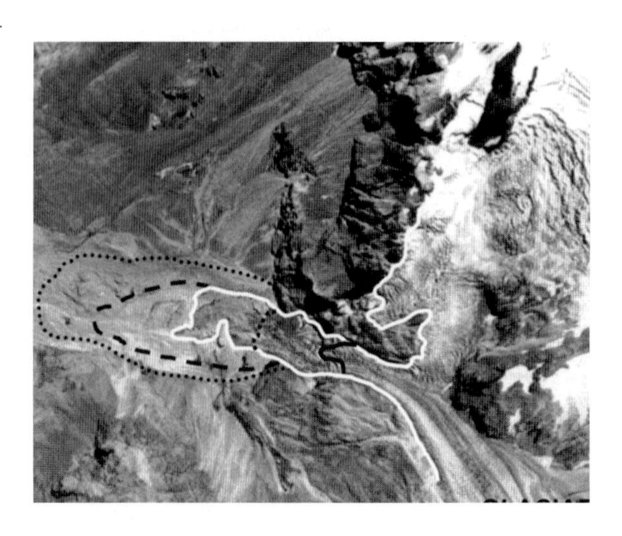

Glaciers in the Chilean central Andes (Fig. 49.1, 32°S–41°S) have shown significant frontal retreat, area shrinkage and ice thinning in an accelerating trend during recent decades, presumably in response to atmospheric warming and reduction of precipitation.

Nearly 1600 glaciers with a total ice area of ca. 1300 km^2 have been inventoried in the Chilean central Andes, which have experienced a total volume loss due to thinning and retreat of 46 ± 17 km^3 of water equivalent between 1945 and 1996 (Rivera *et al.*, 2002), affecting water resources availability for agriculture, mining and human consumption.

This region includes the most populated part of the country (33°S–36°S) and its glaciers have been recognized as a key factor in contributing to late summer runoff in many of the main river basins, especially during summers with severe drought when up to 67% of the water flow is generated by glacier meltwater (Peña & Nazarala, 1987). In spite of the importance of the regional glaciers, very limited glaciological research has been carried out since the pioneering work of Lliboutry (1956). There are still some basins without glacier inventories, especially regarding debris-covered ice, very few ice-thickness data exist (Rivera *et al.*, 2001) and very little is known about the energy and mass balance of the glaciers (Corripio, 2002).

The only systematic mass balance programme is taking place at Glaciar Echaurren Norte (33°35′S, 70°08′W) where results have shown that during warm (cold) phases of ENSO events, high (low) winter precipitation has generated more positive (negative) annual balances (Escobar *et al.*, 1995). As a result, the water resources of central Chile have been under pressure during recent decades owing to increased competition for water allocation as a result of rapid economic growth, and also because the availability of these resources has been stressed by higher interannual variability of the weather system and reduction of the glacier areas.

One of the most dramatic glacier responses has taken place on Glaciar Juncal Sur (33°05′S, 70°06′W), which has experienced an average frontal change of $-50\,m\,yr^{-1}$ between 1955 and 1997, with a total area loss of ca. 10% since 1955.

The glacier with the longest historical record of frontal variations is Glaciar Cipreses (34°33′S, 70°22′W), which has been systematically retreating and shrinking since 1860; in an accelerated trend during recent decades when the retreat rates tripled (Fig. 49.2A). Ice-elevation changes have been measured in this glacier by comparing digital elevation models based upon regular cartography generated from aerial photographs acquired in 1955 and Shuttle Radar Topography Mission (SRTM) data acquired in 2000, yielding an average thinning rate of $1.06 \pm 0.45\,m\,yr^{-1}$ for the ablation area (Fig. 49.2B).

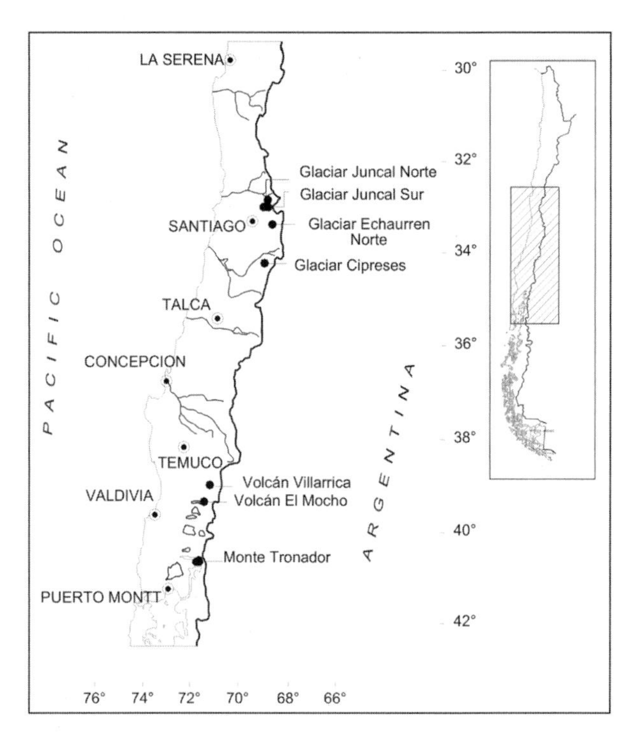

Figure 49.1 Index map showing the location of the glaciers discussed in the text.

Between 36°S and 41°S, most of the glaciers are located on active volcanoes where eruptions have generated damaging lahars owing to sudden melting of snow and ice by ash deposition and lava flows. One of the most active volcanoes in this region is

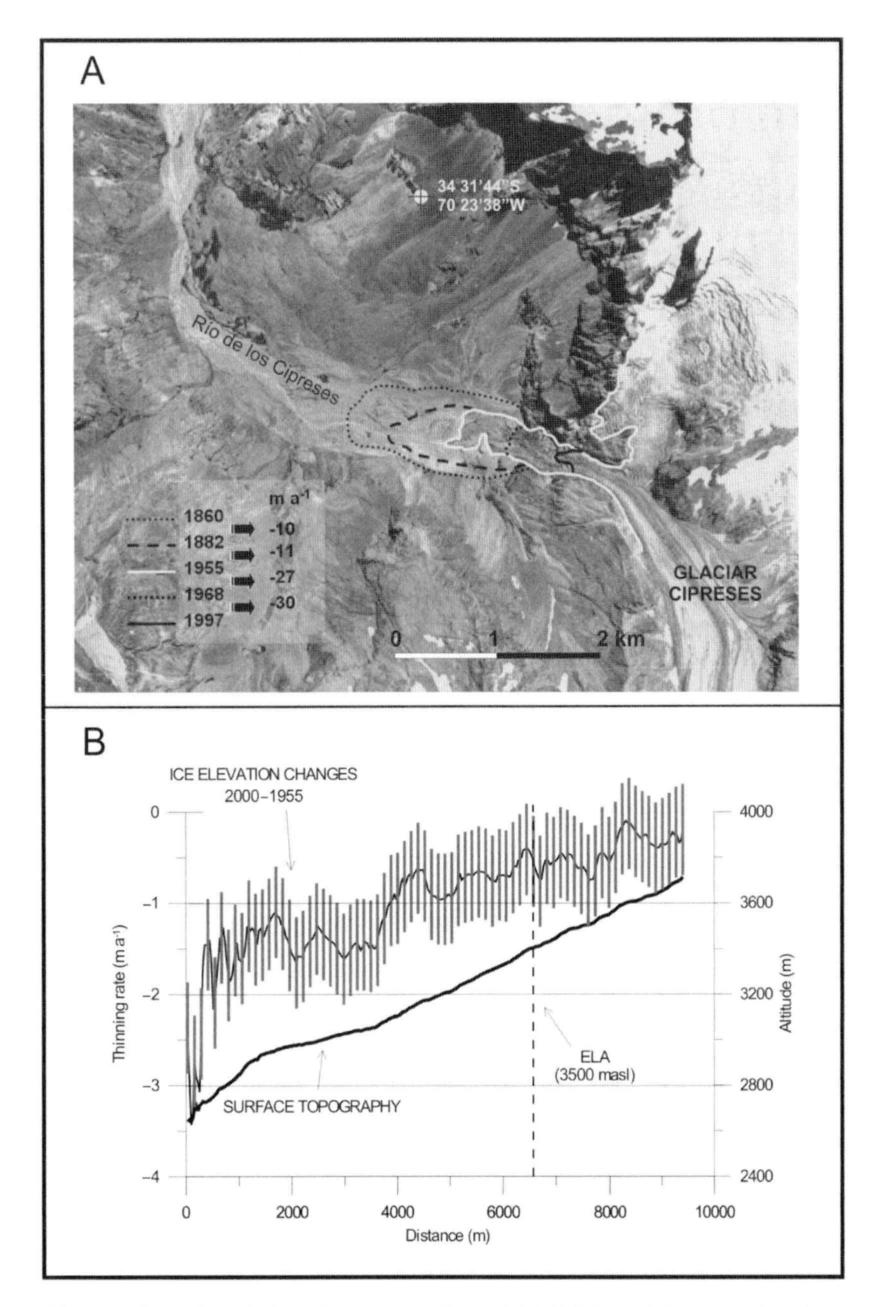

Figure 49.2 Glaciar Cipreses frontal variations between 1860 and 1997 (A) and ice elevation changes between 1955 and 2000 (B).

Volcán Villarrica (39°25′S, 71°55′W), which is ice capped by partially debris-covered glaciers that have been shrinking between 1961 and 2003 at an average rate of $-0.4\,km^2\,yr^{-1}$, with an areal loss of ca. 25% of the 1961 glacier area. Most of these glacier changes have taken place after large eruptions and they are not necessarily related to climate changes.

In spite of having experienced surface atmospheric cooling between the 1950s and the 1980s (Rosenblüth *et al.*, 1997), the region between 38°S and 41°S shows extensive glacier retreat during the second half of the 20th century, presumably in response to both an atmospheric warming at the approximate altitude of the equilibrium line altitude (ELA) of the glaciers (ca. 2000 m a.s.l.) and a significant reduction of precipitation between 1960 and 2000 (Bown & Rivera, in press).

Considering the current glacier behaviour it is possible to presume that the glaciers are out of balance with present climate and further retreat will probably take place.

FIFTY

Palaeoglaciology of the last British–Irish ice sheet: challenges and some recent developments

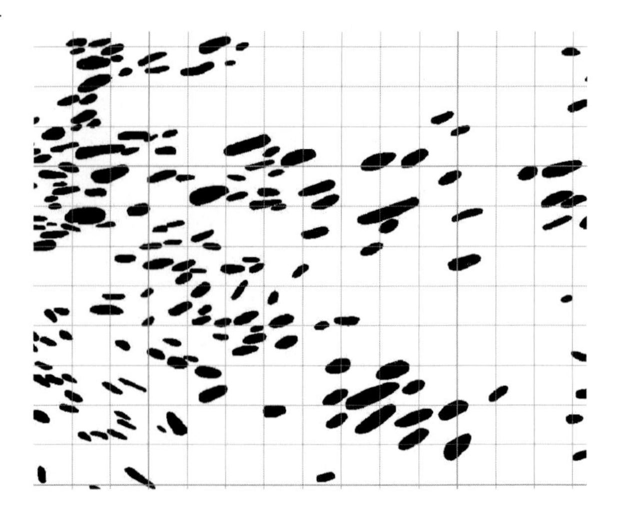

Chris D. Clark*, Sarah L. Greenwood* and David J. A. Evans†

**Department of Geography, University of Sheffield, Sheffield S10 2TN, UK*
†Department of Geography, Durham University, South Road, Durham DH1 3LE, UK

50.1 Introduction

Reconstructing the extent, flow geometry and topography of former ice sheets has recently become more than an academic exercise because of the increasing perception of the importance of the cryosphere in climate change in the earth–ocean–atmosphere system. Of particular note is the discovery that punctuated delivery of freshwater from ice sheets into the oceans (via icebergs and ice-dammed lakes) has the ability to significantly perturb the thermohaline circulation, and as a consequence force extremely abrupt climate flips (e.g. Broecker, 1994). The abruptness of change raises concerns about the robustness of current (global warming) climate predictions and severely challenges our process knowledge of how the climate system works. In addition, a better understanding of palaeo-ice sheets will allow us to assess: (i) the effect of their vertical extent on atmospheric circulation and climate variation (e.g. Shin & Barron, 1989); (ii) the influence of land-ice volumes on global sea level and glacial rebound, leading also to an improved knowledge of mantle viscosity (e.g. Lambeck, 1993a,b); (iii) the effect of ice as an agent of long-term landscape evolution, specifically its impacts on uplift, tectonics and climate (e.g. Sugden *et al.*, 1999); (iv) ice-sheet dynamics (e.g. Boulton *et al.*, 2001); and (v) more national issues related to Quaternary ice-sheet deposits, such as the explanation of landscapes (e.g. Meehan & Warren, 1999), land-use planning, construction engineering, waste disposal and mineral exploitation (e.g. Eyles & Dearman, 1981; DiLabio & Coker, 1989; Merritt, 1992; Gray, 1993; McMillan, 2002).

In order to address the above, we need to work towards palaeogeographicalal reconstructions of ice sheets that include: centres of initial nucleation and subsequent growth; the main centres of mass (ice divides) and margin configuration; migrations of ice divides and readjustments of the overall flow geometry and margins; major ice streams and calving bays; and retreat patterns to the final places of disappearance. Ideally we would be able to constrain the ice thickness (and hence volume) through the growth and decay phases, and know the timing of major events (i.e. maximum extent, ice-divide shifts, successive margin positions, activation of ice streams and major calving events, and chronology of ice retreat). Figures 50.1 & 50.2 schematically illustrate the type of palaeoglaciological reconstructions that we need to work towards.

If we possessed a wide knowledge of the palaeogeography of the last British–Irish Ice Sheet (BIIS) it would contribute considerably to the wider earth system science objectives outlined earlier in which ice-sheet volumes, configurations, discharge events and their timing are prioritized. However, we are far from being able to provide such complete information, even though more than 150 yr of research and publication have uncovered many key fragments of proxy data and the British and Irish landscapes are rich with relevant data. Further field investigations, mapping programmes, more geochronometric dating and developments in the techniques used to map glacial geomorphology from satellite images and digital elevation models (DEMs) will undoubtedly help us to advance our palaeogeographical reconstructions. We argue, however, that the key weakness at present relates to a limited effort at assembling pieces of the jigsaw puzzle. This is not intended as a criticism of existing work, but derives from comparison of what is known about the BIIS compared against reconstructions of other palaeo-ice sheets such as the Laurentide and

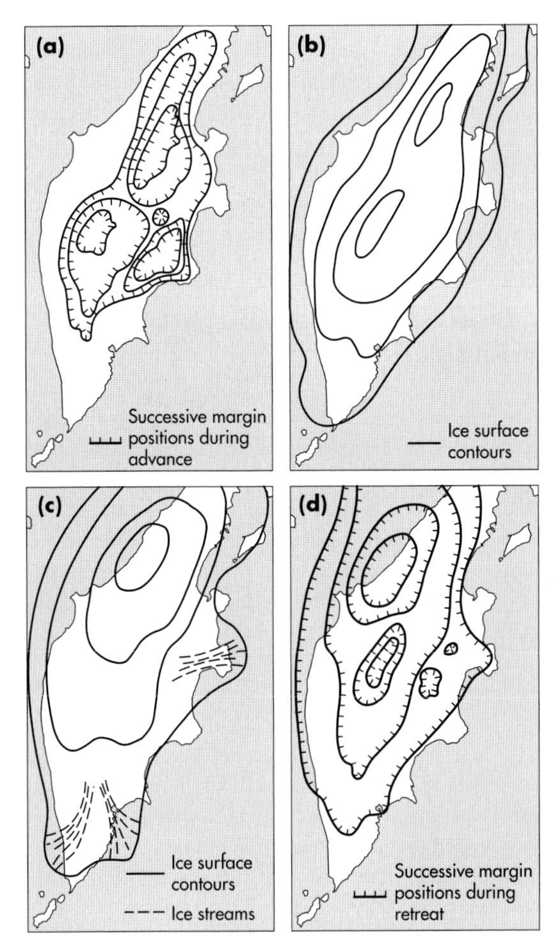

Figure 50.1 A schematic reconstruction of the advance and retreat pattern, and surface topography of a hypothetical ice sheet. (a) If climatic conditions are appropriate ice may first nucleate in the highlands as separate ice caps, which eventually merge to form a large ice sheet. (b) The surface topography of the resulting maximal ice sheet is likely to reflect its growth history, but also might be conditioned by changes in atmospheric circulation and zones of moisture delivery arising from presence of the ice mass. (c) Climatic feed-back effects (moisture starvation in certain sectors), sea-level controls on calving flux, changes in thermal regime or the activation of ice streams might drive the ice sheet's morphology into a different configuration. For example, as illustrated, major ice streams might evacuate ice in sufficient quantities to drive the ice divide away from its original position. (d) The retreat pattern would now reflect such major configuration changes, might not be back towards the high ground, and may be asymmetric and differ considerably from the growth pattern. Retreat might not be to a single ice mass, but may fragment into component ice caps (e.g. consider the incomplete deglaciation of Iceland into its ice caps). We emphasize the hypothetical nature of the above, and use it merely to illustrate the kind of dynamism, migration of ice divides and asymmetry that it is reasonable to expect of an ice sheet's evolution. For convenience the topography of the Kamchatka Peninsula is used as a background, but in no way is the above based on any evidence, indeed glaciation of this region is highly controversial and poorly known.

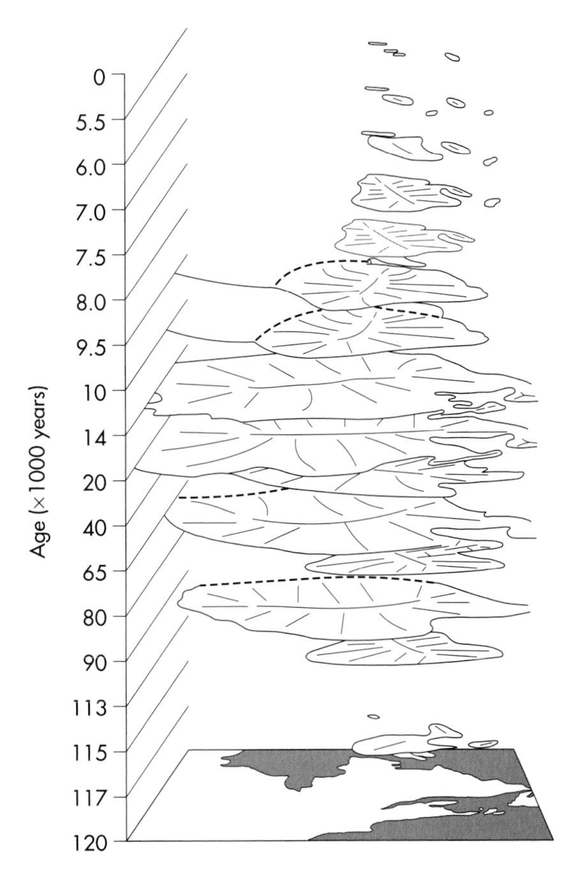

Figure 50.2 An example of the kind of palaeoglaciological reconstruction it should be possible to work towards. Time–space diagram illustrating reconstructed margin positions, flow patterns and ice divide locations of the Quebec–Labrador sector of the Laurentide Ice Sheet. Note how growth and decay is asymmetric and that divide positions migrate. This synthesis is based on an inversion approach using glacial landforms to derive successive flow patterns, combined with information on available dating constraints and moraine positions. Data based on the reconstruction in Clark *et al* (2000). (See www.blackwellpublishing.com/knight for colour version.)

Fennoscandian for which reconstructions of ice-divide locations, flow patterns, ice margins, ice streams and retreat patterns and timing are available (e.g. Dyke & Prest, 1987a; Dyke *et al.*, 2003; Clark *et al.*, 2000; Kleman *et al.*, 1997; Boulton *et al.*, 2001).

It is perhaps unrealistic to expect full reconstructions of the ice sheet and its dynamics through time, based solely on the geomorphological and geological evidence that it left behind. This is because ice sheets do not always inscribe a record of their activity in the landscape, and because we still have incomplete knowledge of, and gaps in, what likely exists. The most profitable route of enquiry is to advance knowledge through combined use of numerical ice-sheet modelling (e.g. Boulton *et al.*, 1985), isostatic inversion modelling (e.g. Lambeck, 1993a,b) and palaeoglaciological reconstruction driven by geomorphological and geological evidence. Intercomparisons between these approaches and hybrids combining them should ultimately yield the palaeoglaciological information required.

The motivation for this chapter is not to review existing knowledge about the BIIS (cf. Evans *et al.*, 2005), but to clarify challenges, point to where efforts could be most usefully directed and highlight some recent developments that might take us towards an empirically realistic BIIS palaeoglaciology.

50.2 The British–Irish Ice Sheet

Presently we have neither a palaeogeographical overview of the regional landform and sedimentary evidence of the BIIS, nor an empirically driven palaeoglaciological reconstruction. This underlines the point made earlier about a lack of synthesis of the existing pieces of the jigsaw puzzle. For such overviews and reconstructions, we need to appeal to the outputs of numerical models, which to varying degrees have utilized field evidence as inputs or as (limited) validation. Such models are useful, because they aim to use the physics of, for example, ice motion or isostatic rebound, to extend our knowledge beyond what is known from the fragmentary geological record. Figure 50.3 illustrates four such reconstructions for the BIIS at its maximum extent in the Late Devensian. It is not sensible to make detailed comparisons between them because they have been derived by different methods of modelling and with different degrees and types of constraining field information. At the broad scale, however, similarities include: centres of mass over Ireland and the Highlands of Scotland and to varying degrees over the Southern Uplands and the Pennines and Wales. The southern extent of ice is similar and all portray ice cover over the Irish Sea. Ice limits elsewhere vary markedly, especially on the west coast of Ireland and Scotland and over the North Sea. Considerable debate has ensued as to whether the BIIS and Fennoscandian Ice Sheet were confluent over the North Sea, with most recent interpretations based on geological evidence confirming confluence at the LGM (Carr *et al.*, 2000; Sejrup *et al.*, 2003). Large differences exist between the models regarding ice thickness and volume. Over Scotland, for example, outputs suggest a range between 1000 and 2000 m for surface elevation. It is difficult to assess which is the most appropriate reconstruction of the ice sheet, and the best way of doing so requires careful comparison with the observational record of geomorphology and geology. For example, Ballantyne *et al.* (1998a,b) reconstructed the surface profile for northwest Scotland based on observed trimlines, and Gray (1995) has constrained isostatic rebound for the Southern Uplands based on former sea-level evidence and comments on the validity of the various models. Examples such as these, where field evidence is brought together to test the robustness of model outputs, are rare. It is clear that advances in knowledge of the palaeoglaciology of the BIIS critically depend on our ability to use field observations to guide or test numerical models. The onus is on modellers to make greater use of the observational record to either drive their models or test them, and the onus is on field researchers to make available their findings at a scale appropriate for this.

Many questions arise from mismatches between modelled simulations and field observations. Is the physics in the model flawed or missing a key ingredient (i.e. ice streaming)? Are the climate drivers inappropriate? Has the observational data been interpreted incorrectly or is the geochronology inaccurate? Advances depend on greater interaction between modelling and empirical approaches, because alone, it is unlikely that either approach can yield the answers. At present, the small number of published numerical models that exist for the BIIS remain largely unrelated to the wider observational record. Moreover, the field evidence remains fragmented into too many small parcels of information, thereby limiting model intercomparisons and coherent syntheses of the regional glacial geomorphology.

50.3 Observational-based palaeoglaciology: the main hurdles

Most investigations of the BIIS have been on a local to regional basis, which makes ice-sheet-wide synthesis difficult, especially where differences in interpretation between areas remain unresolved. Reconstructions of the whole ice-sheet geometry, based on available evidence, have rarely been attempted. The lack of synthesis, or reconstruction, may be attributed to the complexity and scale of the task. Barriers to the production of a coherent description of the BIIS are considered to be:

1. *The fragmented nature of the evidence*, i.e. many spatially separate studies with few links or even gaps between them, and many unresolved contradictions between areas.
2. *The volume of information*. Paradoxically, it might be argued, there is too much evidence. There has been so much written and mapped that it is daunting to attempt a synthesis.
3. Much of the data may be what Rhoads & Thorn (1993) call *theory-laden evidence*, i.e. as information has been collected over a long period of time, and during which glaciological ideas have changed considerably, it is likely that 'evidence' has been tainted by interpretations, some of which may no longer be valid. Some of the theory-laden evidence has likely propagated through the literature to add to the confusion.
4. *Contemporaneity of evidence*. In seeking to reconstruct ice-sheet geometry and extent based on the evidence available it is easiest and most convenient to assume that most evidence was formed penecontemporaneously as this provides maximum information about the ice sheet at a snapshot in time. However, this approach encourages contrived or unrealistic reconstructions that can be falsified in places by evidence that does not match or by implausible ice dynamics. We presume that most evidence is likely to relate to the pattern of deglaciation with underlying palimpsests of maximal or even build-up phases of ice-sheet configuration. Recognition of these multi-temporal aspects and an 'inversion' methodology for making sense of it (Kleman & Borgström, 1996, Clark, 1997; Kleman, this volume, Chapter 38) has led to advances in our understanding of other ice sheets, but has yet to be widely or systematically applied to the British Ice Sheet (cf. Salt, 2001).
5. *Dating control*. Much of the landform and stratigraphical evidence remains undated, and is thus difficult to fix in time and use in dynamic reconstructions.
6. *Incomplete mapping*. Some key parcels of information that could unlock important parts of the glacial history are likely unidentified and unmapped at present.

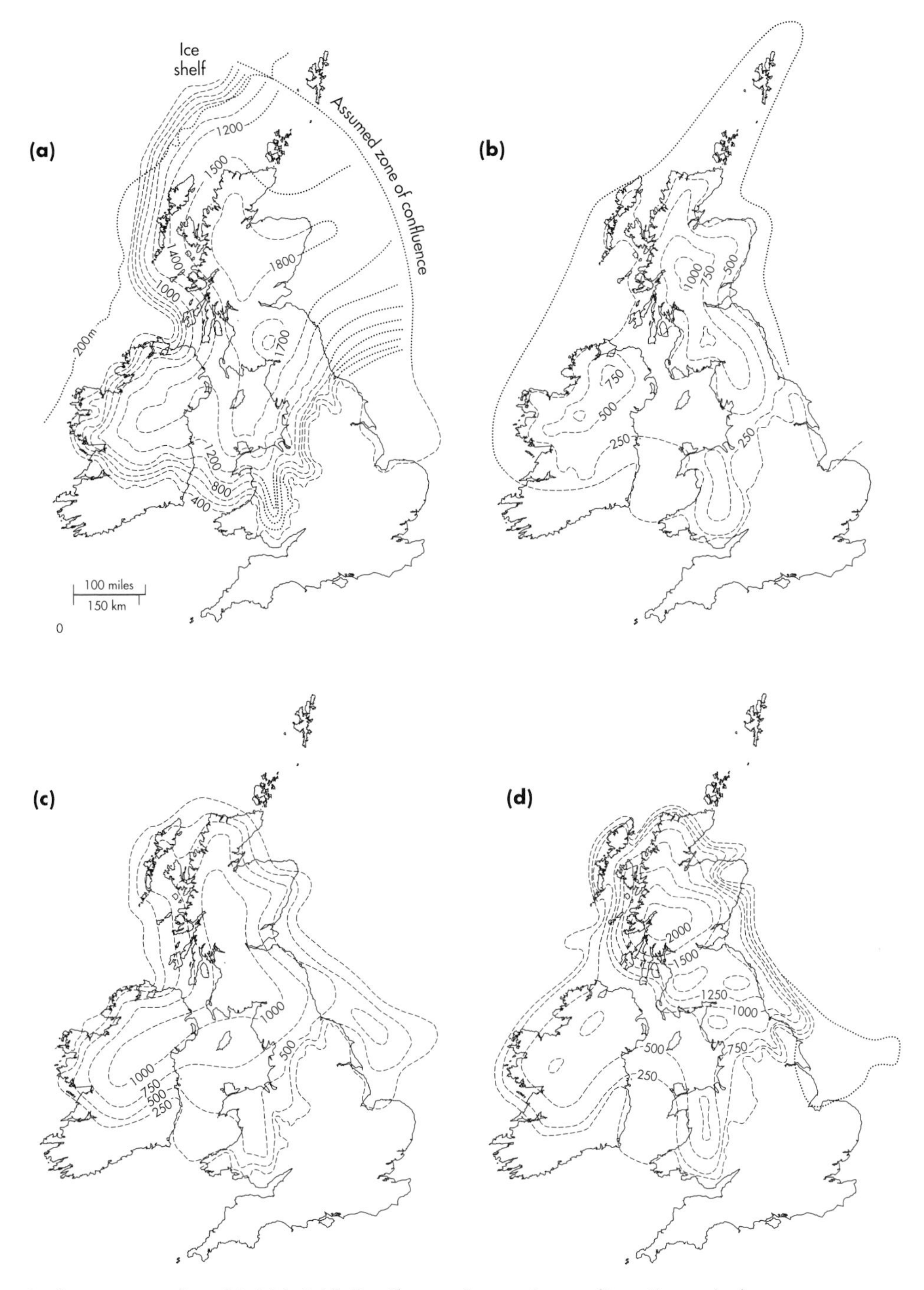

Figure 50.3 Surface topography of British-Irish Ice Sheet at its maximum (Late Devensian) extent, as reconstructed by four numerical models. The first three are ice sheet models partly driven by geomorphological data and the last is produced by isostatic inversion modelling. (a) from Boulton *et al.* (1977), (b) from Boulton *et al.* (1985), (c) from Boulton *et al.* (1991) and (d) Lambeck (1993b). Information redrawn from Gray (1995).

7 *Glaciological plausibility.* It is likely that many evidence-based reconstructions of ice-sheet flow and geometry may in fact be glaciologically implausible. Greater use of modelling may help to eliminate these cases and provide alternatives. In addition to modelling we can use contemporary analogues such as the Antarctic Ice Sheet, but this may be misleading because of its predominantly cold-based thermal regime and its climatic setting.

8 *An objective methodology.* There appears to be little consensus regarding the method by which we utilize observations to build a reconstruction. There are many different approaches taken. At the worst we could (perhaps justifiably) be accused of

'story-building' by attempting to fit patterns of retreat or flow geometry to best explain our own observations and data. To remain objective it is essential that the basis of the methodology is well described, with all assumptions clearly stated, and actual evidence kept distinct from interpretation and interpolation.

50.4 Improved utilization of existing observations

From points 1–3 above and our earlier comments it is clear that it is difficult to make use of published knowledge because there is so much written (at least 2000 papers), and which is also difficult to assess because of propagation of theory-laden evidence, and because of fragmentation with numerous small and separate studies. Critical literature reviews can synthesize and overcome some of these problems, and when performed have significantly helped to advance knowledge (e.g. Sutherland, 1984; Ehlers *et al.*, 1991; Ehlers & Gibbard, 2004). However, for reasons of brevity they are likely to be selective, and focus on the generalized picture rather than on localized details. Also, it is questionable whether the richness in data generated over the past 100 yr can be adequately assessed and presented in a journal article or book chapter when the more appropriate format is a map or map series. Ideally, we require a map of Britain and Ireland and the surrounding continental shelf on which all geomorphological, geological and geochronometric information (i.e. moraines, striae, till limits, dates, etc) relating to the last ice sheet is plotted. This would be the jigsaw puzzle assembled, from which ice-sheet-wide reconstructions could be attempted. Notably, for the Laurentide Ice Sheet such a 'glacial map' has existed since 1968 (Prest *et al.*, 1968; Fulton, 1995) and numerous reconstructions have been derived from it. We argue that a similar approach is required for Britain and Ireland, and would likely yield similar and significant advances in our knowledge.

A recent project, called BRITICE, has partly addressed this issue by reviewing all the relevant literature from academic journals and PhD theses (Evans *et al.*, 2005) and 100 yr of BGS maps of onshore and offshore geology (Clark *et al.*, 2004). Information was extracted and entered into a geographical information system (GIS), from which a 'glacial map' (1 : 625,000 scale, 1 × 1.6 m) has been produced (Clark *et al.*, 2004). This project focused primarily on geomorphological features that inform us about the last ice sheet, and includes the following information: moraines, eskers, drumlins, meltwater channels, tunnel valleys, trimlines, limit of key glacigenic deposits, glaciolacustrine deposits, ice-dammed lakes, erratic dispersal paths and shelf-edge fans. The GIS contains over 20,000 individual features split into thematic layers (as above). Figure 50.4 provides an overview of the database and Fig. 50.5 is an enlarged extract illustrating some of the types of data and their distribution. The task is incomplete in that some information was excluded, such as striae and geochronometric dates, but it does represent the most complete assembly of the BIIS jigsaw puzzle to date. The purpose of the work was to bring together valuable information from the literature in the hope that it can help motivate the following.

1 *Evidence-based reconstruction of the ice sheet.* Meltwater channels for example could be combined with the moraines, ice-dammed lakes and eskers to build a sequential pattern of glacier retreat, to infer palaeo-ice dynamics and inform palaeoclimate reconstructions (cf. landsystems approach; Evans, 2003a). The drumlin and erratic-pathway data could be analysed and enhanced by mapping from digital elevation models (DEMs) and satellite images to derive the ice-flow patterns. These could be used to reconstruct changes in flow geometry and ice-divide positions through time. Ideally a full inversion approach utilizing all the data could be attempted and constrained by available stratigraphical and dating evidence.
2 *Numerical ice-sheet modelling.* Modelling has become increasingly important as an aid to reconstructing ice sheets and for assessing their relationship to other factors such as sea-surface temperatures, climate and sediment discharge (cf. Siegert, 2001). Flow-pattern or ice-extent information can be used to drive the modelled reconstructions or as validation of the modelled result, or in some combination. An example of the former was presented by Boulton *et al.* (1977), who used flow-patterns to constrain British Ice Sheet geometry (Fig. 50.3a) and from this derived a modelled estimate of the surface topography. The alternative is to 'grow' an ice sheet over the topography using climate drivers (i.e. derived from ice-core records) and then assess the plausibility of the modelled ice sheet by comparing ('testing') it with empirically derived evidence of ice-flow configuration. It is frequent to hear field investigators criticize the work of ice-sheet modellers because they sometimes fail to use the wealth of geological information available and because of a lack of testing of their results, against what is known. Such criticism is perhaps unfair given that evidence is so rarely compiled in a consistent format so as to make it useable. It is hoped that the BRITICE GIS compilation, made freely available via the world-wide web, will facilitate increased use of geomorphological data in modelling experiments.
3 *Directing fieldwork.* It is evident from the Glacial Map (Fig. 50.4) that although there is a reasonable spread of information across the ice-sheet bed, there are notable gaps and great variability in data density. This compilation may assist field workers in choosing areas for future investigation, mitigating the tendency within the academic literature to keep re-investigating the same area.

Conflicts and discrepancies exist in the data recorded on the Glacial Map and in the BRITICE GIS database. For example, some confusion is likely to exist with regard to the true age of some of the features. The inclusion of some pre-Devensian and Loch Lomond Stadial features is likely in some locations where dating control is poor. It is hoped that, in addition to the points above, compilations such as BRITICE may encourage greater scrutiny of valuable published work.

50.5 Case study: using meltwater channels to constrain ice-sheet retreat patterns

Here we illustrate how some of the data layers in the BRITICE GIS can be used to build a preliminary reconstruction of the

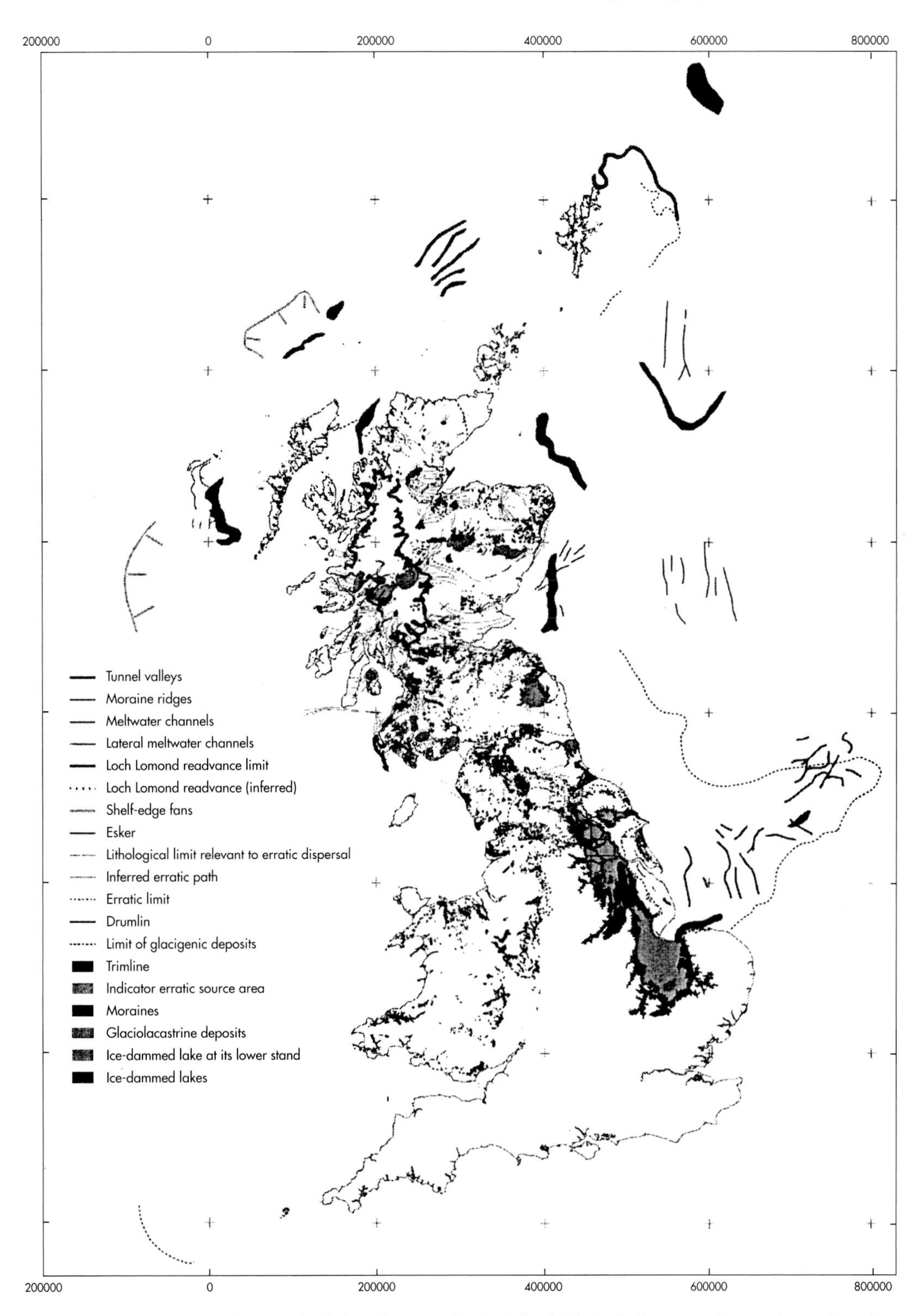

Figure 50.4 The BRITICE compilation of a 'Glacial map of Britain', which includes: moraines, eskers, drumlins, meltwater channels, tunnel valleys, trimlines, limit of key glacigenic deposits, glaciolacustrine deposits, ice-dammed lakes, erratic dispersal paths and shelf-edge fans. The data are not intended to be fully visible at this scale, see Fig. 50.5 for a zoomed extract and Clark *et al.* (2004) where a full scale, 1 : 625,000 map is reproduced, but the illustration does emphasize the spread of information and gaps. (See www.blackwellpublishing.com/knight for colour version.)

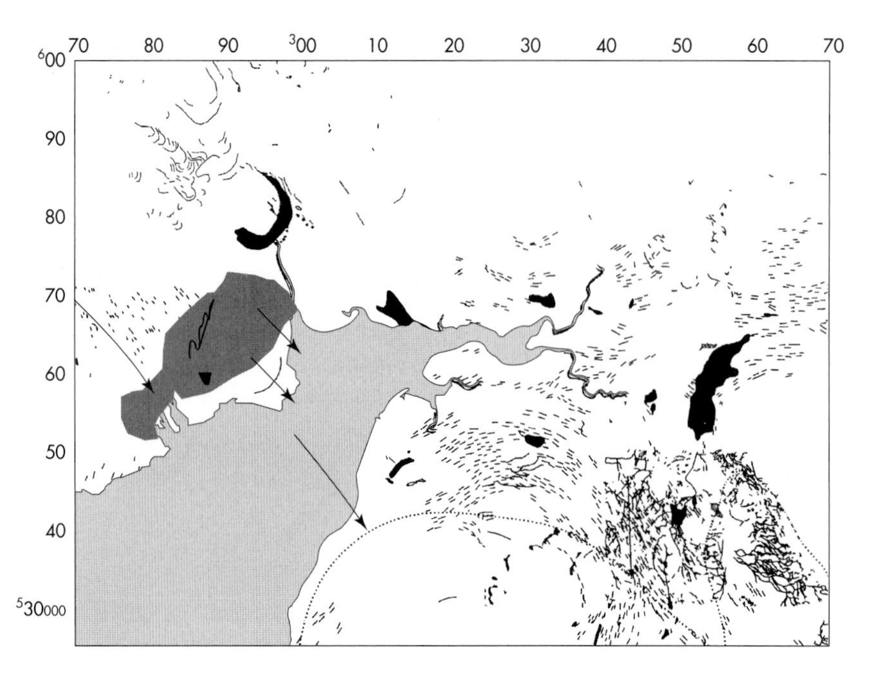

Figure 50.5 Zoomed extract (for northern Lake District–Solway Firth) of the information held in the BRITICE GIS and on the Glacial Map. Drumlins are recorded as short straight black lines; meltwater channels, eskers and small moraines are recorded as curving black lines. Erratic source areas, pathways and limits are also shown. Large black features are moraines. For a colour version of this figure, and indeed map of the whole of Britain, please consult Chris Clark's website at the University of Sheffield. Printed versions of the Glacial Map can also be purchased from the website. Ordnance Survey grid-squares are 10 km.

ice-sheet retreat pattern. Approximately 8000 meltwater channels have been identified and mapped in the published literature and BGS maps and reports, but have never been utilized previously to build an ice-sheet-wide synthesis of retreat.

It is crucial to distinguish between subglacial, ice-marginal and proglacial meltwater channels in order to derive appropriate palaeoglaciological information. Well-defined subglacial channels (i.e. with humped long-profiles) inform us of a warm-based glacial thermal regime and disposition of the hydraulic head, which in turn can be used to assess former ice-surface slope and hence ice-flow direction. Ice-marginal or lateral meltwater channels are formed in positions where the glacier edge meets a slope and are of great value for ice-sheet reconstruction, because they record former ice-marginal positions during recession (e.g. Dyke, 1993; Kleman, 1994; Hättestrand & Clark, this volume, Chapter 39). Proglacial channels are more difficult to differentiate from non-glacial, fluvial channels. Diagnostic features are channel sizes much larger than the modern fluvial catchment could reasonably produce or incongruous locations, for example, on upland cols.

Unfortunately, most field workers have failed to distinguish between different types of meltwater channel and prior to the influence of the Scandinavian school of glacial geomorphology in the 1950s many channels were regarded as ice-dammed-lake spillways (Evans, in press). All channels from the literature and BGS maps were digitized and entered into the BRITICE GIS database (see Clark *et al.*, 2004, for further details) as two data layers. One layer was for channels interpreted as lateral in origin (ca. 1000 examples), but the majority were entered into a generic meltwater channel layer (ca. 7000 examples). We have utilized this database and extended the work by:

1 developing observational criteria to permit subdivision of generic meltwater channels into their appropriate category (i.e. lateral, subglacial etc);
2 applied the criteria to all cases, to establish populations for each type;
3 utilized the above to produce a preliminary reconstruction of ice-retreat pattern.

Based on published accounts of the various types and characteristics of meltwater channels, a list of criteria was developed (Table 50.1). Although all forms of channel possess discriminating characteristics, subglacial and lateral meltwater channels are of most importance in reconstructing ice retreat. The investigation was therefore designed to distinguish primarily between these types. However, not all criteria presented in Table 50.1 are appropriate for application to this study, because the coarse resolution of the DEM prevents assessment of widths for example, and we did not visit the channels for field analysis. Our analysis was based on consideration of channel form and position in relation to local topography. The data layer of channels was superimposed upon a DEM of Britain (50 m grid size) and, using qualitative and quantitative analysis in combination, observed characteristics were matched to the appropriate class of channel as defined by the established criteria. A visual assessment was performed, posing four types of observation: channel position relative to topography (e.g. valley floor/side); orientation relative

Table 50.1 Criteria to define meltwater channel type. Lateral channels are split into two categories, with submarginal being those that approximate the glacier margin but have characteristics that indicate a component of flow that penetrated beneath the ice margin. Numbers refer to publications from which criteria were assembled: 1, Sissons (1961); 2, Clapperton (1968); 3, Sugden *et al.* (1991); 4, Sissons (1960); 5, Benn & Evans (1998); 6, Schytt (1956); 7, Dyke (1993); 8, Price (1960). Italicized criteria indicate those that we were not able to utilize because our analysis was principally from the form of channels and their relation to topography as recorded on a DEM of Britain

Subglacial	Lateral		Proglacial	Supraglacial/englacial
	Marginal	Submarginal		
Undulatory long profile[1]	Parallel with contemporary contours[2]		Regular meander bends[5]	Isolated crescentic meander forms on face of hill[1]
Descent downslope may be oblique[1]	Approximately straight[2]		Occasional bifurcation[5]	Low gradient[8]
Sudden meanders[2]	Absence of networks[2]		Flows directly downslope[8]	Sinuous[8]
Complex systems—bifurcating and anastomosing[3]	Perched on valley sides[5]		*Large dimensions—wide and deep*[5]	*Approximately constant width*[8]
Abandoned loops[2]	May terminate in downslope chutes[6]		*Crater chains*[5]	
Abrupt beginning and end[1]	Gentle gradient[1]	Steeper gradient (oblique downslope)[1]		
Absence of alluvial fans[1]	Parallel for long distance[1]	Sudden change in direction[1]		
Cavity systems and potholes[3]	*May terminate abruptly*[5]			
Ungraded confluences[3]	May be found in isolation from all other glacial features[7]			
Variety of size and form within the same connected system[4]				

to contours; channel form (straight or sinuous); and interchannel relationships (isolated or within a sequence/system). Preliminary classifications made on this basis were supported, where required, using GIS analytical tools. Gradients were calculated where the channel appeared neither gentle nor a chute and where further precision was required to aid identification. Subglacial and lateral channels of 6–28° and 2% gradient are described by Sugden *et al.* (1991) and Sissons (1961) respectively. Longitudinal profiles were taken along channels to establish if undulations occurred (i.e. indicative of subglacial formation). As digitizing errors are inherent in the GIS database, each long profile was taken three times and mean values calculated. The number of undulations was recorded and their gradient calculated, in order to express the degree to which the channel flows up and downhill.

Observations were recorded for all channels in the BRITICE GIS. Each channel type was subdivided into three categories of certainty according to our subjective level of confidence as: 'definite', 'probable' and 'possible'. The process of channel classification is demonstrated in Fig. 50.6 and Table 50.2, for an area on the western flanks of the southern Pennines, east of the Cheshire–Shropshire plain. Data generated by the categorization procedure for these channels is presented in Table 50.2.

Results from the above analysis are presented in Fig. 50.7. Most channels fall into the lateral (including submarginal) or subglacial categories. It is unfeasible that the mapping reported here represents the true population of meltwater channels, and it is clear that much further investigation is required. Indeed when performing the above analysis we sometimes found obvious, but hitherto unreported channels (Fig. 50.8). These were not entered into the database, but could form part of a future project.

Although the identification and mapping of channels is incomplete for Britain, there is sufficient density and coverage to formulate reconstructions for parts of the ice sheet. The logic behind this is simple: lateral channels record successive margin positions as the ice sheet retreated and subglacial channels must be orientated parallel to the direction of the steepest ice-sheet-surface profile, and hence can be used to approximate ice-flow direction. As the channels clearly require meltwater production we infer that they must have been created in the ablation zone of the glacier and presume that they therefore record the configuration of ice close to the termini and during retreat. In building the retreat pattern only our certainty classes of 'definite' and 'probable' were utilized. Figure 50.9 illustrates how retreat patterns are generated from such data. This process was performed for all areas to produce our overall assessment of ice-marginal retreat (Fig. 50.10).

We do not comment in detail here on the patterns and possible timing of retreat. This can be attempted only at a later stage when other data layers are considered and are integrated with timing constraints provided by geochronometric dates. However, a number of general points are self-evident.

1 By combining the wealth of published literature on meltwater channels it is possible to reconstruct large fragments of ice-sheet retreat.
2 Retreat is clearly a three-dimensional phenomenon and in many circumstances the emergence of higher terrain is

Table 50.2 Extract from the data table in which individual channel characteristics were recorded and categorized for all channels. These channels are those depicted in Fig. 50.6

Channel identity	Absence of catchment	Position*	Orientation to contours†	Form‡	Relationship to others§	Gradient¶	Long profile**	Identification	Reference	Comments
1319	Yes	F	Pl	M	Sys	G	S	Subglacial	Knowles 1985	Network as a whole has up and down profile
3017	Yes	F	Pl	M	Sys		S	Subglacial	Knowles 1985	
3013	Yes	F	Pl	Ls	Sys	G	S	Subglacial	Knowles 1985	
3007	Yes	F	Pl	S	Sys	G	S	Subglacial	Knowles 1985	
3009	Yes	F	Pp; Pl	S	Sys		S	Subglacial	Knowles 1985	
3006	Yes	F	O	S	Sys		S	Subglacial	Knowles 1985	
3004	Yes	F	Pl	S	Sys	G	S	Subglacial	Knowles 1985	
3002	Yes	F	Pl	M	Sys	G	S	Subglacial	Knowles 1985	
3000	Yes	F	Pl	S	Sys	G	S	Subglacial	Knowles 1985	
3008	Yes	F	Pl	Ls	Sys	G	S	Subglacial	Knowles 1985	
3001	Yes	F	Pl	Ls	Sys	G	S	Subglacial	Knowles 1985	
2997	Yes	F	Pl	Ls	Sys	G	S	Subglacial	Knowles 1985	
2996	Yes	F	Pl	Ls	Sys	G	S	Subglacial	Knowles 1985	
2992	Yes	F	Pl	S	Sys	G	S	Subglacial	Knowles 1985	
2991	Yes	F	Pl	S	Sys	G	S	Subglacial	Knowles 1985	
2988	Yes	F	Pl	Ls	Sys	G	S	Subglacial	Knowles 1985	
3068	Yes	F	Pl	M	Sys	G	H	Subglacial	Knowles 1985	
2986	Yes	S	O	Ls	Ind		S	Subglacial	Rees & Wilson 1998	
2985	Yes	S	O	S	Ind		S	Subglacial	Rees & Wilson 1998	
3010	Yes	F	Pl	Ls	Sys	G	H	Subglacial	Knowles 1985	
2995	Yes	F	Pl	Ls	Sys	G	S	Subglacial	Knowles 1985	
2994	Yes	F	Pl	Ls	Sys		S	Subglacial	Knowles 1985	
2972	Yes	S	O	S	Ind		S	Prob subglacial	Rees & Wilson 1998	
2973	Yes	S	O	Ls	Ind		S	Prob subglacial	Rees & Wilson 1998	
2977	Yes	S	O	S	Ind	G	S	Prob lateral	Rees & Wilson 1998	
2981	Yes	S	O	S	Ind	G	S	Poss lateral	Rees & Wilson 1998	
3066	Yes	S	O	Ls	Ind	G	S	Poss lateral	Rees & Wilson 1998	
2966	Yes	S	O	S	Ind	G	S	Prob lateral	Rees & Wilson 1998	
1243	Yes	S	O	Ls	N		S	Prob subglacial	Rees & Wilson 1998	
1244	Yes	S	O	Ls	N		S	Prob subglacial	Rees & Wilson 1998	
3067	Yes	S	O	S	N		S	Prob subglacial	Rees & Wilson 1998	
2990	Yes	S	O	S	Ind		S	Prob subglacial	Rees & Wilson 1998	
2987	Yes	F	Pl	M	Sys		S	Subglacial	Knowles 1985	
2974	Yes	S	O	M	Sys		H	Subglacial	Knowles 1985	
2964	Yes	F	Pl	Ls	Sys	G	S	Subglacial	Knowles 1985	
2962	Yes	F	Pl	S	Sys	G	S	Subglacial	Knowles 1985	

490	Yes	S	O	S	Ind		S	Subglacial	Johnson 1965	Original literature suggests humped profile, confirming subglacial identification, though DEM resolution is too coarse to identify humped bed. Digitizing is slightly dis-aligned from channels evident on DEM: observations taken assuming digitized channels fit DEM.
492	Yes	S	O	S	Sys		S	Subglacial	Johnson 1965	
483	Yes	S	O	S	Sys		S	Subglacial	Johnson 1965	
494	Yes	S	O	S	Sys		S	Subglacial	Johnson 1965	
493	Yes	S	Pp	S	Sys		S	Subglacial	Johnson 1965	
495	Yes	S	O	S	Sys	G	S	Subglacial	Johnson 1965	
484	Yes	S	O	S	Sys		S	Subglacial	Johnson 1965	
485	Yes	S	O	S	Sys		S	Subglacial	Johnson 1965	
491	Yes	S	O	S	Sys		S	Subglacial	Johnson 1965	
497	Yes	S	O	S	Sys		S	Subglacial	Johnson 1965	
496	Yes	S	O	S	Sys	G	S	Subglacial	Johnson 1965	
486	Yes	S	O	S	Sys		S	Subglacial	Johnson 1965	
487	Yes	S	O	Ls	Sys		S	Subglacial	Johnson 1965	
498	Yes	S	O	S	Sys	G	S	Subglacial	Johnson 1965	
499	Yes	S	O	S	Sys	G	S	Subglacial	Johnson 1965	
488	Yes	S	O	S	Sys		S	Subglacial	Johnson 1965	
500	Yes	S	O	S	Sys		S	Subglacial	Johnson 1965	
503	Yes	F	Pl	Ls	Sys		S	Subglacial	Johnson 1965	
489	Yes	S	O	S	Sys		S	Subglacial	Johnson 1965	
504	Yes	F	Pl	S	Sys		S	Subglacial	Johnson 1965	
505	Yes	S	Pl	S	Ind	G	S	Lateral	Johnson 1965	
460	Yes	S	O; Pl	S	Ind	G	S	Lateral	Johnson 1965	
506	Yes	S	Pl	S	Ind		S	Lateral	Johnson 1965	
459	Yes	S	O	S	Ind		S	Lateral	Johnson 1965	
507	Yes	S	O	S	Ind		S	Lateral	Johnson 1965	
508	Yes	S	O	S	Ind		S	Lateral	Johnson 1965	
463	Yes	S	O; Pl	Ls	Sr	G	S	Lateral	Johnson 1965	
510	Yes	S	Pl	S	Sr	G	S	Lateral	Johnson 1965	
509	Yes	S	Pl	S	Sr		S	Lateral	Johnson 1965	
511	Yes	S	Pl	S	Sr	G	S	Lateral	Johnson 1965	
512	Yes	S	Pl	S	Sr	G	S	Lateral	Johnson 1965	
513	Yes	S	O	S	Sr	G	S	Lateral	Johnson 1965	
514	Yes	S	O	S	Sr	G	S	Lateral	Johnson 1965	
464	Yes	S	Pl	S	Ind	G	S	Lateral	Johnson 1965	
461	Yes	S	Pl	S	Sr	G	S	Lateral	Johnson 1965	
462	Yes	S	Pl	S	Sr	G	S	Lateral	Johnson 1965	

*F, valley floor; S, valley side; C, col; Fp, flat plateau.
†Pp, perpendicular; O, oblique; Pl, parallel.
‡S, straight; Ls, low sinuosity; M, meanders.
§Ind, individual; Sr, series; N, network; Sys, bifurcating and anastomosing system.
¶G, gentle; S, steep chute.
**S, smooth; H, humped.

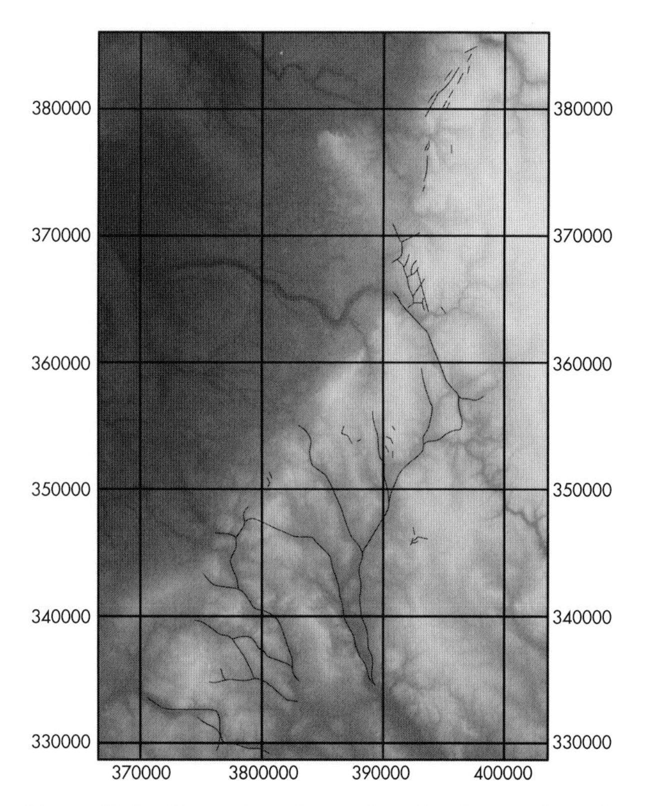

Figure 50.6 Illustration of grouping of meltwater channels into lateral and subglacial categories. In the northeast is a sequence of lateral channels positioned on the slope sides. Running parallel or subparallel to contours, these channels have a gentle gradient, straight form and smooth longitudinal profile. The main network is mostly subglacial channels, flowing across valley sides and floors predominantly oblique to the slope gradient. Many have a sinuous form and develop a complex bifurcating and anastomosing system. Three channels demonstrate an undulating longitudinal profile, lending strong support to a subglacial classification. The area is on the western flanks of the southern Pennines, east of the Cheshire–Shropshire Plain.

evident, for example, in the Yorkshire Dales. The varied topography of Britain complicates what could otherwise have been a simple concentric pattern of ice-sheet shrinkage. Interpretations of glacial history will need to pay close attention to the interaction of ice with topography in order to form coherent syntheses of ice-sheet behaviour. This point is emphasized if we consider how much easier it is to reconstruct retreat patterns for the Laurentide Ice Sheet given the largely flat Canadian Shield.

3 Although the data does not inform the configuration of ice divides and flow patterns at the LGM, it is clear that during ice decay, the ice sheet consisted of multiple centres of outflow (ice domes) and that these interacted or competed with each other. It is unlikely that the retreat pattern presented (Fig. 50.10) records a single, ice-sheet-wide sequential recession pattern, because 'multiple deglaciations' are evident. In the Vale of Eden and adjacent Pennines for example, opposing patterns cannot feasibly be contemporaneous, and thus require an explanation that includes retreat, subsequent readvance and further retreat in a different direction. The possibility remains that some of the meltwater channels are preserved from an earlier glacial stage.

4 The ice sheet did not retreat as a single entity, but rather as it thinned and retreated it separated into component ice caps, which then accomplished their own retreat, such as over the mountains of North Wales and the Southern Uplands of Scotland.

5 Ice retreat was not always to higher terrain. For Scotland this appears to be the case, but in the Pennines for example, ice retreated westwards through the trunk valleys and beyond the main north–south topographic axis. Meltwater channels on the western flank of the southern Pennines and Peak District, along with moraines and channels in the Cheshire–Shropshire Basin, indicate retreat *towards* lower ground and the Irish Sea Basin. Taken together these observations indicate that substantial ice volumes must have existed to the west of the main north–south topographic drainage divide and off the west coast of England, in the Irish Sea. This might have been a result of either significant ice build-up there, perhaps in the confluence between the British and Irish Ice Sheet, or from a late-stage readvance.

6 Given the above-mentioned retreat patterns it is evident that the growth and decay of the British Ice Sheet was asymmetric. According to accepted glaciological principles, growth likely initiated on the higher ground and spread out from here, but the data for England indicates retreat well to the west of the high ground.

Future challenges lie in the compilation of a more complete reconstruction than that outlined above. However, our case study demonstrates the value of an ice-sheet-wide and inversion approach to reconstruction, and highlights the wealth of underutilized geomorphological evidence that exists in the literature.

50.6 Chronological constraints

Prior to our more recent understanding of mobile and dynamic ice sheet behaviour, glacial researchers focused on demarcating and dating ice-sheet maximum extents and the pattern and tempo of retreat. For example, if an ice sheet was initiated in a highland region, expanded to its maximum and then retreated back in a symmetrical manner, then a small number of geochronometric dates (e.g. of deglaciation) distributed across the region could reasonably constrain its behaviour and permit correlations with changes in climate and ocean circulation. Palaeo-ice sheet history, however, is often more complex than this. Interestingly, almost all archaeological discoveries reported in the press seem to take the flavour 'ancient man was actually far more advanced than we had previously thought'. All recent discoveries about palaeo-ice sheets take the same form, in that we now know their behaviour to be much more dynamic, with numerous unexpected surprises and with asymmetry in their growth and decay. Ice sheets are unlikely to have reached their maximum extent synchronously around their margins; ice divides might have shifted during their growth and decay; surges and readvances likely occurred; ice streams

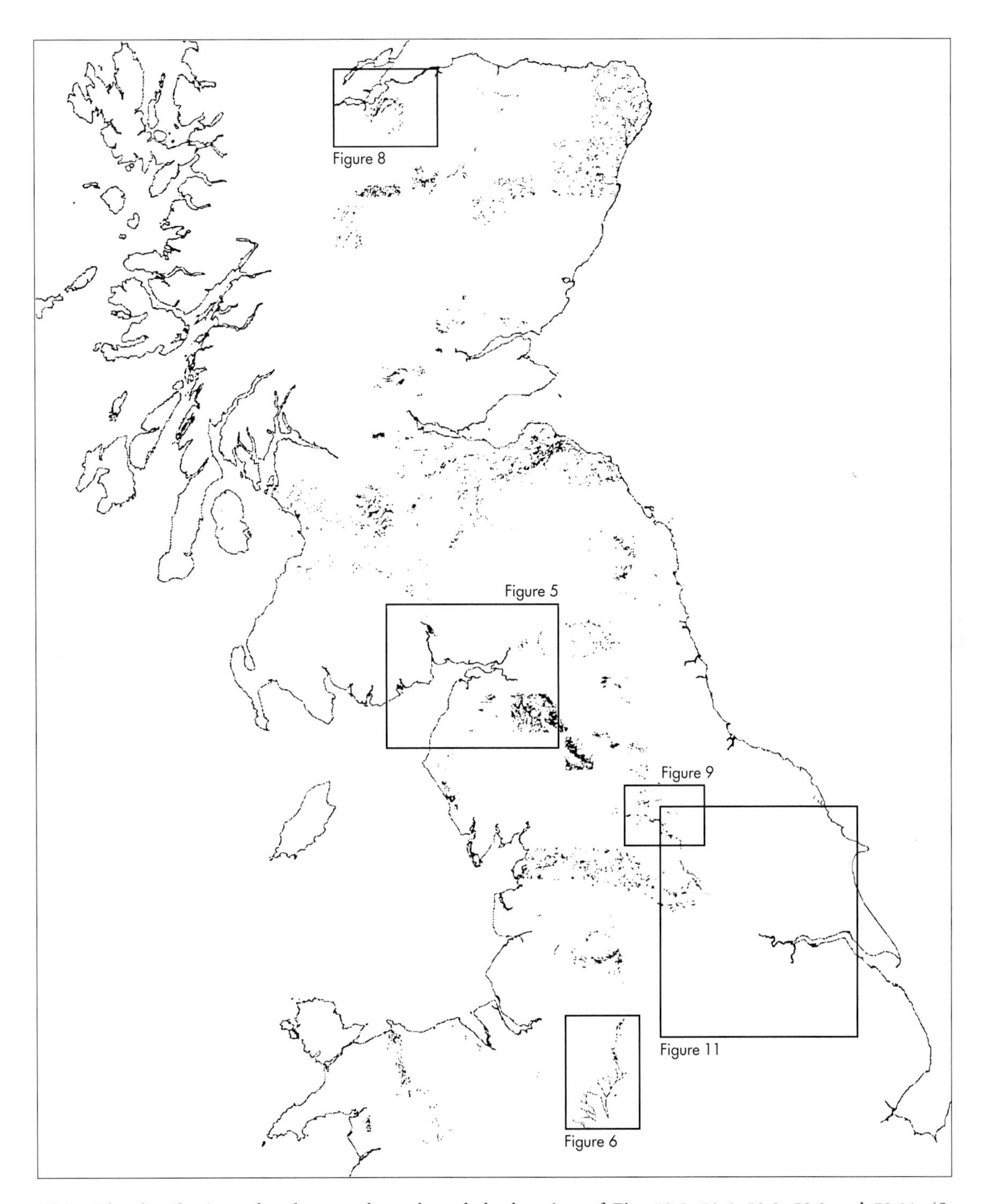

Figure 50.7 The distribution of meltwater channels and the location of Figs 50.5, 50.6, 50.8, 50.9 and 50.11. (See www.blackwellpublishing.com/knight for colour version.)

likely turned on and off like their modern counterparts. This makes the task of constraining ice sheets in time and space much more difficult and requires many more geochronometric dates.

Most dating uses an exposure method, that is, by estimating the length of time elapsed since ice cover disappeared from a location. This permits dating to be used to establish the timing of retreat of ice margins or to bracket, using two dates positioned stratigraphically, a particular advance. Unfortunately it does not allow us to determine major internal changes within the ice sheet, such as when an ice stream turned on or when an ice divide migrated, because there is no method for dating subglacial landforms or deposits. By dating sediments deposited distal to the ice sheet, say in a shelf-edge fan or from ocean cores, it is possible to constrain major depositional events in time (e.g. ice-rafted debris events) and infer that they relate to an ice sheet reorganization such as activation of an ice stream. So for the margins and major

Figure 50.8 Solar-shaded digital elevation model (DEM) of the terrain containing the Flemington eskers and associated kame and kettle topography to the south of the Moray Firth in Scotland. Esker ridges (E) and meltwater channels (M), as depicted by Merritt *et al.* (1995) and on the Glacial Map of Britain (Clark *et al.*, 2004), are highlighted. Note the large lateral meltwater channels slightly northeast of the image centre. These are outside the area mapped by previous researchers and therefore not included on the Glacial Map of Britain. Similarly, the fluted terrain visible in the northwest sector of the image and documenting the flow direction of the former Moray Firth Ice Stream has never been mapped. Image is 45 km across and is derived from a 50-m grid-sized DEM.

events it should be possible to use absolute dating to constrain the timing, but for changes internal to the ice sheet we have to rely on correlations further afield and some relative-age dating methods. Superimposition of landforms for example clearly reveals the relative age of each event and thus can be used to build sequences of ice-sheet changes but that remain unfixed in absolute time.

For the BIIS, dating constraints in the past have been inadequate in both number and quality, and much of the literature was used in making and arguing about tentative correlations between places. Bowen *et al.* (2002) for example comments on the dearth of radiocarbon dates and that the timing of the LGM ice sheet was adduced from only five dates from four localities. Recent developments in dating are beginning to rectify this, in particular by cosmogenic and amino acid methods (Bowen *et al.*, 2002), although problems in the British aminostratigraphy have been identified (McCarroll, 2002). However, when compared with the number of dates that are required to constrain the dynamics or compared with the amount of dates related to other ice sheets (e.g. Laurentide Ice Sheet has some 4000 dates constraining its deglaciation; Dyke *et al.*, 2003), it is clear that dating control is largely inadequate. Major advances in understanding of the BIIS and elucidating its links with climate are likely to accrue from systematic dating programmes. Until such time that more data are collected, we consider it useful for someone with a good knowledge of dating methods and their uncertainties to produce a database of all British–Irish dates. How many are there, what is the spread in time and space, where are the gaps, which ones are reliable? This could include a map of locations, comments regarding their stratigraphical context, age estimates and error bars and some assessment of which are considered reliable. Dyke *et al.* (2003) have produced this for the Laurentide Ice Sheet and it will prove invaluable for those reconstructing ice dynamics of this ice sheet and correlating events further afield. A preliminary version of an assessment of chronological constraints for the British Ice Sheet is available in Evans *et al.* (2005).

50.7 Further mapping and investigation and the role of satellite images and DEMs

Fieldwork aimed at identifying and mapping landforms and deposits is likely to continue to be a valuable means by which we can derive new information pertaining to the dynamics of the ice sheet. However, recent developments in remote sensing and DEMs, and increased availability of these data, are set to revolutionize our approaches to mapping, and permit mapping over much larger areas, allowing the recognition of large-scale patterns that might be invisible at the fieldwork scale.

Satellite images typically cover areas of 100 × 100 km and greater, permitting a single researcher to conduct widespread mapping of glacial geomorphology in a systematic manner and avoiding the persistence of a fragmentary map record. The spatial resolution (related to pixel size) varies between 2.5 and 80 m. The combination of area of coverage and spatial resolution determines the range of appropriate mapping scales. With aerial photographs it is usual to map at scales around 1 : 20,000 but it is hard to work at smaller scales (e.g. 1 : 100,000) without recourse to specially prepared mosaics of photographs, which are often not available. Because most satellite imagery is digital it is easy to work at a wide range of scales, up to the limit imposed by the spatial resolution, such as 1 : 45,000 for Landsat TM. Being able to work across a wide range of scales (e.g. 1 : 45,000 to 1 : 1,000,000) frequently leads to the observation of landforms or patterns that would otherwise have gone undetected. Hitherto unknown landforms or patterns often can be distinguished that are only faintly discernible or even invisible when viewed from ground level or on aerial photography. The utility of satellite images should not be used to justify the abandonment of aerial photograph analysis. In addition to the fact that aerial photographs provide a higher level of detail, the increased digital analysis of images makes it likely that they will soon be available as orthorectified digital mosaics. This will simplify their use in geomorphological mapping and permit synergistic use with satellite imagery.

Although landforms can be identified on satellite images it is not a direct approach in that the interaction of reflected radiation with topography is used to visualize and infer the three-dimensional form of landforms. Digital elevation models are often superior sources as they directly record the shape of the landforms, and a variety of image processing methods, such as illuminated solar shading and perspective viewing, can be used to highlight features. At present, complete DEM coverage of Britain is available from the Ordnance Survey, with a grid size of 50 m. Figures 50.8 & 50.11 illustrate visualizations that can be generated from such data. Large features (hundreds of metres) can be

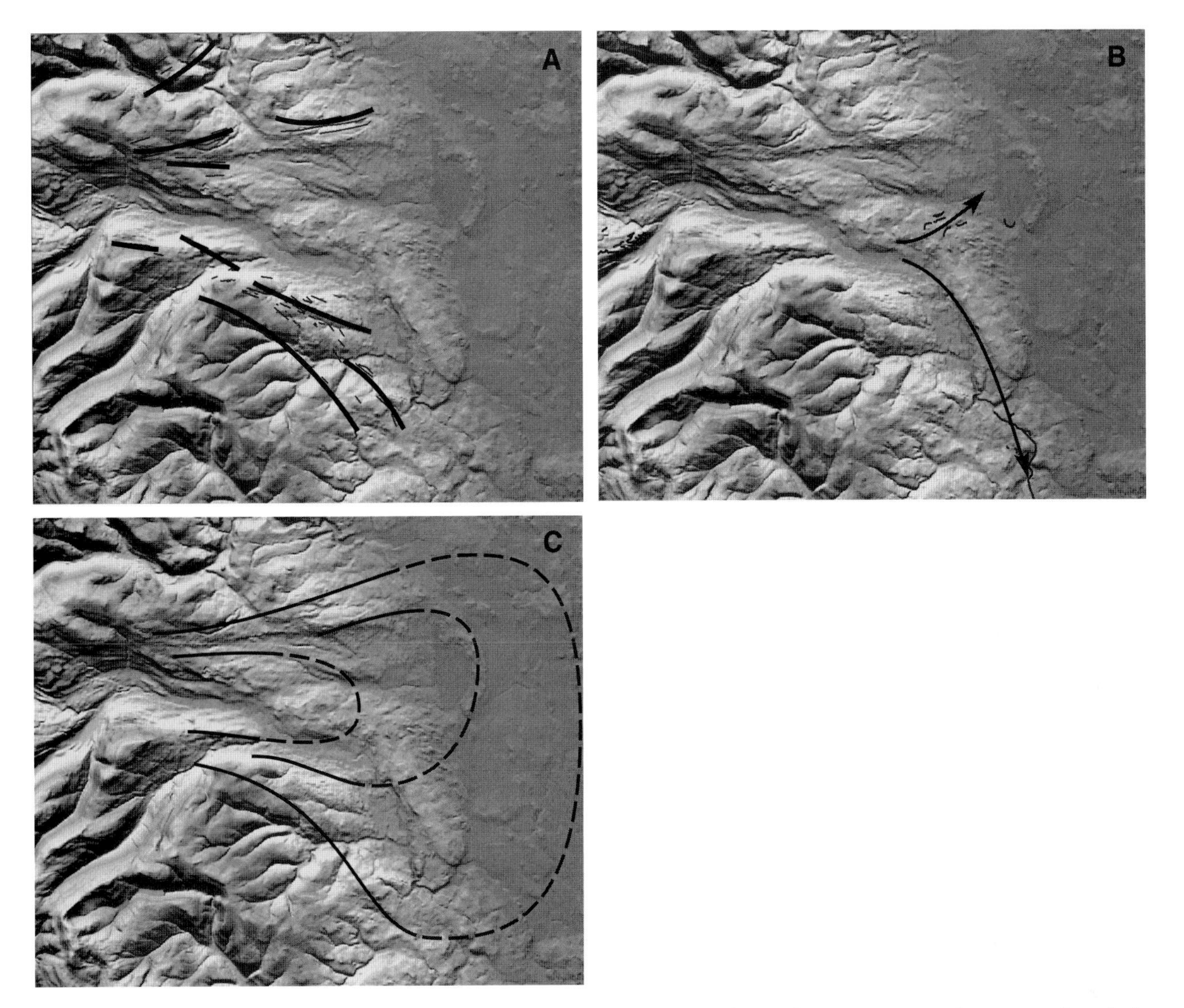

Figure 50.9 Illustration of how meltwater channels, in conjunction with their relation to topography, are used to derive patterns of retreat: (A) lateral channels provide information on successive margin positions; (B) subglacial channels indicate the flow direction; (C) combined, these features permit a reconstruction of retreat. The dashed lines indicate inferred margin positions based on topography. This area is of the eastern end (outlet) of Wensleydale in Yorkshire.

readily recognized and mapped from such data. In the immediate future higher resolution sources of DEM data are likely to become available (from Lidar and airborne radar interferometry), and with grid resolutions of the order of 5 m and less it is expected that an even wider range of landforms will be visible and thus permit complete country-wide mapping of glacial landforms. Data sources and methods for satellite image and DEM analysis are reviewed in Clark (1997).

50.8 Some future challenges

Future challenges lie both in the acquisition of new information recording parts of the ice-sheet's behaviour and in producing ice-sheet-wide syntheses of existing published information. The GIS compilation of evidence and 'Glacial Map' (Clark *et al.*, 2004) requires extending to include Ireland, for which there is a wealth of published data. For both Britain and Ireland data the layers could be extended to include geochronometric constraints, striae, hummocky moraines and stratigraphical information.

The extent of the BIIS onto the continental shelf has always been controversial and yet it is critical to derive correct information as it informs the extent and likely volume of ice. Offshore investigations are hampered by the veneer of Holocene sediments, but with the use of three-dimensional seismics it should be possible to image the palaeosurface beneath these and find bedform patterns, moraines and imprints of ice streams. It might be that there are good geomorphological records off the west coast of Ireland and Scotland and within the North Sea. Given that the dynamics and stability of the BIIS were partly governed by sea level (driven largely by the punctuated demise of the Laurentide Ice Sheet), especially the proclivity towards ice streaming, then we

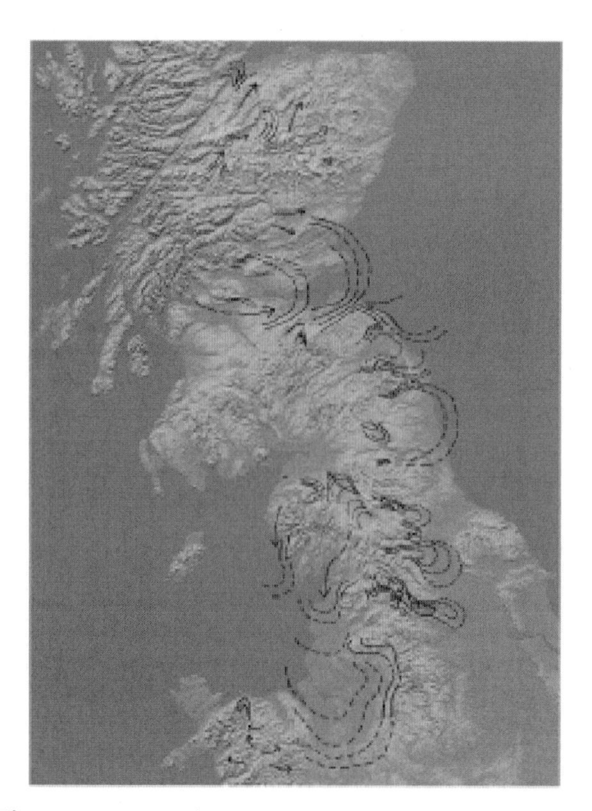

Figure 50.10 Final retreat pattern reconstructed from meltwater channels. Solid lines are supported by actual channel positions and dashed lines are inferred positions based on likely ice configuration given the topography of the area. (See www.blackwellpublishing.com/knight for colour version.)

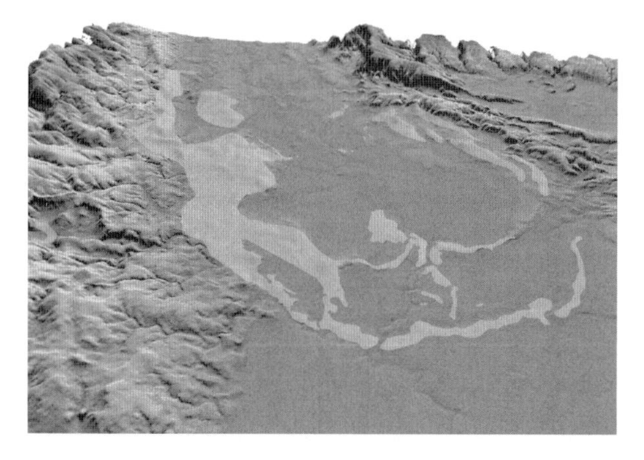

Figure 50.11 Perspective view (looking north) of the topography of the Vale of York. Overlaid (in light grey) on the digital elevation model (DEM) are the York and Escrick moraines and the Linton–Stutton gravels, as mapped by various researchers and depicted on the Glacial Map of Britain (Clark *et al.*, 2004). Note that the DEM picks out sections of the moraine ridges not mapped previously. Image is ca. 60 km across in the foreground and is derived from a 50-m grid-sized DEM.

need to address the possibility of catastrophic break-up of parts of the 'marine sectors' of the ice sheet and consequent streaming behaviour and reorganization of flow patterns and divides. Identification and the dating of any such events will help inform the growing body of research on the impacts of meltwater discharge on ocean circulation.

Compared with other glaciated regions, very little research has been undertaken on the identification of ice-stream imprints in the British glacial record. Based upon the evidence compiled on the Glacial Map of Britain we suggest that the most compelling evidence available so far for palaeo-ice streams occurs in the Vale of York, Vale of Eden, Wensleydale and the Tweed Valley, where subglacial lineations document the impact of former trunk flow. Elsewhere, such as the Irish Sea basin and Moray Firth, ice-stream activity has been proposed based upon topographic constraints and sedimentological evidence (e.g. Merritt *et al.*, 1995; Evans & Ó Cofaigh, 2003). On the continental shelf to the west of Scotland two huge debris fans exist (the Barra and Sula Sgeir Fans; Stoker *et al.*, 1993) and these are known to have been fed by debris during the Devensian. We see it as likely that major ice streams contributed to these fans, and this hypothesis needs to be tested by searching for possible terrestrial imprints. Given the large sediment volumes in these fans the possibility exists that the feeding ice-streams were responsible for Heinrich-style events, and that erosion beneath the source areas significantly controlled landscape evolution in parts of Scotland.

Although it is becoming clear that several generations of ice-flow lineations exist and demonstrably cross-cut each other (e.g. Rose & Letzer, 1977; Mitchell, 1994; Salt, 2001; Smith, 2002; Clark & Meehan, 2001), reconstructions of BIIS palaeo-ice flow dynamics are in their infancy compared with other regions (e.g. Dyke & Morris, 1988; Dyke *et al.*, 1992; Boulton & Clark, 1990b; Boulton *et al.*, 2001; Kleman *et al.*, 1997). We suggest that a systematic subglacial lineation mapping programme, utilizing satellite images and DEMs, is required for Britain and Ireland in order to assess the dynamics of the ice sheet through the last glacial cycle. Based on existing published information on palaeo-ice flow on the 'Glacial Map' (Clark *et al.*, 2004), it appears that ice-divide shifts and competing ice domes were characteristics of the BIIS. The occurrence of ice-flow patterns that were both independent of (i.e. thick ice) and constrained by topography (i.e. thin ice) suggests that the bedform information records changes in ice-sheet configuration during progressive deglacial thinning. The break-up of 'marine sectors' of the ice sheet might have promoted dramatic ice-stream initiation, driving ice divides further inland, thinning the overall surface profile and changing flow configurations. It is only by a more complete mapping and assessment of flow patterns that we can address these speculations.

Recent developments in the cosmogenic isotope dating of mountain summit blockfields in upland Britain (e.g. Ballantyne *et al.*, 1998a,b) has given Quaternary researchers confidence in identifying LGM nunataks and thereby demarcating the upper limits of the last ice sheet. Although alternative interpretations of periglacial trimlines as the former boundaries of ice sheet thermal regimes persist (e.g. Kleman & Borgstrom, 1990), the increasing number of pre-LGM dates on blockfields (e.g. Stone *et al.*, 1998; Bowen *et al.*, 2002) together with independent evidence of prolonged periods of weathering (e.g. gibbsite, Ballantyne, 1994;

Dahl *et al.*, 1996) appear to be confirming the existence of full-glacial unglaciated enclaves. Continued identification and mapping of trimlines, and an ice-sheet-wide synthesis of their distribution is likely to provide information with regard to ice-sheet thickness and surface slopes. This is important as it provides independent data to that more usually collected and can act as tests of palaeoglaciological reconstructions by inversion of geomorphological data or by numerical ice-sheet modelling.

There is a long and controversial history of research on ice-dammed lakes. From a palaeoglaciological perspective such lakes are important as they constrain ice-margin positions and recession patterns. Most research has been conducted on individual ice-dammed lakes rather than syntheses aimed at constraining ice-margin retreat patterns. A preliminary effort has been made to record the most prominent lakes on the 'Glacial Map of Britain' (Clark *et al.*, 2004), but this requires more systematic work to better establish the number and validity of lakes. An alternative line of investigation would be to conduct GIS analysis using a high-resolution DEM, across which the ice margin is backstepped to successively predict where ice-dammed lakes should have occurred and compare this with the geological record. Such an approach has been used for part of the Laurentide Ice Sheet, in which predicted lakes were found to display a good match with the geological record. This approach predicted further lake positions, and importantly was able to verify published accounts of the general ice-margin retreat pattern (Stokes & Clark, 2004).

The basal thermal regime of an ice sheet is important in governing its flow mechanisms (basal sliding, ice creep, sediment deformation) and hence zones of faster or slower flow, overall configuration and thickness. For example, Kleman & Hättestrand (1999) have argued that large parts of the Laurentide Ice Sheet were cold-based and that this necessitates a thick and more stable ice dome than recently envisaged. Thermal regime also affects the nature of any geomorphological activity at the bed, from virtually none under cold-based ice to the production of erosional landscapes and subglacial bedforms under warm-based flow. Considerable advances have been made in using the geomorphological record to infer and map zones of thermal regime, which have been applied to parts of the Fennoscandian and Laurentide Ice Sheets (Kleman *et al.*, 1999; Kleman & Hättestrand, 1999; Hättestrand & Stroeven, 2002). Where a landscape is devoid of any flow traces such as drumlins and flutes or glacial striae and the only recognizable landforms are lateral meltwater channels, it can be argued that ice-sheet retreat across the area was cold-based. An abundance of flow traces, subglacial meltwater channels and eskers indicates that retreat was accomplished under a warm-based regime. Using such an approach, ice-sheet reconstructions can include information on the thermal regime, which is important for considerations of ice thickness (warm-based basal sliding promotes lower surface profiles and thinner ice), and helps us understand ice retreat across areas that are seemingly sparse in glacial landforms. An example of this is illustrated elsewhere in this volume (Hättestrand & Clark, Chapter 39) and inversion methods for thermal regime from the geomorphological evidence are reviewed by Kleman (this volume, Chapter 38). Such an approach has yet to be applied to the BIIS. It is interesting that for the Laurentide, Fennoscandian and Irish ice sheets the majority of the bed (>60%?) is covered by landforms typical of a warm-based thermal regime (i.e. drumlins and eskers), and yet for Britain these are much more sparse (<20%?). In part this certainly arises from incomplete mapping as highlighted earlier, and may also partly be a consequence of upland terrain with a thin sediment cover. However, we suspect that a major cause for this overall difference is that large parts of the ice sheet over Britain were cold-based, and if recognized and established it may help explain the distribution of landforms and guide reconstructions of ice dynamics.

50.9 Conclusions

Our main conclusions are:

1. Deriving a palaeoglaciology of the BIIS is important for wider earth system science objectives of understanding the climate system, ice–ocean interactions and how ice sheets operate.
2. There is a wealth of published information collected over 150 yr that remains considerably underutilized at present. This is especially so for landform evidence.
3. A weakness in existing literature is our limited accomplishment at assembling the palaeo-ice sheet jigsaw puzzle. This is now being addressed, such as the recent synthesis on dating constraints (Bowen *et al.*, 2002), assessment of spatial ice-sheet limits (Ehlers & Gibbard, 2004), elevational limits (Ballantyne *et al.*, 1998a,b) and the compilation of landform evidence in the BRITICE GIS and 'Glacial Map' (Clark *et al.*, 2004; Evans *et al.*, 2005).
4. High-resolution satellite images and DEMs are set to revolutionize our approach to the mapping of glacial geomorphology, and in the coming years may permit elucidation of major ice streams and configuration changes as the ice sheet evolved. Indeed, once sub-5 m grid-sized DEMs become available (and at a cost appropriate to academic research!) it will be possible to complete the mapping of glacial geomorphology of Britain and Ireland in a decade.
5. It is unlikely that it will be possible to capture the full dynamics and behaviour of the BIIS from landform and stratigraphical observations and geochronometric dating alone. Numerical ice-sheet modelling and isostatic inversion modelling will be paramount. At present the small numbers of models that exist remain largely untested or informed by the observational record. We need to be asking:
   - Which elements of the observational record are robust?
   - Are the models compatible with these?
   - Which model best explains the data?
   - Where and why do mismatches occur?
   - What is missing from the modelling?

In the progress of knowledge it is understandable to start at a simple level and gradually build in the complexity. When it was once thought that ice sheets were sluggish to change and we had not fully recognized their sensitivity to climatic and oceanographic variability, focus was on establishing the maximum limits of the ice sheet and elucidating generalized flow patterns and ice sources. It was safe to presume simple symmetric growth to the LGM state and retreat from it, albeit with some readvances. We now know that the BIIS must have been considerably more dynamic, especially given its proximity to the thermohaline conveyor of the North Atlantic. Growth and decay was probably asymmetric, sourced in one area and possibly retreating back to another. Marine sectors may have catastrophically broken-up and discharged Heinrich-style iceberg calving events, ice streams turned on and off, and ice divides migrated. The challenge is to derive new information and make much better use of existing evidence to better constrain some of these dynamics.

FIFTY-ONE

A regional glacial readvance in Ireland: self-promulgating theory, or science-based reality?

Robert T. Meehan

86 Athlumney Castle, Navan, County Meath, Ireland

51.1 Introduction

As with all glacial processes, the processes of ice advance and retreat operate at a variety of scales: that associated with the general growth of an ice sheet over thousands of years and thousands of kilometres, or shunts of the ice margin of a few centimetres on a single day. The measurement of past regional-scale ice advances is achieved by mapping 'first order' glacial geological features (e.g. drumlins and ribbed moraines), which identify concentric zones of advancement of the ice sheet, so the flow becomes the dominant imprint on the landscape over time (Hughes, 1998). These features are important as they inform us of major climatic shifts. 'Second order' glacial geology consists of lineations, forms and sediments that arise near ice margins and weakly overprint first-order glacial geology; these are measured by mapping local-scale features such as flutes, tectonic structures and till units. Second-order glacial geology is being constantly overprinted as the ice sheet advances and retreats at this local scale.

Within this system, most moraines are formed because sediment is constantly being delivered to the ice margin; where the margin pauses a positive moraine feature is created. Therefore, most moraines are built at still-stands during overall retreat, and not by ice 'bulldozing' sediment into a ridge. It is therefore erroneous to equate moraines as indicative of first order ice (re)advance.

51.2 Readvances in Ireland and Scotland

Glacial readvances of first-order magnitude based on moraines as the defining element for ancient ice sheets have, however, been interpreted in Ireland and Scotland (e.g. Charlesworth, 1928; Sissons, 1979). The readvance chronicled in Scotland dates from the Younger Dryas (YD) and is termed the 'Loch Lomond Readvance'; the major readvance in Ireland dates from the end of the last glacial maximum and is termed the 'Drumlin Readvance' (DR). Both are of critical importance in each country's glacial history, as each are attributed to a major climatic deterioration; one easily datable (YD), one less so (DR). In recognition of each, scholars initially used morphostratigraphy in interpreting the position of the ice margin (e.g. Charlesworth, 1928; Sissons, 1979), and 'filled in' more exact details of the glacial process, form–sediment relationships and associations following this (e.g. Charlesworth, 1955; Ballantyne, 2002). From this and the ensuing discussion, the readvances and processes associated with them have been generally accepted but often reinterpreted, repositioned and even partially discounted by various workers as new data have

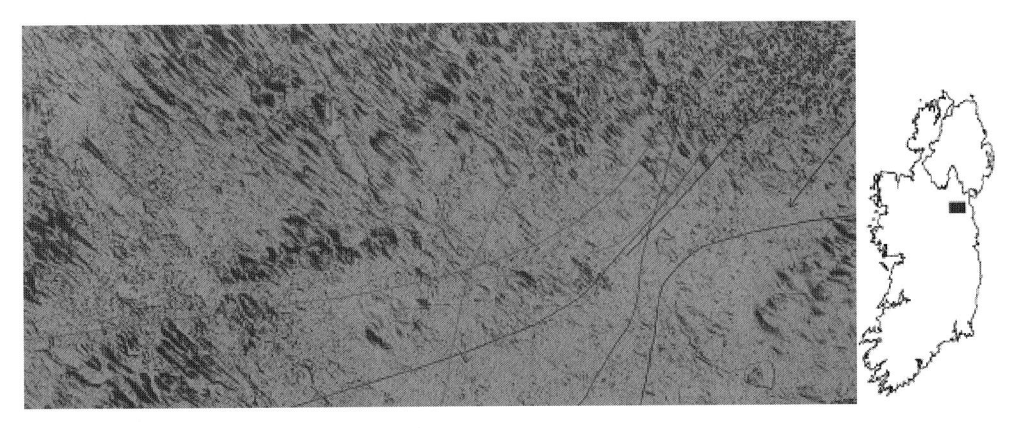

Figure 51.1 Non-azimuth bias shaded DEM of the area around the type-site for the Drumlin Readvance Moraine of Ireland, around Kells in County Meath. The image covers 45 km by 20 km. The lines show the various positions of the moraine according to different authors through the years; eight different positions in all. Note the presence of intact and buried streamlined features south of all moraine lines, and the absence of any definite end-moraine feature.

appeared (e.g. Synge & Stephens, 1960; McCabe, 1985; Warren, 1991; Bennett & Boulton, 1993). It is important to note that most of these supporting data occur at the second-order scale described above.

With the Irish DR, a broad band of hummocky topography in the north-centre of the country was interpreted as a composite landform assemblage 'probably' forming an end-moraine associated with accelerated ice flow, drumlinization and margin readvance (Charlesworth, 1928). There is no glaciological basis for this and indeed most drumlin patterns (e.g. those covering most of Canada; see Boulton & Clark, 1990b) do not conform to this association, and in fact mostly relate to the retreat phase of the ice sheet. Up until the 1990s there had been no mapping of sediments and/or landforms anywhere inland along the feature, notably around the 'type-site' at Kells, County Meath. Morphological and sediment mapping was therefore carried out here over an area of 900 km^2 (Meehan, 1998).

The mapped results of this, which were at the second-order scale of geological mapping (1 : 50,000), were then cross-checked with DEM analysis of the first-order, regional indicators of ice flow. The resultant pattern shows that the moraine feature seemed to be a misnomer as drumlins, crag-and-tails and ribbed moraines associated with the same first-order ice advance occur south of all previously drawn 'moraine' lines (Fig. 51.1). Furthermore, following subsequent work (Clark & Meehan, 2001) it was observed that the end of the features' associated flow is at least 30 km (and possibly more) further south. On the second-order level, results showed that the huge expanse of hummocky terrain, which has been simplified as a single, linear feature, has a more complex process history. Within this area, portions of hummocks comprise areas of bedrock, tectonite, fan and delta sands and gravels, pitted sandur, minor ribbed moraines and even buried drumlins, all of which had distinctly different formational processes and settings (e.g. Meehan, 1998; Clark & Meehan, 2001). The sediment-form geometry suggests a sustained, coherent retreat of ice with the development of a number of broad outwash systems in valleys between moulded crag-and-tail ridges, drumlins and subtle streamlinings. Furthermore, around the areas of the moraine 'feature', nowhere was evidence of a readvance (a 'capping' till) found, even in 49 specially dug trenches and 39 specially drilled boreholes.

51.3 Discussion and conclusions

It seems that earlier literature overlooked the fact that even in a retreating ice sheet, the ice flow is still directed forwards and is capable of producing streamlined landforms. The early idea of equating drumlins with first-order (re)advances has propagated through the literature, but with no basis, and strong contradictions from elsewhere.

In conclusion, none of these morphologically based arguments is valid, and yet they still have influence. The only morphological evidence capable of demonstrating a readvance is an ice marginal feature that has been overridden, leaving a signature of drumlins or flutes on top of it. Stratigraphically, evidence of subglacial sediment (till) or ice-marginal features (moraines) superimposed stratigraphically above proglacial sediments (outwash, lake sediments, organic deposits, soil horizons) is needed. These are both at the second-order scale. As scientific writing has developed over the decades since the 1920s the distinction between description and interpretation has become an overriding tenet. The DR moraine's history proves this was not so in the past. Supposed readvances in Britain, Ireland and elsewhere may require re-examination using the methods outlined above in light of these results.

FIFTY-TWO

Average glacial conditions and the landscape of Snowdonia

Danny McCarroll

Department of Geography, University of Wales Swansea, Swansea, UK

The Snowdonian mountains of North Wales hold a special place in the history of glacial geomorphology. It was here that Darwin (1842) was convinced of the former existence of glaciers and of their role in shaping the landscape. In the 19th century, it was the battleground of the 'glacialists' and 'diluvialists', with evidence such as the high-level shelly drift of Moel Tryfan, in the Snowdonian foothills, playing a key role in the debate (Campbell & Bowen, 1989). The glacial cirques (cwms in Wales) and spectacular U-shaped valleys of the region have become classic examples of glacial erosion and generations of geographers, at all levels, have used the area as a classroom. Recent work, however, calls into question the established interpretation of the landscape.

The evidence for glaciations of Snowdonia was reviewed by Whittow & Ball (1970). They argued that at the maximum extent of the last glaciation, ice from a major ice centre located to the southeast of Snowdonia was sufficiently powerful to push straight through the mountains, carving the spectacular U-shaped valleys of the Pass of Llanberis (Fig. 52.1) and the Ogwen Valley. This interpretation has been promulgated by Addison (e.g. 1990), whose works have become standard fare to the many school and university parties visiting the region. The evidence was first questioned by Gemmell *et al.* (1986), who showed that geomorphological evidence around the triple junction at the southern end of the Pass of Llanberis indicated radial drainage away from the Snowdon (Yr Wyddfa) massif during the last glaciation. Addison (1990, p. 12) countered that although the basal ice might radiate, 'transfluent ice high up in the outflow is believed to have breached the mountains at a number of sites, excavating the major outlet troughs of Nant Ffrancon and Llanberis pass'. This argument requires ice from the southeast to override the mountains and predicts that evidence of ice directions high on the mountains should indicate flow dominantly towards the northwest.

McCarroll & Ballantyne (2000) investigated the evidence for ice thickness in Snowdonia, and for directions of ice movement. A periglacial trimline was mapped, separating areas showing evidence of glacial erosion from palaeonunatak summits with evidence of prolonged weathering. The evidence for former nunataks in Snowdonia proved more spectacular than perhaps anywhere else in Britain. The high peaks of the Glyder range, separating the Pass of Llanberis and the Ogwen Valley, host spectacular blockfields, including the great cantilevered slabs of Glyder Fach, but the lower slopes and cols, including that above Cwm Idwal, are clearly glacially eroded. The lower slopes of the Carneddau, to the east, retain beautifully striated vertical faces to at least 700 m a.s.l., but the high plateau above 850 m a.s.l. is swathed in largely peat-covered blockfield punctured by relict tors, the finest example of which rises 8–12 m above the blockfield on Yr Aryg (865 m a.s.l.). The interpretation of the trimline as the surface of the last ice sheet at its maximum is supported by the presence of gibbsite in relict soils on the weathered summits. Gibbsite is an end product of the weathering of silicates and in mountainous environments is regarded as evidence of a very prolonged period of weathering (Ballantyne *et al.*, 1998a). The trimline cannot, therefore, represent the surface of a later readvance.

The altitude of the ice surface reconstructed by McCarroll & Ballantyne (1990), and the evidence for ice movement high on the mountains, provides no support for the argument that ice from the southeast has ever overridden the mountains of Snowdonia, or that the major valleys represent 'outlet troughs'. On the contrary, all of the evidence points to radial drainage of ice away from Snowdonia, even when the ice was at its thickest (Fig. 52.2). The periglacial trimline reaches 850 m a.s.l. in the main mountain ranges of Snowdonia, including the Snowdon massif, the Glyders and the Carneddau, but only about 820 m a.s.l. to the southeast on Moel Siabod. The ridge of Moel Siabod (Fig. 52.2) lies directly in the path of the ice purported to have overridden Snowdonia, but striae demonstrate that ice crossed it not from the south but from the west.

The geomorphological evidence in Snowdonia clearly demonstrates that the mountains remained an area of ice dispersal throughout the last glaciation, with no evidence of ice from the southeast ever having overridden them. If the Pass of Llanberis and Ogwen Valley were cut by ice flowing northwest through Snowdonia, they would have to have been cut in some previous

Figure 52.1 The Pass of Llanberis, looking to the northwest. Breaching of former ice-sheds, to produce the major valleys cutting through Snowdonia, can be explained by the action of local ice. There is no need to invoke ice from the southeast overriding the mountains. (See www.blackwellpublishing.com/knight for colour version.)

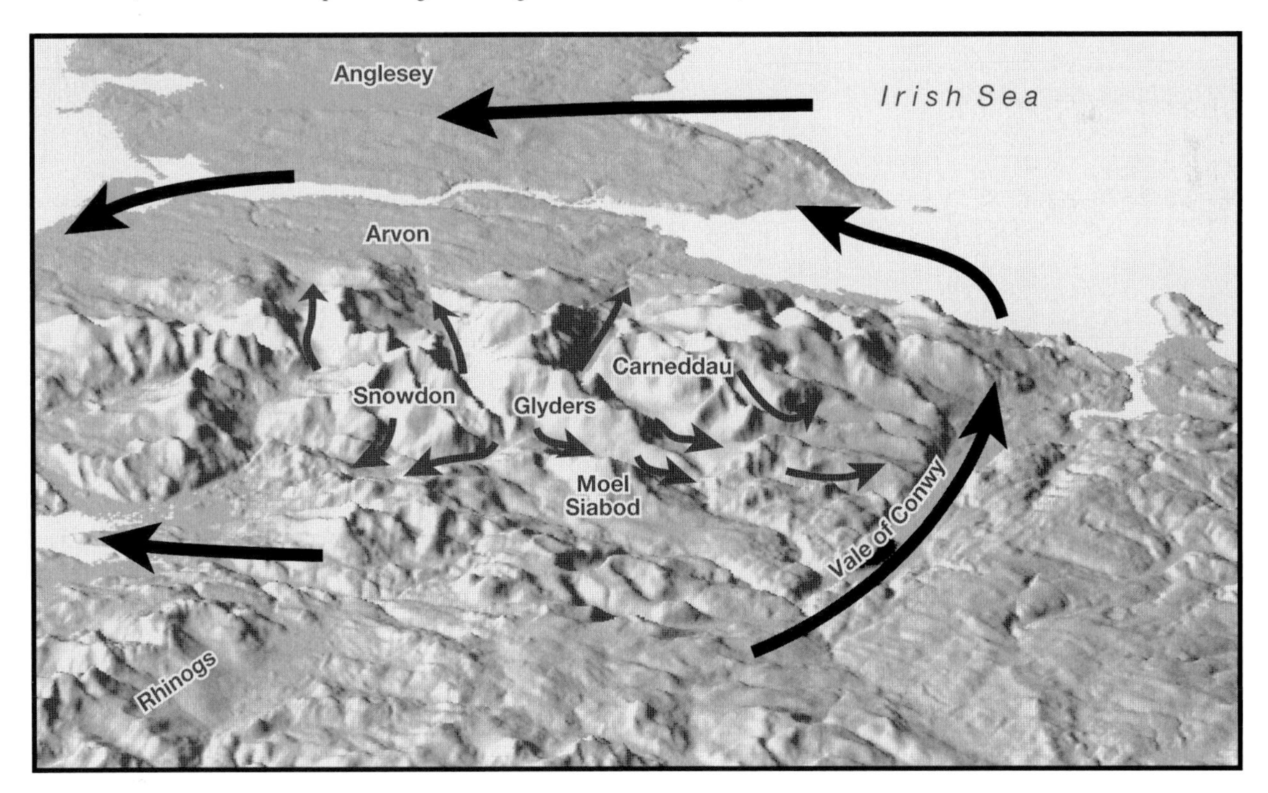

Figure 52.2 Digital elevation model (DEM) of Snowdonia, viewed from the southeast, showing the dominantly radial drainage away from the mountains of Snowdonia during the last glaciation. (DEM courtesy of Rob Davies.)

glaciation, when the ice from the southeast was very much more powerful. However, there is no direct evidence for such a major event and it is possible to explain these valleys as produced by the action of local ice.

It is a mistake to interpret a mountainous landscape like Snowdonia as the product only of major glaciations, with huge ice sheets, and of minor events such as the Younger Dryas (Loch Lomond Stadial) when only the cirques were occupied by ice. For most of Pleistocene time the climate of northern Europe has been much colder than today, but not cold enough to generate major ice sheets. Porter (1989) has stressed that these 'average' glacial conditions may be much more important in shaping landscapes than the infrequent and short-lived extremes. Under these 'average' conditions, Snowdonia's glaciers would radiate out of the cirques and into the valleys, varying in size as the climate fluctuated. In early glaciations the drainage would have been radial, with ice moving away from the mountains in all directions. However, during 'major' glaciations, when ice was able to build up over the hills and plateaux to the southeast of Snowdonia, the southerly drainage of the Snowdonian ice would be constrained. As ice flow to the south was restricted, southerly flowing glaciers would thicken, allowing them to overflow their own ice-sheds, forming the major glacial breaches.

Most of the Snowdonian landscape, therefore, is tuned to 'average' glacial conditions, with ice flowing out of cirques and being evacuated radially via valley glaciers of varying size. The often-spectacular erosional landscapes of the cirques are not the product of minor events such as the Younger Dryas, when the cirques were occupied by small glaciers and glacial erosion was minimal, but have been occupied by ice for much of the Pleistocene. The major through valleys, which are clearly the product of breached ice-sheds, are the product of much thicker local ice constrained by the advance of ice from high ground in the southeast and also southwards down the basin of the Irish Sea.

FIFTY-THREE
Mountain glaciers

Georg Kaser

Tropical Glaciology Group, Institute of Geography, Innrain 52, 6020 Innsbruck, Austria

53.1 Introduction

A glacier forms when the accumulation of ice exceeds its loss over a time span longer than a few years. The appropriate climatic conditions are driven by the general regional free atmosphere conditions but are also affected by the local topography (e.g. Oerlemans, 2001). A glacier's extent is determined not only by climate: glacier-bed conditions also influence the glacier's geometry. Whereas in the highest latitudes climate allows glaciers to cover land up to continental scale and to reach sea level, in lower latitudes glaciers occur only on mountains. The warmer the climate, the higher, generally, are the lowest limits of glaciers. A very rough approximation is that glaciers rise by approximately 1000 m in elevation for each 1000 km in horizontal distance from polar and subpolar regions toward the Equator. Consequently, mountain glaciers cover considerably less area than land ice masses at high latitudes. Because of their small extent, mountain glaciers are rather sensitive to climate variations and changes. They are also 'closest' to human activities and, therefore, attract public and scientific interest:

1 Since human beings first started to extend their activities into the high mountain areas, glaciers have been in their sphere of interest. Since the finding of the Ice Man in the Ötztal Alps we know that this has occurred for more than 5000 yr (Bortenschlager & Öggl, 2000).
2 The variation of mountain glaciers affects the appearance of a landscape in a way that is recognizable within a human's lifetime. Tourism in high mountains is, to a certain extent, also encouraged by glacier landscapes. Recent years of strong glacier melt in the European Alps have affected glacier ski resorts that now seek strategies to protect glaciers from too much loss.

3 In many mountain regions, glaciers substantially control the availability of fresh water. Their growth and shrinkage can have a dramatic impact on the economic, social and cultural activities close to mountains.
4 In conjunction with tectonic and geomorphological conditions of mountain environments, glacier variations can cause hazards. It was, actually, the threat from an ice-dammed lake to which we owe the very first painting of a glacier, the Vernagtferner, damming the runoff from Hintereisferner and Hochjochferner in the Austrian Ötztal Alps in 1601 (Nicolussi, 1990).
5 Since the end of the Little Ice Age, mountain glaciers have lost a considerable amount of mass, which is considered to have contributed significantly to the observed 20th century sea-level rise (e.g. Warrick *et al.*, 1996).

The interest in mountain glaciers and in their fluctuations focuses partly on their impact on the world at lower elevations, but also on the driving forces behind observed changes. Mountain glaciers appear to be sensitive indicators of climate and are comparatively well distributed over the globe. Among climate proxies, glaciers are the only ones that exclusively follow physical laws and these laws are the same, no matter where the glaciers are. Furthermore, these laws are well known and understood, although their application for analytical or numerical solutions in order to describe glacier–climate interaction is complex and calls for a variety of simplifications. Among the advantages of using glaciers as climate indicators is the narrow band of climate conditions in which glaciers exist. Several mechanical and thermodynamic properties, such as glacier ice being impermeable to water and the upper limit of temperature at the pressure melting point, are helpful. Others, such as being partially transparent to solar radiation and the particular way in which glaciers create their own atmospheric boundary layer, are unhelpful to simple solutions for the description of the behaviour of glaciers. In the following, the potential of understanding mountain glaciers as products of a complex climate and of their impacts on human interests, mainly for water supply, is outlined.

Most of today's mountain glaciers are considered to be the products of the Little Ice Age climate. They have grown within a few centuries and they may again diminish or even vanish within similar time periods (e.g. Cogley & Adams, 1998). On geological time-scales they are short-lived phenomena and they have grown and diminished several times throughout the Holocene (e.g. Hormes *et al.*, 2001). Still, as already mentioned, their short-term variation is evident on a human scale, especially as mountain regions occupy about one-fifth of the Earth's surface and provide goods and services to about half of humanity (Messerli & Ives, 1997).

53.2 Mass balance on mountain glaciers

Climate and, in many cases, mechanical processes drive the mass balance of a glacier. The resulting and permanently changing storage of ice, snow and water can be seen at different time-scales (Jansson *et al.*, 2003). The long-term storage of ice and firn on time-scales of 10^0–10^2 yr and longer determines the extent of glaciers: their volume, length and geometry. This relates to the general character of the hydrological regime of the corresponding catchment areas, and affects global sea-level changes. The seasonal variations of snow and water masses can be seen as intermediate-term storage. Glacier mass balance studies usually investigate the intermediate-term storage and provide net variations over monthly to 1-yr time-steps in order to relate them to seasonal climate. Seasonal runoff from glacierized basins is characterized by intermediate-term storage changes. Short-term storage concerns diurnal processes of melt-water production and drainage. Individual storage releases such as drainage from glacier surges and drainage of glacier-dammed water are catastrophic events in many cases.

In a long to intermediate-term, climatological, view, only the storage of firn and ice is of concern. It increases mainly from solid precipitation, which, to a varying extent, is redistributed by drift and avalanches. Mass loss is most effective from melt-water runoff but can be also due to sublimation, which increases with dryness of the atmospheric conditions. Calving and avalanches can contribute to the ablation as well. When mechanical processes such as avalanches and calving are of minor importance for a given mountain glacier, the mass balance can be related directly to weather and climate. In most monitoring and research programmes, attention is focused on the mass balance on the surface of the glacier. Mass changes at the base of a glacier are comparatively small and the refreezing of melt water is usually limited to parts of a glacier and to subseasonal time periods. Basal mass change is only of concern in areas of extraordinarily high geothermal heat. For example, the growth and sudden outburst of subglacial water storages as jökullhlaups in Iceland are the result of marked basal melt (e.g. Björnsson, 2002). Internal accumulation occurs when melt water from the surface penetrates into the firn and refreezes in layers older than from the ongoing year. It is difficult to measure but, particularly at high latitudes where the winter cold wave penetrates into lower and older firn layers, internal accumulation is thought to contribute considerably to the mass balance of glaciers (e.g. Trabant & March, 1999). When snow starts to melt on a glacier tongue during the early ablation season, melt water can also refreeze on the previous summer's glacier-ice surface. Still, this superimposed ice is usually removed during the summer and only in positive mass balance years may some survive in a small area on which winter snow is not removed from the previous year's ice surface.

The surface mass balance of a mountain glacier is different in intensity, timing and effect on different parts of a glacier, but the variation of mass balance with elevation clearly dominates (e.g. Klok & Oerlemans, 2002). Annual net accumulation is highest in the higher parts of a glacier and decreases toward lower elevations until net ablation occurs and increases down-glacier. Drift removal of snow usually leads to little accumulation on the uppermost flanks of a mountain glacier, whereas shading and drift snow deposition may cause smaller ablation amounts on the lowest tips of the tongues. In many cases, glacier tongues are covered by debris that can increase or decrease ablation depending on the type of debris and its thickness. Ablation underneath a debris cover is difficult to measure and monitor and model approaches are used to determine ablation (e.g. Nakawo *et al.*, 2000). The so-called vertical balance profile, VBP, is the mass balance per unit area or specific mass balance (in kg m^{-2} or metres water equiva-

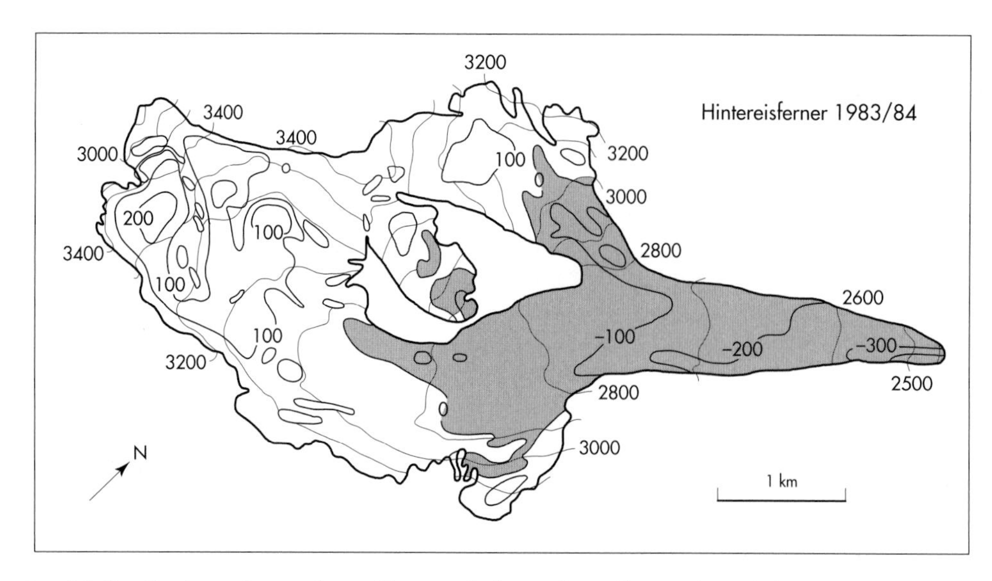

Figure 53.1 Spatial distribution of annual specific mass balance (kg m^{-2}) on Hintereisferner, Austrian Alps in 1984 as an example. (Kindly provided by G. Markl and M. Kuhn, Institute of Meteorology and Geophysics, University of Innsbruck.)

lent, m we) changing with elevation. It is best obtained from direct glaciological mass balance measurements carried out with a well-distributed network of ablation stakes and snow pits (e.g. Fountain *et al.*, 1999) (Fig. 53.1). Traditional techniques as well as more recent high-resolution geodetic techniques also yield spatial distributions of glacier changes by subtracting one terrain model from another. With this method, however, densities of snow and ice must be assumed and the spatial distribution of changes shows not only the climatologically induced variations of mass balance but also variations associated with ice-flow properties. Careful glaciological measurements show that the mass balance not only varies with altitude but also has horizontal differences (see Fig. 53.1). However, they are by far less pronounced and a VBP, if derived as mentioned above, averages the horizontal differences for each altitude step (Fig. 53.2). In an inverse way of application, the dominance of the vertical variation of the specific mass balance is often used to derive a total mass balance (in kg or m^3 we) of a glacier from a series of measuring points lined up, for example, along the central flow line. This, however, ignores all possible horizontal gradients, which cannot be disregarded when looking in detail at an individual glacier's reaction to climate variations (e.g. Oerlemans & Hoogendoorn, 1989). It is obvious that the VBP is the result of a combination of climate parameters and their vertical gradients. This is, in most cases, the basis for glacier mass-balance models of different complexity. Such models are used for calculating glacier mass balance from measured or prognostically modelled climate input, for deriving climate from reconstructed volume and mass changes from former glacier extents (e.g. Oerlemans, 2001), and for modelling glacier runoff (e.g. Hock, 2003; Juen *et al.*, 2003).

The VBP is, in a first step, the result of the vertical distribution of snow accumulation from precipitation and, possibly, redistribution by wind. The latter is of concern under cold and strong wind conditions, namely in the higher latitudes and in very high altitudes. It decreases toward warmer environments where drift is

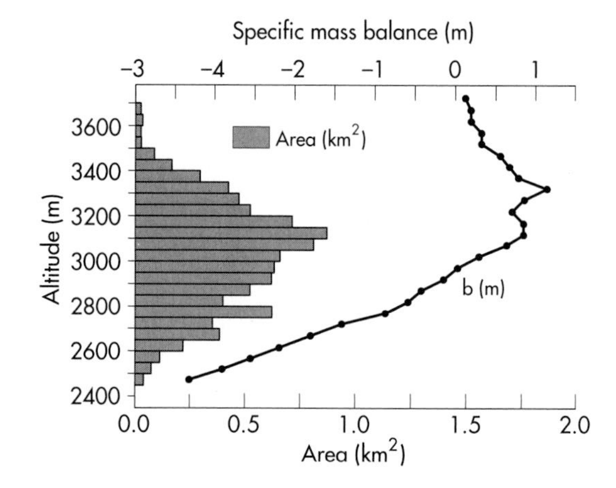

Figure 53.2 The vertical balance profile as well as the area altitude distribution of Hintereisferner (Fig. 53.1) in the hydrological year 1984. (Kindly provided by G. Markl and M. Kuhn, Institute of Meteorology and Geophysics, University of Innsbruck.)

more and more reduced toward a limited small-scale importance on crests and saddles. During the ablation period, climate impact on a glacier and its VBP is much more complex. Different parameters drive the fluxes of energy that are finally used for ablation. Out of all fluxes that determine the processes at the atmosphere–glacier interface, extraterrestrial solar radiation is the only independent one. Its portion reaching the glacier surface is affected by topography, atmospheric moisture content and related cloudiness. The reflection of shortwave radiation and, finally, the available net shortwave radiation is determined by the glacier surface albedo, which is, in turn, the result of previous accumulation and ablation processes. Profiles of wind, temperature and air humidity in the atmospheric boundary above the glacier surface determine the turbulent fluxes of sensible and latent heat.

These fluxes can be positive or negative toward the surface. The latent heat flux is the only term that makes up part of both the energy and, as sublimation or resublimation, the mass balance. Air temperature and atmospheric moisture content also determine the atmospheric longwave emission and stand for the occurrence of solid precipitation which, in turn, interferes via the albedo with the shortwave radiation balance. The resulting energy heats the glacier surface and the layers below whereas the consequent surface temperature determines the emission of longwave radiation from the glacier. If, at the very end, a surplus of energy results, this is used for melting which, together with sublimation, makes up ablation.

In many cases, air temperature and its variation with elevation have a very high statistical correlation with ablation and also with the mass balance (e.g. Ohmura, 2001). Air temperature influences the sensible heat flux, the condensation of moisture in the atmosphere, thus cloudiness and precipitation, the ratio between liquid and solid precipitation, thus the albedo, and the emission of longwave radiation from the atmosphere. Air temperature is itself influenced by solar radiation, surface temperature and sublimation. The so-called temperature index or positive degree-day models (e.g. Hock, 2003) are based on these relations and give good results in many studies on mid- and high-latitude mountain glaciers where temperature follows a pronounced seasonality separating an accumulation from an ablation season (e.g. Braithwaite & Zhang, 2000) and mean hygric conditions vary comparatively little. If applied inversely to retrieve climate scenarios from reconstructed glacier extents, they exclude possible changes in atmospheric moisture content. However, the exclusive use of air temperature and, possibly, precipitation for climate–mass-balance studies becomes insufficient if looking at low-latitude, high-mountain glaciers. In the tropics and subtropics, seasonal variation in air temperature is subdued but the seasonal variation of atmospheric moisture content and all related climate variables is dominant. This fact makes low-latitude glaciers highly sensitive not only to trends in air temperature but also to the fluctuation of moisture-related parameters and their seasonal and interannual variation (e.g. Wagnon *et al.*, 1999). The occurrence of strong sublimation under dry conditions or the time of onset of accumulation at the beginning of the wet season with its effect on albedo are crucial for the mass balance of low-latitude glaciers (Francou *et al.*, 2003). These variations change the VBP markedly and have implications on the equilibrium line altitude, its position in relation to the 0°C level, a possible steady-state extent of a glacier, the ratio of ice loss and, finally, the glacier runoff.

53.3 The runoff from mountain glaciers

From a mid- and high-latitude viewpoint, runoff from mountainous catchment areas basically follows the seasonal variation of air temperature and is dominated by the storage of precipitation as snow cover during the cold season and its progressive release during spring and summer. The higher the ratio of glacierization, the higher and later the summer maximum of runoff (Fig. 53.3). During periods of strong mass loss and consequent retreat of glaciers, runoff is generally increased until the effect is compensated by reduced glacier surface area. When a glacier disappears, runoff follows the melt of the seasonal snow cover and, during the warm periods, liquid precipitation (e.g. Kuhn, 2003). Owing to the reduced or non-existent thermal seasonality, the situation looks different in lower latitudes. There, no seasonal snow cover occurs outside the glaciers and, thus, glacier runoff is almost the only freshwater supply during dry seasons (Fig. 53.3). Glacier runoff is, as in higher latitudes, increased when glaciers retreat but when they vanish, runoff only follows the occurrence of precipitation. It is important to note also that renewed steady-state conditions or glacier advances would reduce the availability of water. Major attention must be paid to this fact in terms of near-future water management not only in the tropical South American Andes, but also in parts of the Himalayas which are, climatically speaking, intermediate between the mid-latitudes and the tropics and seasonality is characterized by both temperature and hygric variations in a complex way (e.g. Ageta & Higuchi, 1984).

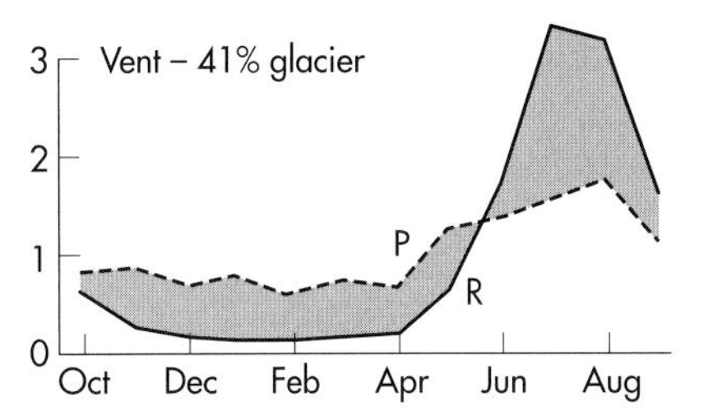

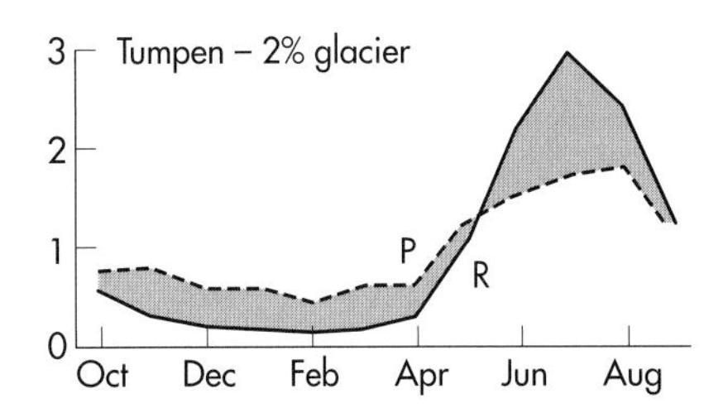

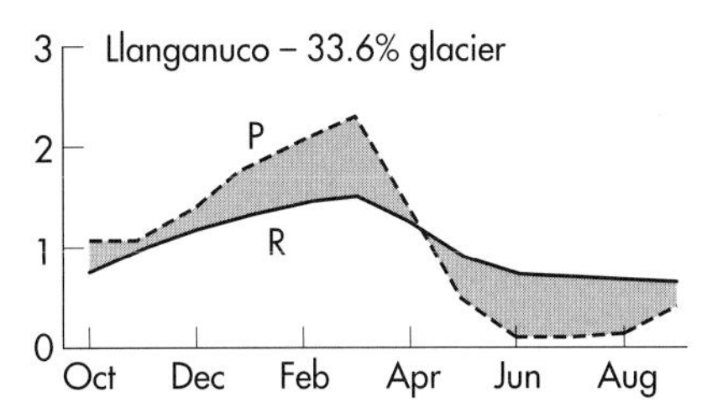

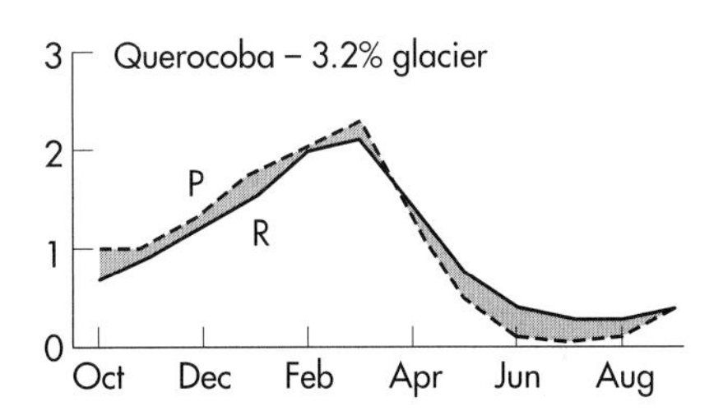

Figure 53.3 Seasonal variations of precipitation, P, and runoff, R (normalized), for differently glacierized catchment areas in the Austrian Alps (left) and the Peruvian Cordillera Blanca (right). (Modified from Kaser *et al.*, 2003.)

FIFTY-FOUR

The Little Ice Age glacial record at Franz Josef Glacier, New Zealand

Krista M. McKinzey

Institute of Geography, School of GeoSciences, The University of Edinburgh Drummond Street, Edinburgh EH8 9XP Scotland, UK

The Franz Josef Glacier (FJG) descends from the western peaks of the Southern Alps of New Zealand (Fig. 54.1A). It is one of the world's lowest lying mid-latitude glaciers and as of March 2001 terminated at 275 m a.s.l. This maritime valley glacier has a terminus response time of 5–20 yr (Oerlemans, 1997), which implies that it responds rapidly to perturbations in temperature, precipitation and, hence, mass balance.

Recently, the Little Ice Age (LIA) history of the FJG was reassessed using mapped tree-ring counts and diameter at breast height (DBH; 1.4 m) measurements of the largest southern rata (*Metrosideros umbellata*) and kamahi (*Weinmannia racemosa*) within moraine limits and trimlines to determine the minimum time elapsed since deglaciation (McKinzey, 2001). This approach differed from previous interpretations because it used rata as the critical indicator species for the time elapsed since deglaciation (rather than kamahi alone, which consistently underestimates the timing of surface exposure) and combined both tree size and age for the reconstruction. Accordingly, the FJG's LIA maximum culminated before AD 1600 (rata > 225 cm DBH; Fig. 54.1B & C), when it terminated 4.5 km down-valley of its 2001 position. Subsequent, but smaller magnitude, readvances occurred by ca. AD 1600 (3.7 km down-valley, rata and kamahi 100–200 cm DBH) and ca. AD 1800 (3.2 km down-valley, rata and kamahi <60 cm DBH). Numerical modelling experiments have delimited possible climatic envelopes for equilibrium lengths of the FJG during the LIA. Results indicate a minimum cooling of −1.15°C or +57% precipitation for the LIA maximum extent (departures relative to 1970–1999 mean temperature and precipitation), −1°C or +47% precipitation for the second LIA advance and −0.8°C or +37% precipitation for the third LIA advance (Anderson, 2004).

Other proxy evidence of late Holocene climate change in New Zealand corroborates an early LIA maximum for the FJG. The speleothem record from northwest Nelson indicates periods of prolonged cold temperatures ca. AD 1400–1450 and ca. AD 1600–1650 (Wilson *et al.*, 1979). Additionally, tree-rings from silver pine (*Lagarostrobos colensoi*) growing at Oroko Swamp reveal that the coldest time in New Zealand during the LIA occurred during the early 16th century when temperatures declined by ca. −1.5°C (Cook *et al.*, 2002). Subsequent cold periods of smaller magnitude occurred around the early 17th and late 18th centuries. Strengthened southwesterly airflow during the LIA (see Schulmeister *et al.*, 2004), which would have brought an influx of cool, moisture-laden air across the Southern Alps, possibly triggered sudden switches in the FJG's mass balance, manifested as a series of rapid fluctuations of the glacier terminus.

Regionally, some glaciers of the Southern Alps reached their LIA maxima at least a century earlier (14th–16th centuries; Fig. 54.2) than the 'classic' period of maximum LIA glacier advances (17th–19th centuries) across large parts of the Northern Hemisphere. Furthermore, glaciers in the North Atlantic were still advancing towards their maxima when New Zealand glaciers, including the FJG, experienced their final and smallest magnitude LIA advance. This suggests that climatic amelioration occurred in New Zealand by the early 19th century. Overall, the New Zealand glacial record presents increasing evidence that both the timing and magnitude of LIA climate change and glacier response between hemispheres were not necessarily synchronous.

Historical records in New Zealand date from ca. AD 1860, and numerous accounts document glacier readvances throughout the late 19th century, although glaciers never regained their volume lost since the end of the LIA. Somewhat periodic, short-term advances of the FJG superimposed on long-term recession during the 20th century suggest complicated relationships between glacier behaviour, climate and synoptic-scale circulation (e.g.

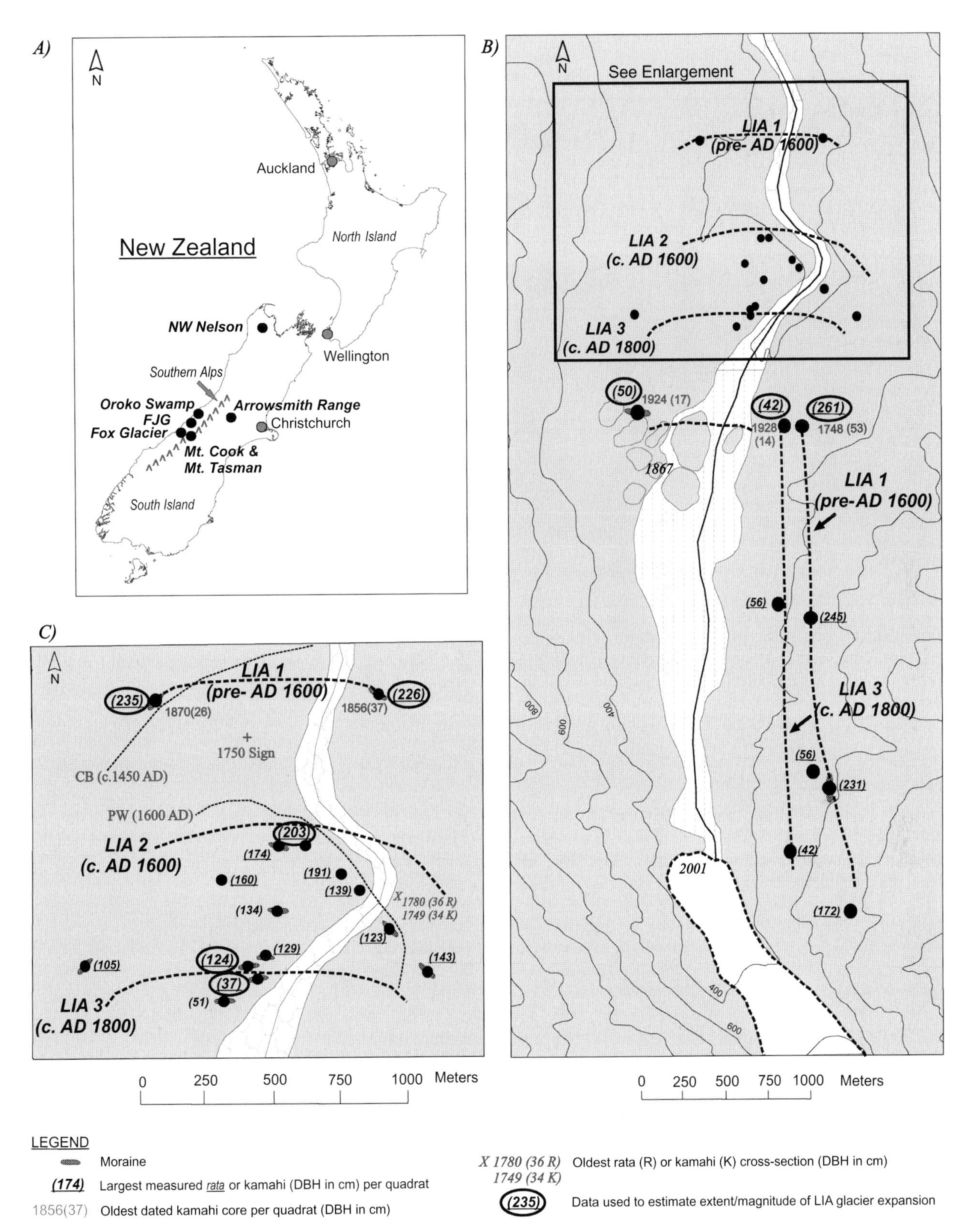

Figure 54.1 (A) The FJG, New Zealand (43°26′S, 170°10′E), and other key South Island locations referred to in text. (B) The reconstructed LIA chronology of the FJG, including prominent trimlines on the eastern valley side, and the first historically dated position of the ice-front as photographed by T. Pringle in AD 1867. Surface contours are drawn every 200 m. (C) Enlargement of the valley floor area with data illustrated as in (B). Previous interpretations of the LIA maximum extent have also been drawn, e.g. the AD 1750 sign (cf. Lawrence & Lawrence, 1965), PW (Wardle, 1973) and CB (Burrows, 1990). (After McKinzey, 2001.)

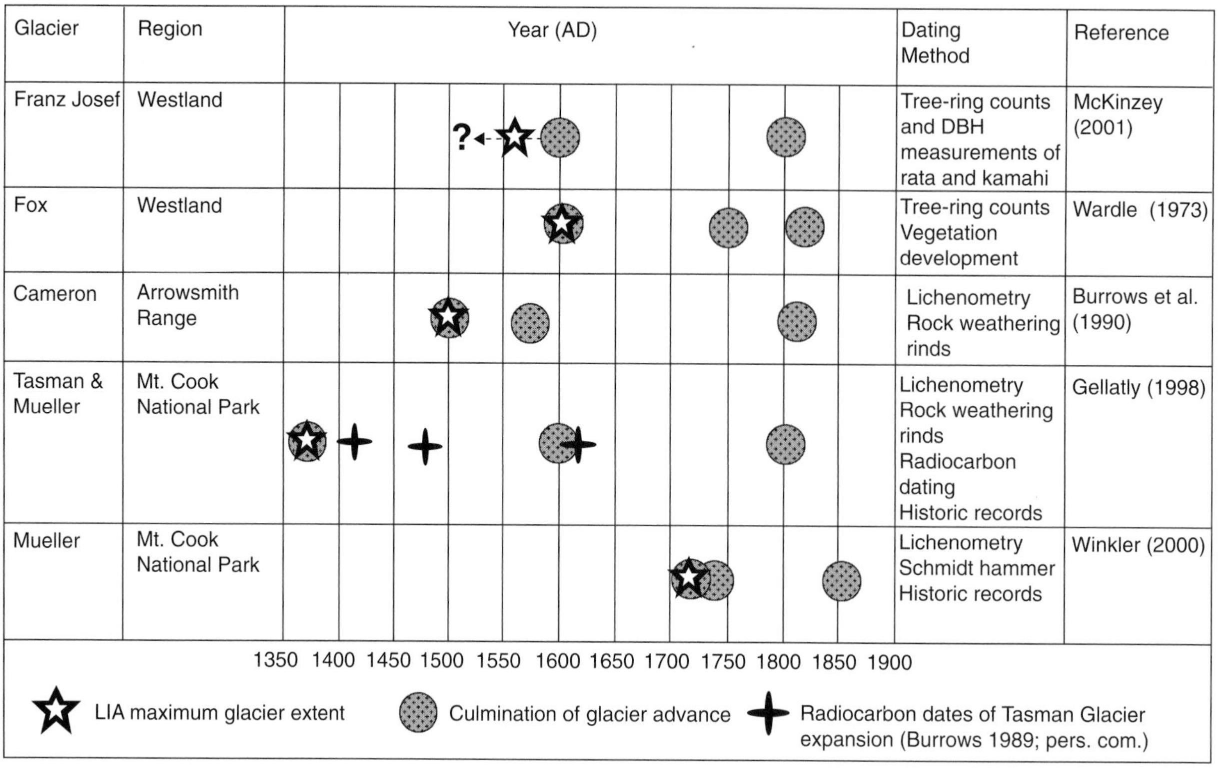

Glacier	Region	Year (AD)	Dating Method	Reference
Franz Josef	Westland	?	Tree-ring counts and DBH measurements of rata and kamahi	McKinzey (2001)
Fox	Westland		Tree-ring counts Vegetation development	Wardle (1973)
Cameron	Arrowsmith Range		Lichenometry Rock weathering rinds	Burrows et al. (1990)
Tasman & Mueller	Mt. Cook National Park		Lichenometry Rock weathering rinds Radiocarbon dating Historic records	Gellatly (1998)
Mueller	Mt. Cook National Park		Lichenometry Schmidt hammer Historic records	Winkler (2000)

1350 1400 1450 1500 1550 1600 1650 1700 1750 1800 1850 1900

LIA maximum glacier extent; Culmination of glacier advance; Radiocarbon dates of Tasman Glacier expansion (Burrows 1989; pers. com.)

Figure 54.2 Comparison of LIA records for selected glaciers of the Southern Alps. The Fox Glacier chronology (cf. Wardle, 1973) is derived mainly from tree-ring counts including kamahi (but not rata). Discrepancies regarding the timing of the LIA maximum at the Mueller Glacier may, in part, be reconciled by independent radiocarbon dates from the nearby Tasman Glacier (Burrows, 1989, personal communication), which support an early 15th century maximum extent. However, along with the reassessed FJG chronology, these examples highlight the necessity of further research across the region. (After McKinzey, 2001.)

Hooker & Fitzharris, 1999). Thus, the reassessed LIA history of the FJG can be used as a key resource with which to refine future predictions about glacier–climate interactions. Moreover, additional multiparameter dating studies are required to further elucidate the Southern Alps glacial record, especially for those climatically sensitive, such as the FJG, so that the nature of interhemispheric connections for recent events, such as the LIA, can be better understood.

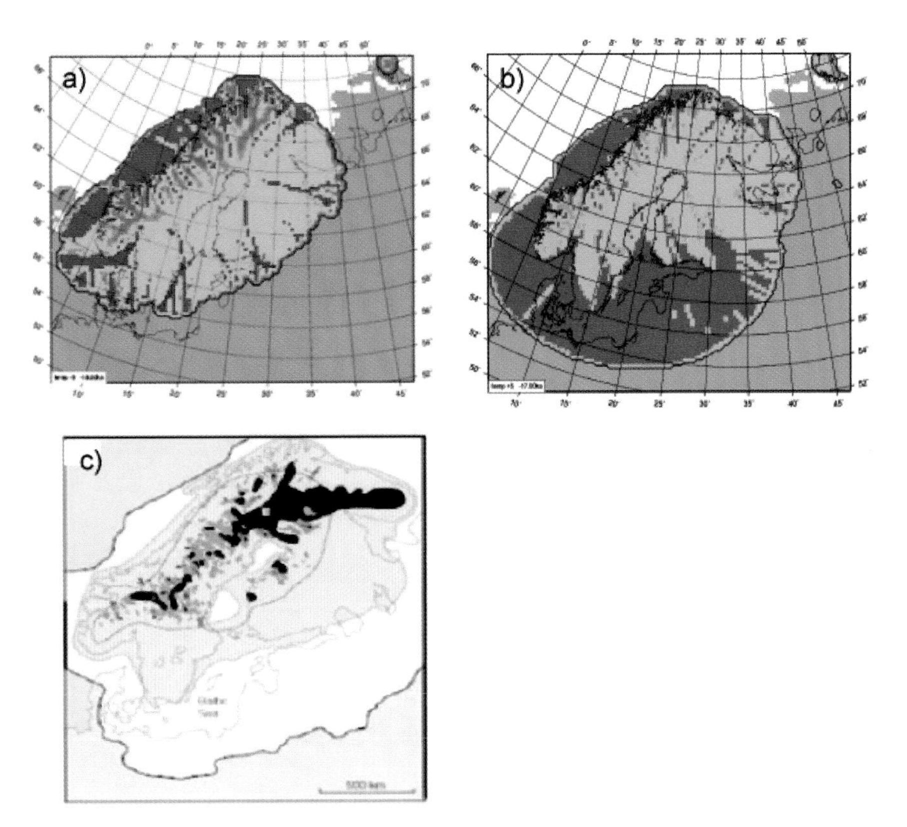

Plate 2.1 (a–c) Simulation of zones of basal freezing and melting beneath the European ice sheet at the LGM, and comparison with geological reconstruction of the basal freezing and melting zones. (a) Basal freezing (blue) and melting (red) for a 'cold' surface temperature simulation. (b) The same for a 'warm' surface temperature simulation. Note the zones of melting along simulated ice streams. (c) The inferred probable maximum zone of basal freezing at the LGM in yellow, and inner zone of sustained freezing in black (from Kleman & Hattestrand, 1999). Clearly the model in (b) is the better fit.

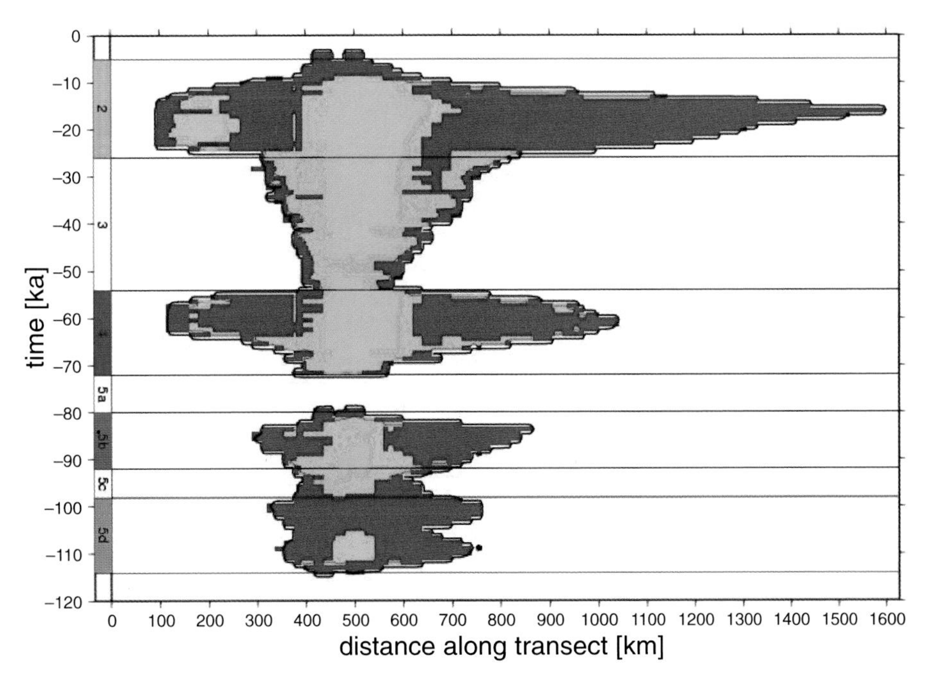

Plate 2.2 A section through the simulation model through the last glacial cycle from the central-western continental shelf of Norway (left) to northern Germany. It shows the zones of basal freezing (turquoise) and zones of basal melting (red).

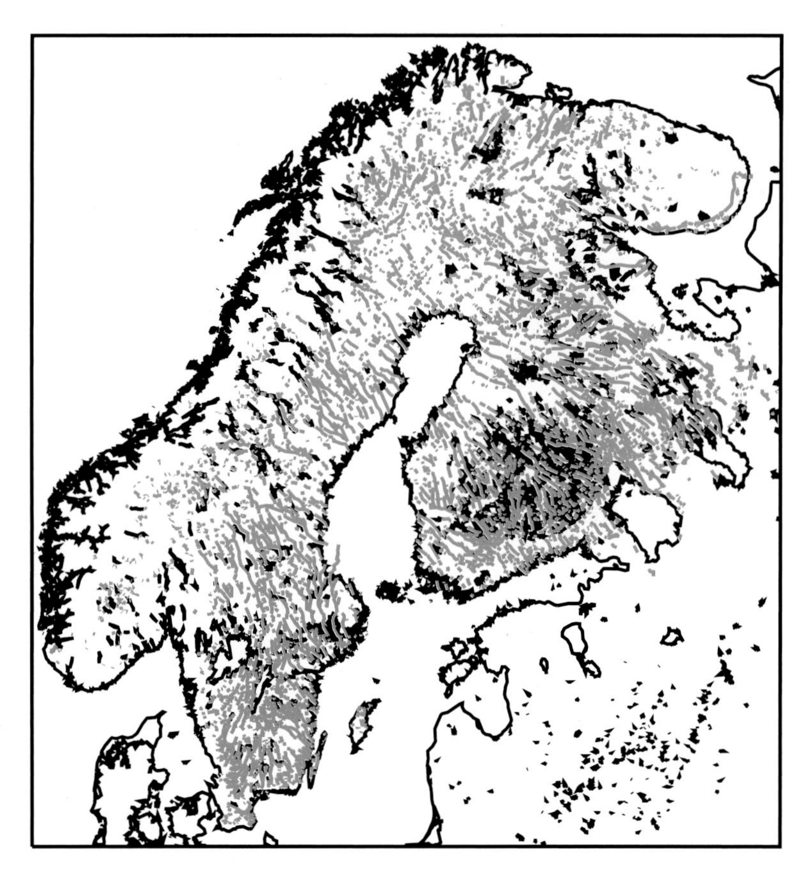

Plate 2.3 The distribution of eskers in the area of the European ice sheet.

Plate 9.1 Hydrofracture caused by pressurized groundwater at the margin of the Saalian ice sheet in northwest Germany. The fracture dissects outwash sediments and is filled with sand injected from below.

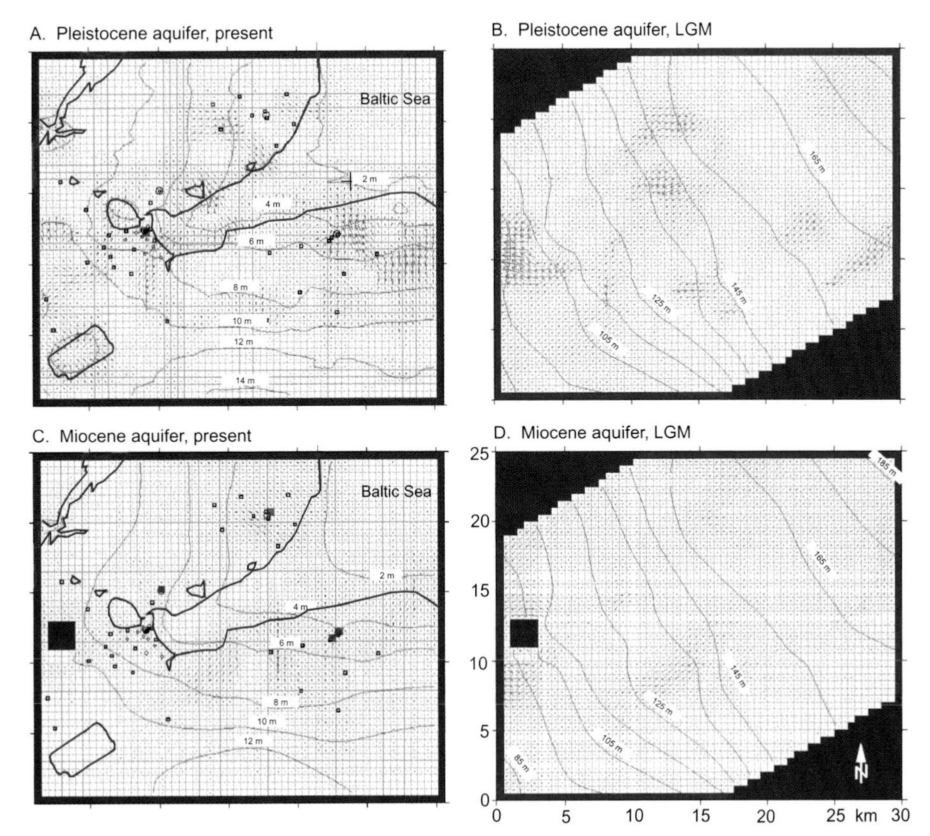

Plate 10.1 Modelled groundwater flow velocity vectors and equipotential lines in the two major aquifers in the study area at present and under the Last Glacial Maximum ice sheet. Ice movement was from northeast to southwest. Inactive cells are marked black. Note the complete reorganization of the groundwater flow field between glacial and interglacial (present) conditions.

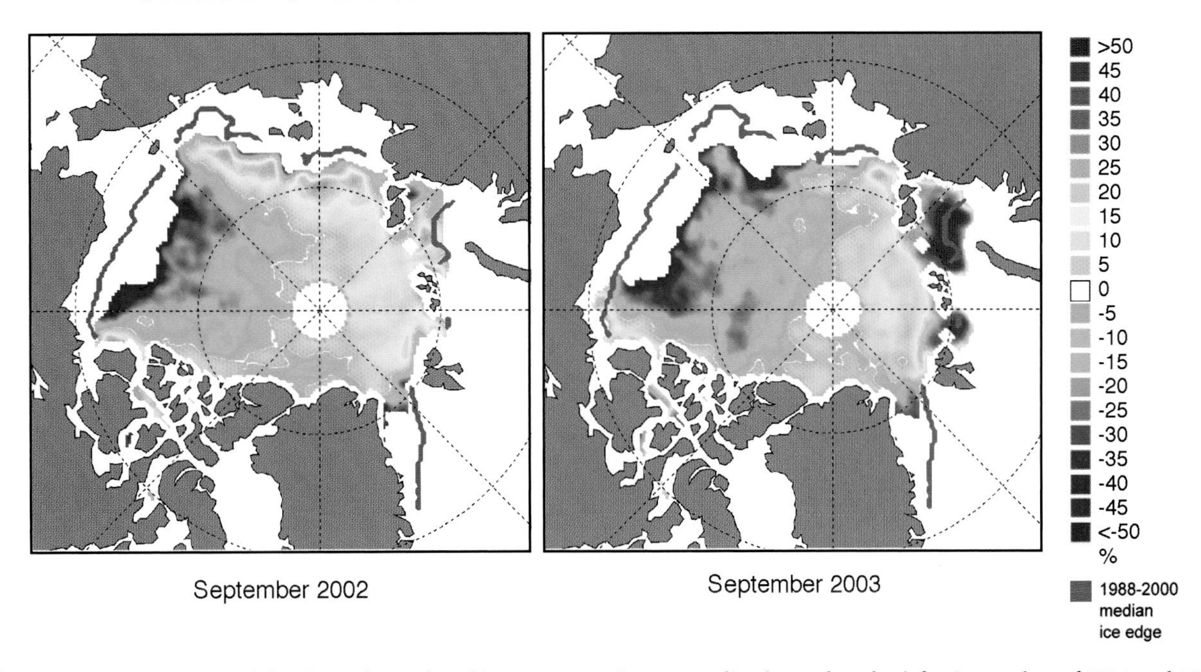

Plate 23.1 Sea-ice extent (all coloured areas) and ice concentration anomalies (see colour bar) for September of 2002 and 2003. Ice concentration anomalies are referenced to means for the period 1988–2000. Median ice extent based on the same period is shown by the red line. (Courtesy of National Snow and Ice Data Center, Boulder, CO.)

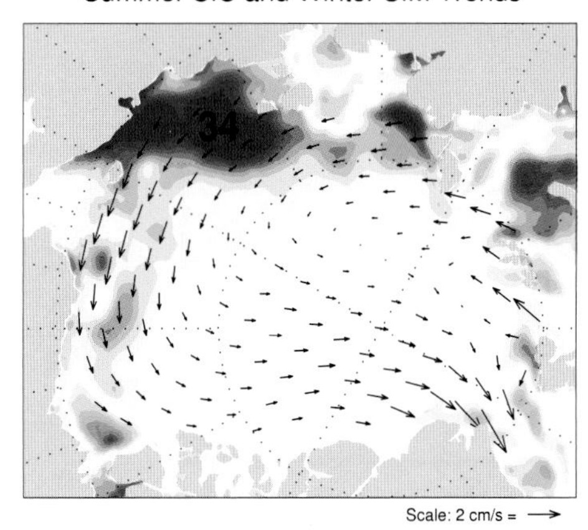

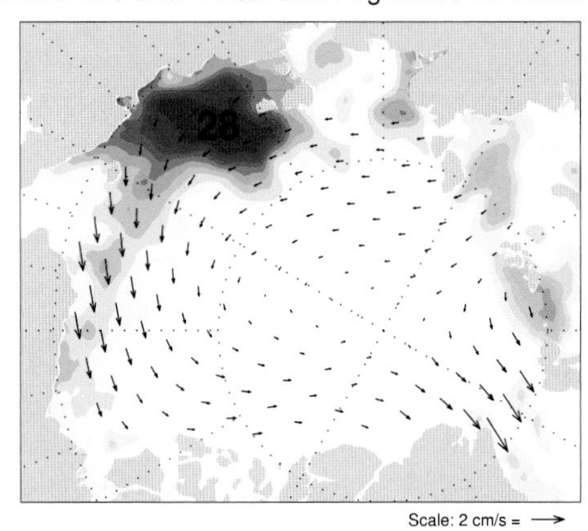

Plate 23.2 Large-scale trends in observed winter sea-ice motion (SIM) and summer sea-ice concentration (SIC) (top) and regressions on the prior winter AO index (bottom). Results are based on the period 1979–1998. Areas with negative and positive SIC trends of at least 5% over the record period are indicated by yellow/red and blue, respectively. The numbers indicate the largest negative trends. (Adapted from Rigor *et al.*, 2002.)

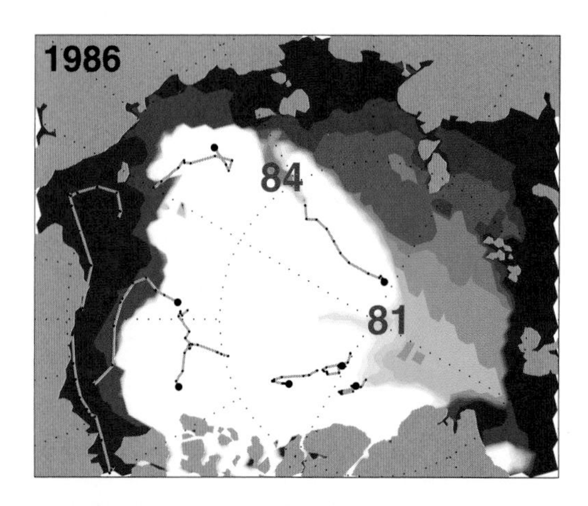

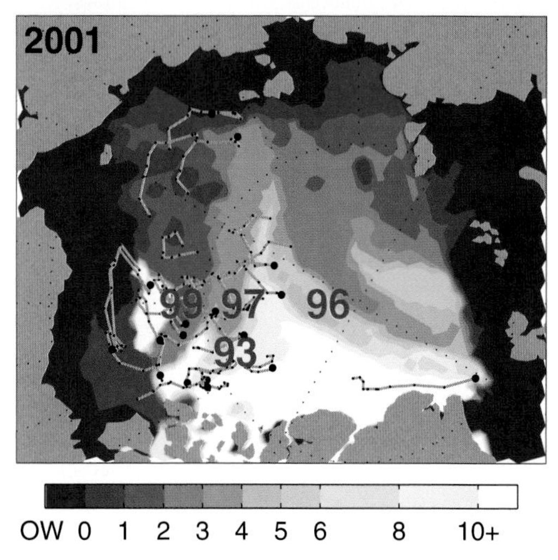

Plate 23.3 Estimated age of sea-ice in September of 1986 and 2001 (in years, see colour scale at bottom). Open water (OW) is shown as dark blue, and the oldest ice is shown as white. The years that younger ice was produced in the Beaufort and Chukchi seas are also shown. (Adapted from Rigor & Wallace, 2004.)

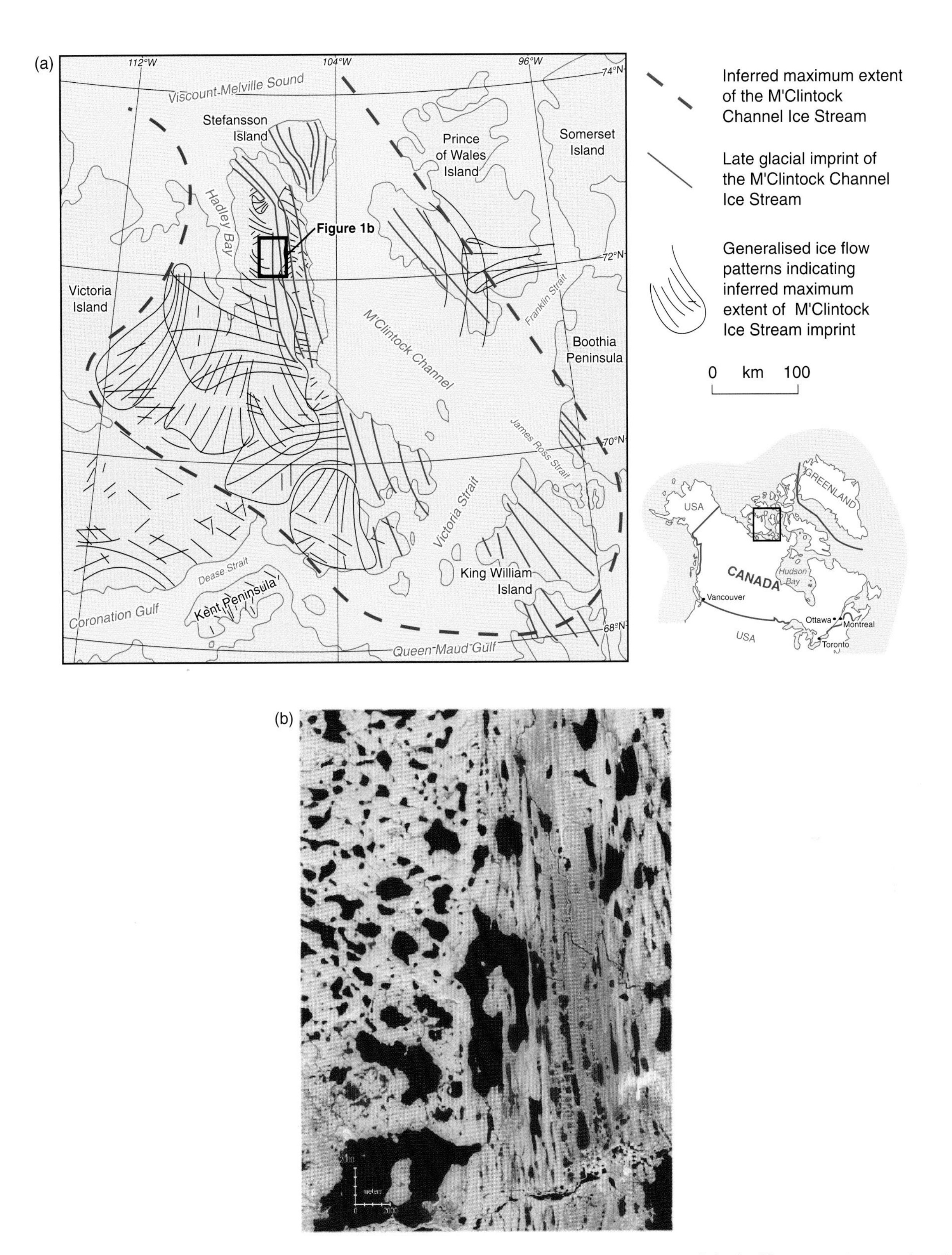

Plate 26.1 Location map of the M'Clintock Channel Ice Stream (a) and satellite imagery of the bedform imprint on Victoria Island (b). The late glacial imprint of the ice stream occupies present day M'Clintock Channel and infringes on western Prince of Wales Island and eastern Storkerson Peninsula (thin red lines). Landsat satellite imagery in (b) shows the margin of the late glacial ice stream imprint on Storkerson Peninsula. However, older flow patterns (thin black lines) indicate that the ice stream may have been much bigger during the Last Glacial Maximum, extending eastward and occupying Hadley Bay.

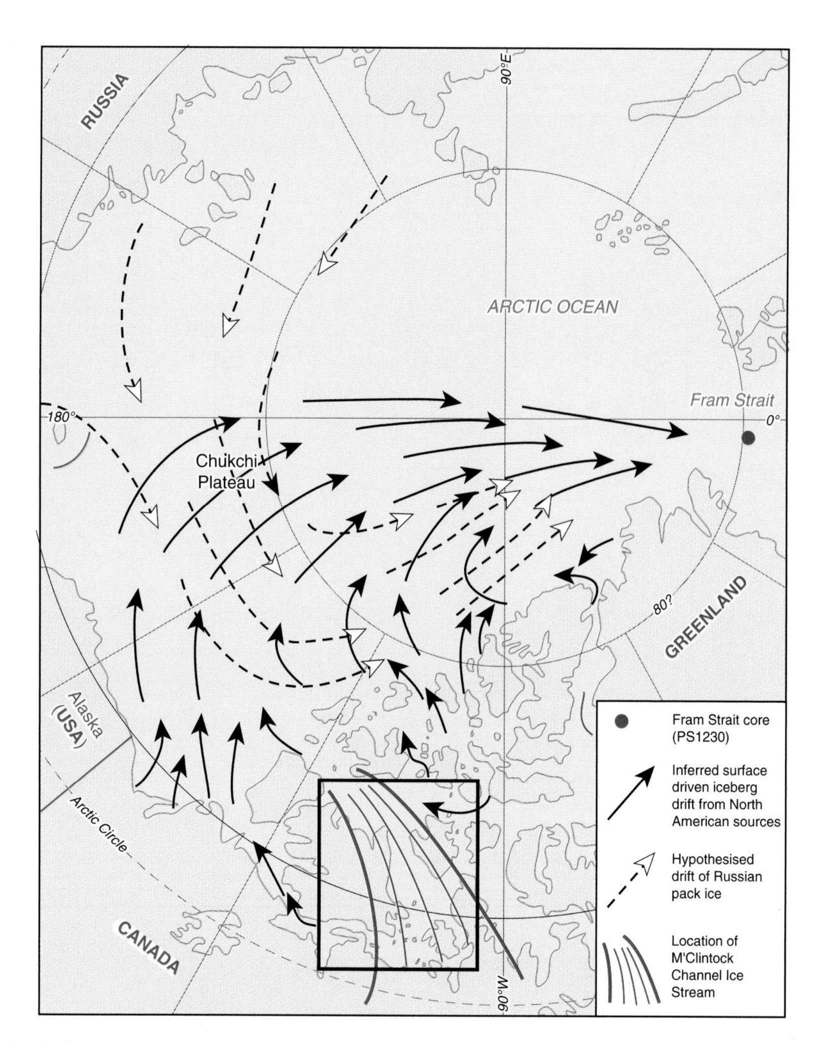

Plate 26.2 Map of the inferred surface current-driven iceberg drift directions from the Canadian Arctic Archipelago (solid arrows) and concurrent hypothesized drift of Russian pack ice (broken arrows) during glacial intervals (modified from Bischof & Darby, 1997). The expanded M'Clintock Channel Ice Stream is shown in red and it can be seen that icebergs issued from this region would enter Fram Strait relatively rapidly compared to present day conditions (Bischof & Darby, 1997). The box indicates the area shown in Plate 26.1a.

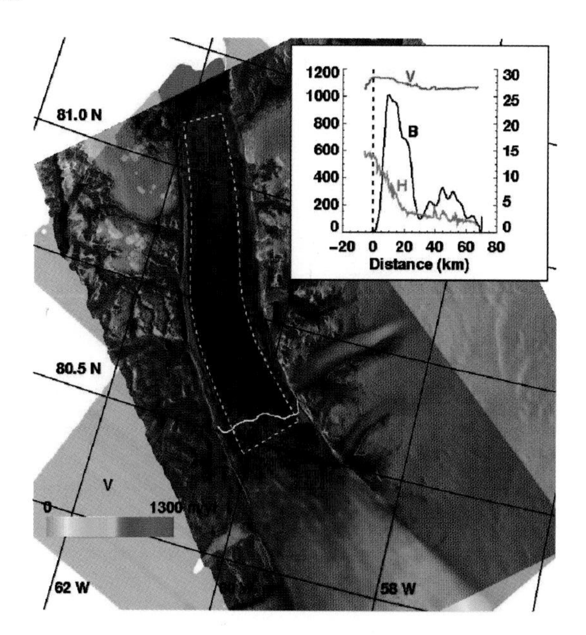

Plate 27.1 Ice velocity of Petermann Gletscher, northwestern Greenland measured from Radarsat-1 interferometric synthetic-aperture radar (InSAR) data. Grounding line inferred from double difference InSAR is white. Ice flow is to the north. Bounding box of calculation of bottom melt rates is dotted white. Inset shows velocity V (red, in $m\,yr^{-1}$, left scale), thickness H (blue, in m, left scale) and bottom melt rate B (black, in $m\,yr^{-1}$, right scale) calculated over the glacier width, versus the distance (in km) from the grounding line.

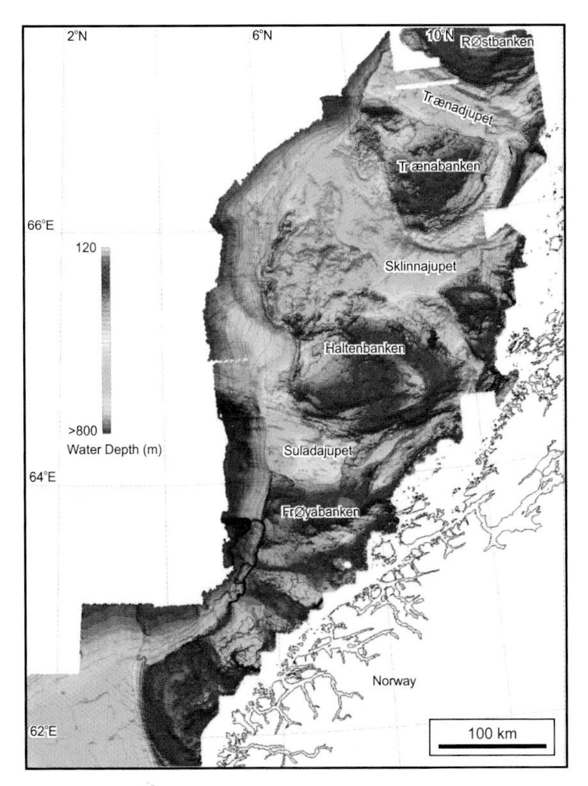

Plate 30.1 Bathymetry of the mid-Norwegian shelf showing cross-shelf troughs and intervening banks. (Modified from Ottesen *et al.*, 2002.)

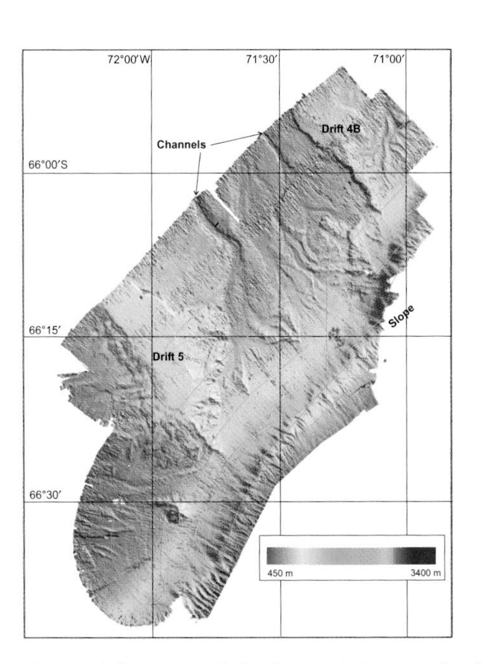

Plate 30.2 Multibeam swath bathymetric image of sediment drifts and intervening channels on the western Antarctic Peninsula continental margin. (Modified from Dowdeswell *et al.*, 2004b.)

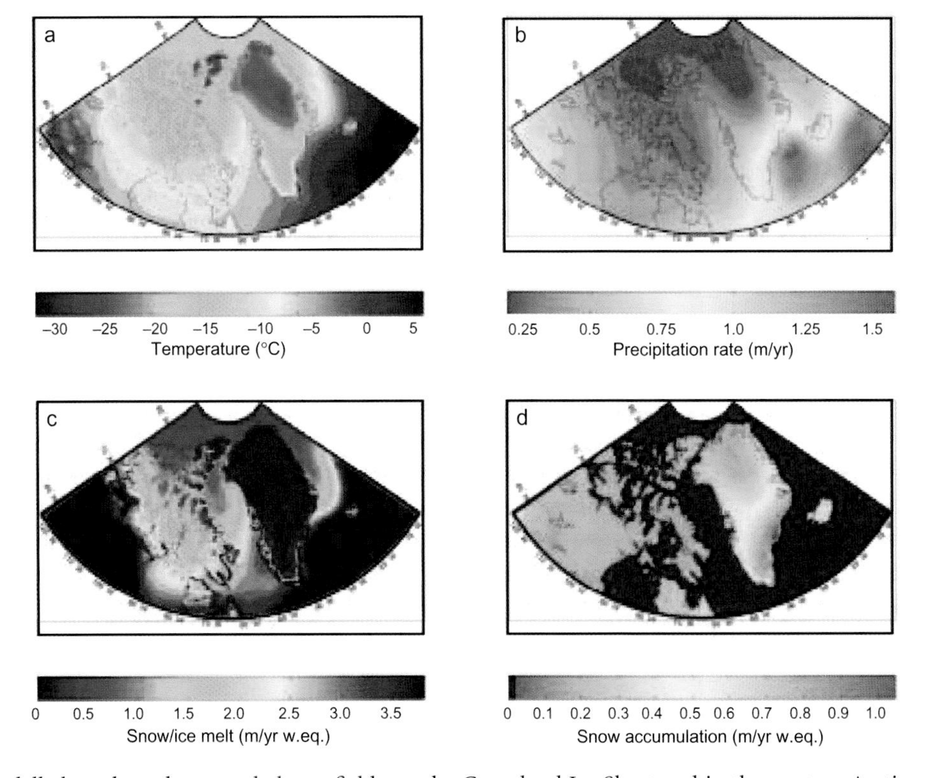

Plate 32.1 Modelled modern-day mass balance fields on the Greenland Ice Sheet and in the western Arctic, using degree-day methodology and climate fields from the NCAR Community Climate System Model (CCSM), v.2.0, with climate fields provided by B. Otto-Bleisner (personal communication, 2003). (a) and (b) show the precipitation and temperature maps that go into the calculation of mass balance fields. (c), (d) and (e) plot annual accumulation, ablation, and mass balance, all in m yr^{-1} water-equivalent. (f) shows ice sheet thickness (m). Model resolution is 1/6° latitude by 1/2° longitude.

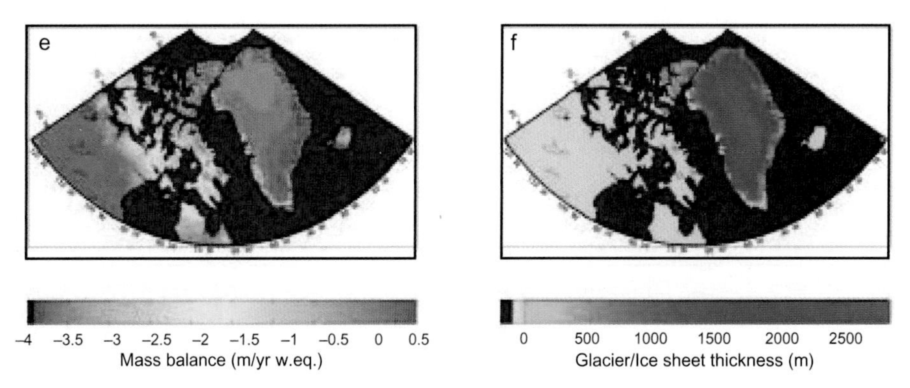

Plate 32.1 *Continued*

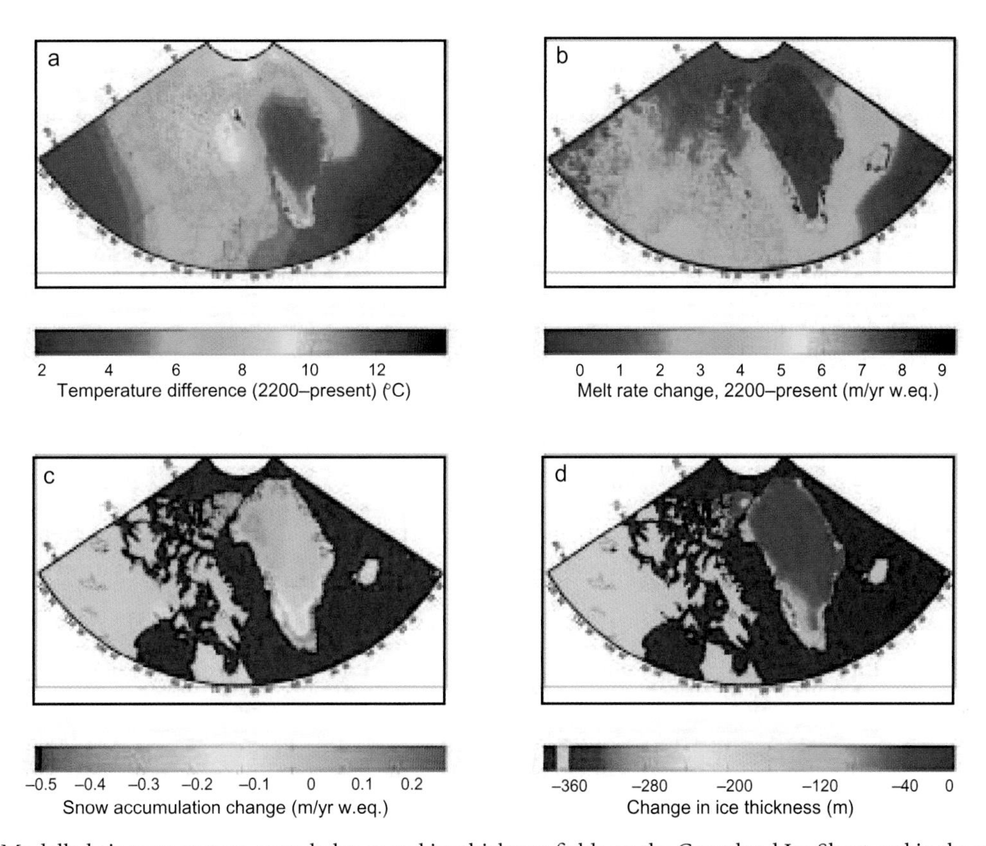

Plate 32.2 Modelled air temperature, mass balance and ice thickness fields on the Greenland Ice Sheet and in the western Arctic for 2200. All plots are shown as difference maps from the reference modern-day (2000) conditions shown in Plate 32.1. (a) Difference in air temperature, 2200 – present (°C). (b & c) Difference in snow/ice ablation and accumulation rates, 2200 – present (m yr^{-1} water-equivalent). (d) Difference in ice sheet thickness, 2200 – present (m). Model resolution is 1/6° latitude by 1/2° longitude.

Plate 38.1 Four ice sheet-scale reconstructions using inversion protocols by (a) Boulton *et al.* (1985), (b) Dyke & Prest (1987a), (c), Boulton & Clark (1990a,b) and (d) Kleman *et al.* (1997). The emphasis in (a) and (b) is on post-LGM configuration changes, whereas in (c) and (d) it is on ice-sheet evolution through the last glacial cycle, with an emphasis on events pre-dating the LGM. (Panel (a) is reproduced with permission from the Geological Society, London. Panel (b) is modified from Dyke & Prest (1987a). Panel (c) is reprinted by permission from Nature, 346, 813-817 (1990), copyright 1990 Macmillan Publishers Ltd. Panel (d) is modified from Kleman et al. (1997).)

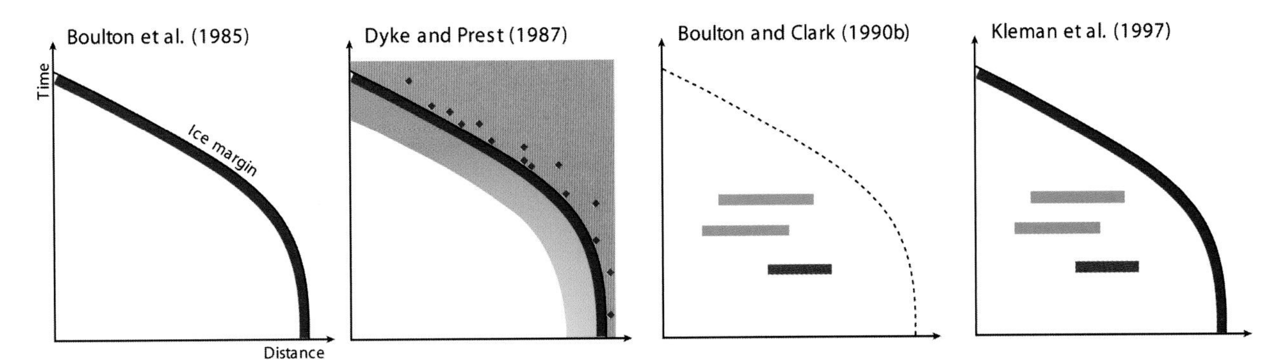

Plate 38.2 A visualization of the differences between four ice-sheet reconstructions, focusing on particular time–space data domains and the data types used in the inversion procedures. Coloured items mark the primary data domains: thick red line marks the deglacial landforms; green, blue and purple mark glacial 'events' reflected by till lineations pre-dating the final decay phase; red diamonds schematically illustrate radiocarbon dates (which always reflect ice-free conditions); orange colour represents a 'stretching' of the deglacial landform record for inferences about older non-deglacial events.

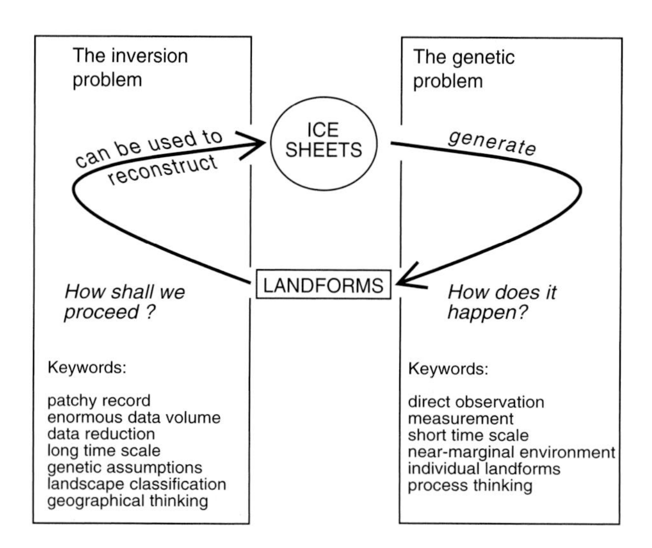

Plate 38.3 The genetic and inversion problems in glacial geomorphology are associated with fundamentally different suites of assumptions, scale and generalization considerations, as well as methodical issues.

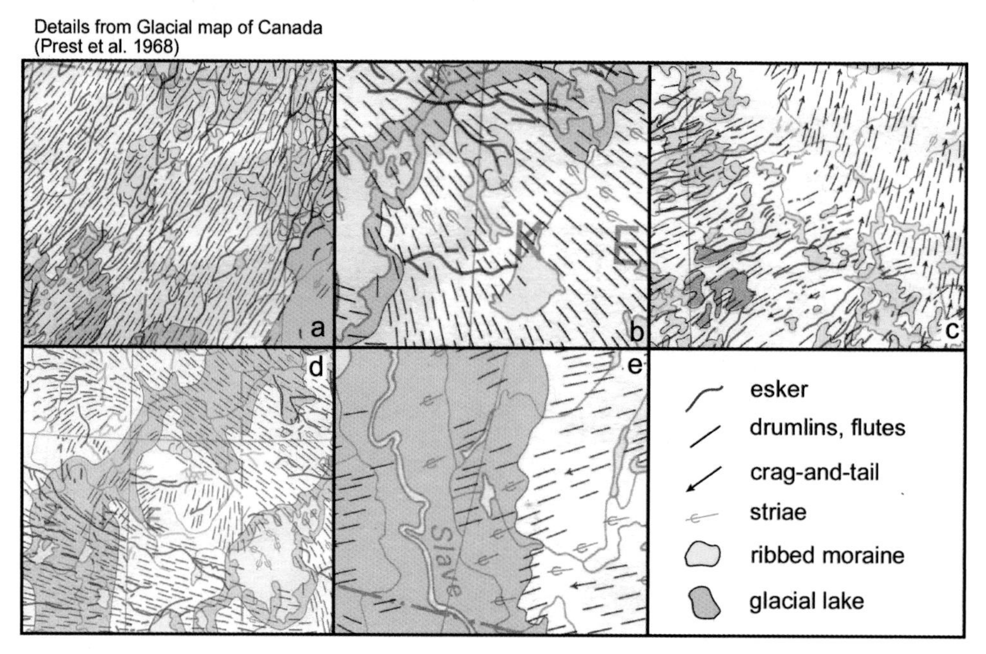

Plate 38.4 Details from the Glacial Map of Canada (Prest *et al.*, 1968). (a) The southwestern sector of Keewatin displays a 'classic' glacial landscape where abundant eskers parallel a single coherent system of till lineations. Fields of ribbed moraine occur in the proximal part of the till lineation swarm, and probably mark areas that changed from cold-based to warm-based (wet-bed) conditions (Hättestrand & Kleman, 1999). This type of landscape is thought to have formed in marginal wet-bed zones of substantial width during ice-sheet decay (Kleman *et al.*, 1997). (b) Eskers cutting obliquely across the convergent head zone of the Dubawnt lineation swarm in Keewatin. The lineations were probably formed by a short-lived ice stream (Stokes & Clark, 1999). The eskers indicate that a major change in flow direction occurred between the ice-stream phase and the deglaciation stage. (c) An intersection zone in central Quebec–Labrador where two different glacial landscapes, with opposing flow directions, occur in close contact. The southwest-orientated landscape in the lower left-hand half of the map displays a full suite of deglacial meltwater features aligned parallel to the lineation system, leading us to regard it as being formed during the last deglaciation. The NNE-orientated Ungava Bay swarm in the northeastern half, in contrast, almost entirely lacks eskers, leading us to believe that the lineations formed underneath central portions of the ice sheet during an earlier flow phase (Jansson *et al.*, 2003). The apparent ice divide is probably entirely fictitious and instead denotes only the up-glacier boundary of wet-bed conditions of the southwastward flow, during the last deglaciation. (d) An isolated patch of relict N–S and NNE–SSW orientated lineations southwest of the Dubawnt lineation swarm. The patch is probably an erosional remnant of a lineation system formed during a glacial event that pre-dated the LGM (Kleman *et al.*, 2002). Hence, its present extent is governed by subsequent Dubawnt ice-stream erosion to the north and sheet-flow erosion to the south. Eskers from the last deglaciation cut the relict north–south trending lineations in the patch at almost right angles. (e) A landscape without eskers on the northwestern flank of the Keewatin sector. Because lineations and aligned striae yield few clues to their age or the duration of flow, these landscapes are difficult to treat in inversion models. If they are part of a cold-based deglaciation landscape (which, typically, lacks eskers), the distribution of glacial-lake shorelines, spillways and drainage channels may give the only solid guidance for decay reconstruction in such areas (Borgström, 1989; Jansson, 2003). Reproduced with the permission of the Minister of Public Works and Government Services Canada, 2004 and Courtesy of Natural Resources Canada, Geological Survey of Canada.

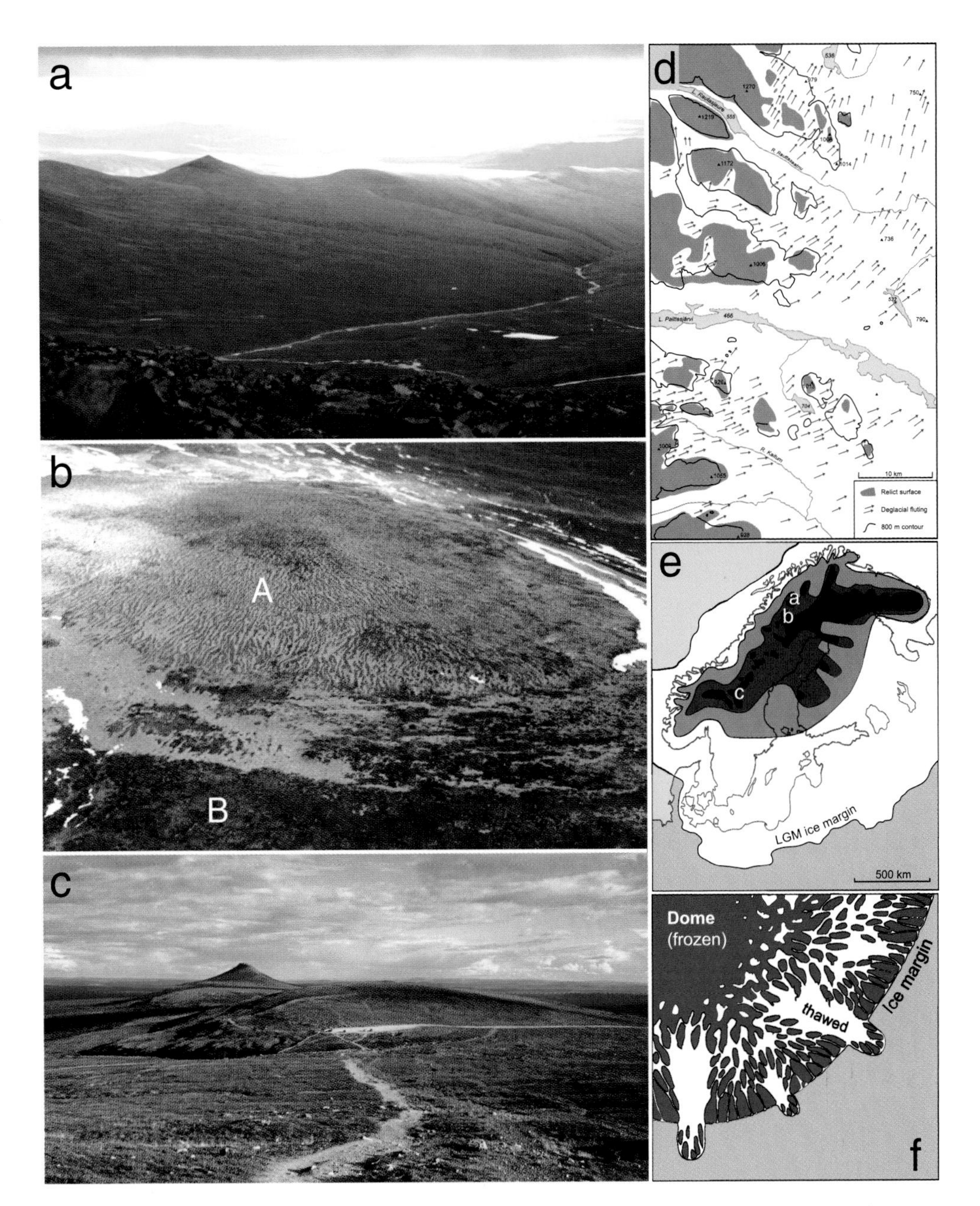

Plate 38.5 (a) Relict surfaces lacking glacial landforms, such as the Tjeuralako Plateau, northern Sweden, are interpreted to mark sustained frozen-bed conditions under one or more successive ice sheets. Cosmogenic dating (^{10}Be) of exposed bedrock on this plateau yielded an exposure age of 45 kyr, indicating inheritance from one or more previous ice-free intervals, and negligible erosion by the last ice sheet (Stroeven *et al.*, 2006) (b) In Fennoscandia, periglacially formed surfaces, such as this striped boulder surface (A) at Tjuolma, The Ultevis plateau, Sweden, occur preferentially on uplands with clear erosional boundaries (Kleman, 1992) to younger glacial landscapes (B) comprising fluting and drumlinization from the last ice sheet. An erratic perched on surface (A) yielded an exposure age (^{10}Be) of 7.4 kyr, whereas bedrock exposed on the same surface yielded exposure ages of 32.7 and 35.2 kyr (Fabel *et al.*, 2002; Stroeven *et al.*, 2006). (c) The Städjan-Nipfjället upland in the southern Scandinavian mountains comprises marginal moraines older than the the last ice sheet, and a >2-m-deep weathering mantle, indicating negligible erosion by the last ice sheet. (d) Relict surfaces and glacially eroded surfaces display an archipelago-like pattern west of Kiruna, northern Sweden. The flow pattern indicated by lineations is consistent with the pattern expected for thick overriding ice and polythermal bed conditions, but inconsistent with the flow pattern expected from a thin-ice scenario comprising nunataks and ice-tongues in valleys. Modified from Kleman *et al.* (1999). (e) Frozen-bed extent under the Fennoscandian Ice Sheet, as inferred for three time periods. Approximate LGM extent, largely based on the distribution of ribbed moraine, inferred to have formed during transition from frozen-bed to thawed-bed conditions, shown as light grey shading. Approximate extent during Younger Dryas, after onset of major ice streams in Finland, is shown as medium grey. Black mark zones with abundant pre-Late Weichselian glacial and non-glacial landforms, and stratigraphic and cosmogenic dating evidence for non-erosive frozen-bed conditions under the last ice sheet. (f) Hughes (1981b) hypothesized that terrestrial core areas would comprise a frozen-bed core, a patchy transition zone to mostly thawed-bed conditions, lenticular frozen-bed patches in ice-stream heads, and ice stream corridors with sharp thermal boundaries to intervening ice-stream ridges. The collective evidence from the Fennoscandian and Laurentide ice sheet areas (Dyke *et al.*, 1992; Kleman *et al.*, 1999) confirms all essential aspects of this hypothesis.

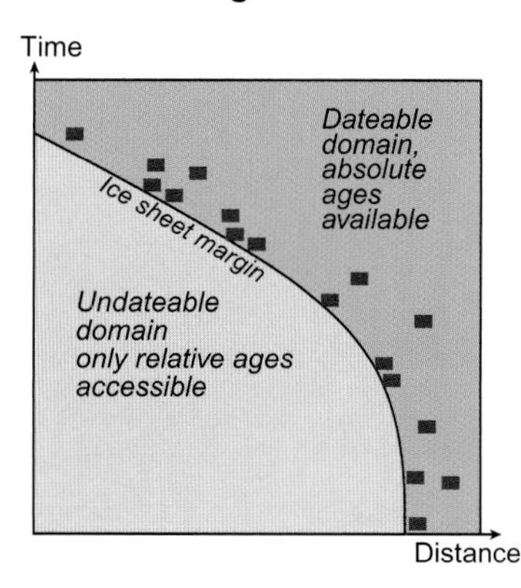

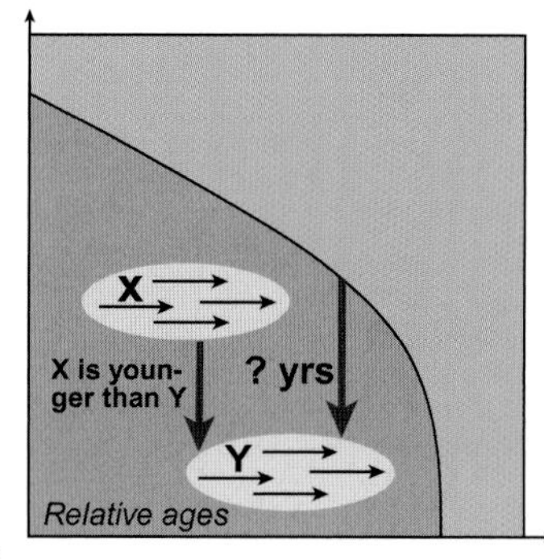

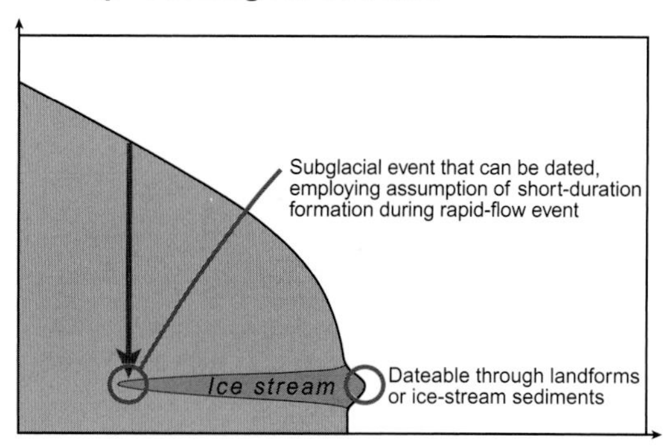

Plate 38.6 (a) Two chronological domains are defined: the extramarginal domain, to which all currently available dating methods pertain (radiocarbon, OSL, cosmogenic and amino-acid racemization techniques), and the subglacial, which is not accessible by any current absolute dating methods. The chronology of the deglacial envelope is currently defined mainly by a scatter of radiocarbon and OSL dates of widely varying spatial density. Through ancillary data, e.g. pollen, a specific dating often can be related to climatic evolution, but only rarely can any direct link to ice-dynamics be established. (b) In the subglacial dating domain, only relative ages are readily available. The relative age of flow events is established through cross-cutting relationships (Clark, 1993). Through analysis of landform assemblages and relative ages, inferences about ice dynamics can be made, but only rarely can any information pertaining to climate be gained. (c) Ice-stream landscapes have a substantial but yet largely unrealized potential for absolute dating of subglacial events, and may prove to be the only realistic tool for that purpose.

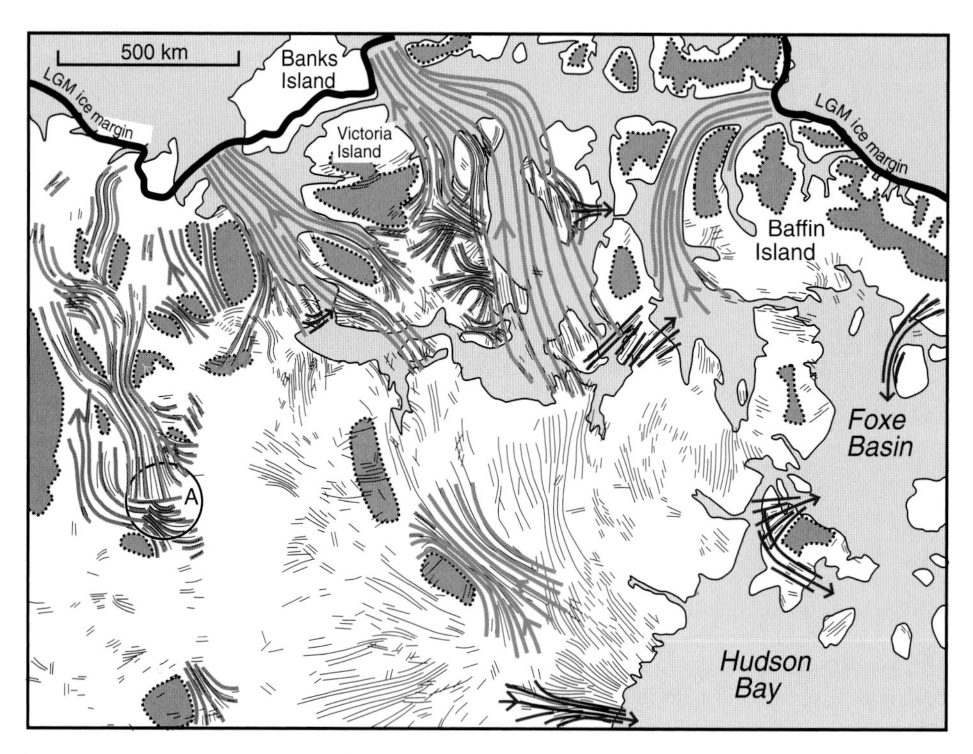

Plate 38.7 Palaeo-ice streams in the northwestern Canadian Arctic. Prime data sources are Prest *et al.* (1968) and morphological mapping using Landsat MSS and TM imagery (Kleman, unpublished data): thin black lines show till lineations mapped from MSS imagery; blue lines show major ice streams of type 1; green lines show type 2 ice streams; red lines show ice streams of type 3; grey lines show unclassified ice streams. See text for description of ice-stream types. Grey shading shows areas displaying relict non-glacial morphology and areas inferred to have been the sites of frozen-bed interstream ridges in the last ice sheet.

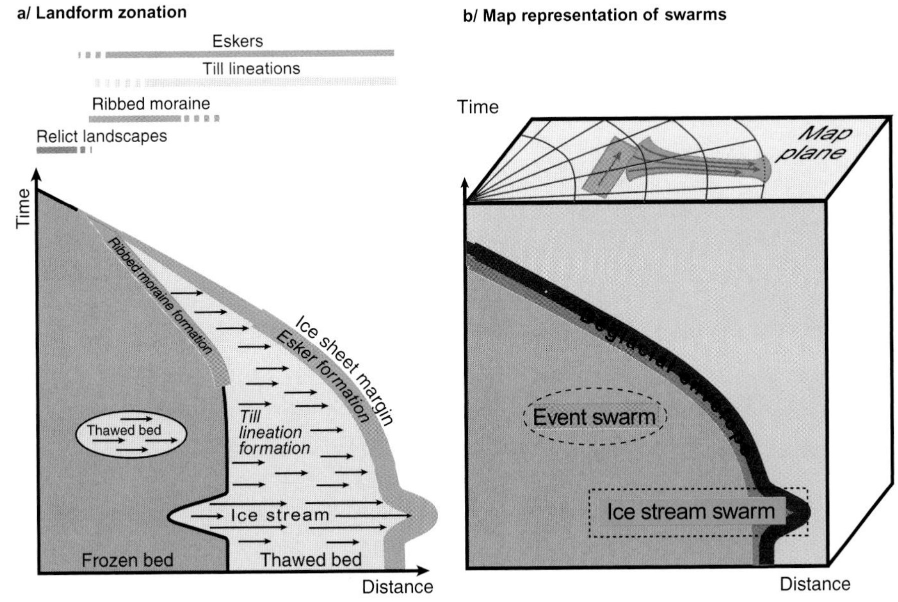

Plate 38.8 (a) Domains of landform formation in a time–distance diagram. Eskers form close to the ice margin in a time-transgressive fashion. Ribbed moraine is inferred to form during transition from a frozen bed to a thawed bed (Kleman & Hättestrand (1999). Glacial lineations form wherever the bed is thawed and subglacial sediments are available. (b) The three swarm types we recognize are event swarms, ice-stream swarms and the deglacial envelope. The latter is defined by eskers and other meltwater landforms and may or may not be associated with till lineations. *Swarms* are simplified and spatially delineated map representations of many individual landforms.

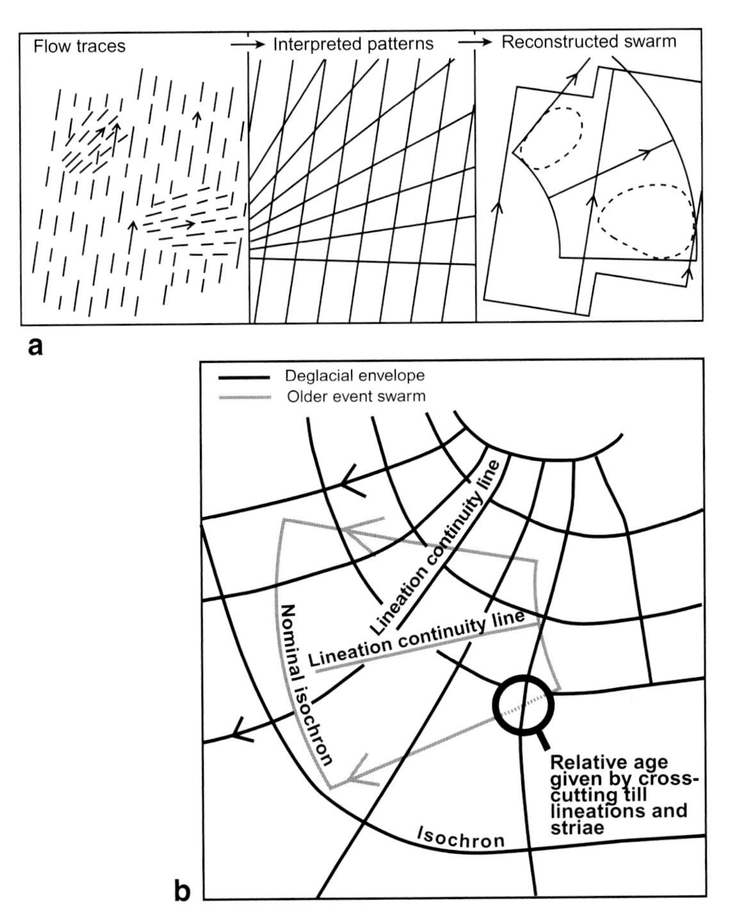

Plate 38.9 (a) A swarm is spatially defined by longitudinal continuity lines, aligned to a visually coherent system of flow traces, and transverse up- and downstream boundaries. The latter are drawn transverse to continuity lines, if necessary in a stepped fashion. Those elements which allow definition of a swarm can be any geological features that reflect ice-flow direction (e.g. striae, flutes, till fabrics, glaciotectonic folds, etc.). (b) An example of an event swarm underlying the deglacial envelope. The angular difference between the flow indicators in the event swarm and the deglacial envelope will differ depending on location, and may be small or non-existent at some locations.

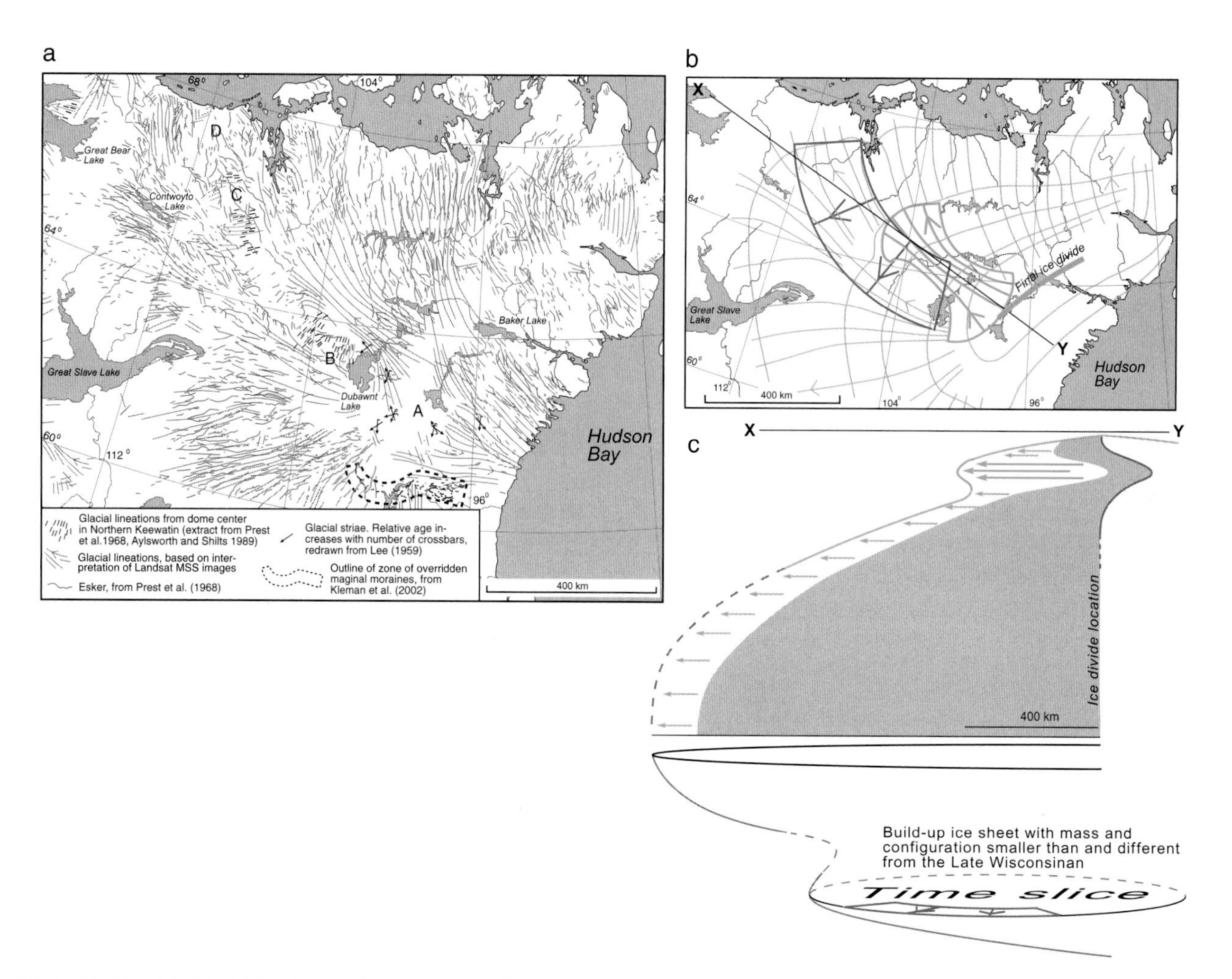

Plate 38.10 (a) Glacial landforms in Keewatin. Till lineations interpreted in Landsat MSS data. Striae observations are from Lee (1959). Eskers are redrawn from the Glacial Map of Canada (Prest *et al.*, 1968). Letters A–D mark patches of lineations, and striae observations, indicative of older ice flow from a dispersal centre in northern Keewatin or the central Arctic. The trunk of the Dubawnt Ice Stream is marked by E. Overridden end moraines in southeastern Keewatin are from Kleman *et al.* (2002). (b) The deglacial envelope is shown by yellow colour and the swarm formed by older flow from the northeast is shown in purple. Green colour marks the Dubawnt ice stream swarm. X–Y marks the location of the transect shown in Plate c. (c) The three Keewatin swarms schematically illustrated in a time–distance diagram. The Dubawnt swarm is interpreted as a short-lived ice-stream event during the Late Wisconsinan deglaciation. The deglacial envelope formed time-transgressively during the entire post-LGM time. The northeast swarm (purple) probably represents a pre-LGM event with small ice volume and a very northerly dispersal centre.

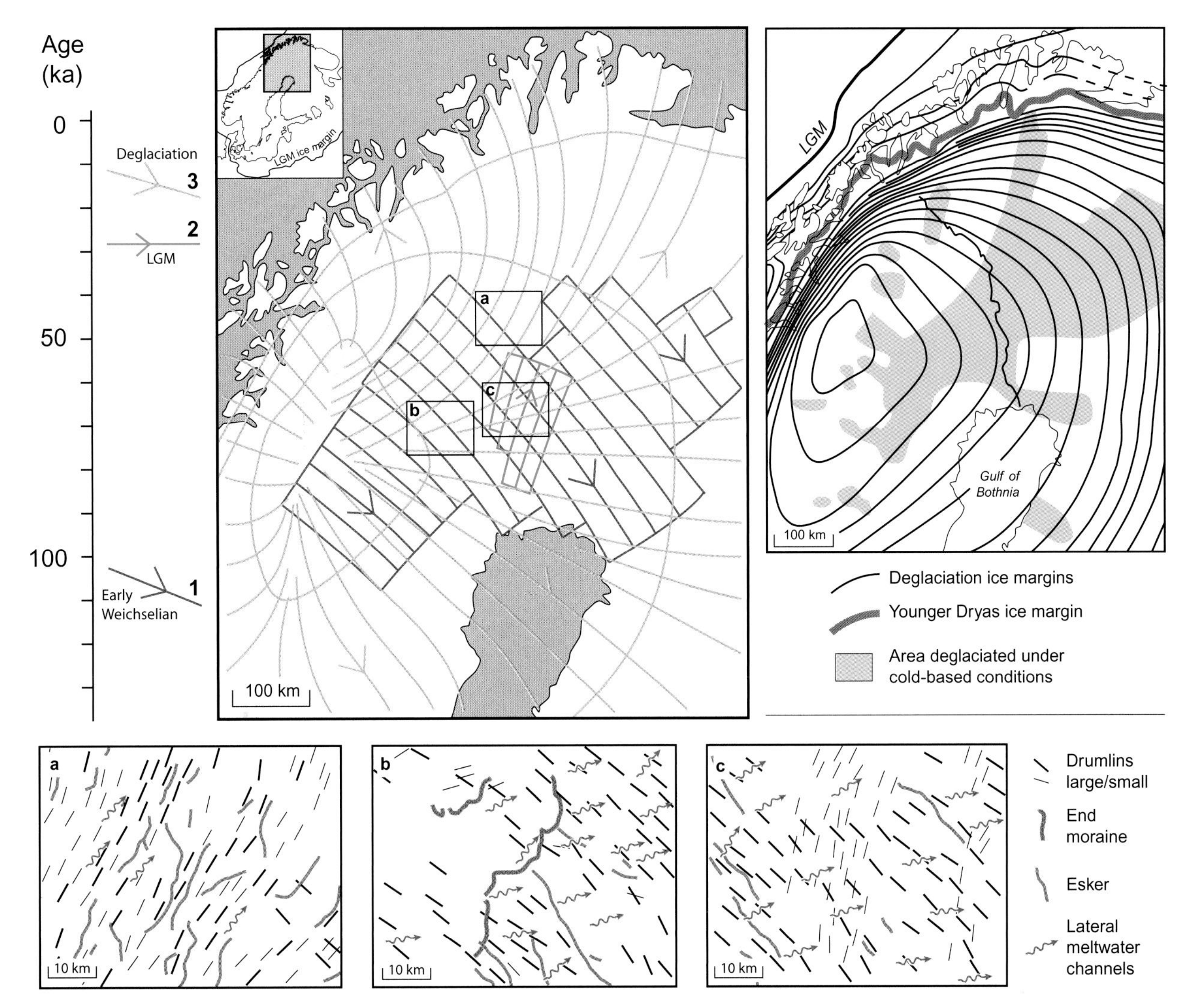

Plate 38.11 Glacial swarms (1–3) in northern Fennoscandia. The spatial distribution is based on Kleman *et al.* (1997). The age assignments are based on Lagerbäck & Robertsson (1988) and Kleman *et al.* (1997). Geomorphological maps (a)–(c) are redrawn from Hättestrand (1997), and show three areas with increasing swarm complexity. Area (a) represents a classic last deglaciation landscape with eskers and lineations (Swarm 3). Only a few large older drumlins with a transverse direction (in the southeastern corner) have survived the deglaciation wet-based ice flow. In area (b), the last deglaciation envelope (Swarm 3) is represented only as a swarm of lateral meltwater channels, and overprints an older wet-bed deglaciation landscape (Swarm 1; with lineations, eskers and end moraines) of an Early Weichselian age. Frozen-bed conditions during the last deglaciation have prohibited formation of subglacial landforms. The third area, (c), is similar to (b), but an event swarm of small drumlins (Swarm 2) overprints the old deglaciation landscape, and is in turn overprinted by the deglacial evelope from the last deglaciation, only manifested by meltwater channels (Swarm 3). The intermediate Swarm 2 lacks any meltwater channels and is therefore interpreted to have formed within the interior of the ice sheet, far from the meltwater system at the ice margin. The area interpreted to have deglaciated under cold-bed conditions is interpreted from the distribution of preserved *pre*-Late Weichselian landforms (indicating a minimum of subglacial erosion), and areas where the deglaciation meltwater system consists of lateral meltwater channels rather than eskers (indicating a minimum of subglacial meltwater (Kleman & Hättestrand, 1999)).

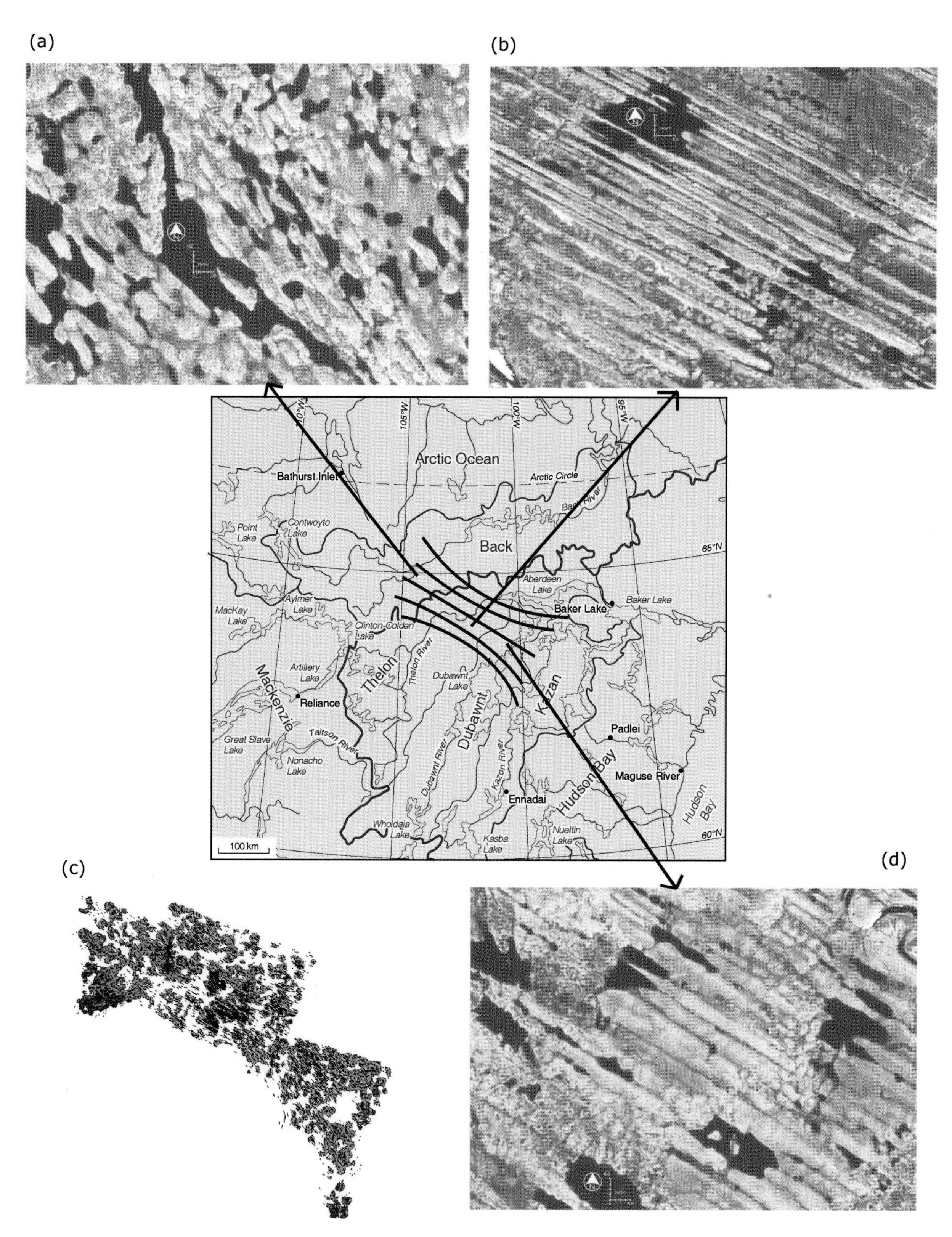

Plate 41.1 Location of the Dubawnt Lake Ice Stream on the north-western Canadian Shield and location of major drainage basins in red (centre). Mapping from digital satellite imagery (Landsat ETM+) indicates that the subglacial bedforms (drumlins and megascale glacial lineations) mimic the expected ice velocity of an ice stream, which speeds up in the onset zone (d), reaches a maximum in the main trunk (b) and decreases at the divergent terminus (a). Each image (a, b, d) is approximately 10 km wide. The overall pattern is shown in (c) where red/orange colours indicate areas of long bedforms and green through to blue show shorter bedforms.

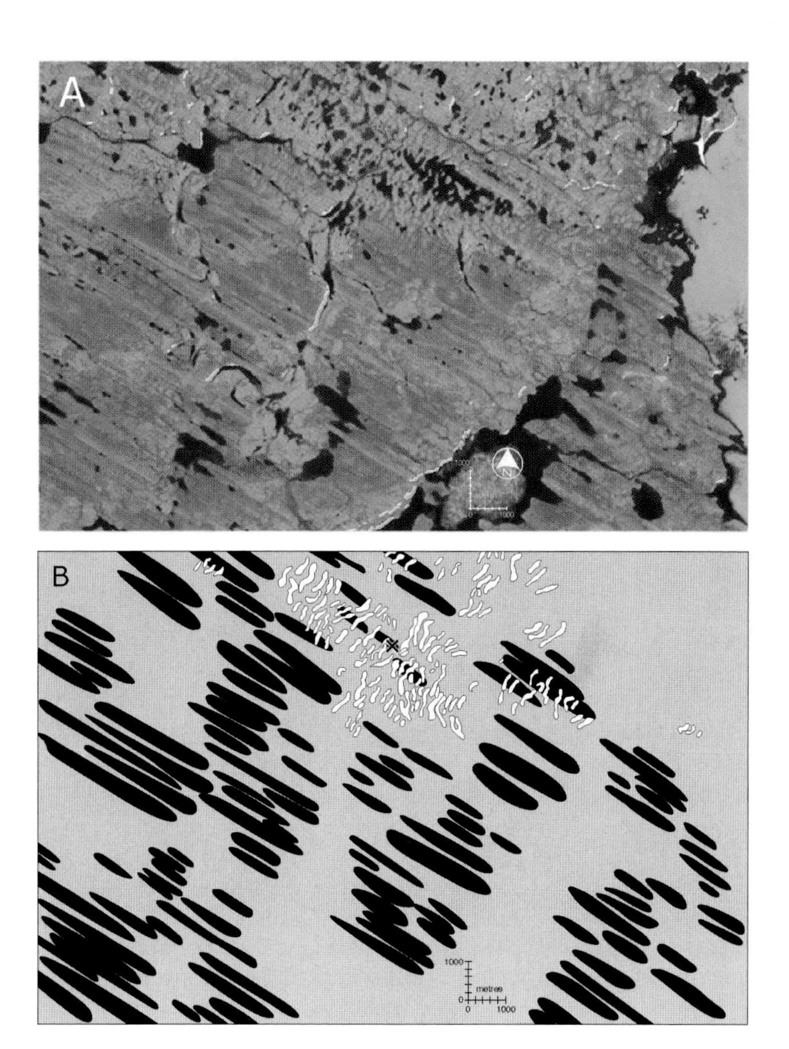

Plate 41.2 (A) Potential sticky spot formed during or in response to ice-stream shut-down. Ice flow is from bottom right to top left, characterized by elongated drumlins and megascale glacial lineations. Bedform mapping in (B) illustrates that the lineaments have been modified by the formation of ribbed moraines. The association between ribbed moraine formation and zones of cold-based ice suggests that this area may be indicative of localized basal freeze-on as the ice-stream shuts down.

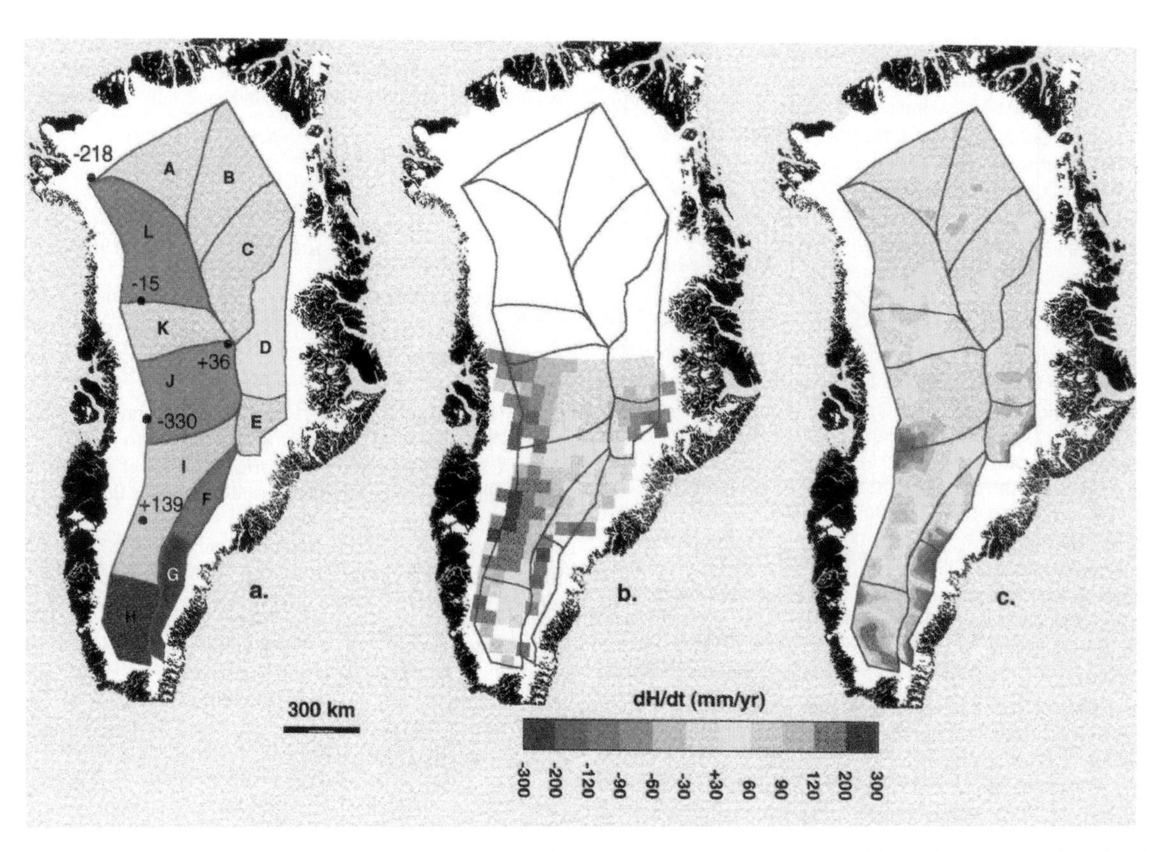

Plate 44.1 Three independent estimates of Greenland Ice Sheet thickening/thinning rates: (a) for the past few decades derived from comparison of ice discharge with snow accumulation, (b) for 1978–1988 derived from comparison of Seasat and Geosat radar-altimeter data, and (c) for 1993/1994–1998/1999 derived from repeated aircraft laser-altimeter surveys. The altimetry data have been corrected for estimated rates (ca. 5 mm yr^{-1}) of isostatic uplift of the underlying bedrock. (From Thomas *et al.*, 2001.)

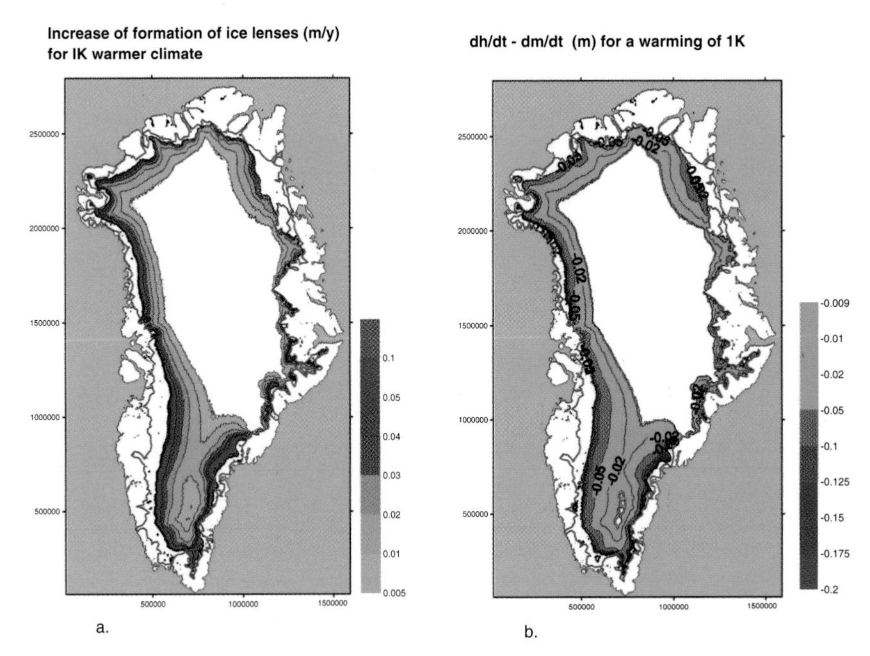

Plate 44.2 (a) Calculated increase of ice lens formation at the surface of the Greenland Ice Sheet for a 1 K warming. (b) Difference between surface-elevation change and local mass change for a 1 K warming (in ice equivalent thickness).

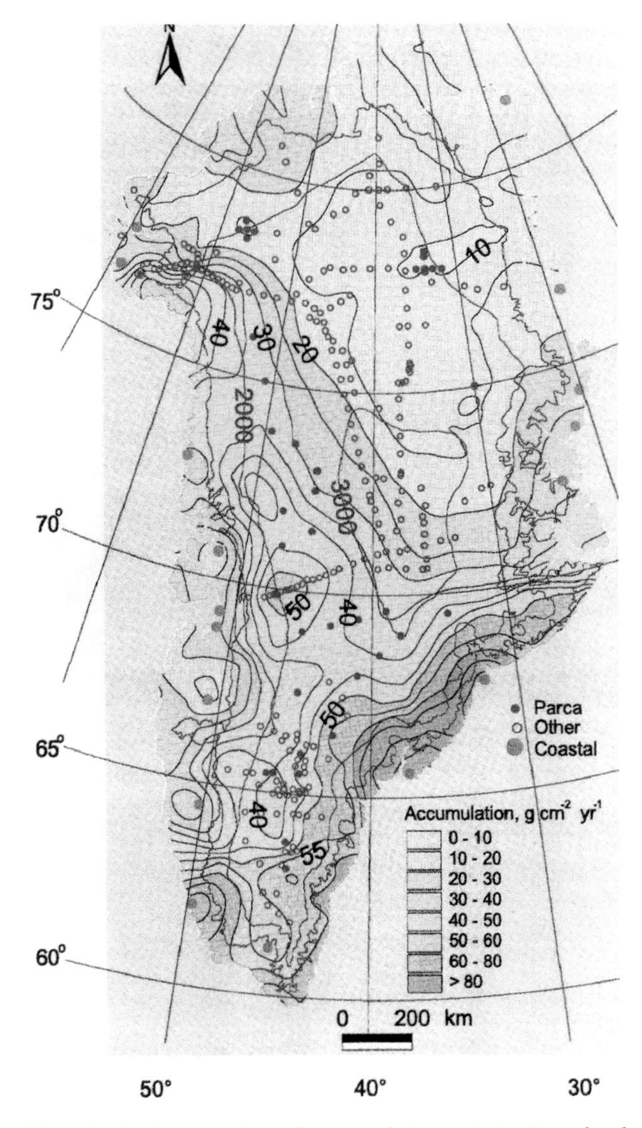

Plate 44.3 Distribution of accumulation rate in Greenland based on snow pit and firn core observations. (From Bales *et al.*, 2001.)

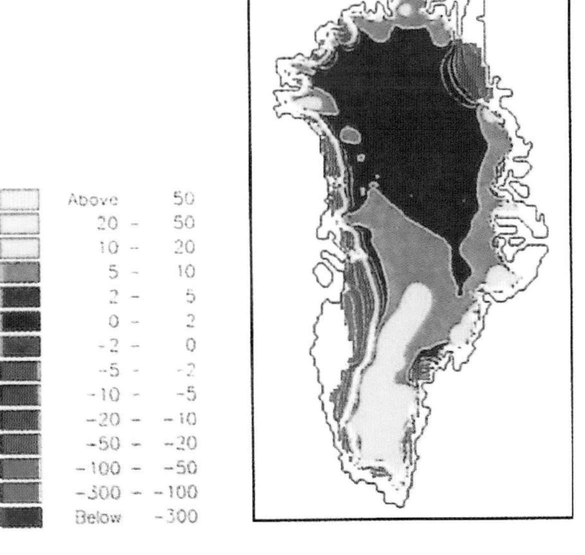

Plate 44.4 Present-day surface-elevation evolution in Greenland (mm yr^{-1}) averaged over the past 200 yr as derived by ice-sheet modelling. (From Huybrechts & LeMeur, 1999.)

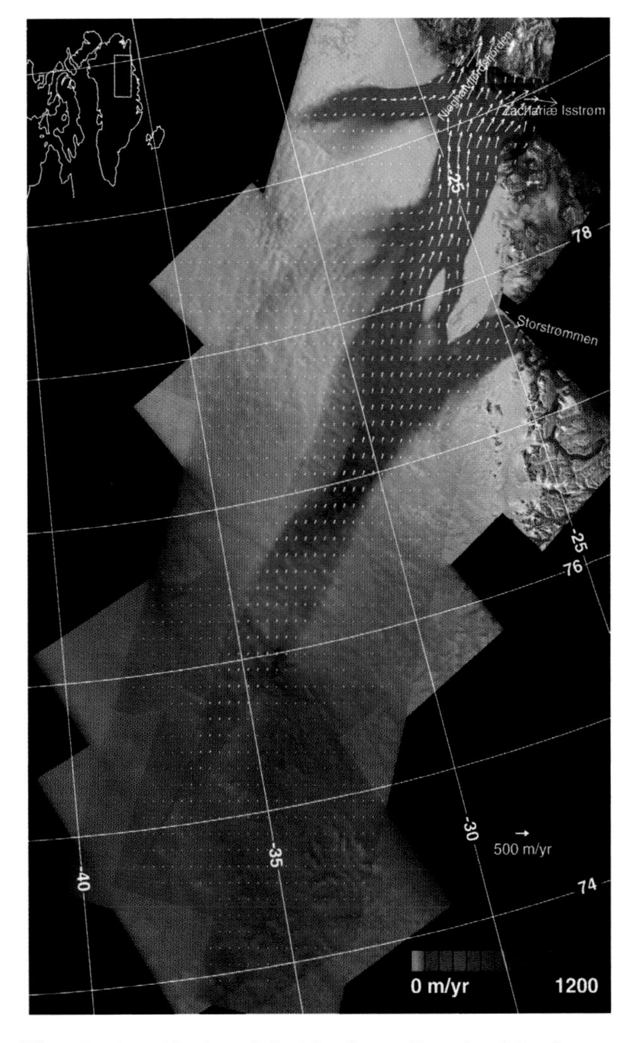

Plate 74.1 Velocity of the Northeast Greenland Ice Stream (colour, vectors) displayed over a SAR image mosaic.

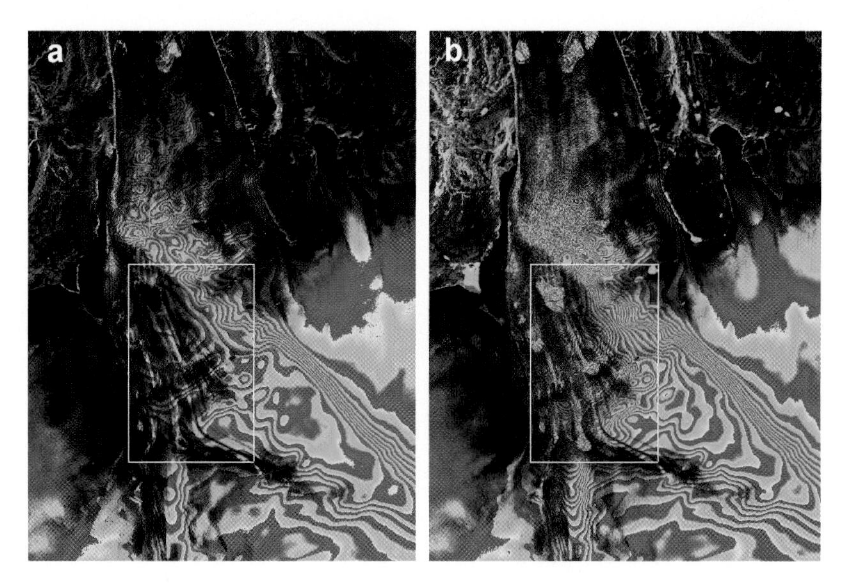

Plate 74.2 Interferograms from the fast moving area of the Ryder Glacier displayed as hue-saturation-value images with value (brightness) determined by the SAR amplitude, hue determined by the interferometric phase, and saturation held constant. Each fringe (yellow–red transition) represents 2.8 cm of displacement directed toward or away from the radar. (a) Interferogram for the interval 21–22 September 1995. (b) Interferogram for the interval 26–27 October 1995. The much denser fringes, particularly on the lower portions of the glacier (white box), indicate a dramatic change in velocity over the September observation.

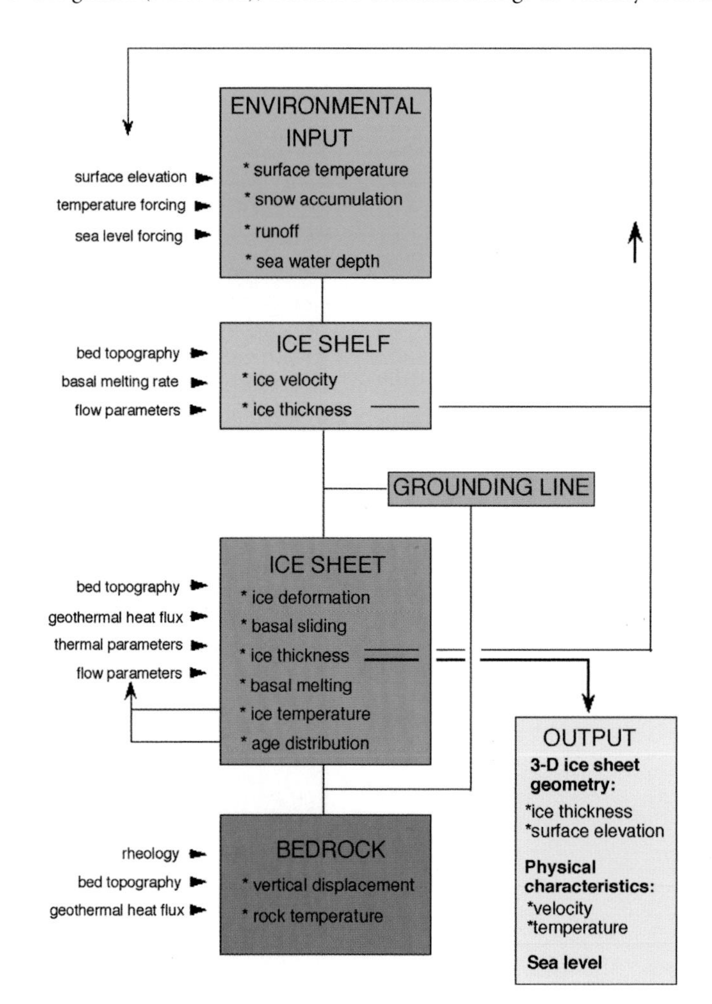

Plate 80.1 Structure of a comprehensive three-dimensional ice-sheet model applied to the Antarctic ice sheet. The inputs are given at the left-hand side. Prescribed environmental variables drive the model, which has ice shelves, grounded ice and bed adjustment as major components. The position of the grounding line is not prescribed, but internally generated. Ice thickness feeds back on surface elevation, an important parameter for the calculation of the mass balance. The model essentially outputs the time-dependent ice-sheet geometry and the coupled temperature and velocity fields. Three-dimensional models applied to the Northern Hemisphere ice sheets are similar, but do not include ice-shelf flow and explicit grounding-line dynamics. (After Huybrechts, 1992.)

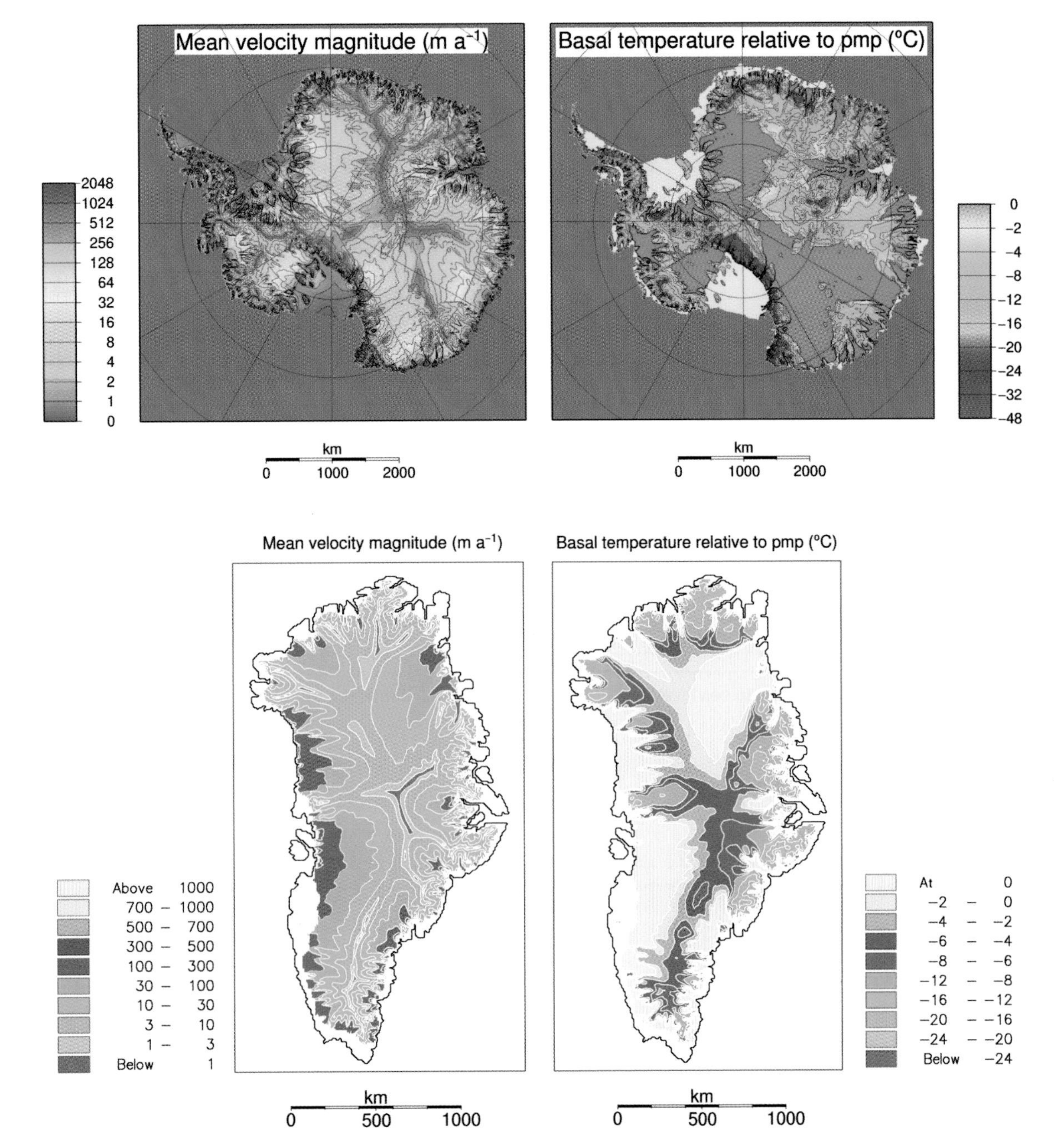

Plate 80.2 Present-day vertically averaged ice velocities and basal temperature field as simulated by three-dimensional models applied to the Antarctic (upper pictures) and Greenland (lower pictures) ice sheets. The orange, respectively yellow, colours in the pictures at the right are areas where the basal ice is at the pressure melting point and basal sliding occurs. These fields were obtained from model versions implemented at 10 km resolution spun up over several glacial cycles. Basal temperatures were obtained for a uniform geothermal heat flux of 50.4 W m^{-2} for Greenland and 54.6 W m^{-2} for Antarctica. Note that despite what the figures may suggest, the Antarctic ice sheet is about eight times larger than the Greenland ice sheet.

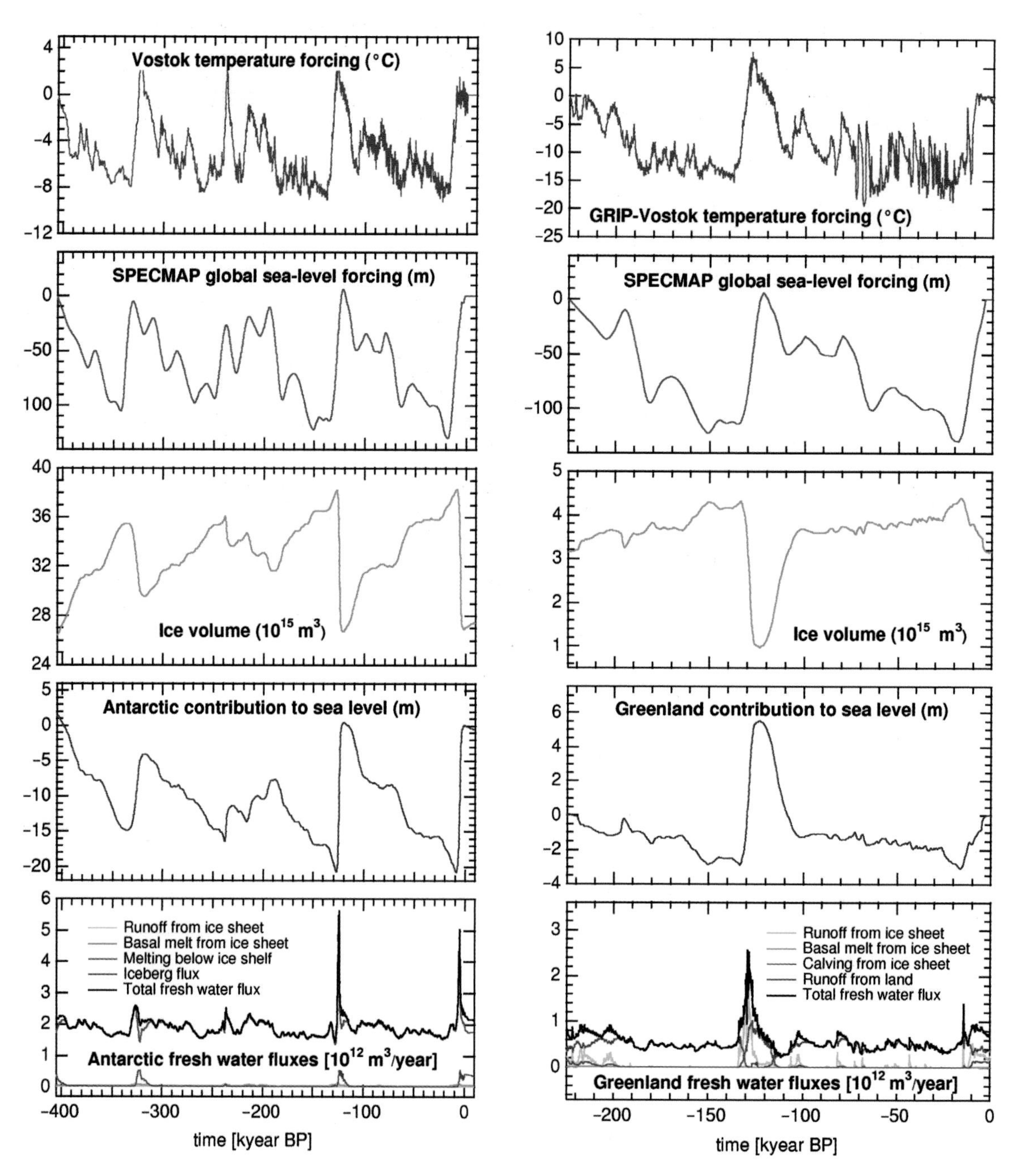

Plate 80.3 Forcing (mean annual air temperature and eustatic sea level) and predicted evolution of key glaciological variables (ice volume, contribution to sea level, freshwater fluxes into the ocean) in typical three-dimensional model experiments over the last few glacial cycles. (Based on the ice-sheet model experiments described in Huybrechts, 2002.)

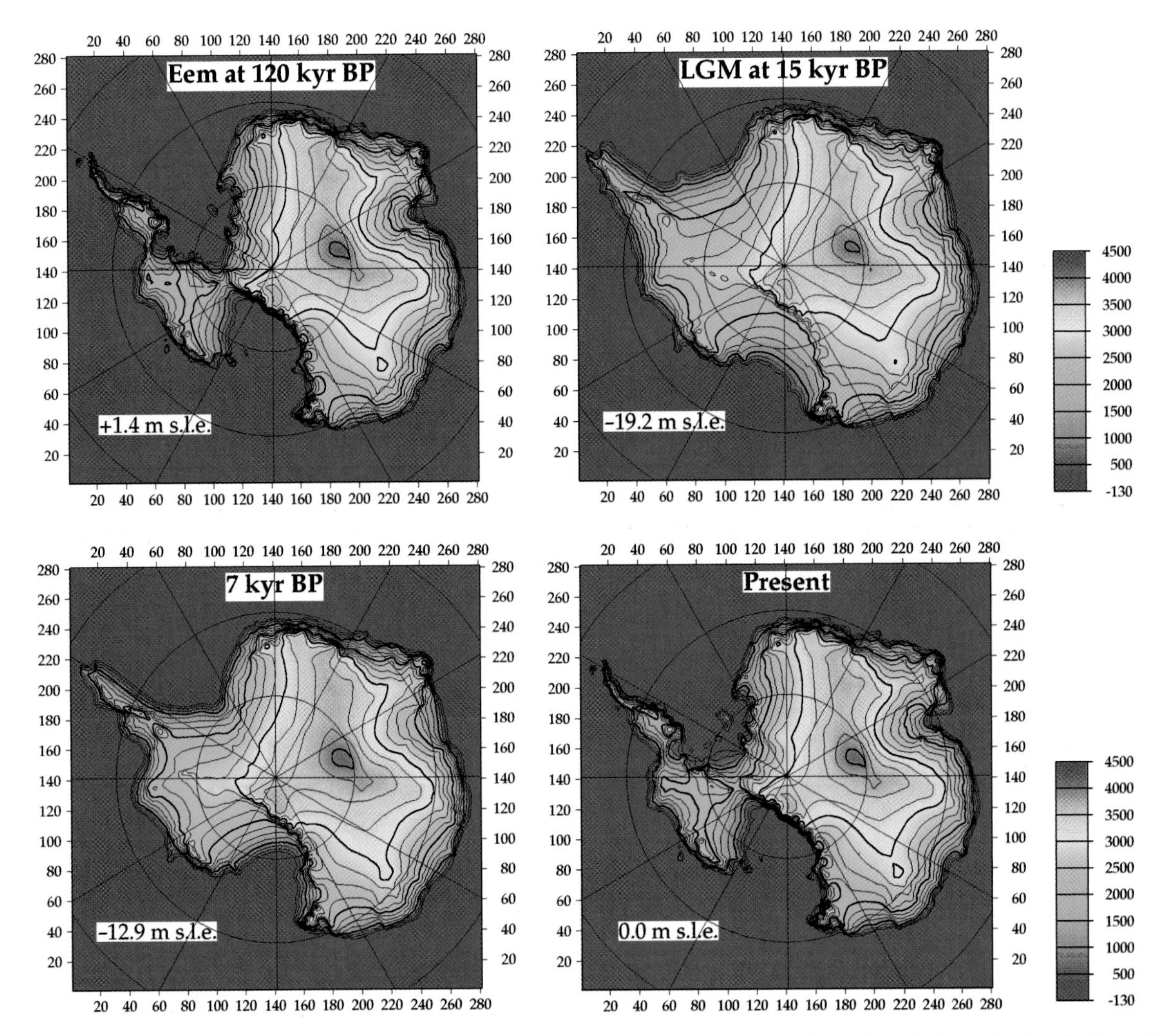

Plate 80.4 Modelled extent and surface topography of the Antarctic ice sheet at a few selected times during the last glacial cycle. In line with glacial–geological evidence, the most pronounced changes take place in the West Antarctic ice sheet. In East Antarctica, variations in ice-sheet geometry are comparably small. A main characteristic of the model is the late Holocene retreat of the grounding line in West Antarctica, still continuing today. Contour interval is 250 m; the lowest contour approximately coincides with the grounding line. (Modified after Huybrechts, 2002.)

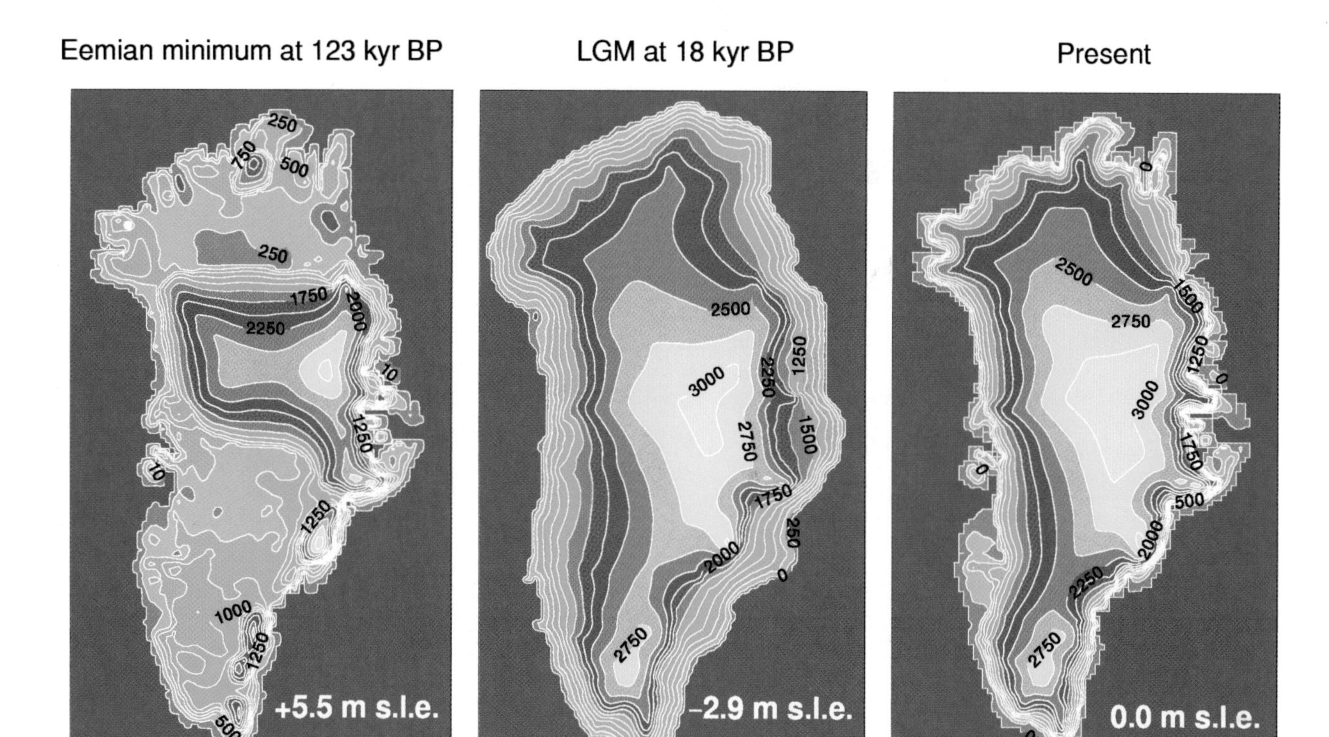

Plate 80.5 Snapshots of Greenland's ice-sheet evolution at three intervals during the last glacial cycle. According to the model, the ice sheet retreated to a small central dome during the Eemian warm period before expanding over most of the continental shelf at the Last Glacial Maximum. Implied global sea-level changes are between −3 m and +6 m. (Modified after Huybrechts, 2002.)

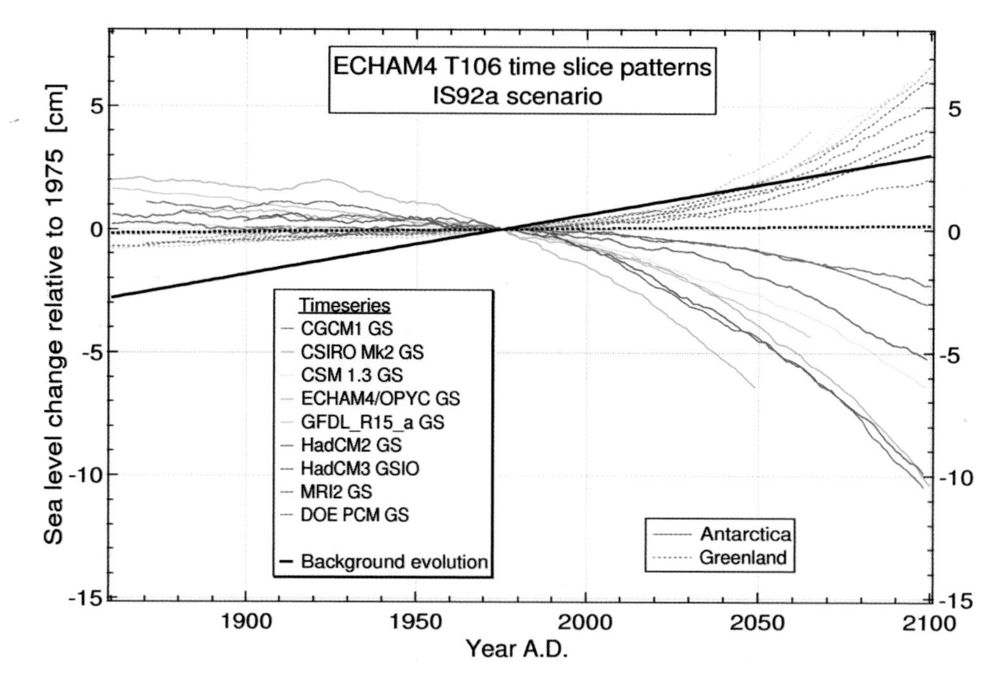

Plate 80.6 Volume changes of the Greenland and Antarctic ice sheets in greenhouse warming experiments expressed in equivalent global sea-level changes. The climatic forcing was derived from scaling time slices from a high-resolution AGCM (ECHAM4) with a suite of lower-resolution AOGCMs. On these short time-scales, the ice-sheet response is entirely dominated by the direct effect of mass-balance changes. The background trend resulting from past environmental changes is shown separately by the thick black lines. The stippled lines refer to the Greenland ice sheet; the full lines are for the Antarctic ice sheet. These experiments were at the base of the polar ice-sheet component to the global sea-level projections of the IPCC Third Assessment Report (Church *et al.*, 2001). (From Huybrechts *et al.*, 2004b.)

FIFTY-FIVE

Changing glaciers in High Asia

Yao Tandong*,† Pu Jianchen*,† and Liu Shiying†,*

**Institute of Tibetan Plateau Research, Chinese Academy of Sciences, Beijing 100029, China*
†Cold and Arid Regions Environmental and Engineering Research Institute, Chinese Academy of Sciences, Lanzhou Gansu 730000, China

55.1 Introduction

There are about 46,298 glaciers in High Asia, the total glacial area is about 59,406 km^2, and total glacial volume about 5590 km^3 (Table 55.1). These glaciers mainly concentrate around the Himalayas, Nyaiaqentanglha, Kunlun, Karakoram and Tienshan mountains. The glaciers in the Tibetan Plateau are the major component of the glaciers in High Asia. These glaciers extend north to the arid and desert regions, and become the main water resources in northwest China. Especially, the large glacier coverage around Tarim basin can supply about 137.7 × 10^8 m^3 glacial meltwater to the lower reaches of the Tarim basin each summer. These glaciers extend south to the warmer, wetter forests and mainly concentrate around the Brahmaputra drainage basin, and form the largest mountain-glacier centre in High Asia. There are about 10,813 glaciers in the Brahmaputra drainage basin, the total glacial area is about 14,491 km^2 and the glacial volume about 1293 km^3.

Most of the inland rivers in norhwest China are supplied with glacial melt water (Table 55.2). According to the studies of Yang & Hu (1992) and Yang (1995), the annual glacial melt water runoff in China was about 56.4 km^3 or 564 × 10^8 m^3, which is close to the total annual runoff of the Yellow River, and is 2% of the total runoff in China, 10% of the total runoff in northwest China and 13% of the total runoff (4431 × 10^8 m^3) of the four provinces in west China (Gansu, Qinghai, Xingjiang and Tibet). In fact, the glacial water resource is very important to the arid inland, mainly including Xingjiang, Qinghai and Gansu province in northwest China. According to the results of Yang (1995), the total glacial melt water runoff in northwest China was about 220.07 × 10^8 m^3.

Some rivers are dependent on glacial melt water in northwest China. The Tarim River is an example. There are 14,285 glaciers in the Tarim River basin, the total glacial area is 23,628.98 km^2, the glacial volume is 2669.435 km^3 and the average glacial depth is 113 m. According to a previous study (Yao *et al.*, 2004), the total annual runoff of six tributaries of the Tarim River is 310 × 10^8 m^3. The runoff of the Tarim River supplied with the glacial melt water reached to about 50% in the past 40 yr.

Glaciers in High Asia are most sensitive to climatic change and fluctuate with climatic cooling and warming. Temperature rise since the termination of the Little Ice Age has had a great impact on glacial distribution, causing a decrease of glacial area and glacial volume (Yao & Shi, 1990). This is particularly the case for the temperate glaciers in the monsoon regions in High Asia. In the 1980s, most of the glaciers had retreated extensively because of climatic warming. Even some previously advanced glaciers had also shifted into retreat phase with rapid climatic warming. In the 1990s, glaciers had retreated more extensively, and the runoff of some rivers had increased largely with the glaciers melting (Shi, 2001; Yao *et al.*, 2004).

From the 1950s, scientists have been studying the glacial fluctuation in High Asia. The study at the very beginning is pure field investigation and aerial photographs. Some long-term monitoring stations were established in the 1960s, 1970s and 1990s, which improved the continuous monitoring of glacial fluctuations. In the 1990s, remote sensing and GIS methods were used to study glacial fluctuation. These studies have provided the base for our discussion here.

55.2 Glacial retreat since the termination of the Little Ice Age

The Little Ice Age (LIA) was a typical cold period over High Asia (Yao *et al.*, 1997). The LIA terminated about AD 1890 (Yao *et al.*, 1996). Large-scale glacial retreat has started since then. The LIA in the High Asia can be divided into three cold stages, which appeared in the 15th century, the 17th century and the 19th century. The coldest stage appeared in the 15th century. The moraines corresponding to the LIA are different for different stages. The largest moraine is the one corresponding to the 15th century cold stage.

In High Asia, there are two types of glaciers: the temperate glaciers and the subpolar glaciers. These two types of glaciers are different in their response to climatic changes.

The Qilian Mountain is a region where subpolar glaciers developed. Some key areas were selected to study glacial fluctuations in the region. As shown in Table 55.3, glacial changes in different periods including the LIA, 1956, 1990 in the Big Snow Mountain in the Qilian Mountain are analysed and the magnitude of glacial fluctuations are calculated. In summary, the decrease of glacial area since the LIA to 1956 is about 4.7%, which is quite small compared with some other regions. However, the decrease from 1956 to 1990 is 4.8%, which is larger than the magnitude of decrease from the LIA to 1956 and indicates an accelerating decrease of glaciers.

Table 55.4 shows glacial area decrease from the LIA to 1956 in another key area in the Qilian Mountain. The magnitude of glacial fluctuation in the area is much larger compared with the Big Snow Mountain. As shown in Table 55.5, the magnitude of glacial fluctuation between the LIA and 1956 is also larger in the Shule River than in the Big Snow Mountain. Two facts are influencing the magnitude of the glacial fluctuations in the two regions: ice temperature and intensity of the monsoon. The Big Snow Mountain is in the western part of the Qilian Mountain. The ice temperature of glaciers is lower (close to polar glaciers) and less influenced by the monsoon. The ice temperature of glac-

Table 55.1 Glaciers in different mountains in High Asia

Mountains	Number of glaciers	Glacial area (km^2)	Glacial volume (km^3)
Altay	403	280	16
Sawuer	21	21	17
Tienshan	9081	9236	1012
Parmir	1289	2696	248
Karokoram	3454	6231	686
Kunlun	7694	12,266	1283
Alkin	235	275	16
Qilian	2815	1931	93
Qiangtang	958	1802	162
Tanggula	1530	2213	184
Gandis	3538	1766	81
Nyainqentanglha	7080	10,701	1002
Hdengduan	1725	1580	97
Himalayas	6475	8412	709
Total	46,298	59,406	5590

Table 55.2 Inland rivers supplied by glaciers in High Asia

Rivers	Number of glaciers	Glacial area (km^2)	Glacial volume (km^3)
Erqisi	403	289.29	16.40
Zhunger	3412	2254.10	137.44
Inland rivers	2385	2048.16	143.71
Tarim	11,711	19,888.81	2313.30
Tu-Ha	446	252.73	12.63
Hexi	2194	1334.75	61.55
Quadam	1581	1865.05	128.53
Yellow	108	40.97	1.25
Total	22,240	27,973.86	2814.81

Table 55.3 Glacial changes during different periods in the Big Snow Mountain of the Qilian Mountains

Basin	Area in Little Ice Age (LIA) (km^2)	Area in 1956 (km^2)	Area in 1990 (km^2)	Changes in LIA–1956		Changes in 1956–1990	
				Area (km^2)	%	Area (km^2)	%
Chaganbulgas	34.61	32.02	28.98	−2.59	−8.1	−3.04	−9.5
Laohugou	98.76	94.80	91.65	−4.43	−4.7	−3.14	−3.3
Tashihe	5.86	5.55	5.26	−0.31	−5.6	−0.32	−5.7
Yemahe	30.66	30.44	29.12	−0.26	−0.9	−1.32	−4.3
Total/average	169.90	162.80	155.00	−7.59	−4.7	−7.82	−4.8

Table 55.4 Glacier-area fluctuations in the Shule River basin in the Qilian Mountains

Basin	Glacier area in 1956 (km^2)	Glacier area in LIA (km^2)	Changes in LIA–1956 (km^2)	(%)
Shule River	440.1	509.0	63.6	14.5
Danghe River	59.3	66.8	9.0	15.2
Beidahe River	66.9	66.9	14.0	20.9
Hala Lake	72.4	97.4	21.0	29.0
Total/average	638.6	740.1	107.6	16.9

iers in the Shule Rivers is higher (close to subpolar glaciers) and is influenced by the monsoon.

The magnitudes of glacial fluctuation of temperate glaciers in the intensive monsoon region are much greater. The temperate glaciers are in the Hengduan Mountains, the east section and on the south slope of the Himalayas, and the east section of the Nyainqentanglha Mountains in the southeast part of the Tibetan Plateau. Su & Shi (2002) studied the glacial fluctuation since the LIA. After investigation of 1139 glaciers from the Yulong, Gongga and Hengduan mountains, it was found that the total glacial area of the measured temperate glaciers has reduced by 30%. In the Yulong Mountains, the glacial area has reduced as much as 60% (He *et al.*, 2003). The study indicates that there is an inverse relationship between glacier size and glacier area reduction. That is, the larger the glacier, the smaller the glacier area reduction, and *vice versa*.

Snowline in the temperate glacial region is a very sensitive indicator of glaciers to climatic changes. Through field investigation and careful comparison of lateral moraines in different periods, we obtained an estimate of snowline fluctuations in temperate glacier regions in High Asia. For example, the snowline of glaciers rose by 150–180 m on the east slope of the Yulong Mountains since the LIA, rose by 100–150 m in the Zayu River basin, and rose by 60–80 m in the Queershan Mountains.

55.3 Glacial retreat in recent years

In the 20th century, the glaciers in High Asia have retreated extensively as a result of climatic warming. The glacial changes can be divided into several stages, as follows:

1 in the first half of the 20th century glaciers had advanced or shifted from advance to retreat;
2 between the 1950s and 1960s many glacial observations were started, and the glaciers in High Asia had begun to retreat extensively (Table 55.6)—according to previous studies (Zhang *et al.*, 1981; Ren, 1988; Shi *et al.*, 2002) about two-thirds of glaciers were retreating and 10% advancing, with the remainder being stable;
3 between the late 1960s and 1970s the glacial mass balance was positive, the snowline dropped and many glaciers advanced, with the proportion of advancing glaciers increasing and that of retreating glaciers decreasing;
4 in the 1980s glaciers again retreated extensively;
5 in the 1990s glacial retreat was more extensive than at any other period in the 20th century.

The glaciological expedition to the Tibetan Plateau in 1989 showed that the glaciers of the southeast Tibetan Plateau retreated extensively, and the Zepu Glacier and Kaqing Glacier were the most obvious examples of glacial retreat (Yao *et al.*, 1991). Some glaciers, however, are still advancing. Detailed research of the Large Dongkemadi Glacier and the Small Dongkemadi Glacier in the Tanggula Mountains and the Meikuang Glacier in the Kunlun Mountains showed that these glaciers were still advancing. However, all these glaciers have shifted from advance to retreat during the 1990s. Now, with very few exceptions of glaciers still advancing, the glaciers in High Asia are retreating.

The glacial retreat since the 1990s has several main features as follows.

1 The magnitude of glacial retreat is increasing. Glacier No.1 in the Urumchi River basin in the Tianshan Mountains is an example. Glacier No.1 comprises two branches (east and west). The ice tongues of the east and west branches were joined together in 1962, but the junction was thinning continuously with the retreat of the two branches. In 1993, the two branches were totally separated from each other and the distance between the two branches reached to more than 100 m in 2001. Figure 55.1 shows the retreat of the Glacier No.1 between the 1960s and 2000. Glacier No.1 retreated most extensively from the early 1960s to the early 1970s when the retreat rate reached to 6 m yr^{-1}. The magnitude of glacial retreat decreased obviously during the mid-1970s and reached a minimum in the early 1980s, but increased again during the

Table 55.5 Glacier fluctuations in the Qilian Mountain in 1956–1990

Basin	Area	Area change	%
Baidahe River	290.76	41.64	14.3
Shule River	589.64	49.56	8.4
Danghe River	259.74	25.04	9.6
Hala Lake	89.27	7.89	8.9
Total/average	1229.41	124.21	10.3

Table 55.6 Proportions of advancing and retreating glaciers in High Asia at different periods

Time	Glaciers	Retreating glaciers (%)	Advancing glaciers (%)	Stable glaciers (%)	Reference
1950–1970	116	53.44	30.17	16.37	He *et al.*, 2003 Jing *et al.*, 2002
1970–1980	224	44.2	26.3	29.5	He *et al.*, 2003 Jing *et al.*, 2002
1980–1990	612	90	10	0	Liu *et al.*, 2003 This paper
1990 to present	612	95	5	0	This paper

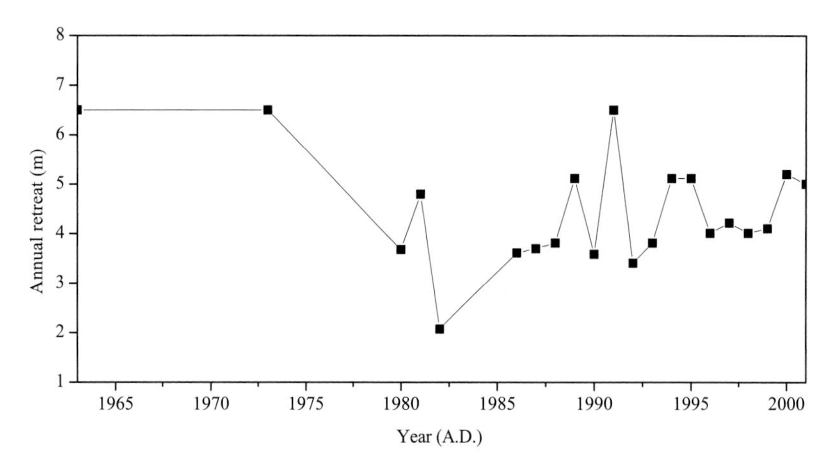

Figure 55.1 Fluctuation of Glacier No.1 in the Urumqi River Basin.

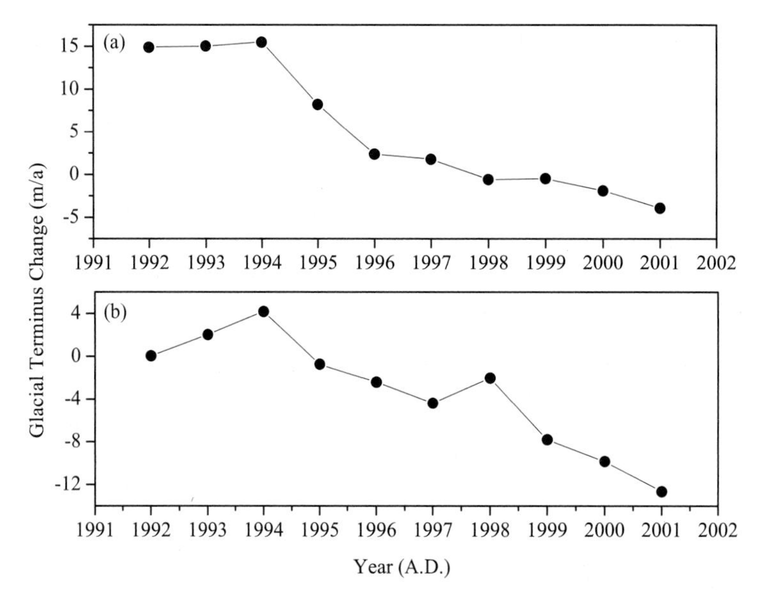

Figure 55.2 Fluctuations of the Larger Dongkemadi Glacier (a) and the Small Dongkemadi Glacier (b) in the Tanggula Mountains.

late 1980s to the 1990s, with the retreat rate reaching a maximum of 6.5 m yr^{-1} between 1990 and 1991.

2 Most of the advancing glaciers gradually shifted to retreat. The Large and Small Dongkemadi glaciers in the Tanggula Mountains are examples. Figure 55.2a & b show the process of the Large and Small Dongkemadi glaciers shifting from advance to retreat. These two glaciers were both advancing when they were first observed in 1991. The total area of the Large Dongkemadi Glacier is about 14.63 km^2, and the Small Dongkemadi Glacier about 1.77 km^2. Based on glaciological theory, there is a lag-time as a glacier responds to climatic change. The lag-time is dependent on glacial size: the larger the glacier, the longer the lag-time. Therefore, the time when the Small Dongkemadi Glacier began to shift from advance to retreat would have been earlier than the Large Dongkemadi Glacier. As shown in Fig. 55.2b, the Small Dongkemadi Glacier advanced about 4 m during the summer in 1992, and then shifted to retreat in 1993, with a retreat rate of only 0.2 m in that year. After 1993, the Small Dongkemadi Glacier kept retreating, and the retreat rate increased year by year and reached to 2.86 m yr^{-1} in 2000. The Large Dongkemadi Glacier had advanced about 15.7 m between 1989 and early 1994, and then shifted to retreat after the summer in 1994. The annual retreat of the Large Dongkemadi Glacier also increased continuously and the retreat rate reached to about 4.56 m yr^{-1} in 2001.

3 At the highest peak of Mount Qomolangma there is also a record of the impact of global warming on the glacial process.

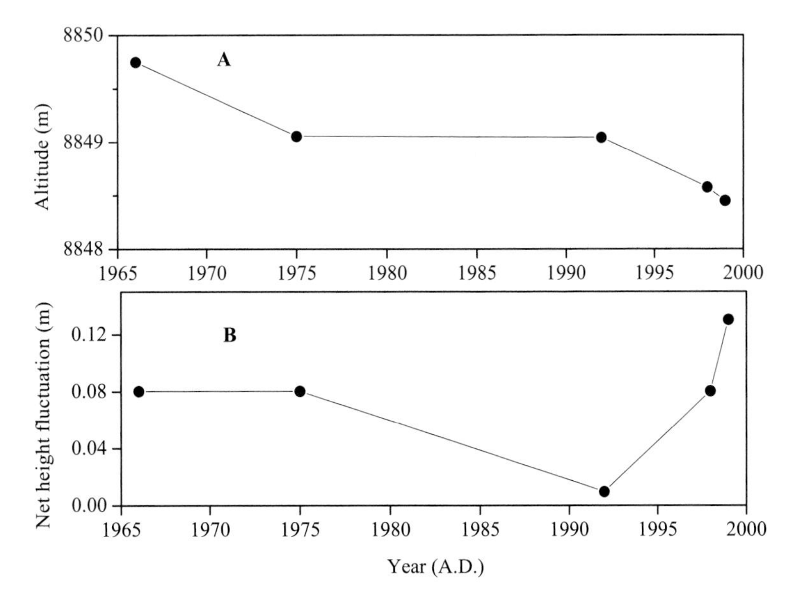

Figure 55.3 Height fluctuations at the top of Mount Qomolangma, where A is the lowering of surface height and B is the lowering rate in different periods.

According to Ren *et al.* (1998), climatic warming caused glacial retreat in Mount Qomolongma. Furthermore, evidence of the impact of climatic warming on glaciers was also found at the top of Mount Qomolangma. According to the observation of Chen *et al.* (2001), the surface height of the snow and ice at the peak of Mount Qomolangma has lowered since 1966. The observation data show a variable lowering rate over the past several decades. As shown in Fig. 55.3A, the surface height at the top of Mount Qomolangma lowered by a total of about 1.3 m (from 8849.75 m to 8848.45 m) between 1966 and 1999. The annual fluctuations of the surface height of snow and ice (in Fig. 55.3B) are as follows: the rate of lowering was very fast between 1966 and 1975, with the annual lowering rate reaching about 0.1 m; the rate of lowering slowed between 1975 and 1992 to only about 0.01 m yr^{-1}, one-tenth of that between 1966 and 1975; it then increased again between 1992 and 1998 to a rate of 0.1 m yr^{-1}, with the annual lowering rate reaching a maximum of 0.13 m yr^{-1} between 1998 and 1999. Such a large lowering of surface height in so short a time confirms that the reduction in height cannot be caused by lithospheric movement, and can be explained only by glacial response to climatic change. Strictly speaking, glacial retreat cannot cause the lowering of glacier surface height at 8848 m a.s.l, but the process of glacial ice formation can induce lowering of the glacier surface height. The depth of snow and ice at the top of Mount Qomolangma is still unclear. A maximum depth of 2.5 m was observed by the Italy Mountaineering Team using a stick, but the true depth of snow and ice cannot be obtained with this method. However, the snow and ice depth at the top of Mount Qomolangma should be deeper than 2.5 m. Prior to global warming, the snow–ice formation process at this altitude is one of very slow densification under gravity. It is similar to that of the Antarctic and Arctic regions. Following global warming, the snow–ice formation process will accelerate as a result of temperature increase, which will cause rapid lowering of the glacier surface height. In fact, the rate of lowering of surface height since 1992 at the top of Mount Qomolangma corresponds to a period of rapid climate warming.

4 The pattern of glacial retreat is different in different regions according to many studies (Su *et al.*, 1996, 1999; Wang & Liu, 2001; Pu *et al.*, 2001; Lu *et al.*, 2002; Jing *et al.*, 2002; Chen *et al.*, 1996; Liu *et al.*, 2000, 2002). Figure 55.4A indicates the observations of the actual retreat of typical glaciers in different regions. These observations show that glacial retreat was extensive in the Karakorum Mountains and southeast Tibetan Plateau, with the annual retreat of Poshu Glacier in the Karakorum Mountains reaching about 50 m. Glacial retreat in the innner Tibetan Plateau was less, at no more than 10 m yr^{-1}. For example, the annual retreat rates of the Puruogangri Glacier and the Malan Ice Cap in the Tibetan Plateau were within 10 m yr^{-1}.

At the plateau scale, the magnitude of glacial retreat is smaller inland and larger at the margin. Figure 55.4B indicates that the magnitude of glacial retreat was large in the southeast Tibetan Plateau and Karakorum Mountains (e.g. the annual glacial retreat in the Karakorum Mountains reached 30 m and 40 m in the southeast Tibetan Plateau), whereas the annual glacial retreat in the Kunlun Mountains and Tanggula Mountains (located in the central Tibetan Plateau) was smaller at no more than 10 m. This kind of regional difference of glacial retreat forms an elliptical shape of the glacial retreat in recent years on the Tibetan Plateau, and the distribution feature was similar to that of the glacial

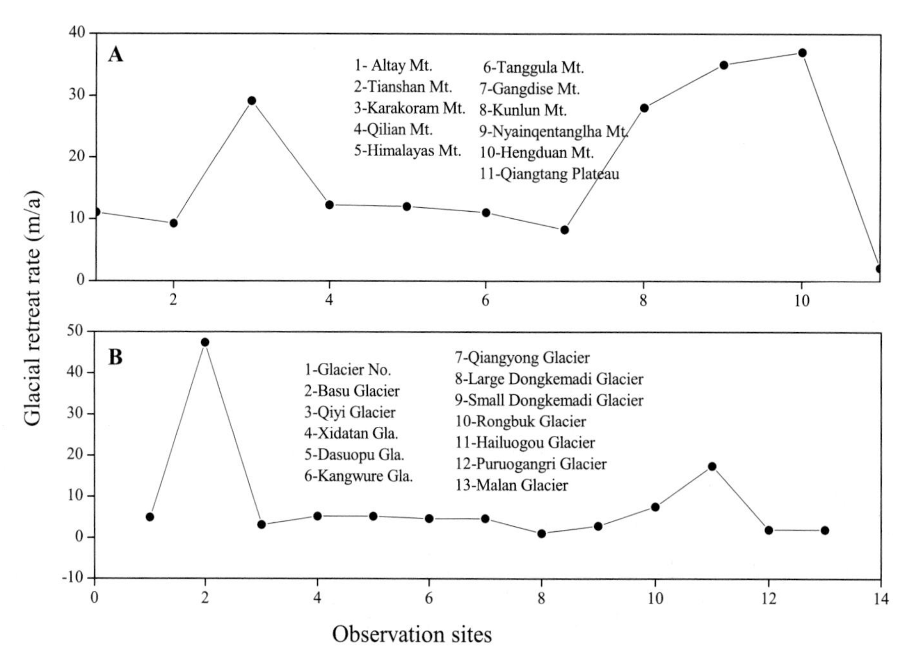

Figure 55.4 Regional features of glacial retreat in High Asia, where (A) shows the annual retreat rate of all the glaciers in a region and (B) shows the annual retreat rate of glacial length observed in different regions.

shrinkages from the maximum of the Little Ice Age to present (Fig. 55.5). The central part of the elliptical regional distribution is located in the Tanggula Mountains, Kunlun Mountains and Qiangtang Plateau in the inner Tibetan Plateau, which record minimum glacial retreat. The glacial retreat increases from inland areas to the margin of the Tibetan Plateau, and reaches a maximum in the southeast Tibetan Plateau and Karakorum Mountains.

55.4 Negative glacial mass balance causes glacial retreat in High Asia

Mass balance is the algebraic sum of glacial mass increase (precipitation on the glacier) and glacial mass loss (glacier melting) in the glacier system. A positive value means a positive glacial mass balance and *vice versa*. The general glacial retreat pattern in High Asia is closely related to the strong negative glacial mass balance of recent years.

Continuous observation sites of glacial mass balance in High Asia include Glacier No.1 in the Urumqi River basin (1956–2001), Small Dongkemali Glacier in the Tanggula Mountains (1990–2001) and Meikuang Glacier (1990–2001) in the Kunlun Mountains. Figure 55.6 shows mass-balance fluctuations of these glaciers in recent years. Obviously, the mass balance of these glaciers cannot reflect the pattern of the whole of High Asia, but studying the characteristics of these fluctuations can help us understand the retreat trends for the whole of High Asia.

There are several different features in the mass balance of these glaciers. Glacier retreat is most extensive in the Tianshan Mountains where the mass balance was strongly negative all the time. The advancing glaciers in the central and north Tibetan Plateau shifted to a retreat phase recently, and their mass balances changed from positive to negative. From Fig. 55.6, the mass balance of Glacier No.1 in the Tianshan Mountains was not only strongly negative all the time, but also its absolute value was the largest among the three glaciers. The mass balance of the Meikuang Glacier was very similar to that of the Small Dongkemadi Glacier, being mostly positive before the 1990s, which coincides with advances of the two glaciers before the 1990s. The mass balances of the Small Dongkemadi Glacier in the Tanggula Mountains and Meikuang Glacier started becoming negative in the mid-1990s. It was a strong signal of general retreat of glaciers in High Asia.

As mentioned above, negative mass balance is the direct cause of glacial retreat in High Asia. Precipitation in most parts of High Asia is increasing, providing the potential for a shift to a positive mass balance. However, many studies have shown obvious temperature rises in High Asia. Therefore, the key cause of general glacial retreat in High Asia is temperature rise due to global warming.

55.5 Conclusions

Glaciers in High Asia are retreating extensively as a result of global warming. The glacial retreat can be divided into several stages during the 20th century. Glaciers advanced, or shifted from an advancing to a stable state, during the early half of the 20th century. From the 1950s to the late 1960s the glaciers in High Asia retreated on a large scale, the retreat slowed down in the 1970s, and became extensive again in the 1980s. The glacial retreat in the

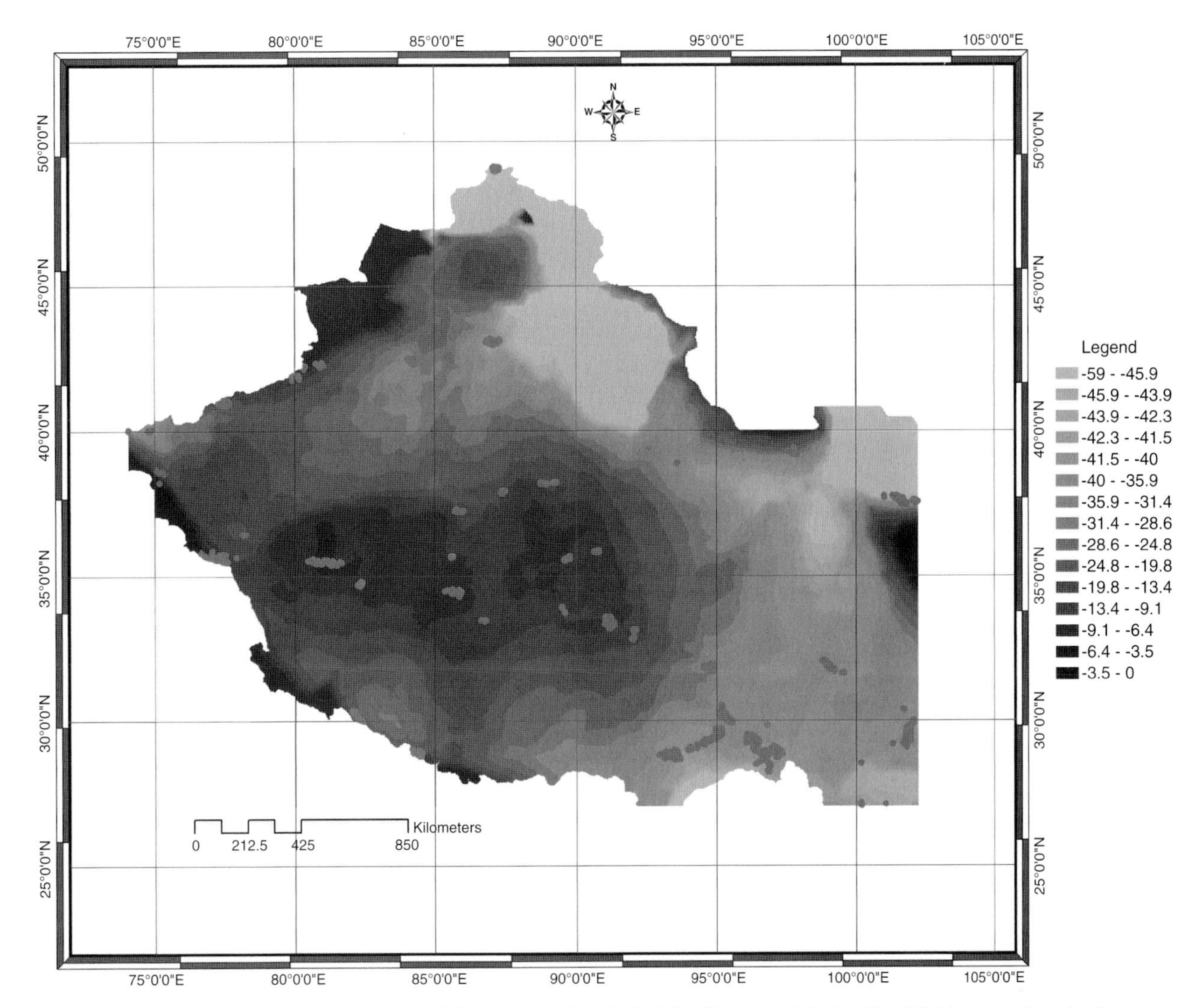

Figure 55.5 Regional features of the glacial fluctuations in High Asia. (See www.blackwellpublishing.com/knight for colour version.)

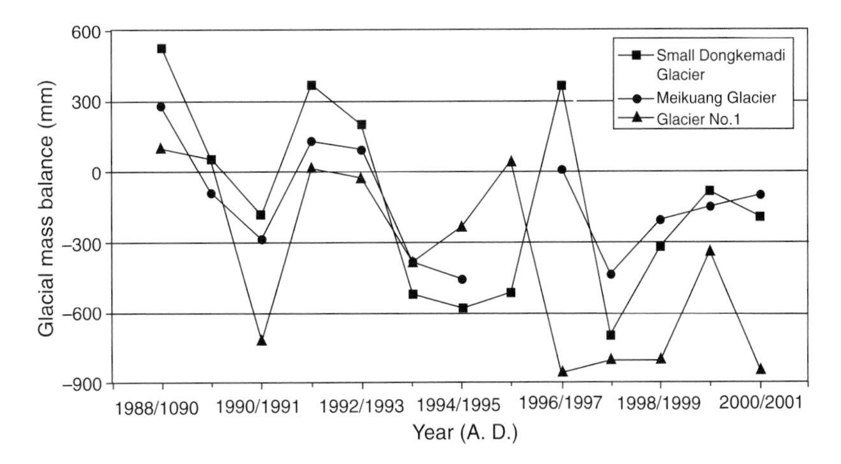

Figure 55.6 Glacial mass balances of some glaciers in High Asia.

1990s is the most extensive. During this period, most glaciers that were advancing are now retreating. The glacial retreat was most extensive on the southeast Tibetan Plateau and in the Karakorum Mountains and least extensive in the central Tibetan Plateau.

Glacial retreat in High Asia is due to negative glacial mass balance as a result of higher temperatures caused by global warming. The long-term data of several glaciers show that the positive mass balance recorded between the end of the 1960s and the late 1970s caused the glacial snowline to fall. At that time, the ratio of advancing glaciers increased and that of retreating glaciers decreased. In the 1980s, glacial mass balance became more negative. In the 1990s, the glaciers with negative mass balance became the most negative and a few glaciers with positive mass balance previously also shifted to a negative balance. The glacial retreat in the 1990s was the most extensive compared with any other period during the 20th century. It is concluded that the cause of the most recent glacial fluctuation in High Asia is increased temperatures due to global warming.

Acknowledgements

This study has been supported by the Innovation Group Fund of the National Natural Science Foundation of China (Grant No. 40121101), the Project KZCX2-SW-339 and the Project KZCX2-SW-118 of the Chinese Academy of Sciences.

PART 4

Glacier composition, mechanics and dynamics

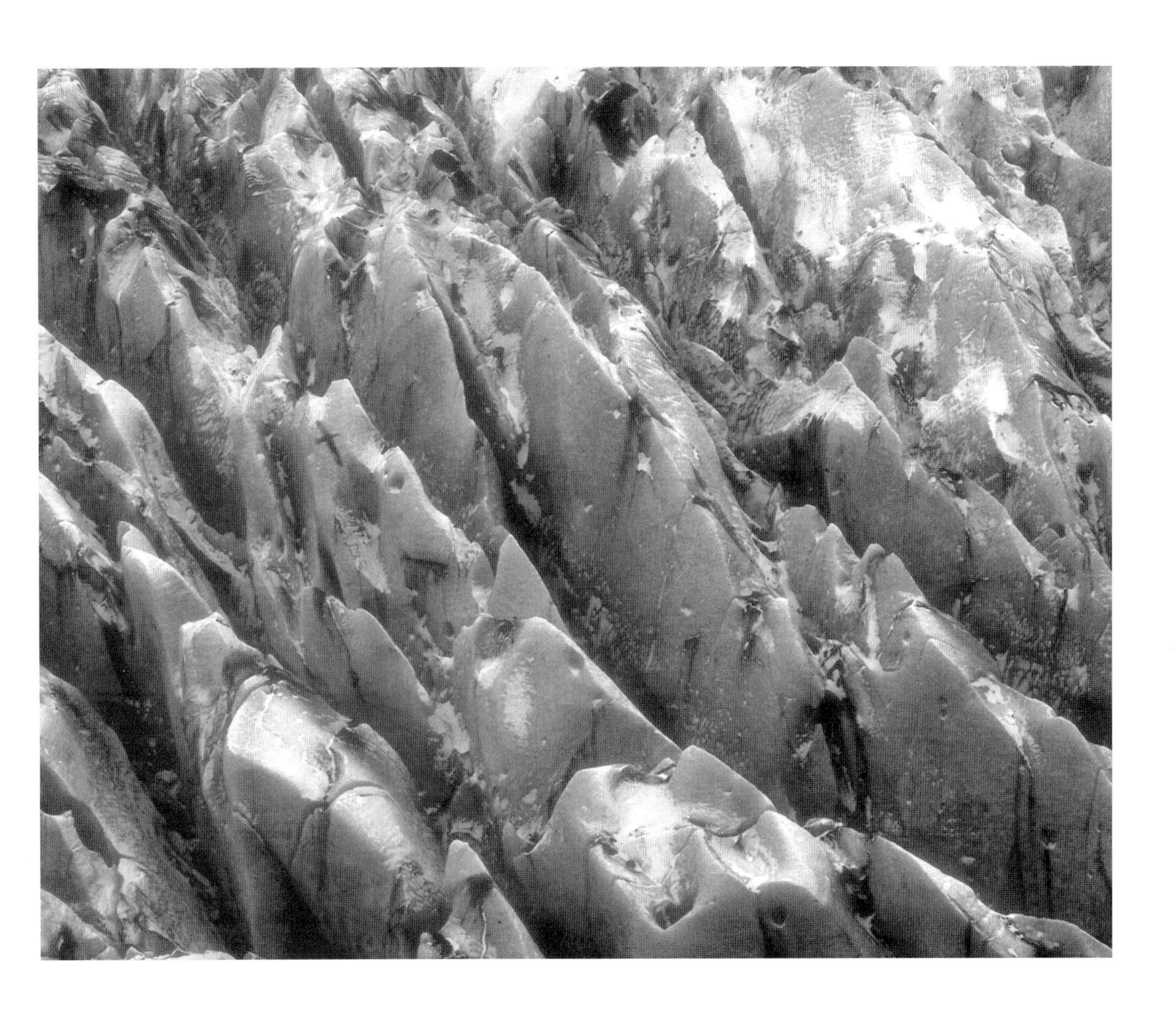

FIFTY-SIX

Glacier composition, mechanics and dynamics

T. H. Jacka

Department of the Environment and Heritage, Australian Antarctic Division, Channel Highway, Kingston, Tasmania, 7150, Australia

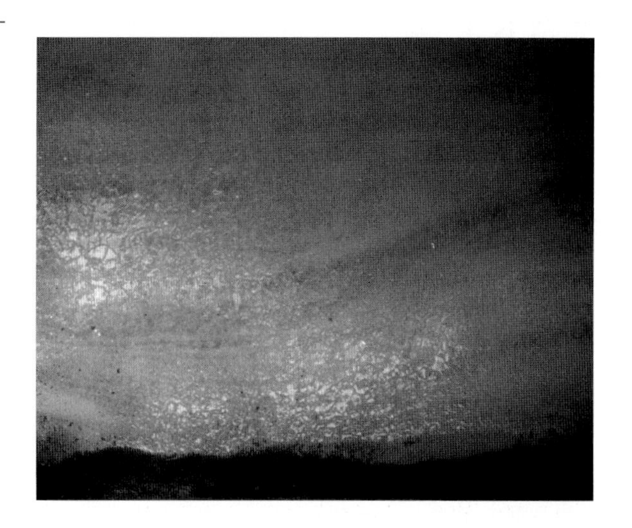

56.1 Introduction

The interrelations between the Earth's glaciers and global change (primarily climate change) are largely due to changes in the size and shape of the glaciers. The largest of these glaciers, by far, are the polar ice sheets of Antarctica and Greenland. Be that as it may, the temperate glaciers are important on time-scales <100 yr because, with the climatic warming rates currently evident across the planet, many of these glaciers are dramatically reducing in size and their contents are adding to sea-level rise. Of course the mass of water released by these glaciers is tiny compared with that trapped in the polar ice sheets. However, because the average temperature of the polar ice sheets is about −30°C, there is no realistic possibility, even with a large increase in atmospheric greenhouse gas concentrations, that they will completely melt within the next millennium (DeConto & Pollard, 2003) (at least without major changes in the orbital and solar parameters controlling Earth's climate). In fact a warmer climate is, on average, a wetter climate, and at temperatures typical of the polar regions, the extra precipitation falls as snow. Thus it is likely, for the Antarctic ice sheet at least, that it will increase in mass with global atmospheric and oceanic warming. Here lies one of our most challenging problems (The ISMASS Committee, 2004). We do not yet know for certain, even whether the polar ice sheets are actually increasing in mass (i.e. have a positive mass balance) or decreasing (i.e. a negative mass balance), but we have seen changes over the past decade that have been much more dramatic than previously anticipated; e.g. rapid glacier retreat and thinning (Rignot, 1998, 2001; Rignot & Thomas, 2002; Shepherd *et al.*, 2002; Zwally *et al.*, 2002a), ice-shelf collapse (Skvarca, 1993, 1994; Rott *et al.*, 1996) and unexpected changes in ice-stream velocities (Stephenson & Bindschadler, 1988; Joughin *et al.*, 2002; Rignot *et al.*, 2002). Satellite- and ground-based remote sensing and over-snow or aerial survey measurements are now beginning to shed more light on the mass balance state of the polar ice sheets, and results indicate some areas of rapid mass loss, whereas other areas indicate mass gain. Rignot & Thomas (2002) estimated a net mass gain for the East Antarctic ice sheet of 22 $km^3 yr^{-1}$, yet a net mass loss of 48 $km^3 yr^{-1}$ for the West Antarctic ice sheet.

Other research that is important for estimating changes in ice mass includes computer modelling. We can never cover the entire Antarctic or Greenland ice sheets, or every glacier with ground-based measurements, but with ice-sheet computer models we can calculate physical characteristics over entire ice sheets. Computer models also can be coupled with ocean models and atmospheric models, leading to a much improved understanding of the feedbacks and relationships between the ice, ocean and atmosphere. Once they are proven (by testing against known observations), ice-sheet computer models can be used to tell us more about the ice sheets and glaciers in the past—useful for understanding past changes—and to forecast what the ice sheets will look like in the future (along with the effects on ocean and atmosphere). To develop accurate and reliable computer models, however, and to properly interpret results from both the computer modelling and remote sensing results, it is essential that we understand the processes involved, i.e. the physics of how ice flows. In turn, as you will read in the following chapters, the physics depends on the composition of the glaciers.

Of course, all glaciers and ice sheets are composed of ice. But it is not that simple. Above the ice is a layer of recently precipitated snow (typically with a surface density, $\rho \approx 0.3\,g\,cm^{-3}$). The snow increases in density as a result of compaction under its own weight (and is usually termed 'firn' at densities greater than ≈ $0.3\,g\,cm^{-3}$), to eventually form ice. In some cases this process is not complete until a depth of >100 m. During compaction, atmospheric air is trapped in the firn and once the firn has reached a density of ca. $0.83\,g\,cm^{-3}$ the air pockets have been isolated from the atmosphere. We call this process 'close-off' and it is a very important concept for understanding ice-core measurements of past gas composition of the atmosphere (see chapters of this book on ice-core and climate studies). Once close-off has occurred, the air pockets are bubbles. Glacier ice is riddled with bubbles under high hydrostatic pressure and they constitute impurities within the ice. The ice flow parameters are affected in proportion to the decrease in density due to the bubbles, because the air has no resistance to deformation and, at densities higher than the close-off density, these bubbles seem to be (and no laboratory experiment has contradicted this) of little consequence in terms of the physics of the ice flow. Although deformation rates for snow are higher than for ice, once the close-off density has been attained, laboratory measurements indicate flow rates similar to those for solid ice (Mellor & Smith, 1966; Jacka, 1994). Below the close-off depth in the ice sheets and glaciers, the ice density continues to increase until it reaches a final density of ca. $0.91\,g\,cm^{-3}$.

For the polar ice sheets, the ice in the upper layers is very clean, although that has not always been the case and deeper layers of the ice sheets (especially the ice sheets of the Northern Hemisphere; e.g. Agassiz and Greenland) show layers of dust from past times. For temperate glaciers, the ice is not so clean. It often contains rocks, from boulder size to pebbles, and dust emanating from surrounding hills and mountains. All glaciers, polar and temperate, also contain atmospheric borne impurities, including fine insoluble impurities (dust, fine enough to be transported in the high atmosphere) and soluble impurities dissolved in the snow that makes up the glacier. Volcanic eruptions, for example, are one source of the soluble and insoluble impurities. Another is dust transported from the non-ice-covered deserts of surrounding land masses. Again at times past, e.g. during the last glacial period when there were greater areas of desert on the planet, there was (on average) a higher concentration of these atmospheric borne impurities blown by the wind over the Greenland and (to a lesser degree) Antarctic ice sheets, and so these impurities are in higher concentrations in deeper layers of the glaciers and ice sheets. Some notable peaks, however, do appear in the not-so-deep layers (e.g. from recent volcanic eruptions). The effect, on the flow of the ice, of these soluble and insoluble impurities needs to be known.

In the basal layers, the ice picks up dirt and rocks as it moves over the bedrock. Complicated flow therefore results in the basal layers that consist of a mix of ice with different impurity types and levels. In addition, the flow is complicated by folding, faults and shear of the ice. The flow of ice is strongly dependent upon temperature. Temperate glaciers are warm (at the melting point) throughout. The temperature just 10 to 20 m below the ice surface in the polar ice sheets is close to the mean annual air temperature. Geothermal heat from the Earth's interior heats the base of the ice, resulting in a temperature profile that is not far from isothermal in the top layer, and increasing approximately linearly in deeper layers, to the basal temperature. The depth of the isothermal layer depends on the snow accumulation rate, and in central Antarctica (e.g Vostok, Dome Fuji, Dome C) where the accumulation rate is extremely low, the isothermal layer is almost non-existent. The basal temperature in many parts of the polar ice sheets is at the pressure melting point, which results from the high hydrostatic pressure, at about −1 to −2°C. So, this dirty basal ice is also relatively warm. There is some uncertainty about the effect on ice flow of the impurities; there is very little question concerning the temperature effects. Certainly, the flow of the basal ice in polar ice sheets has a strong effect on the flow of the ice sheets as a whole.

As mentioned above, the basal ice is warm. Thus it also contains another impurity—water! Water exists on grain boundaries and dislocations within the ice (Wettlaufer, 1999), possibly at temperatures as low as −25°C. Certainly at −10°C, water is present, and it has a profound effect on ice flow at temperatures greater than −5°C (Duval, 1977; Morgan, 1991). This effect can also include sliding of the ice over the bedrock, lubricated by a thin film of water.

For much of the area of the polar ice sheets, the basal ice is frozen to the solid bed; in other areas sliding occurs. There are also some situations in which the bed itself deforms. The best (or at least best known) examples of this are in areas where the West Antarctic ice sheet drains into the Ross Ice Shelf. Even though this ice is not 'channelled' by glacial valleys in the rock below, ice movement rates are sometimes extremely high (up to $20\,km\,yr^{-1}$), and incredibly these high velocity 'pulses' can last for periods of just hours, to years (Tulaczyk, this volume, Chapter 70). These areas are known as 'ice streams' and the rapid accelerations and subsequent retardations are attributed to sliding of the ice and deformation of the under-riding till. Although, as Tulaczyk points out in Chapter 70, some ice streams have been observed even in regions with a hard bed.

Let us now examine in more detail six of the chapters in Part 4. Three of the chapters are concerned, in one way or another, with the flow of glacier and ice-sheet ice generally, and they each (with necessarily, some overlap) are concerned with processes of how ice flows: Duval and Montagnat (Chapter 59) with the crystalline microprocesses leading to an ice flow law; Cuffey (Chapter 57) with a detailed mathematical approach, incorporating the flow law along with equations for the dynamics of glacier flow; and Tulaczyk (Chapter 70) with a more descriptive explanation of fast ice flow in which basal processes are of uppermost consideration. Basal ice observations are then examined in greater detail—not surprising given their importance as outlined earlier—in three other chapters of Part 4. Again in this second group of three chapters, there is much overlap. Hubbard (Chapter 67) provides a background and an approach for development of a model of temperate glacier flow. Fitzsimons (Chapter 65) examines laboratory and field observations in cold-based glaciers of the McMurdo Dry Valleys (Antarctica). He concludes we need to recognize a 'continuum of material properties and deformation properties which is variable in time and space'. Finally, Lawson (Chapter 63) looks at environmental controls on glacier dynamic behaviour.

56.2 Clean ice

In Chapter 59 Duval and Montagnat investigate the physical mechanisms involved in the flow of clean ice, and they also include discussion of the effects of impurities, including water, on the flow. We now have a good understanding, I think Duval and Montagnat will agree, of the rheological properties for clean ice at temperatures above about −15°C and driving stresses higher than 0.2 MPa. For this ice Glen's (modified) ice-flow law (Glen, 1955)

$$\dot{\varepsilon}_0 = EA\tau_0^n$$

which in one form or another is quoted in every one of the following six chapters, suffices well with power exponent $n = 3$. This expression relates strain rate (here the root mean square of the tensor deviators or 'octahedral' value), $\dot{\varepsilon}_0$, to the octahedral (deviator or driving) stress, τ_0, where n is a constant, A is related to temperature through the Arrhenius relation, and E is an enhancement factor due to anisotropic fabric development.

The effect of water entrapped in the ice lattice at the warmer temperatures is already accounted for in the value of A above as this flow law is an empirical law, i.e. based on observations from the laboratory and from the field. The relation A is a function of temperature and of the creep activation energy. Figure 56.1 shows activation energy as a function of the inverse of temperature, based on the results of ice deformation tests at temperatures near the melting point by Morgan (1991). For most materials the activation energy is constant, and we see from Fig. 56.1 that for ice colder than ca. −10°C ($1/T = 0.0038\,\mathrm{K}^{-1}$) this is the case. At higher temperatures, however, the activation energy increases rapidly as a function of temperature as a consequence of the presence of water in the ice lattice.

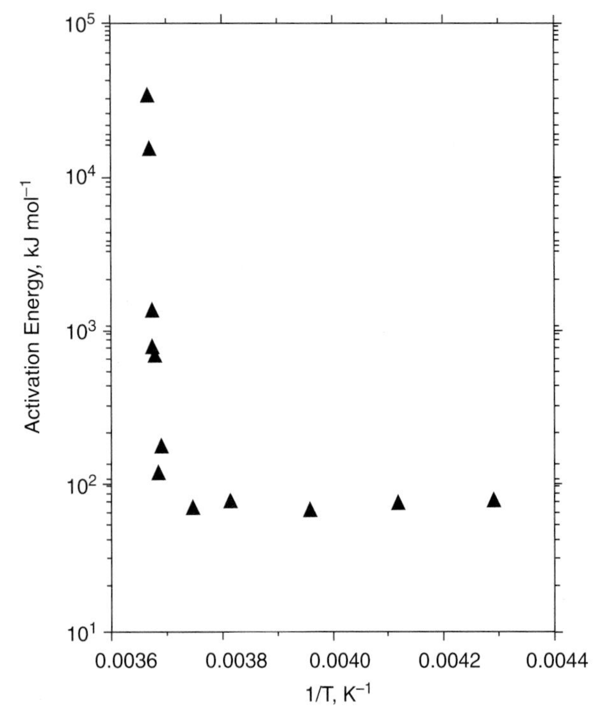

Figure 56.1 Activation energy as a function of the inverse of temperature, based on the results of ice-deformation tests at temperatures near the melting point by Morgan (1991).

Even for anisotropic ice (see Duval and Montagnat's section on 'Fabric development') at relatively high temperatures and stresses, the flow enhancement, E, due to fabric development (ca. 3 in compression; ca. 12 in simple shear) is generally agreed upon. The flow configuration in glaciers and ice sheets is vertical compression near the surface (due to the weight of the accumulation above), and 'bed parallel' simple shear in the deeper layers. Through most of the ice mass then, the predominant stress configuration is a combination (in different ratios) of compression and shear. Thus we need to know the enhancement factor for any combination. Li *et al.* (1996) have carried out laboratory tests aimed at this. The octahedral stress, τ_0, and strain rate, $\dot{\varepsilon}_0$, relations for this stress combination are given by

$$\tau_0 = \sqrt{\frac{2}{3}}\sqrt{\frac{\sigma^2}{4} + \tau^2}$$

and

$$\dot{\varepsilon}_0 = \sqrt{\frac{2}{3}}\sqrt{\dot{\varepsilon}^2 + \dot{\gamma}^2}$$

where σ is the confined normal stress and $\dot{\varepsilon}$ is the compressive strain rate, and where τ is the 'bed parallel' shear stress and $\dot{\gamma}$ the strain rate due to shear. Figure 56.2 shows results from the tests of Li *et al.* (1996), which were all carried out at the same octahedral values. It is seen first that the minimum isotropic strain rates

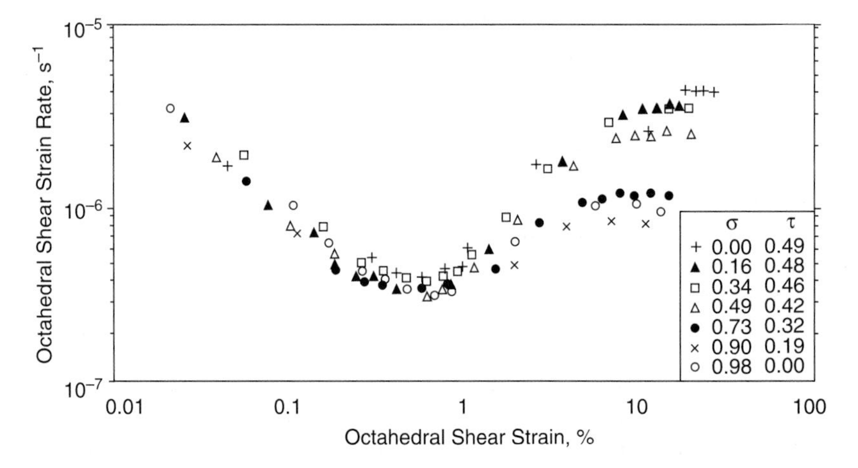

Figure 56.2 Creep curves (log–log plots of octahedral strain rate as a function of octahedral strain) for ice-deformation tests in various combinations (indicated) of compression and shear. The combined octahedral stress for all tests was 0.4 MPa. The test temperature was −2.0°C. (After Li *et al.*, 1996.) Reproduced by permission of the International Glaciological Society.

are the same for compression and shear—the use of root mean square values of the deviator tensors ensures this (Nye, 1953)—and second that the tertiary, steady-state strain rates vary between a factor of 3 and up to 12 times the minimum isotropic strain rates, depending on the ratio of compression to shear. Warner *et al.* (1999) have developed a formulation which fits these data well and which allows a simple modified form of the Glen flow relations to be used. They note for the individual component strain rates that the addition of a compressive stress increases the shear strain rate by less than the Glen flow law would indicate, whereas the addition of a shear stress increases the compression strain rate by more than the Glen flow law provides for. This means that for tertiary (anisotropic) flow, shear stress has more effect on the (normal) compression rate than the compression has on the shear rate. For a shear stress, τ, and compression deviator, S

$$\dot{\varepsilon}_{ij} = AT_0^2 S_{ij}$$

where the octahedral shear stress, τ_0, has been replaced by an 'anisotropic mean' stress given by

$$T_0 = \sqrt{(\alpha^2\tau^2 + \beta^2 S^2)}$$

in which α and β are related to the tertiary strain rate enhancements in shear alone, E_S ($\approx$ 12), and compression alone, E_C ($\approx$ 3) by

$$\alpha = E_S^{1/3} \quad \text{and} \quad \beta = E_C^{1/3}$$

For the temperate glaciers and the warm basal ice of the polar ice sheets, Glen's ice-flow law (as modified above) might suffice, except for the complications imposed by the soluble and insoluble impurities.

For the colder ice (temperatures as low as −60°C in Antarctica) and low driving stresses (<0.2 MPa) relevant to the flow by internal deformation of the bulk of the polar ice masses, there is still some question as to an appropriate n value to use in computations of flow rates. Duval and Montagnac (Chapter 59) argue, with strong emphasis on analysis of crystalline deformation mechanisms, for $n < 2$. Unfortunately, however, there is no direct laboratory confirmation—there cannot be because of the large times (tens to hundreds of years) required to attain minimum isotropic or tertiary steady state (anisotropic) flow rates at these low values.

For the enhancement factor, E, due to the steady-state anisotropic fabric development in clean cold ice with low driving stress, there is now laboratory evidence (at least in uniaxial compression) that as temperature and stress decrease, so does E, from 3 to 1, i.e. no enhancement (Jacka & Li, 2000). This seems to be a result of a reducing capacity, as temperature and stress decrease, for anisotropic fabrics to develop. Figure 56.3a shows a series of creep curves (plots of strain rate as a function of strain) and crystal orientation fabrics from Jacka and Li's laboratory experiments at −19°C in which compressive stress ranges from 0.8 to 0.1 MPa, and Figure 56.3b shows a series at 0.2 MPa in which temperature ranges from −5 to −21°C. These tests suggest E itself may be better described by a function of temperature and deviatoric stress.

56.3 Glacier dynamics, fast flow and ice streaming

Cuffey (Chapter 57) expands the crystal scale to the glacier or ice-sheet scale. He introduces the fundamental glacier mass balance equation that temporal change in ice thickness is governed by the difference between the accumulated mass of snow and the horizontal discharge of the mass (the flux). Cuffey then considers the simplest case in which the bed is mechanically strong, and the ice flow is not affected by impurities. For this case he derives a flat-bed expression for the ice thickness. After a discussion of the consequences of this solution, he examines the added complication of a weak-bed, i.e. applicable to ice streams.

Tulaczyk (Chapter 70) provides us with comprehensive descriptive definitions of fast glacier flow. Of particular concern to him is the ice-stream flow formed in those sections of the West Antarctic Siple Coast ice sheet flowing into the Ross Ice Shelf. The 'definition' after Bindschadler *et al.* (2001a) that slow ice flow is that due to internal deformation alone, whereas fast ice flow is due to rapid basal motion, is useful. It might, for example, be utilized to explain some glacier surges. That is, some glaciers are in slow ice-flow mode most of the time, but as upstream accumulation adds to the thickness basal melt increases, and fast ice flow is activated. Once the additional mass is discharged to the front of the glacier (by a wave-like passage of thicker ice down the glacier) and the glacier is thinned (although longer), basal melting is reduced and slow ice flow is reinstated until, as the front retreats, the up-stream accumulation again builds up.

The West Antarctic ice streams are not surging (in the sense described above). They are (at least some of the time) in fast flow mode but exhibit extraordinarily rapid accelerations and retardations. The mechanisms behind this 'stick-slip' motion are not completely understood, but in some cases seem related to tidally triggered spurts. Tulaczyk (Chapter 70) provides us with a comprehensive account of the theories and arguments that have been proposed. Whatever the mechanisms are, there is no doubt that to properly model ice-stream flow, knowledge of till rheology is also required (Alley *et al.*, 1987a). Tulaczyk presents the case for a power flow law for till, with exponent ca. 1 to 2, and for a Coulomb-plastic flow law, as favoured by laboratory tests of till deformation. As he states, fast ice streaming is enabled by the presence of subglacial water at pressures close to the overburden pressure, and understanding subglacial water generation, storage and transport is key to understanding ice-stream flow and consequently to grasping a full knowledge of ice-sheet mass balance changes. Observations of subglacial processes, however, are particularly difficult (see Alley, this volume, Chapter 72).

56.4 Dirty ice

Lawson (Chapter 63), Fitzsimon (Chapter 65) and Hubbard (Chapter 67) deal with basal ice—ice near the bedrock that is warm, has a high concentration of impurities including a range of dissolved species, insoluble fine dust particles, rocks and boulders, and is folding, faulting, at times shearing and at other times sliding on the bed, which itself may be deforming. To study the flow of dirty ice, we need to consider the effects of (i) insoluble atmosphere-borne dust, (ii) soluble impurities and (iii) basal

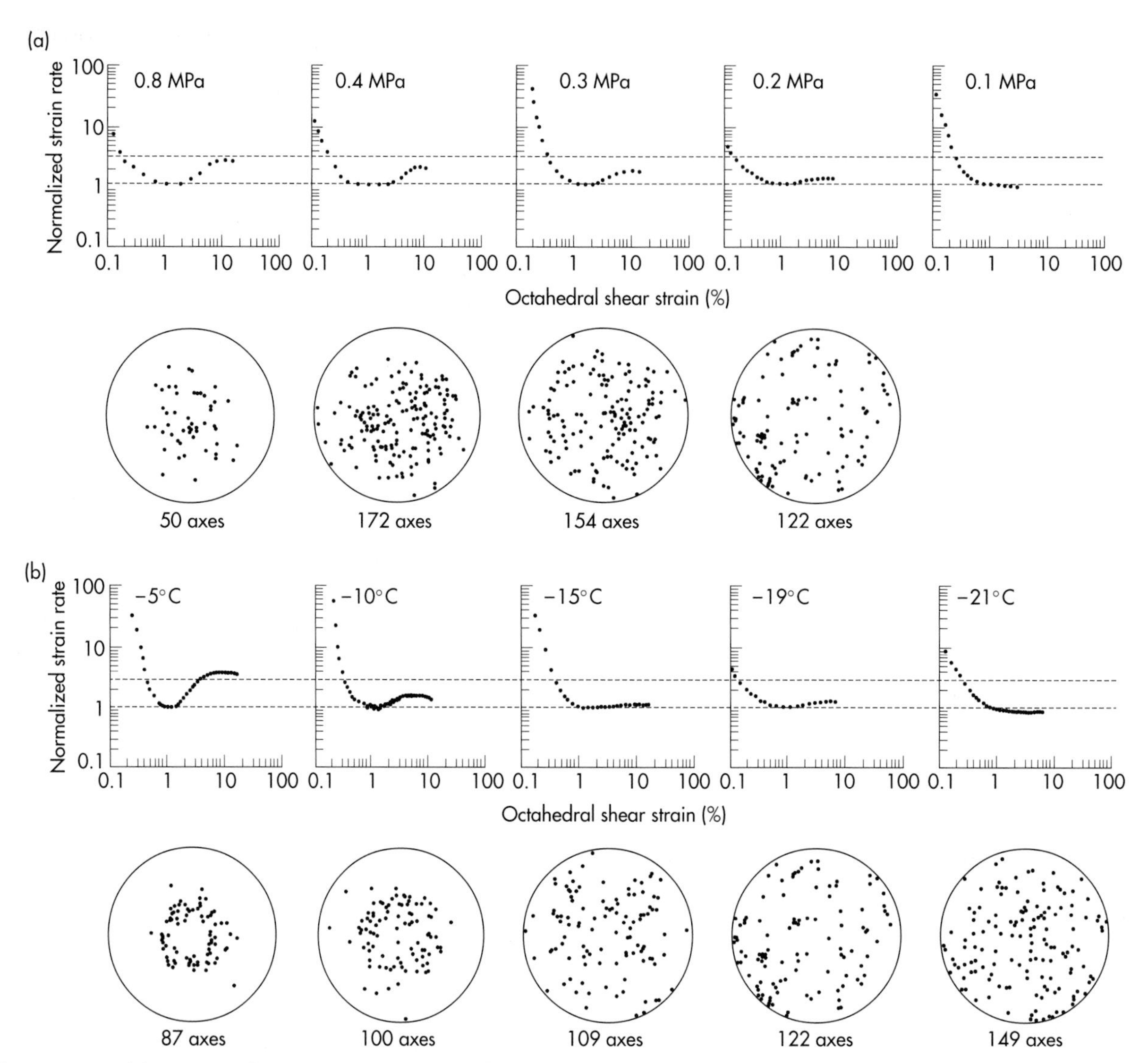

Figure 56.3 (a) A series of creep curves (normalized to a minimum strain rate of 1, and with a dashed line indicating a flow enhancement of 3) and crystal orientation fabrics (after Jacka & Li, 2000) from laboratory experiments at −19°C in which compressive stress ranges from 0.8 to 0.1 MPa. (b) A series of creep curves (normalized to a minimum strain rate of 1, and with a dashed line indicating a flow enhancement of 3) and crystal orientation fabrics (after Jacka & Li, 2000) from laboratory experiments at 0.2 MPa compressive stress in which temperature ranges from −5.0 to −21°C. Reproduced by permission of Hokkaido University Press.

debris. Note that dust delivered to the glaciers and polar ice sheets by atmospheric transport, i.e. in the wind, is necessarily very fine (microparticles) and will be incorporated in the ice within the crystal lattice and at grain boundaries. Similarly, soluble impurities, dissolved in the snow that is deposited to form the ice, are included within the ice lattice and at grain boundaries. Basal debris, however, also includes material that is of the same order of magnitude or much larger than the crystals themselves.

As the presence in the ice of the dust and soluble impurities has an influence on crystal structure (size and orientation fabric development) we need to first consider these interactions and consequences for the ice flow. We know that mean crystal area, in clean ice under little or no stress, increases linearly with time at a rate dependent on temperature through the Arrhenius relation (Stephenson, 1967; Gow, 1969)—often referred to as 'normal crystal growth'. In deforming ice the deformation itself controls the crystal size, which in steady state is a function primarily of the stress (Gow & Williamson, 1976; Jacka & Li, 1994). From ice from the lowest levels of the Greenland, Agassiz and Antarctic ice sheets (e.g. Dahl-Jensen & Gundestrup, 1987; Fisher & Koerner, 1986) and from laboratory tests (Li *et al.*, 1998) we have seen that crystal growth rates seem to be unaffected by dust content, provided that the content does not exceed ca. $10^4\,g^{-1}$. Figure 56.4 (after Li *et al.*, 1998) shows crystal size plotted as a function of microparticle concentration. Data are from ice core DSS, on Law Dome, East Antarctica, from clean ice within the Holocene depth interval, from Last Glacial Maximum (LGM) samples and from samples from the beginning of the last glacial period. One datum is from a Dye 3 (Greenland) LGM sample containing high microparticle

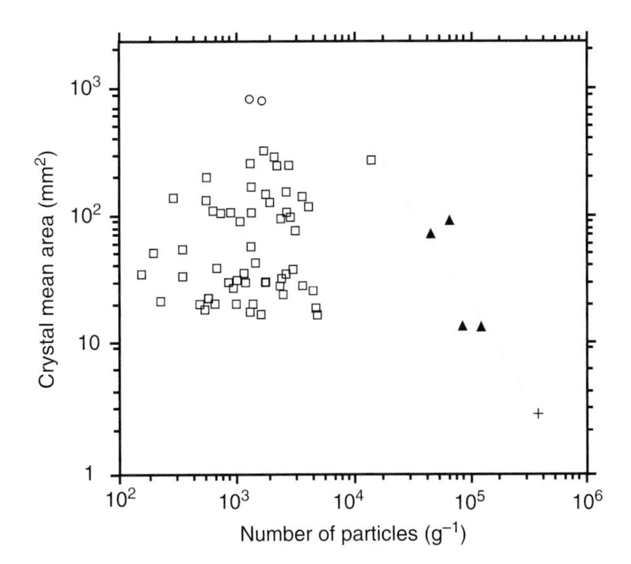

Figure 56.4 Crystal mean area plotted as a function of microparticle concentration. Data represented by squares are from ice core DSS, on Law Dome, East Antarctica, and samples are from within the Holocene depth interval. Solid triangles represent DSS Last Glacial Maximum (LGM) samples. Circles represent samples from the beginning of the last glacial period. The cross is a datum from a Dye 3 (Greenland) LGM sample. (After Li *et al.*, 1998.) Reproduced by permission of the International Glaciological Society.

content. If dust content does exceed ca. $10^4 g^{-1}$, it seems that crystal growth is inhibited as a function (power law exponent ca. −1.5) of microparticle content. This is seen also in the lowest levels of some of the ice cores where crystals are extremely small and dust content very high.

Thin layers, 'spikes', of very small crystals have been located in some ice cores at levels well above the base. These have been associated with higher concentrations of soluble and insoluble impurities, high shear rates and strongly anisotropic crystal orientation fabrics. There is still considerable debate concerning which of these factors are cause, which are effect and which are unrelated. Because of simultaneous spikes in the oxygen isotope ratio, it is clear that some, although not all, of these spikes are related to climatic cool periods—colder, dryer conditions lead to greater amounts of atmospheric borne soluble and insoluble species. It has been argued (e.g. Paterson, 1991) that the higher impurity concentrations lead to smaller crystals and that these smaller crystals will facilitate higher deformation rates and thus development of anisotropic crystal fabrics. On the other hand, as indicated above, insoluble impurities in these layers are not expected to lead to smaller crystals because concentrations are below the threshold of $10^4 g^{-1}$. In addition, Duval & LeGac (1980) and Jacka (1984) found that for the crystal sizes usually found in glaciers and polar ice sheets (ca. 1–10 mm), and within the temperature and stress regime in which $n = 3$, crystal size does not significantly affect ice-flow rates. For much smaller crystals, however, Goldsby & Kohlstedt (1997) do find a crystal size effect on flow rate, and for the fine-grained, dirty ice from Merserve Glacier (Antarctica) and Dye 3 (Greenland), Cuffey *et al.* (2000a,b) argue that there is a grain-size effect on the ice flow.

Another explanation for the shear spikes, as yet not satisfactorily tested in the field or laboratory, may be that some soluble impurities, e.g. F^- (Jones & Glen, 1969) or NH_4^+, which are found along with high insoluble impurity levels because they are generated by the same events, e.g. volcanic eruptions and forest fires (Alley & Woods, 1996), may facilitate accelerated development of anisotropic crystal orientation fabrics in shear. This will lead to even higher shear rates and, as a result, smaller crystals.

Let us now consider the direct effect of the impurities on the ice flow. For the soluble and insoluble atmosphere-borne impurities, concentration levels in Earth's glaciers and polar ice sheets are very low. It seems reasonable to expect that any effect they may have on the ice-flow rates will be by indirect processes such as described above, i.e. by their effect on the ability of the ice to develop crystal anisotropies that accelerate or retard the flow. For the larger impurities, e.g. basal debris and debris originating on valley walls of mountain glaciers, primarily because of the importance of understanding the flow of the basal layers of polar ice sheets, many laboratory studies (Shoji & Langway, 1984; Fitzsimons *et al.*, 1999) and field studies (Holdsworth & Bull, 1970; Anderton, 1973; Fisher & Koerner, 1986; Echelmeyer & Wang, 1987; Tison *et al.*, 1993, 1994; Gow & Meese, 1996; Lawson, 1996) have been carried out. Despite this, we still do not have a full understanding, even of whether the presence of solid debris accelerates or retards ice-flow rates. Several of the authors cited above have found enhanced flow for debris-laden ice, especially at higher temperatures (Lawson, 1996). However, Butkovich & Landauer (1959), Nickling & Bennett (1984) and, at temperatures below −5°C, Lawson (1996) found retarded flow for higher debris content. Hooke *et al.* (1972) and Jacka *et al.* (2003) provided evidence for no significant dependence of flow rate on debris content. The picture of whether debris content accelerates or retards ice-flow rate is confusing and further laboratory and field studies are required. Overall it seems likely that there is a flow rate enhancement at high temperatures, at high stresses, and possibly only for high debris content. Little attention has been paid in the studies so far to the development of crystal anisotropies in debris-laden ice. Only Jacka *et al.* (2003) have examined tertiary flow of debris-laden ice, and their results were similar to those for minimum isotropic ice.

56.5 Conclusions

I have attempted to provide a summary of the 'state of the art' at the time of writing, of our development of an ice-flow law suitable for use in glacier and ice-sheet models. For clean ice, Glen's ice-flow law with $n = 3$, and with modifications for flow enhancement due to the development of crystal anisotropy (varying from ca. 3 to ca. 12, depending on the ratio of compression to shear), suffices well for temperate glaciers and warmer conditions in polar ice sheets. For the colder bulk of the polar ice masses there is still some debate concerning the most appropriate value for n, which possibly may be <2. There is some evidence also that the development of crystal orientation fabrics may be retarded at the lower temperatures and stresses, so that the flow enhancement is reduced as a function of these parameters.

For ice containing soluble impurities, I have argued that some species may accelerate crystal fabric development, causing the spikes in shear and associated very small crystal sizes that have been located in some polar ice sheets. I have argued that insoluble microparticles do not affect ice-flow rates, but are co-located in these high-shear spikes only because they are deposited on the ice sheets along with the soluble species. At the base of the ice sheets, where the concentration of micro- and of larger particles can be very high, crystal growth is retarded by these high concentrations. Our knowledge of the flow of the basal ice due to these larger particles remains the area of greatest doubt. The results from field and laboratory studies are still unclear, and only a very small amount of research has been carried out to understand the steady-state flow of this dirty ice.

Understanding the processes involved in ice-stream flow is of particular importance for modelling ice sheets, and for estimating the mass balance, particularly of the Antarctic ice sheet. Improvements are required in our understanding of till rheology and of subglacial water generation, storage and transport.

FIFTY-SEVEN

Manifestations of ice microphysical processes at the scale of whole ice sheets

K. M. Cuffey

Department of Geography, University of California–Berkeley, 507 McCone Hall, Berkeley, CA 94720-4740, USA

57.1 Introduction

Ice sheets live and die according to a grand contest between climate and ice flow. Specifically, the configuration and evolution of ice sheets is governed by the competition between net accumulation rate and the divergence of ice flux arising from gravitationally induced flow. Ice microphysical processes strongly regulate the character of this flow, especially the dependence of flow rate on ice-mass geometry. Thus the great ice sheets are not only an imprint of climatic forcing, they are also continental-scale manifestations of molecular-scale and grain-scale processes.

This chapter provides a conceptual review of the most important of these manifestations. Ice sheets have a profound impact on the global environment by modifying sea level, planetary albedo and atmospheric circulation. The magnitude of these impacts is related to ice-sheet geographical characteristics: volume, area and topographic form. The first and third of these are most cleanly determined by microphysical processes, if ice-sheet span and climatic forcing are given. The ice-sheet area (and hence span) are determined in a substantially more complex fashion by feed-backs between climatic forcing and ice-sheet topography (and hence microphysical processes), and geographical contingencies such as large-scale influences on climatic forcing, and the distribution of land mass and ocean. The present discussion will focus mostly on ice-sheet volume and profile.

At present, it is not possible to connect microphysics to whole-ice-sheet properties in a manner that is simultaneously direct and convincing. Instead, the connection is made using, as intermediary, a phenomenological description of the constitutive properties of polycrystalline ice derived from laboratory bench-scale experiments and *in situ* measurements in glacier boreholes and tunnels. In this article, I will explore the ice-sheet-scale implications of four primary factors that embody the consequences of microphysics at this phenomenological scale. These are the effective viscosity of ice η, the stress exponent n in the constitutive relation, the temperature-dependence of viscosity (related to the activation energies for ice deformation), and the viscosity variations due to other ice properties (collectively known as 'enhancement', meaning enhancement of ice fluidity). I will finish the article with succinct summaries of how these intermediate-scale properties are thought to arise from molecular and grain-scale processes.

The approach adopted here is to analyse and use one-dimensional, steady-state ice-sheet models to examine controls on ice-sheet character (following Vialov, 1958; Nye, 1960; Haefeli, 1961; Weertman, 1961; most of the relevant physics is summarized in Paterson, 1994). In terms of geographical realism, such a model is inferior to three-dimensional, whole-ice-sheet models but represents a distillation of the relevant physics and geography to their essentials. The aim is to be illustrative but quantitative, with illustration provided both by formulae and numerical results. None of the conceptual results presented here are altered by relaxing the simplifying assumptions used.

57.2 Model framework

57.2.1 Foundations

The formal statement of the governing competition for ice sheets is

$$\frac{\partial H}{\partial t} = \dot{b} - \nabla \cdot \vec{Q} \tag{1}$$

wherein H is the ice thickness, $\dot{b}$ the net accumulation rate (ice-equivalent thickness per time added to the glacier surface) and $\vec{Q}$ the ice flux. The ice flux in a horizontal direction x arises from basal velocity (rate u_b) and internal shear strain rate ($\dot{\varepsilon}_{xz}$)

$$Q_x = \int_b^{H+b} u(z)\,\mathrm{d}z \approx u_b H + 2\int_b^{H+b}\int_b^z \dot{\varepsilon}_{xz}\,\mathrm{d}z'\,\mathrm{d}z \tag{2}$$

where z is the vertical coordinate ($z = b$ at the bed). A massive amount of empirical evidence shows that the deformation of ice is a form of power-law creep (Glen, 1955; Nye, 1957; Barnes *et al.*, 1971; Weertman, 1983; Budd & Jacka, 1989; Goldsby & Kohlstedt, 2001) meaning that $\dot{\varepsilon}_{xz}$ is a function of powers of the deviatoric stress, the deviation of stress from its mean normal value. Most commonly, ice deformation is treated using the phenomenological relation

$$\dot{\varepsilon}_{xz} = EA(T)\tau_{(2)}^{n-1}\tau_{xz} \tag{3}$$

Other strain-rate components are similarly proportional to the corresponding deviatoric stress component. Here $\tau_{(2)}$ is the second invariant of the deviatoric stress tensor, n is the stress exponent (usually estimated as 3 for rapidly deforming ice; data reviewed by Weertman, 1983), $A(T)$ is the temperature-dependent ice softness (or fluidity), and E is the enhancement factor, a correction factor used to account for anisotropic effects and other variables not otherwise included. In ways that are not yet completely understood, microphysical processes determine this functional form, and values for E, $A(T)$ and n. From the compilation due to Paterson (1994, p. 97), the function $A(T)$ is well approximated by

$$A = A_0 \exp\left[\frac{E_a}{R}\left(\frac{1}{263} - \frac{1}{T}\right)\right] \tag{4}$$

where $A_0 = 4.9 \times 10^{-25}\,\mathrm{s}^{-1}\cdot\mathrm{Pa}^{-3}$, and the activation energy is $E_a = 60\,\mathrm{kJ\,mol^{-1}}$ for $T < 263\,\mathrm{K}$ and $E_a = 139\,\mathrm{kJ\,mol^{-1}}$ for $T > 263\,\mathrm{K}$.

In ways that are even less completely understood, microphysical processes also help control u_b, which depends in part on deformation properties of ice, thermal properties of ice and interactions of ice with melt water. High basal water pressure (P_w close to the ice overburden pressure P_i) lubricates basal motion (increases u_b), as does the presence of unconsolidated substrate. Although not a predictive relationship, the statement

$$u_b = \lambda_b\left[1 - \frac{P_w}{P_i}\right]^{-1}\tau_b^m \tag{5}$$

is consistent with our understanding of basal motion and is useful for analyses (Bindschadler, 1983), bearing in mind that variations in the lubrication parameter λ_b can largely determine the variations of u_b in some cases.

Ice sheets flow because gravity induces pressure gradients in the ice, resulting from the ice–air surface slope. The flow gives rise to deformations (including $\dot{\varepsilon}_{xz}$) which are associated with resisting stresses (including τ_{xz}) via the ice constitutive relation approximated by Equation (3). These resisting stresses balance the 'driving stress' τ_d, the net horizontal force-effect of gravity, per unit horizontal area (Van der Veen & Whillans, 1989). The τ_d is a simple function of ice thickness and surface slope, and is (given bed elevation b)

$$\tau_d = \rho g H \frac{d}{\mathrm{d}x}(H + b) \tag{6}$$

57.2.2 The strong-bed ice sheet

Throughout much of the ice sheets the ice–bed interface is mechanically strong and the flow is not confined in narrow channels. Consequently, the driving stress is balanced dominantly by the primary down-flow shear stress τ_{xz} acting at the bed (value τ_b), and the τ_{xz} varies approximately as a linear function of depth

$$\tau_{xz} \approx \tau_d\left[1 - \frac{z-b}{H}\right] \tag{7}$$

Much can be learned by making this and further approximations (Nye, 1959, 1960; Haefeli, 1961). Substitutions into Equation (2) give

$$Q_x = \lambda_b\left[1 - \frac{P_w}{P_i}\right]^{-1}\tau_d^m H + 2\tau_d^n\int_b^{H+b}\int_b^z EA(T)\left(1 - \frac{z-b}{H}\right)^n \mathrm{d}z''\mathrm{d}z' \tag{8}$$

The multiplicative relation of variables in the integrand of Equation (8) has important implications for large-scale ice-sheet flow, in that values for E and A in ice close to the bed where the shear is concentrated are most significant. One can replace the term $E\,A(T)$ with an effective value A_*, a constant defined to yield the equivalent flux as Equation (8), so that

$$Q_x = \lambda_b \left[1 - \frac{P_w}{P_i}\right]^{-1} \tau_b^m H + \frac{2}{n+2} A_* \tau_d^n H^2 = \frac{2}{n+2} A_* \left[1 + \frac{u_b}{\bar{u}_d}\right] \tau_d^n H^2 \quad (9)$$

$$A_* \equiv \frac{n+2}{H^2} \int_b^{H+b} \int_b^{z} EA(T)\left(1 - \frac{z-b}{H}\right)^n dz'' dz' \quad (10)$$

where $\bar{u}_d$ is the depth average of the deformational velocity. The effective temperature T_* is the temperature for which $A(T)$ has the value A_*/E. For a strong-bed system the ratio of u_b to $\bar{u}_d$ typically varies in a range between zero and order one (Paterson, 1994, p. 135), so the factor $1 + u_b/\bar{u}_d$ varies between one and a few. For subsequent use, I define the combined softness as

$$S \equiv A_* \left[1 + \frac{u_b}{\bar{u}_d}\right] \quad (11)$$

If there is non-trivial basal motion ($u_b \neq 0$) the basal temperature is at the bulk melting point, and the T_* will always be greater than −10°C. In this range, a factor of two change in S (as when the u_b changes from being 20% to 140% of $\bar{u}_d$, a large change) is equivalent to an effective temperature change of only approximately 3°C.

Further, defining the fraction of the total ice flux due to internal deformation as

$$q_d \equiv \left[1 + \frac{u_b}{\bar{u}_d}\right]^{-1} \quad (12)$$

and the surface slope as α, the ice-sheet evolution Equation (1) for one-dimensional flow is

$$\frac{\partial H}{\partial t} = \dot{b} + \gamma_1 \frac{\partial^2 b}{\partial x^2} + \left[\gamma_1 \frac{\partial^2 H}{\partial x^2} + \gamma_2 \frac{\partial H}{\partial x}\right] - \frac{Q_x}{S}\frac{\partial S}{\partial x} \quad (13)$$

$$\gamma_1 \equiv \frac{Q_x}{\alpha}[m(1-q_d) + nq_d] \quad (14)$$

$$\gamma_2 \equiv \frac{Q_x}{H}[(m+1)(1-q_d) + (n+2)q_d] \quad (15)$$

Because m and n are similar numbers and the q_d is of order one, the presence of basal motion in a strong-bed ice sheet does not significantly alter its properties.

The steady-state cross-sectional profile of an ice sheet determines the volume of stored ice per length of ice divide. Integrating Equation (9) by specifying the flux as the integrated net accumulation up-flow of location x yields an estimate for this profile, which for an ice-sheet of span L, a flat bed, a spatially uniform S, and a zero thickness at the ice margin ($x = 0$) is

$$H(x) = c_1 \left[\frac{1}{S}\right]^{\frac{1}{2n+2}} \left[\int_0^x \left(\int_{x'}^L \dot{b}\, dx''\right)^{\frac{1}{n}} dx'\right]^{\frac{n}{2n+2}} \quad (16)$$

$$c_1 \equiv \left[2^{n-1}(n+2)\left(\frac{n+1}{n\rho g}\right)^n\right]^{\frac{1}{2n+2}} \approx 0.053 \quad (17)$$

The case of a uniform accumulation rate $\dot{b}$ illustrates the properties of this curve most simply (Vialov, 1958), and in particular gives the ice thickness in the centre of the ice sheet (the divide thickness H_d) as

$$H_d = c_2 \dot{b}^{\frac{1}{2n+2}} S^{\frac{-1}{2n+2}} L^{1/2} \quad (18)$$

$$c_2 \equiv \left[2^{n-1}(n+2)\left(\frac{1}{\rho g}\right)^n\right]^{\frac{1}{2n+2}} \approx 0.048 \quad (19)$$

The ice-sheet volume scales with $\dot{b}$ and S in similar fashion, and with $L^{3/2}$. Values for c_1 and c_2 given here are for $n = 3$.

57.2.2.1 *Perspectives from the strong-bed case*

An ice sheet is a forced diffusive system (Equation 13), with the forcing being net accumulation determined by climate and curvature of underlying topography, modulated by gradients in effective softness S. Ice sheets respond to variable forcings via both diffusive and wave-type behaviours. The variables γ_1 and γ_2 are the non-linear diffusivity and wave speed, respectively (Equations 14 & 15). Both vary spatially. Diffusive behaviour dominates where surface slope is small, in the extensive thick interior regions of ice sheets. Ice-sheet thickness evolves in a manner analogous to temperature evolution in a plate with spatially varying energy source terms.

The non-linearity of the constitutive relation (n,m values) affects the response time of ice sheets to variable forcings; higher non-linearity (larger n,m) enables faster response. This is, in essence, a consequence of the fact that the stress change needed to accommodate a change in ice flux is smaller if the non-linearity is higher. The stress change, in turn, is a function of the redistribution of mass in the glacier body, and this redistribution is less extensive in the higher non-linearity case and thus achieved more rapidly. The non-linearity also directly increases the magnitude of the topographic forcing term for a given bed curvature.

Closely related is that the steady-state ice volume becomes more weakly dependent on the net accumulation rate and softness as the non-linearity increases (Equation 18). For $n = 3$, the value thought to be most appropriate for the ice sheets, a doubling of accumulation rate, or halving of softness, changes the steady-state volume by only 9%, if L is fixed. This is a tremendous insensitivity and largely explains why the modern Greenland and Antarctic ice sheets are only modestly different from their Last Glacial Maximum versions. Also, a uniform increase of the basal motion from zero to $u_b = \bar{u}_d$ would decrease ice volume by only ca. 8%. This explains why ice-sheet models do a very good job of explaining topography of the Greenland and East Antarctic ice sheets even though we have no predictive understanding of basal motion and limited knowledge of its actual spatial distribution in the ice sheets. Similarly, the temperature of deep ice layers need not be known precisely, although it is essential to know it approximately.

57.2.3 The weak-bed, ice-stream-dominated ice sheet

The important case for which the assumptions $\tau_b \approx \tau_d$ and $u_b \sim \bar{u}_d$ are completely wrong is that for which the ice flux is accommodated almost entirely by ice streams that have minimal basal strength (Whillans & Van der Veen, 1997; Raymond *et al.*, 2001). This situation, typified by the modern Siple Coast of West Antarctica and possibly applicable to sectors of the Pleistocene ice sheets, is made possible by high basal water pressures in combination with a weak, unconsolidated, deformable substrate (high λ_b and P_w in Equation 5). In the limiting case, the driving stress is entirely balanced by wall stresses (average magnitude τ_w) in the shear margins bounding the ice streams.

To explore how this situation differs from the previous one, consider an approximation in which the shear stress τ_{xy} varies linearly across the ice stream from a value $+\tau_w$ to a value $-\tau_w$. The ice stream has a width W and half-width φw, and the cross-glacier coordinate is y (zero at the margin). Then

$$\omega\tau_d = H\tau_w \tag{20}$$

$$\tau_w = \rho g\omega\frac{d}{dx}(H+b) \tag{21}$$

The flux per unit width in the ice stream is (Raymond, 2000)

$$Q_s = \frac{2H}{\omega}\int_0^{\omega}\int_0^{y}\dot{\varepsilon}_{xy}dy''dy' = \frac{2H\tau_w^n}{\omega}\int_0^{\omega}\int_0^{y}EA(T)\left(1-\frac{y}{\omega}\right)^n dy''dy' \tag{22}$$

Defining the fractional area (or cross-flow length) of the ice stream within the total ice area as f_s and the number density of ice streams per cross-flow distance as N_s, and defining an effective uniform A_* as before, gives the ice flux per unit width as

$$Q_x = \frac{2}{n+2}f_sA_*\tau_d^nH^{1-n}\omega^{n+1} = \frac{4}{n+2}N_sA_*\tau_d^nH^{1-n}\omega^{n+2} \tag{23}$$

and the governing equation for evolution of the thickness as

$$\frac{\partial H}{\partial t} = \dot{b} + \gamma_1^{(s)}\frac{\partial^2 b}{\partial x^2} + \left[\gamma_1^{(s)}\frac{\partial^2 H}{\partial x^2} + \gamma_2^{(s)}\frac{\partial H}{\partial x}\right] - \frac{Q_x}{A_*}\frac{\partial A_*}{\partial x} - \frac{(n+2)Q_x}{\omega}\frac{\partial \omega}{\partial x} \tag{24}$$

$$\gamma_1^{(s)} \equiv \frac{nQ_x}{\alpha} \tag{25}$$

$$\gamma_2^{(s)} \equiv \frac{Q_x}{H} \tag{26}$$

Analogous calculations to those above yield the thickness profile, given a flat bed, uniform ω and f_s, and a marginal (grounding-line) thickness H_{og}

$$H(x)^{\frac{n+1}{n}} = H_{og}^{\frac{n+1}{n}} + c_3\left[\frac{1}{f_sA_*}\right]^{\frac{1}{n}}\left[\frac{1}{\omega}\right]^{\frac{n+1}{n}}\left[\int_0^x\left(\int_{x'}^L\dot{b}dx''\right)^{\frac{1}{n}}dx'\right] \tag{27}$$

$$c_3 \equiv \left[\frac{n+1}{n}\right]\left[\frac{n+2}{2(\rho g)^n}\right]^{\frac{1}{n}} \approx 0.087 \tag{28}$$

For comparison to the strong-bed case, consider a hypothetical situation for which the ice-stream flow spans the entire ice sheet from divide to margin, and the marginal thickness is zero. The thickness at the ice divide for a constant accumulation rate case would be

$$H_d = c_4\frac{1}{\omega}\left[\frac{\dot{b}}{f_sA_*}\right]^{\frac{1}{n+1}}L \tag{29}$$

$$c_4 = \left[\frac{n+2}{2(\rho g)^n}\right]^{\frac{1}{n+1}} \approx 0.0014 \tag{30}$$

Values for c_3 and c_4 given are again those for $n = 3$.

57.2.3.1 *Perspectives from the weak-bed case*

This type of ice sheet is also a forced diffusive system (Equation 24), with an additional term in the forcing involving the gradient of ice-stream width. This term strengthens the dependence of the forcings on the stress exponent. The effective diffusivity (Equation 25) is identical to that for the strong-bed case, but the new wave speed (Equation 26) is independent of the non-linearity.

The ice volume (above flotation level if $H_{og} \neq 0$) is now dependent on additional variables, the ice-stream width and abundance (ω and f_s, or ω and N_s), with the dependence on width being strong; wider ice streams are substantially more effective conduits for ice transport, so the dependence is inverse. The dependence of steady-state volume on the accumulation rate and ice softness is stronger than in the strong-bed case, but still moderate, with a doubling of accumulation or halving of softness implying a ca. 20% increase of ice volume. Dependence of volume on span is also stronger (ca. L^2).

57.2.3.2 *Perspective on ice microphysics in the weak-bed case*

Of overwhelming importance to the weak-bed case is the ice-stream width and abundance, and more generally both the spatial distribution of weak-bed zones and the extent to which the weak-bed regions truly have no role in balancing the driving stress. The distribution of weak-bed zones depends crucially on the accumulation of basal lubricants (water and sediments) (Alley *et al.*, 1986a; Alley & Whillans, 1991), boundary conditions for the subglacial water system, and on the interactions between interstream ice and the shear margins. The latter certainly depend on ice microphysical processes (e.g. by the dependence of ridge-to-stream flow on stress and temperature; Jacobson & Raymond, 1998) but are not well understood at present. The basal lubrication certainly depends on ice microphysics via the heat of fusion for ice, which essentially specifies how much melt water is produced given a geothermal flux and an overlying ice sheet of given character. Connecting this property of ice to the width and abundance of the ice streams remains a challenge, however, and I will not consider this connection further here (Raymond (2000) would be a good starting point).

As the preceding analysis makes clear, however, the constitutive properties of ice have central importance for determining whole-ice-sheet properties in the weak-bed case, because resistive stresses are still due to ice deformation in this system. In some cases, resistive stresses associated with longitudinal stretching and compressing of the flow will also be important (although they are not found to be important in sections of the ice streams that have been examined fully). As illustrated by the modern West Antarctic ice sheet, weak-bed streaming flow will almost certainly occur only on the flanks of the ice sheet, beyond a strong-bed core region. This arises from the need to concentrate basally produced water for lubrication, and from feed-back between the production of such water and concentration of heat dissipation via concentrated ice discharge. The character of the transition zone (the 'onset' region), and the character of the transition from ice-stream flow to floating ice shelves, will be dependent on ice constitutive properties via longitudinal and wall-stress terms.

57.3 Ice-sheet profiles

The framework outlined in section 57.2 provides the most concise formalism for examining the connections between the phenomenological description of ice and whole-ice-sheet characteristics. I next use it to evaluate whether our conception of these intermediate-scale properties is explanatory at the continental scale, by comparing solutions to Equations (16) and (27) with ice-sheet surface-elevation profiles (following the general idea used by Vialov (1958) and especially Haefeli (1961); cf. Hamley *et al.*, 1985). Specifically, I ask which values for S, A_* and n yield best-fit matches between calculated profiles and data, whether these matches are good, and whether these parameter values are consistent with expectations.

The three major extant ice sheets are all very different. Although each varies from sector to sector, the following generalizations are most broadly true, and help motivate the choice of a representative elevation profile for each. The Greenland Ice Sheet (GIS) is strong-bedded and relatively narrow, and has a moderate to high accumulation rate, a mostly frozen bed, and a marginal ablation zone where melt rates are high. The East Antarctic Ice Sheet (EAIS) is strong-bedded and broad, has a low accumulation rate and a bed that is thawed in much of its interior, and mostly terminates in small floating ice shelves rather than ablation zones. The West Antarctic Ice Sheet (WAIS) is broad and has a weak-bed, ice-stream-dominated flank but a strong-bedded inner core, a moderate accumulation rate and a thawed bed, and terminates in immense ice shelves.

57.3.1 Three representative profiles

The GIS profile used here leads westward down-gradient from the ice divide at Summit (Ohmura & Reeh, 1991). The $\dot{b}$ used in Equation (16) is $0.23\,\mathrm{m\,yr^{-1}}$ at the divide, increases to a pronounced orographic high on the flank, and then decreases sharply to ca. $-6.5\,\mathrm{m\,yr^{-1}}$ at the terminus. This melt rate was chosen to ensure zero net balance along the profile. Given an n value, adjustment of one parameter ($S^{(G)}$) optimizes the model.

The EAIS profile used here is that directly inland from Mirny Station, as used originally by Vialov (also see Hamley *et al.*, 1985). This profile is unusually simple for EAIS, where most gradient lines either cross margin-region mountain ranges or are affected by the Lambert basin. The $\dot{b}$ is $0.03\,\mathrm{m\,yr^{-1}}$ at its head (chosen to be exactly 1000 km inland), and increases sharply near the margin to more than $0.2\,\mathrm{m\,yr^{-1}}$. Much of the ice in this area terminates in ice shelves, so I specify a nominal 200 m grounding line thickness, and this initial thickness is included in the integration giving Equation (16). Given an n value, adjustment of one parameter ($S^{(EA)}$) optimizes the model.

The WAIS profile is an approximate Siple Coast transect from the grounding line up Ice Stream C and to the divide (Alley & Whillans, 1991; Fahnestock & Bamber, 2001). To model this profile, a transition point ($x = 500\,\mathrm{km}$) is chosen that separates the ice-stream section from the inner core. Below this point, Equation (27) is solved, starting with 1 km-thick ice at the grounding line, and using a uniform ice-stream width of 40 km ($\omega = 20\,\mathrm{km}$). Above the transition point, the version of Equation (16) accounting for a non-zero initial thickness is solved. The $\dot{b}$ is maximum at the ice divide ($0.25\,\mathrm{m\,yr^{-1}}$) and decreases linearly to $0.10\,\mathrm{m\,yr^{-1}}$ at the grounding line. The plan-view curvature of the ice streams, the differences between ice streams, the interstream ridges and the mountainous topography in the southern region of the divide all add complexity and variability to Siple Coast transects. Here I am only interested in the most general features of the topography (the gently sloped ice-stream section and the steeper inner core) and ignore these complexities; similar results are obtained for other profiles, given that parameters such as the ice-stream width and the location of transition from streaming flow to creeping flow are here specified, not modelled. Given an n value, two parameters are adjusted to optimize the model: the shear-zone $A_*^{(WA)}$ for the ice stream section, and $S^{(WA)}$ for the inland section.

57.3.2 Profile results

Using a standard value of $n = 3$, the calculated elevation profiles match the topographies very well (Fig. 57.1). To interpret the optimized values, I use the fact that these are essentially thermometers (though imprecise ones). I calculate the equivalent effective temperature T_* as the temperature corresponding to the $\mathcal{A}$ value given by

$$S = \frac{E\mathcal{A}}{q_{\mathrm{d}}} \tag{31}$$

The inverse of $\mathcal{A}$ gives effective temperature via Equation (4).

For Greenland, E is known from borehole studies to be ca. 2 to 4 due to the ice-age ices being soft, and I use the value $E = 2.5$ as suggested by Paterson (1991). For the Antarctic cases, I use $E = 1$. Much of Greenland is frozen at its bed, so the appropriate flux partitioning $q_{\mathrm{d}} = 1$. For East Antarctica and inland West Antarctica I will use both $q_{\mathrm{d}} = 1$ and $q_{\mathrm{d}} = 0.5$, the latter meaning equal partitioning of the flux between deformation and basal motion. Optimized values are shown in Table 57.1.

In column 5 $\hat{T}_*$ is effective temperature calculated independently from direct temperature estimates from boreholes and models (Huybrechts & Oerlemans, 1988; Engelhardt & Kamb, 1993; Paterson, 1994, p. 222; Huybrechts, 1996; Engelhardt, 2005), and using Equation (10). The agreement is very strong between these independent estimates.

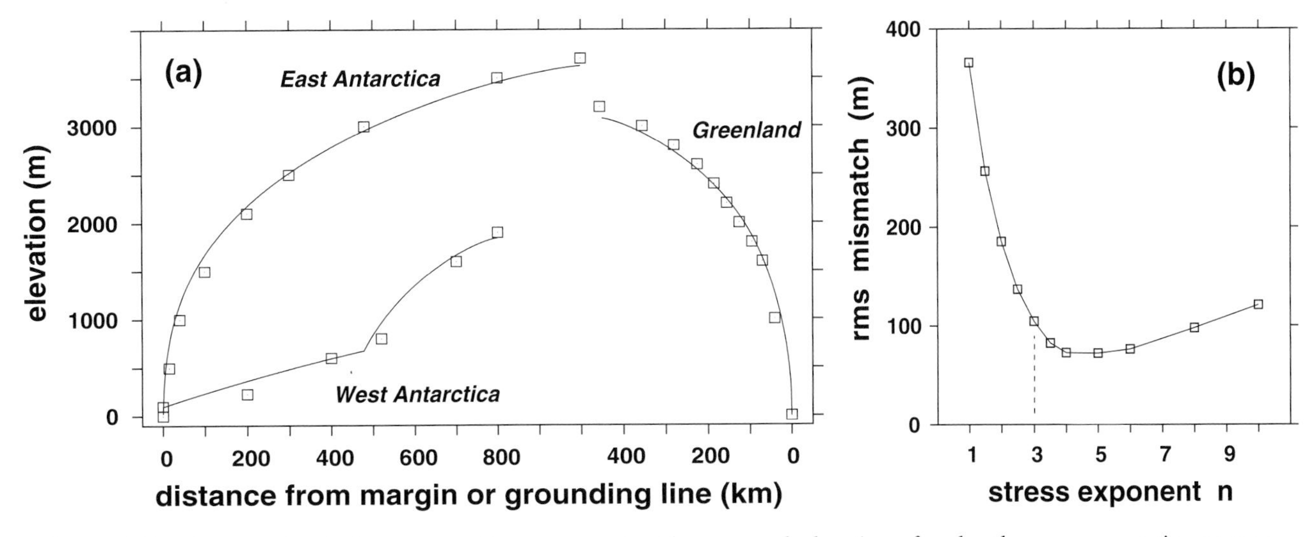

Figure 57.1 (a) Calculated elevation profiles compared with measured elevations for the three representative transects described in the text. (b) Root-mean-square mismatch of calculated and measured elevations, averaged for the three profiles, as a function of stress exponent. For each n value, each profile is least-squares optimized by adjusting softness parameters as described in the text.

Table 57.1 Optimized values used in profile calculations

Parameter	Value	q_d	T_* (°C)	$\hat{T}_*$ (°C)
$S^{(G)}$	1×10^{-24}	1	−12	−7 to −11
$S_{\diamond}^{(G)}$	2×10^{-24}	1	−8	−7 to −11
$S^{(EA)}$	7×10^{-25}	1	−9	−6 to −11
$S^{(EA)}$	7×10^{-25}	0.5	−13	−6 to −11
$S^{(WA)}$	1×10^{-24}	1	−7	−5 to −8
$S^{(WA)}$	1×10^{-24}	0.5	−10	−5 to −8
$A_*^{(WA)}$	2×10^{-25}	N/A	−18	−15 to −20

There is a bit of deception hidden in Fig. 57.1, which is that for Greenland the use of a flat basal topography, as shown, reproduces the elevation profile better than does a more realistic topography, given a spatially uniform S. The second tabulated value for Greenland softness ($S_{\diamond}^{(G)}$) included here is a more realistic case for which the western margin mountain range has been included as basal topography in the profile calculation (abstracted from Bamber *et al.*, 2001b). In this case the profile matches the measured elevations as well as shown in Fig. 57.1 only if S increases toward the ice margin, which is expected to be the case, because the bed becomes warmer toward the margin and some sliding occurs.

The profile shapes depend also on the stress exponent n, and this permits evaluation of which n values are plausible. I specify n, find best-fit values for the four softness parameters, and then calculate the root-mean-square mismatch of the model profile elevations from the known topography. The result of this exercise (Fig. 57.1) clearly indicates that 'average' ice is neither linear viscous nor plastic, and that the non-linearity is pronounced (certainly $n > 2$). This, of course, matches expectation from laboratory experiments, and *in situ* analyses of ice deformation. There is some suggestion here that $n \approx 4$–5 is more appropriate than $n = 3$, but the approximations used here must be remembered. Note that (Equation 16) the sensitivity of profile shape to n becomes progressively smaller as n becomes larger, toward the plastic limit; the broad upper half of the mismatch curve (Fig. 57.1) is inherent in the method.

The topographic profiles of the modern ice sheets thus provide direct confirmation that, for the deep, rapidly deforming ices most relevant here, ice properties have been accurately characterized phenomenologically using laboratory experiments and local *in situ* measurements in active glaciers and ice shelves (Weertman, 1983). This is a major accomplishment.

57.4 General microphysical control on ice-sheet character

Thus emboldened, I use the formalism of section 57.2 to resume the more general exploration (sections 57.2.2.1 and 57.2.3.1) of how the microphysically determined ice properties (effective viscosity, stress exponent, activation energy and enhancement) control ice-sheet characteristics. The accumulation rate patterns, spans and grounding-line thicknesses of the three representative profiles define Greenlandic-type, East Antarctic-type, and West Antarctic-type ice-sheet profiles, conceptual constructs used here for subsequent calculations.

To compare different forms of power-law creep, $\dot{\varepsilon} = A_{(n)}\tau^n$, it is necessary to define an effective viscosity η. Here I do this by using $\eta \equiv A_{(1)}^{-1}$, equating the ice flux in a linear-viscous flow to the ice flux in a flow with stress exponent n (given the same basal stress), and then substituting for τ_b using Q, H and $A_{(n)}$. This gives

$$\eta = \frac{(n+2)^{\frac{1}{n}}}{3} A_{(n)}^{\frac{-1}{n}} \left[\frac{2H^2}{Q}\right]^{\frac{n-1}{n}} \qquad (32)$$

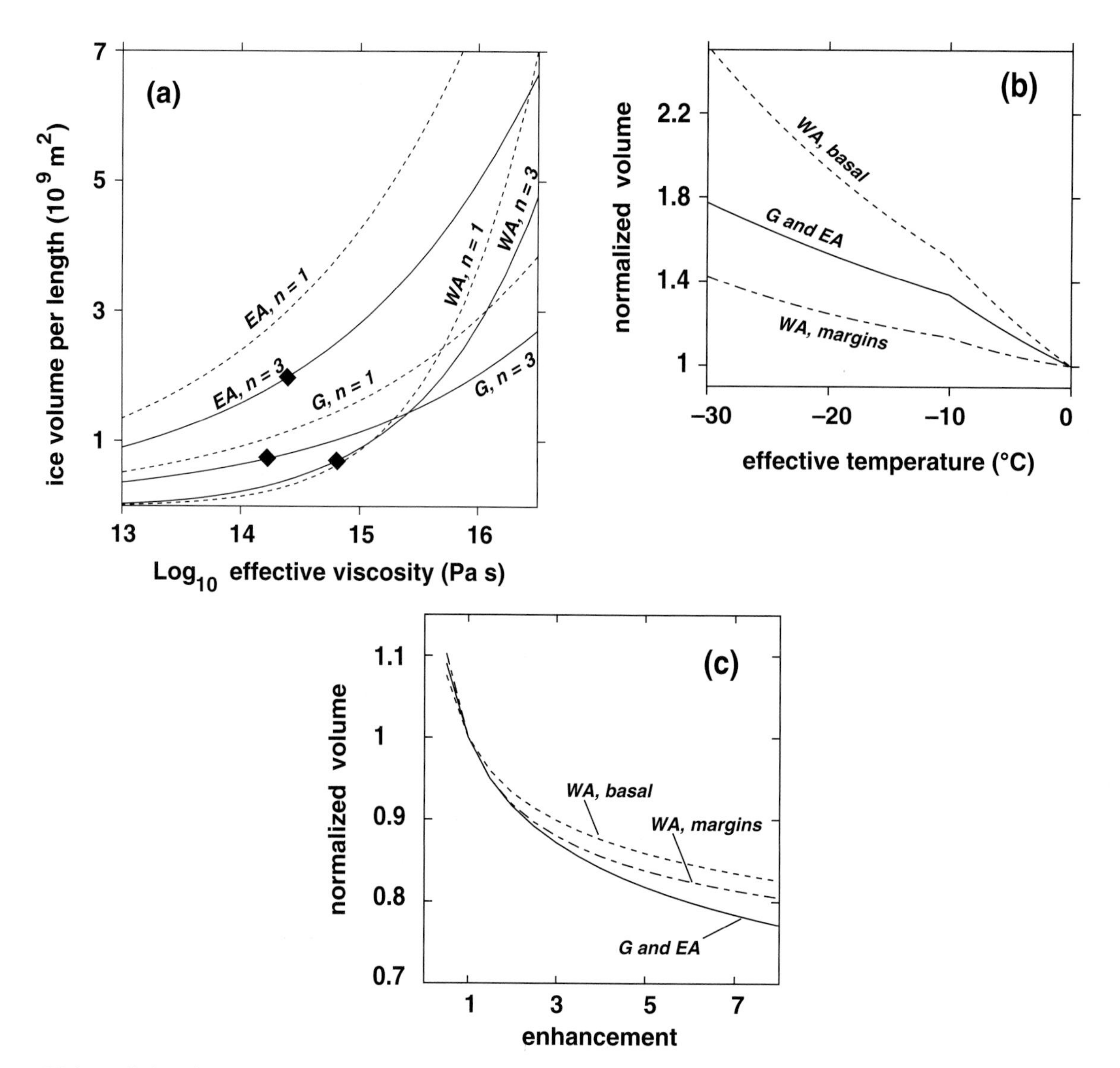

Figure 57.2 Relations between steady-state ice volume per divide length (i.e. area of cross-sectional profiles), effective viscosity, stress exponent and two specific controls on viscosity (temperature and enhancement). (a) Black diamonds indicate estimated positions for the three modern ice sheets, according to representative profiles: EA, East Antarctic-type; WA, West Antarctic-type; G, Greenland-type. West Antarctic volumes are those above the flotation level (hence sea-level equivalent). (b and c) Volumes have been normalized to values at 0°C and at $E = 1$. For 'WA, basal' the effective temperature or enhancement of basal shear has been varied. For 'WA, margins' only the effective temperature or enhancement of the ice-stream shear margins has been varied.

and a similar relation for the ice-stream case. Effective viscosities shown here are averages of η for each profile.

57.4.1 Ice volume, given fixed climatic forcing and span

Ice volumes (per length of ice divide) are calculated by integrating profile solutions. In the case for which both L and $\dot{b}(x)$ are fixed, ice volume depends on the microphysical parameters as shown in Fig. 57.2. Effective viscosity exerts primary control, with ice volume increasing by approximately a factor of two per order-of-magnitude increase in viscosity. This sublinear (roughly logarithmic) dependence of volume on η arises because the flux is fixed by climate, whereas the accommodation of the flux by ice flow relies on ice thickness in at least three multiplicative ways: one through the magnitude of stress, one through the integration to convert strain rate to velocity, and one through the integration to convert velocity to flux.

The consequence of having a stress exponent $n > 1$ is to reduce ice volume. Compared with an ice sheet with $n = 3$, a linear-viscous one would have a flatter surface in the ice-sheet interior, a thicker mid-flank, and a much steeper marginal zone. At the low stresses in the interior of the ice sheet, the linear-viscous ice would be more readily deformed, allowing a smaller surface slope to accommodate the ice flux for a given thickness.

The more pronounced dependence of the WAIS-type ice-sheet volume on viscosity results from steepening of the weak-bed flank

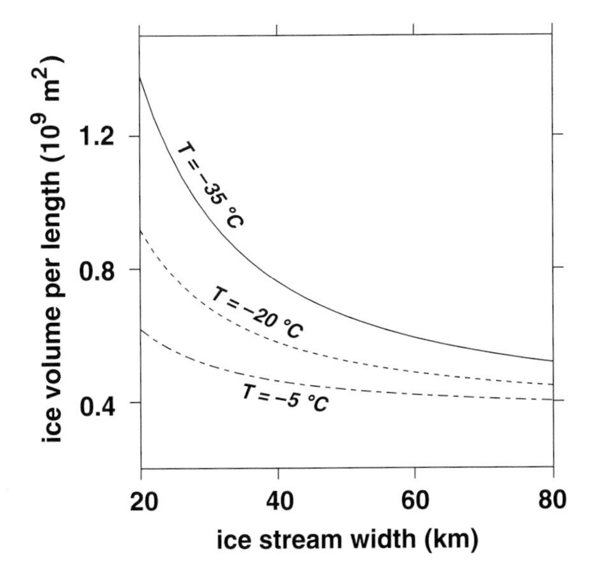

Figure 57.3 Dependence of WAIS-type volume per length on ice stream width (2ω), at three different effective temperatures for the shear margins.

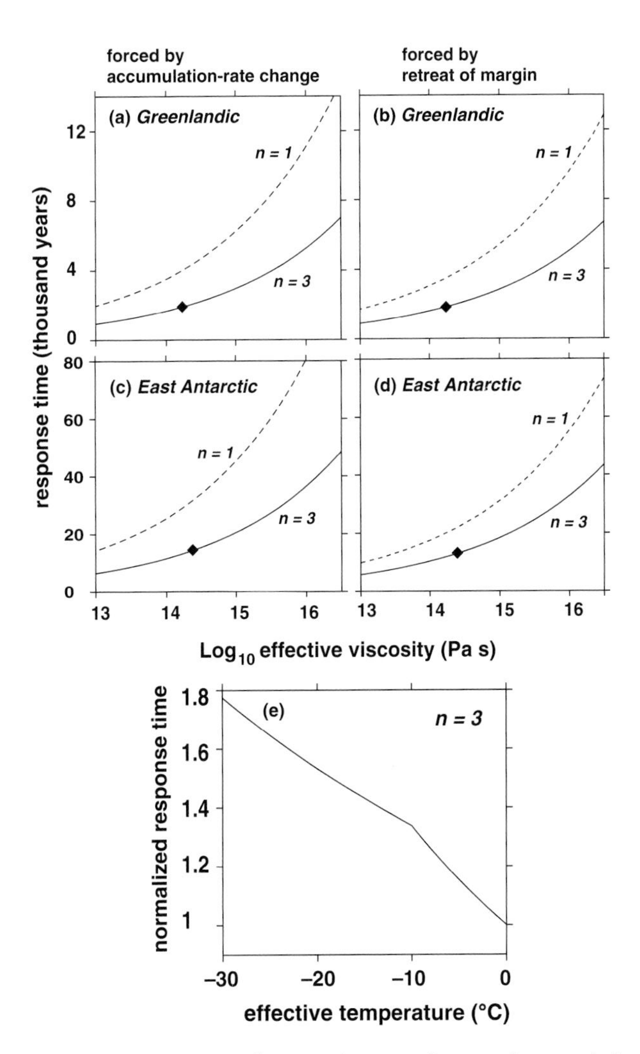

Figure 57.4 Dependence of approximate characteristic response-times on effective viscosity, stress exponent, and one control on viscosity (temperature), shown in panel e. Black diamonds (in a, b, c, and d) are estimates for modern EAIS and GIS. The response-time to accumulation-rate change is that for which ice surface slope does not change, corresponding to a situation wherein the ice margin is free to expand. The comparable response times for fixed margins are smaller, by a factor of $(n+2)/(2n+2)$. These response times neglect covariant change in temperature, which will slow all responses.

above the flotation level as the shear margins become stronger. This type of ice sheet could, in principle, convert from being concave-upward on its flank to being convex-upward, as ice streams stiffen or are replaced by strong-bed regions.

Given the measured activation energies for ice, the full range of effective temperatures for terrestrial glaciers could induce a nearly factor-of-two variation in ice volumes. However, because the thick ice sheets are thermal insulators, it is unlikely that effective temperatures for the ice sheets would vary outside the range of approximately −1 to −15°C, corresponding to a ca. 50% volume variation. The capacity for enhancement variations to induce volume variation is even smaller, but still significant. An enhancement variation of a factor of three, as observed in some of the Arctic ice-cap boreholes, would induce a ca. 15% variation (as can be seen directly from Equation 18).

The ability of temperature decrease to stiffen the margins of ice streams implies that the ice-stream width versus ice-sheet volume relation for a WAIS-type ice sheet is temperature dependent (Fig. 57.3). Thus, despite the vast complexity of the ice-stream-dominated system, it is clearly important to account for the activation energy of ice creep when analysing, for example, the configuration of the ice streams during the Last Glacial Maximum.

57.4.2 Response time

An ice sheet has many different response times, corresponding to different types of external and internal forcings. Most important of the external forcings are changes of interior accumulation rate, changes of marginal position induced by ablation-zone processes or sea-level change, and changes in temperature. These response times decrease as the stress exponent n increases (for reasons explained in section 57.2.2.1). They are also functions of ice thickness through the characteristic time $H/\dot{b}$, which is essentially a residence time and is related to how rapidly mass can be redistributed to approach a new equilibrium configuration.

Two examples of how response times might increase with effective viscosity and with decreased stress exponent are shown here (Fig. 57.4): (i) the linearized estimate for response time of ice thickness to a change in accumulation rate ($H/(n+2)\dot{b}$; see Whillans, 1981); and (ii) the diffusive-limit estimate for response time of interior thinning to retraction of the ice margin (which depends inversely on the diffusivity γ_1; see Cuffey & Clow, 1997). Figure 57.4 also shows the dependence of the response times on effective ice temperature; cold ice sheets respond more slowly. Because this effect arises entirely from the dependence of

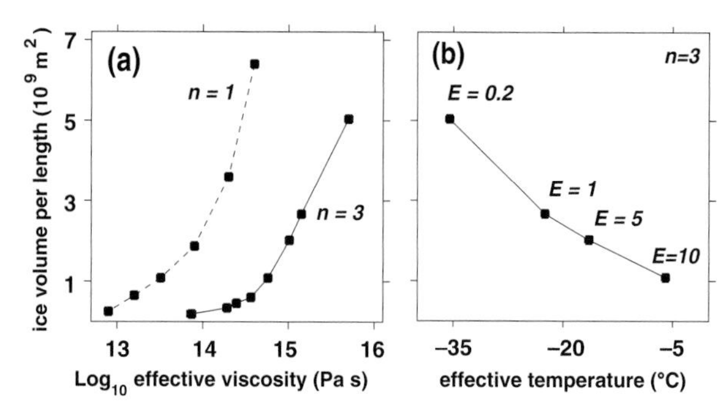

Figure 57.5 Relations between steady-state ice volume per divide length and other variables for a Greenlandic-type ice sheet with a flat bed, allowing for feed-back between accumulation rate and topography. Black squares indicate points for which the calculation was performed. In panel (b), enhancement values are indicated corresponding to the equivalent softness change induced by the temperature change as shown.

ice-sheet geometry on temperature, it is in fact identical to the volume–temperature relation.

57.4.3 Ice volume, given climate feed-back

In the preceding calculations, climatic forcing and ice-sheet span were fixed. In reality, there will be complex feed-backs between ice-sheet topography and the climatic forcing, with melt rates generally being reduced as elevation increases, and with ice-sheet interiors becoming very dry as they are elevated and isolated from atmospheric moisture transported to the margins (Weertman, 1976; Oerlemans, 1980). In some cases, the margins (and hence L) will indeed be fixed, by sea-level or topography. In other cases, as with the GIS and the Pleistocene Laurentide Ice Sheet, the ice sheet is free to expand across the land surface until a warmer climate is reached or a calving margin established.

The effect of such feed-backs is to greatly amplify the dependence of ice volume on viscosity and on the stress exponent. Consider one example (Fig. 57.5), a Greenland-type ice sheet for which the accumulation rate above the equilibrium line is fixed as a function of elevation (above equilibrium line) rather than as a function of position. Further, let the equilibrium line altitude rise linearly with distance from the ice-sheet centre, with ablation rate at zero elevation increasing similarly. This situation corresponds to an ice sheet that is centred in a climatically cold region and that encounters an increasingly hostile warm climate as it expands outward (L increasing). Here the ice-sheet profiles are calculated using a finite-difference solution to Equation (13), and this profile calculation iterated with increasing L values until the steady-state profile is found. Note that I do not include isostatic effects here (see Oerlemans, 1980).

The sensitivity of the steady-state ice volume to effective viscosity is much greater in this case than in those of Fig. 57.2 because the thicker ice also implies a cooler surface climate and therefore a narrower ablation zone. Further, as the span L increases, the total flux through the equilibrium line becomes larger so the ablation needs to be more intense to balance it. Reducing the stress exponent n causes a particularly large increase in L and volume before a new steady configuration is attained because the ice-sheet margin is steeper, as discussed in section 57.4.1.

The exact magnitude of volume change per viscosity change shown in Fig. 57.5 is, of course, dependent on the choice of horizontal ablation-rate gradient ($-0.01\,\mathrm{m\,yr^{-1}\,km^{-1}}$ used here), and so only applies for this one case.

57.5 Connecting to underlying microphysics

Can we understand how microphysics determine the phenomenological constitutive relation for ice? At present, this is only partly possible (Duval *et al.*, 1983; Weertman, 1983; Budd & Jacka, 1989; Alley, 1992; Goldsby & Kohlstedt, 2001).

57.5.1 Why is $\dot{\varepsilon} \propto A\tau^n$?

The bulk constitutive properties are manifestations of the underlying deformation mechanisms. These will vary with conditions of stress, temperature and grain size (Duval *et al.*, 1983; Goldsby & Kohlstedt, 2001). It is first essential to recognize that the constitutive relation for ice analysed and supported in this article (power-law creep, with $n \approx 3$) is applicable specifically to those parts of the ice sheets most important for their large-scale flow and structure: the rapidly deforming layers near the glacier bed on the ice sheet flanks and in ice-stream shear margins. A modified constitutive rule will be needed elsewhere if conditions favour different deformation mechanisms, such as in the low-stress environment of the ice divide region (see Pettit, this volume, Chapter 58).

One important deformation mechanism relevant here is dislocation glide, the motion of crystal lattice defects along the basal planes of the hexagonal ice crystals. Individual ice crystals are like little decks of cards, with easy shear deformation in one plane only; these are the basal planes. This deformation mechanism is certainly dominant in ice, as it quantitatively explains the observed development of crystal c-axis fabrics (the statistics of orientations) and explains the observed dependence of fluidity enhancement on these fabrics (Li *et al.*, 1996; Azuma & Goto-Azuma, 1996; De La Chapelle *et al.*, 1998). It is unlikely to be the rate-limiting mechanism, however. Laboratory measurements of single ice crystals deforming by dislocation glide show the stress exponent is $n \approx 2$–3 for this mechanism in isolation (data are reviewed by Goldsby & Kohlstedt, 2001). In a polycrystal, ice grains cannot deform only by this mechanism because adjacent grains with different shapes and orientations interfere. Other mechanisms must accommodate such incompatible deformations, and these are probably rate-limiting. These may include dislocation motion in stiff directions, diffusion along grain boundaries and grain-boundary sliding (which may also be a consequence of dislocation motion). Rate limitation may also arise from a need to recrystallize or move grain boundaries to eliminate dislocation tangles (Montagnat & Duval, 2000). Uncertainty about the rate-limiting mechanism through the full range of ice-sheet stresses is a major gap in our current knowledge.

Generally, the rate of deformation arising from dislocation mechanisms is proportional to the product of dislocation speed and dislocation density (Weertman, 1983). Both speed and density depend on deviatoric stress, and it is the product of these dependences that determines the exponent n for these mechanisms. The speed of dislocation motion is linear with stress. Weertman (1983) has argued that $n = 3$ occurs when the deformation attains a steady-state, requiring that the external applied stress is balanced by internal stresses associated with lattice distortions around dislocations; in such a condition, theory suggests that dislocation density will be proportional to the square of stress, yielding a cubic product. This sort of steady-state is likely to occur in the rapidly straining deep layers of the ice sheets and smaller glaciers (Alley, 1992), where recrystallization is known to occur.

Clouding of this simple picture (Weertman, 1983) arises from clear data showing a transition to $n \approx 4$ at high stresses in excess of 1 MPa (data reviewed by Goldsby & Kohlstedt, 2001). This is the dislocation creep regime, where the rate-limiting process is dislocation motion in stiff directions. This has led to the idea that the phenomenological result $n \approx 3$ near 1 bar stress is describing a transition regime, where dislocation creep with high n is comparable in magnitude to some other mechanism with lower n that is rate-limiting at lower stress.

Molecular-scale properties that control the ice softness A will also be different depending on which deformation mechanisms are active and rate-limiting. For dislocation-motion mechanisms (Weertman, 1983), the molecular spacing in the lattice is important. Each dislocation jump will move a distance similar to this spacing (the Burgers vector), so the dislocation speed should be proportional to it. Dislocation density, if controlled by the balance of internal and external stresses, will depend inversely on the second power of both the molecular spacing and the shear modulus for the ice lattice. There is presently no calculation of polycrystalline ice viscosity from first principles and known lattice properties.

57.5.2 What controls temperature-dependence?

The temperature-dependence $A(T)$ arises because dislocation jumps are thermally activated (Weertman, 1973). As temperature increases, molecules are increasingly agitated and it becomes more probable that a strained bond adjacent to a dislocation will be broken. Thus the speed of dislocation motion increases with temperature according to statistics of the Boltzmann distribution, giving the Arrhenius relation (Equation 4).

The value for activation energy at $T < -10°C$ is similar to that for self-diffusion of ice (Weertman, 1983). The increase of apparent activation energy above this temperature indicates that the rate-limiting mechanisms are changing qualitatively as the melting temperature is approached (Paterson, 1994). This suggests an increasing accommodation of incompatible deformations by grain-boundary mechanisms, because the grain boundaries are probably significantly broadened and altered by liquid films at these temperatures (see Dash *et al.*, 1995).

57.5.3 What controls enhancement?

Enhancement variations (Budd & Jacka, 1989; Paterson, 1991; Cuffey *et al.*, 2000a) arise from grain-scale processes associated with two physical properties of polycrystalline ice: c-axis fabric strength and grain size. As noted previously, ice crystals deform easily by glide on basal planes, the planes normal to the c-axis. Preferred orientation of c-axes (also referred to as 'strong fabric') makes ice soft for shear resolved on these planes. The preferred orientation itself develops as a function of strain as crystals rotate when deforming to minimize interaction with neighbours. Most of the rapidly shearing deep ice in the ice sheets has been strained to very large values and consequently has strong vertically aligned fabrics, with partial broadening by recrystallization. In a simple shear flow, the capacity for enhancement variation between isotropic ice and ice with a favourably oriented perfect fabric is approximately a factor of ten (Azuma & Goto-Azuma, 1996). Typical actual documented enhancement contrasts that are relevant to large-scale, ice-sheet shear flow are approximately a factor of three (Dahl-Jensen & Gundestrup, 1987). This is much smaller than the potential factor of ten because ice deforming in shear in the high-stress deep layers of the ice sheets is never isotropic.

Both small grain sizes and high soluble impurity contents contribute to much larger enhancements of dirty basal layers of glaciers (Cuffey *et al.*, 2000a). There is no convincing evidence that soluble impurities affect enhancement at the concentrations found in the main body of the ice sheets, but grains are probably fine enough in some cases to induce enhancement variations (Cuffey *et al.*, 2000c). This needs more investigation. Viscosity variations due to grain size could result from grain size dependence of dislocation recovery processes (Montagnat & Duval, 2000), or possibly from deformation by grain-boundary sliding (Goldsby & Kohlstedt, 2001). Laboratory experiments (Jones & Glen, 1969) do show that some acids (such as HF and HCl) have the capacity to soften ice even at low concentrations. The soft ice-age ices of the Northern Hemisphere are alkaline, however, due to the abundance of rock microparticles.

57.6 Concluding summary and perspective

Ice sheets powerfully influence the global environment through their major characteristics: sea-level-equivalent volume, span, and topographic profile. Microphysical processes in ice determine how these major characteristics arise from climatic and sea-level forcings (though this is strongly modulated in some cases by the distribution of weak-bed regions). The most important effects of microphysical processes are embodied at the phenomenological scale by the power-law creep constitutive relation for ice. Although approximate, this has tremendous predictive power, explaining much about the cross-sectional volume and topographic form of the ice sheets, given climatic forcings. The most notable characteristics of this relation are that:

1 the behaviour of rapidly shearing deep ice is intermediate between linear-viscous and perfectly plastic, with $n \approx 3$;
2 the effective viscosity is strongly temperature-dependent;
3 the effective viscosity depends on physical properties of the ice associated with crystal-orientation fabrics and possibly with grain size.

For a given climatic forcing and distribution of weak-bed regions, ice volume increases with increasing effective viscosity and decreasing stress exponent. These dependencies are greatly enhanced if there is full feed-back between climatic forcing and ice sheet surface elevation and span.

The origin of the constitutive relation in terms of the microphysical processes themselves remains unclear, but is strongly affected by the density and motion of flaws (dislocations) in the ice crystal lattice. Deformation is easy by dislocation glide along basal planes. In polycrystalline ice, this gives rise to incompatible deformations that are accommodated by other mechanisms, including dislocation motion in stiff directions and grain-boundary sliding. These other mechanisms are rate-limiting, and their importance varies with conditions of stress, temperature and grain size.

The relationship between microphysical processes and large-scale ice-sheet characteristics discussed in this chapter is a pervasive influence on the response of ice sheets to environmental changes, discussed from numerous perspectives in this volume. As climate warms and equilibrium lines rise, the ice-sheet surface profiles strongly control the expansion of ablation zones and consequent mass losses. Similarly, as sea level rises and forces ice-sheet margins to retreat, the resulting mass loss depends critically on the adjustment of the ice sheet interior toward a new stable profile. Climate changes that cause increased accumulation on an ice sheet interior will induce thickening and mass gain, the magnitude and timing of which are consequences of the dynamics discussed in section 57.2. A sound understanding of the large-scale ice-sheet characteristics and dynamics arising from microphysics is a conceptual key for understanding the response of ice sheets to environmental changes.

FIFTY-EIGHT

Ice flow at low deviatoric stress: Siple Dome, West Antarctica

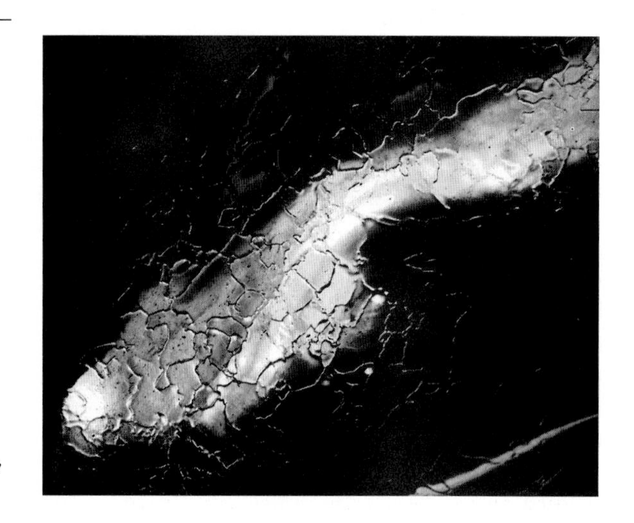

Erin C. Pettit

Department of Earth and Space Sciences, Box 351310, University of Washington, Seattle, Washington 98195, USA

There are situations where Glen's Law is inadequate for modelling ice flow, especially as we ask our models to reflect more details of the natural system. In this case study I focus on the flow of ice in low-deviatoric-stress regimes, such as exist in the divide region of an ice sheet.

In ice, as with other polycrystalline materials, the dominant microscale flow mechanism changes with temperature, deviatoric stress, and grain size. A comprehensive flow law to describe ice deformation should include a term for each flow mechanism. From a microphysical perspective, this kind of flow law is essential for understanding and modelling ice behaviour. Translating this behaviour to the macroscale flow of ice sheets, on the other hand, requires a flow law that is simple to implement in an ice sheet model, yet still captures the effect of shifts in the flow mechanisms. I have modified Glen's Law to accommodate the shift in behaviour at low stress by adding a linear term:

$$\dot{\varepsilon}_{ij} = \left[\underbrace{\frac{E_l A_{o_l}}{d^{p_l}} e^{-\frac{Q_l}{RT}}}_{\text{linear term}} + \underbrace{\frac{E_{nl} A_{o_{nl}}}{d^{p_{nl}}} e^{-\frac{Q_{nl}}{RT}} (\tau_{eff}^2)}_{\text{Glen term}} \right] \tau_{ij} \tag{1}$$

where $\dot{\varepsilon}_{ij}$ is the strain-rate tensor, τ_{ij} is the deviatoric-stress tensor and τ_{eff} is the second invariant of τ_{ij}; E is an enhancement factor (with subscripts referring to linear, l, and non-linear, nl, processes); A_o is the temperature-independent part of the softness parameter defined for clean, isotropic Holocene ice; d is grain size (with exponent p); Q is activation energy; and T is temperature. Although this formulation of the flow law does not account for each separate mechanism, its form can approximate the behaviour observed by laboratory and field experiments. Equation (1) can also be written

$$\dot{\varepsilon}_{ij} = \Gamma[k^2 + \tau_{\mathrm{eff}}^2]\,\tau_{ij} \tag{2}$$

where

$$\Gamma = \frac{E_{\mathrm{nl}}A_{o_{\mathrm{nl}}}}{d^{p_{\mathrm{nl}}}}e^{-\frac{Q_{\mathrm{nl}}}{RT}} \quad \text{and} \quad k = \left[\frac{E_1 A_{o_1}}{E_{\mathrm{nl}}A_{o_{\mathrm{nl}}}}\frac{d^{p_{\mathrm{nl}}}}{d^{p_1}}e^{-\frac{Q_1 - Q_{\mathrm{nl}}}{RT}}\right]^{1/2} \tag{3}$$

where k is the *crossover stress*: the effective deviatoric stress at which the linear and non-linear terms contribute equally to the total strain rate. The variable k is a rheological property of ice; although it may depend on temperature and grain size, it is independent of the geometry and climate of the ice sheet. Prior to this work, the stress at which the dominant flow mechanism changes was poorly known. Here I use conditions at Siple Dome to determine k.

This linear term becomes significant only at low deviatoric stresses. The lowest stresses in an ice sheet are found in the ice-divide region, where both normal and shear stresses approach zero. As Raymond (1983) observed, the non-linear nature of Glen's Law predicts very stiff ice under the divide where stresses are low, when compared with ice on the flanks. This stiff ice affects the flow and, therefore, the internal stratigraphy. Raymond predicted that a steady-state divide would reflect this special flow pattern by forming an arch in the isochrons. If a linear flow mechanism dominates at low stress this special divide flow pattern would be altered.

58.1 The ideal ice-sheet model

The relative importance of the linear deformation mechanisms at a particular divide depends on the stress, temperature and grain-size distributions at that divide. I define the characteristic stress of a divide as: $\tau_{\mathrm{char}} = [2\Gamma H/\dot{b}]^{-1/3}$, which is derived from the characteristic strain rate, $\dot{\varepsilon} \approx \dot{b}/H$, and Glen's Law. If the characteristic stress for a particular divide is equal to or smaller than the crossover stress, k (for ice with similar temperature and grain-size characteristics), then the linear term will play a significant role in the flow pattern. Figure 58.1 shows the isochrons from a model of an idealized ice sheet resembling Siple Dome, West Antarctica, an ice ridge between Bindschadler Ice Stream and Kamb Ice Stream. With 1000-m-thick ice and an accumulation rate of 10 cm yr^{-1}, the characteristic stress for Siple Dome is approximately 0.2 bar. The figure shows isochrons for four different values for the unknown crossover stress. The left-most figure is a divide dominated by Glen-type flow, showing the arch in the isochrons predicted by Raymond (1983). As the unknown crossover stress increases, the arch decreases in magnitude. The arch disappears when the linear term dominates the deformation rate. In ice-penetrating-radar images of Siple Dome, an arch in the isochrons is visible (e.g. Nereson *et al.*, 1998b), implying that the non-linear (Glen) term in the flow law is significant. Due to transient processes such as divide migration (Nereson *et al.*, 1998b), however, this analysis does not identify Siple Dome as a Glen-type divide or a transitional divide.

58.2 The Siple Dome model

Siple Dome is a well studied divide. As it is the site of a recent deep-drilling project by the United States Antarctic Program, Pettit (2003) uses available ice-core data and deformation measurements to solve an inverse problem for flow-law parameters that best describe the flow at Siple Dome utilizing an ice-flow model developed by H.P. Jacobson and T. Thorsteinsson. I focus on four flow-law parameters: the crossover stress (k), which I assume is spatially constant, and a three-layer isotropic enhancement parameterization, describing Holocene ice ($\hat{E}_1$), ice-age ice ($\hat{E}_2$), and the deep recrystallized ice ($\hat{E}_3$), where $\hat{E}_i = \frac{E_{\mathrm{nl}}}{d^{p_{\mathrm{nl}}}}$ is derived from Equation (3) assuming $A_{o_{\mathrm{nl}}}$ is the suggested value from Paterson (1994, p. 91, based on several empirical studies of ice flow). These enhancement factors, $\hat{E}_i$, are for isotropic enhancement only. The model accounts for anisotropic flow explicitly using an analytical formulation for anisotropic deformation developed by Thorsteinsson (2001). Pettit (2003) describes the

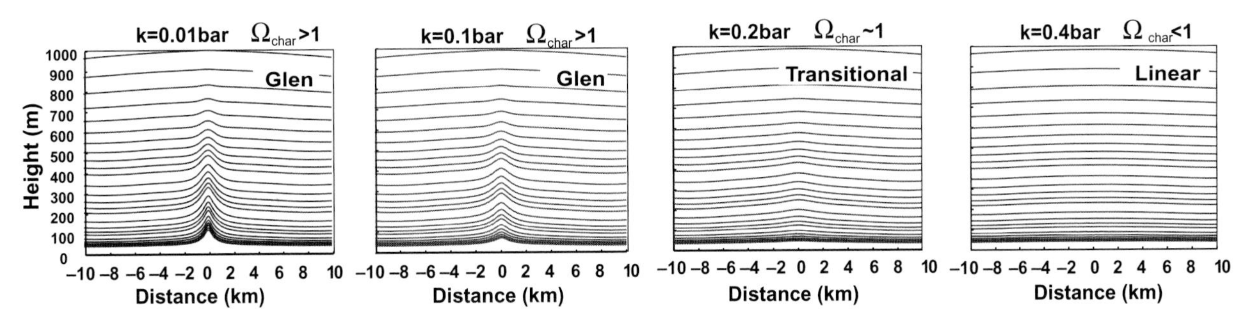

Figure 58.1 Modelled steady-state isochrons for an idealized divide with a thickness of 1000 m and accumulation rate of 0.1 cm yr^{-1}, resembling Siple Dome, West Antarctica. The crossover stress used in the model increases from left to right. The importance of the linear term also increases from left to right, as reflected in the diminishing size of the isochron arch. Ω_{char} is the non-dimensional ratio τ_{char}/k. (From Pettit & Waddington, 2003, reprinted from the Journal of Glaciology with permission of the International Glaciological Society.)

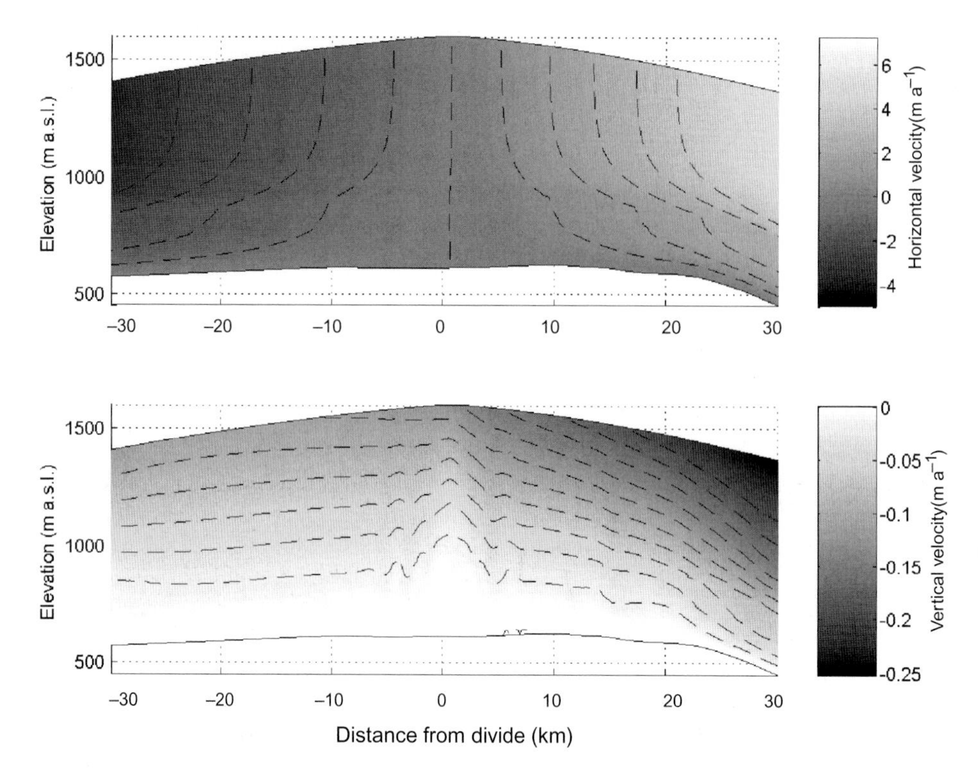

Figure 58.2 Horizontal (top panel) and vertical (bottom panel) velocity fields from the best-fitting model assuming steady state. For both panels, Bindschadler Ice Stream is to the right and Kamb Ice Stream is to the left. The dashed lines are velocity contours.

implementation of Thorsteinsson's formulation in this ice-flow model. The inputs to the model are the surface and bed geometry (from GPS and radar, respectively, Nereson *et al.*, 1998b), the temperature (from hot-water drilled borehole logging; H. Engelhardt, personal communication), and the crystal fabric profile (from borehole sonic log; G. Lamorey, personal communication).

In order to invert for the four unknown flow-law parameters, I compare the model results to measurements of the vertical strain-rate profile at the divide and on the flank (e.g. Zumberge *et al.*, 2002; Elsberg *et al.*, 2004). The vertical strain-rate profile at the divide is sensitive to the relative magnitude of the linear term, whereas that at the flank is not. I chose to compare modelled and measured instantaneous deformation rates rather than internal layer shapes, as the internal layer shapes are a function of the deformational history of the divide.

Our model fits the data best when the crossover stress, k, is 0.22 bar. As the characteristic stress at Siple Dome is about 0.2 bar, I consider it to be a transitional divide, which means that the linear and non-linear terms contribute roughly equally to the overall deformation-rate pattern. The best-fitting enhancement factors are 1.3 for Holocene ice, 0.06 for ice-age ice, and 0.2 for the recrystallized ice near the bed. As the enhancement factor is defined relative to the suggested softness parameter from Paterson (1994), our result of 1.3 for Holocene ice at Siple Dome suggests that our model produces deformation rates close to those expected for clean Holocene ice under similar conditions. The enhancement factors for the two deeper layers are much lower than I expected. This remains a puzzle, although Thorsteinsson *et al.* (1999), who also attempted to separate the softness due to anisotropy from that due to other ice properties, found a similar result at Dye 3 in Greenland. Possibly a poorly understood rate-limiting process such as grain-boundary migration increases the stiffness of the ice.

Using these best-fitting flow-law parameters, Fig. 58.2 shows the horizontal- and vertical-velocity fields for Siple Dome. The horizontal-velocity field shows a sharp change in gradient about 300 m above the bed. This shift corresponds in depth to a shift towards tighter crystal fabric alignment and has been dated to near the transition from ice-age to Holocene ice (Taylor *et al.*, 2004). The tight vertically oriented crystal fabric of the ice-age ice concentrates the shear strain, as anisotropic ice is easy to deform along the basal planes.

Our goal in this case study was to determine the relative importance of linear deformation mechanisms in low-stress regions, particularly near ice divides. Toward this goal, I suggested a two-term flow law where the crossover stress, k, is a rheological property of ice. We found that k is about 0.2 bar for the temperature and grain size of the ice within Siple Dome. Divides worldwide range in characteristic stress from 0.1 bar (Valkyrie Dome, site of Dome Fuji Station) to near 0.4 bar (Quelccaya Ice Cap, Peru). When modelling flow at many divides, therefore, including a linear term in the flow law is likely to be important. Our model of Siple Dome also implies that other processes, in particular crystal-fabric anisotropy, are equally important to successfully capturing the characteristic behaviour of an ice divide.

Acknowledgements

This work resulted from a collaboration among the University of Washington, University of Alaska Fairbanks, and the University of California San Diego. Ed Waddington, Paul Jacobson, Throstur Thorsteinsson, Will Harrison, Dan Elsberg, John Morack, Mark Zumberge and Eric Husmann contributed ideas and data. Greg Lamorey, Richard Alley, Larry Wilen and others from the Siple Dome Deep Drilling Project contributed their data for this analysis.

FIFTY-NINE

Physical deformation modes of ice in glaciers and ice sheets

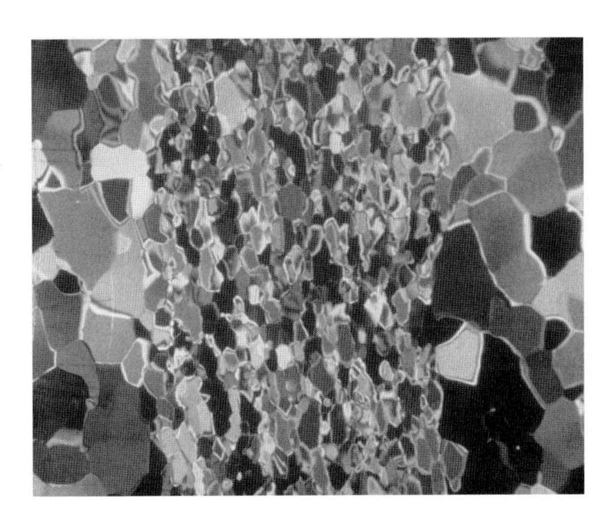

Paul Duval and Maurine Montagnat

Laboratoire de Glaciologie et Géophysique de l'Environnement, B.P. 96, 38402 Saint Martin d'Hères Cedex, France

59.1 Introduction

The slow motion and changes of glaciers and large ice masses are governed by the deformation of polycrystalline ice. The constitutive law of polycrystalline ice for ice-sheet modelling is that of an incompressible, non-linear viscous fluid. Deviatoric stresses in ice sheets are generally lower than 0.1 MPa and strain rates are typically between 10^{-10} and $10^{-13}\,s^{-1}$, but can reach $10^{-7}\,s^{-1}$ in temperate glaciers. It is difficult to obtain valuable information on the ice-flow law under low stress conditions. Laboratory tests take too long to obtain a significant amount of deformation under these conditions and extrapolation from tests performed at higher stresses introduces significant uncertainty. Much progress has been made, however, with the study of the ice structure in deep ice-cores recently retrieved from Antarctica and Greenland. The sensitivity of strain rate to stress in ice sheets is characterized by a stress exponent lower than 2. The rate-controlling processes are not totally clear, but basal slip is the dominant deformation mode (Montagnat & Duval, 2000; Montagnat *et al.*, 2003).

The deformation of polar ice induces the development of lattice-preferred orientations (fabrics) producing a non-random distribution of c-axis orientation. Owing to the very large anisotropy of ice crystals (Duval *et al.*, 1983) and the preponderance of intracrystalline dislocation glide in polar ice (Alley, 1992), initially isotropic ice formed after transformation of snow and firn near the surface becomes anisotropic as fabrics develop. Very large variations of strain rates with the applied stress direction have been found using both laboratory and *in situ* measurements (Russell-Head & Budd, 1979; Pimienta *et al.*, 1987; Budd & Jacka, 1989). Up to now, models describing the evolution of ice sheets have not accounted for the changing rheological properties of ice with time. A first attempt, however, has been made by Mangeney *et al.* (1997) and Salamatin & Malikova (2000).

This work focuses on the deformation modes of polycrystalline ice for conditions prevailing in glaciers and ice sheets. We will show that intracrystalline dislocation glide is compatible with *in situ* deformation measurements and the development of fabrics. This paper is also intended to clarify the relationship between recrystallization and fabrics. Emphasis is placed on the influence of impurities and crystal size on strain rate in the low stress conditions of ice sheets. The role of a liquid phase in the deformation of ice is also discussed.

59.2 Deformation of ice single-crystals

The main feature of the plasticity of an ice crystal is its outstanding anisotropy. For shear stresses corresponding to those found in glaciers and ice sheets, ice deforms by basal slip. Basal slip is caused by the glide motion of basal dislocations with the $\langle 11\bar{2}0\rangle$ Burgers vector. Variations of steady-state strain rates with stress for basal creep are shown in Fig. 59.1. Almost all authors report a stress exponent of about 2 and activation energy close to $63\,kJ\,mol^{-1}$.

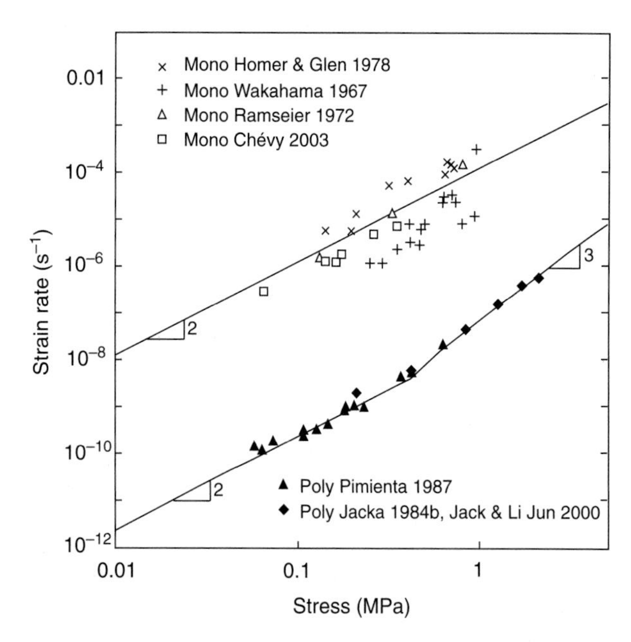

Figure 59.1 Strain rate as a function of stress for basal slip in ice crystals and for creep of isotropic polycrystalline ice at −10°C.

When ice crystals are loaded such that there is no resolved shear stress on the basal plane, creep rates are so small that measurements are uncertain. A slight misorientation of the c-axis with respect to the direction of the applied axial stress induces basal slip. According to Duval *et al.* (1983), under the same prescribed stress, a crystal sheared parallel to its basal plane exhibits a strain rate that can be up to three orders of magnitude greater than that measured when an ice crystal is compressed along or perpendicular to the c-axis. The rapid glide of short-edge dislocations on non-basal planes has been observed by X-ray topography (Higashi *et al.*, 1985; Ahmad & Whitworth, 1988). The fact that basal slip is dominant, in spite of faster movement of such non-basal dislocations, is attributed by these authors to the large difference in respective dislocation density. According to Hondoh *et al.* (1990), the long basal screw dislocations should be dissociated on the basal plane thereby impeding the glide of these dislocations on non-basal planes.

59.3 Deformation of isotropic polycrystalline ice

59.3.1 Viscous behaviour at high stresses

Creep data for isotropic polycrystalline ice are given in Fig. 59.1. They concern the minimum creep rate reached after a strain of about 1%. The stress exponent is close to 3 for stresses between 0.1 and 2 MPa. During primary creep, strain rate decreases by more than two orders of magnitude. This behaviour is directly related to the strong anisotropy of ice crystals (Duval *et al.*, 1983). On initial loading, stress state within the polycrystalline ice sample is almost uniform. Deformation is essentially produced by basal slip. Because grain boundaries act as obstacles to dislocation slip, resolved stress on basal planes is relaxed through load transfer to harder slip systems. As a result, an increasingly non-uniform state of internal stress develops with peaks in amplitude near grain boundaries. Initial creep rate is largely determined by basal slip systems, but the steady strain rate is determined by an appropriate average over the resistance of all slip systems (Ashby & Duval, 1985). Basal slip provides two independent systems. Slip or climb of dislocations on non-basal planes, as discussed above, would give additional independent systems.

Beyond 1 to 2% strain, creep rate increases to a maximum corresponding to about 10% strain (Jacka & Maccagnan, 1984). Tertiary creep with almost constant strain rate is explained by the formation of fabrics and the occurrence of dynamic recrystallization acting as a softening process (Duval *et al.*, 1983). Depending on stress state, tertiary creep rate is between three and ten times higher than the minimum strain rate (Budd & Jacka, 1989).

59.3.2 Viscous behaviour of ice at low stresses; application to polar ice sheets

59.3.2.1 Deformation modes

For conditions prevailing in ice sheets (equivalent stress lower than 0.2 MPa), the stress exponent is slightly lower than 2, a value close to that found in isolated single crystals (Fig. 59.1). This result is supported by densification measurements of bubbly ice at Vostok (Lipenkov *et al.*, 1997). The high difference in strain rate between crystals oriented for basal slip and isotropic ice (Fig. 59.1) cannot be explained by a geometric effect related to the random orientation of grains. As at high stresses, basal slip is the dominant deformation mode, but other deformation modes are required to assure the compatibility of deformation. However, the internal stress field induced by the mismatch of slip at grain boundaries is reduced under the low stress conditions of ice sheets by grain-boundary migration. By sweeping away dislocations located in front of moving grain boundaries, grain-boundary migration associated with grain growth and recrystallization prevents kinematic hardening caused by the incompatibility of deformation between grains. Accommodation of slip by grain-boundary migration in polar ice must be taken into account to explain the low stress exponent observed at low stresses.

A physical deformation model, considering dislocation density within an average grain as an internal variable, was developed by Montagnat & Duval (2000). The increase of dislocation density by work hardening is balanced by grain-boundary migration and by the formation of new grain boundaries. This model accounts for the transition between grain growth and recrystallization in several locations in the Greenland and Antarctic ice sheets. However, the low stress exponent of the flow law cannot be deduced from such a model. It is also difficult to assume that the same value of stress exponent for single crystals and polycrystalline ice at low stresses (Fig. 59.1) implies the same rate-controlling processes.

Grain-boundary sliding (GBS) is also suggested to accommodate basal slip, but the occurrence of such a process in the flow conditions of ice sheets has not been proven. This deformation mode was put forward by Goldsby & Kohlstedt (2001, 2002), who consider polar ice as a superplastic material. It is worth noting that the dominant deformation mode corresponding to the n = 1.8 regime in superplastic materials is GBS (Langdon, 1994; Goldsby & Kohlstedt, 1997). The observed microstructures and the development of preferential orientations of ice crystals in ice sheets, associated with easy basal slip, are clearly not compatible with GBS as the dominant deformation mode in polar ice sheets (Duval & Montagnat, 2002).

59.3.2.2 Dynamic recrystallization in glaciers and ice sheets

The evolution of texture in ice sheets is achieved via three recrystallization processes, which are termed: normal grain growth, rotation recrystallization and migration recrystallization (Alley, 1992; Duval & Castelnau, 1995; De La Chapelle *et al.*, 1998). In the upper layers of ice sheets (several hundred metres), the mean grain size increases with depth (Gow, 1969; Gow & Williamson, 1976). The driving force for the normal grain growth results from the decrease in free energy that accompanies reduction in grain-boundary area. The driving force $3\gamma_{gb}/D$ is less than $100\,\mathrm{J\,m^{-3}}$ because grain size D is larger than 1 mm and the grain-boundary free energy $\gamma_{gb} = 0.065\,\mathrm{J\,m^{-2}}$. The boundary migration velocity is typically $10^{-15}\,\mathrm{m\,s^{-1}}$ at −50°C and can reach a value higher than $10^{-6}\,\mathrm{m\,s^{-1}}$ at the melting point. Grain growth appears to be inhibited by soluble and (or) insoluble impurities (Alley *et al.*, 1986b,c; Fisher & Koerner, 1986; Alley & Woods, 1996; Li *et al.*, 1998). Particles deposited during the Last Glacial Maximum (LGM) and located on grain boundaries seem to reduce the rate of grain growth (Weiss *et al.*, 2002). The pinning pressure due to particles randomly distributed within the ice volume is shown to be too low to explain fine-grained ice in LGM (Alley *et al.*, 1986c).

High-angle boundaries form by the progressive misorientation of sub-boundaries (De La Chapelle *et al.*, 1998). This recrystallization mechanism is termed rotation recrystallization (Poirier, 1985). It counteracts further grain-size increase due to grain growth from 400 m depth in the Byrd ice core (Alley *et al.*, 1995) and 650 m in the GRIP ice core (Thorsteinsson *et al.*, 1997). In this recrystallization regime, grain boundaries migrate in the same low-velocity regime as the one associated with normal grain growth (Duval & Castelnau, 1995).

In the last hundred metres of ice sheets, just above the bedrock, temperature can be higher than −10°C, reaching the melting point at the interface between ice and rock. In this zone, rapid migration of grain boundaries can occur between dislocation-free nuclei and deformed grains. This recrystallization regime, referred to as migration recrystallization, produces coarse and interlocking grains (Gow & Williamson, 1976). The velocity of grain-boundary migration is more than 100 times higher than that associated with rotation recrystallization (Duval & Castelnau, 1995). It is worth noting that migration recrystallization is very active in temperate glaciers and is associated with tertiary creep.

59.3.2.3 Rate-controlling processes in the creep of polar ice; effect of crystal size

The deformation of polar ice is essentially produced by dislocation slip on basal planes. The mismatch of slip at grain boundaries induces lattice distortion and strain gradients. Grain-boundary migration associated with grain growth and recrystallization reduces the dislocation density and the internal stress field. As a consequence, strain rate is higher than that extrapolated from high stress conditions with a stress exponent $n = 3$. The large difference between the basal slip of single crystals and the creep behaviour of polycrystalline ice (Fig. 59.1) shows that the incompatibility of deformation between grains is significant even at low stresses. The amount of non-basal slip or climb of dislocations required for compatibility reasons, however, should be reduced in the flow conditions of ice sheets. With regard to non-basal slip, the bending of basal planes observed by hard X-ray diffraction (Montagnat *et al.*, 2003) must be taken into account when considering deformation along the c-axis. Mechanisms that give the value of the stress exponent $n = 2$ at low stresses cannot be determined on the basis of this analysis alone. The slowest deformation mechanism could control strain rate if it occurs in series with basal slip. It is probably the case at high stresses where non-basal slip or climb of dislocations is the likely rate-controlling process. At low stresses, the amount of these hard deformation systems is reduced by several processes such as the formation of strain heterogeneities within grains and recovery by recrystallization.

The influence of grain size on strain rate has been the subject of discussions during the past few decades. The lack of significant grain-size effect on the minimum creep rate as shown by laboratory tests under deformation conditions corresponding to $n = 3$ (Duval & LeGac, 1980; Jacka, 1984a, 1994) is in accordance with the deformation modes discussed above. Owing to the difficulty of obtaining good laboratory data at low stresses, it is not possible to prove a grain size effect in the deformation conditions of ice sheets on the basis of laboratory experiments alone. The value of the stress exponent lower than 2, unambiguously deduced from *in situ* measurements in polar ice sheets, cannot prove the occurrence of grain-boundary sliding or other mode of deformation. A convincing analysis of deformation measurement data on Meserve Glacier (Antarctica) by Cuffey *et al.* (2000a) suggests a direct dependence of strain rate on grain size. A multi-term flow law with grain-size dependence was adopted by Pettit & Waddington (2003) to simulate the surface morphology of ice divides. A grain-size effect at low stresses, if not yet proven, is likely. It is in accordance with intracrystalline slip accommodated with grain-boundary migration (Montagnat *et al.*, 2003).

59.4 Fabric development

Tertiary creep is associated with migration recrystallization (Steinemann, 1958b). Fabrics are considered to be 'recrystallization fabrics' because the orientation of grains is related to the nucleation and migration of grains (Duval, 1981; Alley, 1992). Plastic strain associated with the formation of recrystallization fabrics is too low to produce important rotation of the lattice. Figure 59.2a shows a typical creep curve obtained at −3°C with the c-axis orientation measured after a strain of about 30%. Recrystallization fabrics are stress-controlled and recrystallized grains are well oriented for basal slip (Duval, 1981). Most of the fabrics that form in temperate glaciers result from migration recrystallization.

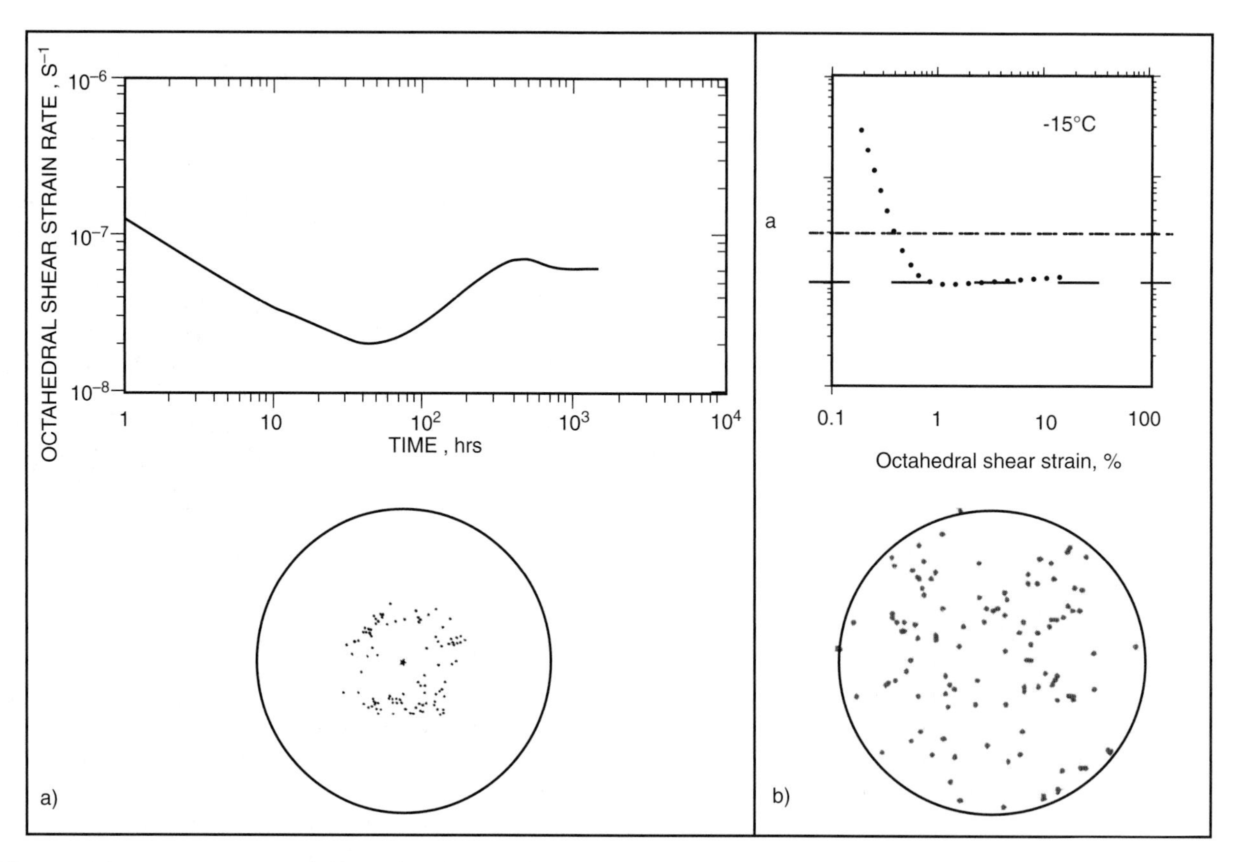

Figure 59.2 Creep curves and fabric data for initially isotropic ice samples deformed in uniaxial compression at 0.2 MPa. (a) Temperature θ = −3°C (from Jacka & Maccagnan, 1984). (b) Temperature θ = −15°C (from Jacka & Li, 2000).

In ice sheets, fabrics develop under the effect of intracrystalline slip (Azuma & Higashi, 1985; Alley, 1988) as long as migration recrystallization does not occur. High finite strain is therefore required to obtain significant preferential c-axis orientation. Figure 59.2b shows a creep curve obtained at −15°C under compression at 0.2 MPa by Jacka & Li (2000). Tertiary creep and, as a consequence, migration recrystallization were not initiated in this laboratory test. The weak fabric pattern is in accordance with this assertion (Fig. 59.2b). Figure 59.3 clearly illustrates the marked difference between fabrics induced by migration recrystallization and those induced by the rotation of the lattice by slip. Small-grained ice is associated with rotation recrystallization, which is stabilized by the presence of fine particles in glacial ice.

In conclusion, migration recrystallization readily occurs in temperate glaciers, but it is involved in ice sheets only at relatively high temperature and if the stored energy is sufficiently high. Microparticles can prevent the transition between rotation and migration recrystallization.

59.5 Effect of impurities and a liquid phase

59.5.1 Enhancement factor in the LGM ice

Analyses of the deformation behaviour of polar ice clearly show that ice deposited during the Last Glacial Maximum (LGM) deforms more readily than Holocene ice (cf. reviews of Budd & Jacka, 1989; Paterson, 1991). The enhanced flow is believed to be due to the high concentration of solid impurities (Dahl-Jensen, 1985; Fisher & Koerner, 1986; Dahl-Jensen & Gundestrup, 1987). Data on the concentration of solid impurities in Holocene and LGM ice in the Antarctic and Greenland ice sheets are given in Table 59.1. There is, indeed, a good correlation between ice viscosity and impurity content.

In order to express the relative fluidity of LGM ice, the flow law for isotropic ice is given by

$$\dot{\varepsilon}_{i,j} = E\frac{B}{2}\tau^{n-1}\sigma'_{i,j} \qquad (1)$$

where $\dot{\varepsilon}_{i,j}$ and $\sigma'_{i,j}$ are the strain rate and stress deviator components, τ is the effective shear stress, B is a parameter which depends on temperature and E is the enhancement factor. From borehole tilting measurements at Dye 3, this factor is equal to unity in the Holocene ice and has a value close to 3 in the LGM ice (Dahl-Jensen, 1985). A similar value was found in other Arctic stations (Paterson, 1991). By using an anisotropic flow model, Azuma & Goto-Azuma (1996) attributed the enhancement factor at Dye 3 to fabric anisotropy. From Thorsteinsson *et al.* (1999), the enhancement factor at Dye 3 is larger than expected from the effect of fabrics alone. These authors assume that impurities and crystal size should be invoked to explain the relative softness of

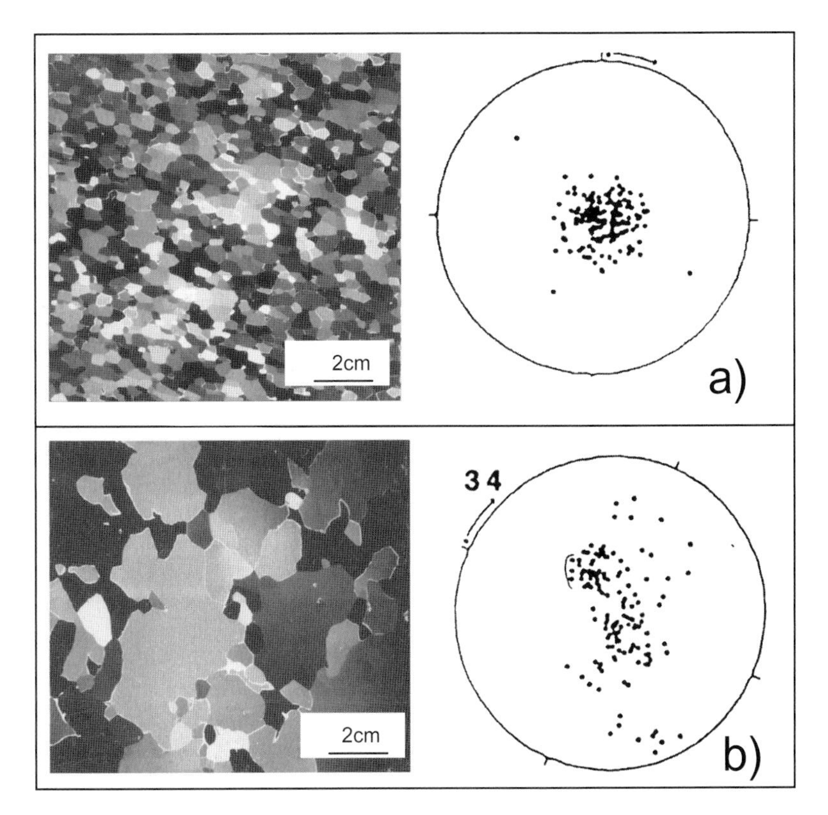

Figure 59.3 Photographs of thin-sections in polarized light and ice fabrics from the Greenland Ice Core at 2806 (top) and 2860 m (bottom) (from De la Chapelle *et al.*, 1998).

Table 59.1 Mean concentration of solid impurities in the Antarctic and Greenland ice sheets during the Holocene and the Last Glacial Maximum (LGM)

Station	Climatic period	Typical concentration ($ng\,g^{-1}$)	Source
Byrd	Holocene	10	Paterson (1991)
	LGM	100	
Vostok	Holocene	15	Petit *et al.* (1999)
	LGM	750	
Camp Century	Holocene	80	Paterson (1991)
	LGM	1200	
Dye 3	Holocene	50	Hammer
	LGM	1300	*et al.* (1985)
GRIP	Holocene	50	Steffensen
	LGM	5000	*et al.* (1997)

LGM ice. It is important to note that the high concentration of dust and other impurities is highly correlated to crystal size, which makes it difficult to assess their respective effect on ice viscosity (Dahl-Jensen & Gundestrup, 1987). Using data on both borehole tilt and closure rate in the Agassiz Ice Cap, Fisher & Koerner (1986) concluded that solid impurities associated with small crystal size are at the origin of the enhancement factor in glacial ice. Cuffey *et al.* (2000c) suggest the enhancement factor in the LGM ice at Dye 3 is totally explained by fabric and crystal size. Based on the analysis developed above on the deformation modes of polar ice at low stresses, the enhancement factor in glacial ice is probably due to the effect of crystal size and crystal orientation. Impurities, by affecting the crystal growth, would indirectly be at the origin of the relative low viscosity of LGM ice. It is, however, worth noting that solid impurities found in basal glacier ice with a much higher concentration can harden ice by impeding dislocation slip (Ashby, 1966, Fitzsimons, this volume, Chapter 65).

59.5.2 The role of a liquid phase in the deformation behaviour of ice

Knowledge of the mechanical response of ice containing a liquid phase is important for modelling the flow of temperate glaciers. Laboratory experiments demonstrate that a few per cent of melt can have a large effect on ice viscosity (De la Chapelle *et al.*, 1999). The role of the liquid phase is analysed by considering the reduction of the internal stress field induced by the mismatch of slip at grain boundaries. The deformation regime with $n < 2$ is extended toward high stresses when the water content increases. A much larger effect was observed in debris-laden ice by Echelmeyer & Zhonxiang (1987) and Cohen (2000). The presence of liquid films around particles even at low temperature would be at the origin of a strong decrease of ice viscosity. In this case, this large effect of interfacial water is not directly related to the volume of the liquid phase but rather to the surface area of particles. The liquid phase would contribute to the accommodation of slip by promoting some grain adjustment and preventing the formation of deformation inhomogeneities.

59.6 Conclusions

The deformation of ice crystals is dominated by the glide of dislocations on the basal plane. Slip or climb of dislocations on non-basal planes cannot be dismissed completely with regard to dislocation multiplication and the required compatibility between grains. The large anisotropy of ice crystals lies at the origin of the development of a non-uniform state of internal stresses during the transient creep of polycrystalline ice. Large strain inhomogeneities within grains are induced by the incompatibility of deformation between grains. For the low stress conditions prevailing in polar ice sheets, grain-boundary migration and dynamic recrystallization are efficient recovery processes. The low stress exponent corresponding to the deformation of ice at low stresses is in accordance with the primordial role of recrystallization in the accommodation of basal slip. Fabrics in glaciers and ice sheets are strain-induced as long as grain growth and rotation recrystallization prevail and stress-induced when migration recrystallization occurs. The enhancement factor in LGM ice appears to be due to the effect of fabric and crystal size. Solid impurities from atmospheric input impede grain growth, but they do not directly affect the ice viscosity.

Acknowledgements

This work was supported by the Départements Sciences Pour l'Ingénieur and Sciences de l'Univers du CNRS. We are very grateful to V. Lipenkov for valuable suggestions during the preparation of the manuscript.

SIXTY

Superplastic flow of ice relevant to glacier and ice-sheet mechanics

David L. Goldsby

Department of Geological Sciences, Brown University, Providence, RI 02912, USA

60.1 Introduction

Understanding the rheological behaviour of ice, particularly at the low stresses relevant to flow within the Earth's cryosphere, is a longstanding problem in glaciology. For purposes of glacier and ice-sheet modelling, the flow of ice is most often described by the Glen flow law

$$\dot{\varepsilon} = B\sigma^n \qquad (1)$$

where $\dot{\varepsilon}$ is strain rate, B is a constant at constant temperature, σ is differential stress, and n is the stress exponent. Glen-law flow is attributed to a single deformation mechanism, dislocation creep, deemed independent of grain size and characterized by a stress exponent n of ca. 3. Despite the widespread adoption of the Glen flow law in glacier and ice-sheet mechanics over the past 50 yr, however, numerous laboratory studies on relatively coarse-grained (≥1 mm) samples have suggested that a transition to a creep mechanism characterized by $n \leq 2$ occurs at stresses lower than investigated in Glen's experiments and relevant to glacier and ice-sheet flow (σ <0.1 MPa) (e.g. Steinemann, 1954, 1958a; Mellor & Smith, 1966; Mellor & Testa, 1969; Colbeck & Evans, 1973; Pimienta & Duval, 1987). Unfortunately, the transition to values of $n \leq 2$ observed in these previous studies occurs at impractically slow strain rates (i.e. $<1 \times 10^{-8}\,s^{-1}$), such that one cannot be certain whether the low values of n result from transient dislocation creep or steady-state creep via a low-n creep mechanism such as diffusion creep (Weertman, 1983).

Within the past decade laboratory experiments have demonstrated conclusively the existence of an extensive 'superplastic flow' regime for ice (Goldsby & Kohlstedt, 1997, 2001; Durham *et al.*, 2001), in which compatible deformation proceeds via intragranular dislocation slip on (primarily) the basal slip system acting in concert with grain-boundary sliding (GBS), with the overall creep rate limited by the rate of GBS. For a given grain size and temperature, superplastic flow (hereafter termed GBS-limited creep) controls the flow of ice at lower stresses than for dislocation creep. The GBS-limited creep of ice is characterized by $n = 1.8$ and, unlike dislocation creep, depends strongly on grain size. The discovery of this creep mechanism was made possible in part by the development of techniques for fabricating samples of fine grain size ($\leq$0.2 mm) (Goldsby & Kohlstedt, 1997, 2001; Stern *et al.*, 1997), which permit investigation of grain-size-sensitive creep mechanisms in ice at practical laboratory strain rates of $\geq 1 \times 10^{-8}\,s^{-1}$.

Most importantly for glaciology, extrapolation of the new laboratory creep data for fine-grained ice to somewhat larger grain sizes provides compelling evidence for the relevance of GBS-limited flow for the dynamics of glaciers and ice sheets. In fact, such extrapolations, coupled with data from previous laboratory and field experiments, strongly suggest that deformation of glaciers and ice sheets, at all but the lowest stresses near the surfaces and the highest stresses near the bases of these structures, occurs via GBS-limited creep.

60.2 Creep of fine-grained ice

The rheological behaviour of materials can be expressed in a general flow law of the form

$$\dot{\varepsilon} = A\frac{1}{d^p}\sigma^n \exp\left(\frac{-Q}{RT}\right) \qquad (2)$$

where A is a material-dependent parameter, d is grain size, p is the grain-size exponent, σ is differential stress, Q is the activation energy for creep, R is the gas constant and T is absolute temperature.

Ambient-pressure experiments on ice samples with uniform grain sizes of 3 μm to 0.2 mm (hereafter referred to as 'fine-grained') unequivocally reveal the existence of three creep regimes, shown schematically in Fig. 60.1, each with characteristic values of n, p and Q: dislocation creep, GBS-accommodated basal slip (i.e. GBS-limited) creep and basal slip-accommodated GBS (i.e. basal slip-limited) creep (Goldsby & Kohlstedt, 1997, 2001). A fourth creep mechanism, diffusion creep, should dominate the flow of ice at the lowest stresses but has not yet been accessed in the laboratory (Goldsby & Kohlstedt, 2001).

At high stresses ($\geq$1 MPa), dislocation creep of fine-grained ice is characterized by a value of $n \approx 4$ and is independent of grain size (i.e. $p = 0$, see fig. 2 of Goldsby & Kohlstedt, 2001). Dislocation creep data for fine-grained ice are in near-perfect agreement

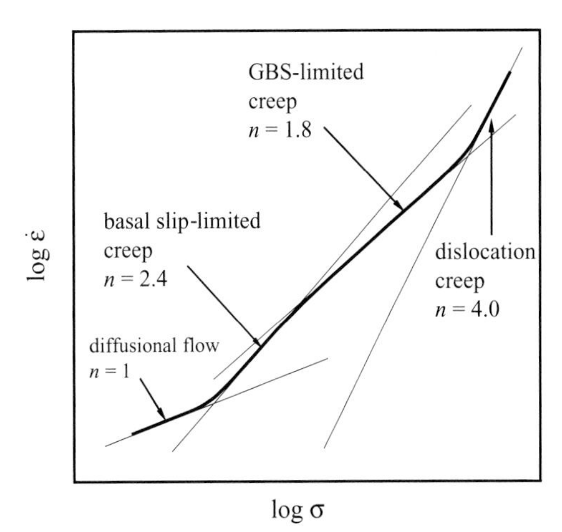

Figure 60.1 Schematic diagram depicting four ice-creep regimes: Dislocation creep, characterized by $n = 4.0$ and $p = 0$, GBS-limited basal-slip creep, with $n = 1.8$ and $p = 1.4$, basal-slip-limited GBS creep, with $n = 2.4$ and $p = 0$, and diffusion creep, with $n = 1$ and $p = 2$ or 3, depending on whether volume or grain-boundary diffusion, respectively, is rate-limiting. Creep data for the $n = 2.4$ creep regime for fine-grained ice are in excellent agreement with creep data for single crystals of ice oriented for basal slip (e.g. Wakahama, 1967). The heavy solid line represents the composite flow behaviour of ice given by Equation (3).

with results of an exhaustive study of dislocation creep of coarser grained ($\geq$0.25 mm) samples deformed at elevated confining pressures (e.g. Durham *et al.*, 1992; 1997). Microcracking, sometimes invoked to explain values of $n > 3$ in ambient pressure tests, is suppressed in elevated pressure experiments. Dislocation creep data for fine-grained ice are also in very good agreement with results of experiments on coarse-grained ($\geq$1 mm) ice samples deformed at ambient pressure and stresses above ca. 1 MPa (e.g. Steinemann, 1958a; Barnes *et al.*, 1971); at a given stress and temperature, strain rates determined at elevated pressure vary from strain rates measured at ambient pressure by less than 10% (see fig. 7 of Goldsby & Kohlstedt, 2001). The activation energy in the $n = 4$ creep regime, below the onset of pre-melting at ca. 255 K (see Goldsby & Kohlstedt, 2001), is equivalent to that for volume diffusion of hydrogen and of oxygen (ca. 60 kJ mol^{-1}). The value of $n = 4$ and the equivalence of the activation energy for creep with that for diffusion of hydrogen and oxygen are consistent with models of climb-limited dislocation creep (e.g. Weertman, 1968).

With decreasing stress, a transition occurs from the dislocation creep regime to the GBS-limited creep regime in which compatible deformation occurs via dislocation slip on primarily the basal-slip system combined with GBS. Although GBS contributes significantly to the strain rate in this regime, the strain rate is probably dominated by basal dislocation slip and rate-limited by GBS (Goldsby & Kohlstedt, 2001). The GBS-limited creep regime is characterized by $n = 1.8$. Unlike dislocation creep, the strain rate due to GBS-limited flow depends strongly on grain size, with $p = 1.4$.

Table 60.1 Constitutive equation parameters (after Goldsby & Kohlstedt, 2001)

Creep regime	A (units)	n	p	Q (kJ mol^{-1})
Dislocation creep, $T < 258$ K	1.2×10^{6} (MPa$^{-4.0}$ s^{-1})*	4.0	0	60
Dislocation creep, $T > 258$ K	6.0×10^{28} (MPa$^{-4.0}$ s^{-1})	4.0	0	181
GBS-limited creep, $T < 255$ K	3.9×10^{-3} (MPa$^{-1.8}$ m$^{1.4}$ s^{-1})	1.8	1.4	49
GBS-limited creep, $T < 255$ K	3.0×10^{26} (MPa$^{-1.8}$ m$^{1.4}$ s^{-1})	1.8	1.4	192

*This value of A for dislocation creep is revised from the original value in Goldsby & Kohlstedt (2001), providing a better fit to the combined data set for dislocation creep from high pressure and ambient pressure tests at $T < 258$ K.

As stress is decreased further, a transition occurs to a creep regime in which the creep rate is probably dominated by GBS but rate-limited by basal slip. The basal slip-limited creep regime is characterized by a value of $n = 2.4$ and the strain rate due to this mechanism is independent of grain size. Both the value of n and the absolute magnitude of the strain rate at a given stress in this regime are in excellent agreement with data for single crystals of ice oriented for basal slip (e.g. Wakahama, 1967), as shown in fig. 2a of Goldsby & Kohlstedt (2001).

At the lowest stresses, a transition is expected to the diffusion creep regime (Mukherjee, 1971), which is characterized by a value of $n = 1$ and a strong dependence of strain rate on grain size (with $p = 2$ or 3 depending on whether volume diffusion or grain-boundary diffusion, respectively, limits the creep rate). Using a value for the unknown parameter in the diffusion-creep equation, the grain-boundary diffusion coefficient, constrained by experiments on fine-grained ice, Goldsby & Kohlstedt (2001) concluded that investigation of diffusion creep in the laboratory would require the maintenance of submicron grain sizes at $T > 248$ K. This requirement may render diffusion creep 'inaccessible' in the laboratory.

Goldsby & Kohlstedt (2001) proposed a composite flow law for ice that includes contributions to the overall creep rate from the four mechanisms described above

$$\dot{\varepsilon} = \dot{\varepsilon}_{\text{diff}} + \left(\frac{1}{\dot{\varepsilon}_{\text{basal}}} + \frac{1}{\dot{\varepsilon}_{\text{GBS}}} \right)^{-1} + \dot{\varepsilon}_{\text{disl}} \quad (3)$$

Each term in the composite flow law is a flow law of the form of Equation (2), where the subscript 'diff' denotes diffusion creep, 'basal' denotes basal slip-limited creep, 'GBS' denotes GBS-limited creep, and 'disl' denotes dislocation creep. Using experimentally determined values of A, n, p and Q for dislocation creep, GBS-limited creep and basal-slip-limited creep, along with estimates of the diffusion-creep rate, Goldsby & Kohlstedt (2001) extrapolated creep data for fine-grained ice via Equation (3) to larger grain sizes (≥1 mm) and to temperatures up to 273 K. Such extrapolations indicate that either GBS-limited creep or dislocation creep is the rate-limiting creep mechanism for ice over a wide range of conditions of glaciological interest, i.e. grain sizes ≥ 1 mm, stresses higher than ca. 0.0001 MPa and temperatures >220 K (Goldsby & Kohlstedt, 2001). Therefore, a simplified form of the composite flow law can be used for the purposes of comparison with previous laboratory data and for glacier and ice-sheet modelling

$$\dot{\varepsilon} = \dot{\varepsilon}_{\text{GBS}} + \dot{\varepsilon}_{\text{disl}} \quad (4)$$

The flow-law parameters A, n, p and Q for the GBS-limited creep and dislocation creep regimes are given in Table 60.1.

60.3 Comparison with previous laboratory data

In the laboratory, both GBS-limited creep and dislocation creep have been observed for samples with grain sizes as large as 0.2 mm and temperatures as high as 268 K (Goldsby & Kohlstedt, 2001). Comparison of creep data for these two mechanisms with creep data for coarse-grained samples ($d \geq 1$ mm) from previous laboratory tests conducted at temperatures at or near 273 K therefore requires comparatively modest extrapolations in grain size (as little as a factor of five) and temperature (no more than 5 K).

In Fig. 60.2, the strain rate versus stress relationship for coarse-grained ice is calculated from Equation (4), using the constitutive parameters for GBS-limited creep and dislocation creep given in Table 60.1, and compared with selected creep data for samples with grain sizes ≥ 0.25 mm from ambient-pressure and high-pressure deformation experiments. The Glen flow law is also shown in Fig. 60.2 for comparison. This straightforward extrapolation of the data for fine-grained ice to larger grain sizes strongly suggests that the Glen flow law represents not one creep mechanism but rather contains contributions from two creep mechanisms, dislocation creep proper ($n = 4$) and GBS-limited creep ($n = 1.8$), yielding an intermediate, transitional value of $n \approx 3$. A fundamental consequence of this conclusion is that the Glen flow law significantly underestimates strain rates in glaciers and ice sheets, which deform at lower stresses than explored in Glen's creep experiments.

60.4 Does GBS-limited flow occur within glaciers and ice sheets?

To provide an overview of the relevance of GBS-limited creep for glacier and ice sheet mechanics, a deformation mechanism map for ice, drawn on axes of grain size and stress, is shown in Fig. 60.3 for $T = 268$ K. The GBS-limited creep regime and the dislocation creep regime are separated by the solid boundary, along which both mechanisms contribute equally to the strain rate. Within each creep regime, strain-rate contours are calculated using the appropriate flow law for that creep regime. As shown in

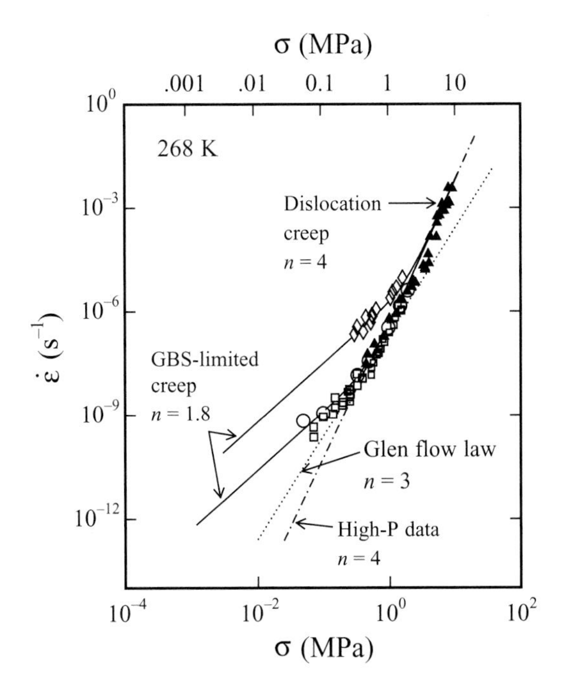

Figure 60.2 Plot comparing the composite flow law of Equation (4) (solid lines) with previous laboratory data for coarse-grained ice samples, for $T = 268$ K. The upper solid line is calculated for d = 0.2 mm, the lower for d = 2 mm. The dotted line represents the Glen flow law; the dotted–dashed line represents data from experiments conducted at high confining pressure (Durham *et al.*, 1992). Data points are from ambient pressure tests: ◊, d = 0.2 mm, Goldsby & Kohlstedt (2001); □, $d \geq 1$ mm, Steinemann (1958a); ○, $d \geq 1$ mm, Mellor & Smith (1966); ▲, $d \geq 1$ mm, Barnes *et al.* (1971).

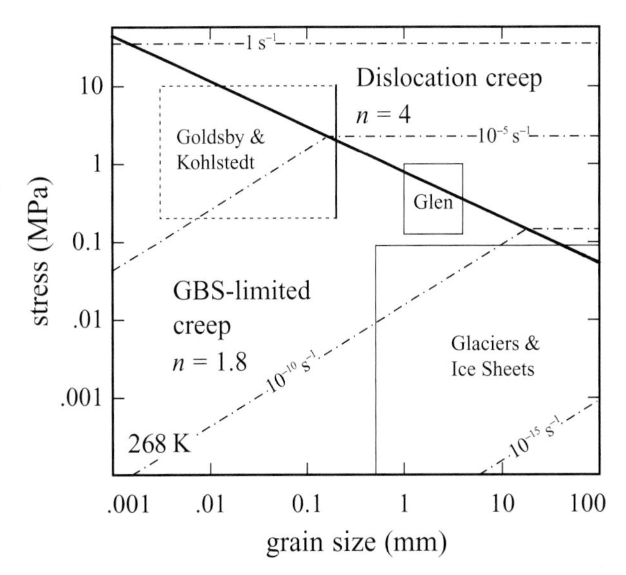

Figure 60.3 Deformation map for ice constructed from the flow laws for GBS-limited creep and dislocation creep for a temperature of 268 K. The heavy solid line of negative slope is the boundary between mechanisms. The dotted–dashed lines are strain-rate contours, calculated using the appropriate flow law for the rate-limiting creep mechanism in each creep regime. The box labelled 'Glen' bounds the approximate σ versus d conditions of Glen's experiments (Glen, 1952, 1955). The box with dashed lines on three sides outlines the full range of stresses and grain sizes in the Goldsby and Kohlstedt experiments (most of which were conducted at temperatures < 268 K); the solid vertical line on the right edge of this box marks the approximate range of stresses for experiments on samples with grain sizes of ca. 0.2 mm at 268 K (see fig. 3 of Goldsby & Kohlstedt, 2001). The approximate range of stresses in glaciers and ice sheets is shown by the large rectangle.

the figure, strain rates in the dislocation creep regime are independent of grain size (horizontal strain-rate contours), whereas strain rates in the GBS-limited creep regime are strongly dependent on grain size (sloping contours).

As shown in Fig. 60.3, the box bounding the conditions explored in Glen's experiments is transected by the boundary between the two creep mechanisms, illustrating that the Glen flow law is really a composite law containing contributions from both GBS-limited creep and dislocation creep. The overlay of the approximate range of grain size versus stress conditions encountered in glaciers and ice sheets in Fig. 60.3 suggests that flow of these bodies is limited by GBS at all but the highest stresses and largest grain sizes near their bases, where a transition to the dislocation creep regime occurs (see below).

60.4.1 Glaciers

That GBS-limited flow occurs within glaciers (and ice sheets) is supported by numerous field observations, including measurements of borehole tilt and cavity (borehole and tunnel) closure. Subsets of the data from such studies are compared in Fig. 60.4 with calculated relationships between strain rate and stress from Equation (4). In most of the cases shown, at sufficiently low stresses, the field studies yield a value of $n \leq 2$. In some cases (e.g. Meier, 1960; Holdsworth & Bull, 1970), a transition to $n > 3$ is observed at the highest stresses (ca. 0.1 MPa), indicating a transition to the dislocation creep (n = 4.0) regime. The agreement between the magnitude of the strain rate in the $n \leq 2$ regime in the field with that calculated from Equation (4) is in some cases very good (e.g. see data of Gerrard *et al.*, 1952; Meier, 1960), although considerable uncertainty exists in the appropriate value of grain size to use in Equation (4), because grain sizes from field studies are often not reported or only typical or 'average' values are reported. In other cases, strain rates observed in the field are significantly higher than predicted by Equation (4). For example, the strain rate at a given stress observed at the Meserve Glacier site of Holdsworth & Bull (1970) and the Barnes Ice Cap site of Hooke (1973) is in excellent agreement with the strain rate calculated via Equation (4) using a temperature of 273 K and a grain size of 1 mm, although the observed temperatures at the Meserve and Barnes Ice Cap sites were ca. 256 and ca. 263 K, respectively, with grain sizes >1 mm at both sites. Strain rates determined in the Holdsworth & Bull (1970) study are higher than those predicted from Equation (4) by about two orders of magnitude. Such

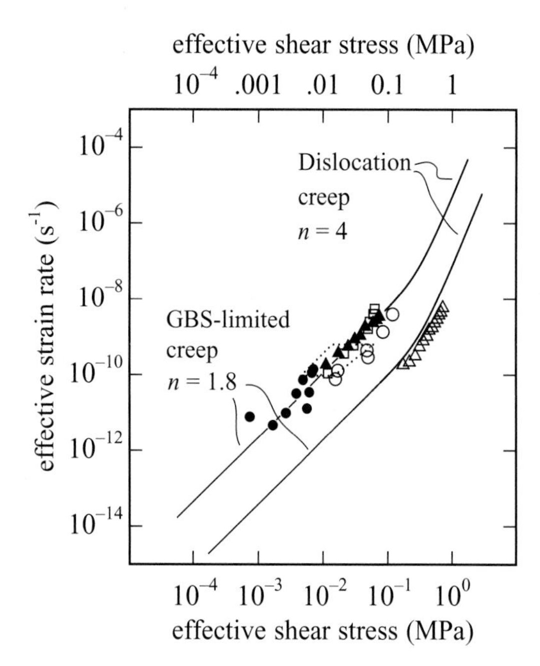

Figure 60.4 Plot comparing the composite flow law of Equation (4) (solid line) with data from field studies on glaciers and ice sheets. The upper solid line is calculated for $d = 2$ mm and $T = 273$ K, the lower solid line for $d = 1$ mm and $T = 255$ K. The dotted-line parallelogram represents the range of conditions of the experiment on the Barnes Ice Cap by Hooke (1973), which yielded $n = 1.65$. Data points are from: ▲, Gerrard *et al.* (1952), with data reanalysed as in Nye (1953); ○, ●, Meier (1960); △, Gow (1963); □, Holdsworth & Bull (1970).

discrepancies between laboratory and field data may reflect relatively large uncertainties in the calculated stresses or measured creep rates in the field studies, the effects of impurities on the creep rate due to GBS-limited creep in nature, and/or the effects of strong *c*-axis fabrics on GBS-limited creep in nature.

60.4.2 Newtonian flow?

Several field studies yield values of $n \leq 1$ at stresses between ca. 0.001 and ca. 0.1 MPa (e.g. Meier, 1960; Marshall *et al.*, 2002), consistent with either diffusion creep ($n = 1, p = 2$–3) or Harper-Dorn creep ($n = 1, p = 0$). Experimentally constrained estimates of the diffusion creep rate, however, suggest that diffusion creep of glaciers and ice sheets is unlikely for grain sizes ≥ 1 mm and stresses > 0.0001 MPa over the entire range of glaciologically significant temperatures (Goldsby & Kohlstedt, 2001). Moreover, stresses in the diffusion creep regime are too low to activate the dislocation sources required to develop strong crystallographic fabrics, which are nearly ubiquitous in glaciers and ice sheets (Frost & Ashby, 1982). The operation of Harper-Dorn creep (a Newtonian dislocation mechanism that is independent of grain size) in glaciers is inconsistent with field studies, which demonstrate that flow of glaciers and ice sheets is grain size-sensitive (e.g. Cuffey *et al.*, 2000a; see below). Finally, I note that values of $n = 1$ from field studies are in many cases only apparent values if grain size increases with depth (stress) in glaciers, as is commonly observed. The low apparent value of n results from the decrease in strain rate with increasing grain size as depth (stress) increases, which is typically not taken into account in analyses of mechanical data from field studies.

60.4.3 Grain size-sensitive flow of glaciers

As grain size, impurity content and *c*-axis fabrics often co-vary in nature, it can be difficult to isolate the effects of each variable on strain rate. Cuffey *et al.* (2000a) measured strain rates in a tunnel in the basal layers of the Meserve Glacier. The temperature and shear stress at the site were essentially constant (ca 256 K and ca. 0.05 MPa) and *c*-axis fabrics varied only slightly, allowing the convolved effects of grain size and impurity content on strain rate to be isolated and semi-quantified via a statistical analysis. The authors concluded that a grain-size-sensitive creep mechanism contributed to the deformation at Meserve. A minor enhancement of strain rate was attributed to segregation of chemical impurities to grain boundaries, which the authors suggest may have resulted in enhanced GBS. Although the statistical analysis used did not allow the contributing grain-size-sensitive mechanism to be identified, one end-member case allowed for a grain-size-sensitive creep mechanism with $p = 1.4$ (i.e. GBS-limited creep) to contribute at least 15% but less than 50% to the strain rate for $d = 1$ mm. Such a scenario is consistent with transitional behaviour between GBS-limited creep and dislocation creep. For the shear stress (ca. 0.05 MPa) and temperature (256 K) of the site, Equation (4) predicts a contribution of 100% from GBS-limited creep to the strain rate for $d = 1$ mm, ca. 90% for $d = 2$ mm and ca. 70% for $d = 5$ mm. Although estimates of the relative contributions of GBS-limited creep and dislocation creep from Equation (4) do not agree precisely with the field data analysis, it nevertheless appears likely that ice at this site deforms in the transitional regime between GBS-limited creep and dislocation creep. Discrepancies between the predicted and observed contributions of GBS-limited flow to the creep rate may reflect errors in the determination of strain rate and/or stress in the field, the effects of impurities present in nature on flow rates, and the effects of *c*-axis fabrics on the relative rates of flow in the GBS-limited and dislocation creep regimes. Transitional flow behaviour via GBS-limited creep and dislocation creep at the Cuffey *et al.* (2000a) site also agrees with tunnel closure and borehole tilt measurements in a nearby tunnel in a previous study on Meserve by Holdsworth & Bull (1970). Temperatures were the same and grain sizes similar in the two studies, but unlike in the Cuffey *et al.* (2000b) study, in which shear stress was nominally constant, the shear stress in the Holdsworth & Bull (1970) study varied from ca. 0.01 to 0.06 MPa. Holdsworth & Bull (1970) observed a transition from $n = 1.9$ to $n = 4.5$ at a shear stress of ca. 0.05 MPa, the same value of shear stress as that determined at the Cuffey *et al.* (2000c) site, and the same transition shear stress as that predicted from Equation (4).

60.4.4 Ice sheets

The Glen flow law underestimates strain rates measured in the ice sheets, particularly for fine-grained ice-age ices (e.g. Thorsteinsson et al., 1999). This discrepancy has led to the introduction of a strain-rate enhancement factor E in the Glen flow law (see Equation (1) in Iverson, this volume, Chapter 91), variously attributed to the effects on flow of c-axis fabrics, the presence of impurities and/or variations in grain size (e.g. Dahl-Jensen, 1985; Thorsteinsson *et al.*, 1999; Cuffey *et al.*, 2000c). Numerous studies have demonstrated that the flow rate in the ice sheets increases with increasing impurity content (both dissolved ions and insoluble particles) and/or decreasing grain size (e.g. Dahl-Jensen, 1985; Thorsteinsson *et al.*, 1999). However, the co-variation of impurity content with inverse grain size in the ice sheets, due to grain-boundary pinning effects of impurities, makes it difficult to quantify the effect of either variable on strain rate. In has been argued that variations of grain size in the ice sheets have a negligible effect on strain rate (e.g. Thorsteinsson *et al.*, 1999), a conclusion which stems in part from the long-held view that grain size is determined by strain rate, rather than being an independent variable that affects the strain rate. Recently, Cuffey *et al.* (2000c) applied results from analysis of flow of Meserve Glacier (Cuffey *et al.*, 2000a) (discussed above) to borehole closure measurements at the Dye 3 borehole in Greenland and in the Agassiz Ice Cap. Noting that chemical and insoluble impurities probably have a negligible effect on strain rate at these sites, Cuffey *et al.* (2000c) demonstrated that residual enhancements in ice-sheet flow (over the enhancement attributed to c-axis fabrics) likely result from reductions in grain size.

60.4.5 A generalized hypothesis for ice sheet flow

At the shallowest depths within an ice sheet (at tens-of-metres depths), where effective stresses are <0.0001 MPa, deformation likely proceeds via diffusion creep (Goldsby & Kohlstedt, 2001). Diffusion creep in this shallow ice is consistent with the lack of c-axis fabrics near the surfaces of the ice sheets, although finite strains may be too small in this region to yield significant fabrics even if creep proceeded via a c-axis-fabric-producing mechanism.

With increasing depth (stress), a transition probably occurs from diffusion creep to GBS-limited flow. At Byrd Station (discussed below), this transition occurs at a depth of ≤135 m. Dislocation slip on the basal-slip system likely dominates the overall strain rate in the GBS-limited creep regime. Basal slip causes a rotation of c-axes toward the compressional axis, eventually forming a single-maximum fabric (Alley, 1992). The tightening of the c-axis pattern with depth results from the progressive rotation of grains (caused by basal slip) with increasing strain. Grain rotation, and hence basal slip, is accommodated and rate-limited by GBS.

Grain size increases via normal grain growth at shallower depths in the ice sheets (Alley, 1992) (e.g. up to ca. 400 m depth at Byrd Station, see below), then remains constant at intermediate depths (e.g. over the range 400–1000 m at Byrd, as shown below) within the GBS-limited creep regime. It has been suggested that relatively constant grain sizes at intermediate depths result from a dynamic balance between normal grain growth and polygonization (subgrain rotation recrystallization) (e.g. Alley, 1992; Thorsteinsson *et al.*, 1997; De La Chapelle *et al.*, 1998). It seems difficult to reconcile a constant grain size in this region, however, with subgrain rotation recrystallization, given the well-known inverse relationship between stress (which, for example, increases by more than a factor of two in the constant grain size regime at Byrd Station) and subgrain size (e.g. Twiss, 1986; Jacka & Li, 1994), suggesting that impurities segregated to grain boundaries may have the greater control on grain size by pinning grain boundaries (e.g. Evans *et al.*, 2001; Barnes *et al.*, 2002; Weiss *et al.*, 2002). The decrease in grain size in the ice-age ices below the constant grain size regime may result from colder accumulation temperatures and the presence of impurities (Weiss *et al.*, 2002). In general, equiaxed polygonal grains and single-maximum c-axis fabrics that become progressively tighter with increasing depth are likely characteristic features of GBS-limited creep.

Temperature increases with increasing depth in the ice sheets due to geothermal heat flux into their bases. With the onset of pre-melting at grain boundaries and at three- and four-grain junctions in ice at ca. 255 K (as indicated by the enhancement in creep rate at that temperature for both GBS-limited and dislocation creep, Goldsby & Kohlstedt, 2001), an abrupt increase in the rate of increase of grain size with depth is observed. This increase in grain size results from the orders-of-magnitude increase in grain-boundary mobility caused by the presence of water on grain boundaries (Duval *et al.*, 1983; Dash *et al.*, 1995), allowing grain boundaries to overcome the pinning forces due to the presence of impurities segregated to the grain boundaries. The depth at the onset of pre-melting (at a temperature of ca. 255 K) corresponds with the first occurrence with increasing depth of multiple maximum c-axis fabrics at both Byrd Station (Gow & Williamson, 1976) and the GRIP site (Thorsteinsson *et al.*, 1997), demonstrating that multiple maximum c-axis fabrics are associated with grain-boundary migration (GBM) recrystallization.

Increases in grain size, temperature and stress with increasing depth conspire to bring the ice closer and closer to the dislocation creep regime (see Fig. 60.5 below). The activation of non-basal slip systems at the very highest stresses within the ice sheets may also contribute to the occurrence of multiple-maximum c-axis fabrics (e.g. Gow & Williamson, 1976; Alley, 1992). A transition from GBS-limited flow to dislocation creep in the basal ice of the ice sheets is consistent with values of $n = 3$ derived from analyses of ice-sheet shape (e.g. Hamley *et al.*, 1985; Cuffey, this volume, Chapter 57), since the shape and velocity of the ice sheet are controlled in large part by deformation of the basal layer (Alley, 1992).

60.4.6 Byrd station example

To illustrate this generalized description of ice-sheet flow, a comparison is made between data from studies on the Byrd Station borehole in Antarctica with the deformation mechanism map for ice. In Fig. 60.5, the measured grain size versus depth profile (Gow *et al.*, 1968) and estimated shear stress versus depth profiles (Frost & Ashby, 1982) for the Byrd Station drill site are plotted. Following Frost & Ashby (1982), shear stresses τ were estimated from $\tau = \rho g h \sin(\alpha)$, where ρ is ice density, g is the acceleration due to gravity, h is the depth in the ice and α is the surface slope. Temperatures increase with increasing depth and are shown at discrete points along the borehole, ranging from ca. 245 K at the surface to ca. 273 K at the base. Superimposed on this plot is the τ versus d deformation mechanism map for ice constructed for $T = 273$ K.

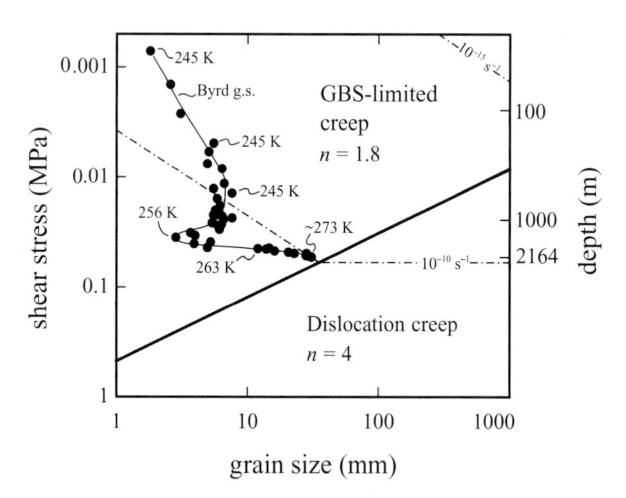

Figure 60.5 Plot of shear stress and depth versus grain size for the deep ice-core from Byrd Station, Antarctica. Shear stresses τ were estimated from the depth h using $\tau = \rho g h \sin(\alpha)$, where ρ is density, g is gravitational acceleration and α is the surface slope (ca. 2.5×10^{-3}, after Frost & Ashby, 1982). Grain-size versus depth data (solid symbols connected schematically with the solid curve) are from Gow *et al.* (1968). Superimposed on the figure is a shear stress versus grain size deformation mechanism map constructed for a temperature of ca. 273 K. The heavy solid line is the boundary between GBS-limited creep and dislocation creep; the dotted–dashed lines are strain-rate contours. Temperatures measured at discrete depths along the borehole (from Gow *et al.*, 1968) are indicated in the figure.

It must be emphasized that the deformation map represents a limiting case, with the boundary between mechanisms and the shear-strain-rate contours within each creep regime calculated only for the highest temperature at Byrd, ca. 273 K. However, the boundary between the GBS-limited creep and dislocation creep regimes shifts downward in stress in Fig. 60.5 only slightly, by a factor of only 1.3, when calculated for a temperature of 245 K.

As suggested in Fig. 60.5, the ice at Byrd Station probably deforms within the GBS-limited creep regime over nearly the entire depth of the ice sheet. At the base of the ice sheet, the ice is expected to deform in a transitional regime between GBS-limited creep and dislocation creep. The observed sharp increase in the rate of grain growth with increasing depth occurs at a temperature of ca. 256 K, in excellent agreement with the temperature of the onset of pre-melting inferred from creep data (ca. 255 K, Goldsby & Kohlstedt, 2001). A similarly sharp increase in the rate of grain growth with increasing depth is observed at the GRIP site in Greenland at a temperature of ca. 258 K (Thorsteinsson *et al.*, 1997). As the ice at 256 K at Byrd deforms well within the GBS-limited creep regime (in which non-basal slip is insignificant), the increased rate of grain growth with depth at $T > 255$ K probably results from enhanced grain-boundary mobility, rather than an increase in driving force due to variations in strain energy (dislocation density) caused by the activation of non-basal slip systems.

60.5 Summary

Understanding the grain-scale deformation of ice is fundamental to the study of glacier and ice-sheet mechanics. The recognition that flow of glaciers and ice sheets is very likely controlled by either GBS-limited creep or dislocation creep provides a unifying working hypothesis for better understanding how glaciers and ice sheets flow. This unifying framework successfully integrates new and existing laboratory creep data for ice with observations from field studies, including mechanical data from field measurements as well as microstructural observations of grain size, grain shape and *c*-axis fabrics.

In spite of the first-order success of the newly obtained constitutive laws for ice in unifying a wide range of laboratory and field observations, however, several outstanding problems remain before the full implications of GBS-limited creep and other creep mechanisms for glacier and ice-sheet mechanics can be realized. Two of the most prominent questions are:

1 How do the creep rates due to GBS-limited basal slip creep and dislocation creep proper ($n = 4.0$) vary with the development of significant *c*-axis fabrics?
2 How does the presence of impurities, both dissolved ions and insoluble particles, present in natural ice bodies, affect the creep rate due to GBS-limited creep?

With regard to the first, nearly all studies of *c*-axis fabric development to date have been conducted on samples deformed within the transitional regime between GBS-limited creep and dislocation creep. Therefore, the effects of *c*-axis fabric development on flow are a (perhaps) complicated convolution of the effects of fabric development on each creep mechanism. With regard to the second, ionic impurities segregated to grain boundaries might significantly enhance the rate of GBS and hence the creep rate due to GBS-limited flow, owing to enhanced grain-boundary disorder, grain-boundary diffusion and pre-melting. Insoluble particles on grain boundaries might impede the motion of grain-boundary dislocations, slowing the rate of GBS-limited creep (Ashby, 1969). The variations of strain rate in the GBS-limited creep regime for ice with variations in both *c*-axis fabrics and impurity content are at present unknown and require further experiments.

SIXTY-ONE

Anisotropy and flow of ice

Throstur Thorsteinsson

Institute of Earth Sciences, University of Iceland, Sturlugata 7, IS-101 Reykjavik, Iceland

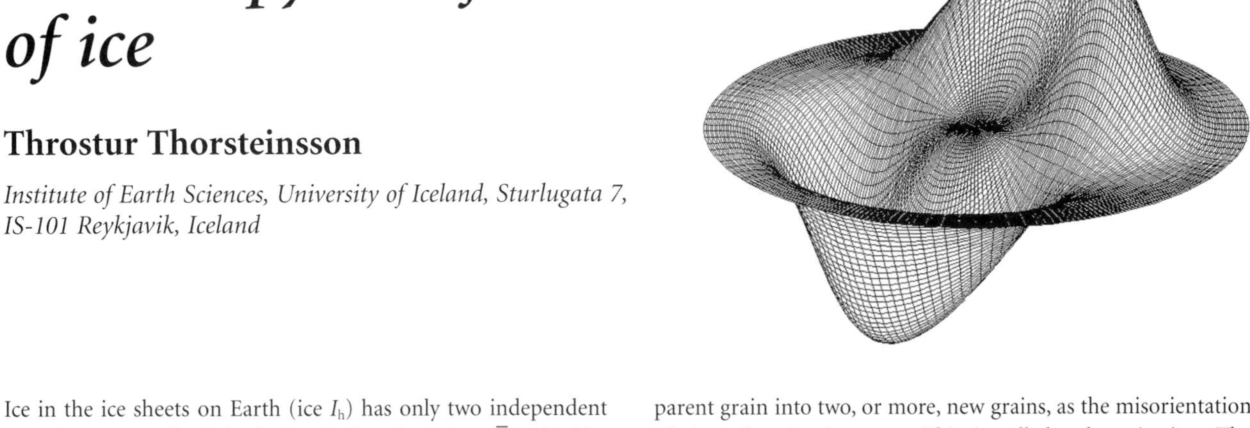

Ice in the ice sheets on Earth (ice I_h) has only two independent easy slip systems, both in the basal plane (0001)$\langle 11\bar{2}0 \rangle$ (Ashby & Duval, 1985). When ice undergoes ductile deformation, it develops fabric (lattice-preferred orientation) due to intracrystalline slip (Wenk *et al.*, 1991). The c-axes rotate towards the principal compression axis (Azuma, 1994), in the absence of recrystallization, as qualitatively known from thin-section measurements (Gow *et al.*, 1997; Thorsteinsson *et al.*, 1997), and from sonic-logging measurements in boreholes (Taylor, 1982). Dynamic recrystallization also induces changes in the fabric; then the recrystallized fabric reflects the symmetry of the current stress state, and memory of prior strain history and fabric evolution can be obscured or lost. As a consequence of fabric development, the bulk physical properties become anisotropic.

To model the flow of ice sheets, we must account for the anisotropy. Models of the fabric evolution and deformation of arrays of crystals as they travel along particle paths can characterize fabric over large parts of the ice sheets, where the flow field changes slowly. Where there are rapid changes in the flow field, a fabric-evolution model following a particular path represents the properties of a smaller volume of ice.

61.1 Fabric evolution model

Fabric development and macroscopic deformation are studied here by examining the effects of nearest-neighbour interaction (NNI) among crystals and dynamic recrystallization (Thorsteinsson, 2002).

The 'strength' of grain interaction can vary from no interaction (no-NNI), with homogeneous stress (Sachs, 1928), to 'strong' interaction (full-NNI), with significant stress redistribution. Increasing the NNI leads to a more homogeneous strain of all the crystals.

There are three dynamic recrystallization regimes for ice: normal grain growth (which has no effect on fabric; Gow, 1971), polygonization (small effect on fabric) and migration recrystallization (significant effect on fabric). As the grains strain, sub-boundaries (dislocation walls) may form due to heterogeneous deformation within grains that relieves stress concentrations. The formation of sub-boundaries can lead to the division of the parent grain into two, or more, new grains, as the misorientation of the subgrains increases. This is called polygonization. The effect on the fabric development is small, because the orientation of the new grains usually deviates by less than 5° from the parent grain, but the effect on grain size is clearly visible in thin-sections (Thorsteinsson *et al.*, 1997).

The formation of sub-boundaries can create small sections of grains that are in a strain shadow. Being strain-energy-free, these small grains can act as seeds for migration recrystallization. The idea adopted in the model is that within the crystal aggregate there are many such 'seeds' that provide potential nucleation sites for new grains. When the temperature is high enough for grain-boundary migration to be efficient (Duval & Castelnau, 1995), and/or the strain energy is greater than the grain-boundary energy, these seeds can quickly consume highly strained crystals, thus reducing the free energy of the system (Montagnat & Duval, 2000).

In studies of high-temperature (−5°C to 0°C) creep of ice, Kamb (1972) found that after only about 0.04 shear strain there was already strong evidence of recrystallization. A totally random orientation for the new grain is not to be expected. In uniaxial compression of ice, for instance, a small-circle girdle fabric forms (Budd & Jacka, 1989). This indicates that new crystals that form in orientations with high resolved shear stress (soft orientations) are favoured to grow. Figure 61.1 shows the model results when polygonization and migration recrystallization are active in uniaxial compression. The results are encouraging because they show both the typical girdle type fabric (Fig. 61.1a), and the nearly constant strain rate beyond 10% strain (Fig. 61.1b).

61.2 Fabric and flow

The effects of anisotropy on deformation are complex. One way to represent fabric is through a cone angle, where the cone angle goes from $\alpha = 0°$ for all crystals aligned, to $\alpha = 90°$ for isotropic ice. In general, stronger anisotropy (smaller cone angle) makes the ice softer in shear parallel to the basal planes, and stiffer in compression normal to the basal planes (think of a deck of cards), as shown in Fig. 61.2.

It has been a common practice to account for the effects of anisotropy with scalar 'enhancement' factors. Where there is one

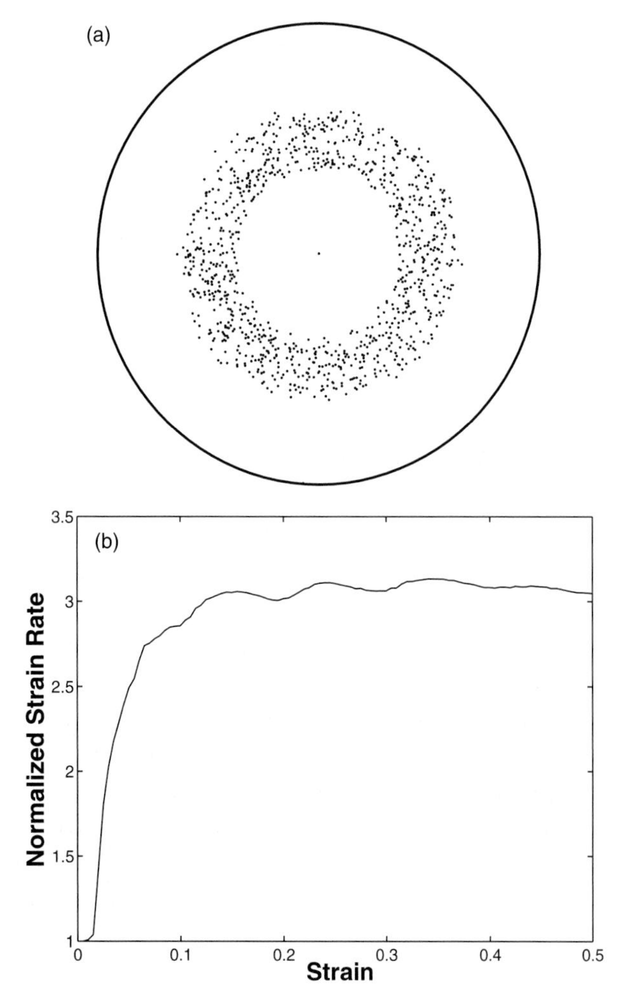

Figure 61.1 Deformation in uniaxial compression where polygonization and migration recrystallization are active: (a) the fabric after 0.5 vertical strain, where the initial fabric was isotropic, and (b) normalized strain rate versus strain.

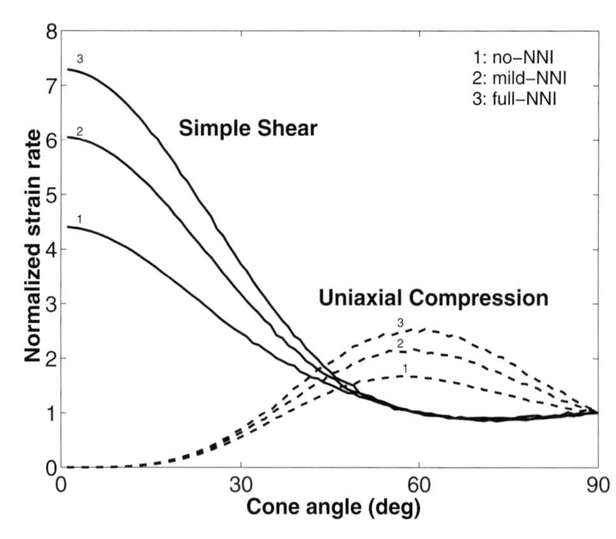

Figure 61.2 Normalized strain rate for different levels of nearest neighbour interaction (NNI) in uniaxial compression and simple shear as a function of cone angle. Isotropic ice has a cone angle of 90° and ice with all crystals aligned has a cone angle of 0°.

dominant stress component (shear or compression) it may be a reasonable approximation; however, in general, a scalar enhancement factor is inadequate, because the enhancement of strain rate due to fabric depends on both stress orientation and deformation direction. Accounting for anisotropy explains most of the 'enhanced' deformation in Greenland, about 75% at Dye 3 for instance (Thorsteinsson *et al.*, 1999).

The vertically symmetric fabric commonly observed in ice-sheets makes horizontal shearing easier (Fig. 61.2). This facilitates folding, i.e. elements with smaller slopes can be overturned. The effects of anisotropy on folding are complicated by the fact that, for a range of cone angles, ice is also softer in compression (Figs 61.2 & 61.3a); this opposes folding (Thorsteinsson & Waddington, 2002). Figure 61.3 shows (a) the vertical strain rate and (b) shear strain rate in combined uniaxial compressive stress and simple shear stress (as when moving away from the centre of a dome). In layers with spatially variable and evolving fabric, the fabric causes localized flow variations, which could create layer disturbances.

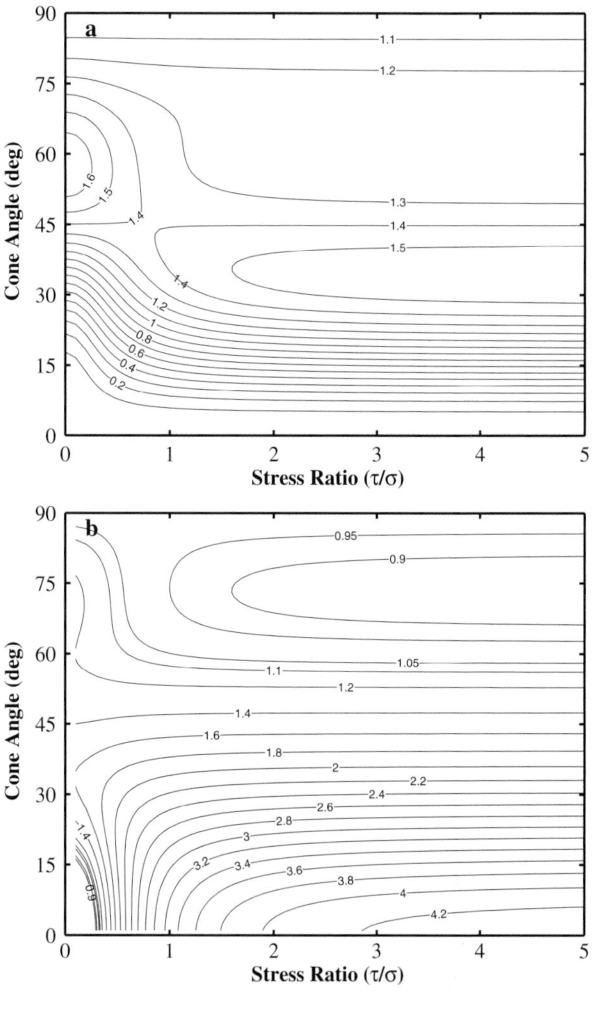

Figure 61.3 Normalized strain rate in combined uniaxial compression (σ) and simple shear (τ) stress state as a function of stress ratio (τ/σ) and cone angle: (a) vertical strain rate and (b) shear strain rate (Thorsteinsson & Waddington, 2002).

Along most particle paths in polar ice sheets, ice experiences a slowly changing local deviatoric-stress pattern, and develops a fabric characteristic of its current stress state, in which there is generally a correspondence between non-zero strain-rate components and non-zero deviatoric-stress components. However, when the stress pattern changes more rapidly than the fabric can evolve, unusual or unexpected deformation patterns can result (Thorsteinsson *et al.*, 2003). The degree to which fabric follows the local stress is determined by the characteristic time-scales for changes in stress, given by the transition-zone width and ice velocity, and for changes in crystal orientation, given, in the absence of recrystallization, by the inverse of the local strain rate due to the principal stress. Recrystallization can significantly reduce the time-scale for fabric adjustment. Stress and fabric tend to be misaligned in ice-stream margins (although recrystallization will also be active) and in flow through a saddle (Thorsteinsson *et al.*, 2003). Non-intuitive deformation patterns can also result if the stress state applied to an ice sample in an experiment is markedly different from the *in situ* stress state in the ice sheet.

SIXTY-TWO

Ductile crevassing

Antoine Pralong

Section of Glaciology, Laboratory of Hydraulics, Hydrology and Glaciology, Swiss Federal Institute of Technology, CH-8092 Zürich, Switzerland

Fracture in ice is a complex competitive process of microcrack nucleation and propagation. The spatial organization of the microcracks during their development determines the brittleness or the ductility of ice. Because of the difficulties in taking the influence of the microcracks into account individually, the fracture is usually considered at a scale (the mesoscale) that homogenizes the microstructures. The analysis of the fracture process is thus based on quantities defined at that scale. In this approach, the brittleness or the ductility of the fracture is frequently related to the stress magnitude, the stress state and the temperature of the ice. Fracture in tension (the usual form of crevassing) becomes ductile at low stress and high temperature.

62.1 Particularities of ductile crevassing

Ductile fracture is characterized by a large fracture process zone (FPZ, a zone where microcracks are active because of stress concentration at the crevasse tip) and a subcritical crevassing (a crevasse that propagates below the fracture toughness; more generally called subcritical crack growth, SCCG).

The size of the FPZ is associated with the stress relaxation at the crevasse tip due to viscoplastic deformations (creep), which are related to dislocation kinetics, and also to microcrack growth (Kachanov, 1999). The stress concentration at the tip is thus attenuated and the stress is redistributed in a zone around the tip. In that zone, microcracks grow. Some microcracks dominate and extend the crevasse. The FPZ influences the ice flow, because the microcracks reduce the mechanical properties of ice. In order to evaluate the size of the FPZ, crevasse patterns observed in nature can be considered. Weiss (2003) analysed the crevasses of a temperate alpine glacier (Glacier d'Argentière, France). He found that the size of the crevasse length follows an exponential distribution, with a minimum crevasse size of 10 m. The spacing between crevasses could be approximated well using a log-normal distribution with a mode of 12.6 m. These results suggest that the size of the FPZ is in the order of magnitude of 10 m in the case of ductile crevassing. This is confirmed by direct simulations of crevasse growth (see below).

Observations of subcritical crevassing are numerous, but the SCCG process has not yet been studied in the case of ice. Figure 62.1 presents a simplified pattern of crack velocity reported as a function of the stress intensity factor K, as observed in ceramics or glass. At the fracture toughness K_c, the crack propagates at a speed close to the body wave speed (critical crack growth). Under the threshold K_{th}, the crack does not progress. For ice, K_{th} can be related to the crack nucleation stress observed at low stress. Between K_{th} and K_c the crack grows with a controlled velocity. This mode of fracture corresponds to the SCCG. For rocks, ceramics or glass, SCCG is usually associated with diffusion processes, chemical reactions and plastic flow (Weiss, 2004). For ice, diffusion processes are known to be very slow. They can,

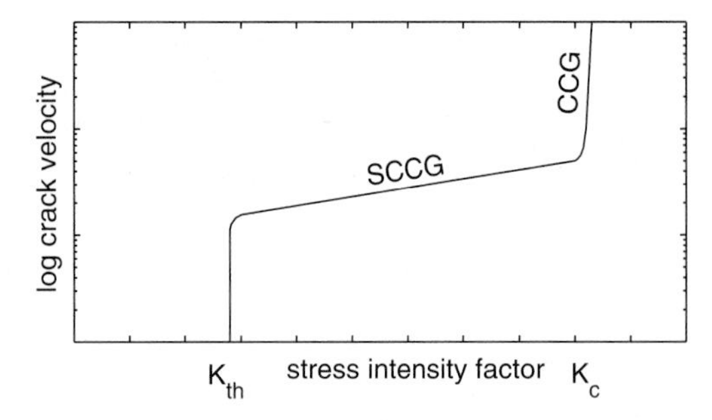

Figure 62.1 Schematic crack velocity versus stress intensity factor. Critical crack growth (CCG) occurs for $K \geq K_c$, where K is the stress intensity factor and K_c is the fracture toughness. Subcritical crack growth (SCCG) arises for $K_{th} \leq K < K_c$, with K_{th} being the threshold of the stress intensity factor.

however, influence the SCCG at a low deformation rate. Chemical reactions due to ice impurities are not required for SCCG, as laboratory experiments performed with pure ice show subcritical effects (Mahrenholtz & Wu, 1992). As asserted above, viscoplastic flow is responsible for the formation of the FPZ. In the FPZ, the development of microcracks is not homogeneous because of the heterogeneity of the ice. In such conditions (e.g. Amitrano *et al.*, 1999), some microcracks progressively dominate and extend the crevasse. As revealed by creep to failure measurements performed in the laboratory (Mahrenholtz & Wu, 1992), the duration of the microcrack accumulation process (until the formation of a crack) depends strongly on the applied load. At low stress (<1 MPa), it is in the order of weeks. Thus, SCCG could be interpreted as a discontinuous process of microcrack nucleation and propagation. At the macroscale, the organization of these microcracks can be regarded as a continuous process of crevasse growth.

62.2 Model overview

Dynamic models of crevasses or rift openings are scarce. They can be classified according to the velocity of the crack they consider. For rapid velocities, i.e. for K close to K_c, elastic effects dominate, and the ice is more brittle. The FPZ is small compared with the system size, and linear elastic fracture mechanics can be applied (Bazant, 1999). This theory is able to describe the opening of rifts in ice-shelfs (Larour *et al.*, 2004). For low crack velocities, i.e. for K close to K_{th}, viscoplastic effects are prevalent. Non-linear fracture mechanics or continuum damage mechanics should be applied. Glacier crevassing has been modelled successfully by using a damage mechanics approach valid for creep materials (Pralong *et al.*, 2003; see also next section), a simple non-linear fracture model (Iken, 1977), or an improved linear fracture mechanics approach, which accounts for subcritical crevassing (Weiss, 2004).

62.3 Continuum damage model

Continuum damage mechanics describes the deterioration of material due to a progressive increase of damage (damage means

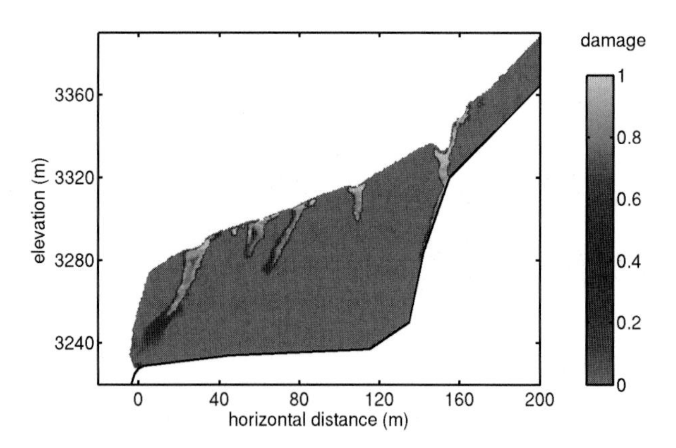

Figure 62.2 Simulation of crevasse formations in the west face hanging glacier of the Eiger, Switzerland. According to Pralong *et al.* (2003), the damage is assumed to be isotropic. The crevasses appear in the model as concentration of damage. The dark grey colour represents ice without microcracks ($D = 0$). Light grey corresponds to broken ice ($D = 1$), i.e., ice with a very high density of microcracks. In the simulation, the frontal crevasse grows. The fracture process zone at its tip is depicted in black and corresponds to intermediate values of D. (See www.blackwellpublishing.com/knight for colour version.)

the density of microcracks, which affect the mechanical behaviour of the material). This approach consists of three elements: (i) quantification of the damage with a variable D (called 'damage variable'), (ii) description of the influence of the damage on the material rheology and (iii) description of the evolution of the damage in time and space. The damage variable D is defined at the mesoscale. As the mesoscale is much smaller than the size of the crevasses, D is, for the modelling of crevassing, a local variable. It has the following meaning: $D = 0$ represents no damage and $D = 1$ corresponds to a fully damaged ice element. The damage variable should adequately describe the level of anisotropy: isotropic damage is represented by a scalar variable and anisotropic damage by a second- or fourth-order tensor. The influence of the microcracks on the ice flow is calculated by assuming a dependence of D on the ice viscosity. For anisotropic damage, the viscosity becomes anisotropic. The increase of damage in material is modelled using a local progressive accumulation law valid for creep materials.

With such an approach, ductile crevassing is handled naturally:

1 In the model, a concentration of damage appears at the crevasse tip and forms the FPZ, thereby taking into account the influence of the FPZ on the ice flow.
2 The subcritical crevassing is reproduced by modelling the local and progressive accumulation of damage.

The crevassing in a hanging glacier is computed (Fig. 62.2) by applying the isotropic damage model developed by Pralong *et al.* (2003). The simulation reproduces the subcritical crevasse growth and the formation of the FPZ.

The physics of ductile crevassing is not well understood at present. The microprocesses of fracture and the spatial organization of microcracks should be analysed in greater depth. Field and laboratory measurements should also be carried out in order to quantify more precisely subcritical crack-growth for ice.

SIXTY-THREE

Environmental conditions, ice facies and glacier behaviour

Wendy Lawson

Department of Geography and Gateway Antarctica, University of Canterbury, Christchurch, New Zealand

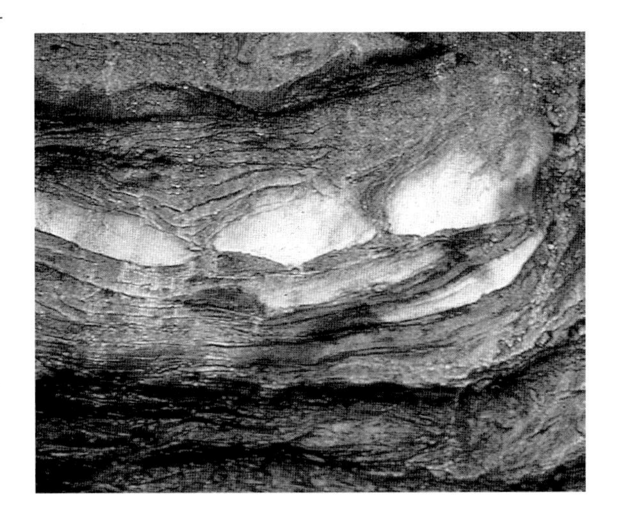

63.1 Introduction

This chapter will explore the way in which the controls on glacier dynamic behaviour exerted by environmental conditions are mitigated by the characteristics of the glacier itself, focusing on the impact of ice-facies variability on stress response. We begin with an overview of the main environmental factors that affect glacier behaviour in order to contextualize the more focused discussion of ice facies and ice flow. In discussion, both in the introductory comments and in the more detailed analysis, we initially attempt to treat each variable or set of variables independently. In reality, many of the variables are strongly interdependent, and cause-and-effect relationships hard to isolate. The issue of causality also varies with scale of analysis, so that a variable that is an independent controlling variable at an individual glacier scale of analysis may become a dependent variable at another scale (Schumm & Lichty, 1965). The proportion of precipitation that falls as snow, for example, is an independent variable at the glacier scale, but a dependent variable at the mountain range scale. The scale of approach in this chapter is essentially that of understanding individual glacier behaviour, in keeping with the majority of glaciological analyses.

63.2 Environmental controls and glacier behaviour: an overview

Glacier behaviour, and therefore the response of glaciers to environmental change, is fundamentally determined by an interdependent combination of climatic regime and the interaction of the glacier with its immediate geological and topographic environment. Climate controls glacier mass balance through its impact on rates, distributions and types of precipitation and ablation. This mass balance regime in turn is the fundamental rate-controlling process for dynamics, because the rate of input and output of mass, sometimes known as the 'activity index' (Meier, 1961), determines the velocity needed to achieve equilibrium. Climatic regime is also the major (but not only: see below) determinant of the thermal regime of the ice mass. Thermal regime has a major impact on ice dynamics through the effect of ice temperature on ice deformation rates, and also because of the importance of the presence of liquid water for basal sliding.

The interaction of a glacier with its immediate geological and topographic environment affects glacier behaviour and response in various ways. In terms of the geological environment, the nature of the glacier substrate has a significant impact on glacier behaviour. The most fundamental parameter is the presence or absence of a layer of unconsolidated sediment at the basal interface (e.g. Boulton, 1986). A layer of basal unconsolidated sediment allows more rapid motion than would otherwise occur, and is thought to be crucial in rapid flow in fast-moving ice streams (e.g. Alley, 1986). More specifically, the lithology of the substrate affects glacier behaviour, and there is a widely known association between the subglacial presence of relatively erodible bedrock, and a propensity for relatively rapid ice flow (e.g. Anandakrishnan *et al.*, 1998; Bell *et al.*, 1998; Jiskoot *et al.*, 2003). The presence of erodible bedrock may provide the source of sediment for a subglacial deforming till layer, thereby allowing enhanced flow rates, as thought to be the case at the Siple Coast ice streams (Bell *et al.*, 1998). On hard rock beds, bedrock type affects bed roughness, which in turn affects sliding rate (Hubbard *et al.*, 2000a), and is also a factor in the subglacial erosion rate (Boulton, 1979).

The thermal regime is also partly determined by geological conditions, and in particular subglacial geothermal heat flux. The variability of geothermal heating can lead to temporally variable effects, such as massive subglacial melting beneath Vatnajokull in Iceland associated with volcanic activity, leading to intermittent jokulhlaups (e.g. Gudmundsson *et al.*, 1997), or spatially variable effects, such as the initiation of rapid flow within the Greenland Ice Sheet over areas of thin crust with high geothermal heat flux rates (Fahnestock *et al.*, 2001).

The topographic environment affects glacier behaviour and response most directly through its effect on glacier surface slope and bed slope (see below). At the mountain-range scale, topography has an indirect effect on behaviour through its role as a determinant of climatic regime of the glacier because of orographic effects on precipitation, and also has an effect through the impact of topographic complexity and glacier shape on behaviour (Jiskoot *et al.*, 2003).

These general comments about environmental controls on glacier behaviour can be brought into focus through the lens of the fundamental relationships governing glacier dynamics, as follows. The fundamental driving stress that determines glacier flow is the vertical shear stress τ_{xy}, which for a glacier confined by valley walls is given by

$$\tau_{xy} = f\rho gh \sin \alpha \quad (1)$$

where h is overlying ice thickness, ρ is ice density, g is acceleration due to gravity, and f is the shape factor, which is a function of the cross-section shape of the valley and broadly its width/depth ratio. Therefore, the two key glacier variables that determine the magnitude of the driving stress are ice thickness, h, and surface slope, α, as well as the cross-sectional topography of the valley. Ice thickness and surface slope, and their spatial distribution, are determined by interactions between the climate and topography, and the way those interactions determine the distribution and rates of precipitation and melting.

The rate of deformation of ice resulting from the applied shear stress τ_{xy} is determined by Glen's flow law for ice, as follows

$$\dot{\varepsilon}_{xy} = A\tau_{xy}^{n} \quad (2)$$

where $\dot{\varepsilon}_{xy}$ is the shear strain rate, n is a constant and A is a flow parameter with a value that is sensitive to a range of characteristics of the ice, which are a function of the environmental interactions outlined above. In particular, A is sensitive to ice temperature, water content and presence of physical and chemical impurities, as well as crystal properties such as size and fabric. The temperature dependence of A is strong, and relatively well understood: it is such that the strain rate at 0°C is about 14 times that at −10°C (Paterson, 1994, p. 97). The sensitivities of A to the other parameters, especially the presence of impurities, are much less well known, although the characteristics of ice can vary significantly through a glacier. The next section addresses this issue.

63.3 Ice-facies formation and types

As indicated above, the characteristics of glacier ice are known to vary significantly through individual glaciers in ways that affect the value of flow parameter A in Equation (2), and hence the response of the ice to applied stress. The most systematic occurrence of significantly different ice within glaciers occurs in the basal zone, because the interaction of the glacier with the bed can lead to the development of layers near the bed of distinctive regelation ice, and to metamorphism of overlying meteoric ice (Weertman, 1957). Regelation ice is formed by melting and refreezing of the ice in contact with the bed, owing to local changes in the melting point resulting from stress concentrations (Fig. 63.1). The melting process allows liquid water to interact with material at the bed, and therefore to acquire a distinctive

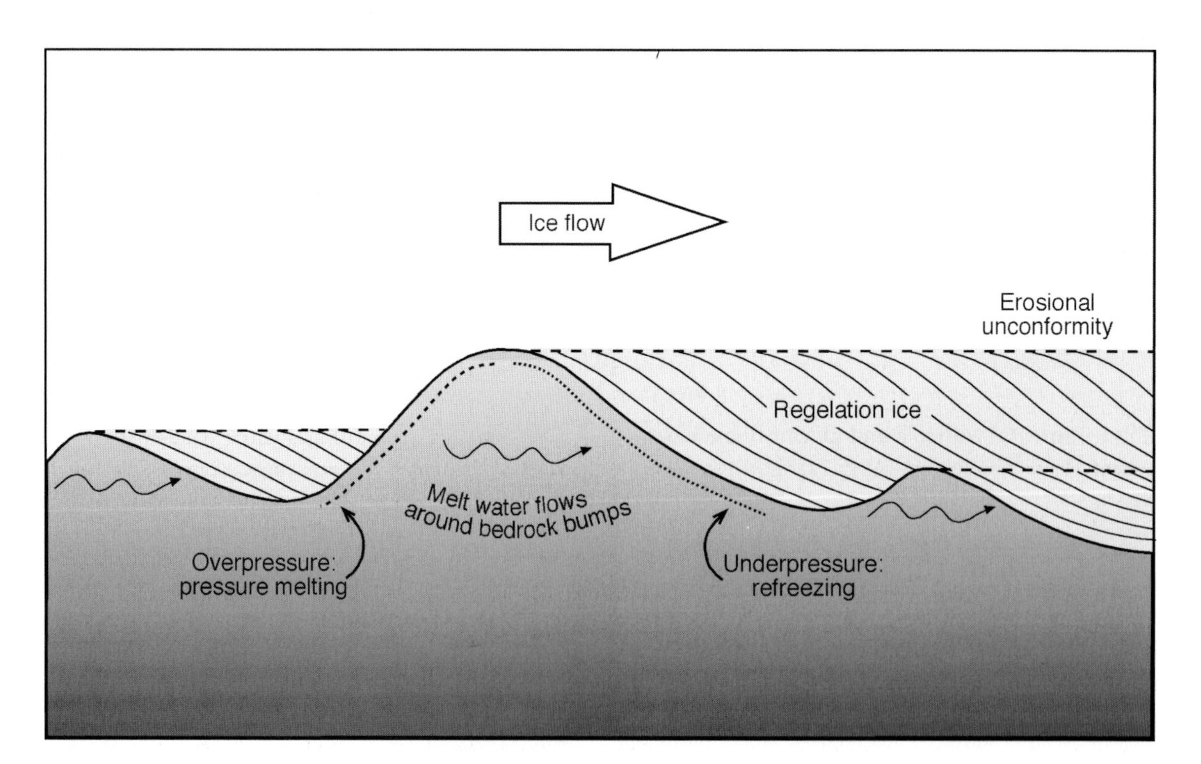

Figure 63.1 Schematic diagram illustrating the process of formation of distinctive basal regelation ice by interaction of the glacier at its basal boundary with bed roughness elements. This process, which contributes to basal motion, is known as Weertman (1957) regelation, and is probably most effective for bed roughness elements with dimensions 10^{-1}–10^{0} m. It produces layers of basal ice with distinctive physical and chemical characteristics.

chemical signature, and the refreezing process allows debris to be entrained into this chemically enriched ice.

As a result of these processes, thick layers of distinctive ice are often found in the basal zones of glaciers and ice sheets in both temperate and cold conditions (Hubbard & Sharp, 1989; Knight, 1997). Temperate glaciers, in particular, can have basal ice layers that make up a significant proportion of the total depth of the ice. Hubbard *et al.* (2000a), for example, found basal ice layers up to 14.5 m thick in ice with a total depth of 45 m at Tsanfleuron Glacier in the Swiss Alps. Fisher & Koerner (1986) identified a basal layer 8 m thick of ice with enhanced solid impurity content in an ice core 340 m deep in the Devon Ice Cap in Arctic Canada. Simoes *et al.* (2002) reported a layer 89 m thick near the base of the Vostok core containing entrained bedrock particles.

In terms of dynamic response, the fact that ice with significantly different characteristics occurs in the basal zone—rather than anywhere else in the glacier system—is particularly important because it is the basal zone where the driving shear stress is greatest, and also where local stress concentrations occur due to bed roughness.

In order to simplify analysis of ice types, and especially in order to characterize complex sequences of apparently chaotic basal ice, various typologies of ice facies have been developed. Lawson's description and classification of ice types at the terminus of the Matanuska Glacier (Lawson, 1979b) was the first application of a systematic stratigraphical facies approach to ice types. He denoted the upper horizons of clean meteoric glacier ice formed by snow metamorphism as englacial ice facies, and distinguished between two basal ice facies. The basal dispersed facies, usually found stratigraphically between upper englacial facies and lower stratified facies, had fewer bubbles and higher sediment concentration than englacial ice. The stratified facies, of which there were three subfacies, was typically strongly layered, and had significantly higher debris concentrations than dispersed-facies ice.

Lawson's non-genetic facies approach to ice classification, and its three basic ice facies, has been widely adopted, with slight variations. Hubbard & Sharp (1995), for example, distinguished seven facies in the western European Alps, although several of their facies might be considered subfacies (Knight, 1997). In some environments, local names have been used: for example, Holdsworth (1974) denoted a basal ice facies with debris and bubble characteristics like those of Lawson's dispersed facies as 'amber ice' because of its distinctive colour, and this term continues to be used (e.g. Fitzsimons *et al.*, 2000).

Recently, there has been a move towards genetic classifications of basal ice facies, although there is often a strong overlap between genetic interpretations and purely stratigraphical typographies. Knight (1997), for example, distinguished two genetic basal ice facies, the first comprising ice-debris mixes entrained at the bed in various ways. In a non-genetic classification, this would include stratified ice and its subtypes. The second comprises ice that has been affected by metamorphic processes near to the bed, and is basically dispersed facies ice. In terms of mechanisms of emplacement, the significance of tectonic effects in determining facies assemblages is also becoming increasingly clear. Waller *et al.* (2000), for example, outlined the way in which post-accretion glaciotectonism had affected basal ice facies at the Russell Glacier in Greenland, such that inferences about subglacial conditions based on a twofold genetic approach to facies classification would be problematic.

Essentially, then, there are three main ice-facies groups. The majority of a glacier is comprised of meteoric englacial ice (Fig. 63.2). This ice may have varying crystallographic and ice-fabric characteristics, but physically and chemically is relatively homogeneous. In the basal zone, there may be ice adjacent to the bed that has been significantly affected by its interactions with the bed in various ways, and this is called 'stratified ice' (Fig. 63.2). Between the englacial ice and underlying stratified ice is ice with some debris but less bubbles than meteoric ice, referred to as dispersed-facies ice.

63.4 *In situ* evidence for ice-facies rheological variability

A range of field evidence from various environments indicates that different ice facies found contiguously, and therefore in identical or similar stress conditions, deform at very different rates in response to similar stresses (Fig. 63.3). Competence-contrast boudinage in the basal zone of temperate Variegated Glacier, in which dispersed facies basal ice was boudinaged within layers of debris-laden stratified facies basal ice (Fig. 63.3a), clearly indicated the relatively rapid deformation of the stratified facies compared with the dispersed facies. At cold conditions, velocity profiles through the basal zone of Suess Glacier in Antarctica indicated that a 0.8 m thick layer of dispersed-facies ice experienced much more rapid deformation than either the overlying englacial ice or the underlying stratified (solid subfacies) ice (Fig. 63.3b; Fitzsimons *et al.*, 2000). Similarly, various types of borehole evidence have shown that ice with enhanced debris concentrations near the base of ice cores deforms more readily than the overlying clean ice (Shoji & Langway, 1984; Fisher & Koerner, 1986; Simoes *et al.*, 2002).

The rapid rate of deformation of debris-laden, stratified-facies ice is such that deformation of these basal layers can account for large proportions of surface motion: for example, Echelmeyer & Wang (1987) found that deformation in a basal debris-laden ice layer 35 cm thick accounted for up to 60% of surface motion at Urumqi Glacier No 1 in the Tienstshan Mountains. Brugman (1983) reported similar importance of deformation across debris-laden ice in motion of glaciers on Mount St Helens.

It is clear then, that debris-laden, basal-ice facies *in situ* have a lower effective viscosity than relatively debris-free ice facies. It is also clear that the processes involved in the formation of the debris-laden ice, and in particular regelation (Fig. 63.1), produce ice with a range of distinctive characteristics as well as the physical presence of debris. In the next section, we attempt to understand the cause of the lower effective viscosity of debris-laden ice facies, by reviewing what is known about the effect of various physical and chemical ice characteristics on ice rheology. These ice facies also have distinctive crystal characteristics. The effects of crystal fabrics on ice deformation are relatively well understood (Budd & Jacka, 1989) and discussion below focuses on impurity effects.

Figure 63.2 Englacial ice facies overlying an erosional unconformity with distinctively layered debris-laden basal stratified-ice facies in the basal zone at Taylor Glacier, Antarctica.

(a)

Figure 63.3 Examples of field evidence for the lower effective viscosity of different ice facies *in situ*: (a) Competence-contrast boudinage at Variegated Glacier, showing the relatively high effective viscosity of the cleaner boudinaged dispersed facies ice, with stratified facies ice layers deformed around it. (b) Velocity profiles in the basal zone at Suess Glacier, showing relatively rapid deformation in the 'amber' (dispersed) facies basal ice (from Fitzsimons *et al.*, 2000). (c) Displacement profile in a marginal cliff at Taylor Glacier, Antarctica, showing rapid displacement rates in the lower part of the debris-laden basal ice section.

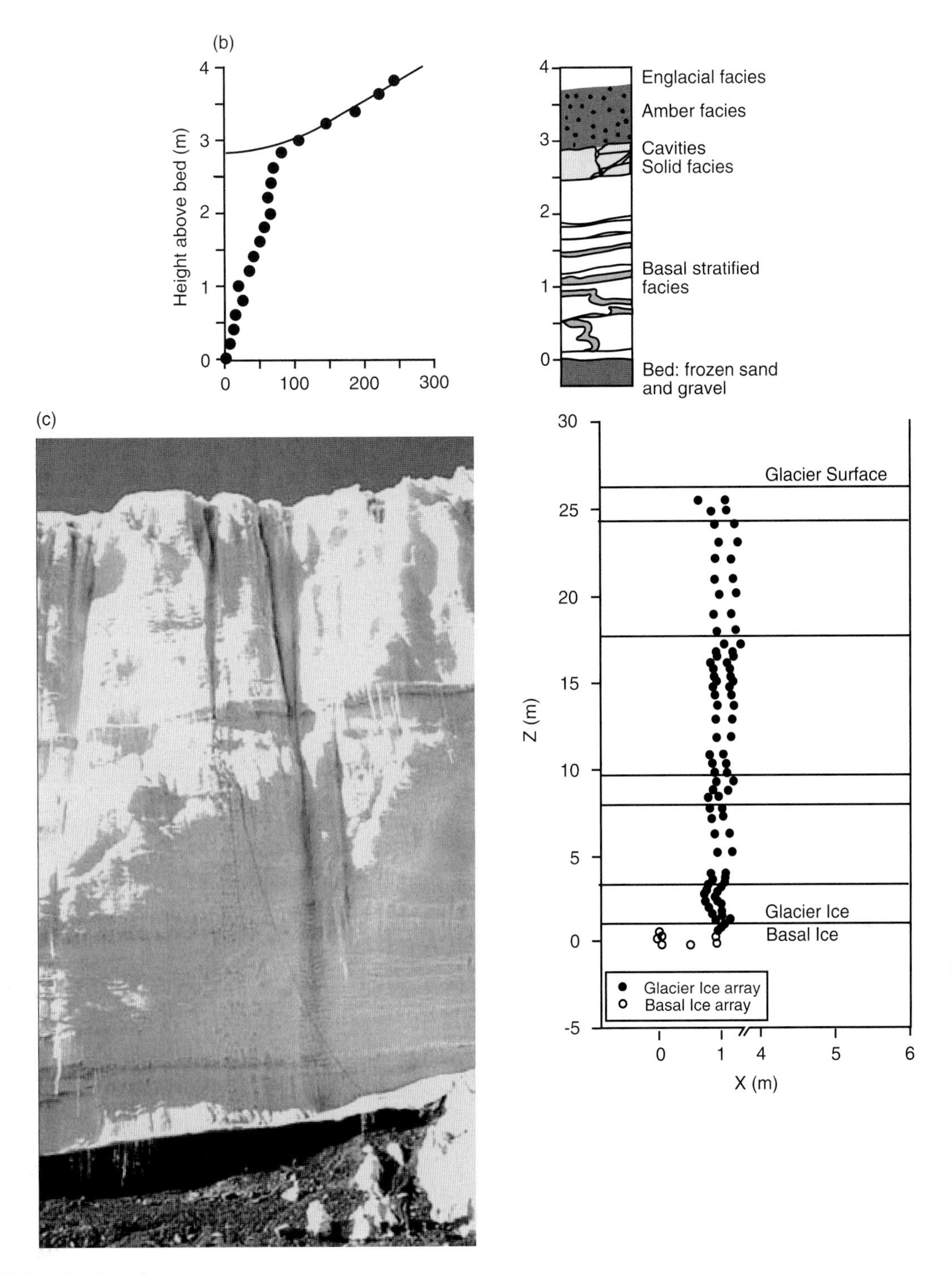

Figure 63.3 *Continued*

63.5 Causes of variability of ice-facies deformation rates

A range of laboratory studies have attempted to determine the effect of the physical presence of debris on ice rheology, by creating ice in the laboratory with embedded debris in carefully controlled conditions. These studies generally have found that the presence of debris decreases deformation rates, and that strain rates decrease with increasing debris-content (Fig. 63.4). For example, Hooke *et al.* (1972) found that the strain rate of ice decreased exponentially with the addition of debris for debris concentrations greater than 10%, although at lower concentrations, results were inconsistent. They attributed this strengthening to crystal-scale processes and, specifically, the inhibition of dislocation motion by the physical presence of debris (cf. Weiss *et al.*, 2002). Nickling & Bennett (1984) similarly found that

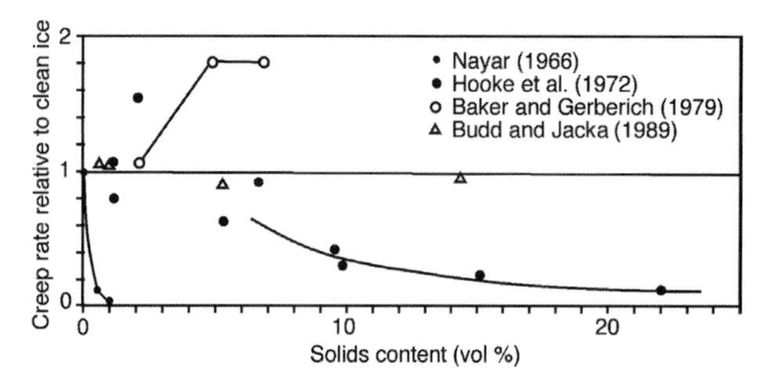

Figure 63.4 The results of various studies showing the effect of the presence of solid impurities on the creep rate of ice, in relation to the creep rate of debris-free ice. (From Budd & Jacka, 1989.)

the strength of debris–ice mixes increased with debris concentration until debris reached 75% by volume, and accounted for this pattern in terms of changes in internal friction and cohesion. Jacka *et al.* (2003), however, found that there was little impact of debris concentration in laboratory ice on minimum strain rates.

Studies of the rheology of debris-laden ice sampled from real glaciers, on the other hand, indicate fairly consistently that the presence of solid impurities leads to an increased flow rate. The results of those few studies, in keeping with the general structural and dynamic observations for relative rheological behaviour of glacier ice *in situ as* outlined above, have found that debris-laden basal ice was weaker than contiguous clean ice sampled from the same glacier. At near zero temperatures, Lawson (1996) found that samples of stratified facies debris-laden basal ice (5–20% debris) from the Taylor Glacier had a mean peak strength only 40% of that of overlying englacial ice. Fisher & Koerner (1986) found enhanced flow rates in basal ice with solid impurities that they correlated with the solid impurity content, rather than any other possible enhancement factor.

The findings for laboratory ice, then, contrast with the various kinds of evidence from more field oriented situations for higher flow rates in ice with solid impurities. This supports the suggestion that the physical presence of debris does not in itself have a substantial effect on the strength of ice (Budd & Jacka, 1989; Durham *et al.*, 1992), although Budd & Jacka suggest that an exception to this may be near the melting point, and we investigate the effect of temperature later in this chapter. In general terms, however, it seems likely that it is indirect effects associated with the presence of debris, for example, in terms of chemical and water content, that cause the observed higher flow rates in debris-laden ice facies.

Basal ice facies in all environments typically have enhanced concentrations of soluble impurities (Cuffey *et al.*, 1999; Hooker *et al.*, 1999; Fitzsimons *et al.*, 2000; Hubbard *et al.*, 2000a, 2004), and the presence of soluble impurities in ice tends to increase the flow rate (e.g. Jones & Glen, 1969; Nakamura & Jones, 1973). Nakamura & Jones, for example, found that a concentration of just 28 ppm of dopant hydrogen fluoride increased the flow rate by a factor of five. The soluble impurities are likely to be acquired during the process of regelation, because solute acquisition takes place rapidly when meltwater interacts with solid impurities (Brown *et al.*, 2001). Soluble impurities may also be concentrated in liquid water found in veins at triple grain junctions and interfacial films (Cuffey *et al.*, 1999).

The rheological effect of bulk soluble impurity content is hard to isolate, however, because it is closely and causally related to bulk water content, such that

$$W = K\frac{M}{\theta} \tag{3}$$

where W is the bulk water concentration, M is the bulk ionic concentration, θ is temperature depression of the ice below its melting point and K is a constant (Hubbard *et al.*, 2004). The bulk water content affects ice-flow rate, even at relatively small water contents. The presence of water, even in small amounts in interfacial films nanometres thick, also has important effects on sliding (Cuffey *et al.*, 1999), so water content is a key variable in terms of glacier dynamic behaviour. Duval (1977) found that strain rate increased rapidly with small water contents, from 5 yr^{-1} for 0.1% water content to 20 yr^{-1} for 1% water content. He found a linear relationship between ice-flow rate and water content (Duval, 1977), such that

$$A = 10W + 4.5 \tag{4}$$

where A is relative hardness of the ice. Given the enhanced soluble impurity content of basal ice facies, it is likely that basal ice generally has relatively high bulk water contents compared with englacial ice facies, although there is little research in this area reported in the literature. There is evidence, however, that water is squeezed out of basal ice during metamorphic type processes (Souchez & Tison, 1981). Hubbard *et al.* (2004) used the two relationships above, in conjunction with measurements of the ionic content of basal ice layers, to estimate that the basal dispersed-facies ice at Tsanfleuron Glacier had an effective viscosity more than an order of magnitude greater than the overlying englacial ice, as a result of differences in bulk water content caused by differences in soluble impurity content.

It is clear that the water content of ice decreases at temperatures further from the melting point (see relationship (3) above), and it is often implicitly assumed that it is only in temperate ice and close to the melting point that liquid water is significant in flow rate enhancement (e.g. Budd & Jacka, 1989). However, the small-scale concentration of stress at asperities on solid impurities can cause melting even at temperatures many degrees below the melting point: for example, in uniaxial compression tests on ice samples from Taylor Glacier at −25°C, meltwater was observed to form (Lawson, 1996). Similarly, Cuffey *et al.* (1999) found that liquid water in interfacial films at the base of Meserve Glacier at −17°C could account for rapid sliding and clean ice lenses within debris-laden ice. They attributed the presence of this water in thicker interfacial films than predicted theoretically to the high solute concentration in the ice.

As indicated above, temperature has a major impact on ice deformation, through its impact on flow parameter A, such that colder ice deforms much less readily. In terms of differential thermal effects on different ice facies, the characteristics of ice facies are such that different facies are likely to have different thermal sensitivities, although this is currently relatively poorly

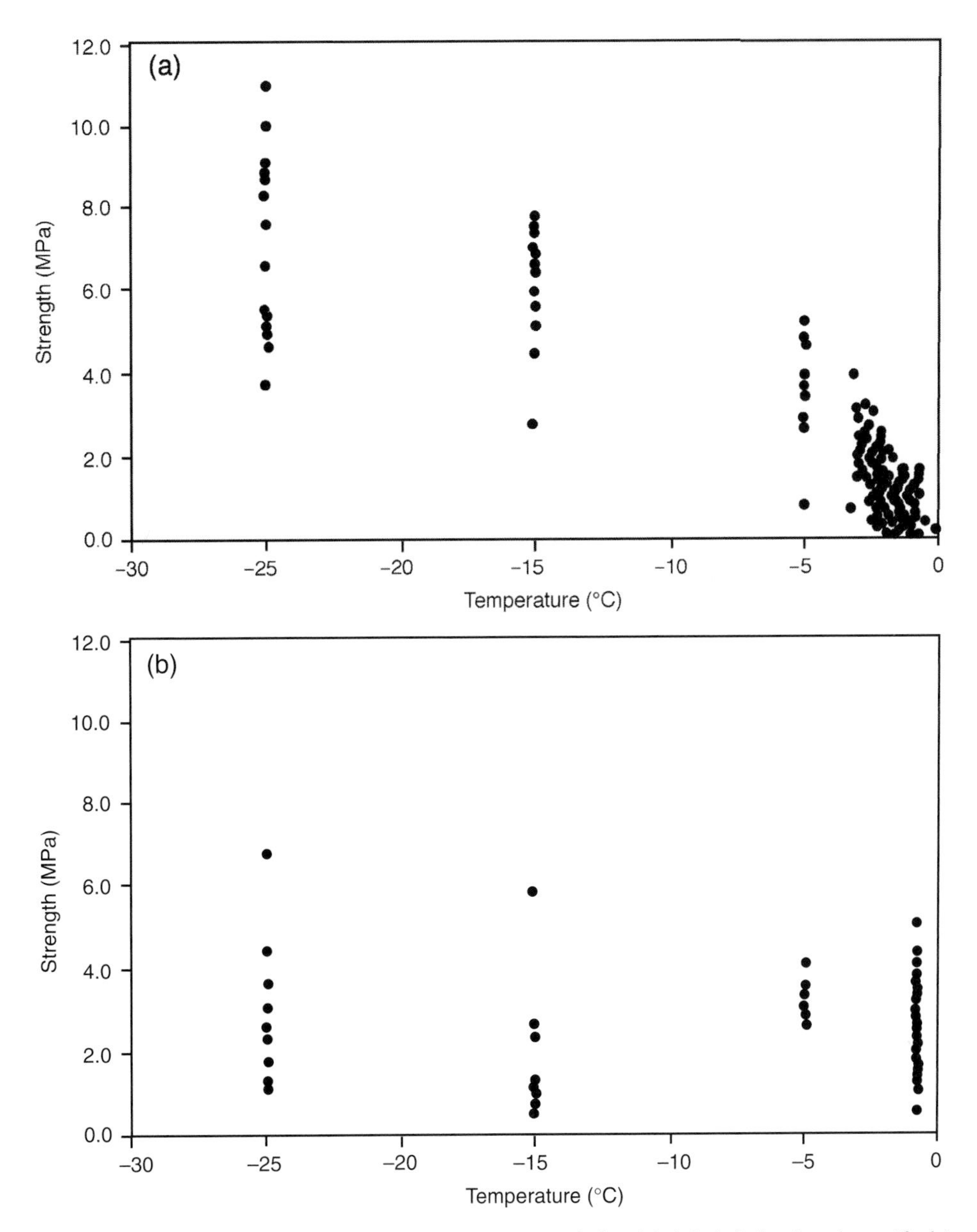

Figure 63.5 Temperature dependence of uniaxial compressive strength for (a) debris-laden basal stratified-ice facies and (b) englacial-ice facies from Taylor Glacier, Antarctica. Note that the envelope of strength defined by the maximum value of strength at each increases systematically as temperature decreases for the debris-laden basal stratified-ice facies.

understood. In rapid strain rate tests, Lawson (1996) found that the strength of debris-laden basal ice facies from Taylor Glacier increased with decreasing temperature, such that the ice was more than twice as strong at −25°C than at −5°C, whereas the strength of the overlying englacial ice did not vary systematically with temperature (Fig. 63.5)

63.6 Effect of presence of basal facies on glacier behaviour

It is clear that the presence of a rheologically different layer at the base of glacier or ice sheet will have a major effect not only on the dynamics of an ice mass, but also, as a result, on its morphology and extent (Boulton & Jones, 1979; Beget, 1986). However, there have been relatively few studies examining the effect of softer basal ice layers on glacier dynamics, although the importance of the issue is often indicated. In a recent study, Hubbard *et al.* (2004) explicitly examined the impact on glacier response behaviour of the presence of two softer layers in the basal zone in a case study at the Tsanfleuron Glacier. They found significant differences between predictions for response using multilayered models. In particular, they found that a multilayered glacier with rheologies constrained by field observations of basal ice was more sensitive in a retreat scenario than a single layered glacier with a rheology constrained by observations of overlying glacier ice.

63.7 Conclusions

Ice in glaciers and ice sheets is not homogeneous. In particular, there are distinctive ice facies found in significant depths in the basal zones that form as a result of the interaction of the glacier with its bed. These layers have a range of characteristics associated with the process of their formation, the net effect of which is that they deform more readily than overlying englacial ice. The cause of this enhanced deformation is hard to isolate, because there are complex and causal interdependencies between variables. The presence of solid impurities *per se* does not soften the ice, but the associated soluble impurities and enhanced water content seem to be causally associated with lower viscosity. There may also be differential thermal effects on different ice facies, so that temperature-sensitivity becomes more complex where different ice types are present. Despite the presence of distinctive basal layers, modellers examining response behaviour often make the simplifying assumption that viscosity is uniform. More research is needed to establish and isolate the causal variables in ice facies rheological variability, to establish the nature and extent of variability, and to more fully elucidate the effects of ice heterogeneity on glacier response behaviour.

SIXTY-FOUR

The behaviour of glaciers on frozen beds: modern and Pleistocene examples

Richard I. Waller* and Julian B. Murton

**School of Physical and Geographical Sciences, Keele University, Keele, Staffordshire, ST5 5BG, UK*
†Department of Geography, University of Sussex, Brighton BN1 9QJ, UK

Recent research has challenged the traditional view that interactions between glaciers and permafrost are limited, suggesting instead that such interactions are both glaciologically significant and potentially widespread (Waller, 2001). Two case studies here illustrate glacier–permafrost interactions in modern and Pleistocene settings.

64.1 Leverett Glacier, west Greenland

The Leverett Glacier is a small outlet glacier in west Greenland (67°06′N, 50°09′W). It has experienced significant oscillations during the past 100 yr, during which the margin appears to have interacted with a proglacial area underlain by continuous permafrost. The proglacial area is dominated by an arcuate moraine complex rising 15–20 m above the surrounding sandur. The complex is characterized by smaller moraine ridges 1–2 m high and steep-sided ponds up to ca. 100 m in diameter.

Erosion of a section through the northern extremity of the moraine revealed its internal composition and structure, which comprises five units separated by unconformities (Table 64.1 & Fig. 64.1). The majority of the sequence (units 2, 4 and 5) is composed of ice-rich diamicton containing pore ice and layers and lenses of debris-poor ice. Close similarity to stratified-facies basal ice exposed at the current margin suggests that it represents relict basal ice. Ice-rich gravel (within unit 3) is interpreted as a proglacial, glaciofluvial deposit on account of its sorted character and the subrounded nature of the constituent clasts. Unit 1 represents a melt-out till, as it is underlain by a thaw unconformity and is similar in thickness to the local active layer. Three principal structures were also observed: a major thrust fault separating units 2 and 3, an underlying drag fold associated with the deformation of unit 3, and an erosional unconformity between units 4 and 5.

Geomorphological similarities to published descriptions of Arctic push moraines suggest a likely origin through proglacial deformation and the stacking of imbricate thrust sheets (Bennett,

2001). However, the internal structure bears little relation to its geomorphology, implying a more complex origin. Instead, its structural characteristics are best explained by successive phases of ice advance and compression, subglacial erosion, and ice retreat. First, ice retreat led to stagnation and burial of basal ice and the aggradation of glaciofluvial sediment. Second, the resulting sequence was frozen as permafrost aggraded downwards. Third, ice readvance resulted in the longitudinal compression of the sequence, the formation of the thrust fault between units 2 and 3, and the associated drag fold deformation of units 3 and 4. Continued ice advance led to the site being overridden and eroded subglacially, producing the erosional unconformity between units 4 and 5. Finally, ice retreat led to its exposure, the deposition of small recessional moraine ridges, and the formation of lakes and melt-out till as the surface layers thawed.

Proglacial permafrost is widely acknowledged to be influential in the generation of push moraines (Bennett, 2001). In this instance, the preservation of a range of ice and frozen sediment facies within the proglacial zone has influenced the rheological response of the materials to subsequent ice advance, and resulted in the formation of a structurally complex landform. In this respect, the lateral continuity of thick permafrost may also have limited the moraine's distal migration and encouraged overriding. Finally, the presence of deformed ice lenses and an erosional unconformity corroborates suggestions that sliding and pervasive deformation can remain active at subfreezing temperatures.

Table 64.1 Key units exposed within the Leverett Glacier section

Unit	Description	Interpretation
1	Layer of sand-rich, ice-free diamicton (≤2 m thick). Caps the entire section	Melt-out till
2	Ice-rich diamicton (ice content ca. 50% by volume). Contains numerous ice layers and lenses that dip up-glacier at ca. 40°; some are deformed into tight folds	Basal ice
3	Folded sequence comprising intercalated layers of ice-rich diamicton, debris-poor ice and ice-rich gravel	Intercalated sequence of basal ice, englacial ice and outwash
4	Ice-rich diamicton containing numerous ice layers and lenses occurring as a series of recumbent, isoclinal folds with strongly attenuated limbs	Basal ice
5	Ice-rich diamicton containing subhorizontal ice layers and lenses	Basal ice

64.2 Tuktoyaktuk Coastlands, western Arctic Canada

The Tuktoyaktuk Coastlands are located on the mainland Beaufort Sea coast of western Arctic Canada (ca. 69–70°N, 129–136°W). During Marine Isotope Stage 2, they were overridden by the north-western Laurentide Ice Sheet (LIS), resulting in widespread glacier–permafrost interactions beneath and, later, in front of its cold-based margin. Investigation of three coastal sequences of glacially deformed permafrost revealed massive ice overlain by ice-rich diamicton and sandy silt (Murton *et al.*, 2004). From these sections, four types of interactions are inferred:

64.2.1 Submarginal deformation

Isoclinal recumbent folds and pinch-and-swell structures within the ice-rich diamicton and frozen sandy silt (Fig. 64.2A) indicate simple shear and extension. Such features are consistent with subglacial rather than proglacial glaciotectonic deformation, and formed within a frozen glaciotectonite.

64.2.2 Submarginal erosion

Blocks of ice up to 15 m long and 1.5 m high, some folded, are dispersed throughout the glaciotectonite (Fig. 64.2B). Where bands in the ice blocks are truncated along the margins of the blocks, *in situ* formation of segregated or intrusive ice is discounted. Instead the ice was eroded from pre-existing ice and thus represents ice clasts. Submarginal erosion of ice also explains the occurrence of an angular unconformity along the top of the

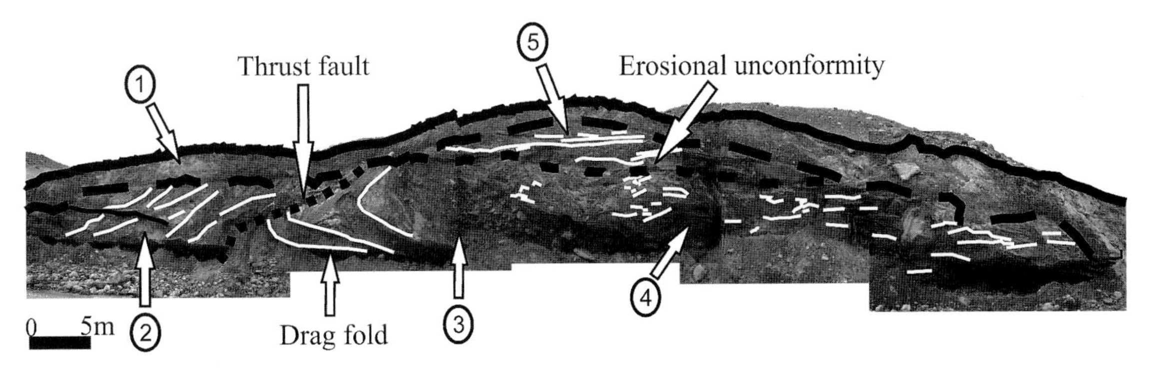

Figure 64.1 Photomontage and structural interpretation of the section through the moraine complex at the Leverett Glacier. Numbered arrows show the location of the key units, thick black lines delineate their boundaries, and thin white lines highlight the orientation and extent of structures within them. Ice flow was from left to right.

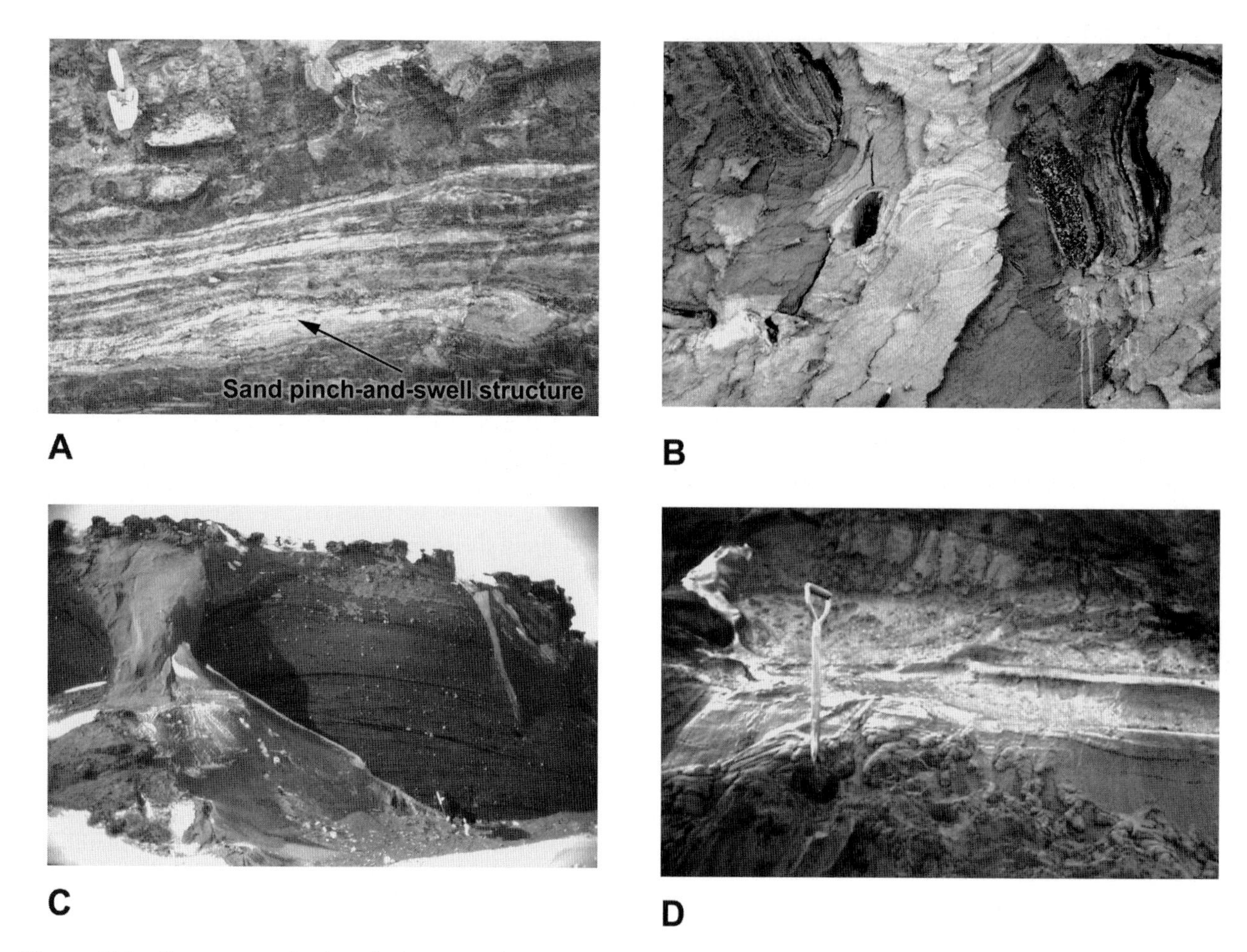

Figure 64.2 Structures reflecting glacier–permafrost interactions in western Arctic Canada: (A) pinch-and-swell structure of sand, North Head (trowel for scale) (69°43′N, 134°26′W); (B) ice clasts within glaciotectonite, Pullen Island (area depicted ca. 1.5 m wide) (69°46′N, 134°25′W); (C) buried basal ice, Mason Bay (face is ca. 4 m high) (69°33′N, 134°02′W); (D) ice dyke–sill truncating a composite wedge, North Head (spade for scale).

underlying massive ice. Had the unconformity resulted from downward thaw, then the ice clasts above it would have thawed. Preservation of the ice clasts and the underlying massive ice indicates that erosion occurred at subfreezing temperatures.

64.2.3 Burial of basal ice

The massive ice (Fig. 64.2C) comprises both buried basal ice and intrasedimental ice (Murton *et al.*, 2005). Burial of basal ice in this region occurred by three processes:

1 glacial thrusting and shearing of frozen sediment (inferred where frozen glaciotectonite overlies basal ice);
2 downward melt of ice previously exposed at the ground surface (inferred where supraglacial melt-out till overlies basal ice);
3 deposition of aeolian, glaciofluvial or lacustrine sediments above basal ice.

64.2.4 Injection of pressurized water into proglacial permafrost

Superimposed on some glacially deformed permafrost are undeformed dykes and sills of intrusive ice and some undeformed massive segregated-intrusive ice (Fig. 64.2D). Such ice clearly post-dates the time of glacial deformation. Because the underlying sedimentary sequence has remained frozen ever since it was deformed, water supplied by pore-water expulsion during permafrost aggradation is unlikely. Instead, an external source of overpressurized water is inferred, probably from glacial meltwater flowing beneath proglacial permafrost during the retreat of the LIS (Murton, 2005).

These interactions illustrate the dynamic coupling of the cold-based margin of an ice sheet, first with subglacial permafrost and later, during deglaciation, with proglacial permafrost.

Acknowledgements

Support from the University of Greenwich, the Quaternary Research Association, the Leverhulme Trust, the Geological Society, the Natural Sciences and Engineering Research Council of Canada, the Geological Survey of Canada, the Polar Continental Shelf Project and the Inuvik Research Centre enabled fieldwork in Greenland and Canada between 1989 and 2002. Dr George Tuckwell (STATS Ltd) undertook the structural interpretation of the Leverett section.

SIXTY-FIVE

Mechanical behaviour and structure of the debris-rich basal ice layer

Sean Fitzsimons

Department of Geography, University of Otago, P.O. Box 56, Dunedin, New Zealand

65.1 Introduction

Basal ice can be defined as a relatively thin layer of ice that forms through a combination of thermal and mechanical processes which link a glacier to its substrate. These thermal and mechanical processes result in several characteristics, including layers and/or lenses of debris entrained from the bed, deformation structures, relatively high solute concentrations and unusual gas composition, that collectively distinguish it from ice that has formed solely by the firnification of snow at the glacier surface (Hubbard & Sharp, 1989; Knight, 1997). Basal ice is thought to be an important component in the rheological behaviour of glaciers.

Most models of glacier behaviour are founded on the assumptions of homogeneous deformation behaviour that is described by Glen's flow law and on the presence of a sharp ice–substrate interface. These assumptions are increasingly at odds with field and laboratory observations that demonstrate that basal ice is characterized by compositional and mechanical properties that distinguish it from glacier ice formed by precipitation alone. These observations suggest that unconsolidated deposits beneath glaciers are common and that these deposits may deform in response to shear stresses imposed by the flowing ice. The separation between modelling efforts and field and laboratory observations is compounded by a tendency to conceptualize the behaviour of basal ice and the glacier substrate as independent entities when such a separation may not be supported by field observations.

Despite a call for enhancement of our understanding of the behaviour of ice at glacier beds by Theakstone in 1967, and restatement of the same imperative in a landmark review of basal ice formation and deformation by Hubbard & Sharp (1986) we have not yet achieved a complete understanding of the formation and deformation of basal ice. Until such comprehension is achieved it seems likely that basal ice will continue to be treated as an isotropic material. The purpose of this chapter is to examine recent developments in the understanding of the deformation at the base of glaciers in the light of field observations and experiments conducted beneath cold-based glaciers in the McMurdo dry valleys.

65.2 Debris-rich basal ice or ice-rich basal sediment?

Early observations of the nature of glacier beds demonstrated that dirty basal ice can rest above frozen sediment and ice. For example Hansen & Langway (1966) and Herron & Langway (1979) described several metres of an ice-laden till-like subice material at the Camp Century core site and Koerner & Fisher (1979) observed a glacier bed that consisted of loose angular rock fragments with 31% rock fragments by weight.

More recently, several studies have sought to understand the distribution of strain at the base of glaciers that have such unconsolidated substrates. For example Truffer *et al.* (2000, 2001) describe a subglacial till layer between 4.5 and 7.5 m thick that was 'clearly separated' from overlying basal ice containing some dirt layers at the base of Black Rapids glacier in Alaska. A sample of the subglacial sediment retrieved from the drill hole yielded a water content of 29% when melted suggesting that the change in material properties at the glacier bed is one of dirty basal ice to ice-rich debris.

In contrast, Echelmeyer & Wang (1987) encountered bands of frozen subglacial till beneath the Urmqui No. 1 Glacier in China and concluded that 10 to 25% of motion occurred within these bands. They reported a sharp contact between the clear glacier ice, which had a debris content of about 1%, and material that they termed the glacier bed, which consisted of 'active ice-laden drift several metres thick'. Samples of the ice-laden drift contained 21–39% debris by volume. These descriptions of the material properties of the glacier substrate overlap with the material properties of what other workers refer to as basal ice.

Examination of the basal zones of glaciers in the McMurdo dry valleys, where most glaciers are cold-based, has demonstrated a wide variety of types of basal ice (Holdsworth & Bull, 1970;

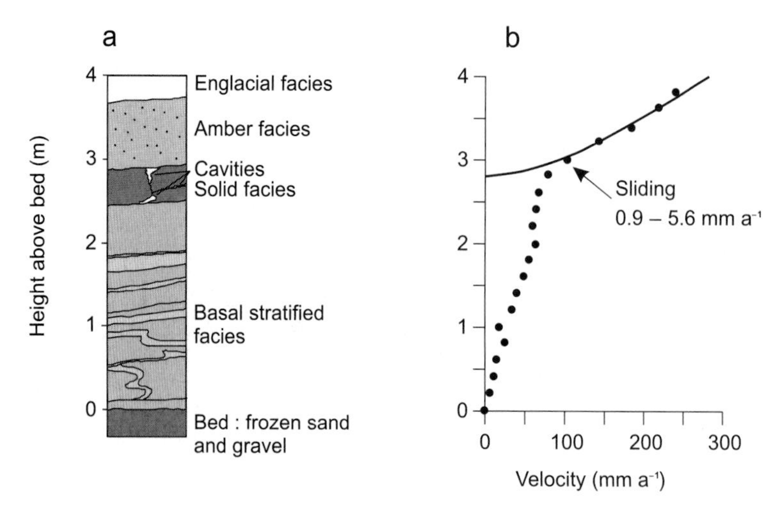

Figure 65.1 Ice composition, structure and deformation at the base of Suess Glacier in the Taylor Valley, Antarctica.

Fitzsimons, 1996; Fitzsimons *et al.*, 1999; Lorrain *et al.*, 1999; Sleewaegen *et al.*, 2003). Holdsworth & Bull (1970) demonstrated that the effective bed of Meserve Glacier (basal temperature −18°C) occurred along the tops of boulders that protruded from the glacier bed into the basal ice. Holdsworth & Bull recorded a debris-rich amber basal ice layer up to 0.6 m thick and demonstrated that salts from the glacier substrate and/or the amber ice extended up to 6 m above the glacier base. These observations, later verified by Cuffey *et al.* (2000c), tend to reinforce the view in the glaciological literature that cold-based glaciers are inefficient agents of erosion.

However, the thickness, composition and structure of basal ice of glaciers in the McMurdo dry valleys that rest on unconsolidated sediments are distinctly different from those confined to the valley sides, such as Meserve Glacier. Glaciers that reach the valley floor are characterized by a thicker basal ice layer, higher debris concentrations, frozen blocks of sediment within the basal ice, and in some cases well-developed ice-marginal moraines (Fitzsimons, 1996; Humphreys & Fitzsimons, 1996). Analysis of the composition of the ice exposed at the margins of some of these glaciers suggests that at least part of the basal ice has been derived from accretion of marginal water and sediment. Tunnels excavated into several of these glaciers have provided an opportunity to closely examine the composition, structure and behaviour of basal ice and glacier beds.

At Suess Glacier the basal ice sequence is approximately 3.8 m thick and can be divided into five main units on the basis of physical appearance, debris concentration and disposition of debris (Fig. 65.1). The layers include clean englacial ice (4–3.7 m), amber ice (*sensu* Holdsworth, 1974) (3.7–2.9 m), a solid layer of frozen sediment (2.9–2.4 m), a basal laminated facies (2.4–0.0 m) and the substrate (Fig. 65.1). The clean englacial ice is underlain by amber ice that contains dispersed particles and fine aggregates, which give the ice a greenish to amber colour. The amber ice is underlain by a broken layer of frozen sand and fine gravel that contains well preserved sedimentary structures. The laminated ice that lies below the main debris band consists of multiple layers of clean bubbly ice to complexly deformed layers of ice with very high debris concentrations (Fig. 65.2). Occasional boulders up to 1.2 m in diameter were observed in the laminated ice up to 2 m above the bed. Ice with relatively low debris concentrations contained abundant ductile structures (Figs 65.3–65.5), whereas structures in the solid layers indicate predominantly brittle defor-

Figure 65.2 Photograph of the stratified ice facies in the basal ice zone of Suess Glacier 26 m into the tunnel. The solid facies occurs at the roof of the tunnel at this location.

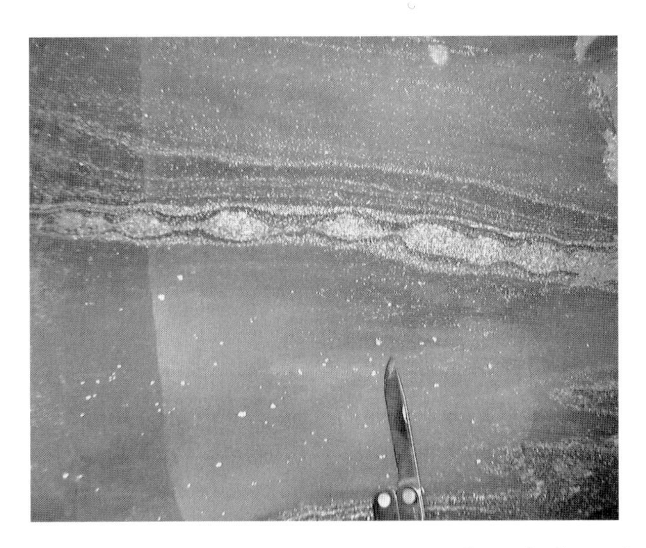

Figure 65.3 Boundinage structures 1.7 m above the base of Suess Glacier. The boudins have formed from a 10-mm-thick layer of sand. The glacier flow direction is from right to left.

Figure 65.5 Sheared recumbent folds 0.8 m above the base of Suess Glacier. The glacier flow direction is from right to left.

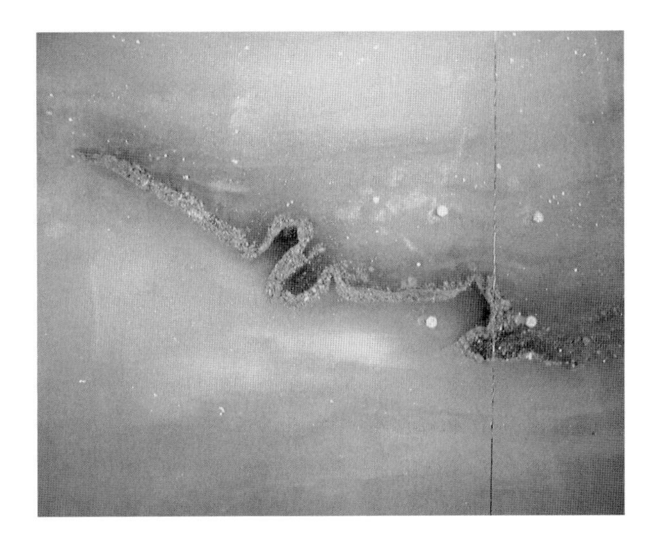

Figure 65.4 A folded layer of gravelly sand 1.0 m above the bed of Suess Glacier. The heads of the bolts are 10 mm in diameter and the glacier flow direction is from right to left.

Figure 65.6 Photograph of a 4 m shaft excavated through part of the basal zone of Wright Lower Glacier showing layers of sand interbedded with ice. The glacier flow direction is from top to bottom.

mation. The presence of large blocks of sediment within which sedimentary structures are well preserved suggests that at least part of the basal ice in the glacier has been formed by erosion and entrainment of the frozen bed of the glacier.

Similar observations made at Wright Lower Glacier demonstrated that large layers of the frozen sand occur within the basal ice (Fig. 65.6). The presence of these layers demonstrates that there is no clear distinction between basal ice and the glacier substrate.

The configuration and composition of basal ice in glaciers in the McMurdo dry valleys that rest on unconsolidated sediment show that there is a continuum of ice–sediment mixtures that range between basal ice and frozen debris. The case for this continuum is supported by studies of glaciated permafrost environments, which have been interpreted as the product of coupling between glaciers and frozen substrates (Astakhov *et al.*, 1996).

65.3 Basal ice deformation

Examination of deformation processes at the base of glaciers is beset by numerous problems, which include limited accessibility, structural complexity and spatial variability in physical properties and temporal variability of deformation processes. Access to subglacial locations is a significant problem because deformation of basal ice and subglacial sediment takes place at the ice–

substrate interface beneath a substantial thickness of ice, which makes direct observation of deformation processes very difficult. Even where access is possible there are considerable uncertainties associated with measurements that are made remotely. If boreholes are used it is often not clear where instruments are located or observations made with respect to the glacier bed. In the case of excavations made to access to the bed they are usually limited to glacier margins where ice is relatively thin and the presence of unsupported walls may result in substantial disturbance of subglacial deformation processes by strain-relief creep. In addition, the flow behaviour of basal ice is likely to vary in time and space in response to both flow perturbations as well as variations in the composition and structure of materials.

Despite these difficulties several important observations concerning the behaviour of basal ice and glacier substrates have been made. Most experiments have been driven by two different concerns: understanding the contribution of subglacial sediment deformation to glacier behaviour, which has been reviewed by Murray (1997), and understanding the rheological behaviour of ice–debris mixtures.

There has been considerable research on the rheology of ice–sediment mixtures, both in the laboratory and in the field. The results of this research have provided evidence to suggest that the behaviour of basal ice is very sensitive to debris and solutes entrained within the ice. However, laboratory experiments conducted on the behaviour of basal ice have produced contradictory results. For example, Goughnour & Andersland (1968) concluded that enhanced deformation was associated with low concentrations of debris, whereas Hooke *et al.* (1992) and Nickling & Bennett (1984) concluded that the creep rate of ice decreased with the addition of fine sand and rock debris respectively.

Empirical studies of the mechanical behaviour of basal ice suggest a strong sensitivity to debris content. Holdsworth (1974) measured deformation of basal ice in a tunnel excavated in Meserve Glacier (−17°C) in the McMurdo dry valleys and found that an amber ice layer experienced enhanced deformation, which he attributed to high solute and debris concentrations. He suggested that the exponent in Glen's flow law for the amber ice was between 5 and 6. More recently a reassessment of the Meserve Glacier basal ice suggests slow but detectable sliding of the amber ice over the substrate (Cuffey *et al.*, 1999) and that the enhanced deformation of the amber layer could be attributed to the small size of ice crystals in the ice (Cuffey *et al.*, 2000c). Brugman (1983) reported preferential deformation along debris-rich ice layers and Lawson (1996) suggested that the presence of debris can enhance creep rates. *In situ* measurements made by Echelmeyer & Wang (1987) in a tunnel in Urumqui No. 1 Glacier in China demonstrated that 60–85% of surface velocity was due to enhanced deformation in ice-laden till, motion in discrete shear planes (10–25% of glacier motion) and sliding at the ice–sediment interface. Similarly Waller & Hart (1999) measured deformation at the margin of an outlet glacier in Greenland and found that most motion could be attributed to sliding or subglacial sediment deformation and relatively little motion could be explained by creep within the basal ice.

In addition to the work described above, several studies have made direct measurements of movement close to glacier beds using tilt meters, plough meters and drag spools deployed in holes drilled from the surface (e.g. Blake *et al.*, 1992, 1994; Fischer & Clarke, 1994). Engelhardt & Kamb (1998) used a tethered stake to measure deformation at the base of Ice Stream B and concluded that the majority of surface velocity could be accounted for by sliding or deformation within a thin (30 mm thick) layer of subglacial till. More recently Porter & Murray (2001) used tilt sensors hammered into the glacier substrate of Bakaninbreen, Svalbard to monitor near-bed deformation. They interpreted data from the drag spool as evidence of movement predominantly in the form of internal deformation and calculated viscosities that suggested the subglacial sediment may have been partly frozen. They concluded that where the bed is cold the sediment-rich basal layer grades into ice-rich sediment without a clear boundary.

Recent excavations of tunnels in the basal zones of several glaciers in the McMurdo dry valleys have provided opportunities for detailed deformation measurements to be made close to the glacier bed. The basal velocity profile, together with tunnel deformation measurements in Suess Glacier, demonstrates that deformation in the lower 4.5 m of the glacier is characterized by progressive simple shear and that strain is heterogeneous. Four distinctive strain domains occur in the basal ice: a high-strain domain associated with the presence of the amber ice where the velocity profile approximates a power law (Fig. 65.1); a low strain domain associated with the presence of the solid facies (2.9–2.4 m); a moderate strain domain associated with the stratified facies; and sliding at the interface between the amber ice and the solid facies (Fig. 65.1). At the boundary between the rapidly deforming amber ice and the underlying frozen sand and gravel numerous air-filled cavities are present, both where the frozen layer protruded into the amber ice and where the contact was relatively flat. At this boundary dial gauges recorded movement from 0.93 to 5.65 mm yr^{-1}. The measurements can be interpreted as sliding velocities or zones of high shear concentrated in a very thin layer. However, the coincidence of the high strain rates with the presence of slickensides on the cavity roofs clearly indicates that sliding has occurred at the interface. The slickensides have formed as ice sliding over the frozen sediment layer has moulded itself around the roughness elements on top of the layer and produce an imprint of the form roughness in the cavity.

The underlying control of the deformation processes and the consequent deformation structures appears to be the strength of the ice and sediment mixture, the averages of which are 0.9 MPa for the amber ice, 1.28 MPa for the stratified ice and 2.53 MPa for the solid facies (Fitzsimons *et al.*, 2001). Material with a low viscosity (low shear strength) supports the highest strain rates and the material with the highest viscosity (high shear strength) supports the lowest strain rates (Table 65.1). Motion within the strat-

Table 65.1 Peak shear strength values from basal ice samples from Suess Glacier

Ice facies	Peak strength (MPa)	Debris content (% volume)
Englacial	1.39	0.06
Amber	0.90	0.87
Stratified	1.28	43.6
Solid	2.53	65

Figure 65.7 The basal ice solid facies with well-preserved planar bedding in Wright Lower Glacier.

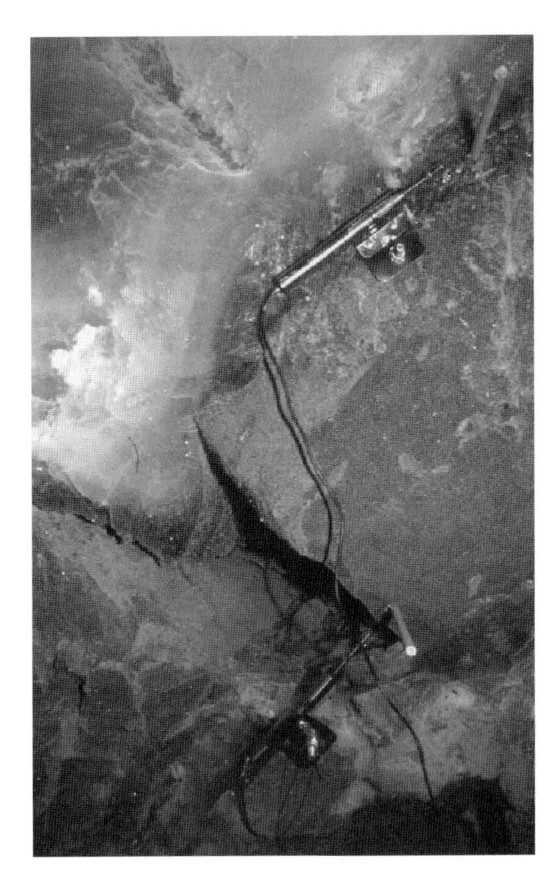

Figure 65.8 Crack in the solid facies in Wright Lower Glacier. The glacier flow direction is from right to left.

ified basal ice is discrete (slickenslides and offsets in velocity profiles measured using strain markers) (see figs 4, 5 & 6 in Fitzsimons *et al.*, 1999).

At the margin of Wright Lower Glacier (basal ice −17°C) plumblines and displacement transducers were used to monitor deformation through layers of frozen sand entrained into the basal zone (Figs 65.6 & 65.7). These measurements show that there was no detectable creep within the frozen sediment layers. However, displacement transducers mounted across cracks in the frozen sediment (Fig. 65.8) demonstrated the cracks were opening at a rate of 22 mm yr^{-1} measured over an 11 month period. The structure of ice adjacent to the cracks suggests that the ice is creeping into the cracks as they open (Fig. 65.9).

Measurements made at the base of Taylor Glacier, which has a basal temperature of −18°C at the margin, have also demonstrated the presence of sliding interfaces within the basal zone. These interfaces are characterized by the development of cavities on the stoss and lee sides of clasts that protruded through the sliding boundary. Sliding velocities up to 167 mm yr^{-1} have been measured using a combination of plumblines, engineering dial gauges and linear variable displacement transducers. Close to the glacier margin the sliding constitutes approximately 10% of glacier motion. Deformation structures that developed in the basal zone over the monitoring period demonstrate that debris-rich ice deformed at considerably greater rates than adjacent clean ice (Figs 65.10 & 65.11).

Taken together, the observations and measurements made in the tunnels in glaciers in the McMurdo dry valleys demonstrate a considerably more complex pattern of deformation than is suggested by the literature on cold-based glaciers. In the cases of the Suess, Wright Lower and Taylor glaciers if observations were made from drill holes from the glacier surface it is very likely that a very different picture of glacier behaviour would have emerged because a drill would not have penetrated the solid facies. Under these circumstances measurements of the first contact between ice and sediment would have been interpreted as the glacier bed. Even

Figure 65.9 Ice from above and below intruding into a crack in the solid facies in Wright Lower Glacier. The glacier flow direction is from right to left.

in the case of measurements and observations made in the tunnels which provide excellent access, our experience in measuring motion at the base of a glacier is that the closer we examine the glacier bed the more problematic the concept of a distinct bed

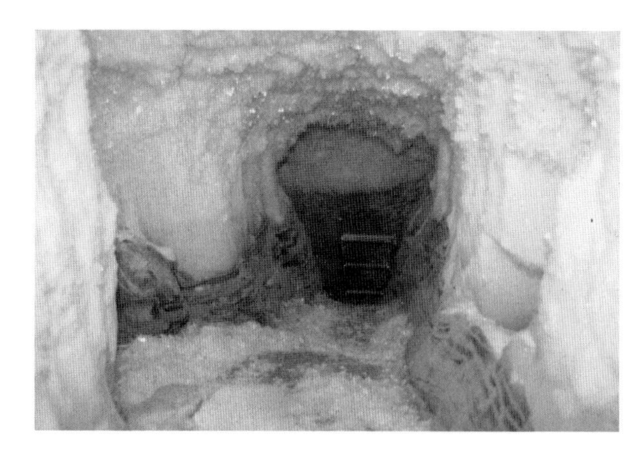

Figure 65.10 Photograph of a tunnel in the left margin of Taylor Glacier showing the 'key-hole' structure produced from more rapid deformation associated with debris-rich ice. The tunnel walls were vertical initially. Photograph taken 11 months after the tunnel was excavated. The glacier flow direction is from the ladder toward the reader.

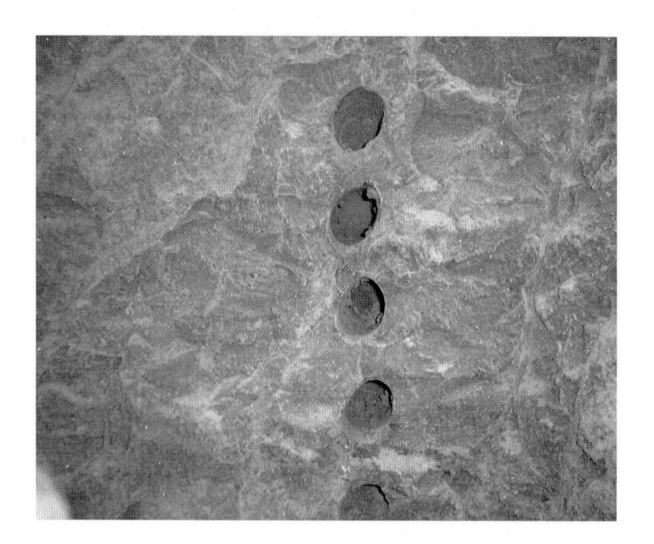

Figure 65.11 Circular strain markers deformed into ellipses after 10 days in debris-rich basal ice in Taylor Glacier. The glacier flow direction is from left to right.

becomes—both in terms of physical characteristics and in terms of deformation patterns (Fig. 65.1).

65.4 Discussion

Although velocity profiles associated with different basal thermal regimes are conceptually useful and widely utilized (e.g. Bennett & Glasser, 1996; Boulton, 1996b; Benn & Evans, 1998) they promote a simplistic view of the ice–sediment interface as a clean boundary between ice and the glacier substrate. From investigations undertaken in the McMurdo dry valleys two problems arise when we attempt to understand the distribution of basal strain with respect to such conceptual velocity profiles.

Firstly, our attempts to identify a glacier-ice–substrate boundary on the basis of a boundary between debris-rich basal ice and ice-rich basal debris ultimately fail because there is no such distinction in any of the glaciers we have examined. In most cases debris-rich ice occurs below what appears to be the zero point in the velocity structure and frozen sediment layers and blocks occur above the zero point. Thus, on the basis of physical characteristics it is frequently not possible to identify the glacier bed.

Secondly, deformation measurements made close to the base of glaciers demonstrate partial coupling between the glacier and the substrate. In Suess Glacier the compound velocity profile (Fig. 65.1), together with measurements of sliding at the ice–solid-facies boundary, demonstrates that the glacier bed is not a simple ice–substrate contact that can be adequately described by a power law. The measured velocity structure is the product of variability in the rheological behaviour of ice with distinctive physical and chemical composition. The form of the profile is similar to theoretical and empirically determined velocity structures associated with subglacial sediment deformation (Boulton, 1996b).

Such a conclusion is consistent with the view expressed by Waller (2001) in a recent review of the basal behaviour of cold glaciers that the assumption of zero-based velocity is simplistic and in reality the situation is likely to be much more complicated. This view has already been demonstrated by Fitzsimons *et al.* (1999), who, as in the accounts of glacier behaviour given above, suggest that rather than a boundary the base of a glacier is a zone. Ultimately the point of zero velocity in any glacier is governed by the thermal and mechanical processes, which are likely to vary in space and time. Accordingly, where a glacier rests on an unconsolidated substrate the base is likely to be a zone rather than a boundary. In the case of the cold-based Antarctic glaciers described above the zone has developed as a result of folding, faulting and shearing that has produced a complex mosaic of ice-rich sediment and debris-rich ice within which many structural characteristics of the substrate are preserved after the material has been entrained. Consequently, drawing distinct boundaries between basal ice and substrate deformation is not supported by the field observations and measurements.

65.5 Summary

1 Examination of the literature suggests that there is confusion concerning the treatment of debris-rich basal ice and ice-rich subglacial sediment. What is described as subglacial permafrost by some researchers is treated as debris-rich basal ice by others.
2 Observations of subglacial conditions in cold-based glaciers in the McMurdo dry valleys demonstrate that in several cases there is no single clear boundary that separates basal ice from a frozen substrate. There is abundant evidence to suggest that deformation of the glacier substrate has resulted in both *en masse* entrainment of the glacier substrate and mixing of material from the glacier substrate with basal ice.
3 Deformation measurements made at the base of glaciers in the McMurdo dry valleys demonstrate that the strain fields close to glacier beds are much more complicated than would be expected from Glen's flow law. Such measurements,

together with the close examination of the structure of basal zones, suggests that treatment of a glacier bed as a simple ice–substrate boundary is flawed.

4 In order to advance our understanding of the behaviour of subglacial materials we need to move away from the persistent separation of basal ice and the glacier substrate. Instead we need to recognize a continuum of material properties and deformation processes that is variable in time and space.

Acknowledgements

This work was supported by the Marsden Fund, Royal Society of New Zealand. Logistical support was provided by Antarctica New Zealand.

SIXTY-SIX

High-resolution time series of basal motion at an Arctic glacier margin

David Chandler

Institute of Geography and Earth Sciences, University of Wales, Aberystwyth, SY23 3DB, UK

66.1 Introduction

Numerous studies of glacier sliding have found that changing subglacial hydrology can explain variations in glacier motion over time-scales of days and longer. Direct measurements of sliding (Fischer & Clarke, 1997a; Hubbard, 2002) and seismicity data (e.g. Deichmann *et al.*, 2000) have provided evidence of unsteady (stick-slip) glacier sliding at much shorter time-scales. These data suggest that friction may be important at the glacier bed, so that elastic strain accumulation and release also form a component of basal motion.

Data collected from the northern margin of Russell Glacier, West Greenland, during September 2002 have allowed investigation of basal motion variations at very short (subhour) time-scales. This paper presents a brief summary of the results and possible implications for basal motion at this and similar sites.

66.2 Field site and methods

Russell Glacier is a drainage lobe of the Greenland Ice Sheet, and terminates ca. 20 km east of Kangerlussuaq. The northern margin is predominantly sediment based, and is characterized by a layer of basal ice up to several metres thick. The structure of the basal ice comprises a debris-poor, dispersed-ice facies overlying a strongly deformed, debris-rich stratified-ice facies (facies descriptions follow Lawson, 1979b). Daily surveys of anchors in the basal ice during late summer 1996 indicated basal motion in the range 20 to 40 m yr^{-1}, with velocities increasing by 100–150% following a period of rainfall (Waller & Hart, 1999).

In this study, the method used to record ice displacement (Fig. 66.1) was based on that of Hubbard (2002). Thin copper wire (diameter 0.2 mm) tied to each anchor was wound on to plastic spools (22 mm diameter) mounted on 100 kΩ multi-turn, conducting-polymer potentiometers. A data logger applied 100 mV across each potentiometer, and recorded the voltage at the central pin. This voltage is proportional to the angle turned through by the spool, and when converted to displacement gives an instrumental resolution of ca. 0.1 mm. At two sites on the northern margin of the glacier, anchors were placed in: (i) nearby sediments, (ii) the stratified ice facies and (iii) the dispersed ice facies. A fourth spool was mounted on the support but not connected to any anchor, to monitor the magnitude of electronic errors. The logger internal temperature was recorded to provide an indication of meteorological conditions. Measurements were made every 5 min at site 1, and every 1 min at site 2. At site 1 there was a ca. 10-cm-thick layer of ice-free debris at the contact between the two ice facies; at site 2 the contact was continuous. At both sites, debris melting out of the stratified-ice facies made the exact position of the bed unclear.

66.3 Results

Table 66.1 summarizes the anchor positions and their mean velocities over the observation period. At site 1 the wire to anchor P2 was disturbed by debris melting out of the ice. Measurements of the relative displacement between the lower anchor and a length of string suspended from the top anchor, however, indicate that all of the velocity of the upper anchor is accommodated within the contact between the two ice types. At site 2, the upper anchor (P3) moved faster than the lower (P2) (Table 66.1). Anchors P1 and P4 at both sites showed negligible overall motion, with short-term fluctuations limited to ±2 logger resolutions (ca. 0.2 mm).

The velocity time series (Fig. 66.2) for anchor P3 at site 1 shows the fastest velocities occurring in the early afternoon, coinciding with higher air temperatures, and long periods of zero motion overnight. At site 2, an increase in velocity occurs during higher air temperatures on 12 September (Fig. 66.3), but the velocities of both anchors are characterized by highly unsteady motion (Fig. 66.4), and lack the long periods of zero motion seen at site 1. Very small, negative velocities in Fig. 66.4 are due to instrumental noise and represent voltage fluctuations of ±2 logger resolutions.

At site 2 the upper anchor moved more smoothly than the lower: this is illustrated by the ratios of the standard deviation to the mean velocity, which are 1.99 and 4.97 for the upper and lower anchors respectively.

66.4 Discussion and conclusions

At diurnal time-scales the results from both sites suggest a link between basal motion and temperature. It is likely that this faster motion is triggered by melt water reaching the bed: either locally,

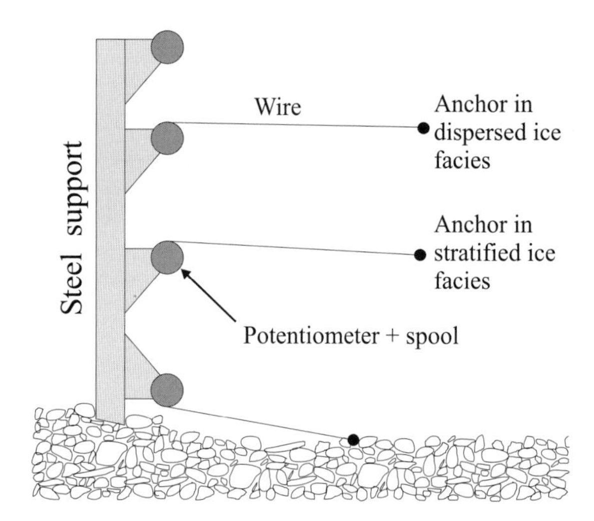

Figure 66.1 Instrument used to record ice displacement. The steel support is ca. 50 cm high.

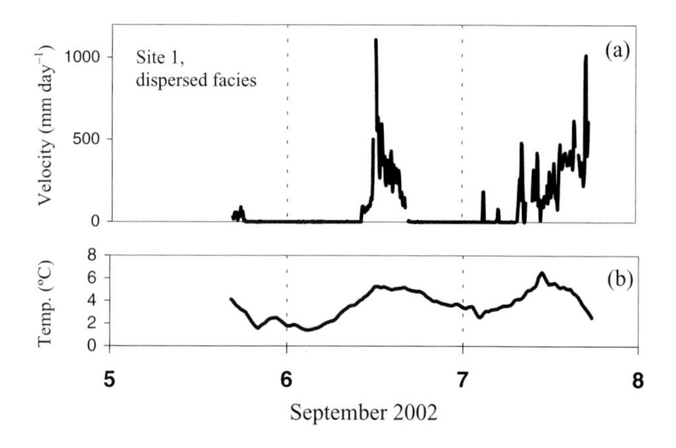

Figure 66.2 Results for site 1: (a) ice velocity for anchor P3; (b) air temperature. Gaps in the record occur after each complete turn of the potentiometer, when the resistance returns to zero.

Table 66.1 Summary of anchor positions and the data recorded

Location (ice thickness)	Description	Height* (cm)	Mean velocity (mm day^{-1})
Site 1 (ca. 5 m)	P1: Sediments near support	0	0.0
	P2: Stratified ice facies	100	—†
	P3: Dispersed ice facies	150	92.1
	P4: Not connected	—	0.0
Site 2 (ca. 15 m)	P1: Sediments near support	0	0.0
	P2: Stratified ice facies	3	53.9
	P3: Dispersed ice facies	60	117.5
	P4: Not connected	—	0.0

*The height above the debris next to the ice; not the height above the bed.
†The wire attached to anchor P2 at site 1 was disturbed by debris falling out of the contact between the two ice facies.

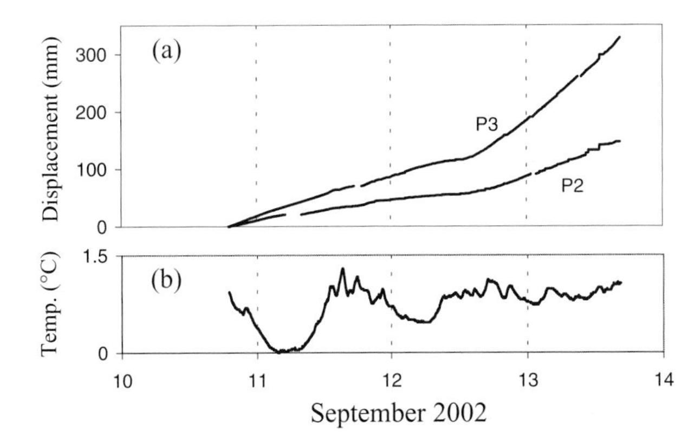

Figure 66.3 Results for site 2: (a) displacements of anchors P2 and P3; (b) air temperature.

or some distance up glacier. In the former case, at least some of the glacier must be warm-based near its margin. In the latter case, diurnal variations in motion up-glacier could be transmitted to the margin via longitudinal stresses. Subdiurnal measurements of surface motion might reveal how far up-glacier these variations occur, and their importance in the dynamics of the glacier over longer time-scales.

Evidence for stick-slip behaviour was seen for all the anchors fixed in the ice and suggests that friction is an important control on the basal motion. Negligible movement of spools P1 and P4 (Table 66.1), along with the varying characteristics of the velocity time-series at different anchors, provides confidence that these slip events are real, and not instrumental artefacts. It is unclear whether these slip events are generated at the bed, or in discrete layers within the ice. At site 2, both these possibilities may occur because some slips are common to both anchors, and others restricted to just one (Fig. 66.4).

The velocity difference between anchors P2 and P3 at site 2 indicates the presence of a high shear strain rate in the ice between the anchors. It is possible that this shear is concentrated within narrow zones, in particular the debris-rich layers. These layers may be associated with the formation of debris bands that occur both at Russell Glacier and at other polythermal glaciers (e.g. Hambrey & Muller, 1996).

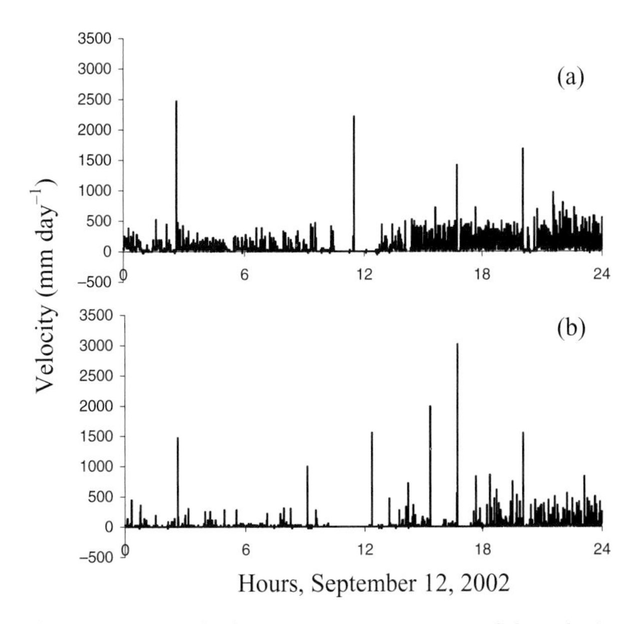

Figure 66.4 Stick-slip motion at site 2. Part of the velocity time series for (a) the upper anchor P3 and (b) the lower anchor P2.

More detailed analysis of the characteristics of these time series may show how friction controls basal motion at this site. Further field studies involving intense spatial coverage using many anchors, in combination with seismic data, might allow investigation of how slip events are generated and how they propagate within the ice.

Acknowledgements

W. Adam organized and supported the field work; R. Waller and B. Hubbard provided many helpful comments.

SIXTY-SEVEN

On the relationships between field data and numerical models of ice-mass motion

Bryn Hubbard

Centre for Glaciology, Institute of Geography and Earth Sciences, University of Wales, Aberystwyth, Ceredigion SY23 3DB, UK

67.1 Introduction

Predictions of ice-mass response to climate change are based on numerical models that replicate the behaviour of ice masses under the influence of changing mass balance. Accurately predicting spatially distributed ice motion is central to the success of these models because it is the glacier's flow field that dictates its pattern of geometrical adjustment to mass-balance change. For this accuracy to be achieved with confidence, models must include close representations of all relevant processes of mass motion, preferably incorporating variations in those processes at as high a temporal and spatial resolution as possible. Field-based glaciologists have a great deal to offer, and to learn from, the modelling community in this context.

67.2 Models of ice-mass motion

Spatially distributed numerical models of glacier and ice-sheet evolution are based on calculating changes in ice thickness (H) through time (t) according to the mass continuity equation

$$\frac{\partial H}{\partial t} = Ac - Ab - b - \nabla(H\bar{u}) \quad (1)$$

where Ac is accumulation, Ab is ablation, b is basal melt, $\bar{u}$ is the vertically integrated two-dimensional horizontal velocity and ∇ is the two-dimensional horizontal divergence operator. It follows that an ice mass' geometric response ($\partial H/\partial t$) to a given change in its surface mass balance boundary condition (Ac–Ab) is critically dependent on ice redistribution by motion ($\bar{u}$). The most advanced of these models are *three-dimensional* and, where necessary, *thermomechanically coupled*, and predominantly, although not exclusively, are solved on fixed, predefined arrays of cells (*finite differences*) where mass-balance, temperature, stress, strain and ultimately the change in ice thickness are computed iteratively. Over the past decade, as numerical time-evolving ice-sheet models have been applied at greater resolutions to a wider variety of scenarios, a significant effort has been expended in developing more sophisticated ice-motion algorithms, which attempt the realistic incorporation of physical processes known to influence specific motion components at appropriate scales and conditions. Such algorithms, however, are often prohibitively expensive in computational resources when incorporated into time-evolving ice-sheet models. Thus, simplified models with reduced ice physics are still widespread, especially in solving problems at large spatial and temporal scales, such as the Antarctic ice sheet, where computational resources are at a premium. For example, a model may include only motion by internal ice deformation, and not that focused near the glacier bed (so-called *basal motion*). Furthermore, in the most simple and widespread case, the *zero-order approximation* considers motion to be driven only by local shear stress (also known as the *driving stress*, τ) calculated from local slope (α), ice thickness (H), density (ρ) and gravity according to

$$\tau = -\rho g H \alpha \quad (2)$$

However, at fine scales where there are large changes in slope or basal motion, significant stresses can be transferred longitudinally or laterally. In order to model these inherited stresses, higher order terms from the Navier–Stokes equations need to be included and solved, requiring more complex derivations and numerical algorithms. The more common of these *higher order* models solve the first-order terms (i.e. those raised to the power of 1, neglecting those raised to higher powers). Such *first-order approximation* models of grounded ice flow thereby account for longitudinally transferred stresses. For the vertical or bridging stresses to be calculated, however, higher order terms again need to be solved. Such *full solution* models are extremely demanding computationally and, to date, have been solvable only on fixed boundary, finite-element glacier geometries to investigate specific small-scale issues (e.g. Gudmundsson, 1999; Cohen, 2000) rather than to long-term studies of ice mass response to climate change, to which the finite-element method is not easily applicable. Neglecting higher order terms is not necessarily a major setback, however, depending on the horizontal resolution of the model in question and the specific conditions under which it is applied. For example, Hindmarsh (personal communication, 2003) has demonstrated, using analytical solutions to a parallel-sided slab geometry, that second-order terms only become important in the ice-flow solution when dealing with grounded ice masses with significant basal motion when the horizontal grid scale approaches the scale of ice thickness.

In reality, inherited stresses are important where variations in traction at the ice margins represent a significant proportion of a glacier's total stress field. Such situations arise at smaller ice masses, where the ratio of boundary area to ice volume is high, and partly- or wholly-temperate ice masses, where marginal traction can vary with the presence of basal meltwater (Clarke & Blake, 1991; Murray *et al.*, 2000b). Thus, inherited stresses are particularly important at temperate valley glaciers. Evidence from Haut Glacier d'Arolla, Switzerland, for example, indicates that such stresses account for up to half of the total stress field in places (Hubbard, 2000). Importantly, ice streams have certain critical features in common with temperate valley glaciers and therefore also demand higher-order modelling to reproduce lateral shear margins and zones of reduced basal traction.

67.3 Ice deformation

67.3.1 Current formalization

Field and laboratory experiments carried out in the 1940s and 1950s led to the initial formulation of a constitutive relation for ice, now known as Glen's flow law (Glen, 1955). In tensor notation with $i, j = x, y, z$, the three axes of the Cartesian coordinate system, it takes the form

$$\dot{\varepsilon}_{ij} = A\tau_e^{n-1}\tau_{ij} \quad (3)$$

where $\dot{\varepsilon}_{ij}$ is the strain rate tensor, A is a rate factor that reflects ice hardness (principally considered as solely temperature-dependent), τ_e is the effective stress, which is a measure of the *total stress state* of the ice under consideration, τ_{ij} is the imposed stress tensor and the exponent n is generally taken to have a value of ca. 3 (Hooke, 1981). Although ice can be modelled alternatively as a Newtonian viscous fluid ($n = 1$, whereby strain rate varies linearly with the applied stress) or as a plastic (whereby ice fails completely, giving infinite strain at a given critical stress of ca. 100 kPa), Glen's non-linear viscous relation remains the most commonly used representation. Yet, owing to its non-linearity and the fact that the calculated strain rate is not just dependent on the imposed stress but on the total stress state, this equation is a challenge to solve at the ice-mass scale.

67.3.2 Field data

Field researchers have long realized that ice masses are characterized by a high degree of spatial variability in their physical composition. Bulk ice properties other than temperature that are known to vary include ice crystallography (size, orientation and shape), and the concentrations of included gas bubbles, debris, water and dissolved impurities. Each of these properties exerts some control over ice flow rate, resulting in large-scale spatial variations in enhancements in flow rate over that of isotropic ice as described by Glen's flow law (Equation 3). Hereafter, such variations in flow rate will be generalized as ice 'softness'. Although quantifying this effect still largely eludes glaciologists, important progress has been made in certain fields. This situation reflects an improving, but still incomplete, understanding of both (i) the exact influence of covarying suites of physical properties on ice softness, and (ii) the precise distribution of those ice properties within real ice masses. Ice cores recovered from large polythermal ice masses have shed some light on this matter, indicating the presence of a basal ice layer, often of Pleistocene age, that is typically between three (Dahl-Jensen, 1985) and 100 (Shoji & Langway, 1984) times softer than the overlying Holocene ice. Analysis of these ice cores indicates that this enhancement probably results from a suite of physical properties, although a preferred ice crystal orientation is central to most explanations (Thorsteinsson *et al.*, 1999). It is also probable that the liquid water content of ice (principally held within the intercrystalline vein system) also plays an important role in this context. Laboratory studies by Duval (1977) indicate that a 1% increase in water content increases a sample's strain rate by ca. 400%.

In contrast to the Earth's larger ice masses, very little is currently known about the internal physical structure of temperate valley glaciers, where ice flow is sufficiently rapid that no ice of Pleistocene age survives, and from where few ice cores have been retrieved. However, recent ice-core evidence suggests that the physical structure of temperate valley glaciers also varies systematically. Analysis of the crystallography, ionic composition, included debris and included gas of a series of ice cores from Tsanfleuron Glacier, Switzerland, led Hubbard *et al.* (2000) and Tison & Hubbard (2000) to argue for the presence of three distinctive zones at this temperate valley glacier. The Basal Zone has a thickness of ca. 1.5 m above the ice–bed interface, the Lower Zone extends for ca. 15 m above this, and the Upper Zone extends from the top of the Lower Zone to the glacier surface (Fig. 67.1).

Finally, it is worth noting that all glaciers are characterized by some degree of discrete brittle failure. Crevasses are common and indicative of complex processes, for example, at the glacier bed, within an ice fall or at a calving ice margin, and represent a boundary across which stresses cannot be transferred. They become a significant stumbling block to the application of continuum mechanics on which Glen's law is based, although void theory is one means by which the effect of crevassing can be incorporated realistically into models of ice motion. Still, very little is known about the integrity of ice at depth. For example, although crevasse depth is generally restricted to some tens of metres, hydrofracturing may extend the failure deeper, which will be critical in controlling iceberg calving as an outlet glacier approaches flotation and basal traction reduces towards zero. Further, healed crevasses can represent planes of weakness for subsequent exploitation. Indeed, numerous recent field studies have focused on the role of thrusting along slip planes within

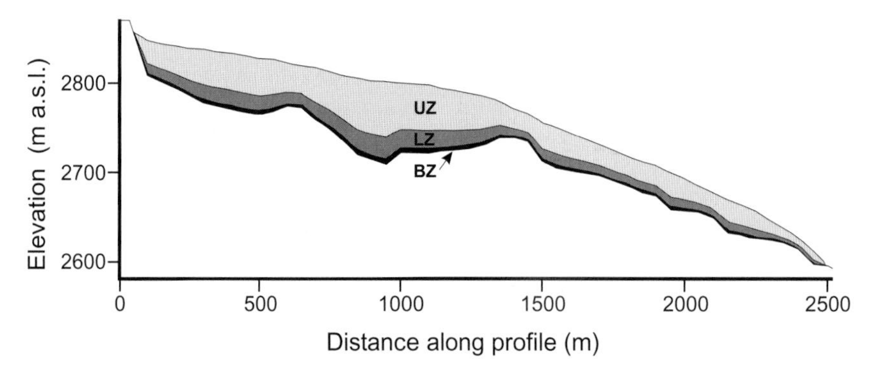

Figure 67.1 Long section of Tsanfleuron Glacier, Switzerland, illustrating its constituent ice zones (UZ = Upper Zone; LZ = Lower Zone; BZ = Basal Zone) as reconstructed from core characteristics. (After Hubbard *et al.* (2003) with the permission of the International Glaciological Society.)

glacier ice, particularly at polythermal glaciers (e.g. Hambrey *et al.*, 1997, 1999).

67.3.3 Incorporation into models of ice-mass motion

The constitutive relation for ice forms the basic building block of all ice-flow models, many of which until recently adopted Glen's law with a spatially uniform value of n and a temperature-dependent value of A in Equation (3). However, ice-sheet models are increasingly adopting a two-layered structure, with a basal layer that is softer than the overlying ice (e.g. Huybrechts *et al.*, 2000). This normally is achieved by introducing an additional multiplier, termed an *enhancement factor*, into Equation (3). This distinction, however, is only one among many potential spatial variations in ice softness, including, for example, those associated with down-glacier changes in crystal fabric and water content that remain unknown and, thus, unaccounted for in ice-sheet models.

Although the binary layering of polythermal ice masses is imperfect, the treatment is markedly better than that of temperate valley glaciers, models of which tend to include no spatial variability in ice softness. One exception is an evaluation of the potential significance of ionic variability within the ice cores recovered from Tsanfleuron Glacier (above). Here, Hubbard *et al.* (2003) (i) used measured ice bulk ionic concentrations to approximate the water content and ice softness of each of the ice zones identified in the cores; (ii) incorporated those rheologies and their host zone geometries (Fig. 67.1) within a two-dimensional first-order flow model, and (iii) ran the model to compare the effects of a homogeneous (uniform rheology) glacier and a three-layered glacier. Results of the ice analysis indicated that the Lower Zone and Basal Zone ice are respectively 1.80 and 10.74 times softer than the Upper Zone ice. Modelling indicated that the imposition of a 75 m rise in equilibrum line altitude (ELA) resulted in a predicted homogeneous glacier that was 65% larger than the predicted layered glacier (Fig. 67.2).

67.4 Basal motion

Periods of enhanced ice-mass velocity, or steady velocities that exceed those induced by ice deformation alone, may be attributed to supplementary basal motion. At hard-bedded glaciers, underlain by bedrock, this component is supplied by basal sliding. At soft-bedded glaciers, underlain by unconsolidated sediments, the motion is generally considered to be achieved by deformation of those sediments. These processes are considered in turn.

67.4.1 Basal sliding

67.4.1.1 Current formalization

Weertman (1957, 1964) considered basal sliding in terms of two discrete processes: *enhanced deformation* and *regelation.*

67.4.1.1.1 Enhanced deformation

Local stress fields are established within basal ice as it passes over a rough bed. Locally enhanced stoss-face stresses induce enhanced local basal drag (τ_b), forcing the ice to deform around bedrock hummocks particularly rapidly. As velocity scales with strain rate times distance, the larger the hummock of a given shape the greater the ice deformation rate U_d, yielding (for a linear flow law) a relation of the form

$$U_d \propto \tau_b a \quad (4)$$

where a is the hummock height (or size in general for a given shape).

67.4.1.1.2 Regelation

Regelation involves melting and refreezing cycles that result from the pressure dependence of the melting point of ice. The process therefore operates most effectively at the pressure melting point. Enhanced stoss-face stresses locally lower the freezing point of ice, inducing melting. The meltwater thereby produced flows along the local pressure gradient to the lee side of the bump, where the local pressure is reduced, the freezing temperature is raised and the water film refreezes. This refreezing releases latent heat, which is conducted back through the rock to the stoss face, where it fuels further melting. Due to this requirement for heat

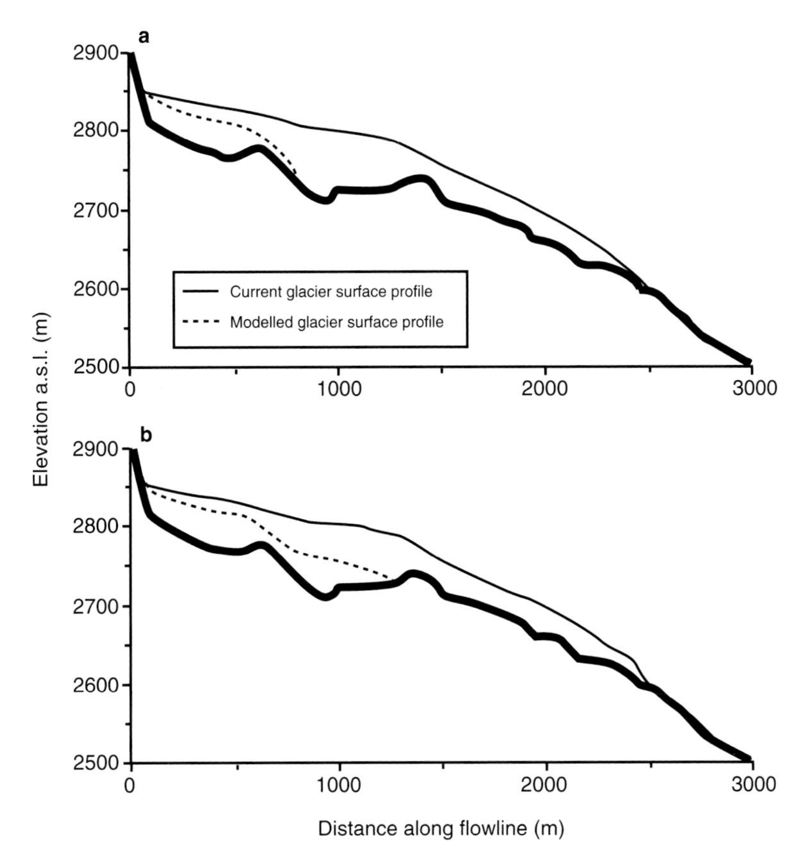

Figure 67.2 Results of the modelled response of Tsanfleuron Glacier to a 75 m rise in ELA. The current measured glacier surface profile is given as a solid line and the modelled steady-state profile is given as a dashed line: (a) multilayer rheology model and (b) single-layer rheology model. (After Hubbard *et al.* (2003) with the permission of the International Glaciological Society.)

flow, regelation sliding velocity U_r is greater around smaller bumps, yielding

$$U_r \propto \tau_b/a \quad (5)$$

67.4.1.1.3 Role of bedrock roughness

As sliding by enhanced ice deformation is predominantly controlled by stresses around large bedrock irregularities and regelation occurs most effectively around small bedrock irregularities, an intermediate, or *controlling*, hummock size may be identified that presents the greatest resistance to basal sliding. Other things being equal, it is the predominance of roughness elements of this size that determines differences in sliding velocity over different bedrock substrates. Making use of this fact to minimize the summed velocities from enhanced deformation and regelation allows total sliding speed U_t to be expressed in terms of the basal drag τ_b and roughness r, defined as the ratio of hummock size to hummock spacing, yielding (for $n = 3$)

$$U_t \propto \tau_b^2/r^4 \quad (6)$$

Further, the value of r may be weighted towards the controlling obstacle size and the analysis may be refined to cater, for example, for hummocks of different sizes and spacing by considering the bed as a spectrum of roughness waveforms (e.g. Nye, 1970).

67.4.1.1.4 Role of basal meltwater

The presence of water-filled cavities at the ice–bedrock interface enhances sliding speed through separating ice from its bed, increasing the shear stress on the locations remaining in contact, and exerting a net down-glacier force on the overlying ice. Although these processes have not been incorporated formally into sliding theory, the most applicable relation expresses sliding speed U_t as an inverse function of effective pressure N (defined as ice pressure minus basal water pressure)

$$U_t \propto \tau_b^p/N^q \quad (7)$$

The values of p and q may be determined empirically.

67.4.1.2 Field data

Studies focusing on periods of enhanced motion generally indicate a positive correlation between rates of measured or inferred basal motion and basal water pressures (Willis, 1995). For example, in their pioneering study Iken & Bindschadler (1986) reported simultaneous records of surface velocity and borehole water pressure (a surrogate for basal water pressure) from Findelengletscher, Switzerland. When plotted against each other these two variables revealed a positive non-linear correlation (Fig.

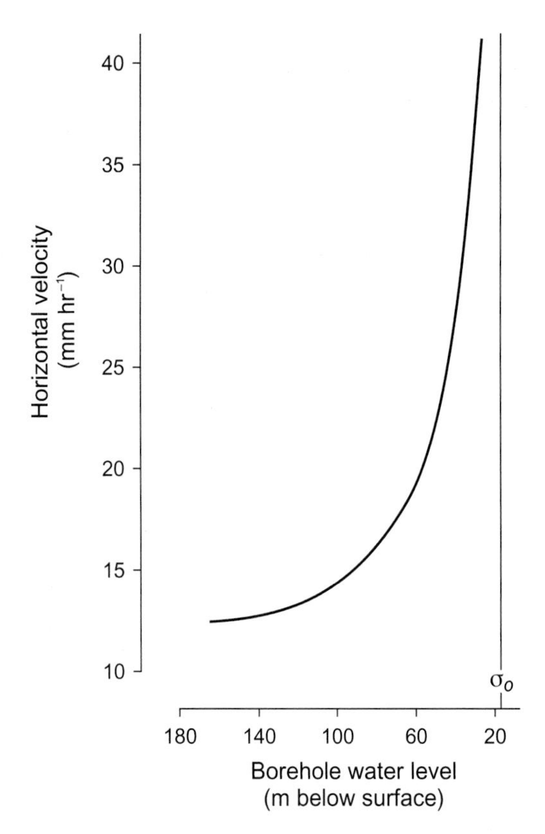

Figure 67.3 Measured relationship between horizontal velocity of a surface stake and water level in a borehole at Findelengletscher, Switzerland. (After Iken & Bindschadler (1986) with the permission of the International Glaciological Society.)

67.3). Jansson (1995) compared these with similar data from Storglaciären, Sweden, both of which matched

$$U_t \propto N^{-0.4} \tag{8}$$

although the multiplier terms in the original equations were an order of magnitude different. Although this is similar to Equation (7) above, simultaneous field data that also allow the basal drag term to be calibrated are rare. One exception was a laboratory-based study by Budd *et al.* (1979) which yielded results that were consistent with Equation (7), indicating values for p and q of 3 and 1 respectively. Basal water pressure, however, could not be varied in these tests.

Despite the general correspondence between basal water pressure and sliding velocity, the precise details of the relationship are unclear. For example, peak velocities may actually coincide with periods of rising subglacial water pressure (e.g. see Sugiyama, this volume, Chapter 68). This raises the possibility that sliding velocity varies in a complex manner in response to a combination of controls, including basal water pressure, rate of (positive) change of pressure, and perhaps even the extent of ice-bed separation.

One means of evaluating rates of basal sliding in detail is to instrument the glacier bed via access provided by boreholes drilled through the ice. Engelhardt *et al.* (1978) used a borehole video camera to record differential ice-bed velocity as a recognizable object, such as a pebble, moved across the field of view. More recently, *dragspools* have been used to record high-resolution time series of sliding (see Fischer & Hubbard, this volume, Chapter 76). Blake *et al.* (1994) first used these instruments, recording a basal sliding component of ≤70% of the total surface speed of Trapridge Glacier, Canada. One drawback with the use of dragspools, however, is that the anchor needs to be inserted into subglacial material, which can itself deform.

Finally, large basal cavities formed naturally beneath thin ice or artificially by hydropower companies provide researchers with direct access to the basal interface. A small number of experiments carried out in such environments indicate that sliding can occur at subfreezing temperatures (Cuffey *et al.*, 2000a) and that sliding may involve significant slip of the ice over bedrock (Cohen *et al.*, 2000; Hubbard, 2002).

67.4.2 Subglacial sediment deformation

Basal motion of ice masses may also be achieved through the deformation of unconsolidated subglacial sediments, the recognition of the importance of which began with the work of Boulton and co-workers only some 20 years ago. Today, subglacial sediment deformation is widely considered to be responsible for the fast flow of many ice masses, key examples being the well-documented ice streams of the Siple Coast, Antarctica.

67.4.2.1 Current formalization

Two general forms of constitutive relation have been advanced for subglacial sediments. The most widely used model is a viscous approximation with an inverse dependence on effective pressure, proposed by Boulton & Hindmarsh (1987)

$$\dot{\varepsilon} \propto \frac{(\tau - \tau_0)^a}{N^b} \tag{9}$$

Here, the yield strength (τ_0) is defined by a Coulomb failure criterion

$$\tau_0 = C_0 + N \tan\theta \tag{10}$$

where C_0 is sediment cohesion and θ is the angle of internal friction.

More recently, several researchers have proposed a highly non-linear rheology for subglacial sediments, arguing that their deformation is closer to plastic failure than flow. Hooke *et al.* (1997), for example, characterized this rheology as

$$\dot{\varepsilon} \propto e^{k\tau} \tag{11}$$

for values of $\tau \geq \tau_0$. The constant k has a value of between 10 and 60.

A less well understood process is that of *ploughing*, whereby clasts protrude from the basal ice into the underlying sediment. Although the effect is not yet incorporated into models of ice-mass motion, Hooyer & Iverson (2002b) considered this effect in terms of the basal stress supported by ploughing clasts.

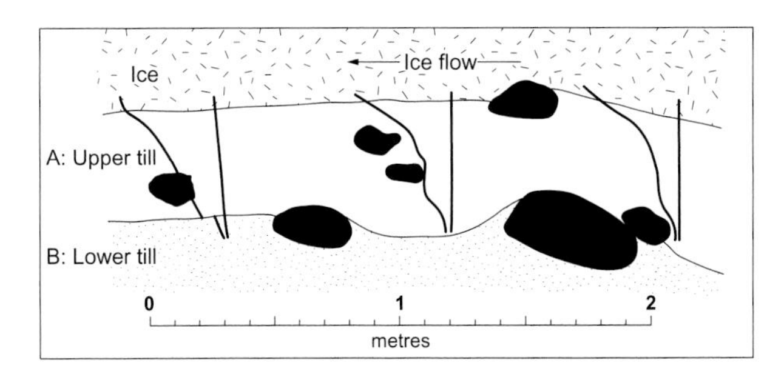

Figure 67.4 Displacement of initially vertical segmented rods emplaced for ca. 5 days in subglacial sediments beneath Breiðamerkurjökull, Iceland. (After Boulton & Hindmarsh (1987) with the permission of the American Geophysical Union.)

67.4.2.2 Field data

The first direct records of sediment deformation were reported by Boulton & Jones (1979) and Boulton & Hindmarsh (1987). This study involved installing vertical arrays of deformation pegs through a hole excavated in the floor of a subglacial tunnel at Breiðamerkurjökull, Iceland. The pegs were left in place for about 5 days, during which time local ice surface velocity and sediment pore-water pressure were measured, and then re-excavated. Over the period, the glacier surface moved ca. 0.42 m, of which ca. 90% was achieved by sediment deformation. The shape of the excavated deformation array revealed that this motion was focused within the uppermost 0.5 m of the sediment layer, and that no motion occurred below this depth (Fig. 67.4). Data from this experiment indicated that the empirical constants a and b in Equation (9) above have values close to 1.

Examination of the deforming sediment layer at Breiðamerkurjökull indicated that the upper 'A' horizon was characterized by a high porosity (generally >55%), whereas the underlying 'B' horizon, which had not deformed, had a porosity of <45%. This substrate structure is compatible with observations of the structure of unconsolidated sediments remaining in other glaciated terrains, providing general support for the viscous model (e.g. Hart, 1995b).

Further direct evidence of the nature of the deformation of subglacial sediments has come from borehole-based instrumentation, including sediment samplers, *penetrometers*, *tilt meters* and *plough meters* (described in Fischer & Hubbard, this volume, Chapter 76). Such studies generally indicate a positive relationship between sediment deformation rate and subglacial water pressure (e.g. Hooke *et al.*, 1997). However, the details of the relationship may be complex. For example, research by Iverson *et al.* (1995) at Storglaciären, Sweden, indicated that the glacier base may decouple from the bed at high subglacial water pressures. Under such conditions, sediment deformation may actually decrease. Further, subglacial sediments retrieved from Ice Stream B were tested in the laboratory and found by Kamb (1991) to be very weak; indeed, it is likely that the basal sediments of such ice streams exert very little restraining influence at all (e.g. Tulaczyk *et al.*, 2000a) and that they are only restrained by occasional basal protrusions (termed *sticky spots*). Sediment beneath Black Rapids Glacier, Alaska, also failed rather than deformed viscously, but in this case at a depth of some metres rather than near the layer's upper surface (Truffer *et al.*, 2000). Together, these studies, along with laboratory-based shear testing by Iverson *et al.* (1997), support a highly non-linear stress–strain relationship for subglacial sediments.

Many valley glaciers may be neither wholly bedrock-based nor wholly sediment-based. In such cases it may not be possible to discriminate at the glacier-wide scale between motion by sliding and motion by deformation. On the other hand, research at such glaciers may provide important information on controls common to both components of basal motion. Borehole-based research at Haut Glacier d'Arolla, for example, points to the key role played by the structure of the basal drainage system in controlling the glacier's overall dynamics. Initially, Harbor *et al.* (1997) noted that a major subglacial channel that develops in a similar location each melt season acts as a corridor of low basal traction (a *slippery zone*). More recently, results from three-dimensional velocity measurements at the glacier indicate that these events generate extrusion flow in the ice located immediately above this preferential drainage axis (Willis *et al.*, 2003). In contrast, deformation profiles either side of the axis are characterized by standard power-functions (Fig. 67.5).

These variations at temperate valley glaciers also have a temporal dimension. Intra-annual velocities vary markedly, with dramatic speed-ups being recorded through the summer melt-season relative to the winter (Willis, 1995). This summer speed-up is caused by the onset of basal motion as large volumes of meltwater reach the glacier bed. Conversely, winter flow fields reflect motion by ice deformation alone. Short-lived early melt-season speed-ups also have been noted at several temperate (Gudmundsson *et al.*, 2000, Gudmundsson, 2002, Mair *et al.*, 2001) and polythermal (Bingham *et al.*, 2003) valley glaciers. These *spring events* reflect the initial delivery of surface-generated meltwater to an underdeveloped basal drainage system. Finally, intradiurnal velocity variations also have been recorded at a few glaciers during the summer melt season, being faster during the afternoon and evening than overnight and during the morning (e.g. Gudmundsson *et al.*, 2000; Sugiyama, this volume, Chapter 68).

Although it is generally accepted that subglacial sediment deformation is only effective where the interface is at the pressure melting point (where liquid water weakens the sediments), researchers who have accessed subfreezing sedimentary beds also report enhanced and localized motion. Echelmeyer & Wang (1987), for example, recorded enhanced deformation and discrete shearing within sediment-rich basal ice that was at a mean temperature of <–4°C. More recently, Fitzsimons *et al.* (1999) reported similar enhanced motion and shearing at the base of Suess Glacier, Antarctica.

67.4.3 Incorporation into models of ice-mass motion

Incorporation of basal motion explicitly into ice-sheet models generally is achieved by introducing a sliding term across those cells where basal temperature attains the pressure melting point according to Equation (9), using the local basal shear stress but

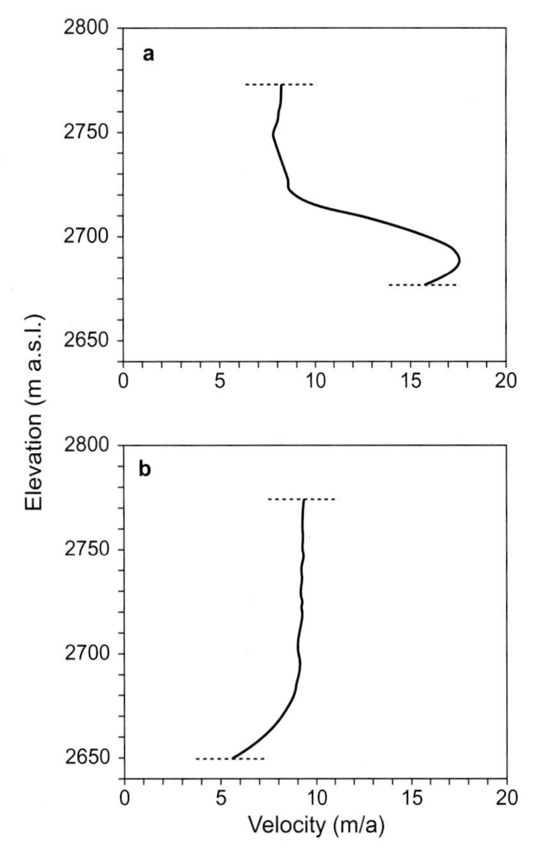

Figure 67.5 Deformation profiles in glacier ice (a) above a basal zone of low traction corresponding to the location of a major melt-season basal channel and (b) above adjacent ice characterized by higher basal traction. (After Willis *et al.* (2003) with the permission of the International Glaciological Society.)

neglecting any effective pressure term (i.e. $N = 1$). For example, Payne *et al.* (2000) utilize a sliding term of the form

$$U_t = \zeta \rho g H \alpha \quad (12)$$

which relates to Equation (7) with $p = 1$, $q = 0$ and local shear stress calculated through Equation (2), with the sliding parameter ζ as a constant. More recently though, in a newly developed first-order thermomechanical model, Payne (personal communication, 2005) allows ζ to vary with the basal melt flux, which goes some distance to incorporating variations in basal water pressure where there is a closed hydrological system (from which water drains ineffectively). Pattyn (2002) uses a sliding term of a similar form to Equation (7), with $p = 3$ and $q = 1$. This follows the work of Bindschadler (1983), who compared different sliding relations against observations and concluded that such a 'Weertman-type' sliding law relation, explicitly corrected for subglacial effective water pressure N, was optimal (although basal water pressure is not predicted by the Pattyn model). Marshall & Clarke (1997b) address the problem slightly differently by introducing a stress-free basal boundary condition (determined by subglacial geology, topography and thermal regime) to represent ice streaming in the Hudson Bay area in their model of the Laurentide Ice Sheet.

In contrast to ice sheets, high resolution velocity data are available to constrain spatial and temporal variations in the basal motion at small valley glaciers. For example, Hubbard *et al.* (1998) used winter (no-sliding) patterns of surface motion measured at Haut Glacier d'Arolla, Switzerland, to direct a three-dimensional, first-order algorithm of the glacier's ice deformation field. This field was then perturbed in a manner consistent with field-based borehole inclinometry data (indicating systematic temporal and spatial patterns of basal sliding) to model the long-term aggregate pattern of glacier motion (presented in more detail by A. Hubbard, this volume, Chapter 69). Similarly, Sugiyama *et al.* (2003) and Sugiyama (this volume, Chapter 68) introduced a slippery zone at the base of their finite-element (full solution) models of Unteraargletscher and Lauteraargletscher, to match variations in the glacier's measured surface velocity field.

To date, slippery zones such as these have been imposed as varying boundary conditions and no truly integrated hydrological-motion model has yet been developed at the ice-mass scale. However, integrated hydrological models such as that developed by Flowers & Clarke (2002a) are moving some way towards this end.

67.5 Research priorities

Numerical models of ice-mass motion should ultimately calculate spatially distributed variations in both softness and basal motion implicitly. To facilitate this, field glaciologists need to supply quantitative relationships that will allow these processes to be expressed in terms of known or modelled variables. Although this task is challenging, it is by no means impossible.

67.5.1 Ice deformation

Three-dimensional, first-order models of glacier flow have the capacity to incorporate fully spatially distributed variations in softness. To achieve this, however, modellers need to be provided with quantitative schemes describing the spatial distribution of ice flow at real ice masses. Thus, the key physical controls over ice softness need to be identified and their influence over softness quantified over a realistic range of temperatures and stress regimes. In parallel with this, schemes describing the spatial evolution of those key ice properties, such as ice crystal size, shape and fabric, at ice masses need to be developed. One means of moving towards such schemes is to acquire sets of ice cores from along flow-lines at ice masses and to compare their physical characteristics and softness variations. In such studies, softness can be measured either directly in the laboratory or inferred from comparing measured three-dimensional motion fields with model output. Although the pilot study by Hubbard *et al.* (2003) summarized above has demonstrated the potential importance of including spatial variations in ice character to the modelling of temperate glaciers, the analysis is limited because it is based solely on the bulk ionic composition of the ice. Future studies need to consider suites of rheologically-important characteristics. Where such schemes may be considered to be too logistically demanding, it may be possible to quantify key properties such as bulk water content by non-intrusive geophysical techniques such as radar (e.g. Murray *et al.*, 2000a).

67.5.2 Basal motion

Field research indicates that basal motion is essentially governed by a small number of factors, including basal shear traction, bed temperature and the presence and pressure of water at the basal interface (whether hard or soft). Basal water pressure is the least-well constrained of these controls, being governed by a balance between water delivery to the glacier base and water removal from it. Although our understanding of spatial and temporal variations in basal water pressure at temperate valley glaciers is now fairly well advanced, little is known about these variables beneath larger and/or thermally complex ice masses. There is therefore a pressing need for field programmes that simultaneously record motion components and their principal controls at these ice masses. Indeed, some work has already begun in this arena. Bingham *et al.* (2003), for example, report systematic annual and seasonal motion variations at polythermal John Evans Glacier in the Canadian Arctic.

This lumped basal motion approach should develop in tandem with further detailed investigations of specific processes of basal motion. For example, the effects of variations in bedrock roughness and the integrity of the basal drainage system are currently neglected in sliding models. The balance between sliding by slip, enhanced deformation and regelation also needs to be formalized, perhaps as functions of stress regime, location and basal temperature. Similarly, controls over processes and rates of subglacial sediment deformation need further investigation. Ultimately, a constitutive relation for subglacial sediments is required. This relation should include terms for processes such as ploughing and perhaps even terms for grain-size evolution and the development of clast fabric that models could calculate in a time-evolving manner. Finally, basal motion at subfreezing temperatures also needs to be investigated more thoroughly and at the glacier-wide scale, both at hard- and soft-bedded glaciers.

SIXTY-EIGHT

Measurements and modelling of diurnal flow variations in a temperate valley glacier

Shin Sugiyama

Section of Glaciology, Versuchsanstalt für Wasserbau Hydrologie und Glaziologie, ETH, Gloriastrasse 37/39, CH-8092 Zürich, Switzerland

68.1 Introduction

During the ablation season, surface melting causes diurnal flow variations in temperate glaciers. Because basal motion plays a key role in short-term velocity fluctuations, measurements of diurnal flow variations provide insight into subglacial conditions and their influence on the glacier flow field.

This contribution reports high-frequency measurements of diurnal flow speed and strain rate variations in a temperate valley glacier in the Swiss Alps. Numerical simulations of the observed flow field using a two-dimensional finite-element flow model indicated that basal motion was non-uniformly distributed along the glacier and its pattern varied in a diurnal manner.

68.2 Field measurements

Measurements were carried out in the ablation area of Lauteraargletscher, Switzerland (Fig. 68.1) from 22 to 27 August 2001. A 6.5-m-long aluminium pole was installed on the surface and continuously surveyed with GPS (Global Positioning System) to obtain hourly records of horizontal flow speed and vertical displacement. Three other poles were also surveyed twice a day at 0600 hours and 1800 hours to compute the horizontal surface strain rate. The precise distance (±5 mm) from the base of a ca. 173-m-long borehole to a surface reference was also measured in order to calculate the vertical strain of ice over the depth of the borehole (Gudmundsson, 2002; Sugiyama & Gudmundsson,

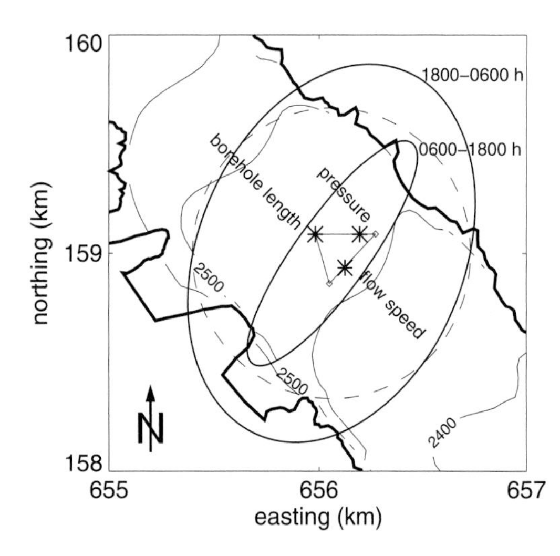

Figure 68.1 Map of the study site on Lauteraargletscher, showing the locations of the surface and borehole measurement sites and the strain grid. The two ellipses are strain rate ellipses during the indicated time periods of the days from 22 to 27 August 2001. These ellipses represent the deformation of the circle (broken line) subjected to the computed surface strain rate in units of 10^{-4} day^{-1}.

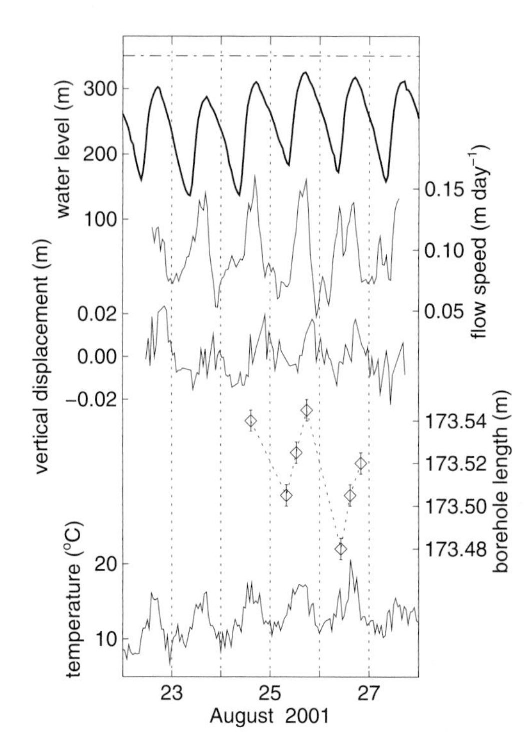

Figure 68.2 Records of water level in a borehole, horizontal flow speed, vertical displacement of the surface relative to the mean elevation, borehole length and air temperature. The dash–dot line at the top indicates the water level corresponding to the ice overburden.

2003). A water-pressure transducer was also installed in a nearby borehole to measure the water level every 10 min.

Results of the measurements are shown together with air temperature recorded at the glacier flank, in Fig. 68.2. During the study period, 40–60 mm w.e. of daily ablation took place, and the borehole water level oscillated more than 100 m diurnally. The surface flow-speed showed clear diurnal variations, reaching a maximum in the afternoon and a minimum in the morning. Diurnal variations are also observed in the vertical displacement and the borehole length. Peaks in flow speed coincide with water-level maxima, suggesting that basal motion was enhanced by high subglacial water pressure. These peaks also coincide with the maximum rate of upward vertical displacement. Surface uplift during fast-flow events has been considered to result from the formation of water-filled basal cavities (Iken *et al.*, 1983). Nevertheless, the observed uplift cannot be attributed totally to basal cavity formation as the borehole length measurements showed vertical tensile strains at times when the surface moved upward. These vertical strain rates reached 5×10^{-4} day^{-1}. The horizontal surface strain data (Fig. 68.1) indicate compression along the flow direction during the daytime (0600–1800 hours), which is consistent with the vertical extension. These changes in the strain regime indicate consistent systematic diurnal variations in the spatial pattern of flow speed along the glacier.

During the same period, we operated two further GPS receivers located 1.5 km up-glacier and down-glacier from the main study site. Although diurnal velocity variations were recorded at all three sites, those up-glacier of the main study site were greatest. This suggests that the measured vertical extension during the daytime was probably caused by a compressive flow established at the study site. Because the seasonal snow line was about 1 km up-glacier, it is plausible that the highest peak diurnal pressure was greater in the upper reach owing to less developed subglacial drainage conditions (Nienow *et al.*, 1998). The compressive flow field dissipated overnight as subglacial water drained down-glacier and flow speed slowed down.

68.3 Numerical modelling

To test the hypothesis proposed above, the englacial flow field was investigated with a finite-element glacier flow model. The aim of the modelling was to prescribe basal conditions to reproduce the measured diurnal variations in flow speed and vertical strain rate.

68.3.1 Model configuration

The model computes flow-velocity fields in a longitudinal cross-section of a slab, based on Stokes' equations and Glen's flow law with $n = 3$ (Sugiyama *et al.*, 2003). A finite-element scheme was used to solve the differential equations for two components of the flow velocity, using quadratic triangle elements with horizontal and vertical grid sizes of 500 m and 50–100 m, respectively. Basal motion was introduced in the form:

$$u_b(x) = c(x)\tau_b^{n'}$$

where $u_b(x)$ is the basal sliding speed at the longitudinal coordinate x, τ_b is the basal shear stress, and n' and $c(x)$ are parameters

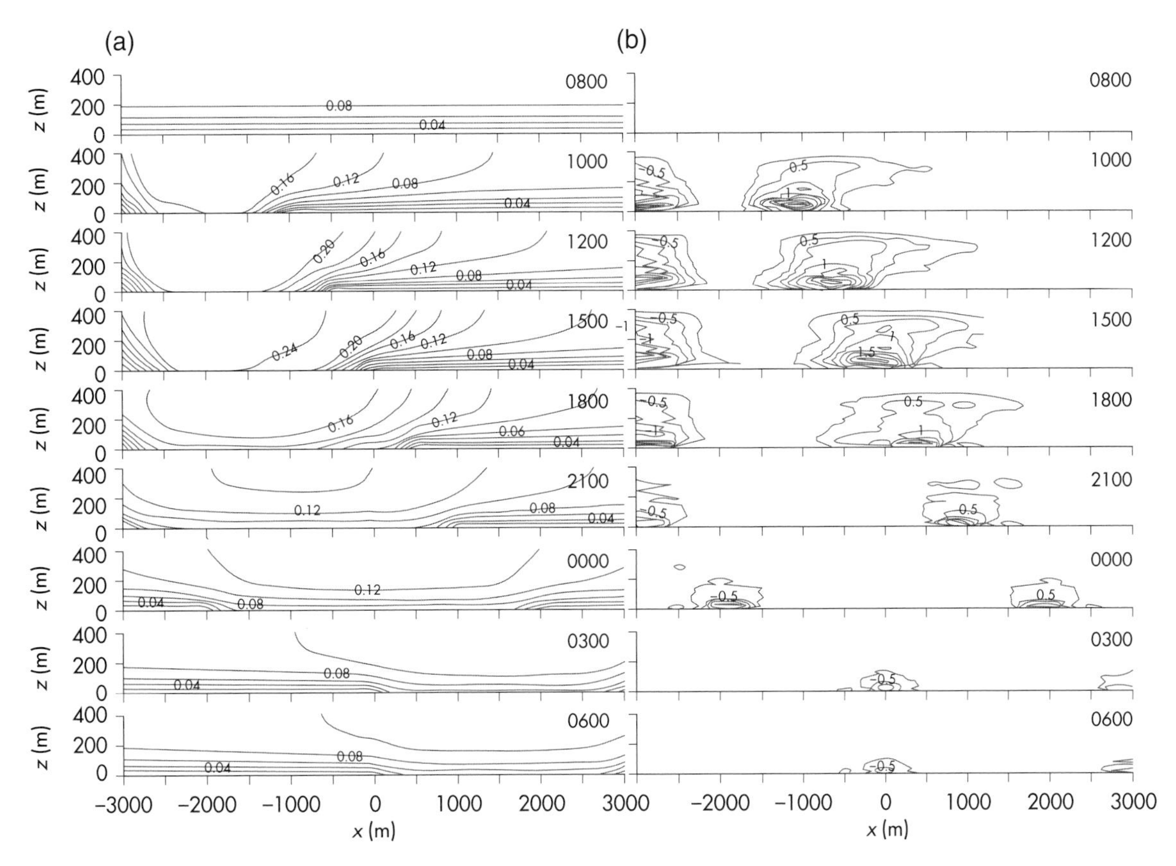

Figure 68.3 Englacial flow fields in a longitudinal cross-section computed for the times indicated in the figures. (a) Horizontal flow speed (m day^{-1}). (b) Vertical strain rate (10^{-4} day^{-1}). Positive value indicates tensile strain rate.

of the sliding law. We assume $n' = 1$, and prescribe $c(x)$ in order to reproduce the observed diurnal flow variations.

68.3.2 Model output

Horizontal flow speed and vertical strain rate were computed for a region between 3 km up-glacier and down-glacier of the study site (Fig. 68.3). Ice thickness at the study site was modelled to be 400 m. No sliding $c(-3000 < x < 3000) = 0$ was assumed at 0800 hours when the surface speed was at a minimum, and then $c(x)$ was determined for each time of day so that the computed surface-flow speeds at $x = 0$ agree to within 10% of the measurements on 25 August (Fig 68.3a). This 10% tolerance was within GPS measurement errors. Based on the hypothesis that the basal motion was higher in the upper reach, a sliding zone with greater value of c was allocated up-glacier of the study site at 1000 hours and its magnitude increased during the daytime. The position of the sliding zone was then gradually moved down-glacier at night, reducing the magnitude of c toward next morning.

Output from the model recreates the principal features of the field measurements:

1 vertical strain rate is tensile during the day from 1000 to 1800 hours;
2 magnitude of the strain rate reaches 10^{-4} day^{-1} at 1500 hours at $x = 0$;
3 diurnal velocity variation is greater up-glacier.

Although the computed strain rate is still smaller than that measured in the borehole, the basal sliding zone established in the upper reach and its temporal evolution was able to explain the diurnal flow variations in Lauteraargletscher.

68.4 Discussion

High temporal resolution measurements of surface velocity and vertical strain rate indicated systematic diurnal flow variations in a temperate valley glacier. These data are consistent with a cause related to spatially and temporally non-uniform basal motion.

These surface flow fluctuations are not simply the results of variations in local basal motion, but also of variations in neighbouring basal and englacial motion, as shown in Fig. 68.3. Thus, the spatial distribution of basal motion and its transfer to the englacial flow field control short-term flow variations in temperate glaciers. High-frequency measurements of the englacial flow field are crucial to improve our understanding in this research field and accurate computation of the glacier flow field is important to infer bed conditions and basal-flow processes from surface observations.

SIXTY-NINE

Using field data to constrain ice-flow models: a study of a small alpine glacier

Alun Hubbard

Department of Geography, University of Edinburgh, Edinburgh EH8 9XP, UK

This case study is concerned with the application of a three-dimensional ice flow model to Haut Glacier d'Arolla, a small temperate glacier in the Swiss Alps. The study illustrates both the potential and certain limitations of using high-resolution, three-dimensional flow modelling constrained by observations to aid our understanding of alpine glacier systems. In particular, the combination of flow modelling with detailed field measurements at Haut Glacier d'Arolla provides a powerful tool to investigate its rheological properties, the interaction between basal hydrology and ice dynamics, the occurrence and formation of surface structures, its past and present response and, ultimately, its future trajectory to climate change.

The model is based on Blatter's (1995) solution of the mass and force balance equations using a non-linear rheology (with the flow law exponent: $n = 3$) applied to the surface and bed topography of Haut Glacier d'Arolla at 70 m horizontal resolution with 40 vertical layers scaled to thickness. The model is steady-state; it computes the instantaneous stress and strain distribution based on Glen's (1958a) flow law:

$$\dot{\varepsilon}_{ij} = A\tau_e^{n-1}\tau_{ij}$$

where $\dot{\varepsilon}_{ij}$ is the strain rate tensor, A is the rate factor reflecting ice hardness, τ_{ij} is the imposed stress tensor and τ_e is the effective stress given by

$$2\tau_e^2 = 2(\tau_{xz}^2 + \tau_{xy}^2 + \tau_{yz}^2) + \tau'^2_{xx} + \tau'^2_{yy} + \tau'^2_{zz}$$

The model computes first-order terms, that is longitudinal and transverse deviatoric stresses (τ'_{xx} and τ'_{yy}) which result in compression and tension throughout the ice mass and which, in extreme circumstances, lead to observed structural failure such as overthrusting, shearing and crevassing. The resulting solution is highly dependent on the basal boundary condition, which can be specified as either a velocity or drag distribution or a combination of both, which enables the spatial interaction of slip/stick patchiness resulting from heterogeneity in the bulk subglacial properties, such as roughness, hydrology, effective pressure and sediment strength, to be modelled. This is achieved through the prescription of zero or reduced basal shear traction to replicate decoupled, low drag zones, whereas zero velocity or increased shear traction can be prescribed to simulate 'sticky' zones. For the purposes of initial model optimization though, the model is first specified with zero sliding across the whole of the bed.

By holding the flow-law exponent (n) constant then the only parameter that requires 'tuning' is the rate factor (A) related to ice viscosity. Assuming Haut Glacier d'Arolla to be temperate with negligible basal motion during winter, then comparison of modelled with observed winter surface velocities provides an effective means for calibrating A. Observations that winter surface velocities are consistently low and, furthermore, that there is virtually no supraglacial melt-water available to drive ice–bed separation, lend tentative support for this assumption. Tuning of the model against surface velocities measured over 10 days in January 1995 yields an optimum value of $A = 0.063\,\text{yr}^{-1}\,\text{bar}^{-3}$, corresponding to an $R^2 = 0.74$ on a bivariate plot of modelled against measured surface velocities (Fig. 69.1). Although this value is half that expected (Paterson, 1994) it does lie within a narrow range of 0.07 $\pm$ 0.01 $\text{yr}^{-1}\,\text{bar}^{-3}$ reported for other temperate glaciers modelled using higher order solutions (e.g. Gudmundsson, 1999). Such consistency lends confidence in the predictive quality of the internal strain component of these models but also suggests that they may be applied to temperate ice masses without significant tinkering with the rate factor.

Computation of the full stress and strain field enables model comparison with additional observables such as the occurrence and orientation of crevassing and measured principal strains. Comparison of zones of maximum computed surface tensile strain and their direction with the actual distribution and orientation of crevassing reveals a good general correspondence (Fig. 69.2a & b). Furthermore, the orientation of modelled principal strain faithfully reproduces the pattern measured from repeat survey of a dense network of strain diamonds from 1994 to 1995 (Fig. 69.3a & b). However, it is apparent that even though their

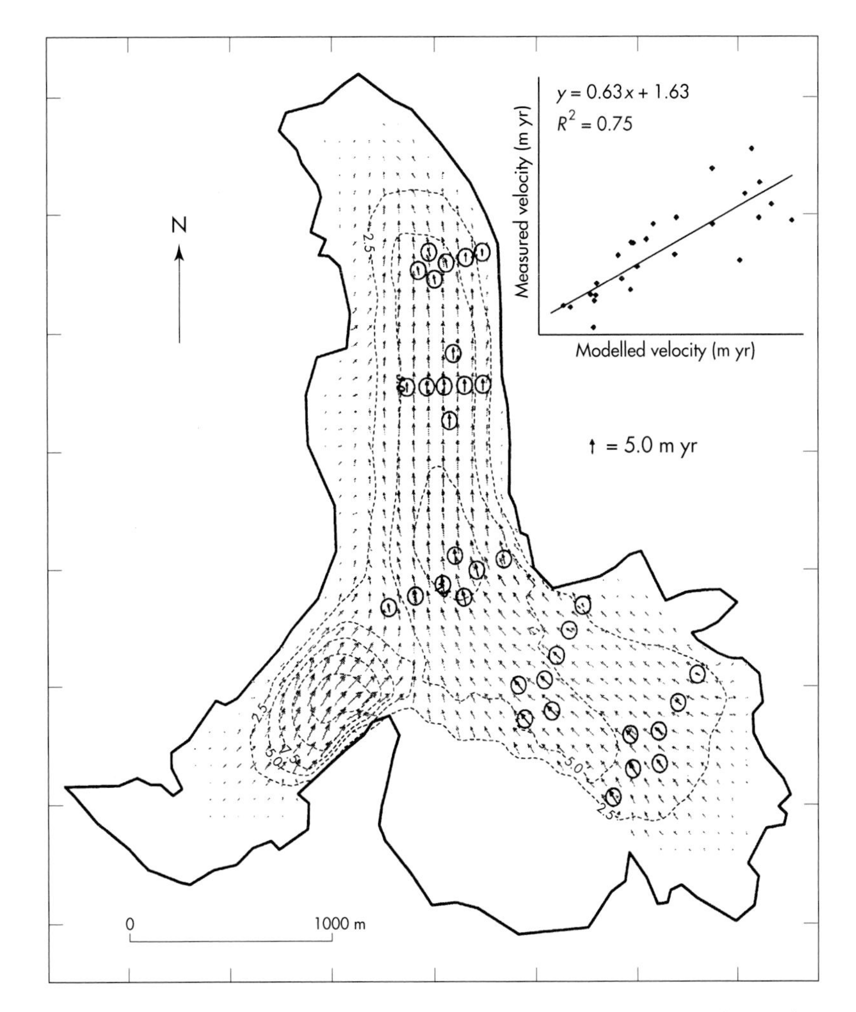

Figure 69.1 Modelled horizontal surface velocity across Haut Glacier d'Arolla at 70 m resolution for $n = 3$ and $A = 0.063\,\text{yr}^{-1}\,\text{bar}^{-3}$. The contours are plotted at $2.5\,\text{m yr}^{-1}$ intervals. Overlain within circles are the winter 1995 velocity vectors and a bivariate analysis of modelled against observed velocities.

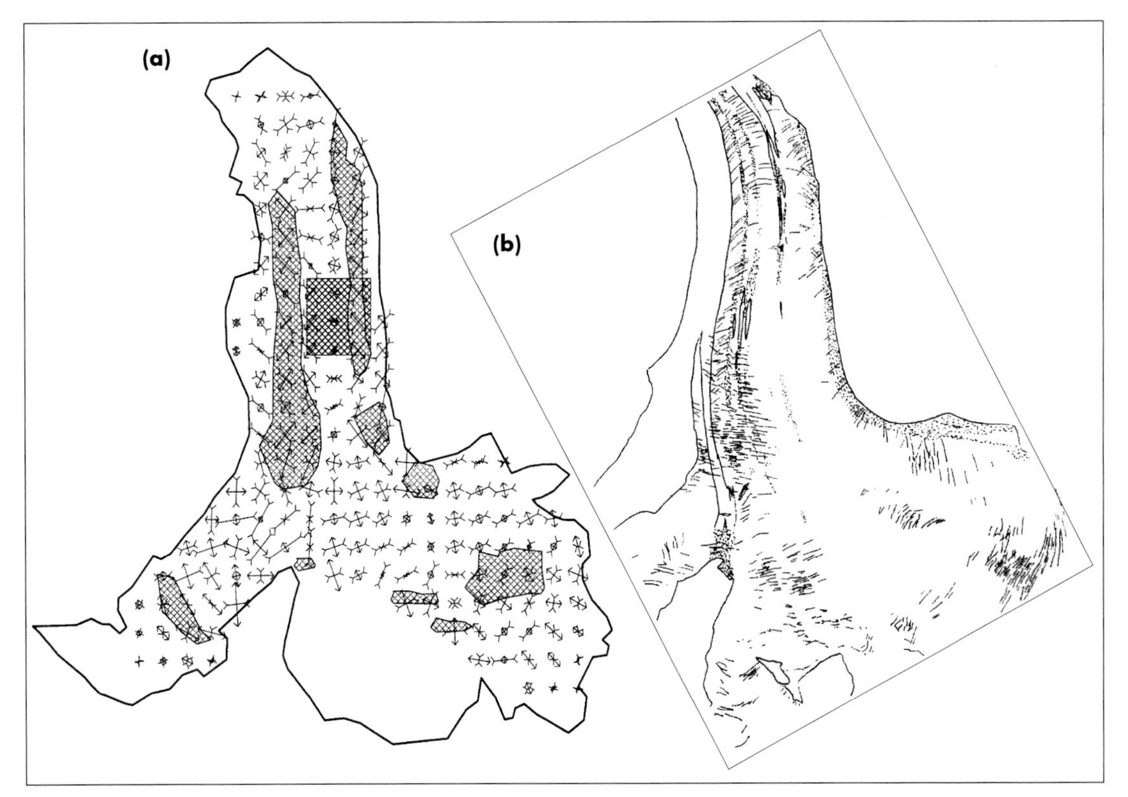

Figure 69.2 (a) The magnitude and orientation of modelled surface-parallel principal stresses at 140 m resolution; inward arrows indicate compression, outward arrows indicate extension. The shaded areas represent zones of maximum surface stress and indicate areas of potential ice failure. (b) The distribution of surface crevasses across Haut Glacier d'Arolla as observed from aerial photography and ground mapping.

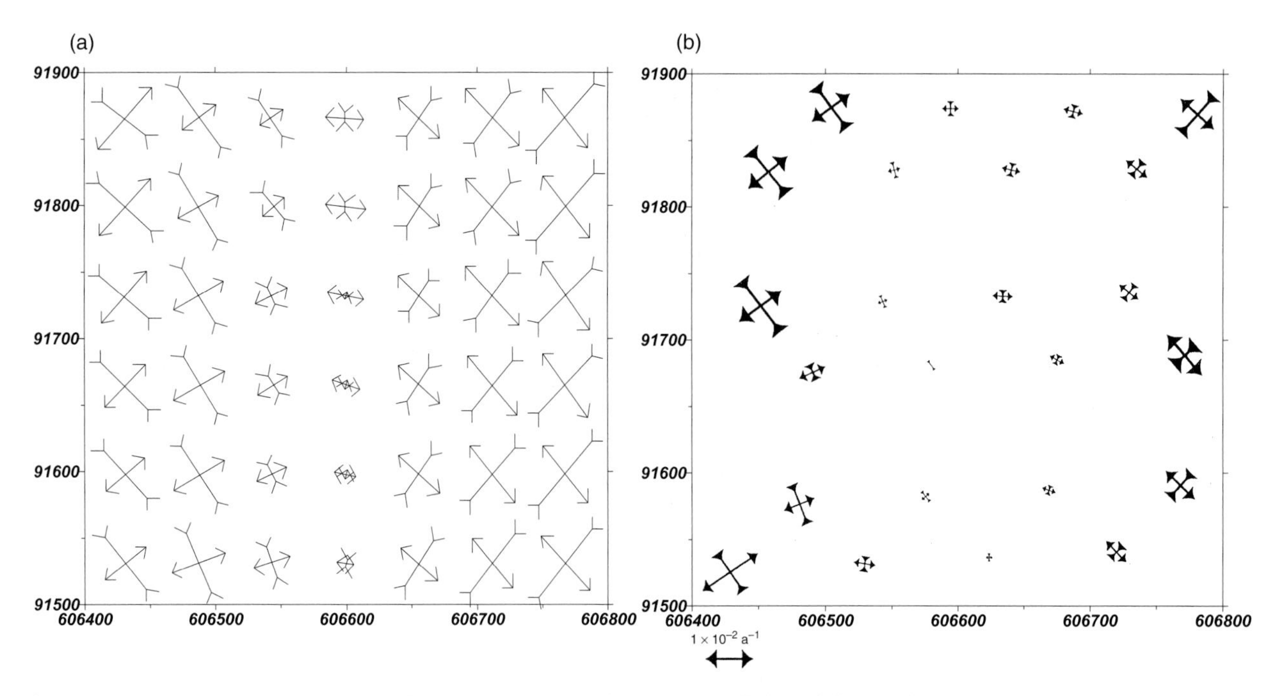

Figure 69.3 (a) The relative magnitude and orientation of surface-parallel modelled and (b) measurement derived principal stresses and strains, respectively, in the region of the high-density strain network indicated in (a).

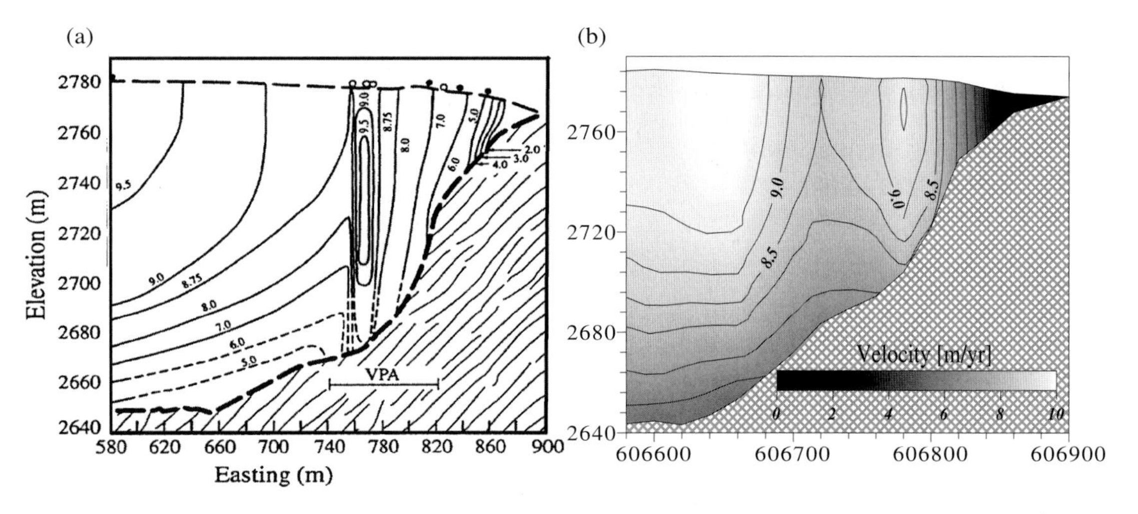

Figure 69.4 (a) Distribution of measured annually averaged (August 1995 to August 1996), horizontal velocity within a half cross-section at Northing 91700. (b) The modelled composite velocity distribution within the glacier cross-section composed of time–weight averages of 20/52 'winter' no sliding, 31/52 'normal summer' sliding and 1/52 enhanced 'spring event' sliding.

orientations correspond, their absolute magnitudes do show some discrepancy away from the centre-line, probably reflecting the effect of large lateral moraines on ice-fabric heterogeneity, combined with the fact that there is a significant component of basal motion influencing the annual flow regime. Both of these effects will significantly alter the relative magnitude of principal strains, but not their orientation.

The model may also be applied to contrasting basal boundary conditions in order to investigate an anomalous pattern of internal strain measured by repeat inclinometry of boreholes between 1995 and 1996 (Harbor *et al.*, 1997) (Fig. 69.4a). The distinctive feature of this cross-sectional strain distribution is the 80-m-wide zone of rapid sliding, low strain 'plug-flow' which coincides with a subglacial pathway predicted on the basis of hydraulic potential analysis (Sharp *et al.*, 1993) and characterized by highly variable water-pressures (often exceeding overburden) during the summer melt-season (Hubbard *et al.*, 1995). To simulate this anomalous velocity distribution, annual flow is modelled as a time-weighted composite of three scenarios reflecting observed seasonal contrasts: (i) non-sliding during winter, (ii) 'normal sliding' during

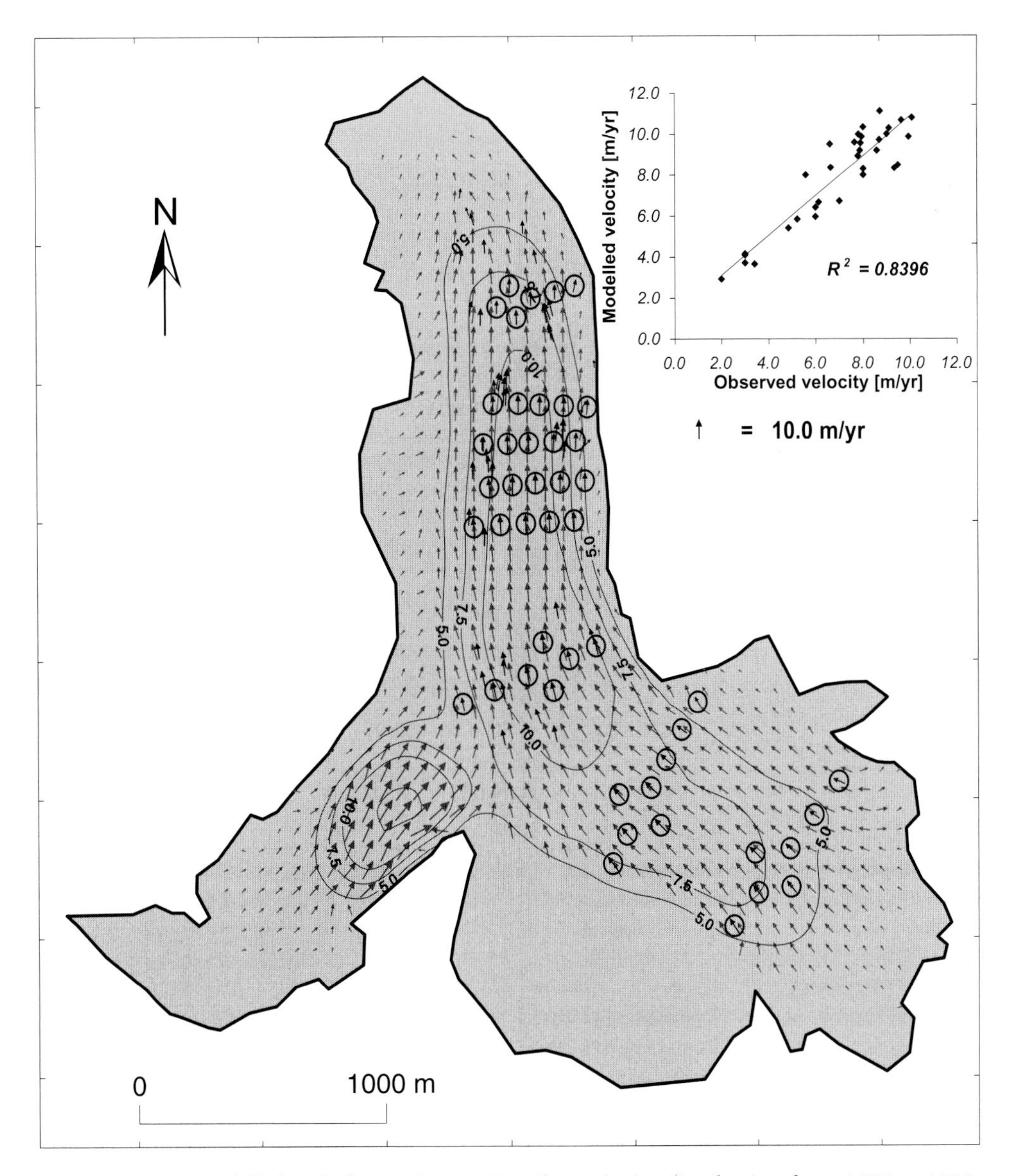

Figure 69.5 Comparison of modelled and observed annual surface velocity distribution from 1995 to 1996 overlain on the subglacial channel network reconstructed from hydraulic potential analysis (Sharp *et al.*, 1993).

the summer melt-season and (iii) 'enhanced' sliding during a week-long spring-event, when observed surface velocities across much of the glacier were an order of magnitude higher than the annual mean. The pattern and magnitude of the modelled basal perturbation are defined with respect to the hydraulic potential analysis. The basal area covered by the main subglacial pathways was decoupled with zero-traction for the 'normal summer' scenario, and an extended area including and adjacent to the two main pathways was decoupled to match the pattern of the measured surface velocity anomaly during the spring-event. Application of these generalized basal scenarios provides a first step towards realistic modelling of melt-season dynamics and enables the reproduction of the key features of both the annual surface velocity distribution (Fig. 69.5) as well as that of the two-dimensional cross-section (Fig. 69.4b). Given the limitations imposed by the 70-m operating resolution, the modelled and observed annual velocity distributions compare well (R^2 = 0.83) and confirm that variations in basal decoupling have a profound impact on the dynamics of Haut Glacier d'Arolla. These variations, however, are spatially and temporally complex, principally reflecting patterns of change in the glacier's basal drainage system.

Finally, in order to investigate the potential long-term response of Haut Glacier d'Arolla to climate change, this steady-state flow model can be coupled to a time-dependent model to yield the evolution of glacier thickness (H) through time (t) determined by the continuity equation:

$$\frac{\partial H}{\partial t} = b - \nabla \cdot (H\overline{u})$$

where b is the net mass balance and $\nabla \cdot (H\overline{u})$ is the two-dimensional flux divergence operator. Using mass-balance

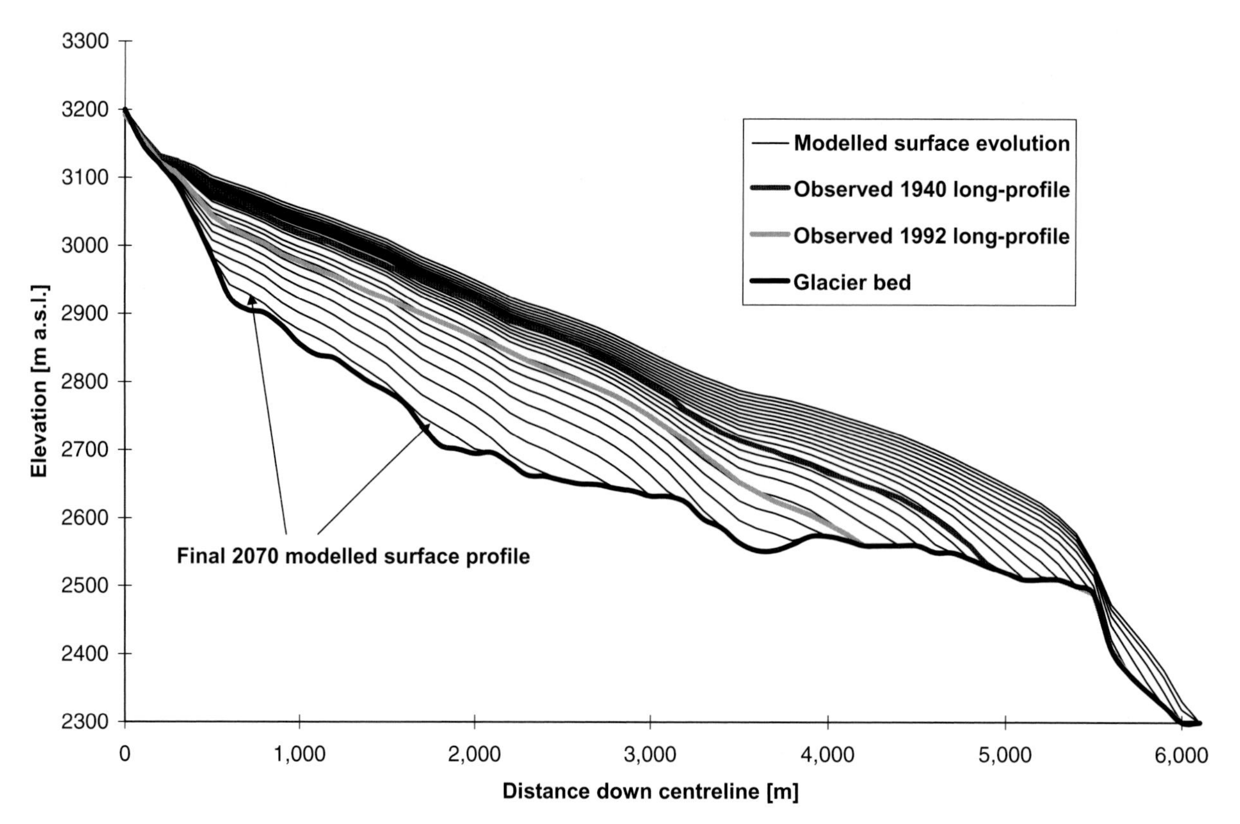

Figure 69.6 Modelled temporal-evolution of Haut Glacier d'Arolla surface centre-line in 10 yr intervals from the mapped 1880 (maximum) extent through to the glacier's predicted demise in 2070. Observed 1940 and 1992 long-profiles are overlain for reference.

measurements from 1989 to 1996 as a benchmark, the model is forced to match the maximum historical position mapped by the Swiss Survey in 1880. The model is then integrated forward in time to correspond to known positions in 1920, 1940 and 1992, and, assuming a constant 1989 to 1996 mass-balance distribution, is used to predict the evolving glacier geometry through to 2100 (Fig. 69.6). This modelling reveals that Haut Glacier d'Arolla has been undergoing accelerating retreat since the Little Ice Age and even under current conditions will not survive the next 70 yr. Such a response is no surprise given that the mean 1989 to 1996 equilibrium line altitude (ELA) was over 3000 m a.s.l., reducing the effective accumulation area to ca. 15% of that required to maintain the present geometry.

SEVENTY

Fast glacier flow and ice streaming

Slawek Tulaczyk

Department of Earth Sciences, University of California, Santa Cruz, CA 95064, USA

70.1 Introduction

Fast ice flow plays a key role in the mass balance and dynamics of glaciers and ice sheets. In West Antarctica, for instance, fast ice streaming accounts for ca. 90% of total ice discharge into the ocean (Bentley, 1987; Vaughan & Spouge, 2002). Fast ice streams are also the most dynamic components of the West Antarctic ice sheet and have been shown to change their velocity by up to several orders of magnitude over time-scales as short as several minutes (Joughin *et al.*, 2002; Bindschadler *et al.*, 2003). On millennial time-scales, variability in flow of the Hudson Bay ice stream has been implicated as the driver of rapid climate changes observed in Greenland ice cores and deep-sea sediments (MacAyeal *et al.*, 1995). Increasing evidence indicates that ice streams also played an important role as drainage pathways in palaeo-ice sheets (Stokes & Clark, 1999; Jansson *et al.*, 2003; Sejrup *et al.*, 2003). Bundles of megascale lineations found on continental shelves ringing Antarctica and other continents suggest that many palaeo-ice streams developed as ice sheets extended toward the shelf break during the Last Glacial Maximum (Shipp *et al.*, 1999; Canals *et al.*, 2000; Wellner *et al.*, 2001; Anderson *et al.*, 2002; Dowdeswell *et al.*, 2004a). Some palaeo-ice streams existed also within terrestrial, rather than marine, portions of Pleistocene ice sheets (e.g. Stokes & Clark, 2003).

Measured glacier velocities range from zero to >10 km yr^{-1} in Jakobshavn Isbrae, Greenland, considered the fastest flowing glacier on Earth. Peak velocities during glacier surges lasting months to years have been recorded to reach ca. 53 m day^{-1} (equivalent to ca. 19 km yr^{-1}) (Kamb *et al.*, 1985, fig. 2C). Recent measurements of stick-slip motion on the ice plain of Whillans Ice Stream, West Antarctica, revealed 10–30 min long bursts of velocity in excess of 1 m h^{-1}, equivalent to ca. 10 km yr^{-1}. There is no single widely agreed velocity threshold above which ice flow can be always classified as 'fast'. Rather, the concept of fast ice flow is a relative one and must be considered in a regional context. It is somewhat easier to define what velocities do not constitute fast ice flow. Ice moving with velocity of a few dozens of metres per year or less is clearly slow moving. Similarly, ice-flow velocities of a few hundred metres per year or more are fast. The difficulties may arise with velocities between ca. 50 and ca. 100 m yr^{-1}. Figure 70.1 illustrates how one can distinguish between slow and fast moving ice using a regional context. The ice velocity data of Joughin *et al.* (1999) revealed that ice flow in West Antarctica is organized into ice streams, their tributaries and inland flow.

The distinction between slow and fast ice flow is often explicitly or implicitly associated with the distinction between motion due to internal ice deformation (slow) and motion caused by basal sliding or till deformation (fast). The inference that slow ice motion is caused by internal deformation whereas fast motion is due to rapid basal motion (a term used here after Blankenship *et al.* (2001) to encompass both basal sliding and till deformation) is frequently justified. For instance, observations made in West Antarctica show that fast ice streams, where the base is at melting point, move through rapid basal motion, whereas slow-moving interstream ridges are frozen to their beds and move through internal deformation (Bentley *et al.*, 1998; Kamb, 2001). It is still uncertain what is the relative role of rapid basal motion and internal deformation in the motion of ice-stream tributaries (Hulbe *et al.*, 2003). Recent force-balance estimates indicate that tributaries may flow by either of these mechanisms and that the predominant mechanism of motion may change along flow (Price *et al.*, 2002; Joughin *et al.*, 2004). However, the relation between fast motion and rapid basal motion is non-unique. For instance, it has been conjectured that Jakobshavn Isbrae attains its high velocity at least partly through deformation of a thick, temperate basal ice layer (e.g. Luthi *et al.*, 2002).

Despite its importance, the phenomena of fast ice flow and ice streaming are still only partly understood and their representation in quantitative models of ice masses is rudimentary. Further observational studies and advances in modelling of fast ice flow are necessary if glaciology is to make a full contribution to understanding the role of the terrestrial cryosphere in past and future changes of global sea level and climate. This chapter is intended to be mainly a review of developments in the understanding of ice streaming that have taken place since the late 1980s. It builds on reviews provided by Clarke (1987a) and Bentley (1987). The considerable scientific interest in ice streaming is illustrated by the fact that a search for the keyword 'ice stream*' in the ISI Web of Science database at the time of this writing returns >500 publications. Other forms of fast ice flow, such as glacier surging, have been covered by recent reviews (e.g. Harrison & Post, 2003).

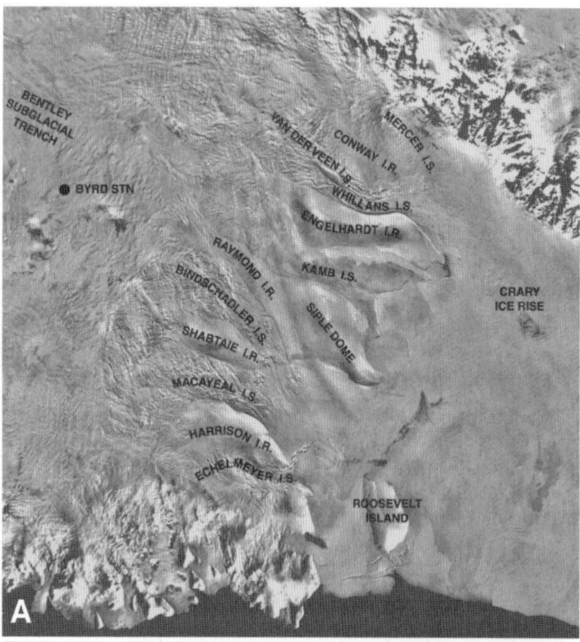

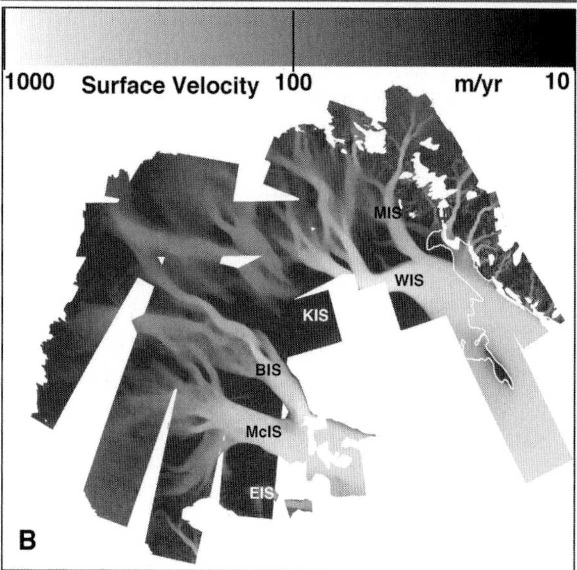

Figure 70.1 Satellite image (AVHRR) of the Siple Coast region in West Antarctica (A) and a velocity map (B) for those parts of this region where RADARMAP interferometry data are available (Joughin *et al.*, 2002). The satellite map shows names of the major ice streams and of other important glaciological features (source: http://igloo.gsfc.nasa.gov/wais/articles/images/ismap2.jpg). Velocity scale is logarithmic. Velocity data provided by Dr I. Joughin (JPL-Caltech).

70.2 Ice streaming

After Bentley (1987), I use Swithinbank's definition of an ice stream as: 'part of an inland ice sheet in which the ice flows more rapidly than . . . the surrounding ice.' A distinction can be drawn between ice streams, which are bounded laterally by ice, and outlet glaciers, which are confined in bedrock valleys. In general this is a useful distinction, although in nature there are many examples of transitional situations (Bentley, 1987, 2003). Remote-sensing observations combined with calculations of balance velocities indicate that within modern polar ice sheets flow is organized into major drainage pathways, which channel most of the ice discharging from the ice sheets (Joughin *et al.*, 1999; Bamber *et al.*, 2000a,b). Ongoing efforts to map ice-surface velocity and bedrock topography will enable delineation of the specific sections of these pathways that can be categorized as either ice streams or outlet glaciers.

Recent applications of satellite radar interferometry provided surface velocity maps for parts of Greenland and Antarctica but the coverage represents still a small portion of the latter continent. Only one large ice stream has been identified in Greenland (Fahnestock *et al.*, 1993). This ice sheet is ringed by mountain ranges and drained by about 30 large outlet glaciers (Bamber *et al.*, 2000b). In Antarctica, there are about three dozen major drainage pathways with pure ice streams concentrated in the Ross Sea sector of the West Antarctic ice sheet (Bamber *et al.*, 2000a; Rignot & Thomas, 2002). These are the Siple Coast ice streams characterized by fast motion, hundreds of metres per year, under low driving stresses, typically <50 kPa (Bindschadler *et al.*, 2001a; Blankenship *et al.*, 2001). The surface slope of these ice streams decreases downstream, resulting in a concave-up shape, so that driving stresses drop to several kPa on the ice plain of the Whillans Ice Stream (Bindschadler *et al.*, 1987). This fast motion under low driving stresses is possible because basal resistance to flow of these ice streams is low and much of the support for driving stress is shifted to the margin (Raymond, 2000; Tulaczyk *et al.*, 2000b). Although these features characterize the 'classic' West Antarctic ice streams, they are not included in Swithinbank's (1954) definition of ice streams.

70.3 Lateral shear margins

West Antarctic ice streams are separated from interstream ridges by lateral shear margins, which are typically several kilometres wide and heavily crevassed (Vornberger & Whillans, 1990). The presence of these features has been long recognized from field observations and satellite images. However, their significant influence on the force balance and dynamics of ice streams has been realized only within the last decade. Past models have made the assumption, common in glaciology, that the gravitational driving stress acting on ice streams is practically completely balanced by basal resistance to flow (e.g. Alley *et al.*, 1989). A priori, there are at least two good reasons to think that such an assumption is reasonable. Firstly, these ice streams have a very large width-to-thickness ratio (10–100; Raymond, 2000; Tulaczyk *et al.*, 2000b), which should favour a relatively small contribution of lateral margins to force balance. Secondly, ice in the shear margins is experiencing high strain rates and strains, which should lead to strong fabric alignment, shear heating and fracturing. All of these factors are expected to weaken ice, leading to an enhancement factor in the flow law of ice of up to ca. 10 (Echelmeyer *et al.*, 1994). Consequently, most ice-stream models published in the early 1990s made the assumption of local stress balance, i.e. that local gravitational driving stress is fully balanced by local basal shear stress,

$\tau_d = \tau_b$ (Lingle, 1984; Alley *et al.*, 1987a, 1989, Fastook, 1987; Kamb, 1991; Lingle *et al.*, 1991). Only the ice-shelf model of MacAyeal adapted to ice streams included the effect of shear margins on ice-stream force balance (MacAyeal, 1989a,b, 1992b).

The presumption that ice in the shear margins should exhibit a large enhancement factor has been undermined by work of Jackson & Kamb (1997), who reported a small enhancement factor (ca. 1–2) from laboratory studies on ice samples recovered from the southern shear margin of Whillans Ice Stream (WIS) in West Antarctica. This is in spite of the fact that at this location the expected total strain accumulation in the shear-margin ice is ca. 20 strains. Measurements of ice-temperature profiles across the same shear margin have demonstrated that a significant portion of the gravitational energy dissipated during motion of this ice stream is expanded in marginal shear heating (Harrison *et al.*, 1998). The latter study indicated that over the past 50 yr the marginal shear averaged 200 kPa there, in good agreement with the independent estimate of Jackson & Kamb (1997) of 220 ± 30 kPa coming from their laboratory tests on ice sampled in the same study area. Other observational studies indicate that marginal shear stresses may support ca. 50–100% of the gravitational driving stress in ice streams and their tributaries (Price *et al.*, 2002; Joughin *et al.*, 2004).

These findings corroborated earlier inferences of Whillans & van der Veen (1993), who calculated that marginal shear stress supports almost all of the driving stress on Whillans Ice Stream. In a subsequent publication, the same authors (Whillans & van der Veen, 2001) analysed the details of stress transfer from the shear margin to a several-kilometres-wide basal zone outside of the ice stream. They have shown that the lateral drag is the highest in the upper 40% of ice thickness, which is colder and more viscous than the lower part of the ice column. The transfer of marginal stress to the ice base outside of the ice stream leads to a peculiar situation, in which ice is propelled forward by stress much greater than the local gravitational driving stress that can be calculated from ice thickness and surface slope. In fact, the magnitude of stress transferred from the marginal shear zone to the bed just outside of the ice stream, ca. 60 kPa, exceeds by a factor of about three to five times the magnitude of the local gravitational driving stress at the site analysed by Whillans & van der Veen (2001). The unusual force balance around ice-stream shear margins is of great significance because:

1 a large fraction of ice-stream velocity is attained outside of the highly lubricated bed of an ice stream;
2 this force balance helps control the position of shear margins and, thus, the velocity of ice-stream motion and its changes (Jacobson & Raymond, 1998).

The location of shear margins of ice streams can be either fixed by subglacial conditions external to ice thermodynamics, e.g. topography and geology, or it can evolve with the thermodynamic state of the ice sheet itself. Anandakrishnan *et al.* (1998) and Bell *et al.* (1998) published results of geophysical studies indicating that, at least in one location, a tributary of Kamb Ice Stream (KIS) has a margin located over a boundary of a subglacial sedimentary basin. This observation provides evidence for the possibility of geological control over ice-stream geometry. However, extensive geophysical studies in the trunk regions of KIS and WIS indicate that there the boundary between slow and fast moving ice corresponds to a transition between frozen and melted bed conditions (Bentley *et al.*, 1998; Gades *et al.*, 2000; Kamb, 2001). There may be a difference in the geological setting of the main ice-stream trunks and the tributary region, which is in general much more dissected by deeply incised subglacial valleys as compared with the gentle topographic troughs associated with ice-stream trunks (Blankenship *et al.*, 2001; Studinger *et al.*, 2001). Where geological control is not the predominant factor, the position of shear margins may evolve through time in response to changes in force balance and thermal energy dissipation in and around the margins (Jacobson & Raymond, 1998; Raymond, 2000; Whillans & van der Veen, 2001). Measurements indicate that in the recent past (ca. 10–100 yr) at least some sections of ice-stream shear margins migrated outward at rates of the order of 1–10 m yr^{-1} (Harrison *et al.*, 1998; Echelmeyer & Harrison, 1999). In one location, inward migration of the southern shear margin of WIS has beeb inferred to have taken place at the rate of ca. 100 m yr^{-1} within the past few hundred years (Clarke *et al.*, 2000).

70.4 Basal processes

The important fact that shear margins support a significant fraction of ice stream gravitational driving stress is caused ultimately by excess basal lubrication of ice streams, which leads to the situation in which $\tau_d < \tau_b$ or even $\tau_d << \tau_b$. (Raymond, 2000; Tulaczyk *et al.*, 2000a,b; Kamb, 2001). Early models of ice-stream mechanics focused on application of the hard-bed sliding theory assuming that the basal lubrication is due to ice–bed separation by a subglacial water film/layer (e.g. Weertman & Birchfield, 1982). However, in the recent decades students of ice streams focused on the soft bedded end-member of basal ice-stream conditions, as exemplified by the trunks of the Siple Coast ice streams (Alley *et al.*, 1986a, 1987a,b, 1989; Blankenship *et al.*, 1986, 1987, 2001; Bentley, 1987; MacAyeal, 1989a,b, 1992b; Kamb, 1991; Whillans & van der Veen, 1993, 1997, 2001; Echelmeyer *et al.*, 1994; Engelhardt & Kamb, 1997; Smith, 1997a,b; Harrison *et al.*, 1998; Hulbe & MacAyeal, 1999; Anandakrishnan *et al.*, 2003; Vaughan *et al.*, 2003b; and many others). Kamb (2001) provides a comprehensive and recent overview of borehole observations of basal conditions beneath three of these ice streams. Although much of the recent work on ice streams focused on the soft-bedded conditions, it is important to remember that ice streams develop in hard-bedded settings as well. This caution is perhaps best underscored by findings of Stokes & Clark (2003), who identified a footprint of a palaeo-ice stream in northern Canada, where the ice stream was flowing over a predominantly hard bed.

70.4.1 Till rheology

To simulate the mechanics and dynamics of soft-bedded ice streams, and glaciers, it is necessary to quantitatively capture till rheology. Early treatments of soft-bedded ice-stream motion focused on the mildly non-linear till model with stress exponent in the till flow law of ca. 1–2 (e.g. Alley *et al.*, 1987b, 1989). This was a convenient choice because such mildly non-linear rheology is in many ways similar to ice rheology. Hence, shear deformation of a till layer was treated quantitatively just as shear deformation of an unusually soft basal ice layer (e.g. MacAyeal, 1989a,b, 1992b). However, laboratory tests on samples of till recovered from beneath WIS have shown that this material behaves as a Coulomb-plastic material whose strength is (nearly) independent of strain rates and depends mainly on effective stress (Kamb, 1991; Tulaczyk *et al.*, 2000a). The values of stress exponents implied by the laboratory data are ca. 50–100. These findings were corroborated by *in situ* and laboratory studies of other tills that also supported the highly non-linear, nearly plastic nature of till rheology (e.g. Fischer & Clarke, 1994; Iverson *et al.*, 1997, 1998; Hooke *et al.*, 1997).

There are still some reservations about applying a rheological till model, based on data derived from small laboratory samples and localized *in situ* studies (spatial scales of ca. 0.1–1.0 m), in simulations of soft-bedded ice masses, for which the length-scales relevant to determining the ice flow rate are several orders of magnitude larger (ca. 100–1000 m) (Hindmarsh, 1997; Fowler, 2002, 2003). One of the most prominent arguments advanced against the (nearly) Coulomb-plastic model of till rheology is based on the fact that ice flow rates in soft-bedded systems appear to be less temporally variable than one would expect if all of the flow resistance were provided by a bed with no significant strain-rate-dependence of strength (Hindmarsh, 1997). However, there are two important problems with this argument. Firstly, much of the flow resistance in soft-bedded ice streams appears to come from sources other than the bed (e.g. shear margins). It is sufficient to look at ice shelves to see ice bodies with no direct velocity control from the base, which, nonetheless, have a relatively narrow and stable range of velocities because the flow rates are controlled ultimately by the mildly non-linear rheology of ice itself (e.g. MacAyeal & Lange, 1988; MacAyeal, 1989a,b). The second important fact, which undermines the arguments against the (nearly) Coulomb-plastic rheology of till, stems from the recent observations of highly variable velocities on the ice plain of Whillans Ice Stream (Bindschadler *et al.*, 2003). There, the ice stream accelerates from a stationary state to velocities greater than ca. 10 km yr^{-1} within minutes and comes back to a stationary state within a similar time period. These wild velocity fluctuations occur in response to small tidally driven stress perturbations. Such behaviour cannot be reconciled easily with the viscous bed model proposed previously for this ice stream (e.g. Alley *et al.*, 1987b, 1989) but is fully consistent with the (nearly) Coulomb-plastic till rheology observed in laboratory tests on samples recovered from beneath the ice stream ca. 200 km upstream of the ice plain (Kamb, 1991; Tulaczyk *et al.*, 2000a). These observations demonstrate that at least in the case of Whillans Ice Stream, small-scale laboratory rheological tests and regional *in situ* observations are both consistent with highly non-linear till rheology and inconsistent with nearly linear viscous till rheology.

70.4.2 Subglacial water

Whether fast ice streaming is accommodated by hard-bed sliding or till deformation, at the most basic level it is enabled by the presence of subglacial water at pressures close to the overburden pressure (e.g. Engelhardt & Kamb, 1997; Engelhardt *et al.*, 1990b; Kamb, 1991, 2001). Theoretical calculations of basal melting/freezing rates indicate that, at least in the Siple Coast region, basal melting predominates inland, particularly beneath ice-stream tributaries, where it takes place at rates of several millimetres per year (Joughin *et al.*, 2003a, 2004; Vogel *et al.*, 2003). Calculations for ice-stream trunks are more ambiguous and suggest that these experience either basal melting or freezing at rates of a few millimetres per year. Basal freezing is particularly likely just upstream of grounding lines, near which ice streams may be susceptible to stoppage unless there is a sufficient supply of water from upstream (Bougamont *et al.*, 2003a,b). The spatial distribution of melting and freezing rates predicted by best available models strongly suggests a need for a regional subice-stream water transport, which should move subglacial water from the inland areas of production toward the marginal zone. This is in agreement with borehole observations, which included the discovery of a water-filled subglacial cavity in the palaeomargin of Kamb Ice Stream (Engelhardt & Kamb, 1997; Kamb, 2001; Carsey *et al.*, 2002). However, these recent results are inconsistent with the proposition of Tulaczyk *et al.* (2000b) that ice-stream beds may have no organized regional drainage (their 'undrained bed model').

It is somewhat unfortunate that understanding subglacial water generation, storage and transport is key to understanding ice-stream flow rates and mass balance because it is difficult to make observations of subglacial water storage and flow at length-scales over which ice flow rates are determined (several ice thicknesses or more). The best direct data, thus far, come from borehole experiments, which sample relatively short scales (fraction of a metre to dozens of metres). Nonetheless, even extensive borehole investigations conducted over a dozen of years on the Whillans, Kamb, and Bindschadler Ice Streams did not produce an unequivocal picture of the physics of water flow beneath soft-bedded ice streams (Engelhardt & Kamb, 1997; Kamb, 2001). What can be positively concluded from these borehole studies is that subglacial water is present beneath ice streams at high pressure, either as till pore water or as free water occurring at the ice–till interface. Although the water pressure is close to the overburden pressure, it is not quite established how close. Borehole water-level measurements suggest effective stress (overburden less water pressure) of less than ca. 100 kPa whereas till properties indicate less than ca. 10 kPa (Tulaczyk *et al.*, 2001a). The presence of a relatively widespread subglacial water film with a thickness of the order of millimetres or more (e.g. Weertman & Birchfield, 1983; Alley *et al.*, 1986a, 1987b, 1989) is not supported by borehole observations (Engelhardt & Kamb, 1997; Kamb, 2001). Inflow of borehole water into the subglacial system was able to create new accommodation space in the form of a water gap, which overprinted the 'pristine' subglacial water system. As has been observed on mountain glaciers, the presence of boreholes perturbs subglacial conditions. In addition, measurements made in an individual borehole may be only locally representative. These complications hinder the use of borehole experiments in constructing models of subice-stream water flow useful in simulations of ice flow (e.g. Harper & Humphrey, 1995; Harper *et al.*, 2002).

However imperfect borehole observations of subglacial water systems are, not too many viable alternatives for studying subglacial water flow and storage exist. A geophysical technique, which could provide regional-scale constraints on subglacial water systems, is desirable. However, thus far, geophysical techniques provide either only qualitative constraints or approximate quantitative constraints on selected properties (e.g. Blankenship *et al.*, 1986, 1987; Anandakrishnan & Alley, 1997a,b). For instance, ice-penetrating radar can successfully map out areas of 'wet' and 'frozen' bed (Bentley *et al.*, 1998) but it is much more difficult to quantify spatial changes in subglacial water abundance within the 'wet' zones from radar surveys alone (Gades *et al.*, 2000; Catania *et al.*, 2003). Seismic surveys have been also valuable in showing spatial variability of subice-stream beds but have difficulty with quantifying properties of subglacial water systems (Anandakrishnan & Bentley, 1993; Atre & Bentley, 1993; Anandakrishnan & Alley, 1997a,b; Smith, 1997a,b; Vaughan *et al.*, 2003b).

Recent advances in ground-based, airborne and satellite geodetic techniques offer the tantalizing possibility that regional-scale subglacial water storage (length-scales of several ice thicknesses or greater) may be monitored through accurate measurements of ice-surface elevation changes. It has been long known that changes in subglacial water storage may result in large (ca. 0.1–1 m) variations in ice-surface elevation on mountain glaciers (e.g. Iken *et al.*, 1996; Fatland & Lingle, 2002). However, no similar observations have been made on ice streams and ice sheets until recently, when Grey *et al.* (2005) reported oval-shaped inflation and deflation of the ice surface on several Siple Coast ice streams based on satellite radar interferometry. These localized elevation changes of between ca. 0.5 and ca. 1 m occurred over a time period of 24 days. Because of their high magnitude and localized nature, the observed ice-surface changes are best explained by filling and drainage of subglacial lakes, which are several kilometres across (Grey *et al.*, 2005). In the upper part of the Kamb Ice Stream, the data imply drainage of ca. 20 million m^3 from the subglacial lake within the observation period. This indicates that a significant focusing of water flow into subglacial lakes occurs in the region because this much water is produced in 24 days over an area ca. 100,000 km^2 if one assumes a reasonable basal melting rate of several milllimetres per year (Joughin *et al.*, 2003a, 2004; Vogel *et al.*, 2003). This has significant implications for models of subice-stream water flow because until now their basic assumption was that no large changes in subice-stream water storage occur over such temporal scales and that subglacial water inputs and outputs are more-or-less in balance at all times (Alley *et al.*, 1989; Kamb, 1991, 2001; Walder & Fowler, 1994; Ng, 2000b). It is as yet unknown whether the observed variability in subglacial water storage and flow has its reflection in changes in ice-stream velocity. Nonetheless, the observations reported by Grey *et al.* (2005) call for a paradigm shift in understanding and quantitative treatment of subice-stream water flow to account for the influence of water storage in subglacial lakes on the rate and physical nature of subice-stream water drainage.

70.4.3 Sticky spots

The term 'sticky spots' has been proposed to describe localized sources of high basal resistance within an otherwise weak till bed (Alley, 1993). Although they represent a pretty logical theoretical construction and most likely do exist, the physical nature of sticky spots and their exact role in controlling soft-bedded ice stream motion are not constrained by observations. The best constraints on spatial distribution and physical nature of sticky spots come from passive seismic experiments described by Anandakrishnan & Bentley (1993) and Anandakrishnan & Alley (1993). Whillans *et al.* (1993) examined the strain rate field on parts of Whillans Ice Stream but failed to find clear evidence of sticky spots. Engelhardt and Kamb (Kamb, 2001) drilled an array of four boreholes in a location where geophysical data suggested the presence of a sticky spot (Rooney *et al.*, 1987) but found that subglacial conditions in these boreholes were not distinguishably different from those revealed by boreholes drilled in other parts of the same ice stream.

Notwithstanding the limited success with direct investigations of sticky spots, analyses of ice-stream velocity data indicate that bed resistance has some level of spatial non-uniformity, at least over spatial scales longer than an ice thickness (MacAyeal *et al.*, 1995; Price *et al.*, 2002; Joughin *et al.*, 2004). However, because ice acts as a low-pass filter with respect to basal inhomogeneities, inversions constrained by ice- surface velocity data do not provide direct constraints on sticky spots, the dimensions of which are much smaller than the ice thickness. In other words, a till bed that is peppered with such small sticky spots is indistinguishable in these inversions from a uniform till bed.

The most geographically extensive inversion of basal shear stress in the Siple Coast region is that found by Joughin *et al.* (2004). It is based on the vertically integrated stress-balance equations for ice streams proposed by MacAyeal (1989a,b). Hence, it yields more realistic quantitative estimates of basal stress when basal shear stress is a small fraction of the gravitational driving stress, $\tau_d > \tau_b$, and internal ice deformation is negligible. However, it provides less reliable results when that condition is not met. The inversion of Joughin *et al.* (2004) is constrained by satellite-derived ice-surface velocity data from autumn of 1997 available at a horizontal resolution of 500 m. Results of this work suggest that over most of the bed beneath ice-stream trunks basal shear stress is weak (ca. 10 kPa or less) but that smaller areas (tens and hundreds of square kilometres) of elevated resistance (up to ca. 100 kPa) do exist. Out of the Siple Coast ice streams, the Whillans Ice Stream has the most uniform and the weakest bed, whereas the MacAyeal Ice Stream has abundant and strong sticky spots. In addition, the work of Joughin *et al.* (2004) and Price *et al.* (2002) shows that the basal resistance beneath ice-stream tributaries tends to be more spatially variable than beneath ice-stream trunks.

70.5 Temporal variability of ice-stream flow

Perhaps the most important aspect of ice-stream flow is the ability of ice streams to change their velocity, and thus alter their contribution to ice-sheet mass balance, on different time-scales (e.g. Bindschadler & Vornberger, 1998; Joughin & Tulaczyk, 2002). This is dramatically illustrated by the recent discovery of stick-slip behaviour on the ice plain of Whillans Ice Stream, which is stationary over most of the time and moves in short, tidally triggered spurts during which ice-flow velocity is equivalent to about 10 km yr^{-1} (Bindschadler *et al.*, 2003). The same ice stream also experienced a continuous slow down in average velocities, at least in the period between 1974 and 1997, with deceleration rates reaching ca. 5 m yr^{-1} on the ice plain (Joughin *et al.*, 2002). If this process continues, the ice stream may shut down in ca. 80 yr (Bougamont *et al.*, 2003b). Another key example of ice-stream flow variability is the shutdown of Kamb Ice Stream, which occurred ca. 150 yr ago (Retzlaff & Bentley, 1993; Smith *et al.*, 2002). This event has widespread implications because it pushed the Siple Coast sector of the ice sheet into a positive mass balance (Joughin & Tulaczyk, 2002), caused the growth of an ice bulge at the confluence of the two tributaries of this ice stream (Joughin *et al.*, 1999; Price *et al.*, 2001), and is already forcing rearrangements in the regional ice flow pattern (Conway *et al.*, 2002).

Observations suggest that the past few hundred years were not overly unusual in terms of ice-stream flow variability when compared with all of the past millennium (Jacobel *et al.*, 1996, 2000; Nereson & Raymond, 1998; Fahnestock *et al.*, 2000; Gades *et al.*, 2000; Nereson *et al.*, 2001). Thus, internal ice-stream processes are able to cause significant changes in ice-flow patterns and rates during time periods when climate forcing is relatively stable. The most important question, however, is whether the documented changes in ice-stream flow are consistent with the idea of a continuing Holocene collapse of the West Antarctic Ice Sheet or they are indicative of an end to the post-glacial retreat of this ice mass (Bindschadler, 1998; Bentley, 1999; Conway *et al.*, 2002).

Joughin & Tulaczyk (2002) speculated that the latter may be the case because the last two major events in the Siple Coast region involved the stoppage of the Kamb Ice Stream and the slowdown of the Whillans Ice Stream. Nonetheless, they acknowledged that these individual events may simply be a part of a century-scale oscillation, which will ultimately return the Siple Coast region to a negative mass balance and enable further retreat of the WAIS. The physical argument is that the Holocene retreat of the ice sheet may have thinned it sufficiently for ice streams to start freezing to their beds, particularly near their grounding lines (Bougamont *et al.*, 2003a,b; Christoffersen & Tulaczyk, 2003a,b). This proposition has been questioned by Parizek *et al.* (2002, 2003) who used a flowline model to infer that there is enough subglacial water beneath Siple Coast ice streams to keep their beds lubricated. Although ultimately the contention of Parizek *et al.* (2002, 2003) may turn out to be correct, their flowline model neglects the dissipation of gravitational energy in lateral shear margins of ice streams, thereby overestimating the basal shear heating and water generation.

The existing thermodynamic models of ice-stream flow evolution suggest that major oscillations in ice-stream flow ('on/off cycles') are longer than hundreds of years and should last a few thousands of years or more (MacAyeal, 1993a; Marshall & Clarke, 1998a,b; Payne, 1998, 1999; Tulaczyk *et al.*, 2000b; Calov *et al.*, 2002; Bougamont *et al.*, 2003a,b; Christoffersen & Tulaczyk, 2003a,b). However, these models make no, or limited, provision for the control of subglacial water dynamics on ice dynamics and they do not incorporate behaviour of ice bulges, such as the one growing on the stopped Kamb Ice Stream. In addition, models with more realistic ice-stream physics tend to consider only single ice streams, neglecting potential interactions between neighbouring ice streams (e.g. Bougamont *et al.*, 2003a; Christoffersen & Tulaczyk, 2003a,b). On the other hand, ice-sheet models represent ice streams in a rudimentary fashion (Marshall & Clarke, 1998a,b; Payne, 1998, 1999; Calov *et al.*, 2002). For these reasons, it is too early to reject the hypothesis that the observed recent changes in ice-stream velocities are just a part of some relatively short, century-scale fluctuation, which will not amount to cessation of the Holocene collapse of WAIS.

Whereas understanding of the dynamics of a system of interacting ice streams is still difficult to achieve, significant advances have been made in the analysis of velocity evolution for a single ice stream (MacAyeal, 1993a; Tulaczyk *et al.*, 2000b; Joughin *et al.*, 2002; Bougamont *et al.*, 2003a,b; Christoffersen & Tulaczyk, 2003a,b). Of greatest interest is the variability in ice-stream velocity over socially relevant time-scales of dozens to a few hundred years. Susceptibility of an ice stream to different perturbations can be illustrated by taking a time derivative of Raymond's (1996, equation 33) analytical equation for centre-line velocity of an ice stream flowing in a rectangular channel. When expressed as a percentage of the original ice-stream velocity, the linear accelerations in response to independent and small perturbations in ice stream width, W, basal shear stress, τ_b, and gravitational driving stress, τ_d, can be expressed through relatively simple equations:

$$\dot{U}_w = 100\frac{1+n}{W}\dot{W} \tag{1a}$$

$$\dot{U}_b = -100\frac{n}{\tau_d - \tau_b}\dot{\tau}_b \tag{1b}$$

$$\dot{U}_d = 100\frac{n}{\tau_d - \tau_b}\dot{\tau}_d \tag{1c}$$

where $\dot{W}$, $\dot{\tau}_b$, $\dot{\tau}_d$ denote the magnitude of the small perturbations and n is the stress exponent in the flow-law of ice (here assumed equal to 3). As equations (1b) and (1c) differ only in a sign, Fig. 70.2 contains just two panels because panel (B) illustrates the magnitude of ice-stream velocity variations in response to changes in either driving stress or basal shear stress. The clear message of Fig. 70.2 is that stress perturbations, such as weakening or strengthening of basal resistance, have much greater impact on ice-stream velocity than changes in ice-stream width. Even relatively large ice-stream widening/narrowing rates of 100 m yr^{-1} (Clarke *et al.*, 2000; Bindschadler & Vornberger, 1998) will result in velocity changes of the order of 0.1–1.0% per year, with only weak dependence on the initial width. However, reasonable perturbations in stresses (order of ca. 0.1 kPa yr^{-1}) can cause ice stream velocity to change by tens, hundreds, or even thousands of per cent per year, particularly where the initial difference between driving stress and basal stress (τ_d–τ_b) is small (of the order of 1 kPa or less). Such small stress differences are characteristic for ice plains, where driving stresses are just a few to several kPa. Hence, the recent observations of high flow-rate variability near the grounding line of Whillans Ice Stream (Bindschadler *et al.*, 2003) are consistent with this simple model. The high sensitivity of ice-stream velocity to changes in stress difference is quite unfortunate because of the difficulty with constraining basal shear stress and quantifying its likely temporal evolution in response to changes in subglacial water flow/storage and other forcings.

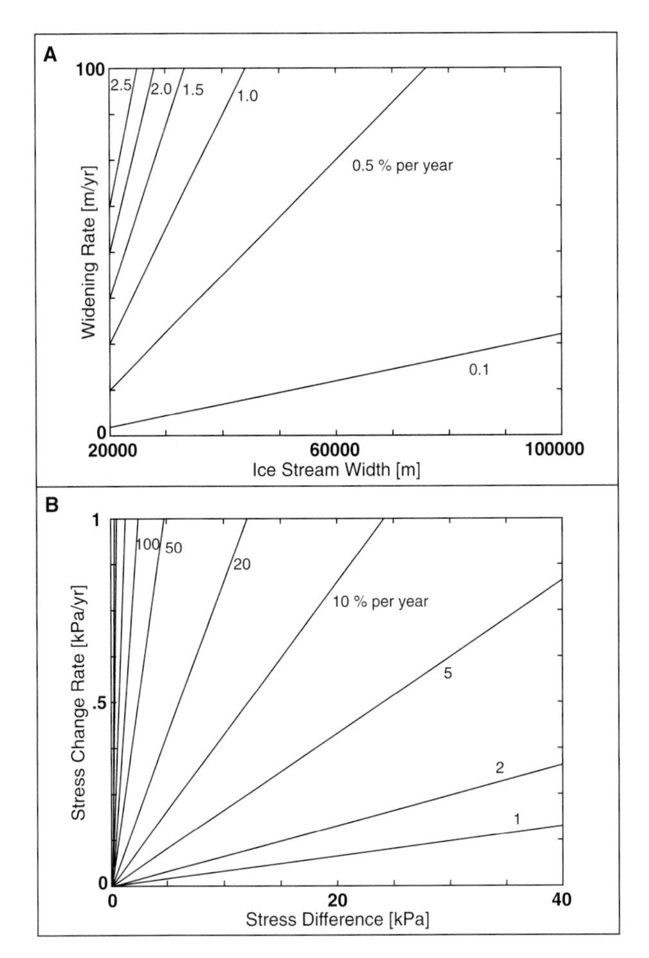

Figure 70.2 Sensitivity of ice-stream velocity to changes in ice-stream width (A) and basal or driving stress (B) based on Equation (1a–c). The range of ice-stream widths in (A) covers tributaries (ca. 20 km), ice plains (ca. 100 km) and ice-stream trunks in between (Joughin *et al.*, 1999; Raymond, 2000; Tulaczyk *et al.*, 2000b). The stress difference in (B) is defined as the difference between driving stress and basal resistance. The stress change rate represents variations in either driving stress or basal resistance. Ice accelerations result from an increase in driving stress or a decrease in basal resistance. Velocity changes are plotted for the case of acceleration (positive per cent per year) but equivalent decelerations can be obtained by simply switching signs. The velocity change isolines are plotted on a linear scale in (A) and on a logarithmic scale in (B).

70.6 Representation of ice streams in ice-sheet models

If, as it is assumed here, the ultimate goal of glaciological research on ice streams is to produce a quantitative model of ice-stream behaviour that can be used to predict future ice-sheet evolution and to reconstruct past ice-sheet flow, then considerable future efforts are needed to achieve this objective. At the present time, treatment of ice streams in numerical ice-sheet models lags behind our understanding of the complexity of ice-stream physics. This is because some of the key physical processes determining ice-stream velocity and its changes occur on relatively short spatial scales and cannot be directly resolved in ice-sheet models, which are mostly vertically integrated and have horizontal resolution of tens of kilometres.

Numerical models based on the shallow-ice approximation are particularly unsuitable for capturing the complex processes taking place in and around lateral ice-stream margins because their force balance equation neglects the very terms which in reality give rise to marginal shear stress and to stress transfer from the margins to the bed (Whillans & van der Veen, 2001). To properly capture these phenomena in a numerical ice-sheet model, a higher order, three-dimensional treatment (e.g. Pattyn, 2003) is necessary with horizontal and vertical resolution of the order of 100 m, at least in and around lateral margins. A considerable increase in computational power would be necessary to achieve such a resolution over the whole ice sheet given that modern ice-sheet models have ca. 100,000 elements and the number of elements in a suitable three-dimensional model would be ca. 1000–100,000 greater. Although these numbers may seem daunting, it is important to note that further advancements in computer processor technology are likely to increase computational power of a single processor by a factor of 1000 in the next decade (Moore, 1965). Additional technological advances, e.g. grid computing, may increase the availability of inexpensive computational time for advanced ice-sheet modelling. Moreover, shallow-ice approximation works well over much of an ice-sheet area (Pattyn, 2003) and development of 'smart' grids, which would apply the three-dimensional treatment only where needed (e.g. lateral shear margins), can further cut down the number of elements needed to achieve realistic representation of ice streams.

In addition to increasing spatial resolution and deploying three-dimensional force balance, ice-sheet models need to take into account that the mass balance of subglacial water is the key factor controlling the distribution and efficiency of lubrication enabling rapid basal motion. Subglacial water is a conserved quantity and its flow and storage can be treated in ice-sheet models (Johnson & Fastook, 2002). Sliding coefficients used to simulate rapid basal motion in areas of melted bed cannot be arbitrary constants but have to be coupled to subglacial water dynamics, even if the exact nature of the relationships between ice velocity, basal shear stress and subglacial water flow/storage is not yet determined. Without such parametrizations ice-sheet models will not be able to predict the future mass balance of modern ice sheets and reconstruct the behaviour of past ice sheets because subglacial water is one of the key factors controlling fast ice-flow rates and their changes (e.g. Zwally *et al.*, 2002a). In spite of the remaining challenges, the rapid progress in ice-stream research observed over the past few decades provides solid foundations for future advances in this important area of glaciological research.

SEVENTY-ONE

Regional basal-thermal budget: implications for ice streaming along the Siple Coast, West Antarctica

Byron R. Parizek

Department of Geosciences and Earth and Environmental Systems Institute, The Pennsylvania State University, University Park, PA

71.1 Introduction

Local changes in the West Antarctic ice sheet (WAIS) have regional effects on ice-sheet mass balance with the potential for global impacts through cryospheric influence on eustatic sea level and ocean circulation patterns. Although a majority of the ice in the WAIS is tied up in the slow-moving (1–10 m yr^{-1}) interior, attention is focused on the fast-moving (100 m yr^{-1}) ice streams that rapidly discharge ice to the ocean through three main outlet systems.

Despite major ice-age advance of grounded ice into the Ross Embayment to near the continental-shelf edge followed by an approximately 1300-km retreat (Conway *et al.*, 1999), the Siple Coast ice streams maintained roughly their same locations and their anomalously low-elevation, low-slope surface profiles (Alley & White, 2000; Anderson & Shipp, 2001). The apparent long-term persistence (Anderson & Shipp, 2001) and short-term variability (e.g. Rose, 1979; Retzlaff & Bentley, 1993; Bindschadler & Vornberger, 1998) of these ice streams are probably tied to regional thermal conditions and local basal lubrication (Parizek *et al.*, 2002, 2003).

Previous research has indicated that available basal water is almost certainly essential to till lubrication and therefore to the past, present and future of these fast-flowing ice streams (e.g. Engelhardt & Kamb, 1998; Tulaczyk *et al.*, 2000b; Kamb, 2001). Furthermore, recent observations and theoretical studies suggest that local basal freeze-on is occurring along stretches of several Siple Coast ice streams (e.g. Tulaczyk *et al.*, 2000b; Raymond, 2000; Kamb, 2001; Joughin *et al.*, 2003a). Significant slow-down of the Kamb Ice Stream occurred ca. 150 yr ago (Rose, 1979; Retzlaff & Bentley, 1993), and more recent deceleration is under-

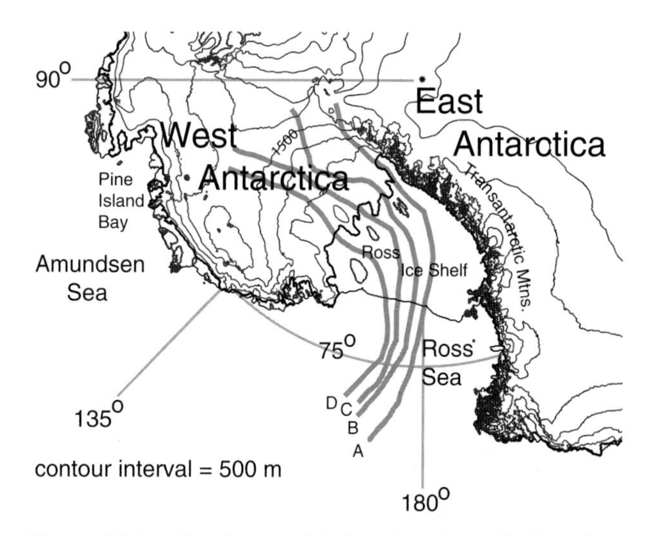

Figure 71.1 Flowline-model domains, from divides down Mercer Ice Stream (A), Whillans Ice Stream (B), Kamb Ice Stream (C) and Bindschadler Ice Stream (D) and across the Ross embayment to the continental-shelf edge. Topographic data from BEDMAP data set (sponsored by the Scientific Committee on Antarctic Research and coordinated by the British Antarctic Survey, Cambridge). (Based on fig. 1 in Parizek *et al.*, 2002, 2003.)

way on Whillans Ice Stream (Bindschadler & Vornberger, 1998; Joughin *et al.*, 2002) and possible Mercer Ice Stream (Joughin & Tulaczyk, 2002; Joughin *et al.*, 2002). It has been suggested that dewatering subglacial tills quickly causes deceleration, indicating impending shutdown of Whillans Ice Stream (e.g. Joughin *et al.*, 2003a).

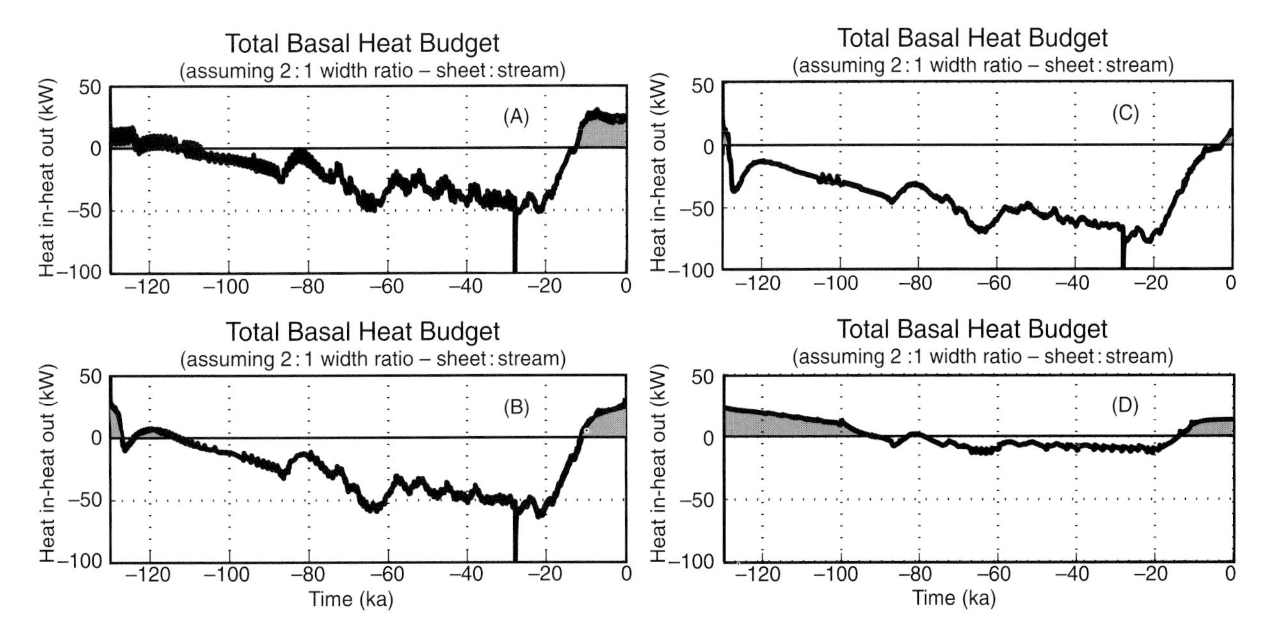

Figure 71.2 History of basal heat budget for flowlines (A), (B), (C) and (D), respectively. Total heat budgets are calculated assuming catchment sheet is twice as wide as stream. Shaded regions indicate a positive budget. (Based on fig. 3 in Parizek *et al.*, 2002, 2003.)

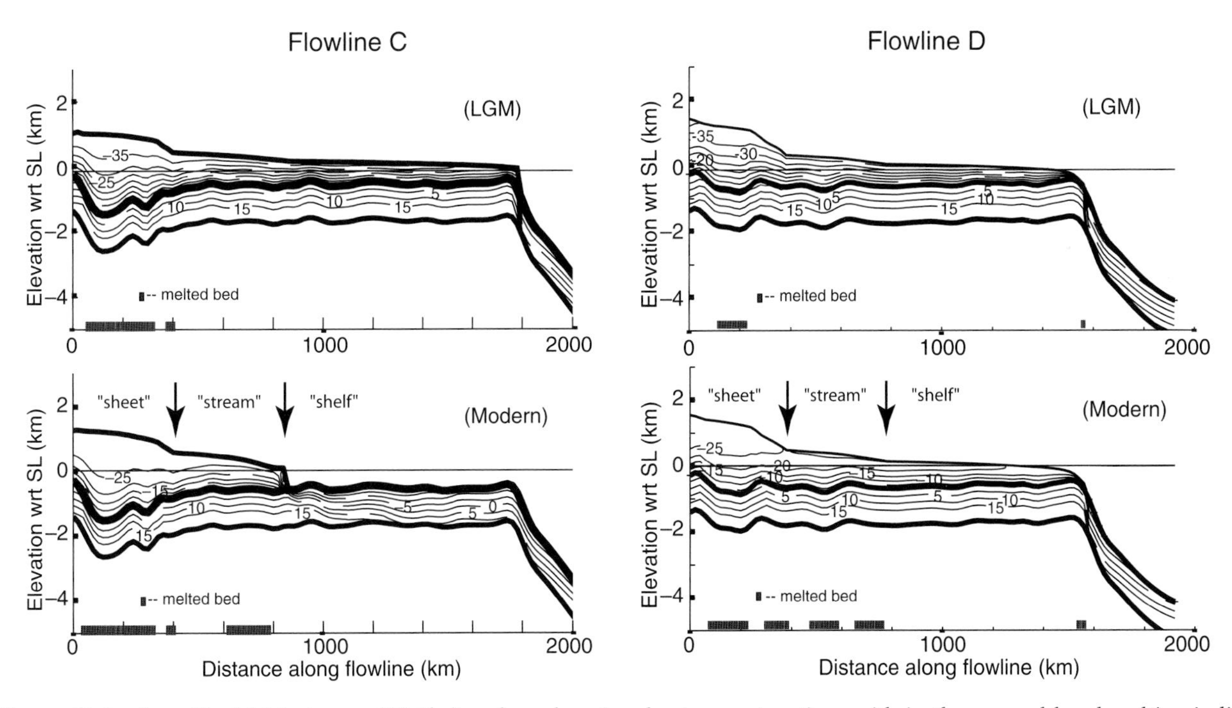

Figure 71.3 Last Glacial Maximum (22.1 ka) and modern ice-sheet reconstructions with isotherms and basal melting indicated for flowlines C (with grounding-line forcing) and D (without grounding-line forcing). (Based on fig. 2 in Parizek *et al.*, 2002, 2003.)

71.2 Numerical Simulations

In an attempt to reconcile localized basal-freezing tendencies amidst ongoing fast-glacier flow, Parizek *et al.* (2002, 2003) conducted numerical studies of the basal heat budget. Using a two-dimensional (vertical and along-flow) thermomechanical flowline model, a wide range of parameter space (sensitivity to accumulation rate, geothermal heat-flux density and basal resistance to flow) was tested along the four flowlines (A–D; down Mercer, Whillans, Kamb and Bindschadler Ice Streams, respectively) displayed in Fig. 71.1. The boundary conditions for the coupled ice-sheet and bedrock model include a temperature gradient from the geothermal heat-flux density, slip/no-slip basal conditions and variable basal-resistance parameterizations for 'sheet', 'stream' and 'shelf' zones (see Fig. 71.3), and forcings for surface temperature from the Byrd ice core, sea level from SPECMAP, accumulation rate following Huybrechts (1992), and specified grounding-line evolution (along flowlines A–C) following Conway *et al.* (1999). Throughout forty-one 130-kyr reconstructions, the basal heat budget was calculated as the difference between the heat flow into the ice–bedrock interface (from geothermal and frictional-heating sources) and the conductive heat flow out of the interface and into the ice. Locally, excess heat generates basal melting, whereas a deficit indicates basal freezing. A positive (negative) 'total' heat budget indicates that there is (not) sufficient up-glacier basal melting to buffer down-glacier freeze-on. Figure 71.2 illustrates four such 'total' balances along the four flowlines assuming the inland 'sheet' region is twice as wide as the 'stream' region based on modern observations (Joughin *et al.*, 1999). Last Glacial Maximum (LGM) and modern reconstructions for the same parameterizations along flowlines C and D are displayed in Fig. 71.3.

71.3 Results

For today, all of the flowlines are simulated to have excess heat (Fig. 71.2). Supporting the observations of Engelhardt & Kamb (1998) and Kamb (2001), meltwater from beneath thick inland ice and in bedrock lows (Fig. 71.3) is driven in a throughgoing hydrological system to the ice streams, where it locally freezes to their cold bases without exhausting the flow. Hence, subglacial tills do not need to supply water for freeze-on, and rapid ice flow can continue.

Data (Conway *et al.* 1999) show that grounding-line retreat of as much as 1300 km began ca. 13 ka after most of the post-glacial sea-level rise was completed. As shown in Fig. 71.2, this is when simulated thermal budgets became positive in several of the model runs. Hence, Parizek *et al.* (2002, 2003) suggested that slightly thicker, colder LGM ice streams were little affected by sea-level rise, that post-glacial warming increased basal lubrication allowing speed-up, thinning, flotation and grounding-line retreat, and that this retreat can continue. The mismatch between the short time constant of the subglacial water and the slower ice response may help explain the short-term variability of the persistent ice streams.

71.4 Discussion and conclusions

These results were generated with a thermal code coupled to an ice-dynamics code that utilizes a diffusion formulation of the thin-ice approximation (Hutter, 1983) in which only the vertical shear stress drives ice deformation. At the scale used in the studies, longitudinal stresses are negligible in the force balance (Whillans & van der Veen, 1993). However, a possible bias is introduced by neglect of ice-stream sides, because modern ice streams dissipate much of their potential energy as heat within the ice at their sides (e.g. Raymond, 2000). The simulated ice streams also have less advection of cold ice than indicated by observations, reducing calculated basal temperature gradients to 5–7% below observed values in ice streams. To offset these unavoidable problems of the reduced-dimensional simulations, basal 'slipperiness' was chosen to produce thinner inland ice and flatter ice streams than observed, reducing, on average, the mechanically generated heating rate per unit area below that modelled in other studies (e.g. Raymond, 2000; Joughin *et al.*, 2002). Taking these difficulties and corrections together, the simulated flowline beds are probably slightly colder than in reality, yet have excess water. Thus, although these two-dimensional studies are not the final word and three-dimensional modelling will certainly improve upon the preliminary results, the regional conclusions indicating continuation of rapid ice flow and thermal ties to the onset of deglaciation are likely robust and highlight the need to account for the basal thermal and water balances when studying the dynamic evolution of the WAIS.

PART 5

The practice of glaciology

SEVENTY-TWO

The practice of glaciology

Richard B. Alley and Sridhar Anandakrishnan

Department of Geosciences and Earth and Mineral Sciences EESI, The Pennsylvania State University, University Park, PA 16802, USA

72.1 Introduction

The practice of glaciology often seems to receive less notice than does theorizing. However, a strong case can be made that our ability to observe what is happening is the most important factor in making glaciological progress, and at least is an indispensable adjunct to theory. We lack first-principles ways to learn the resistance of ice to deformation, the thickness of ice, the flow velocity, etc. We can explain measured values once they are supplied, but any physical framework must be made real using empirically determined quantities.

Furthermore, amazing progress is being made in observational techniques. Although the modern theory of glacial flow is recognizably related to the pioneering postulates of Nye, Weertman and Kamb from several decades ago, satellite-based measurements using global positioning systems (GPS), synthetic aperture radar (SAR) interferometry or laser altimetry (see the papers by Bamber and Joughin, this volume, Chapters 73 & 74) are almost unrecognizably removed from the pressure-altimetry that mapped the Antarctic ice sheet during the early traversing days.

72.2 Some challenges in practicing glaciology

The strong papers in Part 5 review the field well, so we will use our space to speculate on further advancements in the practice of glaciology—What are we now missing? What new developments might advance our field greatly? The topics discussed here are a little idiosyncratic, weighted towards things we work on (or wish we did!).

72.2.1 Surface and borehole geophysics

Ice-penetrating radar has made determination of ice thickness seem routine (e.g. Drewry, 1983b). The ability to map isochrons (Whillans, 1976; Vaughan *et al.*, 1999b; Nereson *et al.*, 2000) allows calibration of ice-flow models using tracer fields as described below (Clarke & Marshall, 2002), and assessment of basal melting (Fahnestock *et al.*, 2001) and flow irregularities (Jacobel *et al.*, 1996). Radar remains reasonably good at distinguishing frozen from thawed beds (e.g. Bentley *et al.*, 1998; Gades *et al.*, 2000), offers the possibility of assessing bed roughness (Doake *et al.*, 2002), can contribute to study of c-axis fabrics (Matsuoka *et al.*, 2003), and can even allow highly precise measurement of the time-rate of change of ice thickness or of basal melting (Corr *et al.*, 2002). Improvements, especially in characterizing glacier beds, would be highly beneficial.

Numerous other geophysical techniques provide additional important insights (e.g. Shabtaie & Bentley, 1995; Behrendt *et al.*, 1998). Of these, seismic studies are especially useful. Active seismic techniques, in which a signal is generated as part of the experiment (Figs 72.1–72.3), and passive techniques, using signals from natural earthquakes, are both of great value.

Active seismic techniques reveal much about the materials and conditions under ice, including both deep geology and the shallow conditions related to ice flow. For example, active seismic studies revealed the existence of soft till under West Antarctic ice streams (Blankenship *et al.*, 1986; also see Anandakrishnan, 2003) and guided the borehole studies that demonstrated deformation (Kamb, 2001). The key role of geological control of those soft tills is also indicated by the co-location of ice streams over sedimentary basins (e.g. Anandakrishnan *et al.*, 1998). Changes over time in basal properties can be monitored using active seismic techniques (Nolan & Echelmeyer, 1999). Important c-axis fabric boundaries detected in ice cores can be mapped across broad areas using seismic techniques (Fig. 72.4; e.g. Bentley, 1971).

Passive seismic surveys similarly produce insight to subglacial geological conditions through receiver-function analyses and identification of earthquake sources (Winberry & Anandakrishnan, 2003). Glaciologically, characterization of earthquake sources in and just under ice is especially relevant (Fig. 72.5; e.g. Anandakrishnan & Bentley, 1993; Anandakrishnan & Alley, 1994; Ekstrom *et al.*, 2003). Moreover, the discovery of strong tidal control of Siple Coast, West Antarctica ice-stream motion was first made through observation of the timing of basal seismicity

Figure 72.1 Hot-water drilling for a seismic shothole, ice stream D, West Antarctica, 2002–2003 field season. (Photograph courtesy of Don Voigt.)

Figure 72.2 Don Voigt, Pennsylvania State University, and seismic blasting controller, ice stream D, West Antarctica, 2002–2003 field season. (Photograph courtesy of Don Voigt.)

of an ice stream (Anandakrishnan & Alley, 1997), and later confirmed and extended by GPS surveying (Anandakrishnan *et al.*, 2003; Bindschadler *et al.*, 2003).

The tidal oscillations of glaciers (Walters & Dunlap, 1987; O'Neel *et al.*, 2001) and ice streams are quite interesting features (also see Kulessa *et al.*, 2003). A highly successful investigative technique in numerous fields is to excite a complex system and then infer processes from the propagation and decay of the response. Specific hypotheses can be elaborated to produce predictions of the outcomes of such experiments, and the outcomes then used to discriminate among the hypotheses. Examples include pump tests or slug tests in hydrogeology, nuclear magnetic resonance (NMR) applications in materials science, high-pressure fluid injection triggering microearthquakes in seismology, and kicking the tyre of a used car to see if the door falls off.

Figure 72.3 Large seismic shot, ice stream D, West Antarctica, 2002–2003 field season. (Photograph courtesy of Don Voigt.)

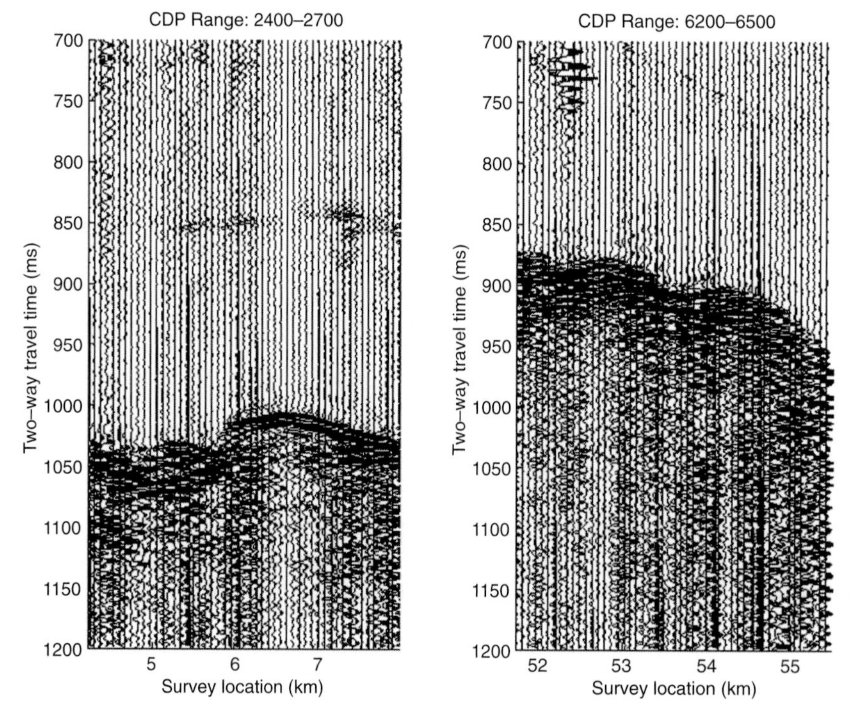

Figure 72.4 Seismic section from West Antarctica. Left panel is from the West Antarctic ice sheet near the onset of streaming flow for ice stream C. A strong reflector from the bottom of the ice stream is visible near 1025 ms two-way travel-time, and a clear but weaker internal reflector is evident near 850 ms; we have interpreted this reflector in the ice as evidence of a change in crystal orientation fabric. The right panel is from farther downstream in ice stream C, and shows a strong basal reflector but less-distinct fabric development, perhaps because recrystallization has affected essentially the entire thickness of the ice stream (Voigt *et al.*, 2003).

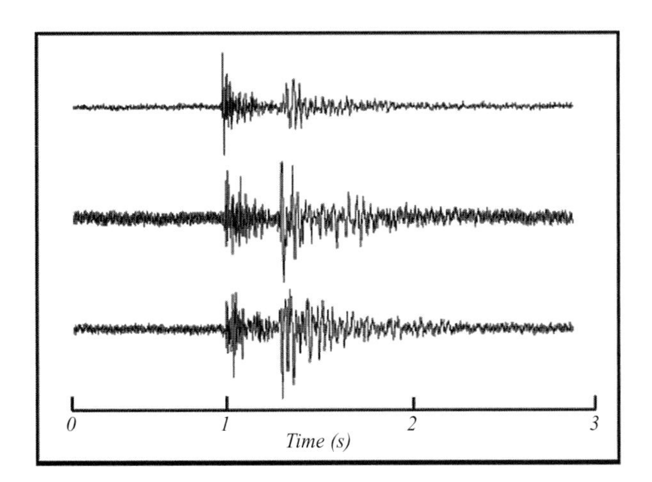

Figure 72.5 Three-component seismogram from site E10 on the ice plain of Ice Stream E (top, vertical; middle, N–S; bottom, E–W). The clear P-wave arrival near 1 s and S-wave arrival near 1.5 s are from a basal event associated with motion of the ice stream.

Ordinarily, earth scientists are limited in their ability to excite a system, and thus must rely on unplanned events (such as high-pressure-injection 'disposal' of toxic wastes, or reservoir filling loading sensitive crustal regions) or natural sources (large earthquakes causing free oscillations of the planet). The tidal signal provides a beautiful excitation in glaciology; nature is kicking the tyres of tidewater glaciers and ice streams for us, allowing testing of hypotheses based on our knowledge and ideas about ice streams. Changes in the motion and seismicity of the ice reveal the response (Anandakrishnan *et al.*, 2003; Bindschadler *et al.*, 2003), and thus are targets for hypotheses.

The tidal data are among the many data sets addressing the basal boundary condition, the most important in controlling ice flow. Uncertainties about ice fabrics (Budd & Jacka, 1989) or temperature introduce errors of safely less than an order of magnitude in most ice-flow modelling, whereas uncertainties about basal conditions can bring order(s)-of-magnitude errors. Knowing more about glacier beds thus is central to our understanding. Borehole access is necessarily limited to restricted places, so surveys such as provided by seismic techniques are essential.

The techniques of seismic surveying are well-known and powerful (e.g. Jarvis & King, 1993). The difficulty remains of achieving extensive surveys—planting and powering geophones, drilling and using shotholes are neither easy nor fast. Numerous technologies have been tried, including 'streamers' containing geophones to avoid planting them (Anandakrishnan *et al.*, 1998; Sen *et al.*, 1998) and 'thumpers' to avoid shothole drilling. These innovations have had successes, but come with trade-offs in quality. Incremental improvements are expected to continue, but improvements comparable to the shift from pressure- to laser-altimetry are not expected at this time. Increased resources to allow more such experiments, and new ideas, will be required to really understand what is going on beneath the world's ice to control its flow.

Downhole surveying is another powerful technique for learning about ice. The possibility of seismic monitoring and shooting in boreholes, and the ability to use borehole seismic techniques to measure preferred orientations of ice crystals, recommend the technique (Thorsteinsson *et al.*, 1999). Presently, many additional possibilities are being explored, including identification of annual layers and volcanic fallout and perhaps even location of biological materials (Price, 2000; Bay *et al.*, 2001; Carsey *et al.*, 2002, 2003; Hawley *et al.*, 2003). It is clear that much more work is possible, with many additional 'payoffs'. Developmental work on the techniques, together with improved access through increased core drilling, borehole milling, or access-hole melting, may allow much progress quickly.

72.2.2 Numerical modelling

Models range from simple and conceptual to highly inclusive and complex, and underlie interpretation of all data sets (Huybrechts, this volume, Chapter 80). Of special note is the rise of the fully coupled, thermomechanical, three-dimensional models, which have grown and expanded greatly since the pioneering work of Mahaffy (1976). A suite of such models is summarized briefly in the European Ice Sheet Modelling Initiative (EISMINT) exercise (Payne *et al.*, 2000; Payne & Baldwin, 2000).

Numerous new directions are being explored now in ice-flow modelling. Older models have largely relied either on the shallow-ice approximation (Hutter, 1983) appropriate for inland regions of ice sheets with spatial averages taken over horizontal distances of many ice thicknesses, or on the depth-averaged longitudinal deviatoric stress balance more appropriate for ice-shelf conditions (MacAyeal, 1992a), sometimes joining these end-to-end (Huybrechts, this volume, Chapter 80). Pioneering work is moving to inclusion of more stress terms, approaching but not yet using the full stress tensor (e.g. Schmeltz *et al.*, 2002; Pattyn, 2003; Vieli & Payne, 2003). Inclusion of the multiple stresses is essential to address the central problems of ice streams, subglacial lakes, etc.

The tie between ice-core interpretation and ice-sheet flow is important, e.g. for interpreting accumulation rates from annual layer thicknesses, surface-temperature histories from borehole temperatures, and climatic temperature histories from surface-temperature histories correcting for thickness change (e.g. Alley *et al.*, 1993; Cuffey *et al.*, 1995). An especially interesting advance in this regard is the inclusion of tracers in ice-flow models, so that such characteristics as the age, latitude, longitude and surface elevation of deposition of ice can be simulated through long times and across space, showing the origin of ice collected in a core (Clarke & Marshall, 2002).

The strong anisotropy of a single ice crystal, together with the strong tendency for development and destruction of coordinated orientations of neighbouring crystals (preferred c-axis fabrics), can have major effects on the flow of ice sheets (e.g. Budd & Jacka, 1989; Thorsteinsson *et al.*, 1999), with smaller effects from grain size and other characteristics of ice (Cuffey *et al.*, 2000a). Attempts have been made and are being made to include these effects into ice-sheet models (e.g. Castlenau *et al.*, 1997; Thorsteinsson, 2002; Faria *et al.*, 2002), although the full evolution equations under the full stresses are not yet available (see 'Forensic glaciology', below).

Improved confidence in interpretation of ice-core palaeoclimatic histories will require improvements in this modelling.

The scale of the associated difficulties in 'handling' full stresses, anisotropy and tracers, together with the surface mass balance, water runoff and basal lubrication, till deformation, etc., is daunting. The main modelling groups are moving vigorously (see Huybrechts, Payne and Gudmundsson, this volume, Chapters 80–82), but a fully integrated ice-flow model appears to be some distance off. The modelling of other key aspects of the Earth system is handled quite differently, with large and often stably funded groups (e.g. the Hadley Centre of the UK Meteorological Office, http://www.meto.gov.uk/research/hadleycentre/models/modeltypes.html, or the U.S. National Center for Atmospheric Research, http://www.ncar.ucar.edu) dedicated to long-term development of complete models and to providing community access to those models or their products.

At present, we lack community ice-flow models with the level of support, intercomparison, etc., expected from the main climate models. Incorporating full stress tensors, anisotropy, fabric development, till deformation, hydrology and more are great challenges. Despite the remarkable progress in ice-flow modelling, the piecemeal funding situation and limited available resources leave us somewhat pessimistic about the ability to really address the major problems rapidly. A few groups, not just one, with a mission and resources more like those of the atmospheric general-circulation-modelling community, would greatly change the situation. Given the major, or even preeminent, importance assumed by sea-level change in assessment of future impacts of changing climate (e.g. IPCC, 2001b), and the dominance of ice sheets as reservoirs of potential sea-level rise (IPCC, 2001a), the value of improvement of ice-flow models should be clear.

72.2.3 Forensic glaciology

The c-axis fabrics, grain sizes and shapes, bubble sizes and shapes, etc., of ice preserve a history of the ice flow and climatic conditions that produced these characteristics, over some 'memory' interval that depends on the active physical processes. In addition, some of these characteristics partially control the rate of ongoing deformation in response to applied stress and temperature. The study of these relations has been a special interest of ours over the years, but the efforts of many other glaciologists show the value of the subject, so we include it despite the possibility that it appears self-serving.

A fundamental difficulty remains that no one has successfully calculated the behaviours of ice physical properties from first principles. (For some idea of one of the many difficulties involved, see Miguel *et al.*, 2001.) Some models (such as that for rotation of existing grains under nearest-neighbour influences; e.g. Azuma & Higashi, 1985; Alley, 1988; van der Veen & Whillans, 1994) have had some success, but are clearly not complete descriptions. Other key quantities, such as conditions for the onset of nucleation and growth of new, strain-free grains, or the rate of grain subdivision by polygonization, remain very poorly known (see review by Alley, 1992).

The 'obvious' solution, that of laboratory experiments, is quite unlikely to work for most of the ice on Earth. This was shown most clearly by the work of Jacka & Li (2000). A great range of experiments, under different stresses and temperatures and with different starting conditions, produced highly consistent results in terms of c-axis fabric, grain size, etc. (also see Budd & Jacka, 1989). All of these consistent results, however, came from experiments that shared the 'problem' of having higher strain rates (usually by order(s) of magnitude) than observed in most of the world's ice, either through higher temperature or stress. Jacka & Li (2000), however, conducted especially heroic experiments approaching strain rates of upper regions of ice sheets, and obtained a very different result: the onset of nucleation and growth of new grains was delayed, consistent with numerous field observations. High-deformation-rate situations develop grain sizes and c-axis fabrics that reflect the stress state over the last 10% strain or so; lower deformation rates allow the physical properties to integrate conditions over much longer times, to more than 100% strain (Alley, 1992).

A rather straightforward explanation is possible. Nucleation and growth of new, strain-free grains, with the attendant loss of 'memory' of earlier conditions and the attendant change in c-axis fabric, grain size, etc., are favoured by accumulation of 'damage' to ice-crystal lattices during deformation (dislocation tangles, etc.). The rate at which damage is produced increases as the deformation rate increases. Various diffusive processes remove or 'heal' damage (recovery). These are not driven by the stress directly, and some at least have a lower activation energy than does the accumulation of damage (grain-boundary versus volume diffusion), so the healing rate increases with increasing temperature more slowly than does the rate of damage accumulation. Above some temperature–stress threshold, the rate of recovery is inconsequential compared with the rate of damage accumulation. Accumulation of sufficient damage will drive nucleation and growth of new grains, and faster deformation simply speeds the attainment of the damage threshold for such nucleation. At slower deformation rates, however, the recovery rate is sufficient to delay onset of nucleation by removing some of the stored strain energy.

Because we lack the ability to calculate the key deformational fields from first principles, and because the natural deformation rates of most of the world's ice are largely out of reach for laboratory experiments, we are left with 'forensic glaciology'—we must look at the ice, and figure out what happened using clues from the situation, the known behaviour of ice and theory from other materials. Through sufficient measurements of ice-core physical properties with different histories of temperature and stress, different impurity loadings, etc., we should be able to work out the key thresholds, and make them available for further ice-core interpretation and ice-flow modelling.

Some possibility of failure exists (if we cannot learn enough from sufficient samples to make a useful 'inversion' for the unknown parameters), but we are optimistic. Early work such as that by Gow & Williamson (1976) points the way, and efforts such as Cuffey *et al.* (2000a) show how successful this can be.

Perhaps the most important step is generation of extensive new data sets, with improving standards of quality. Here, the advent of automatic c-axis-fabric analysers (reviewed by Wilen *et al.*, 2003), of downhole logging and seismic techniques, and perhaps of additional frontiers such as c-axis analyses from radar (Matsuoka *et al.*, 2003) offer the possibility of great advances.

Figure 72.6 Ice-core drilling (Jay Kyne), central Greenland between Summit and Dye 3, 2003.

Improved methods of thin-section analysis, and techniques of analysis that do not utilize thin-section cutting, may be useful. The order-of-magnitude improvement in accuracy afforded by automatic c-axis-fabric analysis compared with human-conducted Rigsby-stage work (Hansen & Wilen, 2002) strongly recommends the new techniques.

72.2.4 Ice-core research

Ice-cores provide the best palaeoclimatic records for guiding many aspects of human behaviour and policy. Recent advances in ice-core research are numerous and impressive. For example, continuous-flow melters feed samples to inductively coupled plasma mass spectrometry (ICPMS) and other instruments (e.g. McConnell, 2002), producing high time-resolution, accurate histories of an amazing range of soluble, insoluble and even gaseous components (Huber *et al.*, 2003). Analyses of important trace gases such as methyl bromide and methyl chloride produce accurate results at the part-per-trillion level (Saltzman *et al.*, 2004). Isotopic analyses of key species produce new and unexpected insights (e.g. Alexander *et al.*, 2003; Sowers *et al.*, 2003). Biological effects are discovered and identified (e.g. Price, 2000; Campen *et al.*, 2003). Electrical characterization now involves multi-technique, multifrequency, multi-track approaches (Moore *et al.*, 1989; Wilhelms *et al.*, 1998; Taylor & Alley, 2004), and physical techniques include x-ray transmission and others (Freitag *et al.*, 2004). And, we are left with the suspicion that the best is yet to come.

In such a promising situation, the most important need is to obtain more ice (Fig. 72.6), from key places, together with the resources to allow technique development and application. Obviously, drilling has advanced greatly (Azuma & Fujii, 2002), but additional rapid improvement in core-recovery speed and quality is unlikely to be easy. Nonetheless, whether the future lies in refinement of existing drills, testing of new techniques, access drills, sidewall corers, or other technologies, better drilling will undoubtedly pay off in better science.

Major ice cores can now rival satellites in lead times. During a 1988 meeting, the U.S. Ice Core Working Group focused on an inland, high-accumulation-rate site in West Antarctica to follow the GISP2 drilling in central Greenland. The report of that meeting (Ice Core Working Group, 1989) stated 'A long-range ice core research plan for the next decade comprises: a deep core in West Antarctica; drilling to start around 1994–95 . . .' (p. 25). Drilling of that core may start slightly more than a decade after the target date if all goes well. Fortunately, during the time that the main project was logistically precluded, intervening coring at Taylor Dome and Siple Dome has proven valuable. The value of having a suite of drilling targets, with site-selection work well advanced, should be clear.

72.2.5 Remotely operated vehicles for under-ice-shelf and grounding-line study

The spectacular side-scan-sonar images and associated core samples enabled by new marine technologies (e.g. Anderson *et al.*, 2002; Canals *et al.*, 2002) have served as the 'satellite photographs' of marine glaciology. The long debate about the glacial versus glacial-marine origin of Antarctic marine diamictons, for example, is easily resolved with a single picture of 100-km-long flutings across the continental shelf. However, that knowledge largely ends at the fronts of the ice shelves. What lies beneath is still poorly sampled at best.

Technologically, this is not a huge hurdle. Grounding-line processes were studied by remotely operated vehicle (ROV) by Powell *et al.* (1996) (also see Dowdeswell & Powell, 1996;

Bruzzone *et al.*, 2003), providing many insights to processes of interest. A long-range ROV with side-scan sonar, seismic capability and perhaps even bottom-sampling capability could open some of the most important parts of the glacial environment to inspection. How 'dirty' are ice-shelf bases, with implications for ice-rafted debris and Heinrich events (Hemming, 2004)? What occurs at their grounding lines? What deposits exist beneath them? How crevassed are they? How warm is the water beneath them? We know how to find out.

72.2.6 Cosmogenic isotope studies of glacial erosion

The presence of abundant cosmogenic radioactive isotopes in rock under one of the large modern ice sheets demonstrates strong non-steadiness—the ice must have gone away to allow cosmic-ray bombardment (Nishiizumi *et al.*, 1996). Mapping such exposure would provide possibly the strongest constraints on the history of ice-sheet existence and stability/instability. Access-hole boring followed by shallow drilling into rock and careful analysis of the samples recovered should be sufficient. This also would allow geothermal-flux determination, providing a critical but poorly known boundary condition for ice-flow modelling.

Glacial erosion of mountain belts is of interest, and figures in some discussions of global biogeochemical cycling, etc. Among the many questions that can be addressed is whether, in partially glaciated basins, the erosion is primarily beneath the ice or also attacks non-glaciated regions nearby. The relative cosmogenic-isotope loading of rocks around the glacier versus those discharged by the glacier should reveal the balance (Perg *et al.*, 2002).

72.2.7 And so much more

We could lengthen this list considerably, because the future of glaciology is so bright and is so tightly linked to the future of glaciological practice. Ice is central in the study of the Earth, through archival of the best climate records, control on freshwater fluxes affecting climate and ocean circulation and sea level, and modification of landscapes, among others. Snow and sea ice exert extreme feed-backs on the climate system. The study of ice has been small and marginal compared with many other fields, yet has produced the spectacular results we see. With appropriate levels of interest and support, much larger progress is likely in the future.

Acknowledgements

We thank numerous colleagues, including those in the WAIS and WAISCORES communities, the Tides project and the Matanuskans. Partial funding was provided by the US National Science Foundation through grants 0229609 and 0229629, and by the Gary Comer Foundation.

SEVENTY-THREE

Remote sensing in glaciology

Jonathan Bamber

Bristol Glaciology Centre, School of Geographical Sciences, University of Bristol, University Road, Bristol, BS8 1SS, UK

73.1 Introduction

The Antarctic Ice Sheet covers an area of some 13 million km^2 and is one of the most inhospitable and remote environments on the planet. Sea ice in the Southern Ocean fluctuates in area between about 4 and 20 million km^2 over the period of a year. By comparison, the area of the USA is 10 million km^2. The number of glaciers is not well known but exceeds 160,000. Monitoring such large, heterogeneous areas, often in remote and inaccessible locations, is ideally suited to satellite-based observations, which provide the only practical means of obtaining synoptic, regional-scale coverage. It should be stressed, however, that, to be of most value, remote-sensing data require thorough validation and calibration with *in situ* measurements. Rather than making field observations redundant, therefore, remote sensing is crucially dependent on them.

The subject of remote sensing of glaciers and ice sheets could fill a book in its own right and, as a consequence, what is presented here is designed to be a primer, rather than a comprehensive review. The aim of this chapter is to introduce the reader to the key relevant concepts of remote sensing and to demonstrate how these concepts have been used to extend our knowledge and understanding of the cryosphere. There are three objectives of this chapter:

1 to present a selective review of what has been achieved, and what advances have taken place, over the past decade;
2 to identify some of the key gaps in satellite/airborne observations;
3 to consider how these gaps may be filled in the near future.

73.2 Basic principles of satellite remote sensing

This section offers a brief refresher in some of the key concepts, sensors and missions that are of particular relevance to glaciology. These concepts are important to understanding the applications discussed but can be safely skipped by any reader with a basic knowledge of satellite remote sensing.

Observations of the Earth from space utilize a relatively small number of wavebands where modulation of the electromagnetic wave by the atmosphere is low. These wavebands are known as atmospheric windows. Three main windows exist in the visible, infra-red and microwave part of the spectrum. For observing the cryosphere, an increasingly useful waveband is the latter, as clouds are transparent in this part of the spectrum and microwave sensors can be used day or night. In addition, since 1991, a continuous record of high-resolution microwave measurements has been available from an instrument known as a synthetic aperture radar (discussed in section 73.3.3).

All of the satellites discussed below are in what are known as low Earth orbits, at altitudes of typically 750–1000 km and are generally in exact repeat cycles. This means that after a certain number of days the pattern of orbit tracks on the ground is repeated. The length of the repeat period varies anywhere from a few days to a month or longer and this clearly affects the temporal resolution with which processes can be observed. The inclination of the orbit (its angle with respect to the Equator) is also an important parameter as it defines the latitudinal limit of the satellite. Typically, for Earth resources missions, this is around 80°, which means that the highest latitude portion of Antarctica and the Arctic ocean may not be covered by a particular sensor. It is also important to remember that the surface of the Earth is, at any time, covered by about 50% cloud and for some areas, such as the marginal sea-ice zone, this value can be substantially higher. In addition, polar regions suffer extended periods of darkness during their winters. Consequently, the revisit time of a satellite sensor operating in the visible or infra-red does not necessarily indicate its temporal sampling rate. Microwave instruments, however, acquire data under almost all conditions (rain can sometimes produce interference). Their temporal resolution is, therefore, directly related to the revisit interval.

Visible sensors receive solar radiation that has been reflected by the surface but also scattered by the atmosphere back to the sensor. This latter component of the signal, known as sky noise or skylight, is an unwanted signal due to scattering by air molecules and more significantly particulate matter, which can produce 'haze' in an image. In general, however, sky noise does not seriously affect discrimination of surface types over glacierized terrain but it can influence, for example, the calculation of the albedo of a surface (Stroeve *et al.*, 1997). Infra-red sensors measure a combination of reflected solar radiation (below a wavelength of about 3 μm) and thermal radiation emitted by the surface of the Earth and the atmosphere.

73.3 Satellites and sensors

There are a number of books on the general principles of remote sensing that cover, in detail, the concepts and the characteristics of the more common and ubiquitous sensors/satellites (Lillesand & Kiefer, 2000; Rees, 2001). These textbooks, however, do not, necessarily, carry details of the sensors relevant to cryospheric applications. Thus, here, we provide only brief details of established sensors and technology with appropriate references where necessary. Greater detail is provided on recently launched and upcoming missions, especially those that have a particular emphasis on polar applications and those instruments that are particularly relevant to mass balance studies of land and sea ice, that are not well represented in the existing literature.

73.3.1 Visible sensors

The relevance of visible and infra-red imaging instruments is primarily for mapping the extent and, to a lesser degree, the characteristics of ice on the planet. Because of the relatively long time series provided by some of these instruments they are particularly valuable for determining variations in areal extent of glaciers, ice caps and sea ice. Surface velocities of land ice have also been obtained using 'feature tracking' methods (section 73.3.4). Given below are some brief details pertaining to the instruments that have been used most extensively for cryospheric monitoring applications.

The Landsat series of satellites has been providing visible and near infra-red (IR) imagery of the Earth's surface (up to a latitudinal limit of ca. ±82.5°) since 1972 and is, at present, the most ubiquitous satellite in Earth resource applications. It merits, therefore, some space here. The original instrument was known as the multispectral scanner (MSS) and had a resolution of 79 m and a repeat cycle of 18 days. The MSS was superseded by the thematic mapper (TM), which has a resolution of 30 m, improved dynamic range in digitization (0–255) and seven channels in the visible, near and thermal infra-red (Table 73.1).

Landsat 7 was launched in April 1999 with the enhanced thematic mapper plus (ETM+) onboard. The ETM+ includes a number of new features that make it a more versatile and effective instrument compared with its predecessors. The primary new features on Landsat 7 are a panchromatic band with 15 m spatial resolution, on board, full aperture, 5% absolute radiometric calibration and a thermal IR channel with 60 m spatial resolution (Lillesand & Kiefer, 2000). A preliminary assessment of the glaciological utility of the ETM+ was undertaken shortly after the data became available (Bindschadler *et al.*, 2001b). A summary of the principal characteristics of the ETM+ is given in Table 73.1.

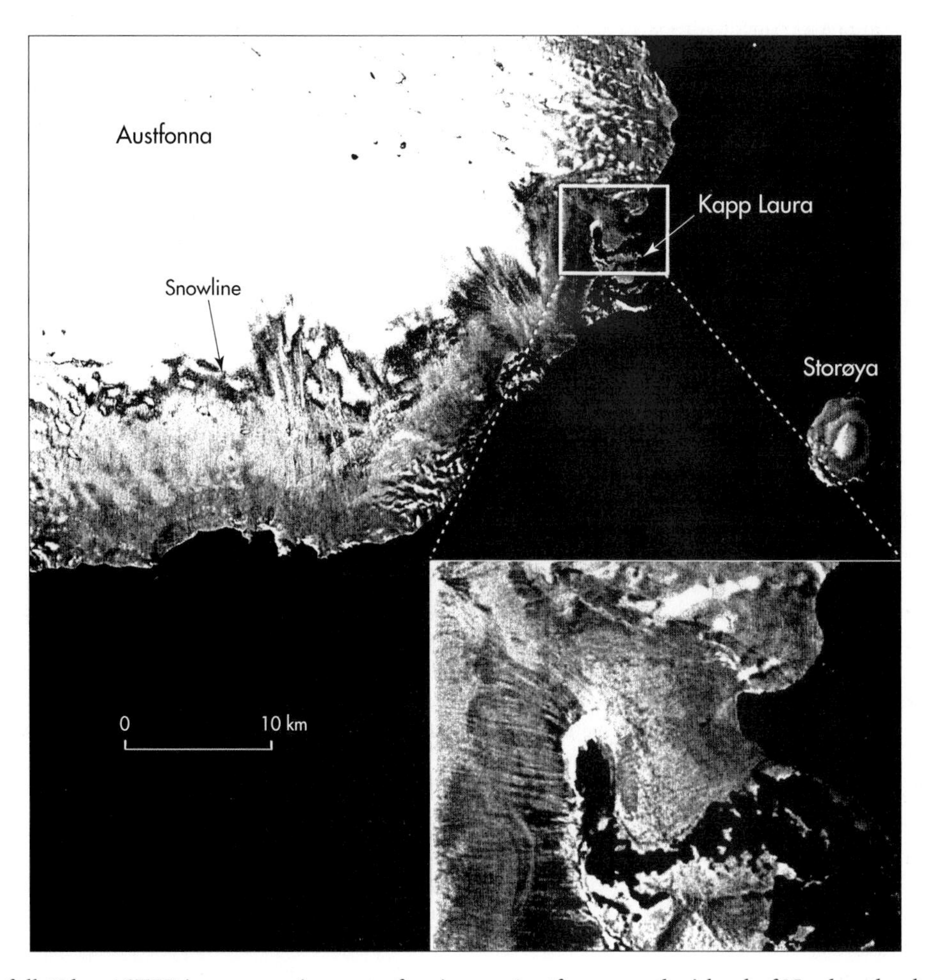

Figure 73.1 A full 60 km ASTER image covering part of an icecap, Austfonna, on the island of Nordaustlandet in the Svalbard archipelago, Norwegian Arctic. The three visible channels were combined to produce a colour composite image. The scene was acquired in June 2001.

Table 73.1 Characteristics of the ETM+ channels

Band number	Spectral range (μm)	Resolution (m)
1	0.45–0.515	30
2	0.525–0.605	30
3	0.63–0.690	30
4	0.75–0.90	30
5	1.55–1.75	30
6	10.40–12.5	60
7	2.09–2.35	30
Pan	0.52–0.90	15

Because of their relatively high spatial resolution and temporal extent, Landsat data have proved useful in mapping and monitoring changes in the areal extent of glacier cover.

A newer generation of visible/IR sensor with several important enhancements compared with the ETM+ is the advanced spaceborne thermal emission and reflection radiometer (ASTER). This comprises a set of three radiometers operating onboard the US Terra satellite, launched in December 1999 as part of NASA's Earth Observing System (EOS). The ASTER is the prime instrument that will be used within a project aimed at mapping global land-ice cover (the GLIMS project) due to its high resolution, low data costs and the capability of producing stereo pairs, allowing the derivation of digital elevation models (DEMs) using photogrammetric techniques. An example of a nadir view ASTER image is shown in Fig. 73.1, which covers part of the ice cap Austfonna on the island of Nordaustlandet in the Arctic Svalbard archipelago. The image shown is a full 60 km scene and to illustrate the level of detail present in the data a smaller subscene is inset. The image was acquired in June and a supraglacial lake has formed near the margin of ice cover. Areas of bare ice are easily identifiable as is the snowline. Also visible are sediment plumes in the ocean emanating from the bed in areas of fast flow.

73.3.2 Infra-red and microwave radiometers

These instruments record the thermally emitted radiation produced by all natural surfaces. This thermal energy is a function of the physical temperature of the surface and a term known as the emissivity, which defines how much of the radiation is emitted at a given wavelength. Emissivity is the converse of the albedo, or reflectivity of a surface. In the case of microwave radiometers, the emissivity is a function of the properties of the subsurface snowpack, allowing valuable information to be obtained not just from the surface, but from a depth down to 20 m or more in dry firn.

The most widely used IR radiometer is the advanced very high resolution radiometer (AVHRR). It views an area that is 2400 km in width (the swath) as it moves along its orbit, with a resolution of about 1.1 km at nadir (directly below). There has been continuous coverage by at least one AVHRR instrument since 1978. Currently four satellites carrying AVHRR are operational and the data are, in general, freely available to the scientific community. The AVHRR provides a useful complement to the higher resolution visible sensors, such as the Landsat ETM, as it has a wide swath width and moderate resolution, providing daily (or better, particularly at high latitudes) coverage of cloud-free areas of sea and land ice. The instrument has been used to examine sea-ice extent, although cloud discrimination is a serious problem for this application and, as a consequence, passive microwave radiometers are preferred for this application, despite their much poorer resolution (see below). The AVHRR has also been used to map the margins, surface characteristics and morphology (such as albedo, flow stripes and surface temperature) of the ice sheets (Fujii *et al.*, 1987; Scambos & Bindschadler, 1991; Steffen *et al.*, 1993; Stroeve *et al.*, 1997). Landsat TM and ETM offer a similar capability for smaller ice masses such as valley glaciers.

As mentioned earlier, clouds are transparent in the microwave part of the spectrum, which means that microwave imaging systems have the advantage over visible or IR sensors (such as Landsat TM, SPOT (Systeme pour l'Observation de La Terre), AVHRR and ASTER) that they can offer day/night coverage even in cloudy conditions. In the polar regions, where cloud cover is often extensive and permanent darkness occurs for many months of the year, they have the potential, therefore, to provide continuous, synoptic observations. Thus passive microwave radiometers (PMRs) provide valuable (and sometimes the only) data over the polar oceans. The first instrument to provide useful data on sea-ice extent was the electrically scanning microwave radiometer (ESMR) launched in 1972, which was later superseded by the special sensor microwave imager, SSM/I. This instrument was first launched in 1987 and is still in operation. The SSM/I data have been used extensively for monitoring sea-ice characteristics and extent in both the Southern and Arctic oceans (as discussed in section 73.4.1).

A problem with PMRs, however, is their relatively poor spatial resolution (typically ca. 40 km). This is due to the fact that spatial resolution is, usually, proportional to wavelength, which is about 10^4 larger in the microwave compared with the visible. It should be noted, however, that a new generation of PMRs are now in operation.

The advanced scanning microwave radiometer (AMSR) has an improved resolution (by a factor of two) compared with SSM/I and utilizes a greater range and number of frequencies. The first modified instrument (AMSR-E) was launched on the US AQUA satellite in May 2002 and the second onboard the ADEOS II satellite in December 2002 (which subsequently failed in October 2003). Table 73.2 details the frequencies and resolutions of the instrument.

Table 73.2 Frequencies and resolutions of the AMSR microwave radiometer onboard the US AQUA satellite

AMSR (2002–present)	Swath width: 1445 km
6.9	76 × 44
10.7	49 × 28
18.7	28 × 16
23.8	22 × 13
36.5	14 × 8
89	6 × 4

73.3.3 Synthetic aperture radar

The relatively poor resolution of PMRs has been overcome for one type of active microwave sensor by using the motion of the satellite to simulate an antenna much larger than can actually be flown in space. To achieve this, both the phase and the amplitude of the radar signal are recorded. The sensor in question is known as a synthetic aperture radar (SAR) and has, since the launch of the first European Remote Sensing satellite (ERS-1) in 1991, become an invaluable tool in studies of both land and sea ice. The typical resolution of current spaceborne SARs is ca. 25 m. Table 73.3 lists recent and planned satellite SAR missions. Although SARs offer good spatial resolution, their use for certain glaciological applications is limited by the fact that the current suite of sensors are single frequency. Thus discrimination of surface characteristics can be hampered because a multispectral interpretation of the signal is not possible. Differentiation between sea ice

Table 73.3 Details of SAR missions suitable for remote sensing of the cryosphere

SAR	Time period	Frequency (GHz)	Incidence angle	Polarization
ERS-1	1991–1995	5.3	23	VV
ERS-2	1995–2002	5.3	23	VV
ENVISAT	2002–present	5.3	23	VV, HH, VH
SRTM	2000	5.3 and 10		
JERS-1	1992–1998	1.2	35	HH
RADARSAT-1	1995–2003	5.3	20–50	HH
RADARSAT-2	2006	5.3	20–50	HH

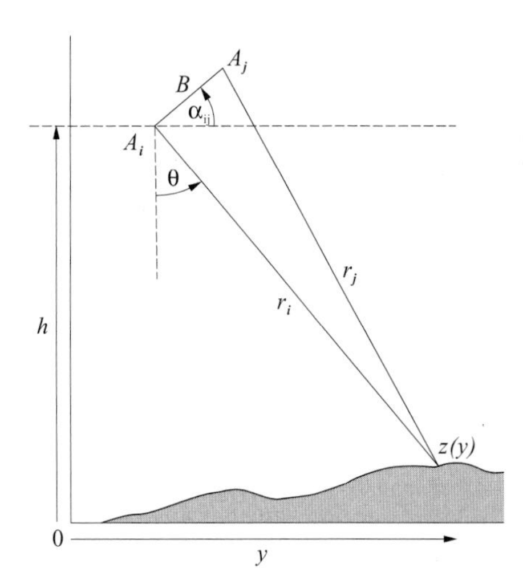

Figure 73.2 Schematic diagram illustrating the geometry of repeat pass synthetic aperture radar interferometry. A_i and A_j are the positions of the satellite at epochs i and j, B is the separation in space, or baseline, between the two measurements and r_i and r_j are the ranges to a target on the surface.

and open water, for example, must rely on backscatter and textural differences, which is not always adequate.

73.3.3.1 Interferometric synthetic aperture radar (InSAR)

The fact that SARs record the phase of the signal has allowed for the possibility of combining images taken at different times and/or locations to produce interference patterns caused by differences in phase in the two images. This is known as interferometric SAR or InSAR and its use has resulted in some remarkable results. Repeat pass interferometry is where pairs of images are combined that have been taken at different times (and from slightly different positions along an orbit) and began to be developed for glaciological applications after the launch of ERS-1 in 1991. Single pass interferometry (as exemplified by the shuttle radar topography mission, SRTM) is where two images are recorded at the same time but from different positions. Differences in path length of a fraction of a wavelength can be measured from the phase offset between the two images allowing, for example, millimetric displacements to be observed. The interference pattern is a function of (i) the topography, (ii) any displacement of the surface that has taken place between the two image acquisitions and (iii) the separation in space (known as the baseline) of the SAR when the two images were acquired (Fig. 73.2). A brief outline of the concept is given below.

For the *i*th and *j*th observations in repeat pass observations, the interferometric phase difference at any point in the interferogram is given by

$$\Delta\phi_{ij} = (4\pi/\lambda)B_{ij}\sin(\theta - \alpha_{ij}) + (4\pi/\lambda)\Delta\rho_{ij} = \phi_{\text{topography}} + \phi_{\text{motion}} \quad (1)$$

where the baseline, B_{ij}, is the distance separating the two points, with baseline B_{ij}, θ is the observation angle of the SAR, and α is the tilt of the baseline with respect to the horizontal (Fig. 73.2). The first term, $\phi_{\text{topography}}$, contains phase information related to the topography of the surface with respect to the baseline. If the targets are displaced by $\Delta\rho_{ij}$ in the range direction between the two observations, then the observed phase will include a second contribution due to this displacement. The sensitivity of the measurements to topography is proportional to the baseline B. For example, if the baseline is zero then $\phi_{\text{topography}}$ is zero and the interference pattern is dependent on displacement only. Variable ϕ_{motion}, however, is independent of B. When the *i*th and jth observations are acquired at the same time (single-pass interferometry), only $\phi_{\text{topography}}$ is non-zero.

Normally, InSAR can provide only relative height information, and ground control points (GCPs) in the form of either GPS data or a pre-existing course resolution DEM are required for three purposes: (i) to provide absolute height control, (ii) to improve the baseline estimate, which is crucial to obtain accurate results, and (iii) to separate $\phi_{\text{topography}}$ from ϕ_{motion}. It is also possible to achieve (iii) by using more than two images to produce two or more interferograms, while making the assumption that ϕ_{motion} is a constant term in all of them. More often, a DEM derived from some other source is used. Perhaps of greater value than the topographic term in Equation (1), however, is the component ϕ_{motion}. Major advances in our understanding of, and ability to monitor ice motion for glaciers, ice caps and sheets, has been achieved over the past decade through the use of InSAR techniques (Joughin *et al.*, 1996, 1998, 2002; Rignot, 1996, 2002b; Dowdeswell *et al.*, 1999a; Luckman *et al.*, 2002). Joughin (this volume, Chapter 74) illustrates the utility and power of interferometry for land-ice applications and this is further discussed in section 73.4.2.

One mission has been flown, however, where deriving topography was the sole objective. The shuttle radar topography mission (SRTM) was fundamentally different from the others listed in Table 73.3 as it was a short, 11-day flight specifically designed to map topography of land surfaces between 60°N and 56°S at a resolution of 30 m and with a vertical accuracy of better than 16 m. The SRTM deployed two different antennae flown on the same platform to achieve simultaneous observations with a fixed, known baseline (Rabus *et al.*, 2003; Smith & Sandwell, 2003). For many regions the accuracy achieved by SRTM was considerably better than 16 m (Rignot *et al.*, 2003), providing the potential for the use of these data to map lower latitude glaciers and ice caps with sufficient accuracy to detect elevation changes over time (see later).

73.3.4 Satellite altimetry

The use of an active ranging system for mapping the elevations of the ice sheets was first postulated as far back as the early 1960s (Robin, 1966). It was not, however, until the launch of Seasat in 1978 that the concept was tested and its value for glaciology explored. There are two primary types of observation that altimetry can provide: (i) absolute elevation and (ii) estimates of elevation change or d*h*/d*t*. The former has been used to produce digital elevation models of the ice sheets (Bamber, 1994a; Ekholm, 1996) and estimates of sea-ice freeboard and, hence, thickness (Laxon *et al.*, 2003). Below, we describe the basic measurement principles and properties of the current fleet of altimeters, which differ

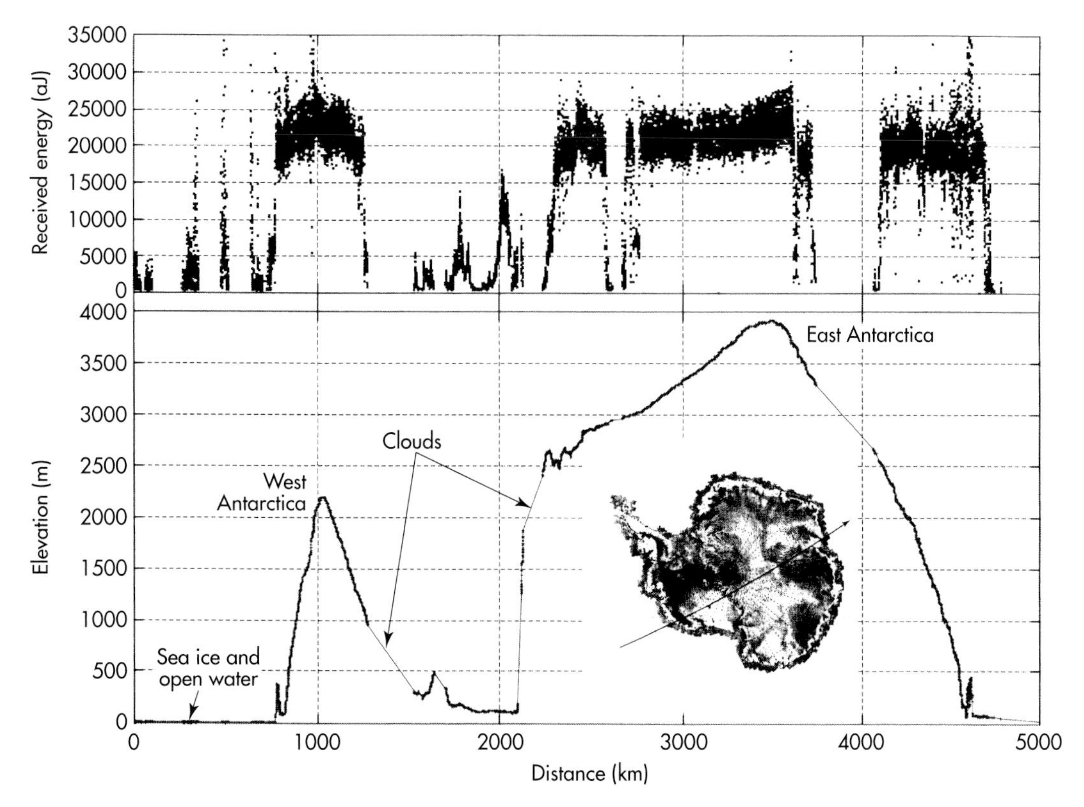

Figure 73.3 Example of data from the Geoscience Laser Altimeter System (GLAS) onboard ICESat. This track was the first to cross Antarctica (20 February 2003) and shows topography from parts of the lower West Antarctic Ice Sheet, the higher East Antarctic Ice Sheet, the steep TransAntarctic Mountains between them as well as sea ice and open water adjacent to the continent in the Southern Ocean (see inset map). Clouds along the track are indicated by reduced or no received energy in the upper panel, for example between 1200 and 1500 km along track. (Image courtesy of Christopher Shuman, NASA/GSFC.)

significantly from other satellite systems described so far, as an altimeter is not an imaging instrument.

There are two distinct categories of altimeter: radar and laser. The first satellite laser (the geoscience laser altimeter system: GLAS) was launched onboard the Ice, Cloud and land Elevation Satellite (ICESat) in January 2001 (Zwally *et al.*, 2002b). The ICESat marks a landmark in Earth observation of the cryosphere: it is the first satellite mission with a specific glaciological remit and is an indication of the increasing importance placed on understanding and monitoring the cryosphere by space agencies and the nations that fund them. An example of GLAS data, covering part of the Antarctic ice sheet, is shown in Fig. 73.3, after initial post-processing of the waveform data. Prior to ICESat, laser altimeter observations of ice have been undertaken from airborne platforms with the primary objective of measuring elevation change over time (dh/dt) and relating this to mass balance (Krabill *et al.*, 2000). Satellite radar altimeter (SRA) data, however, have been obtained, discontinuously, from the ice sheets since 1978. Their use for dh/dt observations has been, therefore, more fully explored compared with satellite laser data.

A thorough review of the use of SRAs over ice sheets can be found elsewhere (Zwally & Brenner, 2001). Presented here is a brief overview of the key issues and problems associated with the use of satellite radar and laser data over ice sheets. The SRAs are active microwave instruments that transmit a microwave pulse to the ground and measure its two-way travel time. They were designed, primarily, for operation over oceans where they can achieve centimetric accuracy (Chelton *et al.*, 2001). Over non-ocean surfaces, such as ice sheets, certain limitations exist. In particular, the current fleet of SRAs are unable to range to surfaces that have a slope significantly greater than the antenna beamwidth of the instrument (which is typically about 0.7°). Thus their useable limit is for slopes less than about 1°. As a consequence, accurate height estimates can be obtained only from larger ice masses with low regional slopes: i.e. over the interiors of Antarctic and Greenland.

There are four satellite missions that satisfied the dual requirement of having accurate enough orbit determination and an orbital inclination that provided substantive coverage of the ice sheets. The first of these was Seasat, launched in 1978. This satellite had a latitudinal limit of 72° providing coverage of the southern half of Greenland and about one-fifth of Antarctica. Although the mission lasted only 100 days, it clearly demonstrated the value of SRA data for mapping the topography of the ice sheets. Geosat, which flew from 1985 to 1989, extended the temporal record but not the spatial coverage. In 1991 the first European remote sensing satellite, ERS-1, was placed in an orbit that provided coverage to 81.5°. In 1995, this satellite was superseded by ERS-2, which had similar characteristics. There is, thus, a continuous record of elevation change for the past decade, covering the whole of the Greenland ice sheet and four-fifths of Antarctica.

Several corrections must be applied/considered when using radar altimeter data over ice sheets. The first of these is a range-estimate refinement procedure known as waveform retracking (Bamber, 1994b). This involves attempting to determine a point within the returned echo that can be related uniquely to some geophysical property of the surface such as the mean elevation within the altimeter footprint on the ground. The retrack correction is typically in the order of a few metres.

The second problem that needs to be addressed is known as the slope-induced error (Brenner *et al.*, 1983) and results from the fact that the altimeter ranges to the nearest point on the ground rather than the nadir point. For a slope of 1° the difference in range between the nadir point and the nearest point is about 120 m. As with waveform retracking there are several methods for correcting for this error, all of which involve using slope information derived from the SRA data. The third problem is that the radar wave can penetrate several metres into dry firn. As a consequence the returned waveform is made up of two components, one from the surface and one from the snow volume beneath. These two components have different ranges and their relative strength depends on the snowpack properties and, therefore, can change with time and location (Davis, 1996). The error this introduces is relatively small (ca. 50 cm) when absolute elevation is the parameter of interest, but is significant and problematic if elevation change measurements (dh/dt) are being estimated (Davis, 1997).

It should be noted that SRAs have also been used successfully to map the extent and characteristics of sea ice (Laxon, 1990). Most recently, a method has been developed to derive sea-ice freeboard and hence, in principle, thickness (Laxon *et al.*, 2003). Sea-ice thickness is an important variable in determining mass balance and for validating models. Snowpack penetration is still a problem for this application, although slope correction is irrelevant and retracking less problematic. It is hoped that GLAS will also be able to determine sea-ice freeboard.

Until the launch of ICESat the only satellite altimeters were radars, with their inherent limitations for cryospheric observations: a large footprint of several kilometres, inability to record and/or obtain accurate data for surface slopes greater than about 1°, and the penetration of the radar wave into dry snowpack, as discussed above. Although the basic concept is the same with a laser (i.e. measurement of a two-way travel time), they overcome these three problems, but introduce a new set, the most important of which is the effect of the atmosphere both through multiple scattering affecting the time delay and cloud cover preventing observations of the surface (Zwally *et al.*, 2002b). Nonetheless, the aim of GLAS is to provide decimetre accuracy elevation data for larger glaciers, ice caps, ice sheets and sea ice. The instrument has a 70 m footprint on the ground and uses a dual frequency laser (green and near infra-red) to enable correction for atmospheric delay effects (Zwally *et al.*, 2002b). The original plan was for a minimum of a 3 yr mission with the satellite placed in a 183-day repeat cycle with an inclination of 94°. Unfortunately, problems encountered with the laser subsystem have reduced the operating lifetime, mission capabilities and scope.

73.3.4.1 CryoSat

The SRAs have proved a useful tool for determining volume changes of the ice sheets and, as a consequence, inferring mass balance. Their use is, as mentioned, not without problems. For example, in a study of the mass balance of the Antarctic Ice Sheet using ERS-1 data, reliable observations were achieved for only 63% of the ice sheet (Wingham *et al.*, 1998) owing to (i) the latitudinal limit of the satellite and (ii) poor accuracy and coverage in the steeper marginal parts of the ice sheet. Similar difficulties were encountered in the marginal zone of the Greenland Ice Sheet, where surface melting produces an additional complication through changes to the microwave properties of the snowpack (Davis *et al.*, 1998).

A new type of SRA has been proposed that could address some of these problems by combining the principles of operation of an altimeter with synthetic aperture processing in the across-track direction (Raney, 1998). The mission is called CryoSat. Among the key objectives of this mission is to derive sea-ice freeboard and improved estimates of elevation changes over the ice sheets. Unlike the current fleet of SRAs, CryoSat will not only provide reliable elevation estimates for the marginal portions of the ice sheets but also for smaller ice masses with an area of ca. 10,000 km^2 or more. The satellite was launched in October 2005 but failed. A second satellite is hoped to be launched in 2009.

73.4 Applications

In this section I present some of the uses of remote sensing and the sensors discussed earlier for understanding and monitoring the behaviour of three distinct components of the cryosphere, namely sea ice, glaciers and ice sheets. The selection of applications is by no means comprehensive but acts to illustrate the capabilities of satellite data for observing, characterizing and monitoring various components of the cryosphere.

One of the most fundamental attributes of an ice mass is its mass balance, and remote sensing can play a key role in measuring this, particularly for the ice sheets and sea ice, where spatially comprehensive *in situ* measurements are not feasible. Although measurement of the mass balance of an individual glacier is tractable through *in situ* observations, it is rarely possible to extend such observations to a whole region, and data from a single glacier cannot be easily extrapolated due to variable local controlling factors. For glaciers, this is where satellite observations are particularly valuable.

Possibly the single most significant advance in the past decade for observing all aspects of the cryosphere has been the continuous operation and availability of synthetic aperture radar data. As mentioned above, this instrument has the resolution required for detailed observations, combined with the all weather, day/night functionality of a microwave instrument. This means, for example, that the normally challenging task of discriminating sea ice from cloud cover is not an issue with SAR data. In addition, the development of interferometric techniques has revolutionized our understanding of land-ice dynamics.

Figure 73.4 Photograph of summer sea ice in the Weddell Sea, Southern Ocean taken in January 1992. A thin layer of grease ice can be seen to be forming in the leads and the floes range in size from a few metres to kilometres in length. Surface roughness is also highly variable.

73.4.1 Sea-ice characteristics

Sea ice is one of the most dynamic components of the cryosphere (alongside seasonal snowcover), varying in area by as much as a factor of five between the summer minimum and winter maximum. Its extent is particularly sensitive to surface air temperature and, as a consequence, has been identified as one of seven key indicators of climate change. Monitoring its behaviour is, therefore, important for climate research but also for many other reasons including operational applications such as ship routing and weather forecasting. This presents a major challenge as (i) it covers a vast area, (ii) it changes rapidly and therefore requires frequent observations and (iii) it is extremely heterogeneous. There are, for example, different types of sea ice (mainly determined by age), which have very different characteristics when viewed from space. In addition, sea-ice cover is never uniform but interspersed with leads and polynyas (areas of open water) that can vary in size from metres to kilometres. This is illustrated in Fig. 73.4, which shows summer sea ice in the Weddell Sea, Antarctica. Grease ice (of a few centimetres in thickness) can be seen to be forming within the leads, while the floes vary greatly in size and ridging is also present within some of the floes. To resolve such features would require high-resolution sensors such as a SAR but the data volumes required to provide complete spatial and temporal coverage are enormous. There is, therefore, a trade-off between spatial and temporal resolution. Synthetic aperture radar can satisfy the former and PMRs offer the latter.

Passive microwave radiometer data have been used to produce a continuous record of ice concentration estimates for the Arctic and Southern Oceans since the launch of ESMR in 1972. Since 1987 the workhorse for routine observations of hemispheric ice coverage has been the SSM/I. The resulting sea-ice-extent time series constitutes one of the longer continuous records of the cryosphere and is, therefore, on climatological grounds alone, of considerable value. Owing to the size of the radiometer footprint, of tens of kilometres (Table 73.2), a mixture of open water and ice is typically contained within each pixel. Retrieval procedures generally attempt to solve for relative coverage based on assumptions of the behaviour of the radiometric signature of ice and water. Ice concentration and extent have been produced routinely using two different approaches known as the NASA team and bootstrap algorithms. The former (Cavalieri *et al.*, 1984) takes advantage of the highly polarized emission of open water compared with ice, and the decreasing contrast between multiyear (MY) and first-year (FY) ice with increasing wavelength. In the bootstrap algorithm (Comiso *et al.*, 1997), the ice concentration is given as the ratio of the difference between the observed brightness temperature (T_b) and the value for open water and the difference between T_b of ice and open water in the multichannel data. This algorithm does not estimate ice type coverage. Comparison between the two algorithms (Comiso *et al.*, 1997) shows only small differences in the central Arctic in the winter whereas larger disagreements can be found in the seasonal ice zones and during the summer (Singarayer & Bamber, 2003). These differences can be as much as 20%, representing an area equivalent to the whole of the Greenland ice sheet. Recently, Hanna & Bamber (2001) reported on the advantages of a new passive microwave ice concentration algorithm that systematically accounts for the effect of changing environmental conditions on the open water and ice tiepoints used in the other two approaches discussed. In addition, an updated version of the NASA Team algorithm has been proposed recently, which copes with low concentration ice more effectively (Markus & Cavalieri, 2000). This algorithm (known as NT2) is being used to generate ice-concentration estimates for the Arctic with AMSR-E data.

Uncertainties in the retrievals are affected by varying surface and atmospheric conditions. As satellite passive microwave radiometers measure emitted radiation at the top of the atmosphere, the observed brightness temperatures, T_b, are contaminated by atmospheric emissions (predominantly from various forms of water vapour). The uncertainties in the ice concentration depend, therefore, on the climatic conditions and season. In the winter Arctic and Antarctic, the uncertainties are about 6% with possible biases of similar size. In the summer, snow wetness, surface melt and melt ponds are additional complications in the retrieval process (Comiso & Kwok, 1996). The uncertainty in summer ice extent is, as a consequence, much higher (at least 10%) and the ice concentration data are of less value. Using the 15% concentration isopleth as the ice edge gives an edge location uncertainty of approximately 12 km. The use of AMSR-E data should, however, reduce this figure with the benefit of the increased resolution and channels that the sensor offers compared with SSM/I.

73.4.1.1 Ice type

As most satellites sense only the surface and near-surface properties of the ice, it is useful to relate the surface properties (e.g. salinity, roughness, etc.) to ice thickness, through the surrogate of 'ice type'. Thus, it can be useful to describe the ice cover in terms of its fractional coverage by MY and FY ice. Multiyear ice is sea ice that has survived at least one summer's melt, whereas FY ice is seasonal and is removed during the summer. Simple ice types, FY and MY ice, as inferred from active or passive microwave (Cavalieri *et al.*, 1984) data sets, can be used as a proxy, albeit crude, indicator of ice thickness because MY ice is generally thicker. The climatic significance of MY ice coverage in the Arctic Ocean is associated with its strong relation to the summer ice concentration (Comiso, 1990; Thomas & Rothrock, 1993). If changes in the climate cause persistent decreases in the summer ice concentration, it would be reflected in decreases in the winter MY ice coverage. This reduction would be due to increased melt during the previous summer or ice export. At the same time, the outflow of thick Arctic MY ice into the Fram Strait also represents a major source of surface fresh water into the Greenland–Iceland–Norwegian Seas, which are source regions for much of the deep water in the world's oceans (Aagaard & Carmack, 1989). An accurate record of the MY ice coverage and its variability is therefore crucial in understanding the relationship between climate and MY ice balance. An adequate description of the sea-ice cover requires the relative proportions of FY and MY ice to be known as a function of time. Even though this distinction between the two ice types is relatively simple, accurate estimates of the relative coverage of the two types in the Arctic Ocean have been difficult to obtain. Nonetheless, assuming that the summer minimum represents the MY ice extent, Comiso (2002) has shown, using data from SSM/I, that MY ice extent has been decreasing by 9% per decade since 1978. This decrease appears to be associated with a concomitant reduction in thickness derived from upward looking sonar data over a somewhat longer time interval (Rothrock *et al.*, 1999).

Synthetic aperture radar data have the advantage that they are relatively free of weather effects that sometimes plague sea-ice type retrievals from PMR data. Under cold dry winter conditions there is significant contrast and stability in the SAR radar signatures of MY ice and FY ice, making it possible to differentiate between the two. The signatures of younger and thinner ice types, however, are more complex and less easy to identify uniquely. Typically, therefore, only the two crude categories, FY and MY, can be readily identified in classification procedures. Ice-type data sets are being produced routinely from the SAR onboard RADARSAT (and hopefully its successor, RADARSAT-2) (Kwok, 1998). Because of the higher resolution of SARs (ca. 25 m), the assumption of a mixture of ice types within a pixel is not necessary.

The MY ice coverage differs between the Arctic and Southern Ocean. Because Antarctic ice is so efficiently advected from the coastal regions north to the marginal areas where it melts, MY ice appears to be relatively scarce around most of Antarctica. However, large areas of Antarctic sea ice remain virtually unstudied (Comiso *et al.*, 1992). Regional-scale ice-type classification using active microwave data has not been attempted to the degree that it has for the Arctic Ocean (Drinkwater, 1998). In both hemispheres, differentiation of ice type cannot be carried out reliably in summer owing to surface melt. The presence of liquid water in the snow layer prevents any penetration of the radio waves into the ice layer and therefore eliminates the contribution of any scattering from the subsurface ice layer which would, otherwise, help in discrimination.

Synthetic aperture radar data for sea-ice mapping and type discrimination clearly have a number of advantages over other data sources. Their high resolution, in particular, reduces the problem of mixed pixels but, at present, there are also significant problems. For hemispheric coverage the data processing issues are substantial and the temporal coverage limited by the relatively narrow swath-width (typically 100 km) and small number of active/planned SAR missions. A multipolarization SAR was launched onboard ENVISAT in 2002 (ASAR, see Table 73.3), which should help with ice-type classification but a multifrequency instrument is some way from being realized.

73.4.1.2 Ice thickness

Sea-ice mass balance depends not only on extent but also on the thickness of the multiyear ice (this is more important in the Arctic where multiyear ice is more pervasive compared with the Southern Ocean). Thickness is influenced by both thermodynamic processes (melting/freezing at the upper and lower surfaces) and mechanical processes (through motion and ice advection). Changes in climate are likely to have an impact on thickness and the measurement of this parameter is, therefore, highly desirable. *In situ* measurements have been obtained from moored and submarine upward-looking sonar but, because of the variability of ice thickness on all scales from metres to kilometres, it is difficult to construct useful averages using limited spatial samples from *in situ* data (Wadhams *et al.*, 1985). Consequently, determination of ice thickness at almost any spatial scale has been a long-term goal for satellite remote sensing. Satellites, however, do not observe the lower surface of the ice. As mentioned earlier, current sensors only receive radiation emitted or scattered from the top surface or upper few tens of centimetres of the ice. Nonetheless, three promising approaches that utilize only surface observations, discussed below, have emerged in recent years that allow us to address different ranges of the thickness distribution. These techniques use measurements from radar altimeters, small-scale ice motion from SAR, and IR imagery. Here, we discuss the first of these to illustrate the potential and the problems, briefly mentioning the other methods.

Sea-ice freeboard (the height of ice above the ocean surface) can be estimated by careful analysis of altimeter waveforms from sea ice and adjacent leads. It has been shown that height differences between quasispecular returns (from smooth leads) and diffuse echoes from the sea-ice surface can be used to estimate ice freeboard (Laxon, 1994). As only 11% of the floating ice is above the ocean surface, freeboard measurement errors are magnified when used to estimate ice thickness. Nonetheless, the best estimate of the uncertainty in the retrieved ice thickness is approximately 0.5 m. The radar altimeter measurements address the mean ice thickness over length scales of perhaps 100 km (Laxon *et al.*, 2003). With the launch of ICESat in 2001, both the spatial and vertical resolution of freeboard measurements should improve as the laser has a footprint of 70 m on the ground compared with kilometres for a radar altimeter.

There are still a number of difficulties associated with understanding the achievable accuracy of the freeboard measurement. Some examples include the possible height biases introduced by the snow layer and the dependence of the height measurement on the location of the dominant scattering surface (snow–ice boundary or within the snow layer). Another issue is the variability in the estimates due to ice advection and deformation, as height estimates within a grid cell are taken from measurements obtained over a month. Nevertheless, this technique represents a significant advancement in the measurement of sea-ice thickness. The impact of this method for climate and sea-ice studies would be enormous if continual long-term direct observations of ice freeboard and thence ice thickness can be realized. In the near term, GLAS and the planned ESA CryoSat radar altimeter missions are tasked to address these observational needs.

The second method relies on measuring the surface temperature of the ice using an IR radiometer (such as AVHRR) and relating this temperature to thickness. The approach appears to have a similar accuracy to moored upward looking sonars for thin ice at least (Yu & Rothrock, 1996). The third method uses estimates of ice motion (from SAR observations) to measure fluxes of ice in Lagrangian cells. These data are combined with a model of ice growth to estimate volume (Kwok & Cunningham, 2002). Both these methods rely on making a measurement that is indirectly related to thickness but, nonetheless, show promise, especially when *in situ* data are available for calibration.

73.4.2 Ice-sheet topography, mass balance and dynamics

Probably the most frequently used method for obtaining topography from remote-sensing data (either space or airborne) is stereo photogrammetry. This approach, however, using visible sensors such as SPOT, has a number of limitations over ice sheets (and to a lesser extent glaciers). First, there is rarely sufficient contrast to carry out stereo matching in the visible part of the electromagnetic (EM) spectrum. Second, cloud is ubiquitous in the polar regions and difficult to discriminate from snow. Third, to obtain absolute height measurements, ground control points are required, which are rarely available. As a consequence, other approaches have often proved more valuable. As mentioned earlier, microwave sensors are particularly useful in the polar regions owing to their all-weather, day/night functioning. The two instruments that have been used most successfully for deriving topography are radar altimeters and synthetic aperture radars (SARs).

Topography is perhaps the most fundamental observation for an ice mass. Land ice, over distances that are ca. 10–20 times the ice thickness, flows downhill, under the force of gravity. Accurate topography provides, therefore, information on both the magnitude and direction of flow. It also allows the identification of ice divides that separate flow in one basin (or glacier) from that in another. Surprisingly, until the launch of ERS-1 in 1991, the topography of Antarctica and Greenland was very poorly constrained, with uncertainties exceeding 200 m over much of the former (Bamber, 1994a).

73.4.2.1 Topography from radar altimetry

A number of groups have used combinations of Seasat, Geosat and ERS SRA data to derive digital elevation models (DEMs) of both Antarctica and Greenland (Zwally *et al.*, 1983; Remy *et al.*, 1989; Bamber, 1994a; Ekholm, 1996; Liu *et al.*, 1999; Bamber *et al.*, 2001). The limit of coverage in Antarctica by satellite altimeter data was, until the launch of ICESat, 81.5°S. This 'hole at the pole' (covering an area of almost 3 million km^2) is still problematic for numerical modelling and mass-balance studies of the ice sheet, as little reliable elevation data exist for this region. The GLAS data has recently provided coverage to 86° and CryoSat-2 (aimed for launch in 2009) should go a further 2° closer to the pole.

The SRA data from ERS-1 covers almost the entirety of Greenland barring the steeper margins of the ice sheet and the most northerly tip. To produce a full coverage DEM of the whole ice sheet and surrounding bedrock, SRA data from ERS-1 and Geosat were supplemented with airborne stereo-photogrammetry, airborne LIDAR (light detecting and ranging) data, InSAR-derived topography and, where no other adequate data were available, digitized cartography (Bamber *et al.*, 2001). The result was a 1 km posting DEM with a root mean square error over the ice sheet of between 2 and 14 m. Over the bedrock the accuracy ranged from 20 to 200 m, dependent on the data source available. Figure 73.5 is a planimetric shaded relief plot of the DEM. Certain large-scale characteristics of flow are clearly visible. Major drainage basins and ice divides (where only vertical motion takes places) can be seen. The latter appear as bright 'ridges' ranging from relatively broad features for much of the northern half of the ice sheet to narrower and 'sharper' in character in the south. Much of the southern half of the large basin in the northeast has a broken appearance, reflecting a 'disturbed' pattern of flow in this area, most probably associated with a rougher bedrock topography and the existence of a fast-flow feature in the basin (Fahnestock *et al.*, 1993; Layberry & Bamber, 2001).

As is evident from Fig. 73.5, the Greenland ice sheet is comprised of a number of drainage basins separated by ice divides. This is also the case for Antarctica. Each drainage basin can be treated, for the purpose of mass-balance studies, as a separate, independent flow unit. Owing to the size of the ice sheets, their

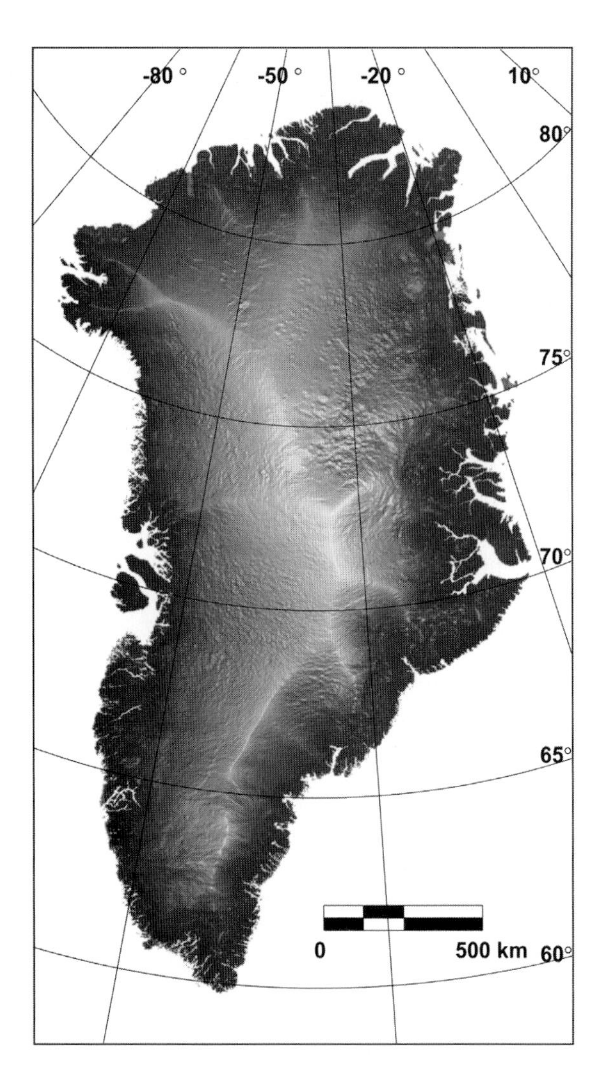

Figure 73.5 Planimetric shaded relief map of a 1 km posting digital elevation model of the Greenland Ice Sheet and surrounding rock outcrops. Lighter shading indicates higher elevation, flatter areas, and darker shading indicates steeper slopes.

non-linear, integrative response to different forcing fields and their long response time, different basins may behave in entirely different ways at any one time. In southern Greenland, for example, one basin, east of the main divide, was observed to be losing mass, and the adjacent basin to the west of the divide was gaining mass by almost the same significant amount (Thomas *et al.*, 2000b). This was a surprising result, with no clear explanation, and highlights the importance of a regional interpretation of mass balance signals. The key to this regional approach is the accurate delineation of ice divides. This has been undertaken for both Antarctica and Greenland using a GIS-based approach by combining slope and aspect, obtained from satellite-derived DEMs (Vaughan *et al.*, 1999a; Hardy *et al.*, 2000). The drainage basin masks were used to estimate the area of each basin and the volume of ice deposited as snow within each basin.

The InSAR has the capability of providing much higher resolution topography (ca. 25 m) than is possible with conventional SRAs. Small-scale topography (roughly equal to the ice thickness) can provide information on flow features that are related to longitudinal stress gradients in the ice and basal topography. In fact, bed topography and lubrication have been inferred from surface topography and motion data using an inverse modelling approach (Thorsteinsson *et al.*, 2004). As mentioned earlier, InSAR, in general, produces relative elevations only and the absolute height control must be provided from another source, such as a course resolution DEM derived from SRA data (Joughin *et al.*, 2001).

73.4.2.2 *Balance velocities*

Over an ice sheet, the gravitational driving force that makes the ice flow is a function of the surface slope and ice thickness. If the slope is estimated over an appropriate distance (typically ca. 20 times the ice thickness) it is reasonable to assume that the ice flows downhill (Paterson, 1994). It is therefore possible to trace particle paths downslope, from an ice divide to the coast. If the net mass balance (accumulation—ablation) is integrated along these flow lines, then the ice flux at any point can be estimated. The depth-averaged velocity at a point is simply this flux divided by the ice thickness. This estimate of velocity is what is required to keep the ice mass in steady-state. Hence the name: balance velocity. This quantity has been calculated for the grounded portions of the Antarctic and Greenland ice sheets using satellite-derived topography, combined with terrestrial measurements of accumulation, ablation and ice thickness (Joughin *et al.*, 1997; Bamber *et al.*, 2000a,b). The results provide the most spatially extensive depiction of flow over the entirety of the ice sheets. Although balance velocities/fluxes cannot be used on their own to determine mass balance, when combined with observations of the present-day velocity field from InSAR data, for example, they can highlight important changes in flow regime that may be related to changing state of balance (Bamber *et al.*, 2000a; Bamber & Rignot, 2002).

73.4.2.3 *Measured land-ice dynamics*

Balance velocities provide a valuable qualitative picture of the pattern of flow but they are an estimated rather than observed quantity. There are two commonly used methods for measuring surface velocities directly. The first is known as feature tracking and uses the motion of 'features' from one or more images (usually from visible sensors) to determine velocity vectors (Scambos *et al.*, 1992; Bindschadler *et al.*, 1996). The main problem with this approach is that much of the interior of the ice sheets (and even some glaciers) is featureless and there is, therefore, nothing to track. In addition, errors are proportional to pixel size, the time interval separating image acquisitions and the actual velocity. As a consequence, for slow-moving ice (less than about $100\,m\,yr^{-1}$) the errors can be unacceptably large.

The second approach, as mentioned earlier, uses SAR interferometry. The advent of repeat-pass InSAR, after the launch of ERS-1, has resulted in a wealth of data on both ice sheet and glacier surface motion. InSAR works well over ice masses ranging in size from small valley glaciers to ice-sheet-basin scale, although there are a number of caveats. The method relies on maintaining coherence between the two or more images used. This means that the scattering properties of the surface must not change significantly. If they do, due to surface melting, snowfall or strong winds, for example, it is difficult to produce an interferogram. This limits the use of InSAR largely to winter months for ice masses that experience melt during the summer. Given that even parts of the ice sheets appear to have seasonal variation in their velocity field (with higher values during the summer melt season) (Zwally *et al.*, 2002a), winter-time data may not provide a representative estimate of the mean annual velocity or ice flux. Perhaps the most fundamental issue surrounding the use of InSAR is the relatively small archive of data and the dearth of planned future missions that might alleviate this. As a consequence using InSAR methods alone to investigate temporal changes in velocity is generally problematic and often impossible. In addition, interferometry only provides motion information in the look direction of the SAR and if this deviates from the flow direction of the ice by more than about 60° then it is unusable. If the ice is flowing too fast then a problem known as phase ambiguity can arise. This is where the relative displacement of a pixel with respect to a neighbour, in the time interval between the image acquisitions, is greater than one wavelength. This results in the fringes becoming disjointed and broken, as can be seen in the downstream portion of Ryder Glacier in Plate 74.2b of Joughin (this volume). A solution to the last two problems has been to apply the principles of feature tracking in visible imagery to SAR data (Gray *et al.*, 1998, 2001). This approach is known variously as speckle or amplitude tracking and complements InSAR and visible feature tracking procedures (Joughin, 2002). It can be used, for example, in areas where no clearly identifiable 'features' exist, such as the interior of the ice sheets, and is well suited to fast flowing features such as ice streams. The method still requires coherence to be maintained between images and the relatively long orbital repeat period of satellites such as ENVISAT (35 days) can be problematic (shorter repeat periods, such as 3 days in the case of ERS-1, reduces the likelihood of changes to surface having taken place). It remains to be seen, therefore, whether the use of SAR data for land-ice motion studies will remain as fruitful as it has been over the past decade.

73.4.2.4 Mass balance

The techniques described above lend themselves to two distinct approaches to determining mass balance. The first, known as the integrative or geodetic approach, is used to infer a mass change from estimates of a volume change. The latter quantity is derived from measurements of dh/dt, often, but not exclusively, from either laser or radar altimeter data. Airborne laser data have been used, for example, to infer the mass balance of the whole of the Greenland ice sheet (Krabill *et al.*, 1999, 2000) and SRA data, covering a different and slightly longer time interval (1978–1988), were used for elevations above the equilibrium line (Davis *et al.*, 2001). The ERS-1 radar altimeter data, obtained between 1991 and 1996, have been used to assess the mass balance of about two-thirds of the Antarctic ice sheet (Wingham *et al.*, 1998). Although these results are of great value, they do have their problems. Perhaps the most profound of these is the relatively short time interval of the measurements, which, when combined with the relatively high natural variability in accumulation rates in Greenland, means that most of the dh/dt signal observed in the SRA data has been attributed to variations in accumulation rate rather than any long-term trend in mass balance (McConnell *et al.*, 2000). Another problem is that variations in climatic conditions, principally precipitation, temperature and wind, affect (i) the ratio of surface to volume scattering of the radar pulse and, hence, the characteristics of the returned signal and (ii) the densification rate of the surface firn layer, which can be some 100–120 m deep. Elevation changes may, therefore, not be related to a change in mass but merely a change in density of the firn layer. With new SRA missions providing an extended time series, and the possibility of combining dh/dt data with gravity measurements from satellites such as GRACE (Wahr *et al.*, 2000), it is hoped that the error in mass-balance determination from this approach will be considerably reduced in the near future (Velicogna & Wahr, 2002; Wahr *et al.*, 2004).

The second approach to mass-balance determination is known as the component or flux divergence approach (Rignot & Thomas, 2002). Here, the flux of ice crossing some line or 'gate', perpendicular to the direction of flow, is compared with the accumulation of ice upstream of this line. A convenient location to determine ice flux, for glaciers with a floating tongue, is the grounding line, GL. The data required for this approach comprise: location of the GL, ice thickness and surface velocity at this location and net mass balance (the accumulation minus ablation, integrated over the upstream catchment area). The location of the GL has been successfully identified using InSAR techniques, which have also been used to derive the velocity field (Rignot, 1996). Ice thickness can be obtained either from *in situ* observations or assuming hydrostatic equilibrium for the floating ice and inverting elevation to give thickness, if the ice and seawater densities are known. Accumulation rates must, in general, be determined from sparse *in situ* data, possibly combined with PMR brightness temperatures (Vaughan *et al.*, 1999a). At present, accumulation rates are one of the main error sources for the flux divergence approach. Nonetheless, the approach has been used on a number of outlet glaciers in Greenland and Antarctica, providing improved estimates of mass balance on an extensive, regional basis (Rignot *et al.*, 1997; Joughin & Tulaczyk, 2002; Rignot, 2002b; Rignot & Thomas, 2002).

73.4.3 Glacier mass balance and the global glacier inventory: GLIMS

The two methods discussed above for determining ice-sheet mass balance can be, and have been, used to determine the mass balance, not only of ice sheets, but also glaciers. Airborne laser altimetry was combined, for example, with cartographic maps, derived from aerial stereo photogrammetry from the 1950s, to estimate dh/dt over a period as long as 40 yr for some 67 Alaskan glaciers (Arendt *et al.*, 2002). The results suggested a mass wastage much higher than previous estimates, equivalent to twice the loss from the Greenland ice sheet over the same period. A similar approach was taken for the Southern Patagonian Icefield but using a DEM derived from SRTM data instead of laser altimetry and comparing this with historical cartography (Rignot *et al.*, 2003). Again, the results showed thinning rates far greater than expected, based on previous mass-balance estimates. In this study, the accuracy of the SRTM data was found to be around ±2 m. Thus, these data could have considerable value for use in elevation change measurements for other low-latitude glaciers, where reasonable historical cartographic data exist. Unfortunately, the northern limit of SRTM was 60°N, missing all of the major glaciated areas of the Arctic such as Svalbard, Greenland, Severnaya Zemlya and so on. For subpolar glaciers, the flux divergence approach has been less useful as there is, in general, no well-defined gate across which the flux can be measured, ice thickness is rarely known and difficult to infer, as is the net mass balance.

Subpolar glaciers are clustered within specific mountain regions. They can range in size from ca. 1 km^2 to ca. 1000 km^2 and can be partially (or even wholly) debris covered, snow covered or bare ice. The problems associated with monitoring the behaviour and mass balance of these types of ice mass are, therefore, often different compared with the ice sheets of Greenland and Antarctica. Thus, although the instruments used may be common, the approach may not be. For example, velocity data can be obtained using either feature tracking or InSAR techniques and these methods have been used to monitor the development of a surge in an Arctic glacier (Luckman *et al.*, 2002). To use these data, however, to determine mass balance requires, as mentioned, the addition of estimates for accumulation, ablation and ice thickness: data that are rarely available for more than a handful of glaciers. Although elevation change estimates have been obtained for some areas, as mentioned above, in many cases no historical cartographic data exist. Consequently, the preferred approach for assessing, indirectly, the mass balance of these 'data sparse' regions has been to monitor, remotely, changes in spatial extent and/or parameters such as the equilibrium line altitude (ELA). This is the principal aim of an international programme known as Global Land Ice Monitoring from Space (GLIMS). The objectives of GLIMS are to establish a global inventory of land-ice characteristics, including surface topography, to measure the changes in extent of glaciers and, where possible, their surface velocities (Paul *et al.*, 2002). The project is designed to use data primarily from the ASTER instrument, supplemented with Landsat ETM+ imagery where necessary/appropriate. The key objective is to establish a baseline digital inventory of glacier extent for comparison with measurements at later dates (Paul *et al.*, 2002). It is the first project of its kind, with a global remit, relying on some 23 regional centres (all separately funded), which have a responsibility to manage the data acquisition and processing for specific areas.

The aim is to achieve annual coverage based on four to five acquisitions during a year. To reduce the influence of seasonal effects, sampling should be achieved at around the same time of year. This problem is illustrated in Fig. 73.1, which shows the variability in surface characteristics common to many glaciers and the issues associated with attempting to undertake a consistent estimate of glacier extent. Highly automated, tailored software is required and is being developed by the GLIMS consortium (Paul *et al.*, 2002) with the aim of allowing automatic classification of terrain into glacier and non-glacierized surfaces.

73.5 Concluding remarks

The small sample of applications discussed above are, inevitably, highly selective, but serve to illustrate the immense capabilities of satellite remote sensing for applications related to glaciology, and also some limitations. Great advances have taken place in the past decade, particularly in two key areas. The first is active radar remote sensing, instigated by the launch of ERS-1 in 1991, which carried both a SAR and radar altimeter onboard. The second major advance has been in the use of laser altimetry, both on airborne and, since 2001, satellite platforms. This is a result of advances in laser technology but also, crucially, in platform navigation and pointing accuracy. Both laser and radar sensors still have much more to offer the glaciological community. ICESat was only launched in 2001 and will, despite problems encountered with the laser subsystem, provide a wealth of new data on land- and sea-ice characteristics. CryoSat-2 should in due course also offer many new insights. Both are altimeter missions with a primary focus on ice, and this is a current area of strength in cryospheric remote sensing. In fact, these are the first satellite missions with a specific focus on the cryosphere, highlighting the importance of understanding and monitoring this sensitive component of the Earth system. Beyond these two missions, however, the future is less clear. There are, for example, currently no planned SAR missions that provide the right sort of conditions for optimum use of interferometry over ice and there still remain gaps in the capabilities of current and planned missions for monitoring changes to the cryosphere. Observations of summer sea-ice, for example, are problematic, with large measurement errors in concentration and extent and no agreed 'gold standard' for the most appropriate algorithm. Mass-balance measurements of land ice have improved but the error budget is still too large to determine whether the Antarctic ice sheet (the largest ice mass on the planet by an order of magnitude) is growing or shrinking (Rignot & Thomas, 2002). There is still no satisfactory approach for remotely determining accumulation rates over the vast expanse of the ice sheets although recent research shows promise in helping to address this (Winebrenner *et al.*, 2001). Glacier mass balance from space still provides many challenges, which have no immediate solution and the space agencies are coming under increasing pressure to support ever more specialized satellite missions. Nonetheless, one thing is certain, the next decade will certainly provide many surprising, important and unexpected advances in our ability to observe and understand the cryosphere from space.

Acknowledgements

I would like to thank Toby Benham for providing the ASTER data used to produce Fig. 73.1 and Ron Kwok for his permission to use some of his material on sea-ice remote sensing and Chris Shuman for providing Fig. 73.3.

SEVENTY-FOUR

Interferometric synthetic aperture radar (InSAR) study of the Northeast Greenland Ice Stream

Ian Joughin

Polar Science Center, Applied Physics Laboratory, University of Washington, Seattle, USA

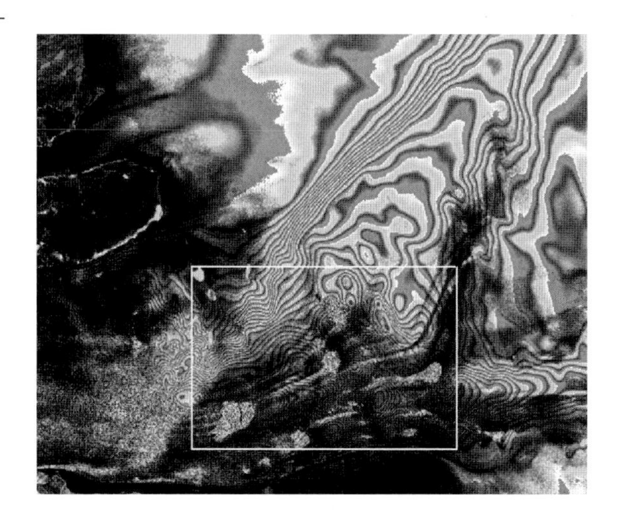

The Northeast Greenland Ice Stream (NEGIS) provides a good illustration of the utility of interferometric synthetic aperture radar (InSAR) for studying ice dynamics and ice-sheet mass balance. This 700-km-long ice stream was discovered only in the mid-1990s when its presence was detected through the analysis of features visible in SAR imagery (Fahnestock *et al.*, 1993). Soon after velocity on a central portion of this ice stream was mapped using InSAR (Kwok & Fahnestock, 1996). This was followed by a complete mapping of the velocity field using many ERS-1/2 ascending and descending passes (Joughin *et al.*, 2000, 2001) as illustrated in Plate 74.1.

The velocity map (Plate 74.1) reveals a pattern of organized flow extending deep into the interior of the ice sheet where the ice stream first manifests itself as a roughly 15-km-wide area of enhanced flow within about 100 km of the ice divide. Although the speed is initially only about 10 m yr^{-1} greater than that of the surrounding ice, the enhanced flow is delineated by well-defined shear margins. A subsequent study has revealed that this flow onset is coincident with an inferred area of enhanced (10–15 cm yr^{-1}) basal melt that may be the result of a geothermal heat flux 15–30 times the continental background average (Fahnestock *et al.*, 2001).

Downstream from the onset, the width of the ice stream increases to approximately 60 km, with flow speeds of 50–60 m yr^{-1}. At this point, the ice stream is visible in ice sheet DEMs as an area of rougher surface topography. At about 300 km downstream from the onset, the flow begins to concentrate on the southeastern side of the ice stream. This narrowing of the main trunk results in an increase in speed to approximately 125 m yr^{-1}. About another 240 km further downstream, the ice-sheet surface flattens out forming a large ice plain with significantly smoother topography. Just upstream of the ice plain, the flow divides with a southern branch feeding Storstrømmen and a northern branch flowing through the ice plain to feed the outlet glaciers of Zachariæ Isstrøm and Nioghalvfjerdsfjorden.

Although InSAR and other spaceborne sensors provide a detailed view of ice flow at the surface, they do not directly reveal the internal dynamical controls on ice streams. Inverse methods can be constrained by velocity and other data sets (e.g. thickness and elevation) to determine the basal shear stress (MacAyeal, 1992b). Such methods have been applied to the velocity data shown in Plate 74.1. The resulting inversions suggest that the bed beneath the ice plain is weak and can only support basal shear stresses in the range of 10 to 15 kPa (Joughin *et al.*, 2001). This weak bed is located in an area where the bed lies several hundred metres below sea level, and as a result, may be an area where weak deformable sediments have collected. Thus, the ice plain, which represents a significant departure from a typical parabolic ice sheet profile, is probably the result of a weak bed without the ability to support larger driving stresses, and hence steeper slopes.

The velocity data shown in Plate 74.1 have also been used in conjunction with ice thickness data from an airborne radio echo sounding system to estimate mass balance at several points along the ice stream. The results suggest that, at present, the ice stream is roughly in balance with its catchment and there is little or no significant thinning of the inland parts of the ice stream. Nearer the coast, Storstrømmen is thickening following a surge in the early 1990s (Mohr *et al.*, 1998). In contrast, laser altimeter results suggest thinning of a few tens of centimetres per year may be occurring on Zachariæ Isstrøm and Nioghalvfjerdsfjorden (Krabill *et al.*, 2000).

Although ice sheets are often considered to have dynamic response times measured in centuries or millennia, InSAR observations have revealed a surprising degree of temporal variability at decadal and subdecadal time-scales (e.g. Mohr *et al.*, 1998).

One example of such variation is the Ryder Glacier minisurge that took place over a few weeks in 1995 (Joughin *et al.*, 1996). Plate 74.2 shows interferograms acquired in late September and October 1995. The much greater density of the fringes on the fast moving areas of the glacier in the October image indicate a more than threefold increase in velocity. Subsequent interferograms (not shown) show that by early November the glacier had slowed down, returning to its normal speed.

Over the past decade, InSAR has evolved rapidly from a tool that could detect motion on isolated regions of a glacier to the point where now vector velocity maps spanning multiple ice streams can be generated easily. Recent acquisition campaigns such as the Modified RADARSAT Antarctic Mapping Mission have provided extensive new data sets. After highly successful missions, many of the early generation of spaceborne SARs have ceased operation or are near the end of their operational life. Future advances will largely rely on the launch of new instruments with the mission parameters favourable for ice-sheet mapping.

SEVENTY-FIVE

An overview of subglacial bedforms in Ireland, mapped from digital elevation data

Mike J. Smith*, Paul Dunlop† and Chris D. Clark‡

**School of Earth Sciences and Geography, Kingston University, Penrhyn Road, Kingston-upon-Thames, Surrey KT1 2EE, UK*
†School of Environmental Sciences, University of Ulster, Coleraine, BT52 1SA, Northern Ireland, UK
‡Department of Geography, University of Sheffield, Winter Street, Sheffield S10 2TN, UK

With the gradual release of the near-global Shuttle Radar Topography Mission and data from other sources, digital elevation maps (DEMs) will revolutionize our approach to the visualization and mapping of glacial landforms. This case study presents reconnaissance mapping of glacial bedforms from DEM data of Ireland, to provide a simple overview and highlight the utility of the approach, guiding future work. It complements and extends detailed bedform mapping completed by Clark & Meehan (2001) and Dunlop (2004).

The DEM used for this case study was created by the Landmap project (Kitmitto *et al*, 2000) using European Remote Sensing (ERS) tandem pairs from space-borne radar (synthetic aperture radar—SAR). With a spatial resolution of 25 m, the DEM is capable of resolving lineaments, although detailed morphology and smaller forms are difficult to distinguish. The DEM visualization was performed using locally based contrast stretches (a non-azimuth based visualization technique), supplemented with traditional azimuth based relief shading (Smith & Clark, 2005). Mapping was conducted by on-screen digitizing.

Figure 75.1 presents a summary map of glacial bedforms for Ireland, incorporating drumlins and ribbed (Rogen) moraine. Some of the landform distributions and patterns conform to existing published accounts based on field mapping, but a number of new and significant observations are apparent. The recognition that the 'Irish Drumlin Belt' is more realistically described as a large suite of ribbed moraine overprinted by drumlins is helpful (Fig. 75.2). Previous results have been extended by DEM mapping (Fig. 75.1) to show the spatial extent of this large ribbed moraine field. Now that these transverse elements have been identified and mapped it is easier to understand the flow patterns in the area which had hitherto appeared complex. Ice flow producing these landforms was towards the southeast and

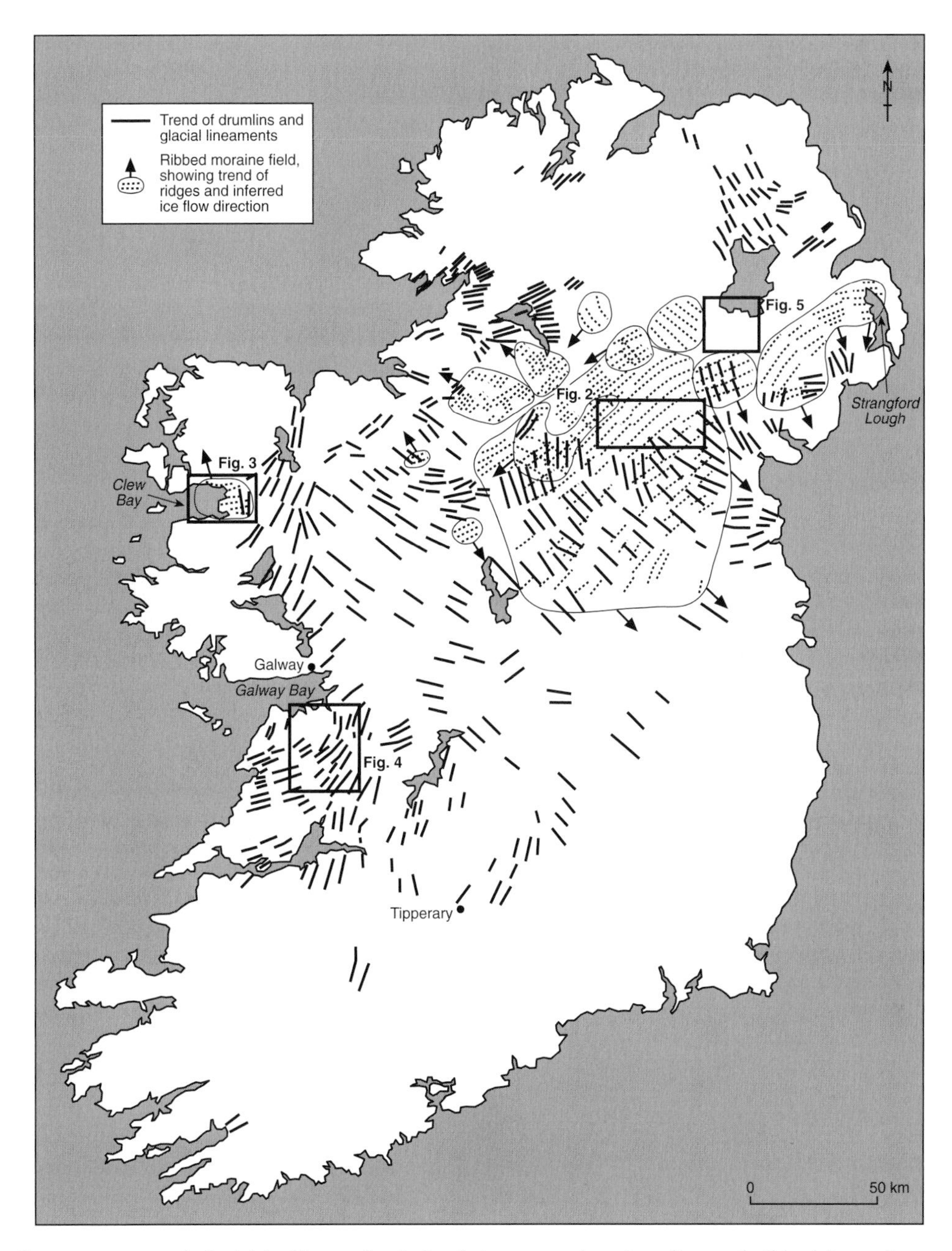

Figure 75.1 Summary map of glacial bedforms for Ireland, incorporating drumlins and ribbed (rogen) moraine including extensive ribbed moraine fields reported by Clark & Meehan (2001). Note the areas of cross-cutting ribbed moraine in north central Ireland indicating shifts in the ice divide.

their distribution over Ireland indicates that a major ice divide must have existed over the northwest Irish Coast with ice extending westwards onto the continental shelf.

Glacial lineaments (drumlins and larger streamlined forms) visible on the DEM conform to many published accounts indicating flow convergence into many of the bays and inlets of Ireland. In addition we have identified strong patterns of streamlining in Tipperary and Galway, far to the south of the main drumlin belt. In Clew Bay (half way up the west coast) a series of wave-like and irregular ridges are apparent (Fig. 75.3) that we interpret as ribbed moraine, possibly forming the western end of the main ribbed moraine belt, created beneath ice flowing towards the north-northwest. Our explanation deviates from the usual interpretation of these landforms as drumlins recording ice flow westwards into the bay (Warren, 1992). Jordan (1997) used satellite imagery to map these landforms, recording them as drumlins formed by ice flowing eastwards onshore. Figure 75.4 depicts some classic drumlins and streamlining south of Galway Bay. Surprisingly, the well known drumlins around Strangford Lough appear very poorly expressed on the DEM. We have mostly

Figure 75.2 Oblique view of the megascale ribbed moraine landscape (after Clark & Meehan, 2001). Note the arcuate planform shape and regularly spaced, anastamosing, ridges. Diagnostic features include consistent size (in relation to neighbours), downstream pointing horns and undulating ridge crests. Image is approximately 30 km across in the foreground and is looking northwards across Co. Monaghan. (Original data used to produce visualization, copyright University of Manchester/University College London, 2001.)

Figure 75.4 Oblique view of classic drumlins and larger scale streamlining just southeast of Galway Bay. Image is approximately 20 km across in the foreground and is looking northwards across Co. Clare. (Original data used to produce visualization, copyright University of Manchester/University College London, 2001.) (See www.blackwellpublishing.com/knight for colour version.)

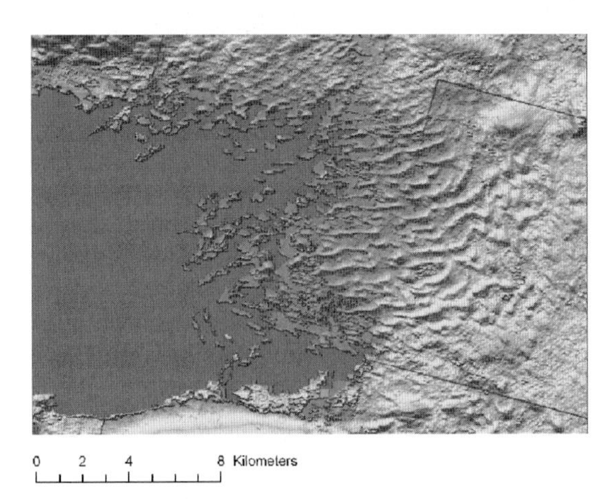

Figure 75.3 Vertical view of bedforms around Clew Bay, Co. Mayo. The DEM data clearly demonstrate that these are not drumlins but are a series of WSW–ENE trending ridges with some sinuosity, which we interpret as ribbed moraine. These exhibit preferential planform concavity towards the north-northwest (usually taken to indicate flow towards this direction). Note the superimposition of drumlins on top of the ribbed moraine on the eastern side of the image. (Original data used to produce visualization, copyright University of Manchester/University College London, 2001.) (See www.blackwellpublishing.com/knight for colour version.)

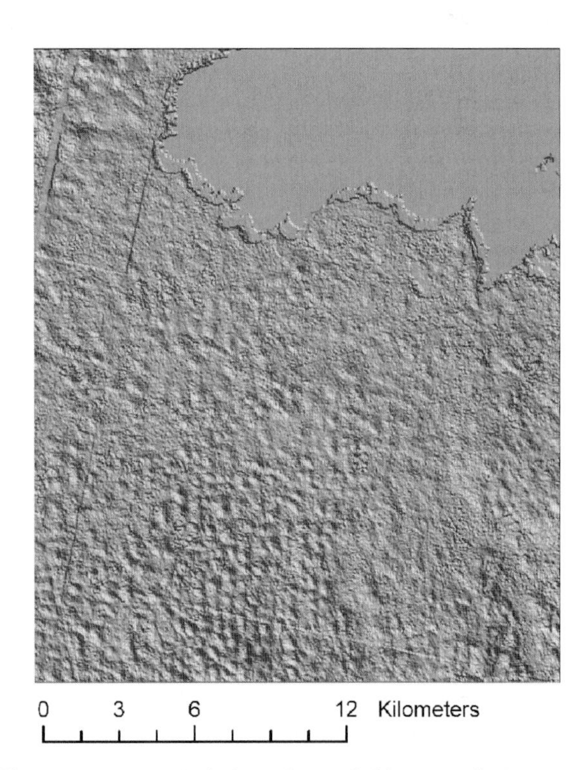

Figure 75.5 Vertical view of rounded hummocks just south of Lough Neagh, Co. Armagh. Note the ovoid nature of the landforms with no preferred orientation. Image is approximately 20 km across. (Original data used to produce visualization, copyright University of Manchester/University College London, 2001.)

interpreted this area as ribbed moraine. It might be that the field-mapped drumlins are below the resolution of the DEM or that they have been confused with ribbed moraine. In many places subglacial bedforms take the form of rounded hummocks with isolated summits (Fig. 75.5). It remains to be ascertained what these are: a new type of bedform or degraded versions of drumlins or ribbed moraine.

This case study demonstrates the use of DEMs for mapping subglacial bedforms. The ready availability, near-global coverage and unbiased visualization (Smith *et al.*, 2001; Smith & Clark, 2005) of DEMs make them suitable for geomorphological mapping. The DEM data can be quickly compiled, at low cost, for reconnaissance mapping or detailed surveys.

SEVENTY-SIX

Borehole-based subglacial instrumentation

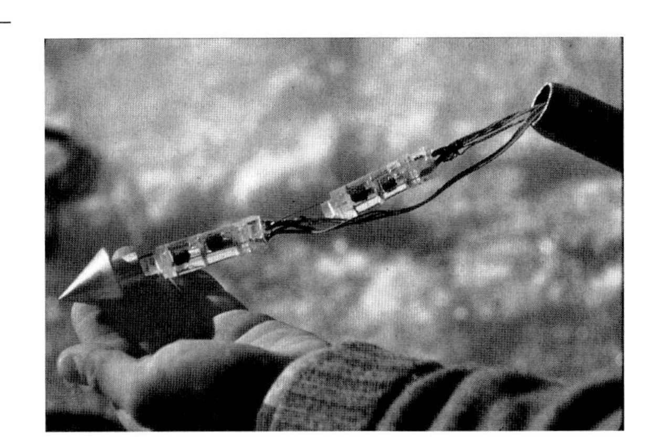

Urs H. Fischer* and Bryn P. Hubbard†

**Antarctic Climate and Ecosystems Cooperative Research Centre and Australian Antarctic Division, GPO Box 252-80, Hobart, Tasmania 7001, Australia*
†*Centre for Glaciology, Institute of Geography and Earth Sciences, University of Wales, Aberystwyth, Ceredigion, SY23 3DB, UK*

76.1 Introduction

Despite the difficulty of making observations at the base of glaciers, numerous studies of processes and conditions at the ice–bed interface have been performed. Direct observations of basal sliding have been carried out in deep, marginal crevasses (Carol, 1947), in natural subglacial cavities (Vivian & Bocquet, 1973; McKenzie & Peterson, 1975; Hubbard, 2002) and in tunnels excavated in the marginal regions of glaciers (Haefeli, 1951; McCall, 1952; Kamb & LaChapelle, 1964; Fitzsimons *et al.*, 1999). Such tunnels were also used to access basal sediments, allowing the measurement of deformation rates and styles (Boulton & Jones, 1979; Boulton & Hindmarsh, 1987). These observations were, however, mostly confined to easily accessible parts of glaciers not flooded by subglacial streams and where the overlying ice was comparatively thin. In a few cases, tunnels excavated in subglacial bedrock for hydroelectric power provide access to the glacier bed. These unique settings have enabled the installation of instruments beneath thick ice to measure simultaneously sliding velocity, shear and normal stresses and temperature on hard rock beds (Boulton *et al.*, 1979; Hagen *et al.*, 1983; Cohen *et al.*, 2000) and shear deformation, pore-water pressure and normal stress in a soft sediment layer (Iverson *et al.*, 2003).

The advent of hot-water drilling has permitted borehole access in central regions of glaciers where basal conditions are probably more characteristic of the bed as a whole. This development has allowed the observation of bed conditions and basal sliding using borehole photography (Harrison & Kamb, 1973; Engelhardt *et al.*, 1978; Koerner *et al.*, 1981) and borehole TV video analysis (Koerner *et al.*, 1981; Pohjola, 1993; Harper & Humphrey, 1995; Copland *et al.*, 1997a,b; Hooke *et al.*, 1997; Harper *et al.*, 2002). With the refinement of hot-water drilling equipment and techniques over the past 20 years, a wide range of instrumentation has been developed that can be deployed through boreholes to retrieve sediment samples and to measure directly the deformation of the substrate, the shear strength of sediments and the sliding of ice over the bed. These developments reflect the efforts of numerous researchers working at a variety of glaciers, including Trapridge Glacier, Canada (e.g. Fischer & Clarke, 2001; Kavanaugh & Clarke, 2001), Storglaciären, Sweden (e.g. Iverson *et al.*, 1995; Hooke *et al.*, 1997; Iverson *et al.*, 1999), Bakaninbreen, Svalbard (Porter *et al.*, 1997; Porter & Murray, 2000; Murray & Porter, 2001), Black Rapids Glacier, Alaska (e.g. Truffer *et al.*, 1999, 2000), Whillans Ice Stream (formerly Ice Stream B), West Antarctica (Engelhardt *et al.*, 1990a,b; Kamb, 1991, 2001; Engelhardt & Kamb, 1998) and Haut Glacier d'Arolla,

Switzerland (e.g. Hubbard *et al.*, 1995; Gordon *et al.*, 1998; Mair *et al.*, 2004).

Many of the more significant developments of down-borehole instrumentation have been published as individual papers (many in the Instruments and Methods section of the Journal of Glaciology). However, routine procedures and numerous relatively minor, but nonetheless significant, improvements to equipment remain unreported. This chapter aims at giving an overview of some of the equipment, procedures and instrumentation as used in the borehole-based investigations at Haut Glacier d'Arolla, Trapridge Glacier and Storglaciären. Such an overview necessarily describes an imperfect snapshot of continually changing technology: new and improved versions of this equipment are being developed and may already be in use elsewhere. Indeed, it is hoped that one consequence of this chapter will be to stimulate such developments further, perhaps particularly by appealing to skilled workers in other fields who may be unaware of the glaciological problems and technology involved.

76.2 Investigations of the glacier substrate

The borehole-based studies undertaken at Haut Glacier d'Arolla, Trapridge Glacier and Storglaciären have involved investigations of glacier hydrology, glacier hydrochemistry and glacier motion. As part of these investigations, we have drilled and instrumented numerous boreholes to study the glacier substrate. Specifically, the sediments accessible at the base of these boreholes have been recovered directly by subglacial sediment sampler, and investigated indirectly by penetrometer, ploughmeter, dragometer, bed tilt cells and drag spools.

76.2.1 Hot-water drilling

The drilling of boreholes through the glacier is achieved by melting ice with a pressurized jet of hot water (e.g. Gillet, 1975; Iken *et al.*, 1977). In principle, the method entails that water drawn from a supraglacial stream, water-filled crevasse or a pit dug into the soaked snowpack is heated to near the boiling point with a unit consisting of a water circulation coil and a burner and subsequently pumped with a pressure of up to 100 bar through a high-pressure hose into a drill stem. The drill stem is a heat-insulated, double-walled steel tube, typically between 1.5 and 3 m long, which ends in a nozzle having an inside diameter of a few millimetres. The hot water jet ejected from the nozzle melts away the ice in front of it at the base of the lengthening borehole.

76.2.2 Water pressure sensor

Owing to the strong interdependence of subglacial hydrological conditions and mechanical processes and properties at the glacier base, the analysis of data from borehole instruments cannot be complete without some means of measuring subglacial water pressures. Boreholes drilled to the bed are often assumed to act as manometers, allowing the water levels within them to be interpreted as subglacial water pressures. Most water-pressure transducers record the resistance of strain gauges bonded to a diaphragm as it flexes in response to the pressure differential imposed across it. The inner cavity of such transducers is commonly sealed at a fixed pressure (often termed 'absolute' or 'sealed-gauge' devices in contrast to 'relative' devices, in which a gland in the supply cable maintains the cavity at atmospheric pressure). Water-pressure transducers, which are typically 4–5 cm long (with the open pressure port at their base) and 2–3 cm in diameter, may be purchased as sealed units with cable attached (expensive but effective) or as open units that may be cast in resin, leaving only the pressure port exposed. Clear resin is recommended for home-potting, such that the cables and contacts may be viewed in the event of a problem. The most common problems encountered in using such sensors are shorting by water due to imperfect sealing, and damage to the conductor strain-gauge contacts within the body of the transducer. These strain-gauge contacts may be particularly fragile and damage may be caused by water ingress, build-up and discharge of static electricity, and rough handling.

We suspend our sensors on their cables in the borehole ca. 0.25–0.5 m above the bed. If a sensor is positioned too low, it may become packed with debris or sheared by glacier motion; if a sensor is positioned too high, the borehole water level might drop below it during periods of low subglacial water pressures.

76.2.3 Subglacial hammer

The subglacial hammer supplies the impulse to any instrument that is designed to be inserted by force into the glacier substrate. Hammer design should minimize the chances of damaging, or becoming entangled with, the cable of the instrument being inserted. The hammer should also be robust, and easy to use and repair. It is also desirable, but not necessary, for the hammer to provide an impulse of consistent magnitude.

Subglacial hammers are controlled from the glacier surface via a single cable that is used to raise and lower an impulse weight (the 'hammer') on to an anvil, to which the subglacial instrument to be inserted is attached. The hammer is free-running relative to the anvil, allowing an impulse to be transferred at the base of both the downstroke and the upstroke. The latter is required, for example, in order to detach a hammer from an inserted instrument to which it is attached by a frictional coupling or to remove an instrument that has been hammered into the subglacial material such as a penetrometer or sediment sampler (see below). The hammer must also be attached to the anvil, allowing the entire assemblage to be lowered down and raised up the borehole on a single cable. Most hammers currently in use are adapted versions of that designed by Blake *et al.* (1992) for use at Trapridge Glacier. According to this design, a cylindrical outer hammer slides up and down an inner rod of approximately twice the hammer's length, being held in place by stoppers at the ends of the rod.

The subglacial hammer used at Haut Glacier d'Arolla (Fig. 76.1) involves one notable development on Blake's original design in that the hammer stroke is restricted and guided by two retaining pins protruding from the inner rod that run along grooves in the outer hammer. This adaptation enables simpler wiring attachments (just one attachment point at the top of the instrument) and a reduction in total instrument length of ca. 50%, as the rod and hammer are now of similar length. The lower tip of the rod houses the anvil and its end is threaded for the

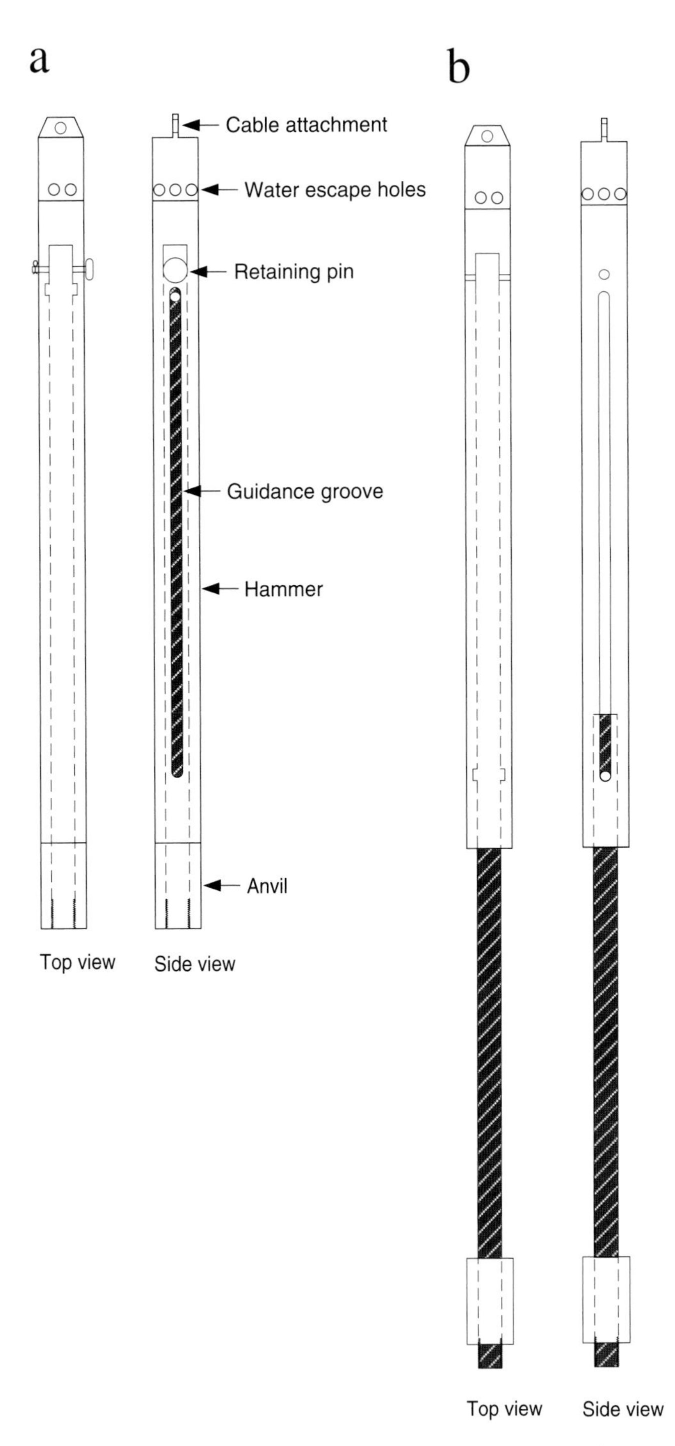

Figure 76.1 Sketch of subglacial hammer: (a) closed position for transport and storage; (b) ready for instrument insertion (open position).

attachment of subglacial tools. The hammer slides freely along the rod, but cannot detach completely from it, being restricted at the base of the stroke by the anvil and at the top of the stroke by two retaining pins. The length of the grooves in the hammer dictates the maximum length of the hammer stroke, ca. 50 cm in the case of the hammer used at Haut Glacier d'Arolla. Hammering, and raising and lowering the hammer along the borehole, requires only a single length of wire: 1.5 or 2.0 mm diameter, multistrand stainless steel wire serves the purpose well. The wire is attached via a loop to the top of the hammer. For transport and storage, the hammer is collapsed to its minimum length and held in place by a retaining pin (Fig. 76.1a). The screw attachment point for subglacial instruments may also be protected from damage during transport by screwing the anvil closed over its length.

The depth to which an instrument has been inserted into the subglacial material can be measured by reference to a mark on the hammer wire (electrical tape suffices for this purpose). The position of the edge of the tape is then noted before and after insertion relative to a fixed point, normally the rim of the borehole at the ice surface, and the depth of insertion is given by the difference between the two readings.

76.2.4 Subglacial sediment sampler

The ideal down-borehole sediment sampler should be capable of recovering large, representative sediment samples from the base of water- or air-filled boreholes. Most current samplers used on temperate-bedded glaciers are based on a design of Hooke *et al.* (1997) for use at Storglaciären. It is formed from a hollow inner barrel with two windows and an outer sleeve that can slide up and down the length of the barrel to reveal or cover the windows (Fig. 76.2). The barrel is 40 mm in diameter and ca. 150 mm long, and its base is conical to ease insertion into the material to be sampled. The opposite end of the barrel screws directly onto the threaded end of the subglacial hammer. The sampler works on the principle that friction with the sediment rising up the sampler during insertion raises the sleeve relative to the barrel, revealing the windows into which sediment moves. Upon removing the instrument the sleeve falls back into place, closing the windows and trapping the sediment inside the barrel as it is raised back up to the glacier surface. On a cautionary note, however, one must bear in mind that the sleeve does not necessarily prevent some loss of fines when the sampler is retrieved through a water-filled borehole.

76.2.5 Penetrometer

A penetrometer can be used to obtain an estimate of the minimum sediment thickness. The method involves the pounding of a pointed, hardened steel rod into the sediment. As described by Hooke *et al.* (1997), a weight that is free to slide up and down along the rod can be raised by a single wire operated from the glacier surface, and then dropped onto a collar welded ca. 0.5 m above the point. At Haut Glacier d'Arolla, we make use of the percussive force provided by the subglacial hammer to drive a ca. 0.3 m long steel rod, with a hardened conical tip that is screwed directly onto the hammer's threaded end, into the sediment. In practice, pounding continues until no further progress in penetration is made. However, there is no way of knowing whether penetration ceases because the penetrometer encounters sediment that is too compact to penetrate, a rock in the sediment, or bedrock.

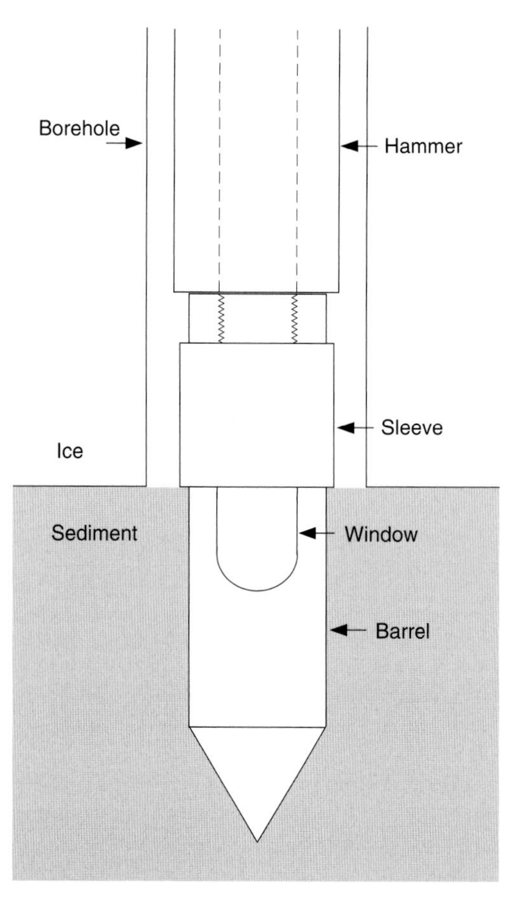

Figure 76.2 Subglacial sediment sampler during insertion. As the sleeve is raised, the windows (one of two shown) are revealed allowing sediment to move into the barrel.

76.2.6 Subglacial tiltmeters

Shear strain of the subglacial substrate may be measured with tiltmeters. As decribed by Blake *et al.* (1992) and Porter & Murray (2001), tiltmeters are built around dual-axis tilt cells. The most robust of these cells takes the form of an arrangement of five wire electrodes that are partially submerged in an electrolyte. As the cell tilts the pattern of resistances between the electrodes changes. Subglacial tilt cells are commonly housed within short resin-filled rigid cylinders, ca. 50 mm long in the case of the cells used at Haut Glacier d'Arolla. They are inserted into the subglacial substrate, either singly or, more commonly, as offset sets of two or three cells. Where inserted as strings of cells, the lowermost cell is attached to the sediment by an anchor and to other cells by a 20–30 mm length of flexible wire (Fig. 76.3). In this way, the cells are positioned at different depths within the sediment, providing information relating to variations in strain rate with depth (Fig. 76.3b). Tiltmeters are inserted via a steel tube which screws onto the threaded end of the subglacial hammer. The cells sit within the tube (Fig. 76.3a), allowing the lip of the anchor to protrude from one end and the cables to exit from the top of the groove. After insertion, the hammer and tube are withdrawn leaving the tiltmeter embedded in the sediment. Tilt cells are calibrated on a special jig that rotates a cell abouts its long axis while the tilt of the axis is fixed at various angles. Tilt cells do not currently record tilt direction, such that the orientation of tilt is therefore commonly assumed to be in a plane defined by the direction of motion of the glacier over the substrate. Under this assumption, the net tilt can be decomposed into down-flow and cross-flow components of tilt.

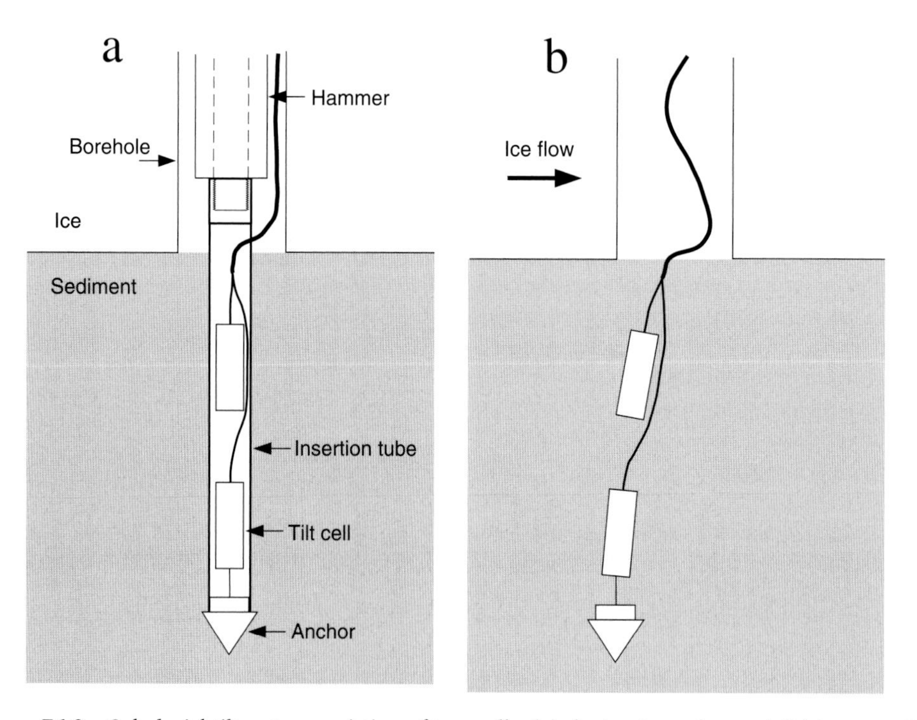

Figure 76.3 Subglacial tiltmeter consisting of two cells: (a) during insertion and (b) in operation.

76.2.7 Ploughmeter and dragometer

The ploughmeter (Fig. 76.4) is described in its current form in some detail by Fischer & Clarke (1994) who developed an earlier concept by Humphrey *et al.* (1993) for use at Trapridge Glacier. The core of the instrument is a ca. 1.5 m long and ca. 20 mm diameter steel rod, the lower end of which is machined into a conical tip for ease of insertion into the subglacial sediment. The rod is sheathed in a clear vinyl tube (internal diameter ca. 25 mm) beneath which strain gauges are bonded by adhesive to polished sections of the rod and the entire intervening space between the rod and sheath is filled with clear resin. Strain gauges are arranged as two sets of two perpendicular to one another near the base of the rod. The rod is hammered into the sediment via an insertion tool, the upper end of which is screwed to the base of the subglacial hammer and the lower end of which fits snugly into a cylindrical hole (ca. 100 mm long and ca. 10 mm wide) drilled into the top of the steel ploughmeter rod. The ploughmeter and hammer are lowered down the borehole together, with the total weight being taken by the ploughmeter cable to prevent the two instruments from becoming detached during lowering. Practice indicates that the most effective way of lowering requires two people—one to direct the instruments into the top of, and down, the borehole, and the other to walk the cables, which are extended down-glacier of the borehole to just over the length of the anticipated borehole depth, to the borehole.

Upon contacting the glacier base, the ploughmeter is hammered into the sediment to a depth that should submerge the strain gauges (Fig. 76.4a). A bending force is then imparted on the ploughmeter, and registered by the strain gauges, as the top of the ploughmeter moves with the borehole into which it protrudes while the base ploughs through the sediment (Fig. 76.4b).

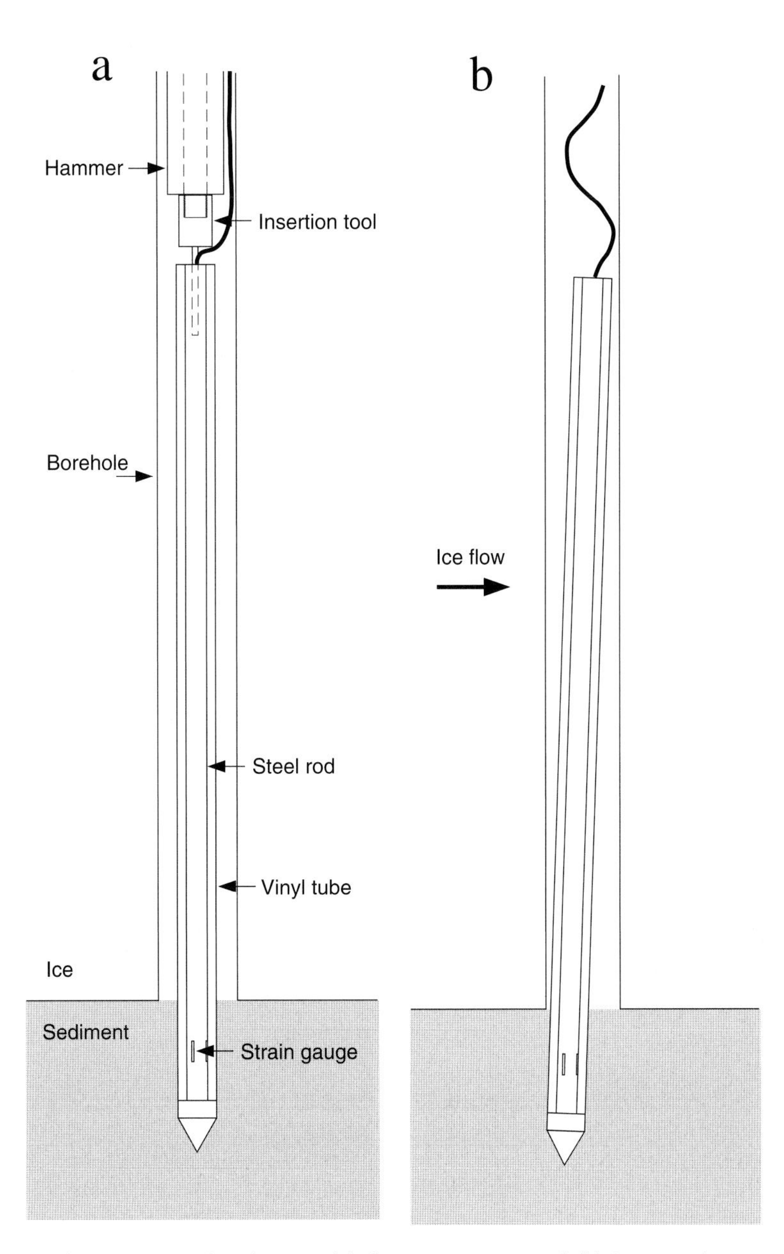

Figure 76.4 Ploughmeter (a) during insertion and (b) in operation.

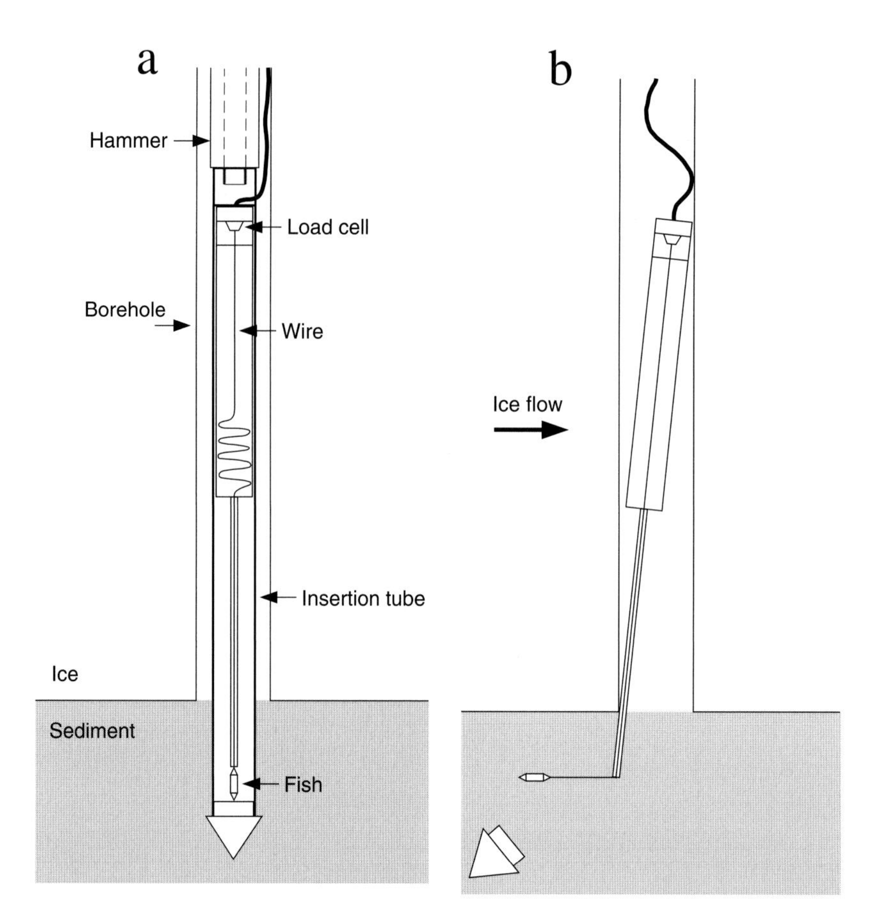

Figure 76.5 Dragometer (a) during insertion and (b) in operation (adapted from Iverson *et al.*, 1994). (Reprinted from the Journal of Glaciology with permission of the International Glaciological Society.)

The forces recorded by the strain gauges, which are pre-calibrated by bending the ploughmeter in the laboratory, can then be interpreted in terms of the strength (Fischer & Clarke, 1994; Fischer *et al.*, 1998a) and sedimentology (Fischer & Clarke, 1997b) of the subglacial sediment layer.

The dragometer (Fig. 76.5), which is an instrument closely related to the ploughmeter, was developed by Iverson *et al.* (1994) to measure sediment strength beneath Storglaciären by dragging an object through the subglacial sediment and measuring the force on the object. The instrument consists of a 100 mm long and 19 mm diameter cylinder with conical ends, dubbed the 'fish', which is connected to a steel wire. The other end of the wire is fed through the bottom end of a 1.9 m long and 19 mm diameter steel pipe and is attached to a load cell which is housed in the upper end of the pipe. Both pipe and fish are lowered down a borehole and driven into the soft substrate with the subglacial hammer and an insertion tube similar to that used for the subglacial tiltmeters (Fig. 76.5a). Eventually, owing to the forward motion of the glacier, the wire becomes taut and the fish begins to be dragged through the sediment (Fig. 76.5b). The length of the wire is chosen such that the fish is positioned beneath the ice rather than beneath the borehole as it is dragged behind the base of the pipe. This configuration has the advantage that—unlike in the case of the ploughmeter—measurements are not made at the base of the borehole where effective normal stresses are likely to be perturbed. Prior to insertion, the dragometer is calibrated by fixing the pipe horizontally and hanging weights on the wire at a right angle to the pipe axis.

Measurement of sediment strength with a ploughmeter or dragometer is only appropriate when subglacial sediment is sufficiently permeable that compression in front of the ploughing object (i.e. base of the rod in the case of the ploughmeter or fish in the case of the dragometer) does not result in pore-water pressure in excess of hydrostatic pressure (Iverson *et al.*, 1994; Fischer *et al.*, 2001; Rousselet & Fischer, 2005). Where perturbations of the ambient pore-water pressure in the sediment by the measuring device cannot be precluded (e.g. beneath glaciers that slide rapidly over impermeable subglacial sediments), the pore pressure should be measured or estimated and factored into the data analysis. Rousselot & Fischer (2003) have experimented with a new version of the ploughmeter which is equipped with an integral pressure sensor. This pressure sensor enables the measurement of pore pressure in the sediment at a pressure port located in the tip of the ploughmeter. A rudder mounted onto the tip ensures that the pressure port rotates into the direction of flow as the glacier slides forward and the tip is dragged through the sediment.

76.2.8 Drag spool

The drag spool, initially described by Blake *et al.* (1994), is designed to record basal sliding over a soft-sediment substrate. The instrument comprises an anchor, which is hammered to a

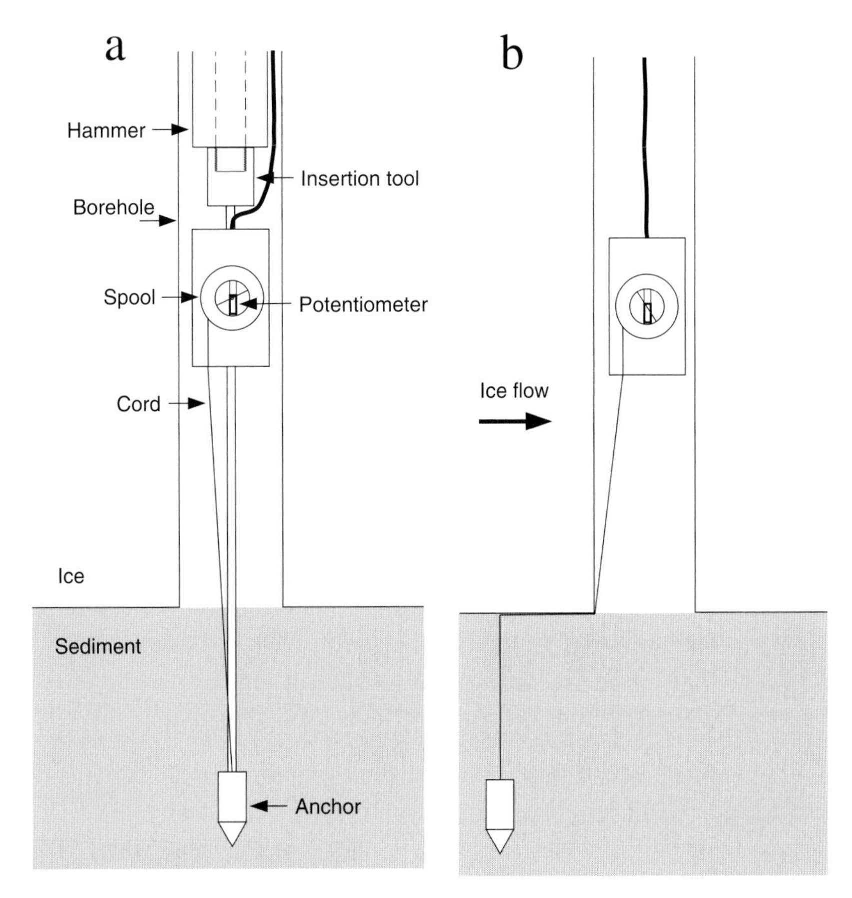

Figure 76.6 Drag spool (a) during insertion and (b) in operation.

known depth into the substrate, a multi-turn potentiometer, which is connected to a spool and housed within a protective case near the base of the borehole, and a (non-stretchable) cord linking the two (Fig. 76.6). The theory behind the instrument is that the anchor remains in place within the sediment and, as the base of the borehole slides over that sediment layer, the cord is spooled out, turning the potentiometer (Fig. 76.6b). The resistance of the potentiometer is then recorded by a datalogger located at the glacier surface, providing high-resolution time series of the length of cord paid out. However, this potential performance is tempered by some ambiguity in interpreting the resulting records. This difficulty arises principally from the precise geometries of (i) initial emplacement and (ii) subsequent motion being unknown at the glacier base. The initial insertion of drag spools is difficult, with loops of electrical insulating tape attached to the outside of the protective housing designed to slide off the insertion tool (a steel rod ca. 7.5 mm diameter) once the anchor has been hammered into the glacier bed. The end of the insertion tool inserts into a cylindrical hole drilled into the top of the anchor such that the two are held together by friction at this join and tension (ca. 1 N) in the cord linking the anchor to the potentiometer housing (Fig. 76.6a). In practice, drag spool insertion is frequently not straightforward, particularly through ice thicknesses greater than ca. 100 m or where sediment is not uniformly soft. Problems can also arise from the housing not sliding off the insertion tool easily enough, in which case the housing will come back up the borehole, paying out its cord, as the hammer and insertion tool are retrieved. The success of insertion can, however, be monitored closely from the glacier surface by measuring the resistance of the potentiometer with a handheld multimeter. Once a successful insertion is suspected and the hammer has been removed from the borehole, the potentiometer housing can be gently raised away from the bed until the potentiometer spools out, registering as a change in resistance on the multimeter. At this point the instrument is ready for use, and the cable is tied-off at the glacier surface and wired into a datalogger. Overcoming such insertion errors is a matter of practice, patience and trial-and-error. However, it is difficult to interpret the resulting data unequivocally, as the cord spooling rate will depend on the path the cord takes through the sediment between the base of the borehole and the anchor, which is unknown. The fundamental problem here is that the instrument, which is designed to measure basal sliding, requires a soft, and therefore potentially deformable, substrate for successful emplacement. This problem cannot be overcome with the current instrument design.

An interesting development based on the operating principle of the drag spool is reported by Boulton & Dobbie (1998) and Boulton *et al.* (2001a) in which four anchors (referred to as strain markers) that are individually attached by cord to spools are installed above one another in the subglacial substrate to measure relative velocities with depth. Installation through the same borehole is accomplished by pre-packing the strain markers in a column of sediment of similar granulometry to that anticipated at the insertion site beneath the glacier. The sediment column is contained in a steel sleeve and inserted into the subglacial substrate. The sleeve is subsequently withdrawn such that the sediment column and the strain markers it contains remain emplaced in the bed. As described above, the principal uncertainty in interpretation is whether the cords linking the spools with the strain markers bend with the pattern of strain within the sediment or cut through the sediment to form a straight path.

76.2.9 Instrument deployment in the glacier substrate

Deployment of instrumentation into subglacial sediment suffers from the uncertainty about the precise position of the instrument with respect to the ice–bed interface, not least because of its expected complexity. Subglacial sediment may squeeze upward into the bottom of boreholes. In this case, instruments, although inserted in sediment, are potentially installed within the body of the glacier. From our experience, this possibility can in most cases be ruled out because the shear-deformation rates implied by tiltmeter measurements exceed the ice-deformation rates that would control sediment deformation within an ice-walled hole. Equally, the high forces usually recorded with ploughmeters can be explained only by their tips being inserted into basal sediment below the ice–bed interface. Another difficulty arises from the hydraulic excavation by the hot-water drill, which is believed to loosen subglacial material to a depth of several decimetres below the ice–bed interface (Blake *et al.*, 1992). It is therefore likely that the assemblage of instrument and hammer, once lowered to the bottom of the borehole, settles into this disturbed layer simply by its own weight. In this case, the measured insertion depth represents the added penetration that results from the hammering procedure. Unfortunately, there is currently no reliable technique for measuring the overall insertion depth.

An assessment of the extent to which the disturbed subglacial material affects the depth of instrument insertion was attempted at Trapridge Glacier by driving two steel rods with the same dimensions as a ploughmeter into the soft bed. Prior to insertion, the lower 50 cm of the rods had been coated with paint. Upon retrieval 10 days later, subhorizontal scratches in the paint suggested that the total insertion depth exceeded the measured depth by ca. 20–30 mm (Fischer, unpublished data).

SEVENTY-SEVEN

Instrumenting thick, active, clast-rich till

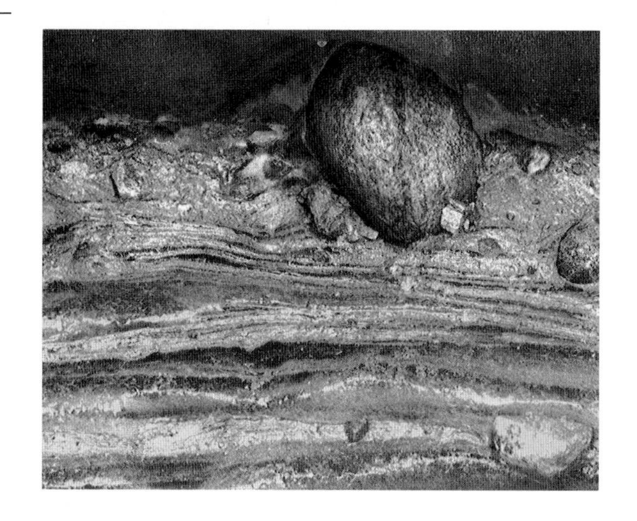

W. D. Harrison and M. Truffer

Geophysical Institute, University of Alaska, Fairbanks, Alaska 99775-7230, USA

77.1 Introduction

The bed of Black Rapids Glacier, a quiescent surge-type glacier in the Alaska Range, offers some interesting challenges. First, it underlies ice which is up to 650 m thick. Second, it consists of a till which is also relatively thick (5 to 7 m where known), it contains large clasts which inhibit penetration, and it is active in the sense that processes deep within it (rather than near the ice–till interface) seem to account for the majority of the observed surface motion (Truffer *et al.*, 2000). This underscores the importance of basal processes deep within the till, both in the quiescent and surge states (Harrison & Post, 2003). The bed can be reached by standard hot-water drilling techniques, and instruments can be placed within it with the help of a large wire-line drill rig (Truffer *et al.*, 1999). However, the expense of the rig, and its difficulties, motivated a search for an alternative.

77.2 The hammer

After considering several possibilities, we decided to place instruments in the till (at a site where the ice is 500 m thick) with the help of a large, commercially available down-hole hammer, which was designed so that its action was not inhibited by the cushioning effect of water. The active weight of the hammer is adjustable; we usually used 360 kg. The most basic problem is to determine

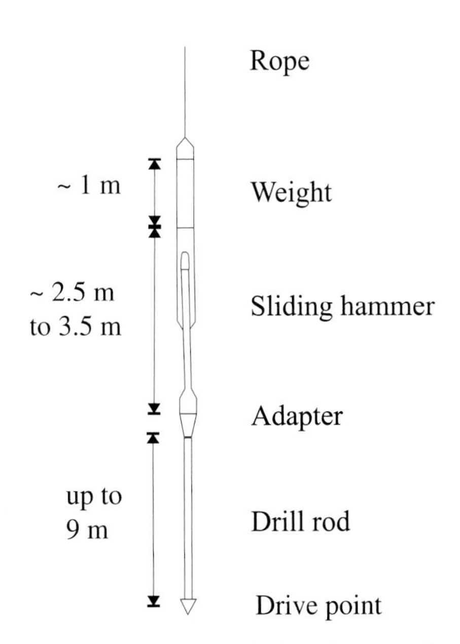

Figure 77.1 Sketch of the borehole tools. Up to three probes can be placed inside the drill rod. More would interfere with the couplings between the 3 m rod sections. After hammering the probes are pulled out by the detachable drive point that acts as anchor. The hammer (shown in extended position) is connected to the drill rod via an adapter. A variable amount of weight is added to increase the active hammering weight. The hammer is accentuated with a composition rope.

whether a light rope or cable can be used to actuate the hammer from the surface. We decided on a 12 mm diameter rope after a full dynamical calculation of the effects of its stretch on hammer operation. Part of the problem is to determine the position of the hammer with respect to its anvils, of which there are two to enable it to hammer either up or down. The surface equipment consists of a tower to assemble down-hole equipment, a heavy tripod to guide the rope down the hole, and a 37 kW engine driving either a drum containing the rope when raising and lowering, or a cathead (a rotating friction drum) to engage the rope (or a short piece of natural fibre rope attached to it) when hammering. The total weight is 2.3 t.

The instrumented probes are placed in a section of 67 mm diameter drill rod (9.5 mm wall thickness, Fig. 77.1), after being tied to each other and to a detachable drive point, all of which is driven into the till by the hammer. Upon reaching the maximum depth, the action of the hammer is reversed, and the drill rod hammered out. The drive point has a larger diameter than the drill rod and remains behind, pulling the probes out of the drill rod as it is retracted. The maximum till depth reached was about 2.5 m.

77.3 The probes

Our probes were shock mounted to survive damage or calibration loss during hammering (Fig. 77.2). They measured pressure

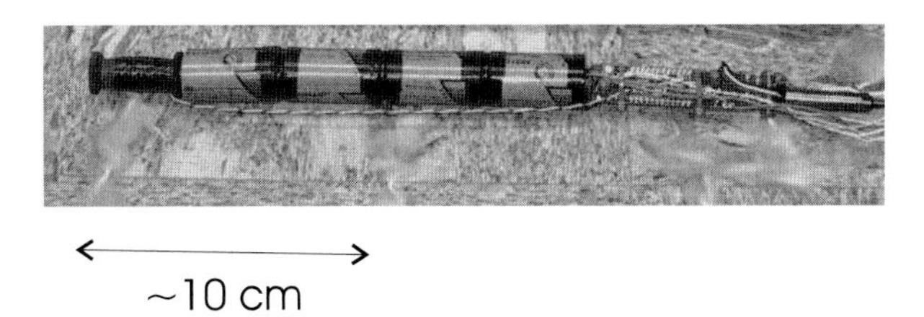

Figure 77.2 Interior of probe, consisting from left to right of transmitting coil, four lithium cells, electronics and tilt sensors, and pressure transducer. This assembly is then cast in epoxy, wrapped in styrofoam and packed into a casing. The casing is 610 mm long and 44 mm in diameter. The wires will be cut after the probe is programmed.

and two axes of tilt, which were sent, without wires, to a receiver just above the ice–till interface using a low-frequency magnetic field. We would have used a relatively high-frequency electromagnetic field, which might have avoided the need for a down-hole receiver, except for the difficulty of transmitting through the heavy metal cases, which we thought might be necessary for survival of the probes in active till. The question of whether plastic casings would survive in such conditions remains open. At any rate, the wireless capability avoided the difficulty of operating a heavy moving hammer in the presence of signal cables, and was essential in our judgement. The probes were designed to transmit data once a day for a year. Data reception was intermittent owing to relatively straightforward electronic problems that could not be fixed in our single season of testing (beginning in April 2002). More details are given by Harrison *et al.* (2004).

77.4 Status

This work showed that a heavy down-hole hammer can be operated from the surface with a long light rope, and that wireless communication from instrumented till probes is feasible. Without a drill rig, a heavy hammer seems at present to be the best tool for inserting instruments into till.

Nevertheless, penetration of clast-rich till is still a problem. We penetrated only a about a third of it, although more progress could probably have been made with more weight on the hammer. A more subtle and perhaps ultimately more difficult problem is to determine the nature of the ice–bed interface ('diffuse' or 'sharp' in the simplest terms), and exactly where the probes are located with respect to it. This is a problem first because hot-water drills can penetrate sediment, and second because loose material sometimes intrudes into a borehole, as is well known in commercial drilling. The mechanical problems of penetration and location of the interface are more difficult than the more glamorous one of wireless communication.

Remembering that till is not a continuum, the ultimate unanswered question is what types of sensors should be in the probes, and what the dimensions of the probes and their separation should be. An iterative process probably will be needed. As deformation mechanisms become better understood, the instrumentation can be optimized to study them.

SEVENTY-EIGHT

Ice-core chronology

Claus U. Hammer

Niels Bohr Institute, University of Copenhagen, Geophysical Department, Juliane Maries Vej 30, DK 2100 Copenhagen, Denmark

78.1 Introduction

The major ice sheets of Antarctica and Greenland play an important and special role in palaeoclimatic research. These kilometre-thick ice deposits owe their existence to precipitated snow. The snow compacts into glacier ice, which deforms during the flow of the ice sheet. The number of years covered by the cores depends on how deep and where the cores were drilled. Drill sites with low annual snow accumulation are also the sites where the oldest ice can be retrieved; the EPICA deep ice core from East Antarctica is presently the ice core that covers the most years back in time, i.e. 740 kyr (EPICA Community Members, 2004). Several other deep ice cores from East Antarctica add to the geographical coverage, i.e. the Dome Fujii, Vostok, Dronning Maudland and Dome C cores.

In West Antarctica the Byrd, Taylor Dome, South Pole and Siple station cores have contributed important palaeoclimatic information: owing to the higher annual snow accumulation in West Antarctica compared with the plateau of East Antarctica the information in West Antarctic ice cores is mainly limited to the last glacial cycle, but they offer a better seasonal resolution and hence a more precise dating by seasonal stratigraphy (see e.g. Hammer *et al.*, 1994). Many more cores to shallow depth have been obtained from Antarctica in order to improve the geographical coverage of more recent times.

In Greenland the Camp Century, Dye 3, GRIP, GISP2 and North GRIP deep ice cores provide the information source for the last glacial cycle on the main ice sheet. Also in Greenland many more ice cores have been retrieved, but they are mainly cores drilled to shallow or intermediate depths.

Palaeoclimatic and environmental information from ice cores is not limited to the polar ice sheets and many important ice cores have been obtained from glaciers and ice caps. I have, however, in the following chosen to concentrate on the chronology of a few deep ice cores as they offer detailed information over the last glacial cycle. This also ensures that the chronology of abrupt climatic changes (DO events, Dansgaard *et al.*, 1993) during the last glacial stage is included.

Before presenting and discussing some of the most up to date chronologies of the polar ice cores it is appropriate to consider briefly some general approaches to the dating of ice cores.

78.2 Dating of ice cores—choice of approach

In the more central parts of the polar ice sheets, except close to the bedrock, the ice flow has left the chronological order of the past snow deposition intact although the original annual surface snow accumulation has been thinned as the ice moved vertically downwards.

78.2.1 Model chronologies

Models for this thinning with depth involve time, and the more realistic the ice flow model is the more realistic the model time-scale for a retrieved ice core will be. The number of annual ice layers from the ice surface to a certain depth z determines the age $t(z)$ of the ice. The annual ice layer thickness $a(z)$ is controlled by the ice flow, which depends on many parameters changing over time. Model time-scales therefore need a number of realistic assumptions in order to be useful. Under relatively simple boundary conditions and reasonable empirical information input, model time-scales can be quite helpful, but to judge their accuracy is not a straightforward matter. In fact a three-dimensional ice-flow model is not necessarily more accurate than a two-dimensional model; the extra information needed can be quite substantial and additional assumptions must be added to the model.

The accuracy of model time-scales does of course depend on the circumstances, i.e.: is the ice flow relatively simple? How is the change of snow accumulation during the glacial cycle treated in the model? What flow law is used for the ice and is the the temperature profile of the ice sheet included? The model can also be confined by adding boundary conditions obtained from other studies. Model chronologies have the advantage that they deliver a date for any depth in the core. Also they can be modified easily if more precise or new input parameters later become available. The disadvantage is that a number of assumptions and simplifications must be included in the model; these eventually will determine the accuracy of the time-scale. The accuracy can in principle be estimated by changing the input parameters and assumptions within the margins determined by our knowledge on the possible errors related to these.

The use of analytic models can be quite helpful in some cases, but such models cannot live up to the present demands of a high dating accuracy as they are based on too simplistic assumptions. The interested reader can find useful information on such models in *The Physics of Glaciers* (Paterson, 1994).

78.2.2 Seasonal stratigraphy chronologies

Another approach to ice-core dating is seasonal stratigraphy. In ideal cases this technique offers a tree-ring-like accuracy. The prerequisite for dating by seasonal stratigraphy is a site with a reasonably high annual snow accumulation, which a priori excludes a whole range of drill sites, for example, the Vostok core on the East Antarctic plateau was dated by a combination of two models, one for the past annual accumulation and one for the ice flow (Lorius *et al.*, 1985).

High accumulation areas do not suffer from underrepresentation of seasonal snowfall or mixing of annual snow layers by wind action, apart from certain marginal ice-sheet sites. Although the high accumulation areas are obvious candidates for dating by seasonal stratigraphy they are, however, also areas where the older layers are thinned dramatically at great depths: a consequence of the ice flow. This is the main reason why cores from the Greenland Ice Sheet can be dated by seasonal stratigraphy only back to approximately 100 ka. The older annual layers are relatively close to the bedrock, which not only makes them very thin, but also increases the probability of breaking the ordered layering or they may simply have melted away at the bedrock.

The seasonal variations can be observed in many of the parameters studied along ice cores and their seasonality may have different explanations: Not all of them can be used as a dating tool to any depth of the core. In the literature some of the most used parameters are, for example, isotopic variation of the ice composition or concentration variations of nitrate, dust, calcium or chloride. It is, however, important to understand why the various parameters are changing with season in order to interprete the validity of the stratigraphical dating; not all apparently seasonal variations may actually turn out to be true seasonal variations. Multiple parameter dating and cross-dating may help to identify the true annual layers, but the accuracy will still be influenced by a number of 'difficult' layers. It is not the intention here to discuss the various methods, but a number of techniques and their limitations can be found in Hammer *et al.* (1978) and Hammer (1989).

The estimation of the accuracy can be judged if events described in historical documents can be identified in the core, for example a well-known volcanic eruption. This will, however, not work for layers older than a few thousand years. For older layers the accuracy of the time-scale can be verified only if it is compared with other well-dated records containing comparable events. Presently this is a controversial subject, because no 'master chronology' exists for palaeorecords (SPECMAP, which is used to compare sea-sediment records is a type of master chronology for sea sediments, but it cannot be used for comparing the more detailed events in ice records, for example, DO events).

78.2.3 Other dating methods

Another approach to the dating of ice cores is the use of radioactive decay methods, but their usefulness is limited by factors such as the half-life, the low concentration of trace substances and gases in the ice as well as complexities concerning sampling and time resolution. The reader is referred to the literature, for example, the review paper by Stauffer *et al.* (1989). Such methods can, however, serve as a rough verification of the model or seasonal chronologies, or can be used where other methods are not applicable. Radioactive dating also can be of use to ice-core chronologies in an indirect way, especially for the older part of the ice cores, which is difficult to date directly. If an event in the ice core can be reconciled with an event in marine or terrestial records and if the latter can be dated by a radiometric method the date can be transferred to the ice core.

This leads us into the use of reference horizons, which holds a great potential for tying together the chronologies of ice cores, terrestial and marine records, for example, by mutual volcanic events. The final goal is to obtain precise and accurate chronologies for all palaeoclimatic records in order to compare their information over time.

The chronological work on ice cores is presently in an interesting and exciting phase, which in my opinion makes it possible to see faintly the future chronological framework of ice cores. The framework could be viewed in the following way.

1. A 'master chronology' spanning the past 100,000 yr is constructed based on the Greenland deep ice cores from Dye 3, GRIP and North GRIP.
2. Based on common major equatorial volcanic eruptive signals in Greenland and Antarctic ice cores the two polar regions are locked into a common chronology representing the past 100,000 yr. The use of volcanic signals is important, because it offers a high number of common events, thus limiting the time periods for which interpolation between reference horizons are needed. If the master chronology has a high precision and accuracy so will the dates of the reference horizons.
3. For ice-core layers older than 100,000 yr it will be necessary to use common reference horizons between ice cores, marine and terrestrial sediments.

In the following I shall concentrate on the first point, because most ice-core chronologies deal with ages younger than 100,000 yr and a common masterchronology for these ages has not yet been agreed upon. All dates given as BP will refer to AD 1950.

78.3 Holocene chronology

During the years 1979–1982 the South Greenland Dye 3 ice core was drilled in a high accumulation area, i.e. some 1.25 m of annual snow deposition corresponding to 0.56 m of ice equivalent. Such a high annual snow deposition combined with a high number of major snowfalls per year ensures that the seasonal isotopic composition of the ice could be used as a dating technique back to approximately 8000 yr BP. The accuracy of the dating was improved by using the continuous acidity profile, electrical conductivity method (ECM) (Hammer, 1980) and in some cases also a detailed dust concentration profile, as a cross-check of the isotopic cycles.

The ECM and dust profiles were measured with a much higher resolution along the core than the isotopic profile and the diffusive processes obliterating the seasonal signal with depth are much smaller than for the corresponding isotopic profile. The isotopic seasonal changes are, however, connected to the seasonal cycle of the atmospheric circulation in a more regular way than the changes in the ECM signal and the dust concentration. Irrespective of this the two profiles were quite helpful in resolving some of the more subtle seasonal variations in the isotopic profile.

The resulting time-scale was checked by historical dated volcanic reference horizons for the past 1000 yr, which indicated an accuracy of ±3 yr for a 1000-yr-old ice layer. Recently verification of the Dye 3 Holocene dating was extended 900 more years back in time when the volcanic signal of the AD 79 Vesuvius eruption was identified in the GRIP ice layer of AD 79 (AD 80 in the Dye 3 core) (Clausen *et al.*, 1997; Barbante *et al.*, in preparation), i.e. only one year from the expected year AD 80. As the stratigraphical dating of the Dye 3 core between AD 80 and 8000 yr BP was not basically different from its dating between AD 1980 and AD 80 it can be concluded that the Dye 3 dating can serve as a master chronology back to some 8000 yr BP with a tree-ring-like precision. This was further confirmed after the GRIP deep drilling, when the major volcanic signals in the Dye 3 and GRIP ice cores were compared, based on the independent stratigraphical dating of the two cores back to 4000 yr BP (Clausen *et al.*, 1997).

The GRIP core was drilled during the years 1989–1992. The annual snow deposition on the GRIP site is only some 40% of the deposition in the Dye 3 area, which makes seasonal isotopic dating more difficult and limits a fairly accurate dating to 4000 yr BP. The Holocene part of the GRIP core, back to 8000 yr BP, therefore has been dated by using the several common volcanic signals, with the Dye 3 core as reference horizons: only unambiguous volcanic signals were used in order to avoid misinterpretations. This could be accomplished because some of the major volcanic signals in Greenland ice cores are characterized by an acid deposition, which is clearly revealed by signal strength (acidity), duration of deposition and depositional pattern over time (see Hammer, 2002). Figure 78.1 shows how the various Greenland deep ice cores were tied together around 8000 yr BP. At deeper strata the Dye 3 core is less relevant for accurate dating owing to the ice flow in the Dye 3 region: fast thinning of the annual layers and problems close to the bedrock. To proceed further back in time we need to consider the GRIP deep ice core from central Greenland.

Between 8000 yr BP and the Younger Dryas (YD) the seasonal stratigraphy of the GRIP core was superior to the Dye 3 core, because the annual layer thicknesses were four to six times larger than in the Dye 3 core. Also the analytical techniques have improved considerably since the Dye 3 drilling; this made the interpretation of the very detailed nitrate and dust concentration profiles fairly simple. The counting of annual years was checked by several independent countings and the end of the YD was established as 11,510 ± 40 yr BP with a deviance of only 5 yr between the various countings over the period 8000 yr BP to the end of the YD. The accuracy of this date is not only a function of the interpretation of seasonal variations in the Dye 3 and GRIP core, but must also include the uncertainty of defining the end of the YD: the latter has been chosen here as the time when the dust concentrations prevailing during the YD fall dramatically. This happens almost at the same time as the isotopic values of the ice are changing, but is seen much more clearly in the dust concentration profile.

The same dating technique was used to take the dating further back in time, but already during the *in situ* measurements of the nitrate and dust concentrations over the YD period it became clear that the nitrate and dust variations were not independent. The two methods therefore could not be used to cross-date the core. Hence the dating of the last glacial stage had to be performed with only one seasonal dating parameter—the dust concentration. The dust concentration was chosen because it was less apt to diffusion and could be measured with a very high resolution along the core, i.e. a few millimetres. The high resolution of the measurements made it in theory possible to date the core as long as the annual layer thicknesses were larger than 1 cm. By using a special technique it even could be used as long as the annual layers were larger than a few millimetres, but in the latter case the measurements were very time consuming.

78.4 Chronology of the last glacial stage

The GRIP and GISP 2 drillings at Summit in central Greenland (1989–1993) retrieved two ice cores to bedrock and 28 km apart, which came to play an essential role in the palaeoclimatic research of the Late Quaternary. Not only were the *in situ* measurements and sample collections more extensive than for any previous ice-core drilling, but for the first time ice deposits covering the last glacial stage became available with an unprecedented time resolution. Hundreds of papers have been published dealing with the results obtained on the Summit ice cores and the readers who want to know more are recommended to seek information in a special issue of the *Journal of Geophysical Research* (1997) on the Greenland Summit Ice Cores. Information on the chronology of the GISP 2 core can be found in, for example, Alley *et al.* (1993, 1997a) and Meese *et al.* (1997).

From a chronological point of view of the last glacial stage the Summit drillings offered 1200-m-long ice-core records covering the time between the end of the Eemian to the end of the YD. Such long records made it in principle possible to date the records by seasonal stratigraphy back to the Eemian. Technically this was done for both ice cores, although the techniques used were different.

Before the Summit drillings the Byrd ice core of West Antarctica had actually been dated by seasonal stratigraphy back to 50 kyr BP (Hammer *et al.*, 1994) using the ECM technique. The successful dating of the Byrd core owed its origin to the seasonal production of sulphur-rich products in the ocean around West Antarctica (the measurements were actually performed during the 1980s). This eventually led to a very strong seasonal acid variation in the ice layers. It was, however, a problem that the Byrd ice-core chronology could not be compared, in a strict sense, with

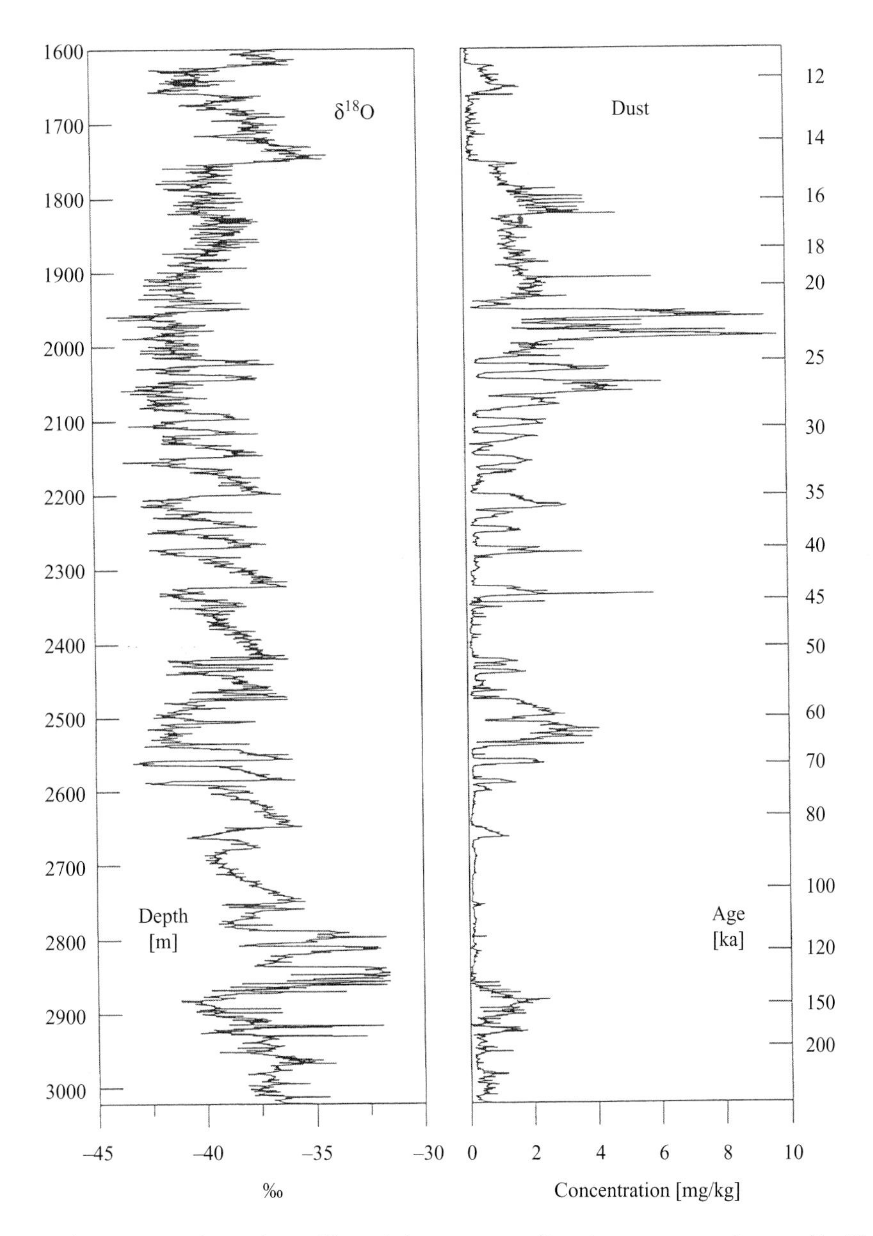

Figure 78.1 The pre-Holocene GRIP isotopic profile and the corresponding dust concentration profile. The time-scale shown to the right is the ss09 (see also main text).

other chronologies of a similar accuracy. Actually the Byrd chronology was not verified until Blunier *et al.* (1998) proved that the abrupt climate changes (DO events) during the last glacial were not in phase between the two hemispheres. Nevertheless the findings of Blunier *et al.* did not depend on absolute dating and it was still a problem that the deep ice cores from Antarctica and Greenland presented various chronologies: no agreement existed on a master chronology.

In 1996 a new deep drilling, North GRIP, started in northeast Greenland and in 1995 the EPICA drilling started in East Antarctica. These two deep drillings have now recovered two new deep ice cores and in the coming years several papers on the results will appear. Both drillings were characterized by an almost comprehensive use of experimental techniques both *in situ* and in the laboratory. The chronology of these cores will extend ice-core chronology far back in time and with new precision. Their chronology will be the result of all the new data and the knowledge already obtained on the previous deep ice cores. Before all these data become available it will be worthwhile to consider the chronologies of several deep ice cores as the situation was in 2002. Schwander *et al.* (2001) compared the various time-scales in order to discuss and show the differences. The result of this comparison over the various chronologies of the last glacial stage clearly illustrated the overall problem: the time-scales deviated up to 5 kyr! This is problematic, because the important climatic events observed in various palaeorecords must be on a more or less accurate time-scale in order to reveal the climatic teleconnections correctly. The time-scales can be criticized, but what we need is not just criticism but verifications or rejections of the various time-scales.

I have in the following chosen to present some data on the GRIP ice core, which in my opinion strongly indicate that it is indeed possible to establish a master chronology for the last glacial stage.

78.4.1 Model time-scales for the GRIP core

The first publication of the entire GRIP isotopic record used a model time-scale—now called ss09 (Dansgaard *et al.*, 1993). The ss09 time-scale was based on a model for thinning of the annual layers and an empirical relation between the past surface snow accumulations and the isotopic values in the core. Further the ss09 time-scale was constrained by fixing the age of the end of YD and the Eemian to 11.5 kyr BP and 110 ka, respectively; the first date was taken from the stratigraphical dating of the GRIP core and the second was based on information from other records.

The ss09 scale was later corrected for the influence of the isotopic composition of the ocean water and was then called $ss09_{sea}$ (Johnsen *et al.*, 2001). Recently a paper by Genty *et al.* (2003) on speleothems in France claims to be more accurate, but in my opinion the speleothem time-scale has not yet been verified sufficiently.

The main problem is that the various time-scales deviate several thousand of years over the last glacial stage, which is unsatisfactory because the abrupt and strong climatic changes demand a better time control in order to analyse the causal relationship between the various records. Whereas the dating of the Holocene is in principle solved with a sufficient accuracy, at least for some ice cores, the glacial chronology still needs to be scrutinized.

78.4.2 Stratigraphic time-scale between 11,510 and 60,000 yr BP

The stratigraphical dating of the GRIP core has hitherto been presented only in an internal report (Hammer *et al.*, 1997), but in the following I will try to show why this dating of the GRIP core is probably the most accurate dating of the last glacial stage back to 60 ka. The most important dates are presented in Table 78.1.

As mentioned earlier only the dust concentration profile was available to date the GRIP core prior to 11,550 yr BP. This immediately raises the questions 'How sure can we be that the interpretations of the seasonal variations are correct during the glacial cycle and can we verify the corresponding time-scale?'.

The interpretations of the seasonal cycles are based on the following:

1 the seasonal variations in the GRIP core are clearly annual to some extent, because there are no other reasonable explanations for the observed consistent variations;
2 a reasonable explanation of the seasonal dust concentration during the last glacial stage is the presence of extreme dryness between winter and spring and more wet summer and autumn conditions;
3 the corresponding ice time-scale can be verified by comparison with other types of well-dated records.

Table 78.1 GRIP time-scale between Preboreal and 60 kyr BP. The dates refer to the start of the stated climate periods. The IS numbers refer to the interstadials as given in Dansgaard *et al.* (1993) and the C numbers to the preceding cold stadials

Climate period	Start of period (before AD 1950)
Preboreal	11510
Younger Dryas	12704
Allerød	14136
Bølling	14872
C1	23615
IS2(a + b)	24089
C2	28130
IS3	28495
C3	29407
IS4	29799
C4	33490
IS5	33853
C5	34936
IS6	35248
C6	36152
IS7	36964
C7	38098
IS8	39846
C8	41442
IS9	41583
C9	42358
IS10	42854
C10	43492
IS11	44482
C11	47072
IS12	49101
C12	51125
IS13	51706
C13	54382
IS14	57319
C14	58400
IS15	58807
C15	59874

The first two points are based on reasonable assumptions, but their correctness cannot be proven without involving the dating of other kinds of records, hence the third point is important. In order to verify the stratigraphical dating of the GRIP core it will be helpful first to consider the transition from the Last Glacial Maximum (LGM) to the end of the YD. This is a period worth special attention as many kinds of palaeoclimatic data cover this period and it represents very abrupt and strong climate changes. The climate changes can be seen in the isotopic profile (Fig. 78.2), but are even more clearly observed as annual layer changes; these changes directly reflect the changing climate. The isotopic profile is only a quantitative indicator of palaeotemperatures if a calibration curve exists. The calibration curve will depend on the seasonal distribution of the snowfall and the height of the old ice surface; information on these parameters is difficult to obtain, although a rough calibration can be made by means of borehole temperatures (see Cuffey *et al.*, 1995; Dahl-Jensen *et al.*, 1998; Johnsen *et al.*, 2001).

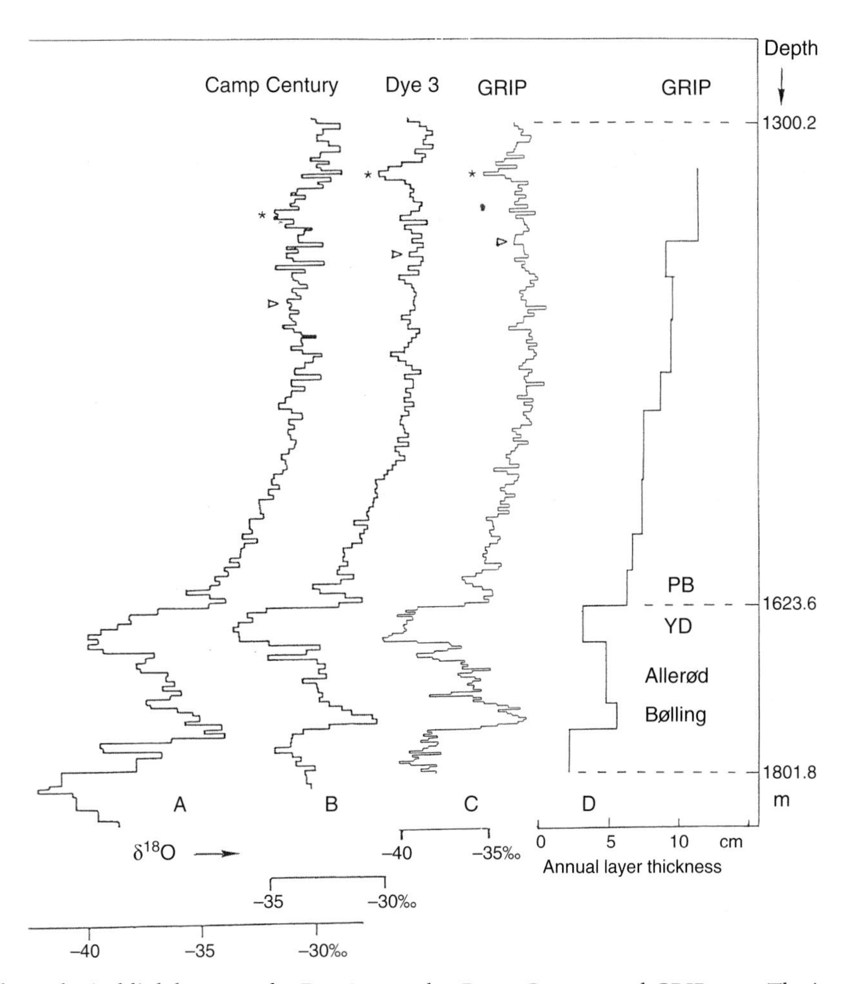

Figure 78.2 The chronological link between the Dye 3 core, the Camp Century and GRIP core. The isotopic minimum in all three records (asterisk) and a common major well-identified volcanic signal (triangle) tie the chronology of the Camp Century (A), Dye 3 (B) and the GRIP core (C) together around 8000 yr BP. All records are plotted on a depth scale, but only the GRIP depth scale is shown (to the right). Profile D indicates the corresponding annual layer thickness of the GRIP core.

Figure 78.3 shows the annual layer thickness $a(z)$ in the GRIP core over the transition from pre-Bølling to a little after the end of the YD. Even though this ice core sequence is 130 m long the change in $a(z)$ due to the ice flow is only around 10%, i.e. the variation in $a(z)$ first of all reflects the relative change in the original surface accumulation $A(t)$. The changes are dramatic both in the values of $a(z)$, nearly a factor of two, and in the rapidity of the changes. Further an accurate tree-ring date exists for the end of the YD (Spurk *et al.*, 1998) and a similar accurate U/Th date for the onset of the Bølling (Bard *et al.*, 1993). The agreement between the dates is quite satisfactory: the ice chronology indicates 11,510 ± 40 yr BP for the end of the YD and the tree-ring date 11,500 ± 30 yr BP. In the case of the onset of the Bølling the ice-core date is 14,872 ± 80 yr BP and the U/Th dating 15,000 ± 100 yr BP. With this verification it can be concluded that the GRIP stratigraphical dating is accurate enough to serve as a master chronology over the transition.

With a GRIP master chronology at hand it is of interest to compare some transitional GRIP events to similar events in other palaeorecords. Some obvious events could be: the icelandic Vedde volcanic ash layer has been dated to ca. 12,000 yr BP (Grønvold *et al.*, 1995) and is observed in the GRIP ice core at 12,008 yr BP; the Laacher See volcanic ash from the Eifel mountains in Germany is generally placed at 12,800 yr BP and a possible ash layer in the GRIP core exists at 12,800 yr BP (not analysed yet): the rapid decline in $a(z)$ around 12,600 yr BP concurs with the estimated date of 11,000 ^{14}C yr BP for the onset of the YD (see W. Broecker's review, 2003). The review by Broecker also discusses the chronological problems associated with the Lake Agassiz catastrophic flood and the onset of the YD. If the Laacher See tephra can be identified in the GRIP core around 12,800 yr BP (GRIP stratigraphical date) it would prove that there was indeed a 200 yr period between the Laacher See eruption and the onset of the YD. In any case a 130-m-long, well dated ice-core sequence exists from which year by year information on atmospheric changes can be inferred over this unusual climatic transition.

The ss09 model time-scale over the transition differs from the stratigraphical time-scale as it places the start of Bølling at 14,700 yr BP. The most likely explanation is that the empirical relation between the isotopic values of the ice and the surface accumula-

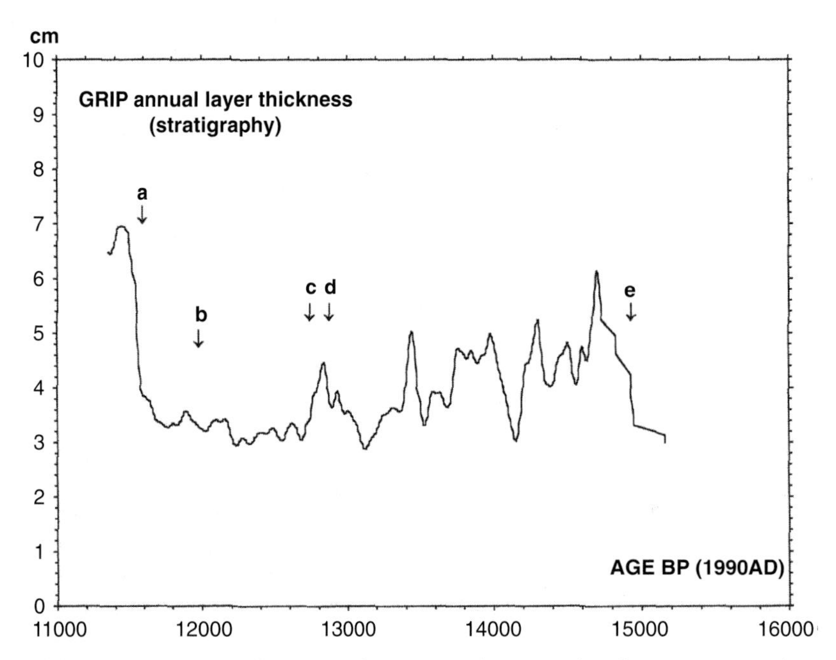

Figure 78.3 The annual layer thickness as a function of time over the Last Glacial Maximum–Holocene transition. The plot is a smoothed version of the more detailed layer counting. The arrows indicate various 'events' that can be used as reference points in order to validate the time-scale by comparing it to other types of dated records: end of Younger Dryas (a); Vedde ash layer (b); Lake Agassiz meltwater pulse (c); Lacher See ash layer? (d); start of Bølling (e).

tion overestimates the surface accumulation; but whatever the reason only the seasonal stratigraphical chronology fits the U/Th date for the start of Bølling.

What about the the pre-Bølling chronology?

78.4.3 Pre-Bølling chronology

The seasonal stratigraphy could in principle have been measured *in situ* as long as $a(z)$ was at least a few millimetres thick, but in practice this was not possible as the measurements became too time consuming as $a(z)$ approached 1 cm; hence only selected sequences of the deeper ice strata were measured with a seasonal resolution. The GRIP continuous seasonal stratigraphy therefore ends around 60 kyr BP. For older layers the sequential analysis can be used to establish a chronology, but the accuracy cannot compete with other palaeoclimatic chronologies. This may change in the future, when the North GRIP core becomes fully analysed. Such an extended ice-core master chronology will be limited to the past 100–130 kyr simply because this seems to be the limit for obtaining old ice with continuous and well resolved seasonal stratigraphy in the north polar region.

A major problem in estimating the stratigraphical chronology between 15 and 60 kyr BP is the LGM period covering 15–30 kyr BP. During this period the dust concentration in the ice is up to 100 times the Holocene values and there is a risk that the interpretation of the seasonal signal is biased. In order to estimate the accuracy back to 60 kyr BP we are presently left with only one possibility for verification: the Z2 volcanic tephra layer. This layer has been traced in several deep-sea sediment records from the North Atlantic and has also marked the GRIP ice with a visible layer (Grønvold *et al.*, 1995). The layer has been dated by interpolation in the DSDP 609 core to 57.5 kyr BP (Bond *et al.*, 1993). In the GRIP core it is found at 2431 m depth with a stratigraphical age of 58,400 yr BP. The ss09 model time-scale dates the Z2 layer to 52 kyr BP. The ss09 time-scale is again underestimating the age, as for the onset of the Bølling, but now the difference is 6 kyr. The ss09 time-scale was later refined by including the change of the isotopic composition of seawater (Johnsen *et al.*, 2001) and the new time-scale was called $ss09_{sea}$. Recently a paper on speleothems in France by Genty *et al.* (2003) claimed a higher precision than the ss09 and $ss09_{sea}$ time-scales. The authors noted that the model chronologies differed by an entire DO event, i.e. indicating that the accuracy of the model time-scale could be out by several thousand years. The claimed precision of the new speleothem time-scale, dated by the U/Th technique, is in itself questionable, because it is based on a not too detailed comparison between DO events in the ice and a speleothem record; the latter record needs to be verified and so does the estimation of the claimed accuracy.

The accuracy of the stratigraphic dating of the GRIP core at 60 kyr BP is estimated to ±1–2 kyr, but it rests on the interpolation of the DSDP 609 core, which can be questioned.

78.5 Final remarks

The continuous Greenland ice-core record reaches some 100 kyr back in time and offers a very high time resolution. The last glacial stage is covered by nearly 1200 m of annual ice deposits containing information on regional and global changes of atmospheric composition and climatic changes. Some of the global changes have their counterpart in Antarctic ice cores and therefore may serve as reference horizons tying the well-dated Greenland record and the Antarctic cores into a tight chronological framework.

An interesting automated layer counting technique is also on the way, but it remains to be seen if such a technique is more accurate than the classic counting by 'eye'.

The chronological accuracy of the pre-100 ka ice layers reached by deep Antarctic ice cores will probably depend on the dating accuracy offered by marine and terrestrial records. The unique information stored in the kilometre-long deep ice cores will, however, add to our understanding of the Earth's changing climate and environment.

SEVENTY-NINE

The 420,000-year climate record from the Vostok ice core

Jean Robert Petit

LGGE-CNRS BP96 38402 Saint Martin d'Hères Cedex, France

The Russian station of Vostok was established in 1957 during the International Geophysical Year at the south geomagnetic pole by the Soviet Antarctic Expeditions. The site (78°28′S, 106°48′E) is 1500 km from the ocean, at an altitude of 3488 m a.s.l. with a continental climate and a mean annual temperature of −55°C. The station overlies the southern end of a giant subglacial lake, Lake Vostok. The total ice thickness is 3750 m, and the snow accumulation rate is only 2.2 cm water equivalent per year. Since 1965, several ice cores have been retrieved by Russian drillers. A 2000-m-deep core was obtained in 1980 and provided the first climatic sequence covering the previous interglacial period (120,000 yr ago) and the preceding glacial period at ca. 150,000 yr ago (Lorius *et al.*, 1985). Since then, a borehole was extended down to 3623 m, and stopped 130 m above the water interface. The palaeoclimatic record deduced from analysis of the first 3310 m was the first long record covering the past 420,000 yr (Petit *et al.*, 1999). Below 3310 m the ice layering is disturbed by glacier dynamics, but the very basal ice from 3539 m to 3750 m is formed by freezing of the lake water (Jouzel *et al.*, 1999).

Here, the main results from the palaeoclimate reconstruction from the first 3310 m of the core are summarized. The past temperature, greenhouse gas content (CO_2, CH_4) and concentrations of aerosols of marine and continental origin have been reconstructed (Fig. 79.1).

The variations of the temperature at the surface of the ice sheet are deduced from the stable isotope composition of the ice. Both simple and more complex isotopic models predict that deuterium (δD) and $\delta^{18}O$ values should vary linearly with temperature in mid- and high latitudes, and this is well documented from observations of modern Antarctic precipitation (Jouzel *et al.*, 2003b). With respect to the present mean annual temperature (−55°C), the reconstructed Vostok temperature profile displays an overall amplitude of about 12°C, although slightly different estimates of climatic changes (up to 18°C) have been proposed (Salamatin *et al.*, 1998). From the mean time separating each warm period (interglacial), the record suggests the presence of cycles with a periodicity of about 100,000 yr. Glacial periods dominate the records, representing almost 90% of cycle duration. The present warm period, referred to as the Holocene, started ca. 12,000 yr ago and appears to be a very stable period with respect to other interglacials. The overall climate record displays a 'saw tooth' structure with temperatures gradually decreasing from the interglacial period to reach the minimum (ca. −67 to −73°C) of the glacial period. This was followed by a more rapid deglaciation, taking place in just a few millennia (ca. 5 to 10 kyr). The overall pattern of the Vostok temperature mimics the reversed global ice volume profile deduced from marine sediment studies, and the reconstructed atmospheric temperature clearly leads the ice volume by a few thousand years.

Spectral analysis of the record indicates the presence of three major periodicities of about 100,000, 40,000 and 20,000 yr. These characterize orbital geometry variations and the Earth's move-

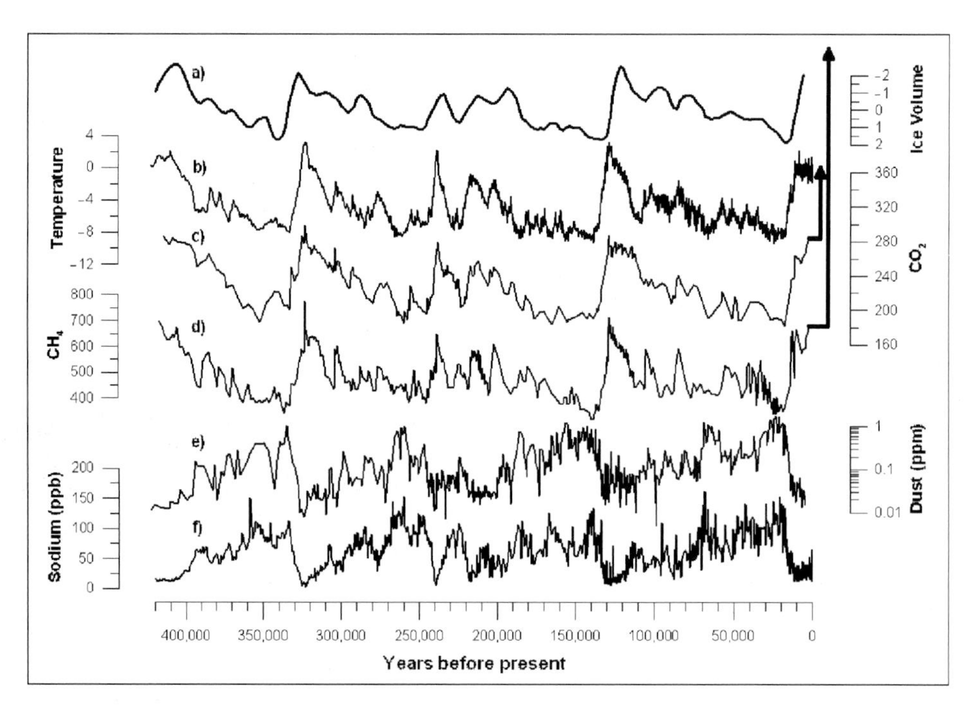

Figure 79.1 The climatic record over the past 420,000 yr deduced from the first 3310 m of the Vostok ice core (adapted from Petit *et al.*, 1999). (a) Global ice volume (in relative units, and inverted scale) as deduced from the marine sediment record (from Bassinot *et al.*, 1994). (b) Temperature (difference °C with the present surface temperature) deduced from the stable isotope composition of the ice. (c & d) Records of greenhouse gases: CO_2 (ppmv) and CH_4 (ppbv) as deduced from entrapped air bubbles. Note the recent increase up to the present level for CO_2 (360 ppmv) and CH_4 (1750 ppbv) reflecting anthropogenic activity since the 1850s. (e) Profile of continental dust concentration (ppm) as plotted in log scale. (f) Profile of sodium concentration (ppb), representative of marine aerosols. (See www.blackwellpublishing.com/knight for colour version.)

ment around the sun, which trigger large climatic variations by modulating solar energy received by the Earth according to latitude and season (Berger, 1978). The Vostok record confirms the astronomical theory of palaeoclimate (i.e. the so-called 'Milankovic' theory) that was firstly supported by marine records (e.g. Bassinot *et al.*, 1994).

Impurities in the Antarctic ice are found at very low concentration ($10^{-9}\,g\,g^{-1}$), mostly comprising the small fraction of primary aerosols emitted by the ocean (sea salt) and the continents (dust). Sulphuric acid is present in ice, resulting from the biogenic emission of dimethyl sulphide (DMS) and gas to particle conversion (Legrand & Mayewski, 1997). Volcanic inputs include discrete events of ash and H_2SO_4 peaks from the SO_2 oxidation, whereas anthropogenic activity is revealed through heavy metals (e.g. Pb) for the most recent period (Boutron *et al.*, 1994).

Sodium (Na) characterizes sea spray whereas dust, for which the mass is mostly represented by particles with sizes between 0.8 and 3 μm, originated mostly from the continents and arid zones. During glacial periods, the colder atmosphere was also drier than during interglacials, reducing the precipitation rate and the hydrological cycle, and therefore the capability of the atmosphere to be naturally washed out by precipitation. This is of importance because it significantly increases the residence time of aerosols and their ability to be disseminated worldwide. For glacial periods, the measured sodium concentration is three to four times higher than throughout the Holocene period, an effect partly due to the reduced precipitation and longer residence time in the atmosphere, but also to a probable more cyclonic activity around Antarctica. These factors, taken together, more than account for the greater sea-ice extent.

It has been hypothesized recently that the Antarctic sea-salt record is mostly influenced by salt blown off fresh sea ice, and could be used as an indicator for ice coverage (Wolff *et al.*, 2003). Indeed, this may be applicable for coastal sites but questionable for inland sites. In fact the sodium records from Vostok and from the new EPICA Dome C appear almost synchronous and closely anticorrelated with isotopic temperature ($r^2 \approx 0.70$ over the past 420,000 yr), supporting the idea that sodium concentrations are firstly linked to atmospheric processes: for example, cyclonic activity around Antarctica, residence time of aerosols, etc. On the other hand, as the sea ice extent is probably also influenced by the heat provided by the deep ocean surrounding Antarctica, some significant leads or lags between a sea-ice indicator and atmospheric temperature are expected, but are not observed for the sodium profile.

For glacial periods, the dust record shows concentrations of up to 750 ppb, 50 times greater than for interglacial periods (15 ppb), a magnitude now documented by several other East Antarctic ice cores (Delmonte *et al.*, 2002, 2004a). Again a drier atmosphere and a probably more efficient meridional transport toward Antarctica are suggested as main factors, to which a source effect is added by the significant extension of continental aridity. Indeed, transport effect is still unclear: general circulation model (GCM) simulations suggest the transport change is less significant by comparison with the source extension effect (Andersen *et al.*, 1998). In addition, recent studies provide evidence for opposite changes in the size distribution of dust between sites during the deglaciation, suggesting that the dust advection to the East Antarctic plateau has a regional character that varies with time (Delmonte *et al.*, 2004b). The source effect and the residence time of aerosols appear, therefore, the most important factors.

The geochemical composition of the dust (isotopes from Sr and Nd) from several East Antarctic ice cores made it possible to determine their geographical origin, mostly in southern South America and in particular Patagonia (Delmonte *et al.*, 2004a). This area is very sensitive to climate change because it is under the influence of sea-ice extension in the Drake passage between South America and the Antarctic Peninsula. Moreover, the Southern Andes, on which ice caps can develop, act as a meteorological barrier to precipitation. Together, this leads to intense periglacial erosion in the mountains, efficient transport of sediment to the plains by seasonal runoff and intense aeolian surface aggradation of the outwash plain during the dry seasons.

Air bubbles entrapped within the ice provide atmospheric air samples that are naturally and continuously collected during snow and firn densification. Vostok ice contains ca. 8% of air by volume. Between the different glacial and interglacial periods, greenhouse gas contents varied similarly as did the temperature. For CO_2 the concentrations varied from 180 to 300 ppmv (parts per million by volume) and for CH_4 from 360 to 700 ppbv. For the recent period, since 1850, CO_2 and CH_4 have both steadily increased and have reached the highest concentrations ever observed for the past 420,000 yr: 360 ppmv and 1750 ppbv respectively. Such changes are the consequence of anthropogenic activity with the use of fossil fuels for energy as well as the intensification of agricultural activity.

Over the 420,000 yr preceding this recent period, the record shows clear correlations between temperature and greenhouse gases, indicating that theses gases participated and contributed to the amplitude of the temperature variations. During the last deglaciation, an initial forcing (insolation?) leading to a small increase in greenhouse gas content (CO_2, but also CH_4 and other gases such as N_2O, etc.) gave first a very small radiative forcing change, which was amplified by positive reactions from the climate system: for example, decrease of sea ice, increase of the atmospheric water content and its own associated greenhouse effect, increase of surface albedo, warming the ocean temperature which releases more CO_2, and warming of soils which increases the biological CH_4 emissions. Finally up to 50% of the global temperature change, which was about 4°C from the glacial to interglacial period, may be due to the resulting greenhouse gas effect (Lorius *et al.*, 1990). Such results and amplification effects are now also supported by climate simulations using general circulation models (GCMs).

Information derived from the ice-core records makes it possible to describe the natural changes of the climate, and is useful to test climatic models with scenarios from the past (e.g. glacial period). The estimate of climate sensitivity to greenhouse gases as deduced from the past and from GCMs is of importance for predicting future global warming and the environmental changes due to the almost inescapable doubling of atmospheric CO_2 by the end of this century.

EIGHTY

Numerical modelling of polar ice sheets through time

Philippe Huybrechts

Alfred-Wegener-Institut für Polar- und Meeresforschung, Postfach 120161, D-27515 Bremerhaven, Germany; and Departement Geografie, Vrije Universiteit Brussel, Pleinlaan 2, B-1050 Brussel, Belgium

80.1 Introduction

Ice sheets respond dynamically to changes in boundary conditions, such as climate variations, basal thermal conditions, and isostatic adjustments of the underlying bedrock. These cause the ice sheets to evolve towards a new equilibrium. Long response time-scales of up to 10^4 years are involved, determined by the ratio of ice thickness to yearly mass turnover, physical and thermal processes at the bed, and processes affecting ice viscosity and mantle viscosity. The response of the ice sheets is further complicated by feedback processes which may amplify or mitigate the ice sheet's adjustment to the forcing or by internal instabilities that may cause rapid changes in ice volume due to changes in the dynamic flow regime. A primary motivation for developing numerical models of ice flow is to gain a better understanding of the spatial and temporal behaviour of ice sheets and glaciers and to predict their response to external forcing. Modelling ice-sheet dynamics presents a powerful framework to investigate the complex interactions between the ice sheets and the climate system in a quantitative way, in past as well as future environments. Ice-flow models are commonly based on fundamental physical laws and assumptions thought to describe glacier flow.

At the top end of the class of ice-sheet models are so-called three-dimensional thermomechanical models, which are able to describe the time-dependent flow and shape of real ice sheets. These models are akin to general circulation models developed in other branches of climate science. Their development closely follows technical progress in such fields as computer power, ice-core and sediment drilling, remote sensing and geophysical dating techniques, which are providing both the required calculating means and the necessary data to feed and validate such models. Models of this type have been applied to the existing ice sheets of Greenland and Antarctica, and to those which covered the continents of the Northern Hemisphere during the Quaternary ice ages. Typical studies have concentrated on mechanisms and thresholds of ice-sheet inception during the Tertiary (Huybrechts, 1994a; DeConto & Pollard, 2003), ice-sheet form and extent during glacial–interglacial cycles (Marshall *et al.*, 2000; Ritz *et al.*, 2001; Charbit *et al.*, 2002; Huybrechts, 2002), and the response of the polar ice sheets to future climatic warming (Huybrechts & de Wolde, 1999; Van de Wal *et al.*, 2001). In this context, the key interactions being investigated are between the effects of a change in climate on the accumulation and ablation fields and the ice sheet's response in terms of changed geometry and flow, including the ice sheet's contribution to the worldwide sea-level stand. Related work has considered the ice sheets as a boundary condition for other components of the Earth's geophysical system, providing changes in surface loading for isostasy and gravity models (Le Meur & Huybrechts, 2001; Tarasov & Peltier, 2004), or providing changes in freshwater fluxes for ocean models, especially to investigate changes of the thermohaline circulation of the North Atlantic Ocean (Schmittner *et al.*, 2002; Fichefet *et al.*, 2003). Three-dimensional thermomechanical ice-sheet models have also been used to investigate the potential for internally generated flow instability (Payne, 1995; Payne & Dongelmans, 1997; Marshall & Clarke, 1997a). In this application, the crucial interactions are between the thermal and flow regimes. In addition, models of the Greenland and Antarctic ice-sheets are in use to assist with the location and dating of ice cores (Greve, 1997b; Huybrechts *et al.*, 2004a), estimating internal distributions of passive tracers such as oxygen isotope ratios (Clarke & Marshall, 2002), yield information about fields that are inaccessible for direct observation such as at the ice-sheet base (Huybrechts, 1996), or assess the component of their present-day evolution due to adjustment to past climate changes (Huybrechts & Le Meur, 1999).

In this chapter the discussion concentrates on three-dimensional whole ice-sheet models applied to the ice sheets of Antarctica and Greenland. This has the advantage that the quality of the model simulations can be assessed against available observations. Also, the range of physical characteristics, climate regimes, and flow mechanisms encountered in both polar ice sheets is probably a good representation of the range of behaviour which occurred in the palaeo-ice-sheets at various times during their evolution. The Antarctic ice sheet is located in a very cold climate, where almost no surface melting occurs and precipitation amounts are limited by low air temperature. Therefore virtually all Antarctic ice is eventually transported into floating ice shelves that experience melting or freezing at their underside and eventually break up to form icebergs. External forcing of such a cold ice sheet is mainly through changes in accumulation rate and basal melting rates below the ice shelves. The ice-sheet extent is limited mainly by the depth of the surrounding ocean and by the capacity of the ocean to float the ice shelves, touching upon the crucial issue of grounding-line dynamics. An important characteristic of the West Antarctic ice sheet is that it is a marine ice sheet, resting on a bed far below sea level. Much of its ice transport towards the coast occurs in ice streams, which are distinct fast-flowing features that rest on smooth sedimentary beds, have very flat surface profiles, and are sharply bordered by relatively stagnant ice at their sides. Their ice flux is dominated by basal flow, mostly through deformation of the underlying water-saturated sediments (Alley *et al.*, 1986a). By contrast, the Greenland ice sheet is situated in a much warmer climate, with a temperature difference of 10–15°C in the annual mean. Summer temperatures are high enough to initiate widespread summer melting. All around the ice-sheet margin, mean annual ablation exceeds the accumulation. A negative surface budget results at elevations below about 1000 m in the north and 1600–1800 m in the southwest. High coastal temperatures do not favour ice shelves, but there are a few along the north and northeast coast. The Greenland ice sheet loses mass by calving of icebergs, mostly at grounding lines in a tidewater environment, and by meltwater runoff from the surface, in roughly equal shares.

80.2 Building a three-dimensional thermomechanical ice-sheet model

Planform time-dependent modelling of ice sheets largely stems from early work by Mahaffy (1976) and Jenssen (1977), extending on the pioneering Antarctic study of Budd *et al.* (1971). These papers develop work by Nye (1957) on what has become known as the shallow-ice approximation (Hutter, 1983). This approximation recognizes the disparity between the vertical and horizontal length scales of ice flow, and implies grounded ice flow by simple shear. This means that the gravitational driving stress is balanced by shear stresses and that transverse and longitudinal strain rate components are neglected. Although the assumption is not valid at all places in the ice sheet, such as at the ice divide or near the ice-sheet margin (Baral *et al.*, 2001), it has shown general applicability in large-scale ice-sheet modelling as long as surface slopes are evaluated over horizontal distances an order of magnitude greater than ice thickness. Under the shallow-ice approximation, both components of the horizontal velocity can be represented as algebraic functions of the local ice geometry (surface slope and ice thickness), which greatly simplifies the numerical solution. The model by Mahaffy (1976) was vertically integrated and was developed as a computer program to find the heights of an arbitrary ice sheet on a rectangular grid. It incorporated Glen's flow law (Glen, 1955) for polycrystalline ice deformation by dislocation creep. That is an empirical relation derived from laboratory tests in analogy with the behaviour of metals at temperatures near to their melting point, and is most commonly used in ice flow modelling. It considers ice as a non-Newtonian viscous fluid, relating strain rates to stresses raised mostly to the third power. However, in ice the rate of deformation for a given stress is also a strong function of temperature. For the range of temperatures found in natural ice masses (−50°C to 0°C), the effective viscosity changes by more than three orders of magnitude. The first model that dealt with the flow–temperature coupling in a dynamic fashion was developed by Jenssen (1977). Jenssen introduced a stretched vertical coordinate, transformed the relevant continuity and thermodynamic equations, and presented a framework to solve the system numerically.

At the heart of three-dimensional thermomechanical models is the simultaneous solution of two evolutionary equations for ice thickness and temperature, together with diagnostic representations of the ice velocity components. These express fundamental conservation laws for momentum, mass and heat, supplemented with a constitutive equation (the flow law). Plate 80.1 shows the structure of one such model as it was described in Huybrechts (1992), and further refined in Huybrechts & de Wolde (1999) and Huybrechts (2002). This model was first developed for the Antarctic ice sheet. It solves the thermomechanically coupled equations for ice flow in three subdomains, namely the grounded ice sheet, the floating ice shelf, and a stress transition zone in between at the grounding line. The flow within the three subdomains is coupled through the continuity equation for ice thickness, from which the temporal evolution of ice-sheet elevation and ice-sheet extent can be calculated by applying a flotation criterion. Grounded ice flow is assumed to result both from internal deformation and from basal sliding over the bed in those areas where the basal temperature is at the pressure melting point and a lubricating water layer is present. Ice deformation in the ice-sheet domain results from vertical shearing, most of which occurs near to the base. For the sliding velocity, a generalized Weertman relation is adopted, taking into account the effect of the subglacial water pressure. Ice shelves are included by iteratively solving a coupled set of elliptic equations for ice-shelf spreading in two dimensions, including the effect of lateral shearing induced by sidewalls and ice rises. At the grounding line, longitudinal stresses are taken into account in the effective stress term of the flow law. These additional stress terms are found by iteratively solving three coupled equations for depth-averaged horizontal stress deviators. Calving of ice shelves is ignored. Instead, the ice shelves extend to the edge of the numerical grid but this has little influence on the position of the grounding line. The temperature dependence of the rate factor in Glen's flow law is represented by an exponential Arrhenius equation.

The distinction between ice-sheet flow and ice-shelf flow was also made in the Antarctic model developed by Ritz *et al.* (2001), but they introduced an additional subdomain representing a 'dragging ice shelf' to incorporate ice-stream dynamics. In their model, inland ice is differentiated from an ice stream zone by the magnitude of basal drag. This is based on the observation that ice stream zones are characterized by low surface slopes and thus low driving stresses, but yet have fast sliding, as is the case at the Siple Coast in West Antarctica. Ritz *et al.* (2001) treat these zones as semi-grounded ice shelves by replacing the shallow ice approximation by the set of ice-shelf equations to which basal drag is added. Gross model behaviour turns out to be quite similar to that of the Huybrechts model, except that the West Antarctic ice sheet has a lower surface slope near the grounding line and that the break in the slope now occurs further upstream at the place where the dragging ice shelf joins the inland ice subject to the shallow ice approximation. One consequence is that grounding-line retreat in the Ritz model occurs more readily in response to rising sea levels, and that the West Antarctic sheet contains less additional ice for an expanded grounding line.

Three-dimensional flow models applied to the Greenland ice sheet have been developed along similar lines, except that ice flow is only considered in the grounded ice domain and ice shelves are not dealt with (Ritz *et al.*, 1997; Greve, 1997a; Huybrechts & de Wolde, 1999). The position of the calving front in these models is not predicted from a self-consistent treatment of calving dynamics, as its physics are poorly understood and a convincing calving relation does not exist. Instead, these models prescribe the position of the coastline, beyond which all ice is removed as calf ice. Another characteristic of most Greenland models concerns the incorporation of variable ice fabric in the flow law to account for different ice stiffnesses as established for Holocene and Wisconsin ice in Greenland ice cores. This is achieved by prescribing a variable enhancement factor in the rate factor of the flow law, and necessitates the simultaneous calculation of ice age to track the depth of the Holocene/Wisconsin boundary. The model developed by Greve (1997a) furthermore incorporates polythermal ice and considers the possibility of a temperate basal ice layer, in which the water content and its impact on the ice viscosity are computed.

In whole ice-sheet models it is necessary to take into account the isostatic adjustment of the bedrock to the varying ice load. Early studies considered a damped return to local isostatic equilibrium (e.g. Oerlemans & van der Veen, 1984) but in most recent models the bedrock model consists of a rigid elastic plate (lithosphere) that overlies a viscous asthenosphere. This means that the isostatic compensation not only considers the local load, but integrates the contributions from more remote locations, giving rise to deviations from local isostasy. For an appropriate choice of the viscous relaxation time, this treatment produces results close to those from a sophisticated self-gravitating spherical visco-elastic earth model, while at the same time being much more efficient in terms of computational overhead (Le Meur & Huybrechts, 1996). Another feature common to most thermomechanical models is the inclusion of heat conduction in the bedrock, which gives rise to a variable geothermal heat flux at the ice-sheet base depending on the thermal history of the ice and rock.

Interaction with the atmosphere and the ocean in large-scale ice-sheet models is effectuated by prescribing the climatic input, consisting of the surface mass-balance (accumulation minus ablation), surface temperature, and, if applicable, the basal melting rate below the ice shelves. Changes in these fields are usually heavily parametrized in terms of air temperature. Precipitation rates are often based on their present distribution and perturbed in different climates according to temperature sensitivities derived from ice cores or climate models. This is principally because of a lack of a better method and implies that interaction between the pattern of precipitation and an evolving ice sheet cannot be accounted for properly. Meltwater runoff, if any, is usually obtained from the positive degree-day method (Braithwaite & Olesen, 1989; Reeh, 1991). This is an index method providing the bulk melting rate depending on air temperature only, but is very efficient in its use and generally gives very acceptable results (Ohmura, 2001). Models of this type are usually driven by time series of regional temperature changes (available from ice-core studies) and by the eustatic component of sea-level change, relative to present values.

Ice-sheet models are typically implemented using finite-difference techniques on a regular grid of nodes in the two horizontal dimensions, and using a stretched co-ordinate system in the vertical. Horizontal grid resolutions are mostly in the range of 20 to 50 km with between 20 and 100 layers in the vertical, concentrated towards the base where the bulk of the velocity shear takes place. Finite element implementations exist (e.g. Hulbe & MacAyeal, 1999) although these are often performed on a regular grid (Fastook & Prentice, 1994). Recent model applications have used much improved compilations of crucial input data such as bed elevation that became available on high-resolution grids from the BEDMAP (Lythe *et al.*, 2001) and PARCA projects (Bamber *et al.*, 2001a; Gogineni *et al.*, 2001).

80.3 Model applications

Three-dimensional models of the Antarctic and Greenland ice sheets have been used to address two main issues: the expansion and contraction of these ice sheets during the glacial–interglacial cycles, and the likely effects of greenhouse-induced polar warming.

80.3.1 Model validation

Before models can have any predictive capabilities, it is necessary to confirm that they are a realistic representation of the real-world system. One often distinguishes between the steps of calibration and validation. The usual practice is to first vary a few adjustable parameters to give a qualitatively best fit with observations. This mainly concerns the multiplier ('enhancement factor') in the rate factor of the flow law and/or the basal sliding parameter, which are chosen to give a good representation of the present-day ice-sheet configuration, preferably after spinning up over the glacial cycles as the ice sheets are currently not in steady state. Often it is also necessary to adjust the values of the degree-day factors in the melt-and-runoff parametrization in order to have the modelled ice margin coincide as closely as possible with its observed location. Another parameter available for tuning is the geothermal heat flux. Its value is not very well defined and therefore can be adjusted to obtain a good fit with measured borehole temperatures where these are available. The problem with using observations of ice thickness and basal temperature to calibrate a model is that these fields are strictly speaking no longer available to validate the model. This problem is difficult to avoid because of the paucity of suitable test data (Van der Veen & Payne, 2004, who prefer the term 'confirmation' rather than 'validation'), leaving only velocity as an independent field for model verification (or confirmation). Another way of confirming time-dependent models is to compare simulations of past behaviour against the geological record but this procedure is also by itself not fully conclusive as different parameter combinations may yield the same result, and the geological record is often ambiguous. Nevertheless, these seem to be the only options at hand to assess the performance of current ice sheet models.

Plate 80.2 shows examples of fundamental output fields available for model testing. As far as can be judged from available data, the predicted fields of vertically averaged velocity and basal temperature look very reasonable. In Greenland most of the base in central areas appears to be frozen to bedrock, with homologous basal temperatures typically between 4 and 8°C below the pressure melting point. Temperate ice is mainly confined to the coastal region and a number of fast-flowing outlet glaciers where dissipation rates are highest. More widespread basal melting also occurs in the northeastern and central-western parts of the Greenland ice sheet. In Antarctica, bottom ice at pressure melting point is widespread in both West and East Antarctica. Both heat dissipation at the base and the insulating effect of thick interior ice in combination with low vertical advection rates play a role here. Also for Antarctica, basal temperatures can be verified in only a few deep boreholes. However, the pattern of pressure melting shown in Plate 80.2 generally can be well correlated with radio-echo sounding data indicative of basal water (Siegert, 2000). The coolest basal layers are found above the Gamburtsev Mountains and the fringing mountain ranges, where the ice is thinnest. It should, however, be kept in mind that a strong control on basal temperatures is exerted by the geothermal heat flux, a parameter already used for tuning the model. Moreover, its value is known to have a large spatial variation and this is not included in current models because of lack of data.

Perhaps the best independent test to confirm models is to compare their velocity fields with observations. This option has not been fully exploited yet. Remote sensing techniques using satellite-derived information make it possible to obtain good representations of surface velocities, but ice-sheet-wide maps have so far not been published. At this stage, modelled velocities can be compared with so-called balance velocities. These are also modelled velocities based on the assumption of stationary downhill flow, which are believed to be correct to within 25% of reality (Bamber *et al.*, 2000a). Gross comparison of the overall patterns of modelled velocities in Plate 80.2 with balance velocities (Budd & Warner, 1996; Joughin *et al.*, 1997; Huybrechts *et al.*, 2000) is certainly favourable, but the details differ. In particular the flow concentration in narrow outlet glaciers and ice streams does not occur to the same degree in the modelled fields. This is a matter of model resolution, but also the diffusive properties of the model physics and the numerical scheme, insufficient basal flow, and the neglect of additional terms in the force balance play a role. Features such as the northeast Greenland ice stream and the flow over Lake Vostok are missing altogether from the fields shown in Plate 80.2, chiefly because the specific mechanisms thought responsible for their formation are not included in the models.

80.4 Glacial cycle simulations

80.4.1 The Antarctic ice sheet

Long integrations of the Antarctic ice sheet during the last glacial cycles were analysed in Budd *et al.* (1998), Huybrechts & de Wolde (1999), Ritz *et al.* (2001) and Huybrechts (2002). Plate 80.3 (left panel) shows the evolution of key glaciological variables over the last four glacial cycles in a typical run with the Huybrechts model, with forcing derived from the Vostok ice core (Petit *et al.*, 1999) and the SPECMAP sea-level stack (Imbrie *et al.*, 1984). In line with the generally accepted view, volume changes are largely concentrated in the West Antarctic and Peninsula ice sheets. These are caused by a repeated succession of areal expansion and contraction of grounded ice close to the continental break during glacial maxima. Around the East Antarctic perimeter, grounding-line advance was limited because of the proximity of the present-day grounding line to the continental shelf edge. In these models, glacial–interglacial fluctuations are mainly controlled by changes in the global sea-level stand and dynamic processes in the ice shelves. This supports the hypothesis that the Antarctic ice sheet basically follows glacial events in the Northern Hemisphere by means of sea-level teleconnections. Typical glacial–interglacial volume changes correspond to global sea-level contributions of about 20 m. Freshwater fluxes originating from the Antarctic ice sheet are an important output because of their role in modulating the deep-water circulation of the ocean. Model predictions displayed in Plate 80.2 show that these are fairly constant in time and are almost entirely dominated by the iceberg flux. During the last two glacial–interglacial transitions meltwater peaks occurred about three times larger than the normal background fluxes. During interglacials, melting from below the ice shelves is also an important contribution but surface runoff always remained negligible.

According to the model, surface elevations over most of West Antarctica and the Antarctic Peninsula were, at the Last Glacial Maximum (LGM), up to 2000 m higher than at present in direct response to the grounding-line advance (Plate 80.4). Over central East Antarctica, surface elevations at the LGM were 100–200 m lower because of the lower accumulation rates (Huybrechts, 2002). A characteristic of this model is that most of the Holocene grounding-line retreat in West Antarctica occurs after 10 kyr BP and lags the eustatic forcing by up to 10 kyr. This behaviour is related to the existence of thresholds for grounding-line retreat, and to the offsetting effect of the late-glacial warming leading to enhanced accumulation rates and a thickening at the margin. The late timing is in line with recent geological evidence (Ingolffson *et al.*, 1998; Conway *et al.*, 1999) and is supported by some interpretations of relative sea-level data (Tushingham & Peltier, 1991), but other inferences have been made. The implication is an ongoing shrinking of the Antarctic ice sheet at the present time equivalent to a global sea-level rise of about 2.5 cm per century (Huybrechts & de Wolde, 1999). An important unknown regarding the glacial history of the West Antarctic ice sheet is whether widespread ice-streaming comparable to the present Siple Coast continued to exist at LGM, in which case surface elevations may have been substantially lower than shown in Plate 80.4, and the contribution to the global sea-level lowering was less by perhaps several metres (Huybrechts, 2002).

80.4.2 The Greenland ice sheet

Similar results from glacial cycle simulations of the Greenland ice sheet are shown in Plate 80.3 (right panel) and Plate 80.5. Here the temperature forcing was assembled from the GRIP $\delta^{18}O$ record (Dansgaard *et al.*, 1993) for the most recent 100 kyr, and from the Vostok record (Petit *et al.*, 1999) for the period before that to circumvent known defects in the GRIP record during the last interglacial. The most conspicuous feature over the last two glacial cycles concerns the fate of the Greenland ice sheet during the Eemian interglacial, when temperatures peaked up to 7°C higher than today and global sea levels are believed to have been 6 m higher. At this time, the model indicates that massive marginal melting caused the ice sheet to shrink to a central-northern dome that existed together with small pockets of residual mountain glaciation over the southeastern highlands. Nevertheless, the ice sheet did not disappear entirely, as evident from the retrieval of pre-Eemian ice from central Greenland ice cores. This behaviour was confirmed by Cuffey & Marshall (2000), although Huybrechts (2002) found that the Greenland minimum is not very well constrained: for plausible combinations of climate conditions and only small shifts in the duration and magnitude of the peak warming, the Eemian ice sheet could have varied from just a little smaller than today to only a small single dome in central-north Greenland. During glacial periods, when melting is unimportant, the Greenland ice sheet expanded beyond the present coastline to cover most of the continental shelf, with an implied contribution to the LGM sea level lowering of about 3 m (Plate 80.5). The evolution of freshwater fluxes displayed in Plate 80.2 shows an interesting contrast with the Antarctic ice sheet. Whereas iceberg calving is also the dominant component during glacial periods, most of the interglacial retreat is caused by surface runoff. A striking feature of the temporal evolution of freshwater fluxes are the recurrent meltwater peaks of up to two times larger than the present-day surface runoff during glacial times. These can be correlated with the warm interstadials punctuating the Dansgaard–Oeschger events and with the warm interval which occurred prior to the Younger Dryas cold period.

80.4.3 Response of the polar ice sheets to future climatic warming

80.4.3.1 Response during the 21st century

Three-dimensional ice-sheet modelling studies all indicate that on time-scales less than a century the direct effects of changes in the surface mass-balance dominate the response. This means that the response is largely static, and thus that the ice flow on this time-scale does not react much to changes in surface mass balance. Greenland studies by Van de Wal & Oerlemans (1997) and Huybrechts & de Wolde (1999) found that ice-dynamics counteract the direct effect of mass-balance changes by between 10 and 20%. The mechanism arises because surface slopes at the margin are steepened in response to the increased melting rates. This causes the ice to flow more rapidly from the accumulaton to the ablation zone, leading to a dynamic thickening below the equilibrium line. The higher surface level of the ablation zone in turn leads to less melting than would be the case if ice dynamics were not included. Because of its longer response time-scales, the Antarctic ice sheet hardly exhibits any dynamic response on a century time-scale, except when melting rates below the ice shelves are prescribed to rise by in excess of 1 m yr^{-1} (O'Farrell *et al.*, 1997; Warner & Budd, 1998; Huybrechts & de Wolde, 1999).

These responses should be considered in addition to the long-term background trend as a result of ongoing adjustment to past environmental changes as far back as the last glacial period. The IPCC Third Assessment Report estimates the latter contribution to be between 0 and 0.5 mm yr^{-1} of equivalent sea-level rise for both polar ice sheets combined (Church *et al.*, 2001a). Three-dimensional modelling studies which analyse the imbalance pattern resulting for the present-day in glacial cycle simulations typically find a long-term sea-level evolution of between 1 and 4 cm per century for Antarctica but a negligible contribution of only a few millimetres per century for Greenland (Huybrechts & de Wolde, 1999; Huybrechts & Le Meur, 1999). Another component to the current and future evolution of ice sheets are the effects of 'unexpected ice-dynamic responses' which may or may not be related to contemporary climate changes, and which find their origin in variations at the ice-sheet base or at the grounding line. Examples are the measured thinning of the Pine Island and Thwaites sectors of the West Antarctic ice sheet (Shepherd *et al.*, 2002), the oscillatory behaviour of the Siple Coast ice streams (Joughin *et al.*, 2002), or the surging behaviour of some Greenland outlet glaciers (Thomas *et al.*, 2000a). Such mechanisms are hard to predict and currently are not incorporated in any large-scale model of the polar ice sheets.

Plate 80.6 shows an example of a series of ice-sheet simulations predicting 20th and 21st century volume changes. Boundary conditions of temperature and precipitation were in these experiments derived by perturbing present-day climatologies according to the geographically and spatially dependent patterns predicted by the T106 ECHAM4 model (Wild *et al.*, 2003) for a doubling of CO_2 under the IS92a scenario. To generate time-dependent boundary conditions, these patterns were scaled with the area-average changes over the ice sheets as a function of time for available AGCM results. Typically, mass-balance changes cause a Greenland contribution to global sea level rise of +2 to +7 cm between 1975 and 2100, and an Antarctic contribution of between −2 and −14 cm. This differential response is because increased marginal melting on Greenland is predicted to outweigh the effect of increased precipitation, whereas a warmer atmosphere over Antarctica is expected to lead to more precipitation, but still negligible surface melting. For the majority of the driving AGCMs, the Antarctic response is larger than for Greenland, so that the combined sea-level contribution from mass-balance changes alone is negative. However, when the background trend is taken into account, the sea-level contribution from both polar ice sheets is not significantly different from zero (Huybrechts *et al.*, 2004b), strengthening earlier conclusions that Antarctica and Greenland may well balance one another on a century time-scale.

The results shown in Plate 80.6 were used as the base for the IPCC TAR projections of sea-level rise from the polar ice sheets. To do that, they were regressed against global mean temperature to enable further scaling to take into account the complete range of IPCC temperature predictions for the most recent SRES emission scenarios. Taking into account the background evolution and various sources of uncertainties, this yielded a predicted Antarctic contribution to global sea-level change between 1990 and 2100 of between −19 and +5 cm, which range can be considered as a 95% confidence interval (Church *et al.*, 2001). For Greenland, the range was −2 to +9 cm. Most of this spread came from the climate sensitivity of the forcing AGCMs, and less from the emission scenario or the uncertainty in the ice-sheet models. These numbers should be compared with the predicted contributions to 21st century sea level rise of between +11 and +43 cm from thermal expansion of the sea water and of between +1 and +23 cm from melting of mountain glaciers and small ice caps, based on the same set of AGCMs. Taking into account all sources and uncertainties, the IPCC TAR predicts a sea-level rise from 1990 to 2100 of between 9 and 88 cm, with a central estimate of 48 cm (Church *et al.*, 2001).

80.4.3.2 *Response during the third millennium and beyond*

Beyond the 21st century, the approximate balance between both polar ice sheets is, however, unlikely to hold. If greenhouse warming conditions were to be sustained after the year 2100, the picture is expected to change drastically. In particular the Greenland ice sheet is very vulnerable to a climatic warming. For an annual average warming over Greenland of more than about 2.7°C, mass-balance models predict that ablation will exceed accumulation (Huybrechts *et al.*, 1991; Janssens & Huybrechts, 2000). Under these circumstances, the ice sheet must contract, even if iceberg production is reduced to zero as it retreats from the coast. For a warming of 3°C, the ice sheet loses mass slowly and may be able to approach a new steady state with reduced extent and modified shape if this results in less ablation. For greater warming, mass is lost faster and the Greenland ice sheet eventually melts away, except for residual glaciers at high altitudes. Two powerful positive feedbacks may accelerate the melting process: lower ice-sheet elevations lead to higher surface temperatures, and land-surface changes from ice to tundra further increase summer temperatures (Toniazzo *et al.*, 2004). Huybrechts & de Wolde (1999) find the Greenland ice sheet to contribute about 3 m of sea level rise by the year 3000 for a sustained warming of 5.5°C. For a warming of 8°C, they calculate a contribution of about 6 m. Greve (2000) reports that loss of mass would occur at a rate giving a sea-level rise of between 1 mm yr^{-1} for a year-round temperature perturbation of 3°C to as much as 7 mm yr^{-1} for a warming of 12°C. Gregory *et al.* (2004) have investigated the development of Greenland's temperature using IPCC scenarios in which atmospheric CO_2 stabilizes at different levels over the next few centuries. They find that the 2.7°C threshold is passed in all but one of 35 combinations of AOGCM and stabilization level; the warming exceeds 8°C in many cases and continues to rise after 2350 for the higher concentrations. The conclusion is that the Greenland ice sheet is likely to be eliminated over the course of the next millennia, unless drastic measures are taken to curb the predicted warming. Even if atmospheric composition and the global climate were to return to pre-industrial conditions, the ice-sheet might not be regenerated, implying that the sea-level rise could be irreversible (Gregory *et al.*, 2004; Toniazzo *et al.*, 2004).

On centennial to millennial time-scales, Antarctic model predictions demonstrate how several mechanisms depending on the strength of the warming come into play. For warmings below about 5°C, runoff remains insignificant and there is hardly any change in the position of the grounding line (Huybrechts & de Wolde, 1999). For larger warmings, however, significant surface melting occurs around the ice-sheet edge and basal melting increases below the ice shelves, causing the ice shelves to thin. When rapid ice-shelf thinning occurs close to the grounding line, grounding-line retreat is induced. In large-scale ice-sheet models, this occurs in two ways: steeper gradients across the grounding zone cause larger driving stresses, and higher deviatoric stress gradients across the grounding zone lead to increased strain rates, and hence a speed-up of the grounded ice and subsequent thinning. In the model studies performed by the Australian group (Budd *et al.*, 1994; O'Farrell *et al.*, 1997; Warner & Budd, 1998), large increases in bottom melting are the dominant factor in the longer-term response of the Antarctic ice sheet, even for moderate climate warmings of a few degrees. Budd *et al.* (1994) found that without increased accumulation, the increased basal melt of 10 m yr^{-1} would greatly reduce ice shelves and contribute to a sea-level rise of over 0.6 m by 500 yr, but no drastic retreat of the grounding line. With a similar model but different climatic forcing, O'Farrell *et al.* (1997) find a sea-level rise of 0.21 m after 500 yr for a transient experiment with basal melt rates evolving up to 18.6 m yr^{-1}. In the study by Warner & Budd (1998), a bottom melt rate of 5 m yr^{-1} causes the demise of WAIS ice shelves in a few hundred years and removal of the marine portions of the West Antarctic ice sheet and a retreat of coastal ice towards more firmly grounded regions elsewhere over a time period of about 1000 years. Predicted rates of sea-level rise in these studies are up to between 1.5 and 3.0 mm yr^{-1} depending on whether accumulation rates increase together with the warming. Although these are large shrinking rates, obtained under severe conditions of climate change, they cannot be considered to support the concept of a catastrophic collapse or strongly unstable behaviour of the WAIS, which is usually defined to mean its demise within several centuries, implying sea-level rises in excess of 10 mm yr^{-1} (Oppenheimer, 1998; Vaughan & Spouge, 2002). It should, however, be noted that the mechanics of grounding-line migration are not fully understood, and that none of these three-dimensional models adequately includes ice streams, which may be instrumental in controlling the behaviour and future evolution of the ice sheet in West Antarctica.

80.5 Conclusions and future outlook

Three-dimensional ice-sheet modelling significantly contributes to a better understanding of the polar ice sheets and their interactions with the climate system. Current models available to the community are able to predict the spatial and temporal ice-sheet response to changes in environmental conditions with increasing confidence. Large-scale models perform best over interior portions of continentally-based ice sheets, where ice deformation is well understood, obeys a simple force balance, and can be reliably modelled taking into account the flow law of ice. In some instances, when the basal ice has developed a strong fabric, making the ice anisotropic, or when crystal properties have introduced gradients in hardness, the resulting effects usually can be handled satisfactorily by prescribing a (variable) enhancement factor in the flow law.

Shortcomings in these models require further investigation in two main fields: incorporation of more appropriate physics and incorporation of improved boundary data. In particular basal sliding, marine ice dynamics and iceberg calving remain problematic. These processes are not easily quantified and are typically highly parametrized. Fast glacier conditions at the base are poorly understood, and so is the development of ice streams in marine-based ice sheets. Processes related to bed roughness, till rheology, and basal water pressure are all thought to be important elements but a realistic basal boundary condition for use in numerical models has not yet been developed. A credible treatment will need to include subglacial hydrology and the geological controls on soft-sediment deformation. The physics of grounding-zone migration is subgrid scale and is also not yet portrayed reliably in current models. In the grounding zone, a change takes place between flow dominated by basal stress to a basal stress-free regime, with flow primarily driven by longitudinal extension rather than vertical shear deformation. The spatial scale over which this transition takes place is unclear, however, and is therefore included in current models in an *ad hoc* way, if at all. The classic example where many of these problems converge is the Siple Coast area of the West Antarctic ice sheet, which is characterized by extensive ice streaming, low surface slopes, and a seemingly smooth transition into a floating ice shelf. Iceberg calving may be an even greater challenge to model in large-scale treatments. Calving at marine margins is related to fracture dynamics and temperature and stress fields in the ice, but the process is not well understood and therefore impossible to model with confidence. A proper treatment of calving is nevertheless warranted because ice-front degradation into bordering marine waters is the dominant means of ablation in Greenland tidewater glaciers and Antarctic ice shelves.

Although ice-sheet evolution is sensitive to several glaciological controls, long-term variability is dictated by climate and mass-balance related boundary conditions. Uncertainties in parametrizations have a large impact, particularly with respect to ice-sheet ablation. The mass-balance calculation is also sensitive to model resolution, as topographic detail is important in high-relief areas and ablation is typically concentrated in a narrow band at the ice-sheet margin. Present-day atmospheric boundary conditions such as mean annual air temperature and snow accumulation are known to a level of accuracy commensurate with that required by ice-sheet models but their patterns of change in past as well as future climates are poorly constrained. Even more troublesome is the melt rate from the underside of the ice shelves, which may affect grounding lines but for which we have very limited data. The same is also true of the geothermal heat warming at the ice-sheet base, which exerts a crucial control on the spatial extent of basal melting, but for which there are very few data.

Current developments in large-scale ice-sheet modelling mainly occur along two lines: incorporation of ice-sheet models in climate or earth system models of varying complexity, and refinements of the ice dynamics at the local scale using higher-order representations of the force balance. Interactive coupling of ice-sheet models with atmosphere and ocean models enables mass-balance changes over the ice sheets to be prescribed more directly and the effects of circulation changes to be dealt with more properly. If the coupling is two-way, the approach can additionally take into account the effect of ice-sheet changes on its own forcing. Such coupled experiments have only just begun but are likely to highlight interesting behaviour. Recent examples include the effects of freshwater fluxes on the circulation of the North Atlantic Ocean (Schmittner *et al.*, 2002; Fichefet *et al.*, 2003), the enhancing effect of ice-dammed lakes on ice-sheet growth (Krinner *et al.*, 2004), or the climate feedbacks resulting from Greenland ice-sheet melting (Ridley *et al.*, 2004). A second line of current research concerns the nesting of detailed higher-order flow models (e.g. Pattyn, 2003) into large-scale models to study the flow at high spatial resolution for which the usual assumptions made in zero-order models are known to break down. First attempts in this direction for limited inland areas near ice divides were presented in Greve *et al.* (1999) and further explored for the purpose of ice-core dating and interpretation in Huybrechts *et al.* (2004a).

EIGHTY-ONE

Ice-flow models

Antony J. Payne and Andreas Vieli

Centre for Polar Observation and Modelling, School of Geographical Sciences, University of Bristol, Bristol BS8 1SS, UK

In this case study we will illustrate the range of models available to study terrestrial ice masses by concentrating on the study of the ice streams of Antarctica. Ice streams are the focus of a great deal of contemporary research because they discharge the majority (more than 90%) of the ice leaving Antarctica, and their dynamics are therefore likely to affect the volume of ice stored in the ice sheet and, hence, global sea levels.

Ice-flow models can best be classified according to the accuracy with which they depict the true force balance within an ice mass. Based on the ratio between a typical ice thickness and the characteristic length of the ice mass (the aspect ratio), three main sets of approximations can be made. These are zero, first and second-order models, which are appropriate at successively finer spatial scales of investigation. The order of such models is indicative of how many effects are incorporated into the modelled force balance.

All models assume that the balance of forces is static (i.e. that acceleration is not significant in the balance). Zero-order models then simply assume that all the gravitational driving stress is balanced locally by vertical shear stresses and basal drag

$$\tau_{xz} = \rho g(s - z)\frac{\partial s}{\partial x} \tag{1}$$

where the symbols are defined in Table 81.1 and a similar expression is used for the force balance in the y dimension.

The other components of a coupled ice-flow model are fairly similar in all three types of model. They consist of (see details in Huybrechts, this volume, Chapter 80): a constitutive relation such as Glen's flow law (Glen, 1955) to convert the predicted stress field to a velocity field; models to predict the evolution of ice thickness and internal temperature field through time; possibly a link between the temperature of ice and its rheology; as well as the use of the continuity equation to diagnose vertical velocity once the horizontal velocity field is known from Equation (1) or similar. Other components may be incorporated to account for processes such as snow accumulation and ablation, basal slip (as a function of gravitational driving stress) and isostasy.

Zero-order models of this type are typically used to study the evolution of the Antarctic Ice Sheet (AIS) over long time-scales of tens to hundreds of thousands of years. They are surprisingly realistic in their depiction of the location of ice streams in Antarctica. Figure 81.1 shows the results from such a model. The positions of many of the major ice streams are depicted correctly and can be compared with their locations derived from the use of balance flux calculations (e.g. Budd & Warner, 1996). The latter assumes that the present-day geometry of Antarctica is in equilibrium and routes the ice flux generated by snow accumulation across the surface of the ice sheet by assuming that ice always flows downhill.

Table 81.1 Definition of symbols used in the text (appropriate units in brackets)

Symbol	Definition
x,y	Horizontal coordinates (m)
z	Vertical coordinate (positive upwards, m)
s	Elevation of upper ice surface (m)
H	Ice thickness (m)
h	Elevation of lower ice surface (bedrock, m)
g	Acceleration due to gravity ($m\,s^{-2}$)
ρ	Ice density ($kg\,m^{-3}$)
τ_{xz},τ_{yz}	Vertical shear stresses (Pa)
τ_{xy}	Horizontal shear stress (Pa)
$\tau_{xz}(h),\tau_{yz}(h)$	Basal tractions (Pa)
σ_x,σ_y	Normal stresses (Pa)
u_x,u_y	Horizontal velocity components ($m\,yr^{-1}$)

The ice streams in Figure 81.1 are generated by the interaction of ice flow, surface topography and temperature. In the model, ice is assumed to slip over its bed only where the bed is predicted to be at melting point, and the amount of slip is a function of the amount of water at the bed. The fact that the location of many ice streams can be predicted accurately by such a simple model implies that many owe their locations to troughs in the underlying bedrock topography. It is likely that the deeper ice in troughs will flow faster (gravitational driving stresses are larger and velocities are further amplified by the non-linear nature of Glen's flow law). In addition, deeper ice is likely to be warmer (and hence softer) because of the enhanced insulation afforded by thicker ice and the larger amounts of frictional heat generated by the flow.

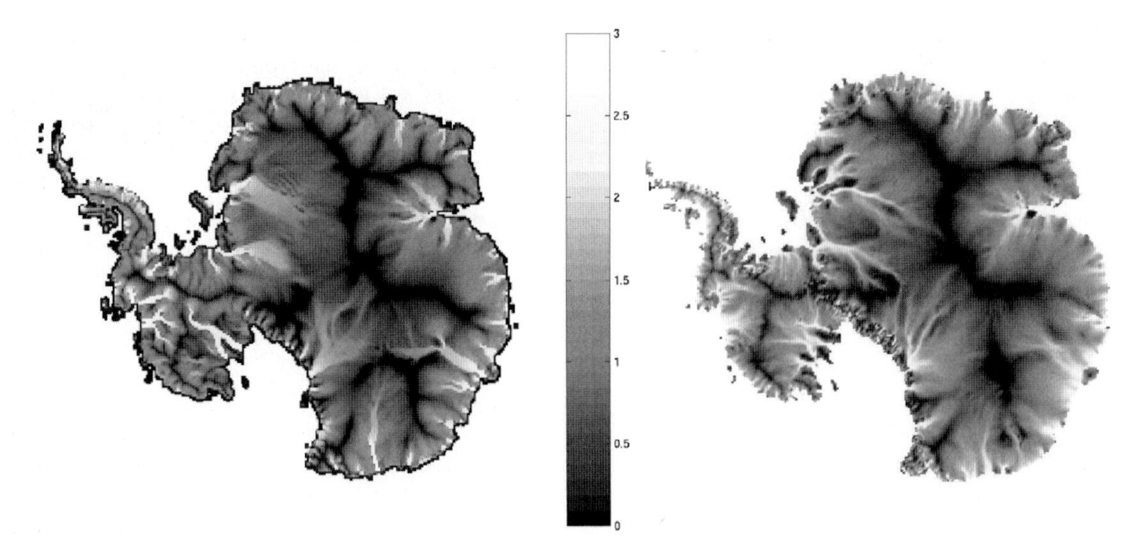

Figure 81.1 A comparison of the predicted ice flux from a numerical model of the present-day Antarctic Ice Sheet (left) with the balance flux determined by routing snow accumulation across the ice sheet (right, after Bamber *et al.*, 2000a). The same scale ($\log_{10}$ m yr^{-1}) is used in both panels. The numerical model is based on Payne (1998) and uses geothermal heat flux of 0.06 W m^{-2} and a slip coefficient B ($u_x(h) = B\tau_{xz}(h)$) of 1×10^{-2} m yr^{-1} Pa^{-1} in West Antarctica and 1×10^{-2} m yr^{-1} Pa^{-1} in East Antarctica. (See www.blackwellpublishing.com/knight for colour version.)

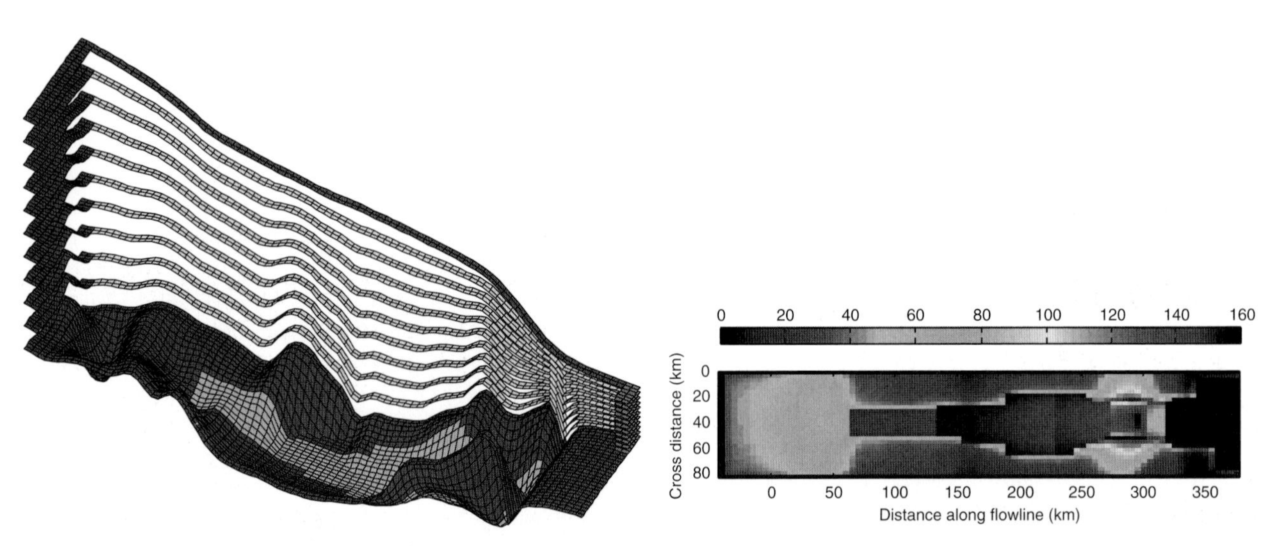

Figure 81.2 Illustration of the numerical domain (upper panel) of a three-dimensional first-order model applied to Pine Island Glacier, West Antarctica. The type of boundary condition applied to each exterior surface is: zero traction along the upper surface (air interface) and the lower surface of the ice shelf (water interface); a force balance with the weight of displaced ocean water at the ice-shelf front; a specified (zero) velocity on the other (upstream and side) lateral boundaries and on the bed outside of the ice stream; and a viscous traction law applied to the bed of the ice stream itself. Four of the many intermediate model layers are shown to illustrate the terrain-following nature of the model. Example results (lower panel) show the calculated basal traction (in Pa). Note the enhanced tractions immediately adjacent to the margins of the ice stream, which arise through the lateral transfer of stress away from the weak bed underlying the ice stream. (See www.blackwellpublishing.com/knight for colour version.)

Although simple, zero-order models may be sufficient to correctly predict the locations of ice streams, it is unlikely that they will be able to predict the flow of ice within the streams accurately. Higher order models are therefore used, which do not make so many restrictive assumptions about the force budget of the ice mass. The mostly commonly used model is the first-order one in which all stress components are considered

$$\frac{\partial \sigma_x}{\partial x} + \frac{\partial \tau_{xy}}{\partial y} + \frac{\partial \tau_{xz}}{\partial z} = \rho g \frac{\partial s}{\partial x} \tag{2}$$

and its vertically integrated form

$$\frac{\partial H \sigma_x}{\partial x} + \frac{\partial H \tau_{xy}}{\partial y} + \tau_{xz}(h) = \rho g H \frac{\partial s}{\partial x} \tag{3}$$

where the third term on the left-hand side represents basal traction and is often parameterized using a linear relation to slip velocity (MacAyeal, 1989). In both cases, similar expressions are used for the force balance in the y dimension and, in practice, the equations are solved by using Glen's flow law to substitute the relevant horizontal velocity components for the stress components.

Models based around Equation (3) have relatively low computational demands because they are two dimensional (x and y). However, Equation (3) is strictly only appropriate to the case of plug flow, where vertical shear (and τ_{xz}) is minimal, and is therefore ideally used to model floating ice shelves that experience no basal traction. However, it is also often used to study the dynamics of ice streams (e.g. Joughin *et al.*, 2001), where the plug flow assumption is more questionable.

More recently, models based on the three-dimensional form of the first-order equations (e.g. Equation (2)) have been developed (e.g. Pattyn, 2003). These models obviously have far higher computational demands but are generally applicable to any flow regime. Figure 81.2 shows an example of the application of this type of model to the study of Pine Island Glacier, West Antarctica.

The final type of flow model (second order) is rarely used in practice and is strictly only required at spatial scales well below the magnitude of the local ice thickness. The derivation of this model differs from that of the first-order ones in that the vertical force balance is explicitly coupled with the horizontal force balances.

In summary, the numerical modelling of ice flow has reached an important milestone and the community now has models that are generally applicable without the need to make a priori assumptions about the nature of the particular flow regime to be modelled. The future challenge lies in integrating these models effectively with the wide range of information now becoming available from satellite sensors.

EIGHTY-TWO

Estimating basal properties of glaciers from surface measurements

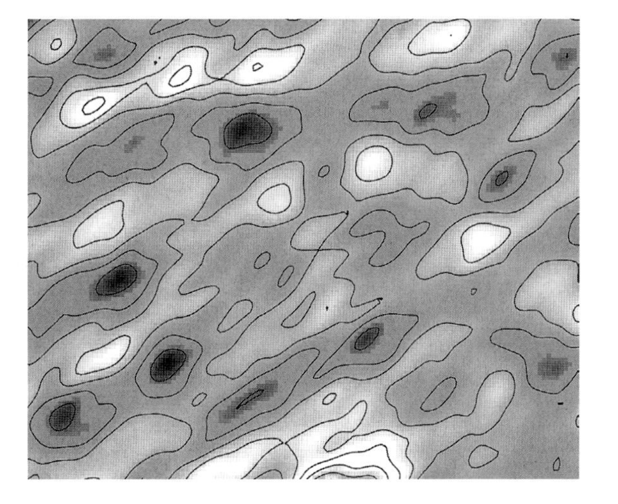

G. Hilmar Gudmundsson

British Antarctic Survey, High Cross, Madingley Rd, Cambridge CB3 0ET, UK

82.1 Introduction

Estimating the bed topography and basal slipperiness indirectly through their effect on surface shape and surface velocity is an example of an inverse problem. If we put all available surface measurements into a vector $\boldsymbol{y}$, denote the basal properties through $\boldsymbol{x}$, and write the relationship between bed and surface as $\boldsymbol{y} = \boldsymbol{f}(\boldsymbol{x})$ where $\boldsymbol{f}$ is the forward model, the inverse problem is that of determining the conditional probability distribution function (PDF; $P(\boldsymbol{x} \mid \boldsymbol{y})$) of a system state given the measurement vector. This PDF also can be written as $P(\boldsymbol{b},\boldsymbol{c} \mid \boldsymbol{s},\boldsymbol{u},\boldsymbol{v},\boldsymbol{w})$ where $\boldsymbol{b}$ and $\boldsymbol{c}$ are the bed topography and the basal slipperiness, respectively, and $\boldsymbol{s}$,

$\boldsymbol{u}$, $\boldsymbol{v}$ and $\boldsymbol{w}$ the surface topography and the three components of the surface velocity vector.

The forward model $\boldsymbol{f}$ may well not have an inverse in the usual sense. We can, for example, expect that for some $\boldsymbol{x}$, $f(\boldsymbol{x}) = 0$, indicating that some properties of the bed have no effect on the surface. Using Bayes' theorem, however, this conditional probability also can be written as

$$P(\boldsymbol{x}|\boldsymbol{y}) = P(\boldsymbol{y}|\boldsymbol{x})P(\boldsymbol{x})/P(\boldsymbol{y})$$

where $P(\boldsymbol{y})$ and $P(\boldsymbol{x})$ are the prior probability density functions of the measurement vector and the system state, respectively, and $P(\boldsymbol{y}|\boldsymbol{x})$ is the conditional PDF of $\boldsymbol{y}$ given $\boldsymbol{x}$. Instead of trying to find the inverse of f, which in general will not exist, we use Bayes' theorem to calculate $P(\boldsymbol{x}|\boldsymbol{y})$ from $P(\boldsymbol{y}|\boldsymbol{x})$. Assuming that the PDFs are Gaussian and the forward model can be linearized so that $\boldsymbol{y} = \boldsymbol{Fx} + \boldsymbol{\varepsilon}$ where $\boldsymbol{F}$ is a matrix and ε the measurement errors, taking the minus logarithm of the above expression and maximizing with respect to $\boldsymbol{x}$ leads to a maximum a posteriori solution $\boldsymbol{x}'$ of the inverse problem given by

$$\boldsymbol{x}' = \boldsymbol{x}_{\mathrm{a}} + \boldsymbol{R}_a\boldsymbol{F}^H(\boldsymbol{F}\boldsymbol{R}_a\boldsymbol{R}^H + \boldsymbol{R}_\varepsilon)^{-1}(\boldsymbol{y} - \boldsymbol{F}\boldsymbol{x}_{\mathrm{a}})$$

where $\boldsymbol{x}_{\mathrm{a}}$ is the a priori value of $\boldsymbol{x}$, $\boldsymbol{R}_{\mathrm{a}}$ and $\boldsymbol{R}_\varepsilon$ are the covariance matrixes of $\boldsymbol{x}_{\mathrm{a}}$ and $\boldsymbol{\varepsilon}$, respectively, and the superscript H denotes the conjugate transpose (Rodgers, 2000). The above equation gives an estimate for the system state ($\boldsymbol{x}$) that can be interpreted as a weighted sum of the a priori ($\boldsymbol{x}_{\mathrm{a}}$) and the new information on the system state gained by the measurement ($\boldsymbol{y}$).

82.2 The forward problem

The forward problem consists of determining the surface shape and surface velocities given the bed topography, basal slipperiness and the form of the constitutive equation. The equations to be solved are $\sigma_{ij,j} + f_i = 0$, and $v_{ij} = 0$ where σ_{ij} are the components of the symmetrical Cauchy stress tensor, v_i are the components of the velocity vector and f_i the components of the volume force. A Cartesian coordinate system is used with the x and y axes spanning the horizontal plane, and with the z axis pointing upwards. The constitutive law is Glen's flow law with a stress exponent n. The boundary conditions are the kinematic boundary conditions at the upper ($s = s(x,y)$) and lower ($b = b(x,y)$) boundaries, free surface condition for the stresses at the upper boundary and a sliding law, $u_b = c(x,y)\tau_b$, along the lower boundary, where u_b is the basal sliding velocity, τ_b the basal shear stress and $c(x,y)$ the basal slipperiness. This problem is non-linear because the constitutive law is non-linear for $n \neq 1$ and because the surface fields react in a non-linear fashion to a finite perturbation in either bed topography ($b(x,y)$) or basal slipperiness ($s(x,y)$).

The long-wavelength sensitivity of the forward solution to changes in basal properties can be determined analytically for non-linear rheology. For arbitrary length scales, analytical solutions exist for linear rheology and numerical solutions for non-linear rheology (Hutter, 1983; Reeh, 1987; Jóhannesson, 1992; Gudmundsson, 2003; Raymond *et al.*, 2003, Hindmarsh, 2004). In either case the perturbations in surface topography ($s(x,y)$) and the surface velocity (u,v,w), are related in frequency space through transfer functions, for which analytical expressions exist (Gudmundsson, 2003). This relation can be written in the form $\boldsymbol{y} = \boldsymbol{Fx}$, where $\boldsymbol{y}$ is a vector containing all surface measurements of surface topography and surface velocities, $\boldsymbol{x}$ is a vector containing the to-be-estimated basal slipperiness and bed topography, and $\boldsymbol{F}$ is a matrix with the transfer functions as elements.

82.3 Inverse formulation

The inverse problem consists in determining the maximum of the conditional probability density $P(\boldsymbol{b},\boldsymbol{c}|\boldsymbol{s},\boldsymbol{u},\boldsymbol{v},\boldsymbol{w})$ and the covariance of that maximum for $\boldsymbol{b}$ and $\boldsymbol{c}$, given surface data and the corresponding covariance matrixes. If the measurements of the surface topography and the measurements of each of the surface velocity components are independent, and the bed topography independent of the slipperiness distribution, $P(\boldsymbol{b},\boldsymbol{c}|\boldsymbol{s},\boldsymbol{u},\boldsymbol{v},\boldsymbol{w})$ can be written as a product of eight PDFs. Each PDF gives the conditional probability of one surface variable ($\boldsymbol{s},\boldsymbol{u},\boldsymbol{v},\boldsymbol{w}$) with regard to one unknown basal variable ($\boldsymbol{b},\boldsymbol{c}$). Surface and basal quantities ($\boldsymbol{s},\boldsymbol{u},\boldsymbol{v},\boldsymbol{w},\boldsymbol{b}$ and $\boldsymbol{c}$) are assumed to be arranged in columns. Assuming all PDFs are Gaussian, maximizing the minus of the logarithm of that product with respect to both $\boldsymbol{b}$ and $\boldsymbol{c}$ leads to

$$\boldsymbol{b} = (\boldsymbol{R}_{\mathrm{b}}')^{-1}(\boldsymbol{F}_{\mathrm{sb}}^H\boldsymbol{R}_{s\varepsilon}^{-1}\boldsymbol{s} + \boldsymbol{F}_{su}^H\boldsymbol{R}_{u\varepsilon}^{-1}\boldsymbol{u} + \boldsymbol{F}_{sv}^H\boldsymbol{R}_{v\varepsilon}^{-1}\boldsymbol{v} + \boldsymbol{F}_{sw}^H\boldsymbol{R}_{w\varepsilon}^{-1}\boldsymbol{w})$$

where the $\boldsymbol{F}$s are the corresponding forward models (for example $\boldsymbol{s} = \boldsymbol{F}_{\mathrm{sb}}\boldsymbol{b}$), $\boldsymbol{b}$ is the vector giving elevation at each grid location and

$$(\boldsymbol{R}_{\mathrm{b}}')^{-1} = \boldsymbol{F}_{\mathrm{sb}}^H\boldsymbol{R}_{s\varepsilon}^{-1}\boldsymbol{F}_{\mathrm{sb}} + \boldsymbol{F}_{\mathrm{ub}}^H\boldsymbol{R}_{u\varepsilon}^{-1}\boldsymbol{F}_{\mathrm{ub}}^{-1} + \boldsymbol{F}_{\mathrm{vb}}^H\boldsymbol{R}_{v\varepsilon}^{-1}\boldsymbol{F}_{\mathrm{vb}} + \boldsymbol{F}_{\mathrm{wb}}^H\boldsymbol{R}_{w\varepsilon}^{-1}\boldsymbol{F}_{\mathrm{wb}} - \boldsymbol{R}_{\mathrm{b}}^{-1}$$

An estimate for the slipperiness distribution ($\boldsymbol{c}'$) is of the same form but with $\boldsymbol{c}$ replacing $\boldsymbol{b}$ in all lower indices. To keep the notation simple it has been assumed that the a priori values of bed and the basal slipperiness are equal to zero.

The above expression gives the solution as a function of x and y. To use the transfer function formulation, the surface data must be transformed into frequency space. This usually requires the surface data to be interpolated onto an equidistant grid. This will give rise to interpolation errors and some spatial correlations between interpolation values, both of which must be estimated. Methods of optimal interpolation can be used for this purpose. For the surface topography, for example, we look for an estimate in the form of $\boldsymbol{s}' = \boldsymbol{Fs} + \boldsymbol{\varepsilon}$, where $\boldsymbol{F}$ is now an unknown matrix to be determined. The unbiased minimum variance interpolator is given by $\boldsymbol{s}' = \boldsymbol{s}_{sa} = \boldsymbol{R}_{sa}\,(\boldsymbol{R}_{sa} + \boldsymbol{R}_\varepsilon)^{-1}$ and the covariance matrix of the interpolated values by $\boldsymbol{R}_s' = \boldsymbol{R}_{sa} - \boldsymbol{R}_{sa}\,(\boldsymbol{R}_{sa} + \boldsymbol{R}_\varepsilon)^{-1}\boldsymbol{R}_{sa}$. The $\boldsymbol{R}_{\mathrm{sa}}$ covariance matrix can be determined from an analysis of the experimental variogram (Kitanidis, 1997). Fourier transform of the interpolated surface fields and the covariance matrices leads to an estimate having the same form as the one given above, but with, for example, $\boldsymbol{s}$ substituted by $\boldsymbol{Ws}$ and $\boldsymbol{R}_s$ by $\boldsymbol{W}\boldsymbol{R}_s\boldsymbol{W}^H$, where $\boldsymbol{W}$ is the discrete Fourier matrix. If the noise and the a priori fields are uncorrelated, the corresponding covariance matrixes are of diagonal form and the expressions given above can be simplified considerably.

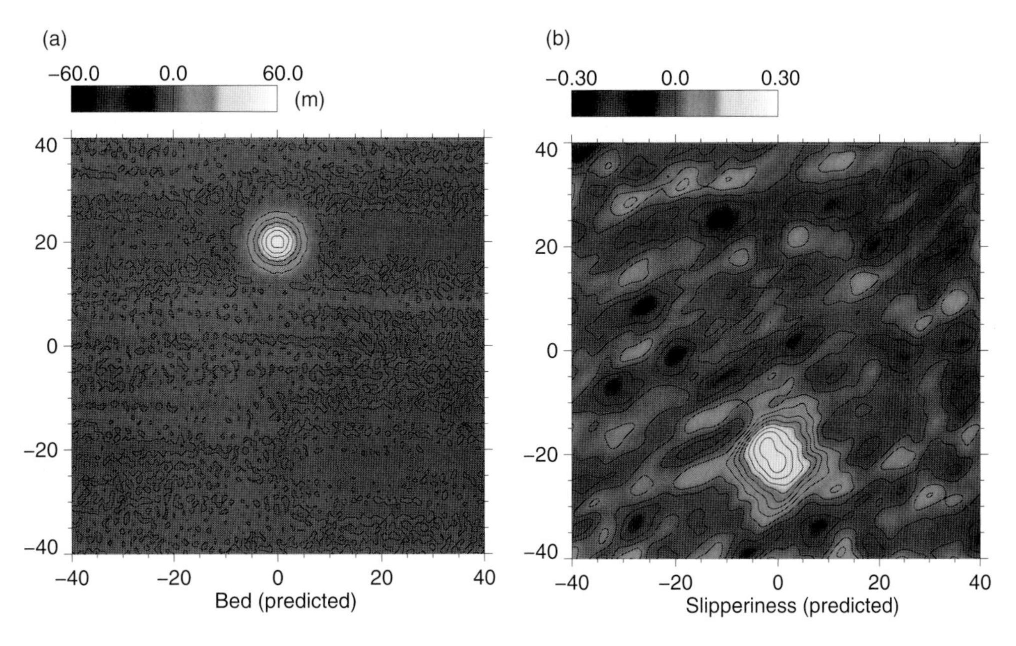

Figure 82.1 Retrieved bed topography and basal slipperiness distribution. The units on the x and y axes are kilometres. The contour interval for the upper figure is 7.5 m and 0.025 m in the lower figure. (See www.blackwellpublishing.com/knight for colour version.)

82.4 Application example

As an example for a simultaneous inversion for bed topography and basal slipperiness, synthetic surface data were generated using a Gaussian shaped perturbation in bed topography and slipperiness distribution. For the forward problem, 30% uncorrelated noise was added to the solution and the surface data were then inverted. Figure 82.1 shows the retrieved bed topography and basal slipperiness distribution. The (total) basal slipperiness perturbation is $C\Delta C$, where C is the mean value of the basal slipperiness and ΔC the (fractional) perturbation. Figure 82.1b shows the contour lines of the retrieved fractional perturbation (ΔC). In Fig. 82.2 a transverse section along the y axis gives a comparison between the amplitude of the initial (dashed lines) and the retrieved (solid lines) perturbations, with the left-hand peak representing the slipperiness perturbation and the right-hand peak the bed perturbation.

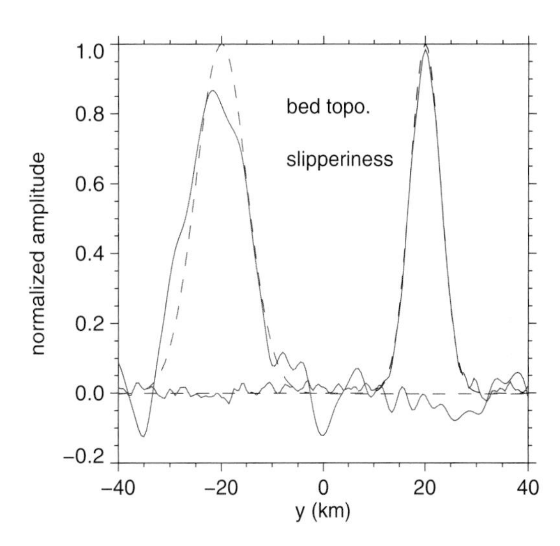

Figure 82.2 Transverse section along the $x = 0$ line in Fig. 82.1 showing the agreement between retrieved (dashed lines) and original bed properties for bed topography (right-hand peaks) and basal slipperiness (left-hand peaks). (See www.blackwellpublishing.com/knight for colour version.)

EIGHTY-THREE

Measuring and modelling the mass balance of glaciers for global change

Roger J. Braithwaite

School of Environment and Development, The University of Manchester, Manchester M13 9PL, UK

83.1 Introduction

Glacier mass balance study is concerned with changes in glacier mass through time, and especially the changes from year to year (Ahlmann, 1948; Meier, 1962; Anonymous, 1969; Østrem & Stanley, 1969; Østrem & Brugman, 1991; Paterson, 1994; Kaser *et al.*, 2002). Glacier mass balance forms the vital link between the changing atmospheric environment and glacier dynamics and hydrology. The most obvious present-day connection between glacier mass balance and global change is the rise in global sea level that will occur if glacier melting increases in response to global warming. Thorarinsson (1940) is usually credited as the first scientist to study the effect of present-day glacier changes on sea level. He stated the problem in terms of the following (slightly paraphrased) questions:

1 Are present-day glaciers in equilibrium?
2 Are glacier changes synchronous, or are the effects of the growth of some offset by the shrinkage of others?
3 Is there a measurable variation in the volume of the oceans due to variations in glacier volume?

These questions are still relevant today but are still difficult to answer. Thorarinsson (1940) established our modern convention of discussing separate sea-level contributions from 'glaciers' (mountain glaciers and ice caps) and 'ice sheets' (Greenland and Antarctica). After a systematic survey of the literature for the major glacial regions, Thorarinsson (1940) concluded that there were widespread and similar amounts of glacier thinning in different regions. He estimated the then-current glacier contribution (late 19th and early 20th Century) to global sea-level rise to be about 0.5 mm yr^{-1}. Thorarinsson (1940) also noted that his estimates of glacier thinning would be improved by '. . . extending to an increased number of districts quantitative investigations into glacier regimes of the kind undertaken by Ahlmann and his assistants round the Norwegian Sea'. He is obviously referring here to the pioneering studies of glacier mass balance, summarized by Ahlmann (1948). Ahlmann's concepts have since been extended to several hundred glaciers in almost every part of the world (Dyurgerov & Meier, 1997; Cogley & Adams, 1998; Haeberli *et al.*, 1998a; Dyurgerov & Meier, 2000; Braithwaite, 2002; Dyurgerov, 2002, 2003; Dowdeswell & Hagen, 2004; Dyurgerov & Meier, 2004; Haeberli, 2004; Hagen & Reeh, 2004).

Interest in glaciers and sea level was revived in the 1980s by a seminal paper (Meier, 1984) and the results of an international workshop on the topic (Polar Research Board of the National Research Council, 1985). Not surprisingly, The Intergovernmental Panel on Climate Change (IPCC) considered sea-level changes in each of its scientific assessments of climate change (Warrick & Oerlemans, 1991; Warrick *et al.*, 1996; Church *et al.*, 2001). These conclude that the major source of sea level rise to 2100 under expected global warming ('business as usual') will be thermal expansion of sea water with significant extra contributions from increased melting of mountain glaciers and ice caps, excluding Greenland and Antarctica. We can surely all agree that glacier mass balance should be seen within the context of global climate change but very few glacier studies were started with this connection in mind.

83.2 Mass-balance measurements

Ahlmann (1948) first developed mass-balance concepts in a series of pioneering measurements in Nordic countries. The longest continuous measurements were started in 1946 on Storglaciären in northern Sweden (Schytt, 1962) and similar measurements started in the 1940s and 1950s on other glaciers in Norway, the Alps, western North America, and on numerous glaciers in the former USSR. The number of glaciers studied and their geographical coverage expanded rapidly in the 1960s under the impetus of the International Hydrological Decade (IHD) 1965–1974. Although the IHD involved an emphasis on the hydrological role of glaciers (Collins, 1984), interest has more recently shifted to the possible increased melting of glaciers as a cause of rises in global sea level, and recent analyses of mass-balance data exclusively stress this aspect of mass-balance study.

Regular tabulations of mass-balance data have been published in hard copy since 1967 as part of The *Fluctuations of Glaciers* series (Kasser, 1967, 1973; Müller, 1977; Haeberli, 1985; Haeberli & Müller, 1988; Haeberli & Hoelzle, 1993; Haeberli *et al.*, 1998b). The World Glacier Monitoring Service (WGMS), sponsored by UNESCO, now publishes this series and also distributes bi-annual summaries of mass balance data over the Internet, e.g. WGMS (2001). However, the latter data are very basic summaries and many prefer the more extensive information (to say nothing of the beautiful free maps!) provided by *Fluctuations of Glaciers*.

The format for the dissemination of data by the World Glacier Monitoring Service (WGMS), and its predecessors, reinforces the impression that 'true' mass-balance data are measured with stakes and snowpits, i.e. using the 'glaciological' method of Ahlmann (1948), and that hydrological and geodetic methods (Hoinkes, 1970) measure something else. The amount of published mass-balance data based on stakes and snowpits has steadily increased over the years, both with respect to number of glaciers studied and length of series for individual glaciers. Collins (1984) could only find annual mass-balance data for 95 glaciers, Dyurgerov & Meier (1997) identify 257 glaciers, Cogley & Adams (1998) identify 231 glaciers and Braithwaite (2002) analyses data for 246 glaciers. Dyurgerov (2002) compiles some previously overlooked data, especially from American and Russian glaciers, and corrects many published mass-balance series to give data for 280 glaciers. There are numerous minor discrepancies between the different data compilations but we can say that annual mass balance has been measured for at least one year on about 300 glaciers. This latter figure is reached by combining data from Braithwaite (2002) and Dyurgerov (2002).

In principle, it should be possible to calculate the present contribution to sea-level rise from glaciers and ice caps by averaging available measurements, but we must recognize the relatively large variability of mass balance in both space and time; see figs 8 & 9 in Braithwaite (2002). Available data for 300 glaciers are strongly biased to western Europe, North America and the former USSR and are further biased to wetter conditions; see figs 6 & 7 in Braithwaite (2002). It therefore seems implausible to me that averages of available data can adequately describe mass-balance variations over the whole globe. For example, the chance inclusion/exclusion of a particular glacier for a particular year will bias the average mass balance for that year.

Despite the above scepticism, results from Dyurgerov & Meier (2000, 2004) and Meier *et al.* (2003) deserve mention. Dyurgerov & Meier (2000) compare average mass balances for 1961–1976 and 1977–1997 with global average temperatures and infer a temperature sensitivity of $-0.37\,\mathrm{m\,yr^{-1}\,K^{-1}}$ for global glacier mass-balance. This estimate of mass-balance sensitivity to global temperature is remarkably close to the global mass-balance sensitivity estimated by Oerlemans (1993c); see below. This cosy agreement seems to be contradicted by table 16.2 in Dyurgerov & Meier (2004) but we might agree that the average of available massbalance measurements (Dyurgerov & Meier, 2000, 2004; Meier *et al.*, 2003) is somehow picking up real variations in global mass balance.

For the future, we should establish new measurements in key areas where feasible but also supplement the 'glaciological' method with modern implementations of the 'geodetic' method, for example, using laser altimetry (Sapiano *et al.*, 1998; Arendt *et al.*, 2002). This will not greatly improve the global coverage of mass balance data, however, and we will always deal with 'a data set that is sparse geographically' in the words of Meier (1993). If existing measurements are not adequate for inferences on a global scale they must be supplemented by modelling.

83.3 Mass-balance modelling

The simplest model for mass balance and climate is the regression model, where an annual series of mean specific balance is correlated with parallel summer temperature and annual, or winter, precipitation series at some nearby climate station. Liestøl (1967), Martin (1975), Braithwaite (1977), Tangborn (1980), Günther & Widlewski (1986), Holmlund (1987), Letréguilly (1988), Laumann & Tvede (1989), Chen & Funk (1990), Trupin *et al.* (1992) and Müller-Lemans *et al.* (1994) have made such studies. The regression coefficient can be interpreted as the sensitivity of mass balance to changes in the corresponding climate variables. The regression model is simple to use and provides a correlation coefficient to give an idea of 'goodness of fit'. The main practical disadvantages are that a few years of observed data (5–10 yr?) are needed to give meaningful results and that a nearby climatic station is needed. The need to have a few years of mass-balance record drastically reduces the number of glaciers available for modelling. Wood (1988) could find only 53 glaciers with at least 7 yr of record and even up to 1999 there are only 143 glaciers with at least 7 yr of record (188 glaciers with at least 5 yr of record).

The fundamental objection to the regression model is that its coefficient reflects conditions prevailing during the period in which the data are measured and will not be valid under a greatly different climate with a changing surface-area distribution. Ultimately there can be no unique relation between glacier mass balance and temperature because, under a constant temperature, glaciers will always tend towards a steady state. Nevertheless, regressions of mass balance on summer temperature yield sensitivities that are comparable, although slightly lower, to those obtained by more advanced models. For examples, see table 6 in Braithwaite & Zhang (2000) and table 1 in Braithwaite *et al.* (2003) for details. High-latitude glaciers have a low regression coefficient for summer temperature, whereas Alpine glaciers have a higher coefficient (Braithwaite, 1977).

If you want a mass-balance model to drive an ice-dynamics model, you have to calculate mass balance at different altitudes, i.e. to use a distributed mass-balance model rather than a lumped model such as the regression model (above). The most obvious models here are degree-day and energy-balance models, where the terms 'degree day' and 'energy balance' refer to the ways in which melting is calculated, although the models must also have routines to calculate the accumulation of snow.

Energy balance models for single glaciers are very demanding in terms of data and are best used for glaciers that are well instrumented. Examples of the approach are by Escher-Vetter (1985),

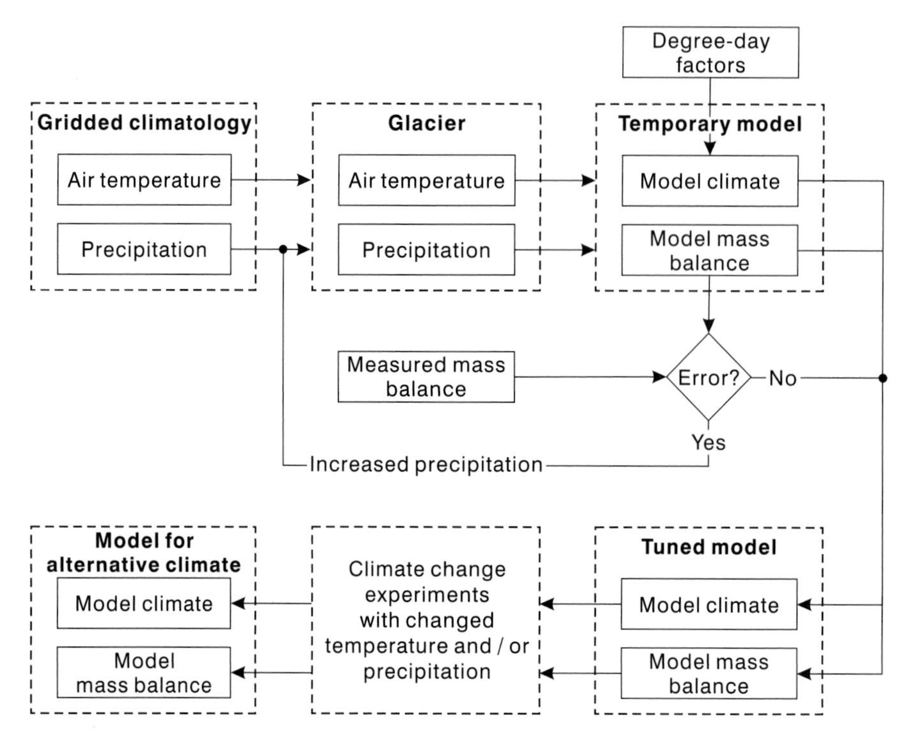

Figure 83.1 Box diagram illustrating the principle of tuning a glacier-climate model to fit observed mass-balance data (Braithwaite *et al.*, 2003).

Braithwaite & Olesen (1990), van de Wal & Russell (1994), Arnold *et al.* (1996), Hock & Holmgren (1996), Hock (1999), van den Broeke (1996) and Braithwaite *et al.* (1998). There is little doubt that the future of improved mass-balance models for single glaciers, e.g. for runoff forecasting, belongs to detailed energy-balance modelling although there are still problems to be solved, especially with respect to turbulent fluxes and albedo variations. By contrast, the energy-balance model of Oerlemans (1993b) is highly parameterized and hardly uses more than temperature and precipitation as input data. The model has been applied to only a few glaciers (Oerlemans & Fortuin, 1992; Fleming *et al.*, 1997), which might suggest that its application is laborious.

For a large-scale assessment of the sensitivity of glacier mass balance to climate change, we can also use the degree-day model. Different examples of this approach are by Braithwaite (1985), Braithwaite & Olesen (1989), Reeh (1991), Huybrechts *et al.* (1991), Laumann & Reeh (1993), Jóhannesson *et al.* (1995), Jóhannesson (1997), Braithwaite & Zhang (1999, 2000) and Braithwaite *et al.* (2003). Although de Woul & Hock (in press) use a degree-day model, their approach is regression based. The following outline is based on the version used by Braithwaite *et al.* (2003).

The accumulation for each month is calculated from monthly precipitation multiplied by the probability of below-freezing temperatures, which is in its turn calculated from monthly mean temperature assuming that temperatures are normally distributed within the month. Monthly melt is assumed proportional to the positive degree-day sum that is calculated from monthly mean temperature (Braithwaite, 1985). The proportionality factors (positive degree-day factors) linking melt to positive degree-day sum depend upon whether the melting refers to ice or snow. There is a range of values in the literature (Braithwaite & Zhang, 2000; Braithwaite, *et al.* 2003; Hock, 2003). Refreezing is estimated for subpolar glaciers by assuming that all annual melt is refrozen within the snow cover up to a definite fraction of the annual accumulation. This factor is taken to be 0.58 if snow density has to change from 375 to 890 $kg\,m^{-3}$ before runoff can occur (Braithwaite *et al.* (1994), but other estimates of initial snow density and ice density for runoff give melt/accumulation ratios of 0.4 to 0.7. The monthly temperature and precipitation are extrapolated to the altitude in question from a nearby weather station, or from a gridded climatology (New *et al.*, 1999). Consistent values of model parameters (degree-day factors for snow and ice and temperature lapse rate) and precipitation are chosen by 'tuning' the model, i.e. varying conditions in the model until the modelled distribution of mass balance with altitude agrees with the field data. Coarse tuning involves establishing suitable degree-day factors and temperature lapse rate for the whole glacier, and fine tuning involves varying the precipitation with altitude to obtain a nearly perfect fit between modelled and observed mass balance as a function of altitude (Fig. 83.1). There is an element of subjectivity in this tuning as several different combinations of model parameters will produce a similar fit of model to data (Braithwaite & Zhang, 2000). The precipitation obtained by fine tuning the model is the effective precipitation over the glacier, e.g. resulting from snow drift and topographic channelling, which is often higher than the precipitation value available in a gridded climatology, or read off a weather map. Oerlemans & Fortuin (1992) tune their energy balance model by a similar variation of model precipitation.

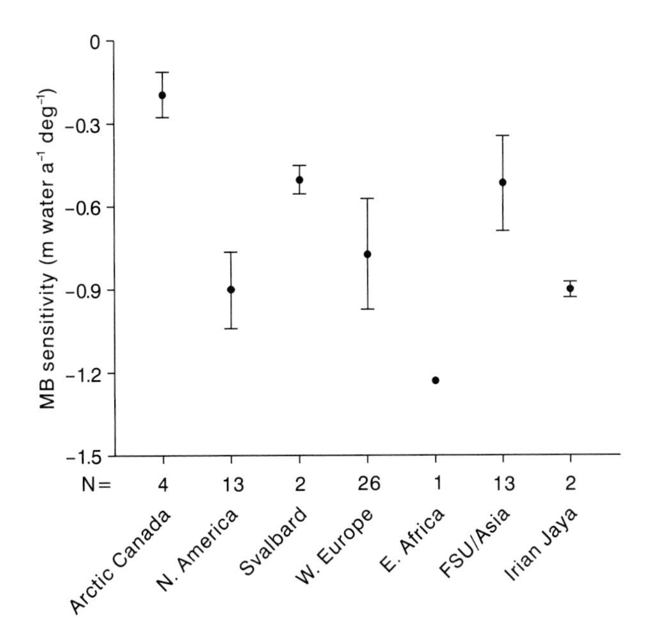

Figure 83.2 Mass-balance sensitivities for 61 glaciers in different glacierized regions. Results from Braithwaite *et al.* (2003).

After you have tuned the degree-day or energy-balance model to agree (more-or-less) with the field data, you can simulate the effects of climate change by changing the temperature and/or precipitation in the model. This necessarily involves the assumption that the model parameters do not change as climate changes. This may not be the case but it is difficult to know how the parameters should be changed.

In the following discussion we consider only the sensitivity of mass balance to a 1 K temperature change that is applied equally to every month in the year. This is a good 'standard' change for comparing results from different glaciers and different regions although real climate changes will involve different changes for different times of the year.

83.4 Upscaling from local to global scale

The simplest way to upscale a glaciological characteristic from the local to global scale is to assume that the global value is equal to the average of the available measurements (local values by definition). This seems unsatisfactory for the temperature sensitivity of glaciers because we know that the sensitivity varies greatly between glaciers (see Fig. 83.2), with lower sensitivities in the Arctic (Canada and Svalbard) and in the former Soviet Union and Asia and higher sensitivities in North America and in the tropics. The most attractive alternative is to correlate the measured temperature sensitivity values with some appropriate climatic element, and then to calculate regional and global values of sensitivity from regional and global values of the chosen climatic element. Oerlemans & Fortuin (1992) and Oerlemans (1993c) pioneered this approach by applying a glacier-climate model to 100 glaciological regions, ranging in size from 80,000 km^2 (Ellesmere Island) down to 1 km^2 (Mount Kenya). They calculated the temperature sensitivity of mass balance for 12 glaciers, correlated the resulting sensitivities with the logarithm (base 10) of precipitation, and then estimated the mass-balance sensitivity of each of their 100 glacierized regions from its estimated precipitation. The area-weighted average of mass balance sensitivity for the 100 regions is -0.39 m yr^{-1} K^{-1}. This represents the mass-balance sensitivity of global glacier cover to a globally uniform temperature change. It is a useful parameter for comparing different models.

The work by Oerlemans & Fortuin (1992) and Oerlemans (1993c) is very important as it informed the 1996 and 2001 assessments of sea-level change by the IPCC (Warrick *et al.*, 1996; Church *et al.*, 2001) and forms the basis for further work. Zuo & Oerlemans (1997) calculate the contribution of glacier melt to sea-level rise since 1865 by applying historic temperature series to the mass-balance sensitivities for the 100 regions, plus four more regions representing sectors of the Greenland Ice Sheet. Gregory & Oerlemans (1998) and van de Wal & Wild (2001) simulate future sea-level rise by applying temperature projections from GCMs to the same 100 glacier regions. Despite the increasing sophistication of these experiments they are still based on results from only 12 glaciers upscaled to 100 regions and their validity rests on the correctness of the upscaling.

The precipitation for each of the 100 regions is based on 'climatological maps and data from climatological stations' (Oerlemans & Fortuin, 1992) or from a 'Climatic Atlas (of Europe, North America, Asia, etc)' (Zuo & Oerlemans, 1997). On the basis of 'a comparison between the precipitation rates obtained from maps and from the mass-balance studies on the 12 selected glaciers', Oerlemans & Fortuin (1992) increase precipitation in all 100 regions by 25%. This takes account of the fact that '. . . glaciers and ice caps tend to form in the wettest parts of terrain with a pronounced topography' but Oerlemans (2001, p. 115) says 'The correction applied, +25%, is just a guess, however'. It is difficult to assess the correctness of these procedures and I prefer a new approach, see below, that combines results from Oerlemans & Fortuin (1992) and Braithwaite *et al.* (2003).

The mass-balance sensitivities for the 12 glaciers common to Oerlemans & Fortuin (1992) and Braithwaite *et al.* (2003) are quite similar across a wide range from the Devon Island ice cap (Arctic Canada) to Ålfotbreen (Norway). Braithwaite *et al.* (2003) clearly distinguish between regional precipitation, in a gridded climatology or on a climatic map, and glacier precipitation on the glacier itself. The latter is generally larger than the former, in agreement with Oerlemans & Fortuin (1992), although their 25% augmentation of precipitation seems a little too small. Despite this, the mass-balance sensitivity from Braithwaite *et al.* (2003) agrees very well with Oerlemans (1993a) when plotted against glacier precipitation (Fig. 83.3). There is therefore no deep contradiction between results of energy-balance and degree-day models.

Aside from the uncertain provenance of the precipitation data in Oerlemans (1993c), it is inconvenient to use 100 glacier regions of widely varying size because climate models are usually based on a regularly spaced grid and climate-model results have to be interpolated to the 100 regions. Furthermore, glaciological conditions can vary widely within the single regions of Oerlemans (1993c). Braithwaite & Raper (2002) update the approach of

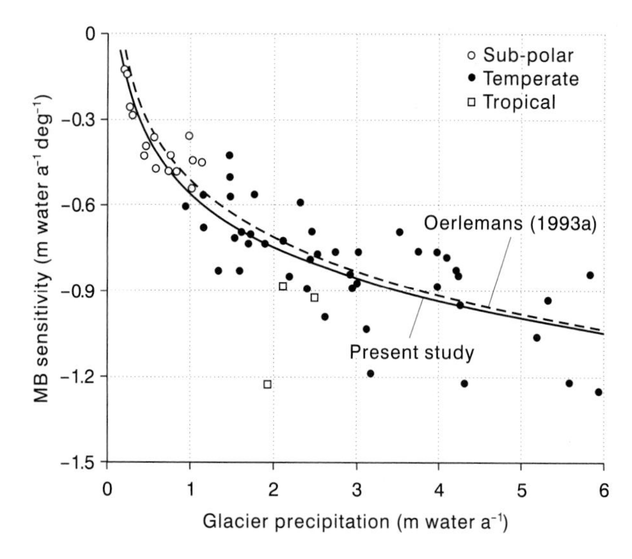

Figure 83.3 Mass-balance sensitivity for 61 glaciers versus glacier precipitation compared with a curve for the 12 glaciers modelled by Oerlemans (1993a). Results for the 61 glaciers are calculated by a degree-day model (Braithwaite *et al.*, 2003).

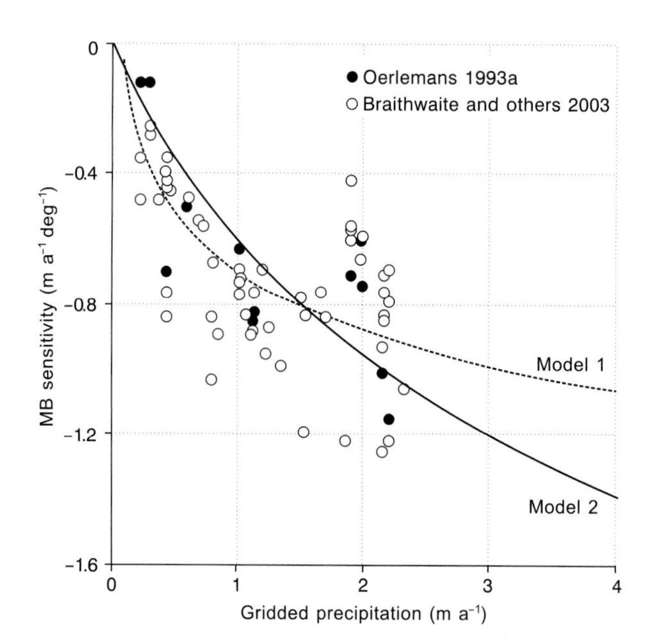

Figure 83.4 Mass-balance sensitivity versus gridded precipitation. Mass-balance sensitivity from Oerlemans (1993c) and Braithwaite *et al.* (2003) and precipitation data from the global gridded climatology of New *et al.* (1999). The two curves represent different logarithmic functions of precipitation.

Oerlemans (1993c) by correlating the mass-balance sensitivity with precipitation from a global gridded climatology with 0.5° resolution (New *et al.*, 1999). They use the resulting regression to estimate mass-balance sensitivity for every 1° latitude by 1° longitude grid square containing glaciers according to the gridded glaciology of Cogley (1998). Their global average of mass-balance sensitivity is $-0.41\,\mathrm{m\,yr^{-1}\,K^{-1}}$ (Braithwaite & Raper, 2002), which is in close agreement with $-0.39\,\mathrm{m\,yr^{-1}\,K^{-1}}$ (Oerlemans & Fortuin, 1992). There appears to be a consensus that the global average of mass-balance sensitivity is rather smaller than sensitivities found in mainland North America and western Europe (Fig. 83.2). This reflects the bias of global glacier cover towards Arctic glaciers with low mass-balance sensitivity.

There is some uncertainty about the area of glaciers, excluding ice sheets, in the world, e.g. somewhere between 0.52 and $0.68 \times 10^6\,\mathrm{km^2}$ (Braithwaite & Raper, 2002). With this range of areas, a mass balance sensitivity of $-0.4\,\mathrm{m\,yr^{-1}\,K^{-1}}$ is equivalent to a sea-level sensitivity of 0.6 to $0.8\,\mathrm{mm\,yr^{-1}\,K^{-1}}$. To answer Thorarinsson's (1940) third question (see Introduction), a sea-level sensitivity of this magnitude should be detectable over a few decades even if it is less than the temperature sensitivity of thermal expansion.

83.5 Future outlook

The apparent consensus above must not blind us to the need for future work. This point can be illustrated by Fig. 83.4 where results from Oerlemans (1993c) and Braithwaite *et al.* (2003) are plotted against gridded precipitation from the global climatology of New *et al.* (1999). Any correlation between mass-balance sensitivity could be used to calculate the global distribution of mass-balance sensitivity. However, two slightly different models achieve widely different results (Fig. 83.4). Model 1 involves the logarithm of precipitation (to base 10) following Oerlemans (1993a), whereas model 2 involves the logarithm of (precipitation + 1). Model 2 is more 'realistic' than model 1 because it passes through the origin, such that mass-balance sensitivity is zero when precipitation is zero, but model 1 gives a better fit to results for low precipitation. The two models diverge at high values of precipitation and we are left wondering whether the increase of mass-balance sensitivity really flattens out for glaciers with high precipitation as indicated by model 1. This is an important question when assessing the contribution to sea-level rise from 'wet' areas such as the Gulf of Alaska.

It is significant that de Woul & Hock (in press) find large mass-balance sensitivity for Icelandic glaciers, i.e. up to $-2.0\,\mathrm{m\,yr^{-1}\,K^{-1}}$ but data from many more glaciers than shown in Fig. 83.4 are needed to obtain a representative picture of mass balance sensitivity across the full precipitation range. Aside from the new data from Iceland, the approaches of Oerlemans (1993a) and Braithwaite *et al.* (2003) are unlikely to yield many more points to add to Fig. 83.4. There is, however, a new possibility based on the concept of an elevation on glaciers where a glaciological variable is similar to the area-weighted average of that variable for the whole glacier. For the special case that the glaciological variable varies linearly with altitude, the elevation is the area-weighted mean elevation of the glacier (Lliboutry, 1965). The mass-balance sensitivity was therefore calculated at the mean elevation for the 61 glaciers and compared with the corresponding sensitivity for the whole glacier. There is a strong correlation (Fig. 83.5) suggesting that mass-balance sensitivity need only be calculated at a

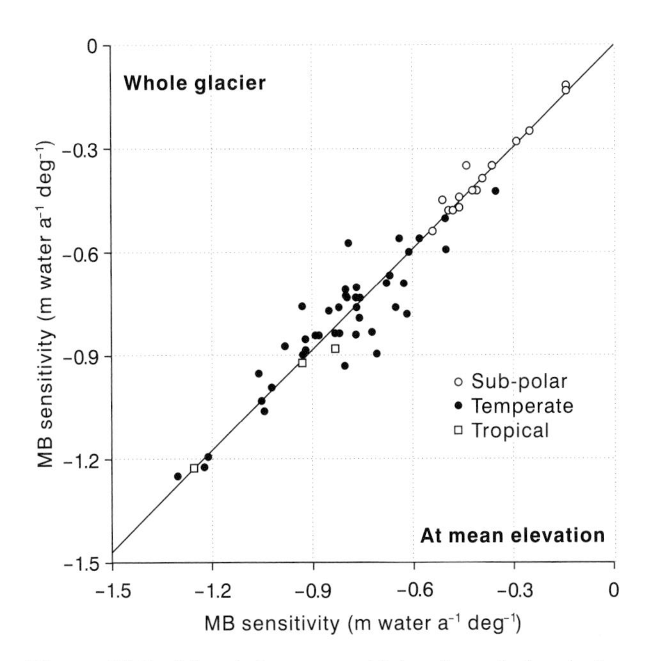

Figure 83.5 Mass-balance sensitivity for whole glaciers compared with mass balance sensitivity at the arithmetic mean elevation of the glacier. Both sensitivities calculated with the degree-day model of Braithwaite *et al.* (2003).

single characteristic altitude on the glacier rather than over the full altitude range. This is an important result because the degree-day model can now be applied to the estimated glacier equilibrium line altitude (ELA) over whole glacier regions. The mass balance at the ELA must be zero so the model can be tuned to give zero mass balance at this altitude and the degree-day model (and probably a simple energy-balance model) can be applied to thousands of glaciers where ELAs are roughly estimated from aerial photographs and topographic maps.

Acknowledgements

My current work on degree-day modelling was originally inspired by grants from NERC (GR9/01777 for 1995-1997) and the EU (ENV4-CT95-0124 for 1996–1999). I have had many stimulating discussions on mass-balance modelling with S. C. B. Raper, Alfred Wegener Institute for Polar and Marine Research, Bremerhaven, Germany.

EIGHTY-FOUR

Integrated perception of glacier changes: a challenge of historical dimensions

Wilfried Haeberli

World Glacier Monitoring Service (WGMS), Glaciology and Geomorphodynamics Group, Geography Department, University of Zurich, Switzerland

84.1 Introduction

Mountain glaciers are key variables for early detection strategies in global climate-related observations. Their changes have been observed systematically in various parts of the world for more than 100 yr. During this time, however, various aspects involved have changed in a most remarkable way. Future perspectives must envisage the possibility of dramatic evolutions, including the rapid deglaciation of entire mountain chains within decades. Worldwide documentation of such developments represents a challenge that must be met using the best available process understanding, methodology and strategy. The Global Terrestrial

Network for Glaciers (GTN-G) recently established as part of the Global Terrestrial Observing System (GTOS/GCOS) and operated by the World Glacier Monitoring Service (WGMS) follows a global hierarchical observing strategy. This integrative approach is based on a combination of **in situ, remote and numerical-modelling** components and consists of observations at several levels which link detailed process studies at one extreme with global coverage by satellite imagery and digital terrain information at the other. The present contribution briefly outlines the historical background and present challenge of worldwide glacier observations, introduces the integrative concept of the applied multilevel monitoring strategy, summarizes some principal aspects of past measurements and potential future scenarios and concludes with recommendations relating to the most urgent needs. It is based on numerous discussions with, and feed-back from, the staff members, national correspondents and principal investigators of the World Glacier Monitoring Service (WGMS) as well as colleagues involved with the ongoing Global Land Ice Measurements from Space (GLIMS) project led by the US Geological Survey.

84.2 Background and perspectives

Fluctuations of glaciers and ice caps have been observed systematically for more than 100 yr in various parts of the world (Haeberli *et al.*, 1998a) and are considered to be highly reliable indications of worldwide warming trends (cf. fig. 2.39a in IPCC, 2001). Mountain glaciers and ice caps are, therefore, key variables for early detection strategies in global climate-related observations. The internationally coordinated collection of information about ongoing glacier changes was initiated in 1894 with the foundation of the International Glacier Commission at the 6th International Geological Congress in Zurich, Switzerland. It was hoped that the long-term observation of glaciers would provide answers to the questions about global uniformity and terrestrial or extraterrestrial forcing of past, ongoing and potential future climate and glacier changes (Forel, 1895). Since then, various aspects involved have changed in a most remarkable way:

1 Concern is growing that the ongoing trend of worldwide and fast, if not accelerating, glacier shrinkage at the 100-yr time-scale is of non-cyclic nature—there is hardly a question any more of the originally envisaged 'variations périodiques des glaciers' (Haeberli *et al.*, 1998a).
2 Under the growing influence of human impacts on the climate system (enhanced greenhouse effect), dramatic scenarios of future developments—including complete deglaciation of entire mountain ranges—must be taken into consideration (Haeberli & Hoelzle, 1995; IPCC, 2001).
3 Such future scenarios may lead far beyond the range of historical/Holocene variability and most likely introduce processes (extent and rate of glacier vanishing, distance to equilibrium conditions) without precedence in the Holocene.
4 A broad and worldwide public today recognizes glacier changes as a key indication of regional and global climate and environment change.
5 Observational strategies established by expert groups within international monitoring programmes build on advanced process understanding and include extreme perspectives.
6 These strategies make use of the fast development of new technologies and relate them to traditional approaches in order to apply integrated, multilevel concepts (*in situ* measurements to remote sensing, local-process oriented to regional and global coverage), within which individual observational components (length, area, volume/mass change) fit together enabling a comprehensive view.

An international network of glacier observations such as the World Glacier Monitoring Service (WGMS) of the International Commission on Snow and Ice (ICSI/IAHS) and the Federation of Astronomical and Geophysical Data Analysis Services (FAGS/ICSU), together with its Terrestrial Network for Glaciers (GTN-G; Haeberli *et al.*, 2000) within the Global Terrestrial Observing System (GTOS) and the Global Climate Observing System (GCOS), is designed to provide quantitative and understandable information in connection with questions about process understanding, change detection, model validation and environmental impacts in a transdisciplinary knowledge transfer to the scientific community as well as to policy makers, the media and the public. This difficult but increasingly important task makes adequate perception of glacier changes a challenge of historical dimensions.

84.3 An integrated observational strategy

Within the framework of the global climate-related terrestrial observing systems, a Global Hierarchical Observing Strategy (GHOST) was developed to be used for terrestrial variables. According to a corresponding system of tiers, the regional to global representativeness in space and time of the records relating to changes in glacier mass and area should be assessed by more numerous observations of glacier length changes as well as by compilations of regional glacier inventories repeated at time intervals of a few decades—the typical dynamic response time of mountain glaciers (Haeberli *et al.*, 2000). The individual tier levels can be described as follows:

Tier 1 (multicomponent system observation across environmental gradients) Primary emphasis is on spatial diversity at large (continental-type) scales or in elevation belts of high-mountain areas. Special attention should be given to long-term measurements. Some of the already observed glaciers (e.g. those in the American cordilleras or in a profile from the Pyrenees through the Alps and Scandinavia to Svalbard) could later form part of Tier 1 observations along large-scale transects.

Tier 2 (extensive glacier mass balance and flow studies within major climatic zones for improved process understanding and calibration of numerical models) Full parameterization of coupled numerical energy-/mass-balance and flow models is based on detailed observations for improved process understanding, sensitivity experiments and extrapolation to areas with less comprehensive measurements. Ideally, sites should be located near the centre of the range of environmental conditions of the zone that they are representing. The actual locations will depend more on existing infrastructure and logistical feasibility rather than on strict spatial guidelines, but there is a need to capture a broad range of climatic zones (such as tropical, subtropical, monsoon-type, mid-latitude maritime/continental, subpolar, polar).

Tier 3 (determination of regional glacier volume change within major mountain systems using cost-saving methodologies) There are numerous sites to reflect regional patterns of glacier mass/volume change within major mountain systems, but they are not optimally distributed (Cogley & Adams, 1998). Observations with a limited number of strategically selected index stakes (annual time resolution) combined with precision mapping at about decadal intervals (volume change of entire glaciers) for smaller ice bodies or with laser altimetry/kinematic GPS (Arendt *et al.*, 2002) for large glaciers constitute optimal possibilities for extending the information into remote areas of difficult access. Repeated mapping and altimetry provide important data at lower time resolution (decades).

Tier 4 (long-term observations of glacier-length-change data within major mountain ranges for assessing the representativeness of mass balance and volume change measurements) At this level, spatial representativeness is the highest priority. Locations should be based on statistical considerations (Meier & Bahr, 1996) concerning climate characteristics, size effects and dynamics ('normal' flow versus effects from calving, surge, debris cover, etc.). Long-term changes of glacier length at a minimum of about 10 sites within each of the mountain ranges should be measured either *in situ* or with remote sensing techniques at annual to multi-annual frequencies.

Tier 5 (glacier inventories repeated at time intervals of a few decades by using satellite remote sensing) Continuous upgrading of preliminary inventories and repetition of detailed inventories using aerial photography or—in most cases—satellite imagery should enable global coverage and the validation of climate models (Beniston *et al.*, 1997). The use of digital terrain information in geographical information systems (GIS) greatly facilitates automated procedures of image analysis, data processing and modelling/interpretation of newly available information (Haeberli & Hoelzle, 1995; Kääb *et al.*, 2002; Paul *et al.*, 2002). Preparation of data products from satellite measurements must be based on a long-term programme of data acquisition, archiving, product generation and quality control.

This integrated and multi-level strategy aims at integrating *in situ* observations with remotely sensed data, process understanding with global coverage and traditional measurements with new technologies. Tiers 2 and 4 mainly represent traditional methodologies which remain fundamentally important for deeper understanding of the involved processes, as training components in environment-related educational programmes and as unique demonstration objects for a wide public. Tiers 3 and 5 constitute wide-open doors for the application of new technologies.

A network of 60 glaciers representing Tiers 2 and 3 is established. This step closely corresponds to the data compilation published so far by the World Glacier Monitoring Service with the biennial Glacier Mass Balance Bulletin and also guarantees annual reporting in electronic form. Such a sample of reference glaciers provides information on presently observed rates of change in glacier mass, corresponding acceleration trends and regional distribution patterns. Long-term changes in glacier length must be used to assess the representativeness of the small sample of mass balance values measured during a few decades with respect to the evolution at a global scale and during previous time periods. This can be done by (i) intercomparison between curves of cumulative glacier length change from geometrically similar glaciers, (ii) application of continuity considerations for assumed step changes between steady-state conditions reached after the dynamic response time (Haeberli & Holzhauser, 2003; Hoelzle *et al.*, 2003), and (iii) dynamic fitting of time-dependent flow models to present-day geometries and observed long-term length change (Oerlemans *et al.*, 1998). New detailed glacier inventories are now being compiled in areas not covered in detail so far or, for comparison, as a repetition of earlier inventories. This task is greatly facilitated by the launching of the ASTER/GLIMS programme (Kieffer *et al.*, 2000). Remote sensing at various scales (satellite imagery, aerophotogrammetry) and GIS technologies must be combined with digital terrain information (Kääb *et al.*, 2002; Paul *et al.*, 2002; Bishop *et al.*, 2004) in order to overcome the difficulties of earlier satellite-derived preliminary inventories (area determination only) and to reduce the cost and time of compilation. In this way, it should be feasible to reach the goals of global observing systems in the years to come.

84.4 Views to the past and to the future

Changes in glacier extent and volume provide important qualitative and quantitative information on pre-industrial variability, rates of change, acceleration tendencies and potential future scenarios relating to energy exchange at the earth–atmosphere interface over various time-scales. A brief examination of the main facts illustrates the key issues related to climate change questions.

Long-term *mass balance* measurements (Fig. 84.1) are available for about 50 glaciers and cover the past few decades. They provide direct (undelayed) signals of climate change and constitute the basis for developing coupled energy-balance/flow models for sensitivity studies. Extensive investigations explore complex feedback effects (albedo, surface altitude, dynamic response) and can be used in conjunction with coupled ocean–atmosphere general circulation models (model validation, hydrological impacts at regional and global scales, etc.; cf. Beniston *et al.*, 1997). Simpler observations using strategically selected index stakes furnish evidence of regional developments. Both types of monitoring combine direct glaciological with geodetic/photogrammetric methods in order to determine changes in volume/mass of entire glaciers (repeated mapping) with high temporal resolution (annual measurements at stakes and pits). Laser altimetry com-

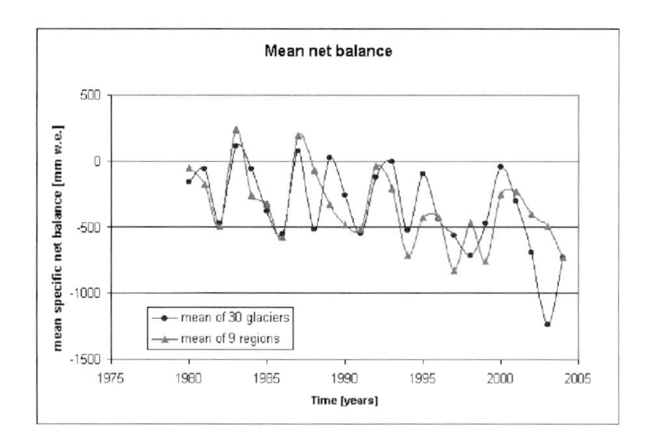

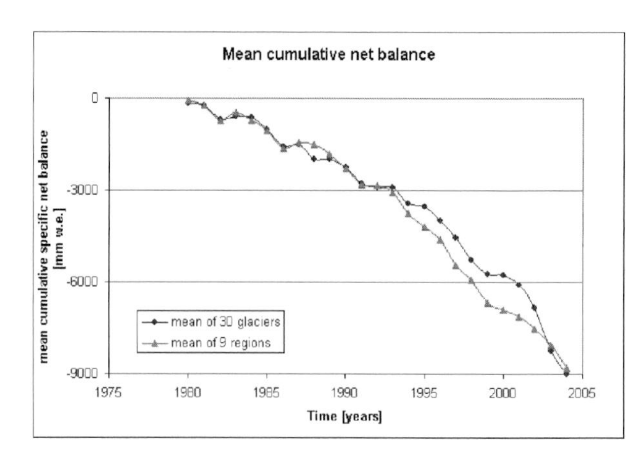

Figure 84.1 Annual (left) and cumulative (right) mass balances based on World Glacier Monitoring Service data.

bined with a kinematic Global Positioning System (GPS) is applied for monitoring thickness and volume changes of very large glaciers, which are the main melt-water contributors to ongoing sea-level rise (Arendt *et al.*, 2002).

The two decades 1980–2000 show a clear trend of increasingly negative balances with average annual ice thickness losses of a few decimetres. Because unchanged climatic conditions would cause mass balances to approach zero values, constantly non-zero mass balances reflect continued climatic forcing. The observed trend of increasingly negative mass balances is consistent with an accelerated trend in global warming and correspondingly enhanced energy flux towards the earth surface. There is considerable spatio-temporal variability over short time periods: glaciers around the North Atlantic, for instance, exhibited considerable mass increase during the recent past and the sensitivity of mass balance and melt-water runoff from glaciers in maritime climates is generally up to an order of magnitude higher than for glaciers in arid mountains (Oerlemans & Fortuin, 1992; Oerlemans, 2001). Statistical analysis indicates that spatial correlations of short-term mass balance measurements typically have a critical range of about 500 km (Cogley & Adams, 1998; Rabus & Echelmeyer, 1998) but tend to increase markedly with increased length of time period under consideration (as it applies to meteorological variables in general): decadal to secular trends are comparable beyond the scale of individual mountain ranges, with continentality of the climate being the main classifying factor (Letréguilly & Reynaud, 1990) in adition to individual hypsometric effects (Furbish & Andrews, 1984; Tangborn *et al.*, 1990).

Cumulative changes in glacier length are strongly enhanced and easily measured but indirect, filtered and delayed signals of climate change (Oerlemans, 2001). They represent an intuitively understood and most easily observed phenomenon to illustrate the reality and impacts of climate change. Trends in long time-series of cumulative glacier length represent convincing evidence of fast climatic change at a global scale, as the retreat of mountain glaciers during the 20th century is striking all over the world (Haeberli *et al.*, 1998a). Total retreat of glacier termini during the 20th century is commonly measured in kilometres for larger glaciers and in hundreds of metres for smaller ones.

Characteristic average rates of glacier thinning (mass loss) can be calculated from cumulative length-change data using a continuity approach over time periods corresponding to the dynamic response time of individual glaciers (Johannesson *et al.*, 1989; Haeberli & Hoelzle, 1995; Haeberli & Holzhauser, 2003; Hoelzle *et al.*, 2003). Characteristic values are a few decimetres per year for larger temperate glaciers in wet coastal climates and centimetres to a decimetre per year for small glaciers and glaciers in dry continental areas with firn areas below melting temperature (Fig. 84.2). At retreating glacier termini, the total secular surface lowering is up to several hundred metres. The apparent homogeneity of the signal at the century time-scale, however, contrasts with great variability at local/regional scales and over shorter time periods of years to decades (Letréguilly & Reynaud, 1990). Intermittent periods of mass gain and glacier readvance during the second half of the 20th century have been reported from various mountain chains (IAHS(ICSI)/UNEP/UNESCO, 1988, 1993, 1998), especially in areas of abundant precipitation such as southern Alaska, Norway and New Zealand. Glaciers in the European Alps, on the other hand, have lost about 50% of their original volume between the middle of the 19th century and 1970–1980 when systematic glacier inventories were compiled (Haeberli & Hoelzle, 1995). Rates of change and acceleration trends comparable to the ones observed during the past 100 yr must have taken place before, within the framework of Holocene glacier fluctuations and, hence, during times of weak anthropogenic forcing. In analogy to the glacier shrinkage documented during the 20th century, the Holocene record of Alpine glacier advance/retreat (Fig. 84.3) mirrors a (regional, hemispherical global?) pre-industrial variability of integrated secular to millennial energy flux towards or from the earth surface. As indicated by the finding of the Oetztal ice man, the 'warm' or 'high-energy' limit of this Holocene variability range may now have been reached in the Alps and possibly in other mountain regions, too. In such cases, continuation or acceleration of the observed trend could soon lead to conditions beyond those occurring during the Holocene precedence.

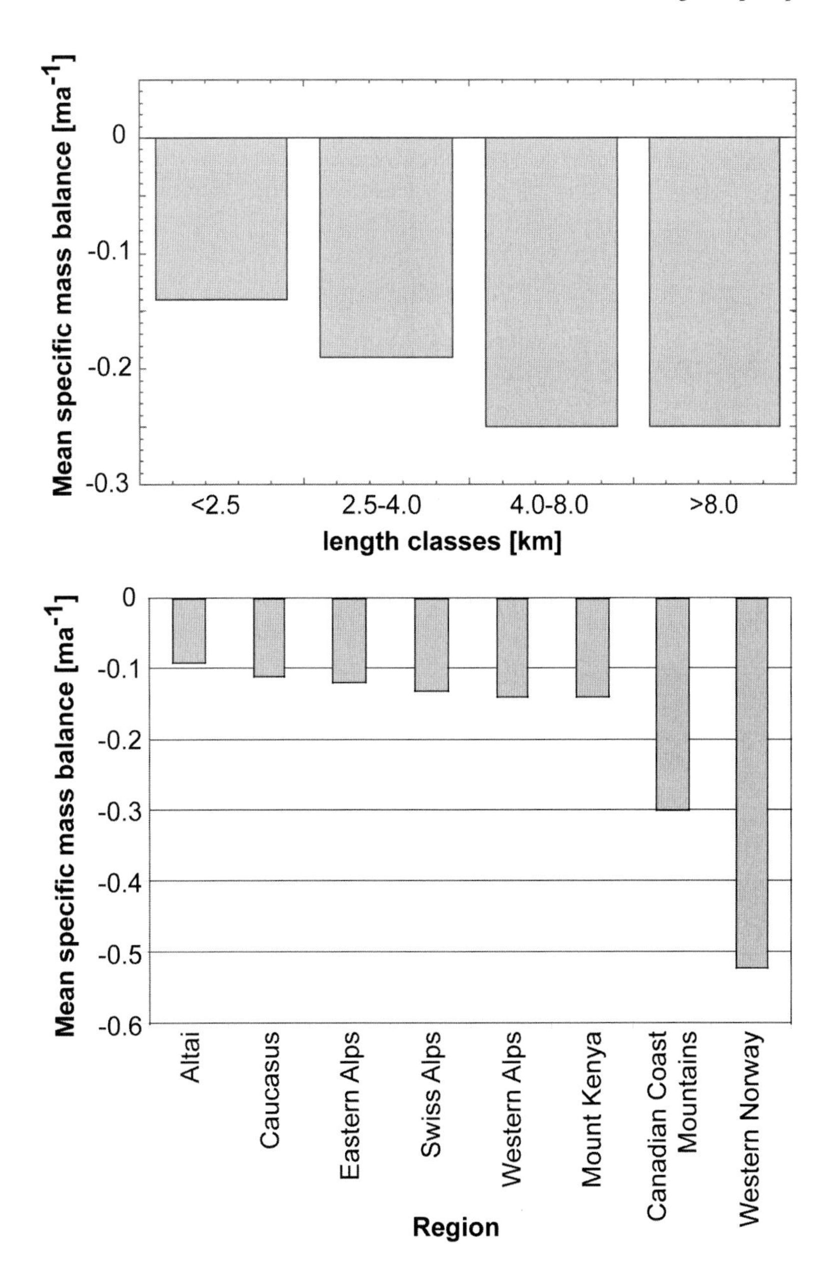

Figure 84.2 Characteristic mass balances during the 20th century as a function of glacier size (top) and region (bottom); values reconstructed from cumulative length change and continuity. (From Hoelzle *et al.*, 2003.)

With the information in detailed glacier inventories (highest and lowest point, area and length), continuity approaches in combination with assumed step changes in mass balance (Haeberli & Hoelzle, 1995) can be applied for calculations of climate-change effects over time periods of a few decades. This rough and simple approach enables realistic quantitative estimates for entire mountain ranges. For the European Alps, a loss in ice volume of at least one-quarter is estimated to have occurred since the 1970s (Haeberli *et al.*, 2002). The extremely hot and dry summer of the year 2003 alone may have caused the melting—within one single year—of some 5% or even more of what is left by now (Frauenfelder *et al.*, in press). With a realistic scenario of future atmospheric warming, almost complete deglaciation could occur within decades, leaving only some ice on the very highest peaks and in thick but downwasting rather than retreating glacier tongues (Table 84.1).

The complex chain of dynamic processes linking glacier mass balance and length changes is at present numerically simulated for only a few individual glaciers that have been studied in great detail (e.g. Greuell, 1992; Oerlemans & Fortuin, 1992; Schmeits & Oerlemans, 1997). A new possibility is to dynamically fit mass-balance histories to present-day geometries and historical length-change measurements of long-observed glaciers using time-dependent flow models (Oerlemans *et al.*, 1998). This approach not only provides important insights concerning mass balances during past periods that are not documented by direct

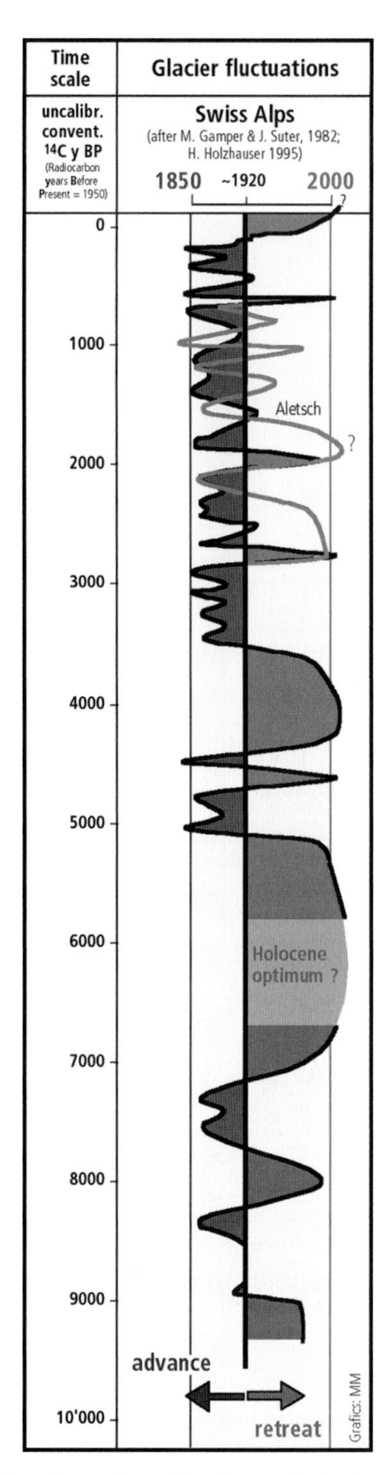

Figure 84.3 Chronology of Holocene glacier length changes in the Swiss Alps. (Compiled by M. Maisch.)

measurements, but also indicates details of potential time-dependent future evolution including feed-backs from effects of flow dynamics. Wallinga & Van de Val (1998) show that Rhonegletscher could disappear within decades if the presently observed trend continues or accelerates. The extensive modelling study by Oerlemans *et al.* (1998) indeed confirms that this could be the case for many if not most other glaciers of the current worldwide mass balance network (Fig. 84.4).

84.5 Needs and recommendations

Worldwide monitoring of glacier changes can build on more than 100 yr of systematic/coordinated observation within the framework of international scientific collaboration, a wealth of excellent—even though still by far incomplete—information, a highly developed understanding of the processes involved and well-reflected integrated/multilevel operational strategies. The task of documenting potentially dramatic developments in remote areas thereby represents a challenge that can be met only by using the best available techniques and concepts. The potential of high-resolution satellite imagery with stereo capacity and in combination with geoinformatics for automated image processing or the use of atmosphere–ocean general circulation models (AOGCMs) for analysis and interpretation of the observed data are prominent examples of avenues for research and development.

The recent availability of high resolution Landsat-7 and ASTER images together with new methods for automated analysis based on GIS techniques in digital inventories of glaciers (Bishop *et al.*, 2004) not only opens new possibilities for upgrading preliminary inventories and repeating earlier inventories in view to assessing regional and global developments but also provides important information on impacts such as rock/ice avalanches or hazards from glacier lakes (Kääb *et al.*, 2003). The interpretation of regional aspects is assisted by the use of statistically downscaled AOGCMs together with seasonal sensitivity characteristics (SSC) on mass-balance models of intermediate complexity. A corresponding study by Reichert *et al.* (2001) demonstrates that mass balances in Norway and Switzerland, respectively, are highly correlated with decadal variations in the North Atlantic Oscillation (NAO). This mechanism, which is entirely due to internal variations in the climate system, can explain the strong contrast between recent mass gains for some Scandinavian glaciers as compared with the marked ice losses observed in the European Alps.

Regional scaling with advanced AOGCM calculations reflects part but not all of current process understanding. In particular, two fundamental physical aspects still await inclusion into simulations and assessments: the *firn/ice temperature effect* and the *size/dynamics effect*. Firn warming relates to latent heat exchange involved with percolation and refreezing of surface melt-water within cold accumulation areas; this process makes the rate of firn warming considerably higher than corresponding air temperature change (Hooke *et al.*, 1983; Haeberli & Alean, 1985). Once the firn becomes temperate, mass loss starts taking place with continued warming of the air. This means that the mass-balance sensitivity of large and still cold firn areas in the Canadian Arctic or in Central Asia, etc., could (i) strongly increase during the coming decades and thereby (ii) reduce the regional differences in sensitivity. The large and relatively flat glaciers around the Gulf of Alaska or in Patagonia, which produce the most important melt-water contribution to sea-level rise, have dynamic response times beyond the century scale and cannot dynamically adjust by tongue retreat to rapid forcing but rather waste down with little area loss. This, in turn, causes the mass balance/altitude feed-back

Table 84.1 Analysis of glacier inventory data for the European Alps (From Haeberli & Hoelzle, 1995, updated.)

Situation 1970–1980		**Simulation (moderate warming scenario)**	
Total glacierized area 1970–1980	2909 km^2	Area reduction 1970–1980 to 2025	ca. 30% of 1970–1980
Total glacier volume 1970–1980	ca. 130 km^3	Mass loss 1970–1980 to 2025	ca. 50% of 1970–1980
Sea-level equivalent	ca. 0.35 mm	Area reduction 1970–1980 to 2100	ca. 90% of 1970–1980
Number of glaciers > 0.2 km^2	1763	Mass loss 1970–1980 to 2100	ca. 95% of 1970–1980
Average mass balance 1850 to 1970–1980	$-0.25\, m\, yr^{-1}$		
Average mass balance 1980–2000	$-0.65\, m\, yr^{-1}$		
Area reduction 1850 to 1970–1980	ca. 40%		
Mass loss 1850 to 1970–1980	ca. 50% of 1850		
Mass loss 1980–2000	>25% of 1970–1980		
Mass loss 2003 alone	ca. 5–10% of 2000		

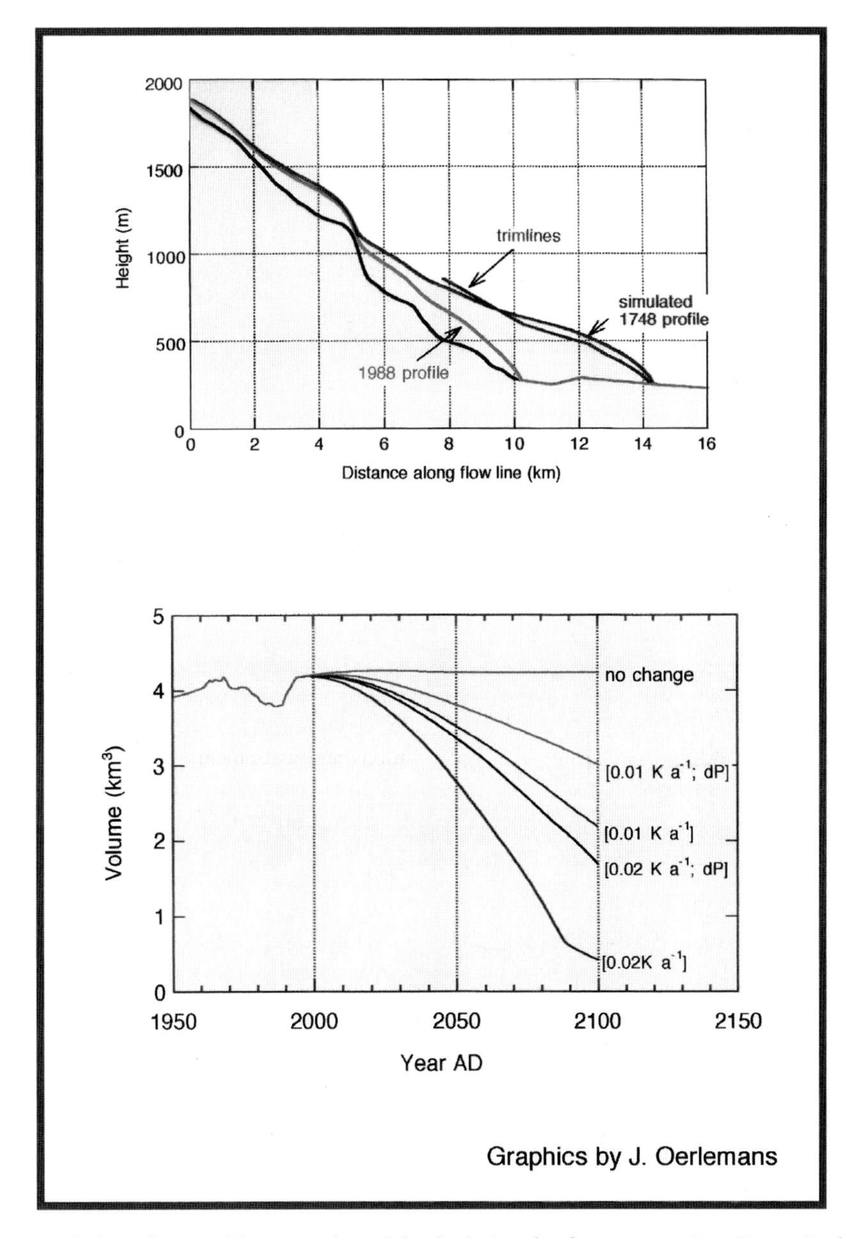

Figure 84.4 Dynamic fitting of Nygardsbreen and model calculation for future scenarios. (From Oerlemans *et al.*, 1998.)

to become important. A cumulative surface lowering of about 50–100 m within a century or so could, indeed, easily increase the mass-balance sensitivity by a factor of two, correspondingly doubling the surface lowering and, hence, leading to a runaway effect. The corresponding growth in size of the ablation area on such glaciers would probably by far overcompensate the effect of shrinking total areas on small glaciers elsewhere. This means that the sensitivity of the main melt-water producers with respect to sea-level change is likely to strongly increase during the coming decades and strengthen regional differences accordingly. The effects on sea level, however, would be reduced to some degree by the fact that important parts of such large maritime melt-water producers are below sea level.

These examples clearly illustrate the key to a successful glacier observation programme: the combination of mapping, monitoring and modelling with advanced process understanding.

EIGHTY-FIVE

The Global Land Ice Measurements from Space (GLIMS) project

A. Kääb*

**Department of Geography, University of Zurich*
**For the GLIMS consortium.*

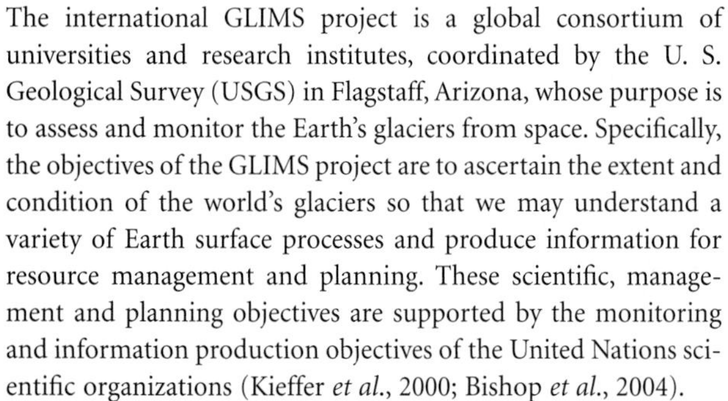

The international GLIMS project is a global consortium of universities and research institutes, coordinated by the U. S. Geological Survey (USGS) in Flagstaff, Arizona, whose purpose is to assess and monitor the Earth's glaciers from space. Specifically, the objectives of the GLIMS project are to ascertain the extent and condition of the world's glaciers so that we may understand a variety of Earth surface processes and produce information for resource management and planning. These scientific, management and planning objectives are supported by the monitoring and information production objectives of the United Nations scientific organizations (Kieffer *et al.*, 2000; Bishop *et al.*, 2004).

The GLIMS project entails:

1 comprehensive satellite multispectral and stereo-image acquisition of land ice;
2 use of satellite imaging data to measure interannual changes in glacier area, boundaries and snowline elevation;
3 measurement of glacier ice-velocity fields;
4 assessment of water resource potential;
5 development of a comprehensive digital database to inventory the world's glaciers, with pointers to other data and relevant scientific publications—the database is developed and located at the National Snow and Ice Data Center (Boulder, CO).

This work and the global image archive at the EROS Data Center (Sioux Falls, SD) will be useful for a variety of scientific and planning applications (Bishop *et al.*, 2004).

The GLIMS project will primarily utilize multispectral imaging from the Landsat TM and ETM+ series, and the new ASTER sensor. Landsat TM and ETM+ data represent a well-established 'working horse' for glacier inventorying and monitoring from space (Kääb *et al.*, 2002; Paul *et al.*, 2002). The ASTER sensor, available since 2000 onboard the NASA Terra spacecraft, opens additional possibilities for glacier observation. The spectral and geometric capabilities of ASTER include three bands in VNIR (visible and near infra-red) with 15 m resolution, six bands in the SWIR (shortwave infra-red) with 30 m resolution, five bands in the TIR (thermal infra-red) with 90 m resolution, and a 15-m resolution NIR along-track stereo-band looking backwards from nadir. The stereo band 3B covers the same spectral range as the nadir band 3N. Of special interest for glaciological studies are the high spatial resolution in VNIR and the stereo and pointing capabilities of ASTER. With topography being a crucial parameter for the understanding of high-mountain phenomena and processes, digital elevation models (DEMs) generated from the ASTER along-track stereo-band are especially helpful (Fig. 85.1; Kääb, 2002). Imaging opportunities by ASTER are governed by Terra's 16-day nadir–track repeat period and the fact that the ASTER

Figure 85.1 Tasman glacier, New Zealand. Depicted terrain section is about 25 × 25 km. The ASTER satellite image was taken on 29 April 2000. North is to the top. Imaging is within the project Global Land Ice Measurements from Space (GLIMS) project. (Satellite data courtesy of NASA/GSFC/METI/ERSDAC/JAROS, and US/Japan ASTER science team; processed by A. Kääb, University of Zurich.) The ASTER satellite sensor carries cameras with different viewing angles. Based on such imagery, the three-dimensional form of the Earth's surface can be computed. The resulting digital terrain models can be used for various analyses to understand glaciers and their environment. The image shows the ASTER satellite image draped over a three-dimensional model of the Earth surface. Mount Cook, the highest peak of New Zealand, is to the middle left of the depicted terrain section. (See www.blackwellpublishing.com/knight for colour version.)

Figure 85.2 Kolka–Karmadon rock and ice avalanche, Caucasus, 20 September 2002. Infra-red false colour composites of satellite imagery of 22 July 2001 (before) and 27 September 2002 (after the event) were taken from the ASTER sensor aboard the NASA TERRA spacecraft. North is to the top. Image size is about. 15 × 15 km. Imaging is within the Global Land Ice Measurements from Space (GLIMS) project. (Images courtesy of NASA/GSFC/METI/ERSDAC/JAROS, and US/Japan ASTER science team; processed by A. Kääb, University of Zurich.) See text for description of the event. (See www.blackwellpublishing.com/knight for colour version.)

VNIR sensor can be pointed cross-track by up to ±24°, which allows for repeat imaging as frequently as every second day in response to urgent priorities.

The ASTER sensor also proved to be very suitable for assessing glacier hazards and managing related disasters. This became particularly evident during the Kolka–Karmadon disaster (Fig. 85.2; Kääb *et al.*, 2003). During the late evening of 20 September 2002, a combined rock and ice avalanche of several million cubic metres started from the Dzimarai-khokh peak in the Kazbek massif, Russian Caucasus (lower left of the images). The large avalanche fell onto the Kolka glacier tongue. The impact of the initial rock and ice avalanche sheared off a major part of the Kolka glacier tongue and started a sled-like rock and ice avalanche of tens of million cubic metres. The Kolka rock and ice avalanche crossed the tongue of the Maili glacier. On its devastating journey northward, the avalanche picked up a large amount of loose sediments in the valley bottom. A few minutes after initiation and 18 km down-valley from the Kolka glacier, the gigantic mass overran the lower parts of the village of Karmadon, killing dozens of inhabitants. Shortly beyond this point, the avalanche was abruptly stopped by the narrowing valley flanks of the Karmadon gorge, and roughly 100 million cubic metres of ice and debris were deposited (middle top of the images). Large amounts of mud were suddenly pressed out of the mass. The resulting mudflow, up to 300 m wide, ravaged the valley bottom below Karmadon, travelling for another 15 km northward from the gorge. In total, the avalanche and subsequent mudflow killed over 120 people. The avalanche deposits at Karmadon soon started to block the rivers entering the gorge. Over a period of approximately 1 month the dammed river water progressively flooded the populated areas near Karmadon that had not been directly affected by the avalanche. The ASTER imagery provided invaluable help in understanding and documenting the event, and in supporting the disaster management.

The GLIMS project closely collaborates with the World Glacier Monitoring Service (WGMS) and the international working group on glacier and permafrost hazards in mountains under the International Commission on Snow and Ice (ICSI) and the International Permafrost Association (IPA).

In addition to spaceborne optical data, the GLIMS project intends to increasingly utilize the integration of solar reflective, thermal and microwave remote sensing to assist in glacier analysis, thereby addressing the limitations of multispectral approaches. This is a key future direction of the GLIMS project.

EIGHTY-SIX

Historical glacier fluctuations

Frank Paul and Max Maisch

Department of Geography, Glaciology and Geomorphodynamics Group, University of Zurich, Winterthurerstrasse 190, CH-8057 Zurich, Switzerland

Glacier changes are among the clearest natural indicators of climatic changes (Haeberli *et al.*, 1998). In particular changes in glacier length are easy-to-follow witnesses of past and ongoing climatic variations and their shifting trends. In contrast to the glacier mass balance, which is the direct and undelayed reaction to the yearly meteorological conditions, length changes display an integrated behaviour, reflecting climatic conditions (mainly temperature and precipitation), in a long-term pattern persisting over many years. Thereby, length variations are an enhanced and, due to glacier motion from the accumulation to the ablation zone, filtered and delayed signal that depend on glacier size and geometry as well as on various other specific topographic factors. Thus, cumulative length changes reflect various frequencies of short-term, decadal or even centennial magnitude (cf. Fig. 86.1).

In Switzerland changes in glacier length have been measured annually since the 1880s, starting with large, long and easily accessible glaciers. At the moment data of about 120 glaciers are collected every year within the framework of the Swiss length measurement network. In particular medium-sized glaciers reflect decadal climate oscillations quite well, as visible from Fig. 86.1. Figure 86.1 also depicts the maximum extent at the end of the Little Ice Age (around 1850–1860) and a general recession period since then, interrupted by two intermittent phases of readvance in the 1920s and 1970s (1965–1985). A similar pattern of glacier fluctuations has been observed around the globe (e.g. Hoelzle *et al.*, 2003). Apart from direct measurements documenting changes in glacier length, much effort has been put into the reconstruction of former glacier extent from indirect evidence, such as lateral moraines (geomorphological approach), or ancient maps, historical paintings and chronicles. In the Alps in recent decades several detailed studies on the historical evolution of individual glaciers and larger mountain groups have been completed. Zängl & Hamberger (2004) illustrate photographically the striking glacier retreat of many Alpine glaciers during the past 100 yr.

By extending the existing Swiss inventory backwards in time a thorough documentation (presenting extensive statistical analyses) of Swiss glacier changes is given by Maisch *et al.* (2000). They have shown that from 1850 to 1973 the entire Swiss glacier area was reduced by more than 25% and glacier volume has shrunk by one-third in total. In most of the mountain regions that typically have a majority of small glaciers, the volume loss was 50% or even more. At the same time a nearly 100 m rise in the equilibrium line altitude (ELA) was calculated, with a regional tendency to higher values in the drier parts of the southern Swiss Alps. The most recent assessment from satellite data has shown that the Alpine glacier area has decreased by another 20% from 1985 to 1999 (Paul *et al.*, 2004).

In order to manage the challenges and needs of worldwide glacier monitoring in the 21st century and with regard to utilizing glaciers as key elements in global climate related observation programmes (Haeberli *et al.*, 1998), modern digital methods are developed for glacier monitoring. These procedures are based on:

1 digitizing of historical and recent glacier outlines from topographical maps in a geographical information system (GIS);
2 combination with a digital elevation model (DEM) to obtain three-dimensional glacier inventory parameters (e.g. slope, aspect);
3 automated delineation of glacier outlines from multispectral, high-resolution (ca. 20 m) satellite imagery (e.g. Paul *et al.*, 2002; Kääb, this volume, Chapter 85).

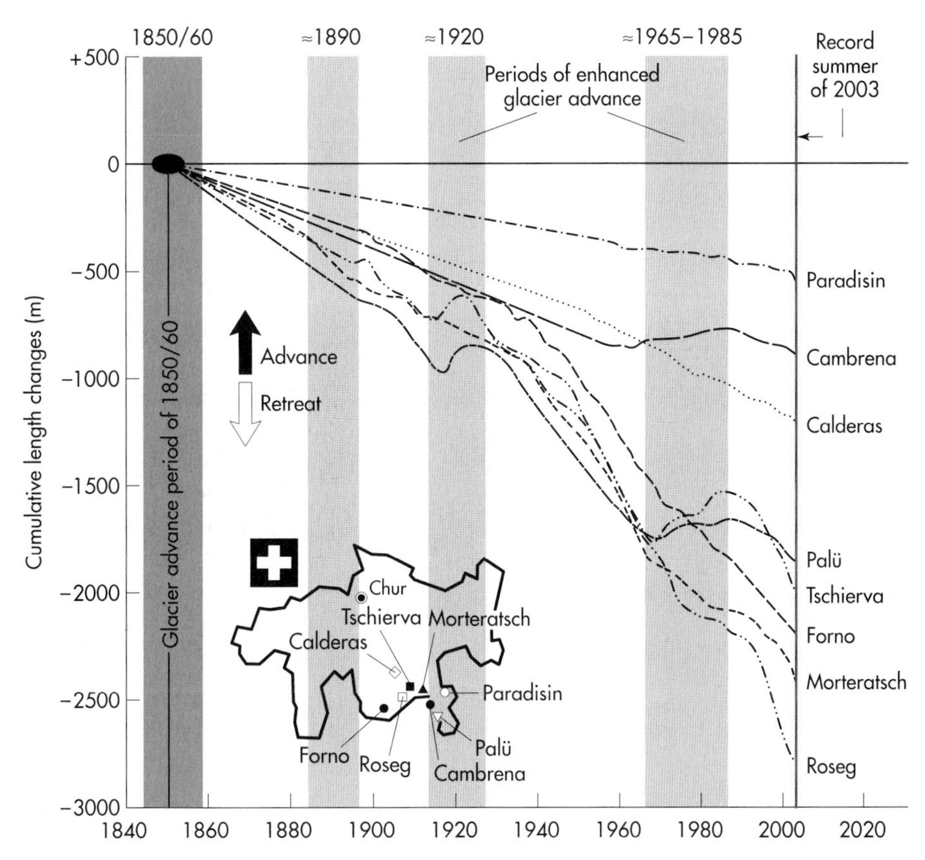

Figure 86.1 Cumulative changes of glacier length measurements from 1850 to 2003 for eight differently sized glaciers in the eastern Swiss Alps. Whereas the largest glaciers depicted (e.g. Morteratsch) reflect only low-pass-filtered and long-term climate changes, medium-sized glaciers indicate decadal fluctuations (e.g. Tschierva) and small glaciers reflect annual or stochastic variations (e.g. Paradisin). (Data source: Swiss Glaciological Commission of the SCNAT (Swiss Academy of Science).)

Figure 86.2 Oblique perspective view of the Bernina region (eastern Swiss Alps), using a high-resolution satellite image from 1997 (IRS-1C) and glacier outlines from 1850 (white) and 1973 (black) draped over a DEM with 25 m spatial resolution. (DEM: © swisstopo (BA 04622).)

The synergetic use of these three techniques (satellite, GIS, DEM) enables impressive visualizations of both the Alpine mountain scenery and the ongoing glacier retreat (Fig. 86.2). Future applications of the digital Swiss glacier inventory and related digital data sets will include: numerical modelling of glacier changes and their relation to climate change, application of distributed glacier mass balance models (e.g. Klok & Oerlemans, 2002) over large mountain regions as well as the forcing of such models with output from regional climate models to determine the impact of past or future climatic conditions on mountain glaciers.

EIGHTY-SEVEN

Interpreting glacial sediments

Doug Benn

School of Geography and Geosciences, University of St Andrews, Fife KY16 9AL, UK

87.1 Introduction

Since the early 1970s there has been a dramatic increase in the scope and sophistication of glacial sedimentology, to the extent that the discipline of today bears little resemblance to that of 30 yr ago. The intervening period has seen the introduction of a wide range of observational and analytical techniques which allow much more detailed characterization of glacigenic sediments than was hitherto possible. Moreover, the same period has witnessed a deep shift in outlook among glacial geologists, who have increasingly come to regard glaciers and their environment as integrated systems, in which glacial sediments and landforms are viewed in the context of temporal and spatial hierarchies of events and processes (e.g. Sugden & John, 1976; Eyles, 1983c; Benn & Evans, 1998). This perspective has encouraged glacial geologists to embrace new developments in glaciology, oceanography, climatology and other disciplines and, conversely, to use glacial sediments to address important cross-disciplinary problems in earth system science.

Several excellent reviews of the current knowledge base of glacial geology have been published in recent years (see e.g. volumes edited by Maltman *et al.*, 2000; Dowdeswell & Ó Cofaigh, 2003; Evans, 2003a). Rather than attempt a necessarily highly selective summary here, this chapter will focus on glacial geology as a scientific *process*, rather than a corpus of knowledge, and will examine a few case studies in detail to explore the workings of the subject as it now stands.

It is useful to think of the evolution of ideas in any science as a process operating on two scales. First, at the scale of the individual research project, the collection, analysis and interpretation of data follows (one hopes) the principles of the scientific method. The degree to which researchers are successful in this is scrutinized by journal editors and reviewers, who aim to prevent poor work from taking up valuable shelf space. Although aggrieved authors may feel otherwise, this process generally works reasonably well without preventing good work from being published, although not necessarily in one's first choice of journal. Once research papers are in the public domain, a larger scale process begins, whereby research work is appraised, criticized and tested by the wider community. Work thus filtered then enters the perceived 'content' of glacial geology at any given time. As with all human activities, this process sounds good in theory but is commonly flawed in practice. For a variety of reasons, some work may be given undue prominence and accepted with relatively little criticism, whereas other work of equal inherent merit may be unfairly ignored, overcriticized or rejected prematurely. In some cases a stable concensus is achieved quickly, whereas in others it may be many years before a satisfactory understanding emerges. Although it is easy to see examples of this with hindsight, it is much harder to gain a reasonably balanced view of the uneven character of the scientific process as it actually happens.

This paper will examine the development of some recent ideas in glacial sedimentology, in an attempt to identify instances where the scientific process (at both scales identified above) has worked effectively, and others where it has not. Three cases will be examined:

1 the problem of reconstructing former subglacial processes from the characteristics of tills;

2 the use of large-scale process–form models as frameworks for generating testable hypotheses;
3 the potential application of glacial sedimentology to reconstructing the deep past, particularly testing the radical 'Snowball Earth' model of Neoproterozoic glaciation.

87.2 Subglacial facies and depositional processes

The interpretation of glacigenic sediments has generated an extraordinarily large literature and numerous classification schemes have been proposed (see e.g. Dreimanis, 1989; Brodzikowski & van Loon, 1991; Benn & Evans, 1998). In large measure, this reflects the myriad ways in which glacigenic sediment can be eroded, transported, deformed, sorted and remixed before it finally comes to rest, and the difficulty of reducing this complexity to a workable yet meaningful system. Nevertheless, considerable progress has been made in identifying criteria and algorithms for reconstructing depositional and deformational processes from field evidence (e.g. Krüger & Kjaer, 1999). An important element of modern glacial sedimentology is the use of modern analogues to guide the interpretation of ancient sediments, and glacial geologists can now draw upon a broad range of observations from different environments (e.g. Evans, 2003a). Additionally, glacial geology has benefited greatly from productive borrowing from other disciplines, such as the use of concepts from structural geology in the study of glacitectonics, and work on subaqueous mass movements in models of glacilacustrine and glacimarine environments. A perennial source of difficulty, however, is the inaccessibility of many glacial environments, and the consequent paucity of unambiguous modern analogues for some ancient sediment facies. This means that the interpretation of some glacial facies is not firmly based on direct observation of modern depositional processes and their products, but must be inferred from the perceived meaning of sediment characteristics. As a result, reconstructions of past processes and environments are subject to revision as ideas and perspectives change. Nowhere is this problem more acute than in the study of subglacial sediments.

Subglacial processes play a fundamental role in glacier dynamics at a wide range of spatial and temporal scales, from short-term glacier fluctuations to millennial-scale interactions between ice sheets, oceans and global climate (e.g. Iverson *et al.*, 1995; Clark *et al.*, 1999). Consequently, it is of great interest to determine whether vanished glaciers and ice sheets were underlain by hard substrata or deforming beds, the degree to which sliding, bed deformation and other processes contributed to glacier motion, and how basal conditions changed through time. Although modelling studies have an important role to play, our knowledge of what former glaciers actually *did* ultimately rests on understanding the geological record, particularly the relationship between measurable characteristics of tills and related sediments, and the depositional and deformational processes that formed them.

For the past 25 yr or so, most subglacial tills have been assigned to one of the triumvirate of melt-out, lodgement and deformation, or some combination of the three (e.g. Hicock, 1990). Which interpretation is applied to a particular till can depend as much on research traditions and shifting ideas as on the sediment properties themselves. One of the most striking examples of this is the rise to prominence of the *deforming bed model* in the mid-1980s, which was wide-reaching enough to be described as a paradigm shift by Boulton (1986). Prior to the 'shift', the majority of researchers regarded glacier beds as rigid, and basal tills as the result of deposition from sliding ice (lodgement) or stagnant ice (melt-out). Then, landmark papers by Alley *et al.* (1986a) and Boulton & Hindmarsh (1987) promoted the idea that many glaciers and ice sheets are underlain by viscous deforming layers up to several metres thick, and that pervasive strain within such layers contributes substantially to glacier motion. Subsequently, vast areas of former 'lodgement' and 'melt-out tills' were reinterpreted as deformation tills. It has been argued that unremoulded melt-out tills should be very rare (e.g. Paul & Eyles, 1990), and in some regions both melt-out and lodgement tills appear to be extinct. In some cases, however, the shift in interpretation was not based on particularly strong evidence, and the widespread applicability of the deforming bed model has been questioned (e.g. Clayton *et al.*, 1989; Ham & Mickelson, 1994; Piotrowski *et al.*, 2001, 2002). Although not all objections are valid (see Boulton *et al.*, 2001a), some are not easily dismissed and it would be unwise to ignore challenges to the prevailing model.

I think it is fair to say that, despite several years of intensive research and a greatly expanded theoretical understanding of glacier beds, definitive criteria for the interpretation of tills remain elusive. One of the reasons for this is that so many tills are macroscopically structureless, or nearly so, and that this characteristic can be interpreted as evidence for pervasive deformation and complete homogenization (e.g. Alley, 1991b) or for no deformation at all (e.g. Piotrowski *et al.*, 2001). Numerous attempts have been made to find clear-cut methods for identifying deformation tills, but many promising candidates have turned out to be considerably more ambiguous than they first appeared. For example, Hooke & Iverson (1995) argued that deformation tills are analogous to fault gouges, and hence should have characteristic fractal grain-size distributions reflecting the minimization of stress concentrations at grain contacts at all scales. However, Khatwa *et al.* (1999) found that the 'fractal dimensions' of glacigenic deposits do not appear to exhibit any simple relationship with depositional processes, and Benn & Gemmell (2002) showed that apparently 'fractal' distributions may be misleading and are not uniquely diagnostic of subglacial shear.

Several authors have argued that clast a-axis fabrics can be used as diagnostic criteria, and that interpretation of tills can be facilitated by comparing till-fabric eigenvalues with those of sediments of known origin (e.g. Dowdeswell & Sharp, 1986; Hart, 1994; Benn, 1994; Benn & Evans, 1996; Hicock *et al.*, 1996). Much of this research concluded that tills formed by the deformation of soft glacier beds have relatively weak a-axis fabrics parallel to glacier flow, and some suggested that transverse fabrics may be indicators of high cumulative strains. It has been argued that weak parallel or transverse fabrics reflect continuous rolling of clasts within pervasively deforming viscous till matrix (e.g. Hicock & Dreimanis, 1992; Hart, 1994), a process explored theoretically by Glen *et al.* (1955). Alternatively, Benn (1995) argued that weak flow-parallel fabrics reflect frequent rapid realignment of clast long axes under unsteady strain regimes within the deforming layer, perhaps as a result of interactions between particles, and that rolling is not an important process.

Recently, however, the association between deformation tills and weak a-axis fabrics has been questioned. On the basis of pioneering ring-shear experiments, Hooyer & Iverson (2000a) showed that clast long axes simply rotate with the strain ellipsoid, and are passive markers of the axis of maximum extension. Thus, they concluded that tills that have undergone significant strain should have strong fabrics aligned in the flow direction. Furthermore, they argued that tills with weak fabrics have either undergone little or no shear strain or have been subjected to some other process that may have disrupted clast alignment, and that the common assumption that deformation tills can be identified by weak a-axis fabrics should be rejected. Support for this view has been provided by Larsen & Piotrowski (2003), who found very strong cluster fabrics in tills for which there is clear independent evidence for deformation.

In view of this work, it is worth asking how the association between deformation tills and weak a-axis fabrics came to be established. According to Piotrowski *et al.* (2002), there is a danger 'of *assuming* a deforming bed, describing characteristics of a "deforming till" [e.g. weak a-axis fabrics], then using these characteristics to "identify" deforming beds'. Although it is undoubtedly true that this kind of circular reasoning has featured in some of the literature, it cannot be invoked to entirely explain away what we might call the *weak fabric hypothesis*. In an attempt to overcome the problem of independently identifying massive deformation tills, Benn (1995) studied tills on the foreland of Breidamerkurjøkull, Iceland, near the site where evidence for till deformation beneath the glacier had been obtained by Boulton & Hindmarsh (1987). The reasoning was, that this site was the one place on earth where *both* subglacial deformation had been demonstrated *and* glacier margin retreat had exposed the till for detailed *in situ* study of its properties. The till was found to have a two-layered structure, consisting of a stiff lower till with consistently strong a-axis fabrics, and a soft upper till with strong fabrics within fluted moraines and weaker fabrics in interflute areas. Numerous fabric elements (a-axes, a–b planes, poles to polished facets, and the orientation of stoss–lee forms) and patterns of striae were used to argue that the relatively weak patterns reflected frequent clast realignment under transient strains, as noted above. At the time, the fabric data were interpreted in terms of pervasive and continuous (if unsteady) deformation. However, the data are also compatible with discontinuous deformation (in space and time), with the locus of shear moving in response to varying pore-water pressure pulses (cf. Tulaczyk, 1999; Boulton *et al.*, 2001a). In addition to the Breiðamerkurjökull study, there is also compelling geological evidence from several sites for deformed sediments with weak fabrics, where deformation is clearly attested by structures such as attenuated augen and streaks of soft inclusions, including chalk.

The laboratory experiments of Hooyer & Iverson (2000) represent a very important new tool in the armoury of techniques for investigating subglacial tills, but it is perhaps premature to assume that the results to date replicate all the important characteristics of deformation tills and related sediments. Borehole studies on modern glaciers show that subglacial conditions are highly variable in time and space and provide evidence for widely varying styles of subglacial deformation (Alley, 2000), and this variability is almost certain to be manifest in till characteristics. Tills also exhibit a very wide variety of grain-size distributions, and interactions between particles are likely to exert a variable influence on till response to stress. It seems probable that deformation tills will exhibit a wide range of fabric types, depending on till stiffness, strain history (e.g. steady or unsteady), granulometry and other factors. Much remains to be learned.

Recently, evidence has been accumulating for an alternative mode of glacier bed behaviour, in which most relative motion is focused near the ice–till interface (e.g. Tulaczyk *et al.*, 2001). According to this *ploughing model*, soft subglacial till does not exert a primary control on glacier velocity, but exhibits a largely passive response to overpassing ice, the velocity of which is determined by 'sticky patches' on the bed or drag at the margins of the glacier or ice stream. The subglacial till is repeatedly ploughed by roughness elements protruding down from the overlying ice, either clasts (Brown *et al.*, 1987) or irregularities in the ice itself ('ice keels': Tulaczyk *et al.*, 2001). The implications of this model for till fabric development remain to be explored, but ploughing should be expected to produce complex (weak?) fabric patterns during repeated disturbance by more or less distant protuberances ploughing through the sediment, clast alignment by drag at the ice–till interface, and other processes. The situation as it now stands presents interesting challenges to glacial sedimentologists. On one hand, we have two contrasting views about the fabric characteristics of deformation tills ('strong' versus 'weak'), and on the other we have a single characteristic ('weak' fabrics) which could be used to support two rival models of till genesis (deformation and ploughing). To progress our understanding of the significance of subglacial tills, these issues need to be resolved.

There is an important general point here. A large amount of sedimentological research does not explictly test hypotheses, but adopts an inductive approach in which evidence is adduced to argue *towards* a conclusion. This approach, which has much in common with classical rhetoric, has a venerable tradition in geology, and has an important role to play in the process of transforming observations into a body of understanding. There is a risk, however, that by adopting an inductive approach we simply construct narratives that 'explain' the observational record to our own satisfaction (and, with luck, that of journal editors and reviewers). Such narratives can appear very persuasive, even self-evident. If we observe characteristics which are compatible with a certain explanation, it seems reasonable to believe that the explanation is correct, particularly if we already believe it to be so. However, induction relies on assumptions, and if these remain unexamined, any conclusions are suspect, as Piotrowski *et al.* (2002) have pointed out. This is why it is useful to be reminded that there may be other models that predict the same characteristics as our favoured model. Apparent contradictions and anomalies are opportunities, because they highlight potential weak points in our understanding, and stimulate the search for ways of finding firmer foundations for knowledge about the world. Thus, where rival models both appear to be compatible with existing data, there is a clear need to design research that will clearly eliminate one or other of the possibilities in any given case.

Our current understanding of tills results from a combination of sedimentology, glaciology, laboratory studies and theory, and it is likely that futher advances will require carefully designed research using multiple methods, including detailed study of the macroscopic and microscopic properties of naturally and artifically deformed samples. Additionally, a thorough understanding of former glacier beds must involve integrating the results of small-scale till properties with larger scale analyses of landform characteristics and distribution. Reconstructing glacier behaviour is not a trivial problem, and no one technique is likely to provide a 'golden ticket' to understanding the genesis of all tills.

87.3 Process–form models

In various incarnations, process–form models have been very widely used as explanatory tools in glacial studies. By making explicit links between sedimentological data and genetic processes at a range of scales, process–form models provide a powerful means of representing past or present glacial environments as integrated systems. Since the 1970s, glacial process–form models have become increasingly sophisticated and comprehensive as glacial geologists have gained experience in an expanding range of environments. The influential paper by Boulton & Paul (1976) recognized just two glacial landsystems: (i) the *subglacial landsystem*, based largely on Icelandic glacier forelands, and (ii) the *supraglacial landsystem*, based on debris-covered glacier margins in Svalbard. Subsequently, many additional landsystems have been described, encompassing a wide range of climatic, glaciological and topographic environments, including surging glaciers, glaciated valleys in high-relief and low-relief mountain landscapes, glacimarine and glacilacustrine settings, and temperate, subpolar and high polar climates (e.g. Boulton & Eyles, 1979; Eyles, 1983a; Benn & Evans, 1998; Evans, 2003a). Further variety arises from the range of temporal and spatial scales considered in process–form models, which encompass both small-scale models of depositional processes and basin-scale models of sedimentation over glacial cycles. The latter are of particular importance, because they provide valuable perspectives on the behaviour of glacial systems on time-scales far beyond those that can be considered in modern process studies.

It has been argued that process–form models are potentially misleading, because they can encourage 'pigeon holing' of data into narrow, preconceived categories (e.g. Kemmis, 1996). At best, however, process–form models are not a set of rigid templates, but provide a liberating framework for structuring studies of glacial environments (Benn & Evans, 1998). Such frameworks have clear pedagogic value, allowing large amounts of information to be presented in easily understood form. More importantly, they have predictive power, and hence can be used to generate hypotheses to be tested against further observations. In other words, process–form models allow us to say, *if* existing observations fit together in this way, *then* we should expect to find particular kinds of supporting evidence if we look *here*. This role of process–form models cannot be overemphasized because, by their very nature, large-scale models of former depositional systems are unlikely to be entirely correct (or at least complete) in their initial form. This is because the reconstruction of depositional environments at larger and larger scales involves the integration of larger and larger amounts of evidence, with a concomitant decrease in the level of detail that can be considered, both by the researchers themselves and within the limits imposed by journal editors. This loss of detail inevitably involves greater degrees of generalization.

Mounting pressures from funding bodies and research assessment exercises have also encouraged researchers to 'think big' and to tackle global-scale questions rather than problems of a perceived local or regional nature. Although this has undoubtedly contributed to the rapid evolution of our science, there is a danger that this approach is skewing research agendas towards 'headline science', to the detriment of the painstaking attention to field evidence that characterized the best work of earlier generations. This is not to detract from the enormous explanatory and predictive power of basin-scale depositional models, but to emphasize that, ultimately, all models arise from and are answerable to field evidence. Without a solid empirical basis, even the most persuasive models are more akin to creation myths than products of scientific rigour.

Although they may be overgeneralized or incomplete, basin-scale depositional models can play an important role in setting research agendas. A good example of this is provided by the debate generated by the ideas of Nick Eyles and Marshall McCabe on the deglaciation of the Irish Sea basin. Eyles & McCabe (1989) proposed that glacigenic deposits exposed around and beneath the Irish Sea can be subdivided into subglacial and glacimarine depositional systems. They argued that the subglacial landsystem, consisting mainly of drumlins and other streamlined landforms, can be correlated stratigraphically in a down-ice direction with subaqueous morainal bank complexes, which in turn pass distally into glacimarine sediments. The proposed stratigraphical relationships between the terrestrial and marine depositional sytems were used to establish an event stratigraphy for the basin. Eyles & McCabe concluded that high relative sea levels caused by glacio-isostatic depression (up to ca. 150 m above present sea level) helped to trigger rapid calving retreat of the ice margin. In turn, this initiated draw-down and fast flow of glacier ice in the lowlands surrounding the northern margins of the basin, resulting in subglacial streamlining of sediment into extensive drumlin fields. The stabilization of ice margins at approximately the present coastline of the Irish Sea Basin resulted in the deposition of morainal banks at the outer edges of the drumlin fields.

This is a bold and imaginative hypothesis, which attempts to integrate sedimentological evidence over a wide geographical area in terms of a single space–time model. Much of the appeal of the model lies in its claim to explain all elements of the landscape (and sea-floor sediments) as linked elements of a basin-scale system. Moreover, it makes clear predictions about what types of evidence should be found in different parts of the basin, and therefore can be used as a framework for future investigations. In the decade following its publication, Eyles & McCabe's paper was criticized by several researchers who argued that the marine limit lies below present sea-level in most of the basin, and that much of the evidence adduced for glacimarine deposition in fact records *terrestrial* environments (e.g. Dackombe & Thomas, 1991; Huddart, 1991; McCarroll & Harris, 1992; Walden, 1994). At the same time, support for the glacimarine hypothesis was presented in several papers (see McCarroll (2001) for a good summary). In the course of this debate, it became clear that in many cases opposing conclusions were being reached using the *same evidence*, prompting a search for unambiguous criteria for differentiating glacimarine and terrestrial glacigenic deposits (e.g. Hart & Roberts, 1994; Ó Cofaigh & Dowdeswell, 2001), and a systematic examination of the predictions of the two models and identification of key tests that could decide between them (e.g. McCarroll, 2001). Using several lines of evidence, McCarroll (2001) concluded that glacigenic deposits around much of the basin were deposited above contemporary sea-level, except in the northern part where raised glacimarine sediments occur at modest altitudes. The debate has not been resolved to the satisfaction of all protagonists (see McCarroll *et al.*, 2001), but McCarroll's position appears to represent the most balanced interpretation at this time. Important issues, such as the timing and significance of glacier readvances and flow events in the millennia following the Last Glacial Maximum, remain the subject of current research.

The Irish Sea example shows how a bold hypothesis can act as a catalyst and a framework for systematic testing of specific hypotheses. A case where this process has worked less well concerns the work of John Shaw and co-workers, which proposes that much of the glacial landscape in North America was produced in subglacial floods of gigantic proportions. Like the Eyles & McCabe model for the Irish Sea, this is a bold idea, offering an integrated framework for interpreting a wide range of evidence. However, there has been a general unwillingness among Quaternary scientists to engage in debate about the model, and instead we have seen the development of two parallel literatures on the glacial record of North America: that in support of the megaflood interpretation (by Shaw and co-workers), and the rest. Although Shaw's ideas are not highly regarded by the majority of glacial sedimentologists, among a wider readership the megaflood interpretation appears to have taken on an aura of truth by dint of the energy and fervour of its proponents. This apparent truth value has developed more by default than as the outcome of a balanced process of hypothesis, prediction and test among specialists, raising the important issue of how the 'content' of a subject is defined, both among specialists and a wider audience. For this reason, Dave Evans and I have attempted to provide a broad critique of the megaflood interpretation (this volume, Chapter 8). We hope that this will spur further debate and comment.

87.4 Long ago and far, far away: the 'Snowball Earth'

One of the most interesting ideas about the Earth's glacial history to emerge in recent years is the proposal that during the Neoproterozoic (Late Precambrian) the Earth was subject to global 'icehouse' conditions on at least two occasions (the Snowball Earth hypothesis; Hoffman *et al.*, 1998; Hoffman & Schrag, 2002). The foundation for this idea is the worldwide occurrence of Neoproterozoic glacigenic deposits, including areas with robust palaeomagnetic evidence for low palaeolatitudes. This evidence presents an intriguing problem: if the Tropics were ice-covered, high surface albedo would tend to keep the Earth in a glacial state, and so, if such a state existed, how could it ever end? Geologists have responded to this problem in several ways. One is to question whether the rocks are glacial in origin, and to argue instead that many supposed 'tillites' are submarine debrites (e.g. Schermerhorn, 1975; Arnaud & Eyles, 2002; Eyles & Januszczak, 2004) or, less plausibly, impact ejecta (Rampino, 1994). A second is to accept that there is evidence for glaciers, but not at low palaeolatitudes. For example, Kent & Smethurst (1998) proposed that the early Earth may have had a non-dipole magnetic field, which would invalidate the assumption that low magnetic inclination implies low geographical latitude. Thirdly, it has been argued that Neoproterozoic climate anomalies may reflect high tilt of the Earth's axis at that time (e.g. Williams, 1993).

The Snowball Earth model proposes a fourth response, which accepts the idea of low-latitude glaciation, and offers a framework that purports to explain all of the evidence (Hoffman *et al.*, 1998; Hoffman & Schrag, 2002). According to this model, global cooling was initiated by drawdown of greenhouse gases, triggered by tectonic factors (e.g. Donnadieu *et al.*, 2004). Ice cover expanded in response to cooling, and once a certain threshold had been crossed, a catastrophic ice-albedo feed-back precipitated the Earth into global icehouse conditions. These persisted until CO_2 outgassing from volcanos initiated warming, which ultimately triggered catastrophic melt-back and a supergreenhouse climate. This framework appears to explain the global distribution of tillites (icehouse phase), and the occurrence of 'cap' carbonates and iron precipitates on top of glacial sediments (melt-back phase).

The model has been criticized on several grounds, but it has been robustly defended (see Hoffman & Schrag (2002) for a good overview of criticism and defence). This debate has focused very welcome attention on the Earth's ancient glacial record, but usually with little emphasis on rigorous analysis of the supposed glacigenic rocks themselves. There has been an unfortunate tendency for supporters of the Snowball Earth to accept any interpretations of the rock record which appear to support the hypothesis, regardless of their inherent plausibility. For example, Neoproterozoic sand-wedge polygons in Australia, apparently indicative of cold but fluctuating temperatures, have been attributed to alternating frozen and thawed bed conditions beneath surging glaciers. To anyone familiar with the sediment–landform associations formed by surging glaciers, this is patently absurd, but the idea has currency nonetheless. Another example is the 'sea-glaciers' proposed by Warren *et al.* (2002) to account for thick Neoproterozoic glacigenic sediment sequences. One of the objections to the Snowball Earth model is that, in a deeply frozen Earth, the hydrological cycle would be too weak to drive significant ice and sediment fluxes, and thus it cannot explain observed sediment thicknesses (e.g. Condon *et al.*, 2002). Warren *et al.* (2002) addressed this problem by invoking theoretical ocean-wide ice shelves that creep equatorward in response to thermally controlled thickness gradients. Although such 'sea glaciers' are physically plausible, their role as efficient engines of sediment transport is much less so, because to erode and entrain significant amounts of sediment, ice must be grounded. This requirement is incompatible with the proposed flow mechanism, which invokes spatial variations in the thickness of *floating* ice. The geological record would thus appear to imply sediment transport by grounded glaciers and an efficient hydrological cycle, but some proponents of the Snowball Earth model appear to prefer such hypothethical and problematic entities as 'sea glaciers' over more prosaic readings of the rock record that raise uncomfortable issues.

Theoretical models now play a justifiably important role in earth science, allowing complex interactions between the atmosphere, hydrosphere, lithosphere and cryosphere to be explored at a high level of detail. Sadly, however, sedimentology is undervalued in the current model-driven research agenda, with the result that models are commonly much less constrained by meaningful assumptions and boundary conditions than they could, or indeed, should be. It is rather ironic that glacigenic sediments have prompted the idea that the whole Earth was glaciated, then have been virtually ignored in the search to discover what kind of cold world it actually was. Detailed, empirical reconstruction of glacial environments has been neglected in favour of geochemical and atmospheric models unfettered by the realities of cold, hard glaciers.

There are some exceptions to this generalization. Condon *et al.* (2002) and Leather *et al.* (2002) have given brief accounts of Neoproterozoic glacigenic facies and discussed their implications for climate change. More recently, Eyles & Januszczak (2004) presented a very lengthy and impressive criticism of the Snowball Earth hypothesis, in which they questioned the evidence for global glaciations during the Neoproterozoic, and argued that 'Neoproterozoic diamictites are overwhelmingly the product of tectonically-related, non-glacial, subaqueous mass-flow processes in marine basins', and that glaciation was diachronous and regional (rather than global) in scope. By attempting to review and interterpret global evidence in a single paper, this work necessarily makes sweeping generalizations, and many geologists would question the conclusion that the majority of Neoproterozoic diamictites are non-glacial. By emphasizing the importance of the rock record, however, Eyles & Januszczak lay down important challenges. First, they challenge the assumptions of modellers, and show that they are not at liberty to assume boundary conditions based on simplistic interpretations of diamictite. Second, by questioning accepted interpretations of Neoproterozoic diamictites, they challenge glacial geologists to test rival models (glacial or non-glacial) in the field. The outcomes of such tests are important, because different interpretations of the rock record imply very different environmental histories, and very different modes of operation of the Earth's climate system.

Glaciologists and glacial geologists have a potentially vital role to play in deciphering the long-term behaviour of the Earth's climate system. They have a wealth of experience of the rich diversity and complexity of modern glacial environments, which is of direct relevance to important and deep questions about the history and future of the Earth. May they go forth and put it to good use.

EIGHTY-EIGHT

Moraine sediment characteristics as indicators of former basal ice layers

William George Adam

School of Physical and Geographical Sciences, Keele University, Keele, Staffordshire ST5 5BG, UK

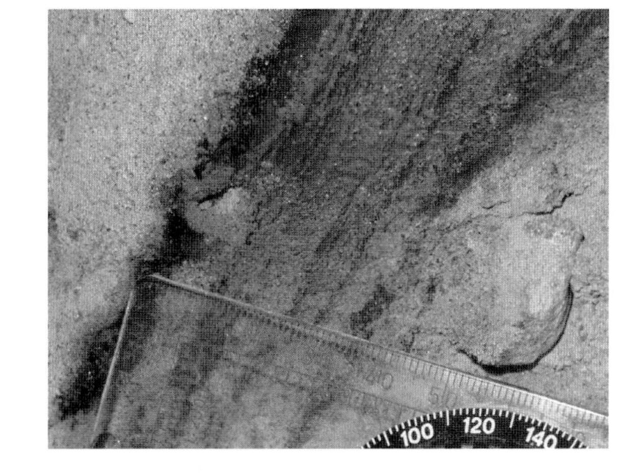

Many interpretations of glacial sediments to reconstruct the basal ice layers of former glaciers have focused upon melt-out tills and glacigenic sediment flows, relying on the preservation of diagnostic glacially derived structures (e.g. Lawson, 1979b; Ham & Mickelson, 1994). This approach can be strengthened if the interpretation did not rely solely upon the preservation of diagnostic glacial structures but was coupled with another technique. For example, Hambrey *et al.* (1999) have provided evidence that a sedimentological difference may exist between basal ice facies at glaciers in Svalbard, using the technique of displaying the RA (aggregate roundness) and C_{40} (shape) indices on a covariant plot (Benn & Ballantyne, 1994). Knight *et al.* (2000) suggested that

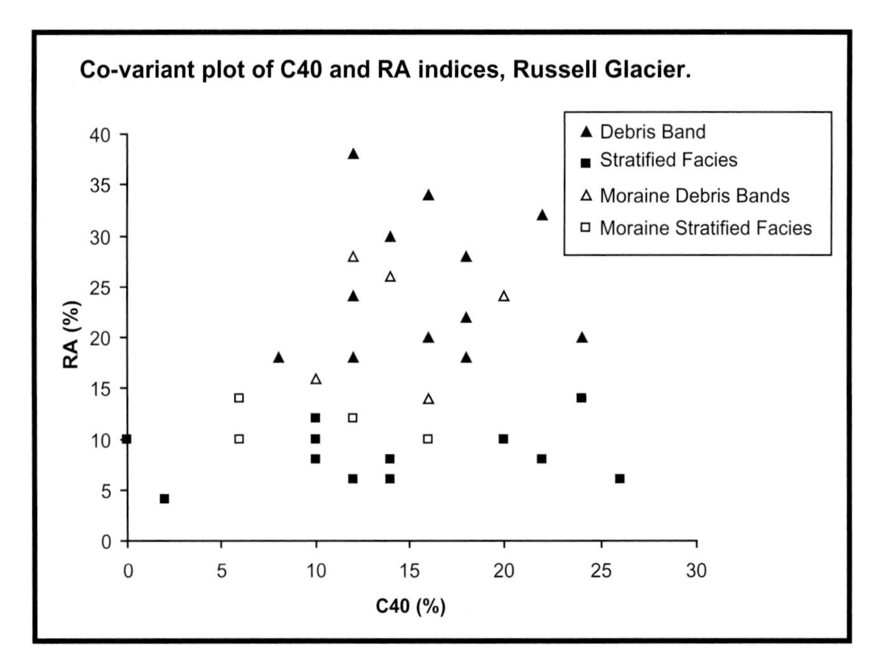

Figure 88.1 Covariant plot of C_{40} and RA indices with samples from stratified facies, debris bands and the ice-marginal moraine, Russell Glacier, western Greenland.

particle-size differences between debris samples from different basal ice facies at Russell Glacier, west Greenland, were transferred to and preserved in the ice-marginal moraine, so that the distribution of particle-size characteristics in moraine sediments could be used to reconstruct patterns of subglacial thermal regime and ice flow of former glaciers.

Adam & Knight (2003) applied Benn & Ballantyne's (1994) method to identify another sedimentological distinction between basal ice facies at the Russell Glacier using covariant plots of aggregate roundness (RA index) and aggregate shape (C_{40} index) of samples from different ice facies and adjacent moraines. Multiple samples of 50 pebbles each were taken from the stratified facies, from debris bands, from the moraine where the stratified facies was the dominant source of sediment supply and from the moraine where the debris bands were the dominant source of supply. Figure 88.1 displays the covariant plot of the C_{40} and RA indices with samples from the stratified facies, debris bands and samples from the moraine at the site referred to by Adam & Knight (2003). The results from the Russell Glacier indicate that there is both a particle-size difference and a morphological difference in clasts between basal facies, which are transferred to and preserved in the ice-marginal moraine.

Observations at Svinafellsjökull, an outlet glacier of the Vatnajökull ice cap, also provide insight into sedimentologically distinctive characteristics within the basal layer. This research was conducted to test Adam & Knight's (2003) conclusions. A covariant plot was built from sample points taken from the stratified facies, from the debris bands and from the moraine. However,the resultant covariant plot shows that there is no significant difference recorded in the pebble-sized debris between the stratified facies and debris bands at Svinafellsjökull. The C_{40} index values of the stratified facies and debris bands range from 8 to 24 and the RA index values range from 4 to 22. Figure 88.2 indicates a complete morphological overlap between the two facies. Figure 88.3 displays the relative proportions of different particle sizes in the stratified facies, debris bands and the dispersed facies basal layers at Svinafellsjökull. The stratified facies is sedimentologically distinctive in the amount of fine-grained material it contains, with almost 60% of its total volume comprising silt-sized debris, but almost 30% of the dispersed facies debris is also of silt-size. Therefore when examining debris from the moraine it is not possible to ascertain simply by the presence/absence of fine-grained material what basal ice layer the sediment originates from.

It is because there is no distinctive particle-size marker in any one facies that the particle-size distribution cannot be used as an interpretive tool at Svinafellsjökull as it was at Russell Glacier. The morphology of the pebble-sized debris at Svinafellsjökull also fails to display the contrast found in the stratified facies and debris bands at the Russell Glacier. Only if distinctive sedimentological characteristics in the basal ice layer are transferred to and preserved in ice-marginal moraines can a sedimentological interpretation of basal ice and hence subglacial environments be built.

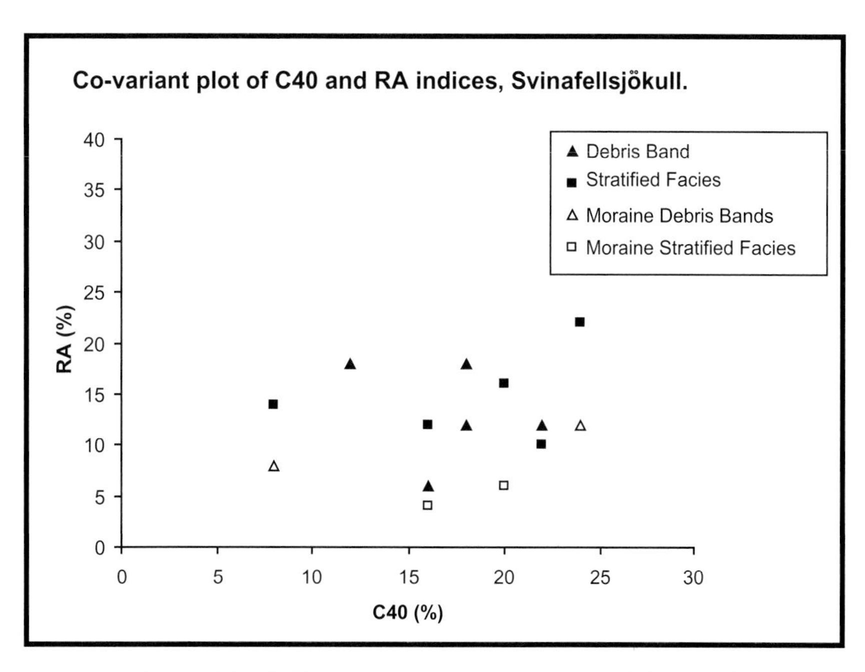

Figure 88.2 Covariant plot of C_{40} and RA indices with samples from stratified facies, debris bands and the ice-marginal moraine, Svinafellsjökull, Vatnajökull ice cap, southern Iceland.

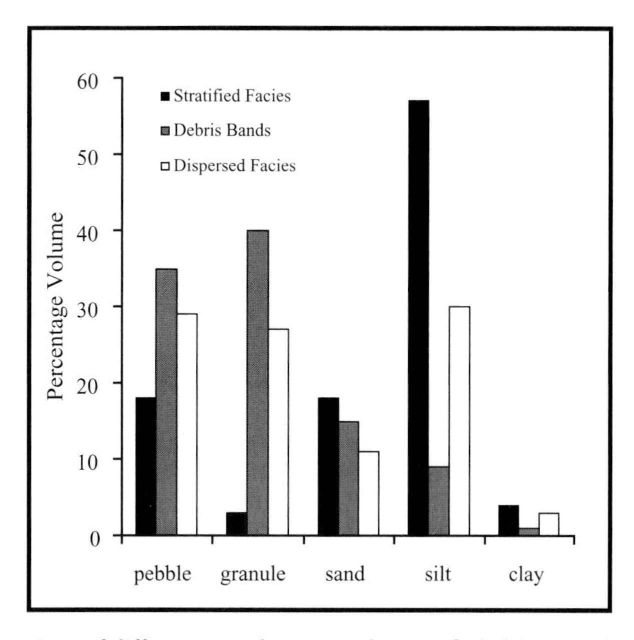

Figure 88.3 Relative proportions of different particle sizes in the stratified, debris band and dispersed basal ice facies.

EIGHTY-NINE

Using cosmogenic isotopes to interpret the landscape record of glaciation: nunataks in Newfoundland?

J.C. Gosse*, T. Bell†, J.T. Gray‡, J. Klein§, G. Yang* and R. Finkel¶

**Earth Sciences, Dalhousie University, Halifax, Canada*
†Geography, Memorial University of Newfoundland, St John's, Canada
‡Geographie, Universitie de Montreal, Montreal, Canada
§Physics, University of Pennsylvania, Philadelphia, USA
¶CAMS, Lawrence Livermore National Laboratory, Livermore, USA

89.1 Introduction

The science of reconstructing temperate palaeo-ice sheet configurations and dynamics is concurrently undergoing two important revolutions. The first is a critical re-evaluation of palaeo-ice sheet records with a renewed appreciation (e.g. Stroeven *et al.*, 2002a,b; Briner *et al.*, 2003; Marquette *et al.*, 2004; Staiger *et al.*, 2005) for non-erosive 'frozen-based' ice conditions (e.g. Sugden, 1968). Landscapes indicative of differential glacial erosion have been recognized in Canada over the past 50 yr (e.g. Sugden & Watts, 1977; Dyke, 1993; Kleman & Hättestrand, 1999; Dredge, 2000). Areas that were covered by ice that did not significantly erode its substrate may have been misinterpreted to have been ice-free during at least the last glaciation (e.g. Batterson & Catto, 2001, Miller *et al.*, 2002). As this is a recent rejuvenation of the cold-based ice concept, there is yet no consensus on the spatial and temporal continuity of non-erosive ice cover during the last glaciation in Canada. Some field evidence points to a patchy but persistent non-erosive ice cover on highlands or in regions of stable ice divides (with zero net flow velocity). Recently improved Antarctica-tuned thermo-mechanical models used to simulate palaeo-ice dynamics suggest that much of the Quebec–Labradorean sector of the Laurentide Ice Sheet (LIS) and peripheral ice domes were frozen-based (no sliding on substrate) for much of the glaciation (e.g. Staiger *et al.*, 2005).

The second reformation is the application of the terrestrial cosmogenic nuclide (TCN) dating method (Lal, 1991, Gosse & Phillips, 2001) to date ice-marginal positions over a much longer time range (hundreds to millions of years) than radiocarbon, particularly where radiocarbon datable material is scarce. The TCN method has also been used to reveal the presence of non-erosive conditions (isotopic ratios indicate that exposure of rock is interrupted) and can be used to estimate the total minimum duration of an exposure history complicated by multiple frozen-based glaciations.

This study is an example of how TCN methods are being used to test two current, long-lived, widely debated conceptual models that explain the genesis of altitudinally distinguishable zones of differential bedrock weathering, or 'weathering zones', which occur along coastal highlands of eastern Canada. The first hypothesis suggests that the boundaries between weathering zones represent the vertical limits of separate glaciations (similar to trimlines). The principal alternative hypothesis suggests that the weathering zones correspond to differences in subglacial erosion during complete ice cover of the region. The TCN data presented here and elsewhere for Labrador support this alternative hypothesis, and are used to suggest that ice cover during the last glaciation was much more extensive than previously thought. The results presented in this case study are an unpublished part of a larger study of the palaeo-ice dynamics of Atlantic Canada (Gosse, 2002). The study area (Fig. 89.1) is in the northern Long Range Mountains of western Newfoundland where evidence supporting both hypotheses is strong.

89.2 Palaeo-ice sheet volume and weathering zones of the Long Range Mountains

In eastern Canada, the extent of glaciation is not easily interpreted from a routine application of glacial geomorphology. This is

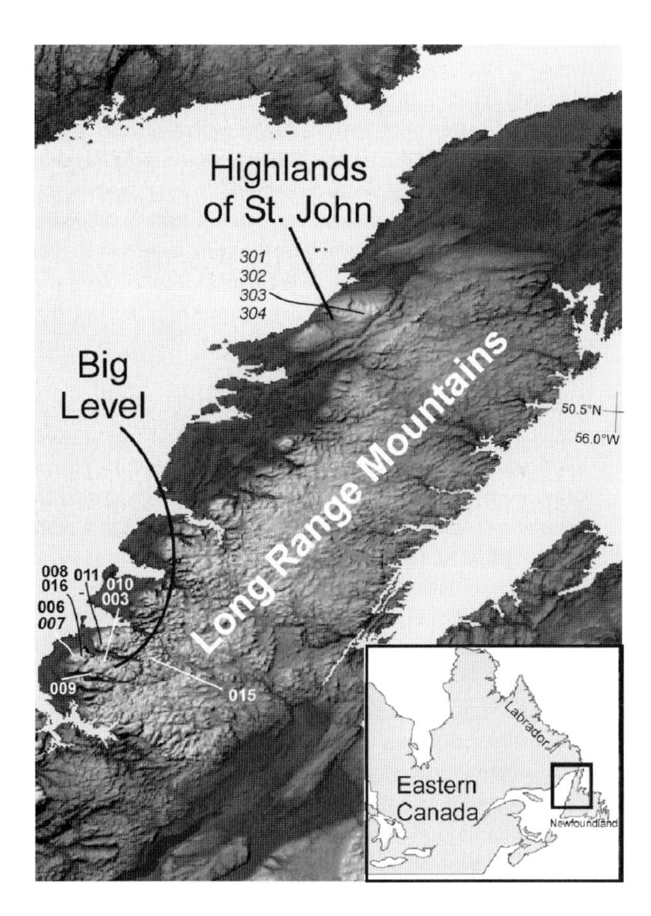

Figure 89.1 Location of the two weathering zone type localities in western Newfoundland of Grant (1977) and the TCN sampling localities (numbers are last three digits of sample IDs in Table 89.1, erratic boulders in italics).

because non-erosive ice may have covered much of the highland summits leaving little or no erosional or depositional evidence of its presence. The few boulders (even perched in places) that have been discovered on bedrock knobs on these summits have been rejected as erratics and considered core stones by previous geomorphologists because their lithology is similar to the underlying bedrock.

An important basis of the 'limited ice cover' paradigm in eastern Canada is the occurrence of altitudinally distinct zones of differential weathering along the coastal highlands (see Marquette *et al.* (2004) for a more complete summary). Like most regions in Eastern Canada, the weathering zones in western Newfoundland have been distinguished according to (i) general appearance of maturity, (ii) presence or preservation of glacial erosional and depositional landforms of various scales, (iii) degree of bedrock weathering, including weathering rinds, rounding of edges, protrusion of relatively resistant veins, and abundance and size of gnammas and rillen, and (iv) geometry of stream cross-sectional profiles. The lowest zones generally have till and ample evidence of glaciation. The upper zones have no evidence of recent glaciation and often comprise felsenmeer (which is a 0.5 to >3 m thick surface layer of porous angular boulders and regolith, particularly in the discontinuous permafrost zones in eastern Canada) and relatively more developed soils. In the 1970s, D. Grant and I. Brookes independently mapped three weathering zones in all high relief regions (800 m) of western Newfoundland (e.g. Brookes, 1977; Grant, 1977, 1986, 1989). They interpreted the boundaries between the zones to represent limits of glaciation. The boundary between the lowest ('A') and intermediate ('B') zone was interpreted to represent the trimline of the last glaciation (letter designations from Grant, 1977). The boundary between 'B' and 'C' zones was the penultimate trimline and zone 'C' areas were either never glaciated (i.e. the summits were nunataks) or were glaciated in the mid- or early Pleistocene. Corroborating the notion that zone boundaries are trimlines is the observation that the boundaries in eastern Canada tend to dip seaward, just as ice in the fjord valleys drained ice from farther inland and spilled onto the coastal piedmont. Early in the development of the nunatak hypothesis were a series of ecological studies that identified over a dozen rare beetle, fish and bryophyte species isolated in the uppermost weathering zones in eastern Canada and Greenland. These disjunct species were believed to have survived in ice-free enclaves along the coastal highlands during the last glaciation (e.g. Fernald, 1911; Belland & Brassard, 1988) (Fig. 89.2a).

An alternative to the nunatak hypothesis is that the weathering zones simply represent regions exhibiting different degrees of glacial erosion, as discussed above and observed by many others elsewhere (Sugden, 1968; Sugden & Watts, 1977; Dyke, 1993; Kleman & Hättestrand, 1999; Dredge, 2000; Stroeven *et al.*, 2002a,b; Marquette *et al.*, 2004; Staiger *et al.*, 2005;). In fjord valleys we should expect the greatest amount of erosion where glacial ice is converging and the ice flux is greatest, with additional sliding due to lubrication from frictional melting. The felsenmeer-capped summits are conducive to frozen-based non-erosive ice cover, particularly where the ice cover is thin due to drawdown into the bounding fjords (Fig. 89.2b).

89.3 Methods

For a recent review of the principles and applications of TCN exposure dating the reader is referred to Gosse & Phillips (2001). Secondary cosmic rays penetrate the upper metres of rock and sediment cover on Earth's surface and interact with nuclei of exposed minerals. For instance, cosmogenic ^{10}Be is produced when a secondary particle breaks apart an oxygen or silicon atom in exposed quartz during an interaction referred to as spallation. Similarly ^{26}Al is produced from ^{28}Si. We used standard sample preparation procedures (e.g. Kohl & Nishiizumi, 1991) and a constant source for Be-carrier (shielded beryl crystal digested by Jeff Klein and Barbara Lawn, University of Pennsylvania) and Al-carrier (commercial inductively coupled plasma mass spectrometry (ICP-MS) standard). The accelerator mass spectrometry (AMS) facilities at University of Pennsylvania (1992 samples) and Lawrence Livermore National Laboratory (2002 samples) were used for the ^{10}Be/^{9}Be and ^{26}Al/^{27}Al measurements, and the inductively coupled plasma atomic emission spectrophotometry (ICP-AES) at EES-1, Los Alamos National Laboratory was used for Al and Be concentrations.

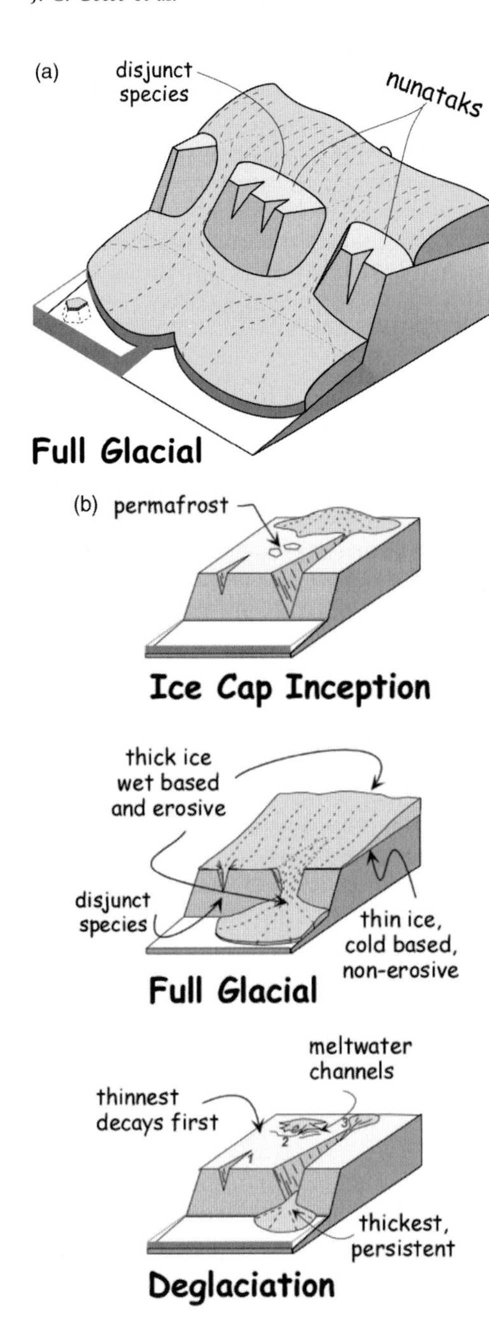

Figure 89.2 Block diagrams of the two opposing hypotheses for the glacial history and ice volume during the last glaciation for western Newfoundland. (a) Nunatak hypothesis, showing extent of Long Range Ice Cap during the Last Glacial Maximum (LGM), after Grant (1977). (b) Complete cover hypothesis, requring thin ice cover on summits and possibly exposed cliffs during the LGM. If the ice cover is from Laurentide Ice Sheet from the west, the cliffs will not be exposed during the LGM.

Our strategy for using TCN to test the weathering zone hypothesis was developed over a decade ago (Gosse *et al.*, 1993). The isotope approach consists of three steps.

1 Using a single TCN, document the timing of last deglaciation in a region by dating erratics strewn throughout a summit plateau or on a moraine. This is a critical step, because it can demonstrate the systematic retreat of ice margins on summits which otherwise appear never to have been glaciated. Unfortunately there are not many large boulders on the highest weathering zones, and the boulders are often similar in lithology to the underlying bedrock because of their short glacial transportation distance. We choose boulders that are perched on bedrock or are on the highest summit to preclude their deposition by mass wasting. If the highest zones were never glaciated, the boulders should have accumulated TCN unhindered by ice cover and therefore should be much older than the timing of the last glacial maximum.
2 Determine the concentration of a single TCN in bedrock surfaces adjacent to the boulders. If the concentrations in the boulders and adjacent bedrock surfaces (typically tor-like features) are equivalent, then the glaciers must have eroded the bedrock more than 2 m to remove any TCN produced in the bedrock prior to glaciation. It is common to have this parity in glaciated valleys where glacial erosion is concentrated (e.g. Labrador and Baffin Island—Marquette *et al.*, 2004, Staiger *et al.*, 2005; elsewhere—Stroeven *et al.*, 2002a,b). If the TCN concentration in bedrock is greater (>2 σ) than that in the adjacent boulders, the glacier that deposited the boulders must have eroded less than 2 m of rock.
3 To substantiate the notion that the bedrock surfaces were actually covered by non-erosive ice (one or more times in the past few million years), a second isotope is measured. The ratio of their concentrations (e.g. $^{26}Al/^{10}Be$) is used to test whether the bedrock surface experienced interruptions in its long exposure history (i.e. due to ephemeral cover by glacier ice, till, ash, water, etc.). A surface that has been eroded more than 2 m deep has very little record of its history of exposure prior to glaciation, so the isotopic ratio will be close to the theoretical production ratio of the isotopes on a continuously exposed surface (upper curve, Fig. 89.3). However, if the surface was buried but not eroded, the measured concentration of TCNs in the rock will record the concentration prior to ice cover (minus any that may have decayed during the ice cover) plus any TCN that has been produced since deglaciation. The ratio of the rapidly:slowly decaying radioisotopes will decrease during periods of burial (below the shaded field, Fig. 89.3).

89.4 Results

89.4.1 Timing of deglaciation of the Long Range Mountains type localities based on TCN in erratics

The exposure ages of boulders in the two type localities of weathering zones show that the summits were deglaciated in the late Pleistocene. On the highest weathering zone of the Highlands of St John (Fig. 89.1), four boulders (>1.5 m height, resting on bedrock or perched on cobbles) have ^{10}Be ages of 13.8 ± 0.4, 13.1 ± 0.6, 13.6 ± 0.5 and 13.9 ± 0.6 ka (±1 σ AMS precision), for an average age of deglaciation of the summit of 13.6 ± 0.7 ka (±2 σ coefficient of variation, i.e. 5%) (Table 89.1). To the south, on Big Level between the fjords of Bakers Pond and Western Brook Pond,

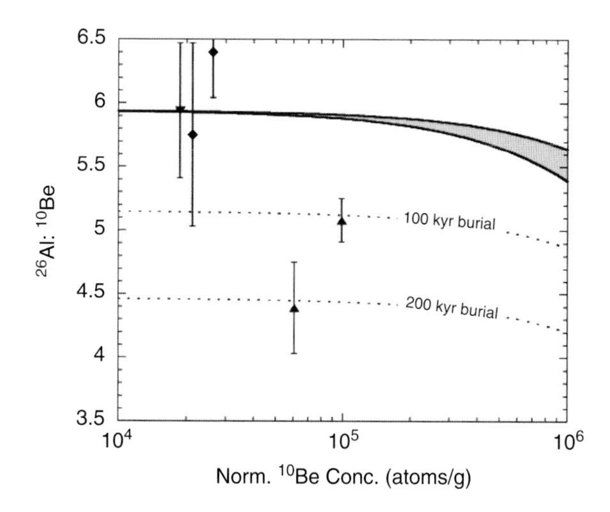

Figure 89.3 $^{26}Al/^{10}Be$ versus ^{10}Be isotope plot for bedrock samples from western Newfoundland. Production rates according to Stone (2000) and interpretation according to Gosse & Phillips (2001). Uncertainties are 1 σ AMS precision. Samples above the shaded field are theoretically impossible and may represent a measurement error. Samples in the field have continuous exposure although they may be gradually eroding. Samples below the field have had their exposures interrupted at least once by shielding cosmic rays, or have experienced recent plucking (episodic erosion >1 m). The latter is unlikely in this case because the samples are from very weathered surfaces (based on quartz vein protrusion and other weathering characteristics). Downward triangles, zone A; diamonds, zone B; upward triangles, zone C.

only one large erratic (80 cm high, but perched on a tor-like ridge) was found and dated with ^{10}Be at 20.9 ± 2.9 ka (±1 σ AMS precision). Significantly, the exposure ages of these erratics place ice cover on the highland summits of western Newfoundland until after the last global glacial maximum. The perched boulder on Big Level was closer to the edge of the fjord and therefore should have become uncovered prior to the deglaciation of the centre of the summits and the adjacent fjords. The ages also agree with the timing of deglaciation in weathering zone A based on bedrock ages.

89.4.2 Concentration of TCN in bedrock surfaces in the Long Range Mountains

The TCN concentration in eight bedrock surfaces in different weathering zones in the Big Level area indicates that even under ice cover the landscapes of the highest weathering zones preserved their mature appearance. Concentrations (normalized to account for differences in production rate due to atmospheric and geomagnetic field effects) from the highest weathering zone ('C' of Grant, 1977) range from 60.8 to 104 × 10^3 atoms g^{-1}. Concentrations in the intermediate weathering zone ('B') range from 21.3 to 125 × 10^3 atoms g^{-1}. In the lowest, least weathered, zone ('A') the bedrock surface has the lowest concentration of 18.7 × 10^3 atoms g^{-1}. The concentration of the erratic in weathering zone 'C' is 21 × 10^3 atoms g^{-1}, which is much lower than the TCN concentrations in the adjacent bedrock. These results show that the bedrock in the lowest and intermediate weathering zones typically have lost more than 2 m of bedrock (and essentially have had their TCN clocks reset about 18 ka). In contrast, the highest weathering zone has retained a memory of exposure prior to glaciation, despite ice cover as recently as 18 ka.

Table 89.1 Sample data (word format)

Sample ID	Elevation (m)	Zone (Grant, 1998)	Description	Normal concentration (atoms g^{-1} qtz)	Age (ka)	Uncertainty (ka)	$^{26}Al/^{10}Be$ (atom $atom^{-1}$)	Uncertainty (atom $atom^{-1}$)
NF-92-015	680	A	Quartz vein, 7 cm W, 0 cm H	18730	18.8	1.0	5.94	0.53
NF-92-009	708	B	Quartz vein 4 cm W, 2 cm H	26291	26.5	1.1	6.40	0.36
NF-92-010	762	B	Pegmatite dyke, 15 cm W, 2 cm H	21296	21.4	1.4	5.75	0.72
NF-92-011	696	B	Granitic gneiss	125031	129	5		
NF-92-008	734	C	Quartz vein, 5 cm W, 3 cm H	98847	101	4	5.08	0.17
NF-92-006	760	C	Quartz vein, 2 cm H	60795	61.7	2.1	4.39	0.36
NF-92-007	760	C	Granitic gneiss erratic	20804	20.9	3.0		
NF-92-003	726	C	Quartz vein, 4 cm W, 3 cm H	99941	102	4		
NF-92-016	641	C	Quartz vein, 3 cm W, 3 cm H	104233	107	4		
NF-02-SJH-301	606	C	Granitic gneiss erratic	13730	13.8	0.4		
NF-02-SJH-302	608	C	Granitic gneiss erratic	13019	13.1	0.6		
NF-02-SJH-303	618	C	Granitic gneiss erratic	13563	13.6	0.5		
NF-02-SJH-304	614	C	granitic gneiss erratic	13867	13.9	0.6		

89.4.3 $^{26}Al/^{10}Be$ to substantiate the previous results for western Newfoundland

The $^{26}Al/^{10}Be$ for weathering zone 'A' is 5.94 ± 0.53 (±1 σ), which is close to the ratio expected for a continuously exposed surface in the late Pleistocene (ca. 6.0 ± 0.1; Gosse & Phillips, 2001) (Fig. 89.3). The intermediate weathering zone 'B' ratio is 5.75 ± 0.72, which, although lower than weathering zone 'A' as expected, is within 1 σ error of the ratio for a continuously exposed surface. The ratio of a second bedrock surface in zone 'B' is 6.40 ± 0.36, which is not a possible ratio based on the production systematics of these two isotopes, and therefore must have an analytical error (Fig. 89.3). The ratios of weathering zone 'C' are substantially lower, at 4.39 ± 0.36 and 5.08 ± 0.17. Collectively, these ratios show that at least some of the bedrock surfaces within the highest weathering zone ('C') have an exposure history that was substantially interrupted by prolonged shielding (200 and 100 kyr minimum total burial durations), that weathering zone 'B' was interrupted by a shorter burial duration or not at all, and weathering zone 'A' surfaces record only continuous exposure since the last glaciation.

89.5 Discussion and conclusion

The results presented here substantiate work elsewhere in Canada and Fennoscandia where data are showing that relict landscapes that lack erosional or depositional evidence of glaciation were actually glaciated as recently as the last global glacial maximum. In the Long Range Mountains of western Newfoundland we use TCN in three ways to document that:

1. deglaciation of the highland summits was approximately coeval or slightly preceded the last deglaciation of thicker ice in the bounding fjord valleys based on radiocarbon age-estimates (on land and offshore, Batterson & Catto, 2001);
2. that despite being glaciated, the summit surfaces have retained an isotopic record of exposure prior to the last glaciation, which means that glacial erosion on the summits was less than 2 m;
3. ratios of $^{26}Al/^{10}Be$ show that the cosmic ray exposure of summit bedrock surfaces was interrupted at least once in the long histories, but that the valleys (zone 'A') and valley walls (many of the zone 'B' surfaces) have apparently been continuously exposed since deglaciation.

Of several explanations that have been posed to counter the interpretation of refugia on ice-free coastal highlands from the biological evidence, there are two that are appropriate for western Newfoundland.

1. The species are isolated today because they are adapted only to the specific ecosystems these summits provide (relative to the adjacent regions, the summits are cold, have wet but evenly distributed annual precipitation, low snow volumes and thin sediment cover).
2. These few species survived the last glaciation(s) in cliff habitats along coasts or between ice caps and valley glaciers that may have remained ice-free during most of the glaciation and subsequently expanded to the summit surfaces.

The latter explanation is not feasible if the entire Long Range Mountains were covered by the Laurentide Ice Sheet flowing from the west (another remaining debate). Clearly additional isotope measurements are needed to substantiate this small dataset for western Newfoundland. More importantly, we need to couple the TCN data with additional soils analyses, biological studies and quantitative geomorphology to provide a more robust test of the nunatak hypothesis.

Acknowledgements

This work benefited and matured from discussions with the late D. Grant, who was invited to be a co-author on an earlier (1996, never submitted) version of this manuscript, but who disagreed with our findings which according to him lacked an adequate means of explaining the distribution of disjunct species of beetles and mosses in the highest weathering zones. B. Lawn completed most of the chemistry for the 1992 samples. The manuscript improved with discussions with I. Brookes and A. Murphy. Field support from C. Gallagher and S. Tubb, in 2002, and field and helicopter support from R.A. Klassen, Geological Survey of Canada, in 1992, were greatly appreciated. Grants ACOA-AIF-1005052 and NSF-OPP-9906280 to JCG and NSERC Discovery Grants to TB and JTG supported this research.

NINETY

Characteristic cosmogenic nuclide concentrations in relict surfaces of formerly glaciated regions

Arjen P. Stroeven*, Jon Harbor†, Derek Fabel‡, Johan Kleman*, Clas Hättestrand*, David Elmore§ and David Fink¶

**Department of Physical Geography and Quaternary Geology, Stockholm University, S-106 91 Stockholm, Sweden*
†Department of Earth and Atmospheric Sciences, Purdue University, West Lafayette, IN 47907, USA
‡Department of Geographical and Earth Sciences, University of Glasgow, Glasgow G12 8QQ, UK
§Purdue Rare Isotope Measurement Laboratory, Purdue University, West Lafayette, IN 47907 USA
¶AMS-ANTARES, Environment Division, ANSTO, PMB1, Menai, NSW 2234, Australia

90.1 Significance of relict surfaces

Many formerly glaciated regions include some areas dominated by glacial landforms (Fig. 90.1A) and other areas with distinct non-glacial features (Fig. 90.1B & C). In addition to a suite of characteristic non-glacial geomorphological features (e.g. tors, mountains with approximate radial symmetry, convex-concave slopes, and well-developed blockfields and other periglacial features), areas with non-glacial features, termed relict areas, are particularly conspicuous in formerly glaciated areas because they often lack features indicative of glacial erosion. Because many relict landscapes in formerly glaciated areas of northern mid-latitude ice sheets occur on highlands close to coasts, and because they often flank deep fjords, in some cases these areas might not have been covered by ice at all (i.e. they were nunataks), and may then have served as glacial refugia for vegetation and animals. Although this contention appears logically plausible along the coasts, where fast flowing ice in fjord systems managed to lower the regional ice surface below the elevation of the interfluves and highest summits, it becomes more problematic when these surfaces occur inland from the current mountain divide. This is the case in many locations in central (Kleman & Borgström, 1990) and northern Sweden (Kleman & Stroeven, 1997). Here we know that ice sheets formerly covered these areas because they include glacial deposits or features from meltwater erosion. In this case the relict areas potentially provide important information about subglacial conditions under these ice sheets. Differentiating between nunataks and areas preserved under ice sheets is important for understanding former ice sheets. The subglacial temperature beneath ice sheets is a key ingredient of the inversion model described in Kleman *et al.* (see this volume, Chapter 38) used for reconstructing the dynamics of former mid-latitude ice sheets.

It remains unclear exactly when relict surfaces were originally formed, and, hence, for how long they were preserved essentially intact despite possible multiple episodes of ice sheet overriding. It is clear that post-glacial Holocene processes were ineffective in some locations and, if this is true of other interglacial periods, then some relict areas may have remained largely intact since the Tertiary.

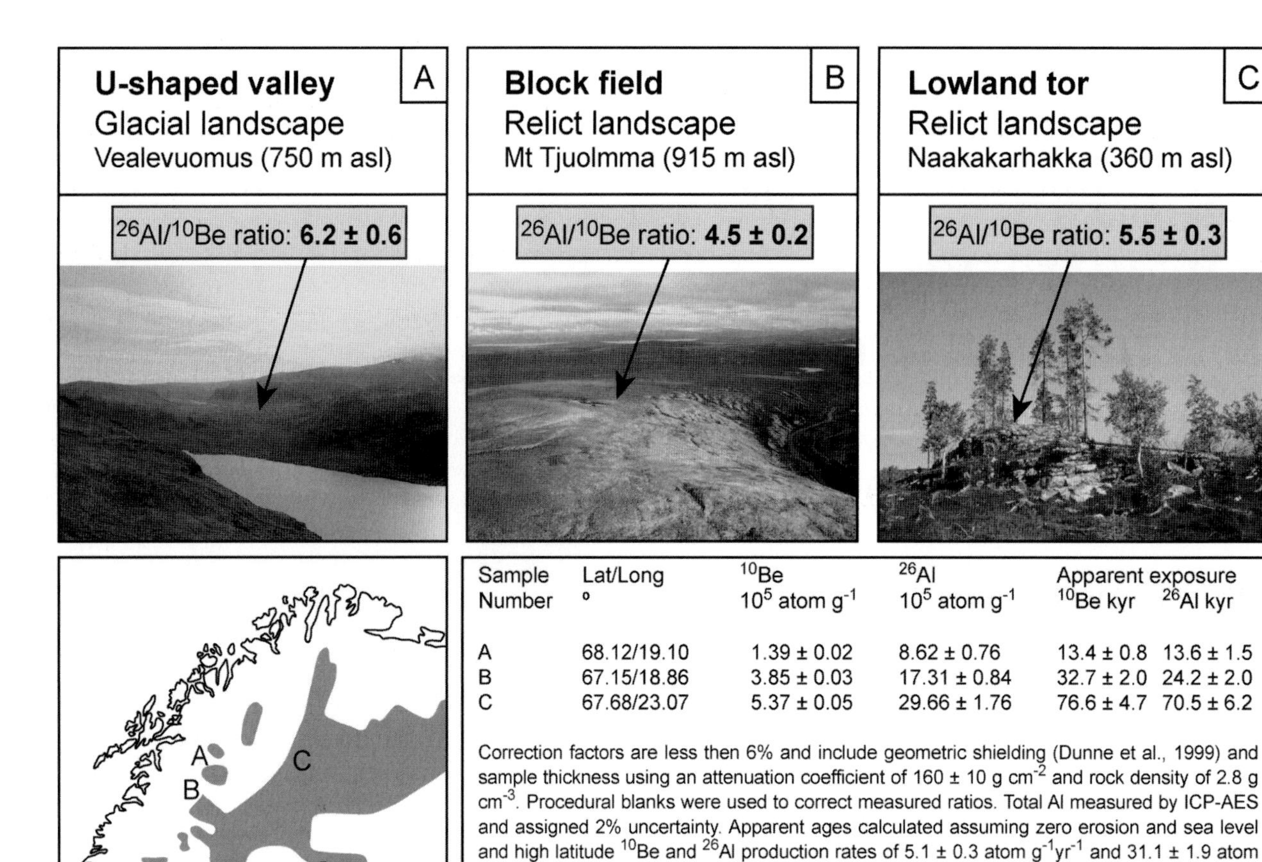

Sample Number	Lat/Long °	^{10}Be 10^5 atom g^{-1}	^{26}Al 10^5 atom g^{-1}	Apparent exposure ^{10}Be kyr	^{26}Al kyr
A	68.12/19.10	1.39 ± 0.02	8.62 ± 0.76	13.4 ± 0.8	13.6 ± 1.5
B	67.15/18.86	3.85 ± 0.03	17.31 ± 0.84	32.7 ± 2.0	24.2 ± 2.0
C	67.68/23.07	5.37 ± 0.05	29.66 ± 1.76	76.6 ± 4.7	70.5 ± 6.2

Correction factors are less then 6% and include geometric shielding (Dunne et al., 1999) and sample thickness using an attenuation coefficient of 160 ± 10 g cm^{-2} and rock density of 2.8 g cm^{-3}. Procedural blanks were used to correct measured ratios. Total Al measured by ICP-AES and assigned 2% uncertainty. Apparent ages calculated assuming zero erosion and sea level and high latitude ^{10}Be and ^{26}Al production rates of 5.1 ± 0.3 atom g^{-1}yr^{-1} and 31.1 ± 1.9 atom g^{-1}yr^{-1} respectively (Stone, 2000). Production rates are scaled to site-specific altitude and latitude using Stone (2000) with 5°C sea level temperature. Uncertainties represent one standard error measurement uncertainty.

Figure 90.1 Apparent cosmogenic isotope ages and ratios of bedrock samples from (A) the bottom of a hanging glacial valley, (B) a relict surface and (C) a lowland tor in northern Sweden. See map insert for approximate locations and table for isotope concentrations and apparent exposure ages. (Data in (B) and (C) are modified from Fabel *et al.* (2002) and Stroeven *et al.* (2002b), respectively.) (See www.blackwellpublishing.com/knight for colour version.)

90.2 Characteristic cosmogenic nuclide concentrations in relict surfaces

The concentrations of cosmogenic nuclides are different in rock surfaces that have been continuously exposed to the cosmic-ray flux, and surfaces that have been shielded by ice for long periods of time. For example the two radionuclides ^{10}Be and ^{26}Al are produced at a fixed ratio in quartz, but have different decay constants; thus if a surface that has been exposed for a time is shielded and nuclide production is halted, ^{26}Al decays faster than ^{10}Be and their ratio diverges from the production ratio of 6.1.

Two issues pertinent to ice-sheet reconstruction were revitalized by recent investigations that have used terrestrial cosmogenic nuclide techniques to examine relict areas within glaciated regions: (i) can these areas help constrain the surface elevation of former ice sheets, i.e. did they survive as nunataks, and (ii) can these areas be used to constrain the subglacial temperature regime, i.e. did they survive underneath cold-based ice?

The ability of the cosmogenic nuclide technique to differentiate between nunataks and areas preserved under ice can be illustrated using data from relict landscapes in northern Sweden. This area has been covered repeatedly by Fennoscandian ice sheets, and includes both relict areas and areas with substantial glacial modification. For example, apparent exposure ages from bedrock samples in U-shaped valleys (Fig. 90.1A) are sometimes consistent with expected deglaciation ages and hence consistent with substantial glacial erosion. Such glacial valleys occur adjacent to relict uplands, where apparent exposure ages from bedrock samples are older than deglaciation (Fig. 90.1B), proving that glacial erosion rates over these surfaces were lower (Fabel *et al.*, 2002). In fact, the presence of tors on many of these uplands and in the northern Swedish lowlands (Fig. 90.1C), and their apparent exposure ages, are consistent with the interpretation of minimal erosion and landscape modification by ice sheets that covered them. In such relict areas the cosmogenic nuclide apparent exposure ages of erratics, erosional scarps and meltwater channels incised into relict surfaces all yield deglaciation ages, reinforcing the conclusion derived from geomorphological mapping that these landscapes survived subglacially (e.g. Fabel *et al.*, 2002, Stroeven *et al.*, 2002a, b). A significant feature of cosmogenic nuclide concentrations in subglacially preserved relict landscapes is that the ratio between cosmogenically produced ^{26}Al and ^{10}Be in relict surfaces is lower (at 1 standard error) than the exposure ratio of 6.1 (Fig. 90.1). Similar results have derived for some relict areas in North America (e.g. Bierman *et al.*, 1999). Thus, in these cases the cosmogenic data allow us to conclude that certain areas were covered repeatedly by ice sheets, and that these sites represent areas where subglacial conditions were not conducive to erosion during growth, maximum and decay phases of multiple glaciations. However, it is important to note that exposure ratios within 1 standard deviation of 6.1 do not prove that a relict area was a nunatak. Unfortunately, the analytical uncertainty in ^{26}Al and ^{10}Be measurements is such that surfaces that have been shielded for up to ca. 150 kyr cannot be confidently distinguished from surfaces that have not been shielded.

90.3 Conclusions

Differentiating between relict areas that were nunataks and relict areas that were preserved as frozen areas under former ice sheets using cosmogenic nuclide studies provides important information for constraining the extent and characteristics of former ice sheets. The data currently available in northern Sweden (e.g. Fabel *et al.*, 2002, Stroeven *et al.*, 2002a, b) indicate that relict landscapes in interior locations of mid-latitude ice sheets probably have survived many consecutive glacial cycles. This provides an important constraint for inversion models used in reconstructing the dynamics of former mid-latitude ice sheets (see Kleman *et al.*, this volume, Chapter 38). It also provides a key boundary condition for ice sheet models, which must be able to indicate non-erosive basal conditions for these areas during growth, maximum and decay phases of repeated ice sheet glaciations. Similar landscapes in coastal locations may have been nunataks or were covered for substantially shorter time spans in total than the interior sites. Differentiating between these two possibilities requires more rigorous testing with multiple nuclides.

NINETY-ONE

Laboratory experiments in glaciology

Neal R. Iverson

Department of Geological and Atmospheric Sciences, Iowa State University, Ames, Iowa 50011, USA

91.1 Introduction

In 1952, John Glen published results of his uniaxial-compression experiments on ice. The resultant flow rule (Glen, 1952, 1955) provided the means to calculate the velocity of a glacier for the first time (Nye, 1952). His experiments helped establish modern glaciology (Clarke, 1987c), and although his flow rule continues to be revised and its limitations better understood, it remains a cornerstone of the discipline.

His results and their lasting impact illustrate the potential utility of laboratory experiments in glaciology. Although experiments are usually viewed as the gold standard for hypothesis testing in physics and chemistry, their perceived importance varies greatly across geoscience disciplines. Critics cite the scale and complexity of open natural systems and conclude, as a result, that Earth processes cannot be fully simulated in the laboratory. These critics are correct but miss the point. Well-designed experiments do not attempt to fully simulate natural conditions. Rather they attempt to isolate and thereby explain important phenomena that cannot be isolated in the field. Through known boundary and initial conditions, complete knowledge of parameter values, control of independent variables and the capacity for true reproducibility, laboratory experiments offer clear advantages that complement full-scale but less controlled field studies.

This review describes laboratory experiments in glaciology, with an emphasis on recent work. It is intended to highlight the contributions of experiments, address their limitations and point to gaps in knowledge that potentially could be filled experimentally. The discussion will be limited to work on glacial processes; for example, experiments that address periglacial processes and sea-ice mechanics will not be discussed.

91.2 Laboratory experiments in glaciology

91.2.1 Ice deformation and structure

The impact of laboratory experiments in glaciology has been greatest in the study of ice rheology and structure. Experiments provide the surest means of determining constitutive relations for ice deformation that when combined with classic conservation rules provide the basis for modelling glacier flow. Relative to rocks in Earth's crust, glaciers are homogeneous and deform at high rates. Laboratory studies are, therefore, more readily applied to glacier flow than to crustal deformation. However, despite the relative homogeneity of glacier ice and its rapid deformation, laboratory studies continue to demonstrate the complexity of ice deformation and to be seriously limited by time-scales for deformation that are too long to be explored in the laboratory. Thus, as was noted by Kamb (1972), laboratory studies of ice deformation, in addition to their relevance to glacier flow, serve as a warning to overly simplistic interpretations of experimental creep results for more complicated geological materials.

Field measurements have also been influential in the study of ice rheology and structure. In such studies, however, the state of stress is usually more complicated than is desirable and usually cannot be measured, requiring estimation of stresses with simplified models. Moreover, because more than one ice property (e.g. temperature, crystal fabric, impurities) commonly changes with depth in glaciers and ice sheets, isolating the effect of a single variable on ice flow is difficult.

Most experiments with polycrystalline ice have been simple in design. Either synthetic ice, usually with initially random c-axis orientations, or glacier ice, usually with some anisotropy, is squeezed. Experiments are normally carried out under a constant stress and under temperatures regulated to 0.1–1.0°C in a cold room or fluid bath. Stress is most commonly applied in uniaxial compression but also in shear and in combined compression and shear, depending upon the objectives of the study. Experiments are conducted over periods ranging from weeks to over 2 yr (e.g. Jacka, 1984b) with the goal of reaching steady-state deformation or at least the minimum strain rate that marks the beginning of tertiary creep.

One of the most important realizations of the last few decades is that no simple flow rule can adequately characterize deformation of ice in glaciers (Lliboutry & Duval, 1985; Alley, 1992). Experiments indicate that the original rule of Glen (1952, 1955) really represents part of an amalgam of flow rules that apply over different ranges of stress and strain rate. A generalization of Glen's flow rule can be written as

$$\dot{\varepsilon} = EA\exp\left(-\frac{Q}{RT}\right)\tau^{n} \tag{1}$$

where $\dot{\varepsilon}$ and τ are the strain rate and differential stress, respectively, A is constant for clean, isotropic ice, E is an enhancement factor that accounts for ice softening due to crystal fabric, impurities and other factors, Q is the activation energy for creep, T is the homologous temperature, R is the universal gas constant and n describes the sensitivity of the strain rate to the stress.

The value of n in Equation (1) is traditionally taken to be a constant equal to 3, based on assessments of both laboratory tests performed primarily at high stresses (>0.1 MPa) and field data (Paterson, 1994). Early laboratory studies conducted at low stresses indicated $n < 2$ (e.g. Mellor & Testa, 1969), but these studies may not have been carried out to sufficient strains to be indicative of steady-state creep, rather than transient creep (e.g. Weertman, 1983). However, strong laboratory evidence has accumulated more recently that, at stresses less than about 0.2 MPa, n is somewhat less than 2 (Pimienta & Duval, 1987; Goldsby & Kohlstedt, 1997, 2001, 2002; De La Chapelle *et al.*, 1999; Duval *et al.*, 2000; Duval & Montagnat, 2002)—a result also supported by some borehole-deformation measurements (Dahl-Jensen & Gundestrup, 1987). This result is important because deviatoric stresses less than 0.1 MPa are typical of ice sheets. Experiments indicate that $n = 4$ at stresses greater than about 0.5 MPa (e.g. Barnes *et al.*, 1971), but such high deviatoric stresses are not usually relevant to glaciers (Paterson, 1994).

Although the lower value of n at low stresses is relatively well accepted, deformation mechanisms responsible for the low value are controversial. Goldsby & Kohlstedt (1997, 2001), aware of the difficulty of reaching steady deformation rates at low stresses, used an innovative sample-preparation procedure to produce very fine-grained ice (3–200 μm). Deformation mechanisms at low stresses commonly result in strain rates that depend inversely on grain size, so reducing grain size better assured that steady-state deformation would be achieved. Goldsby & Kohlstedt found in experiments conducted over a wide range of stress that the low value of n (1.8) was due to basal slip rate-limited by grain-boundary sliding—the deformation mechanism associated with so-called 'superplastic flow'. This differs from the traditional view that dislocation creep is the rate-limiting deformation mechanism. Features indicative of grain-boundary sliding, such as straight grain boundaries, equant grains and four-grain intersections, developed during deformation. High-stress tests yielded results consistent with $n = 4$, as expected for deformation purely by dislocation creep. These data indicate that Glen's rule with $n = 3$ may reflect experiments conducted at stresses near the transition between the grain-boundary-sliding and dislocation-slip regimes (Fig. 91.1). Peltier *et al.* (2000) provocatively assert that an implication of these results is that rates of glacier flow may be underestimated in models by as much as one to two orders of magnitude.

Duval and colleagues (Duval *et al.*, 2000; Montagnat & Duval, 2000; Duval & Montagnat, 2002), however, argue that the low value of n for the coarser-grained ice of glaciers is the result of dislocation creep accommodated by grain-boundary migration. By absorbing dislocations, grain-boundary migration may reduce work hardening that results from local accumulations of dislocations that develop during slip on basal planes. Evidence in support of this viewpoint is that traditional deformation mechanisms, which involve crystal lattice rotation by dislocation slip, have been

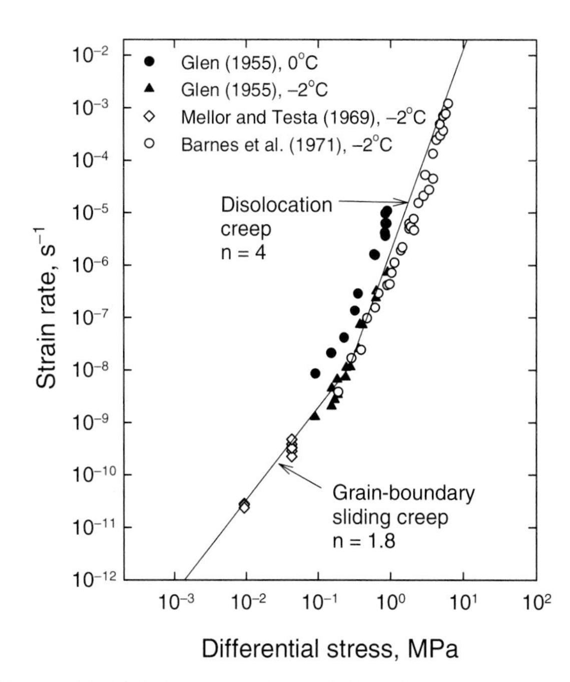

Figure 91.1 Reinterpretation of data from creep experiments on coarse-grained ice at temperatures near the melting point, based on the laboratory results of Goldsby & Kohlstedt (1997, 2001). (Modified from Peltier *et al.*, 2000.)

used to successfully model the development of fabric in ice sheets (e.g. Duval *et al.*, 2000; Montagnat & Duval, 2000). Deformation involving grain-boundary sliding, however, can also cause fabric development because strain that is rate-limited by grain-boundary sliding can accrue primarily from dislocation slip along basal planes (Goldsby & Kohlstedt, 2002). Studies of calcite rocks that develop fabric during superplastic deformation bear this out (e.g. Rutter *et al.*, 1994). However, empirical studies of ice fabrics that result from deformation that is rate-limited by grain-boundary sliding have not been conducted, and, unlike the case for traditional dislocation creep, observed ice-sheet fabrics have not yet been modelled successfully with grain-boundary sliding. Until such studies are conducted, the Goldsby–Kohlstedt hypothesis will probably remain controversial.

This controversy illustrates both the value and limitations of experimental approaches. Use of synthetic fine-grained ice in the experiments of Goldsby & Kohlstedt (1997, 2001) allowed steady-state deformation to be achieved unequivocally, revealing a potentially important deformation mechanism for ice in glaciers. On the other hand, application of the results to ice sheets required an uncertain extrapolation to large grain sizes. Despite this uncertainty, the overall result is positive. Old concepts thought to be quite sturdy are being re-evaluated. For example, there was generally thought to be no influence of ice grain size on deformation rate (Jacka, 1984b; Budd & Jacka, 1989). If grain-boundary sliding, however, is the rate-limiting process, deformation rate should be inversely related to grain size (Goldsby & Kohlstedt, 2001). Thus, this hypothesis provides new context for field studies that demonstrate grain-size dependence (Cuffey *et al.*, 2000c) and/or that seek to infer deformation mechanisms from microstructures.

Other recent laboratory experiments have been less controversial but also innovative and important. De la Chapelle *et al.* (1999) made synthetic ices of different salinities to study the effect of water in ice on its creep rate, extending the work of Duval (1977). Ice consisting of 7% water increased strain rates by more than an order of magnitude relative to pure ice, over stresses ranging from 0.02 to nearly 1 MPa. Softening was attributed to reduction of stress concentrations at grain boundaries that allowed more deformation to occur by easy dislocation slip along basal planes. Thus, by increasing the volume of water at grain boundaries, impurities that are ubiquitous and highly variable in ice cores may help soften ice. These results also may bear on the low effective viscosity of ice near the beds of temperature glaciers, where water contents may exceed 2% (Cohen, 2000).

In experiments conducted in combined compression and shear on initially isotropic ice, Li *et al.* (1996) demonstrated that minimum creep rates prior to the development of flow-induced anisotropy are independent of stress configuration. This result supports the fundamental assumption that deformation of isotropic ice is dependent only on the second invariant of the stress tensor (e.g. Nye, 1953). Not surprisingly, steady-state tertiary creep rates that were attained as ice acquired a fabric depended strongly on stress configuration. For simple shear without compression, enhancement factors of 10 were indicated, consistent with previous experiments (Shoji & Langway, 1988). Subsequent experiments conducted to strains greater than 100% produced single-maximum fabrics with c-axes concentrated perpendicular to the shear plane, similar to fabrics observed deep in polar ice sheets (Li *et al.*, 2000). Most previous simple-shear experiments had produced double-maximum fabrics (Kamb, 1972; Gao *et al.*, 1989), presumably because they were terminated at too low a strain.

91.2.2 Till deformation

Widespread recognition in the 1980s that fast glacier flow might depend on shear deformation of till beneath glaciers provided initial stimulus for experimental work on the mechanical properties of till. At that time, various till rheological rules were beginning to be used in bed-deformation models of glacier flow (e.g. Alley, 1989), usually with the assumption that till obeyed a fluid-like viscous or viscoplastic rheology. Apparent support for this assumption came from deformation profiles measured in subglacial till by Boulton & Hindmarsh (1987) at Breiðamerkurjökull. Since that time, however, deformation profiles measured in that study have been shown to not be uniquely indicative of a particular till rheology (Tulaczyk *et al.*, 2000a; Iverson & Iverson, 2001). This non-uniqueness highlights the difficulty of making definitive interpretations regarding till mechanical properties from measurements of till strain made subglacially, where the state of stress varies spatially and temporally and is difficult to either measure or estimate.

A wide variety of equipment and procedures have been used, to date, in experiments aimed at assessing the rheology of water-saturated till. Kamb (1991) conducted direct-shear tests on fine-grained till collected from the bed of Whillans Ice Stream.

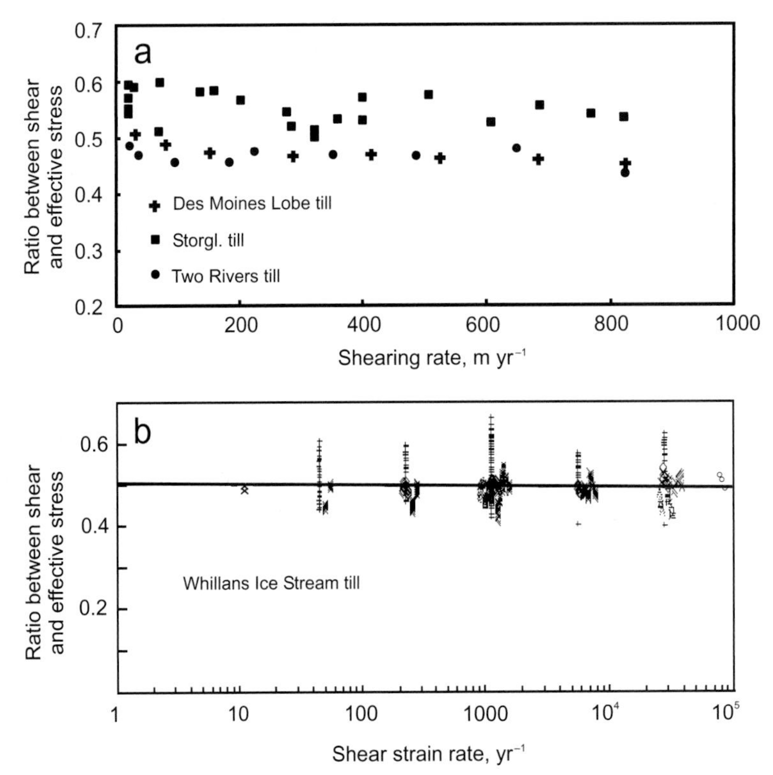

Figure 91.2 (a) Steady-state ratio of shear stress to effective normal stress as a function of shearing rate, as measured in ring-shear experiments on remoulded basal till of Storglaciären (4% clay, 21% silt, 75% sand and gravel), the Two Rivers till of the Lake Michigan lobe (32% clay, 30% silt, 38% sand and gravel), and the basal till of the Des Moines Lobe (16% clay, 36% silt, 48% sand and gravel). Shear rate is the rate of displacement across a shear zone 10–30 mm thick. Effective normal stresses were 20–150 kPa. (Modified from Iverson *et al.*, 1998.) (b) Ratio of shear stress to effective normal stress as a function of shear strain rate, as measured in triaxial experiments on the basal till of the Whillans Ice Stream (35% clay, 23% silt, 42% gravel). Different symbols indicate different effective normal stresses (25–320 kPa) (modified from Tulaczyk *et al.*, 2000a). In both (a) and (b) there is no increase in the ratio of shear stress to effective stress with strain rate that would indicate viscous deformation resistance.

Direct-shear experiments have the great advantage of being standard and easy to perform. Iverson *et al.* (1997, 1998) constructed a large ring-shear device (specimen chamber, 0.6 m o.d., 0.125 m width) and used it to study the mechanical behaviour of basal tills with different grain-size distributions. This device allowed very high shear strains, so that steady-state deformation was ensured, and allowed measurement of all boundary stresses. Tulaczyk *et al.* (2000a) constructed a smaller ring-shear device and also used triaxial and uniaxial testing equipment to study more thoroughly the till tested by Kamb (1991).

Despite the different tills and equipment used in these studies, all results indicate that the steady-state shearing resistance of till is extremely insensitive to strain rate (Fig. 91.2a & b) and that shearing resistance varies linearly with effective normal stress. These are the properties of a Coulomb (frictional) plastic material. Till, therefore, like granular materials in general, does not exhibit intrinsically viscous or Bingham-viscous behaviour, in which shearing resistance increases with strain rate. In contrast with these results, Ho *et al.* (1996) inferred mildly non-linear viscous behaviour from results of stress-controlled direct-shear experiments. Inspection of data from these tests (Vela, 1994), however, indicates that steady strain rates were not attained at a given stress; strain rates were still decreasing when a new larger stress was applied. Use of these data in models of glacier motion on soft beds is dubious (e.g. Licciardi *et al.*, 1998).

Although laboratory results indicate that Coulomb models for till are appropriate for steady deformation (so-called critical-state deformation in which porosity, shearing resistance, and strain rate remain constant), laboratory experiments indicate that transient shearing resistance of till can depend strongly on deformation rate. Moore & Iverson (2002) conducted ring-shear experiments on overconsolidated tills that dilated during the initial stages of shear. Shear stress was held constant and normal stress was reduced until friction within the till was decreased sufficiently to initiate shear deformation. Rapid dilation, as shear rate increased, reduced pore-water pressure, thereby strengthening the till. This apparent viscous response, called dilatant strengthening, occurs when pore-pressure diffusion toward opening voids cannot keep pace with the rate at which porosity increases during shear. Dilatant strengthening eventually slowed deformation, and the consequent decrease in dilation rate allowed pore-water pressure to slowly increase. Pore pressure eventually increased sufficiently to trigger another shearing episode that was again arrested by pore-pressure decline—a cycle that repeated up to 10 times until the

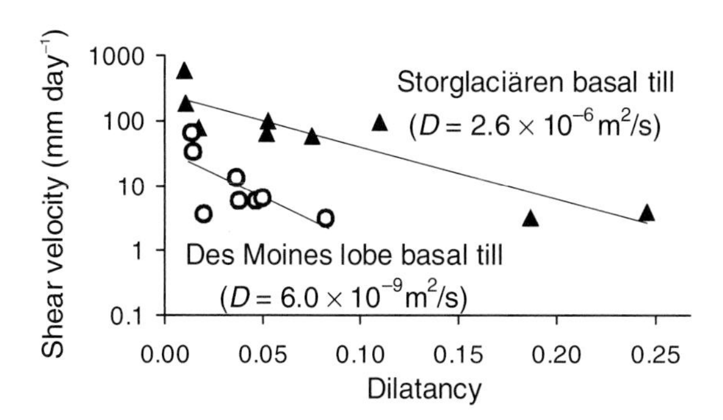

Figure 91.3 Transient shearing velocity as a function of dilatancy averaged for two tills (see Figure 91.2a) over the durations of creep episodes in stress-controlled ring-shear experiments. Dilatancy is the ratio of shear-zone thickening to shearing displacement; this ratio depends inversely on porosity. *D* is hydraulic diffusivity under an effective normal stress of 25 kPa. (From Moore & Iverson, 2002.)

till reached its critical-state porosity and could not dilate further, resulting in catastrophic failure. Time averaged shear velocities prior to catastrophic failure depended inversely on the magnitude of pore dilation with shear and were significantly lower for fine-grained till than for coarse-grained till, owing to different rates of pore-pressure diffusion in the two materials (Fig. 91.3).

These results illustrate that the apparent rheological behaviour of till beneath glaciers will be sensitive to till porosity, hydraulic diffusivity and the rate and magnitude of stress changes. Porosity less than the steady-state value for a given effective normal stress, low hydraulic diffusivity and rapid but small stress changes (limiting frictional equilibrium cannot be sustained by dilatant strengthening if stress changes are too large) favour dilatant strengthening and apparent viscous response. Given the likely variability of these factors beneath glaciers, a reasonable expectation is that the apparent rheological response of till can be highly variable, both temporally and spatially. This conclusion is consistent with recent measurements on Siple Coast ice streams, interpreted as indicating radically different rheological behaviour of tills beneath the Whillans Ice Stream (Coulomb behaviour) and ice streams C and D (apparent viscous behaviour) (Bindschadler *et al.*, 2003).

Till, like ice, develops anisotropy when it deforms. Grains of all sizes tend to become oriented with their long axes parallel to the direction of the largest principal strain. Murray & Dowdeswell (1992) conducted direct shear and triaxial experiments to low strains (<0.12) on various tills and used scanning electron microscopy to evaluate alignment of fine particles. Despite the small strain, alignment of particles was both visually and statistically discernable. A fourfold difference in hydraulic conductivity was estimated parallel and normal to the direction of particle alignment. Shear deformation of a till bed, in the absence of macroscale voids that might collapse during shear (Clarke *et al.*, 1984), should therefore enhance bed-parallel permeability, with likely pore-pressure feed-backs that would affect till mechanical behaviour.

In addition to hydraulic effects, alignment of till particles during deformation can be used to infer the extent to which a basal till of the geological record has been sheared. Hooyer & Iverson (2000a) used a ring-shear device to study the fabric formed by rigid, elongate particles in both till and viscous putty at various total shear strains up to 475. The results with putty agreed with the theory of Jeffery (1922), who predicted that elongate particles in a shearing fluid orbit periodically but reduce their rotation rates when oriented near the shear plane. In tests with till, however, particle behaviour was strikingly different. Rather than rotating continuously, clasts rotated into the shear plane at shear strains of approximately 2 and remained there, resulting in a strong fabric (S_1 eigenvalue of ca. 0.8) that persisted to high strains. This behaviour was caused by slip of the till matrix along clast surfaces, an effect not considered in the theory of Jeffery (1922). These results indicate that if total bed shear strain is high, strong particle fabrics result, contrary to some field inferences (e.g. Hart, 1994).

91.2.3 Sliding

The goal of focusing on recent experimental work cannot be met when considering sliding of glaciers over either a rigid or deformable substrate because there has been no recent experimental work. Indeed, there appear to be no published laboratory studies of glacier sliding since the mid-1980s. Oddly, this situation persists despite growing awareness that accurately modelling processes such as rapid decay of past ice sheets, motion and shutdown of ice streams, Heinrich events and surging of valley glaciers depends on parameterizing basal motion successfully (Clark *et al.*, 1999; Marshall *et al.*, 2000).

Various efforts have been made in the laboratory to study sliding of synthetic ice over rigid roughness elements of very small scale (<0.01 m) (e.g. Barnes *et al.*, 1971) but the study of Budd *et al.* (1979) has been the most influential. Ice, kept at the melting temperature in ice-water baths, was slid across rigid substrates with a wide range of roughnesses. Data collected over the most glacially relevant ranges of shear stress, τ_b, normal stress, N, and sliding speed, v, indicated a relationship of the form, $v = k\,\tau_b^{p}\,N^{-q}$, where k is a constant dependent on bed roughness, $p = 3$ and $q = 1$. Many field studies of sliding have indicated relationships of similar form, if N is replaced by effective normal stress (Paterson, 1994), although values of p and q fitted to field data vary widely. Budd *et al.* (1979) speculated that the value of p from their study reflected the dominance of ice deformation over regelation during sliding because regelation would depend linearly on shear stress. This inference is supported by the likely loss of water from the film at the ice–rock interface to the surrounding ice-water bath (Hooke, 1998). The resultant loss of latent heat would have reduced the heat available for melting on stoss surfaces. Thus, the value of p was undesirably dependent on the proximity of the water film to the bath at atmospheric pressure. Similarly, the value of q may reflect this proximity, which would have controlled the thickness of the water film for a particular value of N. Sliding speed may have been especially sensitive to the thickness of the water film because the amplitudes of roughness elements were small (ca. 0.1–10 mm).

In other experiments, synthetic temperate ice has been slid across larger bed obstacles to study velocity fields and ice rheology. The most ambitious of these studies involved use of a large Couette-type viscometer to slide an annulus of ice at a controlled velocity across two sinusoidal bumps (wavelength, 0.53 m) (Brepson, 1979; Meyssonnier, 1982). The bumps were of low thermal conductivity (epoxy), thereby suppressing regelation. Leeward cavities formed that were comparable in size to the bumps. Using a finite-element model and measured cavity ceilings as a free-surface boundary of the modelling domain, Meyssonnier (1982) calculated steady-state velocity fields in good agreement with measured values using $n = 3$ in the flow law. Hooke & Iverson (1985) pushed streamlined bumps (wavelength, 0.16 m) beneath temperate ice under controlled shear stresses. Motion was accommodated primarily by ice deformation. Comparison of ice deformation measured around bumps with the results of a theoretical model for ice flow past a hemisphere (Lliboutry & Ritz, 1978) indicated $n > 1.5$.

The simplest and possibly most illuminating experiments motivated by the sliding problem have focused on regelation of round wires through ice. Such experiments with temperate ice by Drake & Shreve (1973) indicated that at driving stresses <0.1 MPa, wire speeds were one to two orders of magnitude lower than predicted by simple regelation theory (Nye, 1967). This result agreed in general with previous but less comprehensive laboratory studies (e.g. Townsend & Vickery, 1967). Drake & Shreve's study isolated several seemingly insignificant processes that slowed regelation dramatically. The most relevant of these for glacier sliding was the accumulation of solutes in the rear of wires. This reduced the freezing temperature there and thereby reduced heat flow to the front of the wire. This effect was subsequently incorporated in glacier sliding theories (Hallet, 1976a) and was also used to help interpret the origin of chemical precipitants on deglaciated bedrock surfaces (Hallet, 1976b). Wire-regelation experiments have also been conducted at subfreezing temperatures (Gilpin, 1980) and provide fundamental empirical support for the theory of cold-based glacier sliding (Shreve, 1984). Wire-regelation studies remind us that simple experiments incorporating highly idealized materials and geometries can yield insights that are unlikely to be brought to light by either experiments that are more complex or field studies.

91.2.4 Glacial erosion and sediment transport

Glacial erosion and sediment transport may have had major impacts on uplift in orogenic belts, weathering rates and atmospheric CO_2 (Hallet *et al.*, 1996; Jaeger *et al.*, 2001). This realization has spawned numerous large-scale field and modelling studies of glacial erosion and sediment yields. In comparison, experimental approaches have been unpopular.

Glacial erosion of bedrock occurs principally by abrasion and quarrying. Early experimental efforts to study abrasion of rock by debris-laden sliding ice were performed at temperatures less than −16°C (Lister *et al.*, 1968; Mathews, 1979). Significant abrasion occurred in these experiments, but sliding of cold-based glaciers is too slow to cause much abrasion. Experiments in which a flat rock bed was pushed beneath isolated, gravel-sized particles in temperate ice (Iverson, 1990) have allowed some elements of abrasion theories (Boulton, 1974; Hallet, 1979) to be tested. The glaciological variables in these models responsible for the normal stress in excess of hydrostatic that rock particles exert on the bed have been controversial. These experiments indicated that such stresses depend on the rate of ice movement toward the bed, consistent with the model of Hallet (1979). Stresses associated with ice movement differed from that predicted by regelation and creep theory for an isolated sphere (Watts, 1974) by less than a factor of two, despite the bed as a heat source and rigid boundary. The detachment of rock fragments from the bed by quarrying is probably volumetrically more important than abrasion (Drewry, 1986). However, there have been no laboratory efforts to study this process, and as a result quarrying theories (Iverson, 1991; Hallet, 1996) have not been tested.

Entrainment of sediment from an unlithified bed has been studied in a few experiments. Regelation of temperate ice under pressure through dense arrays of idealized and natural particles has been studied in experiments with and without sliding (Iverson, 1993; Iverson & Semmens, 1995). These experiments indicated that rates of ice motion toward the bed through particles were within a factor of five of that predicted by regelation theory for dense arrays of particles (Philip, 1980). Ice may intrude the glacier substrate by this kind of regelation and entrain particles (Clarke *et al.*, 1999). Particles in experiments ranged from gravel-sized clasts to coarse-grained till (primarily sand and coarse silt). Tightly bound pore water in finer sediments probably impedes or prevents such regelation (Alley *et al.*, 1997c). Knight & Knight (1999) pressed ice at −1°C against a wet sediment bed and observed entrainment of fine sediment (primarily coarse silt) in water at crystal boundaries. In other experiments, they imposed a cold front across the ice–sediment boundary and noted similar movement of fine sediment into ice.

Interactions between grains during basal transport have been studied experimentally to quantify debris communition. In ball-mill experiments, in which rock particles and steel balls were tumbled in a drum (Haldorsen, 1981), two size ranges of particles were produced: a coarse range (0.016–2 mm) due to grain crushing, and a fine range (0.002–0.063 mm) due to abrasion of grains sliding past one another. Momentum exchange between particles in these experiments differed from non-inertial grain interactions expected subglacially. However, ring-shear experiments conducted at glacial rates have revealed similar particle modes (Iverson *et al.*, 1996). Stress concentrations, measured normal to the shearing direction in initially equigranular mudstone, were large at low strains, causing primarily crushing. Consequent production of finer grains reduced stress concentrations by cushioning larger particles, so that abrasion dominated comminution at larger strains.

Mixing between sediment units is another consequence of debris transport in a deforming bed. In ring-shear experiments, Hooyer & Iverson (2000b) studied mixing between equigranular beads of different colours and between lithologically distinct tills. They found that random vertical motions of particles induced by shearing caused linearly diffusive mixing. Mixing coefficients were determined from laboratory measurements. These coefficients were applied to the contact between basal tills of the Des Moines and Superior lobes of the Laurentide Ice Sheet to place an upper bound on the total bed shear strain.

91.2.5 Water in glaciers

The laboratory study of water movement in glaciers is limited by the same scaling considerations that guide river studies (e.g. Peakall *et al.*, 1996), except at the smallest scale in which water moves through microscopic veins at crystal edges. Mader (1992) developed novel methods for making synthetic polycrystalline ice and determining the geometry of such veins photographically. Equilibrium veins—those with no temperature or impurity gradients—were similar in geometry to those predicted theoretically by Nye (1989). Pinched-off veins that help limit ice permeability constituted less than 5% of veins observed.

The flume approaches that have advanced fluvial geomorphology are difficult to extend to englacial and subglacial channels because of the difficulty of incorporating ice melting and deformation at appropriately large scales. Catania & Paola (2001), however, studied braiding of channels at glacier beds in the absence of such melting and deformation to determine how pressure-driven braiding patterns differ from those of subaerial rivers. A stream table was filled with non-cohesive sediment and fitted with piezometers and a transparent, rigid lid to allow direct observation of braiding patterns. The rigid lid, by hindering flow of water over banks and bars, caused lateral gradients in pressure head that were much larger and more variable than elevation-head gradients in rivers. These large, variable pressure gradients resulted in greater channel density and variability of flow direction than in braided rivers. A lid that could both melt and deform like temperate ice would have likely affected this result, but large lateral pressure gradients are certainly characteristic of subglacial hydraulic systems (e.g. Engelhardt & Kamb, 1997).

91.2.6 Field experiments

The manipulation of variables that characterizes experimental work is also clearly a part of some field studies on modern glaciers. For example, borehole water pressure is commonly perturbed to study the nature of subglacial hydraulic systems (Stone & Clarke, 1993; Engelhardt & Kamb, 1997; Stone *et al.*, 1997; Kamb, 2001). Human access to the beds of some glaciers provides opportunities to manipulate the basal environment further. At the Svartisen Subglacial Laboratory in Norway, a conical concrete obstacle (0.2 m high) was installed at the bed of the temperate glacier Engabreen (Cohen *et al.*, 2000). Stresses on the obstacle and sliding speed were measured and used in conjunction with a numerical model of ice flow to estimate A in Glen's flow rule (Equation 1) for basal ice (Cohen, 2000); the mean value of A determined was 5.4 times larger than for 'normal' temperate ice (Paterson, 1994, p. 97). In subsequent experiments beneath Engabreen, a trough (2.5 m long, 2.0 m wide, 0.5 m deep) was blasted in the subglacial bedrock and filled with simulated till (Iverson *et al.*, 2003). After ice closed on the till, water was pumped to it to increase its pore-water pressure. Till sheared beneath the ice only at intermediate values of pore pressure: when pore pressure was sufficiently low or high, ice moved over the till by sliding and ploughing without pervasive shear of the bed, as predicted by some models (Brown *et al.*, 1987; Iverson, 1999).

91.3 Conclusion: future directions

A survey of the *Journal of Glaciology* from 1997 to 2002 indicates that only 4% of papers published during that period involved laboratory experiments. Thus, although such experiments have played an important role in the discipline historically, they are the basis for only a minor fraction of glaciological research. The reasons for this may be more cultural than scientific. The aesthetic appeal of field studies and the increased ease of theoretical approaches made possible by inexpensive and potent computer technology probably contribute to the relative unpopularity of experimental approaches. If so, this is unfortunate; major progress solving longstanding problems in glaciology could be made experimentally.

The most glaring of these problems involves glacier sliding. At present, fundamental aspects of sliding theories that help guide large-scale model parameterizations (e.g. Marshall & Clarke, 1997b) are untested. For example, there are no direct data that unequivocally link basal water pressure to the sizes of basal cavities and cavity size to sliding speed. There are no direct data that link a known basal shear stress, bed roughness and ice-overburden pressure to the different water pressures required for ice–bed separation and unstable sliding. There are no direct data on rates of growth or shrinkage of cavities during their adjustment to variable subglacial water pressure. There are no data to directly compare the contrasting responses of soft- and hard-bedded glaciers to changing basal water pressure. All of these issues could be investigated with an appropriate sliding apparatus; temperatures, boundary stresses and basal water pressure could be controlled, with direct observation of ice–bed separation. The resources required for such experiments would be substantial but no larger than those commonly marshalled for experimental work in disciplines with stronger experimental traditions (e.g. geochemistry).

This review also points to other experimental directions, only a few of which can be mentioned herein. Studies of fabric development in ice deformed by grain-boundary sliding would help clarify key aspects of the debate regarding ice-deformation mechanisms (e.g. Goldsby & Kohlstedt, 2001, 2002; Duval & Montagnat, 2002). The possible dependence of ice effective viscosity on grain size should be revisited experimentally, given the potential importance of grain-boundary sliding and recent field measurements that illustrate grain-size dependence (Cuffey *et al.*, 2000c). Laboratory studies of till deformation should further explore how variables such as till porosity, hydraulic diffusivity and loading rate affect the apparent rheological response of till. Laboratory studies of bedrock quarrying would provide a test of quarrying models and thereby help guide large-scale erosion models (e.g. MacGregor *et al.*, 2000). Laboratory studies of abrasion and associated debris-bed friction would allow subglacial measurements of unexpectedly high debris-bed friction (Iverson *et al.*, 2003) to be critically assessed.

As an example of the potential impact of laboratory experiments, consider progress in understanding a process that is similar in some ways to glacier sliding: slip along crustal faults. Over the past several decades, a rate- and state-dependent friction rule has become widely accepted in the fault-mechanics community (Scholz, 2002) and is used routinely in efforts to model fault-slip dynamics (e.g. Segall & Rice, 1995). This rule did not result primarily from theoretical studies or field monitoring of fault slip. Rather, this rule emerged from multiple long-term efforts to study fault slip in well-controlled laboratory experiments. The experimental ethic responsible for this success story is similar to that which inspired initial laboratory studies of ice rheology (Glen, 1952, 1955). Expansion of this ethic would hasten progress in glaciology.

NINETY-TWO

Laboratory observations of ice formation and debris entrainment by freezing turbid supercooled water

Peter G. Knight and Deborah A. Knight

School of Physical and Geographical Sciences, Keele University, UK

92.1 Introduction

Supercooling of subglacial water by rapid pressure reduction during flow towards the glacier margin has been widely adopted as a hypothesis to explain the origin of anomalously thick basal ice sequences beneath temperate glaciers, and has even been invoked to explain the debris content of North Atlantic Heinrich layers (e.g. Alley *et al.*, 1998; Lawson *et al.*, 1998; Andrews & Maclean, 2003). However, the application of supercooling theory to the problem of basal ice formation remains controversial. For example, whereas Roberts *et al.* (2002) suggested that in southern Iceland the entire basal ice sequence of many glaciers draining the ice cap Vatnajökull can be attributed to freezing of supercooled water, Spedding & Evans (2002), working in the same area, concluded that it is premature to attribute thick sequences of basal ice to supercooling, that other mechanisms remained tenable, and that the evidence for supercooling was unclear. One obstacle to our understanding is the lack of diagnostic physical or chemical criteria that relate the supercooling process with identifiable ice facies. As a contribution to clarifying the characteristics that would be expected of basal ice formed by freezing supercooled subglacial water, we have simulated the supercooling process in the low-temperature laboratory and identified specific facies characteristics associated with the process.

92.2 Methods

We achieved supercooling by two different methods. The first method was to pressurize water to approximately 206 kPa, reduce the water temperature to –0.3°C, and then release the pressure. The second method was to maintain turbulence in the water by means of a bath-recirculating pump, reduce the temperature to –0.3°C, and then switch off the turbulence. Pilot experiments produced similar freezing rates and styles whether the supercooling was achieved by pressure or by turbulence. We adopted the turbulence method for our final experimental procedure because the turbulence also allowed us to keep sediment in suspension within the sample during freezing, which was analogous to the subglacial prototype environment.

Water was supercooled and frozen in 100-L tubs at different ambient laboratory temperatures, with different amounts of clay, silt and fine sand in the water, and in some cases with chemically 'spiked' water. For crystallographic and sedimentological analysis

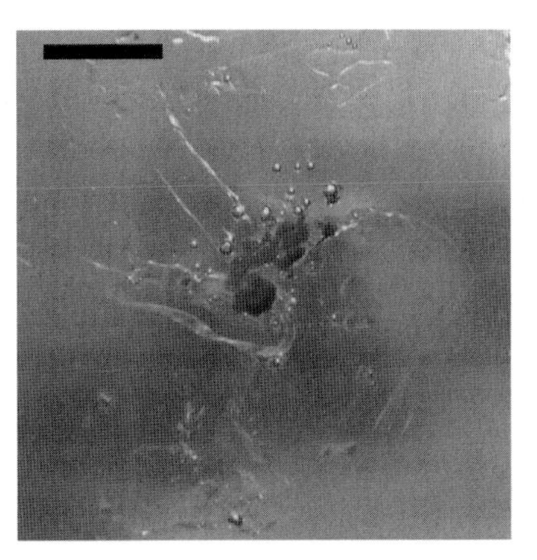

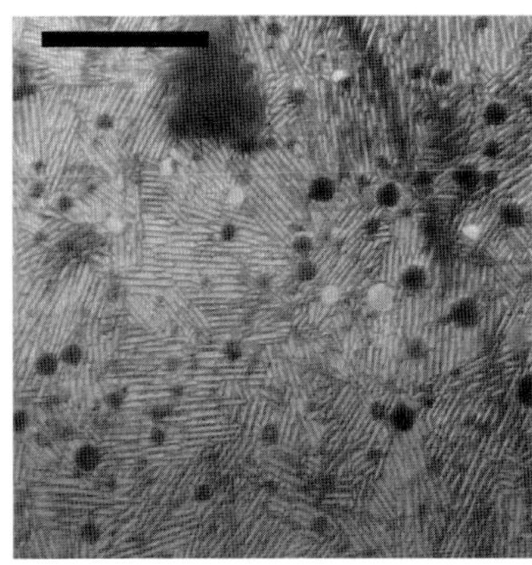

Figure 92.1 Frazil (left), clear (centre) and herringbone (right) facies created during the freezing experiments. Thick sections (0.5 cm) viewed in natural light. Scale bar is 1 cm in each photograph.

the distinctive facies that typically comprised the frozen blocks were sampled by cutting from the block and thin-sectioning. For chemical analysis, samples were taken of water before freezing, of water remaining unfrozen at the end of the experiment, and of each frozen facies.

92.3 Results

Three physically distinctive ice facies, arranged approximately concentrically within the frozen block, formed during the freezing process (Fig. 92.1). While the turbulence was maintained, frazil ice formed at the surface and around the edges of the tub, and created a network of interlocking plates and needles of ice between which debris was trapped. Continued freezing of the turbulent water–sediment mix developed a debris-poor, bubble-poor 'clear' ice facies with suspended aggregates of debris and with a strongly oriented crystal fabric indicating directional freezing towards the centre of the tub. After the turbulence was switched off a third facies, with a distinctive 'herringbone' crystal structure and intracrystalline bubble/debris lineations, was created nearest to the centre of the tub, with a pool of unfrozen water remaining at the tub centre at the end of the experiment. The physical structure of the herringbone ice varied with ambient temperature (Fig. 92.2), with lower temperatures associated with both smaller crystal sizes and finer intracrystal bubble- and debris-lineations. The chemical composition of the three facies was also distinctive. The initial water was spiked with a variety of ionic solutes and the fractionation of the impurities during the experimental freezing is illustrated in Fig. 92.3. Impurities were effectively expelled during the directional freezing of the clear facies but incorporated into the ice during the multidirectional freezing of the frazil ice and, to an even greater extent, of the herringbone facies.

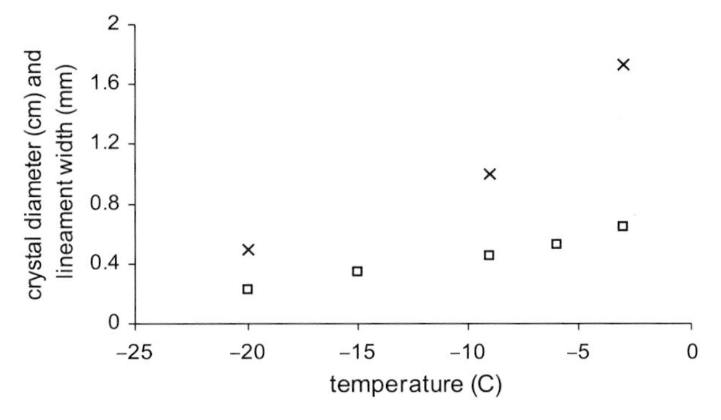

Figure 92.2 Mean crystal diameter (cm, crosses) and mean lineament-width (mm, squares) for samples of herringbone ice facies created by freezing water supercooled to −0.3°C at different ambient laboratory temperatures from −3°C to −20°C.

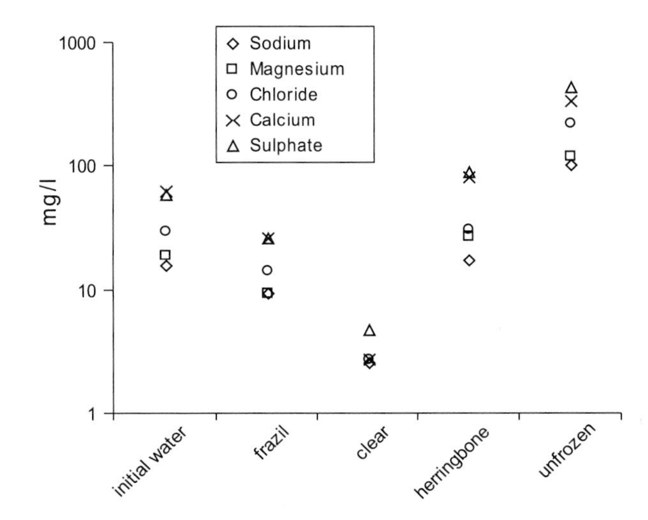

Figure 92.3 Ionic composition of parent water, frazil ice, clear ice, herringbone ice and water remaining unfrozen after the freezing experiment, showing differential expulsion/inclusion of solute in each ice facies.

92.4 Discussion

The ice facies with distinctive physical and chemical characteristics that appear to be created by different styles of freezing at different stages of the experiment are comparable with ice that has been described from the basal ice layer and ice-marginal drainage of glaciers. The frazil ice that developed in the laboratory upon the initial cooling of the turbulent water is similar in structure to that described by Evenson *et al.* (1999) in association with upwelling supercooled subglacial water at the margin of Matanuska glacier. The clear facies with dispersed debris aggregates that developed from directional freezing is similar crystallographically and in debris composition to dispersed and clear facies ice described from many glaciers (e.g. Knight, 1997), although the isotopic composition of these facies described in the field is not always consistent with a simple freezing origin. The herringbone facies associated with the laboratory freezing of supercooled water in a style broadly analogous to that proposed for subglacial supercooling does not resemble previously described basal ice facies from known or proposed supercooling glaciers. However, it does resemble ice grown by supercooling in previous laboratory experiments (e.g. Shibkov *et al.*, 2003) and also resembles ice that we have observed in small quantities at the margin of several Icelandic glaciers where supercooling has been proposed.

92.5 Conclusions

Frazil and clear ice facies physically similar to those observed in the field can be created by the freezing of turbulent water without any pressure-induced supercooling and without any change being imposed on the characteristics of the freezing system during the experiment. Ice with a distinctive herringbone structure was created only when the characteristics of the system were changed during the experiments, either by switching off the turbulence or, in the pilot studies that we conducted using the pressure method, by releasing the pressure. These experiments indicate that freezing water in an environment with varying pressure and/or turbulence can be expected to produce a variety of chemically and physically distinctive facies, the characteristics of which can be used to reconstruct the freezing environment. Further work to elucidate the significance, or otherwise, of supercooling in the formation of basal ice beneath glaciers should consider experimental evidence of how process controls facies character, as well as the field-based descriptions of basal ice that have so far dominated the literature. Experimental work provides a valuable avenue for complementing field observations, and one that has not hitherto been utilized sufficiently in basal ice research.

Acknowledgement

This work was funded by a grant from the Leverhulme Trust.

References

Aagaard, K. & Carmack, E.C. (1989) The role of sea ice and other fresh water in the Arctic circulation. *Journal of Geophysical Research—Oceans*, **94**(C10), 14485–14498.

Aagard, K., Swift, J.H. & Carmack, E.C. (1985) Thermohaline circulation in the Arctic Mediterranean Seas. *Journal of Geophysical Research*, **90**, 4833–4846.

Aharon, P. (2003) Meltwater flooding events in the Gulf of Mexico revisited: Implications for rapid climate changes during the last deglaciation. *Paleoceanography*, **18**, 3-1–3-14.

Aber, J., Croot, D.G. & Fenton, M.M. (1989) *Glaciotectonic Landforms and Structures*. Kluwer Academic Publishers, Dordrecht.

Abdalati, W. & Steffen, K. (2001) Greenland ice sheet melt extent: 1979–1999. *Journal of Geophysical Research*, **106**(D24), 33983–33988.

Ackert, R.P., Barclay, D.J., Borns, H.W., *et al.* (1999) Measurements of past ice sheet elevations in interior West Antarctica. *Science*, **286**(5438), 276–280.

Adam, W.G. & Knight, P.G. (2003) Identification of basal layer debris in ice-marginal moraines, Russell Glacier, West Greenland. *Quaternary Science Reviews*, **22**, 1407–1414.

Addison, K. (1990) Introduction to the Quaternary in North Wales. In: *The Quaternary of North Wales: Field Guide* (Eds K. Addison, M.J. Edge & R. Watkin), pp. 1–19. Quaternary Research Association, Coventry.

Ageta, Y. & Higuchi, K. (1984) Estimation of mass balance components of a summer-accumulation type glacier in the Nepal Himalaya. *Geografiska Annaler*, **66A**(3), 249–255.

Ahlmann, H.W. (1948) *Glaciological Research on the North Atlantic Coasts*. Royal Geographical Society, London.

Ahmad, S. & Whitworth, R.W. (1988) Dislocation motion in ice: a study by X-ray synchrotron topography. *Philosophical Magazine*, **A 57**, 749–766.

Albrecht, O., Jansson, P. & Blatter, H. (2000) Modelling glacier response to measured mass-balance forcing. *Annals of Glaciology*, **31**, 91–96.

Alexander, B., Thiemens, M.H., Farquhar, J., Kaufman, A.J., Savarino, J. & Delmas, R.J. (2003) East Antarctic ice core sulfur isotope measurements over a complete glacial-interglacial cycle. *Journal of Geophysical Research*, **108**(D24). Art. No. 4786.

Allen, J.R.L. (1982) *Sedimentary Structures. Developments in Sedimentology* 30B. Elsevier, Amsterdam, Vol. 2, 663 pp.

Alley, R.B. (1988) Fabrics in polar ice sheets: development and prediction. *Science*, **240**, 493–495.

Alley, R.B. (1989) Water-pressure coupling of sliding and bed deformation: II. Velocity–depth profiles. *Journal of Glaciology*, **35**, 119–129.

Alley, R.B. (1991a) Sedimentary processes may cause fluctuations of tidewater glaciers. *Annals of Glaciology*, **15**, 119–124.

Alley, R.B. (1991b) Deforming bed origin for southern Laurentide till sheets? *Journal of Glaciology*, **37**, 67–76.

Alley, R.B. (1992) Flow-law hypotheses for ice-sheet modeling. *Journal of Glaciology*, **38**, 245–256.

Alley, R.B. (1993) In search of ice-stream sticky spots. *Journal of Glaciology*, **39**, 447–454.

Alley, R.B. (2000) Continuity comes first: recent progress understanding subglacial deformation. In: *Deformation of Glacial Material* (Eds A.J. Maltman, M.J. Hambrey & B. Hubbard), pp. 171–179. Special Publication 176, Geological Society Publishing House, Bath.

Alley, R.B. (2003) Comment on 'When Earth's freezer door is left ajar'. *Eos (Transactions of the American Geophysical Union)*, **84**, 315. Art. No. 2003ES000374.

Alley, R.B. & Ágústsdóttir, A.M. (2005) The 8k event: cause and consequences of a major Holocene abrupt climate change. *Quaternary Science Reviews* **24**, 1123–1149.

Alley, R.B. & Bindschadler, R.A. (Eds) (2001) *The West Antarctic Ice Sheet: Behavior and Environment*. Volume 77, AGU Antarctic Research Series, American Geophysical Union, Washington, DC.

Alley, R.B. & MacAyeal, D.R. (1994) Ice-rafted debris associated with binge–purge oscillations of the Laurentide Ice Sheet. *Paleoceanography*, **9**, 503–511.

Alley, R.B. & Whillans, I.M. (1991). Changes in the West Antarctic Ice Sheet. *Science*, **254**, 959–963.

Alley, R.B. & White, W.C. (2000) Evidence of West Antarctic changes in the Siple Dome ice core (abstract). In *WAIS: The West Antarctic Ice Sheet Initiative*, Seventh Annual Workshop, Algonkian Meeting Center, Sterling, VA.

Alley, R.B. & Woods, G.A. (1996) Impurity influence on normal grain growth in the GISP2 ice core. *Journal of Glaciology*, **42**, 255–260.

Alley, R.B., Blankenship, D.D., Bentley, C.R. & Rooney, S.T. (1986a) Deformation of till beneath ice stream B, West Antarctica. *Nature*, **322**, 57–59.

Alley, R.B., Perepezko, J.H. & Bentley, C.R. (1986b) Grain growth in polar ice, I, theory. *Journal of Glaciology*, **32**, 415–424.

Alley, R.B., Perepezko, J.H., Bentley, C.R. (1986c) Grain growth in polar ice, II, application. *Journal of Glaciology*, **32**, 425–433.

Alley, R.B., Blankenship, D.D., Bentley, C.R. & Rooney, S.T. (1987a) Till beneath Ice Stream B, 3. Till deformation: evidence and implications. *Journal of Geophysical Research—Solid Earth and Planets*, **92**, 8921–8929.

Alley, R.B., Blankenship, D.D., Rooney, S.T. & Bentley, C.R. (1987b) Till beneath Ice Stream-B. 4. A coupled ice–till flow model. *Journal of Geophysical Research—Solid Earth and Planets*, **92**, 8931–8940.

Alley, R.B., Blankenship, D.D., Rooney, S.T. & Bentley, C.R. (1989) Water-pressure coupling of sliding and bed deformation. 3. Application to Ice Stream-B, Antarctica. *Journal of Glaciology*, **35**, 130–139.

Alley, R.B., Meese, D.A., Shuman, C.A., *et al.* (1993) Abrupt increase in snow accumulation at the end of the Younger Dryas event. *Nature*, **362**, 527–529.

Alley, R.B., Anandakrishnan, S., Bentley, C.R. & Lord, N. (1994) A water-piracy hypothesis for the stagnation of Ice Stream C, Antarctica. *Annals of Glaciology*, **20**, 187–194.

Alley, R.B., Gow, A.J. & Meese, D.A. (1995) Mapping *c*-axis fabrics to study physical processes in ice. *Journal of Glaciology*, **41**, 197–203.

Alley, R.B., Shuman, C.A., Meese, D.A., *et al.* (1997a) Visual-stratigraphic dating of the GISP2 ice core; basis, reproducibility, and application. *Journal of Geophysical Research*, **102**(C12), 26367–26381.

Alley, R.B., Mayewski, P.A., Sowers, T., Stuiver, M., Taylor, K.C. & Clark, P.U. (1997b) Holocene climatic instability: a prominent widespread event 8200 years ago. *Geology*, **25**, 483–486.

Alley, R.B., Cuffey, K.M., Evenson, E.B., *et al.* (1997c) How glaciers entrain and transport basal sediment. *Quaternary Science Reviews*, **16**, 1017–1038.

Alley, R.B., Lawson, D.E., Evenson, E.B., Strasser, J.C. & Larson, G.J. (1998) Glaciohydraulic supercooling: a freeze-on mechanism to create stratified debris rich basal ice: II. Theory. *Journal of Glaciology*, **44**(148), 563–579.

Alley, R.B., Clark, P.U., Keigwin, L.D. & Webb, R.S. (1999) Making sense of millennial-scale climate change. In: *Mechanisms of Global Climate Change at Millennial Time Scales* (Eds P.U. Clark, R.S. Webb & Keigwin, L.D.), pp. 301–312. Geophysical Monograph 112, American Geophysical Union, Washington, DC.

Alley, R.B., Anandakrishnan, S. & Jung, P. (2001) Stochastic resonance in the North Atlantic. *Paleoceanography*, **16**, 190–198.

Alley, R.B., Lawson, D.E., Larsen, G.J., Evenson, E.B. & Baker, G.S. (2003) Stabilizing feedbacks in glacier-bed erosion. *Nature*, **424**, 758–760.

Ames, A. (1998) A documentation of glacier tongue variations and lake developments in the Cordillera Blanca. *Zeitschrift für Gletscherkunde und Glazialgeologie*, **34**(1).

Amitrano, D., Grasso, J.-R. & Hantz, D. (1999) From diffuse to localised damage through elastic interaction. *Geophysical Research Letters*, **26**(14), 2109–2112.

Anandakrishnan, S. (2003) Dilatant till layer near the onset of streaming flow of Ice Stream C, West Antarctica, determined by AVO (amplitude vs offset) analysis. *Annals of Glaciology*, **36**, 283–286.

Anandakrishnan, S. & Alley, R.B. (1994) Ice stream C sticky spots detected by microearthquake monitoring. *Annals of Glaciology*, **20**, 183–186.

Anandakrishnan, S. & Alley, R.B. (1997a) Tidal forcing of basal seismicity of ice stream C, West Antarctica, observed far inland. *Journal of Geophysical Research*, **102**(B7), 15183–15196.

Anandakrishnan, S. & Alley, R.B. (1997b) Stagnation of ice stream C, West Antarctica by water piracy. *Geophysical Research Letters*, **24**, 265–268.

Anandakrishnan, S. & Bentley, C.R. (1993) Micro-earthquakes beneath ice stream-B and ice stream-C, West Antarctica—observations and implications. *Journal of Glaciology*, **39**, 455–462.

Anandakrishnan, S., Blankenship, D.D., Alley, R.B. & Stoffa, P.L. (1998) Influence of subglacial geology on the position of a West Antarctic ice stream from seismic observations. *Nature*, **394**, 62–65.

Anandakrishnan, S., Alley, R.B., Jacobel, R. & Conway, H. (2001) The flow regime of Ice Stream C and hypotheses concerning its recent stagnation. In: *The West Antarctic Ice Sheet: Behavior and Environment* (Eds R.B. Alley & R.A. Bindschadler), pp. 283–296. Volume 77. American Geophysical Union, Washington, DC.

Anandakrishnan, S., Voigt, D.E., Alley, R.B. & King, M.A. (2003) Ice stream D flow speed is strongly modulated by the tide beneath the Ross Ice Shelf. *Geophysical Research Letters*, **30**. Art. No. 1361.

Andersen, E.S., Dokken, T.M., Elverhøi, A., Solheim, A. & Fossen, I. (1996) Late Quaternary sedimentation and glacial history of the western Svalbard continental margin. *Marine Geology*, **133**, 123–156.

Andersen, K.K., Armengaud, A. & Genthon, C. (1998) Atmospheric dust under glacial and interglacial conditions. *Geophysical Research Letters*, **25**, 2281–2284.

Anderson, B.M. (2004) *The response of Ka Roimata O Hine Hukatere/Franz Josef Glacier to climate change.* Unpublished PhD thesis, University of Canterbury, New Zealand.

Anderson, J.B. (1999) *Antarctic Marine Geology.* Cambridge University Press, Cambridge, 289 pp.

Anderson, J.B. & Shipp, S.S. (2001) Evolution of the West Antarctic ice sheet. In *The West Antarctic Ice Sheet: Behavior and Environment* (Eds R.B. Alley & R.A. Bindschadler) pp. 45–57. Antarctic Research Series 77, American Geophysical Union, Washington, DC.

Anderson, J.B., Shipp, S.S., Lowe, A.L., Wellner, J.S. & Mosola, A.B. (2002) The Antarctic Ice Sheet during the Last Glacial Maximum and its subsequent retreat history: a review. *Quaternary Science Reviews*, **21**, 49–70.

Anderson, R.S., Anderson, S.P., MacGregor, K.R., *et al.* (2004) Strong feedbacks between hydrology and sliding a small alpine glacier. *Journal of Geophysical Research*, **109**. FO3005, doi:10.1029/2004JF000120.

Anderson, S.P. (2005) Glaciers show direct linkage between erosion rates and chemical weathering fluxes, *Geomorphology*, **67**(1–2), 147–157.

Anderson, S.P., Drever, J.I. & Humphrey, N.F. (1997) Chemical weathering in glacial environments. *Geology*, **25**, 399–402.

Anderson, S.P., Drever, J.I., Frost, C.D. & Holden, P. (2000) Chemical weathering in the foreland of a retreating glacier. *Geochimica Cosmochimica Acta*, **64**(7), 1173–1189.

Anderson, S.P., Longacre, S.A. & Kraal, E.R. (2003) Patterns of water chemistry and discharge in the glacier-fed Kennicott River, Alaska: evidence for subglacial water storage cycles. *Chemical Geology*, **202**, 297–312.

Anderton, P.W. (1973) *Structural Glaciology of a Glacier Confluence, Kaskawulsh Glacier, Yukon Territory.* Report 26, Institute of Polar Studies, Ohio State University, Columbus, OH.

Andrews, J.T. (1970) *A Geomorphological Study of Post-glacial Uplift with Particular Reference to Arctic Canada.* Institute of British Geographers, London.

Andrews, J.T. (1972) Recent and fossil growth rates of marine bivalves, Canadian Arctic, and Late-Quaternary Arctic marine environments. *Palaeogeography, Palaeclimatology, Palaeoecology*, **11**, 157–176.

Andrews, J.T. (1973) The Wisconsin Laurentide Ice Sheet: dispersal centers, problems of retreat, and climatic implications. *Arctic and Alpine Research*, **5**, 185–199.

Andrews, J.T. (1975) *Glacial Systems. An Approach to Glaciers and their Environments.* Duxbury Press, North Scituate, MA.

Andrews, J.T. (1980) Progress in relative sea level and ice sheet reconstructions Baffin Island, N.W.T., for the last 125,000 years. In: *Earth Rheology, Isostasy, and Eustasy* (Ed. N.-A. Morner), pp. 275–200. Wiley, London.

Andrews, J.T. (1987a) Glaciation and sea level: a case study. In: *Sea Surface Studies. A Global View* (Ed. R.J.N. Devoy), pp. 95–126. Croom Helm, London.

Andrews, J.T. (1987b) The Late Wisconsin Glaciation and deglaciation of the Laurentide Ice Sheet. In: *North America and Adjacent Oceans during the Last Deglaciation*, Vol. K-3 (Eds W.F. Ruddiman & H.E.J. Wright), pp. 13–37. Geological Society of America, Boulder, Colorado.

Andrews, J.T. (1989) Quaternary geology of the northeastern Canadian Shield. In: *Quaternary Geology of Canada and Greenland*, Vol. K-1 (Ed. R.J. Fulton), pp. 276–301. Geology of North America Series, Geological Society of America, Boulder, Colorado, and Geology of Canada, No. 1, Geological Survey of Canada, Queen's Printer, Ottawa.

Andrews, J.T. (1990) Fiord to deep-sea sediment transfers along the northeastern Canadian continental margin: models and data. *Géographie Physique et Quaternaire*, **44**, 55–70.

Andrews, J.T. (1991) Association of ice sheets and sea level with global warming: a geological perspective on aspects of global change. In: *Global Changes of the Past* (Ed. R.S. Bradley), pp. 321–339. UCAR/Office for Interdisciplinary Earth Studies, Boulder, CO.

Andrews, J.T. (1998) Abrupt changes (Heinrich events) in late Quaternary North Atlantic marine environments: a history and review of data and concepts. *Journal of Quaternary Science*, **13**, 3–16.

Andrews, J.T. (2000) Icebergs and iceberg rafted detritus (IRD) in the North Atlantic: Facts and Assumptions. *Oceanography*, **13**, 100–108.

Andrews, J.T. (In press) A review: Late Quaternary marine sediment studies of the Iceland shelf, paleoceanography and land–ice sheet–ocean interactions. In: *The Environments of Iceland* (Ed. C. Caseldine). Elsevier, Amsterdam.

Andrews, J.T. & Barber, D.C. (2002) Dansgaard–Oeschger events: Is there a signal off the Hudson Strait Ice Stream? *Quaternary Science Reviews*, **21**, 443–454.

Andrews, J.T. & Giraudeau, J. (2003) Multi-proxy records showing significant Holocene environmental variability: the inner N Iceland Shelf (Hunafloi). *Quaternary Science Reviews*, **22**, 175–193.

Andrews, J.T. & Ives, J.D. (1978) 'Cockburn' nomenclature and the Late Quaternary history of the eastern Canadian Arctic. *Arctic and Alpine Research*, **10**, 617–633.

Andrews, J.T. & Maclean, B. (2003) Hudson Strait ice streams: a review of stratigraphy, chronology, and links with North Atlantic Heinrich events. *Boreas*, **32**, 4–17.

Andrews, J.T. & Mahaffy, M.A.W. (1976) Growth rate of the Laurentide Ice Sheet and sea level lowering (with emphasis on the 115,000 BP sea level low). *Quaternary Research*, **6**, 167–183.

Andrews, J.T. & Miller, G.H. (1972) Quaternary history of northern Cumberland Peninsula, Baffin Island, N.W.T., Canada. Part IV: maps of the present glaciation limits and lowest equilibrium line altitude for north and south Baffin Island. *Arctic and Alpine Research*, **4**, 45–59.

Andrews, J.T. & Miller, G.H. (1979) Glacial erosion and ice sheet divides, northeastern Laurentide Ice Sheet on the basis of limestone erratics. *Geology*, **7**(12), 592–596.

Andrews, J.T. & Tedesco, K. (1992) Detrital carbonate-rich sediments, northwestern Labrador Sea: Implications for ice-sheet dynamics and iceberg rafting (Heinrich) events in the North Atlantic. *Geology*, **20**, 1087–1090.

Andrews, J.T., Guennel, G.K. & Wray, J.L. (1972) An early Tertiary outcrop in north-central Baffin Island, Northwest Territories, Canada: environment and significance. *Canadian Journal of Earth Sciences*, **9**, 233–238.

Andrews, J.T., Davis, P.T. & Wright, C. (1976) Little Ice Age permanent snowcover in the eastern Canadian Arctic: extent mapped from LANDSAT-1 satellite imagery. *Geografiska Annaler*, **58A**, 71–81.

Andrews, J.T., Miller, G.H., Nelson, A.R., Mode, W.N. & Locke, W.W., III (1981) Quaternary near-shore environments on eastern Baffin island, N.W.T. In: *Quaternary Paleoclimate* (Ed. W.C. Mahaney), pp. 13–44. Geo Books, Norwich.

Andrews, J.T., Clark, P. & Stravers, J.A. (1985a) The patterns of glacial erosion across the Eastern Canadian Arctic. In: *Quaternary Environments: Eastern Canadian Arctic, Baffin Bay and Western Greenland* (Ed. J.T. Andrews,), pp. 69–92. Allen and Unwin, Boston, MA.

Andrews, J.T., Stravers, J.A. & Miller, G.H. (1985b) Patterns of glacial erosion and deposition around Cumberland Sound, Frobisher Bay, and Hudson Strait, and the location of ice streams in the Eastern Canadian Arctic. In: *Models in Geomorphology* (Ed. M.J. Woldenburg), pp. 93–117. Allen and Unwin, Boston, MA.

Andrews, J.T., Erlenkeuser, H., Tedesco, K., Aksu, A. & Jull, A.J.T. (1994) Late Quaternary (Stage 2 and 3) Meltwater and Heinrich events, NW Labrador Sea. *Quaternary Research*, **41**, 26–34.

Andrews, J.T., Smith, L.M., Preston, R., Cooper, T. & Jennings, A.E. (1997) Spatial and temporal patterns of iceberg rafting (IRD) along the East Greenland margin, ca. 68°N, over the last 14 cal. ka. *Journal of Quaternary Science*, **12**, 1–13.

Andrews, J.T., Kirby, M.E., Aksu, A., Barber, D.C. & Meese, D. (1998a) Late Quaternary detrital carbonate (DC-) events in Baffin Bay (67°–74°N): Do they correlate with and contribute to Heinrich Events in the North Atlantic? *Quaternary Science Reviews*, **17**, 1125–1137.

Andrews, J.T., Kirby, M.E., Jennings, A.E. & Barber, D.C. (1998b) Late Quaternary stratigraphy, chronology, and depositional processes on the SE Baffin Island slope, detrital carbonate and Heinrich events: implications for onshore glacial history. *Geographie physique et Quaternaire*, **52**, 91–105.

Andrews, J.T., Hardardóttir, J., Helgadóttir, G., *et al.* (2000) The N and W Iceland Shelf: insights into Last Glacial Maximum ice extent and deglaciation based on acoustic stratigraphy and basal radiocarbon AMS dates. *Quaternary Science Reviews*, **19**, 619–631.

Andrews, J.T., Caseldine, C., Weiner, N.J. & Hatton, J. (2001) Late Quaternary (~4 ka) marine and terrestrial environmental change in Reykjarfjördur, N. Iceland: climate and/or settlement? *Journal Quaternary Science*, **16**, 133–144.

Andrews, J.T., Hardardottir, J., Kristjansdottir, G.B., Gronvald, K. & Stoner, J. (2003) A high resolution Holocene sediment record from Húnflóaáll, N. Iceland margin: century to millennial-scale variability since the Vedde tephra. *The Holocene*, **13**, 625–638.

Andruleit, H., Freiwald, A. & Schäfer, P. (1996) Bioclastic carbonate sediments on the southwestern Svalbard shelf. *Marine Geology*, **134**, 163–182.

Aniya, M. (1988) Glacier inventory for the Northern Patagonia Icefield, Chile, and variations 1944/45 to 1985/86. *Arctic and Alpine Research*, **20**(2), 179–187.

Aniya, M. (1999) Recent glacier variations of the Hielo Patagónicos, South America, and their contribution to sea-level change. *Arctic, Antarctic, and Alpine Research*, **31**(2), 165–173.

Aniya, M. & Skvarca, P. (1992) Characteristics and variations of Upsala and Moreno glaciers, southern Patagonia. *Bulletin of Glacier Research*, **10**, 39–53.

Aniya, M., Naruse, R., Shizukuishi, M., *et al.* (1992) Monitoring recent glacier variations in the Southern Patagonia Icefield, utilizing remote sensing data. *International Archives of Photogrammetry and Remote Sensing*, **29**(B7), 87–94.

Aniya, M., Sato, H., Naruse, R., *et al.* (1996) The use of satellite and airborne imagery to inventory outlet glaciers of the Southern Patagonia Icefield, South America. *Photogrammetric Engineering and Remote Sensing*, **62**(12), 1361–1369.

Aniya, M., Sato, H., Naruse, R., *et al.* (1997) Recent glacier variations in the Southern Patagonia Icefield, South America. *Arctic and Alpine Research*, **29**(1), 1–12.

Aniya, M., Park, S., Dhakal, A.M. & Naruse, R. (2000) Variations of some Patagonian glaciers, South America, using RADARSAT and Landsat images. *Science Reports of the Institute of GeoScience, University of Tsukuba, Section A*, **21**, 23–38.

Anonymous (1969) Mass-balance terms. *Journal of Glaciology*, **8**(52), 3–7.

Anslow, F.S., Marshall, S.J. & Shea, J.M. (submitted) A comparison of degree-day and radiation index melt models on the Haig Glacier, Canadian Rockies. *Journal of Applied Meteorology.*

Arendt, A.A., Echelmeyer, K.A., Harrison, W.D., Lingle, C.S. & Valentine, V.B. (2002) Rapid wastage of Alaska glaciers and their contribution to rising sea level. *Science*, **297**(5580), 382–386.

Aristarain, A.J. & Delmas, R.J. (1993) Firn-core study from the southern Patagonia ice cap, South America. *Journal of Glaciology*, **39**(132), 249–254.

Arnaud, E. & Eyles, C.H. (2002) Catastrophic mass failure of a Neoproterozoic glacially influenced margin, the Great Breccia, Port Askaig Formation, Scotland. *Sedimentary Geology*, **151**, 313–333.

Arnold, N.S. & Sharp, M. (2002) Flow variability in Scandinavian ice sheet: modelling the coupling between ice sheet flow and hydrology. *Quarternary Science Reviews*, **21**, 485–502.

Arnold, N.S., Willis, I.C., Sharp, M.J., Richards, K.S. & Lawson, M.J. (1996) A distributed surface energy-balance model for a small valley glacier. I. Development and testing for Haut Glacier d'Arolla, Valais, Switzerland. *Journal of Glaciology*, **42**(140), 77–89.

Arthern, R.J. & Wingham, D.J. (1998) The natural fluctuations of firn densification and their effects on the eodetic determination of ice sheet mass balance. *Climate Change*, **40**, 605–624.

Ashby, M.F. (1966) Work hardening of dispersion-hardened crystals. *Philosophical Magazine*, **14**, 1157–1178.

Ashby, M.F. (1969) On interface-reaction control of Nabarro-Herring creep and sintering. *Scripta Metallurgica*, **3**, 837–842.

Ashby, M.F. & Duval, P. (1985) The creep of polycrystalline ice. *Cold Regions Science and Technology*, **11**, 285–300.

Ashton, G.D. & Kennedy, J.F. (1972) Ripples on the underside of river ice covers. *Journal of the Hydraulics Division, Proceedings of the American Society of Civil Engineers*, **98**, 1603–1624.

Astakhov, V.I., Kaplyanskaya, F.A. & Tarnogradskiy, V.D. (1996) Pleistocene permafrost of West Siberia as a deformable glacier bed. *Permafrost and Periglacial Processes*, **7**, 165–191.

Atkins, C.B., Barrett, P.J. & Hicock, S.R. (2002) Cold glaciers do erode and deposit: evidence from Allan Hills, Antarctica. *Geology*, **30**, 659–662.

Atre, S.R. & Bentley, C.R. (1993). Laterally varying basal conditions beneath Ice Stream-B and Ice Stream-C West Antarctica. *Journal of Glaciology*, **39**(133), 507–514.

Axtmann, E.V. & Stallard, R.F. (1995) Chemical weathering in the South Cascade Glacier Basin, comparison of subglacial and extraglacial weathering. In: *Biogeochemistry of Seasonally Snow-covered Catchments* (Eds K.A. Tonnessen, M.W. Williams & M. Tranter), pp. 431–439. IAHS Publication 228, International Association of Hydrological Sciences, Wallingford.

Aylsworth, J.M. & Shilts, W.W. (1989a) Glacial features around the Keewatin ice divide, Districts of Mackenzie and Keewatin. *Geological Survey of Canada Paper*, **88–24**, 21 pp.

Aylsworth, J.M. & Shilts, W.W. (1989b) Bedforms of the Keewatin ice sheet, Canada. *Sedimentary Geology*, **62**, 407–428.

Aylsworth, J.M. & Shilts, W.W. (1991) Surficial geology of Coats and Mansel Islands, Northwest Territories. *Geological Survey of Canada Paper*, **89–23**, 26 pp.

Azetsu-Scott, K. & Tan, F.C. (1997) Oxygen isotope studies from Iceland to an East Greenland Fjord: behavior of glacial meltwater plume. *Marine Chemistry*, **56**, 239–251.

Azuma, N. (1994) A flow law for anisotropic ice and its application to ice sheets. *Earth and Planetary Science Letters*, **128**(3–4), 601–614.

Azuma, N. & Fujii, Y. (Ed.) (2002) Ice drilling technology 2000. *Memoirs of the National Institute of Polar Research (Japan), Special Issue*, **56**.

Azuma, N. & Goto-Azuma, K. (1996) An anisotropic flow law for ice-sheet ice and its implications. *Annals of Glaciology*, **23**, 202–208.

Azuma, N. & Higashi, A. (1985) Formation processes of ice fabric pattern in ice sheets. *Annals of Glaciology*, **6**, 130–134.

Baker, R.W. & Hooyer, T.S. (1996) Multiple till layers beneath Storglaciären. *Stockholms Universitet Naturgeografisk Institut, Forskningsrapport*, **103**, 25–29.

Baker, V.R. (2001) Water and the Martian landscape. *Nature*, **412**, 228–236.

Baldini, J.U.L., McDermott, F. & Fairchild, I.J. (2002) Structure of the 8200-year cold event revealed by a speleothem trace element record. *Science*, **296**, 2203–2206.

Bales, R.C., McConnell, J.R., Mosley-Thompson, E. & Csatho, B. (2001) Accumulation over the Greenland ice sheet from historical and recent records. *Journal of Geophysical Research*, **106**(D24), 33813–33826.

Ballantyne, C.K. (1994) Gibbsitic soils on former nunataks: implications for ice sheet reconstruction. *Journal of Quaternary Science*, **9**, 73–80.

Ballantyne, C.K. (2002) The Loch Lomond Readvance on the Isle of Mull: glacier reconstruction and paleoclimatic implications. *Journal of Quaternary Science*, **17**(8), 759–772.

Ballantyne, C.K., McCarroll, D., Nesje, A., Dahl, S.O. & Stone, J.O. (1998a) The last ice sheet in north-west Scotland: reconstruction and implications. *Quaternary Science Reviews*, **17**, 1149–1184.

Ballantyne, C.K., McCarroll, D., Nesje, A., Dahl, S.O., Stone, J.O. & Fifield, L.K. (1998b) High-resolution reconstruction of the last ice sheet in NW Scotland. *Terra Nova*, **10**, 63–67.

Bamber, J.L. (1994a) A digital elevation model of the Antarctic ice sheet derived from ERS–1 altimeter data and comparison with terrestrial measurements. *Annals of Glaciology*, **20**, 48–54.

Bamber, J.L. (1994b) Ice Sheet *al*timeter Processing Scheme. *International Journal of Remote Sensing*, **14**(4), 925–938.

Bamber, J.L. & Bindschadler, R.A. (1997) An improved elevation dataset for climate and ice-sheet modelling: validation with satellite imagery. *Annals of Glaciology*, **25**, 439–444.

Bamber, J.L. & Rignot, E. (2002) Unsteady flow inferred for Thwaites Glacier and comparison with Pine Island Glacier, West Antarctica. *Journal of Glaciology*, **48**.

Bamber, J.L., Vaughan, D.G. & Joughin, I. (2000a) Widespread complex flow in the interior of the Antarctic ice sheet. *Science*, **287**, 1248–1250.

Bamber, J.L., Hardy, R.J. & Joughin, I. (2000b) An analysis of balance velocities over the Greenland ice sheet and comparison with synthetic aperture radar interferometry. *Journal of Glaciology*, **46**(152), 67–72.

Bamber, J.L., Ekholm, S. & Krabill, W.B. (2001a) A new, high-resolution digital elevation model of Greenland fully validated with airborne laser altimeter data. *Journal of Geophysical Research*, **106**, 6733–6745.

Bamber, J.L., Layberry, R.L. & Gogenini, S.P. (2001b) A new ice thickness and bed data set for the Greenland ice sheet 1: measurement, data reduction, and errors. *Journal of Geophysical Research*, **106**(D24), 33773–33780.

Banham, P.H. (1977) Glacitectonites in till stratigraphy. *Boreas*, **6**, 101–105.

Baral, D.R., Hutter, K. & Greve, R. (2001) Asymptotic theories of large-scale motion, temperature, and moisture distribution in land-based polythermal ice sheets: a critical review and new developments. *Applied Mechanics Reviews*, **54**, 215–256.

Barber, D.C., Dyke, A., Hillaire-Marcel, C., *et al*. (1999) Forcing of the cold event of 8200 years ago by catastrophic drainage of Laurentide lakes. *Nature*, **400**, 344–348.

Bard, E., Arnold, M., Fairbanks, R.G. & Hamelin, B. (1993) U230,U234 and C14 ages obtained by mass spectrometry on corals. *Radiocarbon*, **35**, 191–199.

Barlow, L.K. (2001) The time period 1400–1980 in central Greenland ice cores in relation to the North Atlantic sector. *Climatic Change*, **48**, 101–119.

Barnes, P., Tabor, D. & Walker, J.C.F. (1971) The friction and creep of polycrystalline ice. *Proceedings of the Royal Society of London, Series A*, **324**, 127–155.

Barnes, P.R.F., Mulvaney, R., Robinson, K. & Wolff, E.W. (2002) Observations of polar ice from the Holocene and the glacial period using the scanning electron microscope. In: *Papers from the International Symposium on Ice Cores and Climate* (Ed. E.W. Wolff). *Annals of Glaciology*, **35**, 559–566.

Barnes, P.W. (1987) Morphological studies of the Wilkes Land continental shelf, Antarctica—glacial and iceberg effects. In: *The Antarctic Continental Margin: Geology and Geophysics of Offshore Wilkes Land* (Eds S.L. Eittreim & M.A. Hampton), pp. 175–194. Earth Science Series 5A, Circum-Pacific Council for Energy and Mineral Resources, Menlo Park, CA.

Barnes, P.W., Reimnitz, E. & Fox, D. (1982) Ice rafting of fine-grained sediment, a sorting and transport mechanism, Beaufort Sea, Alaska. *Journal of Sedimentary Petrology*, **52**, 493–502.

Barnett, D.M. & Holdsworth, G (1974) Origin, morphology, and chronology of sublacustrine moraines, Generator Lake, Baffin Island, Northwest Territories, Canada. *Canadian Journal of Earth Sciences*, **11**, 380–408.

Barnett, P.J. (1992) Quaternary geology of Ontario. In: *Geology of Ontario* (Eds P.C. Thurston, H.R. Williams, R.H. Sutcliffe & G.M. Stott), pp. 1011–1088. Special Volume Part 2, Ontario Geological Survey.

Barrett, P.J. (1996) Antarctic palaeoenvironment through Cenozoic times—a review. *Terra Antarctica*, **3**, 103–119.

Bart, P.J. & Anderson, J.B. (1996) Seismic expression of depositional sequences associated with expansion and contraction of ice sheets on the northwestern Antarctic Peninsula continental shelf. In: *Geology of Siliclastic Shelf Seas* (Eds M. De Batist & P. Jacobs), pp. 171–186. Special Publication 117, Geological Society Publishing House, Bath.

Bassinot, F.C., Labeyrie, L.D., Vincent, E., Quidelleur, X., Shackleton, N.J. & Lancelot, Y. (1994) The astronomical theory of climate and the Bruhes–Matuyama magnetic reversal. *Earth and Planetary Science Letters*, **126**, 91–108.

Batterson, M.J. & Catto, N. (2001) Topographically-controlled Deglacial History of the Humber River Basin, Western Newfoundland. *Geographie Physique et Quaternaire*, **55**(3), 213–228.

Bay, R.C., Price, P.B., Clow, G.D. & Gow, A.J. (2001) Climate logging with a new rapid optical technique at Siple Dome. *Geophysical Research Letters*, **28**, 4635–4638.

Bazant, Z.P. (1999) Size effect on structural strength: a review. *Archive of Applied Mechanics*, **69**(9–10), 703–725.

Beaney, C.L. (2002) Tunnel channels in southeast Alberta, Canada: evidence for catastrophic channelized drainage. *Quaternary International*, **90**, 67–74.

Beaney, C.L. & Hicks, F.E. (2000) Hydraulic modeling of subglacial tunnel channels, south-east Alberta, Canada. *Hydrological Processes*, **14**, 2545–2557.

Beaney, C.L. & Shaw, J. (2000) The subglacial geomorphology of southeast Alberta: evidence for subglacial meltwater erosion. *Canadian Journal of Earth Sciences*, **37**, 51–61.

Beckmann, A., Hellmer, H.H. & Timmermann, R. (1999) A numerical model of the Weddell Sea: large scale circulation and water mass distribution. *Journal of Geophysical Research*, **104**(C10), 23375–23391.

Beget, J.E. (1986) Modeling the influence of till rheology on the flow of the Lake Michigan Lobe. *Journal of Glaciology*, **32**(111), 235–241.

Beget, J. (1987) Low profile of the Northwest Laurentide Ice Sheet. *Arctic and Alpine Research*, **19**(1), 81–88.

Behrendt, J.C., Finn, C.A., Blankenship, D. & Bell, R.E. (1998) Aeromagnetic evidence for a volcanic caldera (?) complex beneath the divide of the west Antarctic ice sheet. *Geophysical Research Letters*, **25**, 4385–4388.

Belchansky, G.I., Douglas, D.C. & Platonov, N.G. (2004) Duration of the Arctic sea ice melt season: regional and interannual variability, 1979–2001. *Journal of Climate*, **17**, 67–80.

Belderson, R.H., Kenyon, N.H. & Wilson, J.B. (1973) Iceberg plough marks in the northeast Atlantic. *Palaeogeography, Palaeoclimatology, Palaeoecology*, **13**, 215–224.

Belkin, I.M., Levitus, S., Antonov, J. & Malmberg, S.-A. (1998) 'Great Salinity Anomalies' in the North Atlantic. *Progress in Oceanography*, **41**, 1–68.

Bell, M. & Laine, E.P. (1985) Erosion of the Laurentide region of North America by glacial and glaciofluvial processes. *Quaternary Research*, **23**(2), 154–174.

Bell, R. (1884) *Observations on Geology, Mineralogy, Zoology and Botany of the Labrador Coast, Hudson's Strait and Bay*. Geological and Natural History Survey of Canada. Report of Progress 1882–83–84.

Bell, R. (1898) *Report on an Exploration on the northern Side of Hudson Strait*. Geological Survey of Canada. Annual Report, Vol. 11, part M. Queen's Printer, Ottawa.

Bell, R.E., Blankenship, D.D., Finn, C.A., *et al*. (1998) Influence of subglacial geology on the onset of a west Antarctic ice stream from aero-geophysical observations. *Nature*, **394**, 58–62.

Belland, R.J. & Brassard, G.R. (1988) The bryophytes of Gros Morne National Park, Newfoundland, Canada: ecology and phytogeography. *Lindbergia*, **14**, 97–118.

Beniston, M., Haeberli, W., Hoelzle, M. & Taylor, A. (1997) On the potential use of glacier and permafrost observations for verification of climate models. *Annals of Glaciology*, **25**, 400–406.

Benn, D.I. (1994) Fabric shape and the interpretation of sedimentary fabric data. *Journal of Sedimentary Research*, **A64**, 910–915.

Benn, D.I. (1995) Fabric signature of till deformation, Breidamerkurjokull, Iceland: *Sedimentology*, **42**, 735–747.

Benn, D.I. (2002) Clast-fabric development in a shearing granular material: implications for subglacial till and fault gouge. Discussion. *Geological Society of America Bulletin*, **114**, 382–383.

Benn, D.I. & Ballantyne, C.K. (1994) Reconstructing the transport history of glacigenic sediments: a new approach based on the co-variance of clast shape indices. *Sedimentary Geology*, **91**, 215–227.

Benn, D.I. & Evans, D.J.A. (1996) The recognition and interpretation of subglacially-deformed materials: *Quaternary Science Reviews*, **15**, 23–52.

Benn, D.I. & Evans, D.J.A. (1998) *Glaciers and Glaciation*. Arnold, London, 734 pp.

Benn, D. & Gemmell, A.M.D. (2002) Fractal dimensions of diamictic particle-size distributions: simulations and evaluation. *Geological Society of America Bulletin*, **114**, 528–532.

Benn, D.I., Kirkbride, M.P., Owen, L.A. & Brazier, V. (2003) Glaciated valley landsystems. In: *Glacial Landsystems* (Ed. D.J.A. Evans), pp. 372–406. Arnold, London.

Bennett, M.R. (2001) The morphology, structural evolution and significance of push moraines. *Earth-Science Reviews*, **53**, 197–236.

Bennett, M.R. (2003) Ice streams as the arteries of an ice sheet: their mechanics, stability, and significance. *Earth-Science Reviews*, **61**, 309–339.

Bennett, M.R. & Boulton, G.S. (1993) A reinterpretation of Scottish 'hummocky moraine' and its significance for the Deglaciation of the Scottish Highlands during the Younger Dryas or Loch Lomond Stadial. *Geology Magazine*, **130**(3), 301–308.

Bennett, M.R. & Glasser, N.F. (1996) *Glacial Geology: Ice Sheets and Landforms*. Wiley, Chichester, 364 pp.

Benson, L., Barber, D., Andrews, J.T., Taylor, H. & Lamothe, P. (2003) Rare-earth elements and Nd and Pb isotopes as source indicators for glacial-marine sediments and North Atlantic 'Heinrich events'. *Quaternary Science Reviews*, **22**, 881–890.

Bentley, C.R. (1971) Seismic anisotropy in the West Antarctic ice sheet. *Antarctic Research Series, American Geophysical Union*, **16**, 131–177.

Bentley, C.R. (1987) Antarctic ice streams—a review. *Journal of Geophysical Research—Solid Earth and Planets*, **92**, 8843–8858.

Bentley, C.R. (1998) Rapid sea-level rise from a West Antarctic ice-sheet collapse: a short-term perspective. *Journal of Glaciology*, **44**, 157–163.

Bentley, C.R., Lord, N. & Liu, C. (1998) Radar reflections reveal a wet bed beneath stagnant Ice Stream C and a frozen bed beneath ridge BC, West Antarctica. *Journal of Glaciology*, **44**, 149–156.

Berger, A. (1978) Long-term variations of daily insolation and Quaternary climatic changes. *Journal of Atmospheric Sciences*, **35**, 2362–2367.

Berger, A. & Loutre, M.F. (1991) Insolation values for the climate of the last 10 million years. *Quaternary Science Reviews*, **10**, 297–318.

Berger, A. & Loutre, M.F. (2002) An exceptionally long interglacial ahead? *Science*, **297**, 1287–1288.

Berger, W.H., Pätzold, J. & Wefer, G. (2002) A case for climate cycles: orbit, sun and moon. In: *Climate Development and History of the North Atlantic Realm* (Eds G. Wefer, W.H. Berger, K.-E. Behre & E. Jansen), pp. 101–123. Springer-Verlag, Berlin.

Bergthorsson, P. (1969) An estimate of drift ice and temperature in Iceland in 1000 years. *Jokull*, **19**, 94–101.

Beyerle, U., Purtschert, R., Aeschbach-Hertig, W., *et al.* (1998) Climate and groundwater recharge during the last glaciation in an ice-covered region. *Science*, **282**, 731–734.

Bianchi, G.G. & McCave, I.N. (1999) Holocene periodicity in North Atlantic climate and deep-ocean flow south of Iceland. *Nature*, **397**, 515–517.

Bierman, R., Marsella, K.A., Patterson, C., *et al.* (1999) Mid-Pleistocene cosmogenic minimum-age limits for pre-Wisconsinan glacial surfaces in southwestern Minnesota and southern Baffin Island: a multiple nuclide approach. *Geomorphology*, **27**(1–2), 25–39.

Bindschadler, R. (1983) The importance of pressurized subglacial water in separation and sliding at the glacier bed. *Journal of Glaciology*, **29**, 3–19.

Bindschadler, R. (1998) GEOSCIENCE: future of the West Antarctic Ice Sheet. *Science*, **282**, 428–429.

Bindschadler, R.A. (2002) History of lower Pine Island Glacier, West Antarctica, from Landsat imagery. *Journal of Glaciology*, **48**, 536–544.

Bindschadler, R. & Vornberger, P. (1998) Changes in the West Antarctic ice sheet since 1963 from declassified satellite photography. *Science*, **279**, 689–692.

Bindschadler, R.A., Stephenson, S.N., Macayeal, D.R. & Shabtaie, S. (1987) Ice dynamics at the mouth of Ice Stream-B, Antarctica. *Journal of Geophysical Research—Solid Earth and Planets*, **92**, 8885–8894.

Bindschadler, R., Vornberger, P., Blankenship, D., Scambos, T. & Jacobel, R. (1996) Surface velocity and mass balance of Ice Streams D and E, West Antarctica. *Journal of Glaciology*, **42**(142), 461–475.

Bindschadler, R., Bamber, J. & Anandakrishnan, S. (2001a) Onset of streaming flow in the Siple Coast region, West Antarctica. In: *The West Antarctic Ice Sheet: Behavior and Environment* (Eds R.B. Alley & R. Bindschadler), pp. 123–136. Volume 77, American Geophysical Union, Washington, DC.

Bindschadler, R., Dowdeswell, J., Hall, D. & Winther, J.G. (2001b) Glaciological applications with Landsat–7 imagery: early assessments. *Remote Sensing of Environment*, **78**(1–2): 163–179.

Bindschadler, R.A., King, M.A., Alley, R.B., Anandakrishnan, S. & Padman, L. (2003) Tidally controlled stick-slip discharge of a West Antarctic ice stream *Science*, **301**, 1087–1089.

Bingham, R.G., Nienow, P.W. & Sharp, M. (2003) Intra-annual and intra-seasonal flow dynamics of a High Arctic polythermal valley glacier. *Annals of Glaciology*, **37**, 181–188.

Bintanja, R. (2001a) Snowdrift sublimation in a katabatic wind region of the Antarctic ice sheet. *Journal of Applied Meteorology*, **40**, 1952–1966.

Bintanja, R. (2001b) Modelling snowdrift sublimation and its effect on the moisture budget of the atmospheric boundary layer. *Tellus*, **53A**, 215–232.

Bintanja, R., Tüg, H. & Lillienthal, H. (2001) Observations of snowdrift over Antarctic snow and ice surfaces. *Annals of Glaciology*, **32**, 168–174.

Birchfield, G.E. & Weertman, J. (1982) A model study of the role of variable ice albedo in the climate response of the Earth to orbital variations. *ICARUS*, **50**, 462–472.

Bischof, J.F. & Darby, D. (1997) Mid- to Late Pleistocene ice drift in the Western Arctic Ocean: evidence for a different circulation in the past. *Science*, **277**, 74–78.

Bishop, M.P., Olsenholler, J.A., Shroder, J.F., *et al.* (2004) Global land ice measurements from space (GLIMS): remote sensing and GIS investigations of the Earth's cyrosphere. *Geocarto International*, **19**(2), 57–84.

Bitz, C.M. & Battisti, D.S. (1999) Interannual to decadal variability in climate and the glacier mass balance in Washington, western Canada, and Alaska. *Journal of Climate*, **12**, 3181–3196.

Björck, S. & Wastegard, S. (1999) Climate oscillations and tephrachronology in eastern middle Sweden during the last glacial-interglacial transition. *Journal of Quaternary Science*, **14**, 399–410.

Björck, S., Kromer, B., Johnsen, S., *et al.* (1996) Synchronized terrestrial–atmospheric deglacial records around the North Atlantic. *Science*, **274**, 1155–1160.

Björnsson, H. (1969) Sea ice conditions and the atmospheric circulation north of Iceland. *Jokull*, **19**, 11–28.

Björnsson, H. (1979) Glaciers in Iceland. *Jökull*, **29**, 74–80.

Björnsson, H. (2002) Subglacial lakes and jökulhlaups in Iceland. *Global and Planetary Change*, **35**, 225–271.

Blake, E.W. (1992) *The deforming bed beneath a surge-type glacier: measurement of mechanical and electrical properties*. PhD thesis, University of British Columbia, Vancouver, 179 pp.

Blake, E.W., Clarke, G.K.C. & Gérin, M.C. (1992) Tools for examining subglacial bed deformation. *Journal of Glaciology*, **38**(130), 388–396.

Blake, E.W., Fischer, U.H. & Clarke, G.K.C. (1994) Direct measurement of sliding at the glacier bed. *Journal of Glaciology*, **40**(136), 595–599.

Blake, W.J. (1966) End moraines and deglaciation chronology in northern Canada with special reference to northern Canada. *Geological Survey of Canada Paper*, **66-26**.

Blanchon, P. & Shaw, J. (1995) Reef-drowning events during the last deglaciation: evidence for catastrophic sea-level rise and ice-sheet collapse. *Geology*, **23**, 4–8.

Blankenship, D.D., Bentley, C.R., Rooney, S.T. & Alley, R.B. (1986) Seismic measurements reveal a saturated porous layer beneath an active Antarctic ice stream. *Nature*, **322**, 54–57.

Blankenship, D.D., Bentley, C.R., Rooney, S.T. & Alley, R.B. (1987) Till beneath Ice Stream-B.1. Properties derived from seismic traveltimes. *Journal of Geophysical Research—Solid Earth and Planets*, **92**, 8903–8911.

Blankenship, D., Morse, D.L., Finn, C.A., *et al.* (2001) Geologic controls on the initiation of rapid basal motion for West Antarctic ice streams. In: *The West Antarctic Ice Sheet: Behavior and Environment* (Eds R.B. Alley & R. Bindschadler), pp. 123–136. Volume 77, American Geophysical Union, Washington, DC.

Blasco, S.M. (2001) Geological history of Fathom Five National Marine Park over the past 15,000 years. In: *Ecology, Culture, and Conservation of a Protected Area: Fathom Five National Marine Park Canada* (Eds S. Parker & M. Munawar), pp. 45–62. Ecovision World Monograph Series, Backhuys Publishers, Leiden.

Blatter, H. (1987) On the thermal regime of an arctic valley glacier: a study of White Glacier, Axel Heiberg Island, N.W.T., Canada. *Journal of Glaciology*, **33**, 200–211.

Blatter, H. (1995) Velocity and stress fields in grounded glaciers: A simple algorithm for including deviatoric stress gradients, *Journal of Glaciology*, **41**(138), 333–343.

Bluemle, J.P. & Clayton, L. (1984) Large-scale glacial thrusting and related processesin North Dakota. *Boreas*, **13**, 279–299.

Blum, J.D. (1997) The effect of late Cenozoic glaciation and tectonic uplift on silicate weathering rates and the marine $^{87}Sr/^{86}Sr$ record. In: *Tectonic Uplift and Climate Change* (Ed. W.F. Ruddiman), pp. 259–288. Plenum Press, New York.

Blunier, T., Chappellaz, J., Schwander, J., *et al.* (1998) Asynchrony of Antarctic and Greenland climate change, during the last glacial period. *Nature*, **394**, 739–743.

Bockheim, J.G. (1990) Soil development rates in the Transantarctic Mountains. *Geoderma*, **47**, 59–77.

Bøggild, C., Reeh, N. & Oerter, H. (1994) Modelling ablation and mass-balance sensitivity to climate change, of Storstrømmen, Northeast Greenland. *Global and Planetary Change*, **9**, 79–90.

Bolduc, A.M. (1992) *The formation of eskers based on their morphology, stratigraphy and lithologic composition, Labrador, Canada*. PhD thesis, Lehigh University.

Bolius, D., Schwikowski, M., Rufibach, B., Jenk, T., Gäggeler, H.W. & Casassa., G. (2004) A shallow ice core record from Mercedario, Argentina. *Geophysical Research Abstracts*, **6**. 01768, SRef-ID: 1607–7962/gra/EGU04-A-01768.

Bond, G.C. & Lotti, R. (1995) Iceberg discharges into the North Atlantic on millennial time scales during the last glaciation. *Science*, **267**, 1005–1009.

Bond, G., Heinrich, H., Broecker, W.S., *et al.* (1992) Evidence for massive discharges of icebergs into the glacial Northern Atlantic. *Nature*, **360**, 245–249.

Bond, G., Broecker, W., Johnsen, S., *et al.* (1993) Correlations between climate records from North Atlantic sediments and Greenland ice. *Nature*, **365**, 143–147.

Bond, G.C., Showers, W., Cheseby, M., *et al.* (1997) A pervasive millennial-scale cycle in North Atlantic Holocene and glacial climates. *Science*, **278**, 1257–1266.

Bond, G.C., Showers, W., Elliot, M., *et al.* (1999) The North Atlantic's 1–2 kyr climate rhythm: relation to Heinrich events, Dansgaard–Oeschger cycles and the Little Ice Age. In: *Mechanisms of Global Climate Change at Millennial Time Scales* (Eds P.U. Clark, R.S. Webb & L.D. Keigwin), pp. 35–58. Geophysical Monograph 112, American Geophysical Union, Washington, DC.

Bond, G., Kromer, B., Beer, J., *et al.* (2001) Persistent solar influence on North Atlantic climate during the Holocene. *Science*, **294**, 2130.

Borgström, I. (1989) *Terrängformerna och den glaciala utvecklingen i södra fjällen*. PhD thesis, Department of Physical Geography, Stockholm Univeristy.

Bortenschlager, S. & öggl, K. (2000) *The Ice Man and its Natural Environment*. Springer-Verlag, Berlin.

Bottrell, S.H. & Tranter, M. (2002) Sulphide oxidation under partially anoxic conditions at the bed of Haut Glacier d'Arolla, Switzerland. *Hydrological Processes*, **16**, 2363–2368.

Bouchard, M.A. (1989) Subglacial landforms and depositis in central and northern Quebec, Canada, with emphasis on Rogen moraines. *Sedimentary Geology*, **62**, 293–308.

Bougamont, M., Tulaczyk, S. & Joughin, I. (2003a) Numerical investigations of the slow-down of Whillans Ice Stream, West Antarctica: is it shutting down like Ice Stream C? *Annals of Glaciology*, **37**, 239–246.

Bougamont, M., Tulaczyk, S. & Joughin, I. (2003b) Response of subglacial sediments to basal freeze-on, 2. Application in numerical modeling of the recent stoppage of Ice Stream C, West Antarctica. *Journal of Geophysical Research—Solid Earth*, **108**, 20-1–20-16.

Boulton, G.S. (1970) On the origin and transport of englacial debris in Svalbard glaciers. *Journal of Glaciology*, **9**(56), 213–229.

Boulton, G.S. (1972). The role of the thermal regime in glacial sedimentation. In: *Polar Geomorphology* (Eds R.J. Price & D.E. Sugden), pp. 1–9. Special Publication 4, Institute of British Geographers, London.

Boulton, G.S. (1974). Processes and patterns of subclagial erosion. In: *Glacial Geomorphology* (Ed. D.R. Coates), pp. 41–87. State University of New York, Binghamton.

Boulton, G.S. (1979). Processes of glacier erosion on different substrata. *Journal of Glaciology*, **23**, 15–37.

Boulton, G.S. (1986) A paradigm shift in glaciology? *Nature*, **322**, 18.

Boulton, G.S. (1987) A theory of drumlin formation by subglacial deformation. In: *Drumlin Symposium* (Eds J. Menzies & J. Rose), pp. 25–80. Balkema, Rotterdam.

Boulton, G.S. (1996a) The origin of till sequences by subglacial sediment deformation beneath mid-latitude ice sheets. *Annals of Glaciology*, **22**, 75–84.

Boulton, G.S. (1996b). Theory of glacier erosion, transport and deposition as a consequence of subglacial sediment deformation. *Journal of Glaciology*, **42**, 43–62.

Boulton, G.S. & Caban, P.E. (1995) Groundwater flow beneath ice sheets: Part II—Its impact on glacier tectonic structures and moraine formation. *Quaternary Science Reviews*, **14**, 563–587.

Boulton, G.S. & Clark, C.D. (1990a) The Laurentide Ice Sheet through the last glacial cycle: drift lineations as a key to the dynamic behaviour of former ice sheets. *Transactions of the Royal Society of Edinburgh, Earth Sciences*, **81**, 327–347.

Boulton, G.S. & Clark, C.D. (1990b) A highly mobile Laurentide Ice Sheet revealed by satellite images of glacial lineations. *Nature*, **346**, 813–817.

Boulton, G.S. & Dent, D.L. (1974) The nature and rates of post-depositional changes in recently deposited till from south-east Iceland. *Geografiska Annaler*, **56A**, 121–134.

Boulton, G.S. & Dobbie, K.E. (1993) Consolidation of sediments by glaciers: relations between sediment geotechnics, soft-bed glacier dynamics and subglacial ground-water flow. *Journal of Glaciology*, **39**(131), 26–44.

Boulton, G.S. & Dobbie, K. (1998) Slow flow of granular aggregates: the deformation of sediments beneath glaciers. *Philosophical Transactions of the Royal Society of London, Series A*, **356**, 2713–2745.

Boulton, G.S. & Eyles, N. (1979) Sedimentation by valley glaciers: a model and genetic classification. In: *Moraines and Varves* (Ed. C. Schluchter), pp. 11–23. Balkema, Rotterdam.

Boulton, G.S. & Hindmarsh, R.C.A. (1987) Sediment deformation beneath glaciers: rheology and geological consequences. *Journal of Geophysical Research*, **92**(B9), 9059–9082.

Boulton, G.S. & Jones, A.S. (1979). Stability of temperate ice caps and ice sheets resting on beds of deformable sediment. *Journal of Glaciology*, **24**, 29–43.

Boulton, G.S. & Paul, M.A. (1976) The influence of genetic processes on some geotechnical properties of tills. *Journal of Engineering Geology*, **9**, 159–194.

Boulton, G.S., Dent, D.L. & Morris, E.M. (1974) Subglacial shearing and crushing, and the role of water pressures in tills from south-east Iceland. *Geografiska Annaler*, **56**(A3–4), 135–145.

Boulton, G.S., Jones, A.S., Clayton, K.M. & Kenning, M.J. (1977) A British ice-sheet model and patterns of glacial erosion and deposition in Britain. In: *British Quaternary Studies, Recent Advances* (Ed. F.W. Shotton), pp. 231–246. Clarendon Press, Oxford.

Boulton, G.S., Morris, E.M., Armstrong, A.A. & Thomas, A. (1979) Direct measurement of stress at the base of a glacier. *Journal of Glaciology*, **22**(86), 3–24.

Boulton, G.S., Smith, G.D., Jones, A.S. & Newsome, J. (1985) Glacial geology and glaciology of the last mid-latitude ice sheets. *Journal of the Geological Society of London*, **142**, 447–474.

Boulton, G.S., Slot, T., Blessing, K., Glasbergen, P., Leijnse, T. & van Gijssel, K. (1993) Deep circulation of groundwater in overpressured subglacial aquifers and its geological consequences. *Quaternary Science Reviews*, **12**, 739–745.

Boulton, G.S., Caban, P.E. & van Gijssel, K. (1995) Groundwater flow beneath ice sheets: part I—large scale patterns. *Quaternary Science Reviews*, **14**, 545–562.

Boulton, G.S., Caban, P.E., van Gijssel, K., Leijnse, A., Punkari, M. & van Weert, F.H.A. (1996) The impact of glaciation on the groundwater regime of Northwest Europe. *Global and Planetary Change*, **12**, 397–413.

Boulton, G.S., Dobbie, K.E. & Zatsepin, S. (2001a) Sediment deformation beneath glaciers and its coupling to the subglacial hydraulic system. *Quaternary International*, **86**(1), 3–28.

Boulton, G.S., Zatsepin, S. & Maillot, B. (2001b). *Analysis of Groundwater Flow Beneath Ice Sheets*. Technical Report TR-0106, Svensk Kärnbränslehantering AB (Swedish Nuclear Fuel and Waste Management), Stockholm, 53 pp.

Boulton, G.S., Dongelmans, P., Punkari, M. & Broadgate, M. (2001c) Palaeoglaciology of an ice sheet through a glacial cycle: the European ice sheet through the Weichselian. *Quaternary Science Reviews*, **20**, 591–625.

Boulton, G.S., Hagdorn, M. & Hulton, N.R.J. (2003) Streaming flow in an ice-sheet through a glacial cycle. *Annals of Glaciology*, **36**, 117–128.

Bourgeois, O., Dauteuil, O. & van Vliet-Lanoe, B. (2000) Geothermal control of flow patterns in the last glacial maximum ice sheet of Iceland. *Earth Surface Processes and Landforms*, **25**, 59–76.

Boutron, C.F., Candelone, J.P. & Hong, S. (1994) Past and recent changes in the large scale tropospheric cycles of lead and other heavy metals as documented in Antarctic and Greenland snow and ice: a review. *Geochemica Cosmoschemica Acta*, **58**, 3217–3225.

Bowen, D.Q., Phillips, F.M., McCabe, A.M., Knutz, P.C. & Sykes, G.A. (2002) New data for the Last Glacial Maximum in Great Britain and Ireland. *Quaternary Science Reviews*, **21**, 89–101.

Bown, F. & Rivera, A. (In press) Climate changes and glacier responses during recent decades in the Chilean Lake District. *Global and Planetary Change*.

Box, J.E., Bromwich, D.H. & Bai, L.-S. (2004) Greenland ice sheet surface mass balance for 1991–2000: application of Polar MM5 mesoscale model and *in-situ* data. *Journal of Geophysical Research*, **109**, No. D16, D16105, 10.1029/2003JD004451.

Bradley, R.S. & Jones, P.D. (1993) 'Little Ice Age' summer temperature variations: their nature and relevance to recent global warming trends. *The Holocene*, **3**, 367–376.

Braithwaite, R.J. (1977) *Air temperature and glacier ablation—a parametric approach*. PhD thesis, McGill University.

Braithwaite, R.J. (1985) Calculation of degree-days for glacier-climate research. *Zeitschrift für Gletscherkunde und Glazialgeologie*, **20**, 1–8.

Braithwaite, R.J. (1995) Positive degree-day factors for ablation on the Greenland ice sheet studied by energy-balance modelling. *Journal of Glaciology*, **137**(41), 153–160.

Braithwaite, R.J. (2002) Glacier mass balance: the first 50 years of international monitoring. *Progress in Physical Geography*, **26**, 1, 76–95.

Braithwaite, R.J. & Olesen, O.B. (1989) Calculation of glacier ablation from air temperature, west Greenland. In: *Glacier Fluctuations and Climatic Change* (Ed. J. Oerlemans), pp. 219–233. Kluwer, Dordrecht.

Braithwaite, R.J. & Olesen, O.B. (1990) A simple energy balance model to calculate ice ablation at the margin of the Greenland ice sheet. *Journal of Glaciology*, **36**(123), 222–229.

Braithwaite, R.J. & Raper, S.C.B. (2002) Glaciers and their contribution to sea level change. *Physics and Chemistry of the Earth*, **27**, 1445–1454.

Braithwaite, R.J. & Zhang, Y. (1999) Modelling changes in glacier mass balance that may occur as a result of climate changes. *Geografiska Annaler*, **81A**(4), 489–496.

Braithwaite, R.J. & Zhang, Y. (2000) Sensitivity of mass balance of five Swiss glaciers to temperature changes assessed by tuning a degree-day model. *Journal of Glaciology*, **46**(152), 7–14.

Braithwaite, R.J., Laternser, M. & Pfeffer, W.T. (1994) Variations of near-surface firn density in the lower accumulation area of the Greenland ice sheet, Pâkitsoq, West Greenland. *Journal of Glaciology*, **40**(136), 477–485.

Braithwaite, R.J., Konzelmann, T., Marty, C. & Olesen, O.B. (1998) Reconnaissance study of glacier energy balance in North Greenland, 1993–94. *Journal of Glaciology*, **44**(147), 239–247.

Braithwaite, R.J., Zhang, Y. & Raper, S.C.B. (2003) Temperature sensitivity of the mass balance of mountain glaciers and ice caps as a climatological characteristic. *Festschrift fur Gletscherkunde und Glazialgeologie*, **38**(1), 35–61.

Braun, H., Christl, M., Rahmstorf, S., *et al.* (2004) Solar forcing of abrupt glacial climate change, in a coupled climate system model. *Eos (Transactions of the American Geophysical Union), Fall Meeting Supplement, Abstract*, **85**(47).

Braun, H., Christl, M., Rahmstorf, S., *et al.* (2005) Possible solar origin of the 1470-year glacial climate cycle demonstrated in a coupled model. *Nature*, **438**, 208–211.

Breemer, C.W., Clark, P.U. & Haggerty, R. (2002) Modeling the subglacial hydrology of the late Pleistocene Lake Michigan Lobe,

Laurentide Ice Sheet. *Geological Society of America Bulletin*, **114**, 665–674.
Brennand, T.A. (2004) Glacifluvial. In: *Encyclopedia of Geomorphology* (Ed. A.S. Goudie), pp. 459–465. Routledge, London.
Brennand, T.A. & Shaw, J. (1994) Tunnel channels and associated landforms: their implications for ice sheet hydrology. *Canadian Journal of Earth Sciences*, **31**, 502–522.
Brennand, T.A., Shaw, J. & Sharpe, D.R. (1995) Regional scale meltwater erosion and deposition patterns, northern Quebec, Canada. *Annals of Glaciology*, **22**, 85–92.
Brenner, A.C., Bindschadler, R.A., Thomas, R.H. & Zwally, H.J. (1983) Slope-induced errors in radar altimetry over continental ice sheets. *Journal of Geophysical Research*, **88**(C3), 1617–1623.
Brepson, R. (1979) Simulated glacier sliding over an obstacle. *Journal of Glaciology*, **23**, 143–156.
Briner, J.P. & Swanson, T.W. (1998) Using inherited cosmogenic Cl-36 to constrain glacial erosion rates of the Cordilleran Ice Sheet. *Geology*, **26**, 3–6.
Briner, J.P., Miller, G.H., Davis, P.T., Bierman, P.R. & Cafee, M. (2003) Last Glacial Maximum ice sheet dynamics in Arctic Canada inferred from young erratics perched on ancient tors. *Quaternary Science Reviews*, **22**, 437–444.
Brodzikowski, K. & van Loon, A.J. (1991) *Glacigenic Sediments*. Elsevier, Amsterdam, 674 pp.
Broecker, W.S. (1987) The biggest chill. *Natural History*, **96**, 74–82.
Broecker, W.S. (1994) Massive iceberg discharges as triggers for global climate change. *Nature*, **372**, 421–424.
Broecker, W.S. (1995). *The Glacial World According to Wally*, 2nd edn. Eldigio Press, Palisades, NY.
Broecker, W.S. (1997) Thermohaline circulation, the Achilles heel of our climate system: will man-made CO_2 upset the current balance? *Science*, **278**, 1582–1588.
Broecker, W.S. (1998) Paleocean circulation during the last glaciation: a bipolar seesaw? *Paleoceanography*, **13**, 119–121.
Broecker, W.S. (2001) Was the medieval warm period global? *Science*, **291**, 1497–1499.
Broecker, W.S. (2003) Does the trigger for abrupt climate change, reside in the ocean or in the atmosphere? *Science*, **300**, 1519–1522.
Broecker, W.S. & Denton, G.H. (1989) The role of ocean–atmosphere reorganizations in glacial cycles. *Geochimica et Cosmochimica Acta*, **53**, 2465–2501.
Broecker, W.S. & van Donk, J. (1979) Insolation changes, ice volumes, and the O^{18} record in deep-sea cores. *Review of Geophysics and Space Physics*, **8**, 169–198.
Broecker, W.S., Peteet, D.M. & Rind, D. (1985) Does the ocean-atmosphere system have more than one stable mode of operation? *Nature*, **315**, 21–26.
Broecker, W.S., Andree, M., Wolfli, W., *et al.* (1988) The chronology of the last deglaciation: implications to the cause of the Younger Dryas event. *Paleoceanography*, **3**, 1–19.
Broecker, W.S., Bond, G., Klas, M., Bonani, G. & Wolfli, W. (1990) A salt oscillator in the glacial North Atlantic? 1. The concept. *Paleoceanography*, **5**, 469–477.
Broecker, W.S., Bond, G., McManus, J., Klas, M. & Clark, E. (1992) Origin of the Northern Atlantic's Heinrich events. *Climatic Dynamics*, **6**, 265–273.
Bromwich, D.H., Chen, B. & Tzeng., R.-Y. (1995) Arctic and Antarctic precipitation simulations produced by the NCAR Community Climate Models. *Annals of Glaciology*, **21**, 117–122.
Bromwich, D.H., Chen, Q., Bai, L., Cassano, E.N. & Li, Y. (2001) Modeled precipitation variability over the Greenland ice sheet, *Journal of Geophysical Research*, **106**(D24), 33891–33908.
Brook, E.J., Kurz, M.D., Ackert, R.P., *et al.* (1993) Chronology of Taylor Glacier advances in Arena Valley, Antarctica, using *in situ* cosmogenic He-3 and Be-10. *Quaternary Research*, **39**, 11–23.
Brookes, I.A. (1977) Geomorphology and Quaternary geology of Codroy lowland and adjacent plateaus, southwest Newfoundland. *Canadian Journal of Earth Sciences*, **14**, 2101–2120.
Brooks, C.E.P. (1926) *Climate Through the Ages: a Study of the Climatic Factors and their Variations*. R.V. Coleman, New York.
Brown, C.S., Meier, M.F. & Post, A. (1982) Calving speed of Alaska tidewater glaciers, with application to Columbia Glacier. *United States Geological Survey, Professional Paper*, **1258-C**, 13 pp.
Brown, G.H. (2002) Glacier meltwater hydrochemistry. *Applied Geochemistry*, **17**, 855–883.
Brown, G.H., Tranter, M., Sharp, M.J. & Gurnell, A.M. (1993) The impact of post-mixing chemical reactions on the major ion chemistry of bulk meltwaters draining the Haut Glacier d'Arolla, Valais, Switzerland. *Hydrological Processes*, **8**, 465–480.
Brown, G.H., Hubbard, B. & Seagren, A.G. (2001) Kinetics of solute acquisition from the dissolution of suspended sediment in subglacial channels. *Hydrological Processes*, **15**, 3487–3497.
Brown, N.E., Hallet, B. & Booth, D.B. (1987) Rapid soft bed sliding of the Pudget Glacial Lobe. *Journal of Geophysical Research*, **92**(B9), 8985–8997.
Brown, P.A. & Kennett, J.P. (1998) Megaflood erosion and meltwater plumbing changes during last North American deglaciation recorded in Gulf of Mexico sediments. *Geology*, **26**, 599–602.
Brown, P.A., Kennett, J.P. & Teller, J.T. (1999) Megaflood erosion and meltwater plumbing changes during last North American deglaciation recorded in Gulf of Mexico sediments. Reply. *Geology*, **27**, 479–480.
Brugman, M.M. (1983) Properties of debris-laden ice: application to the flow response of the glaciers in Mount St Helens. *Annals of Glaciology*, **4**, 297.
Bruzzone, G., Bono, R., Caccia, M. & Veruggio, G. (2003) Internet-based teleoperation of an ROV in Antarctica—Web surfers can remotely operate ROV immersed in the Antarctic Sea by means of a simple connection to the Internet *Sea Technology*, **44**, 47–52.
Bryson, R.A., Wendland, W.M., Ives, J.D. & Andrews, J.T. (1969) Radiocarbon isochrones on the disintegration of the Laurentide Ice Sheet. *Arctic and Alpine Research*, **1**, 1–14.
Budd, W.F. & Jacka, T.H. (1989) A review of ice rheology for ice sheet modeling. *Cold Regions Science and Technology*, **16**, 107–144.
Budd, W.F. & Jenssen, D. (1975) Numerical modelling of glacier systems. In: *Hydrology of Marsh-Ridden Areas. Proceedings of a Symposium held at Minsk, June 1972*, pp. 257–291. IAHS Publication 104, International Association of Hydrologic Sciences, Wallingford.
Budd, W.F. & Smith, I.N. (1981) The growth and retreat of ice sheets in response to orbital radiation changes. In: *Sea Level, Ice, and Climatic Change. Proceedings of a Symposium held during the XVII General Assembly of the IUGG at Canberra, December 1979*, pp. 369–409. IAHS Publication 131, International Association of Hydrologic Sciences, Wallingford.
Budd, W.F. & Warner, R.C. (1996) A computer scheme for rapid calculations of balance-flux distributions. *Annals of Glaciology*, **23**, 21–27.
Budd, W.F., Jenssen, D. & Radok, U. (1971) *Derived Physical Characteristics of the Antarctic Ice Sheet*. ANARE Interim Report, Series A (IV), Glaciology Publication, Australian National Antarctic Research Expedition, Melbourne, 178 pp.
Budd, W.F., Kleage, P.L. & Blundy, N.A. (1979) Empirical studies of ice sliding. *Journal of Glaciology*, **23**, 157–170.

Budd, W.F., Jenssen, D., Mavrakis, E. & Coutts, B. (1994) Modelling the Antarctic Ice Sheet changes through time. *Annals of Glaciology*, **20**, 291–297.

Budd, W.F., Coutts, B. & Warner, R.C. (1998) Modelling the Antarctic and Northern-Hemisphere ice-sheet changes with global climate through the glacial cycle. *Annals of Glaciology*, **27**, 153–160.

Bulat, J. & Long, D. (2001) Images of the sea bed in the Faroe–Shetland Channel from commercial 3D seismic data. *Marine Geophysical Researches*, **22**, 345–367.

Burrows, C.J. (1989) Aranuian radiocarbon dates from moraines in the Mount Cook region, New Zealand. *New Zealand Journal of Geology and Geophysics*, **32**, 205–216.

Burrows, C.J. (1990) *Processes of Vegetation Change*. Unwin Hyman, London.

Burrows, C.J., Duncan, K.W. & Spence, J.R. (1990) Aranuian vegetation history of the Arrowsmith Range, Canterbury II. Revised chronology for moraines of the Cameron Glacier. *New Zealand Journal of Botany*, **28**, 455–466.

Butkovich, T.R. & Landauer, J.K. (1959) *The Flow Law for Ice*. Research Report 56, Cold Regions Research and Engineering Laboratory (CRREL), Hanover, New Hampshire.

Cacho, I., Grimalt, J.O., Pelejero, C., *et al.* (1999) Dansgaard–Oeschger and Heinrich event imprints in Alboran Sea paleotemperatures. *Paleoceanography*, **14**(6), 968–705.

Calkin, P.E. & Feenstra, B.H. (1985) Evolution of the Erie-basin Great Lakes. In: *Quaternary Evolution of the Great Lakes* (Eds P.F. Karrow & P.E. Calkin), pp. 149–170. Special Paper 30, Geological Association of Canada, St John's, NF.

Calov, R., Ganopolski, A., Petoukhov, V. & Claussen, M. (2002) Large-scale instabilities of the Laurentide Ice Sheet simulated in a fully coupled climate-system model. *Geophysical Research Letters*, **29**, 69-1–69-4. doi:10.1029/2002GL016078.

Campbell, S. & Bowen, D.Q. (1989) *Quaternary of Wales*. Geological Conservation Review, Nature Conservancy Council, Peterborough, 238 pp.

Campen, R.K., Sowers, T. & Alley, R.B. (2003) Evidence of microbial consortia metabolizing within a low-latitude mountain glacier. *Geology*, **31**, 231–234.

Canadian Journal of Earth Sciences (1969) Seminar on the causes and mechanics of glacial surges, and Symposium on surging glaciers. *Canadian Journal of Earth Sciences*, **6**.

Canals, M., Urgeles, R. & Calafat, A.M. (2000) Deep sea-floor evidence of pasrt ice streams off the Antarctic Peninsula. *Geology*, **28**, 31–34.

Canals, M., Casamor, J.L., Urgeles, R., *et al.* (2002) Seafloor evidence of a subglacial sedimentary system off the northern Antarctic Peninsula. *Geology*, **30**, 603–606.

Cappa, C.D., Hendricks, M.B., DePaolo, D.J. & Cohen, R.C. (2003) Isotopic fractionation of water during evaporation. *Journal of Geophysical Research*, **108**(D16). 4525, 10.1029/2003JD003597.

Carol, H. (1947) The formation of roches moutonnées. *Journal of Glaciology*, **1**(2), 57–59.

Carr, S.J. (2004) Micro-scale features and structures. In: *A Practical Guide to the Study of Glacial Sediments* (Eds D.J.A. Evans & D.I. Benn), pp. 115–144. Arnold, London.

Carr, S.J., Haflidason, H. & Sejrup, H.P. (2000) Micromorphological evidence supporting Late Weichselian glaciation of the northern North Sea. *Boreas*, **29**, 315–328.

Carrasco, J., Casassa, G. & Rivera, A. (2002) Meteorological and Climatological aspects of the Southern Patagonia Icefield. In: *The Patagonian Icefields. A Unique Natural Laboratory for Environmental and Climate Change Studies* (Eds G. Casassa, F. Sepúlveda & R. Sinclair) pp. 29–41. Kluwer Academic/Plenum Publishers, New York.

Carsey, F., Behar, A., Lane, A.L., Realmuto, V. & Engelhardt, H. (2002) A borehole camera system for imaging the deep interior of ice sheets *Journal of Glaciology*, **48**, 622–628.

Carsey, F., Mogensen, C.T., Behar, A., Engelhardt, H. & Lane, A.L. (2003) Science goals for a Mars polar-cap subsurface mission: optical approaches for investigations of inclusions in ice. *Annals of Glaciology*, **37**, 357–362.

Casassa, G., Espizua, L.E., Francou, B., *et al.* (1998) *Glaciers in South America. Into the Second Century of Worldwide Glacier Monitoring—Prospects and Strategies*. Studies and Reports in Hydrology 56, UNESCO Publishing, Paris, 227 pp.

Casassa, G., Rivera, A., Aniya, M. & Naruse, R. (2002a) Current knowledge of the Southern Patagonia Icefield. In: *The Patagonian Icefields: a Unique Natural Laboratory for Environmental and Climate Change Studies* (Eds G. Casassa, F. Sepúlveda & R. Sinclair), pp. 67–83. Series of the Centro de Estudios Científicos, Kluwer Academic/Plenum Publishers, New York.

Casassa, G., Smith, K., Rivera, A., *et al.* (2002b) Inventory of glaciers in isla Riesco, Patagonia, Chile, based on aerial photography and satellite imagery. *Annals of Glaciology*, **34**, 373–378.

Cassau, C. & Terray, L. (2001) Dual influence of Atlantic and Pacific SST anomalies on the North Atlantic/Europe winter climate. *Geophysics Research Letters*, **30**, 3195–3198.

Castaneda, I.S., Smith, L.M., Kristjansdottir, G.B. & Andrews, J.T. (2004) Temporal changes in Holocene ∡^{18}O records from the northwest and central North Iceland shelf. *Journal of Quaternary Science*, **19**, 321–334.

Castelnau, O., Canova, G.R., Lebensohn, R.A. & Duval, P. (1997) Modeling viscoplastic behavior of anisotropic polycrystalline ice with a self-consistent approach. *Acta Materialia*, **45**, 4823–4834.

Catania, G. & Paola, C. (2001) Braiding under glass. *Geology*, **29**, 259–262.

Catania, G.A., Conway, H.B., Gades, A.M., Raymond, C.F. & Engelhardt, H. (2003) Bed reflectivity beneath inactive ice streams in West Antarctica. *Annals of Glaciology*, **36**, 287–291.

Cavalieri, D.J., Gloersen, P. & Campbell, W.J. (1984) Determination of sea ice parameters with the Nimbus-7 Smmr. *Journal of Geophysical Research—Atmospheres*, **89**(ND4), 5355–5369.

Cavalieri, D.J., Gloersen, P., Parkinson, C.L., Comiso, J.C. & Zwally, H.J. (1997) Observed hemispheric asymmetry in global sea ice changes. *Science*, **278**, 1104–1106.

Cayre, O., Lancelot, Y. & Vincent, E. (1999) Paleoceanographic reconstructions from planktonic foraminifera off the Ibeian margin: temperature, salinity, and Heinrich events. *Paleoceanography*, **14**, 3, 384–396.

Cazorzi, F. & Dalla Fontana, G. (1996) Snowmelt modelling by combining temperature and a distributed radiation index. *Journal of Hydrology*, **181**, 169–187.

Chalmers, A.F. (1976) *What is this Thing called Science?* University of Queensland Press.

Chamberlain, T.C. (1895) Glacial phenomena of North America. In: *The Great Ice Age* (Ed. J. Geikie), pp. 724–775. D. Appleton, New York.

Chapman, M.R. & Shackleton, N.J. (1999) Global ice-volume fluctuations, North Atlantic ice-rafting events, and deep-ocean circulation changes between 130 and 70 ka. *Geology*, **27**, 795–798.

Chapman, W.L. & Walsh, J.E. (1993) Recent variations of sea ice and air temperature in high latitudes. *Bulletin American Meteorological Society*, **74**, 33–47.

Chappell, J. (2002) Sea level changes forced ice breakouts in the Last Glacial cycle: new results from coral terraces. *Quaternary Science Reviews*, **21**, 1229–1240.

Charbit, S., Ritz, C. & Ramstein, G. (2002) Simulations of northern hemisphere ice-sheet retreat: sensitivity to physical mechanisms involved during the last deglaciation. *Quaternary Science Reviews*, **21**, 243–265.

Charlesworth, J.K. (1928) The glacial retreat from central and southern Ireland. *Quarterly Journal of the Geological Society of London*, **84**, 293–344.

Charlesworth, J.K. (1955) The Carlingford Readvance between Dundalk Co. Louth, and Kingscourt and Lough Ramor, Co. Cavan. *Irish Naturalists' Journal*, **2**, 299–302.

Chelton, D.B., Ries, J.C., Haines, B.J., Fu, L. & Callahan, P.S. (2001) Satellite altimetry. In: *Satellite Altimetry and the Earth Sciences* (Eds L.-L. Fu & A. Cazenave), pp. 1–132. Academic Press, New York.

Chen, J. & Funk, M. (1990) Mass balance of Rhonegletscher during 1982/83–1986/87. *Journal of Glaciology*, **36**(123), 199–209.

Chen, J., Liu, C. & Jin, M. (1996) Application of the repeated aerial photogrammetry to monitoring glacier variation in the drainage area of the Urumchi River. *Journal of Glaciology and Geocryology*, **18**(4), 331–336.

Chen, J., Pang, S., Zhang, Y., *et al.* (2001) Height of snow top on the Mt. Everest and global warming. *Advances in Earth Sciences*, **16**(1), 12–14.

Chen, Q., Bromwich, D.H. & Bai, L. (1997) Precipitation over Greenland retrieved by a dynamic method and its relation to cyclonic activity. *Journal of Climate*, **10**, 839–870.

Chevy, J. (2003) *Deformation of Single Crystals of Ice in Torsion*. Internal Report, Laboratoire de Glaciologie et Géophysique de l'Environnement, Grenoble.

Chillrud, S.N., Pedrozo, F.L., Temporetti, P.F. & Planas, H.F. (1994) Chemical weathering of phosphate and germanium in glacial meltwater streams: effects of subglacial pyrite oxidation. *Limnology and Oceanography*, **39**, 1130–1140.

Christiansen, E.A. (1971). *Geology and groundwater resources of the Melville Area 62K, L. Saskatchewan*. Map 12, Geology Division, Saskatchewan Research Council, Saskatoon.

Christoffersen, P. & Tulaczyk, S. (2003a) Thermodynamics of basal freeze-on: Predicting subglacial signatures of interstream ridges and stopped ice streams. *Annals of Glaciology*, **36**, 233–243.

Christoffersen, P. & Tulaczyk, S. (2003b) Response of subglacial sediments to basal freeze-on, 1. Theory and comparison to observations from beneath the West Antarctic Ice Sheet. *Journal of Geophysical Research-Solid Earth*, **108**.

Christoffersen, P. & Tulaczyk, S. (2003c) Signature of palaeo-ice-stream stagnation: till consolidation induced by basal freeze-on. *Boreas*, **32**, 114–129.

Church, J.A., Gregory, J.M., Huybrechts, P., *et al.* (2001) Changes in sea level. In: *Climate Change 2001: the Scientific Basis. Contribution of Working Group I to the Third Assessment Report of the Intergovernmental Panel on Climate Change* (Eds J.T. Houghton, Y. Ding, D.J. Griggs, *et al.*), pp. 639–693. Cambridge University Press, Cambridge.

Ciais, P. & Jouzel, J. (1994) Deuterium and oxygen 18 in precipitation: isotopic model, including mixed cloud processes. *Journal of Geophysical Research*, **99**(D8), 16793–16803.

Clapperton, C.M. (1968) Channels formed by the superimposition of glacial meltwater streams, with special reference to the East Cheviot Hills, North-East England. *Geografiska Annaler*, **50A**, 207–220.

Clark, C.D. (1993) Mega-scale glacial lineations and cross-cutting ice-flow landforms. *Earth Surface Processes and Landforms*, **18**, 1–29.

Clark, C.D. (1994) Large-scale ice-moulding: A discussion of genesis and glaciological significance. *Sedimentary Geology*, **91**, 253–268.

Clark, C.D. (1997) Reconstructing the evolutionary dynamics of former ice sheets using multi-temporal evidence, remote sensing and GIS. *Quaternary Science Reviews*, **16**, 1067–1092.

Clark, C.D. (1999) Glaciodynamic context of subglacial bedform generation and preservation. *Annals of Glaciology*, **28**, 23–32.

Clark, C.D. & Meehan, R.T. (2001) Subglacial bedform geomorphology of the Irish Ice Sheet reveals major configuration changes during growth and decay. *Journal of Quaternary Science*, **16**(5), 483–496.

Clark, C.D. & Stokes, C.R. (2001) The extent and basal characteristics of the M'Clintock Channel ice stream. *Quaternary International*, **86**, 81–101.

Clark, C.D., Knight, J.K. & Gray, J.T. (2000) Geomorphological reconstruction of the Labrador Sector of the Laurentide Ice Sheet. *Quaternary Science Reviews*, **19**, 1343–1366.

Clark, C.D., Evans, D.J.A. & Piotrowski, J.A. (Eds) (2003a) Palaeo-Ice Streams. Special Issue. *Boreas*, **32**, 1–280.

Clark, C.D., Tulaczyk, S.M., Stokes, C.R. & Canals, M. (2003b) A groove-ploughing theory for the production of mega-scale glacial lineations, and implications for ice stream mechanics. *Journal of Glaciology*, **49**(165), 240–256.

Clark, C.D., Evans, D.J.A., Khatwa, A., *et al.* (2004) Map and GIS database of landforms and features related to the last British Ice Sheet. *Boreas*, **33**(4), 359–375.

Clark, I.D., Douglas, M., Raven, K. & Bottomley, D. (2000) Recharge and preservation of Laurentide glacial melt water in the Canadian Shield. *Ground Water*, **38**, 735–742.

Clark, J.A. (1977) An inverse problem in glacial geology: the reconstruction of glacier thinning in Glacier Bay, Alaska between AD 1910 and 1960 from relative sea-level data. *Journal of Glaciology*, **18**, 481–503.

Clark, P.U. (1994) Unstable behavior of the Laurentide Ice Sheet over deforming sediment and its implications for climate change. *Quaternary Research*, **41**, 19–25.

Clark, P.U. & Pollard, D. (1998) Origin of the middle Pleistocene transition by ice sheet erosion of regolith. *Paleoceanography*, **13**, 1–9.

Clark, P.U. & Mix, A.C. (2002) Ice sheets and sea level of the Last Glacial Maximum. *Quaternary Science Reviews*, **21**, 1–7.

Clark, P.U. & Walder, J.S. (1994) Subglacial drainage, eskers, and deforming beds beneath the Laurentide and Eurasian ice sheets. *Geological Society of America Bulletin*, **106**, 304–314.

Clark, P.U., Alley, R.B., Keigwin, L.D., Licciardi, J.M., Johnsen, S.J. & Huaxiao Wang (1996a) Origin of the first global freshwater pulse following the last glacial maximum. *Paleoceanography*, **11**(5), 563–577.

Clark, P.U., Licciardi, J.M., MacAyeal, D.R. & Jenson, J.W. (1996b) Numerical reconstruction of a soft-bedded Laurentide Ice Sheet during the last glacial maximum. *Geology*, **24**, 679–682.

Clark, P.U., Alley, R.B. & Pollard, D. (1999) Northern Hemisphere ice-sheet influences on global climate change. *Science*, **286**, 1104–1111.

Clark, P.U., Marshall, S.J., Clarke, G.K.C., Hostetler, S.W., Licciardi, J.M. & Teller, J.T. (2001). Freshwater forcing of abrupt climate change, during the last glaciation. *Science*, **293**, 283–287.

Clark, P.U., Mitrovica, J.X., Milne, G.A. & Tamisiea, M.E. (2002a) Sea-level fingerprinting as a direct test for the source of Global Meltwater Pulse 1A. *Science*, **295**, 2438–2441.

Clark, P.U., Pisias, N.G., Stocker, T.F. & Weaver, A.J. (2002b). The role of the thermohaline circulation in abrupt climate change. *Nature*, **415**, 863–869.

Clark, P.U., Brook, E.J., Raisbeck, G.M., Yiou, F. & Clark, J. (2003) Cosmogenic 10Be ages of the Saglek Moraines, Torngat Mountains, Labrador. *Geology*, **31**, 617–620.

Clarke, G.K.C. (1987a) Fast glacier flow—ice streams, surging, and tidewater glaciers. *Journal of Geophysical Research—Solid Earth and Planets*, **92**, 8835–8841.

Clarke, G.K.C. (1987b) Subglacial till: a physical framework for its properties and processes. *Journal of Geophysical Research*, **92**(B9), 9023–9036.

Clarke, G.K.C. (1987c) A short history of scientific investigations on glaciers. *Journal of Glaciology*, **Special Issue**, 4–24.

Clarke, G.K.C. & Blake, E.W. (1991) Geometric and thermal evolution of a surge-type glacier in its quiescent state: Trapridge Glacier, Yukon Territory, Canada, 1969–89. *Journal of Glaciology*, **37**(125), 158–169.

Clarke, G.K.C. & Marshall, S.J. (2002) Isotopic balance of the Greenland ice sheet: modelled concentrations of water isotopes from 30000 BP to present. *Quaternary Science Reviews*, **21**, 419–430.

Clarke, G.K. & Prairie, I.L. (2001) Modelling iceberg drift and ice-rafted sedimentation. In: *Continuum Mechanics and Applications in Geophysics and the Environment* (Eds B. Straughan & R. Greve), pp. 182–200. Springer-Verlag, New York.

Clarke, G.K.C., Collins, S.G. & Thompson, D.E. (1984) Flow, thermal structure, and subglacial conditions of a surge-type glacier. *Canadian Journal of Earth Sciences*, **21**, 232–240.

Clarke, G.K.C., Marshall, S.J., Hillaire-Marcel, C., Bilodeau, G. & Veiga-Pires, C. (1999) A glaciological perspective on Heinrich events. In: *Mechanisms of Global Climate Change at Millennial Time Scales* (Eds P.U. Clark, R.S. Webb & L.D. Keigwin), pp. 243–262. Geophysical Monograph 112, American Geophysical Union, Washington, DC.

Clarke, G., Leverington, D., Teller, J. & Dyke, A. (2003) Superlakes, megafloods, and abrupt climate change. *Science*, **301**, 922–923.

Clarke, G.K.C., Leverington, D.W., Teller, J.T. & Dyke, A.S. (2004) Paleohydraulics of the last outburst flood from glacial Lake Agassiz and the 8200 BP cold event. *Quaternary Science Reviews*, **23**, 389–407.

Clarke, T.S., Liu, C., Lord, N.E. & Bentley, C.R. (2000) Evidence for a recently abandoned shear margin adjacent to Ice Stream B2, Antarctica, from ice-penetrating radar measurements. *Journal of Geophysical Research—Solid Earth*, **105**, 13409–13422.

Clausen, H.B., Hammer, C.U., Hvidberg, C.S., *et al.* (1997) A comparison of the volcanic records over the past 4000 years from the Greenland Ice Core Project and Dye 3 Greenland ice cores. *Journal of Geophysical Research*, **102**, 26707–26723.

Clayton, L. & Moran, S.R. (1974) A glacial process–form model. In: *Glacial Geomorphology* (Ed. D.R. Coates), pp. 89–119. State University of New York, Binghamton.

Clayton, L. & Moran, S.R. (1982) Chronology of late Wisconsinan glaciation in middle North America. *Quaternary Science Reviews*, **1**, 55–82.

Clayton, L., Mickelson, D.M. & Attig, J.W. (1989) Evidence against pervasively deformed bed material beneath rapidly moving lobes of the southern Laurentide Ice Sheet. *Sedimentary Geology*, **62**, 203–208.

Clayton, L., Attig, J.W. & Mickelson, D.M. (1999) Tunnel channels in Wisconsin. In: *Glaciers Past and Present* (Eds D.M. Mickelson & J.W. Attig). *Geological Society of America Special Paper*, **337**, 69–82.

Clement, A.C., Cane, M.A. & Seager, R. (2001) An orbitally driven tropical source for abrupt climate change. *Journal of Climate*, **14**(11), 2369–2375.

Cline, D.W. (1997a) Effect of seasonality of snow accumulation and melt on snow surface energy exchanges at a continental alpine site. *Journal of Applied Meteorology*, **36**, 22–41.

Cline, D.W. (1997b) Snow surface energy exchanges and snowmelt at a continental, midlatitude Alpine site. *Water Resources Research*, **33**(4), 689–701.

Cogley, J.G. (1998) GGHYDRO Release 2.2. ftp://ftp.trentu.ca/pub/gghydro. Accessed June 1998.

Cogley, J.G. & Adams, W.P. (1998) Mass balance of glaciers other than the ice sheets. *Journal of Glaciology*, **44**(147), 315–325.

Cogley, J.G., Adams, W.P., Ecclestone, M.A., Jung-Rotthenhaaausler, F. & Ommanney, C.S.L. (1995) *Mass Balance of Axel Heiberg Island Glaciers 1960–1991*. NHRI Science Report No. 6, National Hydrology Research Institute, Saskatoon, 168 pp.

Cohen, D. (2000) Rheology of ice at the bed of Engabreen, Norway. *Journal of Glaciology*, **46**, 611–621.

Cohen, D., Hooke, R. Le, B., Iverson, N.R. & Kohler, J. (2000) Sliding of ice past an obstacle at Engabreen, Norway. *Journal of Glaciology*, **46**(155), 599–610.

Colbeck, S.C. & Evans, R.J. (1973) A flow law for temperate glacier ice. *Journal of Glaciology*, **12**, 71–86.

Colgan, P.M., Bierman, P.R., Mickelson, D.M. & Caffee, M. (2002) Variation in glacial erosion near the southern margin of the Laurentide Ice Sheet, south-central Wisconsin, USA: implications for cosmogenic dating of glacial terrains. *Geological Society of America Bulletin*, **114**, 1581–1591.

Colgan, P.M., Mickelson, D.M. & Cutler, P.M. (2003) Ice-marginal terrestrial landsystems: southern Laurentide ice sheet margin. In: *Glacial Landsystems* (Ed. D.A. Evans), 111–142. Edwin Arnold, London.

Colle, B.A., Westrick, K.J. & Mass, C.F. (1999) Evaluation of MM5 and Eta-10 precipitation forecasts over the Pacific Northwest during the cold season. *Weather Forecasting*, **14**, 137–154.

Collins, D.N. (1979) Hydrochemistry of meltwaters draining from an alpine glacier. *Arctic and Alpine Research*, **11**, 307–324.

Collins, D.N. (1984) Water and mass balance measurements in glacierised drainage basins. *Geografiska Annaler*, **66A**(3), 197–213.

Comiso, J.C. (1990) Arctic Multiyear Ice Classification and Summer Ice Cover Using Passive Microwave Satellite Data. *Journal of Geophysical Research—Oceans*, **95**(C8), 13411–13422.

Comiso, J.C. (2002) A rapidly declining perennial sea ice cover in the Arctic. *Geophysical Research Letters*, **29**(20), 10.1029.

Comiso, J. (2003) Warming trends in the Arctic from clear-sky satellite observations. *Journal of Climate*, **16**, 3498–3510.

Comiso, J.C. & Kwok, R. (1996) Surface and radiative characteristics of the summer Arctic sea ice cover from multisensor satellite observations. *Journal of Geophysical Research—Oceans*, **101**(C12), 28397–28416.

Comiso, J.C., Grenfell, T. C., Lange, M.A., Lohanick, W., Moore, R.K. & Wadhams, P. (1992) Microwave remote sensing of the Southern Ocean ice cover. In: *Microwave Remote Sensing of Sea Ice* (Ed. F.D. Carsey), pp. 243–259. American Geophysical Union, Washington, DC.

Comiso, J.C., Cavalieri, D.J., Parkinson, C.L. & Gloersen, P. (1997) Passive microwave algorithms for sea ice concentration: a comparison of two techniques. *Remote Sensing of Environment*, **60**(3), 357–384.

Condon, D.J., Prave, A.R. & Benn, D.I. (2002) Neoproterozoic glacial-rainout intervals: Observations and implications. *Geology*, **30**, 35–38.

Conway, H.W., Hall, B.L., Denton, G.H., *et al.* (1999) Past and future grounding-line retreat of the West Antarctic ice sheet. *Science*, **286**, 280–286.

Conway, H., Catania, G., Raymond, C.F., Gades, A.M., Scambos, T.A. & Engelhardt, H. (2002) Switch of flow direction in an Antarctic ice stream. *Nature*, **419**, 465–467.

Cook, E.R., Palmer, J.G. & D'Arrigo, R.D. (2002) Evidence for a 'Medieval Warm Period' in a 1,100 tree-ring reconstruction of past austral summer temperatures in New Zealand. *Geophysical Research Letters*, **29**, 1–4.

Cooper, A.P.R. (1997) Historical observations of Prince Gustav Ice Shelf. *Polar Record*, **33**, 285–294.

Cooper, R.J., Wadham, J.L., Tranter, M., Hodgkins, R. & Peters, N.E. (2002) Groundwater hydrochemistry in the active layer of the proglacial zone, Finsterwalderbreen, Svalbard. *Journal of Hydrology*, **269**, 208–223.

Copland, L., Harbor, J. & Sharp, M. (1997a) Borehole video observation of englacial and basal ice conditions in a temperate valley glacier. *Annals of Glaciology*, **24**, 277–282.

Copland, L., Harbor, J., Gordon, S. & Sharp, M. (1997b) The use of borehole video in investigating the hydrology of a temperate glacier. *Hydrological Processes*, **11**(2), 211–224.

Corr, H.F.J., Jenkins, A., Nicholls, K.W. & Doake, C.S.M. (2002) Precise measurement of changes in ice-shelf thickness by phase-sensitive radar to determine basal melt rates. *Geophysical Research Letters*, **29**(8), 73-1–73-4.

Corripio, J. (2002) *Modelling the energy balance of high altitude glacierised basins in the Central Andes*. PhD thesis, University of Edinburgh.

Cowdery, S. (2004) *Average Quaternary extent of the West Antarctic Ice Sheet*. Unpublished MS thesis, University of Washington, Seattle, WA.

Craig, B.G. (1964) Surficial geology of east-central District of Mackenzie. *Geological Survey of Canada Bulletin*, **99**, 1–41.

Craig, H. (1961a) Isotopic variations in meteoric waters. *Science*, **133**, 1702–1703.

Craig, H. (1961b) Standard for reporting concentrations of deuterium and oxygen-18 in natural water. *Science*, **133**, 1833–1834.

Cuffey, K.M. (2001) Interannual variability of elevation on the Greenland ice sheet: effects of firn densification, and establishment of a multy-century benchmark. *Journal of Glaciology*, **47**(158), 369–377.

Cuffey, K.M. & Alley, R.B. (1996) Erosion by deforming subglacial sediments: Is it significant? (Toward till continuity.) *Annals of Glaciology*, **22**, 17–24.

Cuffey, K.M. & Clow, G.D. (1997) Temperature, accumulation, and ice sheet elevation in central Greenland through the last deglacial transition. *Journal of Geophysical Research*, **102**(C12), 26383–26396.

Cuffey, K.M. & Marshall, S.J. (2000) Substantial contribution to sea-level rise during the last interglacial from the Greenland ice sheet. *Nature*, **404**, 591–594.

Cuffey, K.M. & Vimeux, F. (2001) Covariation of carbon dioxide and temperature from the Vostok ice core after deuterium-excess correction. *Nature*, **412**, 523–527.

Cuffey, K.M., Clow, G.D., Alley, R.B., Stuiver, M., Waddington, E.D. & Saltus, R.W. (1995) Large Arctic temperature change at the glacial–Holocene transition. *Science*, **270**, 455–458.

Cuffey, K.M., Conway, H., Hallet, B., Gades, A.M. & Raymond, C.F. (1999) Interfacial water in polar glaciers and glacier sliding at −17°C. *Geophysical Research Letters*, **26**(6), 751–754.

Cuffey, K.M., Conway, H., Gades, A., Hallet, B., Raymond, C.F. & Whitlow, S. (2000a) Deformation properties of subfreezing glacier ice: role of crystal size, chemical impurities, and rock particles inferred from *in situ* measurements. *Journal of Geophysical Research*, **105**(B12), 27895–27915.

Cuffey, K.M., Conway, H., Gades, A.M., *et al.* (2000b) Entrainment at cold glacier beds. *Geology*, **28**, 351–354.

Cuffey, K.M., Thorsteinsson, T. & Waddington, E.D. (2000c) A renewed argument for crystal size control of ice sheet strain rates. *Journal of Geophysical Research*, **105**, 27889–27894.

Cunjak, R.A., Prowse, T.D. & Parrish, D.L. (1998) Atlantic salmon (*Salmo salar*) in winter: 'the season of parr discontent'? *Canadian Journal of Fisheries and Aquatic Sciences*, **55**, 161–180.

Cutler, P.M., Macayeal, D.R., Mickelson, D.M., Parizek, B.R. & Colgan, P.M. (2000) A numerical investigation of ice-lobe—permafrost interaction around the southern Laurentide ice sheet. *Journal of Glaciology*, **46**, 311–325.

Cutler, P.M., Mickelson, D.M., Colgan, P.M., MacAyeal, D.R. & Parizek, B. (2001) Influence of the Great Lakes on the dynamics of the Laurentide ice sheet: numerical experiments. *Geology*, **29**(11), 1039–1042.

Cutler, P.M., Colgan, P.M. & Mickelson, D.M. (2002) Sedimentologic evidence for outburst floods from the Laurentide Ice Sheet margin in Wisconsin, U.S.A.: implications for tunnel-channel genesis. *Quaternary International*, **90**(1), 23–40.

Dackombe, R.V. & Thomas, G.S.P. (1991) Glacial deposits and Quaternary stratigraphy of the Isle of Man. In: *Glacial Deposits in Great Britian and Ireland* (Eds J. Ehlers, P.L. Gibbard & J. Rose), pp. 333–344. Balkema, Rotterdam.

Dahl, R. (1965) Plastically sculpted detail form on rock surfaces in northern Nordland, Norway, *Geografiska Annaler, Series A*, **47**, 3–140.

Dahl, S.-O., Ballantyne, C.K., McCarroll, D. & Nesje, A. (1996) Maximum altitude of Devensian glaciation on the Isle of Skye. *Scottish Journal of Geology*, **32**, 107–115.

Dahl-Jensen, D. (1985) Determination of the flow properties at Dye 3, South Greenland, by bore-hole-tilting measurements and perturbation modelling. *Journal of Glaciology*, **31**, 92–98.

Dahl-Jensen, D. & Gundestrup, N. (1987) Constitutive properties of ice at Dye 3, Greenland. In: *The Physical Basis of Ice Sheet Modelling* (Eds E.D. Waddington & J.S. Walder), pp. 31–43. Publication 170, International Association of Hydrological Sciences, Wallingford.

Dahl-Jensen, D., Mosegaard, K., Gundestrup, N., *et al.* (1998) Past temperatures directly from the Greenland Ice Sheet. *Science*, **282**, 268–271.

Daly, C., Neilson, R.P. & Philips, D.L. (1994) A statistical-topographic model for mapping climatological precipitation over mountainous terrain. *Journal of Applied Meteorology*, **33**, 140–158.

Dansgaard, W. (1964) Stable isotopes in precipitation. *Tellus*, **16**, 436–468.

Dansgaard, W., Johnsen, S.J., Moller, J. & Langway, C.C. (1969) One thousand centuries of climatic record from Camp Century on the Greenland Ice Sheet. *Science*, **166**, 377–381.

Dansgaard, W., Johnsen, S.J., Clausen, H.B., *et al.* (1993) Evidence for general instability of past climate from a 250-kyr ice-core record. *Nature*, **364**, 218–220.

Darby, D.A., Bischof, J.F., Spielhagen, R.F., Marshall, S.A. & Herman, S.W. (2002) Arctic ice export events and their potential impact on global climate during the late Pleistocene. *Paleoceanography*, **17**(2), 1–17.

Darwin, C. (1842) Notes on the effects produced by the ancient glaciers of Caernarvonshire, and on the boulders transported by floating ice. *Philosophical Magazine*, **21**, 180–188.

Dash, J.G., Fu, H.Y. & Wettlaufer, J.S. (1995) The premelting of ice and its environmental consequences. *Reports of Progress in Physics*, **58**, 115–167.

Davies, M.L. & Fitzsimons, S. (2004) Selected case studies of cold-based glaciers, South Victoria Land, Antarctica. *Quaternary Newsletter*, **104**, 30–44.

Davis, C.H. (1996) Temporal change in the extinction coefficient of snow on the Greenland ice sheet from an analysis of seasat and geosat altimeter data. *IEEE Transactions on Geoscience and Remote Sensing*, **34**(5), 1066–1073.

Davis, C.H. (1997) A robust threshold retracking algorithm for measuring ice-sheet surface elevation change from satellite radar altimeters. *IEEE Transactions on Geoscience and Remote Sensing*, **35**(4), 974–979.

Davis, C.H., Kluever, C.A. & Haines, B.J. (1998) Elevation change of the southern Greenland ice sheet. *Science*, **279**(5359), 2086–2088.

Davis, C.H, McConnell, J.R., Bolzan, J., Bamber, J.L., Thomas, R.H. & Mosley-Thompson, E. (2001) Elevation change of the southern Greenland ice sheet from 1978 to 1988: interpretation. *Journal of Geophysical Research—Atmospheres*, **106**(D24), 33743–33754.

Davis, P.T. (1985) Neoglacial moraines on Baffin Island. In: *Quaternary Environments: Eastern Canadian Arctic, Baffin Bay and Western Greenland* (Ed. J.T. Andrews), pp. 682–718B. Allen and Unwin, London.

Davis, W.M. (1926) The value of outrageous geological hypotheses. *Science*, **63**, 463–468.

Davison, S. & Stoker, M.S. (2002) Late Pleistocene glacially-influenced deep-marine sedimentation off NW Britain: implications for the rock record. In: *Glacier-Influenced Sedimentation on High-latitude Continental Margins* (Eds J.A. Dowdeswell & C. Ó Cofaigh), pp. 129–147. Special Publications 203, Geological SocietyPublishing House, Bath.

De Angelis, H. & Skvarca, P. (2003) Glacier surge after ice shelf collapse. *Science*, **299**, 1560–1562.

De La Chapelle, S., Castelnau, O., Lipenkov, V. & Duval, P. (1998) Dynamic recrystallization and texture development in ice as revealed by the study of deep ice cores in Antarctica and Greenland. *Journal of Geophysical Research*, **103**(B5), 5091–5105.

De La Chapelle, S., Milsch, H., Castlenau, O., *et al.* (1999) Compressive creep of ice containing a liquid intergranular phase: rate-controlling processes in the dislocation creep. *Geophysical Research Letters*, **26**, 251–254.

De Mora, S.J., Whitehead, R.F. & Gregory, M. (1994) The chemical composition of glacial melt water ponds and streams on the McMurdo Ice Shelf, Antarctica. *Antarctic Science*, **6**, 17–27.

De Vernal, A. & Hillaire-Marcel, C. (2000) Sea-ice cover, sea-surface salinity and halo-/thermocline structure of the northwest North Atlantic: modern versus full glacial conditions. *Quaternary Science Reviews*, **19**, 65–85.

De Woul, M. & Hock, R. In press. Static mass balance sensitivity of Arctic glaciers and ice caps using a degree-day approach. *Annals of Glaciology*, **42**.

Dean, W.E., Forester, R.M. & Bradbury, J.P. (2002) Early Holocene change in atmosphere circulation in the Northern Great Plains: an upstream view of the 8.2 ka cold event. *Quaternary Science Reviews*, **21**, 1763–1775.

DeConto, R.M. & Pollard, D. (2003) Rapid Cenozoic glaciation of Antarctica induced by declining atmospheric CO_2. *Nature*, **421**, 245–249.

De Geer, G. (1940) Geochronologia Suecica Principles. *Kungliga Svenska Vetenskapsakademiens Handlingar*, **III**(18:6), 367 pp.

Deichmann, N., Ansorge, J., Scherbaum, F., Aschwanden, A., Bernardi, F. & Gudmundsson, G.H. (2000). Evidence for deep icequakes in an Alpine glacier. *Annals of Glaciology*, **31**, 85–90.

Delmonte, B., Petit, J.-R. & Maggi, V. (2002) Glacial to Holocene implications of the new 27,000 year dust record from the EPICA Dome C (East Antarctica) ice core. *Climate Dynamics*, **18**, 647–660.

Delmonte, B., Basile-Doelsch, I., Petit, J.R., *et al.* (2004a) Comparing the Epica and Vostok dust records during the last 220,000 years: stratigraphical correlation and origin in glacials periods. *Earth Science Reviews*, **66**, 63–87.

Delmonte, B., Petit, J.R., Andersen, K.K, *et al.* (2004b).Opposite regional atmospheric circulation changes over east Antarctica during the last climatic transition evidenced by dust size distributions changes. *Climate Dynamics*, **23**, 427–438.

Delworth, T.L. & Knutson, T.R. (2000) Simulation of early 20th century global warming. *Science*, **287**, 2246–2250.

Delworth, P.E. & Dixon, K.W. (2001) Implications of the recent trend in the Arctic/North Atlantic Oscillation for the North Atlantic thermohaline circulation. *Journal of Climate*, **13**, 3721–3727.

Demuth, M.N. & Keller, R. (1997) *An Assessment of the Mass Balance of Peyto Glacier (1966–1995) and its Relation to Recent and Past-century Climatic Variability*. Contribution Series CS–97007, National Hydrology Research Institute, Saskatoon, 43 pp.

Denby, B., Greuell, J.W. & Oerlemans, J. (2002) Simulating the Greenland atmospheric boundary layer. Part I: model description and validation. *Tellus*, **A54**, 512–528.

Denton, G.H. & Hendy, C.H. (1994) Younger Dryas age advance of Franz Josef Glacier in the Southern Alps of New Zealand. *Science*, **264**, 1434–1437.

Denton, G.H. & Hughes, T.J. (Eds) (1981) *The Last Great Ice Sheets*. Wiley, New York.

Denton, G.H., Hughes, T.J. & Karlen, W. (1986) Global ice sheet system interlocked by sea level. *Quaternary Research*, **26**, 3–26.

Denton, G.H., Sugden, D.E., Marchant, D.R., Hall, B.L. & Wilch, T.I. (1993) East Antarctic ice sheet sensitivity to Pliocene climatic change from a Dry Valleys perspective. *Geografiska Annaler*, **75A**, 155–204.

Denton, G.H., Heusser, C.J., Lowell, T.V., *et al.* (1999) Interhemispheric linkage of paleoclimate during the last glaciations. *Geografiska Annaler*, **81A**, 107–153.

Déry, S.J., Taylor, P.A. & Xiao, J. (1998) The thermodynamic effects of sublimating snow in the atmospheric boundary layer. *Boundary-layer Meteorology*, **89**, 251–283.

Deser, C., Walsh, J.E. & Timlin, M.S. (2000) Arctic sea ice variability in the context of recent Atmospheric Circulation Trends. *Journal of Climate*, **13**, 617–633.

Deser, C., Magnusdottir, G., Saravanan, R. & Phillips, A. (2004) The effects of North Atlantic SST and sea ice anomalies on the winter circulation in CCM3. Part II: direct and indirect components of the response.*Journal of Climate*, **17**, 877–889.

DETR and The Meteorological Office (Eds) (1997) *Climate Change and its Impacts: a Global Perspective*. The Meteorological Office, Bracknell.

Dickson, R.R., Meincke, J., Malmberg, S. & Lee, A. (1988) The 'Great Salinity Anomaly' in the northern North Atlantic 1968–1982. *Progress in Oceanography*, **20**, 103–151.

Dickson, R.R., Osborn, T.J., Hurrell, J.W., *et al.* (2000) The Arctic Ocean response to the North Atlantic Oscillation.*Journal of Climate*, **13**, 2671–2696.

DiLabio, R.N.W. & Coker, W.B. (1989) Drift prospecting. *Geological Survey of Canada Paper*, **89-20**.

Doake, C.S.M. & Vaughan, D.G. (1991) Rapid disintegration of the Wordie Ice Shelf in response to atmospheric warming. *Nature*, **350**, 328–330.

Doake, C.S.M., Corr, H.F.J., Jenkins, A., *et al.* (2001) Rutford Ice Stream, Antarctica. In: *The West Antarctic Ice Sheet: Behavior and Environment* (Eds R.B. Alley & R.A. Bindschadler), pp. 221–235. Volume 77, American Geophysical Union, Washington, DC.

Doake, C.S.M., Corr, H.F.J. & Jenkins, A. (2002) Polarization of radio waves transmitted through Antarctic ice shelves. *Annals of Glaciology*, **34**, 165–170.

Dokken, T.E. & Jansen, E. (1999) Rapid changes in the mechanism of ocean convection during the last glacial period. *Nature*, **401**, 458–451.

Domack, E.W. (1988) Biogenic facies in the Antarctic glacimarine environment: basis for a polar glacimarine summary. *Palaeogeography, Palaeoclimatology, Palaeoecology*, **63**, 357–372.

Domack, E.W., Jacobson, E.A., Shipp, S. & Anderson, J.B. (1999) Late Pleistocene–Holocene retreat of the West Antarctic Ice-sheet system in the Ross Sea: Part 2—Sedimentologic and stratigraphic signature. *Geological Society of America Bulletin*, **111**, 1517–1536.

Domack, E.W., Leventer, A., Dunbar, R., *et al.* (2001) Chronology of the Palmer Deep site, Antarctic Peninsula: a Holocene palaeoenvironmental reference for the circum-Antarctic. *The Holocene*, **11**, 1–9.

Donnadieu, Y., Godderis, Y., Ramstein, G., Nedelec, A. & Meert, J. (2004) A 'snowball earth' climate triggered by continental break-up through changes in runoff. *Nature*, **428**, 303–306.

Donnelly, J.P., Driscoll, N.W., Uchupi, E., *et al.*, 2005. Catastrophic meltwater discharge down the Hudson Valley: a potential trigger for the Intra-Allerod cold period. *Geology*, **33**, 89–92.

Douglas, B. (1997) Global sea rise: a redetermination. *Surveys in Geophysics*, **18**, 279–292.

Dowdeswell, E.K. & Andrews, J.T. (1985) The fiords of Baffin Island: description and classification. In: *Quaternary Environments: Eastern Canadian Arctic, Baffin Bay, and Western Greenland* (Ed. J.T. Andrews), pp. 93–121. Allen and Unwin, Boston.

Dowdeswell, J.A. & Elverhøi, A. (2002) The timing of initiation of fast-flowing ice streams during a glacial cycle inferred from glacimarine sedimentation. *Marine Geology*, **188**, 3–14.

Dowdeswell, J.A. & Hagen, J.O. (2004) Arctic ice caps and glaciers. In: *Mass Balance of the Cryosphere* (Eds J.L. Bamber & A.J. Payne), pp. 527–578. Cambridge University Press, Cambridge.

Dowdeswell, J.A. & Ó Cofaigh, C.O. (Eds) (2003) *Glacier-influenced Sedimentation on High-latitude Continental Margins*. Special Publication 203, Geological Society Publishing House, Bath.

Dowdeswell, J.A. & Powell, R.D. (1996) Submersible remotely operated vehicles (ROVs) for investigations of the glacier ocean–sediment interface. *Journal of Glaciology*, **42**, 176–183.

Dowdeswell, J.A. & Sharp, M.J. (1986) Characterization of pebble fabrics in modern terrestrial glacigenic sediments. *Sedimentology*, **33**, 699–710.

Dowdeswell, J.A. & Siegert, M.J. (1999) Ice-sheet numerical modeling and marine geophysical measurements of glacier-derived sedimentation on the Eurasian Arctic continental margins. *Geological Society of America Bulletin*, **111**, 1080–1097.

Dowdeswell, J.A., Whittington, R.J. & Hodgkins, R. (1992) The sizes, frequencies and freeboards of East Greenland icebergs observed using ship radar and sextant. *Journal of Geophysical Research*, **97**, 3515–3528.

Dowdeswell, J.A., Villinger, H., Whittington, R.J. & Marienfeld, P. (1993) Iceberg scouring in Scoresby Sund and on the East Greenland continental shelf. *Marine Geology*, **111**, 37–53.

Dowdeswell, J.A., Whittington, R.J. & Marienfeld, P. (1994) The origin of massive diamicton facies by iceberg rafting and scouring, Scoresby Sund, East Greenland. *Sedimentology*, **41**, 21–35.

Dowdeswell, J.A., Maslin, M.A., Andrews, J.T. & McCave, I.N. (1995) Iceberg production, debris rafting, and the extent and thicknes of Heinrich layers (H-1, H-2) in North Atlantic sediments. *Geology*, **23**, 301–304.

Dowdeswell, J.A., Kenyon, N.H., Elverhøi, A., *et al.* (1996) Large-scale sedimentation on the glacier-influenced Polar North Atlantic margins: long-range side-scan sonar evidence. *Geophysical Research Letters*, **23**, 3535–3538.

Dowdeswell, J.A., Kenyon, N.H. & Laberg, J.S. (1997) The glacier-influenced Scoresby Sund Fan, East Greenland continental margin: evidence from GLORIA and 3.5 kHz records. *Marine Geology*, **143**, 207–221.

Dowdeswell, J.A., Elverhøi, A. & Spielhagen, R. (1998) Glacimarine sedimentary processes and facies on the Polar North Atlantic margins. *Quaternary Science Reviews*, **17**, 243–272.

Dowdeswell, J.A., Unwin, B., Nuttall, A.-M. & Wingham, D.J. (1999a) Velocity structure, flow instability and mass flux on a large Arctic ice cap from satellite radar interferometry. *Earth and Planetary Science Letters*, **167**, 131–140.

Dowdeswell, J.A., Elverhøi, A., Andrews, J.T. & Hebbeln, D. (1999b) Asynchronous deposition of ice-rafted layers in the Nordic seas and North Atlantic Ocean. *Nature*, **400**, 348–351.

Dowdeswell, J.A., ó Cofaigh, C., Andrews, J.T. & Scourse, J.D. (2001) Debris transported by icebergs and paleoceanographic implications. *Eos (Transactions of the American Geophysical Union)*, **82**, 382–386.

Dowdeswell, J.A., Bassford, R.P., Gorman, M.R., *et al.* (2002a) Form and flow of the Academy of Sciences Ice Cap, Severnaya Zemlya, Russian High Arctic. *Journal of Geophysical Research*, **107**. 10.1029/2000/JB000129.

Dowdeswell, J.A., Ó Cofaigh, C., Taylor, J., Kenyon, N.H., Mienert, J. & Wilken, M. (2002b) On the architecture of high-latitude continental margins: the influence of ice-sheet and sea-ice processes in the Polar North Atlantic. In: *Glacier-influenced Sedimentation on High-latitude Continental Margins* (Eds J.A. Dowdeswell & C. Ó Cofaigh), pp. 33–54. Special Publication 203, Geological Society-Publishing House, Bath.

Dowdeswell, J.A., Ó Cofaigh, C. & Pudsey, C.J. (2004a) Thickness and extent of the subglacial till layer beneath an Antarctic paleo-ice stream. *Geology*, **32**, 13–16.

Dowdeswell, J.A., Ó Cofaigh, C. & Pudsey, C.J. (2004b) Continental slope morphology and sedimentary processes at the mouth of an Antarctic palaeo-ice stream. *Marine Geology*, **204**, 203–214.

Drake, L.D. & Shreve, R.L. (1973) Pressure melting and regelation of ice by round wires. *Proceedings of the Royal Society of London, Series A*, **332**, 51–83.

Dredge, L.A. (2000) Age and origin of upland block fields on Melville Peninsula, eastern Canadian Arctic. *Geografiska Annaler*, **82A**, 443–454.

Dreimanis, A. (1989) Tills: their genetic terminology and classification. In: *Genetic Classification of Glacigenic Deposits* (Eds R.P. Goldthwait & C.L. Matsch), pp. 17–84. Balkema, Rotterdam.

Drever, J.I. (1988) *The Geochemistry of Natural Waters*, 2nd edn. Prentice Hall, New Jersey, 437 pp.

Drever, J.I. (Ed.) (2003) *Surface and Ground Water, Weathering, Erosion and Soils*, Vol. 5, *Treatise on Geochemistry* (Eds H.D. Holland & K.K. Turekian). Elsevier-Pergamon, Oxford.

Drewry, D.J. (1983a) *The Surface of the Antarctic Ice Sheet*. Scott Polar Research Institute, Cambridge.

Drewry, D.J. (Ed.) (1983b) *Antarctica: Glaciological and Geophysical Folio*. Scott Polar Research Institute, University of Cambridge.

Drewry, D.J. (1986) *Glacial Geologic Processes*. Edward Arnold, London.

Drewry, D.J. & Morris, E.M. (1992) The response of large ice sheets to climatic change. *Philosophical Transactions of the Royal Society of London*, **338**, 235–242.

Drewry, D.J., Jordan, S.R. & Jankowski, E. (1983) Measured properties of the Antarctic Ice Sheet: surface configuration, ice thickness, volume and bedrock characteristics. *Annals of Glaciology*, **3**, 83–91.

Drinkwater, M.R. (1998) Satellite microwave observations of Antarctic sea ice. *Analysis of SAR data of the Polar Oceans: Recent Advances* (Eds C. Tsatsoulis & R. Kwok), pp. 145–187. Springer-Verlag, Berlin.

Dugdale, R.E. (1972) A statistical analysis of the state of a glacier's 'health'. *Journal of Glaciology*, **11**, 73–80.

Dunlop, P. (2004) *The characteristics of ribbed moraine and assessment of theories for their genesis*. PhD thesis, Department of Geography, University of Sheffield.

Dunne, J., Elmore, D. & Muzikar, P. (1999) Scaling factors for the rates of production of cosmogenic nuclides for geometric shielding and attenuation at depth on sloped surfaces. *Geomorphology*, **27**(1–2), 3–11.

Durham, W.B., Kirby, S.H. & Stern, L.A. (1992) Effects of dispersed particulates on the rheology of water ice at planetary conditions. *Journal of Geophysical Research*, **97**(20), 20883–20897.

Durham, W.B., Kirby, S.H. & Stern, L.A. (1997) Creep of water ices at planetary conditions; a compilation. *Journal of Geophysical Research*, **102**, 16293–16302.

Durham, W.B., Kirby, S.H. & Stern, L.A. (2001) Rheology of ice I at low stress and elevated confining pressure. *Journal of Geophysical Research*, **106**, 11031–11042.

Duval, P. (1977) The role of water content in the creep rate of polycrystalline ice. In: *Proceedings of the Grenoble Symposium, Isotopes and Impurities in Snow and Ice, August–September 1975*, pp. 29–33. IAHS Publication 118, International Association of Hydrological Sciences, Wallingford.

Duval, P. (1981) Creep and fabrics of polycrystalline ice under shear and compression.*Journal of Glaciology*, **27**, 129–140.

Duval, P. & Castelnau, O. (1995) Dynamic recrystallization of ice in polar ice sheets. *Journal de Physique C3*, **5**, C3-197–C3-205.

Duval, P. & Le Gac, H. (1980) Does the permanent creep rate of polycrystalline ice increase with crystal size? *Journal of Glaciology*, **25**(91), 151–157.

Duval, P. & Montagnat, M. (2002) Comment on 'Superplastic deformation of ice: Experimental observations' by D.L. Goldsby & D.L. Kohlstedt. *Journal of Geophysical Research*, **107**(B5). ECV4 1–2. 10.1029/2001JB000946.

Duval, P., Ashby, M.F. & Anderman, I. (1983) Rate-controlling processes in the creep of polycrystalline ice. *Journal of Physical Chemistry*, **87**, 4066–4074.

Duval, P., Arnaud, L. & Brissaud, O. (2000) Deformation and recrystallization processes of ice from polar ice sheets. *Annals of Glaciology*, **30**, 83–87.

Dyke, A.S. (1993). Landscapes of cold-centred Late Wisconsinan ice caps, Arctic Canada. *Progress in Physical Geography*, **17**, 223–247.

Dyke, A.S. (2004) An outline of North American deglaciation with emphasis on central and northern Canada. In: *Quaternary Glaciations—Extent and Chronology, Part II: North America* (Eds J. Ehlers & P.L. Gibbard), pp. 373–424. Developments in Quaternary Science, Vol. 2b, Elsevier, Amsterdam.

Dyke, A.S. & Evans, D.J.A. (2003) Ice-marginal terrestrial landsystems: northern Laurentide and Innuitian ice sheet margins. In: *Glacial Landsystems* (Ed. D.A. Evans), pp. 143–165. Edwin Arnold, London.

Dyke, A.S. & Morris, T.F. (1988) Canadian landform examples—drumlin fields, dispersal trains, and ice streams in Arctic Canada. *The Canadian Geographer*, **32**, 86–90.

Dyke, A.S. & Peltier, W.R. (2000) Forms, response times and variability of relative sea-level curves, glaciated North America. *Geomorphology*, **32**, 315–333.

Dyke, A.S. & Prest, V.K. (1987a) Late Wisconsinan and Holocene history of the Laurentide ice sheet. *Géographie Physique et Quaternaire*, **41**, 237–263.

Dyke, A.S. & Prest, V.K. (1987b) *Paleogeography of Northern North America, 18000–5000 Years Ago*. Map 1703A, Geological Survey of Canada, Queen's Printer, Ottawa.

Dyke, A.S. & Savelle, J.M. (2000) Major end moraines of Younger Dryas age on Wollaston Peninsula, Victoria Island, Canadian arctic: implications for paleoclimate and for formation of hummocky moraine. *Canadian Journal of Earth Sciences*, **37**, 601–619.

Dyke, A.S., Dredge, L.A. & Vincent, J.-S. (1982) Configuration and dynamics of the Laurentide Ice Sheet during the last glacial maximum. *Géographie Physique et Quaternaire* **36**, 5–14.

Dyke, A.S., Morris, T.F., Green, D.E.C. & England, J. (1992) Quaternary geology of Prince of Wales Island Arctic Canada. *Geological Survey of Canada Memoir*, **433**, 142 pp.

Dyke, A.S., Dale, J.E. & McNeely, R.N. (1996) Marine molluscs as indicators of environmental change in glaciated North America and Greenland during the last 18 000 years. *Geographie physique et Quaternaire*, **50**, 125–184.

Dyke, A.S., Andrews, J.T., Clark, P.U., *et al.* (2002) The Laurentide and Innuitian ice sheets during the Last Glacial Maximum. *Quaternary Science Reviews*, **21**, 9–31.

Dyke, A.S., Moore, A. & Robertson, L. (2003) *Deglaciation of North America*. Geological Survey of Canada Open File1574, 2 map sheets, 1 CD-ROM. Queen's Printer, Ottawa.

Dyurgerov, M. (2001) Mountain glaciers at the end of the twentieth century: Global analysis in relation to climate and water cycle. *Polar Geography*, **25**(4), 241–337.

Dyurgerov, M. (2002) *Glacier Mass Balance and Regime: Data of Measurements and Analysis*. INSTAAR Occasional Paper No. 55 (Eds M. Meier & R. Armstrong), Institute of Arctic and Alpine Research, University of Colorado, Boulder, CO. Distributed by National Snow and Ice Data Center, Boulder, CO.

Dyurgerov, M. (2003) Mountain and subpolar glaciers show an increase in sensitivity to climate warming and intensification of the water cycle. *Journal of Hydrology*, **282**, 164–176.

Dyurgerov, M. & Dwyer, J. (2001) The steepening of glacier mass balance gradients with Northern Hemisphere warming. *Zeitschrift fur Gletscherkunde und Glazialgeologie*, **36**, 107–118.

Dyurgerov, M.B. & Meier, M.F. (1997) Year to year fluctuation of global mass balance of small glaciers and their contribution to sea level change. *Arctic and Alpine Research*, **29**, 392–401.

Dyurgerov, M. & Meier, M. (2000) Twentieth century climate change: evidence from small glaciers. *Proceedings National Academy of Science*, **97**, 1406–1411.

Dyurgerov, M.B. & Meier, M.F. (2004) Glaciers and study of climate and sea-level change. In: *Mass Balance of the Cryosphere* (Eds J.L. Bamber & A.J. Payne), pp. 579–622. Cambridge University Press, Cambridge.

Dzulinski, S. & Walton, E.K. (1965) *Sedimentary Features of Flysch and Greywackes*. Developments in Sedimentology, Vol. 7, Elsevier, Amsterdam, 274 pp.

Echelmeyer, K. & Harrison, W.D. (1990) Jakobshavn Isbræ, West Greenland: Seasonal variation—or lack thereof. *Journal of Glaciology*, **36**, 82–88

Echelmeyer, K.A. & Harrison, W.D. (1999) Ongoing margin migration of Ice Stream B, Antarctica. *Journal of Glaciology*, **45**, 361–369.

Echelmeyer, K. & Wang, Z. (1987). Direct observation of basal sliding and deformation of basal drift at subfreezing temperatures. *Journal of Glaciology*, **33**, 83–98.

Echelmeyer, K.A., Harrison, W.D., Larsen, C. & Mitchell, J.E. (1994) The Role of the Margins in the Dynamics of an Active Ice Stream. *Journal of Glaciology*, **40**, 527–538.

Egloff, J. & Johnson, G.L. (1979) Erosional and depositional structures of the southwest Iceland insular margin: thirteen geophysical profiles. In: *Geological and Geophysical Investigations of Continental Margins* (Eds J.S. Watkins, L. Montadert & P.W. Dickerson), pp. 43–63. Memoir 29, American Association of Petroleum Geologists, Tulsa, Oklahoma.

Ehlers, J. & Gibbard, P. (Eds) (2004) *Quaternary Glaciations—Extent and Chronology*, Vol. 1, *Europe*. Developments in Quaternary Science, Elsevier, Amsterdam.

Ehlers, J., Gibbard, P.L. & Rose, J. (1991) *Glacial Deposits in Great Britain and Ireland*. Balkema, Rotterdam, 580 pp.

Einarsson, M.A. (1991) Temperature conditions in Iceland 1901–1990. *Jökull*, **41**, 1–20.

Ekholm, S. (1996) A full coverage, high-resolution, topographic model of Greenland computed from a variety of digital elevation data. *Journal of Geophysical Research—Solid Earth*, **101**(B10), 21961–21972.

Ekman, I. & Iljin, V. (1991) Deglaciation, the Younger Dryas end morianes and their correlation in the Karelian A.S.S.R. & adjacent areas. In: *Eastern Fennoscandian Younger Dryas End Moraines, Field Conference Excursion Guide* (Eds H. Rainio & M. Saarnisto), pp. 73–99. OPAS-guide 32, Geological Survey of Finland: Helsinki.

Ekstrom, G., Nettles, M. & Abers, G.A. (2003) Glacial earthquakes. *Science*, **302**, 622–624.

Elias, S.A., Short, S.K., Nelson, C.H. & Birks, H.H. (1996) Life and times of the Bering land bridge. *Nature*, **382**, 60–63.

Elsberg, D.H., Harrison, W.D., Echelmeyer, K.A. & Krimmel, R.M. (2001) Quantifying the effects of climate and surface change on glacier mass balance. *Journal of Glaciology*, **47**(159), 649–658.

Elsberg, D.H., Harrison, W.D., Zumberge, M.A., Morack, J.L., Pettit, E.C., Waddington, E.D. & Husmann, E. (2004) Strain rates and short term strain events measured at Siple Dome, Antarctica. *Journal of Glaciology*, **50**(171), 511–521.

Engelhardt, H. (2005) Ice temperature and high geothermal flux at Siple Dome, West Antarctica, from borehole measurements. *Journal of Glaciology*, **50**(169), 251–256.

Engelhardt, H. & Kamb, B. (1993) Vertical temperature profile of Ice Stream B. *Antarctic Journal of the US*, **28**, 63–66.

Engelhardt, H. & Kamb, B. (1997) Basal hydraulic system of a West Antarctic ice stream: constraints from borehole observations. *Journal of Glaciology*, **43**, 207–230.

Engelhardt, H. & Kamb, B. (1998) Basal sliding of Ice Stream B, West Antarctica. *Journal of Glaciology*, **44**, 223–230.

Engelhardt, H.F., Harrison, W.D. & Kamb, B. (1978) Basal sliding and conditions at the glacier bed as revealed by bore-hole photography. *Journal of Glaciology*, **20**(84), 469–508.

Engelhardt, H.F., Humphrey, N. & Kamb, B. (1990a) Borehole geophysical observations on Ice Stream B, Antarctica. *Antarctic Journal of the United States*, **25**(4), 80–82.

Engelhardt, H., Humphrey, N., Kamb, B. & Fahnestock, M. (1990b) Physical conditions at the base of a fast moving Antarctic ice stream. *Science*, **248**, 57–59.

England, J. (1999) Coalescent Greenland and Innuitian ice during the Last Glacial Maximum: revising the Quaternary of the Canadian High Arctic. *Quaternary Science Reviews*, **18**, 421–456.

Engquist, P., Olsson, T. & Svensson, T. (1978) Pumping and recovery tests in wells sunk in till. *Nordic Hydrology Conference*, Hanasari, Finland, pp. 134–142.

Environment Canada (1999) *Water Survey of Canada*. Hydat CD-ROM v.99–2.00, Environment Canada, Ottawa.

Environment Canada (2003) *Canadian Climate Normals 1971–2000*. http://www.climate.weatheroffice.ec.gc.ca/climate_normals/index_e.html.

EPICA Community Members (2004) Eight glacial cycles from an Antarctic ice core. *Nature*, **429**, 623–628.

Erxleben, J., Elder, K. & Davis, R. (2002) Comparison of spatial interpolation methods for estimating snow distribution in the Colorado Rocky Mountains. *Hydrological Processes*, **16**, 3627–3649.

Escher-Vetter, H. (1985) Energy balance calculations for the ablation period 1982 at Vernagtferner, Oetztal Alps. *Annals of Glaciology*, **6**, 158–160.

Escobar, F., Vidal, F. & Garín, C. (1992) Water balance in the Patagonia Icefield. In: *Glacier Researches in Patagonia 1990* (Eds R. Naruse & M. Aniya), pp. 109–119. Japanese Society of Snow and Ice, Sapporo.

Escobar, F., Casassa, G. & Pozo, V. (1995) Variaciones de un glaciar de montaña en los Andes de Chile central en las últimas dos décadas. *Bulletin de l'Institut Francais d'études Andines*, **24**(3), 683–695.

Evans, B., Renner, J. & Hirth, G. (2001) A few remarks on the kinetics of static grain growth in rocks. *International Journal of Earth Science*, **90**, 88–103.

Evans, D.J.A. (1990) The effect of glacier morphology on surficial geology and glacial stratigraphy in a high arctic mountainous terrain. *Zeitschrift fur Geomorphologie*, **34**, 481–503.

Evans, D.J.A. (1993) High latitude rock glaciers: a case study of forms and processes in the Canadian arctic. *Permafrost and Periglacial Processes*, **4**, 17–35.

Evans, D.J.A. (Ed.) (2003a) *Glacial Landsystems*. Arnold, London, 532 pp.

Evans, D.J.A. (2003b) Ice-marginal terrestrial landsystems: active temperate glacier margins. In: *Glacial Landsystems* (Ed. D.A. Evans), pp. 12–43. Edwin Arnold, London.

Evans, D.J.A. (In press) Glacial depositional processes and forms. In: *The History of the Study of Landforms or the Development of Geomorphology*, Vol. 4, *Quaternary and Recent Processes and Forms (1890–1965) and the Mid-century Revolutions* (Eds T.P. Burt, R.J. Chorley, D. Brunsden, A.S. Goudie & N.J. Cox). Routledge, London.

Evans, D.J.A. & England, J. (1991) Canadian landform examples 19: high arctic thrust block moraines. *Canadian Geographer*, **35**, 93–97.

Evans, D.J.A. & Ó Cofaigh, C. (2003) Depositional evidence for marginal oscillations of the Irish Sea Ice Stream in southeast Ireland during the last glaciation. *Boreas*, **32**, 76–101.

Evans, D.J.A. & Rea, B.R. (1999) Geomorphology and sedimentology of surging glaciers: a landsystems approach. *Annals of Glaciology*, **28**, 75–82.

Evans, D.J.A. & Rea, B.R. (2003) Surging glacier landsystem. In: *Glacial Landsystems* (Ed. D.A. Evans), pp. 259–288. Edwin Arnold, London.

Evans, D.J.A. & Twigg, D.R. (2000) *Breiðamerkurjökull 1998*. 1:30,000 scale map. University of Glasgow and Loughborough University.

Evans, D.J.A. & Twigg, D.R. (2002) The active temperate glacial landsystem: a model based on Breidamerkurjokull and Fjallsjokull, Iceland. *Quaternary Science Reviews*, **21**, 2143–2177.

Evans, D.J.A., Lemmen, D.S. & Rea, B.R. (1999) Glacial landsystems of the southwest Laurentide Ice Sheet: modern Icelandic analogues. *Journal of Quaternary Science*, **14**, 673–691.

Evans, D.J.A., Rea, B.R., Hansom, J.D. & Whalley, W.B. (2002) The geomorphology and style of plateau icefield glaciation in a fjord terrain, Troms-Finnmark, north Norway. *Journal of Quaternary Science*, **17**, 221–239.

Evans, D.J.A., Clark, C.D. & Mitchell, W.A. (2005) The Last British Ice Sheet: a review of the evidence utilised in the compilation of the Glacial Map of Britain. *Earth Science Reviews*, **70**, 253–312.

Evans, D.J.A., Rea, B.R., Hiemstra, J.F. & Ó Cofaigh, C. submitted. Genesis of glacial sediments and landforms in south-central Alberta, Canada: mega-floods and alternative interpretations. *Quaternary Science Reviews*.

Evans, J. & Pudsey, C.J. (2002) Sedimentation associated with Antarctic Peninsula ice shelves: implications for palaeoenvironmental reconstructions of glacimarine sediments. *Journal of the Geological Society, London*, **159**, 233–238.

Evans, J., Dowdeswell, J.A. & ó Cofaigh, C. (2004) Late Quaternary submarine bedforms and ice-sheet flow in Gerlache Strait and on the adjacent continental shelf, Antarctic Peninsula. *Journal of Quaternary Science*, **19**, 397–407.

Evans, J., Pudsey, C.J., Ó Cofaigh, C., Morris, P. & Domack, E. (2005) Late Quaternary glacial history, flow dynamics and sedimentation along the eastern margin of the Antarctic Peninsula Ice Sheet. *Quaternary Science Reviews*, **24**, 741–774.

Evenson, E.B., Lawson, D.E. Strasser, J.C., *et al.* (1999) Field evidence for the recognition of glaciohydrologic supercooling. *Geological Society of America Special Paper*, **337**, 23–35.

Eyles, C.H. (1988) Glacially and tidally influenced shallow marine sedimentation of the late Precambrian Port Askaig Formation. *Palaeogeography, Palaeoclimatology, Palaeoecology*, **68**, 1–25.

Eyles, C.H. & Eyles, N. (1984) Glaciomarine sediments of the Isle of Man as a key to late Pleistocene stratigraphic investigations in the Irish Sea Basin. *Geology*, **12**, 359–364.

Eyles, N. (1983a) Glacial geology: a landsystems approach. In: *Glacial Geology* (Ed. N. Eyles), pp. 1–18. Pergamon, Oxford.

Eyles, N. (1983b) The glaciated valley landsystem. In: *Glacial Geology* (Ed. N. Eyles), pp. 91–110. Pergamon, Oxford.

Eyles, N. (Ed.) (1983c) *Glacial Geology*. Pergamon, Oxford, 409 pp.

Eyles, N. & Dearman, W.R. (1981) A glacial terrain map of Britain for engineering purposes. *Bulletin of the International Association of Engineering Geology*, **24**, 173–184.

Eyles, N. & Januszczac, N. (2004) 'Zipper-rift': a tectonic model for Neoproterozoic glaciations during the break-up of Rodinia after 750 Ma. *Earth Science Reviews*, **65**, 1–73.

Eyles, N. & McCabe, A.M. (1989) The Late Devensian <22,000 BP Irish Sea Basin: the sedimentary record of a collapsed ice sheet margin. *Quaternary Science Reviews*, **8**, 307–351.

Eyles, N. & Menzies, J. (1983) The subglacial landsystem. In: *Glacial Geology* (Ed. N. Eyles), pp. 19–70. Pergamon, Oxford.

Eyles, N., Dearman, W.R. & Douglas, T.D. (1983a) The distribution of glacial landsystems in Britain and North America. In: *Glacial Geology* (Ed. N. Eyles), pp. 213–228. Pergamon, Oxford.

Eyles, N., Eyles, C.H. & Miall, A.D. (1983b) Lithofacies types and vertical profile models: an alternative approach to the description and environmental interpretation of glacial diamict and diamictite sequences. *Sedimentology*, **30**, 393–410.

Eyles, C.H., Eyles, N. & Miall, A.D. (1985) Models of glaciomarine sedimentation and their application to the interpretation of ancient glacial sequences. *Palaeogeography, Palaeoclimatology, Palaeoecology*, **51**, 15–84.

Eythorsson, J. (1935) On the variations of glaciers in Iceland: Some studies made in 1931. *Geografiska Annaler*, **17**, 121–137.

Fabel, D. & Harbor, J. (1999) The use of *in-situ* produced cosmogenic radionuclides in glaciology and glacial geomorphology. *Annals of Glaciology*, **28**, 103–110.

Fabel, D., Stroeven, A.P., Harbor, J., *et al.* (2002) Landscape preservation under Fennoscandian ice sheets determined from *in situ* produced ^{10}Be and ^{26}Al. *Earth and Planetary Science Letters*, **201**, 397–406.

Fahnestock, M. & Bamber, J. (2001) Morphology and surface characteristics of the West Antarctic Ice Sheet. In: *The West Antarctic Ice Sheet: Behavior and Environment* (Eds R.B. Alley & R.A. Bindschadler), pp. 13–27. Volume 77, American Geophysical Union, Washington, DC.

Fahnestock, M., Bindschadler, R., Kwok, R. & Jezek, K. (1993). Greenland ice sheet surface properties and ice dynamics from ERS-1 SAR imagery. *Science*, **262**(5139), 1530–1534.

Fahnestock, M.A., Scambos, T.A., Bindschadler, R.A. & Kvaran, G. (2000) A millennium of variable ice flow recorded by the Ross Ice Shelf, Antarctica. *Journal of Glaciology*, **46**, 652–664.

Fahnestock, M., Abdalati, W., Joughin, I., Brozena, J. & Gogineni, P. (2001) High geothermal heat flow, basal melt, and the origin of rapid ice flow in central Greenland. *Science*, **294**(5550), 2338–2342.

Fairbanks, R.G. (1989) A 17,000-year glacio-eustatic sea level record: influence of glacial melting rates on the Younger Dryas event and deep-ocean circulation. *Nature*, **342**, 637–642.

Fairchild, I.J., Killawee, J.A., Sharp, M.J., Hubbard, B., Lorrain, R.D. & Tison, J.-L. (1999) Solute generation and transfer from a chemically reactive alpine glacial-proglacial system. *Earth Surface Processes and Landforms*, **4**(13), 1189–1211.

Falconer, G., Ives, J.D., Loken, O.H. & Andrews, J.T. (1965) Major end moraines in eastern and central Arctic Canada. *Geographical Bulletin*, **7**, 137–153.

Fanning, A.F. & Weaver, A.J. (1997) Temporal–geographical meltwater influences on the North Atlantic conveyor: implications for the Younger Dryas. *Paleoceanography*, **12**, 307–320.

Faria, S.H., Ktitarev, D. & Hutter, K. (2002) Modelling evolution of anisotropy in fabric and texture of polar ice. *Annals of Glaciology*, **35**, 545–551.

Farmer, G.L., Barber, D.C. & Andrews, J.T. (2003) Provenance of Late Quaternary ice-proximal sediments in the North Atlantic: Nd, Sr and Pd isotopic evidence. *Earth and Planetary Science Letters*, **209**, 227–243.

Farrand, W.R. & Drexler, C.W. (1985). Late Wisconsinan and Holocene history of the Lake Superior basin. In: *Quaternary Evolution of the Great Lakes* (Eds P.F. Karrow & P.E. Calkin), pp. 18–32. Special Paper 30, Geological Association of Canada, St John's, NF.

Fastook, J.L. & Prentice, M.L. (1994) A Finite-element Model of Antarctica: Sensitivity Test for Meteorological Mass Balance Relationship. *Journal of Glaciology*, **40**, 167–175.

Fatland, D.R. & Lingle, C.S. (2002) InSAR observations of the 1993–95 Bering Glacier (Alaska, USA) surge and a surge hypothesis. *Journal of Glaciology*, **48**, 439–451.

Felzer, B., Oglesby, R., Hyman, D. & Webb, T. III (1994) Sensitivity of the global climate system to changes in northern hemisphere ice sheet size using the NCAR CCM1 (nonlinear sensitivity analysis based on perpetual season model runs). In: *6th Conference on Climate Variations, Nashville, TN*, 23–28 January, American Meteorological Society, pp. 202–206.

Fenton, M.M., Moran, S.R., Teller, J.T. & Clayton, L. (1983) Quaternary stratigraphy and history in the history inn the southern part of the lake Agassiz basin. In: *Glacial Lake Agassiz* (Eds J.T. Teller & L. Clayton), pp. 49–74. Special Paper 26, Geological Association of Canada, St John's, NF.

Fernald, M.L. (1911) An expedition to Newfoundland and Labrador. *Rhodora*, **XIII**(151), 108–162.

Fichefet, T., Poncin, C., Goosse, H., Huybrechts, P., Janssens, I. & Treut, H.L. (2003) Implications of changes in freshwater flux from the Greenland ice sheet for climate of the 21st century. *Geophysical Research Letters*, **30**, 81–84. doi:10.1029/2003GL017826.

Fischer, U.H. & Clarke, G.K.C. (1994) Ploughing of subglacial sediment. *Journal of Glaciology*, **40**(134), 97–106.

Fischer, U.H. & Clarke, G.K.C. (1997a) Stick-slip sliding behaviour at the base of a glacier. *Annals Glaciology*, **24**, 390–396.

Fischer, U.H. & Clarke, G.K.C. (1997b) Clast collision frequency as an indicator of glacier sliding rate. *Journal of Glaciology*, **43**(145), 460–466.

Fischer, U.H. & Clarke, G.K.C. (2001) Review of subglacial hydro-mechanical coupling: Trapridge Glacier, Yukon Territory, Canada. *Quaternary International*, **86**(1), 29–43.

Fischer, U.H., Clarke, G.K.C. & Blatter, H. (1998a) Evidence for temporally varying 'sticky spots' at the base of Trapridge Glacier, Yukon Territory, Canada. *Journal of Glaciology*, **45**(150), 352–360.

Fischer, U.H., Iverson, N.R., Hanson, B., Hooke, R.L. & Jansson, P. (1998b) Estimation of hydraulic properties of subglacial till from ploughmeter measurements. *Journal of Glaciology*, **44**, 517–522.

Fischer, U.H., Porter, P.R., Schuler, T., Evans, A.J. & Gudmundsson, G.H. (2001) Hydraulic and mechanical properties of glacial sediments beneath Unteraargletscher, Switzerland: implications for glacier basal motion. *Hydrological Processes*, **15**(18), 3525–3540.

Fisher, D.A. & Koerner, R.M.J (1986) On the special rheological properties of ancient microparticle-laden northern hemisphere ice as derived from bore-hole and core measurements. *Journal of Glaciology*, **32**(112), 501–510.

Fisher, D.A., Reeh, N. & Langley, K. (1985) Objective reconstruction of the late Wisconsinan Laurentide Ice Sheet and the significance of deformable beds. *Geographie physique et Quaternaire*, **39**, 229–238.

Fisher, T.G. (2003) Chronology of glacial Lake Agassiz melt water routed to the Gulf of Mexico. *Quaternary Research*, **59**, 271–276.

Fisher, T.G. & Shaw, J. (1992) A depositional model for Rogen moraine, with examples from the Avalon Peninsula, Newfoundland. *Canadian Journal of Earth Sciences*, **29**, 669–686.

Fisher, T.G., Smith, D.G. & Andrews, J.T. (2002) Preboreal oscillation caused by a glacial Lake Agassiz flood. *Quaternary Science Reviews*, **21**, 873–878.

Fitzsimons, S.J. (1996) Formation of thrust block moraines at the margins of dry-based glaciers, south Victoria Land, Antarctica. *Annals of Glaciology*, **22**, 68–74.

Fitzsimons, S.J. (2003) Ice-marginal terrestrial landsystems: polar continental glacier margins. In: *Glacial Landsystems* (Ed. D.A. Evans), pp. 89–110. Edwin Arnold, London.

Fitzsimons, S.J., McManus, K.J. & Lorrain, R.D. (1999) Structure and strength of basal ice and substrate of a dry-based glacier: evidence for substrate deformation at sub-freezing temperatures. *Annals of Glaciology*, **28**, 236–240.

Fitzsimons, S., Lorrain, R. & Vandergoes, M. (2000) Behaviour of subglacial sediment and basal ice in a cold glacier. In: *Deformation of Glacial Material* (Eds A.J. Maltman, M.J. Hambrey & B. Hubbard), pp. 181–190. Special Publication 176, Geological Society Publishing House, Bath.

Fitzsimons, S.J., McManus, K.J., Sirota, P. & Lorrain, R. (2001) Direct shear tests of materials from a cold glacier: implications for landform development. *Quaternary International*, **88**, 1–9.

Fleming, K. & Lambeck, K. (2004) Constraints on the Greenland Ice Sheet since the Last Glacial Maximum from sea-level observations and glacial rebound models. *Quaternary Science Reviews*, **23**(9–10), 1053–1077.

Fleming, K.M., Dowdeswell, J. & Oerlemans, J. (1997) Modelling the mass balance of northwest Spitsbergen glaciers and responses to climate change. *Annals of Glaciology*, **24**, 203–210.

Fleming, S.W. & Clark, P.U. (2000) Investigation of water pressure transients beneath temperate glaciers using numerical groundwater flow experiments. *Journal of Quaternary Science*, **15**, 567–572.

Fleming, S.W. & Clarke, G.K.C. (2003) Glacial control of water resource and related environmental responses to climatic warming: empirical analysis using historical streamflow data from northwestern Canada. *Canadian Water Resources Journal*, **28**, 69–86.

Flint, R.F. (1943) Growth of North American ice sheet during the Wisconsin age. *Geological Society of America Bulletin*, **54**, 325–362.

Flint, R.F. (1971) *Glacial and Quaternary Geology*. Wiley, New York.

Flower, B.P., Hastings, D.W., Hill, H.W. & Quinn, T.M. (2004) Phasing of deglacial warming and Laurentide Ice Sheet meltwater in the Gulf of Mexico. *Geology*, **32**, 597–600.

Flowers, G.E. & Clarke, G.K.C. (2002a) A multicomponent coupled model of glacier hydrology 1. Theory and synthetic examples. *Journal of Geophysical Research*, **107**(B1). 2287: doi:10.1029/2001JB001122.

Flowers, G.E. & Clarke, G.K.C. (2002b) A multicomponent coupled model of glacier hydrology 2. Application to Trapridge Glacier, Yukon, Canada. *Journal of Geophysical Research*, **107**(B11). 2288: doi:10.1029/2001JB001124.

Flowers, G.E., Björnsson, H. & Pálsson, F. (2003) New insights into the subglacial and periglacial hydrology of Vatnajökull, Iceland, from a distributed physical model. *Journal of Glaciology*, **49**(165), 257–270.

Folger, D.W., Hathaway, R.A., Christopher, R.A., *et al.* (1978) Stratigraphic test well, Nantucket Island, Massachusetts. *U.S Geological Survey Circular*, **773**.

Fookes, P.G., Gordon, D.L. & Higginbottom, I.E. (1978) Glacial landforms, their deposits and engineering characteristics. In: *The Engineering Behaviour of Glacial Materials*. Proceedings of Symposium, University of Birmingham, pp. 18–51.

Forel, F.-A. (1895 Les variations périodiques des glaciers. Discours préliminaire. *Archives des Sciences physiques et naturelles, Genève*, **XXXIV**, 209–229.

Fortuin, J.P.F. & Oerlemans, J. (1990) Parameterization of the annual surface temperature and mass balance of Antarctica. *Annals of Glaciology*, **14**, 78–84.

Fountain, A.G. (1994) Borehole water-level variations and implications for the subglacial hydraulics of South Cascade Glacier, Washington State, U.S.A. *Journal of Glaciology*, **40**, 293–304.

Fountain, A.G. & Walder, J.S. (1998) Water flow through temperate glaciers. *Reviews of Geophysics*, **36**, 299–328.

Fountain, A.G., Jansson, P., Kaser, G. & Dyurgerov, M. (Eds) (1999) Methods of mass balance measurements and modelling. *Geografiska Annaler*, **81**A(4), special issue.

Fowler, A.C. (2002) Rheology of subglacial till. *Journal of Glaciology*, **48**, 631–632.

Fowler, A.C. (2003) On the rheology of till. *Annals of Glaciology*, **37**, 55–59.

Fox, A.J. & Cooper, A.P.R. (1998) Climate-change indicators from archival aerial photography of the Antarctic Peninsula. *Annals of Glaciology*, **27**, 636–642.

Fox, A.J. & Vaughan, D.G. (In press) The retreat of Jones Ice Shelf, Antarctic Peninsula. *Journal of Glaciology*.

Francou, B., Ramirez, E., Caceres, B. & Mendoza, J. (2000) Glacier evolution in the tropical Andes during the last decades of the 20th century: Chacaltaya, Bolivia and Antizana, Ecuador. *Ambio*, **29**(7), 416–422.

Francou, B., Vuille, M., Wagnon, P., Mendoza, J. & Sicart, J.E. (2003) Tropical climate change recorded by a glacier in the central Andes during the last decades of the 20th century: Chacaltaya, Bolivia, 16°S. *Journal of Geophysical Research—Atmospheres*, **108**(D5), 4154–4165. doi: 10.1029/2002JD002959.

Frauenfelder, R., Zemp, M., Hoelzle, M. & Haeberli, W. (In press) Worldwide glacier mass balance measurements: general trends and first results of the extraordinary year 2003 in Central Europe. Data of Glaciological Studies.

Freeze, R.A. & Cherry, J.A. (1979) *Groundwater*. Prentice-Hall, New Jersey.

Freitag, J., Wilhelms, F. & Kipfstuhl, S. (2004) Microstructure dependent densification of polar firn derived from X-ray microtopography. *Journal of Glaciology*, **50**(169), 243–250.

French, H.M. (1996) *The Periglacial Environment*, 2nd edn. Addison Wesley Longman, Reading, 341 pp.

Frost, H.J. & Ashby, M.F. (1982) *Deformation Mechanism Maps*. Pergamon Press, New York, 167 pp.

Fujii, Y., Yamanouchi, T., Suzuki, K. & Tanaka, S. (1987) Comparison of surface conditions of the inland ice sheet, Dronning Maud Land, Antarctica, derived from NOAA AVHRR data with ground observations. *Annals of Glaciology*, **9**, 72–75.

Fujiyoshi, Y., Kondo, H., Inoue, J. & Yamada, T. (1987) Characteristics of precipitation and vertical structure of air temperature in northern Patagonia. *Bulletin of Glacier Research*, **4**, 15–23.

Fulton, R.J. (Ed.) (1989) *Quaternary Geology of Canada and Greenland*, Vol. K-1. Geology of North America Series, Geological Society of America, Boulder, Colorado, and Geology of Canada, no. 1, Geological Survey of Canada, Queen's Printer, Ottawa, 839 pp.

Fulton, R.J. (1995) *Surficial materials of Canada*, Map 1880A, scale 1 : 5,000,000. Geological Survey of Canada. Queen's Printer, Ottawa.

Funder, S. (1989) Quaternary geology of the ice-free areas and adjacent shelves of Greenland. In: *Quaternary Geology of Canada and Greenland*, Vol. K-1 (Ed. R.J. Fulton), pp. 743–792. Geology of North America Series, Geological Society of America, Boulder, Colorado, and Geology of Canada, No. 1, Geological Survey of Canada, Queen's Printer, Ottawa.

Funder, S. & Hansen, L. (1996) The Greenland ice sheet—a model for its culmination and decay during and after the last glacial maximum. *Bulletin of the Geology Society of Denmark*, **42**, 137–152.

Funk, M. & Röthlisberger, H. (1989) Forecasting the effects of a planned reservoir which will partially flood the tongue of Unteraargletscher in Switzerland. *Annals of Glaciology*, **13**, 76–81.

Furbish, D.J. & Andrews, J.T. (1984) The use of hypsometry to indicate long-term stability and response of valley glaciers to changes in mass transfer. *Journal of Glaciology*, **30**(105), 199–211.

Gades, A.M., Raymond, C.F., Conway, H. & Jacobel, R.W. (2000) Bed properties of Siple Dome and adjacent ice streams, West Antarctica, inferred from radio-echo sounding measurements. *Journal of Glaciology*, **46**, 88–94.

Ganopolski, A. & Rahmstorf, S. (2001) Rapid changes of glacial climate simulated in a coupled climate model. *Nature*, **409**, 153–158.

Gao, X., Jacka, T.H. & Budd, W.F. (1989) The development of ice crystal anisotropy in shear and comparisons of flow properties in shear and compression. In: *Proceedings of the International Symposium on Antarctic Research* (Ed. G. Kun), pp. 32–40. China Ocean Press, Beijing.

Gellatly, A.F. (1982) Lichenometry as a relative-age dating method in Mount Cook National Park. *New Zealand Journal of Botany*, **20**, 343–353.

Gellatly, A.F., Gordon, J.E., Whalley, W.B. & Hansom, J.D. (1988) Thermal regime and geomorphology of plateau ice caps in northern Norway: observations and implications. *Geology*, **16**, 983–986.

Gemmell, C., Smart, D. & Sugden, D. (1986) Striae and former ice flow directions in Snowdonia, North Wales. *Geographical Journal*, **152**, 19–29.

Genty, D., Blamart, D., Ouahdi, R., *et al.* (2003) Precise dating of Dansgaard–Oeschger climate oscillations in western Europe from stalagmite data. *Nature*, **421**, 833–837.

Gerber, R.E., Boyce, J.I. & Howard, K.W.F. (2001) Evaluation of heterogeneity and field-scale groundwater flow regime in leaky till aquitard. *Hydrogeology Journal*, **9**, 60–78.

Gerrard, J.A.F., Perutz, M.F. & Roch, A. (1952) Measurement of the velocity distribution along a vertical line in a glacier. *Proceedings of the Royal Society of London*, **A213**, 546–558.

Gilbert, R. (2000) The Devil Lake pothole (Ontario): Evidence of subglacial fluvial processes. *Géographie physique et Quaternaire*, **54**, 245–250.

Gillet, F. (1975) Steam, hot-water and electrical thermal drills for temperate glaciers. *Journal of Glaciology*, **14**(70), 171–179.

Gillett, N.P., Baldwin, M.P. & Allen, M.R. (2003) *Climate Change*, and the North Atlantic Oscillation. In: *The North Atlantic Oscillation: Climate Significance and Environmental Impact*, pp. 193–209. Geophysical Monograph 134, American Geophysical Union, Washington, DC.

Gilpin, R.R. (1980) Wire regelation at low temperatures. *Journal of Colloid and Interface Science*, **77**, 435–448.

Ginot, P., Schwikowski, M., Gäggeler, H.W., *et al.* (2002) First results of a palaeoatmospheric chemistry and climate study of Cerro Tapado glacier, Chile. In: *The Patagonian Icefields: a Unique Natural Laboratory for Environmental and Climate Change Studies* (Eds G. Casassa, F. Sepúlveda & R. Sinclair), pp. 157–167. Series of the Centro de Estudios Científicos. Kluwer Academic/Plenum Publishers, New York.

Giorgi, F. & Marinucci, M.R. (1996) An investigation of the sensitivity of simulated precipitation to model resolution and its implications for climate studies. *Monthly Weather Review* **124**, 148–166.

Giorgi, F., Hostetler, S.W. & Brodeur, C.S. (1994) Analysis of the surface hydrology in a regional climate model. *Quarterly Journal of the Royal Meteorological Society* **120**, 161–183.

Giorgi, F., Means, L.O., Shields, C. & Mayer, L. (1996) A regional model study of the importance of local versus remote controls of the 1988 drought and the 1993 flood over the central United States. *Journal of Climate* **9**, 1150–1162.

Giorgi, F., Francisco, R. & Pal, J. (2003) Effects of a subgrid-scale topography and land-use scheme on the simulation of surface climate and hydrology. Part I: Effects of temperature and water vapour disaggregation. *Journal of Hydrometeorology*, **4**, 317–333.

Giraudeau, G., Cremer, M., Manthe, S., Laberrie, L. & Bond, G. (2000) Coccolith evidence for instabilities in surface circulation south of Iceland during Holocene times. *Earth and Planetary Science Letters*, **179**, 257–268.

Glasser, N.F. & Hambrey, M.J. (2002) Sedimentary facies and landform genesis at a temperate outlet glacier: Soler Glacier, North Patagonian Icefield. *Sedimentology* **49**, 43–64.

Glen, J.W. (1952) Experiments on the deformation of ice. *Journal of Glaciology*, **2**, 111–114.

Glen, J.W. (1955) The creep of polycrystalline ice. *Proceedings of the Royal Society of London Series B*, **228**, 519–538.

Glen, J.W. (1958a) The flow law of ice. In: *Symposium of Chamonix. Physics of the Motion of Ice*, pp. 169–170. IAHS Publication 47, International Association of Hydrological Sciences, Wallingford.

Glen, J.W. (1958b) The flow law of ice: A discussion of the assumptions made in glacier theory, their experimental foundations and consequences. In: *Symposium of Chamonix. Physics of the Motion of Ice*, pp. 171–183. IAHS Publication 47, International Association of Hydrological Sciences, Wallingford.

Glen, J.W., Donner, J.J. & West, R.G. (1957) On the mechanism by which stones in till become oriented. *American Journal of Science*, **255**, 194–205.

Glidor, H. (2003) When Earth's freezer door is left ajar. *Eos* (*Transactions of the American Geophysical Union*), **84**, 215.

Glover, R.W. (1999) Influence of spatial resolution and treatment of orography on GCM estimates of the surface mass balance of the Greenland Ice Sheet. *Journal of Climate* **12**, 551–563.

Glynn, P.D., Voss, C.I. & Provost, A.M. (1999) Deep penetration of oxygenated meltwaters from warm based ice-sheets into the Fennoscandian Shield. *Proceedings of a SKB workshop 'Use of Hydrochemical Information in Testing Groundwater Flow Models' in Borgholm, Sweden*, pp. 201–241 Organization for Economic Cooperation and Development/Nuclear Energy Agency, Paris.

Godoi, M.A., Casassa, G. & Shiraiwa, T. (2001) Reseña de estudios paleoclimáticos mediante testigos de hielo: potencialidades y evidencia obtenida en el cono sur de Sudamérica. *Anales Instituto Patagonia, Serie Ciencias Naturales*, **29**, 45–54.

Gogineni, S.P., Tammana, D., Braaten, D., Leuschen, C., *et al.* (2001) Coherent radar ice thickness measurements over the Greenland ice sheet. *Journal of Geophysical Research*, **106**, 33761–33772.

Goldsby, D.L. & Kohlstedt, D.L. (1997) Grain boundary sliding in fine-grained ice I. *Scripta Materialia*, **37**(9), 1399–1406.

Goldsby, D.L. & Kohlstedt, D.L. (2001) Superplastic deformation of ice: Experimental observations. *Journal of Geophysical Research*, **106**(B6), 11017–11030.

Goldsby, D.L. & Kohlstedt, D.L. (2002) Reply to comment by P. Duval and M. Montagnat on 'Superplastic deformation of ice: Experimental observations'. *Journal of Geophysical Research*, 107, 10.1029/2002JB001824.

Goldthwait, R.P. (1960) Study of ice cliff in Nunatarssuaq, Greenland. *Tech. Rep. Snow Ice Permafrost Res. Establ.* **39**, 1–103.

Goldthwait, R.P. (1979) Giant grooves made by concentrated basal ice streams. *Journal of Glaciology*, **23**, 297–307.

Gordon, S., Sharp, M., Hubbard, B., Smat, C., Ketterling, B., Willis, I. (1998) Seasonal reorganization of subglacial drainage inferred from measurements in boreholes. *Hydrological Processes*, **12**(1), 105–133.

Gosse, J. (2002) Report on the Atlantic Canada Glacier Ice Dynamics Workshop, May 22–24, 2002. *Geoscience Canada*, **29**(4), 183–185.

Gosse, J.C. & Phillips, F.M. (2001) Terrestrial *in situ* cosmogenic nuclides: Theory and application. *Quaternary Science Reviews*, **20**, 1475–1560.

Gosse, J.C., Grant, D.R, Klein, J., Middleton, R., Lawn, B., Dezfouly-Arjomandy, B. (1993) Significance of altitudinal weathering zones in Atlantic Canada, inferred from *in situ* produced cosmogenic radionuclides. *Geological Society of America, Abstracts and Program*, **25**(6), 394.

Goughnour, R.R. & Andersland, O.B. (1968) Mechanical properties of a sand–ice system. *Journal of Soil Mechanics and Foundations Division, Proceedings of the American Society of Civil Engineers*, **SM4**, 923–950.

Gow, A.J. (1963) Results of measurements in the 309 meter borehole at Byrd Station, Antarctica. *Journal of Glaciology*, **4**, 771–784.

Gow, A.J. (1969) On the rates of growth of grains and crystals in south polar firn, *Journal of Glaciology*, **8**(53), 241–252.

Gow, A.J. (1971) *Depth–Time–Temperature Relationships of Ice Crystal Growth in Polar Glaciers.* Research Report 300, Cold Regions Research and Engineering Laboratory, Hanover, New Hampshire, pp. 1–19.

Gow, A.J. & Meese, D. (1996) *Nature*, of basal debris in the GISP2 and Byrd ice cores and its relevance to bed processes. *Annals of Glaciology*, **22**, 134–140.

Gow, A.J. & Williamson, T. (1976) *Rheological Implications of the Internal Structure and Crystal Fabrics of the West Antarctic Ice Sheet as Revealed by Deep Core Drilling at Byrd Station.* Report 76-35, U.S. Army Cold Regions Research and Engineering Laboratory, Hanover, NH, 25 pp.

Gow, A.J. & Williamson, T. (1976) Rheological implications of the internal structure and crystal fabrics of the West Antarctic ice sheet as revealed by deep core drilling at Byrd Station. *Geological Society of America Bulletin*, **87**, 1665–1677.

Gow, A.J., Ueda, H.T. & Garfield, D.E. (1968) Antarctic ice sheet: preliminary results of first core hole to bedrock. *Science*, **161**, 1011–1013.

Gow, A.J., Epstein, S. & Sheehy, W. (1979) On the origin of stratified debris in ice cores from the bottom of the Antarctic ice sheet. *Journal of Glaciology*, **23**, 185–192.

Gow A.J., Meese, D.A., Alley, R.B., *et al.* (1997) Physical and structural properties of the Greenland Ice Sheet Project 2 ice core: a review. *Journal of Geophysical Research*, **102**(C12), 26559–26575.

Grant, D.R. (1977) Altitudinal weathering zones and glacial limits in western Newfoundland, with particular reference to Gros Morne National Park. *Geological Survey of Canada Paper*, **77-1A**, 455–463.

Grant, D.R. (1986) *Surficial Geology, Port Saunders, Newfoundland.* Geological Survey of Canada, Map 1622A (1 : 250,000).

Grant, D.R. (1989) *Surficial Geology, Sandy Lake–Bay of Islands, Newfoundland.* Geological Survey of Canada, Map 1664A (1 : 250,000). Queen's Printer, Ottawa.

Grasby, S., Osadetz, K., Betcher, R. & Render, F. (2000) Reversal of the regional-scale flow system of the Williston basin in response to Pleistocene glaciation. *Geology*, **28**, 635–638.

Gravenor, C.P. (1975) Erosion by Continental Ice Sheets. *American Journal of Science*, **275**, 594–604.

Gray, A.L., Mattar, K.E., Vachon, P.W., *et al.* (1998) InSAR results from the RADARSAT Antarctic Mapping Mission data: Estimation of glacier motion using a simple registration procedure. In: *International GeoScience and Remote Sensing Symposium (IGARSS 98) on Sensing and Managing the Environment*, Seattle, WA, 6–10 July 1. IEEE Service Center, 445 Hoes Lane, Po Box 1331, Piscataway, NJ 08855–1331.

Gray, A.L., Short, N., Mattar, K.E. & Jezek, K.C. (2001) Velocities and Flux of the Filchner Ice Shelf and Its Tributaries Determined from Speckle Tracking Interferometry. *Canadian Journal of Remote Sensing* **27**(3), 193–206.

Gray, J.M. (1993) Quaternary geology and waste disposal in south Norfolk, England. *Quaternary Science Reviews*, **12**, 899–912.

Gray, J.M. (1995) Influence of Southern Upland ice on glacio-isostatic rebound in Scotland: the Main Rock Platform in the Firth of Clyde. *Boreas*, **24**, 30–36.

Gray, L., Joughin, I., Tulaczyk, S., Spikes, V.B., Bindschadler, R. & Jezek, K. (2005). Evidence for subglacial water transport in the West Antarctica Ice Sheet through three-dimensional satellite radar interferometry. *Geophysical Research Letters*, **32**(3). Art. No. L03501.

Gregory, J.M. & Oerlemans, J. (1998) Simulated future sea-level rise due to glacier melt based on regionally and seasonally resolved temperature changes. *Nature*, **391**, 474–476.

Gregory, J.M., Huybrechts, P. & Raper, S.C.B. (2004) Threatened loss of the Greenland ice sheet. *Nature*, **428**, 616.

Greuell, W. (1992) Hintereisferner, Austria: mass balance reconstruction and numerical modelling of historical length variation. *Journal of Glaciology*, **38**(129), 233–244.

Greve, R. (1997a) Application of a polythermal three-dimensional ice sheet model to the Greenland ice sheet: Response to steady-state and transient climate scenarios. *Journal of Climate*, **10**, 901–918.

Greve, R. (1997b) Large-scale ice-sheet modelling as a means of dating deep ice cores in Greenland. *Journal of Glaciology*, **43**, 307–310.

Greve, R. (2000) On the response of the Greenland ice sheet to greenhouse climate change. *Climatic Change*, **46**, 289–303.

Greve, R. & MacAyeal, D.R. (1996) Dynamic/thermodynamic simulations of Laurentide ice-sheet instability. *Annals of Glaciology*, **23**, 328–335.

Greve, R., Mügge, B., Baral, D.R., Albrecht, O., *et al.* (1999) Nested high-resolution modelling of the Greenland Summit Region. In: *Advances in Cold-Region Thermal Engineering and Sciences* (Eds K. Hutter, Y. Wang & H. Beer), pp. 285–306. Springer Verlag, Berlin.

Grønvold, K., Oskarsson, N., Johnsen, S.J., *et al.* (1995) Ash layers from Iceland in the Greenland GRIP ice core correlated with oceanic and land sediments. *Earth and Planetary Science Letters*, **135**, 149–155.

Grootes, P.M. & Stuiver, M. (1997) Oxygen 18/16 variability in Greenland snow and ice with 10^{-3} to 10^{5}-year time resolution. *Journal of Geophysical Research*, **102**(C12), 26455–26470.

Grousset, F., Pujol, C., Labeyrie, L., Auffret, G. & Boelaert, A. (2000) Were the North Atlantic Heinrich events triggered by the behaviour of the European ice sheets. *Geology*, **28**(2), 123–126.

Groussett, F.E., Cortijo, E., Huon, S., *et al.* (2001) Zooming in on Heinrich layers. *Paleoceanography*, **16**, 240–259.

Grove, J.M. (2001) The initiation of the 'Little Ice Age' in regions round the North Atlantic. *Climatic Change*, **48**, 53–82.

Gudmundsson, G.H. (1999) A three-dimensional numerical model of the confluence area of Unteraargletscher, Bernese Alps, Switzerland. *Journal of Glaciology*, **45**(150), 219–230.

Gudmundsson, G.H. (2002) Observations of a reversal in vertical and horizontal strain-rate regime during a motion event on Unteraargletscher, Bernese Alps, Switzerland. *Journal of Glaciology*, **48**(163), 566–574.

Gudmundsson, G.H. (2003) Transmission of basal variability to a glacier surface. *Journal of Geophysical Research*, **108**(B5), 2253. doi:10.1029/2002JB002107.

Gudmundsson, G.H., Bassi, A., Vonmoos, M., Bauder, A., Fischer, U.H. & Funk, M. (2000) High-resolution measurements of spatial and temporal variations in surface velocities of Unteraargletscher, Bernese Alps, Switzerland. *Annals of Glaciology*, **31**, 63–68.

Gudmundsson, H.J. (1997) A review of the Holocene environment history of Iceland. *Quaternary Science Reviews*, **16**, 81–92.

Gudmundsson, M.T., Sigmundsson, F. & Bjornsson, H. (1997) Ice-volcano interaction of the 1996 subglacial eruption, Vatnajokull, Iceland. *Nature*, **389**, 954–957.

Günther, R. & Widlewski, D. (1986) Die Korrelation verschiedener Klimaelemente mit dem Massenhaushalt alpiner und skandinavischer Gletscher. *Zeitschrift für Gletscherkunde und Glazialgeologie*, **22**(2), 125–147.

Gurnell, A.M. (1982). The dynamics of suspended sediment concentration in a proglacial stream. In: *Hydrological Aspects of Alpine and High Mountain Areas. Proceedings of the Exeter Symposium, July 1982* (Ed. J.W. Glen,), pp. 319–330. Publication 138, International Association of Hydrological Sciences, Wallingford.

Gustafson, G., Liedholm, M., Lindbom, B. & Lundblad, K. (1989). *Groundwater Calculations on a Regional Scale at the Swedish Hard Rock Laboratory*. SKB SHRL Progress Report 25-88-17, Swedish Nuclear Power Inspectorate, Stockholm.

Gustavson, T.C. & Boothroyd, T.C. (1987) A depositional model for outwash, sediment sources, and hydrologic characteristics, Malaspina Glacier, Alaska: A modern analog of the southeastern margin of the Laurentide ice sheet. *Geological Society of America Bulletin*, **99**, 187–200.

Haeberli, W. (Ed.) (1985) Fluctuations of glaciers 1975–1980 (Vol. IV). IAHS(ICSI)–UNESCO, Paris.

Haeberli, W. (2004) Glaciers and ice caps: historical background and strategies of world-wide monitoring. In: Mass balance of the cryosphere (Eds J.L. Bamber & A.J. Payne), pp. 559–578. Cambridge University Press, Cambridge.

Haeberli, W. & Alean, J. (1985). Temperature and accumulation of high altitude firn in the Alps. *Annals of Glaciology*, **6**, 161–163.

Haeberli, W. & Hoelzle, M. (Eds) (1993) *Fluctuations of glaciers 1985–1990* (Vol. VI). IAHS(ICSI)–UNEP–UNESCO, Paris.

Haeberli, W. & Hoelzle, M. (1995) Application of inventory data for estimating characteristics of and regional climate-change effects on mountain glaciers: a pilot study with the European Alps. *Annals of Glaciology*, **21**, 206–212. (Russian Translation in *Data of Glaciological Studies, Moscow*, **82**, 116–124.)

Haeberli, W. & Holzhauser, H. (2003) Alpine glacier mass changes during the past two millennia. *Pages News*, **1**(11), 13–15.

Haeberli, W. & Müller, P. (Eds) (1988) *Fluctuations of glaciers 1980–1985* (Vol. V). IAHS(ICSI)–UNEP–UNESCO, Paris.

Haeberli, W., Hoelzle, M. & Suter, S. (Eds) (1998a) *Into the Second Century of Worldwide Glacier Monitoring: Prospects and Strategies*. A contribution to the International Hydrological Programme (IHP) and the Global Environment Monitoring System (GEMS). UNESCO Studies and Reports in Hydrology 56, UNESCO Publishing, Paris.

Haeberli, W., Hoelzle, M., Suter, S. & Fraunfelder, R. (Eds) (1998b) *Fluctuations of Glaciers 1990–1995*, Vol. VII. IAHS(ICSI)–UNEP–UNESCO, Paris.

Haeberli, W., Barry, R. & Cihlar, J. (2000) Glacier monitoring within the Global Climate Observing System. *Annals of Glaciology*, **31**, 241–246.

Haeberli, W., Maisch, M. & Paul, F. (2002) Mountain glaciers in global climate-related observation networks. *World Meteorological Organization Bulletin*, **51**(1), 18–25.

Haefeli, R. (1951) Some observations of glacier flow. *Journal of Glaciology*, **1**(9), 469–500.

Haefeli, R. (1961) Contribution to the movement and the form of ice sheets in the Arctic and Antarctic. *Journal of Glaciology*, **3**, 1133–1150.

Hagen, J.O. & Reeh, N. (2004) *In situ* measurement techniques: land ice. In: *Mass Balance of the Cryosphere* (Eds J.L. Bamber & A.J. Payne), pp. 11–41. Cambridge University Press, Cambridge.

Hagen, J.O., Wold, B., Liestøl, O., Østrem, G., Sollid, J.L. (1983) Subglacial processes at Bondhusbreen, Norway: preliminary results. *Annals of Glaciology*, **4**, 91–98.

Hald, M. & Aspeli, R. (1997) Rapid climatic shifts of the northern Norwegian Sea during the last deglaciation and the Holocene. *Boreas*, **26**, 15–28.

Hald, M. & Hagen, S. (1998) Early Preboreal cooling in the Nordic seas region triggered by meltwater. *Geology*, **26**, 615–618.

Haldorsen, S. (1981) Grain size distributions of subglacial till and its relation to glacial crushing and abrasion. *Boreas*, **10**, 91–105.

Haldorsen, S., Heim, M. & Lauritzen, S.-E. (1996) Subpermafrost groundwater, Western Svalbard. *Nordic Hydrology*, **27**, 57–68.

Hall, A.M., Peacock, J.D. & Connell, E.R. (2003) New data for the Last Glacial Maximum in Great Britain and Ireland: a Scottish perspective on the paper by Bowen *et al.* (2002). *Quaternary Science Reviews*, **22**, 1551–1554.

Hall, V.A. & Pilcher, J.R. (2002) Late-Quaternary Icelandic tephras in Ireland and Great Britain: detection, characterization and usefulness. *The Holocene*, **12**, 223–232.

Hallet, B. (1976a) The effect of subglacial chemical processes on sliding. *Journal of Glaciology*, **17**, 209–221.

Hallet, B. (1976b) Deposits formed by subglacial precipitation of $CaCO_3$. *Geological Society of America Bulletin*, **87**, 1003–1015.

Hallet, B. (1979) A theoretical model of glacial abrasion. *Journal of Glaciology*, **23**, 39–50.

Hallet, B. (1996) Glacial quarrying: a simple theoretical model. *Annals of Glaciology*, **22**, 1–9.

Hallet, B., Hunter, L. & Bogen, J. (1996). Rates of erosion and sediment evacuation by glaciers: A review of field data and their implications. *Global and Planetary Change*, **12**, 213–235.

Ham, N.R. & Mickelson, D.M. (1994) Basal till fabric and deposition at Burroughs Glacier, Glacier Bay, Alaska. *Geological Society of America Bulletin*, **106**, 1552–1559.

Hambrey, M.J. & Muller, F. (1987) Structures and ice deformation in the White Glacier, Axel Heiberg Island, Northwest Territories, Canada. *Journal of Glaciology*, **20**(82), 41–66.

Hambrey, M.J., Dowdeswell, J.A., Murray, T.A. & Porter, P.A. (1996) Thrusting and debris-entrainment in a surging glacier: Bakaninbreen, Svalbard. *Annals of Glaciology*, **22**, 241–248.

Hambrey, M.J., Huddart, D., Bennett, M.R. & Glasser, N.F. (1997) Genesis of 'hummocky moraine' by thrusting in glacier ice: evidence from Svalbard and Britain. *Journal of the Geological Society, London*, **154**, 623–632.

Hambrey, M.J., Bennett, M.R., Dowdeswell, J.A., Glasser, N.F. & Huddart, D. (1999) Debris entrainment and transfer in polythermal valley glaciers. *Journal of Glaciology*, **45**(149), 69–86.

Hamley, T.C., Smith, I.N. & Young, N.W. (1985) Mass-balance and ice-flow-law parameters for East Antarctica. *Journal of Glaciology*, **31**, 334–339.

Hammer, C.U. (1980) Acidity of polar ice cores in relation to absolute dating, past volcanism and radio-echoes. *Journal of Glaciology*, **25**, 359–372.

Hammer, C.U. (1989) Dating by Physical and Chemical Seasonal Variations and Reference Horizons. In: *The Environmental Record in Glaciers and Ice Sheets* (Eds H. Oeschger & C.C. Langway Jr), pp. 99–121. Dahlem Konferenzen, Wiley, Chichester.

Hammer, C.U. (2002) Holocene Climate and Past Volcanism: Greenland-Northern Europe. In: *Climate Development and History of the North Atlantic Realm* (Eds G. Wefer, W. Berger, K. Behre & E. Jansen), pp. 149–163. Springer-Verlag, Berlin.

Hammer, C.U., Clausen, H.B., Dansgaard, W., *et al.* (1978) Dating of Greenland ice cores by flow models, isotopes, volcanic debris and continental dust. *Journal of Glaciology*, **20**(82), 3–26.

Hammer, C.H., Clausen, H.B., Dansgaard, W., *et al.* (1985) Continuous impurity analysis along the Dye 3 deep core. In: *Greenland ice core: Geophysics, Geochemistry and the Environment* (Eds C.C. Langway, H. Oeschger & W. Dansgaard), pp. 90–94. Geophysical Monograph 33, American Geophysical Union, Washington, DC.

Hammer, C.U., Clausen, H.B. & Langway, C.C. Jr. (1994) Electrical conductivity method (ECM) stratigraphic dating of the Byrd Station ice core, Antarctica. *Annals of Glaciology*, **20**, 115–120.

Hammer, C.U., Andersen, K.K., Clausen, H.B., *et al.* (1997) The Stratigraphic Dating of the GRIP Ice Core. Special report of the Geophysical Department, Niels Bohr Institute for Astronomy, Physics and Geophysics, University of Copenhagen.

Hanna, E. & Bamber, J. (2001) Derivation and optimization of a new Antarctic sea-ice record. *International Journal of Remote Sensing*, **22**(1), 113–139.

Hansel, A.K., Mickelson, D.M., Schneider, A.F. & Larsen, C.E. (1985) Late Wisconsinan and Holocene history of the Lake Michigan basin. In: *Quaternary Evolution of the Great Lakes* (Eds P.F. Karrow & P.E. Calkin), pp. 39–53. Special Paper 30, Geological Association of Canada, St John's, NF.

Hansen, B.L. & Langway, C.C. jr. (1966) Deep core drilling in ice and core analysis at Camp Century, Greenland, 1961–1966. *Antarctic Journal of the United States*, **1**, 207–208.

Hansen, D.P. & Wilen, L.A. (2002) Permofmance and applications of an automated c-axis ice-fabric analyzer. *Journal of Glaciology*, **48**, 159–170.

Hansen, J.E. (2003) *http://www.giss.nasa.gov/data/update/gistemp/*. Goddard Institute for Space Studies, Washington.

Harbor, J.M. (1982) Numerical modelling of the development of U-shaped valleys by glacial erosion. *Geological Society of America Bulletin*, **104**, 1364–1375.

Harbor, J.M., Sharp, M., Copland, L., Hubbard, B., Nienow, P. & Mair, D. (1997) The influence of subglacial drainage conditions on the velocity distribution within a glacier cross-section. *Geology* **25**, 739–742.

Hardy, R.J., Bamber, J.L. & Orford, S. (2000) The delineation of major drainage basins on the Greenland Ice sheet using a combined numerical modelling and GIS approach. *Hydrological Processes*, **14**(11–12), 1931–1941.

Harper, J.T. & Humphrey, N.F. (1995) Borehole video analysis of a temperate glacier's englacial and subglacial structure: implications for glacier flow models. *Geology*, **23**(10), 901–904.

Harper, J.T., Humphrey, N.F. & Greenwood, M.C. (2002) Basal conditions and glacier motion during winter/spring transition, Worthington Glacier, Alaska, U.S.A. *Journal of Glaciology*, **48**(160), 42–50.

Harper, J.T., Humphrey, N.F. & Pfeffer, W.T. (2003) Glacier-scale variability of the subglacial drainage system: year-long records from boreholes at sites along the length of a glacier. *Eos* (*Transactions of the American Geophysical Union*), **84**, Fall Meeting Supplement, Abstract C22A-04.

Harrison, W. (1958) Marginal zones of vanished glaciers reconstructed from pre-consolidation pressure values of overridden silts. *Journal of Geology*, **66**, 72–95.

Harrison, W.D. & Kamb, B. (1973) Glacier bore-hole photography. *Journal of Glaciology*, **12**(64), 129–137.

Harrison, W.D. & Post, A.S. (2003) How much do we really know about glacier surging? *Annals of Glaciology*, **36**, 1–6.

Harrison, W.D., Echelmeyer, K.A. & Larsen, C.F. (1998) Measurement of temperature in a margin of Ice Stream B, Antarctica: implications for margin migration and lateral drag. *Journal of Glaciology*, **44**, 615–624.

Harrison, W.D., Elsberg, D.H., Echelmeyer, K.A. & Krimmel, R.M. (2001) On the characterization of glacier response by a single time scale. *Journal of Glaciology*, **47**(159), 659–664.

Harrison, W.D., Truffer, M., Echelmeyer, K.A., Pomraning, D.A., Abnett, K.A. & Ruhkick, R.H. (2004) Probing the till beneath Black Rapids Glacier, Alaska. *Journal of Glaciology*, **50**(171), 605–614.

Hart, J.K. (1994) Till fabric associated with deformable beds, *Earth Surface Processes and Landforms*, **19**, 15–32.

Hart, J.K. (1995a) Recent drumlins, flutes and lineations at Vestari-Hagafellsjökull, Iceland. *Journal of Glaciology*, **41**, 596–606.

Hart, J.K. (1995b) Subglacial erosion, deposition and deformation associated with deformable beds. *Progress in Physical Geography*, **19**, 173–191.

Hart, J.K. & Roberts, D.H. (1994) Criteria to distinguish between subglacial glaciotectonic and glaciomarine sedimentation, I. Deformation styles and sedimentology. *Sedimentary Geology*, **91**, 191–213.

Hart, J.K. & Watts, R.J. (1997) A comparison of the styles of deformation associated with two recent push moraines, south Van Keulenfjorden, Svalbard. *Earth Surface Processes and Landforms*, **22**, 1089–1107.

Harvey, L.D.D. (1980) Solar variability as a contributing factor to Holocene climatic change. *Progress in Physical Geography*, **4**, 487–530.

Hasnain, S.I., Subramanian, V. & Dhanpal, K. (1989) Chemical characteristics and suspended sediment load of meltwaters from a Himalayan glacier in India. *Journal of Hydrology*, **106**, 99–108.

Hastenrath, S. & Ames, A. (1995a) Diagnosing the imbalance of Yanamarey Glacier in the Cordillera Blanca of Peru. *Journal of Geophysical Research*, **100**(D3), 5105–5112.

Hastenrath, S. & Ames, A. (1995b) Recession of Yanamarey Glacier in Cordillera Blanca, Peru, during the 20th century. *Journal of Glaciology*, **41**(137), 191–196.

Hathaway, J.C., Poag, C.W., Valentine, P.C., *et al.* (1979) U.S. Geological Survey Core Drilling on the Atlantic Shelf. *Science*, **206**(4418), 515–527.

Hättestrand, C. (1997) Ribbed moraines in Sweden—distribution pattern and palaeoglaciological implications. *Sedimentary Geology*, **111**, 41–56.

Hättestrand, C. & Kleman, J. (1999) Ribbed moraine formation. *Quaternary Science Reviews*, **18**, 43–61.

Hättestrand, C. & Stroeven, A.P. (2002) A relict landscape in the centre of Fennoscandian glaciation: geomorphological evidence of minimal Quaternary glacial erosion. *Geomorphology*, **44**, 127–143.

Hawley, R.L., Waddington, E.D., Alley, R.B. & Taylor, K.C. (2003) Annual layers in polar firn detected by borehole optical stratigraphy. *Geophysical Research Letters*, **30**, 10.1029/2003GL017675.

Hays, J.D., Imbrie, J. & Shackleton, N.J. (1976) Variations in the Earth's Orbit: Pacemaker of the Ice Ages. *Science*, **194**, 112–1132.

He, Y., Zhang, Z., Theakstone, W., *et al.* (2003) Changing features of the climate and glaciers in China's monsoonal temperature glacier region. *Journal of Geophysical Research*, **108**(D17), 4530–4536.

Hebrand, M. & Åmark, M. (1989) Esker formation and glacier dynamics in eastern Skåne and adjacent areas, southern Sweden. *Boreas*, **18**, 67–81.

Heinrich, H. (1988) Origin and consequences of cyclic ice rafting in the Northeast Atlantic Ocean during the past 130,000 years. *Quaternary Research*, **29**, 143–152.

Hellmer, H.H. & Olbers, D. (1989) A two-dimensional model for the thermohaline circulation under an ice shelf. *Antarctic Science*, **1**(4), 325–336.

Hemming, S.R. (2004) Heinrich events: massive Late Pleistocene detritus layers of the North Atlantic and their global climate imprint. *Reviews of Geophysics*, **42**, 1–43. Art. No. RG1005.

Hemming, S.R., Bond, G.C., Broecker, W.S., Sharp, W.D. & Klas-Mendelson, M. (2000) Evidence from ^{40}AR/^{39}Ar ages of individual hornblende grains for varying Laurentide sources of Iceberg discharges 22,000 to 10,500 yr B.P. *Quaternary Research*, **54**, 372–383.

Hemming, S., Vorren, T. & Kleman, J. (2002a) Provinciality of ice rafting in the North Atlantic: application of ^{40}Ar/^{39}Ar dating of individual ice rafted hornblende grains. *Quaternary International*, **95–96**, 75–85.

Hemming, S.R., Hall, C.M., Biscaye, P.E., *et al.* (2002b) ^{40}Ar/^{39}Ar ages and ^{40}Ar concentrations of fine-grained sediment Fractions from North Atlantic Heinrich Layers. *Chemical Geology*, **182**, 583–603.

Herron, S. & Langway, CC. jr. (1979) The debris-laden ice at the bottom of the Greenland ice sheet. *Journal of Glaciology*, **23**, 193–207.

Hesse, R. (1995) Continental slope and basin sedimentation adjacent to an ice-margin: a continuous sleeve-gun profile across the Labrador Slope, Rise and Basin. In: *Atlas of Deep Water Environments: Architectural Style in Turbidite Systems* (Eds K.T. Pickering, R.N. Hiscott, N.H. Kenyon, F.R. Lucchi & R.D.A. Smith), pp. 14–17. Chapman and Hall, London.

Hesse, R. & Khodabakhsh, S. (1998) Depositional facies of late Pleistocene Heinrich events in the Labrador Sea. *Geology*, **26**, 103–106.

Hesse, R., Klaucke, I., Ryan, W.B.F. & Piper, D.J.W. (1997) Ice-sheet Sourced Juxtaposed Turbidites Systems in Labrador Sea. *Geoscience Canada*, **24**, 3–12.

Hesse, R., Rashid, H. & Khobabakhsh, S. (2004) Fine-grained sediment lofting from meltater-generated turbidity currents during Heinrich events. *Geology*, **32**, 449–452.

Hicock, S.R. (1990) Genetic till prism. *Geology*, **18**, 517–519.

Hicock, S.R. & Dreimanis, A. (1992) Deformation till in the Great Lakes region: implications for rapid flow along the south-central margin of the Laurentide Ice Sheet. *Canadian Journal of Earth Sciences*, **29**, 1565–1579.

Hicock, S.R., Goff, J.R., Lian, O.B. & Little, E.C. (1996) On the interpretation of subglacial till fabric. *Journal of Sedimentary Research*, **66**, 928–934.

Higashi, A., Fukuda, A., Hondoh, T., *et al.* (1985) Dynamical dislocation processes in ice crystal. *Proceedings of Yamada Conference IX* (Eds T. Susuki, K. Ninomiya & S. Takeuchi), pp. 511–515. University of Tokyo Press, Tokyo.

Hildes, D.H.D., Clarke, G.K.C., Flowers, G.E. & Marshall, S.J. (2004) Subglacial erosion and englacial sediment transport modelled for North American ice sheets. *Quaternary Science Reviews*, **23**, 409–430.

Hillaire-Marcel, C., Occhietti, S. & Vincent, J.-S. (1981) Sakami moraine, Quebec: A 500-km-long moraine without climatic control. *Geology*, **9**, 210–214.

Hindmarsh, R.C.A. (1993) Modelling the dynamicsof ice sheets. *Progress in Physical Geography*, **17**, 391–412.

Hindmarsh, R.C.A. (1997) Deforming beds: viscous and plastic scales of deformation: *Quaternary Science Reviews*, 16, 1039–1056.

Hindmarsh, R.C.A. (1999) Pore-water signal of marine ice sheets. *Global and Planetary Change*, **23**, 197–211.

Hindmarsh, R.C.A. (2004) A numerical comparison of approximations to the Stokes equations used in ice sheet and glacier modeling. *Journal of Geophysical Research*, **109**. F01012, doi:10.1039/2003JF000065.

Hindmarsh, R.C.A. & Le Meur, E. (2001) Dynamical processes involved in the retreat of marine ice sheets. *Journal of Glaciology*, **47**, 271–282.

Hindmarsh, R.C., van der Wateren, F.M. & Verbers, A.L. (1998) Sublimation of ice through sediment in Beacon Valley, Antarctica. *Geografiska Annaler*, **80A**, 209–219.

Hiscott, R.N., Aksu, A.E. & Nielsen, O.B. (1989) Provenance and dispersal patterns, Pliocene-Pleistocene section at Site 645, Baffin Bay. In: *Proceedings ODP, Scientific Results, Vol. 105* (Ed. S.K. Stewart), pp. 31–52. Ocean Drilling Program, College Station, TX.

Hjulström, F. (1935) Studies of the morphological activities of rivers as illustrated by the River Fyris. *Bulletin of the Geological Institute University of Uppsala*, **25**, 221–527.

Ho, C.L., Vela, J.C., Jenson, J.W., *et al.* (1996) Evaluation of long-term time-rate parameters of subglacial till. In: *Measuring and Modeling Time Dependent Soil Behavior* (Eds T.C. Sheehan & V.N. Kaliakin), pp. 122–136. Special Publication 61, American Society of Civil Engineers, New York.

Hock, R. (1999) A distributed temperature-index ice- and snowmelt model including potential direct solar radiation. *Journal of Glaciology*, **45**(149), 101–111.

Hock, R. (2003) Temperature index melt modelling in mountain areas. *Journal of Hydrology*, **282**(1–4), 104–115.

Hock, R. & B. Holmgren. (1996) Some aspects of energy balance and ablation of Storglaciären, Northern Sweden. *Geografiska Annaler*, **78A**(2–3), 121–131.

Hodge, S.M. & Doppelhammer, S.K. (1996) Satellite imagery of the onset of streaming flow of Ice Streams C and D, West Antarctica. *Journal of Geophysical Research*, **101**, 6669–6677.

Hodge, S.M., Trabant, D.C., Krimmel, R.M., Heinrichs, T.A., March, R.S. & Josberger, E.G. (1999) Climate variations and changes in mass of three glaciers in western North America. *Journal of Climate*, **11**(9), 2161–2179.

Hodgkins, R., Tranter, M. & Dowdeswell, J.A. (1997) Solute provenance, transport and denudation in a high Arctic glacierised catchment. *Hydrological Processes*, **11**, 1813–1832.

Hodgson, D.A. (1994) Episodic ice streams and ice shelves during retreat of the northwesternmost sector of the Late Wisconsinan Laurentide Ice Sheet over the central Canadian Arctic archipelago. *Boreas*, **23**, 14–28.

Hodson, A.J., Tranter, M. & Vatne, G. (2000) Contemporary rates of chemical weathering and atmospheric CO_2 sequestration in glaciated catchments: an Arctic perspective. *Earth Surface Processes and Landforms*, **25**, 1447–1471.

Hodson, A.J., Mumford, P. & Lister, D. (2004) Suspended sediment and phosphorus in proglacial rivers: bioavailability and potential impacts upon the P status of ice-marginal receiving waters. *Hydrological Processes*, **18**, 2409–2422.

Hodson, A.J., Mumford, P.N., Kohler, J. & Wynn, P.N. (2005). The High Arctic glacial ecosystem: new insights from nutrient budgets. *Biogeochemistry*, **72**, 233–256.

Hoelzle, M., Haeberli, W., Dischl, M. & Peschke, W. (2003) Secular glacier mass balances derived from cumulative glacier length changes. *Global and Planetary Change*, **36**(4), 77–89.

Hoerling, M.P., Hurrell, J.W. & Xu, T. (2001) Tropical origins for recent North Atlantic climate change. *Science*, **292**, 90–92.

Hoerling, M.P., Hurrell, J.W., Xu, T, Bates, G.T. & Phillips, A. (2004) Twentieth century North Atlantic climate change. Part II: Understanding the effect of Indian Ocean warming. *Climate Dynamics.* doi: 10.1007/s00382-004-0433-x

Hoffman, P.F. & Schrag, D.P. (2002) The snowball Earth hypothesis: testing the limits of global change: *Terra Nova*, 14, 129–155.

Hoffman, P.F., Kaufman, A.J., Halverson, G.P. & Schrag, D.P. (1998) A Neoproterozoic snowball Earth: *Science*, 281, 1342–1346.

Hoffmann, G., Werner, M. & Heimann, M. (1998) The Water Isotope Module of the ECHAM Atmospheric General Circulation Model—A study on Time Scales from Days to Several Years. *Journal of Geophysical Research*, **103**, 16871–16896.

Hoinkes, H. (1970) Methoden und Möglichkeiten von Massenhaushaltsstudien auf Gletschern: Ergebnisse der Messreihe Hintereisferner (Ötztaler Alpen) 1953–1968. *Zeitschrift für Gletscherkunde und Glazialgeologie*, **6**(1–2), 37–89.

Holdsworth, G. (1974) *Meserve Glacier, Wright Valley, Antarctica: part 1. Basal Processes.* Report No. 37, Institute of Polar Studies, Ohio State University, Columbus.

Holdsworth, G. & Bull, C. (1970) The flow of cold ice: investigations on Merserve Glacier, Antarctica. In: *International Symposium on Antarctic Glaciological Exploration (ISAGE). Hanover, New Hampshire, 3–7 September 1968*, pp. 204–216. International Association of Hydrological Sciences, Wallingford.

Holland, H.D. (1978) *The Chemistry of the Atmosphere and Oceans.* Wiley, New York, 351 pp.

Holland, M.M. & Bitz, C.M. (2003) Polar amplification of climate change, in coupled models. *Cliate. Dynamics*, **21**, 221–232.

Holloway, G. & Sou, T. (2002) Has Arctic sea ice rapidly thinned? *Journal of Climate*, **15**, 1691–1701.

Holloway, J.M. & Dahlgren, R.A. (2002) Nitrogen in rock: Occurrences and biogeochemical implications. *Global Biogeochemical Cycles*, **16**, 1118, doi:10.1029/2002GB001862.

Holloway, J.M., Dahlgren, R.A., Hansen, B. & Casey, W.H. (1998) Contribution of bedrock nitrogen to high nitrate concentrations in stream water. *Nature*, **395**, 785–788.

Holmlund, P. (1987) Mass balance of Storglaciären during the 20th century. *Geografiska Annaler*, **69A**(3–4), 439–447.

Holmlund, P. & Fuenzalida, H. (1995) Anomalous glacier responses to 20th century climatic changes in Darwin Cordillera, southern Chile. *Journal of Glaciology*, **41**(139), 465–473.

Holtedahl, H. (1958) Some remarks on the geomorphology of continental shelves off Norway, Labrador and southeast Alaska. *Journal of Geology*, **66**, 461–471.

Homer, D.R. & Glen, J.W. (1978) The creep activation energies of ice. *Journal of Glaciology*, **21**, 429–444.

Hondoh, T., Iwamatsu, H. & Mae, S. (1990) Dislocation mobility for non basal glide in ice measured by in situ X-ray topography. *Philosophical Magazine*, **A62**, 89–102.

Hooke, R. & Le B. (1998) *Principles of Glacier Mechanics.* Prentice Hall, Upper Saddle River, New Jersey.

Hooke, R., Le B. & Iverson, N.R. (1985) Experimental study of ice flow around a bump: comparison with theory. *Geografiska Annaler*, **67A**, 187–198.

Hooke, R., Le B. & Iverson, N.R. (1995) Grain-size distribution in deforming subglacial tills: role of grain fracture. *Geology*, **23**, 57–60.

Hooke, R., Le B., Dahlin, B.B. & Kauper, M.T. (1972) Creep of ice containing dispersed fine sand. *Journal of Glaciology*, **11**(63), 327–336.

Hooke, R., Le B., Gould, J.E. & Brzozowski, J. (1983) Near-surface temperatures near and below the equilibrium line on polar and subpolar glaciers. *Zeitschrift für Gletscherkunde und Glazialgeologie*, **19**, 1–25.

Hooke, R., Le B., Hanson, B., Iverson, N.R., Jansson, P. & Fischer, U.H. (1997) Rheology of till beneath Storglaciären, Sweden. *Journal of Glaciology*, **43**, 172–179.

Hooke, R.L. (1973) Structure and flow in the margin of the Barnes Ice Cap, Baffin Island, N.W.T., Canada. *Journal of Glaciology*, **12**, 423–438.

Hooke, R.L. & Clausen, H.B. (1982) Wisconsin and Holocene $\partial^{18}O$ variations, Barnes Ice Cap, Canada. *Geological Society of America Bulletin*, **93**, 784–789.

Hooke, R.L., Pohjola, V.A., Jansson, P. & Kohler, J. (1992) Intraseasonal changes in deformation profiles revealed by borehole studies, Storglaciaren, Sweden. *Journal of Glaciology*, **38**, 348–358.

Hooker, B., Fitzsimons, S. & Morgan, R. (1999) Chemical characteristics and origin of clear basal ice facies in dry-based glaciers, South Victoria Land, Antarctica. *Global and Planetary Change*, **22**, 29–38.

Hooker, B.L. & Fitzharris, B.B. (1999) The correlation between climatic parameters and the retreat and advance of Franz Josef Glacier, New Zealand. *Global and Planetary Change*, **22**, 39–48.

Hooyer, T.S. & Iverson, N.R. (2000a) Clast-fabric development in a shearing granular material: implications for subglacial till and fault gouge. *Geological Society of America Bulletin*, **112**, 683–692.

Hooyer, T.S. & Iverson, N.R. (2000b) Diffusive mixing between shearing granular materials: constraints on bed deformation from till contacts. *Journal of Glaciology*, **46**, 641–651.

Hooyer, T.S. & Iverson, N.R. (2002) Flow mechanisms of the Des Moines lobe of the Laurentide ice sheet. *Journal of Glaciology*, **48**, 575–586.

Hormes, A., Müller, B.U. & Schlüchter, C. (2001) The Alps with little ice: evidence for eight Holocene phases of reduced glacier extent in the Central Swiss Alps. *The Holocene*, **11**(3), 255–265.

Hosein, R. (2002) *Biogeochemical weathering processes in the glacierised Rhône and Upper Oberaar catchments, Switzerland, and the Apure catchment, Venezuela.* Unpublished PhD thesis, University of Neuchâtel.

Hostetler, S.W., Bartlein, P.J., Clark, P.U., Small, E.E. & Solomon, A.M. (2000) Simulated influences of Lake Agassiz on the climate of central North America 11,000 years ago. *Nature*, **405**, 334–337.

Houghton, J.T., Ding, Y., Griggs, D.J., *et al.* (2001) *Climate Change, 2001: The Scientific Basis.* Cambridge University Press, Cambridge.

Houmark-Nielsen, M. (1999) A lithostratigraphy of Weichselian glacial and interstadial deposits in Denmark. *Bulletin of the Geological Society of Denmark*, **41**, 181–202.

Hubbard, A. (2000) The verification and significance of three approaches to longitudinal stresses in high-resolution models of glacier flow. *Geografiska Annaler Series A–Physical Geography*, **82A**, 471–487.

Hubbard, B. (2002) Direct measurement of basal motion at a hard-beddEd. temperate glacier: Glacier de Tsanfleuron, Switzerland. *Journal of Glaciology*, **48**(160), 1–8.

Hubbard, B. & Maltman, A.J. (2000) Laboratory investigations of the strength, static hydraulic conductivity and dynamic hydraulic conductivity of glacial sediments. In: *Deformation of Glacial Material* (Eds A.J. Maltman, M.J. Hambrey & B. Hubbard), pp. 231–242. Special Publication 176, Geological Society Publishing House, Bath.

Hubbard, B. & Sharp, M. (1989) Basal ice formation and deformation: a review. *Progress in Physical Geography*, **13**, 529–558.

Hubbard, B. & Sharp, M. (1993) Weertman regelation, multiple refreezing events, and the isotopic evolution of the basal ice layer. *Journal of Glaciology*, **39**, 275–291.

Hubbard, B. & Sharp, M. (1995) Basal ice facies and their formation in the Western Alps. *Arctic and Alpine Research*, **27**(4), 301–310.

Hubbard, B.P., Sharp, M.J., Willis, I.C., Nielsen, M.K. & Smart, C.C. (1995) Borehole water level variations and the structure of the subglacial hydrological system of Haut Glacier d'Arolla, Valais, Switzerland. *Journal of Glaciology*, **41**(139), 572–583.

Hubbard, B., Sharp, M. & Lawson, W.J. (1996) On the sedimentological character of Alpine basal ice facies. *Annals of Glaciology*, **22**, 187–193.

Hubbard, A., Blatter, H., Nienow, P., Mair, D. & Hubbard, B. (1998) Comparison of a three-dimensional model for glacier flow with field data from Haut Glacier d'Arolla, Switzerland. *Journal of Glaciology*, **44**(147), 368–378.

Hubbard, B., Siegert, M. & McCarroll, D. (2000a) Spectral roughness of glaciated bedrock surfaces: implications for glacier sliding. *Journal of Geophysical Research*, **105**(B9), 21295–21303.

Hubbard, B., Tison, J., Janssens, L. & Spiro, B. (2000b) Ice-core evidence of the thickness and character of clear-facies basal ice: Glacier de Tsanfleuron, Switzerland. *Journal of Glaciology*, **46**(152), 140–150.

Hubbard, B., Hubbard, A., Mader, H.M., Tison, J.-L., Grust, K. & Nienow, P. (2003) Spatial variability in the water content and rheology of temperate glaciers: Glacier de Tsanfleuron, Switzerland. *Annals of Glaciology*, **37**, 1–6.

Hubbard, B., Hubbard, A., Mader, H., Tison, J.-L., Grust, K. & Nienow, P. (2004) Spatial variability in the water content and rheology of temperate glaciers: Glacier de Tsanflueron, Switzerland. *Annals of Glaciology*, **37**, 1–6.

Huber, C., Leuenberger, M. & Zumbrunnen, O. (2003) Continuous extraction of trapped air from bubble ice or water for on-line determination of isotope ratios. Analytical Chemistry **75**, 2324–2332.

Huddart, D. (1991) The glacial history and glacial deposits of the North and West Cumbrian lowlands. In: *Glacial Deposits of Britain and Ireland* (Eds J. Ehlers, P.L. Gibbard & J. Rose), pp. 151–168. Balkema, Rotterdam.

Hughen, K.A., Overpeck, J.T., Trumbore, S. & Peterson, L.C. (1996) Rapid climate changes in the tropical Atlantic region during the last deglaciation. *Nature*, **380**, 51–54.

Hughes, M.K. & Diaz, H.F. (1994) Was there a 'Medieval Warm Period', and if so, where and when? In: *Medieval Warm Period* (Eds M.K. Hughes & H.F. Diaz), pp. 109–142. Kluwer, Boston.

Hughes, T. (1977) West Antarctic Ice Streams. *Reviews of Geophysics and Space Physics*, **15**, 1–46.

Hughes, T. (1992) Abrupt climatic change related to unstable ice-sheet dynamics: toward a new paradigm. *Palaeogeography, Palaeoclimatology, Palaeoecology*, **97**, 203–234.

Hughes, T.J. (1981a) The weak underbelly of the West Antarctic Ice Sheet. *Journal of Glaciology*, **27**(97), 518–525.

Hughes, T.J. (1981b) Numerical reconstruction of paleo-ice sheets. In: *The Last Great Ice Sheets* (Eds G.H. Denton & T.J. Hughes), pp. 221–261. Wiley, New York.

Hughes, T.J. (1998) *Ice sheets*. Oxford University Press, New York. 343 pp.

Hulbe, C.L. & MacAyeal, D.R. (1999) A new numerical model of coupled inland ice sheet, ice stream, and ice shelf flow and its application to the West Antarctic ice sheet. *Journal of Geophysical Research*, **104**(B11), 25349–25366.

Hulbe, C.L., Wang, W.L., Joughin, I.R. & Siegert, M.J. (2003) The role of lateral and vertical shear in tributary flow toward a West Antarctic ice stream. *Annals of Glaciology*, **36**(36), 244–250.

Hulbe, C.L., MacAyeal, D.R., Denton, G.H., Kleman, J. & Lowell, T.V. (2004) Catastrophic ice shelf breakup as the source of Heinrich event icebergs. *Paleoceanography*, **19**, 1. PA1004 10.1029/2003PA000890.

Hulton, N.R.J. & Sugden, D.E. (1995) Modelling mass balance on former maritime ice caps: a Patagonian example. *Annals of Glaciology*, **21**, 304–310.

Humlum, O. (1981) Observations on debris in the basal transport zone of Myrdalsjokull, Iceland. *Annals of Glaciology*, **2**, 71–77.

Humphrey, N., Kamb, B., Fahnestock, M. & Engelhardt, H. (1993) Characteristics of the bed of the lower Columbia Glacier, Alaska. *Journal of Geophysical Research*, **98**(B1), 837–846.

Humphreys, K.A. & Fitzsimons, S.J. (1996) Landform and sediment associations of dry-based glaciers in polar arid environments. *Zeitschrift fur Geomorphologie*, **105**, 21–33.

Hurrell, J.W. (1995) Decadal trends in the North Atlantic Oscillation: regional temperatures and precipitation. *Science*, **269**, 676–679.

Hurrell, J.W. (1996) Influence of variations in extratropical wintertime teleconnections of Northern Hemisphere temperature. *Geophysics Research Letters*, **23**, 665–668.

Hurrell, J.W., Kushnir, Y., Ottersen, G. & Visbeck, M. (2003) An overview of the North Atlantic Oscillation. In: *The North Atlantic Oscillation: Climate Significance and Environmental Impact* (Eds J.W. Hurrell, Y. Kushnir, G. Ottersen & M. Visbeck), pp. 1–35. Geophysical Monograph 134, American Geophysical Union, Washington, DC.

Hurrell, J.W., Hoerling, M.P., Phillips, A. & Xu, T. (2004) Twentieth century North Atlantic climate change. Part I: assessing determination. *Climate Dynamics*. doi:10.1007/s00382-004-0432-x.

Hutter, K. (1983) *Theoretical Glaciology; Material Science of Ice and the Mechanics of Glaciers and Ice Sheets*. D. Reidel, Dordrecht.

Huuse, M. & Lykke-Andersen, H. (2000) Overdeepened Quaternary valleys in the eastern Danish North Sea: morphology and origin. *Quaternary Science Reviews*, **19**, 1233–1253.

Huybrechts, P. (1992) *The Antarctic Ice Sheet and Environmental Change: a Three-dimensional Modeling Study*. Berichte zur Polarfoschung, 99, Alfred-Wegener-Institut für Polar- und Meeresforschung, Bremerhaven, 241 p.

Huybrechts, P. (1993) Glaciological modelling of the late Cenozoic East Antarctic ice sheet: stability or dynamism? *Geografiska Annaler*, **75A**, 221–238.

Huybrechts, P. (1994a) Formation and disintegration of the Antarctic ice sheet. *Annals of Glaciology*, **20**, 336–340.

Huybrechts, P. (1994b) The present evolution of the Greenland ice sheet: an assessment by modelling. *Global and Planetary Change*, **9**, 39–51.

Huybrechts, P. (1996) Basal temperature conditions of the Greenland ice sheet during the glacial cycles. *Annals of Glaciology*, **23**, 226–236.

Huybrechts, P. (2002) Sea-level changes at the LGM from ice-dynamic reconstructions of the Greenland and Antarctic ice sheets during the glacial cycles. *Quaternary Science Reviews*, 21, 203–231.

Huybrechts, P. & de Wolde, J. (1999) The dynamic response of the Greenland and Antarctic ice sheets to multiple-century climatic warming. *Journal of Climate*, **12**, 2169–2188.

Huybrechts, P. & Le Meur, E. (1999) Predicted present-day evolution patterns of ice thickness and bedrock elevation over Greenland and Antarctica. Polar Research **18**(2), 299–308.

Huybrechts, P. & Oerlemans, J. (1988) Thermal regime of the East Antarctic Ice Sheet: a numerical study on the role of the dissipation-strain rate feedback with changing climate. *Annals of Glaciology*, **11**, 52–59.

Huybrechts, P. & T'siobbel, S. (1995) Thermomechanical modelling of Northern Hemisphere ice sheets with a two-level mass-balance parameterization. *Annals of Glaciology*, **21**, 111–116.

Huybrechts, P., Letreguilly, P. & Reeh, N. (1991) The Greenland ice sheet and global warming. *Palaeogeography, Plaeoclimatology, Palaeoecology (Global and Planetary Change Section)*, **89**, 399–412.

Huybrechts, P., Steinhage, D., Wilhelms, F. & Bamber, J.L. (2000) Balance velocities and measured properties of the Antarctic ice sheet from a new compilation of gridded datasets for modeling. *Annals of Glaciology*, **30**, 52–60.

Huybrechts, P., Janssens, I., Poncin, C. & Fichefet, T. (2002) The response of the Greenland ice sheet to climate changes in the 21st century by interactive coupling of an AOGCM with a thermomechanical ice-sheet model. *Annals of Glaciology*, **35**, 409–415.

Huybrechts, P., Rybak, O.O., Pattyn, F. & Steinhage, D. (2004a) Age and origin of the ice in the EPICA DML ice core derived from a nested Antarctic ice sheet model. *Geophysical Research Abstracts*, **6**, 06533 (CD-ROM).

Huybrechts, P., Gregory, J.M., Janssens, I. & Wild, M. (2004b) Modelling Antarctic and Greenland volume changes during the 20th and 21st centuries forced by GCM time slice integrations. *Global and Planetary Change*, **42**, 83–105, doi:10.1016/j.gloplacha.2003.11.011.

IAHS(ICSI)–UNEP–UNESCO (1988). *Fluctuations of Glaciers 1980–1985* (Eds W. Haeberli & P. Müller). World Glacier Monitoring Service, Paris.

IAHS(ICSI)–UNEP–UNESCO (1989) *World Glacier Inventory–Status 1988–* (Eds W. Haeberli, H. Bosch, K. Scherler, G. Ostrem & C.C. Wallen). World Glacier Monitoring Service, Paris. 368 pp.

IAHS(ICSI)–UNEP–UNESCO (1993) *Fluctuations of Glaciers 1985–1990* (Eds W. Haeberli & M. Hoelzle). World Glacier Monitoring Service, Paris.

IAHS(ICSI)–UNEP–UNESCO (1998) *Fluctuations of Glaciers 1990–1995.* (Eds W. Haeberli, M. Hoelzle, S. Suter & R. Frauenfelder). World Glacier Monitoring Service, Paris.

Ibarzabal y Donangelo, T., Hoffmann, J.A.J. & Naruse, R. (1996) Recent climate changes in southern Patagonia. *Bulletin of Glacier Research*, **14**, 29–36.

Ice Core Working Group (U.S.) (1989) *U.S. Global Ice Core Research Program: West Antarctica and Beyond* (Ed. P. Grootes). Quaternary Research Center, University of Washington, Seattle.

Iken, A. (1977) Movement of large ice mass before breaking off. *Journal of Glaciology*, **19**(81), 595–605.

Iken, A. & Bindschadler, R.A. (1986) Combined measurements of subglacial water pressure and surface velocity of Findelengletscher, Switzerland: conclusions about drainage system and sliding mechanism. *Journal of Glaciology*, **32**(110), 101–119.

Iken, A., Röthlisberger, H. & Hutter, K. (1977) Deep drilling with a hot water jet. *Zeitschrift für Gletscherkunde und Glazialgeologie*, **12**(2), 143–156.

Iken, A., Röthlisberger, H., Flotron, A. & Haeberli, W. (1983) The uplift of Unteraargletscher at the beginning of the melt season—a consequence of water storage at the bed? *Journal of Glaciology*, **29**(101), 28–47.

Iken, A., Fabri, K. & Funk, M. (1996) Water storage and subglacial drainage conditions inferred from borehole measurements on Gornernletscher, Valais, Switzerland. *Journal of Glaciology*, **42**, 233–248.

Imbrie, J.Z., Hays, J.D., Martinson, D.G., MacIntyre, A., *et al.* (1984) The orbital theory of Pleistocene climate: support from a revised chronology of the marine d18O record. In: *Milankovitch and Climate* (Eds A. Berger, J.Z. Imbrie, J.D. Hays, *et al.*), pp. 269–305. D. Reidel, Dordrecht.

Imbrie, J., Boyle, E.A., Clemens, S.C., *et al.* (1992) On the structure and origin of major glaciation cycles, 1, Linear responses to Milankovitch forcing. *Paleoceanography*, **7**, 701–738.

Ingolfsson, O., Hjort, C., Berkman, P.A., Björck, S., *et al.* (1998) Antarctic glacial history since the Last Glacial Maximum: an overview of the record on land. *Antarctic Science*, **10**, 326–344.

IPCC (2001a) *Climate Change 2001: the Scientific Basis* (Eds J.T. Houghton, Y. Ding, D.J. Griggs, *et al.*). Contribution of Working Group I to the Third Assessment Report of the Intergovernmental Panel on Climate Change, Cambridge University Press, Cambridge, 881 pp.

IPCC (2001b) *Climate Change 2001: Impacts, Adaptation, and Vulnerability* (Eds J.J. McCarthy, O.F. Canziani, N.A. Leary, D.J. Dokken & K.S. White). Contribution of Working Group II to the Third Assessment Report of the Intergovernmental Panel on Climate Change, Cambridge University Press, Cambridge, 1000 pp.

IPCC (2001c) *Summary for Policymakers—Climate Change 2001: the Scientific Basis.* Intergovernmental Panel on Climate Change, Geneva.

Iverson, N.R. (1990) Laboratory simulations of glacial abrasion: comparison with theory. *Journal of Glaciology*, **32**, 304–314.

Iverson, N.R. (1991) Potential effects of subglacial water-pressure fluctuations on quarrying. *Journal of Glaciology*, **37**, 27–36.

Iverson, N.R. (1993) Regelation of ice through debris at glacier beds: Implications for sediment transport. *Geology*, **21**, 559–562.

Iverson, N.R. (1999) Coupling between a glacier and a soft bed: II. Model results. *Journal of Glaciology*, **45**, 41–53.

Iverson, N.R. & Hooyer, T.S. (2002) Clast-fabric development in a shearing granular material: Implications for subglacial till and fault gouge: Reply. *Geological Society of America Bulletin*, **114**, 383–384.

Iverson, N.R. & Iverson, R.M. (2001) Distributed shear of subglacial till due to Coulomb slip. *Journal of Glaciology*, **47**, 481–488.

Iverson, N.R. & Semmens, D. (1995) Intrusion of ice into porous media by regelation: A mechanism of sediment entrainment by glaciers. *Journal of Geophysical Research*, **100**, 10219–10230.

Iverson, N.R., Jansson, P. & Hooke, R.L. (1994) In-situ measurements of strength of deforming subglacial till. *Journal of Glaciology*, **40**, 497–503.

Iverson, N.R., Hanson, B., Hooke, R.LeB. & Jansson, P. (1995) Flow mechanism of glaciers on soft beds. *Science*, **267**(5194), 80–81.

Iverson, N.R., Hooyer, T.S. & Hooke, R. LeB. (1996) A laboratory study of sediment deformation: Stress heterogeneity and grain-size evolution. *Annals of Glaciology*, **22**, 167–175.

Iverson, N.R., Baker, R.W. & Hooyer, T.S. (1997) A ring-shear device for the study of sediment deformation: tests on tills with contrasting clay contents. *Quaternary Science Reviews*, **16**, 1057–1066.

Iverson, N.R., Hooyer, T.S. & Baker, R.W. (1998) Ring-shear studies of till deformation: Coulomb-plastic behaviour and distributed strain in glacier beds. *Journal Glaciology*, **44**(148), 634–642.

Iverson, N.R., Baker, R.W., Hooke, R.LeB., Hanson, B., Jansson, P. (1999) Coupling between a glacier and a soft bed: I. A relation between effective pressure and local shear stress determined from till elasticity. *Journal of Glaciology*, **45**(149), 31–40.

Iverson, N.R., Cohen, C., Hooyer, T.S., *et al.* (2003) Effects of basal debris on glacier flow. *Science*, **301**(5629), 81–84.

Ives, J.D. (1957) Glaciation of the Torngat Mountains Northern Labrador. Arctic **10**, 66–87.

Ives, J.D. (1962) Indications of recent extensive glacierization in north-central Baffin Island, N.W.T. *Journal of Glaciology*, **4**, 197–205.

Ives, J.D. & Andrews, J.T. (1963) Studies in the physical geography of north central Baffin Island. *Geographical Bulletin*, **19**, 5–48.

Ivy-Ochs, S., Schlüchter, C., Kubik, P., Dittrich-Hannen, B. & Beer, J. (1995) Minimum ^{10}Be exposure ages of early Pliocene for the Table Mountain plateau and the Sirius Group at Mount Fleming, Dry Valleys, Antarctica. *Geology*, **23**, 1007–1019.

Jacka, T.H. (1984a) Laboratory studies on relationships between ice crystal size and flow rate. *Cold Regions Science and Technology*, **10**(1), 31–42.

Jacka, T.H. (1984b) The time and strain required for development of minimum strain rates in ice. *Cold Regions Science and Technology*, **8**, 261–268.

Jacka, T.H. (1994) Investigations of discrepancies between laboratory studies of the flow of ice: density, sample shape and size, and grain-size. *Annals of Glaciology*, **19**, 146–154.

Jacka, T.H. & Li, J. (1994) The steady-state crystal size of deforming ice. *Annals of Glaciology*, **20**, 13–18.

Jacka, T.H. & Li, J. (2000) Flow rates and crystal orientation fabrics in compression of polycrystalline ice at low temperatures and stresses. In: *Physics of Ice Core Records* (Ed. T. Hondoh), pp. 83–102, Hokkaido University Press, Sopporo.

Jacka, T.H. & Maccagnan, M. (1994) Ice crystallographic and strain rate changes with strain in compression and extension. *Cold Regions Science and Technology*, **8**, 269–286.

Jacka, T.H., Donoghue, S., Li, J., Budd, W.F. & Anderson, R.M. (2003) Laboratory studies on the flow rates of debris-laden ice. *Annals of Glaciology*, **37**, 108–112.

Jackson, M. & Kamb, B. (1997) The marginal shear stress of Ice Stream B, West Antarctica. *Journal of Glaciology*, **43**, 415–426.

Jacobel, R.W., Scambos, T.A., Raymond, C.F. & Gades, A.M. (1996) Changes in the configuration of ice stream flow from the West Antarctic ice sheet. *Journal of Geophysical Research*, **101**(B3), 5499–5504.

Jacobel, R.W., Scambos, T.A., Nereson, N.A. & Raymond, C.F. (2000) Changes in the margin of Ice Stream C, Antarctica. *Journal of Glaciology*, **46**, 102–110.

Jacobs, S.S., Helmer, H.H., Doake, C.S.M., Jenkins, A. & Frolich, R.M. (1992) Melting of ice shelves and the mass balance of Antarctica. *Journal of Glaciology*, **38**, 375–387.

Jacobs, S.S., Hellmer, H.H. & Jenkins, A. (1996) Antarctic ice sheet melting in the Southeast Pacific. *Geophysical Research Letters*, **23**(9), 957–960.

Jacobson, H.P. & Raymond, C.F. (1998) Thermal effects on the location of ice stream margins. *Journal of Geophysical Research*, **103**(B6), 12111–12122.

Jaeger, J.B., Hallet, B., Pavlis, T., *et al.* (2001) Orogenic and glacial research in pristine southern Alaska. *Eos* (*Transactions of the American Geophysical Union*), **82**, 213–216.

Jansen, E. & Sjøholm, J. (1991) Reconstruction of glaciation over the past 6 Myr from ice-borne deposits in the Norwegian Sea. *Nature*, **349**, 600–603.

Janssens, I. & Huybrechts, P. (2000) The treatment of meltwater retention in mass-balance parameterizations of the Greenland ice sheet. *Annals of Glaciology*, **31**, 133–140.

Jansson, K.N. (2003) Early Holocene glacial lakes and ice marginal retreat pattern in Labrador/Ungava, Canada. *Palaeogeography, Palaeoclimatology, Palaeoecology*, **193**, 473–501.

Jansson, K.N., Stroeven, A.P. & Kleman, J. (2003) Configuration and timing of Ungava Bay ice streams, Labrador-Ungava, Canada. *Boreas*, **32**, 252–262.

Jansson, P. (1995) Water pressure and basal sliding on Storglaciären, northern Sweden. *Journal of Glaciology*, **41**(138), 232–240.

Jansson, P., Hock, R. & Schneider, Th. (2003) The concept of glacier storage: a review. *Journal of Hydrology*, **282**, 116–129.

Jarvis, E.P. & King, E.C. (1993) The seismic wavefield recorded on an Antarctic ice shelf. *Journal of Seismic Exploration*, **134**, 69–86.

Jeffery, G.B. (1922) The motion of ellipsoidal particles immersed in a viscous fluid. *Proceedings of the Royal Society of London, Series A*, **102**, 169–179.

Jellinek, H.H.G. (1959) Adhesive properties of ice. *Journal of Colloid Science*, **14**, 268–280.

Jenkins, A., Vaughan, D., Jacobs, S.S., Hellmer, H.H. & Keys, H. (1997) Glaciological and oceanographic evidence of high melt rates beneath Pine Island Glacier, West Antarctica, *Journal of Glaciology*, **43**(143), 114–121.

Jennings, A.E. (1993) The Quaternary History of Cumberland Sound, Southeastern Baffin Island: the Marine Evidence. *Geographie physique et Quaternaire*, **47**, 21–42.

Jennings, A.E., Tedesco, K.A., Andrews, J.T. & Kirby, M.E. (1996) Shelf erosion and glacial ice proximity in the Labrador Sea during and after Heinrich events (H-3 or 4 to H-0) as shown by formanifera. In: *Late Quaternary Palaeoceanography of the North Atlantic Margins* (Eds J.T. Andrews, W.E.N. Austin, H. Bergsten & A.E. Jennings), pp. 29–49. Special Publication 111, Geological Society Publishing House, Bath.

Jennings, A.E., Manley, W.F., MacLean, B. & Andrews, J.T. (1998) Marine evidence for the last glacial advance across Eastern Hudson Strait, Eastern Canadian Arctic. *Journal of Quaternary Science*, **13**, 501–514.

Jennings, A.E., Grönvold, K., Hilberman, R., Smith, M. & Hald, M. (2002a) High resolution study of Icelandic tephras in the Kangerlussuaq Trough, SE East Greenland, during the last deglaciation. *Journal of Quaternary Science*, **17**, 747–757.

Jennings, A.E., Knudsen, K.L., Hald, M., Hansen, C.V. & Andrews, J.T. (2002b) A mid-Holocene shift in Arctic sea ice variability on the East Greenland shelf. *The Holocene*, **12**, 49–58.

Jenssen, D. (1977) A three-dimensional polar ice sheet model. *Journal of Glaciology*, **18**, 373–389.

Jing, Z., Ye, B., Jiao, K., *et al.* (2002) Surface velocity on the Glacier No.51 at Haxilegen of the Kuytun River, Tianshan Mountains. *Journal of Glaciology and Geocryology*, **24**(5), 563–566.

Jiskoot, H., Murray, T. & Luckman, A. (2003 Surge potential and drainage basin characteristics in East Greenland. *Annals of Glaciology*, **36**, 142–148.

Jóhannesson, T. (1992) *Landscape of temperate ice caps*. PhD thesis, University of Washington.

Jóhannesson, T. (1997) The response of two Icelandic glaciers to climatic warming computed with a degree-day glacier mass-balance model coupled to a dynamic glacier model. *Journal of Glaciology*, **43**(144), 321–327.

Jóhannesson, T., Raymond, C.F. & Waddington, E.D. (1989) Time-scale for adjustment of glaciers to changes in mass balance. *Journal of Glaciology*, **35**(121), 355–369.

Jóhannesson, T., Sigurdsson, O., Laumann, T. & Kennett, M. (1995) Degree-day glacier mass-balance modelling with application to glaciers in Iceland, Norway, and Greenland. *Journal of Glaciology*, **41**, 345–358.

Johnsen, S.J. & White, J.W.C. (1989) The origin of Arctic precipitation under present and glacial conditions. *Tellus*, **41B**, 452–468.

Johnsen, S.J., W.Dansgaard, W. & White, J.W.C. (1989) The origin of Arctic precipitation under present and glacial conditions. *Tellus*, **41B**, 452–468.

Johnsen, S.J., Clausen, H.B., Dansgaard, W., *et al.* (1992) Irregular glacial interstadials recorded in a new Greenland ice core. *Nature*, **359**, 311–313.

Johnsen, S.J., Dahl-Jensen, D., Gundestrup, N., *et al* (2001) Oxygen isotope and paleotemperature records from six Greenland ice-core stations: Camp Century, Dye–3, GRIP, GISP2, Renland, and North GRIP. *Journal of Quaternary Science*, **16**(4), 299–307.

Johnson, J. & Fastook, J.L. (2002) Northern Hemisphere glaciation and its sensitivity to basal melt water. *Quaternary International*, **95–96**, 65–74.

Johnson, R.G. (1997) Ice age initiation by an ocean-atmospheric circulation change in the Labrador Sea. *Earth and Planetary Science Letters*, **148**, 367–379.

Johnson, R.H. (1965) Glacial geomorphology of the west Pennine slopes between Cliviger and Congleton. In: *Essays in Geography for A.A. Miller* (Eds J.B. Whittow & P.D. Wood), pp. 58–94. Reading University Press, Reading.

Jones, G.A. & Keigwin, L.D. (1988) Evidence from the Fram Strait (78 N) for early deglaciation. *Nature*, **336**, 56–59.

Jones, I.W., Munhoven, G., Tranter, M., Huybrechts, P. & Sharp, M.J. (2002) Modelled glacial and non-glacial HCO_3^-, Si and Ge fluxes since the LGM: little potential for impact on atmospheric CO_2 concentrations and the marine Ge:Si ratio. *Global Planetary Change*, **33**, 139–153.

Jones, S. & Glen, J.W. (1969) The effect of dissolved impurities on the mechanical properties of ice crystals. *Philosophical Magazine*, **8**(19), 13–24.

Jónsson, S. (1992) Variability of convective conditions in the Greenland Sea. *ICES Marine Science Symposium*, **195**, 32–39.

Jordan, C. (1997) Quaternary geology mapping in the Republic of Ireland—how much can be added through remote sensing? In: *Proceedings of the Twelfth International Conference and Workshop on Applied Geologic Remote Sensing17–19/11/97, Denver, Colorado, USA*, Vol. 2, pp. 37–44.

Jørgensen, F. & Piotrowski, J.A. (2003) Signature of the Baltic Ice Stream on Funen Island, Denmark during the Weichselian glaciation. *Boreas*, **32**, 242–256.

Joughin, I. (2002) Ice-sheet velocity mapping: a combined interferometric and speckle-tracking approach. *Annals of Glaciology*, **34**, 195–201.

Joughin, I. & Tulaczyk, S. (2002) Positive mass balance of the Ross Ice Streams, West Antarctica. *Science*, **295**(5554), 476–480.

Joughin, I., Tulaczyk, S., Fahnestock, M. & Kwok, R. (1996) A mini-surge on the Ryder Glacier, Greenland, observed by satellite radar interferometry. *Science*, **274**(5285), 228–230.

Joughin, I., Fahnestock, M., Ekholm, S. & Kwok, R. (1997) Balance velocities of the Greenland ice sheet. *Geophysical Research Letters*, **24**, 3045–3048.

Joughin, I.R., Kwok, R. & Fahnestock, M.A. (1998) Interferometric estimation of three-dimensional ice-flow using ascending and descending passes. *IEEE Transactions on GeoScience and Remote Sensing*, **36**(1), 25–37.

Joughin, L., Gray, L., Bindschadler, R., *et al.* (1999) Tributaries of West Antarctic ice streams revealed by RADARSAT interferometry. *Science*, **286**(5438), 283–286.

Joughin, I.R., Fahnestock, M.A. & Bamber, J.L. (2000) Ice flow in the northeast Greenland ice stream. *Annals of Glaciology*, **31**, 141–146.

Joughin, I., Fahnestock, M., MacAyeal, D., Bamber, J.L. & Gogineni, P. (2001) Observation and analysis of ice flow in the largest Greenland ice stream. *Journal of Geophysical Research—Atmospheres* **106**(D24), 34021–34034.

Joughin, I., Tulaczyk, S., Bindschadler, R.A. & Price, S. (2002) Changes in west Antarctic ice stream velocities: observation and analysis. *Journal of Geophysical Research*, **107**, 2289. doi:10.1029/2001JB001029.

Joughin, I.R., Tulaczyk, S. & Engelhardt, H.E. (2003a) Basal melt beneath Whillans Ice Stream and Ice Streams A and C. *Annals of Glaciology*, **36**, 257–262.

Joughin, I., Rignot, E., Rosanova, C.E., Lucchitta, B.K. & Bohlander, J. (2003b) Timing of recent accelerations of Pine Island Glacier, Antarctica. *Geophysical Research Letters*, **30**(13). art. no.–1706.

Joughin, I., Macayeal, D.R. & Tulaczyk, S. (2004) Melting and freezing beneath the Ross ice streams, Antarctica. *Journal of Glaciology*, **50**, 96–108.

Journal of Geophysical Research (1997) Greenland Summit Ice Cores. **102**. NO. C12.

Jouzel, J. & Koster, R.D. (1997) A reconsideration of the initial conditions used for stable water isotope models. *Journal of Geophysical Research*, **101**, 22933–22938.

Jouzel, J. & Merlivat, L. (1984) Deuterium and oxygen 18 in precipitation: modeling of the isotopic effects during snow formation. *Journal of Geophysical Research*, **89**(D7), 11749–11757.

Jouzel, J., Merlivat, L. & Lorius, C. (1982) Deuterium excess in an East Antarctic ice core suggests higher relative humidity at the oceanic surface during the last glacial maximum. *Nature*, **299**, 688–691.

Jouzel, J., Russell, G.L., Suozzo, R.J., Koster, R.D., White, J.W.C. & Broecker, W.S. (1987) Simulations of the HDO and ^{18}O atmospheric cycles using the NASA GISS general circulation model: the seasonal cycle for present-day conditions. *Journal of Geophysical Research*, **92**, 14739–14760.

Jouzel, J., Petit, J.R., Souchez, R., *et al.* (1999) More than 200 meters of lake ice above subglacial Lake Vostok, Antarctica.. *Science*, **286**, 2138–2141.

Jouzel, J., Vimeux, F., Caillon, N., *et al.* (2003a) Temperature reconstructions from Antarctic ice cores. *Journal of Geophysical Research*, **108**(D12), 4361, doi:10.1029/2002JD002677.

Jouzel, J., Vimeux, F., Caillon, N., *et al.* (2003b) Magnitude of isotope/temperature scaling for interpretation of central Antarctic ice cores. *Journal of Geophysical Research*, **108**(D12), 4361. doi:10.1029/2002JD002677.

Juen, I., Kaser, G. & Georges, C. (2003) Modelling observed and future runoff from a glacierized tropical catchment (Cordillera Blanca, Perú). In: *Proceedings of the Symposium on Andean Mass Balance*, Valdivia, March.

Jull, M. & McKenzie, D. (1996) The effect of deglaciation on mantle melting beneath Iceland. *Journal of Geophysical Research*, **101**(B10), 21815–21828.

Kääb, A. (2002) Monitoring high-mountain terrain deformation from air- and spaceborne optical data: examples using digital aerial imagery and ASTER data. *ISPRS Journal of Photogrammetry and Remote Sensing*, **57**(1–2), 39–52.

Kääb, A., Paul, F., Maisch, M., Hoelzle, M. & Haeberli, W. (2002) The new remote-sensing-derived Swiss glacier inventory: II. First results. *Annals of Glaciology*, **34**, 362–366.

Kääb, A., Wessels, R., Haeberli, W., *et al.* (2003) Rapid ASTER imaging facilitates timely assessment of glacier hazards and disasters. *Eos (Transactions of the American Geophysical Union)*, **84**(13), 117–124.

Kachanov, L.M. (1999) Rupture time under creep conditions (translated from Russian, 1957). *International Journal of Fracture*, **97**(1–4), xi–xviii.

Kalnay, E., Kanamitsu, M., Kistler, R., *et al.* (1996) The NCEP/NCAR 40-year reanalysis project. *Bulletin of the American Meteorological Society*, **77**, 437–471.

Kamb, B. (1972) Experimental recrystallization of ice under stress. In: *Flow and Fracture of Rocks* (Eds H.C. Heard, I.Y. Borg, N.L., Carter & C.B. Raileigh), pp. 211–241.AGU Monograph Series 16, American Geophysical Union, Washington, DC.

Kamb, B. (1987) Glacier surge mechanism based on linked cavity configuration of the basal water conduit system. *Journal of Geophysical Research*, **92**, 9083–9100.

Kamb, B. (1991) Rheological nonlinearity and flow instability in the deforming bed mechanism of ice stream motion. *Journal of Geophysical Research*, **96**(B10), 16585–16595.

Kamb, B. (2001) Basal zone of the West Antarctic ice streams and its role in lubrication of their rapid motion. In: *The West Antarctic Ice Sheet: Behavior and Environment* (Eds R.B. Alley & R. Bindschadler), pp. 157–199. Volume 77, American Geophysical Union, Washington, DC.

Kamb, B. & LaChapelle, E. (1964). Direct observation of the mechanism of gacier sliding over bedrock. *Journal of Glaciology*, **5**, 159–172.

Kamb, B., Raymond, C.F., Harrison, W.D., *et al.* (1985) Glacier surge mechanism—1982–1983 surge of Variegated Glacier, Alaska. *Science*, **227**, 469–479.

Kaplan, M.R., Miller, G.H. & Steig, E.J. (2001) Low-gradient outlet glaciers (ice streams?) drained the Laurentide Ice Sheet. *Geology*, **29**, 343–346.

Karl, D.M. & Tien, G. (1992) MAGIC: A sensitive and precise method for measuring dissolved phosphorus in aquatic environments. *Limnology and Oceanography*, **37**, 105–115.

Karlen, W. (1988) Scandinavian glacial and climatic fluctuations during the Holocene. *Quaternary Science Reviews*, **7**, 199–209.

Karlen, W. & Denton, G.H. (1976) Holocene glacial variations in Sarek National Park, northern Sweden. *Boreas*, **5**, 25–56.

Karrow, P.F. & Calkin, P.E. (Eds) (1985) *Quaternary Evolution of the Great Lakes*. Special Paper 30, Geological Association of Canada, St John's, NF, 258 pp.

Kaser, G., Ames, A. & Zamora, M. (1990) Glacier fluctuations and climate in the Cordillera Blanca, Peru. *Annals of Glaciology*, **14**, 136–140.

Kaser, G., Fountain, A. & Jansson, P. (2002) *A Manual for Monitoring the Mass Balance of Mountain Glaciers*. International Hydrological Programme, Technical Documents in Hydrology 59, Unesco, Paris.

Kaser, G., Juen, I., Georges, C., Gomez, J. & Tamayo, W. (2003) The impact of glaciers on the runoff and the reconstruction of mass balance history from hydrological data in the tropical Cordillera Blanca, Peru. *Journal of Hydrology*, **282**, 130–144.

Kaspi, Y., Sayag, R. & Tziperma, E (2004) A 'triple sea-ice state' mechanism for the abrupt warming and synchronous ice sheet collapses during Heinrich events. *Paleoceanography*, **19**. PA3004, doi:10.1029/2004PA001009.

Kasser, P. (1967) *Fluctuations of Glaciers 1959–1965*. IAHS(ICSI)–UNESCO, Paris.

Kasser, P. (1973) *Fluctuations of Glaciers 1965–1970*. IAHS(ICSI)–UNESCO, Paris.

Kaufman, D.S., Ager, T.A., Anderson, N.J., *et al.* (2004) Holocene thermal maximum in the western Arctic (0–180°W). *Quaternary Science Reviews*, **23**, 529–560.

Kavanaugh, J. & Cuffey, K.M. (2003) Sapace and time variation of d18O and dD in Antarctic precipitation revisited. *Global Biogeochemical Cycles*, **17**(1), 1017–1031.

Kavanaugh, J.L., Clarke, G.K.C. (2001) Abrupt glacier motion and reorganization of basal shear stress following the establishment of a connected drainage system. *Journal of Glaciology*, **47**(158), 472–480.

Kehew, A.E. & Teller, J.T. (1994) History of late glacial runoff along the southwestern margin of the Laurentide Ice Sheet. *Quaternary Science Reviews*, **13**, 859–877.

Kelly, M., Funder, S., Houmark-Nielsen, M., *et al.* (1999) Quaternary glacial and marine environmental history of northwest Greenland: a review and reappraisal. *Quaternary Science Reviews*, **18**, 373–392.

Kemmis, T.J. (1996) Lithofacies associations for terrestrial glacigenic successions. In: *Past Glacial Environments: Sediments, Forms and Techniques* (Ed. J. Menzies), pp. 285–300. Butterworth-Heinemann, Oxford.

Kempton, J.P., Johnson, W.H., Heigold, P.C. & Cart-wright, K. (1991) Mahomet bedrock valley in east-central Illinois; topography, glacial drift stratigraphy, and hydrogeology. In: *Geology and Hydrogeology of the Teays–Mahomet Bedrock Valley System* (Eds W.N. Melhorn & J.P. Kempton). *Geological Society of America Special Paper*, **258**, 91–124.

Kenneally, J.P. & Hughes, T.J. (1995–96) The calving constraints on inception of Quaternary ice sheets. *Quaternary International*, **95–96**, 43–53.

Kent, D. & Smethurst, M. (1998) Shallow bias of paleomagnetic inclinations in the Paleozoic and Precambrian. *Earth Planetary Science Letters*, **160**, 391–402.

Kenyon, N.H. (1986) Evidence from bedforms of a strong poleward current along the upper continental slope of Northwest Europe. *Marine Geology*, **72**, 187–198.

Khatwa, A., Hart, J.K. & Payne, A.J. (1999) Grain textural analysis across a range of glacial facies. *Annals of Glaciology*, **28**, 111–117.

Kieffer, H., Kargel, J.S., Barry, R., Bindschadler, R., *et al.* (2000) New eyes in the sky measure glaciers and ice sheets. *Eos* (*Transactions of the American Geophysical Union*), **81**(24), 265 + 270–271.

King, D.A. (2004) *Climate Change*, science: Adapt, mitigate, or ignore? *Science*, **303**, 176–177.

King, E.L., Sejrup, H.P., Haflidason, H., Elverhøi, A. & Aarseth, I. (1996) Quaternary seismic stratigraphy of the North Sea Fan: glacially-fed gravity flow aprons, hemipelagic sediments, and large submarine slides. *Marine Geology*, **130**, 293–316.

King-Clayton, L.M., Kautsky, F., Chapman, N.A., Svensson, N.-O., de Marsily, G. & Ledoux, E. (1997) The Central Climate Change Scenario: SKI's SITE-94 project to evaluate the future behaviour of a deep repository for spent-fuel. In: *Glaciation and Hydrogeology* (Eds L. King-Clayton, N. Chapman, L.O. Ericsson & F.Kautsky), p. A33. SKI Report 13, Swedish Nuclear Power Inspectorate.

Kirby, M.E. & Andrews, J.T. (1999) Mid-Wisconsin Laurentide Ice Sheet Growth and decay: implications for Heinrich events–3 and –4. *Paleoceanography*, **14**, 211–223.

Kirkbride, M.P. (1995) Processes of transportation. In: *Modern Glacial Environments* (Ed. J. Menzies), pp. 261–292. Butterworth-Heinemann, Oxford.

Kitanidis, A.K. (1997) *Introduction to Geostatistics: Applications to Hydrogeology*. Cambridge University Press, Cambridge.

Kitmitto, K., Cooper, M., Venters, C.C., *et al.* (2000) LANDMAP: serving satellite imagery to the UK academic community. In: *RSS2000 Adding Value to Remotely Sensed Data*. University of Leicester.

Kjær, K.H. & Krüger, J. (2001) The final phase of dead-ice moraine development: processes and sediment architecture, Kötlujökull, Iceland. *Sedimentology*, **48**, 935–952.

Klassen, R.W. (1989) Quaternary geology of the southern Canadian Interior Plains. In: *Quaternary Geology of Canada and Greenland*, Vol. K-1 (Ed. R.J. Fulton), pp. 138–173. Geology of North America

Series, Geological Society of America, Boulder, Colorado, and Geology of Canada, No. 1, Geological Survey of Canada, Queen's Printer, Ottawa.

Klein, A.G., Seltzer, G.O. & Isacks, B.L. (1999) Modern and last local glacial maximum snowlines in the Central Andes of Peru, Bolivia, and Northern Chile. *Quaternary Science Reviews,* **18**, 63–84.

Kleman, J. (1992) The palimpsest glacial landscape in northwestern Sweden: Late Weichselian deglaciation landforms and traces of older west-centred ice sheets. *Geografiska Annaler,* **74A**, 305–325.

Kleman, J. (1994) Preservation of landforms under ice sheets and ice caps. *Geomorphology*, **9**, 19–32.

Kleman, J. & Borgström, I. (1990) The boulder fields of Mt. Fulufjället, west-central Sweden—Late Weichselian boulder blankets and interstadial periglacial phenomena *Geografiska Annaler,* **72A** (1), 63–78.

Kleman, J. & Borgström, I. (1994) Glacial landforms indicative of a partly frozen bed. *Journal of Glaciology*, **135**, 255–264.

Kleman, J. & Borgström, I. (1996) Reconstruction of paleo-ice sheets: the use of geomorphological data. *Earth Surface Processes and Landforms,* **21**, 893–909.

Kleman, J. & Hätterstrand, C. (1999) Frozen-bed Fennoscandian and Laurentide ice sheets during the last Glacial Maximum. *Nature,* **402**(6757), 63–66.

Kleman, J. & Stroeven, A.P. (1997) Preglacial surface remnants and Quaternary glacial regimes in northwestern Sweden: *Geomorphology*, **19**(1–2), 35–54.

Kleman, J., Bergstrom, I. & Hattestrand, C. (1994) Evidence for a relict glacial landscape in Quebec–Labrador. *Palaeogeography, Palaeoclimatology, Palaeoecology*, **111**, 217–228.

Kleman, J., Hättestrand, C., Borgström, I. & Stroeven, A. (1997) Fennoscandian paleoglaciology reconstructed using a glacial geological inversion model. *Journal of Glaciology*, **43**, 283–299.

Kleman, J., Hättestrand, C. & Clarhäll, A. (1999) Zooming in on frozen-bed patches—Scale-dependent controls on Fennoscandian Ice Sheet basal thermal zonation. *Annals of Glaciology*, **28**, 189–194.

Kleman, J., Fastook, J. & Stroeven, A. (2002) Geologically and geomorphologically constrained numerical model of Laurentide Ice Sheet inception and build-up. *Quaternary International*, **95–96**, 87–98.

Klitgaard-Kristensen, D., Sejrup, H.-P., Haflidason, H., Johnsen, S. & Spurk, M. (1998) A regional 8200 cal. yr BP cooling event in northwest Europe, induced by final stages of the Laurentide ice sheet deglaciation? *Journal of Quaternary Science*, **13**, 165–169.

Klok, E.J. & Oerlemans, J. (2002) Model study of the spatial distribution of the energy and mass balance of Morteratschgletscher, Switzerland. Journal of Glaciology, **48**(163), 505–518.

Knight, P.G. (1997) The basal ice layer of glaciers and ice sheets. *Quaternary Science Reviews*, **16**, 975–993.

Knight, P.G. (1999) *Glaciers*. Stanley Thornes, Cheltenham, 261 pp.

Knight, P.G. & Knight, D.A. (1999) Experimental observations of subglacial entrainment into the vein network of polycrystalline ice. *Glacial Geology and Geomorphology*, http://boris.qub.ac.uk/ggg/papers/full/1999/rp051999/rp05.html.

Knight, P.G., Patterson, C.J., Waller, R.I., Jones, A.P. & Robinson, Z.P. (2000) Preservation of basal-ice sediment texture in ice-sheet moraines. *Quaternary Science Reviews*, **19**, 1255–1258.

Knowles, A. (1985) The Quaternary history of north Staffordshire. In: *The Geomorphology of North West England* (Ed. R.H. Johnson), pp. 222–236. Manchester University Press, Manchester.

Kobayashi, S. (1979) Studies on interaction between wind and dry snow surface. *Contributions to the Institute of Low Temperature Science, Hokkaido University, Series A,* **29**, 64 pp.

Koerner, R.M. (1977) Ice thickness measurements and their implications to past and present ice volumes in the Canadian high Arctic. *Canadian Journal of Earth Sciences*, **14**, 2697–2705.

Koerner, R.M. & Fisher, D.A. (1979) Discontinuous flow, ice structure and dirt content in the basal layers of the Devon Island ice cap. *Journal of Glaciology*, **23**, 209–220.

Koerner, R.M., Fisher, D.A. & Parnandi, M. (1981) Bore-hole video and photographic cameras. *Annals of Glaciology*, **2**, 34–38.

Kohl, C.P. & Nishiizumi, K. (1992) Chemical isolation of quartz for measurement of *in situ* produced cosmogenic nuclides. *Geochimica et Cosmochimica Acta*, **56**, 3583–3587.

Kohout, F.A., Hathaway, J.C., Folger, D.W., *et al.* (1977) Fresh Groundwater Stored in Aquifers under the Continental Shelf, Implications from a Deep Test, Nantucket Island, Massachusetts. *Water Resources Bulletin*, **13**(2), 373–386.

Kor, P.S.G., Shaw, J. & Sharpe, D.R. (1991) Erosion of bedrock by subglacial meltwater, Georgian Bay, Ontario: a regional view. *Canadian Journal of Earth Sciences,* **28**, 623–642.

Krabill, W., Frederick, E., Manizade, S., *et al.* (1999) Rapid thinning of parts of the southern Greenland Ice Sheet. *Science*, **283**, 1522–1524.

Krabill, W., Abdalati, W., Frederick, E., *et al.* (2000) Greenland ice sheet: high-elevation balance and peripheral thinning. *Science*, **289**, 428–430.

Krauskopf, K.B. (1967) *Introduction to Geochemistry*. McGraw-Hill, New York, 721 pp.

Krinner, G., Mangerud, J., Jakobsson, M., Crucifix, M., Ritz, C. & Svendsen, J.L. (2004) Enhanced ice sheet growth in Eurasia owing to adjacent ice-dammed lakes. *Nature*, **427**, 429–432.

Krüger, J. (1993) Moraine ridge formation along a stationary ice front in Iceland. *Boreas*, **22**, 101–109.

Krüger, J. (1994) Glacial processes, sediments, landforms and stratigraphy in the terminus region of Myrdalsjökull, Iceland. *Folia Geographica Danica*, **21**, 1–233.

Krüger, J. & Kjaer, K.H. (1999) A data chart for field description and genetic interpretation of glacial diamicts and associated sediments—with examples from Greenland, Iceland and Denmark. *Boreas*, **28**, 386–402.

Kuhn, M. (2003) Redistribution of snow and glacier mass balance from a hydrological model. *Journal of Hydrology*, **282**(2003) 95–103.

Kulessa, B. & Murray, T. (2003) Slug-test derived differences in bed hydraulic properties between a surge-type and non-surge-type Svalbard glacier. *Annals of Glaciology*, **36**, 103–109.

Kulessa, B., Hubbard, B., Brown, G.H. & Becker, J. (2003) Earth tide forcing of glacier drainage *Geophysical Research Letters*, **30**. Art. No. 1011.

Kulig, J.J. (1985) A sedimentation model for the deposition of glacigenic deposits in west-central Alberta: a single (Late Wisconsinan) event. *Canadian Journal of Earth Sciences*, **26**, 266–274.

Kutzbach, J.E., Bartlein, P.J., Foley, J.A., *et al.* (1996) Potential role of vegetation feedback in the climate sensitivity of high-latitude regions: a case study at 6000 years B.P. *Global Biogeochemical Cycles*, **10**, 727–736.

Kutzbach, J., Gallimore, R., Harrison, S., Behling, P., Selin, R. & Laarif, F. (1998) Climate and biome simulations for the past 21,000 years. *Quaternary Science Reviews*, **17**, 473–506.

Kuvaas, B. & Leitchenkov, G. (1992) Glaciomarine turbidite and current controlled deposits in Prydz Bay, Antarctica. *Marine Geology*, **108**, 365–381.

Kwok, R. (1998) The RADARSAT geophysical processor system. In: *Analysis of SAR Data of the Polar Oceans: Recent Advances* (Eds C. Tsatsoulis & R. Kwok), pp. 235–257. Springer-Verlag, Berlin.

Kwok, R. & Cunningham, G.F. (2002) Seasonal ice area and volume production of the Arctic Ocean: November 1996 through April

1997. *Journal of Geophysical Research—Oceans*, **107**(C10). art. no.–8038.

Kwok, R. & Fahnestock, M.A. (1996) Ice sheet motion and topography from radar interferometry. *IEEE Transactions of Geoscientific Remote Sensing*, **34**(1), 189–200.

Kwok, R. & Rothrock, D.A. (1999) Variability of Fram Strait ice flux and North Atlantic Oscillation. *Journal of Geophysical Research*, **104**(C3), 5177–5189.

Kwok, R., Siegert, M.J. & Carsey, F.D. (2000) Ice motion over Lake Vostok, Antarctica: constraints on inferences regarding accreted ice. *Journal of Glaciology*, **46**, 689–694.

Laaksoharju, M. & Rhén, I. (1999) Äspö project—Hydrogeology and hydrochemistry used to indicate present flow dynamics. *Proceedings of a SKB workshop 'Use of Hydrochemical Information in Testing Groundwater Flow Models' in Borgholm, Sweden*, pp. 65–78. Organization for Economic Co-operation and Development/Nuclear Energy Agency, Paris.

Laberg, J.S. & Vorren, T.O. (1993) A Late Pleistocene submarine slide on the Bear Island Trough Mouth Fan. *Geo-Marine Letters*, **13**, 227–234.

Lacasse, C. (2001) Influence of climatic variability on the atmospheric transport of Icelandic tephra in the subpolar North Atlantic. *Global and Planetary Change*, **29**, 31–56.

Lagerbäck, R. (1988) The Veiki moraines in northern Sweden—widespread evidence of an early Weichselian deglaciation. *Boreas*, **17**, 469–486.

Lagerbäck, R. & Robertsson, A.-M. (1988) Kettle holes—stratigraphical archives for Weichselian geology and palaeoenvironment in northernmost Sweden. *Boreas*, **17**, 439–468.

Laine, E.P. (1980) New evidence from beneath the western North Atlantic for the depth of glacial erosion in Greenland and North America. *Quaternary Research*, **14**, 188–198.

Lal, D. (1991) Cosmic-ray labeling of erosion surfaces—in situ nuclide production-rates and erosion models. *Earth and Planetary Science Letters*, **104**, 424–439.

Lamb, H.H. (1979) Climatic variations and changes in the wind and ocean circulation: The Little Ice Age in the Northeast Atlantic. *Quaternary Research*, **11**, 1–20.

Lambeck, K. (1990) Glacial rebound, sea-level change and mantle viscosity. *Quarterly Journal Royal Astronomical Society*, **31**, 1–30.

Lambeck, K. (1993a) Glacial rebound of the British Isles-I. Preliminary model results. *Geophysical Journal International*, **115**,941–959.

Lambeck, K. (1993b) Glacial rebound of the British Isles-II. A high-resolution, high-precision model. *Geophysical Journal International*, **115**, 960–990.

Lambeck, K., Smither, C. & Johnston, P. (1998) Sea-level change, glacial rebound and mantle viscosity for northern Europe. *Geophysical Journal International*, **134**, 102–144.

Lambeck, K., Yokoyama, Y., Johnston, P. & Purcell, A. (2000) Global ice volumes at the Last Glacial Maximum and early Lateglacial. *Earth and Planetary Science Letters*, **181**, 513–527.

Lambrecht, A., Mayer, C., Oerter, H. & Nixdorf, J. (1999) Investigations of the mass balance of the southeastern Ronne Ice Shelf. *Annals of Glaciology*, **29**, 250–254.

Landais, A., Barnola, J.M., Masson-Delmotte, V., Jouzel, J., Chappellaz, J., Caillon, N., Huber, C., Leuenberger, M. & Johnsen, S.J. (2004) A continuous record of temperature evolution over a sequence of Dansgaard-Oeschger eventw during Marine Stage 4 (76 to 62 kyr BP) *Geophysical Research Letters*, **31**, L22211. doi: 10.1029/2004GL021193, 1–4.

Langdon, T.G. (1994) A unified approach to grain boundary sliding in creep and superplasticity. *Acta Metallurgica et Materialia*, **42**, 2437–2443.

Larsen, C.E. (1985) Lake level, uplift, and outlet incision, the Nipissing and Algoma Great Lakes. In: *Quaternary Evolution of the Great Lakes* (Eds P.F. Karrow & P.E. Calkin), pp. 63–77. Special Paper 30, Geological Association of Canada, St John's, NF.

Larsen, C.E. (1987) Geological history of Glacial Lake Algonquin and the Upper Great Lakes. *United States Geological Survey Bulletin*, **1801**, 36 pp.

Larsen, E. & Mangerud, J. (1992) Subglacially formed clastic dikes. *Sveriges Geologiska Undersökning Series Ca*, **81**, 163–170.

Larsen, H.C., Saunders, A.D., Clift, P.D., *et al.* (1994) Seven million years of glaciation in Greenland. *Science*, **264**, 952–955.

Larsen, N.K. & Piotrowski, J.A. (2003) Fabric pattern in a basal till succession and its significance for reconstructing subglacial processes. *Journal of Sedimentary Research*, **73**, 725–734.

Larter, R.D. & Vanneste, L.E. (1995) Relict subglacial deltas on the Antarctic Peninsula outer shelf. *Geology*, **23**, 33–36.

Larour, E., Rignot, E. & Aubry, D. (2004). Modelling of rift propagation on Ronne Ice Shelf, Antarctica, and sensitivity to climate change. *Geophysical Research Letters*, **31**(16). art. no. L16404.

Laumann, T. & Reeh, N. (1993) Sensitivity to climate change, of the mass balance of glaciers in southern Norway. *Journal of Glaciology*, **39**(133), 656–665.

Laumann, T. & Tvede, A.M. (1989) *Simulation of the Effects of Climate Changes on a Glacier in Western Norway*. Meddelse fra Hydrologisk Avdeling 72, Norges Vassdrags- og Energiverk, Oslo.

Lawrence, D.B. & Lawrence, E.G. (1965) Glacier studies in New Zealand. *Mazama*, **47**, 17–27.

Lawson, D.E. (1979a) Characteristics and origins of the debris and ice, Matanuska Glacier, Alaska. *Journal of Glaciology*, **23**, 437–438.

Lawson, D.E. (1979b) *Sedimentological Analysis of the Western Terminus Region of the Matanuska Glacier, Alaska*. Report. 79–9, Cold Regions Research and Engineering Laboratory, Hanover, New hampshire.

Lawson, D.E, Strasser, J.C., Evenson, E.B., Alley, R.B., Larson, G.J. & Arcone, S.A. (1998) Glaciohydraulic supercooling: a freeze-on mechanism to create stratified. debris-rich basal ice: I. Field evidence. *Journal of Glaciology*, **44**(148), 547–562.

Lawson, W. (1996) The relative strengths of debris-laden basal ice and clean glacier ice: some evidence from Taylor Glacier, Antarctica. *Annals of Glaciology*, **23**, 270–276.

Laxon, S. (1990) Seasonal and inter-annual variations in Antarctic sea ice extent as mapped by radar altimetry. *Geophysical Research Letters*, **17**(10), 1553–1556.

Laxon, S. (1994) Sea-Ice Altimeter Processing Scheme at the EODC. *International Journal of Remote Sensing*, **15**(4), 915–924.

Laxon, S., Peacock, N. & Smith, D. (2003) High interannual variability of sea ice thickness in the Arctic region. *Nature*, **425**(6961), 947–950.

Layberry, R.L. & Bamber, J.L. (2001) A new ice thickness and bed data set for the Greenland ice sheet 2. Relationship between dynamics and basal topography. *Journal of Geophysical Research—Atmospheres*, **106**(D24), 33781–33788.

Laymon, C.A. (1992) Glacial geology of western Hudson Strait, Canada, with reference to Laurentide Ice Sheet dynamics. *Geological Society of America Bulletin*, **104**, 1169–1177.

Le Meur, E. & Huybrechts, P. (1996) A comparison of different ways of dealing with isostasy: examples from modeling the Antarctic ice sheet during the last glacial cycle. *Annals of Glaciology*, **23**, 309–317.

Le Meur, E. & Huybrechts, P. (2001) A model computation of surface gravity and geoidal signal induced by the evolving Greenland ice sheet. *Geophysical Journal International*, **145**, 1–21.

Lean, J., Beer, J. & Bradley, R. (1995) Reconstruction of solar irradiance since 1610: implications for climate change. *Geophysical Research Letters*, **22**, 3195–3198.

Lean, J.L., Wang, Y.-M. & Shelley, N.R.J. (2002) The effect of increasing solar activity on the Sun's total and open magnetic flux during multiple cycles: Implications for solar forcing of climate. *Geophysical Research Letters*, **29**. doi: 10.1029/2002GL015880, 2002.

Leather, J., Allen, P.A., Brasier, M.D. & Cozzi, A. (2002) Neoproterozoic snowball earth under scrutiny: Evidence from the Fiq glaciation of Oman. *Geology*, **30**, 891–894.

Lebreiro, S.M., Moreno, J.C., McCave, I.N. & Weaver, P.P.E. (1996) Evidence for Heinrich layers off Portugal (Tore Seamount: 39N, 12W). *Marine Geology*, **131**, 47–56.

Lee, H.A. (1959) Surficial geology of southern district of Keewatin and the Keewatin ice divide, Northwest Territories. *Geological Survey of Canada Bulletin*, **51**.

Lefebre, F., Gallée, H., van Ypersele, J.-P. & P. Huybrechts, P. (2002) Modelling of large-scale melt parameters with a regional climate model in South-Greenland during the 1991 melt season. *Annals of Glaciology*, **35**, 391–397.

Legates, D.R. & Willmott, C.J. (1990a) Mean seasonal and spatial variations in global surface air temperature. *Theoretical and Applied Climatology*, **41**, 11–21.

Legates, D.R. & Willmott, C.J. (1990b) Mean seasonal and spatial variations in gauge-corrected global precipitation. *International Journal of Climatology*, **10**, 111–127.

Legrand, M. & Mayewski, P. (1997) Glaciochemistry of polar ice cores: a review, *Review of Geophysics*, **35**, 217–243.

Leiva, J.C. (1999) Recent fluctuations of the Argentine glaciers. *Global and Planetary Change*, **22**, 169–177.

Lemmon, D.S., Duk-Rodkin & A, Bednarski, J. (1994) Late glacial drainage systems along the northwestern margin of the Laurentide Ice Sheet. *Quaternary Science Reviews*, **13**, 805–828.

Leonard, E.M. (1986) Varve Studies at Hector Lake, Alberta, Canada, and the Relationship between Glacial Activity and Sedimentation. *Quaternary Research*, **25**, 199–214.

Letréguilly, A. (1988) Relation between the mass balance of western Canadian mountain glaciers and meteorological data. *Journal of Glaciology*, **34**, 11–17.

Letréguilly, A. & Reynaud, L. (1990) Space and time distribution of glacier mass balance in the northern hemisphere. *Arctic and Alpine Research*, **22**(1), 43–50.

Leung, L.R. & Qian, Y. (2003) The sensitivity of precipitation and snowpack simulations to model resolution via nesting in regions of complex terrain. *Hydrometeorology*, **4**, 1025–1043.

Leventer, A., Williams, D. & Kennet, J.P. (1982) Dynamics of the Laurentide ice sheet during the last glaciation: evidence from the Gulf of Mexico. *Earth and Planetary Science Letters*, **59**, 11–17.

Leverington, D.W. & Teller, J.T. (2003) Paleotopographic reconstructions of the eastern outlets of glacial Lake Agassiz. *Canadian Journal of Earth Sciences*, **40**, 1259–1278.

Leverington, D.W., Mann, J.D. & Teller, J.T. (2000) Changes in the bathymetry and volume of glacial Lake Agassiz between 11,000 and 9300 ^{14}C yr BP. *Quaternary Research*, **54**, 174–181.

Leverington, D.W., Mann, J.D. & Teller, J.T. (2002) Changes in the bathymetry and volume of glacial Lake Agassiz between 9200 and 7700 ^{14}Cyr B.P. *Quaternary Research*, **57**, 244–252.

Lewis, C.F.M. (1969) Late Quaternary history of lake levels in the Huron and Erie basins. *Proceedings 12th Conference Great Lakes Research, The University of Michigan, Ann Arbor MI, May 5–7, 1969*, International Association for Great Lakes Research, pp. 250–270.

Lewis, C.F.M. & Anderson, T.W. (1989) Oscillations of levels and cool phases of the Laurentian Great Lakes caused by inflows from glacial Lakes Agassiz and Barlow-Ojibway. *Journal of Paleolimnology*, **2**, 99–146.

Lewis, C.F.M., Moore Jr., T.C., Rea, D.K., Dettman, D.L., Smith, A.J. & Mayer, L.A. (1994) Lakes of the Huron basin: their record of runoff from the Laurentide Ice Sheet. *Quaternary Science Reviews*, **13**, 891–922.

Li, J., Jacka, T.H. & Budd, W.F. (1996) Deformation rates in combined compression and shear for ice which is initially isotropic and after the development of strong anisotropy. *Annals of Glaciology*, **23**, 247–252.

Li, J., Jacka, T.H. & Morgan, V.I. (1998) Crystal size and microparticle record in the ice core from Dome Summit South, Law Dome, East Antarctica. *Annals of Glaciology*, **27**, 343–348.

Li, J., Jacka, T.H. & Budd, W.F. (2000) Strong single-maximum crystal fabrics developed in ice undergoing shear with unconstrained normal deformation. *Annals of Glaciology*, **30**, 88–92.

Licciardi, J.M., Clark, P.U., Jenson, J.W. & Macayeal, D.R. (1998) Deglaciation of a soft-bedded Laurentide Ice Sheet. *Quaternary Science Reviews*, **17**, 427–448.

Licciardi, J.M., Teller, J.T. & Clark, P.U. (1999) Freshwater routing by the Laurentide Ice Sheet during the last deglaciation. In: *Mechanisms of Global Climate Change at Millennial Time Scales* (Eds P.U. Clark, R.S. Webb & L.D. Keigwin), pp. 177–201. Geophysical Monograph 112, American Geophysical Union, Washington, DC.

Licht, K.J. (2004) The Ross Sea's contribution to eustatic sea level during meltwater pulse 1A. *Sedimentary Geology*, **165**, 343–353.

Liestøl, O. (1967) Storbreen glacier in Jotunheimen, Norway. *Norsk Polarinstitut Skrifter*, **141**.

Lillesand, T.M. & Kiefer, R.W. (2000) *Remote Sensing and Image Interpretation*. New York, John Wiley.

Lingle, C.S. (1984) A numerical-model of interactions between a polar ice stream and the ocean—application to Ice Stream-E, West Antarctica. *Journal of Geophysical Research—Oceans*, **89**, 3523–3549.

Lingle, C.S. & Brown, T.J. (1987) A subglacial aquifer bed model and water pressure dependent basal sliding relationship for a West Antarctic ice stream. In: *Dynamics of the West Antarctic Ice Sheet* (Eds C.J. van der Veen & J. Oerlemans), pp. 249–285. D. Reidel, Norwell.

Lingle, C.S. & Covey, D.N. (1998) Elevation changes on the East Antarctic ice sheet, 1978–93, from satellite radar altimetry: a preliminary assessment. *Annals of Glaciology*, **27**, 7–18.

Lingle, C.S., Schilling, D.H., Fastook, J.L., Paterson, W.S.B. & Brown, T.J. (1991) A flow band model of the Ross ice shelf, Antarctica: response to CO_2-induced climatic warming. *Journal of Geophysical Research—Solid Earth and Planets*, **96**, 6849–6871.

Linton, D.L. (1963) The forms of glacial erosion. Transactions of the Institute of British Geographers, **33**, 1–28.

Lipenkov, V. Ya, Salamatin, A. & Duval, P. (1997) Bubbly-ice densification in ice sheets: II. Applications. *Journal of Glaciology*, **43**, 397–407.

Lister, H., Pendlington, A. & Chorlton, J. (1968) Laboratory experiments on abrasion of sandstones by ice. In: *Snow and Ice. Reports and Discussions*, pp. 98–106. IAHS Publication 79, International Association of Hydrologic Sciences, Wallingford.

Liu, H.X., Jezek, K.C. & Li, B.Y. (1999) Development of an Antarctic digital elevation model by integrating cartographic and remotely sensed data: A geographic information system based approach. *Journal of Geophysical Research-Solid Earth*, **104**(B10), 23199–23213.

Liu, S., Xie, Z. & Liu, C. (2000) Mass balance and fluctuations of glaciers. In: *Glaciers and their Environments in China—the Present, Past and Future* (Eds Shi, Y., Huang, M., Yao, T., *et al.*), pp. 101–131. Science Press, Beijing.

Liu, S., Shen, Y., Sun, W., *et al.* (2002) Glaciers variation since the Maximum of the Little Ice Age in the western Qilian Mountains, Northwest China. *Journal of Glaciolog, and Geocryology*, **24**(3), 227–233.

Livingstone, D.A. (1963) Chemical compositions of rivers and lakes, *U.S. Geological Survey Professional Paper*, **440-G**, 64 pp.

Ljunger, E. (1930) Spaltektonik und morphologie der schwedishen-Skaggerakk—kuste. Tiel III. Die erosienformen. *Bulletin of the Geological Institutions of the University of Uppsala*, **21**, 255–475.

Lliboutry, L. (1956) *Nieves y Glaciares de Chile. Fundamentos de Glaciología*. Ediciones de la Universidad de Chile, Santiago, 471 p.

Lliboutry, L. (1965) *Traité de glaciologie. Tome II Glaciers, variations du climat, sols gelés*. Masson, Paris.

Lliboutry, L. (1976) Physical processes in temperate glaciers. *Journal of Glaciology*, **16**, 151–158.

Lliboutry, L. & Duval, P. (1985) Various isotropic and anisotropic ices found in glaciers and polar ice sheets and their corresponding rheologies. *Annales Geophysicae*, **3**, 207–224.

Lliboutry, L. & Ritz, C. (1978) Ecoulement permanent d'un fluide visquesux non linéaire (corps de Glen) autour d'une sphère parfaitement lissé. *Annals of Geophysics*, **34**, 133–146.

Lønne, I. (1995) Sedimentary facies and depositional architecture of ice-contact glacimarine systems. *Sedimentary Geology*, **98**, 13–43.

Lønne, I. (2001) Dynamics of marine glacier termini read from moraine architecture. *Geology*, **29**, 199–202.

Lorius, C., Jouzel, J., Ritz, C., *et al.* (1985) A 150,000-year climatic record from Antarctic ice. *Nature*, **316**, 591–596.

Lorius, C., Jouzel, J., Raynaud, D., Hansen, J. & Le Treut, H. (1990) Greenhouse warming, climate sensitivity and ice core data. *Nature*, **347**, 139–145.

Lorrain, R.D., Fitzsimons, S.J., Vandergoes, M.J. & Stievenard, M. (1999) Ice compositio evidence for the formation of basal ice from lake water beneath a cold-based Antarctic glacier. *Annals of Glaciology*, **28**, 277–281.

Low, A.P. (1893) Notes on the glacial geology of western Labrador and northern Quebec. *Geological Society of America Bulletin*, **4**, 419–421.

Lowe, A.L. & Anderson, J.B. (2002) Reconstruction of the West Antarctic ice sheet in Pine Island Bay during the Last Glacial Maximum and its subsequent retreat history. *Quaternary Science Reviews*, **21**, 1879–1897.

Lowe, A. & Anderson, J.B. (2003) Evidence of abundant subglacial meltwater beneath the paleo-ice sheet in Pine Island Bay, Antarctica. *Journal of Glaciology*, **49**, 125–138.

Lowell, T.V., Heusser, C.J., Andersen, B.G., *et al.* (1995) Interhemipheric Correlation of Late Pleistocene Glacial Events. *Science*, **269**, 1541–1549.

Lu, A., Yao, T., Liu, S., *et al.* (2002) Glacier change in the Geladandong area of the Tibet Plateau monitored by remote sensing. *Journal of Glaciology and Geocryology*, **24**(5), 559–562.

Lucchitta, B.K. (2001) Antarctic ice streams and outflow channels on Mars. *Geophysical Research Letters*, **28**, 403–406.

Luchitta, B.K. & Rosanova, C.E. (1998) Retreat of northern margins of George VI and Wilkins ice shelves, Antarctic Peninsula. *Annals of Glaciology*, **27**, 41–46.

Luckman, A., Murray, T. & Strozzi, T. (2002) Surface flow evolution throughout a glacier surge measured by satellite radar interferometry. *Geophysical Research Letters*, **29**(23). art. no.–2095.

Ludwig, W., Amiotte-Suchet, P., Munhoven, G. & Probst, J.-L. (1998) Atmospheric CO_2 comsumption by continental erosion: present-day controls and implications for the last glacial maximum. *Global and Plantentary Change*, **16–17**, 107–120.

Luthi, M., Funk, M., Iken, A., Gogineni, S. & Truffer, M. (2002) Mechanisms of fast flow in Jakobshavn Isbrae, West Greenland: Part III. Measurements of ice deformation, temperature and cross-borehole conductivityin boreholes to the bedrock. *Journal of Glaciology*, **48**, 369–385.

Lythe, M., Vaughan, D.G. & Consortium, B. (2001) BEDMAP: a new ice thickness and subglacial topographic model of Antarctica. *Journal of Geophysical Research*, **106**, 11335–11351.

MacAyeal, D.R. (1989a) Large-scale ice flow over a viscous basal sediment—theory and application to Ice Stream-B, Antarctica. *Journal of Geophysical Research*, **94**, 4071–4087.

MacAyeal, D.R. (1989b) Ice-shelf response to ice-stream discharge fluctuations. 3. The effects of ice-stream imbalance on the Ross Ice Shelf. *Journal of Glaciology*, **35**, 38–42.

MacAyeal, D.R. (1992a) Irregular oscillations of the West Antarctic ice sheet. *Nature*, **359**, 29–32.

MacAyeal, D.R. (1992b). The basal stress-distribution of Ice Stream-E, Antarctica, inferred by control methods. *Journal of Geophysical Research—Solid Earth*, **97**(B1), 595–603.

MacAyeal, D.R. (1993a) Binge/purge oscillations of the Laurentide Ice Sheet as a cause of North Atlantic's Heinrich events. *Paleoceanography*, **8**, 775–784.

MacAyeal, D.R. (1993b) A low-order model of growth/purge oscillations of the Heinrich-event cycle. *Paleoceanography*, **8**, 767–773.

MacaYeal, D.R. & Lange, M.A. (1988) ice-shelf response to ice-stream discharge fluctuations. 2. Ideal rectangular ice shelf. *Journal of Glaciology*, **34**, 128–135.

MacAyeal, D.R., Bindschadler, R.A. & Scambos, T. (1995) Basal friction of Ice Stream E, West Antarctica. *Journal of Glaciology*, **41**(138), 247–262.

MacGregor, K.R. (2002) *Modeling and field constraints on glacier dynamics, erosion, and alpine landscape evolution*. PhD thesis, Earth Sciences University of California, Santa Cruz, Santa Cruz, pp. 277.

MacGregor, K.R., Anderson, R.S., Anderson, S.P., *et al.* (2000) Numerical simulations of glacial-valley longitudinal profile evolution. *Geology* **28**, 1031–1034.

Mackay, J.R. & Mathews, W.H. (1964). The role of permafrost in ice-thrusting. *Journal of Geology*, **72**, 378–380.

MacLean, B. (1985) Geology of the Baffin Island Shelf. In: *Quaternary Environments: Eastern Canadian Arctic, Baffin Bay, and Western Greenland. Geology of the Baffin Island Shelf* (Ed. J.T. Andrews), pp. 154–177. Allen and Unwin, Boston.

MacLean, B. (2001a) Introduction: geographic setting and studies. In: *Marine Geology of Hudson Strait and Ungava Bay, Eastern Arctic Canada: Late Quaternary Sediments, Depositional Environments, and Late Glacial–Deglacial History Derived from Marine and Terrestrial Studies* (Ed. B. MacLean), pp. 65–69. Geological Survey of Canada Bulletin 566, Queen's Printer, Ottawa.

MacLean, B. (Ed.) (2001b) *Marine Geology of Hudson Strait and Ungava Bay, Eastern Arctic Canada: Late Quaternary Sediments, Depositional Environments, and Late Glacial–Deglacial History Derived from Marine and Terrestrial Studies*. Geological Survey of Canada Bulletin 566, Queen's Printer, Ottawa.

MacLean, B., Williams, G.L., Jennings, A.E. & Blakeney, C. (1986) Bedrock and surficial geology of Cumberland Sound, N.W.T. *Geological Survey of Canada Paper*, **86-1B**, 605–615.

MacLean, B., Andrews, J.T., Gray, J.T., *et al.* (2001) Hudson Strait Quaternary sediments and late glacial and deglaciation history: a discussion and summary. In: *Marine Geology of Hudson Strait and Ungava Bay, Eastern Arctic Canada: Late Quaternary Sediments, Depositional Environments, and Late Glacial–Deglacial History Derived from Marine and Terrestrial Studies* (Ed. B. MacLean), pp. 181–192. Geological Survey of Canada Bulletin 566, Queen's Printer, Ottawa.

Maclennan, J., Jull, M., McKenzie, D., Slater, L. & Grönvold, K. (2002) Link between volcanism and deglaciation in Iceland. *Geochemistry, Geophysics and Geosystems*, **3**(1), 1–25.

Mader, H.M. (1992) Observations of the water-vein system of polycrystalline ice. *Journal of Glaciology*, **38**, 333–348.

Madsen, V. (1921). Terrainformerne på kovbjerg Bakkeø. *Danmark geologische Undersoekelse*, **4**, 1(12).

Mahaffy, M.A.W. (1976) A three-dimensional numerical model of ice sheets: tests on the Barnes ice cap, Northwest Territories. *Journal of Geophysical Research*, **81**, 1059–1066.

Mahrenholtz, O. & Wu, Z. (1992) Determination of creep damage parameters for polycrystalline ice. *Advances in Ice Technology, 3rd International Conference on Ice Technology*, Cambridge, MA, pp. 181–192.

Mair, D., Nienow, P., Willis, I. & Sharp, M. (2001) Spatial patterns of glacier motion during a high-velocity event: Haut Glacier d'Arolla, Switzerland. *Journal of Glaciology*, **47**(156), 9–20.

Mair, D., Willis, I., Fischer, U.H., Hubbard, B., Nienow, P. & Hubbard, A. (2003) Hydrological controls on patterns of surface, internal and basal motion during three spring events, Haut Glacier d'Arolla, Switzerland. *Journal of Glaciology*, **49**, 555–567.

Maisch, M., Wipf, A., Denneler, B., Battaglia, J. & Benz, C. (2000) Die Gletscher der Schweizer Alpen. Gletscherhochstand 1850, Aktuelle Vergletscherung, Gletscherschwund-Szenarien. vdf Hochschulverlag, Zurich.

Malmberg, S.-A. (1985) The water masses between Iceland and Greenland. *Journal Marine Research Institute*, **9**, 127–140.

Malmberg, S.-A. & Jonsson, S. (1997) Timing of deep convection in the Greenland and Iceland Seas. *Journal of Marine Science*, **54**, 300–309.

Maltman, A.J., Hubbard, B. & Hambrey, M.J. (Eds) (2000) *Deformation of glacial Materials.* Special Publication No. 176, Geological Society Publishing House, Bath.

Manabe, S. & Stouffer, R.J. (1988) Two stable equilibria of a coupled ocean-atmosphere model. *Journal of Climate*, **1**, 841–866.

Manabe, S. & Stouffer, R.J. (1995) Simulation of abrupt climate change, induced by freshwater input to the North Atlantic Ocean. *Nature*, **378**, 165–167.

Manabe, S. & Stouffer, R.J. (1997) Coupled ocean-atmosphere model response to freshwater input: comparison to Younger Dryas event. *Paleoceanography*, **12**, 321–336.

Mandle, R.J. & Kontis, A.L. (1992) Simulation of regional ground water flow in the Cambrian–Ordovician aquifer system in the northern Midwest, United States. *U.S. Geological Survey Professional Paper*, **1405-C**.

Mangeney, A., Califano, F. & Hutter, K. (1997) A numerical study of anisotropic, low Reynolds number, free surface flow for ice sheet flow modeling. *Journal of Geophysical Research*, **102**, 22749–22764.

Manley, W.F. & Miller, G.H. (2001) Glacial-geological record on southern Baffin Island reflecting late glacial ice-sheet dynamics in the eastern Hudson Strait region. In: *Marine Geology of Hudson Strait and Ungava Bay, Eastern Arctic Canada: Late Quaternary Sediments, Depositional Environments, and Late Glacial–Deglacial History Derived from Marine and Terrestrial Studies* (Ed. B. MacLean), pp. 19–30. Geological Survey of Canada Bulletin 566, Queen's Printer, Ottawa.

Mann, D.H. (1986) Reliability of a fjord glacier's fluctuations for paleoclimatic reconstructions. *Quaternary Research*, **25**, 10–24.

Mann, G.W., Anderson, P.S. & Mobbs, S.D. (2000) Profile measurements of blowing snow at Halley, Antarctica. *Journal of Geophysical Research*, **105**, 24491–24508.

Mann, M.E. (2000) Lessons for a New Millennium. *Science*, **290**, 253–254.

Mann, M.E. & Jones, P. (2003) Global surface temperatures over the past two millennia. *Geophysics Research Letters*, **30**(15), 5-1–5-4. doi: 10.1029/2003GL017814.

Mann, M.E. & Lees, J.M. (1996) Robust estimation of background noise and signal detection in climatic time series. *Climatic Change*, **33**, 409–445.

Mann, M.E., Gille, E., Bradley, R.S., *et al.* (1999) Global temperature patterns in past centuries: an interactive presentation. www.ngdc.noaa.gov/paleo/ei/ei_cover.html.

Marchant, D., Lewis, A., Phillips, W., Moore, E., Souchez, R., Denton, G., Sugden, D. & Laudis, G. (2002) Formation of patterned-ground and sublimation till over Miocene glacier ice in Beacon Valley, southern Victoria Land, Antarctica. *Geological Society of American Bulletin*, **114**(6), 718–730.

Marczinek, S. (2002) *Zur Hydrogeologie und Paläohydrogeologie zum Zeitpunkt des Weichselhochglazials im Einzugsgebiet der Eckernförder Bucht.* PhD dissertation, University of Kiel, 108 pp.

Marczinek, S. & Piotrowski, J.A. (2002) Grundwasserströmung und –beschaffenheit im Einzugsgebiet der Eckernförder Bucht, Schleswig-Holstein. *Grundwasser*, **2**, 101–110.

Mark, B.G. (2002) Hot ice: glaciers in the tropics are making the press. *Hydrological Processes*, **16**, 3297–3302.

Mark, B.G. & Seltzer, G.O. (2003) Tropical glacier meltwater contribution to stream discharge: a case study in the Cordillera Blanca, Peru. *Journal of Glaciology*, **49**(165), 271–281.

Marks, D., Domingo, J., Susong, D., Link, T. & Garen, D. (1999) A spatially distributed energy balance snowmelt model for application in mountain basins. *Hydrological Processes*, **16**, 1935–1959.

Markus, T. & Cavalieri, D.J. (2000) An enhancement of the NASA Team sea ice algorithm. *Ieee Transactions on GeoScience, and Remote Sensing* **38**(3), 1387–1398.

Marlin, C., Ginidis, P., van Gijssel, K. & Boulton, G.S. (1997) Geochemistry of porewaters in Tertiary clays at Muhlenrade, Schleswig-Holstein, Germany. In: *Simulation of the Effects of Long-term Climatic Change on Groundwater Flow and the Safety of Geological Disposal Sites* (Eds G.S. Boulton & F. Curle), pp. 238–256. Report EUR 17793 EN, Nuclear Science and Technology, European Community.

Marquette, G.C., Gray, J.T., Gosse, J.C., Courchesne, F., Stockli, L., Macpherson, G. & Finkel, R. (2004) Felsenmeer persistence through glacial periods in the Torngat and Kaumajet Mountains, Quebec–Labrador, as determined by soil weathering and cosmogenic nuclide exposure dating. *Canadian Journal of Earth Sciences*, **41**, 19–38.

Marsella, K.A., Bierman, P.R., Davis, P.T. & Caffee, M.W. (2000) Cosmogenic 10Be and 26Al ages for the Last Glacial Maximum, eastern Baffin Island, Arctic Canada. *Geological Society of America Bulletin*, **112**, 1296–1312.

Marshall, H.P., Harper, J.T., Pfeffer, W.T. & Humphrey, N.F. (2002) Depth-varying constitutive properties observed in an isothermal glacier. *Geophysical Research Letters*, **10**.1029/2002GL015412.

Marshall, S. & Oglesby, R.J. (1994) An improved snow hydrology for GCMs. Part I: snow cover fraction, albedo, grain size, and age. *Climate Dynamics*, **10**, 21–37.

Marshall, S.J. (2002) Modelled nucleation centres of the Pleistocene ice sheets from an ice sheet model with subgrid topographic and glaciologic parameterizations. *Quaternary International*, **95–96**, 125–137.

Marshall, S.J. & Clark, P.U. (2002) Basal temperature evolution of North America ice sheets and implications for the 100-kyr cycle. *Geophysical Research Letters*, **29**, 67-1–67-4. doi: 10.1029/2002GL015192.

Marshall, S.J. & Clarke, G.K.C. (1996) Geologic and topographic controls on fast flow in the Laurentide and Cordilleran Ice Sheets. *Journal of Geophysical Research*, **101**, 17827–17839.

Marshall, S.J. & Clarke, G.K.C. (1997a) A continuum mixture model of ice stream thermomechanics in the Laurentide Ice Sheet 1. Theory. *Journal of Geophysical Research*, **102**, 20599–20613.

Marshall, S.J. & Clarke, G.K.C. (1997b) A continuum mixture model of ice stream thermomechanics in the Laurentide Ice Sheet 2. Application to the Hudson Strait Ice Stream. *Journal of Geophysical Research*, **102**, 20615–20637.

Marshall, S.J. & Clarke, G.K.C. (1999) Ice sheet inception: subgrid hypsometric parameterization of mass balance in an ice sheet model. *Climate Dynamics*, **15**, 533–550.

Marshall, S.J. & Cuffey, K.M. (2000) Peregrinations of the Greenland Ice Sheet divide in the last glacial cycle: implications for central Greenland ice cores. *Earth and Planetary Science Letters*, **179**, 73–90.

Marshall, S.J., Clarke, G.K.C., Dyke, A.S., *et al.* (1996) Geological and topographical controls on fast flow in the Laurentide and Cordilleran Ice Sheets. *Journal of Geophysical Research*, **101**, 17827–17839.

Marshall, S.J., Tarasov, L., Clarke, G., K.C. & Peltier, W.R. (2000) Glaciological reconstruction of the Laurentide Ice Sheet: physical processes and modeling challenges. *Canadian Journal of Earth Sciences*, **37**, 769–793.

Marshall, S.J., James, T.S. & Clarke, G.K.C. (2002) North American Ice Sheet reconstructions at the time of the Last Glacial Maximum. *Quaternary Science Reviews*, **21**, 175–192.

Martin, S. (1975) Corrélation bilans de masse annuels—facteurs météorologiques dans les Grandes Rousses. *Zeitschrift für Gletscherkunde und Glazialgeologie*, **10**(1–2), 89–100.

Martin, S., Munoz, E.A. & Drucker, R. (1997) Recent observations of a spring-summer surface warming over the Arctic Ocean. *Geophysics Research Letters*, **24**, 1259–1262.

Maslanik, J.A., Serreze, M.C. & Agnew, T. (1998) On the record reduction in 1998 western Arctic sea-ice cover. *Geophysics Research Letters*, **26**, 1905–1908.

Maslowski, W., Newton, B., Schlosser, P., Semtner, A. & Martinson, D. (2000) Modeling recent climate variability in the Arctic Ocean. *Geophysical Research Letters*, **27**, 3743–3746.

Maslowski, W., Marble, D.C. & Walczowski (2001) Recent trends in Arctic sea ice. *Annals of Glaciology*, **33**, 545–550.

Mathews, W.H. (1974) Surface profiles of the Laurentide ice sheet in its marginal areas. *Journal of Glaciology*, **13**, 37–43.

Mathews, W.H. (1979) Simulated glacial abrasion. *Journal of Glaciology*, **23**, 51–56.

Matsuoka, K. & Naruse R. (1999) Mass balance features derived from a firn core at Hielo Patagonico Norte, South America. *Arctic, Antarctic, and Alpine Research*, **31**(4), 333–340.

Matsuoka, K., Furukawa, T., Fujita, S., Maeno, H., Uratsuka, S., Naruse, R. & Watanabe, O. (2003) Crystal orientation fabrics within the Antarctic ice sheet revealed by a multipolarization plane and dual-frequency radar survey. *Journal of Geophysical Research*, 108, Art. No. 2499.

Matthews, J.A. (1977) A lichenometric test of the 1750 end-moraine hypothesis: Storbreen gletchervorfeld, southern Norway. *Norsk Geologisk Tidsskrift*, **31**, 129–136.

Matthews, W.H. (1974) Surface profiles of the Laurentide Ice Sheet in its marginal areas. *Journal of Glaciology*, **13**, 37–43.

Mayer, C., Reeh, N., Jung-Rothenhausler, F., Huybrechts, P. & Oerter, H. (2000) The subglacial cavity and implied dynamics under Nioghalverdsfjorden Glacier, NE Greenland. *Geophysical Research Letters*, **27**(15), 2289–2292.

Mayewski, P.A., Lyons, W.B., Ahmad, N., Smith, G. & Pourchet, M. (1984) Interpretation of the chemical and physical time series retrieved from Sentik Glacier, Ladakh Himalaya, India. *Journal of Glaciology*, **30**, 66–76.

Mayewski, P.A., Meeker, L.D., Twickler, M.S., Whitlow, S., Yang, Q. & Prentice, M. (1997) Major features and forcing of high-latitude Northern Hemisphere atmospheric circulation Over the Last 110,000 Years using a 110,000 long glaciochemical series. *Journal of Geophysical Research*, **102**, 26345–26366.

McCabe, A.M. (1985) Geomorphology. In: *The Quaternary History of Ireland* (Eds K.J. Edwards & W.P. Warren), pp. 67–93. Academic Press, Dublin.

McCall, J.G. (1952) The internal structure of a cirque glacier: report on studies of the englacial movements and temperatures. *Journal of Glaciology*, **2**(12), 12–131.

McCarroll, D. (2001) Deglaciation of the Irish Sea basin: a critique of the glaciomarine hypothesis. *Journal of Quaternary Science*, **16**, 393–404.

McCarroll, D. (2002) Amino-acid geochronology and the British Pleistocene: secure stratigraphical framework or a case of circular reasoning? *Journal of Quaternary Science*, **17**, 647–651.

McCarroll, D. & Ballantyne, C.K. (2000) The last ice sheet in Snowdonia. *Journal of Quaternary Science*, **15**, 765–778.

McCarroll, D. & Harris, C. (1992) The glacigenic deposits of western Lleyn, north Wales: terrestrial or marine? *Journal of Quaternary Science*, **7**, 19–29.

McCarroll, D., Knight, J. & Rijsdijk, K. (2001) The glaciation of the Irish Sea basin. *Journal of Quaternary Science*, **16**, 391–392.

McConnell, J.R. (2002) Continuous ice-core chemical analyses using inductively Coupled Plasma Mass Spectrometry. *Environmental Science and Technology*, **36**, 7–11.

McConnell, J.R., Arthern, R.J., Mosley-Thompson, E., *et al.* (2000b) Changes in Greenland ice sheet elevation attributed primarily to snow accumulation variability. *Nature*, **406**(6798), 877–879.

McDonald, M.G. & Harbaugh, A.W. (1988) *MODFLOW: A Modular Three-dimensional Finite-difference Ground-water Flow Model.* Techniques of Water-Resources Investigations, Book 6, U.S. Geological Survey.

McIntosh, J.C., Walter, L.M. & Martini, A.M. (2002) Pleistocene recharge to midcontinent basins: Effects on salinity structure and microbial gas generation. *Geochimica et Cosmochimica Acta*, **66**, 1681–1700.

McKenzie, G.D. & Peterson, D.N. (1975) Correspondence. Subglacial cavitation phenomena: comments on the paper by R. Vivian and G. Bocquet. *Journal of Glaciology*, **14**(71), 339–340.

McKinzey, K.M. (2001) *A New Little Ice Age chronology of the Franz Josef Glacier, West Coast, New Zealand.* Unpublished MSc thesis, University of Canterbury, New Zealand.

McMillan, A.A. (2002) Onshore Quaternary geological surveys in the 21st century—a perspective from the British Geological Survey. *Quaternary Science Reviews*, **21**, 889–899.

Meehan, R.T. (1998) *The Quaternary sedimentology and last deglaciation of northwest County Meath and adjacent parts of Counties Westmeath and Cavan, Ireland.* Unpublished PhD thesis, National University of Ireland, 504 pp.

Meehan, R.T. & Warren, W.P. (1999) *The Boyne Valley in the Ice Age: a Field Guide to some of theValley's most Important Glacial Geological Features.* Meath County Council and Geological Survey of Ireland, Dublin, 84 pp.

Meeker, L.D. & Maywekski, P.A. (2002) A 1400-year high-resolution record of atmospheric circulation over the North Atlantic and Asia. *The Holocene*, **12**, 257–266.

Meese, D.A., Gow, A.J., Grootes, P., *et al.* (1994) The accumulation record from the GISP2 core and as indictor of climate change, throughout the Holocene. *Science*, **266**, 1680–1682.

Meese, D.A., Gow, A.J., Alley, R.B., *et al.* (1997) The Greenland Ice Sheet Project 2 depth-age scale; methods and results. *Journal of Geophysical Research*, **102**(C12), 26411–26423.

Meier, M.A. (1965) Glaciers and climate. In: *The Quaternary of the United States* (Eds H.E.J. Wright & D.G. Frey), pp. 795–805. Princeton University Press, Princeton, NJ.

Meier, M.F. (1960) Mode of flow of Saskatchewan Glacier Alberta, Canada. *US Geological Survey Professional Paper*, **351**, 70 pp.

Meier, M.F. (1961) Mass budget of South Cascade Glacier. *US Geological Survey Professional Paper*, **424B**, 206–211.

Meier, M.F. (1962) Proposed definitions for mass budget terms. *Journal of Glaciology*, **4**(33), 252–265.

Meier, M. F. (1984) Contributions of small glaciers to global sea level. *Science*, **226**(4681), 1419–1421.

Meier, M.F. (1993) Ice, climate, and sea level; do we know what is happening? In: *Ice in the Climate System* (Ed. W.R. Peltier), pp. 141–160. Springer-Verlag, Berlin and Heidelberg.

Meier, M.F. & Bahr, D.B. (1996) Counting glaciers: use of scaling methods to estimate the number and size distribution of the glaciers on the world. *Glaciers, Ice Sheets and Volcanoes: a Tribute to Mark F. Meier* (Ed. S.C. Colbeck), 1–120. Special Report 96–27, Cold Regions Research and Engineering Laboratory (CRREL), Hanover, New Hampshire.

Meier, M. & Dyurgerov, M. (2002) How Alaska affects the world. *Science*, **297**, 350–351.

Meier, M.F. & Post, A. (1969) What are glacial surges? *Canadian Journal of Earth Sciences*, **6**, 807–817.

Meier, M.F. & Post, A. (1987) Fast tidewater glaciers. *Journal of Geophysical Research*, **92**(B9), 9051–9058.

Meier, M., Lundstrom, S., Stone, D., *et al.* (1994) Mechanical and hydrologic basis for the rapid motion of a large tidewater glacier 1. Observations. *Journal of Geophysical Research*, **99**, 15219–15229.

Meier, M.F., Dyurgerov, M.B. & McCabe, G.J. (2003) The health of glaciers: recent changes in glacier regime. *Climatic Change*, **59**, 123–135.

Meincke, J., Rudels, B. & Friedrich, H.J. (1997) The Arctic Ocean-Nordic Seas thermohaline system. *Journal of Marine Science*, **54**, 283–299.

Mellor, M. & Smith, J.H. (1966) Creep of snow and ice. *Cold Regions Research and Engineering Laboratory Report*, **220**, 13 pp.

Mellor, M. & Testa, R. (1969) Creep of ice under low stress. *Journal of Glaciology*, **8**, 147–152.

Menzies, J., Zaniewski, K. & Dreger, D. (1997) Evidence, from microstructures, of deformable bed structures in drumlins, Chimney Bluffs, New York State. *Sedimentary Geology*, **111**, 161–175.

Mercer, J.H. (1961) The response of fjord glaciers to changes in the firn limit. *Journal of Glaciology*, **3**, 850–866.

Mercer, J.H. (1978) West Antarctic ice sheet and CO_2 greenhouse effect: a threat of disaster. *Nature*, **271**, 321–325.

Mercer, J.H. (1981) West Antarctic ice volume: the interplay of sea level and temperature, and a strandline test for absence of the ice sheet during the last interglacial. In: *Sea Level, Ice, and Climatic Change. Proceedings of a symposium held during the XVII Assembly of the IUGG at Canberra, December 1979* (Ed. I. Allison), pp. 323–329. IAHS Publication 131, International Association of Hydrological Sciences, Wallingford.

Merlivat, L. & Jouzel, J. (1979) Global climatic interpretation of the deuterium-oxygen-18 relationship for precipitation. *Journal of Geophysical Research*, **84**(C8), 5029–5033.

Merritt, J.W. (1992) A critical review of the methods used in the appraisal of onshore sand and gravel resources in Britain. *Engineering Geology*, **32**, 1–9.

Merritt, J.W., Auton, C.A. & Firth, C.R. (1995) Ice-proximal glaciomarine sedimentation and sea-level change in the Inverness area, Scotland: A review of the deglaciation of a major ice stream of the British Late Devensian Ice Sheet. *Quaternary Science Reviews*, **14**, 289–329.

Messerli, B. & Ives, J.D. (1997) *Mountains of the World: a Global Priority; a Contribution to Chapter 13 of Agenda 21*. Parthenon Publishing Group, London.

Meyssonnier, J. (1982) Sliding of ice over a bump: numerical computation assuming Norton–Hoff's law and experimental values. In: *International Symposium on Numerical Models in Geomechanics* (Eds R. Dungar, C.G. Pande & S.A. Studer), pp. 344–352. A.A. Balkema, Salem, NH.

Michel, F.A. & van Everdingen, R.O. (1994) Changes in hydrogeologic regimes in permafrost regions due to climatic change. *Permafrost and Periglacial Processes*, **5**, 191–195.

Mickelson, D.M., Clayton, L., Fullerton, D.S. & Borns, H.W., Jr. (1983) The late Wisconsin glacial record of the Laurentide ice sheet in the United States. In: *Late Quaternary Environments of the United States*, Vol. 1, *The Late Pleistocene* (Ed. S.C. Porter), pp. 3–37. University of Minnesota Press, Minneapolis.

Mienert, J., Kenyon, N.H., Thiede, J. & Hollender, F.-J. (1993) Polar continental margins: studies off East Greenland. *Eos (Transactions of the American Geophysical Union)*, **74**, 225–236.

Miguel, M.C., Vespignani, A., Zapperi, S., Weiss, J. & Grasso, J.R. (2001) Intermittent dislocation flow in viscoplastic deformation. *Nature*, **410**, 667–671.

Miller, G.H. & de Vernal, A. (1992) Will greenhouse warming lead to northern hemisphere ice sheet growth? *Nature*, **355**, 244–246.

Miller, G.H., Wolfe, A.P., Steig, E.J., Sauer, P.E., Kaplan, M.R. & Briner, J.P. (2002) The Goldilocks dilemma: big ice, little ice, or 'just-right' ice in the Eastern Canadian Arctic. *Quaternary Science Reviews*, **21**, 33–48.

Miller, L. & Douglas, B.C. (2003) Mass and volume contributions to twentieth-century global sea level rise. *Nature*, **25**, 406–409.

Mingram, B. & Brauer, K. (2001) Ammonium concentration and nitrogen isotope composition in metasedimentary rocks from different tectonometamorphic units of the European Variscan Belt. *Geochimica et Cosmochimica Acta*, **65**, 273–287.

Mitchell, W.A. (1994) Drumlins in ice sheet reconstructions with special reference to the western Pennines. *Sedimentary Geology*, **91**, 313–332.

Mitrovica, J. (1996) Haskell [1935] revisited. *Journal of Geophysical Research*, **101**(B1), 555–569.

Mohr, J.J., Reeh, N. & Madsen, S.N. (1998) Three dimensional glacial flow and surface elevations measured with radar interferometry. *Nature*, **391**, 273–276.

Möller, D. (1996) Die Höhen und Höhenänderungen des Inlandeises. Die Weiterfürung der geodätischen Arbeiten der Internationalen Glaziologischen Grönland-Expedition (EGIG) durch das Institut für Vermessungskunde der TU Braunschweig 1987–1993. Deutsche Geodätische Kommission bei der Bayrischen Akademie der Wissenschaften, Reihe B, Angewandte Geodäsie, Heft Nr. 303. Verlag der Bayrischen Akademie der Wissenschaften, 49–58.

Montagnat, M. & Duval, P. (2000) Rate-controlling processes in the creep of polar ice, influence of grain boundary migration associated with recrystallization. *Earth and Plantetary Science Letters*, **183**, 179–186.

Montagnat, M., Duval, P., Bastie, P., *et al.* (2003) Lattice distortions in ice crystals from the Vostok core (Antarctica) revealed by hard X-ray diffraction; implication in the deformation of ice at low stresses. *Earth and Plantetary Science Letters*, **214**, 369–378.

Mooers, H.D. (1997) Terrestrial record of Laurentide Ice Sheet reorganization during Heinrich events. *Geology*, **25**, 987–990.

Moore, G.E. (1965) Cramming more components onto integrated circuits. *Electronics*, **38**(8).

Moore, J.C., Mulvaney, R. & Paren, J.G. (1989) Dielectric stratigraphy of ice—a new technique for determining total ionic concentrations in polar ice cores. *Geophysical Research Letters*, **16**, 1177–1180.

Moore, P.L. & Iverson, N.R. (2002) Slow episodic shear of granular materials regulated by dilatant strengthening. *Geology*, **30**, 843–846.

Moran, S.R. (1971) Glaciotectonic structures in drift. In: *Till: a Symposium* (Ed. R.P. Goldthwait), pp. 127–148. Ohio State University Press, Columbus.

Moran, S.R., Clayton, L., Hooke, R.L, Fenton, M.M. & Andriashek, L.D. (1980). Glacier-bed landforms of the prairie region of North America. *Journal of Glaciology*, **25**, 457–476.

Morgan, V.I. (1991) High temperature ice creep tests, *Cold Regions Science and Technology*, **19**(3) 295–300.

Moros, M., Kuijpers, A., Snowball, I., *et al.* (2002) Were glacial iceberg surges in the North Atlantic triggered by climatic warming? *Marine Geology*, **192**, 393–417.

Morris, E.M. & Mulvaney, R. (1995) Recent changes in surface elevation of the Antarctic Peninsula ice sheet. *Zeitschrift fur Gletscherkunde und Glazialgeologie*, **31**, 7–15.

Morris, E.M. & Vaughan, D.G. (2003) Spatial and temporal variation of surface temperature on the Antarctic Peninsula and the limit of viability of ice shelves. In: *Antarctic Research Series*, Vol. 79 (Eds E. Domack, A. Burnett & A. Leventer), pp. 61–68. American Geophysical Union, Washington, DC.

Morris, E.M. & Vaughan, D.G. (In press) Spatial and temporal variation of surface temperature on the Antarctic Peninsula and the limit of viability of ice shelves. In: *Antarctic Research Series* (Ed. E. Domack). American Geophysical Union, Washington, DC.

Morse, J.W. & Arvidson, R.S. (2002) The dissolution kinetics of major sedimentary carbonate minerals. *Earth-Science Reviews*, **58**, 51–84.

Mote, T.L. (2000) Ablation rate estimates over the Greenland ice sheet from microwave radiometric data. *Professional Geographer*, **52**, 322–331.

Motyka, R., Hunter, L., Echelmeyer, K.A. & Connor, C. (2003) Submarine melting at the terminus of a temperate tidewater glacier, Leconte Glacier, Alaska, U.S.A., *Annals of Glaciology*, **36**, 57–65.

Mukherjee, A.K. (1971) The rate-controlling mechanism in superplasticity. *Materials Science and Engineering*, **8**, 83–89.

Müller, F. (Ed.) (1977) *Fluctuations of Glaciers 1970–1975*, Vol. III. IAHS(ICSI)–UNESCO, Paris.

Müller-Lemans, H., Funk, M., Aellen, M. & Kappenberger, G. (1995) Langjährige Massenbilanzreihen von Gletschern in der Schweiz. *Zeitschrift für Gletscherkunde und Glazialgeologie*, **30**(1994), 141–160.

Munro, M. & Shaw, J. (1997) Erosional origin of hummocky terrain, south-central Alberta, Canada. *Geology*, **25**, 1027–1030.

Munro-Stasiuk, M.J. (2000) Rhythmic till sedimentation: evidence for repeated hydraulic lifting of a stagnant ice mass. *Journal of Sedimentary Research*, **70**, 94–106.

Munro-Stasiuk, M.J. (2003) Subglacial Lake McGregor, south-central Alberta, Canada. *Sedimentary Geology*, **160**, 325–350.

Munro-Stasiuk, M.J. & Shaw, J. (2002) The Blackspring Ridge Flute Field, south-central Alberta, Canada: evidence for subglacial sheetflow erosion. *Quaternary International*, **90**, 75–86.

Murray, E.A. (1988) *Subglacial erosional marks in the Kingston, Ontario, Canada, region*. MSc. thesis, Queen's University, Canada, 171 pp.

Murray, T. & Clarke, G.K.C. (1995) Black-box modelling of the subglacial water system. *Journal of Geophysical Research*, **100**(B7), 10231–10245.

Murray, T. & Dowdeswell, J.A. (1992) Water throughflow and physical effects of deformation on sedimentary glacier beds. *Journal of Geophysical Research*, **97**(B6), 8993–9002.

Murray, T. & Porter, P.R. (2001) Basal conditions beneath a soft-bedded polythermal surge-type glacier: Bakaninbreen, Svalbard. *Quaternary International*, **86**(1), 103–116.

Murray, T. (1997) Assessing the paradigm shift: deformable glacier beds. *Quaternary Science Reviews*, **16**, 995–1016.

Murray, T., Stuart, G.W., Fry, M., Gamble, N.H. & Crabtree, M.D. (2000a) Englacial water distribution in a temperate glacier from surface and borehole radar velocity analysis. *Journal of Glaciology*, **46**(154), 389–398.

Murray, T., Stuart, G. W., Miller, P. J., *et al.* (2000b) Glacier surge propagation by thermal evolution at the bed. *Journal of Geophysical Research–Solid Earth*, **105**, 13491–13507.

Murton, J.B. (2005) Ground-ice stratigraphy and formation at North Head, Tuktoyaktuk Coastlands, western Arctic Canada: a product of glacier-permafrost interactions. *Permafrost and Periglacial Processes*, **16**, 31–50.

Murton, J.B., Waller, R.I., Hart, J.K., Whiteman, C.A., Pollard, W.H & Clark, I.D. (2004) Stratigraphy and glaciotectonic structures of permafrost deformed beneath the northwest margin of the Laurentide Ice Sheet, Tuktoyaktuk Coastlands, Canada. *Journal of Glaciology*, **50**(170), 399–412.

Murton, J.B., Whiteman, C.A., Waller, R.I, Pollard, W.H., Clark, I.D. & Dallimore, S.R. (2005) Basal ice facies and supraglacial melt-out till of the Laurentide Ice Sheet, Tuktoyaktuk Coastlands, western Arctic Canada. *Quaternary Science Reviews*, **24**, 681–708.

Mysak, L.A. & Power, S.B. (1992) Sea-ice anomalies in the western Arctic and Greenland-Iceland Sea and their relation to an interdecadal climate cycle. *Climatological Bulletin*, **26**, 147–176.

Naish, T.R., Woolfe, K.J., Barrett, P.J., *et al.* (2001) Orbitally induced oscillations in the East Antarctic ice sheet at the Oligocene/Miocene boundary. *Nature*, **413**, 719–723.

Naito, N., Ageta, Y., Nakawo, M., Waddington, E.D., Raymond, C.F. & Conway, H. (2001) Response sensitivities of a summer-accumulation type glaciers to climate changes indicated with a glacier fluctuation model. *Bulletin of Glaciological Research*, **18**, 1–8.

Nakamura, T. & Jones, S.J. (1973 Mechanical properties of impure ice crystals. In: *Physics and Chemistry of Ice* (Eds E. Whalley, S.J. Jones & L.W. Gold), pp. 365–369. Royal Society of Canada, Ottawa.

Nakawo, M., Raymond, C.F. & Fountain, A.G. (Eds) (2000) *Debris-covered Glaciers*. IAHS Publication 264, International Association of Hydrological Sciences, Wallingford.

Naruse, R. & Skvarca, P. (2000) Dynamic features of thinning and retreating Glaciar Upsala, a lacustrine calving glacier in Southern Patagonia. *Arctic, Antarctic, and Alpine Research*, **32**(4), 485–491.

Naruse, R., Aniya, M., Skvarca, P. & Casassa, G. (1995) Recent variations of calving glaciers in Patagonia, South America, revealed by ground surveys, satellite-data analyses and numerical experiments. *Annals of Glaciology*, **21**, 297–303.

Naruse, R., Skvarca, P. & Takeuchi, Y. (1997) Thinning and retreat of Glaciar Upsala, and an estimate of annual ablation changes in southern Patagonia. *Annals of Glaciology*, **24**, 38–42.

Nereson, N.A. & Raymond, C.F. (2001) The elevation history of ice streams and the spatial accumulation pattern along the Siple Coast of West Antarctica inferred from ground-based radar data from three inter-ice-stream ridges. *Journal of Glaciology*, **47**, 303–313.

Nereson, N.A., Hindmarsh, R.C.A. & Raymond, C.F. (1998a) Sensitivity of the divide position at Siple Dome, West Antarctica, to boundary forcing. *Annals of Glaciology*, **27**, 207–214.

Nereson, N.A., Raymond, C.F., Jacobel, R.W. & Waddington, E.D. (1998b) Migration of the Siple Dome ice divide, West Antarctica. *Journal of Glaciology*, **44**(148), 643–652.

Nereson, N.A., Raymond, C.F., Jacobel, R.W. & Waddington, E.D. (2000) The accumulation pattern across Siple Dome, West Antarctica, inferred from radar-detected internal layers. *Journal of Glaciology*, **46**, 75–87.

Nesje, A., Kvamme, M., Lovlie, R. & Rye, N. (1991) Holocene Glacial and Climate History of the Jostedalsbreen Region, Western Norway; Evidence from Lake Sediments and Terrestrial Deposits. *Quaternary Science Reviews*, **10**, 87–114.

Nesje, A., Dahl, S.O. & Bakker, J. (2004) Were abrupt lateglaciall and early-Holocene climatic changes in northwest Europe linked to freshwater outbursts to the North Atlantic and Arctic Oceans? *The Holocene*, **14**, 299–310.

New, M., Hulme, M. & Jones, P. (1999) Representing twentieth century space-time variability. I. Development of a 1961–1990 mean monthly terrestrial climatology. *Journal of Climate*, **12**, 829–856.

Ng, F.S.L. (2000a) Canals under sediment-based ice sheets. *Annals of Glaciology*, **30**, 146–153.

Ng, F.S.L. (2000b) Coupled ice-till deformation near subglacial channels and cavities. *Journal of Glaciology*, **46**, 580–598.

Nickling, W.G. & Bennett, L. (1984) The shear strength characteristics of frozen coarse granular debris. *Journal of Glaciology*, **30**, 348–357.

Nicolussi, K. (1990) Bilddokumente zur Geschichte des Vernagtferners im 17. Jahrhundert. *Zeitschrift für Gletscherkunde und Glazialgeologie*, **26**(2), 97–119.

Niemelä, J., Ekman, I. & Lukashov, A. (Eds) (1993) *Quaternary Deposits of Finland and Northwestern Part of Russian Federation and their Resources*. Map at 1 : 1,000,000, Geological Survey of Finland and Institute of Geology, Karelian Science, Centre of the Russian Academy of Sciences.

Nienow, P., Sharp, M. & Willis, I. (1998) Seasonal changes in the morphology of the subglacial drainage system, Haut Glacier d'Arolla, Switzerland. *Earth Surface Processes and Landforms*, **23**, 825–843.

Nishiizumi, K., Finkel, R.C., Ponganis, K.V., Graf, T., Kohl, C.P. & Marti, K. (1996) *In situ* produced cosmogenic nuclides in GISP2 rock core from Greenland Summit (Abstract). *Eos* (*Transactions of the American Geophysical Union*), Fall Meeting 1996, **77**(46) Supplement, F428, Abstract OS41B–10.

Nogami, M. (1972) The snowline and climate during the last glacial period in the Andes Mountains. *Quaternary Research*, **11**, 71–80.

Nolan, M. & Echelmeyer, K. (1999) Seismic detection of transient changes beneath Black Rapids Glacier, Alaska, U.S.A.: II. Basal morphology and processes. *Journal of Glaciology*, **45**, 132–146.

Norddahl, H. (1983) Late Quaternary stratigraphy of Fnjóskadalur central North Iceland; a study of sediments, ice-lake strandlines, glacial isostacy and ice-free areas. *Lunqua*, **12**, 78.

Norddahl, H. (1991) Late Weichselian and Early Holocene deglaciation of Iceland. *Jökull*, **40**, 27–50.

Norddahl, H. & Pétursson, G.P. (In press) Relative sea level changes in Iceland. New aspect of the Weichselian deglaciation of Iceland. In: *The Environments of Iceland* (Ed. C. Caseldine). Amsterdam: Elsevier.

Norman, G.W.H. (1938) The last Pleistocene ice-front in Chibougamau District, *Quebec. Transactions of the Royal Society of Canada*, Series 3, section IV, pp. 69–86.

Normark, W.R. & Reid, J.A. (2003) Extensive deposits on the Pacific Plate from Late Pleistocene North American glacial lake outbursts. *Journal of Geology*, **111**, 617–637.

Nye, J.F. (1952) The mechanics of glacier flow. *Journal of Glaciology*, **2**, 82–93.

Nye, J.F. (1953) The flow law of ice from measurements in glacier tunnels, laboratory experiments, and the Jungfraufirn borehole experiment. *Proceedings of the Royal Society of London, Series A*, **219**, 477–489.

Nye, J.F. (1957) The distribution of stress and velocity in glaciers and ice sheets. *Proceedings of the Royal Society of London Series A*, **239**, 113–133.

Nye, J.F. (1959) The motion of ice sheets and glaciers. *Journal of Glaciology*, **3**, 493–507.

Nye, J.F. (1960) The response of glaciers and ice sheets to seasonal and climatic changes. *Proceedings of the Royal Society of London, Series A*, **256**, 559–584.

Nye, J.F. (1967) Theory of regelation. *Philosophical Magazine*, **16**, 1249–1266.

Nye, J.F. (1970) Glacier sliding without cavitation in a linear viscous approximation. *Proceedings of the Royal Society of London, Series A*, **315**, 381–403.

Nye, J.F. (1973) Water at the bed of a glacier. In: *Symposium on the Hydrology of Glaciers, Cambridge 1969*, pp. 189–194. IAHS Publication 95, International Association of Hydrologic Sciences, Wallingford.

Nye, J.F. (1989) The geometry of water veins and nodes in polycrystalline ice. *Journal of Glaciology*, **35**, 17–22.

Nye, J.F. (2000) A flow model for the polar caps of Mars. *Journal of Glaciology*, **46**, 438–444.

Ó Cofaigh, C. (1996) Tunnel valley genesis. *Progress in Physical Geography*, **20**(1), 1–19.

Ó Cofaigh, C. & Dowdeswell, J.A. (2001) Laminated sediments in glacimarine environments: diagnostic criteria for their interpretation. *Quaternary Science Reviews*, **20**, 1411–1436.

Ó Cofaigh, C. & Evans, D.J.A. (2001) Sedimentary evidence for deforming bed conditions associated with a grounded Irish Sea glacier, southern Ireland. *Journal of Quaternary Science*, **16**, 435–454.

Ó Cofaigh, C., Lemmen, D.S., Evans, D.J.A. & Bednarski, J. (1999) Glacial landform/sediment assemblages in the Canadian High Arctic and their implications for late Quaternary glaciation. *Annals of Glaciology*, **28**, 195–201.

Ó Cofaigh, C., Pudsey, C.J., Dowdeswell, J.A. & Morris, P. (2002a) Evolution of subglacial bedforms along a paleo-ice stream, Antarctic Peninsula continental shelf. *Geophysical Research Letters*, **29**. 10.1029/2001GL014488.

Ó Cofaigh, C., Taylor, J., Dowdeswell, J.A., *et al.* (2002b) Sediment reworking on high-latitude continental margins and its implications for palaeoceanographic studies: insights from the Norwegian-Greenland Sea. In: *Glacier-influenced Sedimentation on High-latitude Continental Margins* (Eds J.A. Dowdeswell & C. ó Cofaigh), pp. 325–348. Special Publication 203, Geological Society Publishing House, Bath.

Ó Cofaigh, C., Taylor, J., Dowdeswell, J.A. & Pudsey, C.J. (2003) Palaeo-ice streams, trough mouth fans and high latitude continental slope sedimentation. *Boreas*, **32**, 37–55.

Ó Cofaigh, C., Dowdeswell, J.A., Kenyon, N.H., Evans, J., Taylor, J., Mienert, J. & Wilken, M. (2004) Timing and significance of glacially-influenced mass wasting in the submarine channels of the Greenland Basin. *Marine Geology*, **207**, 39–54.

Ó Cofaigh, C., Dowdeswell, J.A., Allen, C.S., Hiemstra, J., Pudsey, C.J., Evans, J. & Evans, D.J.A. (2005) Flow dynamics and till genesis associated with a marine-based Antarctic palaeo-ice stream. *Quaternary Science Reviews*, **24**, 709–740.

O'Grady, D.B. & Syvitski, J.P.M. (2002) Large-scale morphology of Arctic continental slopes: the influence of sediment delivery on slope form. In: *Glacier-influenced Sedimentation on High-latitude*

Continental Margins (Eds J.A. Dowdeswell & C. Ó Cofaigh), pp. 11–31. Special Publication 203, Geological SocietyPublishing House, Bath.

Oerlemans, J. (1980) Model experiments on the 100,000-yr glacial cycle. *Nature*, **287**, 430–432.

Oerlemans, J. (1982) Glacial cycles and ice sheet modelling. *Climatic Change*, **4**, 353–374.

Oerlemans, J. (Ed.) (1989) *Glacier Fluctuations and Climatic Change*. Kluwer, Dordrecht, 417 pp.

Oerlemans, J. (1993a) Evaluating the role of climate cooling in iceberg production and Heinrich events. *Nature*, **364**, 783–786.

Oerlemans, J. (1993b) A model for the surface balance of ice masses: part I. Alpine glaciers. *Zeitschrift für Gletscherkunde und Glazialgeologie*, **27/28**, 63–83.

Oerlemans, J. (1993c) Modelling of glacier mass balance. In: *Ice in the Climate System* (Ed. W.R. Peltier), pp. 101–116. Springer-Verlag, Berlin and Heidelberg.

Oerlemans, J. (1997) Climate sensitivity of Franz Josef Glacier, New Zealand, as revealed by numerical modeling. *Arctic and Alpine Research*, **29**, 233–239.

Oerlemans, J. (2001) *Glaciers and Climate Change*. A.A. Balkema, Rotterdam.

Oerlemans, J. & Grisogono, B. (2002) Glacier wind and parameterisation of the related surface heat flux. *Tellus*, **A54**, 440–452.

Oerlemans, J. & Fortuin, J.P.F. (1992) Sensitivity of glaciers and small ice caps to greenhouse warming. *Science*, **258**, 115–118.

Oerlemans, J. & Hoogendoorn, N.C. (1989) Mass-balance gradients and climate change. *Journal of Glaciology*, **35**(121), 399–405.

Oerlemans, J. & Klok, E.J. (2002) Energy balance of a glacier surface: analysis of AWS data from the Morteratschgletscher, Switzerland. *Arctic, Antarctic and Alpine Research*, **34**(123), 115–123.

Oerlemans, J. & Knap, W.H. (1998) A 1 year record of global radiation and albedo in the ablation zone of Morteratschgletscher, Switzerland. *Journal of Glaciology*, **44**(147), 231–238.

Oerlemans, J. & Reichert, B.K. (2000) Relating glacier mass balance to meteorological data using a seasonal sensitivity characteristic (SSC). *Journal of Glaciology*, **46**(152), 1–6.

Oerlemans, J. & van der Veen, C.J. (1984) *Ice Sheets and Climate*. D. Reidel, Dordrecht.

Oerlemans, J., Anderson, B., Hubbard, A., Huybrechts, P., *et al.* (1998) Modelling the response of glaciers to climate warming. *Climate Dynamics*, **14**, 267–274.

Oerter, H., Kipfstuhl, J., Determann, J., *et al.* (1992) Evidence for basal marine ice in the Filchner-Ronne Ice Shelf, *Nature*, **358**, 399–401.

O'Farrell, S.P., McGregor, J.L., Rotstayn, L.D., Budd, W.F., *et al.* (1997) Impact of transient increases in atmospheric CO2 on the accumulation and mass balance of the Antarctic ice sheet. *Annals of Glaciology*, **25**, 137–144.

Ogilvie, A.E.J. (1991) Climatic change in Iceland A.D. c.865 to 1598. *Acta Archaeologica*, **61**, 233–251.

Ogilvie, A.E.J. (1992) Documentary evidence for changes in the climate of Iceland, A.D. 1500 to 1800. In: *Climate Since A.D. 1500* (Eds R.S. Bradley & P.D. Jones), pp. 92–117. Routledge, London.

Ogilvie, A.E.J. (1997) Fisheries, Climate and Sea Ice in Iceland: an historical perspective. In: *Marine Resources and Human Societies in the North Atlantic Since 1500* (Ed. D. Vickers), pp. 69–87. Institute of Social and Economic Research, Memorial University, St Johns.

Ogilvie, A.E.J. & Jónsson, T. (2001) 'Little Ice Age' research: A perspective from Iceland. *Climatic Change*, **48**, 9–52.

Ogilvie, A.E., Barlow, L.K. & Jennings, A.E. (2000) North Atlantic Climate c.A.D. 1000: Millennial Reflections on the Viking Discoveries of Iceland, Greenland and North America. *Weather*, **55**, 34–45.

Ohmura, A. (2001) Physical basis for the temperature-based melt-index method. *Journal of Applied Meteorology*, **40**, 753–761.

Ohmura, A. & Reeh, N. (1991) New precipitation and accumulation maps for Greenland. *Journal of Glaciology*, **37**, 140–148.

Ohmura, A., Kasser, P. & Funk, M. (1992) Climate at the equilibrium line of glaciers. *Journal of Glaciology*, **38**(130), 397–411.

Ohmura, A., Wild, M. & Bengtsson, L. (1996) A possible change in mass balance of Greenland and Antarctic ice sheets in the coming century. *Journal of Climate* **9**, 2124–2135.

Ólafsdóttir, Th. (1975) Jökulgardur á sjávarbotni út af Breidafirdi (English summary: A moraine ridge on the Iceland shelf, west of Breidafjördur). *Náttúrufrædingurinn*, **45**, 247–271.

Olafsson, J. (1999) Connections between oceanic conditions off N-Iceland, Lake Myvatn temperature, regional wind direction variability and the North Atlantic Oscillation. *Rit Fiskideildar*, **16**, 41–57.

Olesen, O.B. & Reeh, N. (1969) Preliminary report on glacier observations in Nordvestfjord, East Greenland. *Grønlands Geoeologiske Undersøgelse Rapport*, **21**, 41–53.

O'Neel, S., Echelmeyer, K.A. & Motyka, R.J. (2001) Short-term flow dynamics of a retreating tidewater glacier: LeConte Glacier, Alaska, USA *Journal of Glaciology*, **47**, 567–578.

O'Neill, P.O. (1985) *Environmental Chemistry*. George Allen and Unwin, London, 232 pp.

Oppenheimer, M. (1998) Global warming and the stability of the West Antarctic ice sheet. *Nature*, **393**, 325–332.

Østrem, G. (1964) Ice-cored moraines in Scandinavia. *Geografiska Annaler*, **46**, 282–337.

Østrem, G. (1965) Problems of dating ice-cored moraines. *Geografiska Annaler*, **47A**, 1–38.

Østrem, G. & Brugman, M. (1991) *Glacier Mass Balance Measurements: a Manual for Field and Office Work*. National Hydrology Research Institute, Saskatoon, Canada, and the Norwegian Water Resources and Electricity Board, Oslo, Norway.

Østrem, G. & Stanley, A. (1969) *Glacier Mass Balance Measurements: a Manual for Field and Office Work*. The Canadian Department of Energy, Mines and Resources, Ottawa, and The Norwegian Water Resources and Electricity Board, Oslo.

Østrem, G., Haakensen, N. & Eriksson, T. (1981) The glaciation level in southern Alaska. *Geografiska Annaler*, **63A**, 251–260.

Ottesen, D., Dowdeswell, J.A., Rise, L., Rokoengen, K. & Henriksen, S. (2002) Large-sacle morphological evidence for past ice-stream flow on the mid-Norwegian continental margin. In: *Glacier-influenced Sedimentation on High-latitude Continental Margins* (Eds J.A. Dowdeswell & C. ó Cofaigh), pp. 245–258. Special Publication 203, Geological SocietyPublishing House, Bath.

Ottesen, D., Dowdeswell, J.A. & Rise, L. (2005) Submarine landforms and the reconstruction of fast-flowing ice streams within a large Quaternary ice sheet: the 2,500 km-long Norwegian–Svalbard margin (57° to 80°N) *Geological Society of America Bulletin*, **117**, 1033–1050.

Overland, J.E., Spillane, M.C., Percival, D.B., Wang, M. & Mofjeld, H.O. (2004) Seasonal and regional variation of pan-Arctic surface air temperature over the instrumental record. *Journal of Climate*, **17**, 3263–3282.

Owen, E.B. (1967) Northern hydrogeological region. In: *Groundwater in Canada* (Ed. I.C. Brown), pp. 173–194. GSC Economic Geology Report No. 24, Geological Survey of Canada, Queen's Printer, Ottawa.

Owen, L.A. & Derbyshire, E. (1989) The Karakoram glacial depositional system. *Zeitschrift für Geomorphologie* **76**, 33–73.

PARCA (2001) *Journal of.Geophysical.Research*, Special Issue **106**(D24).

Parizek, B.R., Alley, R.B., Anandakrishnan, S. & Conway, H. (2002) Sub-catchment melt and long-term stability of Ice Stream D, West Antarctica. *Geophysical Research Letters*, **29**(8). 10.1029/2001GL014326.

Parizek, B.R., Alley, R.B. & Hulbe, C.L. (2003) Subglacial thermal balance permits ongoing grounding-line retreat along the Siple Coast of West Antarctica. *Annals of Glaciology*, **36**, 251–256.

Paterson, W.S.B. (1969) *The Physics of Glaciers*. Pergamom, Oxford, 250 pp.

Paterson, W.S.B. (1972) Laurentide Ice Sheet: estimated volumes during the late Wisconsin. *Reviews of Physics and Space Physics*, **10**, 885–917.

Paterson, W.S.B. (1981) *The Physics of Glaciers*, 2nd edn. Pergamon Press, Oxford.

Paterson, W.S.B. (1991) Why ice-age ice is sometimes 'soft'. *Cold Regions Science and Technology*, **20**, 75–98.

Paterson, W.S.B. (1994) *The Physics of Glaciers*, 3rd edn. Pergamon, Oxford, 480 pp.

Paterson, W.S.B. & Reeh, N. (2001) Thinning of the ice sheet in northwest Greenland over the past forty years. *Nature*, **414**, 60–62.

Patterson, C.J. (1994) Tunnel-valley fans of the St Croix moraine, east-central Minnesota, USA. In: *Formation and Deformation of Glacial Deposits* (Eds W.P. Warren & D.G. Croot), pp. 69–87. Balkema, Rotterdam.

Pattyn, F. (2002) Transient glacier response with a higher-order numerical ice-flow model. *Journal of Glaciology*, **48**(162), 467–477.

Pattyn, F. (2003) A new three-dimensional higher-order thermomechanical ice sheet model: Basic sensitivity, ice stream development, and ice flow across subglacial lakes. *Journal of Geophysical Research*, **108**(B8). Art. No. 2382. doi: 10.102972002JB002329.

Paul, F., Kääb, A., Maisch, M., Kellenberger, T. & Haeberli, W. (2002) The new remote sensing-derived Swiss Glacier Inventory: I. Methods. *Annals of Glaciology*, **34**, 355–361.

Paul, F., Kääb, A., Maisch, M., Kellenberger, T.W. & Haeberli, W. (2004) Rapid disintegration of Alpine glaciers observed with satellite data. *Geophysical Research Letters*, **31**. L21402, doi: 10.1029/2004GL020816.

Paul, M.A. (1983) The supraglacial landsystem. In: *Glacial Geology* (Ed. N. Eyles), pp. 71–90. Pergamon, Oxford.

Paul, M.A. & Eyles, N. (1990) Constraints on the preservation of diamict facies (melt-out tills) at the margins of stagnant glaciers. *Quaternary Science Reviews*, **9**, 51–69.

Payne, A.J. (1995) Limit cycles in the basal thermal regimes of ice sheets. *Journal of Geophysical Research*, **100**(B3), 4249–4263.

Payne, A.J. (1998) Dynamics of the Siple Coast ice streams, West Antarctica: Results from a thermomechanical ice sheet model. *Geophysical Research Letters*, **25**(16), 3173–3176.

Payne, A.J. (1999) A thermomechanical model of ice flow in West Antarctica. *Climate Dynamics*, **15**(2), 115–125.

Payne, A.J. & Baldwin, D.J. (2000) Analysis of ice-flow instabilities identified in the EISMINT intercomparison exercise. *Annals of Glaciology*, **30**, 204–210.

Payne, A.J. & Dongelmans, P.W. (1997) Self organization in the thermomechanical flow of ice sheets. *Journal of Geophysical Research*, **102**, 12219–12234.

Payne, A.J., Huybrechts, P., Abe-Ouchi, A., *et al.* (2000) Results from the EISMINT model intercomparison: the effects of thermomechanical coupling. *Journal of Glaciology*, **46**, 227–238.

Payne, A.J., Vieli, A., Shepherd, A., Wingham, D.J. & Rignot, E. (2004) Recent dramatic thinning of largest West Antarctic ice stream triggered by oceans. *Geophysical Research Letters*, **31**(L23401). doi: 10.1029/204GL021284.

Peakall, J., Ashworth, P. & Best.J. (1996) Physical modeling in fluvial geomorphology: Principles, applications and unresolved issues. In: *The Scientific Nature, of Geomorphology* (Eds B.L. Rhoads & C. Thorn), pp. 221–253. Wiley, Chichester.

Peltier, W.R. (1994) Ice Age paleotopography. *Science*, **265**, 195–201.

Peltier, W.R. (1996) Mantle viscosity and Ice-Age ice sheet topography. *Science*, **273**, 1359–1364.

Peltier, W.R. (1998) Postglacial variations in the level of the sea: Implications for climate dynamics and solid-earth geophysics. *Reviews of Geophysics*, **36**, 603–689.

Peltier, W.R. & Andrews, J.T. (1976) Glacial-Isostatic Adjustment–I. The Forward Problem. *Geophysical Journal of the Royal Astronomical Society*, **46**, 605–646.

Peltier, W.R., Goldsby, D.L., Kohlstedt, D.L., *et al.* (2000) Ice-age ice-sheet rheology: constraints on the Last Glacial Maximum form of the Laurentide ice sheet. *Annals of Glaciology*, **30**, 163–176.

Peña, H. & Nazarala, N. (1987) Snowmelt-runoff simulation model of a central Chile Andean basin with relevant orographic effects. In: *Large Scale Effects of Seasonal Snow Cover* (Eds B.E. Goodison, R.G. Barry & J. Dozier), pp. 161–172. IAHS Publication 166, International Association of Hydrological Sciences, Wallingford.

Perg, L.A., Von Blackenburg, F., Kubik, P. (2002) Cosmogenic nuclide budget in a glaciated mountain range (W. Alps). *Geochimica et Cosmochimica Acta*, **66**(15A), A591–A591 (Supplement).

Person, M., Dugan, B., Swenson, J.B., *et al.* (2003) Pleistocene hydrogeology of the Atlantic continental shelf, New England. *Geological Society of America Bulletin*, **115**, 1324–1343.

Peteet, D. (1995) Global Younger Dryas? *Quaternary International*, **28**, 93–104.

Peterson, B.J.B., Holmes, R.M., McClelland, J.W., *et al.* (2002) Increasing river discharge to the Arctic Ocean. *Science*, **298**, 2171–2173.

Petit, J., White, J., Young, N., Jouzel, J. & Korotkevich, Y. (1991) Deuterium excess in recent Antarctic snow. *Journal of Geophysical Research*, **96**(D3), 5113–5122.

Petit, J.-R., Jouzel, J., Raynaud, D., *et al.* (1999) Climate and atmospheric history of the past 420,000 years from the Vostok ice core, Antarctica. *Nature*, **399**(6735), 429–436.

Pettit, E.C. (2003) *Unique dynamic behaviors of ice divides: Siple Dome and the rheological properties of ice*. PhD thesis, University of Washington.

Pettit, E.C. & Waddington, E.D. (2003) Ice flow at low deviatoric stress. *Journal of Glaciology*, **49**(166), 359–369.

Pfeffer, W.T., Dyurgerov, M., Kaplan, M., *et al.* (1997) Numerical modeling of late glacial Laurentide advance of ice across Hudson Strait: Insights into terrestrial and marine geology, mass balance, and calving flux. *Paleoceanography*, **12**, 97–110.

Pfeffer, W.T., Cohn, J., Meier, M.F & Krimmel, R. (2000) Alaskan glacier beats a dramatic retreat. *Eos (Transactions of the American Geophysical Union)*, **81**(48), 577–584.

Philip, J.R. (1980) Thermal fields during regelation. *Cold Regions Science and Technology*, **3**, 193–203.

Pimienta, P. (1987) Etude du comportement mecanique des glaces polycristallines aux faibles contraintes; applications aux glaces des calottes polaires. PhD thesis, University of Grenoble, France.

Pimienta, P. & Duval, P. (1987) Rate controlling processes in the creep of polar glacier ice. *Journal de Physique*, **48**, 243–248.

Pimienta, P., Duval, P. & Lipenkov, V. Ya. (1987) Mechanical behaviour of anisotropic polar ice. In: *Physical Basis of Ice Sheet Modelling* (Eds E.D. Waddington & J.S. Walder), pp. 57–66. Proceedings of a symposium held during the XIX General Assembly of the IUGG at Vancouver, August 1987. IAHS Publication 170, International Association of Hydrologic Sciences, Wallingford.

Piotrowski, J.A. (1993) Salt diapirs, pore-water traps and permafrost as key controls for glaciotectonism in Kiel area, northwestern Germany. In: *Glaciotectonics and Mapping Glacial Deposits* (Ed. J.S. Aber), pp. 86–98. Hingnell Printing, Winnipeg.

Piotrowski, J.A. (1994) Tunnel-valley formation in northwest Germany—geology, mechanisms of formation and subglacial bed conditions for the Bornhöved tunnel valley. *Sedimentary Geology*, **89**, 107–141.

Piotrowski, J.A. (1997a) Subglacial hydrology in northwestern Germany during the last glaciation: groundwater flow, tunnel valleys, and hydrological cycles. *Quaternary Science Reviews*, **16**, 169–185.

Piotrowski, J.A. (1997b) Subglacial groundwater flow during the last glaciation in northwestern Germany. *Sedimentary Geology*, **111**, 217–224.

Piotrowski, J.A. & Kraus, A. (1997) Response of sediment to ice sheet loading in northwestern Germany: effective stresses and glacier bed stability. *Journal of Glaciology*, **43**, 495–502.

Piotrowski, J.A. & Tulaczyk, S. (1999) Subglacial conditions under the last ice sheet in northwest Germany: ice-bed separation and enhanced basal sliding Germany? *Quaternary Science Reviews*, **18**, 737–751.

Piotrowski, J. A., Bartels, F., Salski, A. & Schmidt, G. (1996) Geostatistical regionalisation of glacial aquitard thickness in Northwestern Germany, based on fuzzy kriging. *Mathematical Geology*, **28**, 437–452.

Piotrowski, J.A., Geletneky, J. & Vater, R. (1999) Soft-bedded subglacial meltwater channel, from the Welzow-Sued open-cast lignite mine, Lower Lusatia, eastern Germany. *Boreas*, **28**, 363–374.

Piotrowski, J.A., Mickelson, D.M., Tulaczyk, S., Krzyszkowski, D. & Junge, F.W. (2001) Were deforming subglacial beds beneath ice sheets really widespread? *Quaternary International*, **86**, 139–150.

Piotrowski, J.A., Mickelson, D.M., Tulaczyk, S., Krzyszkowski, D. & Junge, F.W. (2002) Reply to the comments by G.S. Boulton, K.E. Dobbie, S. Zatsepin on: Deforming soft beds under ice sheets: how extensive were they? *Quaternary International*, **97–98**, 173–177.

Piotrowski, J.A., Hermanowski, P., Wspanialy, A., Lulek, A., Munck, F., Kronborg, C. & Rattas, M. (2004) Modelling subglacial groundwater flow at the southern margin of the Scandinavian Ice Sheet: a key to understanding ice sheet behaviour? *Workshop on Groundwater Dynamics and Global Change, Agricultural University of Norway, 14–16 April*, Ås, Abstract.

Piper, D.J.W., Mudie, P.J., Fader, G.B., Josenhans, H.W., MacLean, B. & Vilks, G. (1991) Quaternary geology. In: *Geology of the Continental Margin of Eastern Canada*, Vol. I-1 (Eds M.J. Keen & G.L. Williams), pp. 475–607. Geological Society of America, Boulder, Colorado.

Pohjola, V.A. (1993) TV-video observations of bed and basal sliding on Storglaciären, Sweden. *Journal of Glaciology*, **39**(131), 111–118.

Poirier, J.P. (1985) *Creep of Crystals*. Cambridge Earth Science Series, Cambridge University Press, Cambridge.

Polar Research Board of the National Research Council (1985) *Glaciers, ice sheets, and sea level: effect of a CO2-induced climatic change.* Report of a workshop held in Seattle, Washington, 13–15 September 1984, US Department of Energy Publication DOE/EV/60235–1.

Pollard, A., Wakarini, N. & Shaw, J. (1996) Genesis and morphology of erosional shapes associated with turbulent flow over a forward-facing step. In: *Coherent Flow Structures in Open Channels* (Eds P.J. Ashworth, S.L. Bennet, J.L. Best & S.J. McLelland), pp. 249–265. Wiley, New York.

Pollard, D. & Ingersoll, A.P. (1980) Response of a zonal climate-ice sheet model to the orbital perturbations during the Quaternary ice age. *Tellus*, **32**, 301–319.

Pollard, D. & Thompson, S. (1997) Driving a high-resolution dynamic ice-sheet model with GCM climate: ice-sheet initiation at 116,000 BP. *Annals of Glaciology*, **25**, 296–304.

Polyak, L., Edwards, M.H., Coakley, B.J. & Jakobsson, M. (2001) Ice shelves in the Pleistocene Arctic Ocean inferred from glaciogenic deep-sea bedforms. *Nature*, **410**, 453–457.

Polyak, L., Curry, W.B., Darby, D., Bischof, J. & Cronin, T.M. (2004) Contrasing glacial/interglacial regimes in the western Arctic Ocean as exemplified by a sedimentary record from the Mendeleev Ridge. *Palaeogeography, Palaeoclimatology, Palaeoecology*, **203**, 73–93.

Polyakov, I.V. & Johnson, M.A. (2000) Arctic decadal and interdecadal variability. *Geophysics Research Letters*, **27**, 4097–4100.

Polyakov, I.V., Alekseev, G.V., Bekryaev, R.V., *et al.* (2002) Observationally based assessment of polar amplification of global warming. *Geophysics Research Letters*, **29**. doi: 10.1029/2001GL011111.

Pomeroy, J.W. & Essery, R.L.H. (1999) Turbulent fluxes during blowing snow: field tests of model sublimation predictions. *Hydrological Processes*, **13**, 2963–2975.

Poore, R.Z., Osterman, L., Curry, W.B. & Phillips, R.L. (1999) Late Pleistocene and Holocene meltwater events in the western Arctic Ocean. *Geology*, **27**, 759–762.

Popovnin, V., Danilova, T. & Petrakov. D. (1999) A pioneer mass balance estimate for a Patagonian glacier: Glaciar de los Tres, Argentina. *Global and Planetary Change*, **22**, 255–267.

Porter, P.R. & Murray, T. (2001) Mechanical and hydraulic properties of till beneath Bakaninbreen, Svalbard. *Journal of Glaciology*, **47**, 167–175.

Porter, P.R., Murray, T. & Dowdeswell, J.A. (1997) Sediment deformation and basal dynamics beneath a glacier surge front: Bakaninbreen, Svalbard. *Annals of Glaciology*, **24**, 21–26.

Porter, S.C. (1981) Glaciological evidence of Holocene climatic change. In: *Climate and History* (Eds T.M.L. Wigley, M.J. Ingram & G.L. Farmer), pp. 82–110. Cambridge University Press, London.

Porter, S.C. (1989) Some geological implications of average Quaternary glacial conditions. *Quaternary Research*, **32**, 245–261.

Porter, S.C. & Denton, G.H. (1967) Chronology of neoglaciation in the North American Cordillera. *American Journal Science*, **265**, 177–210.

Porter, S.J. (1977) Present and Past Glaciation Threshold in the Cascade Range, Washington, U.S.A.: Topographic and climatic controls and paleoclimatic implications. *Journal of Glaciology*, **18**, 101–116.

Potter, J.R. & Paren, J.G. (1985) Interaction between ice shelf and ocean in George VI Sound, Antarctica. In: *Oceanology of the Antarctic Continental Ice Shelf* (Ed. S.S. Jacobs), pp. 35–58. Volume 43, American Geophysical Union, Washington, DC.

Powell, R.D. (1981) A model for sedimentation by tidewater glaciers. *Annals of Glaciology*, **2**, 129–134.

Powell, R.D. (1984) Glacimarine processes and inductive lithofacies modelling of ice shelf and tidewater glacier sediments based on Quaternary examples. *Marine Geology*, **57**, 1–52.

Powell, R.D. (2003) Subaquatic landsystems: fjords. In: *Glacial Landsystems* (Ed. D.A. Evans), pp. 313–347. Edwin Arnold, London.

Powell, R.D., Dawber, M., McInnes, J.N. & Pyne, A.R. (1996) Observations of the grounding line area at a floating glacier terminus. *Annals of Glaciology*, **22**, 217–223.

Pralong, A., Funk, M. & Lüthi, M.P. (2003) A description of crevasse formation using continuum damage mechanics. *Annals of Glaciology*, **37**, 77–82.

Prest, V.K. (1969) *Retreat of Wisconsin and Recent ice in North America.* Map 1257A, Geological Survey of Canada, Queen's Printer, Ottawa.

Prest, V.K. (1990) Laurentide ice-flow patterns: a historical review, and implications of the dispersal of Belcher Island Erratics. *Geographie physique et Quaternaire*, **44**, 113–136.

Prest, V.K., Grant, D.R. & Rampton, V.N. (1968) *Glacial Map of Canada.* Map 1253A, Geological Survey of Canada. Queen's Printer, Ottawa.

Price, P.B. (2000) A habitat for psychrophiles in deep Antarctic ice. *Proceedings of the National Academy of Sciences of the United States*, **97**, 1247–1251.

Price, R.J. (1969) Moraines, sandar, kames and eskers near Breiðamerkurjökull, Iceland. *Transactions of the Institute of British Geographers*, **46**, 17–43.

Price, R.J. (1970) Moraines at Fjallsjökull, Iceland. *Arctic and Alpine Research*, **2**, 27–42.

Price, S.F., Bindschadler, R.A., Hulbe, C.L. & Joughin, L.R. (2001) Post-stagnation behavior in the upstream regions of Ice Stream C, West Antarctica. *Journal of Glaciology*, **47**, 283–294.

Price, S.F., Bindschadler, R.A., Hulbe, C.L. & Blankenship, D.D. (2002) Force balance along an inland tributary and onset to Ice Stream D, West Antarctica. *Journal of Glaciology*, **48**, 20–30.

Principato, S.M. (2003) *The late Quaternary history of eastern Vestfirdir, NW Iceland.* PhD dissertation, Department of Geological Sciences, University of Colorado, Boulder, 258 pp.

Pu, J., Yao, T., Wang, N., *et al.* (2001) Recently variation of Malan glacier in Hoh Xil Region, Center of Tibetan Plateau. *Journal of Glaciology and Geocryology*, **23**(2), 189–192.

Pudsey, C.J. (1992) Late Quaternary changes in Antarctic Bottom Water velocity inferred from sediment grain size in the northern Weddell Sea. *Marine Geology*, **107**, 9–33.

Pudsey, C.J. (2000) Sedimentation on the continental rise west of the Antarctic Peninsula over the last three glacial cycles. *Marine Geology*, **167**, 313–338.

Punkari, M. (1995) Glacial flow systems in the zone of confluence between the Scandinavian and Novaya Zemlya ice sheets. *Quaternary Science Reviews*, **14**, 589–603.

Pusch, R., Borgesson, L. & Knuttson, S. (1990) Origin of silty fracture fillings in crystalline bedrock. *Geologisk Forening Stockholm*, **112**, 209–213.

Rabus, B.T. & Echelmeyer, K.A. (1998) The mass balance of McCall Glacier, Brooks Range, Alaska, U.S.A.; its regional relevance and implications for climate change in the Arctic. *Journal of Glaciology*, **44**(147), 333–351.

Rabus, B.T. & Lang, O. (2003) Interannual surface velocity variations of Pine Island Glacier, West Antarctica. *Annals of Glaciology*, **36**, 205–214.

Rabus, B., Eineder, M., Roth, A. & Bamler, R. (2003) The shuttle radar topography mission—a new class of digital elevation models acquired by spaceborne radar. *ISPRS Journal of Photogrammetry and Remote Sensing*, **57**(4), 241–262.

Rahmstorf, S. (1995a) Bifurcations of the Atlantic thermohaline circulation in response to changes in the hydrological cycle. *Nature*, **378**, 145–149.

Rahmstorf, S. (1995b) Multiple convection patterns and thermohaline flow in an idealized OGCM. *Journal of Climate*, **8**, 3028–3039.

Rahmstorf, S. (2000) The thermohaline ocean circulation: A system with dangerous thresholds? *Climatic Change*, **46**, 247–256.

Rahmstorf, S. (2002) Ocean circulation and climate during the past 120,000 years. *Nature*, **419**, 207–214.

Rahmstorf, S. (2003) Timing of abrupt climate change: a precise clock. *Geophysical Research Letters*, **30**, 10,1510. doi: 10, 1029/2003GL.017115, 1–17.

Rahmstorf, S. & Alley, R. (2002) Stochastic resonance in glacial climate. *Eos (Transactions of the American Geophysical Union)*, **83**, 129–135.

Rahmstorf, S. & Ganopolski, A. (1999) Long-term global warming scenarios computed with an efficient coupled climate model. *Climate Change*, **43**, 353–367.

Rahmstorf, S., Archer, D., Ebel, D.S., *et al.* (2004) Cosmic rays, carbon dioxide, and climate. *Eos (Transactions of the American Geophysical Union)*, **85**, 38–41.

Rains, R.B., Shaw, J., Skoye, R., Sjogren, D. & Kvill, D. (1993) Late Wisconsinan subglacial megaflood paths in Alberta. *Geology*, **21**, 323–326.

Rains, R.B., Shaw, J., Sjogren, D.B, *et al.* (2002) Subglacial tunnel channels, Porcupine Hills, southwest Alberta, Canada. *Quaternary International*, **90**, 57–65.

Raiswell, R. (1984) Chemical models of solute acquisition in glacial meltwaters. *Journal of Glaciology*, **30**, 49–57.

Raiswell, R. & Thomas, A.G. (1984) Solute acquisition in glacial meltwaters I. Fjällsjökull (south-east Iceland): bulk meltwaters with closed system characteristics. *Journal of Glaciology*, **30**, 35–43.

Ramirez, E., Francou, B., *et al.* (2001) Small glaciers disappearing in the tropical Andes: a case-study in Bolivia: Glacier Chacaltaya (16′S). *Journal of Glaciology*, **47**(157), 187–194.

Rampino, M.L. (1994) Tillites, diamictites, and ballistic ejects of large impacts. *Journal of Geology*, **102**, 439–456.

Ramseier, R.O. (1972) *Growth and mechanical properties of river and lake ice.* PhD thesis, Laval University, Quebec.

Raney, R.K. (1998) The delay/Doppler radar altimeter. *Ieee Transactions on GeoScience, and Remote Sensing*, **36**(5), 1578–1588.

Rasch, M., Elbering, B., Jakobsen, B.H. & Hasholt, B. (2000) High-resolution measurements of water discharge, sediment, and solute transport in the River Zackenbergelven, Northeast Greenland. *Arctic, Antarctic and Alpine Research*, **32**, 336–345.

Rashid, H., Hesse, R. & Piper, D.J.W. (2003) Distribution, thickness and origin of Heinrich layer 3 in the Labrador Sea. *Earth and Planetary Science Letters*, **205**, 281–293.

Rasilainen, K., Suksi, J., Ruskeeniemi, T., Pitkänen, P. & Poteri, A. (2003) Release of uranium from rock matrix—a record of glacial meltwater intrusions? *Journal of Contaminant Hydrology*, **61**, 235–246.

Raymo, M.E. (1992) Global climate change: a three million year perspective. In: *Start of a Glacial. Proceedings of the Mallorca NATO ARW* (Eds G. Kukla & E. Went), pp. 207–223. NATO ASI Series I, Vol. 3, Springer-Verlag, Heidelberg.

Raymo, M.E., Backman, J., Clement, B.M., Martinson, D.G. & Ruddiman, W.F. (1989) Late Pliocene Variation in Northern Hemisphere Ice Sheets and North Atlantic Deep Water Circulation. *Paleoceanography*, **4**, 413–446.

Raymond, C. (1996) Shear margins in glaciers and ice sheets. *Journal of Glaciology*, **42**, 90–102.

Raymond, C.F. (1983) Deformation in the vicinity of ice divides. *Journal of Glaciology*, **29**(103), 357–373.

Raymond, C.F. (2000) Energy balance of ice streams. *Journal of Glaciology*, **46**, 665–674.

Raymond, C.F., Echelmeyer, K.A., Whillans, I.M. & Doake, C.S.M. (2001) Ice stream shear margins. In: *The West Antarctic Ice Sheet: Behavior and Environment* (Eds R.B. Alley & R.A. Bindschadler), pp. 137–156. Volume 77, American Geophysical Union, Washington, DC.

Raymond, M., Gudmundsson, G.H. & Funk, M. (2003) Non-linear finite amplitude transfer of basal perturbations to a glacier surface. *Geophysical Research Abstracts,* **5**, 04031.

Rea, B.R. & Evans, D.J.A. (2003) Plateau icefield landsystems. In: *Glacial Landsystems* (Ed. D.A. Evans), pp. 407–431. Edwin Arnold, London.

Rea, B.R., Whalley, W.B., Evans, D.J.A, Gordon, J.E. & McDougall, D.A. (1998) Plateau icefields: geomorphology and ice dynamics. *Journal of Quaternary Science*, **13** (Supplement 1), 35–54.

Rea, B.R., Whalley, W.B., Dixon, T. & Gordon, J.E. (1999) Plateau icefields as contributing areas to valley glaciers and the potential impact on reconstructed ELAs: a case study from the Lyngen Alps, North Norway. *Annals of Glaciology*, **28**, 97–102.

Rebesco, M., Larter, R.D., Camerlenghi, A. & Barker, P.F. (1996) Giant sediment drifts on the continental rise west of the Antarctic Peninsula. *Geo-Marine Letters*, **16**, 65–75.

Reeh, N. (1985) Greenland ice-sheet mass balance and sea-level change. In: *Glaciers, Ice Sheets, and Sea Level: Effect of a CO_2 Induced Climatic Change*, pp. 155–171. United States Department of Energy, Seattle, WA.

Reeh, N. (1987) Steady-state three-dimensional ice flow over an undulating base: first-order theory with linear ice rheology. *Journal of Glaciology*, **33**(114), 177–185.

Reeh, N. (1989, erschienen 1991) Parameterization of melt rate and surface temperature on the Greenland ice sheet. *Polarforschung*, **59**(3), 113–128.

Reeh, N. (1994) Calving from Greenland Glaciers: Observations, balance estimates of calving rates, calving laws. In: *Report on the Workshop on the Calving Rate of West Greenland Glaciers in Response to Climate Change* (Ed. N. Reeh), pp. 85–102. Danish Polar Center, Copenhagen.

Reeh, N. (1999) Mass balance of the Greenland ice sheet: can modern observation methods reduce the uncertainty? Proceedings from the Workshop on Methods of Mass Balance Measurements and Modelling, Tarfala Sweeden, August 10–12, 1998. *Geografiska. Annaler*, **81A**(4), 735–742.

Reeh N., Boggild, C.E. & Oerter, H. (1994) Surge of Storstrommen, a large outlet glacier from the inland ice of north-east Greenland. *Grønlands Geologiske Undersøgelse Rapport*, **162**, 201–209.

Reeh, N., Mayer, C., Miller, H., Thomsen, H.H. & Weidick, W. (1999) Climate control on fjord glaciations in Greenland: implications for IRD-deposition in the Sea. *Geophysical Research Letters*, **26**(8), 1039–1042.

Reeh, N., Mohr, J.J., Krabill, W.B., *et al.* (2002) Glacier specific ablation rate derived by remote sensing measurements. *Geophysical Research Letters*, **29**(16), 10-1–10-4.

Rees, W.G. (2001) *Physical Principles of Remote Sensing*. Cambridge University Press, Cambridge.

Rees, J.G. & Wilson, A.A. (1998) *Geology of the Country around Stoke-on-Trent*. Memoir of the British Geological Survey, Keyworth.

Reichert, B.K., Bengtsson, L. & Oerlemans, J. (2001) Midlatitude forcing mechanisms for glacier mass balance investigated using General Circulation Models. *Journal of Climate*, **14**, 3767–3784.

Reimnitz, E, Dethleff, D. & Nürnberg, D. (1994) Contrasts in Arctic shelf sea-ice regimes and some implications: Beaufort Sea versus Laptev Sea. *Marine Geology*, **119**, 215–225.

Remenda, V.H., Cherry, J.A. & Edwards, T.W.D. (1994) Isotopic composition of old ground water from Lake Agassiz: implications for Late Pleistocene climate. *Science*, **266**, 1975–1978.

Remy, F., Mazzega, P., Houry, S., Brossier, C. & Minster, J.F. (1989) Mapping of the topography of continental ice by inversion of satellite-altimeter data. *Journal of Glaciology*, **35**(119), 98–107.

Ren, B. (1988) Recent fluctuation of glaciers in China. In: *An Introduction to the Glaciers in China*, pp. 171–186. Science Press, Beijing.

Ren, J., Qin, D. & Jing, Z. (1998) Climatic warming causes the glacier retreat in Mt. Qomolangma. *Journal of Glaciology and Geocryology*, **20**(2), 184–185.

Retzlaff, R. & Bentley, C.R. (1993) Timing of stagnation of Ice Stream C, West Antarctica, from short-pulse radar studies of buried surface crevasses. *Journal of Glaciology*, **39**, 553–561.

Reynaud, L. (1987) The 1986 survey of the Grand Moulin on Mer de Glace, Mont Blanc Massif, France. *Journal of Glaciology*, **33**, 130–131.

Reynolds, J.B. (1997) Ecology of overwintering fishes in Alaskan freshwaters. In: *Freshwater of Alaska: Ecological Syntheses* (Eds A.M. Milner & M.W. Oswood), pp. 281–302. Springer-Verlag, New York.

Rhoads, B.L. & Thorn, C.E. (1993) Geomorphology as science: the role of theory. *Geomorphology*, **6**, 287–307.

Richard, P.J.H. & Occhietti, S. (2005) ^{14}C chronology for ice retreat and inception of Champlain Sea in the St Lawrence Lowlands, Canada. *Quaternary Research*, **63**, 353–358.

Ridley, J.K., Huybrechts, P., Gregory, J.M. & Lowe, J.A. (2004) Behaviour of an interactive Greenland ice sheet in HADCM3. *Geophysical Research Abstracts*, **6**, 02003 (CD-ROM).

Rignot, E. (1996) Tidal motion, ice velocity and melt rate of Petermann Gletscher, Greenland, measured from radar interferometry. *Journal of Glaciology*, **42**(142), 476–485.

Rignot, E. (1998) Fast recession of a West Antarctic Glacier. *Science*, **281**(5376), 549–551.

Rignot, E. (2001) Evidence of rapid retreat and mass loss of Thwaites Glacier, West Antarctica. *Journal of Glaciology*, **47**(157), 213–222.

Rignot, E. (2002a) Ice-shelf changes in Pine Island Bay, Antarctica, 1947–2000. *Journal of Glaciology*, **48**, 247–256.

Rignot, E. (2002b) Mass balance of East Antarctic glaciers and ice shelves from satellite data. *Annals of Glaciology*, **34**, 217–227.

Rignot, E. & Jacobs, S. (2002) Rapid bottom melting widespread near Antarctic Ice Sheet grounding lines. *Science*, **296**, 2020–2023.

Rignot, E. & Thomas, R.H. (2002) Mass balance of polar ice sheets. *Science*, **297**(5586), 1502–1506.

Rignot, E.J., Gogineni, S.P., Krabill, W.B. & Ekholm, S. (1997) North and northeast Greenland ice discharge from satellite radar interferometry. *Science*, **276**(5314), 934–937.

Rignot, E.J., Buscarlet, G., Csatho, B., Gogineni, S., Krabill, W. & Schmeltz, M. (2000) Mass balance of the northeast sector of the Greenland ice sheet: a remote-sensing perspective. *Journal of Glaciology*, **46**(153), 265–273.

Rignot, E., Gogineni, J., Joughin, I. & Krabill, W (2001) Contribution to the glaciology of northern Greenland from satellite radar interferometry. *Journal of.Geophysical Research* (Special Issue), **106**(D24), 34007–34019.

Rignot, E.J., Vaughan, D.G., Schmeltz, M., Dupont, T. & MacAyeal, D.R. (2002) Acceleration of Pine Island and Thwaites Glaciers, West Antarctica. *Annals of Glaciology*, **34**, 189–194.

Rignot, E., Rivera, A. & Casassa, G. (2003) Contribution of the Patagonia icefields of South America to sea level rise. *Science*, **302**, 434–437.

Rigor, I.G. & Wallace, J.M. (2004) Variations in the age of Arctic sea-ice and summer sea-ice extent. *Geophysics Research Letters*, **31**. L09401, doi: 10.1029/2004GL019492.

Rigor, I.G., Colony, R.L. & Martin, S. (2000) Variations in surface air temperature observations in the Arctic, 1979–1997. *Journal of Climate*, **13**, 896–914.

Rigor, I.G., Wallace, J.M. & Colony, R.L. (2002) Response of sea ice to the Arctic Oscillation. *Journal of Climate*, **15**, 2648–2663.

Riihimaki, C.A. (2003) *Quantitative Constraints on the Glacial and Fluvial Evolution of Alpine Landscapes*. PhD thesis, Earth Sciences Department, University of California, Santa Cruz, 160 pp.

Rind, D., deMenocal, P., Russell, G., *et al.* (2001a) Effects of glacial meltwater in the GISS coupled atmosphere-ocean model 1. North Atlantic Deep water response. *Journal of Geophysical Research*, **106**(D21), 27335–27353.

Rind, D., deMenocal, P., Russell, G., *et al.* (2001b) Effects of glacial meltwater in the GISS coupled atmosphere-ocean model, 2. A bipolar seesaw in Atlantic deep water production. *Journal of Geophysical Research*, **106**(D21), 27355–27365.

Ritz, C., Fabre, A. & Letréguilly, A. (1997) Sensitivity of a Greenland ice sheet model to ice flow and ablation parameters: consequences for the evolution through the last glacial cycle. *Climate Dynamics*, **13**, 11–24.

Ritz, C., Rommelaere, V. & Dumas, C. (2001) Modeling the evolution of Antarctic ice sheet over the last 420,000 years: implications for altitude changes in the Vostok region. *Journal of Geophysical Research*, **106**, 31943–31964.

Rivera, A. (2004) *Mass balance investigations at Glaciar Chico, Southern Patagonia Icefield, Chile*. PhD thesis, University of Bristol.

Rivera, A. & Casassa, G. (1999) Volume changes on Pío XI glacier, Patagonia: 1975–1995. *Global and Planetary Change*, **22**, 233–244.

Rivera, A., Casassa, G. & Acuña, C. (2001) Mediciones de espesor en glaciares de Chile centro-sur. *Revista Investigaciones Geográficas*, **35**, 67–100.

Rivera, A., Acuña, C., Casassa, G. & Bown, F. (2002) Use of remote sensing and field data to estimate the contribution of Chilean glaciers to the sea level rise. *Annals of Glaciology*, **34**, 367–372.

Rivera, A., Bown, F., Casassa, G., Acuña, C. & Clavero, J. (2005) Glacier shrinkage and negative mass balance in the Chilean Lake District. *Hydrological Sciences Journal*, **50**(6), 963–974.

Roberts, M.J., Tweed, F.S., Russell, A.J., Knudson, O., Lawson, D.E., Larson, G.J., Evenson, E.B. & Bjornsson, H. (2002) Glaciohydraulic supercooling in Iceland. *Geology*, **30**(5), 439–442.

Robin, G. de Q. (1966) Mapping the Antarctic Ice Sheet by Satellite Altimetry. *Canadian Jnl. Earth Sciences*, **3**(6), 893–901.

Robin, G. de Q. (1977) Ice cores and climatic change. *Philosphical Transactions of the Royal Society of London, Series B*, **280**, 143–168.

Robin, G. de Q. (1979) Formation, flow, and disintegration of ice shelves. *Journal of Glaciology*, **24**, 259–265.

Robin, G. de Q., Drewry, D.J. & Meldrum, D.T. (1977) International studies of ice sheet and bedrock. *Philosophical Transactions of the Royal Society of London*, **279**(B), 185–196.

Rodbell, D.T. (1992) Late Pleistocene equilibrium-line reconstructions in the northern Peruvian Andes. *BOREAS*, **21**, 43–52.

Rodgers, C.D. (2000) *Inverse Methods for Atmospheric Sounding: Theory and Practice*. Series on Atmospheric, Oceanic and Planetary Physics, Vol. 2. World Scientific, London.

Rodwell, M.J., Rowell, D.P. & Folland, C.K. (1999) Oceanic forcing of the wintertime North Atlantic Oscillation and European climate. *Nature*, **398**, 320–323.

Roe, G.H. (2002) Modeling orographic precipitation over ice sheets: an assessment over Greenland. *Journal of Glaciology*, **48**, 70–80.

Roe, G.H. & Steig, E.J. (2004) Characterization of millennial-scale climate variability. *Journal of Climate*, **17**, 1929–1944.

Rogers, J.C. & van Loon, H. (1979) The seesaw in winter temperatures between Greenland and northern Europe. part II: some oceanic and atmospheric effects in middle and high latitudes. *Monthly Weather Review*, **107**, 509–519.

Rogers, J.C., Wang, C.-C. & McHugh, M.J. (1998) Persistent cold climatic episodes around Greenland and Baffin Island: Links to decadal-scale sea surface temperature anomalies. *Geophysical Reaserch Letters*, **25**, 3971–3974.

Romm, E.S. (1966) *Flow Characteristics of Fractured Rocks*. Nedra, Moscow.

Rooney, S.T., Blankenship, D.D., Alley, R.B. & Bentley, C.R. (1987) Till beneath Ice Stream-B. 2. Structure and continuity. *Journal of Geophysical Research—Solid Earth and Planets*, **92**, 8913–8920.

Rooth, C. (1982) Hydrology and ocean circulation. *Progress in Oceanography*, **11**, 131–149.

Rose, J. (1987) Drumlins as part of a subglacial bedform continuum. In: *Drumlin Symposium* (Eds J. Menzies & J. Rose), pp. 25–80. Balkema, Rotterdam.

Rose, J. & Letzer, J.M. (1977) Superimposed drumlins. *Journal of Glaciology*, **18**, 471–480.

Rose, K.E. (1979) Characteristics of ice flow in Marie Byrd Land, Antarctica. *Journal of Glaciology*, **24**(90), 63–75.

Rosenblüth, B., Casassa, G. & Fuenzalida, H. (1995) Recent climatic changes in western Patagonia. *Bulletin of Glacier Research*, **13**, 127–132.

Rosenblüth, B., Fuenzalida, H. & Aceituno, P. (1997) Recent temperature variations in southern South America. *International Journal of Climatology*, **17**, 67–85.

Röthlisberger, H. (1972) Water pressure in intra- and subglacial channels. *Journal of Glaciology*, **11**, 177–203.

Rothrock, D.A., Yu, Y. & Maykut, G.A. (1999) Thinning of the Arctic sea-ice cover. *Geophysical Research Letters*, **26**(23), 3469–3472.

Rothrock, D.A., Zhang, J. & Yu, Y. (2003) The Arctic ice thickness anomaly of the 1990s: A consistent view from observations and models. *Journal of Geophysical Research*, **108**(C3), 3083, doi: 10.1029/2001JC001208.

Rott, H., Skvarca, P. & Nagler, T. (1996) Rapid collapse of northern Larsen Ice Shelf, Antarctica. *Science*, **271**(5250), 788–792.

Rott, H., Rack, W., Skvarca, P. & de Angelis, H. (2002) Northern Larsen Ice Shelf, Antarctica: further retreat after collapse. *Annals of Glaciology*, **34**, 277–282.

Rousselot, M. & Fischer, U.H. (2003) A combined field and laboratory study of ploughing. *Geophysical Research Abstracts*, **5**, 30-1-2003.

Rousselot, M. & Fischer, U.H. (2005) Evidence for excess pore-water pressure generated in subglacial sediment: implications for clast ploughing. *Geophysical Research Letters*, **32**, L11501, doi:10.1029/2005GL022642.

Roy, M., Clark, P.U., Barendregt, R.W., Glasmann, J.R. & Enkin, R.J. (2004) Glacial stratigraphy and paleomagnetism of late Cenozoic deposits of the north-central United States. *Geological Society of America Bulletin*, **116**, 30–41.

Royer, D.L., Berner, R.A., Montanez, I.P., Tabor, N.J. & Beerling, D.J. (2004) CO2 as a primary driver of Phanerozoic climate. *Geological Society of America Today*, **14**, 4–10.

Ruddiman, W.F. (1977) Late Quaternary deposition of ice-rafted sand in the subpolar North Atlantic (lat. 40° to 65°N). *Geological Society of America Bulletin*, **88**, 1813–1827.

Ruddiman, W.F. (2003a) The Anthropogenic greenhouse era began thousands of years ago. *Climatic Change*, **61**(3), 261–293.

Ruddiman, W.F. (2003b) Orbital insolation, ice volume, and greenhouse gases. *Quaternary Science Reviews*, **22**, 1597–1629.

Ruddiman, W.F. (2004) The role of greenhouse gases in orbital-scale climatic changes. *Eos (Transactions of the American Geophysical Union)*, **85**(1), 6–7.

Ruddiman, W.F. & Kutzbach, J.E. (1991) Plateau uplift and climatic change. *Scientific American*, **264**, 66–75.

Ruddiman, W.F., Raymo, M.E., Martinson, D.G., Clement, B.M. & Backman, J. (1989) Mid-Pleistocene evolution of Northern Hemisphere climate. *Paleoceanography*, **4**, 353–412.

Russell, H.A.J., Arnott, R.W.C. & Sharpe, D.R. (2002) Evidence for rapid sedimentation in a tunnel channel, Oak Ridges Moraine, southern Ontario, Canada. *Sedimentary Geology*, **3134**, 1–23.

Russell, H.A.J., Sharpe, D.R., Brennand, T.A., Barnett, P.J. & Logan, C. (2003) Tunnel channels of the Greater Toronto and Oak Ridges Moraine Areas, Southern Ontario. *Geological Survey of Canada Open File* 4485, scale 1:250,000. Queen's Printer, Ottawa.

Russell-Head, D.S. & Budd, W.F. (1979) Ice-sheet flow properties derived from bore-hole shear measurements combined with ice-core studies. *Journal of Glaciology*, **24**, 117–130.

Rutter, E.H., Casey, M. & Burlini, L. (1994) Preferred crystallographic orientation development during the plastic and superplastic flow of calcite rocks. *Journal of Structural Geology* **16**, 1431–1446.

Sachs, G. (1928) Zur Ableitung einer Fliessbedingung. *Zeitschrift des Vereines Deutscher Ingenieure*, **72**(22), 734–736.

Sættem, J. (1990) Glaciotectonic forms and structures on the Norwegian continental shelf: observations, processes and implications. *Norsk Geologisk Tidsskrift*, **70**, 81–94.

Sakai, K. & Peltier, W.R. (1997) Dansgaard-Oeschger Oscillation in a Coupled Atmosphere-Ocean Climate Model. *Journal of Climate*, **10**, 949–970.

Salamatin, A.N. & Malikova, D.R. (2000) Structural dynamics of an ice sheet in changing climate. *Data of Glaciological Studies*, **89**, 112–128.

Salamatin, A.N., Lipenkov, V.Y., Barkov, N.I., Jouzel, J., Petit, J.R. & Raynaud, D. (1998) Ice-core age dating and palaeothermometer calibration based on isotope and temperature profiles from deep boreholes at Vostok Station (East Antarctica). *Journal of Geophysical Research*, **103**, 8963–8977.

Salt, K.E. (2001) *Palaeo ice sheet dynamics and depositional settings of the Late Devensian Ice sheet in South West Scotland*. Unpublished PhD thesis, University of Glasgow.

Saltzman, E.S., Aydin, M., De Bruyn, W.J., King, D.B. & Yvon-Lewis, S.A. (2004) Methyl bromide in preindustrial air: Measurements from an Antarctic ice core. *Journal of Geophysical Research*, 109(D5), Art. No. D05301.

Sapiano, J.J., Harrison, W.D. & Echelmeyer, K.A. (1998) Elevation, volume and terminus changes of nine glaciers in North America. *Journal of Glaciology*, **146**(44), 119–135.

Sawagaki, T. & Hirakawa, K. (1997) Erosion of bedrock by subglacial meltwater, Soya Coast, East Antarctica. *Geografiska Annaler (Series A)*, **79**, 223–238.

Scambos, T.A. & Bindschadler, R.A. (1991) Feature maps of ice streams C, D and E, West Antarctica. *Antarctic Journal of the US*, **26**(5), 312–313.

Scambos, T.A., Dutkiewicz, M.J., Wilson, J.C. & Bindschadler, R.A. (1992) Application of Image Cross-Correlation to the Measurement of Glacier Velocity Using Satellite Image Data. *Remote Sensing of Environment* **42**(3), 177–186.

Scambos, T.A., Hulbe, C., Fahnestock, M. & Bohlander, J. (2000) The link between climate warming and break-up of ice shelves in the Antarctic Peninsula. *Journal of Glaciology*, **46**(154), 516–530.

Schelkes, K., Klinge, H., Vogel, P. & Wollrath, J. (1999) Aspects of the use and importance of hydrochemical data for groundwater flow modelling at radioactive waste disposal sites in Germany. *Proceedings of a SKB workshop 'Use of Hydrochemical Information in Testing Groundwater Flow Models' in Borgholm, Sweden*, pp. 151–162. Organization for Economic Co-operation and Development/ Nuclear Energy Agency, Paris.

Schermerhorn, L.J.G. (1975) Tectonic framework of Late Precambrian supposed glacials, *in*, Wright, A.E. & Moseley, F., eds., *Ice Ages: Ancient and Modern*: Seel House, Liverpool, pp. 241–274.

Schmeits, M.J. & Oerlemans, J. (1997) Simulation of the historical variations in length of Unterer Grindelwaldgletscher, Switzerland. *Journal of Glaciology*, **43**(143), 152–164.

Schmeltz, M., Rignot, E., Dupont, T.K. & MacAyeal, D.R. (2002) Sensitivity of Pine Island Glacier, West Antarctica, to changes in ice-shelf and basal conditions: a model study. *Journal of Glaciology*, **48**, 552–558.

Schmidt, R.A. (1972) *Sublimation of Wind-transported Snow—a Model*. Research Report RM–90, Rocky Mountains Forestry and Range Experimental Station, For Services, U.S. Department of Agriculture, Fort Collins, Colorado, 245 pp.

Schmittner, A., Yoshimori, M. & Weaver, A.J. (2002) Instability of glacial climate in a model of the Ocean-Atmosphere-Cryosphere System. *Science*, **295**, 1489–1493.

Schneeberger, C., Albrecht, O., Blatter, H., Wild, M. & Hock, R. (2001) Modelling the response of glaciers to a doubling in atmospheric CO_2: a case study of Störglaciaeren, northern Sweden. *Climate Dynamics*, **17**(11), 825–834.

Scholz, C.H. (2002) *The Mechanics of Earthquakes and Faulting*, 2nd edn. Cambridge University Press, Cambridge.

Schomacker, A., Kroger, J. & Larsen, G. (2003) An extensive late Holocene glacier advance of Kotlujokull, central south Iceland. *Quaternary Science Reviews*, **22**, 1427–1434.

Schulmeister, J., Goodwin, I., Renwick, J., *et al.* (2004) The Southern Hemisphere westerlies in the Australasian sector over the last glacial cycle: a synthesis. *Quaternary International*, **118–119**, 23–53.

Schulz, M., Berger, W.H., Sarnthein, M. & Grootes, P.M. (1999) Amplitude variations of 1470-year climate oscillations during the last 100,000 years linked to fluctuations of continental ice mass. *Geophysical Research Letters*, **26**, 3385–3388.

Schulz, M., Paul, A. & Timmermann, A. (2002) Relaxation oscillators in concert: a framework for climate change at millennial timescales during the late Pleistocene. *Geophysical Research Letters*, **29**(24), 46-1–46-4.

Schumm, S.A. & Lichty, R.W. (1965) Time, space and causality in geomorphology. *American Journal of Science*, **263**, 110–119.

Schwander, J., Jouzel, J., Hammer, C.U., *et al.* (2001) A tentative chronology for the EPICA Dome Concordia. *Geophysical Research Letters*, **28**, NO.22, 4243–4246.

Schwarb, M. (2000) The Alpine precipitation climate. Evaluation of a high-resolution analysis scheme using comprehensive rain-gauge data. Diss. ETH No. 13911, Institute for Climate Research, ETH Zürich.

Schwikowski, M., Brütsch, S., Saurer, M., Casassa, G. & Rivera, A. (2003) First shallow firn core record from Gorra Blanca, Patagonia. *Geophysical Research Abstracts*, **5**, 01427.

Schytt, V. (1956) Lateral drainage channels along the northern side of the Moltke glacier, Northwest Greenland. *Geografiska Annaler*, **38**, 64–77.

Schytt, V. (1962) Mass balance studies in Kebnekasje. *Journal of Glaciology*, **4**, 33, 281–289.

Scourse, J.D., Hall, I.R., McCave, N., Young, J.R. & Sugdon, C. (2000) The origin of Heinrich layers: evidence from H2 for European precursor events. *Earth and Planetary Science Letters*, **182**, 187–195.

Segall, P. & Rice, J.R. (1995) Dilatancy, compaction, and slip instability of a fluid-infiltrated fault. *Journal of Geophysical Research*, **100**, 22155–22171.

Sejrup, H.-P., Larsen, E., Haflidason, H., *et al.* (2003) Configuration, history and impact of the Norwegian Channel Ice Stream. *Boreas*, **32**, 18–36.

Semenov, V.A. & Bengtsson, L. (2003) Modes of wintertime Arctic temperature variability. *Geophysics Research Letters*, **30**, 1781, doi: 10.1029/2003GLO17112.

Sen, V., Stoffa, P.A., Dalziel, I.W.D., Blankenship, D.D., Smith, A.M. & Anandakrishnan, S. (1998) Seismic surveys in central West Antarctica: Data and processing examples. *Terra Antarctica*, **5**, 761–772.

Serreze, M.C. & Francis, J. (In press) The Arctic amplification debate. *Climate Change*.

Serreze, M.C., Maslanik, J.A., Barry, R.G. & Demaria, T.L. (1992) Winter Atmospheric Circulation in the Arctic Basin and Possible Relationships to the Great Salinity Anomaly in the Northern North Atlantic. *Geophysical Research Letters*, **19**, 293–296.

Serreze, M.C., Walsh, J.E., Chapin, F.S. III, *et al.* (2000) Observational evidence of recent change in the northern high latitude environment. *Climatic Change*, **46**, 159–207.

Serreze, M.C., Maslanik, J.A., Scambos, T.A., *et al.* (2003) A record minimum Arctic sea ice extent an area in 2002. *Geophysics Research Letters*, **30**, doi: 10.1029/2002GL016406.

Shabtaie, S. & Bentley, C.R. (1995) Electrical-resistivity sounding of the East Antarctic ice-sheet. *Journal of Geophysical Research*, **100**(B2), 1933–1954.

Shackleton, N.J. (1973) Attainment of isotopic equilibrium between ocean water and the benthonic foraminifera genus uvigerina: isotopic changes in the ocean during the last glacial. *Colloques Internationaux du C.N.R.S.* #219.

Shackleton, N.J. (1974) Attainment of isotopic equilibrium between ocean water and the benthic foraminifera genus Uvigerina: isotopic changes in the ocean during the last glacial. *CNRS Collogues Internationaux*, **219**, 203–209.

Shackleton, N.J. (2000) The 100,000-Year Ice-Age Cycle Identified and Found to Lag Temperature, Carbon Dioxide, and Orbital Eccentricity. *Science*, **289**, 1897–1902.

Shackleton, N.J. & Opdyke, N.D. (1973) Oxygen isotope and paleomagnetic stratigraphy of equatorial Pacific core V28–238: oxygen isotope temperatures and ice volumes on 10^4 and 10^6 year scale. *Quaternary Research*, **3**, 39–55.

Shackleton, N.J., Wiseman, J.D.H. & Buckley, H.A. (1973) Non-equilibrium Isotopic Fractionation between Seawater and Planktonic Foraminiferal Tests. *Nature*, **242**, 177–179.

Shackleton, N.J., Fairbanks, R.G., Chiu, T. & Parreinin, F. (2004) Absolute calibration of the Greenland time scale: implications for Antarctic time scale and for $\delta^{14}C$. *Quaternary Science Reviews*, **23**(14–15), 1513–1522. doi: 10.1016/j.quascirev.2004.03.006.

Sharp, M.J. (1984) Annual moraine ridges at Skalafellsjökull, southeast Iceland. *Journal of Glaciology*, **30**, 82–93.

Sharp, M.J. (1988) Surging glaciers: geomorphic effects. *Progress in Physical Geography*, **12**, 533–559.

Sharp, M.J., Dowdeswell, J.A. & Gemmell, J.C. (1989) Reconstructing past glacier dynamics and erosion from glacial geomorphic evidence: Snowdon, North Wales. *Journal of Quaternary Science*, **4**, 115–130.

Sharp, M., Richards, K., Willis, I., Nienow, P., Lawson, W. & Tison, J.-L. (1993) Geometry, bed topography and drainage system structure of the Haut Glacier d'Arolla, Switzerland. *Earth Surface Process & Landforms*, **18**, 557–572.

Sharp, M., Tranter, M., Brown, G.H. & Skidmore, M. (1995) Rates of chemical denudation and CO_2 drawdown in a glacier-covered alpine catchment. *Geology*, **23**, 61–64.

Sharp, M., Parkes, J., Cragg, B., Fairchild, I.J., Lamb, H. & Tranter, M. (1999) Bacterial populations at glacier beds and their relationship to rock weathering and carbon cycling. *Geology*, **27**, 107–110.

Sharp, M., Creaser, R.A. & Skidmore, M. (2002) Strontium isotope composition of runoff from a glaciated carbonate terrain. *Geochimica et Cosmochimica Acta*, **66**, 595–614.

Sharpe, D.R. (1987) The stratified nature of drumlins from Victoria Island and southern Ontario, Canada. In: *Drumlin Symposium* (Eds J. Menzies & J. Rose), pp. 185–214. Balkema, Rotterdam.

Sharpe, D.R. & Shaw, J. (1989) Erosion of bedrock by subglacial meltwater, Cantley, Quebec. *Geological Society of America Bulletin*, **101**, 1011–1020.

Sharpe, D.R., Pugin, A., Pullan, S. & Shaw, J. (2004) Regional unconformities and the sedimentrary architecture of the oak Ridges Moraine area, southern Ontario. *Canadian Journal of Earth Sciences*, **41**, 183–198.

Shaviv, N.J. & Veizer, J. (2003) Celestial driver of Phanerozoic climate? *Geological Society of America Today*, **13**, 4–10.

Shaw, J. (1979) Genesis of the Sveg tills and Rogen moraines of central Sweden a model of basal meltout. *Boreas*, **8**, 409–426.

Shaw, J. (1983) Drumlin formation related to inverted melt-water erosional marks. *Journal of Glaciology*, **29**, 461–479.

Shaw, J. (1988) Subglacial erosional marks, Wilton Creek, Ontario. *Canadian Journal of Earth Sciences*, **25**, 1442–1459.

Shaw, J. (1989) Drumlins, subglacial meltwater floods and ocean responses. *Geology*, **17**, 853–856.

Shaw, J. (1993) Geomorphology. In: *Edmonton Beneath Our Feet: A Guide to the Geology of the Edmonton Region* (Ed. J.D. Godfrey), pp. 21–32. Edmonton Geological Society, University of Alberta.

Shaw, J. (1994) Hairpin erosional marks, horseshoe vortices and subglacial erosion. *Sedimentary Geology*, **91**, 269–283.

Shaw, J. (1996) A meltwater model for Laurentide subglacial landscapes. In: *Geomorphologie sans Frontières* (Eds S.B. McCann & D.C. Ford), pp. 181–236. Wiley, Chichester.

Shaw, J. (2002) The meltwater hypothesis for subglacial bedforms. *Quaternary International*, **90**, 5–22.

Shaw, J. & Gilbert, R. (1990) Evidence for large-scale subglacial meltwater flood events in southern Ontario and northern New York State. *Geology*, **18**, 1169–1172.

Shaw, J. & Healy, T.R. (1977) The formation of the labyrinth, Wright Valley, Antarctica. *New Zealand Journal of Geology, and Geophysics*, **20**, 933–947.

Shaw, J. & Kvill (1984) A glaciofluvial model for drumlins of the Livingstone Lake area, Saskatchewan. *Canadian Journal of Earth Sciences*, **21**, 1442–1459.

Shaw, J. & Sharpe, D.R. (1987) Drumlin formation by subglacial meltwater erosion. *Canadian Journal of Earth Sciences*, **24**, 2316–2322.

Shaw, J., Kvill, D. & Rains, B. (1989) Drumlins and catastrophic floods. *Sedimentary Geology*, **62**, 177–202.

Shaw, J., Rains, R.B., Eyton, J.R. & Weissling, L. (1996) Laurentide subglacial outburst floods: landform evidence from digital elevation models. *Canadian Journal of Earth Sciences*, **33**, 1154–1168.

Shaw, J., Munro-Stasiuk, M., Sawyer, B., Beaney, C., Lesemann, J.-E., Musacchio, A., Rains, B. & Young, R.R. (1999) The Channeled Scabland: back to Bretz? *Geology*, **27**, 605–608.

Shaw, J., Faragini, D.M., Kvill, D.R. & Rains, R.B. (2000) The Athabasca fluting field: implicationsfor the formation of large-scale fluting (erosional lineations). *Quaternary Science Reviews*, **19**, 959–980.

Shea, J.M., Anslow, F.S. & Marshall, S.J. (In press) Hydrometeorological relationships on the Haig Glacier, Alberta, Canada. *Annals of Glaciology*, **40**.

Shepherd, A., Wingham, D.J., Mansley, J.A.D. & Corr, H.F.J. (2001) Inland thinning of Pine Island Glacier, West Antarctica. *Science*, **291**, 862–864.

Shepherd, A., Wingham, D.J. & Mansley, J.A.D. (2002) Inland thinning of the Amundsen Sea sector, West Antarctica. *Geophysical Research Letters*, **29**, doi: 10.1029/2001GL014183.

Shepherd, A., Wingham, D.J. & Rignot, E. (2004) Warm ocean is eroding West Antarctic Ice Sheet. *Geophysical Research Letters*, **31**(L23402). doi: 10.1029/2004GL021106.

Shi, Y. (2001) Estimation of the water resources affected by climatic warming and glacier shrinkage before 2050 in West China. *Journal of Glaciology and Geocryology*, **23**(4), 333–341.

Shi, Y., Shen, Y. & Hu, R. (2002) Preliminary study on signal, impact and outlook of climatic shift from warm-dry to warm-humid in Northwest China. *Journal of Glaciology and Geocryology*, **24**(3), 219–225.

Shibkov, A.A., Goloviv, Y.I., Zheltov, M.A., Korolev, A.A. & Leonov, A.A. (2003) Morphology diagram of nonequilibrium patterns of ice crystals growing in supercooled water. *Physica A*, **319**, 65–79.

Shilts, W.W. (1980) Flow patterns in the central North American ice sheet. *Nature*, **286**, 213–218.

Shin, R.A. & Barron, E. (1989) Climate sensitivity to continental ice sheet size and configuration. *Journal of Climate*, **2**, 1517–1537.

Shindell, D.T., Schmidt, G.A., Mann, M.E., Rind, D. & Waple, A. (2001) Solar forcing of regional climate change during the Maunder Minimum. *Science*, **294**, 2149–••.

Shipp, S.S., Anderson, J.B. & Domack, E.W. (1999) Late Pleistocene-Holoceneretreat of the West Antarctic ice-sheet system in the Ross Sea: Part 1—geophysical results. *Geological Society of America Bulletin*, **111**, 1486–1516.

Shipp, S.S., Welner, J.S. & Anderson, J.B. (2002) Retreat signature of a polar ice stream: sub-glacial geomorphic features and sediments from the Ross Sea, Antarctica. In: *Glacier-influenced Sedimentation on High-latitude Continental Margins* (Eds J.A. Dowdeswell & C. Ó Cofaigh), pp. 277–304. Special Publication 203, Geological Society Publishing House, Bath.

Shiraiwa, T., Kohshima, S., Uemura, R., *et al.* (2002) High net accumulation rates at Campo de Hielo Patagónico Sur, South America, revealed by analysis of a 45.97 m long ice core. *Annals of Glaciology*, **35**, 84–90.

Shoemaker, E.M. (1986) Subglacial hydrology for an ice sheet resting on a deformable aquifer. *Journal of Glaciology*, **32**, 20–30.

Shoemaker, E.M. (1991) On the formation of large subglacial lakes. *Canadian Journal of Earth Sciences*, **28**, 1975–1981.

Shoemaker, E.M. (1992a) Subglacial floods and the origin of low-relief ice-sheet lobes. *Journal of Glaciology*, **38**, 105–112.

Shoemaker, E.M. (1992b) Water sheet outburst floods from the Laurentide Ice Sheet. *Canadian Journal of Earth Sciences*, **29**, 1250–1264.

Shoemaker, E.M. & Leung, H.K.N. (1987) Subglacial drainage for an ice sheet resting upon a layered deformable bed. *Journal of Geophysical Research*, **92**(B6), 4935–4946.

Shoji, H. & Langway, C.C. (1984) Flow behavior of basal ice as related to modelling considerations. *Annals of Glaciology*, **5**, 141–148.

Shoji, H. & Langway, C.C. (1988) Flow-law parameters of the Dye 3, Greenland, deep ice core. *Annals of Glaciology*, **10**, 146–150.

Shreve, R.L. (1972) Movement of water in glaciers. *Journal of Glaciology*, **11**, 205–214.

Shreve, R.L. (1984) Glacier sliding at subfreezing temperatures. *Journal of Glaciology*, **30**, 341–347.

Siegel, D.I. (1991) Evidence for dilution of deep, confined ground water by vertical recharge of isotopically heavy Pleistocene water. *Geology*, **19**, 433–436.

Siegert, M.J. (2000) Antarctic subglacial lakes. *Earth-Science Reviews*, **50**, 29–50.

Siegert, M.J. (2001) *Ice Sheets and Late Quaternary Environmental Change*. Wiley, Chichester, 231 pp.

Siegert, M.J. & Dowdeswell, J.A. (2002) Late Weichselian iceberg, meltwater and sediment production from the Eurasian Ice Sheet: results from numerical ice-sheet modelling. *Marine Geology*, **188**, 109–127.

Siegert, M.J., Ellis-Evans, J.C., Tranter, M., *et al.* (2001) Physical, chemical and biological processes in Lake Vostok and other Antarctic subglacial lakes. *Nature*, **414**, 603–609.

Siegert, M.J., Dowdeswell, J.A., Svendsen, J.-I. & Elverhoi, A. (2002) The Eurasian Arctic during the last Ice Age. *American Scientist*, **90**, 32–39.

Sigurdsson, F. (1990) Groundwater from glacial areas in Iceland. *Jökull*, **40**, 119–145.

Sigurdsson, O. (1991) Glacier variations 1930–1960, 1960–1990 og 1989–1990. *Jökull*, **41**, 88–96.

Sigurdsson, O. & Jonsson, T. (1995) Relation of glacier variations to climate change in Iceland. *Annals of Glaciology*, **21**, 263–270.

Simoes, J., Petit, J., Souchez, R., Lipenkov, V., de Angelsi, M., Lui, L., Jouzel, J. & Duval, P. (2002) Evidence of glacial flour in the deepest 89 m of the Vostok ice core. *Annals of Glaciology*, **35**, 341–346.

Singarayer, J.S. & Bamber, J.L. (2003) EOF analysis of three records of sea-ice concentration spanning the last 30 years. *Geophysical Research Letters*, **30**(5), art. no.–1251.

Sissons, J.B. (1960) Some aspects of glacial drainage channels in Britain: Part I. *Scottish Geographical Magazine*, **76**, 131–146.

Sissons, J.B. (1961) Some aspects of glacial drainage channels in Britain: Part II. *Scottish Geographical Magazine*, **77**, 15–36.

Sissons, J.B. (1979) The Loch Lomond Stadial in the British Isles. *Nature*, **280**, 199–203.

Sjogren, D.B. (1999) *Formation of the Viking Moraine, east-central Alberta: geomorphic and sedimentary evidence*. Unpublished PhD thesis, University of Alberta.

Sjogren, D.B. & Rains, R.B. (1995) Glaciofluvial erosional morphology and sediments of the Coronation-Spondin scabland, east-central Alberta. *Canadian Journal of Earth Sciences*, **32**, 565–578.

Sjogren, D.B., Fisher, T.G., Taylor, L.D., Jol, H.M. & Munro-Stasiuk, M.J. (1990) Incipient tunnel channels. *Quaternary International*, **90**, 41–56.

SKI (1997) *Deep Repository Performance Assessment Project, Site–94*. SKI Report 5 (Summary), Swedish Nuclear Power Inspectorate, Stockholm, 90 pp.

Skidmore, M.L. & Sharp, M.J. (1999) Drainage system behaviour of a High-Arctic polythermal glacier. *Annals of Glaciology*, **28**, 209–215.

Skidmore, M.L., Foght, J.M. & Sharp, M.J. (2000) Microbial life beneath a High Arctic glacier. *Applied and Environmental Microbiology*, **66**, 3214–3220.

Skusi, J., Rasilainen, K., Casanova, J., Ruskeeniemi, T., Blomqvist, R. & Smellie, J. (2001) U-series disequilibria in a groundwater flow route as an indicator of uranium migration processes. *Journal of Contaminant Hydrology*, **47**, 187–196.

Skvarca, P. (1993) Fast recession of the northern Larsen Ice Shelf monitored by space images. *Annals of Glaciology*, **17**, 317–321.

Skvarca, P. (1994) Changes and surface features of the Larsen Ice Shelf, Antarctica, derived from Landsat and Kosmos mosaics. *Annals of Glaciology*, **20**, 6–12.

Skvarca, P. & Naruse, R. (1997) Dynamic behavior of Glaciar Perito Moreno, southern Patagonia. *Annals of Glaciology*, **24**, 268–271.

Skvarca, P., Rack, W., Rott, H. & Ibarzabal y Donangelo, T. (1998) Evidence of recent climatic warming on the eastern Antarctic Peninsula. *Annals of Glaciology*, **27**, 628–632.

Skvarca, P., Angelis, H., Naruse, R., *et al.* (2002) Calving rates in fresh water: new data from southern Patagonia. *Annals of Glaciology*, **34**, 379–384.

Skvarca, P., Naruse, R. & Angelis, H. (2004) Recent thickening trend of Perito Moreno Glacier, southern Patagonia. *Bulletin of Glaciological Research*, **21**, 45–48.

Sleewaegen, S., Samyn, D., Fitzsimons, Z.S. & Lorrain, R. (2003) Equifinality of basal ice facies from an Antarctic cold-based glacier. *Annals of Glaciology*, **37**, 257–262.

Smellie, J.A.T. (1985) Uranium series disequilibrium studies od drillcore KM3 from the Kamlunge test-site, northern Sweden. *Mineralogical Magazine*, **49**, 271–279.

Smellie, J.A.T. & Frape, S. (1997) Hydrochemical aspects of glaciation. In: *Glaciation and Hydrogeology* (Eds L. King-Clayton, N. Chapman, L.O. Ericsson & F. Kautsky), pp. 45–51. SKI Report 13, Swedish Nuclear Power Inspectorate, Stockholm.

Smellie, J.A.T. & Laaksoharju, M. (1992) The Äspö Hard Rock Laboratory: final evaluation of the hydrochemical pre-investigations in relation to existing geologic and hydraulic conditions. *Svensk Kärnbränslehantering AB Technical Report*, **92-31**, 239 pp.

Smith, A.M. (1996) Ice shelf basal melting at the grounding line, measured from seismic observations. *Journal of Geophysical Research*, **101**, 22749–22755.

Smith, A.M. (1997a) Variations in basal conditions on Rutford Ice Stream, West Antarctica. *Journal of Glaciology*, **43**, 245–255.

Smith, A.M. (1997b) Basal conditions on Rutford Ice Stream, West Antarctica, from seismic observations. *Journal of Geophysical Research—Solid Earth*, **102**, 543–552.

Smith, A.M., Vaughan, D.G., Doake, C.S.M. & Johnson, A.C. (1999) Surface lowering of the ice ramp at Rothera Point, Antarctic Peninsula, in response to regional climate change. *Annals of Glaciology*, **27**, 113–118.

Smith, B. & Sandwell, D. (2003) Accuracy and resolution of shuttle radar topography mission data. *Geophysical Research Letters*, **30**(9), art. no.–1467.

Smith, B.E., Lord, N.E. & Bentley, C.R. (2002) Crevasse ages on the northern margin of Ice stream C, West Antarctica. *Annals of Glaciology*, **34**, 209–216.

Smith, D.G. (1994) Glacial Lake McConnell: paleogeography, age, duration, and associated river deltas, Mackenzie River basin, western Canada. *Quaternary Science Reviews*, **13**, 829–843.

Smith, I.R. (2000) Diamictic sediments within high arctic lake sediment cores: evidence for lake ice rafting along the lateral glacial margin. *Sedimentology*, **47**, 1157–1179.

Smith, L.M., Miller, G.H., Otto-Bliesner, B. & Shin, S.-I. (2003) Sensitivity of the Northern Hemisphere Climate System to extreme changes in Arctic sea ice. *Quaternary Science Reviews*, **22**, 645–658.

Smith, L.M., Andrews, J.T., Castaneda, I.S., Kristjansdottir, G.B., Jennings, A.E. & Sveinbjjronsdottir, A.E. (2005) Temperature reconstructions from SW and N Iceland waters over the last 10,000 cal yr B.P. based on ^{18}O records from planktonic and benthic foraminifera. *Quaternary Science Reviews*, **24**, 1723–1740.

Smith, M.J. (2002) *Technical developments for the geomorphological reconstruction of palaeo ice sheets from remotely sensed data.* Unpublished PhD thesis, University of Sheffield. 278 pp.

Smith, M.J. & Clark, C.D. (2005) Methods for the visualisation of digital elevation models for landform mapping. *Earth Surface Processes and Landforms*, **30**(7), 885–900.

Smith, M.J., Clark, C.D. & Wise, S.M. (2001) Mapping glacial lineaments from satellite imagery: an assessment of the problems and development of best procedure. *Slovak Geological Magazine*, **7**, 263–274.

Solheim, A. (1991) The depositonal environment of surging sub-polar tidewater glaciers. *Norsk Polarinstitutt Skrifter*, **194**, 97 pp.

Sollid, J.L. & Sörbel, L. (1984) Distribution of glacial landforms in southern Norway in relation to the thermal regime of the last continental ice sheet. *Striae*, **20**, 63–67.

Souchez, R. & Jouzel, J. (1984) On the isotopic composition in δD and $\delta^{18}O$ of water and ice during freezing. *Journal of Glaciology*, **30**(106), 369–372.

Souchez, R.A. & Tison, J.-L. (1981) Basal freezing of squeezed water: its influence on glacier erosion. *Annals of Glaciology*, **2**, 63–66.

Souchez, R., Khazendar, A., Ronveaux, D. & Tison, J.-L. (1998) Freezing at the grounding line in East Antarctica: possible implications for sediment export efficiency. *Annals of Glaciology*, **27**, 316–320.

Souchez, R., Jouzel, J., Lorrain, R., Sleewaegen, S., Stievenard, M. & Verbeke, V. (2000) A kinetic isotopic effect during ice formation by water freezing. *Geophysical Research Letters*, **27**(13), 1923–1926.

Souchez, R., Petit, J.R., Jouzel, J., Simões, J., de Angelis, M., Barkov, N., Stievenard, M., Vimeux, F., Sleewaegen, S. & Lorrain, R. (2002) Highly deformed basal ice in the Vostok core, Antarctica. *Geophysical Research Letters*, **29**(7), 40/1–40/4.

Sowers, T., Alley, R.B. & Jubenville, J. (2003) Ice core records of atmospheric N2O covering the last 106,000 years. *Science*, **301**, 945–948.

Spedding, N. & Evans, D.J.A. (2002) Sediments and landforms at Kvíárjökull, southeast Iceland: a reappraisal of the glaciated valley landsystem. *Sedimentary Geology*, **149**, 21–42.

Speed, R.C. & Cheng, H. (2004) Evolution of marine terraces and sea level in the last interglacial, Cave Hill, Barbados. *Geological Society of America Bulletin*, **116**, 219–232.

Speight, J.G. (1963) Late Pleistocene historical geomorphology of the Lake Pukaki area, New Zealand. *New Zealand Journal of Geology, and Geophysics* **6**, 160–188.

Splettoesser, J. (1992) Antarctic global warming? *Nature*, **355**, 503.

Spurk, M., Friedrich, M., Hofmann, J., *et al.* (1998) Revisions and Extension of the Hohenheim Oak and Pine Chronologies: New Evidence about the Timing of the Younger Dryas/Preboreal Transition. *Radiocarbon*, **40**, 1107.

Srivastava, S.P., Arthur, M.A., Clement, B., *et al.* (1989) *Scientific Results Baffin Bay and Labrador Sea.* Ocean Drilling Program, Vol. 105, Texas A & M University, Washington, DC, 1038 pp.

Staiger, J.W., Gosse, J.C., Johnson, J., Fastook, J., Gray, J.T., Stockli, D., Stockli, L. & Finkel, R. (2005) Quaternary relief generation by polythermal glacier ice: a field calibrated glacial erosion model: *Earth Surface Processes and Landforms*. Accepted.

Stauffer, B. (1989) Dating of ice by radioactive isotopes. In: *The Environmental Record in Glaciers and Ice Sheets* (Eds H. Oeschger & C.C.Langway Jr), pp. 123–139. Dahlem Konferenzen, Wiley, Chichester.

Stauffer, B., Neftel, A., Oeschger, H. & Schwander, J. (1985) CO_2 concentration in air extracted from Greenland ice samples. In: *Greenland ice core: Geophysics, Geochemistry and the Environment* (Eds C.C. Langway, H. Oeschger & W. Dansgaard), pp. 85–89. Geophysical Monograph 33, American Geophysical Union, Washington, DC.

Steele, M. & Boyd, T. (1998) Retreat of the cold halocline layer in the Arctic Ocean. *Journal of Geophysical Research*, **103**, 10,419–10,435.

Steffen, K., Bindschadler, R., Casassa, C., *et al.* (1993) *Snow and Ice Application of AVHRR in Polar Regions. Annals of Glaciology*, Boulder, Colorado, 20 May 1992.

Steffen, K., Cuillen, N., Huff, R., Stewart, C. & Rignot, E. (2004) Petermann Gletscher's floating tongue in northwestern Greenland: peculiar surface features, bottom melt channels, and mass balance assessment. In: *34th International Arctic Workshop, 10–13 March*, pp. 158–160. Institute of Arctic and Alpine Research, University of Colorado, Boulder, CO.

Steffensen, J.P. (1997) The size distribution of microparticles from selected segments of the Greenland Ice Core Project ice core representing different climatic periods. *Journal of Geophysical Research*, **102**(C12), 26755–26763.

Steig, E.J. & Alley, R. (2002) Phase relationships between Antarctic and Greenland climate records. *Annals of Glaciology*, **35**, 451–456.

Steig, E.J., Wolfe, A.P. & Miller, G.H. (1998) Wisconsinan refugia and the glacial history of eastern Baffin Island, Arctic Canada: Coupled evidence from cosmogenic isotopes and lake sediments. *Geology*, **26**, 835–838.

Steinemann, S. (1954) Results of preliminary experiments on the plasticity of ice crystals. *Journal of Glaciology*, **2**, 404–413.

Steinemann, S. (1958a) Resultats experimentaux sur la dynamique de la lgace et leurs correlations ave le mouvement et la petrographie des glaciers. In: *Symposium of Chamonix. Physics of the Motion of Ice*, pp. 184–198. IAHS Publication 47, International Association of Hydrological Sciences, Wallingford.

Steinemann, S. (1958b) Experimentelle Untersuchungen zur Plastizität von Eis. *Beiträge Zur Geologie der Schweiz, Hydrologie*, **10**, 72.

Steinþórsson, S. & Óskarsson, N. (1983) Chemical monitoring of Jökulhaup water in Skeidþará and the geothermal system in Grimsvötn, Iceland. *Jökull*, **33**, 73–86.

Stenni, B., Masson, V., Johnsen, S., Jouzel, J., Longinelli, A., Monin, E., Röthlisberger, R. & Selmo, E. (2001) An oceanic cold reversal during the last deglaciation. *Science*, **293**, 2074–2077.

Stephenson, P.J. (1967) Some considerations of snow metamorphism in the antarctic ice sheet in the light of the ice crystal studies. In: *Physics of Snow and Ice*, Vol. 1 (Ed. H. Oura), pp. 725–740. Institute of Low Temperature Science, Hokkaido University.

Stephenson, S.N. & Bindschadler, R.A. (1988) Observed velocity fluctuations on a major Antarctic ice stream. *Nature*, **334**(6184), 695–697.

Stern, L.A., Durham, W.B. & Kirby, S.H. (1997) Grain-sized-induced weakening of H2O ices I and II and associated anisotropic recrystallization. *Journal of Geophysical Research*, **102**, 5313–5325.

Stewart, M.K. (1975) Stable isotope fractionation due to evaporation and isotopic exchange of falling waterdrops: Applications to atmospheric processes and evaporation of lakes. *Journal of Geophysical Research*, **80**, 1133–1146.

Stocker, T.F. (1996) The ocean in the climate system: observing and modeling its variability. In: *Topics in Atmospheric and Interstellar Physics and Chemistry* (Ed. C.F. Boutron), pp. 39–90. European Research Course on Atmospheres, Vol. 2, Les Editions de Physique, Les Ulis.

Stocker, T.F. & Wright, D.G. (1991) Rapid transitions of the ocean's deep circulation induced by changes in surface water fluxes. *Nature*, **351**, 729–732.

Stoker, M.S. (1995) The influence of glacigenic sedimentation on slope-apron development on the continental margin off Northwest Britain. In: *The Tectonics, Sedimentation and Palaeoceanography of the North Atlantic Region* (Eds R.A. Scrutton, M.S. Stoker, G.B. Shimmield & A.W. Tudhope), pp. 159–177. Special Publication 90, Geological Society Publishing House, Bath.

Stoker, M.S. (2002) Late Neogene development of the NW UK Atlantic margin. In: *Exhumation of the North Atlantic Margin: Timing, Mechanisms and Implications for Petroleum Exploration* (Eds A.G. Doré, J.A. Cartwright, M.S. Stoker, J.P. Turner & N. White), pp. 313–329. Special Publication 96, Geological Society Publishing House, Bath.

Stoker, M.S. & Holmes, R. (1991) Submarine end-moraines as indicators of Pleistocene ice-limits off northwest Britain. *Journal of the Geological Society, London*, **148**, 431–434.

Stoker, M.S., Hitchin, K. & Graham, C.C. (1993) *The Geology of the Hebrides and West Shetland Shelves*. HMSO, London.

Stokes, C.R. (2000) *The geomorphology of paleo-ice streams: identification, characterisation and implications for ice stream functioning*. PhD thesis, Department of Geography, University of Sheffield.

Stokes, C.R. & Clark, C.D. (1999) Geomorphological criteria for identifying Pleistocene ice streams. *Annals of Glaciology*, **28**, 67–74.

Stokes, C.R. & Clark, C.D. (2001) Palaeo-ice streams. *Quaternary Science Reviews*, **20**, 1437–1457.

Stokes, C.R. & Clark, C.D. (2003a) Laurentide ice streaming on the Canadian Shield: A conflict with the soft-bedded ice stream paradigm. *Geology*, **31**, 347–350.

Stokes, C.R. & Clark, C.D. (2003b) The Dubawnt Lake paleo-ice stream: evidence for dynamic ice sheet behaviour on the Canadian Shield and insights regarding the controls on ice-stream location and vigour. *Boreas*, **32**, 263–279.

Stokes, C.R. & Clark, C.D. (2004) Evolution of Late Glacial ice-marginal lakes on the north-western Canadian Shield and their influence on the location of the Dubawnt Lake Palaeo-Ice Stream. *Palaeogeography, Palaeoclimatology, Palaeoecology*, **215**(1/2), 155–171. doi:10.1016/j.palaeo.2004.09.006.

Stone, J.O. (2000) Air pressure and cosmogenic isotope production. *Journal of Geophysical Research*, **105**(B10), 23753–23759.

Stone, D.B. & Clarke, G.K.C. (1993) Estimation of subglacial hydraulic properties from induced changes in basal water pressure: A theoretical framework for borehole response tests. *Journal of Glaciology*, **39**, 327–340.

Stone, D.B., Clarke, G.K.C. & Ellis, R.G. (1997) Inversion of borehole-response test data for estimation of subglacial hydraulic properties. *Journal of Glaciology*, **43**, 103–113.

Stone, J.O., Ballantyne, C.K. & Fifield, L.K. (1998) Exposure dating and validation of periglacial weathering limits, northwest Scotland. *Geology*, **26**, 587–590.

Stone, J.O.H., Balco, G., Sugden, D.E., *et al.* (2003) Holocene deglaciation of Marie Byrd Land, West Antarctica. *Science*, **299**, 99–102.

Stoner, J.S. & Andrews, J.T. (1999) The North Atlantic as a Quaternary magnetic archive. In: *Quaternary Climates, Environments and Magnetism* (Eds B. Maher & R. Thompson), pp. 49–80. Cambridge University Press, Cambridge.

Stoner, J.S., Channell, J.E.T. & Hillaire-Marcel, C. (1996) The magnetic signature of rapidly deposited detrital layers from the deep Labrador Sea: Relationship to North Atlantic Heinrich Layers. *Paleoceanography*, **11**, 309–326.

Stott, L., Poulsen, C., Lund, S. & Thunell, R. (2002) Super ENSO and global climate oscillations at Millennial time scales. *Science*, **297**, 222–226.

Stotter, J., Wastl, M., Caseldine, C. & Haberle, T. (1999) Holocene paleoclimatic reconstructions in Northern Iceland: Approaches and Results. *Quaternary Science Reviews*, **18**, 457–474.

Stroeve, J. (2001) Assessment of Greenland albedo variability from the advanced very high resolution radiometer Polar Pathfinder data set. *Journal of Geophysical Research, Special Issue*, **106**(D24), 33,989–34,006.

Stroeve, J., Nolin, A. & Steffen, K. (1997) Comparison of AVHRR-derived and in situ surface albedo over the Greenland ice sheet. *Remote Sensing of Environment*, **62**(3), 262–276.

Stroeven, A.P., Fabel, D., Harbor, J., Hättestrand, C. & Kleman, J.U. (2002a) Quantifying the erosional impact of the Fennoscandian ice sheet in the Tornetrask-Narvik corridor, northern Sweden. *Geografiska Annaler*, **88A**(3–4), 275–287.

Stroeven, A.P., Fabel, D., Hättestrand, C. & Harbor, J. (2002b) A relict landscape in the centre of Fennoscandian glaciation: cosmogenic radionuclide evidence for tors preserved through multiple glacial cycles. *Geomorphology*, **44**, 145–154.

Stroeven, A., Fabel, D., Harbor, J., *et al.* (2002c) Reconstructing pre-glacial landscapes and erosion history of glaciated passive margins: Applications of *in-situ* produced cosmogenic nuclide techniques.

In: *Exhumation of the North Atlantic Margin* (Eds A.G. Dore, J.A. Cartwright, M.S. Stoker, J.P. Turner & N.J. White), pp. 53–168. Special Publication 196, Geological Society Publishing House, Bath.

Stroeven, A.P., Harbor, J., Fabel, D., *et al.* (2006) Slow, patchy landscape evolution in northern Sweden despite repeated ice sheet glaciation. In: *Tectonics, Climate and Landscape Evolution* (Eds S. Willett, N. Hovius, M. Brandon & D. Fisher), 387–396. Geological Society of America Special Paper 398, Tulsa, OK.

Studinger, M., Bell, R.E., Blankenship, D.D., *et al.* (2001) Subglacial sediments: a regional geological template for ice flow in West Antarctica. *Geophysical Research Letters*, **28**, 3493–3496.

Stueber, A.M. & Walter, L.M. (1994) Glacial recharge and paleohydrologic flow systems in the Illinois basin: Evidence from chemistry of Ordovician carbonate (Galena) formation waters. *Geological Society of America Bulletin*, **106**, 1430–1439.

Stuefer, M. (1999) *Investigations on mass balance and dynamics of Moreno Glacier based on field measurements and satellite imagery*. PhD thesis, University of Innsbruck, Austria.

Stuiver, M., Braziunas, T.F., Becker, B. & Kromer, B. (1991) Climatic, solar, oceanic and geomagnetic influences of Late Glacial and Holocene $^{14}C/^{12}C$ change. *Quaternary Research*, **35**, 1–14.

Su, Z. & Shi, Y. (2002) Response of monsoonal temperate glaciers to global warming since the Little Ice Age. *Quaternary International*, **97–98**, 123–131.

Su, Z., Song, G. & Cao, Z. (1996) Maritime characteristics of Hailougou Glacier in the Gongga Mountains. *Journal of Glaciology and Geocryology*, **18**(supplement), 51–59.

Su, Z., Liu, Z., Wang, W., *et al.* (1999) Glacier response to the climatic change and its trend forecast in Qinghai-Tibetan Plateau. *Advance in Earth Sciences*, **14**(6), 607–612.

Sugden, D.E. (1968) The selectivity of glacial erosion in the Cairngorm mountains, Scotland. *Transactions of the Institute of British Geographers*, **45**, 79–92.

Sugden, D.E. (1974) Landscapes of glacial erosion in Greenland and their relationship to ice, topographic and bedrock conditions. *Institute of British Geographers, Special Publication*, **7**, 177–195.

Sugden, D.E. (1976) A case against deep erosion of shields by ice sheets. *Geology*, **4**, 580–582.

Sugden, D.E. (1977) Reconstruction of the morphology, dynamics and thermal characteristics of the Laurentide Ice Sheet at its maximum. *Arctic and Alpine Research*, **9**, 21–47.

Sugden, D.E. (1978) Glacial erosion by the Laurentide Ice Sheet. *Journal of Glaciology*, **20**, 367–392.

Sugden, D.E. & Denton, G.H. (2004) Cenozoic landscape evolution of the Convoy Range to Mackay Glacier area, Transantarctic Mountains: onshore to offshore synthesis. *Geological Society of America Bulletin*, **116**, 840–857.

Sugden, D.E. & John, B.S. (1976) *Glaciers and Landscape*. Arnold, London.

Sugden, D.E. & Watts, S.H. (1977) Tors, felsenmeer, and glaciation in northern Cumberland Peninsula, Baffin Island. *Canadian Journal of Earth Sciences*, **14**, 2817–2823.

Sugden, D.E., Denton, G.H. & Marchant, D.R. (1991) Subglacial meltwater channel systems and ice sheet overriding, Asgard Range, Antarctica. *Geografiska Annaler*, **73A**, 109–121.

Sugden, D.E., Marchant, D.R., Potter, N., *et al.* (1995) Preservation of Miocene glacier ice in East Antarctica. *Nature*, **376**, 412–414.

Sugden, D.E., Summerfield, M.A., Denton, G.H., *et al.* (1999) Landscape development in the Royal Society Range, southern Victoria Land, Antarctica: stability since the mid-Miocene. *Geomorphology* **28**, 181–200.

Sugden, D.E., Balco, G., Cowdery, S.G., Stone, J.O. & Sass, L.C. III. (2005) Selective glacial erosion and weathering zones in the coastal mountains of Marie Byrd Land, Antarctica. *Geomorphology*, **67**, 317–334.

Sugiyama, S. & Gudmundsson, G.H. (2003) Diurnal variations in vertical strain observed in a temperate valley glacier. *Geophysical Research Letters*, **30**(2), 1090.

Sugiyama, S., Gudmundsson, G.H. & Helbing, J. (2003) Numerical investigation of the effects of temporal variations in basal lubrication on englacial strain-rate distribution. *Annals of Glaciology*, **37**, 49.

Summerfield, M.A., Sugden, D.E., Denton, G.H., Marchant, D.R., Cockburn, H.A.P. & Stuart, F.M. (1999) Cosmogenic data support previous evidence of extremely low rates of denudation in the Dry Valleys region, southern Victoria Land, Antarctica. In: *Uplift, Erosion & Stability: Perspectives on Long-term Landscape Evolution* (Eds B.J. Smith, W.B. Whalley & P.A. Warke), pp. 255–267. Special Publication 162, Geological Society Publishing House, Bath.

Sutherland, D.G. (1984) The Quaternary deposits and landforms of Scotland and the neighbouring shelves—a review. *Quaternary Science Reviews*, **3**, 157–254.

Svendsen, J.I., Mangerud, J., Elverhøi, A., Solheim, A. & Schüttenhelm, R.T.E. (1992) The Late Weichselian glacial maximum on western Spitsbergen inferred from offshore sediment cores. *Marine Geology*, **104**, 1–17.

Swift, D.A., Nienow, P.W., Hoey, T.B. & Mair, D.W.F. (2005a) Seasonal evolution of runoff from Haut Glacier d'Arolla, Switzerland and implications for glacial geomorphic processes. *Journal of Hydrology*, **309**(1–4), 133–148. doi:10.1016/j.hydrol.2004.11.016

Swift, D.A., Nienow, P.W. & Hoey, T.B. (2005b) Basal sediment evacuation by subglacial meltwater: suspended sediment transport from Haut Glacier d'Arolla, Switzerland. *Earth Surface Processes and Landforms*, **30**(7), 867–883.

Swithinbank, C.W.M. (1954) Ice streams. *Polar Record*, **7**, 185–186.

Synge, F.M. & Stephens, N. (1960) The Quaternary period in Ireland-an assessment. *Irish Geography*, **4**, 121–130.

Syvitski, J.P.M., Burrell, D.C. & Skei, J.M. (1987) *Fjords: Processes and Products*. Springer-Verlag, New York, 379 pp.

Syvitski, J.P.M., Andrews, J.T. & Dowdeswell, J.A. (1996) Sediment deposition in an iceberg-dominated Glacimarine Environment, East Greenland: Basin Fill Implications. *Global and Planetary Change*, **12**, 251–270.

Syvitski, J.P., Jennings, A.E. & Andrews, J.T. (1999) High-Resolution Seismic Evidence for Multiple Glaciation across the Southwest Iceland Shelf. *Arctic and Alpine Research*, **31**, 50–57.

Talbot, C.J. (1999) Ice ages and nuclear waste isolation. *Engineering Geology*, **52**, 177–192.

Tangborn, W. (1980) Two models for estimating climate-glacier relationships in the North Cascades, Washington, U.S.A. *Journal of Glaciology*, **25**(91), 3–22.

Tanner, V. (1914) Studier öfver Kvartärsystemet i Fennoskandias nordliga delar III. Om landisens rörelser och afsmältning i finska Lapland och angränsande trakter. *Bulletin de la géologique de Finlande*, **38**.

Tarasov, L. & Peltier, W. (2002) Greenland glacial history and local geodynamic consequences. *Geophysical Journal International*, **150**, 198–229.

Tarasov, L. & Peltier, W.R. (2004) A geophysically constrained large ensemble analysis of the deglacial history of the North-American ice-sheet complex. *Quaternary Science Reviews*, **23**, 359–388.

Taylor, J., Dowdeswell, J.A. & Kenyon, N.H. (2000) Canyons and Late Quaternary sedimentation on the North Norwegian margin. *Marine Geology*, **166**, 1–9.

Taylor, J., Dowdeswell, J.A., Kenyon, N.H. & Ó Cofaigh, C. (2002) Late Quaternary architecture of trough-mouth fans: debris flows and suspended sediments on the Norwegian margin. In: *Glacier-influenced Sedimentation on High-latitude Continental Margins* (Eds J.A. Dowdeswell & C. Ó Cofaigh), pp. 55–71. Special Publication 203, Geological SocietyPublishing House, Bath.

Taylor, K.C. (1982) *Sonic logging at Dye 3, Greenland.* MS thesis, University of Wisconsin, pp. 1–64.

Taylor, K.C. & Alley, R.B. (2004) Two dimensional electrical stratigraphy of the Siple Dome, Antarctica ice core. *Journal of Glaciology*, **50**(169), 231–235.

Taylor, K.C., Alley, R.B., Meese, D.A., *et al.* (2004) Dating the Siple Dome (Antarctica) ice core by manual and computer interpretation of annual layering. *Journal of Glaciology*, **50**(170) 453–461.

Teller, J.T. (1987) Proglacial lakes and the southern margin of the Laurentide Ice Sheet. In: *North America and Adjacent Oceans During the Last Deglaciation*, Vol. K-3 (Eds W.F. Ruddiman & H.E. Wright), pp. 39–69. The Geology of North America Series, Geological Society of America, Boulder, Colorado.

Teller, J.T. (1990a) Volume and routing of late glacial runoff from the southern Laurentide Ice Sheet. *Quaternary Research*, **34**, 12–23.

Teller, J.T. (1990b) Meltwater and precipitation runoff to the North Atlantic, Arctic, and Gulf of Mexico from the Laurentide Ice Sheet and adjacent regions during the Younger Dryas. *Paleoceanography*, **5**, 897–905.

Teller, J.T. (1995a) History and drainage of large ice-dammed lakes along the Laurentide Ice Sheet. *Quaternary International*, **28**, 83–92.

Teller, J.T. (1995b) The impact of large ice sheets on continental palaeohydrology. In: *Global Continental Palaeohydrology* (Eds K.J. Gregory, L. Starkel & V.R. Baker), pp. 109–129. Wiley, Chichester.

Teller, J.T. (2001) Formation of large beaches in an area of rapid differential isostatic rebound: the three-outlet control of Lake Agassiz. *Quaternary Science Reviews*, **20**, 1649–1659.

Teller, J.T. (2003) Subaquatic landsystems: large proglacial lakes. In: *Glacial Landsystems* (Ed. D.A. Evans), pp. 348–371. Edwin Arnold, London.

Teller, J.T. (2004) Controls, history, outbursts, and impacts of large late Quaternary proglacial lakes in North America. In: *The Quaternary Period in the United States*, Vol. 1, *Developments in Quaternary Science* (Eds A. Gilespie, S. Porter & B. Atwater), pp. 45–61. Elsevier, Amsterdam.

Teller, J.T. & Clayton, L. (Eds) (1983) *Glacial Lake Agassiz.* Special Paper 26, Geological Association of Canada, St John's, NF.

Teller, J.T. & Kehew, A.E. (Eds) (1994) Late Glacial History of Large Proglacial Lakes and Meltwater Runoff Along the Laurentide Ice Sheet. *Quaternary Science Reviews*, **13**(9–10), 795–981.

Teller, J.T. & Leverington, D.W. (2004) Glacial Lake Agassiz: a 5000-year history of change and its relationship to the $\delta^{18}O$ record of Greenland. *Geological Society of America Bulletin*, **116**, 729–742.

Teller, J.T. & Thorleifson, L.H. (1983) The Lake Agassiz—Lake Superior connection. In: *Glacial Lake Agassiz* (Eds J.T. Teller & L. Clayton), pp. 261–290. Special Paper 26, Geological Association of Canada, St John's, NF.

Teller, J.T., Leverington, D.W. & Mann, J.D. (2002) Freshwater outbursts to the oceans from glacial Lake Agassiz and their role in climate change during the last deglaciation. *Quaternary Science Reviews*, **21**, 879–887.

Teller, J.T., Boyd, M., Yang, Z., Kor, P.S.G. and Fard, A.M. (2005) Alternative routing of Lake Agassiz overflow during the Younger Dryas: new dates, paleotopography and a re-evaluation. *Quaternary Science Reviews*, **24**, 1890–1905.

The ISMASS Committee (2004) Recommendations for the collection and synthesis of Antarctic Ice Sheet mass balance data. *Global and Planetary Change*, **42**, 1–15.

Theakstone, W.H. (1967) Basal sliding and movement near the margin of the glacier Osterdalsisen, Norway. *Journal of Glaciology*, **6**, 805–816.

Theil, H. (1950) A rank invariant method of linear and polynomial regression analysis, I, II, III. *Proceedings, Koninklijke Nederlandse Akademie van Wetenschappen*, **53**, 386–392, 521–525, 1397–1412.

Thomas, A.G. & Raiswell, R. (1984) Solute acquisition in glacial meltwaters II. Argentiere (French Alps): bulk meltwaters with open system characteristics. *Journal of Glaciology*, **30**, 44–48.

Thomas, D.R. & Rothrock, D.A. (1993) The Arctic-Ocean Ice Balance—a Kalman Smoother Estimate. *Journal of Geophysical Research—Oceans*, **98**(C6), 10053–10067.

Thomas, R.H. (1977) Calving Bay dynamics and Ice Sheet retreat up the St Lawrence Valley System. Geographie physique et Quaternaire **31**, 347–356.

Thomas, R.H. (1979a) West Antarctic Ice Sheet: present-day thinning and Holocene retreat of the margins. *Science*, **205**, 1257–1258.

Thomas, R.H. (1979b) The dynamics of marine ice sheets. *Journal of Glaciology*, **24**, 167–177.

Thomas, R.H., Sanderson, T.J.O. & Rose, K.E. (1979) Effect of climatic warming on the West Antarctic ice sheet. *Nature*, **277**, 355–358.

Thomas, R.H., Abdalati, W., Akins, T., Csatho, B.M., *et al.* (2000a) Substantial thinning of a major east Greenland outlet glacier. *Geophysical Research Letters*, **27**, 1291–1294.

Thomas, R.H., Akins, T., Csatho, B., Fahnestock, M., Gogineni, P., Kim, C. & Sonntag, J. (2000b) Mass Balance of the Greenland Ice Sheet at high elevations. *Science*, **289**, 426–428.

Thomas, R., Csatho, B., Davis, C., *et al.* (2001) Mass balance of higher-elevation parts of the Greenland ice sheet. *Journal of Geophysical Research*, **106**(D24), 33707–33716.

Thomas, R.H., Abdalati, W., Frederick, E., Krabill, W.B., Manizade, S. & Steffen, K. (2003) Investigation of surface melting and dynamic thinning on Jakobshavn Isbrae, Greenland. *Journal of Glaciology*, **48**(165), 231–239.

Thomas, R.N. (2001) Program for Arctic Regional Climate Assessment (PARCA): Goals, key findings, and future directions. *Journal of Geophysical Research*, **106**(D24), 33,691–33,705.

Thompson, D.W.J. & Wallace, J.M. (1998) The Arctic Oscillation signature in the wintertime geopotential height and temperature fields. *Geophysics Research Letters*, **25**, 1297–1300.

Thompson, D.W.J. & Wallace, J.M. (2000) Annular modes in the extratropical circulation. Part I: Month-to-month variability. *Journal of Climate*, **13**, 1000–1016.

Thompson, D.W.J., Wallace, J.M. & Hegerl, G. (2000) Annular modes in the extratropical circulation. Part II: Trends. *Journal of Climate*, **13**, 1018–1036.

Thompson, S.L. & Pollard, D. (1997) Greenland and Antarctic mass balances for present and doubled CO_2 from the GENESIS version–2 global climate model. *Journal of Climate* **10**, 871–900.

Thorarinsson, S. (1940) Present glacier shrinkage, and eustatic changes of sea-level. *Geografiska Annaler*, **22**(1940), 131–159.

Thorarinsson, S. (1953) Oscillations of Iceland glaciers during the last 250 years. *Geografiska Annaler*, **25**, 1–54.

Thorleifson, L.H. (1996) Review of Lake Agassiz history. In: *Sedimentology, Geomorphology and History of the Central Lake Agassiz Basin* (Eds J.T. Teller, L.H. Thorleifson, G. Matile & W.C. Brisbin), pp. 55–84. Field Trip Guidebook B2, Annual Meeting 27–29 May, Winnipeg, MB. Geological Association of Canada.

Thorsteinsson, T. (2001) An analytical approach to deformation of anisotropic ice crystal aggregates. *Journal of Glaciology*, **47**(158), 507–516.

Thorsteinsson, T. (2002) Fabric development with nearest-neighbor interaction and dynamic recrystallization. *Journal of Geophysical Research*, **107**(B1), Art. No. 2014.

Thorsteinsson, T. & Waddington, E.D. (2002) Folding in strongly anisotropic layers near ice-sheet centers. *Annals of Glaciology*, **35**, 480–486.

Thorsteinsson, T., Kipfstuhl, J. & Miller, H. (1997) Textures and fabrics in the GRIP ice core. *Journal of Geophysical Research*, **102**(C12), 26583–26599.

Thorsteinsson, T., Waddington, E.D., Taylor, K.C., Alley, R.B. & Blankenship, D.D. (1999) Strain-rate enhancement at Dye 3, Greenland. *Journal of Glaciology*, **45**(150), 338–345.

Thorsteinsson, T., Waddington, E.D. & Fletcher, R.C. (2003) Spatial and temporal scales of anisotropic effects in ice-sheet flow. *Annals of Glaciology*, **37**, 40–48.

Thorsteinsson, T., Raymond, C.F., Gudmundsson, G.H., Bindschadler, R.A., Vornberger, P. & Joughin, I. (2004) Bed topography and lubrication inferred from surface measurements on fast-flowing ice streams. *Journal of Glaciology*, **49**(167), 481–490.

Tinner, W. & Lotter, A.F. (2001) Central European vegetation response to abrupt climate change at 8.2 ka. *Geology*, **29**, 551–554.

Tippett, C.R. (1985) Glacial dispersal train of Paleozoic erratics, central Baffin Island, N.W.T., Canada. *Canadian Journal Earth Sciences*, **22**, 1818–1826.

Tison, J.-L. & Hubbard, B. (2000) Ice crystallographic evolution at a temperate glacier: Glacier de Tsanfleuron, Switzerland. In: *Deformation of Glacial Materials* (Eds A.J. Maltman, B. Hubbard & M.J. Hambrey), pp. 23–38. Special Publication 176, Geological Society Publishing House, Bath.

Tison, J.-L., Petit, J.R., Barnola, J.M. & Mahaney, W.C. (1993) Debris entrainment at the ice-bedrock interface in sub-freezing temperature conditions (Terre Adélie, Antarctica). *Journal of Glaciology*, **39**(132), 303–315.

Tison, J.-L., Thorsteinsson, Th., Lorrain, R.D. & Kipfstuhl, J. (1994) Origin and development of textures and fabrics in basal ice at Summit, central Greenland. *Earth and Planetary Science Letters*, **125**, 421–437.

Tockner, K., Malard, F., Uehlinger, U. & Ward, J.V. (2002) Nutrients and organic matter in a glacial river-floodplain system (Val Roseg, Switzerland). *Limnology and Oceanography*, **47**, 266–277.

Toniazzo, T., Gregory, J.M. & Huybrechts, P. (2004) Climatic impact of a Greenland deglaciation and its possible irreversibility. *Journal of Climate*, **17**, 21–33.

Torell, O. (1873) Undersökningar öfver istiden del II. Skandinaviska landisens utsträckning under isperioden. *Öfversigt af Kungliga Vetenskapsakademiens Förhandlingar 1873*, **1**, 47–64.

Torinesi, O., Fily, M. & Genthon, C. (2003) Interannual variability and trend of the Antarctic Ice Sheet summer melting period from 20 years of spaceborne microwave data. *Journal of Climate*, **16**, 1047–1060.

Townsend, D.W. & Vickory, R.P. (1967) An experiment in regelation. *Philosophical Magazine*, **18**, 1275–1280.

Trabant, D.C. & March, R.S. (1999) Mass-balance measurements in Alaska and suggestions for simplified observation programs. *Georgrafiska Annaler*, **81A**(4), 777–789.

Tranter, M. (1982) *Controls on the chemical composition of Alpine glacial meltwaters*. Unpublished PhD thesis, University of East Anglia.

Tranter, M. (2003) Chemical weathering in glacial and proglacial environments. In: *Treatise on Geochemistry* (Eds H.D. Holland & K.K. Turekian), Vol. 5, *Surface and Ground Water, Weathering, Erosion and Soils* (Ed. J.I. Drever), pp. 189–205. Elsevier-Pergamon, Oxford.

Tranter, M., Sharp, M.J., Brown, G.H., Willis, I.C., Hubbard, B.P., Nielsen, M.K, Smart, C.C., Gordon, S., Tulley, M. & Lamb, H.R. (1997) Variability in the chemical composition of *in situ* subglacial meltwaters. *Hydrological Processes*, **11**, 59–77.

Tranter, M., Huybrechts, P., Munhoven, G., Sharp, M.J., Brown, G.H., Jones, I.W., Hodson, A.J., Hodgkins, R. & Wadham, J.L. (2002a) Glacial bicarbonate, sulphate and base cation fluxes during the last glacial cycle, and their potential impact on atmospheric CO_2. *Chemical Geology*, **190**, 33–44.

Tranter, M., Sharp, M.J., Lamb, H.R., Brown, G.H., Hubbard, B.P. & Willis, I.C. (2002b) Geochemical weathering at the bed of Haut Glacier d'Arolla, Switzerland—a new model. *Hydrological Processes*, **16**, 959–993.

Truffer, M., Motyka, R.J., Harrison, W.D., Echelmeyer, K.A., Fisk, B. & Tulaczyk, S. (1999) Subglacial drilling at Black Rapids Glacier, Alaska, U.S.A.: drilling method and sample descriptions. *Journal of Glaciology*, **45**(151), 495–505.

Truffer, M., Harrison, W.D. & Echelmeyer, K.A. (2000) Glacier motion dominated by processes deep in underlying till. *Journal of Glaciology*, **46**(153), 213–221.

Truffer, M., Echelmeyer, K.A. & Harrison, W.D. (2001) Implications of till deformation on glacier motion. *Journal of Glaciology*, **47**, 123–134

Trupin, A.S., Meier, M.F. & Wahr, J.M. (1992) Effect of melting glaciers on the Earth's rotation and gravitational field: 1965–1984. *Geophysical Journal International*, **108**(1992), 1–15.

Tschudi, S., Schafer, J.M., Borns, H.W., Ivy-Ochs, S., Kubik, P.W. & Schluchter, C. (2003) Surface exposure dating of Sirius Formation at Allan Hills nunatak, Antarctica: New evidence for long-term ice-sheet stability. *Eclogae Geologicae Helvetiae*, **96**, 109–114.

Tsui, P.C., Cruden, D.M. & Thomson, S. (1989) Ice-thrust terrains in glaciotectonic settings in central Alberta. *Canadian Journal of Earth Science*, **6**, 1308–1318.

Tulaczyk, S. (1999) Ice sliding over weak, fine-grained tills: dependence of ice–till interactions on till granulometry. In: *Glacial Processes Past and Present* (Eds D.M. Mickelson & J.V. Attig), pp. 159–177. Special Paper 337, Geological Society of America, Boulder, Colorado.

Tulaczyk, S., Kamb, B. & Engelhardt, H. (2000a) Basal mechanics of Ice Stream B. I. Till mechanics. *Journal of Geophysical Research*, **105**, 463–481.

Tulaczyk, S., Kamb, W.B. & Engelhardt, H.F. (2000b) Basal mechanics of ice stream B, West Antarctica. II. Undrained plastic bed model. *Journal of Geophysical Research*, **105**(B1), 483–494.

Tulaczyk, S., Kamb, B. & Engelhardt, H.F. (2001a) Estimates of effective stress beneath a modern West Antarctic ice stream from till preconsolidation and void ratio. *Boreas*, **30**, 101–114.

Tulaczyk, S.M., Scherer, R.P. & Clark, C.D. (2001b) A ploughing model for the origin of weak tilols beneath ice streams: a qualitative treatment. *Quaternary International*, **86**(1), 59–70.

Tullborg, E.-L. (1989) *Fracture Fillings in the Drillcores KAS05-KAS08 from Äspö, Southeastern Sweden*. SKB Progress Report 25-89-16, Swedish Nuclear Power Inspectorate, Stockholm.

Tullborg, E.-L. (1997a) How do we recognize remnants of glacial water in the bedrock? In: *Glaciation and Hydrogeology* (Eds L. King-Clayton, N. Chapman, L.O. Ericsson & F. Kautsky), pp. A75–A76. SKI Report 13, Swedish Nuclear Power Inspectorate, Stockholm.

Tullborg, E.-L. (1997b) Assessment of redox conditions based on fracture mineralogy. *Abstract, OECD/NEA Group Workshop on the Use of Hydrochemical Information in Testing Groundwater Flow Models, Borgholm, Sweden.*

Tushingham, A.M. & Peltier, W.R. (1991) Ice–3G: a new global model of late Pleistocene deglaciation based upon geophysical predictions of post-glacial relative sea level change. *Journal of Geophysical Research*, **96**, 4497–4523.

Twiss, R.J. (1986) Variable sensitivity piezometric equations for dislocation density and subgrain diameter and their relevance to olivine and quartz. In: *Mineral and Rock Deformation, Laboratory Studies; the Paterson Volume* (Eds B.E. Hobbs & H.C. Heard), pp. 247–261. Geophysical Monograph 36, American Geophysical Union, Washington, DC.

Tyrell, J.B. (1897) Report on the Dubawnt, Kazan and Ferguson rivers, and the northwest coast of Hudson Bay. *Geological Survey of Canada Annual Report*, **9**, 1–218. Queen's Printer, Ottawa.

Tyrrell, J.B. (1898) The glaciation of north-central Canada. *Journal of Geology*, **6**, 147–160.

Tziperman, E. (1997) Inherently unstable climate behaviour due to weak thermohaline ocean circulation. *Nature*, **386**, 592–595.

Vaikmäe, R., Vallner, L., Loosli, H.H., Blaser, P.C. & Juillard-Tardent, M. (2001) Palaeogroundwater of glacial origin in the Cambrian-Vendian aquifer of northern Estonia. *Geological Society of London Special Publication*, **189**, 17–27.

Van de Wal, R.S.W. & Oerlemans, J. (1994) An energy balance model for the Greenland ice sheet. *Global and Planetary Change*, **9**, 115–131.

Van de Wal, R.S.W. & Oerlemans, J. (1997) Modelling the short term response of the Greenland ice sheet to global warming. *Climate Dynamics*, **13**, 733–744.

Van de Wal, R.S.W. & Russell, A.J. (1994) A comparison of energy balance calculations, measured ablation and meltwater runoff near Søndre Strømfjord, West Greenland. *Global and Planetary Change*, **9**(1–2), 29–38.

Van de Wal, R.S.W. & Wild, M. (2001) Modelling the response of glaciers to climate change by applying the volume-area scaling in combination with a high resolution GCM. *Climate Dynamics*, **18**, 359–366.

Van de Wal, R.S.W., Wild, M. & de Wolde, J. (2001) Short-term volume changes of the Greenland ice sheet in response to doubled CO_2 conditions. *Tellus*, **53B**, 94–102.

Van den Broeke, M. (1996) Characteristics of the lower ablation zone of the West Greenland ice sheet for energy-balance modelling. *Annals of Glaciology*, **23**, 160–166.

Van der Meer, J.J.M. (2004) *Spitsbergen push moraines.* Elsevier, Amsterdam.

Van der Meer, J.J.M., Kjaer, K.H. & Krüger, J. (1999) Subglacial water-escape structures and till structures, Sléttjökull, Iceland. *Journal of Quaternary Science*, **14**, 191–205.

Van der Veen, C.J. (1986) Numerical modeling of ice shelves and ice tongues. *Annals of Geophysics Series B*, **4**, 45–54.

Van der Veen, C.J. (1996) Tide-water calving. *Journal of Glaciology*, **42**(141), 375–385.

Van der Veen, C.J. (2002) Polar ice sheets and global sea level: how well can we predict the future? *Global and Planetary Change*, **32**, 165–194.

Van der Veen, C. & Payne, A.J. (2004) Modelling land-ice dynamics. In: *Mass Balance of the Cryosphere: Observations and Modelling of Contemporary and Future Changes* (Eds J.L. Bamber & A.J. Payne), pp. 169–225. Cambridge University Press, Cambridge.

Van der Veen, C.J. & Whillans, I.M. (1989) Force budget: 1. Theory and numerical methods. *Journal of Glaciology*, **35**, 53–60.

Van der Veen, C.J. & Whillans, I.M. (1994) Development of fabric in ice. *Cold Regions Science and Techology*, **22**, 171–195.

Van Kreveld, S., Sarthein, M., Erlenkeuser, H., *et al.* (2000) Potential links between surging ice sheets, circulation changes, and the Dansgaard-Oeschger cycles in the Irminger Sea, 60–18 ka. *Paleoceanography*, **15**, 425–442.

Van Tatenhove, F., van der Meer, J. & Koster, E. (1996) Implications for deglaciation chronology from new AMS age determinations in central West Greenland. *Quaternary Research*, **45**, 245–253.

Van Weert, F.H.A., van Gijssel, K., Leijnse, A. & Boulton, G.S. (1997) The effects of Pleistocene glaciations on the geohydrological system of Northwest Europe. *Journal of Hydrology*, **195**, 137–159.

Vaughan, D.G. & Doake, C.S.M. (1996) Recent atmospheric warming and retreat of ice shelves on the Antarctic Peninsula. *Nature*, **379**(6563), 328–331.

Vaughan, D.G. & Spouge, J. (2002) Risk estimation of collapse of the West Antarctic ice sheet. *Climatic Change*, **52**, 65–91.

Vaughan, D.G., Bamber, J.L., Giovinetto, M., Russell, J. & Cooper, A.P.R. (1999a) Reassessment of net surface mass balance in Antarctica. *Journal of Climate*, **12**(4), 933–946.

Vaughan, D.G., Corr, H.F.J., Doake, C.S.M. & Waddington, E.D. (1999b) Distortion of isochronous layers in ice revealed by ground-penetrating radar. *Nature*, **398**, 323–326.

Vaughan, D.G., Marshall, G.J., Connolley, W.M., King, J.C. & Mulvaney, R. (2001) Climate change: Devil in the detail. *Science*, **293**, 1777–1779.

Vaughan, D.G., Marshall, G.J., Connolley, *et al.* (2003a) Recent rapid regional climate warming on the Antarctic Peninsula. *Climatic Change*, **60**, 243–274.

Vaughan, D.G., Smith, A.M., Nath, P.C. & Le Meur, E. (2003b) Acoustic impedance and basal shear stress beneath four Antarctic ice streams. *Annals of Glaciology*, **36**, 225–232.

Veillette, J.J., Dyke, A.S. & Roy, M. (1999) Ice-flow evolution of the Labrador Sector of the Laurentide Ice Sheet: a review, with new evidence from northern Quebec. *Quaternary Science Reviews*, **18**, 993–1019.

Vela, J.C. (1994) *Rheological testing of subglacial till material.* M.Sc. Thesis, Washington State University.

Velicogna, I. & Wahr, J. (2002) Postglacial rebound and Earth's viscosity structure from GRACE. *Journal of Geophysical Research-Solid Earth*, **107**(B12), art. no.–2376.

Vettoretti, G. & Peltier, W.R. (2004) Sensitivity of glacial inception to orbital and greenhouse gas climate forcing. *Quaternary Science Reviews*, **23**, 499–519.

Vialov, S.S. (1958) Regularities of glacial shields movement and the theory of plastic viscous flow. In: *Symposium of Chamonix. Physics of the Motion of Ice*, pp. 266–275. IAHS Publication 47, International Association of Hydrological Sciences, Wallingford.

Vieli, A. & Payne, A.J. (2003) Application of control methods for modelling the flow of Pine Island Glacier, West Antarctica. *Annals of Glaciology*, **36**, 197–204.

Vimeux, F., Masson, V., Jouzel, J., Stievenard, M. & Petit, J.R. (1999) Glacial-interglacial changes in ocean surface conditions in the Southern hemisphere. *Nature*, **398**, 410–413.

Vimeux, F., Masson, V., Delaygue, G., Jouzel, J., Petit, J.-R. & Stievenard, M. (2001) A 420,000 year deuterium excess record from East Antarctica: information on past changes in the origin of precipitation at Vostok. *Journal of Geophysical Research*, **106** (D23), 31863–31873.

Vimeux, F., Cuffey, K.M. & Jouzel, J. (2002) New insights into Southern Hemisphere temperature changes from Vostok ice core using deuterium excess correction. *Earth and Planetary Science Letters*, **203**, 829–843.

Visser, K., Thunell, R. & Stott, L. (2003) Magnitude and timing of temperature change in the Indo-Pacific warm pool during deglaciation. *Nature*, **421**(9), 152–155.

Vivian, R. & Bocquet, G. (1973) Subglacial cavitation phenomena under the Glacier d'Argentière, Mont Blanc, France. *Journal of Glaciology*, **12**(66), 439–451.

Voelker, A.H.L. (1999) *Zur Deutung der Dansgaard–Oeschger Ereignisse in ulta-hochauflosenden Sedimentprofilen aus dem Europaischen Nordmeer*. Universitat Kiel, Kiel, 271 pp.

Voelker, A. (2000) Potential links between surging ice sheets, circulation changes, and the Dansgaard-Oeschger cycles in the Irminger Sea, 60–18 kyr (2000) *Paleoceanography*, **15**, 4, 425–442.

Voelker, A.H.L. (2002) Global distribution of centennial-scale records for Marine Isotope Stage (MIS) 3: a database. *Quaternary Science Reviews*, **21**(10), 1185–1212.

Vogel, S.W., Tulaczyk, S. & Joughin, I.R. (2003) Distribution of basal melting and freezing beneath tributaries of Ice Stream C: implication for the Holocene decay of the West Antarctic ice sheet. *Annals of Glaciology*, **36**, 273–282.

Voigt, D.E., Alley, R.B., Anandakrishnan, S. & Spencer, M.K. (2003) Ice-core insights into the flow and shut-down of Ice Stream C, West Antarctica. *Annals of Glaciology*, **37**, 123–128.

Von Grafenstein, U., Erlenkeuser, H., Müller, J., Jouzel, J. & Johnsen, S. (1998) The cold event 8200 years ago documented in oxygen isotope records of precipitation in Europe and Greenland. *Climate Dynamics*, **14**, 73–81.

Vornberger, P.L. & Whillans, I.M. (1990) Crevasse deformation and examples from Ice Stream B, Antarctica. *Journal of Glaciology*, **36**, 3–10.

Vorren, T. (2003) Subaquatic landsystems: continental margins. In: *Glacial Landsystems* (Ed. D.A. Evans), pp. 289–312. Edwin Arnold, London.

Vorren, T. & Laberg, J.S. (1997) Trough mouth fans—palaeoclimate and ice sheet monitors. *Quaternary Science Reviews*, **16**, 865–881.

Vorren, T.O., Hald, M., Edvardsen, M. & Lind-Hansen, O.W. (1983) Glacigenic sediments and sedimentary environments on continental shelves: general principles with a case study from the Norwegian shelf. In: *Glacial Deposits in North-West Europe* (Ed. J. Ehlers), pp. 61–73. Balkema, Rotterdam.

Vorren, T.O., Lebesbye, E., Andreassen, K. & Larsen, K.B. (1989) Glacigenic sediments on a passive continental margin as exemplified by the Barents Sea. *Marine Geology*, **85**, 251–272.

Vorren, T.O., Laberg, J.S., Blaume, F., *et al.* (1998) The Norwegian–Greenland Sea continental margins: morphology and Late Quaternary sedimentary processes and environment. *Quaternary Science Reviews*, **17**, 273–301.

Vuille, M., Bradley, R.S., Werner, M. & Keimig, F. (2003) 20th century climate change in the tropical Andes: Observations and model results. *Climate Change*, **59**(1–2), 75–99.

Waddington, B.S. & Clarke, G.K.C. (1995) Hydraulic properties of subglacial sediment determined from the mechanical response of water-filled boreholes. *Journal of Glaciology*, **41**, 112–124.

Wadham, J.L., Hodson, A.J., Tranter, M. & Dowdeswell, J.A. (1997) The rate of chemical weathering beneath a quiescent, surge-type, polythermal based glacier, southern Spitsbergen. *Annals of Glaciology*, **24**, 27–31.

Wadham, J.L., Hodson, A.J., Tranter, M. & Dowdeswell, J.A. (1998) The hydrochemistry of meltwaters during the ablation season at a high Arctic, polythermal-based glacier, South Svalbard. *Hydrological Processes*, **14**, 1767–1786.

Wadham, J.L., Cooper, R.J., Tranter, M. & Hodgkins, R. (2001a) Enhancement of glacial solute fluxes in the proglacial zone of a polythermal glacier. *Journal of Glaciology*, **47**(157), 378–386.

Wadham, J.L., Hodgkins, R., Cooper, R.J. & Tranter, M. (2001b) Evidence for seasonal subglacial outburst events at a polythermal-based high Arctic glacier, Finsterwalderbreen, Svalbard. *Hydrological Processes*, **15**, 2259–2280.

Wadham, J.L., Bottrell, S., Tranter, M. & Raiswell, R. (2004) Stable isotope evidence for microbial sulphate reduction at the bed of a polythermal high Arctic glacier. *Earth and Planetary Science Letters*, **219**, 341–355.

Wadhams, P., McLaren, A.S. & Weintraub, R. (1985) Ice Thickness Distribution in Davis Strait in February from Submarine Sonar Profiles. *Journal of Geophysical Research—Oceans*, **90**(NC1), 1069–1077.

Wagnon, P., Ribstein, P., Francou, B. & Pouyaud, B. (1999) Annual cycle of energy balance of Zongo glacier, Cordillera Real, Bolivia. *Journal of Geophysical Research*, **104**, 3907–3923.

Wagnon, P., Sicart, J.E., Berthier, E. & Chazarin, J.P. (2003) Wintertime high-altitude surface energy balance of a Bolivian glacier, Illimani, 6340 m above sea level. *Journal of Geophysical Research—Atmospheres*, **108**(D6), 4177. doi: 10.1029/2002JD002088.

Wahr, J., Wingham, D. & Bentley, C. (2000) A method of combining ICESat and GRACE satellite data to constrain Antarctic mass balance. *Journal of Geophysical Research-Solid Earth*, **105**(B7), 16279–16294.

Wahr, J., van Dam, T., Larson, K. & Francis, O. (2001) GPS measurements of vertical crustal motion in Greenland. *Journal of Geophysical Research*, **106**(D24), 33755–33759.

Wahr, J., Swenson, S., Zlotnicki, V. & Velicogna, I. (2004) Time-variable gravity from GRACE: First results. *Geophysical Research Letters*, **31**(11). art. no.-L11501.

Wakahama, G. (1967) On the plastic deformation of single crystal of ice. In: *Proceedings of the International Conference on Low Temperature Science, 1966*, Vol. 1, pp. 292–311. Institute of Low Temperature Science, Hokkaido University, Sapporo.

Walden, J. (1994) Late Devensian sedimentary environments in the Irish Sea basin: glacioterrestrial or glaciomarine? In: *The Quaternary of Cumbria: Field Guide* (Eds J. Boardman & J. Walden), pp. 15–18. Quaternary Research Assocication, Oxford.

Walder, J.S. (1982) Stability of sheet flow of water beneath temperate glaciers and implications for glacier surging. *Journal of Glaciology*, **28**, 273–293.

Walder, J.S. & Fowler, A. (1994) Channelized subglacial drainage over a deformable bed. *Journal of Glaciology*, **40**, 3–15.

Walker, G.P.L. (1965) Some aspects of Quaternary volcanism in Iceland. *Transactions of the Leicester Literary and Philosophical Society*, **59**, 25–40.

Walland, D.J. & Simmonds, I. (1996) Sub-grid scale topography and the simulation of Northern Hemisphere snow cover. *International Journal of Climatology*, **16**, 961–982.

Waller, R.I. (2001) The influence of basal processes on the dynamic behaviour of cold-based glaciers. *Quaternary International*, **86**, 117–128.

Waller, R.I. & Hart (1999) Mechanisms and patterns of motion associated with the basal zone of the Russell Glacier, South-West Greenland. *Glacial Geology and Geomorphology* (http://boris.qub.ac.uk/ggg/papers/full/1999/rp021999/rp02.html)

Waller, R., Hart, J. & Knight, P. (2000) The influence of tectonic deformation on facies variability in stratified debris-rich basal ice. *Quaternary Science Reviews*, **19**(8), 775–786.

Wallin, B. (1992) *Sulphur and Oxygen Isotope Evidence from Dissolved Sulphates in Groundwater and Sulphide Sulphur in Fissure Fillings at Äspö, Southeastern Sweden*. SKB Progress Report 25-92-08, Svensk Kärnbränslehantering AB (Swedish Nuclear Fuel and Waste Management), Stockholm, 44 pp.

Wallin, B. (1995) *Paleohydrological Implications in the Baltic Area and its Relation to the Groundwater at Äspö, South-Eastern Sweden—a Literature Survey.* SKB Technical Report 95-06, Svensk Kärnbränslehantering AB (Swedish Nuclear Fuel and Waste Management), Stockholm, 68 pp.

Wallinga, J. & Van de Wal, R.S.W. (1998) Sensitivity of Rhonegletscher, Switzerland, to climate change: experiments with a one-dimensional flowline model. *Journal of Glaciology*, **44**(147), 383–393.

Walsh, J.E., Vinnikov, K. & Chapman, W.L. (1999) On the use of historical sea ice charts in assessments of century-scale climatic variations. In: *World Climate Research Arctic Climate System Study (ACSYS), Proceedings of the Workshop in Sea Ice Charts on the Arctic, Seattle, WA, 5–7 August 1998*, WMO/TD No. 949, IAPO Publication No. 3, pp. 1–3.

Walters, R.A. & Dunlap, W.W. (1987) Analysis of time series of glacier speed: Columbia Glacier, Alaska. *Journal of Geophysical Research*, **92**(B9), 8969–8975.

Wang, Z. & Liu, C. (2001) Geographical characteristics of the distribution of glaciers in Chian. *Journal of Glaciology and Geocryology*, **23**(3), 231–237.

Ward, C.G. (1995) The mapping of ice front changes on Muller Ice Shelf, Antarctic Peninsula. *Antarctic Science*, **7**, 197–198.

Wardle, P. (1973) Variations of the glaciers of Westland National Park and the Hooker Range, New Zealand. *New Zealand Journal of Botany*, **11**, 349–388.

Warner, R.C. & Budd, W.F. (1998) Modelling the long-term response of the Antarctic ice sheet to global warming. *Annals of Glaciology*, **27**, 161–168.

Warner, R.C., Jacka, T.H., Li, J. & Budd, W.F. (1999) Tertiary flow relations for compression and shear in combined stress tests on ice. In: *Advances in Cold-Region Thermal Engineering and Sciences* (Eds K. Hutter, Y. Wang & H. Beer), pp. 259–270. Technological, Environmental, and Climatological Impact. Springer Lecture Notes in Physics, Berlin.

Warren, C.R. (1992) Iceberg calving and the glacioclimatic record. *Progress in Physical Geography*, **16**, 253–282.

Warren, C.R. (1993) Rapid recent fluctuations of the calving San Rafael Glacier, Chilean Patagonia: climatic or non-climatic? *Geografiska Annaler*, **75A**(3), 111–125.

Warren, C.R. & Sugden, D.E. (1993) The Patagonia icefields: A glaciological review. *Arctic and Alpine Research*, **25**, 316–331.

Warren, C.R., Greene, D.R. & Glasser, N.F. (1995) Glaciar Upsala, Patagonia: rapid calving retreat in fresh water. *Annals of Glaciology*, **21**, 311–316.

Warren, C.R., Benn, D.I., Winchester, V. & Harrison, S. (2001) Buoyancy-driven lacustrine calving, Glaciar Nef, Chilean Patagonia. *Journal of Glaciology*, **47**, 135–146.

Warren, W.P. (1991) Fenitian (Midlandian) glacial deposits and glaciation in Ireland and the adjacent offshore regions. In: *Glacial Deposits in Great Britain and Ireland* (Eds J. Ehlers, P.L. Gibbard & J. Rose), pp. 79–88. Balkema, Rotterdam.

Warren, W.P. (1992) Drumlin orientation and the pattern of glaciation in Ireland. *Sveriges Geologiska Undersökning*, **81**, 359–366.

Warrick, R.A. & Oerlemans, J. (1990) Sea level rise. In: *Climate Change—the IPCC Scientific Assessment* (Eds J.T. Houghton, G.J. Jenkins & J.J. Ephraums), pp. 358–405. Cambridge University Press, Cambridge.

Warrick, R.A., Le Provost, C., Meier, M.F., Oerlemans, J. & Woodworth, P.L. (1996). Changes in sea level. In: *Climate Change, 1995—The Science of Climate Change* (Eds J.T. Houghton, L.G. Meira Filho, B.A. Callander, N. Harris, A. Kattenberg & K. Maskell), pp. 257–281. Cambridge University Press, Cambridge.

Wastegård, S. (2002) Early to middle Holocene silicic tephra horizons from the Katla volcanic system, Iceland: new results from the Faroe Islands. *Journal of Quaternary Science*, **17**, 723–730.

Wastl, M., Stotter, J. & Caseldine, C. (2001) Reconstruction of Holocene variations of the upper limit of tree or shrub birch growth in Northern Iceland based on evidence from Vesturardalur-Skidadalur, Trollaskagi. *Arctic, Antarctic and Alpine Research*, **33**, 191–203.

Watts, P.A. (1974) *Inclusions in ice.* PhD thesis, University of Bristol.

Weaver, A.J. & Hughes, T.M.C. (1994) Rapid interglacial climate fluctuations driven by North Atlantic Ocean circulation. *Nature*, **367**, 447–450.

Weaver, A.J., Eby, M., Wiebe, E.C., *et al.* (2001) The Uvic Earth System Climate Model: model description, climatology and application to past, present and future climates. *Atmosphere-Ocean*, **39**, 361–428.

Weaver, A.J., Saenko, O.A., Clarck, P.U. & Mitrovica, J.X. (2003) Meltwater pulse 1A from Antarctica as a trigger of the Bolling-Allerod warm interval. *Science*, **299**, 1709–1713.

Weaver, T.R., Frape, S.K. & Cherry, J.A. (1995) Recent cross-formational fluid flow and mixing in the shallow Michigan Basin. *Geological Society of America Bulletin*, **107**, 697–707.

Webb, P.-N., Harwood, D.M., McKelvey, B.C., Mercer, J.H. & Stott, L.D. (1984) Cenozoic marine sedimentation and ice-volume variation on the East Antarctic craton. *Geology*, **12**, 287–291.

Weertman, J. (1957) On the sliding of glaciers. *Journal of Glaciology*, **3**, 33–38.

Weertman, J. (1961a) Stability of ice-age ice sheets. *Journal of Geophysical Research*, **66**, 3783–3792.

Weertman, J. (1961b) Equilibrium profile of ice caps. *Journal of Glaciology*, **3**, 953–964.

Weertman, J. (1964) The theory of glacier sliding. *Journal of Glaciology*, **5**(39), 287–303.

Weertman, J. (1968) Dislocation climb theory of steady-state creep. *Transactions of the American Society of Metals*, **61**, 681–694.

Weertman, J. (1972) General theory of water flow at the base of a glacier or ice sheet. *Rewiews of Geophysics and Space Physics*, **10**, 287–333.

Weertman, J. (1973) Creep of ice. In: *Physics and Chemistry of Ice* (Eds E. Whalley, S.J. Jones & L.W. Gold), pp. 320–337. Royal Society of Canada, Ottawa.

Weertman, J. (1974) Stability of the junction of an ice sheet and an ice shelf, *Journal of Glaciology*, **13**, 3–11.

Weertman, J. (1976) Milankovitch solar radiation variations and ice age ice sheet sizes. *Nature*, **261**, 17–20.

Weertman, J. (1983) Creep deformation of ice. *Annual Reviews in Earth and Planetary Sciences*, **11**, 215–240.

Weertman, J. & Birchfield, G.E. (1982) Subglacial water flow under ice streams and West Antarctic ice-sheet stability. *Annals of Glaciology*, **3**, 316–318.

Weidick, A. (1968) Observation on some Holocene glacier fluctuations in West Greenland. *Meddelelser om Grønland*, **165**(6), 202 pp.

Weidick, A. (1991) Present-day expansion of the southern part of the Inland Ice. *Report Grønlands Geologiske Undersøgelser*, **152**, 73–79.

Weidick, A. (1995) *Satellite Image Atlas of Glaciers of the World, Greenland.* U.S. Geological Survey Professional Paper 1386-C, United States Government Printing Office, Washington.

Weidick, A. (1996) Neoglacial changes of ice cover and sea level in Greenland—a classical enigma. In: *The Paleo-Eskimo Cultures of Greenland* (Ed. B. Gronnow), pp. 257–270. Danish Polar Center, Copenhagen.

Weiss, J. (2003) Scaling of fracture and faulting of ice on earth. *Surveys in Geophysics*, **24**(2), 185–227.

Weiss, J. (2004) Subcritical crack propagation as a mechanism of crevasse formation and iceberg calving. *Journal of Glaciology*, **50**(168), 109–115.

Weiss, J., Vidot, J., Gay, M., Arnaud, L., Duval, P. & Petit, J.R. (2002) Dome Concordia ice microstructure; impurities effect on grain growth. In: *Papers from the International Symposium on Ice Cores and Climate* (Ed. E.W. Wolff). *Annals of Glaciology*, **35**, 552–558.

Wellner, J.S., Lowe, A.L., Shipp, S.S. & Anderson, J.B. (2001) Distribution of glacial geomorphic features on the Antarctic continental shelf and correlation with substrate: implications for ice behavior. *Journal of Glaciology*, **47**, 397–411.

Wenk, H.-R., Bennett, K., Canova, G. & Molinari, A. (1991) Modelling plastic deformation of peridotite with the self-consistent theory. *Journal of Geophysical Research*, **96**(B5), 8337–8349.

Wettlaufer, J.S. (1999) Ice Surfaces: Macroscopic effects of microscopic structure. *Philosophical Transactions of the Royal Society, Series A*, **357**, 3403–3425.

WGMS (2001) *The World Glacier Inventory*. World Glacier Monitoring System, University of Zürich, Zürich, http://nsidc.org/NOAA/wgms_inventory [Accessed July 2001]

Whillans, I.M. (1976) Radio-echo layers and the recent stability of the West Antarctic ice sheet. *Nature*, **264**, 152–155.

Whillans, I.M. (1981) Reaction of the accumulation zone portions of glaciers to climatic change. *Journal of Geophysical Research*, **86**(C5), 4274–4282.

Whillans, I.M. (1987) Force budget of ice sheets. In: *Dynamics of the West Antartic Ice Sheet* (Eds. C.J. van der Veen & J. Oerlemans), pp. 17–36. D. Reidel, Dordrecht.

Whillans, I.M. & van der Veen, C.J. (1993) New and improved determinations of velocity of Ice Streams B and C, West Antarctica. *Journal of Glaciology*, **39**(133), 483–490.

Whillans, I.M. & van der Veen, C.J. (1997) The role of lateral drag in the dynamics of ice stream B, Antarctica. *Journal of Glaciology*, **43**, 231–237.

Whillans, I.M. & van der Veen, C.J. (2001) Velocity pattern in a transect across Ice Stream-B, Antarctica. *Journal of Glaciology*, **47**, 433–440.

Whillans, I.M., Jackson, M. & Tseng, Y.H. (1993) Transmission of stress between an ice stream and interstream ridge. *Journal of Glaciology*, **39**, 562–572.

Whillans, I.M., Bentley, C.R. & van der Veen, C.J. (2001) Ice streams B and C. In: *The West Antarctic Ice Sheet: Behavior and Dynamics* (Eds R.B. Alley & R.A. Bindschadler), pp. 257–281. Volume 77, American Geophysical Union, Washington, DC.

White, A.F. & Brantley, S.L. (2003) The effect of time on the weathering of silicate minerals: why do weathering rates differ in the laboratory and field? *Chemical Geology*, **202**, 479–506.

White, A.F., Blum, A.E., Bullen, T.D., Vivit, D.V., Schulz, M. & Fitzpatrick, J. (1999) The effect of temperature on experimental and natural chemical weathering rates of granitoid rocks. *Geochimica et Cosmochimica Acta*, **63**, 3277–3291.

White, A.F., Bullen, T.D., Schulz, M.S., Blum, A.E., Huntington, T.G. & Peters, N.E. (2001) Differential rates of feldspar weathering in granitic regoliths. *Geochimica et Cosmochimica Acta*, **65**, 847–869.

White, W.A. (1972) Deep erosion by continental ice sheets. *Geological Society of America Bulletin*, **83**, 1037–1056.

White, W.A. (1988) More on deep erosion by continental ice sheets and their tongues of distributary ice. *Quaternary Research*, **30**, 137–150.

Whittow, J.B. & Ball, D.F. (1970) North-west Wales. In: *The Glaciations of Wales and Adjoining Areas* (Ed. C.A. Lewis), pp. 21–58. Longman, London.

Wild, M. & Ohmura, A. (2000) Change in mass balance of polar ice sheets and sea level from high-resolution GCM simulations of greenhouse warming. *Annals of Glaciology*, **30**, 197–203.

Wild, M., Calanca, P., Scherrer, S.C. & Ohmura, A. (2003) Effects of polar ice sheets on global sea level in high-resolution greenhouse scenarios. *Journal of Geophysical Research*, **108**, 4165. doi:10.1029/2002JD002451.

Wilen, L.A., Diprinzio, C.L., Alley, R.B. & Azuma, N. (2003) Development, principles, and applications of automated ice fabric analyzers. *Microscopy Research and Technique*, **62**, 2–18.

Wilhelms, F., Kipfstuhl, J., Miller, H., Heinloth, K. & Firestone, J. (1998) Precise dielectric profiling of ice cores: a new device with improved guarding and its theory. *Journal of Glaciology*, **44**, 171–174.

Williams, G.E. (1993) History of the Earth's obliquity. *Earth Science Reviews*, **34**, 1–45.

Williams, K.M., Andrews, J.T., Jennings, A.E., Short, S.K., Mode, W.N. & Syvitski, J.P.M. (1995) The Eastern Canadian Arctic at ca. 6 ka: A time of transition. *Geographie physique et Quaternaire* (Canadian Global Change issue), **49**, 13–27.

Williams, L.D. (1975) The variation of corrie elevation and equilibrium line altitude with aspect in Eastern Baffin Island, N.W.T., Canada. *Arctic and Alpine Research*, **7**, 169–181.

Williams, L.D. (1978a) Ice-sheet initiation and climatic influence of expanded snow cover in Arctic Canada. *Quaternary Research*, **10**, 141–149.

Williams, L.D. (1978b) The little ice age glaciation level on Baffin Island, Arctic Canada. *Palaeogeography, Palaeoclimatology, Palaeoecology*, **25**, 199–207.

Williams, P.J. & Smith, M.W. (1991) *The Frozen Earth*. Cambridge University Press, Cambridge.

Williams, R.S. & Ferrigno, J.G. (Eds) (1998) *South America. Satellite Image Atlas of Glaciers of the World. U.S Geological Survey Professional Paper*, **1386-I**, 206 pp.

Willis, I.C. (1995) Intra-annual variations in glacier motion: a review. *Progress in Physical Geography*, **19**, 61–106.

Willis, I.C., Mair, D.W.F., Hubbard, B., Nienow, P.W., Fischer, U.H. & Hubbard, A.L. (2003) Seasonal variations in ice deformation and basal motion across the tongue of Haut Glacier d'Arolla, Switzerland. *Annals of Glaciology*, **36**, 157–167.

Wilson, A.T., Hendy, C.H. & Reynolds, C.P. (1979) Short-term climate change and New Zealand temperatures during the last millennium. *Nature*, **279**, 315–317.

Winberry, J.P. & Anandakrishnan, S. (2003) Seismicity and neotectonics of West Antarctica. *Geophysical Research Letters*, **30**. Art. No. 1931.

Winebrenner, D.P., Arthern, R.J. & Shuman, C.A. (2001) Mapping Greenland accumulation rates using observations of thermal emission at 4.5-cm wavelength. *Journal of Geophysical Research—Atmospheres*, **106**(D24), 33919–33934.

Wingfield, R.T.R. (1990). The origin of marine incisions within the Pleistocene deposits on the North Sea. *Marine Geology*, **91**, 31–52.

Wingham, D.J., Ridout, A.J., Scharroo, R., Athern, R.J. & Shum, C.K. (1998) Antarctic elevation change from 1992 to 1996. *Science*, **282**(5388), 456–458.

Winkler, S. (2000) The 'Little Ice Age' maximum in the Southern Alps, New Zealand: preliminary results at Mueller Glacier. *The Holocene*, **10**, 643–647.

Wolff, C.F., Rankin, A.M. & Rothlisberger, R. (2003) An ice core indicator of Antarctic sea ice production? *Geophysical Research Letters*, **30**(22), 2158. doi:10.1029/2003GL018454.

Woo, M. (1990) Consequences of climate change for hydrology in permafrost zones. *Journal of Cold Regions Engineering*, **4**, 15–20.

Wood, F.B. (1988) Global alpine glacier trends, 1960s to 1980s. *Arctic and Alpine Research*, **20**(4), 404–413.

Woodworth, P. & Player, R. (2003) The Permanent Service for Mean Sea Level: An update to the 21st Century. *Journal of Coastal Research*, **19**(2), 287–295.

Wunsch, C. (2000) On sharp spectral lines in the climate record and the millennial peak. *Paleoceanography*, **15**, 417–424.

Wunsch, C. (2004) Quantitative estimate of the Milankovitch forced contribution to observed Quaternary climate change. *Quaternary Science Reviews*, **23**, 1001–1012.

Wynne, R.H., Magnuson, J.J., Clayton, M.K., Lillesand, T.M. & Rodman, D.C. (1996) Determinants of temporal coherence in the satellite-derived 1987–1994 ice breakup dates of lakes on the Laurentian Shield. *Limnology and Oceanography*, **41**, 832–838.

Yamada, T. (1987) Glaciological characteristics revealed by 37.6-m deep core drilled at the accumulation area of San Rafael Glacier, the Northern Patagonia Icefield. *Bulletin of Glacier Research*, **4**, 59–67.

Yang, Z. (1995) Glacier meltwater runoff in China and its nourishment to river. *Chinese Geographical Science*, **5**, 66–76.

Yang, Z. & Hu, X. (1992) Study of glacier meltwater resources in China. *Annals of Glaciology*, **16**, 141–145.

Yang, Z.-L., Dickinson, R.E., Hahmann, A.N., *et al.* (1999) Simulation of snow mass and extent in global climate models. *Hydrological Processes*, **13**(12–13), 2097–2113.

Yao, T. & Shi, Y. (1990) Fluctuations and future trend of climate, glaciers and discharge of Urumqi River in Xingjiang. *Science in China, Series B*, **33**(4), 504–512.

Yao, T., Ageta, Y., Ohata, T., *et al.* (1991) Preliminary results from China–Japan Glaciological Expedition in Tibet Plateau in 1989. *Journal of Glaciology and Geocryology*, **13**(1), 1–8.

Yao, T., Jiao, K., Tian, L., *et al.* (1996) Climatic variations since the Little Ice Age recorded in the Guliya ice core, *Science in China, Series D*, **39**, 588–596.

Yao, T., Thompson, L., Qin, D., *et al.* (1997) Variations in temperature and precipitation in the past 2000 years on the Xizang (Tibet) plateau—Guliya ice core record. *Science in China, Series D*, **39**, 425–433.

Yao, T., Liu, S., Pu, J., Shen, Y. & Lu, A. (2004) Recent glacial retreat in High Asia and its impact on water resource in Northwest China. *Science in China*, **34**(6), 535–543.

Yarnal, B. (1984) Relationships between synoptic-scale atmospheric circulation and glacier mass balance in southwestern Canada during the International Hydrological Decade, 1965–74. *Journal of Glaciology*, **30**, 188–198.

Yevteyev, S.A. (1959) Opredeleniye kolichestva morennogo materiala, perenosimogo lednikami vostochnogo poberezh'ya Antarktidy. *Informatsionnyy Byulleten' Sovetskoy Antarticheskoy Ekspedititsii*, **11**, 14–16.

Young, R.R. (2000) Glacial landforms. In: *Oxford Companion to the Earth* (Eds P.L. Hancock & B.J. Skinner). Oxford University Press, Oxford.

Yu, Y. & Rothrock, D.A. (1996) Thin ice thickness from satellite thermal imagery. *Journal of Geophysical Research—Oceans*, **101**(C11), 25753–25766.

Zängl, W. & Hamberger, S. (2004) *Gletscher im Treibhaus Eine fotografische Zeitreise in die alpine Eiswelt.* Tecklenborg Verlag, Steinfurth.

Zhang, X., Zheng, B. & Xie, Z. (1981) Recent variations of existing glaciers on the Qinghai-Xizang (Tibet) Plateau. In: *Geological and Ecological Studies of Qinghai-Xizang Plateau*, pp. 1625–1629. Science Press, Beijing.

Zielinski, G.A., Mayewski, P.A., Mekker, L.D., *et al.* (1994) Record of Volcanism since 7000 B.C. from the GISP2 Greenland ice core and implications for the Volcano-Climate system. *Science*, **264**, 948–952.

Zuber, A., Weise, S.M., Motyka, J., Osenbrück, K. & Różański, K. (2004) Age and flow pattern of groundwater in a Jurassic limestone aquifer and related Tertiary sands derived from combined isotope, noble gas and chemical data. *Journal of Hydrology*, **286**, 87–112.

Zumberge, M.A., Elsberg, D.H., Harrison, W.D., *et al.* (2002) Measurement of vertical velocity and strain at Siple Dome by optical sensors. *Journal of Glaciology*, **46**(161), 217–225.

Zuo, Z. & Oerlemans, J. (1997) Contribution of glacier melt to sea-level rise since AD 1865: a regionally differentiated calculation. *Climate Dynamics*, **13**, 835–845.

Zwally, H.J. & Brenner, A.C. (2001) Ice sheet dynamics and mass balance. In: *Satellite Altimetry and the Earth Sciences* (Eds L.-L. Fu & A. Cazenave), pp. 351–370. Academic Press, New York.

Zwally, J.H. & Giovinetto, M.B. (1997) Accumulation in Antarctica and Greenland derived from passive-microwave data: a comparison with contoured compilations. *Annals of Glaciology*, **21**, 123–130.

Zwally, H.J., Bindschaler, R.A., Brenner, A.C., Martin, T.V. & Thomas, R.H. (1983) Surface elevation contours of Greenland and Antarctic ice sheets. *Journal of Geophysical Research*, **88**(C3), 1589–1596.

Zwally, H.J., Brenner, A.C., Major, J.A., Bindschadler, R.A. & Marsh, J.G. (1989) Growth of Greenland Ice Sheet: measurement. *Science*, **246**, 1587–1591.

Zwally, H.J., Abdalati, W., Herring, T., Larson, K. & Steffen, K. (2002a) Surface-melt induced acceleration of Greenland ice-sheet flow. *Science*, **297**, 218–222.

Zwally, H.J., Schutz, B., Abdalati, W., *et al.* (2002b) ICESat's laser measurements of polar ice, atmosphere, ocean, and land. *Journal of Geodynamics*, **34**(3–4), 405–445.

Zweck, C. & Huybrechts, P. (2003) Modeling the marine extent of Northern Hemisphere ice sheets during the last glacial cycle. *Annals of Glaciology*, **37**, 173–180.

Index

Colour plate numbers are in **bold**

F

G

J

K

L

R

S

T